HIGH ALTITUDE

LUNG BIOLOGY IN HEALTH AND DISEASE

Executive Editor

Claude Lenfant
Director, National Heart, Lung and Blood Institute
National Institutes of Health
Bethesda, Maryland

1. Immunologic and Infectious Reactions in the Lung, *edited by C. H. Kirkpatrick and H. Y. Reynolds*
2. The Biochemical Basis of Pulmonary Function, *edited by R. G. Crystal*
3. Bioengineering Aspects of the Lung, *edited by J. B. West*
4. Metabolic Functions of the Lung, *edited by Y. S. Bakhle and J. R. Vane*
5. Respiratory Defense Mechanisms (in two parts), *edited by J. D. Brain, D. F. Proctor, and L. M. Reid*
6. Development of the Lung, *edited by W. A. Hodson*
7. Lung Water and Solute Exchange, *edited by N. C. Staub*
8. Extrapulmonary Manifestations of Respiratory Disease, *edited by E. D. Robin*
9. Chronic Obstructive Pulmonary Disease, *edited by T. L. Petty*
10. Pathogenesis and Therapy of Lung Cancer, *edited by C. C. Harris*
11. Genetic Determinants of Pulmonary Disease, *edited by S. D. Litwin*
12. The Lung in the Transition Between Health and Disease, *edited by P. T. Macklem and S. Permutt*
13. Evolution of Respiratory Processes: A Comparative Approach, *edited by S. C. Wood and C. Lenfant*
14. Pulmonary Vascular Diseases, *edited by K. M. Moser*
15. Physiology and Pharmacology of the Airways, *edited by J. A. Nadel*
16. Diagnostic Techniques in Pulmonary Disease (in two parts), *edited by M. A. Sackner*
17. Regulation of Breathing (in two parts), *edited by T. F. Hornbein*
18. Occupational Lung Diseases: Research Approaches and Methods, *edited by H. Weill and M. Turner-Warwick*
19. Immunopharmacology of the Lung, *edited by H. H. Newball*
20. Sarcoidosis and Other Granulomatous Diseases of the Lung, *edited by B. L. Fanburg*
21. Sleep and Breathing, *edited by N. A. Saunders and C. E. Sullivan*
22. *Pneumocystis carinii* Pneumonia: Pathogenesis, Diagnosis, and Treatment, *edited by L. S. Young*
23. Pulmonary Nuclear Medicine: Techniques in Diagnosis of Lung Disease, *edited by H. L. Atkins*
24. Acute Respiratory Failure, *edited by W. M. Zapol and K. J. Falke*
25. Gas Mixing and Distribution in the Lung, *edited by L. A. Engel and M. Paiva*

26. High-Frequency Ventilation in Intensive Care and During Surgery, *edited by G. Carlon and W. S. Howland*
27. Pulmonary Development: Transition from Intrauterine to Extrauterine Life, *edited by G. H. Nelson*
28. Chronic Obstructive Pulmonary Disease: Second Edition, *edited by T. L. Petty*
29. The Thorax (in two parts), *edited by C. Roussos and P. T. Macklem*
30. The Pleura in Health and Disease, *edited by J. Chrétien, J. Bignon, and A. Hirsch*
31. Drug Therapy for Asthma: Research and Clinical Practice, *edited by J. W. Jenne and S. Murphy*
32. Pulmonary Endothelium in Health and Disease, *edited by U. S. Ryan*
33. The Airways: Neural Control in Health and Disease, *edited by M. A. Kaliner and P. J. Barnes*
34. Pathophysiology and Treatment of Inhalation Injuries, *edited by J. Loke*
35. Respiratory Function of the Upper Airway, *edited by O. P. Mathew and G. Sant'Ambrogio*
36. Chronic Obstructive Pulmonary Disease: A Behavioral Perspective, *edited by A. J. McSweeny and I. Grant*
37. Biology of Lung Cancer: Diagnosis and Treatment, *edited by S. T. Rosen, J. L. Mulshine, F. Cuttitta, and P. G. Abrams*
38. Pulmonary Vascular Physiology and Pathophysiology, *edited by E. K. Weir and J. T. Reeves*
39. Comparative Pulmonary Physiology: Current Concepts, *edited by S. C. Wood*
40. Respiratory Physiology: An Analytical Approach, *edited by H. K. Chang and M. Paiva*
41. Lung Cell Biology, *edited by D. Massaro*
42. Heart–Lung Interactions in Health and Disease, *edited by S. M. Scharf and S. S. Cassidy*
43. Clinical Epidemiology of Chronic Obstructive Pulmonary Disease, *edited by M. J. Hensley and N. A. Saunders*
44. Surgical Pathology of Lung Neoplasms, *edited by A. M. Marchevsky*
45. The Lung in Rheumatic Diseases, *edited by G. W. Cannon and G. A. Zimmerman*
46. Diagnostic Imaging of the Lung, *edited by C. E. Putman*
47. Models of Lung Disease: Microscopy and Structural Methods, *edited by J. Gil*
48. Electron Microscopy of the Lung, *edited by D. E. Schraufnagel*
49. Asthma: Its Pathology and Treatment, *edited by M. A. Kaliner, P. J. Barnes, and C. G. A. Persson*
50. Acute Respiratory Failure: Second Edition, *edited by W. M. Zapol and F. Lemaire*
51. Lung Disease in the Tropics, *edited by O. P. Sharma*
52. Exercise: Pulmonary Physiology and Pathophysiology, *edited by B. J. Whipp and K. Wasserman*
53. Developmental Neurobiology of Breathing, *edited by G. G. Haddad and J. P. Farber*
54. Mediators of Pulmonary Inflammation, *edited by M. A. Bray and W. H. Anderson*
55. The Airway Epithelium, *edited by S. G. Farmer and D. Hay*

56. Physiological Adaptations in Vertebrates: Respiration, Circulation, and Metabolism, *edited by S. C. Wood, R. E. Weber, A. R. Hargens, and R. W. Millard*
57. The Bronchial Circulation, *edited by J. Butler*
58. Lung Cancer Differentiation: Implications for Diagnosis and Treatment, *edited by S. D. Bernal and P. J. Hesketh*
59. Pulmonary Complications of Systemic Disease, *edited by J. F. Murray*
60. Lung Vascular Injury: Molecular and Cellular Response, *edited by A. Johnson and T. J. Ferro*
61. Cytokines of the Lung, *edited by J. Kelley*
62. The Mast Cell in Health and Disease, *edited by M. A. Kaliner and D. D. Metcalfe*
63. Pulmonary Disease in the Elderly Patient, *edited by D. A. Mahler*
64. Cystic Fibrosis, *edited by P. B. Davis*
65. Signal Transduction in Lung Cells, *edited by J. S. Brody, D. M. Center, and V. A. Tkachuk*
66. Tuberculosis: A Comprehensive International Approach, *edited by L. B. Reichman and E. S. Hershfield*
67. Pharmacology of the Respiratory Tract: Experimental and Clinical Research, *edited by K. F. Chung and P. J. Barnes*
68. Prevention of Respiratory Diseases, *edited by A. Hirsch, M. Goldberg, J.-P. Martin, and R. Masse*
69. *Pneumocystis carinii* Pneumonia: Second Edition, *edited by P. D. Walzer*
70. Fluid and Solute Transport in the Airspaces of the Lungs, *edited by R. M. Effros and H. K. Chang*
71. Sleep and Breathing: Second Edition, *edited by N. A. Saunders and C. E. Sullivan*
72. Airway Secretion: Physiological Bases for the Control of Mucous Hypersecretion, *edited by T. Takishima and S. Shimura*
73. Sarcoidosis and Other Granulomatous Disorders, *edited by D. G. James*
74. Epidemiology of Lung Cancer, *edited by J. M. Samet*
75. Pulmonary Embolism, *edited by M. Morpurgo*
76. Sports and Exercise Medicine, *edited by S. C. Wood and R. C. Roach*
77. Endotoxin and the Lungs, *edited by K. L. Brigham*
78. The Mesothelial Cell and Mesothelioma, *edited by M.-C. Jaurand and J. Bignon*
79. Regulation of Breathing: Second Edition, *edited by J. A. Dempsey and A. I. Pack*
80. Pulmonary Fibrosis, *edited by S. Hin. Phan and R. S. Thrall*
81. Long-Term Oxygen Therapy: Scientific Basis and Clinical Application, *edited by W. J. O'Donohue, Jr.*
82. Ventral Brainstem Mechanisms and Control of Respiration and Blood Pressure, *edited by C. O. Trouth, R. M. Millis, H. F. Kiwull-Schöne, and M. E. Schläfke*
83. A History of Breathing Physiology, *edited by D. F. Proctor*
84. Surfactant Therapy for Lung Disease, *edited by B. Robertson and H. W. Taeusch*
85. The Thorax: Second Edition, Revised and Expanded (in three parts), *edited by C. Roussos*

86. Severe Asthma: Pathogenesis and Clinical Management, *edited by S. J. Szefler and D. Y. M. Leung*
87. *Mycobacterium avium*–Complex Infection: Progress in Research and Treatment, *edited by J. A. Korvick and C. A. Benson*
88. Alpha 1–Antitrypsin Deficiency: Biology • Pathogenesis • Clinical Manifestations • Therapy, *edited by R. G. Crystal*
89. Adhesion Molecules and the Lung, *edited by P. A. Ward and J. C. Fantone*
90. Respiratory Sensation, *edited by L. Adams and A. Guz*
91. Pulmonary Rehabilitation, *edited by A. P. Fishman*
92. Acute Respiratory Failure in Chronic Obstructive Pulmonary Disease, *edited by J.-P. Derenne, W. A. Whitelaw, and T. Similowski*
93. Environmental Impact on the Airways: From Injury to Repair, *edited by J. Chrétien and D. Dusser*
94. Inhalation Aerosols: Physical and Biological Basis for Therapy, *edited by A. J. Hickey*
95. Tissue Oxygen Deprivation: From Molecular to Integrated Function, *edited by G. G. Haddad and G. Lister*
96. The Genetics of Asthma, *edited by S. B. Liggett and D. A. Meyers*
97. Inhaled Glucocorticoids in Asthma: Mechanisms and Clinical Actions, *edited by R. P. Schleimer, W. W. Busse, and P. M. O'Byrne*
98. Nitric Oxide and the Lung, *edited by W. M. Zapol and K. D. Bloch*
99. Primary Pulmonary Hypertension, *edited by L. J. Rubin and S. Rich*
100. Lung Growth and Development, *edited by J. A. McDonald*
101. Parasitic Lung Diseases, *edited by A. A. F. Mahmoud*
102. Lung Macrophages and Dendritic Cells in Health and Disease, *edited by M. F. Lipscomb and S. W. Russell*
103. Pulmonary and Cardiac Imaging, *edited by C. Chiles and C. E. Putman*
104. Gene Therapy for Diseases of the Lung, *edited by K. L. Brigham*
105. Oxygen, Gene Expression, and Cellular Function, *edited by L. Biadasz Clerch and D. J. Massaro*
106. Beta$_2$-Agonists in Asthma Treatment, *edited by R. Pauwels and P. M. O'Byrne*
107. Inhalation Delivery of Therapeutic Peptides and Proteins, *edited by A. L. Adjei and P. K. Gupta*
108. Asthma in the Elderly, *edited by R. A. Barbee and J. W. Bloom*
109. Treatment of the Hospitalized Cystic Fibrosis Patient, *edited by D. M. Orenstein and R. C. Stern*
110. Asthma and Immunological Diseases in Pregnancy and Early Infancy, *edited by M. Schatz, R. S. Zeiger, and H. N. Claman*
111. Dyspnea, *edited by D. A. Mahler*
112. Proinflammatory and Antiinflammatory Peptides, *edited by S. I. Said*
113. Self-Management of Asthma, *edited by H. Kotses and A. Harver*
114. Eicosanoids, Aspirin, and Asthma, *edited by A. Szczeklik, R. J. Gryglewski, and J. R. Vane*
115. Fatal Asthma, *edited by A. L. Sheffer*
116. Pulmonary Edema, *edited by M. A. Matthay and D. H. Ingbar*
117. Inflammatory Mechanisms in Asthma, *edited by S. T. Holgate and W. W. Busse*
118. Physiological Basis of Ventilatory Support, *edited by J. J. Marini and A. S. Slutsky*

119. Human Immunodeficiency Virus and the Lung, *edited by M. J. Rosen and J. M. Beck*
120. Five-Lipoxygenase Products in Asthma, *edited by J. M. Drazen, S.-E. Dahlén, and T. H. Lee*
121. Complexity in Structure and Function of the Lung, *edited by M. P. Hlastala and H. T. Robertson*
122. Biology of Lung Cancer, *edited by M. A. Kane and P. A. Bunn, Jr.*
123. Rhinitis: Mechanisms and Management, *edited by R. M. Naclerio, S. R. Durham, and N. Mygind*
124. Lung Tumors: Fundamental Biology and Clinical Management, *edited by C. Brambilla and E. Brambilla*
125. Interleukin-5: From Molecule to Drug Target for Asthma, *edited by C. J. Sanderson*
126. Pediatric Asthma, *edited by S. Murphy and H. W. Kelly*
127. Viral Infections of the Respiratory Tract, *edited by R. Dolin and P. F. Wright*
128. Air Pollutants and the Respiratory Tract, *edited by D. L. Swift and W. M. Foster*
129. Gastroesophageal Reflux Disease and Airway Disease, *edited by M. R. Stein*
130. Exercise-Induced Asthma, *edited by E. R. McFadden, Jr.*
131. LAM and Other Diseases Characterized by Smooth Muscle Proliferation, *edited by J. Moss*
132. The Lung at Depth, *edited by C. E. G. Lundgren and J. N. Miller*
133. Regulation of Sleep and Circadian Rhythms, *edited by F. W. Turek and P. C. Zee*
134. Anticholinergic Agents in the Upper and Lower Airways, *edited by S. L. Spector*
135. Control of Breathing in Health and Disease, *edited by M. D. Altose and Y. Kawakami*
136. Immunotherapy in Asthma, *edited by J. Bousquet and H. Yssel*
137. Chronic Lung Disease in Early Infancy, *edited by R. D. Bland and J. J. Coalson*
138. Asthma's Impact on Society: The Social and Economic Burden, *edited by K. B. Weiss, A. S. Buist, and S. D. Sullivan*
139. New and Exploratory Therapeutic Agents for Asthma, *edited by M. Yeadon and Z. Diamant*
140. Multimodality Treatment of Lung Cancer, *edited by A. T. Skarin*
141. Cytokines in Pulmonary Disease: Infection and Inflammation, *edited by S. Nelson and T. R. Martin*
142. Diagnostic Pulmonary Pathology, *edited by P. T. Cagle*
143. Particle–Lung Interactions, *edited by P. Gehr and J. Heyder*
144. Tuberculosis: A Comprehensive International Approach, Second Edition, Revised and Expanded, *edited by L. B. Reichman and E. S. Hershfield*
145. Combination Therapy for Asthma and Chronic Obstructive Pulmonary Disease, *edited by R. J. Martin and M. Kraft*
146. Sleep Apnea: Implications in Cardiovascular and Cerebrovascular Disease, *edited by T. D. Bradley and J. S. Floras*
147. Sleep and Breathing in Children: A Developmental Approach, *edited by G. M. Loughlin, J. L. Carroll, and C. L. Marcus*

148. Pulmonary and Peripheral Gas Exchange in Health and Disease, *edited by J. Roca, R. Rodriguez-Roisen, and P. D. Wagner*
149. Lung Surfactants: Basic Science and Clinical Applications, *R. H. Notter*
150. Nosocomial Pneumonia, *edited by W. R. Jarvis*
151. Fetal Origins of Cardiovascular and Lung Disease, *edited by David J. P. Barker*
152. Long-Term Mechanical Ventilation, *edited by N. S. Hill*
153. Environmental Asthma, *edited by R. K. Bush*
154. Asthma and Respiratory Infections, *edited by D. P. Skoner*
155. Airway Remodeling, *edited by P. H. Howarth, J. W. Wilson, J. Bousquet, S. Rak, and R. A. Pauwels*
156. Genetic Models in Cardiorespiratory Biology, *edited by G. G. Haddad and T. Xu*
157. Respiratory-Circulatory Interactions in Health and Disease, *edited by S. M. Scharf, M. R. Pinsky, and S. Magder*
158. Ventilator Management Strategies for Critical Care, *edited by N. S. Hill and M. M. Levy*
159. Severe Asthma: Pathogenesis and Clinical Management, Second Edition, Revised and Expanded, *edited by S. J. Szefler and D. Y. M. Leung*
160. Gravity and the Lung: Lessons from Microgravity, *edited by G. K. Prisk, M. Paiva, and J. B. West*
161. High Altitude: An Exploration of Human Adaptation, *edited by T. F. Hornbein and R. B. Schoene*

ADDITIONAL VOLUMES IN PREPARATION

IgE and Anti-IgE in Asthma and Allergic Disease, *edited by R. B. Fick, Jr., and P. M. Jardieu*

Drug Delivery to the Lung, *edited by H. Bisgaard, C. O'Callaghan, and G. C. Smaldone*

Inhaled Steroids in Asthma: Optimizing Effects in the Airways, *edited by R. P. Schleimer, P. M. O'Byrne, S. J. Szefler, and R. Brattsand*

Clinical Management of Chronic Obstructive Pulmonary Disease, *edited by T. Similowski, W. A. Whitelaw, and J.-P. Derenne*

HIGH ALTITUDE

An Exploration of Human Adaptation

Edited by

Thomas F. Hornbein
Robert B. Schoene

University of Washington School of Medicine
Seattle, Washington

MARCEL DEKKER

MARCEL DEKKER, INC. NEW YORK • BASEL

This cover illustration was featured on the cover of the *Journal of Applied Physiology* on the April through June issues 2000 (88:4, 88:5, and 88:6). We acknowledge Steve Graepel for the cover design. This illustration is copyrighted by the Mayo Foundation and reproduced with permission.

ISBN: 0-8247-0313-8

This book is printed on acid-free paper.

Headquarters
Marcel Dekker, Inc.
270 Madison Avenue, New York, NY 10016
tel: 212-696-9000; fax: 212-685-4540

Eastern Hemisphere Distribution
Marcel Dekker AG
Hutgasse 4, Postfach 812, CH-4001 Basel, Switzerland
tel: 41-61-261-8482; fax: 41-61-261-8896

World Wide Web
http://www.dekker.com

The publisher offers discounts on this book when ordered in bulk quantities. For more information, write to Special Sales/Professional Marketing at the headquarters address above.

Current printing (last digit):
10 9 8 7 6 5 4 3 2 1

PRINTED IN THE UNITED STATES OF AMERICA

INTRODUCTION

Adaptation has always been the subject of considerable interest and debate. Philosophers, students of religion, and scientists of many disciplines have devoted their energies to investigating and explaining where our planet (our world) and we (as human beings) came from, and how we have responded to terrestrial changes.

The history of life is indeed a fascinating and mysterious subject. Great names have been associated with it. Among many, two can be singled out: Charles Darwin, who aimed at understanding biological variations and natural selection, and Claude Bernard, who introduced the concept of "internal environment," which postulates that human beings can achieve a life that is independent of the endless changes in external conditions.

To a point, the truth about life remains an open question, but most likely it is an equilibrium between the external, environmental influences and forces and the relative stability of the "milieu interieur." Perhaps the best illustration of how such an equilibrium has been reached by the human species is found in its adaptation to high altitude, which is the subject of this volume.

The reader need only focus on the first chapter to recognize the complexity of the adaptive processes that permit us to "function" at high altitude. It shows that questions about living in thin air have been asked for centuries, even millennia. Because of our great desire to conquer living at high altitude, an understanding of how we adapt to that environment has been achieved, but much still remains to be uncovered.

Ever since the conquest of the highest Himalayan peaks in the second half of the last century, the biology of adaptation to higher altitude has been of considerable interest. Great scientists from all biological disciplines have devoted years of work to these studies. All have worked in academe, some have explored and conquered peaks never reached before. Many of them have contributed to this book. Their scientific achievements and their first-hand experiences present the reader with an exclusive opportunity to explore a biological/medical area of unique dimension.

The Lung Biology in Health and Disease series of monographs has presented many new and challenging topics to its readership, always with the goal of reaching new peaks! Both clinical and basic research subjects have been included: we even explored the human response to the undersea world. *High Altitude: An Exploration of Human Adaptation*, edited by Thomas F. Hornbein and Robert B. Schoene, truly takes us to one of the highest summits. For this, and for the contributions included in this volume, I am extremely grateful to the authors and editors.

Claude Lenfant, M.D.
Bethesda, Maryland

PREFACE

Great things are done when men and mountain meet.
This is not done by jostling in the street.
—William Blake

Maybe it is the proximity to the heavens or perhaps the distance from where most of us live out our lives. The peaks and steppes of high altitude harbor mystery. Something about those regions high above the level of the sea attracts some of us lowlanders to venture and to explore. This book is the creation of just such individuals, those who find in high places not only beauty but also the uncertainty that catalyzes the search for understanding about how humans (and other organisms) adapt in order to survive and even thrive where the pressure of air and hence the partial pressure of its oxygen are diminished.

This volume is about how humans respond acutely to, adapt to, and sometimes fail to adapt to high altitude, in particular to the hypoxia of high altitude. We define high altitude as elevations as low as 2000 meters (6500 feet) to those as high as 8850 meters (approximately 29,000 feet).

First, this book is not a comprehensive review of the literature about our human affair with thin air. Rather, you will find here a series of essays written by individuals at the forefront of studying the effects of high altitude on the human body. One major goal has been to review critically and selectively, rather than comprehensively, the existing knowledge that is most important to our current understanding. These reviews are freely seasoned with the authors' (as well as others') views, hypotheses, and fantasies about how things might work. Our other major intent is to identify some of the black holes in our current knowledge, those areas wanting further exploration.

The thinning of the air over our heads as we climb higher above the sea has a number of consequences: the air grows colder and the rays from the sun become more intense. But most germane to the intent of this volume is that the weight of the air above lessens and with it the pressure of oxygen in the air we breathe. At the summit of Mount Everest, the barometric and oxygen pressures are approximately one-third of what we breathe at sea level.

The natural laboratory of high altitude enables us to study the impact of exogenous hypoxia on healthy humans to better distinguish the effect of hypoxia from that of cardiopulmonary and other diseases that can also cause a person to be hypoxic.

The 25 chapters comprising this volume encompass four domains:

1. *The stage.* Chapters 1 to 3 touch on the history of our inquiry into human existence at high altitude, the nature of our atmosphere and how it came to be, and an anthropological perspective on those who dwell permanently at great heights.
2. *Organism defense.* Chapters 4 and 5 explore how the cells within the body detect that the oxygen they need is in short supply (oxygen sensing) and how they defend themselves against this stress.
3. *Oxygen's journey from air to mitochondrion.* The physiological journey to high altitude is expressed by the traverse of molecular oxygen down a gradient from the air we breathe to the mitochondria, where oxygen fuels aerobic metabolism. This path comprises two conductive systems, one that moves gas (the pulmonary system) and one that transports oxygen-containing blood (the circulatory system) and a diffusive component that fills in the gaps (from alveoli to blood, from capillary blood to tissue mitochondria). In addition to lungs, heart, vessels, and blood, many organs play a role in sensing, facilitating, or modulating this oxygen flux. Chapters 6 through 21 explore the role of individual organs as well as their integrated function in enabling physical and mental performance at high altitude.
4. *Maladaptation.* Chapters 22 to 25 describe failure to thrive at high altitude: acute mountain sickness and high-altitude cerebral and pulmonary edema, chronic mountain sickness, and the challenges faced by lowland dwellers venturing to high altitude who have preexisting medical conditions.

So, where do we go from here? Even during the not-so-brief gestation of this volume, we have watched the nature of high-altitude research evolve, changing in a manner predictable from the directions in which biomedical research in general is unfolding.

As is apparent from many of the chapters in this book, one inevitable evolution is toward understanding not just *what* is happening, but *how*.

- *How* is a change in the partial pressure of oxygen sensed by individual cells in multiple tissues such as the carotid body, renal EPO fibroblasts, pulmonary and other vessels, and a wide array of other organs?
- *How* is this sensed signal genetically expressed (HIF, etc.?) and then transcribed into biochemical messages (VEGF, NO, and myriad more) within cells that tell organs how to defend themselves against hypoxia?
- *How* is the magnitude of the response modulated and why does the response sometimes become inappropriate to the organism's best interests? Are there genetic antecedents contributing to the diversity—both advantageous and disadvantageous—of response?

But the other side of the coin—the *what*—is still far from complete and much information is needed simply to describe adequately what is going on in acclimatizing lowlanders and permanent high-altitude inhabitants. Geographical, ethnic, gender, and aging effects are incompletely characterized. With increasing numbers of lowlanders traveling to high altitudes for play or work, some with regular intermittent exposures, we need to know how they fare. Studies will require epidemiological approaches abetted by use of genetic markers and applied to larger populations than have been studied thus far. Elucidation of mechanisms at a fundamental level may help connect these explorations of hypoxia and high altitude with the other facets of the molecular machinery that underpins our existence.

In the prescient words of John Muir, "When we try to pick out anything by itself, we find it hitched to everything else in the universe."

Thomas F. Hornbein
Robert B. Schoene

CONTRIBUTORS

Alberto Arregui, M.D., Ph.D. Professor, Department of Medicine, Cayetano Heredia University, Lima, Peru

Peter Bärtsch, M.D. Professor, Division of Sports Medicine, Department of Internal Medicine, University Hospital, Heidelberg, Germany

George A. Brooks, Ph.D. Professor, Department of Integrative Biology, University of California, Berkeley, California

Gail E. Butterfield, Ph.D., R.D.* Director, Nutritional Studies, Nutritional Research, GRECC, Palo Alto VA Medical Center, Palo Alto, California

Neil S. Cherniack, M.D. Professor, UMDNJ–New Jersey Medical School, Newark, New Jersey

Kevin P. Davy, Ph.D. Assistant Professor, Department of Health and Exercise Science, Colorado State University, Fort Collins, Colorado

* Deceased.

Jerome A. Dempsey, Ph.D. Professor, The John Rankin Laboratory of Pulmonary Medicine, Department of Preventive Medicine, University of Wisconsin, Madison, Wisconsin

Diana Depla Newcastle Upon Tyne, England

Henry Gautier, M.D. Professor, Atelier de Physiologie Respiratoire, Faculté de Médecine Saint-Antoine, Paris, France

Howard J. Green, Ph.D. Professor, Department of Kinesiology, University of Waterloo, Waterloo, Ontario, Canada

Robert F. Grover Professor Emeritus, University of Colorado, Denver, and Arroyo Grande, California

Peter H. Hackett, M.D., F.A.C.E.P. President, International Society for Mountain Medicine, Ridgway, Colorado

Peter W. Hochachka, O.C., Ph.D., D.Sc., F.R.S.C. Professor, Departments of Zoology, Radiology, and Medicine, University of British Columbia, Vancouver, British Columbia, Canada

Thomas F. Hornbein, M.D. Professor and Chairman Emeritus, Department of Anesthesiology, and Professor of Physiology and Biophysics, University of Washington School of Medicine, Seattle, Washington

Charles S. Houston, M.D. Professor Emeritus, University of Vermont, Burlington, Vermont

Herbert N. Hultgren, M.D.* Professor Emeritus, Department of Medicine, Stanford University School of Medicine, Palo Alto, California

Pamela Parker Jones, Ph.D. Research Assistant Professor, Department of Kinesiology and Applied Physiology, University of Colorado, Boulder, Colorado

Bengt Kayser, M.D., Ph.D. Faculty of Medicine, University of Geneva, Geneva, Switzerland

Sukhamay Lahiri, D.Phil. Professor, Department of Physiology, University of Pennsylvania School of Medicine, Philadelphia, Pennsylvania

---

* Deceased.

Fabiola León-Velarde, D.Sc. Professor, Department of Physiology, Cayetano Heredia University, Lima, Peru

Benjamin D. Levine, M.D. Director, Institute for Exercise and Environmental Medicine and University of Texas Southwestern Medical Center, Dallas, Texas

Richard T. Meehan, M.D., F.A.C.P., F.A.C.R. Department of Medicine, National Jewish Medical Research Center, Denver, Colorado

Joseph Milic-Emili, M.D. Professor Emeritus, Departments of Physiology and Medicine, McGill University, Montreal, Quebec, Canada

Carlos C. Monge Professor Emeritus, Department of Physiology, Cayetano Heredia University, Lima, Peru

Lorna G. Moore, Ph.D. Professor, Department of Anthropology, University of Colorado, and Department of Medicine, University of Colorado Health Sciences Center, Denver, Colorado

Susan Niermeyer, M.D. Associate Professor, Department of Pediatrics, University of Colorado Health Sciences Center, Denver, Colorado

Marcus E. Raichle, M.D. Professor, Departments of Radiology and Neurology, Washington University Medical Center, St. Louis, Missouri

John T. Reeves, M.D. Professor, Departments of Medicine and Pediatrics, University of Colorado Health Sciences Center, Denver, Colorado

Jean-Paul Richalet, M.D., D.Sc. Professor, Faculté de Medicine de Bobigny, Université Paris 13, Bobigny, France

Robert Roach, Ph.D. Research Associate Professor, Department of Life Sciences, New Mexico Highlands University, Las Vegas, New Mexico

Clarence F. Sams, Ph.D. Research Biochemist, Life Science Research Laboratories, NASA–Johnson Space Center, Houston, Texas

Robert B. Schoene, M.D. Professor, Department of Medicine, University of Washington School of Medicine, Seattle, Washington

Douglas R. Seals, Ph.D. Professor, Department of Kinesiology and Applied Physiology, University of Colorado, Boulder, Colorado

John W. Severinghaus, M.D., F.R.C.A. Professor Emeritus, Department of Anesthesiology, University of California, San Francisco, California

Curtis A. Smith, Ph.D. Senior Scientist, The John Rankin Laboratory of Pulmonary Medicine, Department of Preventive Medicine, University of Wisconsin, Madison, Wisconsin

Kurt R. Stenmark, M.D. Professor, Department of Pediatrics, University of Colorado Health Sciences Center, Denver, Colorado

John R. Sutton, M.D., F.R.C.P.* Professor, Exercise Research Centre, University of Sydney, Sydney, Australia

Erik R. Swenson, M.D. Associate Professor, Department of Medicine, University of Washington School of Medicine, Seattle, Washington

Peter N. Uchakin, Ph.D. Lead Research Scientist, Division of Basic Medical Sciences, Mercer University School of Medicine, Macon, Georgia

Peter D. Wagner, M.D. Professor, Department of Medicine, University of California, San Diego, La Jolla, California

John B. West, M.D., Ph.D., D.Sc., F.R.C.P., F.R.A.C.P. Professor, Department of Medicine, University of California, San Diego, La Jolla, California

John V. Weil, M.D. Director, CVP Research Lab, and Professor, Department of Medicine, University of Colorado Health Sciences Center, Denver, Colorado

David P. White, M.D. Circadian, Neuroendocrine, and Sleep Disorder Section, Brigham and Women's Hospital, Boston, Massachusetts

Eugene E. Wolfel, M.D. Professor, Division of Cardiology, Department of Medicine, University of Colorado Health Sciences Center, Denver, Colorado

Stacy Zamudio, Ph.D. Assistant Professor, Department of Anthropology, University of Colorado, and Department of Anesthesiology, University of Colorado Health Sciences Center, Denver, Colorado

---

* Deceased.

CONTENTS

TO THREE WITH WHOM WE SHARED THIS ADVENTURE, WHO FINISHED THEIR OWN JOURNEY TOO SOON:

GAIL BUTTERFIELD, scientist, teacher, mentor, and role model, particularly for young women who would follow the path she traveled;

HERB HULTGREN, clinician, scholar, gentle guide with impish curiosity, who taught so many of us the joy of exploration;

JOHN SUTTON, cauldron of ideas, bubbling energy and enthusiasm, who never understood the phrase "It can't be done."

1

The Growth of Knowledge About Air, Breathing, and Circulation as They Relate to High Altitude

CHARLES S. HOUSTON

University of Vermont
Burlington, Vermont

I. Introduction

Many thousands of years ago, humans realized that an uninterrupted supply of good air was necessary for most life. From accidental as well as deliberate experiments it became clear that humans would die if kept from breathing for only a few minutes. Some philosophers asked why this should be; the answer was many centuries in arriving.

Meanwhile, they pondered what happened to air when it was breathed: where did it go, and how did it get there? Soon the ingenious network of lungs and blood vessels was explored and various theories described. Not for many centuries would the wonderful machinery of respiration and circulation begin to be understood.

At first only a few imaginative scholars believed that invisible, impalpable air actually had substance and weight, and it took a thousand years to prove this—and to show that air weighed less the higher one climbed. That portion of air that was necessary for life was described and eventually its chemistry defined. From there it was a shorter journey to discover how this vital stuff was transported and eventually used, and its residue discarded.

While these voyages of discovery were in process, curious adventurers were pushing across the world, on the oceans, the deserts and jungles, and even onto high mountains, where they described the illnesses that befell them. After all that had gone before, it was only a small leap from these observations to the composition of air and the variability of its weight, and bingo! Thin air, lacking some vital stuff, caused mountain sickness. This connection was possible only because some explorers studied what their forebears had learned of respiration and circulation and the physics of the invisible atmosphere they were penetrating.

Clearly, the history of altitude illness is part of the great fabric of knowledge describing the characteristics of air and blood, of transporting and using a necessity of life. To really comprehend altitude sickness, we must understand how we learned about the physiology of life.

II. First Experiences on High Mountains

The story of humans and altitude extends far back into the past (1), as our feelings changed from awe, worship, and fear to curiosity and love. Two thousand years ago Chinese poets and artists rapturously described the beauty, the peace, the majesty of mountains. Not until some of the adventuresome went high did they become aware of sickness on high mountains.

The ancient Greeks also revered high places: Zeus and his consorts lived on Mt. Olympus; occasionally they might descend for a little earthly dalliance. In the Andes shrines were built on high peaks because the souls of the dead lived there and protected humans, animals, and crops. The remains of stone shelters indicate that people lived for a time and did manual work as high as 22,000 feet, where perfectly preserved mummies have been found (2).

In ancient China a high official of the Western Pan Dynasty, Too Kin, wrote Emperor Chung ti (37–32 BC) describing the dangers encountered when crossing the western mountains and deserts to Kashmir (3): ''Next one comes to the Big Headache and Little Headache Mountains as well as the Red Earth and Swelter Hills. They make a man so hot his face turns pale, his head aches, and he begins to vomit. Even the swine react this way.''

A more explicit description of one form of mountain sickness was written by Buddhist missionary Fa-Hsien (334–420 AD) while crossing a pass some 12,000–14,000 feet high (4):

> Fa-Hsien and the two others proceeding southwards, crossed the Little Snowy Mountains. On them the snow lies accumulated both winter and summer. On the north side of the mountains, in the shade, they suddenly encountered a cold wind which made them shiver and unable to speak. Hwuy-Ring could not go any farther. A white froth came from his mouth and he said to Fa-Hsien, ''I cannot live any longer. Do you immediately go away, that we do not all die here''; and with these words he died.

Hwuy-Ring probably died from a combination of hypothermia and what we now recognize as high altitude pulmonary edema. The party was crossing either the Kilik pass or the Ulagh Rabat pass, both higher than 14,000 feet. Three centuries later another Buddhist, Xuan Zang (602–664 AD), described crossing the mighty Tien Shan, Kun Lun, and Karakoram ranges (5):

> The journey is arduous and dangerous and the wind dreary and cold. Travellers are often attacked by fierce dragons so that they should neither wear red garments nor carry gourds with them, nor shout loudly. Even the slightest violation of these rules will invite disaster.

The dragon theme recurs in other countries. In Europe many centuries later, reputable scientists took notarized statements from people who had been attacked by an extraordinary variety of dragons (6). From 1559 to 1654 a distinguished naturalist, Ulisse Aldrovandi, and his heirs published 13 volumes of beautiful drawings of these monsters. Whatever they were, dragons seem to have experienced an abrupt extinction less than 300 years ago!

Other travelers went high enough to experience mountain sickness. Father Alonzo Ovalde, in an account of his travels in the Andes at the end of the sixteenth century, wrote (7):

> When we come to ascend the highest point of the mountain, we feel an aire so piercing and subtile that it is with much difficulty we can breathe, which obliges us to fetch our breath quick and strong and to open our mouths wider than ordinary, applying to them likewise our handkerchiefs to protect our mouth and break the extreme coldness of the air and to make it more proportionable to the temperature which the heart requires, not to be suffocated; this I have experienced every time I have passed this mighty mountain.

His better known contemporary, Father Jose Acosta, described his misery while crossing a pass called Pariacaca (15,750 feet) in the Peruvian Andes (8):

> [W]hen I climbed the Escaleras (de Pariacaca) . . . I felt such a deadly pain I was ready to hurl myself from the horse onto the ground. . . . And almost immediately there followed so much retching and vomiting that I thought I would lose my soul, because after what I ate and the phlegm, there followed bile and more bile both yellow and green so that I brought up blood from the violence I felt in my stomach. . . .
>
> [I]t is not only . . . the Paricaca pass which produces this effect but also . . . the entire mountain range . . . and much more for those who ascend from the sea coast to the mountain than for those who return from the mountain to the plains . . . *

In another translation, Acosta is quoted as writing, like Ovalde: ''I therefore perswade myselfe that the element of the aire there is so subtile and delicate as it is not proportionable with the breathing of man'' (9).

In the the third century before the Christian era, Xenophon and Alexander had experienced the synergistic effects of altitude, cold, hunger, and dehydration, although they did not recognize the effects of altitude alone.

---

* Gilbert explains how corrupted translations have altered this, the original text, to imply that it is the first description of acclimatization.

In the fourteenth and fifteenth centuries, the Mongol Hordes rampaged across Central Asia, high Tibet, and into Europe, crossing deserts and high mountains. Mirza Muhammad Haider, one of the Mongol chieftains, described the hazards of altitude in perceptive detail (10):

> Another peculiarity of Tibet is the dam-giri which the Moghuls call yas and which is common to the whole country, though less prevalent in the region of forts and villages. The symptoms are a feeling of severe sickness (nakhushi) and in every case one's breath so seizes him that he becomes exhausted, just as if he had run up a steep hill with a heavy burden on his back. On account of the oppression it causes, it is difficult to sleep. Should, however, sleep overtake one, the eyes are hardly closed before one is awake with a start caused by the oppression of the lungs and chest. . . . When overcome by this malady the patient becomes senseless, begins to talk nonsense, and sometimes the power of speech is lost, while the palms of the hands and the soles of the feet become swollen. Often, when this last symptom occurs, the patient dies between dawn and breakfast time; at other times he lingers on for several days. . . .
>
> This malady only attacks strangers; the people of Tibet know nothing of it, nor do their doctors know why it attacks strangers. Nobody has ever been able to cure it. The colder the air, the more severe is the form of the malady.

Haider's account is notable in several respects: he described several important signs and symptoms of mountain sicknesses, he noted that it also affects horses, and he clearly recognized that something protected life-long altitude residents—perhaps the first unequivocal mention of acclimatization.

III. Something in Mountain Air Was Responsible

Mountain dwellers attributed the sickness in newcomers to emanations from ores bearing antimony or to noxious fumes from a variety of plants and shrubs. But a few later explorers suggested that changes in the atmosphere were responsible.

One of these was Horace Benedict de Saussure, a broadly educated philosopher-scientist who made the second ascent of Mont Blanc (15, 771 feet) in 1787. During the next few years, impressed by his sensations on the summit, he studied his pulse, respirations, temperature, and symptoms on that and other mountains. Using a torricellian barometer he found that air weighed less at altitude and attributed his symptoms to the decreased density (11):

> The sort of weariness which proceeds from the rarity of the air is absolutely insurmountable; when it is at its height, the most imminent peril will not make you move a step faster. . . . Since the air (on the summit of Mont Blanc) had hardly more than half of its usual density, compensation had to be made for the lack of density by the frequency of inspirations. . . . That is the cause of the fatigue that one experiences at great heights. For while the respiration is accelerating, so also is the circulation.

de Saussure may not have heard of the discovery of oxygen 10 years earlier, but he clearly recognized that the difference in air was at least part of the cause of his symptoms.

Another vivid picture of mountain sickness is that by Friedrich von Tschudi, who described his nasty experience while exploring the Andes in 1838–1842 (12):

> My panting mule slackened his pace, and seemed unwilling to mount a rather steep ascent which we had now arrived at. To relieve him I dismounted, and began walking at a rapid pace. But I soon felt the influence of the rarefied air, and I experienced an oppressive sensation which I had never known before. I stood still for a few moments to recover myself, and then tried to advance. My heart throbbed audibly; my breathing was short and interrupted. A world's weight seemed to lie upon my chest; my lips swelled and burst; the capillaries of my eyes gave way, and blood flowed from them. In a few moments my senses began to leave me. I could neither see, hear, nor feel distinctly. A grey mist floated before my eyes, and I felt myself involved in that struggle between life and death which, a short time before, I fancied I could discern on the face of nature. Had all the riches of earth, or the glories of heaven, awaited me a few hundred feet higher, I could not have stretched out my hand toward them. In that half senseless state I lay stretched on the ground until I felt sufficiently recovered to remount my mule.

Such severe symptoms struck some but spared others and was said to be more common in some areas than in others. It was all very puzzling. Pieces of the puzzle had been lying about for centuries and others were being discovered, but it would be necessary to assemble them—together with parts of a greater puzzle—before the picture could become more clear. Ovalde, Acosta, De Saussure, and Tschudi all came close to the truth when they wrote of ''subtle air'' or its thinness or lack of density. But a key piece was missing.

IV. How Did the Atmosphere Relate to Lungs and Heart?

The greater puzzle was the intricate relationship between air, breathing, blood, and life. What might there be (or not be) on high mountains that so affected travelers? Was there really a poison in mountaintop air? Or was some stuff lacking up there?

The importance of the invisible, impalpable stuff in which we live like fish in water had been recognized by the Sumerians several thousand years before the Christian era. The ancient Chinese taught that proper breathing (*lien ch'i*) changed air into the soul substance or ''vital essence'' (1,13), but they did not know what this was or how it entered the body.

Although dissection of the human body was forbidden during the First Dynasty (3000 BC), the Egyptians gained an approximate idea of human anatomy from embalming the dead. Gradually they identified the heart as the center of an elaborate system of passages or vessels, some of which carried blood, some air. The Egyptian Ebers Papyrus (1700 BC) clearly describes how air enters the nose, flows to the lungs and from there to the heart (13). Perhaps from such a papyrus came the beginning of ''western'' cardiopulmonary physiology.

During a few incredibly creative centuries shortly before Christ, a brilliant group of philosopher-scientists struggled to understand the relationships between four elements of which all life consisted. These were the same elements described in the ancient Hindu Vedas, or sacred scripts: air, earth, water, and fire. Of them, only air could not be detected, but not surprisingly the *idea* of air was commingled with the *reality* of blood: both were essential for life (13).

Anaximenes of Miletus in 455 BC believed that air was formed from water as vapor or mist and was alive. As the ancients had, he suggested that in its invisible form air sustains the soul, which animates life, even though it could not be seen or felt. During a discussion of what filled empty space, Empedocles (490–430 BC) used a clepsydra (a vessel with one small hole in the top and several small holes in the bottom) to show that when the vessel was held under water with the hole in the top closed, water could not enter until the top was opened. Conversely, once filled with the top closed, water would not flow out. This led him to conclude that air had substance; he proposed that it entered the body not only thorough the nose and mouth to the lungs, but also through the skin of the entire body. He also suggested that blood ebbed and flowed through the body, an idea put forward by the Chinese many centuries earlier.

Hippocrates accepted Empedocles' theory of the circulation and believed that the purpose of respiration was to cool the heart with air pumped out of the lungs by the auricles into the right side of the heart via the pulmonary artery and also through the pulmonary vein to the left auricle. Diogenes of Appolonia wrote the first systematic description of the cardiovascular system in which some vessels carried air as well as blood. Plato and his student Aristotle accepted Empedocles' belief that air was carried from the lungs to the heart in both veins and arteries. Erasistratus was greatly influenced by Strato, who argued that all substance included some empty space and from this hypothesized the existence of a vacuum. Erasistratus concluded that when the heart relaxed and enlarged in diastole, air would rush through the pulmonary artery into the left ventricle to prevent a vacuum. There air was somehow changed or combined with blood into a "vital spirit," which was distributed in the arteries (14).

Written early in the first century BC, the Chinese *Huan-ti-nei-ching* medical texts seemed to describe the circulation of the blood, (although various translations differ), and portions of the text suggest that air flowed in the "conduits" mixed together with blood, much like what the Greeks later described as *pneuma*: ". . . the heart regulates all the blood in the body . . . the blood current flows continuously in a circle and never stops. . . . a circle with no beginning and no end. Blood flows six inches with one respiration making a complete circuit of the body about 50 times in 14 hours . . . " (13,15).

Galen (130–199 AD), arguably the best and most influential physician of the Greco-Roman school, whose dictates dominated medicine for a thousand years, wrongly concluded from animal vivisection that air flowed from lungs into the left side of the heart, where it both fuelled and cooled the "innate heat" which was life itself. There it combined with blood and formed the pneuma or "vital spirit" (14). This was pumped through the arteries, which "in the whole body . . . communicate with the veins and exchange air and blood with them through extremely fine invisible openings.'' To complete the circuit, Galen had to postulate tiny pores between the two ventricles, an hypothesis that persisted for 10 centuries (16).

In mid-thirteenth century Ibn al Nafis (1210–1288) more correctly described the anatomy of the pulmonary circulation and reaffirmed that respiration served to cool the heart and to add air to blood and to remove ''fulginous'' vapors. This important work in Arabic was probably not widely known (17,18).

Michael Servetus broke Galen's iron grip on orthodox physiology by his accurate description of the whole circulation. In 1553 he was burned at the stake for his heretical disbelief in the Trinity, and because most copies of his revolutionary book were burned with him, it is unlikely that Harvey saw one. Then Vesalius also challenged Galen, writing: ''We are driven to wonder at the handiwork of the Almighty, by means of which blood from the right into the left ventricle sweats through passages which escape human vision.'' Vesalius too implied that air did not flow in blood vessels, but instead was fixed to blood like the old concept of *pneuma*, but he never fully grasped the importance of respiration (13).

In 1628 William Harvey published *De Motu Cordis*. His genius lay in combining the works of his many predecessors with some simple arithmetic and some careful observations of living humans. From these he derived an accurate description of the motions of the heart and circulation, lacking only knowledge of the pulmonary capillaries. Forty years later Robert Hooke made a beautiful compound microscope like those already in use in Italy; with this he could have seen the capillaries in the lungs and confirmed Harvey's hypothesis that such connections had to exist. But he did not. That step was left for Malpighi around 1689.

V. What Was the Atmosphere Made Of?

The question of what the atmosphere consisted of remained a troublesome mystery. Strato's concept of a vacuum was forgotten until the seventeenth century, when it was revived by Giovanni Baliani, who respectfully suggested it to Galileo. To support Baliani, a young mathematician named Gaspar Berti around 1638–1640 set up an ungainly water-filled lead pipe 34 feet long, against the wall of his house. With this he demonstrated the vacuum beneath the sealed top, incidentally making the first crude barometer and showing that air had weight! Evangelista Torricelli quickly converted this long pipe to a small mercury-filled instrument with which the weight of the atmosphere could be observed (20).

Very soon thereafter in France, Florin Perier took a torricellian barometer up a small mountain (leaving an identical instrument at the base for control observations) and showed that the atmosphere weighed less the higher he went. In Austria Otto von Guericke built the first vacuum pump and dramatically demonstrated how atmospheric pressure held together the two halves of an evacuated sphere. These experiments fitted a major piece into the puzzle (20).

Robert Boyle asked Robert Hooke to make a copy of von Guericke's ''aire pump'' in 1658, and together the two showed that mice died in an evacuated bell jar, not because their exhaled air poisoned them, but because animals needed something in air in order to live (21). But what was it? Once more their predecessors seemed to have known!

In 756 AD a Chinese scientist Nao Hoa had written that the *yin* in air was not pure, but could be purified by heating potassium nitrate (which suggests that his *yin* was oxygen) (13). Leonardo Da Vinci recognized that air consisted of at least two parts and that air in which fire would not burn would not support life either (1). Centuries earlier Pliny the Elder had written in his *Natural History* that Roman well-diggers would lower a lighted lamp into the well, deciding that if the lamp went out the air was dangerous for them.

Alchemists in the sixteenth and seventeenth centuries released oxygen by heating mercuric oxide or potassium nitrate, or, as Boyle did in 1673, by heating lead oxide, almost anticipating the ''discovery'' of oxygen a century later (1,10). But the alchemists were trying to make gold from dross and failed to appreciate the importance of the gas they had liberated! Although they saw that this gas supported fire and was necessary for life, few asked why.

There were false leads. A Danish chemist, Georg Stahl, around 1690 proposed that all combustible materials contained an impalpable gas which he called phlogiston, or ''fire substance.'' To explain why some materials gained rather than lost weight by burning, Stahl's followers proposed that phlogiston had a property the opposite of weight—levity—which made the substance heavier when it escaped. This seems rather absurd today, but it captured the minds of many scientists for a century.

Sixty years earlier, Jean Rey had interpreted similar observations differently, and he would prove to be right. Puzzled by finding that some matter (he used tin) heated in air gained weight, he wrote: ''This increase in weight comes from air which in the vessel has been rendered denser, heavier, and in some measure adhesive . . . (to the metal) (19). About the same time Ole Borch (aka Borrichius) (1626–1690) joined those who had isolated oxygen, but he too has been forgotten. Finally, between 1770 and 1773 a Swedish pharmacist Carl Scheele heated silver carbonate and obtained a special gas: ''Since this air is necessarily required for the origination of fire, and makes up about one third of our common air, I shall call it for the sake of shortness ''Fire Air'' (18). Scheele noted that this new gas supported life as well as combustion. On September 30, 1774, he wrote Antoine Laurent Lavoisier in Paris thanking him for a book and added a description of his experiment, asking Lavoisier to repeat it so that he ''Will see how much air is produced by this reduction and whether a lighted candle can carry on its flame, and animals live in it'' (18).

A British clergyman, Joseph Priestley, who had been studying a gas formed from fermentation, visited Lavoisier the following month and the two may have discussed Scheele's letter. Lavoisier had been looking at oxidation of inorganic substances, perhaps interested in phlogiston, and in 1772 he had sent a letter to the

French Academy describing this work but asking that the letter be sealed until he was ready to publish. Both men seemed to sense that something important was close to hand. After their meeting both hurried to isolate the new gas.

On Saturday, November 19, 1774, Priestley set up an experiment, but on Sunday he was occupied at church and not until Monday did he actually produce oxygen and take physiology a giant leap forward (22):

> In this air as I had expected, a candle burned with a valid flame. In this air a mouse lived for a full half hour. My reader will not wonder, that, having ascertained the superior goodness of dephlogisticated air by living mice . . . I should have the curiosity to taste it myself. . . . Who can tell but that, in time, this pure air may become a fashionable article in luxury.

On April 26, 1775 Lavoisier read his paper to the French Academy, 5 weeks after Priestley had read his own to the Royal Society. The existence of oxygen had been proven. Debating who was first to do so is irrelevant. There is ample honor for both, but their unsung predecessors should not be forgotten.

One more piece of the puzzle of air, blood, fire, and life was now on the table, but a few more pieces were needed before the relationship of oxygen to altitude sickness would be understood.

VI. Why Breathe—and How?

Long ago the Chinese recognized that an uninterrupted supply of "good" air was necessary to support most life. During the age of the Pharoahs, Egyptian physicians described the respiratory tract through which air entered the chest. The Greco-Roman philosopher-scientists, following these observations, believed that the purpose of breathing was to cool the "innate heat" of the heart; they developed various theories about how air entered the blood to form "pneuma," which could be found in both arteries and veins.

Galen's writings about the mechanics of breathing are sometimes opaque, but basically he believed that air was drawn into the lungs by expansion of the chest, supporting this concept by showing that the lungs collapsed and motion of the chest ceased when the spinal cord was severed (14,17). For the next 1500 years his theory was disputed by those who believed that the lungs expanded actively, sucking in air, thus expanding the thorax.

After Gaspar Berti first demonstrated a vacuum in 1640, Torricelli had made the first barometer, proving that the atmosphere had weight. In turn, this revived Galen's theory of respiratory motion: during inspiration it was the thorax that actively expanded and air was pushed into the lungs by the pressure of the atmosphere. When the thoracic muscles and diaphragm relaxed, the elastic thorax returned to its relaxed position, and air was pressed out. Franciscus Sylvius de la Boe wrote in 1660: "The lungs do not move naturally of their own motion but they follow the motion of the thorax and the diaphragm. . . . The lungs are not expanded because they are filled with air, they are filled with air because they are expanded'' (17).

Ten years later Robert Boyle, Richard Lower, and John Mayow agreed, showing that the chest cavity was enlarged by stimulation of the diaphragm and intercostal muscles. Mayow repeated the work of Vesalius a century earlier and found that the lungs did not expand when the chest was opened; he wrote: ''From this we conclude that the lungs are distended by air rushing in and that they do not expand of themselves.'' But even such work did not persuade the skeptics, and these fundamentals were not fully accepted for another hundred years!

But what stimulated the chest to expand and contract? Galen had shown that control of chest and diaphragm motion came through the spinal cord. After his hypothesis of breathing had been resuscitated, it seemed that the stimulus came from the brain itself. Between 1760 and 1837 a number of workers localized the principal control of breathing in a special respiratory center or centers in the medulla of the brain. This control center was responsive to carbon dioxide and later shown to be equally responsive to acidifaction of blood. But the theory did not explain the hyperventilation due to hypoxia that decreased arterial carbon dioxide and made the blood alkaline.

Two decades of brilliant experimentation with isolated dog heads perfused with normal blood, and with cross transfusions, made it clear that oxygen also controlled respiration. Gesell, Heymans, and Winterstein showed that this was accomplished by a small collection of cells (the carotid body), richly provided with capillaries direct from the caroid artery in each side of the neck. By 1930 it was clear that respiration was under dual control: by carbon dioxide in the respiratory center and by oxygen in the carotid bodies. This became, incidentally, a nice explanation for Cheyne-Stokes breathing with its fluctuating levels of each gas! Bohr, Henderson, and Hasselbach put the icing on the cake by showing that changes in carbon dioxide significantly altered the affinity of hemoglobin for oxygen.

VII. How Does Oxygen Pass from Lungs into Blood?

Once the nature of the atmosphere had been established and the laws that govern gases formulated, oxygen was proven essential for life. The next challenge was to determine how oxygen passed from the atmosphere, through the alveolar structure of the lungs, and into blood. Many had been trying to do this for a long time. Galen and his predecessors thought air somehow entered the heart from the lungs and mixed with blood to form ''pneuma,'' which penetrated throughout the body.

Giovanni Borelli, inspired by Galileo in 1681, applied his imagination and talent as Professor of Mathematics at Pisa to physiology, and wrote perceptively that ''air taken in by breathing is the chief cause of life in animals,'' although of course the ''vital spirit'' or oxygen was still to be isolated. Borelli made the new and important observation that air dissolved in water could pass through certain membranes, and thus pioneered the principles of diffusion.

Soon carbon dioxide would be identified. Jean Baptiste Van Helmont (1577–1644) collected the gas that pooled in the bottom of a famous cave in Italy where dogs perished but their taller masters survived. He showed that this was the same as the gas formed by adding acid to limestone that would extinguish fire and would not support life (17). But it was still too early to relate this to combustion.

A century later Joseph Black found that what he christened ''fixed aire'' was formed by burning charcoal. Furthermore, in an ingenious experiment, he arranged for the air exhaled by 1500 people during 10 hours inside a church at a religious gathering to pass through a ceiling vent and over rags saturated with limewater. By weighing the calcium carbonates thus formed, he showed that the expired gases contained ''fixed aire'' or carbon dioxide. Thus, even before oxygen was isolated, the gas exhaled during respiration, presumably generated by the bodily functions, had been identified (17,23). Equally important was the growing recognition that something in air was essential to life.

Later, after Borelli had demonstrated that air could pass through a membrane and after oxygen had been proven essential to life, a new question arose: If carbon dioxide was carried in the blood and oxygen inspired into the lung, could the two gases pass the lung membrane in opposite directions? John Dalton, after much study, provided the answer in 1808: ''When a vessel contains a mixture of such elastic fluids (gases), each acts independently on the vessel . . . just as if the other were absent'' (24). This was the crucial law of partial pressures, which said that the passage of a gas or gases through a membrane was dependent (among other things) on the difference in the pressure of each gas on the opposite sides of the membrane. It seemed probable that Borelli's observation should apply to the thin membranes of the alveolar walls: oxygen could pass into blood at the same time carbon dioxide was passing out of it.

In 1892 Christian Bohr confirmed earlier work by Baptiste Biot in 1807 (and later extended by Moreau) proving that the swim bladders of fish contained a very high concentration of oxygen; this appeared to violate Dalton's law and could be explained only by active oxygen secretion (25).

Haldane and Lorrain Smith, studying how oxygen moved from the lung into blood, visited Bohr in 1894 and were intrigued by his suggestion that similar secretion might occur through the mammalian alveolar walls. To test this hypothesis, Haldane and Smith measured the oxygen pressure in alveolar air, calculated the arterial oxygen pressure after breathing a low concentration of carbon monoxide for 20 minutes, and found the calculated alveolar oxygen pressure always lower than the arterial (26).

Bohr's students Auguste and Marie Krogh devised a more accurate method for estimating arterial blood oxygen by a single breath of carbon monoxide. With this they were unable to confirm their professor's findings. Reluctantly they published a series of brilliant papers in 1910 and started a famous controversy by their unequivocal statement: ''The passage of (oxygen) and the elimination of carbon dioxide in the lungs takes place by diffusion and by diffusion alone'' (27).

Haldane and Smith's experiments led them to dispute the diffusion theory. In

1910 at a meeting in Vienna, Haldane met young Yandell Henderson, who suggested that the summit of Pikes Peak in Colorado would be an ideal location to detect oxygen secretion if it took place. The following year Haldane, Henderson, Douglas, and Schneider went to Pikes Peak to study acclimatization, including measurement of oxygen pressure in alveolar air and using an improved carbon monoxide method to estimate oxygen in arterial blood at rest and after exertion.

They found the calculated arterial oxygen pressure always higher than the alveolar, which convinced Haldane that active oxygen secretion was a specific function of the lung, which enabled humans to tolerate hypoxia. He proposed that differences between individuals in their susceptibility to mountain sickness were due to different degrees of secretion. His later studies did not shake this belief, even when confronted with contrary evidence collected by his former colleague, Joseph Barcroft (26).

Barcroft had begun his scientific career by studying the metabolism of salivary glands, and in 1901, together with Haldane, he developed a method for measuring gases in small amounts of liquid. He found that the partial pressure of oxygen in saliva was higher than in the capillary blood in the salivary gland. At first he could explain this observation only by active oxygen secretion! Several years later, however, his study of factors affecting the shape of hemoglobin dissociation curves and thus the oxygen content of blood led him to reexamine his earlier data and to question oxygen secretion in the salivary gland circulation.

Barcroft was called up in World War I to treat pulmonary edema due to chlorine gas, but soon after the Armistice he returned to studies of other causes of hypoxia—like high altitude. Renouncing the attractive secretion theory, he challenged Haldane's data.

To do so in 1920 he spent 10 days in a sealed glass room where the oxygen was gradually decreased to the partial pressure equivalent to 18,000 feet. Using a method developed by Stadie a few years earlier, Barcroft had a large cannula tied into his radial artery (which, he wrote matter of factly, "of course had to be sacrificed"). Through this cannula he drew arterial blood and was able to measure oxygen directly. His data showed that his arterial oxygen pressure was lower than the alveolar (28).

Haldane countered that Barcroft had studied only himself, and when he was sick at that. Haldane repeated and corrected his own studies and argued that although secretion might not occur during rest or at normal atmospheric pressure, it was one of the means by which humans adjusted or acclimatized to hypoxia. He supported his belief with clinical observations of those who had been sick or well on Pikes Peak, but again, he did not measure arterial oxygen directly (29).

The argument raged, of course in gentlemanly terms. Barcroft and a strong team went to Cerro de Pasco in Peru (14,200 feet) and repeated the alveolar-arterial studies along with much other work. Once again the alveolar air always contained a higher oxygen pressure than the arterial blood, even in the well-adapted natives. Haldane was wrong. This ended the oxygen secretion theory of acclimatization.

VIII. How Is Oxygen Carried in Blood?

How, once diffused from the lungs into the blood, is oxygen transported to the tissues that are dependent on an uninterrupted supply?

Twenty-one hundred years ago Erasistratus suggested that air somehow combined with blood, and in 1668, long before oxygen was isolated (30), Lower wrote that something in air ''penetrated'' dark venous blood changing it to the bright scarlet of that in the arteries (19):

> . . . in the open chest, so long as the lungs were inflated with a bellows, the blood leaving was bright red; when the bellows stopped the blood became dark. . . . This red color is entirely due to the penetration of particles of air into the blood. . . . May we not concede an inward passage of this foodstuff into the blood through similar tiny pores (in the lungs)?

Robert Hooke added experiments that showed that as long as the lungs received fresh air, whether they moved or not, the blood going through them became red. Once again we see the Greco-Roman concept of ''pneuma'' as a combination of air and blood.

In 1910, two millenia after Erasistratus, Christian Bohr described the S-shaped hemoglobin dissociation curve we know today, a more precise map of the curve defined by Bert in 1872. Bohr and others also measured the effect of carbon dioxide and changes in acidity on this dissociation curve and therefore on oxygen transport. They used these data to explain how entry of carbon dioxide into blood in the tissue capillaries ''helped'' the offloading of oxygen, and how the loss of carbon dioxide into alveolar air enhanced the loading of oxygen in the lungs (17). Such transfer of gases depended on the motions of the chest—the process of breathing.

IX. What Was the Relationship of Oxygen to Mountain Sickness?

Twenty years after Priestley's paper, Thomas Beddoes showed his awareness of the significance of oxygen in mountain sickness when he wrote (31):

> Now in ascending these rugged heights the muscular exertion must expend a great deal of oxygene which the rarefied atmosphere will supply but scantily. . . . The experiments of Mr. Saussure, Pini, and Reboul, concur in shewing that, independent of its rarefaction, the atmosphere of very elevated mountains contains a far smaller proportion of oxygene than that of lower regions, especially than that of the high vallies of the Alps.

In 1802 Alexander von Humboldt climbed to 18,800 feet on Chimborazo in the Andes and, like Beddoes, suggested that lack of oxygen was responsible for his symptoms (32):

> [W]e all became by degrees much distressed; a constant desire to vomit, together with vertigo, were the most prominent symptoms, and proved far more trying than the difficulty of breathing which we likewise suffered from. The distress, the weakness, and the desire to vomit certainly came as much from the lack of oxygen in these regions as from the rarity of the air.

There were other theories: Stanhope Speer picked up Haller's ''hydraulic theory'' but ignored oxygen in his book written in 1853, although some of his concepts were correct (33):

> These symptoms (of mountain sickness) may be referred to a threefold source, viz, a gradually increasing congestion of the deeper portions of the circulatory apparatus, increased venosity of the blood, and loss of equilibrium between the pressure of the external air and that of the gases existing within the intestines. . . . The causes of mountain sickness are themselves the result of a change from a given atmospheric pressure and temperature, to one in which both are greatly and suddenly diminished.

Conrad Meyer-Ahrens, a leading Zurich physician, in 1854 came closer to the truth, though he had an odd explanation for the weakness experienced at altitude: ''Mountain sickness is due to (a) decrease in the absolute quantity of oxygen, (b) rapidity of evaporation (c) intensity of light, (d) expansion of intestinal gases and (e) weakening of the coxo-femoral articulation'' (34).

Studies of altitude illness languished for more than 50 years after de Saussure, during which time 57 ascents of Mont Blanc were made, 2 by women. Then, starting in 1852, a London physician, Albert Smith, gave 2000 illustrated lectures describing his adventures on Mont Blanc so vividly that hundreds took up mountaineering and began the Golden Age of alpine climbing. More than half of the budding alpinists had symptoms, piquing medical curiosity and prompting serious studies of mountain physiology.

X. Finally—The Cause(s) of Mountain Sickness

Lower had shown that dark venous blood changed color when exposed to air, but the relationship between this observation and the symptoms of mountain sickness was not recognized for a long time. Then came a man who, as Harvey had done, was able to put together the work of his predecessors to make a coherent whole.

After studying law, Paul Bert took his degree in medicine and studied respiration under Claude Bernard. He met a Paris physician, Denis Jourdanet, who had traveled among the mountains of Mexico, and the two became friends and fellow-workers. Jourdanet had built decompression chambers for therapeutic purposes and had formulated the hypothesis that blood contained less oxygen on high mountains because the atmospheric pressure was lower; he called this ''barometric anoxemia'' (35). Jourdanet provided the means for the expensive equipment and supported Bert in a series of studies to examine the puzzle of mountain sickness.

Bert used his chamber to decompress animals to different pressures with different oxygen percentages in the chamber and found that there was no consistency in either the *pressure* or the *percentage of oxygen* at which they died. When he kept the percentage of oxygen constant but varied the chamber pressure, animals died more rapidly the lower the pressure. Then he took the imaginative step of multiplying the barometric pressure in the chamber by the percentage of oxygen, and this number gave consistent results (35). He had applied Dalton's law of partial pressure to show that it was the *partial pressure of oxygen* rather than barometric pressure or percentage alone that was decisive—a major advance in physiology and another piece of the puzzle put in place.

As a clincher he had himself rapidly "taken up" in his chamber to 18,000 feet and experienced symptoms we would today call acute hypoxia; then he breathed oxygen and obtained immediate relief. Finally he breathed oxygen while being "taken up" and developed no symptoms. From repeated tests he concluded that mountain sickness was due to lack of oxygen, and that by breathing oxygen even at decreased pressures, he could prevent or alleviate the symptoms.

These careful studies would soon be of great importance, when newly developing warplanes struggled toward heights that would endanger pilots without supplementary oxygen, and would be even more relevant much later, when both oxygen and increased pressure would be necessary in very advanced aircraft.

Bert also collected anecdotal experiences at altitude from hundreds of travelers and explorers who had been on high mountains all over the world, as Boyle had done in a smaller way. He used instruments which he and Jourdanet designed to measure the amount of oxygen combined with hemoglobin under different conditions. He extracted gases from the blood of animals exposed to decreased pressures and defined a portion of the oxy-hemoglobin dissociation curve we know today. He encouraged his young colleagues to go to high altitude in balloons, until two of three died when they went too high after their meager supply of oxygen was exhausted; this ended Bert's interest in altitude studies.

His *La Pression Barometrique*, an extraordinary encyclopedia of all that was then known or believed about barometric pressure, high altitude, and mountain sickness, was published in 1878 (36). His altitude studies have overshadowed his pioneering work with hemoglobin and oxygen transport and with increased barometric pressures as in diving. He and Jourdanet had shown beyond a reasonable doubt that lack of oxygen was the main, if not the only cause of the symptoms experienced on high mountains. But one of their contemporaries disagreed (37).

Angelo Mosso was already an experienced mountaineer and physiologist when he read the works of German poet-physiologist Albrecht von Haller, who argued in 1761 that the physical effect of decreased atmospheric pressure on the blood vessels was solely responsible for altitude sickness. This idea intrigued Mosso, and he began to study altitude in a decompression chamber and later in a laboratory built for him by Queen Margherita of Italy on one of the summits of Monte Rosa (15,300 feet). He concluded that the decrease in carbon dioxide in the blood that resulted from overbreathing at altitude, rather than lack of oxygen, caused mountain sickness, and he coined the word "acapnia" for this phenomenon (37).

Tom Longstaff, one of the great Himalayan pioneers and a practicing physician seemed to disagree with Mosso in 1906 in his doctoral thesis (38):

> With the reduced atmospheric pressure there is necessarily a reduction in the amount of CO_2 per unit of volume of air breathed. The increased pulmonary ventilation already referred to also tends to lower the alveolar CO_2 tension. Nevertheless it must not be forgotten that the main source of CO_2 in the alveolar air is not the atmosphere, and Mosso's theory that the diminished supply of CO_2 at altitude is the cause of mountain sickness will not bear close inspection.

In the last three lines of the preceding quotation, Longstaff misquoted Mosso, who believed the loss of CO_2 by hyperventilation, not the lower CO_2 in the atmosphere, caused mountain sickness. Mosso was mistaken, but his book *Life of Man on the High Alps* is rich in laboratory experiments and mountain experiences.

Other scientists were also turning to high summits. Joseph Vallot was doing botany and geology when he was attracted to Mont Blanc. First he sought to determine whether or not mountain sickness was caused by a decrease in body temperature, at that time a popular theory. In 1887 he spent three nights on the summit of Mont Blanc in a tent and was so impressed that he decided to build a laboratory high on the mountain.

A year later he returned to Chamonix with 19 cases containing three duplicate sets of instruments, one to be read in the valley, one halfway up the mountain, and one in a small building a little below the summit. Among the many studies he made was the change in exercise capacity experienced at altitude by squirrels! He taught them to run on a caged wheel and counted the number of turns in Chamonix, during a week in his high laboratory, and again in Chamonix; squirrels lost half of their exercise capacity at altitude and did not fully regain it after descent (39). After many other experiments and observations his building was abandoned and a new one erected in 1898; it has been enlarged and is an active research laboratory today.

Soon after Vallot, a distinguished astronomer, Jules Janssen built his own observatory near the summit in August 1891. Believing that exertion was a major cause of the mountain sickness he suffered, he later had himself pulled up the mountain on a sled and felt none of the unpleasant symptoms that affected the 12 men who dragged him (40).

Dr. Egli-Sinclair, the leader of Janssen's party, had to depart, and a young Chamonix doctor, Etienne Henri Jacottet, took his place. Jacottet climbed up rapidly, went on to the summit, and on returning to the laboratory he became ill. As he was dying he wrote a farewell letter to his brother describing his harrowing symptoms (37). An autopsy by Dr. Wizard read in part (42): "Poumon: couleur violet, gonfle, fonce, congestion bilaterale, oedeme considerable muqueuse bronchique injectee fortement. Le liquide de la coupe est ecumeneux. Congestion egale partout. Foie, rate, reins normaux. Pas d'oedeme des jambes'' (42). This is the first autopsied case of high altitude pulmonary edema.

Shortly after his expedition to K2 (second highest mountain in the world) in 1902, mountaineer Guy Knowles wrote: ''Pfannl is really seriously ill. . . . What he has is oedema of one lung which is, of course, very serious'' (43). The leader of the party, Aleister Crowley, added: ''Pfannl has oedema of the lungs and his brain is gone.''

In 1893 Hugo Kronecker, consultant for the railroad proposed to reach the 13,645 foot Jungfrau summit, studied the symptoms of mountain sickness not only on mountaineering climbs, but also in a small decompression chamber (a ''bell''). After studying many cases, he was satisfied that ''mountain sickness does exist'' and noted that even slight exertion exacerbated symptoms. To decide whether passengers on the train would be made sick by the altitude, he arranged for seven men and women to be carried in chairs up to the pass (11,385 feet) to which the railroad was eventually built. Like Janssen, they did not have symptoms. Kronecker doubted Bert's evidence that hypoxia was the cause of mountain sickness. Instead, perhaps influenced by Haller's mistaken hydraulic theory, Kronecker believed that diminished atmospheric pressure was the primary cause. He argued that decreased atmospheric pressure dilated the pulmonary blood vessels, straining the heart and causing edema of the lungs. His description of six cases suggests that these were in fact high altitude pulmonary edema (44).

In 1870 Dr. Nathan Zuntz left a country practice for teaching and research in physiology. He made scientific expeditions to Monte Rosa and Pikes Peak and invited Barcroft, Douglas, and Durig to join an expedition to the peak of Teneriffe (12,200 ft) in 1909, specifically to look at Mosso's acapnia theory. Using equipment he had developed, he studied exercise metabolism at sea level and at altitude in a balloon. Before World War I Zuntz was a pioneer in aviation medicine, going himself to 17,500 feet, breathing oxygen from an apparatus he designed, and planning the first pressurized cabin for use in balloon flights above 35,000 feet. With colleague Adolph Loewy and assistants Mueller and Caspari, he wrote another of the classics in altitude physiology: *Hohenklima und Bergwanderung* (45).

These were the beginnings of a great surge of interest in altitude physiology. Soon leading scientists were making expeditions to mountains all over the world.

Half a century earlier, Hennessey, a member of the survey of India, had calculated that Mount Everest was the highest mountain on earth, and in a few decades mountaineers became interested (41). By the end of the century, an attempt to climb the mountain seemed possible as Tibet became more accessible. The summit even seemed attainable as climbers pushed their ceiling higher and higher. In 1892 a distinguished surgeon-mountaineer, Clinton Dent, predicted the summit would be reached (47):

> Selected men will have to work for a year or more with the one definite object before them. . . . We may agree with Mr Whymper that the effects on respiration will impose

> limitations on the range of man, but it does not seem conceivable that this limitation is below the level of the highest point on the earth's crust. . . . It is a tremendous undertaking but a magnificent possibility.

Sixty one years later two men reached the summit, and since then several hundred have done so without supplementary oxygen. But, perhaps influenced by Haller and Meyer-Ahrens, Dent was mistaken about some aspects of altitude illness (47):

> A far more important factor (than abdominal distention) is the effect of diminished pressure on the portion of the spinal marrow which is concerned with the nutrition of the locomotive agents, the lower limbs; greatly increased pressure also produces much the same symptoms. This effect has no relation to the absence of oxygen.

Everest was one thing, the Alps quite another. During the last half of the nineteenth century hundreds of men and a few women swarmed over the Alps and noted that, even on the same climb under similar conditions, some fell victim to mountain sickness, others did not, and the same individual might be sick one day and not another. This could not easily be explained by acclimatization, and Longstaff, who had climbed many high Himalayan peaks, was not impressed: ''As to the general benefit to mountaineers of 'acclimatization' produced by a prolonged residence at an altitude of half an atmosphere or less, I am altogether sceptical, not withstanding that its efficacy has been frequently urged by many writers'' (38). Instead Longstaff insisted that training and physical fitness were protective, claiming that this was why he and others had made some high altitude climbs without symptoms. Today, despite many careful studies, we have neither proved nor disproved the fitness argument, and the ultimate secrets of acclimatization elude us.

Joseph Barcroft, after his Peruvian work with local residents, concluded (28):

> The acclimatized man is not the man who has attained to bodily and mental powers as great in Cerro de Pasco as he would have in Cambridge (whether that town be situated in Massachusetts or in England). Such a man does not exist. All dwellers at high altitude are persons of impaired physical and mental powers. The acclimatized man is he who is least impaired.''

The distinguished Peruvian physiologist, Carlos Monge, Sr., violently disagreed, claiming that Barcroft's observations were flawed because he had himself been affected by the altitude. Both were partly right; we know today that even full acclimatization does not restore sea level ability above 12,000–14,000 feet (48).

Following the precept of Claude Bernard that the constancy of the internal environment is the condition of an active life. Barcroft thought acclimatization was the result of changes in many physiological processes. Drawing on this in 1946, Houston and Riley described acclimatization as ''—a series of integrated adaptations which tend to restore the oxygen pressure of the tissues toward normal sea level values despite the lowered PO_2 of the atmosphere'' (49) (note the careful word ''toward'').

The early Himalayan climbers made no physiological observations, and the first serious studies of man's ability to summit Everest ''without adventitious aids'' were those of mountaineer-chemist Alexander Kellas. He himself did no laboratory work, but from others' data calculated that a well-acclimatized and trained climber could reach the summit of Everest. He was puzzled that the height where Acute Mountain Sickness (AMS) becomes a problem differs between individuals and from place to place. He accepted the belief that facing into the wind increases alveolar oxygen by "packing air into the lungs" (50).

Kellas knew that alkalosis resulted from the hyperventilation response to high altitude, but, lacking accurate measurements, he hedged his bet: "At high altitude . . . the quantity of carbon dioxide in the blood is lowered, but the acidity increases (or rather the alkalosis produced by removal of carbon dioxide is diminished correspondingly) and the respiratory center remains adequately stimulated" (51). Kellas died in 1924 en route to Everest with the expedition on which Mallory and Irvine disappeared near the summit and Norton reached 28,000 feet, where the atmospheric oxygen pressure is only a few torr more than on the summit. Not for more than 50 years would Dent's and Kellas's prediction be confirmed when that last thousand feet would be climbed breathing only air.

During the 1924 expedition Major R. W. G. Hingston, doctor to the party, recorded pulse rates, blood pressure, breath-holding time, and red blood cell counts for eight of the climbers up to 21,500 feet, the first such data collected so high. He noted that resting pulse rate and blood pressure in these acclimatized men did not change greatly from sea level, but breath-holding time decreased and hemoglobin content increased (52).

Access to Everest was strictly controlled by the Tibetans, and only a few British parties were allowed, so those who wished to study the effects of altitude turned to other peaks, lower but no less formidable. In the 1930s German mountaineers made several attempts to climb Nanga Parbat, the seventh highest summit in the world. Although their goal was to reach the top, in 1934 Ulrich Luft joined the party as medical officer and collected data up to 19,000 feet, as he did later at the Jungfraujoch laboratory and in Berlin. His definitive thesis, "Acclimatization to Altitude," was published in 1941, but was only recently translated into English and made generally available (53).

Climbing mountains is not the only way to get high in the air: men have always longed to fly like birds. Daedalus and Icarus and Leonardo failed, but others persisted. Laurenco de Guzmao demonstrated a small hot air balloon to the king of Portugal in 1686, and in November 1782 Joseph and Etienne Montgolfier in Paris secretly flew a tethered hot air balloon. After a few more tethered flights, a larger balloon carried a sheep and a rooster over 1000 feet into the sky. In late October 1783 Pilatre de Rozier and the Marquis d'Arlandes entered the wicker basket of a huge balloon and rose above 3000 feet, landing safely a few miles away. The age of flight had begun. More hot air balloon flights, and soon balloons filled with the newly discovered hydrogen, followed (54).

A century later in March 1874 two of Paul Bert's young associates Croce-Spinelli and Sivel flew to 24,000 feet in a hydrogen balloon, breathing oxygen from leather bags on Bert's insistence. Emboldened by this, on the next flight they were joined by Tissandier, and the trio were lifted even higher. But they ran out of oxygen and when the balloon later descended Croce-Spinelli and Sivel were dead, having succumbed to hypoxia (36).

Then came powered flight, and by 1916 aircraft could fly high enough to endanger an aircrew. This prompted urgent studies of how to select pilots who best tolerated altitude (which proved futile), and rudimentary oxygen systems evolved. The research continued after World War I on mountains where laboratory studies could be done.

As World War II loomed it was obvious that very high altitude might be decisive in an air war. Programs to increase tolerance for hypoxia by residence in the mountains proved effective but impractical. Instead, better oxygen equipment and soon pressurized breathing equipment enabled men to go as high as their aircraft. Since then oxygen equipment has been rendered moot by pressurization of aircraft.

Under the pressure of World War II many studies of altitude physiology opened even more doors to new fields of interest not only to climbers and aviators but to clinicians as well. New equipment, new concepts, and new techniques now build on the work of previous centuries. Although the original puzzle has been assembled, clearly many new pieces must be added to the periphery—and quite probably to the center as well!

Not surprisingly, the large decompression chambers, mostly idle as the war ended, were a tempting opportunity to look at how well man might tolerate the thin air on the summit of Mount Everest, if he could get there! An ambitious study, suitably called Operation Everest, was sponsored by the U.S. Navy in 1946. During a 35-day stay in a large decompression chamber, four men were slowly taken to a simulated altitude of 25,000 feet from which, in one day, they made a "dash for the summit." Two of the four required oxygen, but the other two were able to rest quietly for 20 minutes and to ride a bicycle briefly before descending. They were not fully acclimatized by the relatively short ascent, but the data provided new insights into acclimatization. Most welcome to mountaineers, Operation Everest proved that men could survive breathing only the surrounding air on the highest point on earth (49).

Today, as described in this book by others, sophisticated instrumentation makes possible a fantastic array of studies, even on the actual summit of Everest.

XI. Summary

In this chapter I have touched on only a few of the advances made in the twentieth century; there have been far too many for such a summary survey. Many of the most important have been described elsewhere (51,57,58).

In the last part of our century it has become clear that we have as much to learn today as our predecessors had in centuries past. Only the parameters are different. They were looking at gross anatomy and physiology; we use refined instruments to study minutiae—cells, enzymes, neurotransmitters, and molecular relationships as yet unknown. Although we have begun to understand a great deal about the acquisition and utilization of essential oxygen, the ultimate mechanisms elude us still—and may for many years.

Acknowledgment

I am very grateful to Dr. Ralph Kellogg for many suggestions—and for saving me from embarrassing mistakes.

References

1. Gilbert DL. Oxygen and Living Processes. An Interdisciplinary Approach. New York: Springer-Verlag, 1981.
2. MacNeish RS. Early man in the Andes. Sci Am 19XX; 38–46.
3. Gilbert DL. The first documented report of mountain sickness: the China or Headache Mountain story. Resp Physiol 1983; 52:315–326.
4. Hsien F. The Travels of Fa-Hsien (399–414 AD) or a Record of the Buddhist Kingdoms. Westport, CT: Greenwood, 1981.
5. Zheng Z, Zhenkai L. Footprints on the Peaks: Mountaineering in China. Seattle: Cloudcap, 1995.
6. Gribble F. The Early Mountaineers. London: T Fisher Unwin, 1899.
7. Pinkerton J. A General Collection of the Best and Most Interesting Voyages and Travels in All Parts of the World. London: Longman, 1808.
8. Gilbert DL. The first documented description of mountain sickness; the Andean or Pariacaca story. Respir Physiol 1983; 52:327–347.
9. Personal communication from Kellogg citing Grimstone.
10. Haider MM, Dughlat. The Tarikh-i-Rashid: History of the Mongols in Central Asia. Lahore: Book Traders, 1894:412–413.
11. de Saussure HB. In: Pinkerton J. A General Collection of the Best and Most interesting Voyages and Travels in all Parts of the World. London: Longman, 1808.
12. Tschudi JJ. Travels in Peru. London: Bogue, 1847.
13. Gordon EL. Medicine Throughout Antiquity. Philadelphia, Davis: 1949.
14. Wilson, LG. Erasistratus, Galen and the pneuma. Bull Hist Med 1959; 33:293–314.
15. Unschuld PU. Medicine in China, a History of Ideas. Berkeley: University of California, 1985.
16. Fleming D. Galen on the motion of the blood in the heart and lungs. Isis 1955; 46:14–21.
17. Perkins JF. Historical development of respiratory physiology. In: Fenn WO, Rahn R, eds. Handbook of Physiology. Washington, DC: American Physiological Society, 1964.
18. Houston CS. Going Higher: Oxygen, Man and Mountains. Seattle: Mountaineers, 1998.
19. Fulton JF. Selected Readings in the History of Physiology. Springfield, IL: Charles C Thomas, 1966.
20. Middleton WEK. The History of the Barometer. Baltimore: Johns Hopkins, 1964.
21. Hooke R. History of the Royal Society of London. 28:539–540, 1667.

22. Priestley J. The Discovery of Oxygen. Edinburgh: Livingston, 1961.
23. Guerlac H. Joseph Black and fixed air. A bicentenary retrospective with some new or little known material. Isis 1957; 48:124–151.
24. Dalton J. Experimental enquiry into the proportion of the several gases or elastic fluids, constituting the atmosphere. Manchester: Mem Lit Phil Soc, 1805.
25. Bohr C. The influence of section of the vagus nerve on the disengagement of gases in the swim-bladder of fishes. J Physiol 1894; 15:494–500.
26. Haldane JS, Priestley JG. Oxygen secretion in the lungs. In: West JB, ed. High Altitude Physiology. Stroudsburg: Hutchinson-Ross, 1981.
27. Krogh A. On the mechanism of gas exchange in the lungs. Acta Scand 1910; 23:248–278.
28. Barcroft J. The Respiratory Function of the Blood; Pt 1: Lessons from High Altitude. London: Cambridge University Press, 1925.
29. West JB, Milledge J. The Haldane-Barcroft debate: Resolved that the lungs secrete oxygen. In: Sutton JR, Houston CS, Coates G, eds. Hypoxia and the Brain. Burlington: Queen City Press, 1995.
30. Wilson LG. The problem of the discovery of the pulmonary circulation. J Hist Med 1962; 16:229–244.
31. Beddoes T. Observations on the Nature and Cure of Calculus, Sea Scurvy, Consumption, Cafarrh, and Fever; Together with Conjectures upon Several Other Subjects of Physiology and Pathology. Edinburgh: Alembic Club, 1902.
32. von Humboldt A. Travels to the equinoctial regions of the new continent during the years 1799–1804. In: Bert P. Barometric Pressure. Bethesda, MD: Undersea Medical Society, 1978.
33. Speer ST. On the physiological basis of mountain sickness. London: Richards, 1853.
34. Meyer-Ahrens C. Die Bergkrankheit. Cited in: Bert P. Barometric Pressure. Hitchcock MA, Hitchcock FA, trans. Columbus: College Book Co, 1943.
35. Kellogg RH. La Pression Barometrique: Paul Bert's hypoxia theory and its critics. Respir Physiol 1978; 34:1–28.
36. Bert P. Barometric Pressure. Hitchcock MA, Hitchcock FA, trans Columbus: College Book Co, 1943.
37. Mosso A. Life of Man in the High Alps. London: T Fisher Unwin, 1898.
38. Longstaff TG. Mountain Sickness and Its Probable Cause. London: Spottiswoode, 1906.
39. Vallot J, Bayeux, R. Exeriments faites au Mont Blanc, en 1991–93, sur l'activite spontannee aux tres hautes altitudess. Notes presented at Academie des Science, Francais, 1914.
40. Whymper E. A Guide to Chamonix and the Range of Mont Blanc. London: John Murray, 1896.
41. Egli-Sinclair. Sur le Mal de Montagne. Ann Obs Vallot 1893; 1:110–130.
42. Foray J. Personal communication.
43. Knowles G. Unpublished typescript 1902, courtesy Nick Clinch.
44. Kronecker H. Mountain-sickness. Med Mag 1895; 4:651–666.
45. Gunga H-C, Kirsch KA. Nathan Zuntz (1847–1920)—a German pioneer in altitude physiology and aviation medicine, part 1: Biography. Aviat Space Environ Med 1995; 66:168–171.
46. Ward MP. Mount Everest. Cartographic J 1994; 31:33–34.
47. Dent CT. Can Mount Everest Be Ascended? The Nineteenth Century 1892; 32:604–613.

48. Monge C. Acclimatization in the Andes. Baltimore: Johns Hopkins University, 1948.
49. Houston CS, Riley RL. Respiratory and circulatory changes during acclimatization to high altitude. Am J Physiol 1947; 149:565–588.
50. West JB. Alexander Kellas and the physiological challenge of Mt. Everest. J Appl Physiol 1987; 63:9–11.
51. West JB. High Life: A History of High Altitude Physiology and Medicine. New York: Oxford, 1998.
52. Norton EF. The Fight for Everest 1924. London: Arnold, 1925:243–261.
53. Luft UC. Acclimatization to Altitude. Berlin 1941. Luft FC, trans. Albuquerque: Lovelace Foundation, 1993.
54. Gillespie CC. The Montgolfier Brothers and the Invention of Aviation. Princeton, NJ: Princeton University Press, 1983.
55. Houston CS, Sutton JR, Cymerman A, Reeves JT. Operation Everest II man at extreme altitude. J Appl Physiol 1987; 63:877–882.
56. Kellogg RH. Altitude acclimatization: a historical introduction emphasizing the regulation of breathing. Physiologist 1968; 11:37–57.
57. Hultgren HN. High Altitude Medicine. Stanford, CA: Hultgren Publications, 1997.
58. Houston CS, Going Higher: Oxygen, Man and Mountains. Seattle: Mountaineers, 1998.

2

The Atmosphere

JOHN B. WEST

University of California, San Diego
La Jolla, California

I. Evolution of the Atmosphere

In the beginning God created the heavens and the earth, but only now are the first direct data on the composition of an outer planet atmosphere being transmitted to Earth. It is remarkable that although Earth was formed about 4.5 billion years ago, exploration of the solar system with measurements of the planets' atmospheres has only occurred in the last 45 years, i.e., in 0.000001% of Earth's lifetime.

Earth and its primitive atmosphere were probably formed by the accretion of material from early solar nebulae by gravitational forces. The material is thought to have had the same composition as the sun, i.e., mostly hydrogen and helium, but with small quantities of heavier elements. Oxides and hydrides of the heavier elements condensed into particles and came together to form the planets.

The initial atmosphere mainly consisted of hydrogen and helium, but this atmosphere was lost because of Earth's weak gravitational field. If we consider the masses and distances of the planets and some of their larger satellites from the sun, it transpires that Mercury is too small to retain an atmosphere, and the same is true, or nearly so, of our moon and Mars. However, the outer planets such as Jupiter and Saturn have abundant atmospheres, and the first data from Galileo's atmospheric probe of Jupiter are presently being analyzed in the Jet Propulsion Laboratory in Pasadena, CA. Generally the surface temperature of a planet decreases with increasing distance from the sun, although this is modified by the greenhouse effect of the atmosphere, which is why the surface temperature of Venus exceeds that of Mercury. In addition, Jupiter and Saturn are so massive that they generate and radiate heat. However, only Earth has a surface temperature that allows water to be in the liquid phase, and therefore it is difficult to see how advanced life as we know it could exist elsewhere in the solar system.

The primary atmosphere of hydrogen and helium was replaced by a secondary atmosphere formed by degassing of various constituents of the primitive Earth as a result of the increasing temperature (1). The temperature rose mainly because of the kinetic energy of impacting bodies and radioactive decay. Volcanoes and fumaroles also contributed to this secondary atmosphere, and active volcanoes can still be observed elsewhere in the solar system, for example, on Io, a satellite of Jupiter. Liberation of water of crystallization produced large amounts of water vapor, carbonates decomposed to form carbon dioxide, nitrides and ammonium compounds gave rise to nitrogen and ammonia, radioactive decay released hydrogen, helium, and argon, and sulfur compounds gave rise to hydrogen sulfide and sulfur dioxide.

This secondary atmosphere was anaerobic, but as Miller (2) showed when he heated a similar gas mixture with water and exposed it to electrical discharges, this primordial soup could produce a remarkable range of organic compounds including many amino acids, nucleotides, lipids, and sugars. Such events were not confined to Earth's atmosphere. Radioastronomy has detected over 80 different complex molecular species, including a variety of organic molecules, in space between and around distant stars. Some of Earth's lipids apparently became organized into membranes to encapsulate material as "protocells."

Presumably this was the first stage in the evolution of life on Earth. Although the necessary steps to produce replicating molecules are still poorly understood, some biologists believe that random sequencing of ribonucleic acid (RNA) eventually produced by chance a template for the formation of an enzyme that conferred some biochemical advantage on a "protocell." This enabled it to compete effectively with its neighbors as well as to replicate its RNA. Later the transmission of the genetic code was taken over by deoxyribonucleic acid (DNA) in organisms with the greatest potential for survival.

These primitive forms of life were a far cry from the advanced aerobic life forms that we know today. A dramatic development apparently occurred about 3–3.5 billion years ago when there was the first utilization of visible light as an energy source with the production of oxygen by photosynthesis. Initially the oxygen was a waste product, and it is not certain how rapidly it accumulated in the atmosphere (Fig. 1). There is some evidence that the major change from a reducing to an oxidizing atmosphere occurred about 2 million years ago (3). However, many biologists believe that the oxygen concentration also rose steeply just before the beginning of the abundant fossil records of the Cambrian period, i.e., about 570 million years ago. It is possible that the oxygen concentration peaked in the late Carboniferous

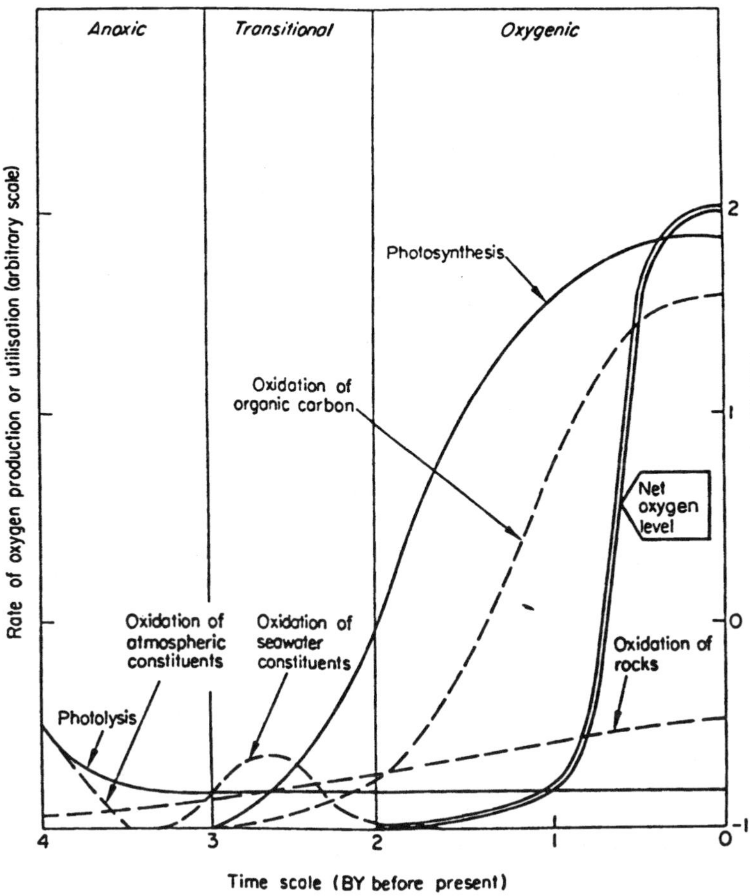

Figure 1 Possible scheme for the sequence of events leading to the oxygen concentration in the present atmosphere. Solid lines indicate oxygen-generating processes and broken lines show oxygen-utilizing processes (left-hand scale). The double line represents the resulting accumulation of oxygen in the atmosphere (right-hand scale). Some authorities believe that the net oxygen level rose earlier than depicted here. BY: billion years. PAL: present atmospheric level. (From Ref. 38.)

period, about 290 million years ago, as a result of production of oxygen by plants. At this time there was a dramatic increase in the evolution of both animals and plants with the development of many amphibia and a large vertebrate invasion of the land. Although some scientists argue that the oxygen concentration may have subsequently declined somewhat, there is other evidence that it has remained nearly constant since the Carboniferous period.

There is evidence that the evolution of life forms has not been a continuous process because of changes in the atmosphere. The extinction of the dinosaurs at the end of Mesozoic era, about 65 million years ago, was probably caused by the impact of an asteroid in the Yucatan, which resulted in an enormous dust cloud that interfered with photosynthesis and the food chain. Apparently an even greater extinction occurred at the end of the Permian era, about 250 million years ago, which decimated marine genera (4).

Two other aspects of the evolution of the atmosphere should be mentioned because of their importance today. Oxygen screens out much ultraviolet radiation, but ozone (O_3) is much more effective. Recently the ozone layer has been thinned in part as a result of the action of man-made contaminants, particularly chlorofluorocarbons, which are used as propellants and refrigerants. These are now being replaced, but a colorful reminder of the danger is seen in Australia where schoolchildren are required to wear legionnaire's hats.

Another contemporary aspect of atmospheric evolution is the apparent warming of Earth as a result of the greenhouse effect caused in part by man-made carbon dioxide emissions. The extent of this human contribution is controversial, but, in any event, carbon dioxide production is an almost inevitable consequence of an increased standard of living in many countries.

II. Altitude and Barometric Pressure

A. History

History is always fun, but in this context it is also important because of recent misunderstandings. Torricelli (5) was probably the first to recognize that the air has weight ("We live submerged at the bottom of an ocean of the element air" he wrote to his friend Michelangelo Ricci), and shortly thereafter Pascal (6) persuaded his brother-in-law, Perier, to take a barometer to the top of the Puy de Dôme (1463 m), thus demonstrating the fall in pressure at increased altitude. A few years later Boyle (7) formulated his well-known law, which states that gas volume and pressure are inversely related (at constant temperature).

In the nineteenth century, many travelers went to high altitude and measured barometric pressure. Typical of their results was a table reproduced in Appendix 1 of Paul Bert's classical *La Pression Barométrique*, published in 1878 (8). Bert acknowledged Jourdanet (9) as the source of much of the data. An interesting feature of Bert's table is that he gives the height and barometric pressure at the summit of Mt. Everest as 8840 m and 248 mmHg, both of which are very nearly correct. Everest had been shown to be the world's highest mountain in 1852.

Zuntz et al. (10) gave the following relationship for determining barometric pressure at any altitude:

$$\log b = \log B - \frac{h}{72(256{\cdot}4 + t)}$$

where h is the altitude difference in meters, t is the mean temperature (°C) of the air column of height h, B is the barometric pressure (mmHg) at the lower altitude, and b is the barometric pressure at the higher altitude. Note that if temperature were constant, log$b$ would be proportional to negative altitude, i.e., the pressure would decrease exponentially as altitude increased.

The expression proposed by Zuntz et al. (which itself was derived from other sources) was extensively used by high altitude physiologists. For example, FitzGerald (11) employed it in her study of alveolar P_{CO_2} and hemoglobin concentrations in residents of the Colorado mountains. Kellas (12) used the same expression to predict barometric pressures in the Himalayan ranges, obtaining the correct value of 251 mmHg for the summit of Mt. Everest, assuming a mean temperature of 0°C. The Zuntz et al. formula was also used by Haldane and Priestley (13).

All would have been well had it not been for the introduction of the Standard Atmosphere by the aviation industry in the 1920s. It had been recognized for many years that the relationship between pressure and altitude was temperature dependent, as shown by the expression of Zuntz et al. (10) above. This meant that the relationship was also latitude dependent. However, there were clear advantages to the aviation industry in having a model atmosphere that applied approximately to mean conditions over the surface of Earth and could be used for calibrating altimeters and low-pressure chambers and assessing the performance of aircraft under well-defined conditions. This was referred to as the ICAO Standard Atmosphere (14) or the U.S. Standard Atmosphere (15), the two being identical up to altitudes of interest to high-altitude physiologists. The assumptions of the Standard Atmosphere were a sea-level pressure of 760 mmHg (1013 millibars), sea-level temperature of 15°C, and linear decrease in temperature with altitude (lapse rate) of 6.5°C/km up to an altitude of 11 km.

It should be emphasized that the model atmosphere was never meant to be used to predict the actual barometric pressure at a particular location. In fact, it was developed by people who clearly recognized that there would be substantial local variations caused by latitude, season, and other factors. Unfortunately the Standard Atmosphere was taken up by many respiratory physiologists in the 1940s and inappropriately used to predict the pressure at various points on Earth's surface, particularly on high mountains. The most glaring errors were apparent when the Standard Atmosphere was used for the pressure on Mt. Everest (16–19). The barometric pressure calculated in this way for the Everest summit (altitude 8848 m) is 236 mmHg, which is far too low as Bert clearly recognized in 1878 (8). Unfortunately the essentially correct relationships that had been used by Jourdanet (9), Bert (8), Zuntz et al. (10), FitzGerald (11), Kellas (12), and Haldane and Priestley (13) were disregarded.

Perhaps the most dramatic consequence of the inappropriate use of the Standard Atmosphere was in 1949 during Operation Everest I. At that time four naval recruits were gradually decompressed to a pressure of 236 mmHg (thought to be the pressure on the summit), and their alveolar P_{O_2} fell to as low as 21 mmHg (18), about 14 mmHg less than that of a well-acclimatized climber on the actual summit (20).

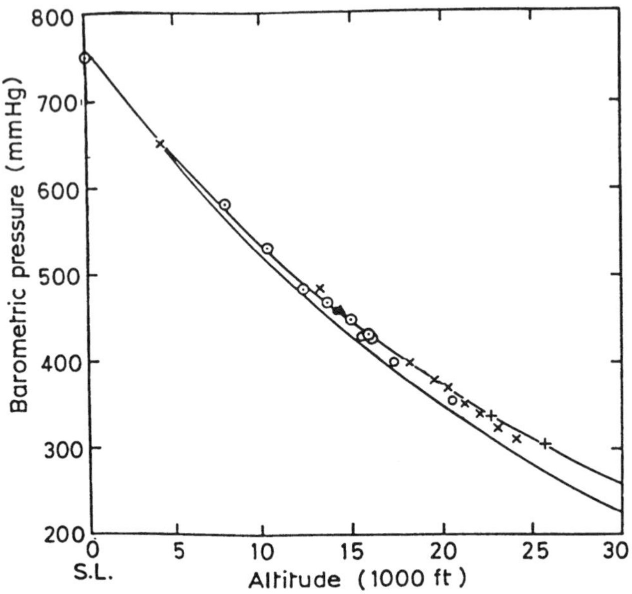

Figure 2 Relationship between barometric pressure and altitude. The upper curve is calculated from the formula of Zuntz et al. (10) using a mean air column temperature of 15°C. The lower curve follows the Standard Atmosphere (14). The plotted points represent observations made on Mt. Everest and elsewhere, mainly in the Andes. They clearly fit the upper curve better than the lower. (From Ref. 21.)

When expeditions to high altitude resumed after World War II, it soon became apparent that barometric pressures in the Himalayas and Andes exceeded those predicted from the Standard Atmosphere. This was clearly shown by Pugh (21) (Fig. 2) and clinched by the first direct measurement of barometric pressure above 8000 m and on the Everest summit during the course of the 1981 American Medical Research Expedition to Everest (22) (Fig. 3). The measured value of 253 mmHg was 17 mmHg higher than predicted from the Standard Atmosphere and this difference almost certainly made it possible to climb Mt. Everest without supplementary oxygen (23).

B. Physical Principles

The reason why barometric pressure decreases with increasing altitude is, as Torricelli recognized, the decrease in the weight of the air above us. If the atmosphere were incompressible, as is nearly the case in a liquid, pressure would decrease linearly with altitude, just as it does in a liquid. However, because the air at lower altitude is compressed, barometric pressure decreases more rapidly with height near Earth's surface. If the temperature of the gas were constant, the decrease would be

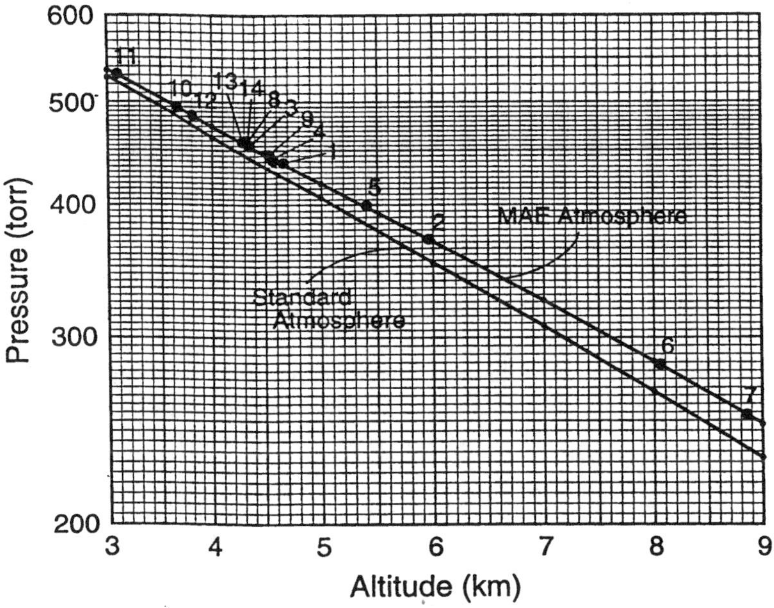

Figure 3 Barometric pressure-altitude relationships for two model atmospheres. The upper line was drawn using the model atmosphere equation (MAE) as described in the text. It shows how well this model atmosphere predicts the barometric pressure at 13 well-known locations where high altitude studies have been carried out, and the barometric pressures accurately measured. In numerical order the sites are: Collahuasi mine, north Chile; Aucanquilcha mine, north Chile; Vallot Observatory, Mont Blanc; Capanna Margherita, Monte Rosa; base camp, Mt. Everest; Camp 5, Mt. Everest; Summit, Mt. Everest; Cerro de Pasco, Peru; Morococha, Peru; Lhasa, Tibet; Crooked Creek, White Mountain, California; Barcroft Laboratory, White Mountain, California; Pike's Peak, Colorado; White Mountain Summit, California. The lower line shows the Standard Atmosphere. (Modified from Ref. 24.)

exponential with respect to altitude [see the expression above from Zuntz et al. (10)]. However, because the temperature decreases, the pressure falls more rapidly than the exponential law predicts.

The relationship between pressure, volume, and temperature is given by the ideal gas law:

$$PV = nRT$$

where P is pressure, V is volume, n is the number of gram molecules of the gas, R is the gas constant, and T is the absolute temperature. Both Boyle's law, pressure times volume is a constant (at constant temperature), and Charles' law, volume is proportional to absolute temperature (at constant pressure), follow from the ideal gas law.

Dalton's law states that each gas in a mixture exerts a pressure according to its own concentration, independently of the other gases present. That is, each component behaves as though it were present alone. This follows from the kinetic theory of gases. The pressure of each gas is referred to as its partial pressure and is given by the total pressure times the fractional concentration of the gas:

$$P_x = PF_x$$

where P_x is the partial pressure, P is the total pressure, and F_x is the fractional concentration of gas x. The fractional concentration always refers to dry gas.

Graham's law states that the rate of diffusion of a gas in the gas phase is inversely proportional to the square root of its density. This follows from the kinetic theory of gases, which states that the kinetic energy ($0.5\ mv^2$ where m is mass and v is velocity) of all gases is the same at a given temperature and pressure. One might therefore expect that very light gases such as hydrogen and helium would be lost from the upper atmosphere, and as we saw earlier, this did occur in the evolution of the atmosphere. However, up to altitudes of about 10 km, which are those of interest to us, convective mixing maintains the composition of the atmosphere constant.

As altitude increases, the atmosphere can be divided on the basis of temperature into the troposphere, stratosphere, and the gas above. The troposphere is the region of interest to us and extends to an altitude of about 19 km at the equator but only about 9 km at the poles. In it, the temperature decreases approximately linearly with altitude down to about −60°C. Above the troposphere, the temperature in the stratosphere remains nearly constant at about −60°C for some 10–12 km of altitude.

C. Model Atmospheres and Pressures on High Mountains

As explained in the historical summary above, the introduction of the Standard Atmosphere caused great confusion in high-altitude physiology because the pressures it predicts on high mountains are generally too high. However, other model atmospheres have been developed by geophysicists. The critical variable is the change of air temperature with altitude, and therefore model atmospheres have been constructed for different latitudes and seasons of the year. These different models give a large range of pressures at a given altitude. For example, the maximum difference of pressure at an altitude of 9 km is from 206 to 248 torr, i.e., about 20%.

However, it has been shown that the mean of the model atmospheres for latitude 15° (in all seasons) and 30° (in the summer) predicts the barometric pressure (P_B) at many locations of interest at high altitude very well (24), with predictions being within 1%. This model atmosphere equation (MAE) is P_B (torr) $= \exp(6.63268 - 0.1112\ h - 0.00149\ h^2)$, where h is the altitude in km. The predictions are good because many high mountain sites are within 30° of the equator, and also many studies are made during the summer. Figure 3 shows the relationship between barometric pressure and altitude for this equation (top line) and demonstrates how successfully it predicts the barometric pressure of 13 well-known sites where high-altitude studies have been carried out. Figure 3 also shows the U.S. Standard Atmosphere, which predicts pressures that are too low. The equation is

$$P_B \text{ (torr)} = \exp(6.63266 - 0.1176\,h - 0.00163\,h^2).$$

D. Effects of Latitude

Figure 4 shows the barometric pressure at different altitudes plotted against latitude for summer and winter. It can be seen that the pressure at Earth's surface and at an altitude of about 24 km is essentially independent of altitude. However, in the altitude range of about 6–16 km there is a pronounced bulge in barometric pressure near the equator in both summer and winter. Since the latitude of Mt. Everest is

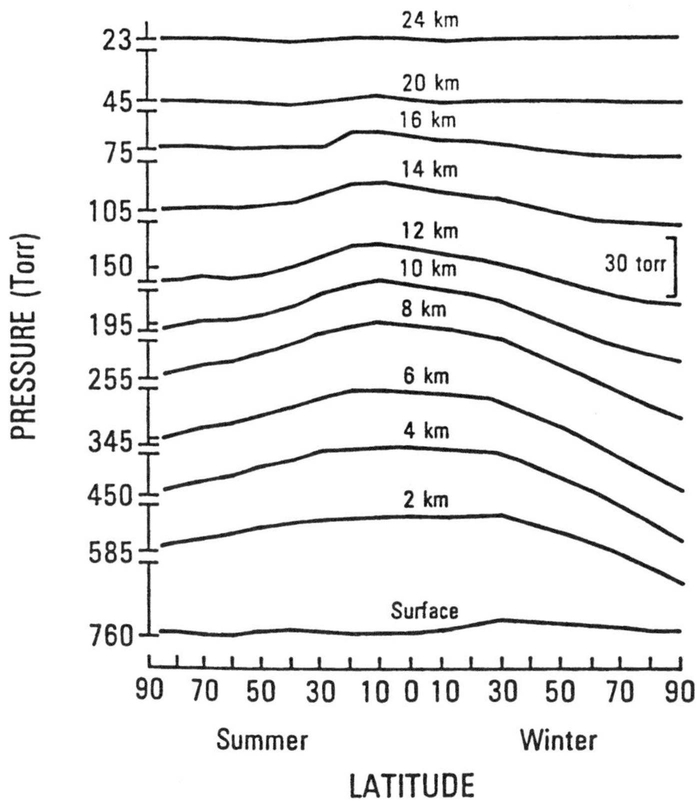

Figure 4 Increase of barometric pressure near the equator at various altitudes in both summer and winter. Vertical axis shows the pressure increasing upward according to the scale on the right. The numbers on the left show the barometric pressures at the poles for various latitudes. The altitude of Mt. Everest is 8848 m. (From Ref. 25.)

28°N, the pressure at its summit (8848 m) is considerably higher than would be the case for a hypothetical mountain of the same altitude near one of the poles (see also Fig. 6).

The cause of the increase in barometric pressure near the equator is a very large mass of very cold air in the stratosphere above the equator (25). Paradoxically, the coldest air in the atmosphere is above the equator. This is brought about because the heat of the sun near the equator causes expansion of the air and a large upwelling of the atmosphere. For the same reason, the height of the tropopause is much greater (19 km) near the equator than near the poles (9 km), as pointed out previously.

E. Effects of Season of the Year

Summer and winter greatly affect the relationship between barometric pressure and altitude because of the temperature effects. Figure 5 shows the mean monthly pressures for an altitude of 8848 m (Everest summit) as obtained from radio-sonde balloons released from New Delhi, India, over a period of 15 years (Delhi is at about the same latitude as Everest.) Note that the mean pressures in the winter months of January and February (243.0 and 243.7 mmHg, respectively) were substantially lower than in the summer months of July and August (254.5 mmHg for both

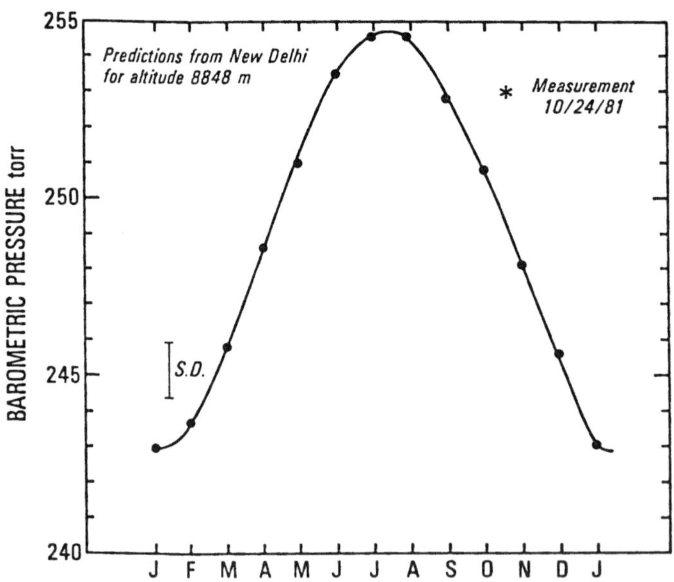

Figure 5 Mean monthly pressures for 8848 m altitude as obtained from weather balloons released from New Delhi, India. Note the increase during the summer months. The mean monthly standard deviation (S.D.) is also shown. The barometric pressure measured on the Everest summit on October 24, 1981 (*) was unusually high for that month. (From Ref. 22.)

months). This 11 mmHg difference of pressure will certainly affect $\dot{V}O_{2max}$ near the summit where the slope of the $\dot{V}O_{2max}$-pressure line is about 13 mL/min/torr (23). Therefore, a fall of barometric pressure of 11 mmHg will reduce $\dot{V}O_{2max}$ at the Everest summit by about 143 mL/min, i.e., by about 13%.

Figure 5 also shows the direct measurement of barometric pressure on the Everest summit made by Christopher Pizzo on October 24, 1981. This was unusually high for the month because Pizzo was fortunate enough to have an exceptionally fine day for his summit climb. The temperature on the summit was measured as −9°C, some 20–30°C higher than expected for that altitude.

Another direct measurement of barometric pressure on the Everest summit was made by David Breashears at 7:00 a.m. local time on May 23, 1997, during the course of an expedition sponsored by the television station WGBH. The hand-held barometer (Pretel Alti Plus K2) was calibrated in our laboratory after the expedition, and the pressure on the summit was within a mmHg or so of the value of 253 mmHg obtained by Pizzo. Data from radio-sonde balloons were also obtained for the same day and time, and the interpolated pressure at an altitude of 8848 m agreed well with the measured value. An interesting point was that Everest was near the center of a zone of high atmospheric pressure. As Figure 5 shows, the pressures for May and October are expected to be similar. Over 2000 measurements of barometric pressure were made on the South Pole (altitude 7986 m) in 1998 and the results agreed well with the pressure–altitude relationship for the summit (25a).

The effects of latitude and season are combined in Figure 6. All the data are from radio-sonde balloons, which are released from meteorological stations all over the world twice a day. The values in Figure 6 are for an altitude of 8848 and are the means from all longitudes (26). The dramatic effects of altitude and season are clearly seen. It is interesting that in midsummer the pressure reaches a maximum near the latitude of Mt. Everest (28°N). Figure 6 also shows that if Everest were at the latitude of Mt. McKinley (63°N), the difficulties of reaching the summit without supplementary oxygen would be enormously increased.

It is possible to obtain the relationship between barometric pressure and altitude all over the world on any particular day from the radio-sonde balloon measurements. Details on how to extract these data are given in West (27). The calculations show that when Messner and Habeler made the first ascent of Everest without supplementary oxygen in 1978, the pressure on the summit was 251 mmHg. When Messner made the first solo ascent without supplementary oxygen in August 1980, he was fortunate that the barometric pressure was unusually high at 256 mmHg. The only winter ascent so far was on December 22, 1987, by Ang Rita Sherpa when the pressure was only 247 mmHg.

F. Physiological Significance of Barometric Pressure at High Altitude

How important are these differences of barometric pressure in determining physical performance at extreme altitude? An analysis of the factors determining maximal exercise during extreme hypoxia has been carried out using data from the 1981

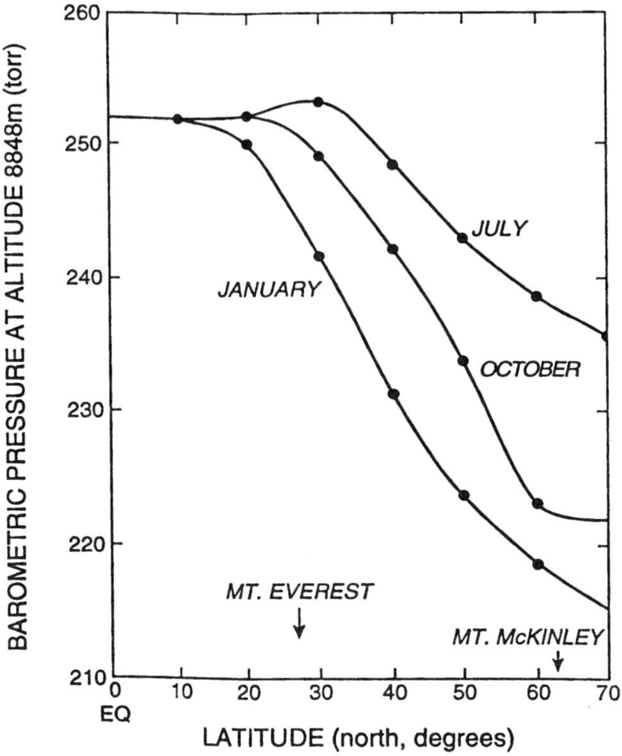

Figure 6 Barometric pressure at the altitude of Mt. Everest plotted against latitude in the northern hemisphere for midsummer, midwinter, and a preferred month for climbing in the postmonsoon period (October). Note the considerably lower pressures in the winter. The arrows show the latitudes of Mt. Everest and Mt. McKinley. (Modified from Ref. 22.)

American Medical Research Expedition to Everest as a basis (23). The relationship between $\dot{V}O_{2max}$ and inspired PO_2 found on Operation Everest II was very similar (28). The analysis showed that although many factors influence $\dot{V}O_{2max}$ at extreme altitudes including the level of alveolar ventilation, lung diffusing capacity, cardiac output, hemoglobin concentration, P_{50} of the blood, and base excess of the blood, the most critical factor is the barometric pressure (Fig. 7). The reason for this is that the pressure determines the inspired PO_2. One of the conclusions of the analysis was that it would be impossible to climb the mountain without supplementary oxygen if the pressure on the summit conformed to the Standard Atmosphere value. In addition, the variations of barometric pressure with season shown in Figures 5 and 6 indicate that it would be considerably more difficult to reach the summit without supplementary oxygen in the winter as a result of the reduced inspired PO_2, quite apart from the obvious difficulties of very low temperatures. It is significant that,

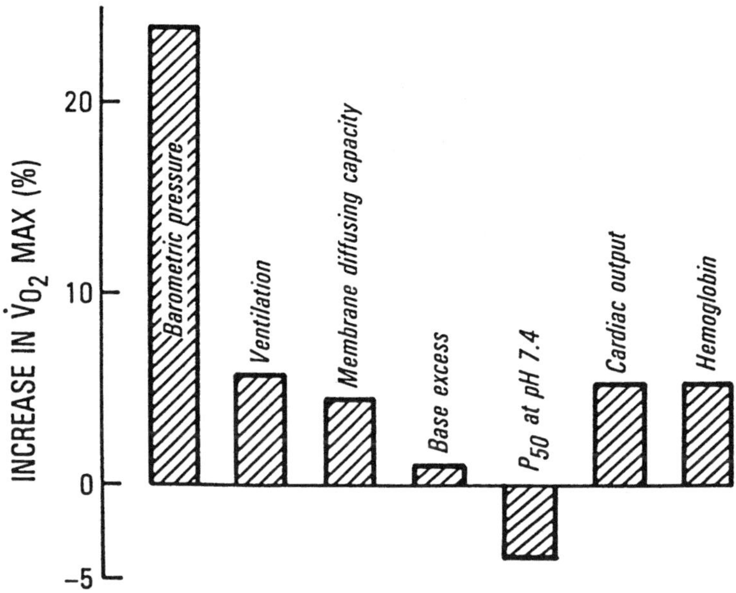

Figure 7 Percentage increase in maximal oxygen consumption for a climber on the summit of Mt. Everest when various parameters are increased by 5% leaving all the other variables constant. Note that the $\dot{V}O_{2max}$ is most sensitive to barometric pressure. See original article for details. (From Ref. 23.)

although there have been many ascents of Everest without supplementary oxygen in the premonsoon and postmonsoon seasons, only one person has made a winter ascent without supplementary oxygen, and this was on the first day of winter, December 22, 1987.

III. Factors Other Than Barometric Pressure at High Altitude

Temperature falls with increasing altitude at the rate of about 1°C per 150 m, and this lapse rate is essentially independent of latitude. As a result, the average temperature near the summit of Mt. Everest throughout the year is about −40°C. Of course most climbers choose the pre- and postmonsoon seasons when the temperature is higher. Wind chill factors are even more important. Wind velocities on Himalayan peaks have been estimated to reach in excess of 150 km/h, although there are few direct measurements.

Absolute humidity, i.e., the amount of water vapor per unit volume of gas at the prevailing temperature, is extremely low at high altitude because water vapor pressure is reduced at low temperatures. For example, while the water vapor pressure at +20°C is 17 mmHg, it is only 1 mmHg at −20°C. The result is that climbers at high altitude are almost always dehydrated. The insensible water loss caused by ventilation is very high because of the dryness of the inspired air and the high levels of ventilation, especially on exercise. Water loss also occurs through sweating, and because of the dryness of the air, sweat rapidly evaporates. Water is difficult to provide because snow must first be melted. Pugh (29) showed that climbers at altitudes over 6000 m require 3–4 L of fluid per day to maintain a urine output of 1.5 L per day. Even so, it is likely that many people at high altitude are in a state of chronic volume depletion. Members of the 1981 American Medical Research Expedition to Everest who were living at an altitude of 6300 m had a significantly increased serum osmolarity compared with sea level in spite of ample fluids being available (30).

Solar radiation is extremely intense at high altitude because of the thinner atmosphere and reflection of the radiation from snow. Ionizing radiation is also greater at high altitude because of less absorption by the atmosphere.

IV. Artificial Atmosphere at High Altitude

The hypoxia of high altitude results from the low barometric pressure, which, because the oxygen concentration of the air is independent of altitude, causes a low inspired P_{O_2}. One way to relieve the hypoxia is to artificially increase the pressure by placing the subject in a lightweight rubberized canvas bag and pressurizing this with a foot pump. Increasing the pressure by about 100 mmHg gives a reduction of equivalent altitude of about 1500 m at altitudes of 4000–5000 m. At least two bags are commercially available: the Gamow from the United States and Certec from France. There have been numerous accounts of their effective use in treating high-altitude pulmonary edema and high-altitude cerebral edema (31).

The other potential mechanism for reducing the hypoxia is to raise the oxygen concentration of the atmosphere. There is great deal of interest in this at the present time partly because of the increase in commercial and scientific activities at altitudes of 4000–5500 m. One example is a new mine in north Chile at an altitude of 4500–4600 m. This mine will employ up to 2000 people, and most of these workers will have their homes at sea level and be bused up to the mine for 7 days and then down to join their families for a further 7 days, with the cycle continuing indefinitely. Another example is a radiotelescope that is planned for an altitude of 5000 m in north Chile. The personnel will live at an altitude of about 2400 m and drive up to the facility every day. These altitudes result in severe hypoxia, with impairment of central nervous system (CNS) function, quality of sleep, and physical work capacity.

The use of oxygen-enriched atmospheres in selected parts of these facilities shows great promise (32). Relatively small degrees of oxygen enrichment result in large improvements in inspired P_{O_2}. For example, every 1% increase in oxygen concentration (e.g., from 21 to 22%) is equivalent to a reduction of altitude of about 300 m (Fig. 8). Thus, increasing the oxygen concentration of the atmosphere to 26%

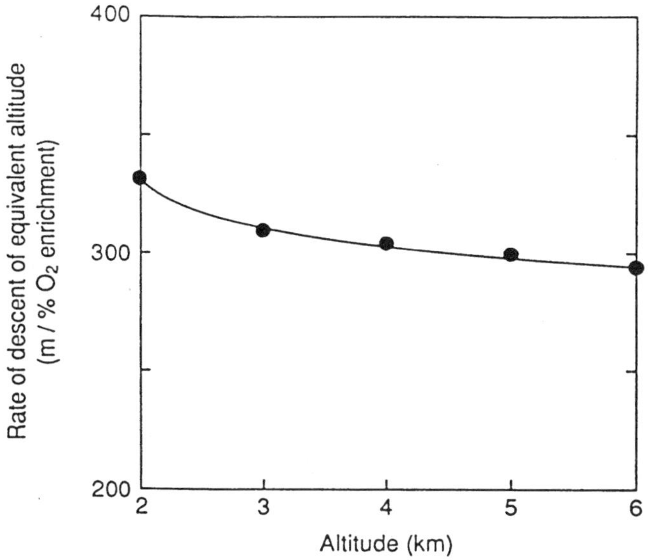

Figure 8 The vertical axis shows the meters of equivalent altitude descent when the oxygen concentration of the atmosphere is increased by 1% at the altitude shown on the horizontal axis. Note that at altitudes up to about 5000 m, each 1% of oxygen enrichment, results in an altitude reduction of more than 300 m. (From Ref. 32.)

at an altitude of 4500 m reduces the equivalent altitude to 3000 m, which is much more easily tolerated. Places where oxygen enrichment has been considered include dormitories, cabins of heavy equipment such as mechanical shovels and trucks, offices, conference rooms, and laboratories.

One of the reasons why oxygen enrichment has become more feasible is that large amounts of oxygen-enriched air can now be produced relatively cheaply. Liquid oxygen is being used in pilot studies in north Chile. A less expensive source of large amounts of oxygen is an oxygen concentrator, which uses a molecular sieve. Thousands of these are now in use in the homes of patients with severe lung disease. The sieve consists of a nonflammable ceramic material, for example, synthetic zeolite. Air is pumped through the sieve at high pressure and nitrogen is preferentially adsorbed, with the result that the effluent gas is oxygen-enriched. Periodically the sieve has to be regenerated by pumping air through it at low pressure to wash out the excess nitrogen, and this is easily accomplished by having two sieves, one of which is undergoing regeneration at any one time. These oxygen concentrators have an indefinite life and can easily produce large volumes of 90% oxygen, which is just as useful in the present application as pure oxygen.

Recently a pilot study of oxygen enrichment was carried out at the Tambo mine in north Chile, altitude 4300 m. Sixteen dormitory rooms had the oxygen concentration raised to 24–25% using a liquid oxygen source and appropriate ducting. The sleep quality of the miners using oxygen enrichment was compared with another group sleeping in ambient air. Measurements showed that the oxygen-enriched group slept better. They had fewer arousals (waking up), respiratory movements were less abnormal with fewer apneas and less total apneic time, and the EEG showed longer periods of deep phase 4 sleep. Psychometric tests performed during the day following the test night showed small but consistent improvement in six psychometric tests, including attention, concentration, and short-term memory. There was also evidence of increased arterial oxygen saturations during the day. These preliminary results are consistent with those obtained by treating patients with obstructive sleep apnea at sea level with continuous positive airway pressure (CPAP). Such patients have improved psychometric function, higher arterial Po_2 values, and a change in CO_2 sensitivity as a result of treatment (33–35).

A gratifying feature of the study was how easy it was to accomplish reliable oxygen enrichment in this relatively large dormitory facility. This pilot study strongly suggests that oxygen enrichment on a large scale in such a commercial facility may well improve the well-being of the workers and therefore the productivity of the mine.

Possible disadvantages of oxygen enrichment should also be considered. Some people immediately think of a possible fire hazard, but in fact the flammability of paper and cotton in 26% oxygen at an altitude of 4500 m is less than in air at sea level (36). Flammability of an atmosphere depends on two variables (37): (1) the partial pressure of oxygen, which is much lower in the enriched air at high altitude than at sea level, and (2) the quenching effect of the inert components, i.e., nitrogen, of the atmosphere. This quenching is slightly reduced at high altitude, but the net effect is still a lower flammability.

Loss of acclimatization to high altitude is sometimes cited as a possible disadvantage of oxygen enrichment. However, there is no basic difference between entering a room with an oxygen-enriched atmosphere and driving down to a lower altitude. Everybody would sleep at a lower altitude if they could. It is true that frequent exposure to a lower altitude will result in less acclimatization to the higher altitude, other things being equal. However, the ultimate objective is effective working at high altitude, and this can be enhanced using oxygen enrichment.

Another argument is that altering the atmosphere in this way might increase the legal liability of the facility if some kind of hypoxia-related illness develops. Actually, the opposite view makes more sense. It is possible that a worker who develops a myocardial infarction while working at high altitude could claim that the altitude was a contributing factor. Any procedure that reduces the equivalent altitude makes altitude-induced illnesses less likely.

Oxygen enrichment of room air at high altitude could be a major advance in the development of commercial and scientific facilities at altitudes over 4000 m. Everybody now expects that the ventilation of a room will provide a comfortable temperature and humidity. Control of the oxygen concentration can be regarded as a further logical step in humans' control of their environment.

Acknowledgment

This work was supported by NIH grant R01 HL 46910.

References

1. Oró J. Early chemical stages in the origin of life. In: Bengtson S, ed. Early Life on Earth, Nobel Symposium No. 84. New York: Columbia University Press, 1994:48–59.
2. Miller SL. A production of amino acids under possible primitive earth conditions. Science 1953; 117:528–529.
3. Kasting JF. Earth's early atmosphere. Science 1993; 259:920–926.
4. Chapman CR, Morrison D. Cosmic Catastrophes. London: Plenum Press, 1989:97.
5. Torricelli E. Letter of Torricelli to Michelangelo Ricci. 1644. English translation of relevant pages in: West JB, ed. High Altitude Physiology. Stroudsburg, PA: Hutchinson Ross Publishing Co., 1981:60–63.
6. Pascal B. Story of the great experiment on the equilibrium of fluids. 1648. English translation of relevant pages in: West JB, ed. High Altitude Physiology. Stroudsburg, PA: Hutchinson Ross Publishing Co., 1981:64–69.
7. Boyle R. New Experiments Physico-Mechanical, Touching the Air. 2nd ed. Oxford: Thomas Robinson, 1662. Relevant pages reprinted in: West JB, ed. High Altitude Physiology. Stroudsburg, PA: Hutchinson Ross Publishing Co., 1981:70–75.
8. Bert P. La Pression Barométrique. Paris: Masson, 1878. English translation by Hitchcock MA, Hitchcock FA. Columbus, OH: College Book Co., 1943.
9. Jourdanet D. Influence de la pression de l'air sur la vie de l'homme. Paris: Masson, 1875.
10. Zuntz N, Loewy A, Muller F, Caspari W. Atmospheric pressure at high altitudes. In: Höhenklima und Bergwanderungen in ihrer Wirkung auf den Menschen. Berlin: Bong, 1906:37–39. Translation of relevant pages in: West JB, ed. High Altitude Physiology. Stroudsburg, PA: Hutchinson Ross Publishing Co., 1981:78–80.
11. FitzGerald MP. The changes in the breathing and the blood of various altitudes. Phil Trans R Soc Lond Ser B 1913; 203:351–371.
12. Kellas AM. A consideration of the possibility of ascending the loftier Himalaya. Geogr J 1917; 49:26–47.
13. Haldane JS, Priestley JG. Respiration. 2nd ed. London: Oxford University Press (Clarendon), 1935.
14. ICAO. Manual of the ICAO Standard Atmosphere. 2nd ed. Montreal, Quebec: International Civil Aviation Organization, 1964.
15. NOAA. U.S. Standard Atmosphere. Washington, DC: National Oceanic and Atmospheric Administration, 1976.
16. Houston CS, Riley RL. Respiratory and circulatory changes during acclimatization to high altitude. Am J Physiol 1947; 149:565–588.
17. Houston CS, Sutton JR, Cymerman A, Reeves JT. Operation Everest II: man at extreme altitude. J Appl Physiol 1987; 63:877–882.
18. Riley RL, Houston CS. Composition of alveolar air and volume of pulmonary ventilation during long exposure to high altitude. J Appl Physiol 1951; 3:526–534.
19. Rahn H, Fenn WO. A Graphical Analysis of the Respiratory Gas Exchange. Washington, DC: American Physiology Society, 1955.

20. West JB, Hackett PH, Maret KH, Milledge JS, Peters RM, Jr., Pizzo CJ, Winslow RM. Pulmonary gas exchange on the summit of Mt. Everest. J Appl Physiol 1983; 55:678–687.
21. Pugh LGCE. Resting ventilation and alveolar air on Mount Everest: with remarks on the relation of barometric pressure to altitude in mountains. J Physiol (London) 1957; 135:590–610.
22. West JB, Lahiri S, Maret KH, Peters RM, Jr., Pizzo CJ. Barometric pressures at extreme altitudes on Mt. Everest: physiological significance. J Appl Physiol 1983; 54:1188–1194.
23. West JB. Climbing Mt. Everest without oxygen: an analysis of maximal exercise during extreme hypoxia. Respir Physiol 1983; 52:265–279.
24. West JB. Prediction of barometric pressures at high altitudes using model atmospheres. J Appl Physiol 1996; 81:1850–1854.
25. Brunt D. Physical and Dynamical Meterology. 2nd ed. Cambridge, UK: Cambridge University Press, 1952:379.
25a. West JB. Barometric pressures on Mt. Everest: new data and physiological significance. J Appl Physiol 1999; 86:1062–1066.
26. Oort AH, Rasmusson EM. Atmospheric Circulation Statistics. Rockville, MD: U.S. Dept. of Commerce, NOAA, 1971:84–85.
27. West JB. Acclimatization and tolerance to extreme altitude. J Wilderness Med 1993; 4:17–26.
28. Sutton JR, Reeves JT, Wagner PD, Groves BM, Cymerman A, Malconian MK, Rock PB, Young PM, Walter SD, Houston CS. Operation Everest II: oxygen transport during exercise at extreme simulated altitude. J Appl Physiol 1988; 64:1309–1321.
29. Pugh LGCE. Animals in high altitudes: man above 5000m—mountain exploration. In: Dill DB, Adolph EF, Wilber CG, eds. Handbook of Physiology, Section 4. Washington, DC: American Physiological Society, 1964:861–868.
30. Blume FD, Boyer SJ, Braverman LE, Cohen A. Impaired osmoregulation at high altitude. J Am Med Assoc 1984; 252:1580–1585.
31. Robertson JA, Shlim DR. Treatment of modern acute mountain sickness with pressurization in a portable hyperbaric (Gamow) bag. J Wilderness Med 1991; 2:268–273.
32. West JB. Oxygen enrichment of room air to relieve the hypoxia of high altitude. Respir Physiol 1995; 99:225–232.
33. Engleman HM, Martin SE, Deary IJ, Douglas NJ. Effect of continuous positive airway pressure treatment on daytime function in sleep apnoea/hypopnoea syndrome. Lancet 1994; 343:572–575.
34. Leech JA, Onal E, Lopata M. Nasal CPAP continues to improve sleep-disordered breathing and daytime oxygenation over long-term follow-up of occlusive sleep apnea syndrome. Chest 1992; 102:1651–1655.
35. Berthon-Jones M, Sullivan CE. Time course of change in ventilatory response to CO_2 with long-term CPAP therapy for obstructive sleep apnea. Am Rev Respir Dis 1987; 135:144–147.
36. West JB. Fire hazard in oxygen-enriched atmospheres at low barometric pressures. Aviat Space Environ Med 1997; 68:159–162.
37. Roth EM. Space Cabin Atmospheres: Part II, Fire and Blast Hazards. Washington, DC: NASA Report SP-48, 1964.

3

The People

SUSAN NIERMEYER

University of Colorado Health Sciences Center
Denver, Colorado

STACY ZAMUDIO and LORNA G. MOORE

University of Colorado
University of Colorado Health Sciences Center
Denver, Colorado

I. Introduction

Nearly 140 million people reside at high altitude, defined as elevations above 2500 m (8000 ft) (Table 1). While this is a small fraction of the world's population, it comprises a sizable proportion of the population in certain countries or regions. The numbers of high-altitude residents are likely to increase as the result of population growth and active migration in many of these regions.

The fundamental adaptive challenge at high altitude is hypoxia. Hypoxia results from the lower barometric pressure, and hence partial pressure of oxygen, at elevations above sea level. There are other distinctive characteristics of the high-altitude environment, including increased solar radiation, greater diurnal temperature fluctuation, aridity, low biomass, and limitations on energy production. Even though the hypoxia of high altitude cannot be readily modified by culture for at least the majority of persons, cultural factors can exert important influences by conditioning biological responses.

Here and throughout, we use "adaptation" to refer to a feature of structure, function, or behavior that enables an organism to live and reproduce in a given environment. These features may be biological or behavioral in nature, genetic or developmental in origin. In most instances, a combination of factors is involved.

Table 1 Estimated Numbers of Persons Residing >2500 m in 1995

Region / Country / Province or state	Total population	Annual population growth (%)	Estimated % > 2500 m[a]	Estimated no. > 2500 m
Africa				24,301,950
Ethiopia	55,053,000	2.9	25	13,763,250
Kenya	28,261,00	2.1	10	2,826,100
Rwanda	7,952,000	2.6	15	1,192,800
Uganda	21,297,000	2.9	10	2,129,700
Zaire	43,901,000	2.1	10	4,390,100
Asia				78,677,965
Afghanistan	20,141,000	5.6	10	2,014,100
Bhutan	1,638,000	2.4	45	737,100
China	1,221,462,000	1.0	~2	22,094,700
Inner Mongolia	8,960,000		20	*1,792,000*
Qinghai	4,120,000		40	*1,648,000*
Sichuan	104,070,000		5	*5,203,500*
Tibet	2,150,000		80	*1,720,000*
Yunnan	31,920,000		20	*6,384,000*
Xinjian Uygur Zizhiqu	13,368,000		40	*5,347,200*
India	935,744,000	1.8	~3	26,815,115
Himchal Pradesh	4,269,569		30	*1,280,871*
Jammu and Kashmir	6,981,600		40	*2,792,640*
Sikkim	425,000		40	*170,000*
Utar Pradesh	112,858,019		20	*22,571,604*
Kazakstan	14,984,100	2.3	20	2,996,820
Kirghizistan	4,698,000	2.5	10	469,800
Nepal	21,918,000	2.5	35	7,671,300
Pakistan	140,497,000	2.8	10	14,049,000
Tajikistan	6,101,000	2.7	30	1,830,030
Central and South America				35,821,750
Argentina	34,587,000	1.2	5	1,729,350
Bolivia	7,414,000	2.3	40	2,965,600
Chile	14,262,000	1.4	10	1,426,200
Colombia	35,101,000	1.5	20	7,020,200
Ecuador	11,460,000	2.0	15	1,719,000
Guatemala	10,621,000	2.8	10	1,062,100
Mexico	93,674,000	1.8	15	14,051,100
Peru	23,780,000	1.8	20	4,756,000
Venezuela	21,844,000	2.0	5	1,092,200
North America				383,900
United States	263,250,000	0.9	<1	383,900
Colorado	3,068,000		10	*306,800*
Utah	1,542,000		5	*77,100*
Total				139,185,565

[a] Population estimates >2500 m for each country, province, or state were made using the total population size, geographic area >2500 m, population size of the largest cities, and population density.
Source: Refs. 1,2.

By "genetic adaptation" we mean a heritable characteristic whose presence reflects the operation of natural selection or other processes modifying gene frequencies over time (e.g., mutation, genetic drift, gene flow). We use "developmental adaptation" to refer to physiological traits that are acquired in response to prolonged exposure to stress(ors). While developmental processes are often considered distinct from genetic ones, the capacity to acquire the trait may be genetically based. "Acclimatization" is used to refer to the time-dependent physiological responses that occur following exposure to high altitude. The distinction between these terms is that adaptations confer evolutionary benefit (enhanced fitness), whereas the physiological responses of acclimatization may or may not be adaptive.

A great deal has been learned in recent years about the mechanisms underlying physiological responses to hypoxia. We begin with a consideration of the extent to which the geographical and historical circumstances differ among current high-altitude populations in order to judge whether some high-altitude groups have lived longer at altitude than others. We also review some of these group's cultural practices that condition their biological responses. The major portion of the chapter is a survey of recent studies of the physiological responses to hypoxia in Himalayan, Andean, and Rocky Mountain high-altitude residents. Our consideration excludes East African highland residents simply because of the lack of recent study in this region. We examine each phase of the life cycle; namely, pregnancy and fetal development, infancy, childhood, adolescence, adulthood, and old age. We end with a consideration of the possible future directions for research on long-term high-altitude residents. In particular, we address recent developments in evolutionary theory and genetic methodology that bear on our ability to detect and interpret genetic differences between populations. Incorporating such perspectives into future studies of high-altitude populations will advance our understanding of the specific ways in which genetic traits interact with physiological and cultural processes.

II. History of Human High-Altitude Habitation

A. Duration of High-Altitude Occupation by Region

World high-altitude populations likely vary in their duration of high-altitude residence and degree of genetic admixture with lowland groups. The comparison of these groups is of interest from an evolutionary perspective, since populations that have spent the longest time at high altitude and have the least degree of admixture from lowland populations can be expected to be the best adapted.

The Himalayan (Tibetan) Plateau is the largest and most geographically isolated of the high-altitude regions. It is roughly oval in shape, stretching 2400 km (1500 miles) east to west and 1100 km (700 miles) along its north-south axis and encompassing over 200 million hectares (800,000 square miles). The distance to the nearest sea coast, the Bay of Bengal, ranges from 800 to 2400 km (500 to 1500 miles). On the south, it is flanked by the world's highest mountains and by peaks reaching 7600 m on the north, west, and east. Only at the northeastern edge does it descend gradually to the low-altitude region drained by the Huang (Yellow) River.

Hominids (the taxonomic family to which humans, not apes, belong) have been present in Asia for a million or more years. This is longer than the duration of hominid occupation of North and South America but not as long as in Africa. Paleoliths and microliths consistent with northern Asian tool cultures of the Upper Paleolithic, dating to approximately 25,000–50,000 years ago, have been found at 4500–5200 m sites on the northern Tibetan Plateau (3,4). Even older material from the late Pliocene (approx. 2 million years ago) has been reported from the Tibetan Plateau in northern Pakistan (5). While these data cannot provide assurance of genetic continuity with current inhabitants or when permanent occupation of the Plateau began, they support the possibility that humans have been on the Tibetan Plateau for upward of 50,000 years.

Genetic and linguistic studies support a long period of residence for the Tibetan population in its current location. Dental morphology, mitochondrial, and nuclear genetic markers show Tibetans to be related to Korean, Siberian, and Mongolian populations, likely to be descendants of the early inhabitants of Asia, and to differ considerably from Han (''Chinese'') and other southeast Asian populations (6–10). The Tibetans' membership in the Tibeto-Burmese language group differentiates them from southeast and north central Asian (including Mongolian) populations and indicates that Tibetans have resided in their south central location long enough for linguistic separation to have occurred. Contact has occurred between Tibetans and other populations via trade and, in the thirteenth and fourteenth centuries, conquest by Mongolians. But the major trading routes (e.g., the Silk Road) avoided the Tibetan Plateau, and the period of Mongolian domination was relatively brief (~100 years) and more in the form of patronage than subjugation (11). Therefore, occupation of Tibet by foreigners has been relatively limited until recently in terms of geographic penetration, impact on survival, and well-being. Genetic divergence increases when populations are separated by physical, linguistic, and other cultural barriers, and the western and eastern boundaries of Tibet show such genetic divergence (12,13). Thus, the expectation would be that Tibetans are genetically distinct from adjacent populations except those to the northeast. However, no specific genetic admixture estimates are available between Tibetans and other populations, including the Sherpas and Mongolians. Such information would be useful for determining admixture rates and the genetic history for this region of the world.

The Andean Altiplano extends nearly 4800 km (3000 miles) along nearly the whole of South America, averaging 200 km (125 miles) wide and encompassing nearly 100 million hectares (400,000 square miles). The Amazon River basin extends to the east and broad grassland plains flank its southernmost portion. The Pacific coastline parallels the altiplano 75–150 km (50–100 miles) to the west. The region between the altiplano and Pacific coast is generally dry but is punctuated by littoral plains, irrigated valleys, and broad harbors.

Native Americans derived from north central Asian populations, who migrated in several waves over the Beringian land bridge which was exposed periodically between approximately 65,000 and 15,000 years ago (14). There is comparatively little mitochondrial DNA (mtDNA) variation among the indigenous populations of the Americas. Existing variation can be accounted for by as few as four mtDNA lineages (15–17). This may reflect a small number of initial migrants (e.g., as few as four women plus some men) or simply limited mtDNA variation among a larger number of persons. Archaeological evidence suggests that humans were present in South America as early as 9,000–12,000 years ago (18–20). Permanent residence of the Andean region required domestication of the potato, quinoa and other grains, camelids (llamas, alpacas), and guinea pigs, which began about 6,000 years ago and was complete by 4,000 years ago (18–20). Somewhat surprisingly, the majority of Quechua and Aymara belong to separate mtDNA lineages (15) and linguistic groups, even though they reside next to each other on the altiplano and have much in common with each other culturally. Such patterns of mtDNA variation may indicate considerable stability and separation of women (maternal lineages) from the two language groups or may be the result of stochastic, genetic drift within small populations (21).

The Pacific Coast was the point of contact between the indigenous civilizations of the Andean and European powers in the 1500s. Contact had devastating consequences for the Andean civilizations. Not only were its rulers killed and cities looted, but within 100 years following conquest population size declined from 12 million to 675,000 as the result of pandemics of infectious disease, malnutrition, and forced resettlement (22). This loss of 95% of the indigenous population meant that the surviving fraction underwent an evolutionary "bottleneck," and, as a result, it is by no means certain that altitude-related traits happened to be preserved. Today, estimates of admixture in Andean populations follow a gradient wherein coastal populations show the most European admixture, sierra (intermediate altitude) populations are intermediate, and altiplano (high altitude) populations show the least admixture. In Ecuador, Peru, and Bolivia the estimates of European admixture vary from 5 to 30% (21,23,24) (R. E. Ferrell, personal communication). There is good correlation between genetic data and surname (23). In groups with Aymara patronyms and matronyms, 89% of the genes surveyed were of Amerindian origin. Spanish-surnamed individuals had fewer Amerindian genes but nonetheless a substantial degree of Indian ancestry, with 67% of their genes being of Amerindian origin. However, given the likelihood that contemporary Andeans are derived from populations whose genetic variation had been severely curtailed, the Amerindian genes present may or may not be of Andean origin and/or contain traits related to high-altitude existence.

The Rocky Mountain Plateau encompasses an oval region, approximately 1,200 km (750 miles) long and 400 km (250 miles) wide or nearly 40 million hectares (150,000 square miles) in the western United States in Wyoming, Colorado, Utah, and New Mexico, with the highest segment being in Colorado. Amerindians lived seasonally in this region, but it was not inhabited permanently until 150 years ago. Its current residents are genetically heterogeneous, having descended from low-altitude European, Amerindian, and Hispanic populations. Thus, the high-altitude populations of the Rocky Mountain region are, by virtue of their short period of residence, not genetically adapted to high altitude.

In summary, it is likely that persons have lived the longest on the Tibetan Plateau, for an intermediate length of time on the Andean altiplano, and for the shortest length of time in the Rocky Mountain region. The greater size, geographic isolation, remoteness from coastal regions, and absence, until recently, of conquest by low-altitude groups also suggests that the residents of the Tibetan Plateau have been subject to less reduction in genetic variation and less genetic admixture with low-altitude groups than the Andean residents.

B. Consideration of Cultural Influences

High-altitude residents employ cultural practices that modify environmental stressors and thus condition biological responses. Indigenous cultural practices affecting food sources, energy expenditure, and population movement provide examples of the ways in which such conditioning takes place.

Plants and animals genetically adapted to high-altitude environments have been traditionally relied upon as food sources. In Tibet, barley is the staple, with green vegetables being added in the summer and dried vegetables and root crops such as potato and turnip eaten in the winter (25). The staple foods in Andean diets consist of numerous tubers (over 2500 kinds of potato, ulloco, oca, mashua), quinoa, and other grains (kiwicha, tarwi, cannihua) (26). The Quechua practice of freeze-drying potatoes (chuño) and meats (charqui) preserves them for long periods and reduces their weight, facilitating their transport by groups using widely dispersed resources (27). Other crops (peanuts, beans, fruits, and coca) from low-altitude regions supplement the Andean diet and, in the case of coca, have nutritional benefits as well as narcotic effects (26). Animals indigenous to the high-altitude environment—the llama and guinea pig of the Andes and the yak of the Himalayas—are good sources of food, clothing, and fuel. The llama and yak are particularly so, providing a source of transport, meat, wool, rope, leather, dung for fuel and fertilizer, and, in the case of the yak, milk for butter, cheese, and yogurt as well as labor for pulling the plow.

Exchange of resources between altitudes plays an important part in the availability of food in high-altitude regions. In the Andes, an extensive network of Inca roads or wide footpaths links the highland and lowland areas. These have been joined more recently by highways, railroads, and air travel. Animal resources (wool, hides, meat) from the Andean altiplano are exchanged for wheat and other foods grown at lower elevations (27). In Tibet, most exchange of crops and animal products occurs between farmers and pastoralists within a region (25). Between regions, trade routes and relationships between monasteries served historically to move products over larger distances. Until the 1950s, nearly all trade was by foot or pack animal because wheeled vehicles were prohibited by Buddhist beliefs and railroads have yet to penetrate the Tibetan Plateau. Today, highways and air travel link the Tibetan Plateau with lowland regions.

Concentrated deposits of minerals with high monetary value are exposed in the mountainous environment and contribute importantly to economic exchanges in the Andean and Rocky Mountain regions. Mining has also been practiced on the Tibetan Plateau but to a lesser extent, at least until recently. The monetary value justifies high levels of expenditure for employing large numbers of persons and purchasing, in the case of Rocky Mountain residents, virtually all the required food stuffs from low-altitude regions. Recreational tourism provides an increasingly important source of income for Rocky Mountain residents.

Exchanges within households are important in the production and consumption of resources. In traditional subsistence economies, food is generally produced by adults and adolescents and distributed to the young and older-aged members of the household. This distribution pattern serves to minimize seasonal change in caloric consumption. For example, in rural highland Peru, preharvest household caloric consumption was less than half that present postharvest (28). However, preferential distribution of food to children, the reduction of household consumption by changes to less energy-intensive activities, and the temporary out-migration of adolescent and adult males protected children from seasonal shortages (29). The use of children or adolescents in herding serves to reduce household caloric consumption since a 12-year-old child can complete the herding work of an adult with 30% fewer calories (27). In the Lhasa valley, men and women eat the same kinds and generally the same amounts of food. The relationship of food need to energy expenditure is recognized by Tibetan farmers; those who do the most work eat the most, regardless of sex. Children do not eat as much as adults at meal times but eat whenever hungry between meals (25).

Conservation of energy is accomplished by the use of housing and clothing with properties that minimize heat loss and maximize heat gain. Houses made of adobe, thick mortared stone, or sod bricks on the Andean and the Himalayan plateaus offer some insulating value and effectively store radiant heat gained during the almost universally sunny days. Houses of piled stone construction provide little buffering against cold but represent a lesser energy and resource investment for mobile pastoralist families in both locations. Clothing also serves as a primary barrier against cold, creating a warm, portable living environment. Clothing typically consists of multiple layers of insulating fiber (usually wool). The outer layer is often dark, tightly woven, and water-resistant and serves to maximize solar heat absorption and prevent convective heat loss. Covering for the head and face provides shielding and helps maintain a warmed, humidified microenvironment around the face. The limited availability of wood or other fuel for heating houses makes the strategy of fully clothed family members sleeping together an important means of conserving body temperature without increased expenditure of caloric energy or fuel resources (30).

Specific cultural practices afford additional protection from environmental stresses around periods of vulnerability in the life cycle. For example, in both Peru and Tibet, the infant or young child sleeps with the mother in the early months of life, is nursed in the warmest location, remains swaddled even while indoors, and is placed in the sunniest areas when outside. The infant is wrapped in multiple layers—diapering, leggings, an inner garment, and a sweater—and wears a knitted hat (31; S. Niermeyer, unpublished observations). In Tibet, infants are carried inside the traditional outer garment (chuba) where nursing can occur within protective layers of clothing. Quechua and Aymara women in the Andes carry their infants using a carrying cloth (manta) worn across the mother's back and fully enclosing the infant, who is also encased in a blanket and swaddled by a cloth belt. The insulating value of this manta pouch is sufficient to raise relative humidity and temperature 12°C from the first layer of the pouch to the infant's skin, but inspired PO_2 is 8–16 mmHg below ambient (32). The adoption of Western-style dress by mothers in the Lhasa valley appears associated with a higher incidence of cold injury in their infants, suggesting that the abandonment of long-standing clothing and carrying practices may be maladaptive (S. Niermeyer, unpublished observations).

Another example of a cultural practice thats affords protection from the high-altitude environment is permanent or temporary outmigration, perhaps the ultimate behavioral solution to a biological problem. Such practices in the Andes were recorded by the seventeenth-century historian Antonio de la Calancha, who observed that pregnant women of Spanish origin would descend to give birth at lower altitudes and not return until the child was more than a year old (33). A similar practice occurs today among pregnant Han women in Tibet. They typically descend to their home districts at or near sea level and remain there or leave their infants with extended family until the infant is approximately 2 years of age before returning to high altitude (34).

In summary, high-altitude residents engage in cultural practices that modify the effects of hypoxia on energy availability as well as the other attributes of the high-altitude environment. Historical as well as present-day practices of Himalayan and Andean residents rely on well-adapted indigenous plants and animals to produce calorically dense food. By seasonal adjustment in energy expenditure and food consumption within households, resources are distributed in ways that protect infants and children at times of shortage. House construction and clothing aid in the conservation of energy. In the Andes and Rocky Mountains, trade between regions, highly valued mineral resources, and, particularly in North America, recreational opportunities augment the resources available. Close proximity to different climatic zones provided by the vertical layering of highland environment in the relatively equatorial Andean region facilitates such resource exchange.

III. High-Altitude Adaptation Across the Life Cycle

The stages of the human life cycle provide a convenient organizing framework for reviewing the physiological responses to hypoxia and demonstrating how these responses influence the ability of one generation to successfully reproduce the next (adaptation). In this section, we consider pregnancy and fetal life; infancy, childhood, and adolescence; adulthood and old age. Mortality risk is unequally distributed across these phases; it is greatest during fetal life, intermediate during infancy and childhood, and lowest in adolescents and adults (except for women during child-bearing years) until the oldest ages. Thus, the early phases of the life cycle are of the greatest interest from an evolutionary point of view. At each phase, we compare high-altitude and sea-level residents or high-altitude residents with recently acclimatized newcomers. The underlying question is the extent to which lifelong high-altitude residents of the three world regions reviewed above differ in their physiological responses to hypoxia and whether such differences support the likelihood that some populations are better adapted than others.

A. Pregnancy and Fetal Life

From an evolutionary perspective, pregnancy represents a critical overlap between the generations. While the health risk to the fetus and neonate is greater than to the mother, the evolutionary effect of a maternal death is magnified since it results in the loss of two individuals, mother and fetus, and curtails the father's genetic contribution as well.

Fetal Wastage/Pregnancy Loss

Fetal wastage is usually assessed by careful monitoring of completed fertility (total number of births per women aged 15–44 years), spontaneous abortions, or miscarriages. In practice, such information is difficult to obtain. Fertility has been reported to be less in the Andean region than at sea level as judged by smaller completed family size, a shortening of the reproductive span by later menarche and earlier menopause, and an increase in completed fertility in persons who migrated from high to low altitudes (35). Other studies in South America and Nepal do not support an altitude-associated reduction in fertility and, in fact, suggest higher completed fertility at high than at low altitudes (36–38). Delayed menarche and earlier menopause do not limit fertility; higher fertility is achieved in highland Peru and Chile by shorter intervals between births and increased frequency of conception during lactation. Births are likely to be underenumerated in the Andean and Himalayan regions where fewer than half the women give birth in hospitals (39). High neonatal (birth to 28 day) or infant (birth to 1 year) mortality can serve either to increase or decrease fertility by prompting a greater number of births to assure surviving offspring or by reducing the number of infants tallied in periodic censuses. Cultural practices affecting exposure to intercourse (e.g., proportion of the population living as celibate nuns) as well as the contribution and costs of children also influence childbearing patterns. In rural highland Peru, children generate more resources than they consume, making high completed fertility desirable (27).

Endocrinological studies suggest alterations in reproductive function at high altitude. Andean highland women of reproductive age showed similar levels of luteinizing hormone but lower prolactin levels compared to low-altitude women (40). Urinary ovarian hormone levels were essentially the same, but serum progesterone was elevated and estrogens (estradiol and estriol) reduced during high- compared to low-altitude pregnancy in Colorado and Peru (41,42). Measures of testicular function were similar, but testosterone levels were lower at 4340–4500 m than at 150 m in Peru. Sherpa males at high compared with low altitude have lower serum luteinizing hormone and a trend for lower follicle-stimulating hormone levels, suggesting less stimulation of the pituitary gland (43). Thus, there may be some level of impairment of reproductive function at high altitude, but it is not sufficient to limit or even reduce fertility.

Intrauterine Growth and Pregnancy Duration

One of the best-documented effects of high altitude is a progressive reduction in birth weight. Birth weights decline an average of 100 g per 1000 m altitude gain in studies conducted over a 40-year period (Fig. 1). In addition to the lower mean birth weight, the percentage of low birth weight babies (<2500 g) is fourfold greater at high (>2700 m) than low altitude in a U.S. population-based study (44). The reduction in birth weight is due to direct effects of high altitude and not to interactive effects with other risk factors such as maternal age, parity, body size, or prenatal care (45). Similar birth weight reductions under other circumstances of reduced fetal-

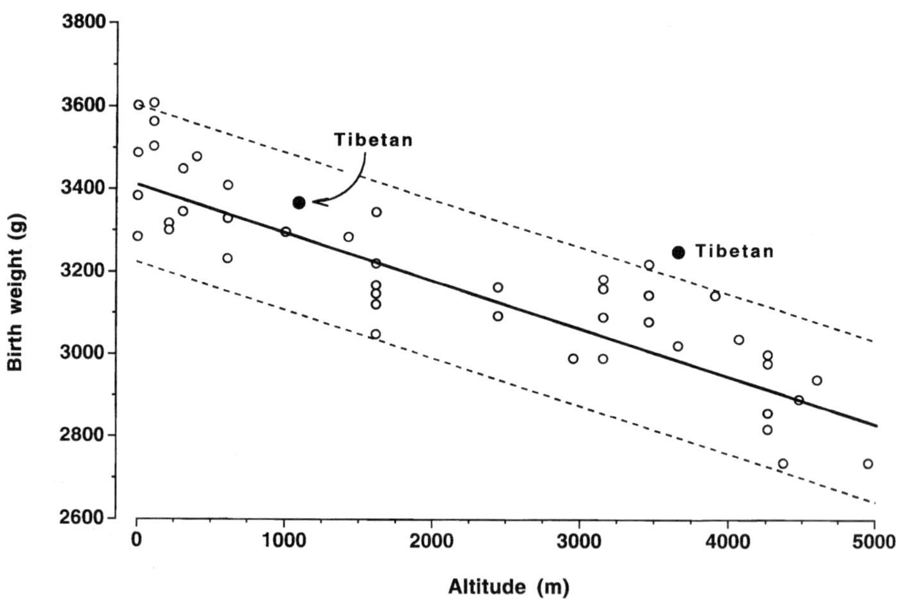

Figure 1 Mean birth weights from previously published studies in North America, South America, and Tibet (labeled) with the upper and lower 90% confidence limits (dotted lines). (From Ref. 222.)

placental oxygen supply implicate the hypoxia of high altitude as causative. The reduction in birth weight at high altitude has historical significance; the first recognition that fetal growth and length of gestation were separable influences on birth weight was made at high altitude (46).

The primary cause of the reduction in birth weight is retardation of intrauterine growth rather than shortened gestation. Babies are both lighter and shorter for gestational age, conforming to a model of growth retardation throughout pregnancy. The greatest absolute reduction in fetal size occurs in the mid to late third trimester, as demonstrated by a progressive reduction in growth rate after 32 weeks at high compared with low altitudes in the Colorado population (47,48). Average gestational ages at high versus low altitude are generally similar (Table 2), but accurate gestational age information is difficult to obtain, particularly in developing countries where standardized reporting of birth weight and gestational age is available for only a small proportion of the population.

The magnitude of fetal growth retardation appears to vary in relation to the duration of high-altitude residence, with the longest resident populations experiencing the least decline and the shortest resident groups demonstrating the most reduction in birth weight. Comparing well-matched samples collected by the same investigator at or near sea level and at greater than 3000 m, the reduction in birth weight is greatest in North Americans (−352 g, $p < 0.001$), intermediate in South Americans (−270 g in Peru, −282 g in Bolivia, $p < 0.001$), and least in Tibetans (−72 g, $p =$ NS) (50). Comparing women of different ancestry at the same altitude (3600 m), women from long-resident high-altitude populations give birth to heavier weight infants than women of low-altitude population ancestry. This is particularly true in Tibet, where birth weights averaged 294–650 g heavier in Tibetan than Han women (34,55), but also in Aymara women whose infants weighed 143 gm more than babies born to women of European or mestizo ancestry (53). The protection from altitude-associated reductions in birth weight has also been observed for Nepalese Sherpa (56) but not Ladakhis (49). The Ladakhi women were smaller than the other samples surveyed, complicating the interpretation of the cause of the lower birth weights observed.

Thus, several factors appear involved in protecting long-term, native high-altitude residents from intrauterine growth retardation (IUGR). Developmental factors likely contribute to the differences in birth weights between high-altitude newcomers and lifelong residents. Genetic factors are involved in the determination of birth weight; genetic factors account for as much as 70% of the variation in birth weight among offspring of monozygotic twins at sea level (57–59). Thus, the protection from altitude-associated IUGR afforded Tibetans compared to other high-altitude natives may result from the possession of different genetic variants. Nutritional, behavioral, and health-related characteristics during pregnancy are also likely important in each location. The particular maternal physiological characteristics that might be affected by genetic, developmental, or pregnancy-specific factors are described in the section below.

Table 2 Effect of High Altitude on Birth Weight, Gestational Age, and Neonatal or Infant Mortality

	Low altitude					High altitude				
Area (Ref.)	Altitude (m)	Weight (g)	Gest. age (wk)	% preterm	Mortality rate	Altitude (m)	Weight (g)	Gest. age (wk)	preterm %	Mortality rate
Rocky Mountains										
Lichty, 1957 (46)	1600	3035			23.4[a]	3100	2655[b]			41.6[a,b]
McCullough, 1977 (47)	<2130	3166	37.0	18.2	11.9[a]	>2740	2962[b]	39.0	19.2	18.5[a,b]
Unger, 1988 (48)	<2130	3235	40.0	11.6	6.0[a]	>2740	3058[b]	39.5	11.5	6.5[a]
Jensen, 1997 (45)	<2130	3297	39.5	11.2		>2740	3056[b]	39.0[b]	14.2	
Andes										
Mazess 1965 (54)	338				28.6[a]	3030				52.8[b]
Beall 1981 (51)	600	3410				3860	3140[b]			
Haas, 1980 (53)	400	3427	39.0		10.6[c]	3600	3165[b]	39.0		9.3[c]
Gonzales, 1993 (40)	150	3178	39.8			4340	2982[b]	38.2*		
Carmen Torres, 1993 (52)	150	3180	39.2	9		4340	2835[b]	38.6	12*	
Himalayas										
Zamudio, 1993 (50)	1200	3299	39.7	7		3600	3236	39.8	7	
Wiley, 1994 (49)						3600	2764	37.8		144[a]

[a] Neonatal mortality rate = deaths within first 28 days/1000 live births.
[b] Comparison with low altitude, $p < 0.05$.
[c] Infant mortality rate = deaths within first year/1000 live births.

Maternal Oxygen Transport Responses to Pregnancy

Because not all women at high altitude deliver growth-retarded babies, we have conducted a series of studies to test the hypothesis that altitude-associated IUGR is due to maternal oxygen transport insufficient to meet fetal-placental demands. Alternate possibilities are that fetal growth is impaired by limitations of placental diffusing capacity for oxygen, other nutrients, or fetal-specific factors. Early reports indicated that the placenta at high altitude was similar in absolute size and larger in relation to fetal size than at low altitude (41). Recent studies show that the high-altitude placenta is more vascularized and has a greater diffusing capacity than at low altitude (60,61). Thus, impaired placental oxygen diffusing capacity is unlikely to be the source of altitude-associated IUGR.

During pregnancy, elevated levels of progesterone and estrogen in combination with increased metabolic rate raise peripheral (carotid body) and central nervous system ventilatory sensitivity to chemosensory stimuli and result in increased resting ventilation (62–64). At low altitude, the increase in ventilation does not appreciably raise arterial oxygen saturation, since it is already nearly maximal, but at high altitude arterial oxygen saturation rises with pregnancy (Table 3). Studies in Peru and Colorado demonstrated that the magnitude of rise in a woman's ventilation, hypoxic ventilatory sensitivity, and arterial oxygen saturation during pregnancy related positively to the birth weight of her infant (72–74). The rise in maternal arterial oxygen saturation among Tibetans appears diminished relative to that seen in Colorado or Peruvian highlanders (Table 3). However, different samples of Tibetan women were compared in the pregnant versus the nonpregnant condition, whereas the same women were studied while pregnant and postpartum in Colorado and Peru. Hence it is unclear whether interindividual variation or differing effects of pregnancy are involved.

These studies suggest that factors that raise maternal arterial oxygenation help preserve fetal growth at high altitude. Since the pregnancy-associated rise in arterial oxygen saturation and higher hemoglobin concentration of high-altitude residents result in levels of arterial oxygen content that are similar to low-altitude values, the question becomes what is responsible for the IUGR observed. One possibility is that the oxygen tension gradient between maternal and fetal blood is more important than maternal arterial oxygen content for deciding fetal-placental oxygen delivery. Maternal arterial PO_2 and, to a lesser extent, uterine venous PO_2 are reduced in pregnant ewes chronically exposed high altitude (75–77), and hence the maternal-fetal oxygen diffusion gradient is likely reduced. Another possibility is that uteroplacental oxygen delivery is reduced as a result of decreased uterine blood flow.

In studies conducted at low altitude, we demonstrated that approximately 1 L/min or 15% of the total cardiac output is directed toward the uteroplacental circulation near term (78). The rise in uteroplacental blood flow was accomplished, in approximately equal part, by a doubling of uterine artery diameter by mid-gestation and a rise in uterine artery flow velocity that was progressive throughout pregnancy. The increase in uterine artery diameter in pregnant Colorado women at 3100

Table 3 Maternal Oxygen Transport During Pregnancy

	USA Low Altitude (63,67,70,71[a])		USA High Altitude (65,66,70[a])		Peru High Altitude (69,70)		Tibetans High Altitude (68,69[a])		Han newcomers High Altitude (68[a])	
	nonpreg.	preg.	nonpreg.	preg.	nonpreg.	preg.	nonpreg	preg.	nonpreg.	preg.
Altitude, m	<2500	<2500	3100	3100	4300	4300	3658	3658	3658	3658
V_E	7.1 ± 0.4	9.6 ± 0.4	8.8 ± 0.3	11.4 ± 0.3	9.5 ± 0.4	12.0 ± 0.7	10.1 ± 0.5	11.7 ± 0.3	9.1 ± 0.5	10.2 ± 0.5
VO_2	196 ± 8	254 ± 9	199 ± 5	244 ± 5	—	—	255 ± 11	267 ± 7	233 ± 18	244 ± 18
VO_2/kg	3.1 ± 0.1	3.5 ± 0.2	3.1 ± 0.8	3.3 ± 0.1	—	—	5.2 ± 0.2	4.7 ± 0.1	5.0 ± 0.3	4.4 ± 0.3
HVR	124 ± 13	237 ± 26	244 ± 20	322 ± 28	23 ± 8	87 ± 17	45 ± 8	134 ± 19	44 ± 11	134 ± 16
HVR/kg	2.1	3.2	3.6 ± 0.3	4.3 ± 0.3	0.4	1.4	1.0 ± 0.2	2.3 ± 0.3	0.8 ± 0.3	2.5 ± 0.4
$PaCO_2$	38	32	31 ± 1	27 ± 1	31 ± 1	26 ± 1	31 ± 1	27 ± 1	31 ± 1	29 ± 1
SaO_2	94.6 ± 0.5	95.4 ± 0.4	90.9 ± 0.2	92.2 ± 0.2	82.9 ± 1.2	87.4 ± 0.4	89.0 ± 0.5	89.8 ± 0.3	86.7 ± 0.6	89.6 ± 0.5
Hgb	13.9	12.6	15.1 ± 0.2	13.8 ± 0.2	14.0 ± 0.4	13.1 ± 0.3	14.9 ± 0.2	12.6 ± 0.3	15.2 ± 0.3	14.4 ± 0.4
CaO_2	17.5	14.5	18.7 ± 0.2	17.3 ± 0.2	15.9 ± 0.4	15.6 ± 0.4	18.1 ± 0.3	15.5 ± 0.3	17.9 ± 0.2	17.3 ± 0.6
Blood vol	66.6	79.7	58.3 ± 1.2	69.9 ± 1.9	—	—	—	—	—	—
Plasma vol	42.6	65.9	33.7 ± 0.8	42.9 ± 1.3	—	—	—	—	—	—
C.O.	5.3	6.5	—	—	5.2 ± 0.5	5.8 ± 0.1	—	—	—	—
MAP	84	85	87 ± 1	85 ± 1	—	—	76 ± 2	85 ± 2	80 ± 2	89 ± 4
HR	74	84	81 ± 1	93 ± 2	77 ± 2	79 ± 2	76 ± 2	81 ± 2	81 ± 3	88 ± 2
UA/CI vel	1.0 ± 0.2	4.3 ± 0.4	0.9 ± 0.1	4.6 ± 0.2	—	—	—	5.5 ± 0.7	—	2.9 ± 0.4
UA vel	9 ± 2	61 ± 3	10 ± 1	69 ± 2	—	—	—	56 ± 3	—	47 ± 4
UA flow	6 ±2	312 ± 22	8 ± 2	203 ± 48						

Mean ± SEM nonpregnant and pregnant values were obtained in the same woman ≥3 months postpartum and the third trimester, respectively, in the United States and eru but from different women in Tibet.

V_E = Ventilation, 1 BTPS/min; VO_2 = O_2 consumption, ml STPD/min; HVR = hypoxic ventilatory response, shape parameter A; HVR/kg = HVR/kg body weight; $PaCO_2$ = arterial or end-tidal PCO_2, mm Hg; SaO_2 = arterial O_2 saturation, %; Hgb = g/100 mL blood; CaO_2 = arterial O_2 content, mL O_2/100 ml blood; blood vol = total blood volume, mL/kg; plasma volume, mL/kg; C.O. = cardiac output, L/min; MAP = mean arterial blood pressure, mmHg; HR = heart rate, bpm; UA/CI vel = ratio of uterine artery/common iliac artery blood flow velocity, cm/sec; UA vel = uterine artery blood flow velocity, cm/sec; UA flow = uterine artery blood flow, mL/min.

[a] L.G. Moore, unpublished data.

m was only about half as great as that seen at low altitude (1600 m), resulting in one-third lower uterine artery blood flow (Table 3). We also found that pelvic blood flow distribution to the uterine artery was diminished at high versus low altitude (70,79), In the same study, women who developed preeclampsia at high altitude had even less redistribution of pelvic blood flow to favor the uterine circulation and hence no increase in uterine artery flow velocity near term, unlike normal pregnant women. These differences were present before the onset of hypertension, suggesting that lowered uterine artery blood flow may be a cause rather than an effect of preeclampsia (79). The reduction in uterine artery blood flow at high altitude was consistent with the decrease in birth weight observed in these and previous experimental animal studies (Fig. 2).

Because babies born to Tibetan women weighed more that those of acclimatized Han newcomers at 3658 m, we asked whether Tibetan preservation of fetal growth could be linked to increased arterial oxygen content or augmented uterine artery blood flow. Pregnancy increased maternal ventilation, but arterial oxygen saturation was unchanged in the Tibetan women. Whereas the increased arterial oxygen saturation and elevated hemoglobin concentration preserved arterial oxygen content at sea-level values in the high-altitude Colorado, Peru, and Han women, the pregnant Tibetan women had lower hemoglobin and lower arterial oxygen content values than the other groups (Table 3). The Tibetans' lower hemoglobins were likely due

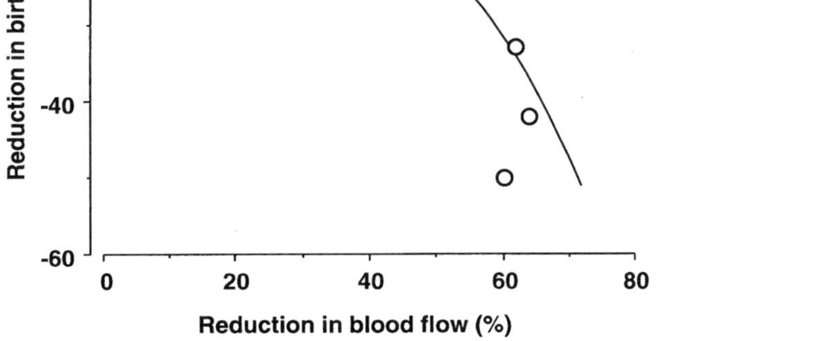

Figure 2 Birth weight, expressed as a percentage of sea-level or low-altitude values, falls as uterine, placental, or uteroplacental blood flow diminishes in experimental animal or human studies. Open circles are previously published values and closed circles are from normotensive or preeclamptic women studied at 3100 m. (Adapted from Ref. 70.)

to greater plasma volume expansion and may, in turn, have facilitated their directing a larger portion of their pelvic (common iliac) blood flow toward the uterine artery (68). This greater redistribution of lower extremity flow to the uterine artery was associated with heavier birth weights in the Tibetan versus the Han samples. Taken together, these data suggested that the Tibetans employed a strategy that maximized the increase in uterine blood flow, not arterial oxygen content (68).

In the aggregate, these data suggest that the ways in which adequate oxygen delivery is maintained to the fetus vary among individuals and populations, with Tibetans emphasizing an increase in uterine blood flow and the other groups protecting arterial oxygen content. Since all the Andean and Tibetan women were born and raised at high altitude and there were no discernible differences between Colorado women born at high altitude versus those moving there as adults, genetic factors appear implicated in the sample differences observed. In addition, because the Tibetan women have oxygen transport characteristics close to sea-level values and give birth to infants with the least reduction in birth weight, those factors preserving oxygen delivery closest to sea-level values appear to be the best adapted.

Neonatal, Infant, and Maternal Mortality

Birth weight is a major determinant of infant mortality. Since mortality increases at both the lower and upper extremes of the birth weight distribution, birth weight is considered to be the classic example of the evolutionary process of stabilizing selection. Given the lower birth weights generally present at high altitudes, the expectation would be that neonatal and, by extension, infant mortality would be increased. This is supported by older studies in Colorado in which mortality rates at high altitudes were higher than those observed at lower elevations (Table 2) (46,47). However, current data demonstrate that mortality rates at high altitudes in Colorado have declined to nationwide levels in association with a modest increase in birth weight, fall in percent preterm births, and likely improvements in the detection and management of complicated pregnancies (88,80). Today, as well as for the past 20 years, Peru and Bolivia have had the highest infant mortality rates in South America. In both countries, mortality rises with increasing elevation when all infants or only urban infants are compared (39). The Colorado data have the advantage of complete population enumeration and reasonably accurate data reporting. In South America, neonatal and infant deaths are almost certainly underreported since registration of a birth or death requires payment and only about one third are certified by a physician (39). No reliable infant mortality data, to our knowledge, are available from Tibet. When comparing infant mortality between regions or countries, it is important to recall that factors other than hypoxia may be involved, including political factors (since infant mortality has taken on broader meaning as a yardstick for social and economic development), infectious disease, and other environmental influences.

Both growth retardation and preterm delivery affect birth weight and mortality risk. Within the range of values commonly observed, premature delivery has a much greater impact on neonatal survival than a reduction in birth weight alone (81).

Accordingly, the increased neonatal mortality observed previously at high altitude in Colorado was due chiefly to deaths among babies who were both growth-retarded and premature (47). In a more recent large U.S. study, there was no difference in birth weight–specific mortality between low and high altitudes when fetal growth retardation and preterm delivery were taken into account (80). In South America, the lack of complete population enumeration and gestational age information in some studies prevents determination of whether birth weight–specific mortality is altered at high versus low altitude. Beall reports a 170 g lower optimal birth weight (i.e., associated with the lowest mortality rate) in a 3860 m compared with 600 m sample from Peru (51). However, since no gestational age information was available, it is possible that the lower optimal birth weight was due to a greater contribution of growth retardation than prematurity to mortality risk at high versus low altitude. Also, the low-altitude optimal birth weight values are considerably higher than in other studies (81,82). A recent study from Ladakh at 3600 m yielded an average birth weight of 2674 g, well below Tibetan and Andean values at similar elevations, which carried with it an even greater birth weight–specific mortality risk (49). Approximately 6% were Tibetans; the remainder were of Indian and other genetic background. It is likely that prematurity, parentage, and other factors such as small maternal body size elevated infant mortality. Thus, in short, "small" is not better at high (or any) altitude.

The IUGR and greater frequency of preterm deliveries in some studies at high altitude may be related to an increased frequency of preeclampsia, placental abruptions, and other complications of pregnancy. Preeclampsia is defined clinically as elevation in blood pressure (>140/90 mmHg, a systolic rise >30 mmHg or a diastolic rise >15 mmHg) after week 20 accompanied by significant proteinuria (>0.3 g/L in a 24-hour collection) in a woman who is normotensive when nonpregnant. Abnormalities of liver function, coagulation, and the central nervous system are sometimes observed as well. Preeclampsia is the leading cause of maternal and fetal mortality in the industrialized world and the third most frequent cause of very low birth weight in the United States (83–85). In three separate studies, we have observed a three- to fourfold increase in the incidence of preeclampsia at high compared to low altitude in Colorado (45,73,86). Data from South America are equivocal as to whether the incidence of preeclampsia is increased (40,88), and no studies have been conducted in the Himalayan region. Interestingly, we found higher blood pressures in Han than in Tibetan pregnant women living at 3658 m in Lhasa (L. G. Moore, unpublished observations). A review of all deliveries in La Oroya, Peru (3750 m), over a 15-year period indicated that placental abruptions were three times more common than at sea level and occurred in 6.8% of women over 40 years and 3.4% with parity greater than 4 (89). An increased incidence of maternal complications during pregnancy may prompt preterm deliveries and raise maternal as well as infant mortality. Maternal mortality in Peru and Bolivia is more than twice the South American average (39) and rises from 13.2 maternal deaths per 10,000 live births at the coast, to 21.5 in the 2000 to 3000 m region, and to 43.1 at elevations above 3000 m in Peru (40). Maternal mortality is not increased at high altitude in the United States, where there is ready access to medical care. Data from the Himalayan region are lacking.

Impaired placentation and uteroplacental ischemia may be a common pathway whereby pregnancy complications are increased and intrauterine growth retarded at high altitude. Uteroplacental ischemia has been associated with IUGR and/or preeclampsia at low altitude (90,91). A growing body of evidence suggests that there is impaired trophoblast invasion of maternal spiral arteries in preeclamptic pregnancies. Trophoblasts are the specialized epithelial cells of the placenta, which ultimately comprise the physical connection between the embryo (fetus) and uterus. Impaired trophoblast invasion results in shallower placentation and maternal spiral arteries that retain their muscular portions in the myometrial (nondecidual) segment, preserve their contractile sensitivity to local and circulating vasopressors and, as a result, reduced uterine blood flow (reviewed in Ref. 92). Trophoblast differentiation and invasion may be oxygen regulated; when grown in the presence of physiological oxygen tensions, trophoblast stem cells differentiate and rapidly invade extracellular matrix. The same cells grown under hypoxic conditions failed to invade extracellular matrix, partially due to inhibition of their transition to an invasive phenotype, marked by switching of integrin cell-extracellular matrix receptors (93).

In summary, fertility is not reduced at high altitude but intrauterine growth is retarded, birth weight is lowered, and neonatal and infant survival are impaired except under conditions of accessible medical care. The cause of the IUGR is likely to be insufficient oxygen delivery to meet fetal-placental demands. This is supported by our findings of consistent associations between infant birth weight and maternal ventilation, arterial oxygenation, and uterine blood flow at high altitude as well as by others' findings that placental diffusing capacity is not impaired relative to low-altitude values. The degree of IUGR is least in the longest resident high-altitude populations. Whether there is a corresponding reduction in birth weight–specific mortality is unclear and will likely remain so, given the many factors affecting mortality rates. The parallels between the magnitude of IUGR and the extent of maternal oxygen transport responses to pregnancy are suggestive that the adaptive strategies and degrees of success differ between and within regions. That is, Tibetans appear distinguished by lesser reductions in birth weight, lower hemoglobin concentrations, and greater blood flow redistribution to the uterine circulation. The higher birth weights in native Aymara than acclimatized newcomer women may be accompanied by differences in oxygen transport, but these remain, to our knowledge, unstudied. Of interest is that uterine blood flow as well as hemoglobin characteristics of the Tibetan women resemble those of healthy low-altitude Colorado residents, whereas the Han women are more like Colorado high-altitude women with preeclampsia.

B. Infant, Childhood, and Adolescent Development

Remarkable changes in oxygen transport take place at birth. At high altitude, these are complicated by the lower ambient oxygen tensions as well as the IUGR and other conditions stemming from prenatal life. Effects of high altitude continue to influence growth and development, exercise performance, and occurrence of illness during infancy and childhood. Differences among regional populations and between groups residing at the same altitude will be considered below to determine the extent to which interpopulation differences exist in infant, childhood, and adolescent development.

Oxygen Transport Characteristics

In the transition following birth, the lungs change from fluid- to air-filled, pulmonary blood flow increases dramatically, and vascular shunts reverse in direction and close. Oxygen plays a critical role in the postnatal transition because of its function as a pulmonary vasodilator.

At sea level, arterial oxygen saturation is modestly lower during the first week of life than the 94–98% values present during the neonatal period and infancy (94). At 1610 m in Denver, arterial oxygen saturation in healthy, term infants was reduced relative to sea level values, averaging 92–94% from 24–48 hours to age 3 months (95). At 3100 m in Leadville, Colorado, arterial oxygen saturation was lower still, especially during the first week of life (Fig. 3) (96). Unlike the pattern at low altitude, arterial oxygen saturation fell during the first week and then rose gradually to attain near-birth values at 2 and 4 months of age. The fall in arterial oxygen saturation at 1 week was consistent with clinical observations that babies who develop signs of hypoxemia (e.g., cyanosis, irritability, poor feeding, and failure to gain weight) often become symptomatic around 1 week of age. At all ages, values were higher during wakefulness than during active or quiet sleep and intermediate during feeding.

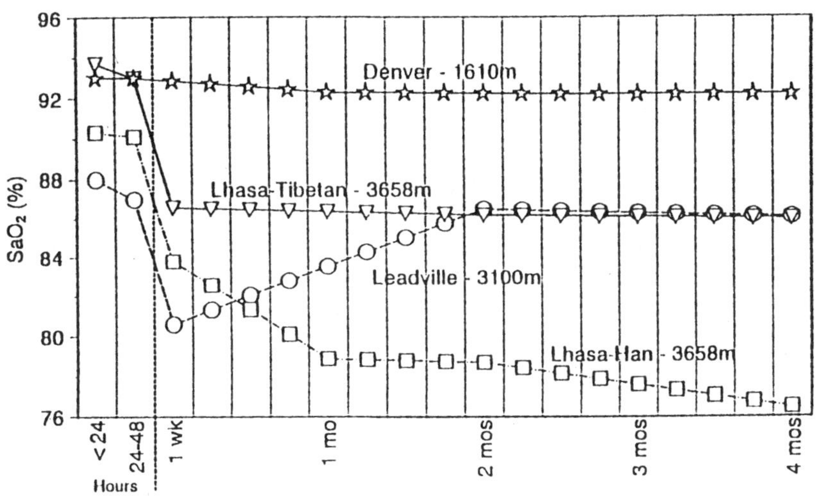

Figure 3 Arterial oxygen saturation in infants during quiet sleep falls with increasing altitude. (Adapted from Refs. 34, 95, and 96.)

At 3658 m in Lhasa, arterial oxygen saturation was higher in Tibetan than in Han infants from birth through 4 months of age (Fig. 3) (34). The neonates were similar in gestational age and Apgar scores; however, the Han had lower birth weights and higher hemoglobin and hematocrit at birth than the Tibetans. In both groups, arterial oxygen saturation was highest in the first 2 days after birth and then declined. Whereas arterial oxygen saturation in the Tibetan infants stabilized at 4 months to values within the range of Leadville infants, arterial oxygen saturation in the Han declined progressively (Fig. 3).

Arterial oxygen saturation in 2- to 5-month-old Quechua infants at 3750 m in Peru averaged 88 ± 3% (97), similar to Tibetan values. At 4540 m in Peru, directly measured arterial oxygen saturation ranged from 57 to 75% in newborn infants 30 minutes to 72 hours old (98) and remained in the range of 74–80% throughout infancy (99).

The cause of the dramatic differences in arterial oxygen saturation between Tibetan and Han infants residing at the same elevation but differing in their high-altitude ancestry likely reflects differences in ventilation and/or pulmonary blood flow, but these have not been measured. Increased ventilation would serve to decrease the oxygen pressure gradient from the atmosphere to the alveoli and to raise arterial oxygen tensions. The accompanying fall in arterial carbon dioxide tensions may also increase pH and left shift the oxyhemoglobin dissociation curve to raise saturation at a given oxygen pressure, but hemoglobin-oxygen dissociation curve position is unknown.

Ventilatory patterns during the neonatal period and infancy change in ways that differ at high compared with low altitude. At sea level, periodic breathing occurs during sleep in the overwhelming majority (78%) of full-term neonates (100). Its prevalence declines perhaps as early as 1 month and definitely by 5–6 months (101) in response to developmental changes in central and peripheral chemoreceptors (101–104). Early studies in healthy infants born in Denver (1610 m) suggested that periodic breathing was "much more frequent and less transitory . . . than has been previously described," but the prevalence reported (65%) was similar to sea level (105,106). At higher altitudes, the prevalence clearly increases; 100% of Leadville neonates demonstrated a repetitive pattern of 4–6 breaths over 6–7 seconds followed by a pause of equal length (105). Current investigations in Leadville confirm the occurrence of periodic breathing in all infants older than 48 hours and its association with cyclic declines in arterial oxygen saturation (S. Niermeyer, unpublished observations). Perhaps periodic breathing is responsible for the decrease in arterial oxygen saturation at 1 week of age at high altitude and the consistently lower values observed during sleep compared with wakefulness throughout the first 4 months of life.

A rapid postnatal fall in pulmonary artery pressure (P_{PA}) occurs in the first days of life at sea level. This is important for achieving first functional and then anatomic closure of the atrial and ductal shunts (107). At 3100 m in Leadville, Colorado, P_{PA} indices obtained by echocardiography in healthy, term infants were normal to moderately elevated during the first week of life and fully normal at 2–

4 months (96). However, all neonates received supplemental oxygen during the first 24–48 hours. At very high altitudes without supplemental oxygen, the postnatal fall in P_{PA} is prolonged or fails to occur. Newborns at 4540 m in Peru with an alveolar PO_2 of approximately 50 mmHg showed persistence of near-systemic P_{PA} for several days following birth (Fig. 4). Administration of 100% oxygen to three infants at 72 hours resulted in a dramatic fall in P_{PA} to near sea level values. Elevated P_{PA} and pulmonary vascular resistance were confirmed by right heart cardiac catheterization in infants and children under 5 years of age at 4330 m and 4540 m in Peru (99). Further support for delayed or, in some cases, absent postnatal regression of the fetal pulmonary vascular pattern comes from histological observations in South American children dying at high altitudes. These infants and children demonstrated peripheral extension and hypertrophy of the muscular layer of the pulmonary arteries and aretioles (108). Elevated right-sided pressure contributes to an increased prevalence of patent ductus arteriosus (109,110). An increased prevalence of atrial septal defect and patent ductus arteriosus has also been observed in Han and Tibetan infants in Qinghai Province of China (the northern portion of the Tibetan Plateau), rising from near zero at sea level to more than 5% at 4500 m (111).

A syndrome of subacute infantile mountain sickness has been described in Lhasa (3658 m) in 15 infants or children who died between 3 and 16 months of age with signs of pulmonary hypertension and right heart failure (112). All were male; 14 were Han and one was Tibetan. All but two were born at low altitude and brought to Lhasa at an average of 2 months before the onset of disease symptoms. Clinical signs and symptoms included dyspnea, cough, cyanosis, sleeplessness, and

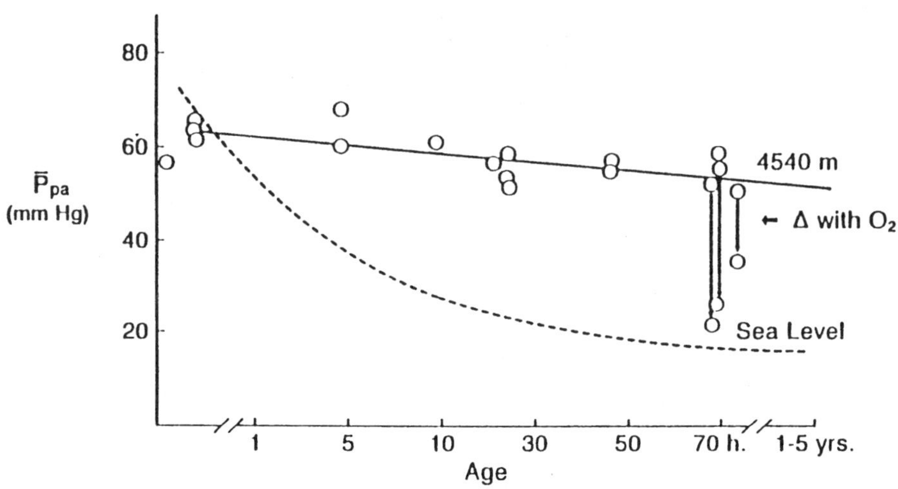

Figure 4 The fall in pulmonary arterial pressure (P_{pa}) in neonates at sea level is greater than at 4540 m in Peru. Supplemental oxygen restores values in high-altitude neonates to sea-level values. (Adapted from Ref. 98.)

irritability, facial edema, hepatomegaly, and oliguria. Histological findings were medial hypertrophy of small pulmonary arteries, muscularization of pulmonary arterioles, and severe right ventricular hypertrophy and dilation. A control group of native Tibetan infants and children who died of noncardiopulmonary causes showed normal thin-walled pulmonary arteries and arterioles after 4 months of age. A similar clinical syndrome was previously reported in five infants and six older children in Leadville, Colorado (3100 m), three of whom underwent cardiac catheterization and demonstrated pulmonary hypertension and one of whom showed medial hypertrophy and intimal thickening at autopsy (113).

In summary, the newborn at high altitude experiences a slower transition from fetal to mature patterns of cardiopulmonary function. Arterial oxygen saturation is lower than at sea level, particularly in samples derived from populations that have recently migrated to high altitude. Periodic breathing appears more common and may contribute to lower arterial oxygen saturation during sleep. The postnatal fall in P_{PA} is markedly slower at very high altitudes but only modestly delayed at more intermediate altitudes where supplemental oxygen is used routinely. Whereas fetal cardiovascular patterns persist into childhood in South America and Han migrant populations in Tibet, histological data in native Tibetan infants suggest a more rapid involution of fetal patterns. A possible reason for this difference between Tibetan and Han children is their better-sustained arterial oxygen saturation during early infancy.

Growth and Nutrition

Growth at high altitude is a product of genetic and/or developmental factors acting in concert with nutrition, levels of habitual activity, and other socioeconomic and environmental characteristics. The relative contributions of these factors to growth has been actively investigated, particularly in the Andean region.

Generally, the decreased growth in utero is sustained postnatally at high altitude. In Andean samples, a consistent reduction in length and weight from birth to 2 years has been observed at high compared with low altitude (114). A comparison of highland and coastal children in Ecuador found that shorter stature in the highlands was due to diminished linear growth velocity within the first 6 months (115). Delayed growth in Andean highlanders continues during childhood and adolescence, resulting in a 1- to 2-year lag in height and a less pronounced adolescent growth spurt. Adolescents grow for about 2 years longer, or until 22 years of age, but adult stature remains shorter than at low altitude (116–121).

The growth failure at high altitude in South America is consistent with that seen among impoverished populations worldwide (114), suggesting the involvement of nutritional factors. This is supported by observations that upper socioeconomic status (SES) children in Nuñoa, Peru, were taller and heavier than lower SES children (28). Over the past two decades, the age at which peak growth velocity was attained and growth completed declined in above-average SES adolescents but remained unchanged in lower SES adolescents in the southern Peruvian Andes (122). Comparisons in Bolivia found no difference in anthropometric characteristics between highland and lowland boys when the comparisons were controlled for SES and nutritional status. Boys from a low SES background at high and low altitude were marginally undernourished and showed a growth delay of approximately 2 years (123).

The pattern of growth retardation in the Himalayas also suggests an influence of nutritional or other socioeconomic factors (25). There do not appear to be any appreciable differences in growth retardation among Tibetan, Han, and Hui male adolescents at 3200 m and 4300 m in Qinghai Province, the northeastern section of the Tibetan Plateau, but in all groups the average body weights were at or below the National Center for Health Statistics (NCHS) 5th percentile and average height was at or slightly above the 5th percentile (124). Rural Tibetans fell below the NCHS standards in height and weight by 4–6 months in both sexes (25). In the urban Lhasa valley, weight for height was within the normal range, but weight and height for age were at the lower range of the NCHS standards. Weight for height in Lhasa was noted to drop below NCHS standards at two time points: one at around 6 months, perhaps reflecting the increased importance of supplemental foods at this age, and a second time point at 1–2 years, when children begin to walk and to depend more on adult foods (25). High-altitude adolescent Bod girls of Ladakh, Jammu, and Kashmir, India, were significantly lighter and shorter than their low-altitude counterparts (125). Smaller body sizes were not found in boys of Tibetan ancestry residing at high compared with low altitude in Nepal, but high-altitude girls over the age of 12 were smaller than low-altitude girls (126). Both high- and low-altitude groups evidenced growth retardation.

As mentioned above, Han infants often remain with extended family in lowland China until about 2 years of age, at which time they are brought back to Tibet. Newly arrived Han children showed poor appetite, diarrhea, and intestinal malabsorption, all of which can adversely affect growth (25). The Han children had markedly lower body weights by age 3 than the native Tibetan children, both rural and urban, despite the generally favored economic circumstances of the Han compared to the Tibetan population.

Enlarged chest dimensions and accelerated chest development, despite generally smaller body size, has been reported in numerous studies at high altitude. Chest shape appears influenced by altitude in both Andean and Himalayan populations such that chest depth increases more than chest width. The contribution of genetic versus developmental factors has received a great deal of study. Since persons of Quechua ancestry born and raised at low altitude also have enlarged chest dimensions (127,128), larger chest dimensions may be a fixed genetic trait in the Quechua. Chest dimensions have significant heritability (i.e., the extent to which the expression of a trait is due to inherited characteristics in a particular environment) among the Quechua and Aymara, and heritability was greater at high altitude (129). This has been interpreted as being due to natural selection acting to increase the representation of genes increasing chest dimensions at high altitudes (130). Phenotypic plasticity may also be involved since genes coding for accelerated chest development might be preferentially expressed in the high-altitude environment.

It is important to recall that lung volumes, not chest dimensions, have functional significance; greater lung volumes are associated with increased surface area for gas exchange and decreased alveolar-arterial oxygen diffusion differences, which help maintain arterial oxygen saturation during exercise at high altitude. While an equivalence is often assumed, chest dimensions are not the same as lung volumes since the chest contains structures other than the lung and larger lung volumes may be accommodated by a lower diaphragm, not a larger rib cage. Despite the large number of studies of chest dimensions, there is a paucity of complete lung volume measurements at high altitude. However, existing data clearly demonstrate that lung volumes are increased worldwide at high compared with low altitude, whether the high-altitude groups are compared with low-altitude U.S. standards or with samples drawn from newcomers living at the same altitude (Fig. 5). Residual volume increases the most, but vital capacity and total lung capacity are also enlarged at high altitude (Fig. 5). Adolescents of Aymara and mestizo ancestry (a mixture of Quechua, Aymara, and Spanish) have modestly (~4%) larger vital capacities when adjusted for differences in body size than similarly aged adolescents of European ancestry born and raised at high altitude (131).

Debate has centered on whether lung volumes (and chest dimensions) are equally enlarged in Himalayan and Andean populations. As demonstrated in Figure

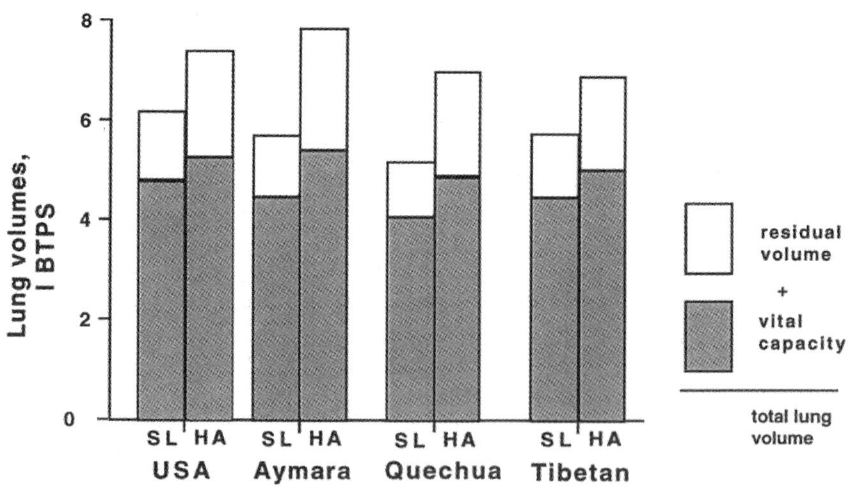

Figure 5 Total lung volume (RV + VC), residual volume (RV), and vital capacity (VC) (all 1, BTPS) in Colorado residents of 3100 m (223,224), Aymara and Quechua South American residents of 3600–4540 m (128,225), and Tibetan residents of 3658 m (226) are greater than sea-level predicted values.

5, Tibetan high-altitude residents clearly have enlarged lung volumes. Other studies have shown that lung volumes are increased in young Bod girls at high altitude in Ladakh, Jammu, and Kashmir, India, compared to their lowland counterparts (125) and in boys between 5 and 18 years at high altitude in Himachal Pradesh (132). Rigorous determination as to whether the magnitude of increase varies among the Andean, Himalayan, and Rocky Mountain regions is hampered by substantial inter-populational variation in body size and disagreement regarding the best reference population. The information summarized in Figure 5 demonstrates that if differences are present, they are of modest proportion.

Larger lung volumes result from exposure to high altitude during growth and development, as they cannot be acquired in adult animals spending an equivalent period of time at high altitude (133). The increase in each lung volume is present by early adolescence (age 11), with residual volume enlarging progressively in 11- to 19-year-old La Paz adolescents (128). A recent, comprehensive review of environmental, developmental, and genetic influences on lung volumes among Bolivian high-altitude residents of varying population ancestry, altitude of birth, and duration of high-altitude residence found that growth and development at high altitude accounts for approximately 25% of the increase in vital capacity and residual volume among males but, interestingly, not among females (134). Genetic factors accounted for an additional 25% of the variability. Occupational characteristics associated with rural high-altitude residence further raised vital capacity but not residual volume. Thus, Frisancho and coworkers concluded that environmental factors (occupation, body composition) exerted greater influence on increasing vital capacity than developmental characteristics, whereas both developmental and genetic factors raised residual volume, at least in males (134).

Taken together, available studies indicate that childhood growth at high altitude is characterized by enlarged lung dimensions—which develop universally during infancy, childhood, and adolescence—and by consistent growth retardation compared to sea-level standards. Nutritional stress plays an important role, especially in rural populations. Malnutrition, energy requirements, intercurrent illness, and genetic factors may variably affect growth of regional populations. However, even when nutritional factors are optimized, some degree of growth retardation appears present at high altitude.

Exercise Performance

Exercise performance in children at high altitude demonstrates the influences of hypoxia as well as nutritional status, developmental, and genetic factors. Greksa et al. (135) studied 11- to 12-year-old healthy, well-nourished Aymara boys in La Paz. VO_2max was slightly lower than that of upper SES European boys in La Paz, but both groups had about a 10% reduction in VO_2max when compared with their low-altitude counterparts. Obert et al. (136) found that VO_2max per kg body weight in highland boys averaged 11% lower than lowland boys. At both high and low altitude, the VO_2max of boys from a high SES, mestizo background was greater than that of boys from a low SES, predominantly Quechua and Aymara background. The high SES group was significantly taller, heavier, and fatter, but when VO_2max was normalized per kg lean body mass, differences disappeared between the high and low SES groups at a given altitude but remained present between altitudes within a given SES group. Maximal power during a force-velocity test and mean power during a 30-second Wingate test were higher in upper SES than lower SES boys, probably reflecting the presence of mild malnutrition in the latter. The authors concluded that a marginal state of malnutrition did not affect VO_2max but led to lower power in prepubertal males at both high and low altitude (136). A similarly designed study of prepubertal Bolivian girls supported these conclusions (137). Girls of low SES had lower VO_2max and anaerobic power than high SES girls. Anaerobic power but not VO_2max remained lower after normalization by body weight in the low SES girls.

Frisancho and coworkers demonstrated that persons born and raised at high altitude (3750 m) in La Paz had greater aerobic capacity than subjects who acclimatized to high altitude during adulthood (138). Younger age of arrival related to greater maximal exercise capacity, particularly when arrival was before age 10. Overall, developmental influences could account for 25% of the variability in maximal exercise capacity. Genetic factors were also influential and could account for an additional 25% of the variability in aerobic capacity among high-altitude residents.

In summary, high-altitude exposure reduces maximal exercise capacity in children in a fashion that is relatively independent of nutritional and other SES influences. However, nutrition and other SES-related characteristics are more important than hypoxia in accounting for altitude-associated reductions in maximal and mean anaerobic power. Being born and raised at high altitude as well as genetic factors serve to restore maximal aerobic capacity toward sea-level values.

Illnesses

In addition to the effects of hypoxia and nutrition, the presence of childhood diseases can exert important influences on exercise performance and growth. There are several ways in which the environmental conditions associated with high altitude interact with childhood illnesses.

Some health problems in children are uniquely related to high altitude. One such problem is reascent high-altitude pulmonary edema (HAPE), which occurs, albeit rarely, in children returning to high altitude after temporary descent to low altitude (139). Reascent HAPE is often preceded by upper respiratory infection (140). Repeated episodes of HAPE can occur in the same child and tend to run in families. Cardiac catheterization in 7 Leadville, Colorado, children (3100 m) after recovery from HAPE showed greater P_{PA} response to hypoxia than in control children and right ventricular hypertrophy by ECG criteria (140). Reascent HAPE has been reported in 4- to 19-year-old residents of Bogotá, Columbia, at 2640 m (141). Data from the Tibetan Plateau suggest an incidence of 1.5% in Han children compared to approximately 0.2% in Tibetans (142). The protection afforded the Tibetan children may be due to diminished pulmonary vasoconstrictor response, better oxygenation, less frequent descent to lower altitudes, and/or other factors reducing susceptibility to altitude-related illness.

Other health problems in children are influenced by cultural responses to the high-altitude environment. Rickets reach a peak incidence of 21% in the 18- to 35-month-old group in the Lhasa valley. Though there are ample hours of sunshine, the cultural responses to the climate (multiple layers of clothing and thick-walled houses with small windows) limit sunlight exposure. Lower prevalence in young children likely reflects the protective effects of breast milk, and older children play outside even during winter. Other cultural factors such as exposure to smoke from cooking fires and crowded living conditions are likely to promote respiratory diseases and exaggerate hypoxemia. Upper respiratory tract infections and pneumonia are more prominent in the colder months, and diarrheal disease is more prevalent in warmer periods (25). Malnutrition and lack of access to health care make diarrheal and respiratory illnesses major contributors to child mortality in high-altitude communities (143).

Finally, some child health "problems" may in fact result from inappropriately applied definitions, despite efforts to altitude-adjust the criteria. The World Health Organization's standard for evaluation of childhood anemia adds 0.57 g of hemoglobin for every 1000 m to the sea-level lower limit of normal (13.03 g/100 mL whole blood) (25). Using these criteria, 49% of Tibetan and 25% of Han children are anemic. It is likely, however, that these criteria overestimate anemia in the Tibetan group. Tibetan hemoglobins are normally distributed in most age groups with mean values being very close to the corresponding sea-level norm. Using sea-level criteria, the overall prevalence of anemia becomes 12% in Tibetan children from 6 months to 7 years of age. This information supports findings in adults and suggests that the Tibetan population does not respond to altitude with as great an increase in hemoglobin as do the Han (25). Consequently, it may be inappropriate to apply the same disease criteria to children from different populations even though they reside at the same altitude.

In summary, the effects of hypoxia on infant, childhood, and adolescent development vary among and within high-altitude regions. Common to all regions is lower arterial oxygen saturation, retention of fetal cardiopulmonary characteristics, delayed growth, enlarged lung volume, and diminished exercise capacity. Inadequate nutritional resources exaggerate growth retardation. Tibetan infants have higher arterial oxygen saturations, less muscularized pulmonary arterioles, and more rapid involution of fetal patterns than Han living at the same altitude. Growth retardation may be less marked in Tibetan than Andean highlanders, but definitive interpretation requires information regarding energy expenditure and substrate utilization. Existing data are insufficient to determine whether differences exist between Tibetan and Andean infants in arterial oxygen saturation, cardiopulmonary characteristics, and exercise capacity. Hypoxia prompts diseases such as reascent HAPE and exacerbates other diseases such as acute respiratory infection and diarrheal diseases, particularly when combined with the aridity, smoke, dust, food shortages, and limited access to health care frequently encountered at high altitude.

C. Adulthood and Old Age

During adulthood, members of one generation are confronted with the need to function as the primary producers of food and other resources required for their own, their children's, and perhaps their parents' and even grandparents' survival. This requires work. Hence, exercise performance is often used to evaluate adult high-altitude adaptation. Work capacity of high-altitude residents is diminished relative to low-altitude values. In this section, we examine exercise performance in acclimatized newcomer and lifelong European, Andean, and Himalayan high-altitude residents and compare the parameters of oxygen transport influencing exercise performance in these groups. We conclude with a consideration of chronic mountain sickness and other diseases that can substantially impair an adult's ability to work and survive at high altitude.

Exercise Performance

Maximal exercise capacity is an index of the integrated functioning of the oxygen transport system. It is reduced at high altitude by approximately 11% per 1000 m altitude gain (144). As a result, exercise capacity is lower upon ascent and after acclimatization to high altitude than it is at low altitude (Fig. 6). The reduction in work capacity is accompanied by reduced cardiac output and decreased brain but not muscle oxygen delivery. Despite extensive study, the cause(s) of the reduction in exercise capacity remain elusive (see Chapter 21).

Several lines of evidence suggest that maximal exercise capacity is greater in high-altitude natives than in acclimatized newcomers (Fig. 6). Averaging values from published studies, exercise capacity was 8% higher in lifelong natives of European ancestry, 16% higher in native Andean residents, and 12% higher in Himalayan natives than in acclimatized newcomers. The differences between natives and newcomers were most consistent when the two groups were well matched for physical training, age, body size, and lean body mass (145,146). Comparing the change in exercise capacity with ascent versus descent also indicates that native high-altitude residents are less affected than acclimatized newcomers; VO_2max decreased 15% with ascent in sea-level residents whereas descent increased VO_2max only half as much (8%) in high-altitude natives (147,148). Variability among high-altitude residents in maximal exercise capacity has been related to developmental, genetic, as well as occupational (e.g., training) characteristics (138), which collectively tend to increase exercise capacity. Based on current data, maximal exercise capacity at high altitude appears to be nearer sea-level values in Andean than in European high-altitude natives, with values being intermediate in Himalayan highlanders (Fig. 6). Relatively few studies have been conducted in Himalayan high-altitude natives, and existing data are variable.

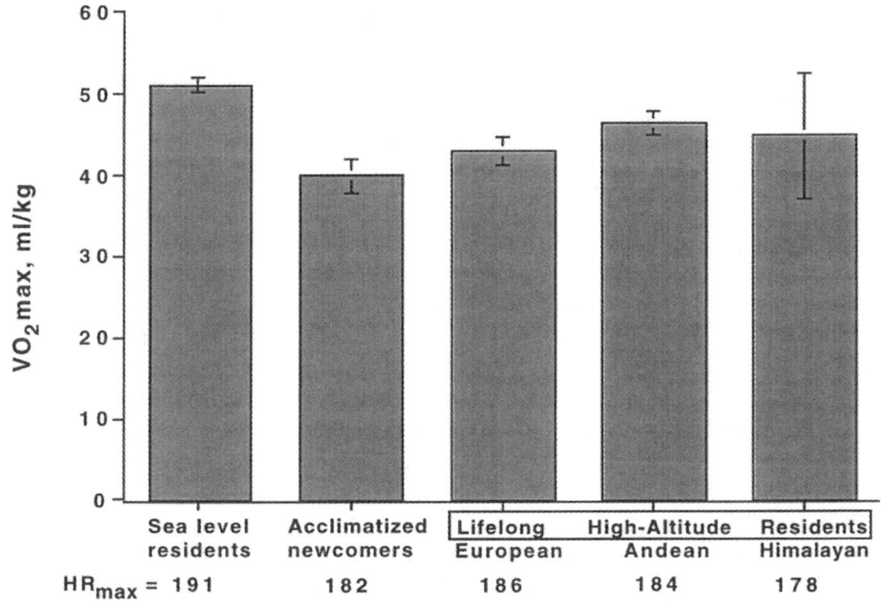

Figure 6 In studies meeting the criterion for maximal oxygen consumption (VO_2max, mL BTPS/min) of a less than 150 mL increase in VO_2 with an at least 25 watt increase in workload, exercise capacity was reduced in acclimatized newcomers as well as lifelong high-altitude residents when compared to values from a large series of normal sea level residents (227). Lifelong high-altitude residence appears to confer some recovery in VO_2max. Relative to acclimatized newcomer values, exercise capacity was increased 8% in lifelong European high-altitude residents, 16% in Andean residents, and 12% in Himalayan natives. Average maximal heart rates (HR_{max}, beats/min) are also shown. (Data from lifelong European high-altitude residents of 3100–3830 m from Refs. 153, 201, and 228. Data on Andean residents of 3400–4540 m are taken from Refs. 146, 147, 228–232. Data from Himalayan residents of 3658–4700 m were reported in Refs. 145, 149, and 150.)

Exercise efficiency or the work that can be performed at a given level of oxygen consumption is increased in some but not other studies of Andean and Himalayan high-altitude natives compared with acclimatized newcomers. The amount of work accomplished at a given VO_2 was nearly 30% greater in native Quechua highlanders than acclimatized lowlanders (148). Increased exercise efficiency has also been seen among higher-altitude Tibetans in some but not all studies (149–151).

Characteristics of Oxygen Transport

A series of recent studies has addressed the mechanisms by which oxygen transport is restored toward sea-level values in high-altitude natives. The components of the oxygen transport system are reviewed below to determine whether such mechanisms differ among regional populations and can explain the greater exercise capacity and/or efficiency of high-altitude natives compared with acclimatized newcomers.

Ventilatory acclimatization, commonly defined as an hypoxia-induced rise in ventilation that overcomes the inhibitory effects of hypocapnia, occurs over days among newcomers at high altitude. It is commonly measured as a fall in arterial PCO_2 at a given PO_2, indicating a rise in effective ventilation (i.e., alveolar ventilation per unit carbon dioxide production) (Fig. 7). Older studies of lifelong Andean (152), Colorado (153,154), and Himalayan (155–157) high-altitude residents found lower effective alveolar ventilation in natives than newcomers. Because infants had similar ventilatory responses at low and high altitude (103) and duration of high-altitude residence correlated with the magnitude of fall in arterial PCO_2, it appeared that ventilatory acclimatization was lost after decades of high-altitude residence. However, most of the earlier Andean and Himalayan studies were based on small samples of uncertain parentage. The newcomer groups were also small and, in some cases, had been exposed to a range of altitudes or severity of hypoxic stimuli. More recent studies have compared larger samples of lifelong high-altitude residents with acclimatized newcomers. Some (158–160) but not all (157) demonstrate levels of effective alveolar ventilation in lifelong high-altitude natives equivalent to those of acclimatized newcomers (Fig. 7). In general, a higher proportion of Himalayan studies lie closer to the newcomer ''after-acclimatization'' curve than did studies of lifelong Andean residents, suggesting that Himalayans have higher effective alveolar

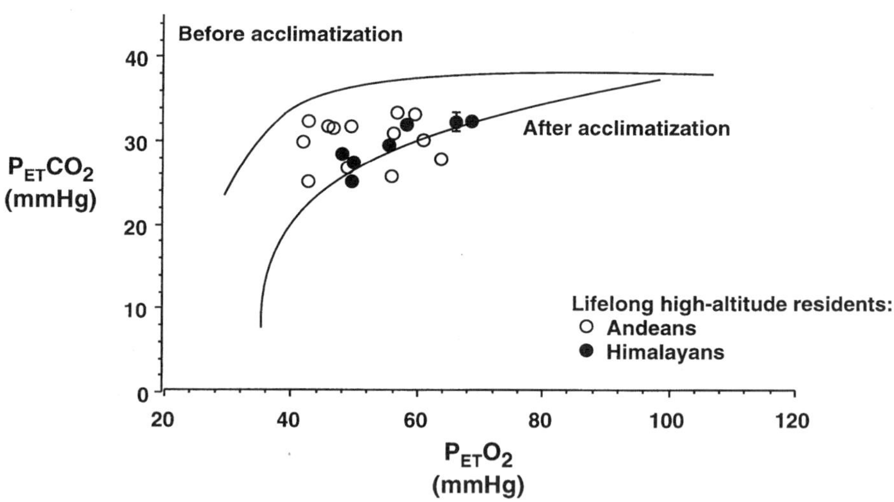

Figure 7 End-tidal PCO_2 ($P_{ET}CO_2$) falls with decreasing end-tidal PO_2 ($P_{ET}O_2$) at high altitude in acclimatized newcomers (after acclimatization line) and lifelong Andean and Himalayan high-altitude residents (From Ref. 158.)

ventilation than Andeans (158). This conclusion is supported by direct comparisons by the same investigators of large numbers of Andeans and Tibetans living at the same altitude (161).

The rise in effective alveolar ventilation is accompanied by an increase in ventilatory sensitivity to hypoxia (162). Conversely, the loss of ventilatory acclimatization has been attributed to a loss of hypoxic ventilatory responsiveness (152,154,163,164). However, it is important to recall that hypoxic ventilatory response is but one factor influencing resting ventilation. For example, Curran et al. (165) showed that Tibetans native to altitudes above 4400 m had levels of resting ventilation similar to Tibetans born and raised at 3658 m, but the higher-altitude Tibetans had lower levels of hypoxic ventilatory response. Likewise, lifelong residents of Leadville, Colorado, with blunted ventilatory drives maintain levels of ventilation equivalent to persons with higher ventilatory responses to hypoxia (154). Consistent with the maintenance of similar levels of effective alveolar ventilation, we found that lifelong Tibetan residents of 3658 m have hypoxic ventilatory responses at least as great as Han acclimatized newcomers and clearly greater than Han who migrated to high altitude as children (158). This agrees with previous reports in which higher or similar levels of hypoxic ventilatory sensitivity were found in Sherpas or Tibetans when compared with acclimatized newcomers (159,160). Comparing all published values, we concluded that Tibetan residents of 3658 m had higher levels of hypoxic ventilatory responsiveness than Andean residents of similar altitudes (158). Recent comparisons in large numbers of Tibetan and Aymara highlanders at the same altitude show that hypoxic ventilatory drives were, on the average, higher and more variable in the Tibetans (161). The higher hypoxic ventilatory drives in Tibetans than Andeans are likely due to innate, possibly genetic factors, although developmental alterations resulting from differences in intrauterine or neonatal oxygenation have not been excluded. Twin studies at low altitude demonstrate that a significant portion of the variation in hypoxic ventilatory response is due to genetic factors (166,167). At high altitude, the studies of Beall, Blangero, and coworkers demonstrate significant heritability for hypoxic ventilatory drives and, interestingly, a higher heritability in Tibetans (46%) than Andeans (12%) (168). The interpretation of heritability is complex, and lower heritabilities do not necessarily mean lesser degrees of genetic involvement. For example, the lower heritability among Andeans may indicate that there is insufficient variation in the genes involved to demonstrate a heritable component or it may mean that the genes influencing hypoxic ventilatory response were lost from the Andean gene pool. Since heritability estimates are derived for a given population, strictly speaking, they should not be compared between populations.

In addition to alveolar ventilation, arterial oxygen saturation is influenced by the alveolar-arterial oxygen diffusion difference [$(A\text{-}a)DO_2$], hemoglobin-oxygen affinity, and alveolar/end-capillary diffusion disequilibrium. Diffusion limitation is important during heavy levels of exercise at sea level (169). It is also important at high altitude where the oxygen driving pressure is low and arterial oxygen tensions are on the steep part of the dissociation curve, although the lower absolute cardiac outputs attained at maximal exercise would be expected to lessen diffusion limitation by lengthening transit time. Reduction in (A-a)DO_2 permits high-altitude natives versus acclimatized newcomers to better maintain arterial oxygen saturation at a given workload (158) or level of oxygen consumption (Fig. 8). The smaller (A-a)DO_2 is likely the result of greater surface area for gas exchange due, in turn, to larger lung volumes (Fig. 5). Of interest is that (A-a)DO_2 is equally narrowed in Tibetan, Andean, and Rocky Mountain high-altitude residents (Fig. 8).

Hemoglobin-oxygen affinity varies as the result of genetic variation in hemoglobin structure as well as variation in the effectors of hemoglobin-oxygen binding (pH, temperature, red blood cell concentrations of 2,3-diphosphoglycerate and adenosine triphosphate). There are no reports of electrophoretic or DNA sequence hemoglobin variants in human high-altitude populations, but if present, they do not appear to affect hemoglobin-oxygen affinity as measured by the position of the hemoglobin-oxygen dissociation curve. Existing Andean and Himalayan data indicate that hemoglobin-oxygen affinity is fully normal by sea-level standards (170,171). Generally, studies have been performed only at rest; in Tibetans at near-maximal exercise, standard P50 does not change and in vivo P50 rises modestly to 28.4 ± 0.5 mmHg as the result of mild acidosis (arterial pH of 7.38 ± 0.01) (L. G. Moore, unpublished observations).

Resting arterial oxygen saturation appears influenced by heritable factors in lifelong Tibetan but not in Andean high-altitude residents (172,173). In Tibetans at 4850–5450 m and 3800–4065 m, complex segregation analysis of familial variation in arterial oxygen saturation revealed a pattern of inheritance consistent with a simple, single, autosomal locus with dominance (172). It is not clear what this gene (or

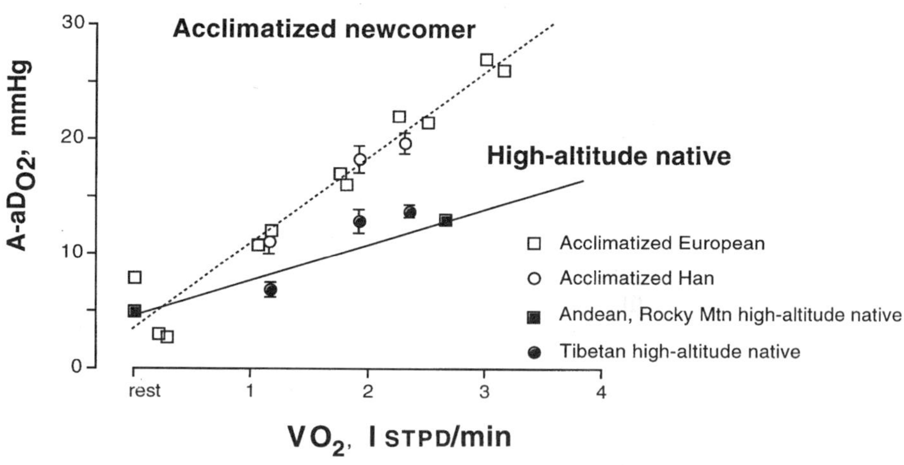

Figure 8 The alveolar-arterial oxygen difference [(A-a)DO_2] is lower in high-altitude natives than acclimatized lowlanders. Andean, Rocky Mountain, and Himalayan values are similar in relation to levels of oxygen consumption (VO_2). (From Ref. 233.)

group of closely linked genes) is. The variation in arterial oxygen saturation was not related to ventilation, ventilatory response to hypoxia, hemoglobin, age, or sex (34,174). It is important to recall, again, that the lack of significant heritability among Andean highlanders does not indicate that genetic influences were absent but may simply be due to insufficient variation to demonstrate a heritable component.

Hemoglobin concentration at a given altitude is lower in Himalayan than Andean highlanders. This is evident in large-scale surveys (175) and carefully matched comparisons of Sherpa and Aymara from similar, rural regions (176). Sherpas demonstrate lower and more normally distributed hemoglobin and hematocrit values than the Aymara (Fig. 9). The lower hemoglobin may result from a lesser hypoxic stimulus, due perhaps to better-maintained ventilation, as described above. Alternatively or additionally, higher erythropoietin levels in Aymara than Sherpa residents at the same altitude suggest that Andeans may have greater erythropoietic response to a given level of hypoxic stimulus (176). Other factors may be involved, including developmental regulation of hemoglobin production (177) or variation in red blood cell destruction. Cultural and environmental characteristics are also important; lower hemoglobin concentrations in rural than urban highlanders suggest that dust and other factors increasing the risk of chronic lung disease may magnify the erythropoietic effects of ambient hypoxia (178).

Cardiac output rises with progressive exercise, serving to increase oxygen and other nutrient delivery to the exercising muscle. The greater exercise capacities in native highlanders than acclimatized newcomers suggest higher cardiac outputs,

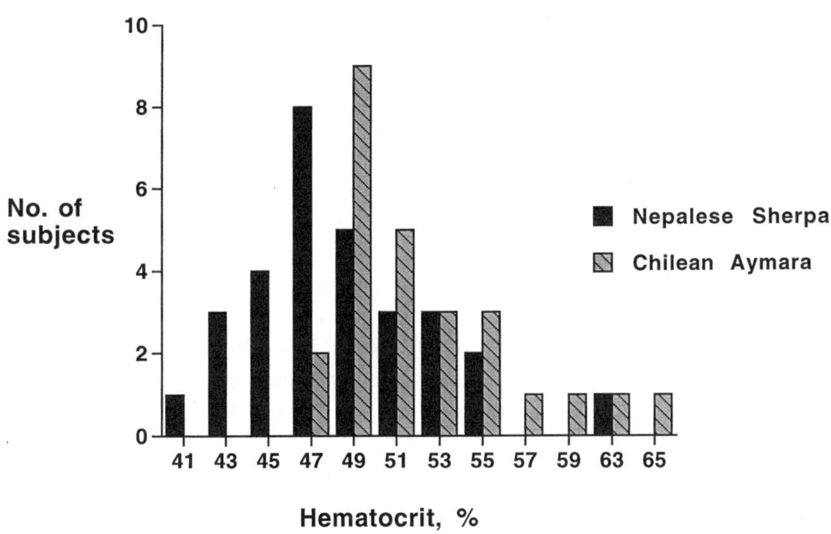

Figure 9 The distribution of hematocrit (%) in adult, rural male residents of equivalent altitudes is shifted to lower values in Nepalese Sherpa (filled bars) than in Chilean Aymara (hatched bars). (From Ref. 176.)

which could be due, in turn, to higher heart rates and/or stroke volume. Existing data do not support higher heart rates at maximal exercise but rather suggest lower values in high-altitude natives than acclimatized newcomers (Fig. 6). Several studies support the concept that natives are more reliant on parasympathetic tone, whereas newcomers depend more on beta-sympathetic stimulation to raise heart rate at maximal effort (179–181). The possibility of increased stroke volume is supported by greater resting heart volumes in young Tibetan men compared with similarly sized and aged, healthy Han residents of Lhasa (604 ± 21 vs. 550 ± 24 cm^3 or 377 ± 11 versus 347 ± 12 cm^3/m^2 body surface area) (L. G. Moore, unpublished data). A rise in stroke volume can be accomplished by an increase in preload, greater myocardial contractility, or a reduction in afterload. Few studies have measured all three factors. In native Tibetans at 3658 m, we found that preload (right and left heart filling pressures) was low and rose normally with exercise (182). Stroke volume rose with increasing filling pressure in a fashion that was normal by sea-level standards and greater than previously reported for Colorado or Peruvian high-altitude natives (183,184). Furthermore, stroke volume at a given right atrial mean or pulmonary capillary wedge pressure was not elevated by the addition of 100% oxygen to the inspired air, implying no hypoxic depression of myocardial contractility. Pulmonary arterial pressure and resistance were remarkably low and unresponsive to added hypoxia when compared with previous studies in lifelong Rocky Mountain or Andean high-altitude natives (182). Thus, particularly during exercise, Tibetans demonstrated a lower pulmonary arterial pressure gradient across the lung than Andean high-altitude natives (Fig. 10) and a decrease in systemic vascular resistance, enabling the Tibetans to achieve a more than threefold elevation in cardiac output with good cardiac functional reserve. Consistent with the low pulmonary arterial pressures and absence of hypoxic vasoconstriction, Gupta et al. have reported a lack of smooth muscle in the small pulmonary arteries of native Ladakhi men (185). The absence of pulmonary hypoxic vasoconstriction and vascular remodeling in these human studies is similar to that which has been reported in the yak (186,187), snow pig (188), and llama (189)—all species considered to be genetically adapted to high altitude. The Tibetans' well-maintained myocardial contractility and low vascular resistance in both the pulmonary and systemic circulations would be expected to permit greater cardiac output during exercise than in Andean or Rocky Mountain high-altitude residents.

During exercise, the preservation of cerebral blood flow appears to be favored in Tibetan high-altitude natives compared to acclimatized newcomers. Using internal carotid artery blood flow as an index of cerebral blood flow, young Tibetan men maintained higher flow and better preserved cerebral oxygen delivery at peak effort than Han acclimatized newcomers (190). A comparison of Himalayan natives with European elite climbers revealed fewer MRI changes indicative of cortical atrophy and brain damage in the high-altitude natives, despite more frequent exposure to extreme high-altitude without supplemental oxygen (191). As reviewed above, the greater redistribution of lower extremity blood flow to favor the uteroplacental circu-

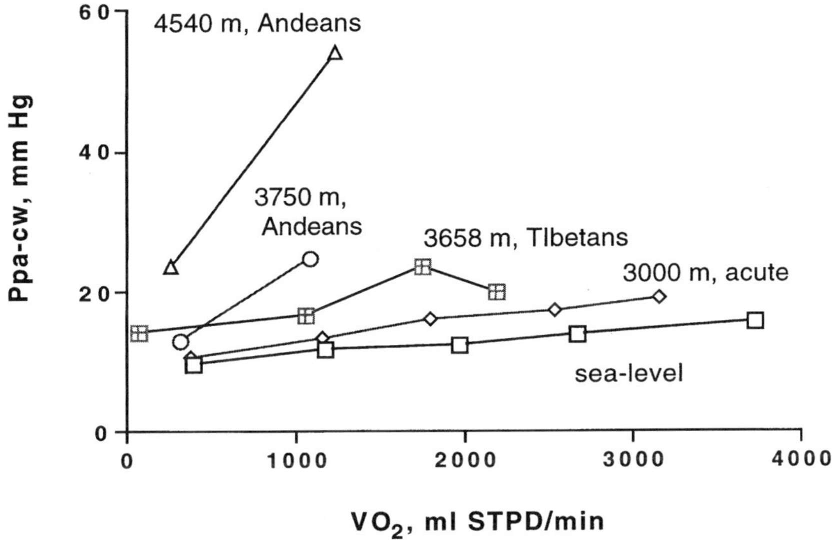

Figure 10 The pressure gradient across the lung (P_{pa}—P_{cw}, pulmonary arterial pressure—pulmonary capillary wedge pressure, mmHg) changes little with increasing oxygen consumption (VO_2, mL STPD/min) in sea-level residents (open squares) (169,234), sea-level residents after acute exposure to 3000 m (open diamonds) (169,234), Tibetan residents of 3658 m (squares with crosses) (182) but rises markedly in Andean lifelong residents of 3750 m and 4540 m (open triangles) (184,235).

lation in Tibetan than Han pregnant women also indicates good blood flow distribution to organs with high oxygen demand.

Tissue oxygen utilization can be increased by an augmentation of arterial oxygen content, blood flow, or tissue oxygen extraction. As a measure of tissue oxygen extraction, we plotted all published data for the differences in arterial-venous oxygen content versus cardiac output in male sea-level residents, high-altitude natives, and after acute or more prolonged exposure to high altitude (Fig. 11). Acute hypoxia lowered the arterial-venous oxygen content per unit blood flow, probably as the result of decreased arterial oxygen saturation. After acclimatization, the slope of the line relating arterial-venous oxygen content to blood flow left-shifted toward sea-level values, probably as the result of a rise in arterial oxygen content. The slopes of the lines in Rocky Mountain, Andean, and Tibetan high-altitude natives were close to that of sea-level natives (Fig. 11), suggesting that further time-dependent changes raised extraction, perhaps by lowering mixed venous oxygen content. Whether regional high-altitude populations differ in terms of oxygen extraction cannot be determined from the available data because only our Tibetan subjects exercised across a sufficient range of exercise loads.

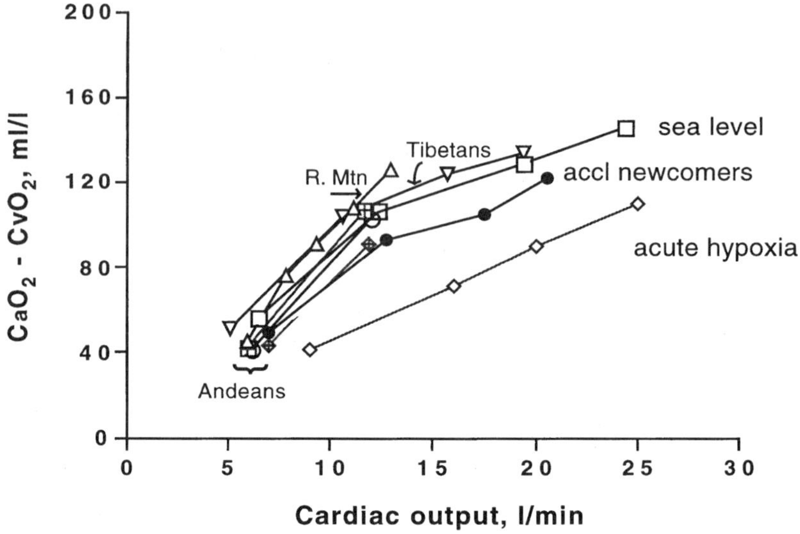

Figure 11 The arterial-venous difference in oxygen content (C_aO_2—C_vO_2, mL O_2/L whole blood) rises with increasing cardiac output during exercise in sea-level residents (open squares) (169,234). The difference is narrowed for a given cardiac output in sea-level residents after acute exposure to 3000 m (open diamonds) (169,234) and in sea-level residents after 14 days acclimatization to 3800 m (open circles) (234). Lifelong high-altitude residents of Tibet (downward pointing triangles) (182), the Andes [squares with crosses (236); diamonds with crosses (235); circles with crosses (184)], and Colorado (upward pointing triangles) (183) more closely resemble sea-level residents.

Tissue oxygen extraction can be influenced by differences in skeletal muscle capillarity, mitochondrial volume density, oxidative enzyme capacity, or reliance on anaerobic metabolism. Capillary density in Sherpa mountaineers was within the range of acclimatized European values. The muscle fiber area perfused by each capillary was greater, not reduced, in Sherpas compared to European climbers (192). The relative or absolute enzyme mitochondrial oxidative capacity was also diminished (144,192). Lower lactate levels at a given workload in lifelong high-altitude natives than acclimatized newcomers (or sea-level residents) suggest that there is not greater reliance on anaerobic metabolism (148,149). Hochachka and coworkers have advanced the idea that closer coupling of ATP supply and demand permitted more efficient mitochondrial oxygen utilization and, in turn, lowered glycolysis and lactate production (144,148). In support of this, NMR studies of muscle metabolism conducted at sea level demonstrated that Quechua highlanders accomplished greater calf muscle work with similar or reduced perturbations in phosphorylation potential, phosphocreatine or ATP concentration than sedentary, endurance, or power-trained athletes. Alternatively, Kayser and coworkers have suggested that shorter oxygen diffusion path, greater oxygen conductance, or increased tissue oxygen stores as the result of elevated myoglobin or other oxygen-binding pigments may be involved (192). Interestingly, the low mitochondrial volume/specific VO_2 peak ratio appears to be an inborn feature of Tibetans (193).

In summary, maximal exercise capacity appears to be restored toward sea level values after generations among Andean and Himalayan high-altitude residents compared with acclimatized newcomers. Several factors are likely to contribute to this recovery. One factor is the maintenance of increased alveolar ventilation and hypoxic ventilatory response, a significant factor for increasing ventilation during exercise as well as at rest (194). Another is the decreased $(A\text{-}a)DO_2$. One of the most decisive differences observed among high-altitude populations is the Tibetans' low pulmonary arterial pressure (182). The consequent reduction in afterload to the right heart may permit the Tibetans to achieve a greater rise in cardiac output during exercise than acclimatized newcomers or other high-altitude residents. Other tissue-related characteristics serving to increase tissue oxygen utilization should receive additional investigation.

Chronic Mountain Sickness and Other Chronic Diseases

Comparatively little attention has been paid to the occurrence of chronic diseases during adulthood and old age at high altitude with the notable exception of chronic mountain sickness (CMS), also called Monge's disease (33). CMS occurs in persons who have lost their ventilatory acclimatization to high altitude after seemingly normal adjustment to residence at high altitude. It is characterized by symptoms of clubbing, facial edema, conjunctival congestion, headache, dizziness, paresthesias, fatigue, loss of memory, insomnia, poor appetite, and dyspnea. CMS is usually diagnosed on the basis of hemoglobin (or hematocrit) concentrations above the normal range in the absence of clear evidence of chronic lung or left-sided heart disease (195,196). A symptoms score is sometimes used (197).

Several hypotheses have been advanced to account for the occurrence of CMS. One hypothesis is that CMS is due to an age-associated loss of ventilatory sensitivity to hypoxia and consequent reduction in ventilation, arterial oxygen tension, and rise in hemoglobin concentration (198,199). Persons with CMS have lower levels of ventilatory responsiveness to hypoxia, effective alveolar ventilation, and arterial oxygen tensions than healthy persons at the same elevation (200). However, persons with low hypoxic drives live successfully at high altitude for years, and some persons with CMS have hypoxic responses within the normal range, suggesting that an age-associated decline in hypoxic ventilatory sensitivity is not the only factor. Other abnormalities of ventilatory control may be involved. For example, we have found that ventilatory depression with hypoxia was greater in individuals with CMS than in age-matched controls (196,200). While an age-associated increase in hemoglobin or hematocrit in Leadville has been reported, it is important to recall that there is considerable variation in this age-related pattern (201).

Another view is that CMS represents subclinical chronic obstructive lung disease or left-sided heart disease, the symptoms of which are altered by residence at high altitude (202). In support of this, approximately half the persons with CMS in Leadville, Colorado, demonstrated some impairment of lung function (202). We observed that CMS patients in Lhasa with lower arterial oxygen saturations had smaller vital and total lung capacities than the better-oxygenated CMS patients, suggesting that impaired lung function and widening of the alveolar-arterial oxygen gradient contributed to the severity of the disease (200). Mining is an important economic activity in the North American and many of the South American communities where CMS has been described, creating the possibility that occupational factors increasing the risk of obstructive lung disease contribute to its etiology. In support of this are the generally lower hemoglobin values in rural than in urban regions of South America (178) and the high prevalence of cigarette smoking among persons with CMS (202,203).

It has been suggested that persons with CMS have an excessive bone marrow response to a given level of hypoxia, but greater serum immunoreactive erythropoietin levels have not been observed in CMS patients compared with healthy controls (204). Additional investigation of factors influencing erythropoiesis is warranted.

CMS may be a manifestation of disordered breathing and brain blood flow regulation during sleep. Kryger and coworkers (205) found sporadic episodes of arterial desaturation in CMS patients in Leadville, Colorado (3100 m). Following treatment with the respiratory stimulant, medroxyprogesterone acetate, episodes of arterial desaturation became less common, hemoglobin fell, and symptoms overall improved. We documented an increased frequency of sleep-disordered breathing (apneas and hypopneas) and prolonged episodes of hypoventilation in CMS patients compared with healthy, age-matched control residents of Lhasa, Tibet (3658 m) (206). While higher hemoglobin concentration prevented lower arterial oxygen saturation from decreasing arterial oxygen content during the day, the nighttime disturbances in breathing reduced arterial oxygen content well below control values. Cerebral oxygen delivery during sleep appeared to be further reduced by a blunted brain blood flow response to hypercapnia and hypoxia; CMS patients had less rise in internal carotid artery blood flow than controls during and immediately following episodes of sleep-disordered breathing (206). The fall in arterial oxygen tensions during the night would be expected to stimulate erythropoiesis and augment pulmonary vasoconstriction. Since metabolic alterations have been implicated in blunting hypoxic ventilatory sensitivity and augmenting hypoxic ventilatory depression, it is possible that similar mechanisms mediate the lack of brain blood flow response and ventilatory characteristics of CMS. Whatever the cause(s), it is likely that the reduction in brain oxygen delivery contributes to the cognitive impairment and neurological symptoms of CMS (206).

The prevalence of CMS appears to vary among and within regions as well as between men and women in the same region. Prevalence data, ironically (considering its proximity), are the least complete for the Rocky Mountain region. Kryger et al. remarked that 60 men and 2 women in Leadville, Colorado, at 3100 m were under treatment that would yield a minimal prevalence estimate of only 3% (196). While the mobile nature of the Rocky Mountain population complicates the conduct of surveys, the altitude at which persons sleep may be less variable and of greater physiological importance.

A comprehensive prevalence study has been undertaken in 2875 male and 152 female residents of 4300 m in Cerro de Pasco, Peru (197,199,207). Population ancestry was not reported, but most persons in this community are of mestizo parentage. In men, diagnosis of excessive polycythemia was based on hemoglobin concentration greater than 21.3 g/dL in combination with an elevated symptom score. This hemoglobin cutoff was defined as 2 SD above the average value for 20 to 39-year-old men. In women, diagnosis of excessive polycythemia was based on a hematocrit greater than 56%, defined as greater than 2 SD above the average hematocrit for women aged 30–40 years. The prevalence of excessive polycythemia was greater in men than women, occurring in 16% of adult males and 9% of women between the ages 30 and 54 years. Prevalence rose with age for both sexes. There was a uniform age-associated increase in hematocrit in men, whereas in women the relationship was relatively flat until after the age of the menopause (197). Among men, 7% aged 20–29 years versus 34% aged 60–69 years had excessive polycythemia (207). Among the 9% of women with excessive polycythemia, 77% were between 45 and 54 years (197).

Prevalence of CMS in the Himalayan region is greater in Han migrants than in Tibetan lifelong residents. In an epidemiological survey of 3201 persons at 3600–4700 m in Tibet, CMS was much more frequent among Han than Tibetans and more common in men than in women (Table 4). Han had a nearly 10-fold excess of CMS compared with Tibetans in a survey of 25,618 persons living at 2261 to 5225 m in

Table 4 Prevalence (%) of Excessive Polycythemia[a] in the Tibet Autonomous Region in 3201 Male (n = 1749) and Female (n = 1452) Migrants (Han) and Native Highlanders (Tibetans) More Than 15 Years Old

	Males			Females		
Region (alt, m)	Migrant workers	Native workers	Native farmers, herders	Migrant workers	Native workers	Native farmers, herders
Lhasa (3658 m)	12.97	1.05	0	1.64	0	0
Gyangze (4040 m)	31.5	4.8	1.5	3.8	0	0.3
Nagqu (4500–4700 m)	38.4	14.4	6.6	7.2	6.5	2.7

[a]Excessive polycythemia at 3658 and 4040 m was defined as red blood cell (RBC) counts > 6.5 × 10^6/μL blood and hemoglobin (hgb) > 20 g/dL blood, or RBC counts > 7.15 × 10^6/μL, or hgb > 22 g/dL. At 4500–4700 m, excessive polycythemia was considered as RBC > 7.0 × 10^6/μL and hgb > 21 g/dL, or RBC > 7.7 × 10^6/μL, or hgb > 23 g/dL.
Source: Ref. 208.

Qinghai Province (209). Most cases occurred in cigarette-smoking Han men (none of whom were miners) who emigrated from lowland regions as adults (200,203). Tibetans are not fully protected from CMS, but all the cases we have seen were in persons born in lower altitude regions of Tibet after extended residence at much higher elevations (200,206).

It is difficult to compare prevalence estimates among the Himalayan, Andean, and Rocky Mountain regions, given the lack of population ancestry data for Andeans, age-specific prevalence estimates for Himalayans, and complete prevalence information from the Rocky Mountain region. Existing data suggest that the prevalence among the Han is within the range of Andean values and that both the Han and Andean values are greater than the Tibetans'. The Tibetans' lower prevalence may, in turn, be due to their lower hemoglobin concentrations, higher ventilations, and/or greater hypoxic ventilatory drives. Lower hemoglobin concentrations appear to be present lifelong in Tibetans. Unknown is whether the effects of aging on ventilation and arterial oxygenation, particularly during sleep, differ in Tibetans and other high-altitude residents. We did not observe any differences in the occurrence of sleep-disordered breathing in young Han and Tibetan men at 3658 m (206).

Information about other chronic cardiovascular or respiratory diseases in high-altitude residents is limited. The prevalence of systemic hypertension is reported to be low in Peruvian high-altitude communities (210) but increased among lifelong Tibetan residents of high altitude compared with Han newcomers (211). In the Peruvian study, both diastolic and particularly systolic pressures were lower (210), whereas in Tibet higher diastolic and systolic pressures were seen (211). In the Peruvian study, genetic, dietary, and other similarities between the high- and low-altitude regions led the investigators to postulate that the blood pressure reduction was due to chronic hypoxia acting to lower peripheral vascular resistance and/or cardiac output (210). This is consistent with the decline in blood pressure observed in men after years of high-altitude residence (212), by which time the increase in blood pressure resulting from sympathetic stimulation (213) has presumably been reversed. The high salt intake in the Tibetan diet may be a factor in the high prevalence of hypertension observed.

Among Rocky Mountain residents, mortality from emphysema and chronic bronchitis is increased at high compared with low altitudes (214,215). Persons died after a shorter duration of illness and more commonly from right heart failure, suggesting that the hypoxia of high altitude rather than some other factor associated with high-altitude residence worsened survival. There are also fewer elderly, measured as the proportion of the population over the age of 65, at high than at low altitudes in Colorado (216). This is not the result of increased mortality but rather increasing outmigration after the age of 50 years in combination with an influx of younger-aged residents. The most frequent cause for relocation cited by persons who migrated from high altitude was poor health, whereas family-related concerns were the principal cause of migration for persons moving from one low-altitude location to another. The health problems cited were, overwhelmingly, complications of heart and lung disease (216).

In summary, CMS is a clear instance of adaptive failure at high altitude. Persons who die from CMS (or its associated conditions, including right heart failure and strokes) are often elderly, but it affects persons in the early years of adulthood as well. Susceptibility to CMS appears to be greater among migrants to high altitude than among natives. Lower prevalence in Andean compared with Tibetan high-altitude residents suggest that generational factors and/or the degree of genetic admixture may be involved. The protection afforded Tibetans may be due to their higher ventilation and hypoxic ventilatory drives as well as to other factors. Further research is needed to provide more consistent, age-specific diagnostic criteria for CMS. Not only altitude but also age and gender should be taken into account for defining cutoff values for what constitutes ''normal'' hemoglobin. In the case of women, such standards need to take account of other factors influencing hemoglobin concentration, such as menstruation and pregnancy.

IV. Summary and Conclusions

Recent studies using an expanded range of techniques have revealed differences between lifelong high-altitude residents and acclimatized newcomers as well as among resident high-altitude populations. These differences are of physiological and anthropological importance insofar as they imply that generational factors influence adaptive processes. Here we try to summarize what has been learned, indicating some of the complexity influencing the interpretation of such studies and suggesting future studies that might be undertaken to resolve at least some of the issues remaining.

As reviewed above, geographical and historical circumstances differ among high-altitude populations. The Tibetan Plateau is larger, more geographically remote, and has been occupied by humans for a longer period of time than the Andean Altiplano or Rocky Mountain Plateaus. Given these distinctions, the Tibetan population is likely to have been the longest resident at high altitude. Compared with Andeans, the Tibetan gene pool is less likely to have been constricted by small numbers of initial migrants and/or severe population decline. Tibetans may also have been subject to less genetic admixture with lowland groups, although this may be changing today with the influx of large numbers of Han migrants. Admixture with lowlanders is not an issue for Rocky Mountain inhabitants as they are persons of low-altitude ancestry. While viewed by some investigators as a limitation, the fact that Rocky Mountain residents are of low-altitude ancestry has the advantage of providing a control group with which other high-altitude populations can be compared.

Future studies are required to better document the population history, extent of genetic admixture, and relationship of genetic to physiological traits in high-altitude populations. For such admixture and population history studies, the important but often overlooked sex-specific nature of gene flow needs to be taken into account. Population history and admixture studies are performed using mitochondrial DNA (mtDNA), y-chromosomal nuclear DNA (yDNA), and/or autosomal nuclear DNA (nDNA) genetic systems. MtDNA and yDNA offer special advantages since they do not undergo recombination (by virtue of being inherited solely through the mother or father, respectively). In addition, mtDNA appears to accumulate variation (mutations) more rapidly than nuclear DNA (nDNA) (217). However, admixture estimates vary, depending on what kind of DNA is employed. For example, in the San Luis Valley, Colorado, intermarriage between Native Americans and Spanish-Americans occurs disproportionately between Native American women and Hispanic men. As a result, 87% of Hispanic persons had mtDNA of Native American origin, whereas in the same individuals the estimate of Native American admixture using nDNA was 48% (15). In an analogous situation, a greater degree of European admixture was inferred in Central American tribes when yDNA rather than mtDNA was evaluated, indicating that the European contribution to Central American gene pools was predominantly male in origin (218). The relevance for high-altitude populations is that admixture with Europeans may not be detected in Andean populations using mtDNA alone. If contact between Tibetan and other populations was sex-specific, admixture estimates would likely vary according to the source of the genetic material. For example, Mongolian conquest and movements of male traders would be expected to yield higher yDNA than mtDNA admixture estimates. The possibility exists that such differences in admixture estimates might be useful for inferring patterns of contact, although such analyses quickly become very complicated. Other cultural practices can also influence mtDNA variation; for example, polygyny (one man with several wives) would be expected to increase mtDNA variation and female infanticide to decrease mtDNA variation.

Several differences in adaptive success between natives and newcomers have been identified. Lifelong high-altitude residents of the Andes and/or Himalayas have the following distinctions when compared with acclimatized newcomers:

- Less intrauterine growth retardation
- Better neonatal oxygenation and involution of fetal cardiopulmonary characteristics
- Enlarged lung volumes and decreased alveolar-arterial oxygen diffusion gradients
- Higher maximal exercise capacity (VO_2 max)
- Better maintained increase in cerebral blood flow during exercise (Tibetans only)
- Lower hemoglobin concentrations (Tibetans only)
- Less susceptibility to CMS (Tibetans only)

Among high-altitude native populations, several differences in adaptive success and physiological strategy have emerged from recent comparisons. According to an evolutionary hypothesis, the greater duration and genetic isolation of the Tibetan population would be expected to result in better adaptation, as judged by the existence of attributes associated with improved chance of reproductive success. Ordering the comparison in relation to the Tibetan population, Tibetans demonstrate:

Less intrauterine growth retardation
Greater reliance on redistribution of blood flow than elevated arterial oxygen content to increase uteroplacental oxygen delivery during pregnancy
Higher levels of resting ventilation and hypoxic ventilatory responsiveness
Less hypoxic vasoconstriction, lower pulmonary arterial pressure and resistance
Lower hemoglobin concentration
Less susceptibility to CMS

Several of the distinctions demonstrated by Tibetans parallel the differences between natives and newcomers, although the degree of protection or adaptive benefit appears enhanced for the Tibetans. However, many of these observations are based on limited information, and the requisite comparative data from other regions are not always available or collected with the same methodologies. For example, direct comparisons of cardiac output and pulmonary arterial pressure are needed in larger numbers of Tibetan and Andean high-altitude residents at similar levels of exercise intensity. Another feature of the above list is that characteristics which are likely to be closely related to mortality risk (e.g., IUGR, uteroplacental blood flow, susceptibility to CMS) demonstrate clearer differences between natives and newcomers or among native populations than other characteristics such as exercise performance. This may be the result of difficulties in accurately measuring exercise performance, given the influences of training, motivation, etc., or it may indicate that differences in exercise performance are less affected by the processes of natural selection operative at high altitudes. The higher frequency of IUGR and preeclampsia at high altitude provides a useful opportunity to generate insights into pathophysiology. Finally, the above list demonstrates that selective pressures differ between the sexes. These differences are particularly apparent during pregnancy. The apparently lower susceptibility to CMS among women than men raises the intriguing possibility that sex differences in adaptive processes are of lifelong significance.

There are also several points of similarity. One is the seemingly equal retardation of postnatal growth among highland infants, children, and adolescents when compared with well-validated, low-altitude NCHS standards. Since it is likely that insufficient nutrition and recurrent illnesses contribute to the growth retardation observed, additional studies are warranted that examine the separate and combined effects of hypoxia, nutrition, and illness on growth. Such studies could be usefully conducted in the Rocky Mountain region where the effects of hypoxia alone should be more readily apparent. Such studies of the interactive effects of altitude, poor nutrition, and intercurrent illness are important for regional and health policy planners. Another related issue concerns the impact of IUGR. As reviewed above, information on gestational age along with birth weight is essential for comparing birth weight–specific mortality. This, in turn, requires complete population enumeration or, at least, large and demonstrably representative samples. With the efforts to improve health and heath care in many regions of the world, such information may be forthcoming and will be important for determining the relative influences of prematurity and IUGR on mortality risk at high altitude. Studies in the neonatal period and infancy are required to address the consequences of preterm delivery, IUGR, poor or delayed cardiopulmonary transition, and intercurrent respiratory illness for mortality and morbidity at high altitude.

Lung volumes appear universally increased in lifelong high-altitude residents. While subtle variations may be important, the decrease in alveolar-arterial oxygen tension differences appears essentially the same in Rocky Mountain, Andean, and Himalayan highland residents. Additional studies are warranted to determine what kinds of growth signals selectively stimulate the lungs and the mechanisms by which such an increase diminishes alveolar-arterial oxygen tension differences at high altitude.

These observations suggest several conclusions regarding convergence and complexity of adaptive events. In terms of convergence, the Himalayan Plateau has not only been the longest occupied and most remote, its occupants demonstrate several physiological advantages that are consistent with their hypothesized evolutionary advantage. Of interest is that the overall pattern of successful adaptation appears to result in establishing the physiological and functional attributes of sea-level residents. For example, Tibetans more closely resemble healthy persons at sea level in terms of their characteristics of intrauterine growth and birth weights, uterine artery blood flow, hemoglobin concentrations, and pulmonary arterial pressures. This supports the view that some of the physiological responses commonly observed in sojourners and North or South American high-altitude residents—e.g., decreased birth weight, polycythemia, diminution of ventilation, moderate elevation in pulmonary arterial pressure—are not adaptive. It also raises the presently unanswered question as to what underlying factor(s) has permitted Tibetans to adapt.

In terms of complexity, it is important to recall that not just the directional forces of natural selection but also the chance-driven processes of genetic drift (i.e., the loss or fixation of alleles as a result of small population size), gene flow, and mutation are involved in evolutionary change. These processes are exemplified by the possibly small numbers of initial migrants to the Americas, severe reductions in population size at the time of sixteenth-century European conquest, and present-day gene flow in the Andean and Himalayan regions. In addition, natural selection is not a simple process. One source of complexity is that a change in adaptation may not alter fitness (i.e., the net result of all adaptations affecting differential fertility and mortality). For example, an adaptation that increases fertility may increase susceptibility to a particular disease and therefore be "canceled out" by increased mortality. Fitness, not adaptation, directs evolutionary change. Neither organisms nor environments remain static and hence an individual's state of adaptation must be evaluated with respect to both historic and present-day environmental conditions. Traits seemingly adaptive today may have arisen under a different set of selective pressures, making the evolutionary scenarios we reconstruct from present-day utility nothing more than "just so" stories (219,220). Other possible explanations of seemingly adaptive traits include close linkage with other traits undergoing selective pressure, random events such as genetic drift, or allometric relationships involved in growth. There may be multiple solutions to a given problem, not a single best adaptive solution.

Future progress will likely be achieved in identifying the genes involved in governing physiological responses to hypoxia and determining whether genetic variants (alleles) contribute to differences between natives and newcomers and/or among native populations. The methods of statistical genetics are useful for determining whether the pattern of transmission in a large group of biologically related individuals (a family set) suggests a single gene with multiple alleles, multiple genes, differential penetrance, or non-Mendelian factors (e.g., imprinting, mosaicism and uniparental disomy) (22). Statistical genetic techniques do not stand alone but must be accompanied by linkage analysis to determine the location of the gene(s). This is accomplished by comparing the co-segregation of the trait of interest with traits that have already been genetically mapped, given that the frequency with which they segregate together is determined by their chromosomal location. While a full genomic search is possible, candidate genes are usually chosen on the basis of being plausibly related to the trait of interest. This is facilitated by the observation that traits that are related in function are often located in contiguous gene blocks on the chromosome, perhaps the result of evolution by gene duplication (221). Of interest will be whether a discrete, common set of hypoxia-sensitive genes are identified or whether there are multiple genetic factors involved that vary broadly among populations.

Future progress can also be anticipated in achieving a more integrated view of high-altitude adaptation. Part of this integration will derive from an increased understanding of the ways in which levels of biological organization are articulated and influence each other. Assistance will be provided by the increasing availability of new (and often minimally invasive) techniques for evaluating cellular metabolism, energy expenditure, and body composition. Another path toward achieving integration is likely to stem from the inclusion of a broader range of subjects, including women and men, young and old, native and newcomer. The detailed physiology of adaptation and long-term acclimatization that has been worked out in men is being extended to women in ways that consider the influences of the menstrual cycle, hormone replacement, oral contraceptives, menopause, and other gender-related influences. Studies of populations residing at high altitude in the twenty-first century will consider not only historically defined groups native to high-altitude regions but heterogeneous groups who form new settlements at high altitude.

References

1. The Hammond Medallion World Atlas. Maplewood, NJ: Hammond, Inc., 1984.
2. World Health Statistics Annual. Geneva: World Health Organization, 1996.
3. Sensui Z. Uncovering prehistoric Tibet. China Reconstructs 1981; 1:64–65.
4. Zhimin A. Paleoliths and microliths from Shenja and Shuanghu, Northern Tibet Curr Anthropol 1982; 23:493–499.

5. Denell RW, Rendell HM, Hialwood E. Late Pliocene artifacts from northern Pakistan. Curr Anthropol 1988; 29:495–498.
6. Matsumoto H. Characteristics of the Mongloid and neighboring populations on the basis of the genetic markers of immunoglobins. J Anthro Soc Nippon 1987; 95:291–304.
7. Torroni A, Miller JA, Moore LG, Zamudio S, Zhuang J, Droma T, Wallace DC. Mitochondrial DNA analysis in Tibet: implications for the origin of the Tibetan population and its adaptation to high altitude. Am J Phys Anthropol 1994; 93:189–199.
8. Turner CG. Late Pleistocene and Holocene population history of East Asia based on dental variation. Am J Phys Anthropol 1987; 73:305–321.
9. Lee TD, Zhao TM, Mickey R, Sun YP, Lee G, Song CX, Cheng DZ, Zhou S, Ding SQ, Cheng DX. The polymorphism of HLA antigens in the Chinese. Tissue Antigens 1988; 32:188–208.
10. Zhao TM, Lee TD. Gm and Km allotypes in 74 Chinese populations: a hypothesis of the origin of the Chinese nation. Hum Genet 1989; 83:101–110.
11. Avedon JF. Exile in the Land of Snows. New York: Vintage Books, 1986:3–33.
12. Farabegoli A, Barbujani G. Diversity of some gene frequencies in European and Asian populations. VI. Geographic patterns of PGM and ACP. Human Hered 1990; 40:313–321.
13. Zei G, Barbujani G, Lisa A, Fiorani O, Menozzi P, Siri E, Cavalli-Sforza LL. Barriers to gene flow estimated by surname distribution in Italy. Ann Hum Genet 1993; 57: 123–140.
14. Neel JV, Biggar RJ, Sukernik RI. Virologic and genetic studies relate Amerind origins to the indigenous people of the Mongolia/Manchuria/southeastern Siberia region. Proc Natl Acad Sci USA 1994; 91:10737–10741.
15. Merriwether DA, Rothhammer F, Ferrell RE. Distribution of the four founding lineage haplotypes in Native Americans suggests a single wave of migration for the New World. Am J Phys Anthropol 1995; 98:411–430.
16. Chen YS, Torroni A, Excoffier L, Santachiara-Benerecetti AS, Wallace DC. Analysis of mtDNA variation in African populations reveals the most ancient of all human continent-specific haplogroups. Am J Hum Genet 1995; 57:133–149.
17. Torroni A, Schurr TG, Cabell MF, Brown MD, Neel JV, Larsen M, Smith DG, Vullo CM, Wallace DC. Asian affinities and continental radiation of the four founding Native American mtDNAs [see comments]. Am J Hum Genet 1993; 53:563–590.
18. MacNeish RS, Berger RP. Megafauna and man from Ayacucho, highland Peru. Science 1970; 166:8975–977.
19. Lynch TF. The paleo-indians. In: Jennings JD, ed. Ancient South Americans. San Francisco: W.H. Freeman and Co., 1978:87–137.
20. Nunez L. Paleoindian and archaic cultural periods in the arid and semiarid regions of northern Chile. Adv World Archaeol 1983; 2:161–203.
21. Blanco R, Chakraborty R. Genetic distance analysis of twenty-two South American Indian populations. Human Hered 1975; 25:177–193.
22. Cook ND. Demographic Collapse, Indian Peru, 1520–1620. New York: Cambridge University Press, 1981.
23. Chakraborty R, Barton SA, Ferrell RE, Schull WJ. Ethnicity determination by names among the Aymara of Chile and Bolivia. Hum Biol 1989; 61:159–177.
24. Post RH, Neel JV, Schull WJ. Tabulation of phenotype and gene frequency for 11 different genetic systems studied in the American Indians. In: Biomedical Challenges

Presented by the American Indian. Washington, DC: Pan American Health Organization, 1968:141–185.
25. Kolsteren P, Atkinson S, Maskall K. Nutrition, Health, Water and Sanitation Assessment in the Lhasa Valley. London: Save the Children Fund/Lhasa Health Bureau, 1990.
26. Richardson JB. People of the Andes. Washington, DC: Smithsonian Books, 1994.
27. Thomas RB. Energy flow at high altitude. In: Baker PT, Little MA, eds. Man in the Andes. A Multidisciplinary Study of High Altitude Quechua. Stroudsburg, PA: Dowden, Hutchinson and Ross, Inc., 1976:379–404.
28. Leonard WR, Thomas RB. Biosocial responses to seasonal food stress in highland Peru. Hum Biol 1989; 61:65–85.
29. Leonard WR. Household-level strategies for protecting children from seasonal food scarcity. Soc Sci Med 1991; 33:1127–1133.
30. Hanna JM. Natural exposure to cold. In: Baker PT, Little MA, eds. Man in the Andes. A Multidisciplinary Study of High-Altitude Quechua. Stroudsburg, PA: Dowden, Hutchinson and Ross, Inc., 1976.
31. Baker TS. Child care, child training and environment. In: Baker PT, Little MA, eds. Man in the Andes. Stroudsburg, PA: Dowden, Hutchinson and Ross, Inc., 1976:85–97.
32. Tronick EZ, Thomas RB, Daltabuit M. The Quechua manta pouch: a caretaking practice for buffering the Peruvian infant against the multiple stressors of high altitude. Child Dev 1994; 65:1005–1013.
33. Monge CM. Acclimatization in the Andes. Baltimore, MD: Johns Hopkins Press, 1948: 36–37.
34. Niermeyer S, Yang P, Shanmina, Drolkar, Zhuang J, Moore LG. Arterial oxygen saturation in Tibetan and Han infants born in Lhasa, Tibet. N Engl J Med 1995; 333:1248–1252.
35. Abelson AE. Altitude and fertility. Hum Biol 1976; 48:83–92.
36. Gonzales GF. Determinates biomedicos de la fertilidad humana en la altura. In: Gonzales GF, ed. Reproduction Humana en la Altura. Lima, Peru: Consejo Nacional de Ciencia y Tecnologica, 1993:73–87.
37. Goldstein MC, Tsarong P, Beall CM. High altitude hypoxia, culture and human fecundity/fertility: a comparative study. Am Anthropol 1983; 85:28–49.
38. Cruz-Coke R, Crisoffanini A, Aspillaga M, Biancani F. Evolutionary forces in human populations in an environmental gradient in Arica, Chile. Hum Biol 1966; 4:421–438.
39. Pan American Health Organization. Health Conditions in the Americas. Scientific publication No. 549. Washington, DC: World Health Organization, 1994.
40. Gonzales GF, Guerra-Garcia R. Caracteristicas hormonales y antropometricas del embarazo y del recien nacido en la altura. In: Gonzales GF, ed. Reproduccion Humana en la Altura. Lima, Peru: Consejo Nacional de Ciencia y Tecnologica, 1993:125–141.
41. Sobrevilla LA, Cassinelli MT, Carcelen A, Malaga JM. Human fetal and maternal oxygen tension and acid-base status during delivery at high altitude. Am J Obstet Gynecol 1971; 111:1111–1118.
42. Zamudio S, Leslie K, White M, Hagerman D, Moore LG. Low serum estradiol and high serum progesterone concentrations characterize hypertensive pregnancies at high altitude. J Soc Gynecol Invest 1994; 1:197–205.

43. Bangham CRM, Hackett PH. Effects of high altitude on endocrine function in the sherpas of Nepal. J Endocrinol 1978; 79:147–148.
44. Yip R. Altitude and birth weight. J Pediatr 1987; 111:869–876.
45. Jensen GM, Moore LG. The effect of high altitude and other risk factors on birthweight: independent or interactive effects? Am J Publ Health 1997.
46. Lichty JL, Ting R, Bruns PD, Dyar E. Studies of babies born at high altitude. Am J Dis Child 1957; 93:666–669.
47. McCullough RE, Reeves JT, Liljegren RL. Fetal growth retardation and increased infant mortality at high altitude. Obstet Gynecol Surv 1977; 32:596–598.
48. Unger C, Weiser JK, McCullough RE, Keefer S, Moore LG. Altitude, low birth weight, and infant mortality in Colorado. J Am Med Assoc 1988; 259:3427–3432.
49. Wiley AS. Neonatal size and infant mortality at high altitude in the western Himalaya. Am J Phys Anthropol 1994; 94:289–305.
50. Zamudio S, Droma T, Norkyel KY, Acharya G, Zamudio JA, Niermeyer SN, Moore LG. Protection from intrauterine growth retardation in Tibetans at high altitude. Am J Phys Anthropol 1993; 91:215–224.
51. Beall CM. Optimal birthweights in Peruvian populations at high and low altitudes. Am J Phys Anthropol 1981; 56:209–216.
52. Carmen-Torres D, Gonzales GF. Edad gestacional al parto a diferentes altitudes del Peru. In: Gonzales GF, ed. Reproduccion Humana en la Altura. Lima, Peru: Consejo Nacional de Ciencia y Tecnologica, 1993:143–151.
53. Haas LD, Frongillo EJ, Stepcik, C., Beard J, Hurtado L. Altitude, ethnic and sex differences in birth weight and length in Bolivia. Hum Biol 1980; 52:459–477.
54. Mazess RB. Neonatal mortality and altitude in Peru. Am J Phys Anthropol 1965; 23: 209–213.
55. Zhoma ZX, Sun SF, Zhang JG, Huang SY, Moore LG. Fetal growth and maternal oxygen supply in Tibetan and Han residents of Lhasa (3658 m). FASEB J 1989; 3: A987.
56. Smith CF. Comparative birthweight data for Sherpa women living at high and low altitudes in Nepal (abstr). Am J Phys Anthropol 1993; 16(suppl):183.
57. Nance WE. Genetic studies of the offspring of identical twins: a model for the analysis of quantitative inheritance in man. Acta Genet Med Gemellol 1976; 25:103–113.
58. Magnus P. Causes of variation in birth weight: a study of offspring of twins. Clin Genet 1984; 25:15–24.
59. Vlietinck R, Derom R, Neale MC, Maes H, van Loon H, Derom C, Thiery M. Genetic and environmental variation in the birth weight of twins. Behav Genet 1989; 19:151–161.
60. Reshetnikova OS, Burton GJ, Milovanov AP. Effects of hypobaric hypoxia on the fetoplacental unit: the morphometric diffusing capacity of the villous membrane at high altitude. Am J Obstet Gynecol 1994; 171:1560–1565.
61. Mayhew TM, Jackson MR, Haas JD. Oxygen diffusive conductances of human placentae from term pregnancies at low high altitudes. Placenta 1990; 11:493–503.
62. Hannhart B, Pickett CK, Weil JV, Moore LG. Influence of pregnancy on ventilatory and carotid body neural output responsiveness to hypoxia in cats. J Appl Physiol 1989; 67:797–803.
63. Moore LG, McCullough RE, Weil JV. Increased HVR in pregnancy: relationship to hormonal and metabolic changes. J Appl Physiol 1987; 62:158–163.
64. Tatsumi K, Hannhart B, Moore LG. Influences of sex steroids on ventilation and venti-

latory control. In: Dempsey JA, Pack AI, eds. Regulation of Breathing. 2d ed. New York: Marcel Dekker, 1995:829–864.
65. Zamudio S, Palmer SK, Dahms TE, Berman JC, Young DA, Moore LG. Alterations in uteroplacental blood flow precede hypertension in preeclampsia at high altitude [see comments]. J Appl Physiol 1995; 79:15–22.
66. Zamudio S, Palmer SK, Dahms TE, Berman JC, McCullough RG, McCullough RE, Moore LG. Blood volume expansion, preeclampsia, and infant birth weight at high altitude. J Appl Physiol 1993; 75:1566–1573.
67. Goodlin RC. Care of the Fetus. Masson Publishing USA, Inc., 1979.
68. Moore LG. Maternal O2 transport and fetal growth in Colorado, Peru and Tibet high-altitude residents. Am J Hum Biol 1990; 2:627–637.
69. Zamudio S, Sun S, Moore LG. Ventilatory characteristics of Peruvian and Tibetan women and men native to high altitude. In: Leon-Velarde F, Arrigui A, eds. Hipoxia: Investgaciones Basicas y Clinicas-Homenaje a Carlos Monge Cassinelli. Lima, Peru: Universidad Perunana Cayetano Heredia, 1993:55–70.
70. Zamudio S, Palmer SK, Stamm E, Coffin C, Moore LG. Uterine blood flow at high altitude. In: Sutton JR, Houston CS, eds. Hypoxia and the Brain. Burlington, VT: Queen City Press, 1995:112–124.
71. Circulatory function during pregnancy: man. In: Altman PL, Dittmer DS, eds. Biological Handbooks, Respiration and Circulation. Bethesda, MD: Federation of American Societies for Experimental Biology, 1971:443–449.
72. Moore LG, Brodeur P, Chumbe O, D'Brot J, Hofmeister S, Monge C. Maternal hypoxic ventilatory response, ventilation, and infant birth weight at 4,300 m. J Appl Physiol 1986; 60:1401–1406.
73. Moore LG, Hershey DW, Jahnigen D, Bowes W Jr. The incidence of pregnancy-induced hypertension is increased among Colorado residents at high altitude. Am J Obstet Gynecol 1982; 144:423–429.
74. Moore LG, Jahnigen D, Rounds SS, Reeves JT, Grover RF. Maternal hyperventilation helps preserve arterial oxygenation during high-altitude pregnancy. J Appl Physiol 1982; 52:690–694.
75. Kitanaka T, Gilbert RD, Longo LD. Maternal responses to long-term hypoxia in sheep. Am J Physiol 1989; 256:R1340–R1347.
76. Jacobs R, Robinson JS, Owens JA, Falconer J, Webster MED. The effect of prolonged hypobaric hypoxia on growth of fetal sheep. J Dev Physiol 1988; 10:97–112.
77. Makowski EL, Battaglia FC, Meschia G, Behrman RE, Schreuffer J, Seeds AE, Bruns PD. Effect of maternal exposure to high altitude upon fetal oxygenation. Am J Obstet Gynecol 1968; 100:852–861.
78. Palmer SK, Zamudio S, Coffin C, Parker S, Stamm E, Moore LG. Quantitative estimation of human uterine artery blood flow and pelvic blood flow redistribution in pregnancy [review]. Obstet Gynecol 1992; 80:1000–1006.
79. Zamudio S, Palmer SK, Droma T, Stamm E, Coffin C, Moore LG. Effect of altitude on uterine artery blood flow during normal pregnancy. J Appl Physiol 1995; 79:7–14.
80. Wilcox AJ. Birth weight and perinatal mortality: the effect of maternal smoking. Am J Epidemiol 1993; 137:1098–1104.
81. Williams RL, Creasy RK, Cunningham GC, Hawes WE, Norris FD, Tashiro M. Fetal growth and perinatal viability in California. Obstet Gynecol 1982; 59:624–632.
82. Karn MN, Penrose LS. Birth weight and gestational time in relation to maternal age, parity and infant survival. Eugenics 1957; 16:147–164.

83. Ales KL, Norton ME, Druzin ML. The clinical significance of predictions based on screening second trimester mean arterial pressure: adverse maternal [corrected] and infant outcomes [published erratum appears in J Clin Epidemiol 1990; 43(5):537]. J Clin Epidemiol 1990; 43:117–124.
84. Ales KL, Frayer W, Hawks G, Auld PM, Druzin ML. Development and validation of a multivariate predictor of mortality in very low birth weight. J Clin Epidemiol 1988; 41:1095–1103.
85. Lehmann DK, Mabie WC, Miller JM, Jr., Pernoll ML. The epidemiology and pathology of maternal mortality: Charity Hospital of Louisiana in New Orleans, 1965–1984. Obstet Gynecol 1987; 69:833–840.
86. Palmer SK, Moore LG, Berman JC, Cregger B, Zamudio S. Altitude and hypertensive complications of pregnancy (abstr). Am J Phys Anthropol 1994; 18:158.
87. Zamudio S, Palmer SK, Regensteiner JG, Moore LG. High altitude and hypertension during pregnancy. Am J Hum Biol 1995; 7:182–192.
88. Lopez-Jaramillo P, y de Felix M. Uso de calcio en la prevencion inducida por el embarazo. Bol Sanit Panam 1991; 110.
89. Quintana D, Briceno G, Axel E. Evaluacion del desprendimento prematuro de placenta en la altura. In: First World Congress of High-Altitude Medicine and Physiology. La Paz, Bolivia: 1994:87.
90. Lunell NO, Sarby B, Lewander R, Nylund L. Comparison of uteroplacental blood flow in normal and in intrauterine growth-retarded pregnancy. Measurements with Indium-113m and a computer-linked gammacamera. Gynecol Obstet Invest 1979; 10:106–118.
91. Lunell NO, Lewander R, Mamoun I, Nylund L, Sarby S, Thornstrom S. Uteroplacental blood flow in pregnancy induced hypertension. Scand J Clin Lab Invest Suppl 1984; 169:28–35.
92. Robertson WB, Khong TY, Brosens I, De Wolf F, Sheppard BL, Bonnar J. The placental bed biopsy: review from three European centers [review]. Am J Obstet Gynecol 1986; 155:401–412.
93. Genbacev O, Joslin R, Damsky C, Polliotti B, Fisher S. Hypoxia alters early gestation human cytotrophoblast differentiation/invasion in vitro and models the placental defects that occur in preeclampsia. J Clin Invest 1996; 97:540–550.
94. Mok JY, McLaughlin FJ, Pintar M, Hak H, Amaro-Galvez R, Levison H. Transcutaneous monitoring of oxygenation: What is normal? J Pediatr 1986; 108:365–371.
95. Thilo EH, Park-Moore B, Berman ER, Carson BS. Oxygen saturation by pulse oximetry in healthy infants at an altitude of 1610 m (5280 ft). What is normal? [see comments]. Am J Dis Child 1991; 145:1137–1140.
96. Niermeyer S, Shaffer EM, Thilo E, Corbin C, Moore LG. Arterial oxygenation and pulmonary arterial pressure in healthy neonates and infants at high altitude. J Pediatr 1993; 123:767–772.
97. Reuland DS, Steinhoff MC, Gilman RH, Bara M, Olivares EG, Jabra A, Finkelstein D. Prevalence and prediction of hypoxemia in children with respiratory infections in the Peruvian Andes. J Pediatr 1991; 119:900–906.
98. Gamboa R, Marticorena E. Pulmonary arterial pressure in newborn infants in high altitude [in Spanish]. Arch Inst Biol Andina 1971; 4:55–66.
99. Sime F, Banchero N, Penaloza D, Gamboa R, Cruz J, Marticorena E. Pulmonary hypertension in children born and living at high altitudes. Am J Cardiol 1963; 11:143–149.

100. Kelly DH, Stellwagen LM, Kaitz E, Shannon DC. Apnea and periodic breathing in normal full-term infants during the first twelve months. Pediatr Pulmonol 1985; 1: 215–219.
101. Glotzbach SF, Ariagno RL. Periodic Breathing. In: Beckerman RC, Brouillette RT, Hunt CE, eds. Respiratory Control Disorders in Infants and Children. Baltimore, MD: Williams & Wilkins, 1992:142–160.
102. Matsuoka T, Mortola JP. Effects of hypoxia and hypercapnia on the Hering-Breuer reflex of the conscious newborn rat. J Appl Physiol 1995; 78:5–11.
103. Lahiri S, Brody JS, Motoyama EK, Velasquez TM. Regulation of breathing in newborns at high altitude. J Appl Physiol 1978; 44:673–678.
104. Barrington KJ, Finer NN, Wilkinson MH. Progressive shortening of the periodic breathing cycle duration in normal infants. Pediatr Res 1987; 21:247–251.
105. Lubchenco LO, Ashby BL, Markarian M. Periodic breathing in newborn infants in Denver and Leadville, Colorado (abstr). Soc Pediatr Res 1964; 50.
106. Deming J, Washburn AH. Respiration in infancy. I. A method of studying rates, volume and character of respiration with preliminary report of results. Am J Dis Child 1935; 49:108–124.
107. Emmanouilides GC, Moss AJ, Duffie ER, Adams FH. Pulmonary artery pressure changes in human newborn infants from birth to 3 days of age. J Pediatr 1964; 65: 327–333.
108. Arias-Stella J, Saldana M. The muscular pulmonary arteries in people living at high altitudes. Fifth Annual Conference on Research in Emphysema, Aspen CO, 1962 (abstract).
109. Alzamora-Castro V, Battilana G, Abugattas R, Sialer S. Patent ductus arteriosus and high altitude. Am J Cardiol 1960; 5:761.
110. Gamboa R, Marticorena E. The ductus arteriosus in the newborn infant at high altitude. Vasa 1972; 1:192–195.
111. Miao CY, Zuberbuhler JS, Zuberbuhler JR. Prevalence of congenital cardiac anomalies at high altitude. J Am Coll Cardiol 1988; 12:224–228.
112. Sui GJ, Liu YH, Cheng XS, Anand IS, Harris E, Harris P, Heath D. Subacute infantile mountain sickness. J Pathol 1988; 155:161–170.
113. Khoury GH, Hawes CR. Primary pulmonary hypertension in children living at high altitude. J Pediatr 1963; 62:177–185.
114. Haas JD. Prenatal and infant growth and development. In: Baker PT, Little MA, eds. Man in the Andes. A Multidisciplinary Study of High-Altitude Quechua. Stroudsburg, PA: Dowden, Hutchinson and Ross, Inc., 1976:161–178.
115. Leonard WR, DeWalt KM, Stansbury JP, McCaston MK. Growth differences between children of highland and coastal Ecuador. Am J Phys Anthropol 1995; 98:47–57.
116. Beall CM, Baker PT, Baker TS, Haas JD. The effects of high altitude on adolescent growth in southern Peruvian Amerindians. Hum Biol 1977; 49:109–124.
117. Beall CM. A comparison of chest morphology in high altitude Asian and Andean populations. Hum Biol 1982; 54:145–163.
118. Frisancho AR, Baker PT. Altitude and growth: a study of the patterns of physical growth of a high altitude Peruvian Quechua population. Am J Phys Anthropol 1970; 32:279–292.
119. Frisancho AR. Human growth and pulmonary function of a high altitude Peruvian Quechua population. Hum Biol 1969; 41:365–379.
120. Mueller WH, Yen F, Rothhammer F, Schull WJ. Multinational Andean genetic and

health program: VII. Lung function and physical growth—multivariate analyses in high- and low-altitude populations. Aviat Space Environ Med 1978; 49:1188–1196.

121. Mueller WH, Schull VN, Schull WJ, Soto P, Rothhammer F. A multinational Andean genetic and health program: growth and development in an hypoxic environment. Ann Hum Biol 1978; 5:329–352.
122. Leatherman TL, Carey JW, Thomas RB. Socioeconomic change and patterns of growth in the Andes. Am J Phys Anthropol 1995; 97:307–321.
123. Obert P, Fellmann N, Falgairette G, Bedu M, Van Praagh E, Kemper H, Post B, Spielvogel H, Tellez V, Qintela A. The importance of socioeconomic and nutritional conditions rather than altitude on the physical growth of prepubertal Andean highland boys. Ann Hum Biol 1994; 21:145–154.
124. Weitz CA, Garruto RM, Zhang Z, Chen JH, Liu RL, Liu JZ, Yuan KF, He X, Chin CT. Comparison of morphological and physiological characteristics of Han, Hui and Tibetans born and raised at high altitude in western China (abstr). Am J Hum Biol 1995; 7:Abs 155.
125. Malik SL, Pandey AK. Respiratory adaptation to high altitude in adolescent Bod girls of the Western Himalayas. Ann Hum Biol 1993; 20:575–581.
126. Pawson IG. Growth characteristics of populations of Tibetan origin in Nepal. Am J Phys Anthropol 1977; 47:473–482.
127. Melton TW. Comparison of growth and development in two Peruvian populations of high altitude ancestry. In: Eckhardt RB, Melton TW, eds. Population Studies on Human Adaptation and Evolution in the Peruvian Andes. 1992:192–241.
128. Greksa LP, Spielvogel H, Caceres E. Total lung capacity in young highlanders of Aymara ancestry. Am J Phys Anthropol 1994; 94:477–486.
129. Kramer AA. Heritability estimates of thoracic skeletal dimensions of a high-altitude Peruvian population. In: Eckhardt RB, Melton TW, eds. Population Studies on Human Adaptation and Evolution in the Peruvian Andes. 1992:25–49.
130. Ahn YI. Heritability estimates of four anthropometric measurements on the thorax in a high altitude Peruvian population. In: Eckhardt RB, Melton TW, eds. Population Studies on Human Adaptation and Evolution in the Peruvian Andes. 1992:113–137.
131. Greksa LP, Spielvogel H, Paz-Zamora M, Caceres E, Paredes-Fernandez L. Effect of altitude on the lung function of high altitude residents of European ancestry. Am J Phys Anthropol 1988; 75:77–85.
132. Malhotra R. Thoracic adaptation to high altitude. Anthropol Anzeig 1986; 44:355–359.
133. Johnson RL, Jr., Cassidy SS, Grover RF, Schutte JE, Epstein RH. Functional capacities of lungs and thorax in beagles after prolonged residence at 3,100 m. J Appl Physiol 1985; 59:1773–1782.
134. Frisancho AR, Frisancho HG, Albalak R, Villain M, Vargas E, Soria R. Developmental, genetic, and environmental components of lung volumes at high altitude. Am J Hum Biol 1997; 9:191–203.
135. Greksa LP, Spielvogel H, Paredes-Fernandez L. Maximal exercise capacity in adolescent European and Amerindian high-altitude natives. Am J Phys Anthropol 1985; 67:209–216.
136. Obert P, Bedu M, Fellmann N, Falgairette G, Beaune B, Quintela A, Van Praagh E, Spielvogel H, Kemper H, Post B. Effect of chronic hypoxia and socioeconomic status on VO_{2max} and anaerobic power of Bolivian boys. J Appl Physiol 1993; 74:888–896.

137. Blonc S, Fellmann N, Bedu M, Falgairette G, Dejonge R, Obert P, Beaune B, Spielvogel H, Tellez W, Quintela A, Sanmiguel JL, Coudert J. Effect of altitude and socioeconomic status on VO2max and anaerobic power in prepubertal Bolivian girls. J Appl Physiol 1996; 80:2002–2008.
138. Frisancho AR, Frisancho HG, Milotich M, Brutsaert T, Albalak R, Spielvogel H, Villena M, Vargas E, Soria R. Developmental, genetic, and environmental components of aerobic capacity at high altitude. Am J Phys Anthropol 1995; 96:431–442.
139. Scoggin CH, Hyers TM, Reeves JT, Grover RF. High-altitude pulmonary edema in the children and young adults of Leadville, Colorado. N Engl J Med 1977; 297:1269–1272.
140. Fasules JW, Wiggins JW, Wolfe RR. Increased lung vasoreactivity in children from Leadville, Colorado, after recovery from high-altitude pulmonary edema. Circulation 1985; 72:957–962.
141. Ucros S, Posada E, Llano G. High altitude pulmonary edema in children living in Bogota. In: First World Congress of High-Altitude Medicine and Physiology. La Paz, Boliva: 1994:60.
142. Wu TY, Ki WS, Li Y. High altitude disease on the Tibetan plateau. In: First World Congress of High-Altitude Medicine and Physiology. La Paz, Bolivia, 1994:55.
143. Luerssen JS. Landlessness, health and the failures of reform in the Peruvian highlands. Human Organ 1994; 53:380–387.
144. Matheson GO, Allen PS, Ellinger DC, Hanstock CC, Gheorghiu D, McKenzie DC, Stanley C, Parkhouse WS, Hochachka PW. Skeletal muscle metabolism and work capacity: a 31P-NMR study of Andean natives and lowlanders. J Appl Physiol 1991; 70:1963–1976.
145. Sun SF, Droma TS, Zhang JG, Tao JX, Huang SY, McCullough RG, McCullough RE, Reeves CS, Reeves JT, Moore LG. Greater maximal O_2 uptakes and vital capacities in Tibetan than Han residents of Lhasa. Respir Physiol 1990; 79:151–161.
146. Frisancho AR, Martinez C, Velasquez T, Sanchez J, Montoye H. Influence of developmental adaptation on aerobic capacity at high altitude. J Appl Physiol 1973; 34:176–180.
147. Favier R, Spielvogel H, Desplanches D, Ferretti G, Kayser B, Hoppeler H. Maximal exercise performance in chronic hypoxia and acute normoxia in high-altitude natives. J Appl Physiol 1995; 78:1868–1874.
148. Hochachka PW, Stanley C, Matheson GO, McKenzie DC, Allen PS, Parkhouse WS. Metabolic and work efficiencies during exercise in Andean natives. J Appl Physiol 1991; 70:1720–1730.
149. Ge RL, Chen QH, Wang LH, Gen D, Yang P, Kubo K, Fujimoto K, Matsuzawa Y, Yoshimura K, Takeoka M. Higher exercise performance and lower VO_{2max} in Tibetan than Han residents at 4,700 m altitude. J Appl Physiol 1994; 77:684–691.
150. Curran-Everett L, Zhang JG, Droma TS, Moore LG. Superior work performance in lifelong Tibetan residents of 4,400 m compared with 3,658 m (abstr). Physiologist 1992; 35.
151. Niu W, Youan W, Li B, Chen N, Song S. Effects of long-term acclimatization in lowlanders migrating to high altitude: comparison with high altitude residents. Eur J Appl Physiol 1995; 71:543–548.
152. Severinghaus JW, Bainton CR, Carcelen A. Respiratory insensitivity to hypoxia in chronically hypoxic man. Respir Physiol 1966; 1:308–334.
153. Dempsey JA, Reddan WG, Birnbaum ML, Forster HV, Thoden JS, Grover RF, Rankin

J. Effects of acute through life-long hypoxic exposure on exercise pulmonary gas exchange. Respir Physiol 1971; 13:62–89.

154. Weil JV, Byrne-Quinn E, Sodal IE, Filley GF, Grover RF. Acquired attenuation of chemoreceptor function in chronically hypoxic man at high altitude. J Clin Invest 1971; 50:186–195.

155. Lahiri S. Alveolar gas pressures in man with life-time hypoxia. Respir Physiol 1968; 4:373–386.

156. Lahiri S, Milledge JS. Acid-base in Sherpa altitude residents and lowlanders at 4880 m. Respir Physiol 1967; 2:323–334.

157. Santolaya RB, Lahiri S, Alfaro RT, Schoene RB. Respiratory adaptation in the highest inhabitants and highest Sherpa mountaineers. Respir Physiol 1989; 77:253–262.

158. Zhuang J, Droma T, Sun S, Janes C, McCullough RE, McCullough RG, Cymerman A, Huang SY, Reeves JT, Moore LG. Hypoxic ventilatory responsiveness in Tibetan compared with Han residents of 3,658 m. J Appl Physiol 1993; 74:303–311.

159. Huang ZR, Zhu SC, Ba ZF, Hu ST. Ventilatory control in Tibetan highlanders. In: Liu DS, ed. Geological and Ecological Studies of Qinghai-Xizang Plateau. Beijing, PRC: Science Press, 1981:1363–1369.

160. Hackett PH, Reeves JT, Reeves CD, Grover RF, Rennie D. Control of breathing in Sherpas at low and high altitude. J Appl Physiol 1980; 49:374–379.

161. Beall CM, Brittenham GM, Strohl KP, Blangero J, Williams-Blangero S, Almasy LA, Decker MJ, Worthman CM, Goldstein MC, Vargas E, Villena M, Soria R, Alaracon AM, Gonzales C. Ventilation and hypoxic ventilatory response of Tibetan and Aymara high-altitude natives. Am J Phys Anthropol 1997.

162. Huang SY, Alexander JK, Grover RF, Maher JT, McCullough RE, McCullough RG, Moore LG, Sampson JB, Weil JV, Reeves JT. Hypocapnia and sustained hypoxia blunt ventilation on arrival at high altitude. J Appl Physiol 1984; 56:602–606.

163. Lahiri S, Kao FF, Velasquez T, Martinez C, Pezzia W. Irreversible blunted respiratory sensitivity to hypoxia in high altitude natives. Respir Physiol 1969; 6:360–374.

164. Milledge JS, Lahiri S. Respiratory control in lowlanders and Sherpa highlanders at altitude. Respir Physiol 1967; 2:310–322.

165. Curran LS, Zhuang J, Droma T, Land L, Moore LG. Hypoxic ventilatory responses in Tibetan residents of 4400 m compared with 3658 m. Respir Physiol 1995; 100: 223–230.

166. Collins DD, Scoggin CH, Zwillich CW, Weil JV. Hereditary aspects of decreased hypoxic response. J Clin Invest 1978; 62:105–110.

167. Kawakami Y, Yoshikawa T, Shida A, Asanuma Y, Murao M. Control of breathing in young twins. J Appl Physiol 1982; 52:537–542.

168. Strohl K, Beall CM. Ventilatory response to experimental hypoxia in adult male and female natives of the Tibetan and Andean Plateaus. In: Houston C, Coates J, eds. Hypoxia 97: Women at Altitude. Toronto: B.C. Decker, Inc., 1997.

169. Wagner PD, Gale GE, Moon RE, Torre-Bueno JR, Stolp BW, Saltzman HA. Pulmonary gas exchange in humans exercising at sea level and simulated altitude. J Appl Physiol 1986; 61:260–270.

170. Moore LG, Curran-Everett L, Droma TS, Groves BM, McCullough RE, McCullough RG, Sun SF, Sutton JR, Zamudio S, Zhuang JG. Are Tibetans better adapted? Int J Sports Med 1992; 13(suppl 1):S86–S88.

171. Samaja M, Veicsteinas A, Cerretelli P. Oxygen affinity of blood in altitude Sherpas. J Appl Physiol 1979; 47:337–341.

172. Beall CM, Blangero J, Williams-Blangero S, Goldstein MC. Major gene for percent of oxygen saturation of arterial hemoglobin in Tibetan highlanders. Am J Phys Anthropol 1994; 95:271–276.
173. Beall CM, Strohl KP, Blangero J, Williams-Blangero S, Brittenham GM, Goldstein MC. Quantitive genetic analysis of arterial oxygen saturation in Tibetan highlanders. Am J Hum Biol 1997;
174. Crausman RS, Lewin JM. Pulmonary arterial hypertension in nonagenarians [letter]. J Am Geriatr Soc 1995; 43:1177.
175. Beall CM, Brittenham GM, Macuaga F, Barragan M. Variation in hemoglobin concentration among samples of high-altitude natives in the Andes and the Himalayas. Am J Hum Biol 1990; 2:639–651.
176. Winslow RM, Chapman KW, Monge CC. Ventilation and the control of erythropoiesis in high-altitude natives of Chile and Nepal. Am J Hum Biol 1990; 2:653–662.
177. Hsu SL, Marks J, Shaw JP, Tam M, Higgs DR, Shen CC, Shen CK. Structure and expression of the human theta 1 globin gene. Nature 1988; 331:94–96.
178. Frisancho AR. Origins of differences in hemoglobin concentration between Himalayan and Andean populations. Respir Physiol 1988; 72:13–18.
179. Zhuang J, Droma T, Sutton JR, McCullough RE, McCullough RG, Groves BM, Rapmund G, Janes C, Sun S, Moore LG. Autonomic regulation of heart rate response to exercise in Tibetan and Han residents of Lhasa (3,658 m). J Appl Physiol 1993; 75: 1968–1973.
180. Savard GK, Kayser B, Binzoni T. Cardiac parasympathetic control in Tibetan newcomers to altitude (5050 m). In: Sutton JR, Houston CS, eds. Hypoxia and the Brain. Burlington, VT: Queen City Press, 1995.
181. Hartley LH, Vogel JA, Cruz JC. Reduction of maximal exercise heart rate at altitude and its reversal with atropine. J Appl Physiol 1974; 36:362–365.
182. Groves BM, Droma T, Sutton JR, McCullough RG, McCullough RE, Zhuang J, Rapmund G, Sun S, Janes C, Moore LG. Minimal hypoxic pulmonary hypertension in normal Tibetans at 3,658 m. J Appl Physiol 1993; 74:312–318.
183. Hartley LH, Alexander JK, Modelski M, Grover RF. Subnormal cardiac output at rest and during exercise in residents at 3,100 m altitude. J Appl Physiol 1967; 23:839–848.
184. Banchero N, Sime F, Penaloza D, Cruz J, Gamboa R, Marticorena E. Pulmonary pressure, cardiac output and arterial oxygen saturation during exercise at high altitude and at sea level. Circulation 1966; 33:249–262.
185. Gupta ML, Rao KS, Anand IS, Banerjee AK, Boparai MS. Lack of smooth muscle in the small pulmonary arteries of the native Ladakhi. Is the Himalayan highlander adapted? Am Rev Respir Dis 1992; 145:1201–1204.
186. Durmowicz AG, Hofmeister S, Kadyraliev TK, Aldashev AA, Stenmark KR. Functional and structural adaptation of the yak pulmonary circulation to residence at high altitude. J Appl Physiol 1993; 74:2276–2285.
187. Heath D, Williams D, Dickinson J. The pulmonary arteries of the yak. Cardiovasc Res 1984; 18:133–139.
188. Sun SF, Sui GZ, Liu YH, Cheng XS, Anand IS, Harris P, Heath D. The pulmonary circulation of the Tibetan snow pig (marmota himalaya hodgson). J Zoology 1989; 217:85–91.
189. Banchero N, Grover RF, Will JA. High altitude-induced pulmonary arterial hypertension in the llama (Lama glama). Am J Physiol 1971; 220:422–427.

190. Huang SY, Sun S, Droma T, Zhuang J, Tao JX, McCullough RG, McCullough RE, Micco AJ, Reeves JT, Moore LG. Internal carotid arterial flow velocity during exercise in Tibetan and Han residents of Lhasa (3,658 m). J Appl Physiol 1992; 73:2638–2642.
191. Garrido E, Segura R, Capdevilla A, Pujol J, Javierre C, Ventura JL. Are Himalayan Sherpas better protected against brain damage associated with extreme altitude climbs? Clin Sci 1996; 90:81–85.
192. Kayser B, Hoppeler H, Claassen H, Cerretelli P. Muscle structure and performance capacity of Himalayan Sherpas. J Appl Physiol 1991; 70:1938–1942.
193. Kayser B, Hoppeler H, Desplanches D, Marconi C, Broers B, Cerretelli P. Muscle ultrastructure and biochemistry of lowland Tibetans. J Appl Physiol 1996; 81:419–425.
194. Martin BJ, Weil JV, Sparks KE, McCullough RE, Grover RF. Exercise ventilation correlates positively with ventilatory chemoresponsiveness. J Appl Physiol 1978; 45:557–564.
195. Monge-C C, Arregui A, Leon-Velarde F. Pathophysiology and epidemiology of chronic mountain sickness. Int J Sports Med 1992; 13(suppl 1):S79–S81.
196. Kryger M, McCullough RE, Collins D, Scoggin CH, Weil JV, Grover RF. Treatment of excessive polycythemia of high altitude with respiratory stimulant drugs. Am Rev Respir Dis 1978; 117:455–464.
197. Leon-Velarde F, Ramos MA, Hernandez JA, Deidiaquez D, Munoz LS, Gaffo A, Cordova S, Durand D, Monge C. The role of menopause in the development of chronic mountain sickness. Am J Physiol 1997; 41:R90–R94.
198. Sime F, Monge C, Whittembury J. Age as a cause of chronic mountain sickness (Monge's disease). Int J Biometeorol 1975; 19:93–98.
199. Leon-Velarde F, Arregui A, Monge C, Ruiz y Ruiz H. Aging at high altitudes and the risk of chronic mountain sickness. J Wilderness Med 1993; 4:183–188.
200. Sun SF, Huang SY, Zhang JG, Droma TS, Banden G, McCullough RE, McCullough RG, Cymerman A, Reeves JT, Moore LG. Decreased ventilation and hypoxic ventilatory responsiveness are not reversed by naloxone in Lhasa residents with chronic mountain sickness. Am Rev Respir Dis 1990; 142:1294–1300.
201. Grover RF, Reeves JT, Grover EB, Leathers JE. Muscular exercise in young men native to 3,100 m altitude. J Appl Physiol 1967; 22:555–564.
202. Kryger M, McCullough R, Doekel R, Collins D, Weil JV, Grover RF. Excessive polycythemia of high altitude: role of ventilatory drive and lung disease. Am Rev Respir Dis 1978; 118:659–666.
203. Pei SX, Chen XJ, Si Ren BZ, Liu YH, Cheng XS, Harris EM, Anand IS, Harris PC. Chronic mountain sickness in Tibet. Q J Med 1989; 71:555–574.
204. Leon-Velarde F, Monge CC, Vidal A, Carcagno M, Criscuolo M, Bozzini CE. Serum immunoreactive erythropoietin in high altitude natives with and without excessive erythrocytosis. Exp Hematol 1991; 19:257–260.
205. Kryger M, Glas R, Jackson D, McCullough RE, Scoggin C, Grover RF, Weil JV. Impaired oxygenation during sleep in excessive polycythemia of high altitude: improvement with respiratory stimulation. Sleep 1978; 1:3–17.
206. Sun SF, Oliver-Pickett C, Ping Y, Micco AJ, Droma TS, Zamudio S, Zhang JG, Huang SY, McCullough RG, Cymerman A, Moore LG. Breathing and brain blood flow during sleep in patients with chronic mountain sickness. J Appl Physiol 1996; 81:611–618.

207. Monge-C C, Leon-Velarde F, Arregui A. Increasing prevalence of excessive erythrocytosis with age among healthy high-altitude miners (letter). N Engl J Med 1989; 321: 1271.
208. Xie CF, Pei SX. Some physiological data of sojourners and native highlanders at three different altitudes in Xizang. In: Liu DS, ed. Geological and Ecological Studies of Qinghai-Xizang Plateau. New York: Gordon and Breach, 1981:1449–1452.
209. Wu TY, Zhang Q, Jin B, Xu F, Cheng Q, Wan X. Chronic mountain sickness (Monge's Disease): an observation in Qinghai-Tibet plateau. In: Ueda G, ed. High Altitude Medicine. Matsumoto, Japan: Shinshu University Press, 1992:314–324.
210. Ruiz L, Penaloza D. Altitude and hypertension. Mayo Clin Proc 1977; 52:442–445.
211. Sun SF. Epidemiology of hypertension on the Tibetan plateau. Hum Biol 1986; 58: 507–515.
212. Marticorena E, Ruiz L, Severino J, Galvez J, Penaloza D. Systemic blood pressure in white men born at sea level: changes after long residence at high altitudes. Am J Cardiol 1969; 23:364–368.
213. Wolfel EE, Selland MA, Mazzeo RS, Reeves JT. Systemic hypertension at 4,300 m is related to sympathoadrenal activity. J Appl Physiol 1994; 76:1643–1650.
214. Moore LG, Rohr AL, Maisenbach JK, Reeves JT. Emphysema mortality is increased in Colorado residents at high altitude. Am Rev Respir Dis 1982; 126:225–228.
215. Renzetti AD, McClement JH, Litt BD. The Veteran's Administration Cooperative Study of Pulmonary Function III. Mortality in relation to respiratory function in chronic obstructive pulmonary disease. Am J Med 1966; 41:115–129.
216. Regensteiner JG, Moore LG. Migration of the elderly from high altitudes in Colorado. J Am Med Assoc 1985; 253:3124–3128.
217. Giles RE, Blanc H, Can HM, Wallace DC. Maternal inheritance of human mitochondrial DNA. Proc Natl Acad Sci USA 1980; 77:6715–6719.
218. Torroni A, Chen YS, Semino O, Santachiara-Beneceretti AS, Scott CR, Lott MT, Winter M, Wallace DC. mtDNA and Y-chromosome polymorphisms in four Native American populations from southern Mexico [see comments]. Am J Hum Genet 1994; 54: 303–318.
219. Gould SJ, Lewontin RC. The Spandrels of San Marcos and the Panglossian Paradigm: A Critique of the Adaptationist Program. In: Sober E, ed. Conceptual Issues in Evolutionary Biology. Cambridge, MA: MIT Press, 1979.
220. Lewontin RC. Adaptation. Sci Am 1978; 239:212–8, 220, 222.
221. Weiss KM. Genetic Variation and Human Disease: Principles and Evolutionary Approaches. Cambridge, MA: Cambridge University Press, 1993.
222. Moore LG, Zamudio S, Curran-Everett L, Torroni A, Jorde LB, Shohet RV, Thupten, Drolkar T. Genetic adaptation to high altitude. In: Wood SC, Roach RC, eds. Sports and Exercise Medicine. New York: Marcel Dekker, 1994:225–262.
223. Cerny FC, Dempsey JA, Reddan WG. Pulmonary gas exchange in non-native residents of high altitude. J Clin Invest 1973; 52:2993–2999.
224. DeGraff AC, Grover RF, Johnson RL,Jr., Hammond MW,Jr., Miller JM. Diffusing capacity of the lung in Caucasians native to 3,100 m. J Appl Physiol 1970; 29:71–76.
225. Hurtado A. Animals in high altitudes: resident man. In: Handbook of Physiology. Section 4: Adaptation to the Environment. Washington, DC: American Physiological Society, 1964:843–860.
226. Droma T, McCullough RG, McCullough RE, Zhuang JG, Cymerman A, Sun SF, Sut-

ton JR, Moore LG. Increased vital and total lung capacities in Tibetan compared to Han residents of Lhasa (3,658 m). Am J Phys Anthropol 1991; 86:341–351.
227. Vogel JA, Patton JF, Mello RP, Daniels WL. An analysis of aerobic capacity in a large United States population. J Appl Physiol 1986; 60:494–500.
228. Mazess RB. Exercise performance at high altitude in Peru. Fed Proc 1969; 28:1301–1306.
229. Kollias J, Buskirk ER, Akers RF, Prokop EK, Baker PT, Picon-Reategui E. Work capacity of long-time residents and newcomers to altitude. J Appl Physiol 1968; 24: 792–799.
230. Vogel JA, Hartley LH, Cruz JC. Cardiac output during exercise in altitude natives at sea level and high altitude. J Appl Physiol 1974; 36:173–176.
231. Elsener RW, Bolstad A, Forno C. Maximum oxygen consumption of Peruvian Indians native to high altitude. In: Weihe WH, ed. Physiological Effects of High Altitude. New York: Macmillan, 1964:217–233.
232. Schoene RB, Roach RC, Lahiri S, Peters RM,Jr., Hackett PH, Santolaya R. Increased diffusion capacity maintains arterial saturation during exercise in the Quechua Indians of Chilean Altiplano. Am J Hum Biol 1990; 2:663–668.
233. Zhuang J, Droma T, Sutton JR, Groves BM, McCullough RE, McCullough RG, Sun S, Moore LG. Smaller alveolar-arterial O_2 gradients in Tibetan than Han residents of Lhasa (3658 m). Respir Physiol 1996; 103:75–82.
234. Bebout DE, Story D, Roca J, Hogan MC, Poole DC, Gonzalez-Camarena R, Ueno O, Haab P, Wagner PD. Effects of altitude acclimatization on pulmonary gas exchange during exercise. J Appl Physiol 1989; 67:2286–2295.
235. Lockhart A, Zelter M, Mensch-Dechene J, Antezana G, Paz-Zamora M, Vargas E, Coudert J. Pressure-flow-volume relationships in pulmonary circulation of normal highlanders. J Appl Physiol 1976; 41:449–456.
236. Sime F, Penaloza D, Ruiz L. Bradycardia, increased cardiac output, and reversal of pulmonary hypertension in altitude natives living at sea level. Br Heart J 1971; 33: 647–657.

4

Cellular and Molecular Mechanisms of O_2 Sensing with Special Reference to the Carotid Body

SUKHAMAY LAHIRI

University of Pennsylvania
School of Medicine
Philadelphia, Pennsylvania

NEIL S. CHERNIACK

UMDNJ–New Jersey Medical School
Newark, New Jersey

I. Introduction

A decrease in the O_2 in the air brings into play an array of compensatory mechanisms, which include an increase in ventilation, changes in the distribution of blood flow to organs in the body, an increase in the O_2-carrying capacity of the blood, and shifts in metabolic pathways. The exact configuration of the response varies with the severity of hypoxia, with its duration, and with the maturity of the human or animal exposed to hypoxia.

Even the intensity of a particular response waxes and wanes with time. For example, the initial increase in breathing caused by hypoxic stimulation of arterial chemoreceptors diminishes in a few minutes because of the release of neurotransmitters in the brain that inhibit respiration, but then ventilation increases if hypoxia is prolonged sufficiently, producing the phenomenon of acclimatization seen with adjustment at altitude.

The defense against hypoxia occurs at different levels. The arterial chemoreceptors are the first line of defense against hypoxia. They are responsible for the rapidly augmented breathing that occurs even with relatively small decreases in Po_2. The rise in Po_2 that occurs with greater ventilation feedback helps adjust the arterial chemoreceptor signal to appropriate levels. This ventilatory increase is supported by changes in the circulation, constriction of the pulmonary vasculature, and dilation in the systemic circulation. Later the formation of erythropoetin by renal tissue in the adult and by the liver in the immature increases the ability of the blood to transport O_2 to the tissues. In addition, all cells that consume oxygen probably have intracellular compensatory mechanisms that are called into play when they suffer from sufficiently severe and/or prolonged hypoxia.

Much current research is devoted to identifying the O_2-sensing molecules in the arterial chemoreceptor and other tissues and organs that recognize deficiencies in O_2 supply and initiate compensatory processes.

Studies in the culture of model cell lines derived from tumor tissue suggest that the rapid response to hypoxia involves molecules in either the cell membrane or the mitochondria. In either case a heme compound seems to be involved in signal recognition. Superoxide anions like peroxide are produced by intracellular oxidative processes. These naturally occurring free radicals can damage cells when they are present in excess, but recently it was proposed that they participate in O_2 sensing. Ultimately, compensation involves the synthesis of increased amounts of existing or entirely new proteins caused by activation and inactivation of specific genes.

The molecular mechanisms responsible for O_2 sensing are probably somewhat different in excitable tissues like the carotid body than in nonexcitable tissues like the erythropoietin-producing cells of the kidney. The ability of cells to respond to hypoxia may also vary with age, surroundings, and the level of metabolic activity. Although many exciting insights into the mechanisms of O_2 sensing and its cellular consequences have been made, important details are missing. Even more important, findings in cell lines have not as yet been related to the physiological sequence of events that occurs in the whole animal faced with hypoxia.

This chapter focuses on the arterial chemoreceptors and reviews data and theories concerning cellular mechanisms of O_2 sensing in the carotid body and in erythropoietin-producing tissues and, to round out the picture, in vascular smooth muscle and neurons as well. The complex cellular and systemic changes that occur with prolonged data are described in later sections. Finally, we summarize the new studies on altered gene expression triggered by hypoxia in model cell systems.

II. Physiological Clues to the Identity of O_2-Sensing Molecules

Although major advances have been made in understanding the biological response to hypoxia, the molecular mechanisms involved in the very first step of O_2 sensing remain elusive.

Although physiological responses offer important clues, tissues differ in their speed of response to hypoxia. The arterial chemoreceptors respond most quickly, followed by the smooth muscles of the pulmonary arteries, which respond in only a few seconds. Erythropoietin-producing cells require more prolonged hypoxia before reacting. A quick response probably involves only membranes and cytoplasmic processes and does not require the synthesis of new protein. If hypoxia is unrelieved for hours, gene activity can be altered, but in some genes changes can begin in minutes.

It would be intellectually quite satisfying if a single molecule or class of molecules was responsible for O_2 sensing in all tissues. Heme proteins do appear to be involved in many tissues, and specific heme proteins, such as the cytochromes, have been suggested as the molecular sensors of O_2 (1–5).

Since different potential O_2-sensing molecules (such as cytochrome oxidase) are transformed from fully oxidized to fully reduced states over different ranges of Po_2, tissue levels of O_2 could be another important clue as to the nature of the actual molecular sensor. In the case of cytochrome oxidase, full oxygenation is achieved at rather low Po_2 levels—$<$10 torr (1–5). It is important to remember that the critical Po_2 is at the site of the molecule and not arterial or venous Po_2. It is possible that there are different molecules involved in O_2 sensing which have different Po_2 response ranges, perhaps organized in some sort of a hierarchical structure, allowing the organism to respond in a graded fashion to Po_2 changes of different duration and severity.

Also, the effects of hypoxia may not be direct but rather involve an altered level of some other molecule (e.g., H_2O_2) whose concentration depends on the Po_2 level.

III. Tissue Po_2 and O_2 Sensing

Very low levels of tissue Po_2 are probably required to stimulate erythropoietin or cause pulmonary artery vasoconstriction. For erythropoietin production, incubation of hepatic cells (Hep3B) with 7 torr Po_2 for 24 hours is necessary (6). A systemic arterial Po_2 below 60–80 torr is needed to raise pulmonary arterial pressure, but the tissue Po_2 in pulmonary smooth muscle is likely to be far less (7). Tissue Po_2 was found to be less than 10 torr when arterial segments were perfused with solutions containing Po_2 at 100 torr (8). At lower perfusate O_2 tension, tissue levels would be even lower and hence compatible with cytochrome oxidase being the molecular sensor (1–5).

However, the tissue Po_2 at which carotid body stimulation occurs is controversial. Arterial Po_2 of the order of 60–70 torr stimulates carotid body nerve activity (5). Acker et al. (9) found the tissue Po_2 in the carotid body to be 7–20 torr at an arterial Po_2 of 100. But Whalen and Nair (10) and Buerk et al. (11) reported a value of 70–80 torr in carotid body tissue when the perfusate Po_2 was 100–110 torr. At a perfusate Po_2 of 60–70, the carotid body tissue Po_2 would be near 2–3 Torr according to Acker, whereas according to Whalen and Nair it would be around 30 torr [and perhaps even higher, since in carotid body, O_2 consumption decreases with hypoxia (9,10)].

Assuming the Vo_2 of the carotid body equals 1.5 mL/min/100 g, flow (Q) equals 1.5 L/min/100 g and the (a-v) O_2 difference is 1.0 mL/L, Gonzalez et al.

(12) calculated the venous Po_2 of the carotid body to be 90 at an arterial Po_2 of 100 torr. Since tissue Po_2 of the carotid body is about the same whether it is perfused with blood or cell-free fluids (10) and the relationships between Po_2 and chemosensory activity were also found to be the same with either type of perfusate (10,13), the influence of red blood cells on tissue Po_2 would seem to be minimal. In an in vitro carotid body preparation, Fay (14) found an (a-v) O_2 difference of 10 torr when the Po_2 of the inflowing perfusate was 100 torr, so that the effluent Po_2 was 90 torr. This suggests that tissue Po_2 is relatively high even when hypoxia produces obvious ventilatory stimulation (which occurs at an arterial Po_2 of about 60 torr).

However, Lahiri et al., using phosphorescence quenching by O_2, found that the microvascular Po_2 in the carotid body was 40–50 torr in vivo when the perfusate Po_2 was 100 (5) and 23 torr in vitro when the perfusate Po_2 was about 80 torr. At a lower arterial Po_2 of 60 torr, the microvascular Po_2 would be even less. After a correction for the gradient required for O_2 diffusion, the tissue Po_2 would be less than 10 torr (a level compatible with cytochrome oxidase being the O_2 sensor) (15).

The complexity of the regulation of carotid body flow and the geometry of its circulation (15,16) further complicates the assessment in vivo of tissue levels of Po_2 in the carotid body and of determining how Po_2 is coupled to chemosensory discharge. The microcirculation of the carotid body is very sensitive to various agents. Vasodilators decreased the O_2 (a-v) Po_2 difference to 5 torr from the normal 10 torr (16). Hemorrhage is also known to excite the chemosensory discharge (17), not by decreasing tissue Po_2 (which seems to be independent of red blood cell numbers), but by vasoconstriction.

IV. The Arterial Chemoreceptors—Carotid and Aortic Bodies

O_2 sensing in higher organisms occurs at the carotid body and aortic bodies, and the information is sent to the brain stem, producing an array of reflex actions, which include breathing. The aortic body and particularly the carotid body also detect changes in pH and Pco_2 (18–24), but this aspect is not unique to the chemoreceptors. Other receptors detect changes in Pco_2 and acidity (25). Not only are the transmission pathways different, but carotid and aortic bodies, unlike taste buds, have some special mechanism for detecting O_2 changes.

Carotid and aortic bodies are located near the carotid artery bifurcation and aortic arch, where baroreceptors are also strategically located. The two types of receptors, chemo and baro, together reflexively help coordinate breathing and cardiovascular responses. The carotid body consists of 40,000–60,000 glomus cells in rats and cats, while the aortic bodies are smaller and contain fewer glomus cells (26) and have a weaker response to acid than the carotid body (27). Both have rich networks of fenestrated capillaries, around which two types of cells are clustered (Fig. 1). The major cell type, Type I glomus cells, is organized in groups of 5–10 encased by 2–3 cells of the other type (Type II). Type I cells are innervated and distinguished by the presence of dense-cored vesicles containing various neurotrans-

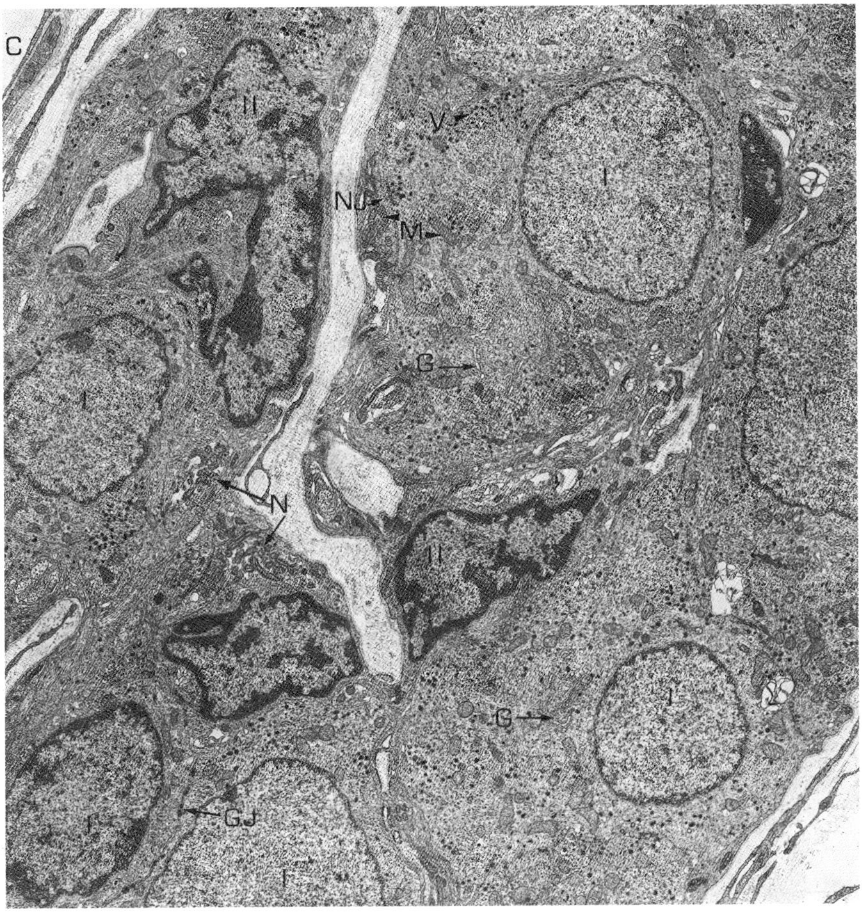

Figure 1 Electron micrograph (× 10,000) of chinchilla carotid body. I and II, type I and type II cells; C, capillary; G, Golgi apparatus; GJ, gap junction; M, mitochondria; N, nerve endings; v, dense-cored vesicles.

mitters. Since neural signals are the hallmark of the chemoreceptor cell, it is only natural that these innervated Type I cells have been targeted as chemoreceptor cells (see Fig. 1).

Recently, Type I cells of the carotid body (but not as yet the aortic body) have been isolated, cultured, and studied. However, investigation of the cellular responses of Type I cells have been handicapped by the paucity of these cells. Hence, studies of O_2 sensing have often used other cells as substitutes. For example, neuroepithelial bodies, which are scattered in the mucosa of the airway, have been isolated and their O_2-sensing properties have been examined (28). They secrete serotonin, but their physiological role is not known. The carotid body cells contain tyrosine hydroxylase, which is the rate-limiting enzyme for the synthesis of dopamine and/or epinephrine (29). Tyrosine hydroxylase metabolism is O_2 sensitive. The PC-12 cell line established from chromaffin cells of the adrenals (like the Type I cells) produce tyrosine hydroxylase that increases with hypoxia. PC-12 cells have been extensively used to study the effects of hypoxia under the assumption that they behave like Type I cells, but glomus cells differ from chromaffin cells in many respects (29,30).

The O_2 sensitivity observed in the binding of transcription factors to promoter regions of genes in the PC-12 nucleus (31) may be an example of a generalized mechanism found in many cells, but it may have nothing to do with the specific O_2 sensor mechanisms that operate in the carotid body that give rise to increased ventilation.

In vivo responses of the carotid body can be influenced by factors that play no role in vitro, e.g., changes in blood flow. Blood flow in the carotid body has been shown to be exquisitely sensitive to O_2 and various other agents, such as nitric oxide, a vasodilator (32–36). Tissue Po_2 diminishes with constant perfusate Po_2, as chemosensory activity is increased by the inhibition of the enzyme nitric oxide synthase.

In addition, in vivo changes in Po_2 and Pco_2 often occur together. Changes in Pco_2 affect the activity of the carotid body even without appreciably affecting its circulation (24). There are similarities between the interaction that occurs with respect to CO_2 and O_2 binding to hemoglobin and the interactive effects of O_2 and CO_2 on chemoreceptor discharge, which support the hypothesis that some hemoglobin-like molecules in the chemoreceptor receptor cell are responsible for sensing O_2 and CO_2/H^+ (13,20).

Single fiber studies have shown that carotid chemoreceptors respond to CO_2 linearly, while hypoxia increases the slope and decreases the intercept of the CO_2 response line (20). In general, these results conform to those obtained using multifiber preparations in vivo (37) and in vitro (38). The mechanism for the interactive effects of CO_2 and hypoxia is not clear. Both are reported to affect K^+ channels and give rise to increases in intracellular Ca^{2+} although one can abolish O_2 sensitivity without appreciably altering the CO_2/H^+ sensitivity of the carotid body (4).

Changes in intracellular pH seem to affect the carotid body response to CO_2. The carotid body discharge responds to CO_2 with an overshoot, which disappears after inhibition of carbonic anhydrase, an enzyme present in glomus cells (39–41).

The carotid body is supplied with sympathetic and parasympathetic fibers, whose activity may be modified by hypoxia and which may alter its activity. Efferent activity, in turn, changes with carotid sinus nerve activity. The targets of these efferent nerve fibers are primarily blood vessels.

Petrosal ganglion cells are activated by sensory afferents, and they in turn influence the firing of the ganglion glomerular nerve through cells in sympathetic ganglia and the activity of parasympathetic fibers through cells in the nodose ganglia (34). The petrosal ganglion is not simply a relay station for sensory discharge, since different ganglion cells contain different neurotransmitters, which certainly modify the output of the cells. However, exactly what role these nerve cells play is not known. Generally the autonomic innervation of the carotid body exerts an inhibitory effect on its sensory discharge (33,34), although there are reports to the contrary. By affecting blood vessel size, efferent control can alter tissue Po_2 in Type I cells. Recent observations reveal that the blood vessels stain positively for nitric oxide synthase (NOS), supporting the idea that nitric oxide is a normal inhibitory transmitter involved in vasomotor control in the carotid body (33–35).

Nitric oxide has also been reported to decrease sensory response in experimental situations in which its vascular effect would be minimized (34). During acute hypoxia, nitric oxide production has been found to decrease, which could be in part responsible for the increase of chemosensory discharge. Nitric oxide increases cGMP so that the rise in carotid body activity may be due to a reduction in cGMP (35). Acute hypoxia also suppresses CO production (66), which also lowers cGMP. Another endogenous gas, H_2O_2, is also believed to cause inhibitory effects on the carotid body and carotid body vasculature.

V. Effects of Hypoxia on the Type I Cell Membrane

There are two leading theories of O_2 sensing by the carotid body: one idea is that hypoxia acts on molecules in the cell membrane, and the other is that it acts on molecules in the mitochondria. Hypoxia depolarizes Type I cells and increases carotid body nerve discharge, perhaps by suppressing K^+ channels in the cell membrane. The O_2-sensing K^+ channel is believed by some investigators to be the O_2-sensing molecule. This is based on the following lines of evidence.

In the late 1980s, patch clamping began to be used to measure the ionic currents across glomus cells. Some types of K^+ currents were found to be suppressed by hypoxia (42–50). A single K^+ channel in an isolated membrane patch could be closed by low Po_2 (48), and it was proposed that K^+ (51) channel O_2 interaction in vivo occurs at the cell membrane (48–51). The sequence proposed was that K^+ current suppression depolarized Type I cells, opened voltage-gated Ca^{2+} channels, and Ca^{2+} entry into the cell triggered neurotransmitter release and increased neural discharge.

Gonzalez et al. suggested that the O_2 sensor might be a molecule independent of the K^+ channel but connected to it and not the channel itself (51). However, no experimental verification was offered. Speculation was that the process might involve a heme-linked NADPH oxidase (52,53). Hypoxia could, for example, decrease the amount of H_2O_2 generated by the oxidase, resulting in increased glutathione levels, and the ratio of GSH to GSSG might decrease cGMP and increase cAMP levels. This might alter membrane proteins, which in turn could alter ion channel potency and membrane ionic conductances (52).

There is considerable evidence that CO_2/H^+ stimulation of carotid chemoreceptors occurs by a mechanism separate from that responsible for O_2 sensing (4), but it has been suggested that H^+ can affect K^+ channels and that, like hypoxia, K^+ current suppression is responsible for H^+ chemoreception (42). Another idea proposed by Gonzalez et al. is that a rise in intracellular $[H^+]$ exchanges with extracellular Na^+, which in turn leads to increased Na^+-Ca^{2+} exchange with Ca^{2+} entry (12). The rise in intracellular $[Ca^{2+}]$ is responsible for neurotransmitter release and neural discharge. But blocker of voltage-gated Ca^{2+} channel inhibited the neural response of the carotid body to hypoxia (53a) and hypercapnia (53b), supporting the membrane channel rather than the exchanger hypothesis.

There are problems with the idea that O_2 acts at the cell membrane (12). For example, voltage-dependent K^+ current inhibition has a threshold for activation at around -40 mV, whereas the resting potential of glomus cells is around -55 mV. As a result, O_2-sensitive K^+ channels would be already closed in normoxic glomus cells and could not be further closed by lowering Po_2. However, if normoxic glomus cells fire spontaneously, then O_2-sensitive K^+ channels can become susceptible to low Po_2 inhibition, leading to more Ca^{2+} entry and more neurotransmitter release and nerve firing. Fiber and McCloskey (54) reported that -20 mV or more are needed for glomus cell depolarization, although Buckler and Vaughan-Jones (55) showed that Ca^{2+} entry occurs as a result of smaller depolarization of glomus cells, around -40 mV. Also, as pointed out by Lahiri (13), the range of K^+ channel sensitivity to hypoxia (80–150 torr) is quite different from the much lower levels of Po_2 (< 85) needed to excite the carotid body appreciably (48–49a).

Others researchers have expressed doubts as to the precise role of oxygen-sensitive K^+ channels in controlling the afferent activity of the carotid body. Wyatt et al. (56) showed that the glomus cells of rats, which have blunted ventilatory responses to hypoxia, do not depolarize in response to hypoxia, although they have O_2-sensitive K^+ channels. Also, Cheng and Donnelly compared carotid sinus nerve responses with the changes in the outward K^+ current produced by hypoxia in the rat carotid body (57). They found that hypoxia failed to alter the outward current in most cells, even when the carotid body showed a brisk nerve response.

K^+ channel blockade with tetraethylammonium (TEA) does not ablate the carotid sinus nerve response to hypoxia. Charybdotoxin (20 nM), which blocks Ca^{2+}-sensitive K+ channels, inhibited the outward K^+ current but caused no change in nerve activity (58). Also, charybdotoxin did not alter the hypoxic response of carotid bodies (58). Buckler and Vaughan-Jones (59) found that charybdotoxin (20 nM), tetraethylammonium (10 nM), and 4 aminopyridine (AP) (1 nM) had no effect on intracellular $[Ca^{2+}]$ i.

Thus, these results are not consistent with the idea that voltage-gated outward currents and nerve response of the carotid body are related. The modulation of O_2-sensitive K^+ currents by hypoxia does not appear to be the primary step in initiating carotid body response to hypoxia in the carotid body (57,60–60b), but this conclusion has not been reached by all (61).

VI. Metabolic Effects of Hypoxia on Type I Cells

Hypoxia might be sensed by its metabolic rather than its membrane effects in the carotid body. Hypoxia might interfere with ATP formation at mitochondria. The resulting decrease in cellular phosphate potential ([ATP]/[Pi] [ADP]) could influence Ca^{2+}-ATPase activity, leading to intracellular [Ca^{2+}] rising and the release of neurotransmitter (30,55,62). Evidence supporting this hypothesis comes from several sources. For example, oligomycin (an inhibitor of oxidative phosphorylation) blocks O_2 chemoreception without preventing CO_2 chemoreception (4,9,62). Oligomycin, which inhibits the formation of ATP and the energy-linked uptake of calcium by mitochondria, also blocks responses to cyanide, antimycin A, and metabolic uncouplers. When oligomycin is first administered, chemoreceptor discharge rises to a peak (4,9,62) but subsequently decreases to a new steady-state level. This sequence of events can be explained as follows. Upon administration of oligomycin, ATP production rapidly falls so that chemoreceptor activity becomes maximal. Then the sensory discharge decreases to a new steady state as glycolysis increases.

Other explanations are possible. It has been shown (2,3) that oligomycin can affect the cell membrane hyperpolarizing glomus cells. Some believe that the effects of oligomycin could result from nonspecific action on K^+ channels (63–65).

Experiments with carbon monoxide also support the metabolic hypothesis. Carbon monoxide is generated in small amounts in vivo by the enzyme hemeoxygenase II and can have many biological effects (66). CO reacts with guanylate cyclase, increasing cGMP, which in turn causes a cascade of reactions such as relaxation in smooth muscle by inhibiting Ca^{2+} entry (66–69). Also, CO at low levels (Pco, 70

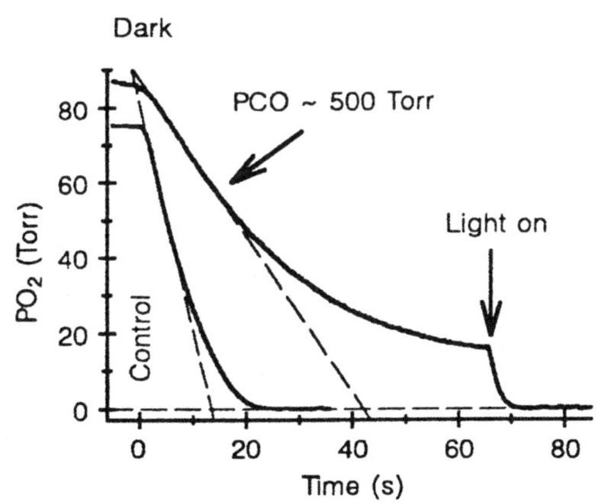

Figure 2 High Pco (500 torr) at Po_2 of 120 torr diminished the O_2 disappearance rate, which was restored by light exposure, compared to control. The O_2 disappearance rate was measured by Po microelectrode upon perfusate flow interruption.

torr) can bind with membrane hemeprotein and can reverse hypoxia-induced membrane depolarization in carotid body glomus cells (70) and decrease chemosensory discharge (71,72). Thus, low amounts CO take part in high-affinity reactions, binding to heme protein and mimicking the effect of O_2.

High concentrations of CO ($P_{CO} > 300$ torr) can bind with a lower affinity to reduced cytochrome a_3, a unit in the cytochrome oxidase complex in mitochondria, displacing O_2 from cytochrome oxidase so that the cells will be hypoxic even in the presence of a P_{O_2} of 100 torr (71–73). High P_{CO} (>300 torr), even without hypoxia, excites the carotid body in the dark, but not if exposed to white light (Fig. 2). This suppression of excitation by light can be duplicated at wavelengths (73) that match exactly those that affect cytochrome oxidase (74–76). These findings strongly support the idea that cytochrome oxidase is the O_2-sensing molecule that initiates carotid body activity. The excitation by CO is associated with a decrease of oxygen consumption by the carotid body, an inhibition that is reversed by light (77,78).

Also, CO interacts with CO_2 in the chemoreceptor in the dark, imitating the effects of hypoxia on P_{CO_2} (18). This supports the idea that high P_{CO} (>300 torr) operates like hypoxia in the dark. Dopamine is released by hypoxia and should be released by high P_{CO} in the dark even in the absence of hypoxia (79–83). The fact that low levels of CO interfere with hypoxic responses, acting like added oxygen, suggests that there is another heme-containing complex with which CO combines in the membrane, which can influence O_2 sensing in the carotid body. The importance of this membrane-combining site in the physiological response of the carotid body is unknown.

VII. Models of Chemoreception in the Carotid Body

The two major hypotheses explaining O_2 sensing in the carotid body are a nonrespiratory membrane hypothesis and a respiratory metabolic hypothesis.

According to the membrane hypothesis, O_2 sensing begins at the membrane of Type I cells with O_2 suppressing directly or indirectly the opening of K^+ channels, which leads to cell depolarization. The K^+ channels that are closed by hypoxia may be special O_2-sensitive channels or various other types of K^+ channels, which can be influenced nonspecifically by hypoxia. For example, both hypoxia and acid suppress certain K^+ and open Ca^{2+} channels in Type I cells (12,42–45,47–51).

It may be that patency of K^+ channels depends on the level of H_2O_2 and only indirectly on P_{O_2} (85–90). Acker has postulated that O_2 radicals are formed by the actions of the ubiquitous membrane protein NADPH oxidase on O_2 (9). These radicals are dismutated to H_2O_2, which then acts as a second messenger according to Acker. The H_2O_2 can then either act on genes (e.g., to increase erythropoietin in erythropoietin-producing cells) or be scavenged by catalase or glutathione. Catalase scavenging leads to the formation of a complex, which activates heme containing guanylate cyclase and enhances the level of cGMP. Alternatively, glutathione scavenging changes the ratio of GSH to GSSG. In this scheme, hypoxia by altering cGMP and/or GSH GSSG ratio (reducing compounds) closes K^+ channels and leads to Ca^{2+} influx.

This hypothesis provides a mechanism for O_2 sensing in both excitable and nonexcitable tissue. It is of interest that sulfhydryl agents that are also reducing agents are known to excite chemoreceptors (91). Drugs that alter H_2O_2 metabolism might help further evaluate the role of O_2 radicals in chemoreception. In addition, the immune deficiency disorder chronic granulomatous disease, in which there is a genetic defect in H_2O_2 production, offers a unique possibility to evaluate its significance in O_2 sensing. Murine models in which portions of NADPH oxidase are knocked out (92) may help determine the role of NADPH in O_2 sensing.

According to the respiratory metabolic hypothesis, O_2 sensing takes place in mitochondria of the cytochrome oxidase complex, reducing energy metabolism and ATP formation. This changes the phosphate potential [ATP]:[ADP] [P1] ratio, allowing intracellular calcium to rise by release of Ca from intracellular stores, which in turn releases neurotransmitters. Figure 3 summarizes the membrane and respiratory metabolic theories of O_2 sensing in the carotid body (1,11,15,18,62,77).

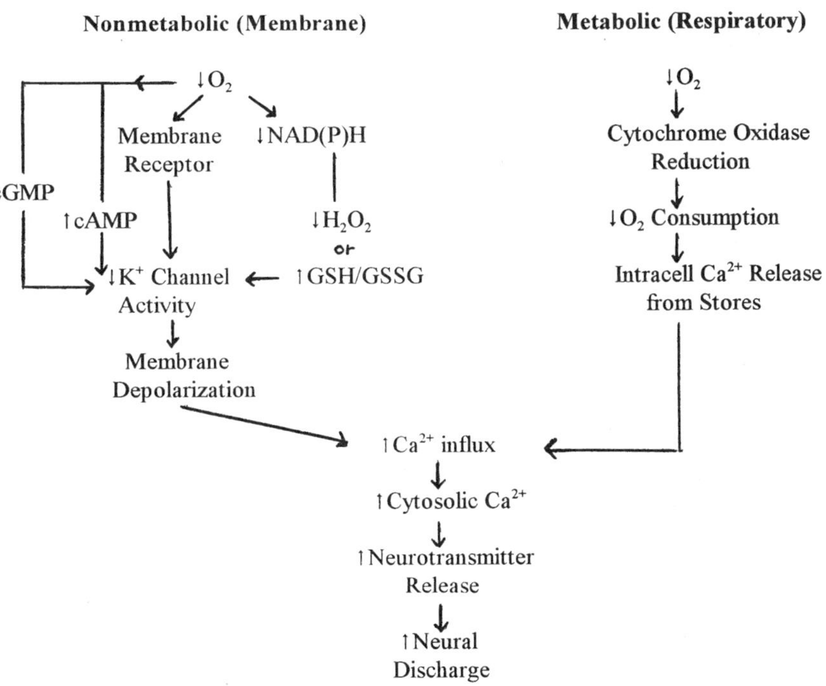

Figure 3 Postulated mechanisms of hypoxic chemoreception in type I carotid body cells.

There is evidence that Ca^{2+} is released from an internal store by hypoxia (93–95). Thapsigargin, a potent inhibitor for Ca-ATPase, resulted in a persistent elevation of intracellular [Ca^{2+}], which increased chemosensory discharge.

While it is now agreed that chemoreception in the carotid body takes place in Type I rather than Type II cells, it is still controversial whether respiratory metabolic or the nonrespiratory membrane models best portray O_2 sensing. O_2 sensing may take place in more than one site in Type I cells. The results of studies of the effects of high and low levels of CO on carotid body activity, which show inhibition with low levels of CO and excitation by higher levels, would be consistent with the idea that at least two different sites (two different molecular mechanisms) are triggered by oxygen lack and produce increasing chemoreceptor activity. Reactions at both sites could give rise to increases in intracellular Ca^{2+} (72). It is of interest that hypercapnia as well as hypoxia (54,58) is followed by a rise in intracellular [Ca^{2+}], which corresponds to the increase in chemosensory discharge. In both models, the rise in intracellular Ca^{2+} produces neurotransmitter release.

Dopamine seems to be an important neurotransmitter involved in modulating carotid sinus nerve discharge. Tyrosine hydroxylase is the rate-limiting enzyme for dopamine and norepinephrine synthesis and is activated within a few hours of exposure of the carotid body to hypoxia (92,97), but it can also be activated by other stimuli (83). Hypoxia releases dopamine from carotid body glomus cells as sensory discharge increases (79–82). Quantal secretion of dopamine also occurs from single glomus cells and is dependent on Ca^{2+} entry (82). Although both sensory discharge and dopamine release are Ca^{2+} dependent (82), other neurotransmitter may be important as well since dopamine stores can be exhausted in the rat CB and release stopped without a reduction in chemosensory response (83). While dopamine is certainly one neurotransmitter that is released by the carotid body, many other agents have been implicated as mediators of carotid sinus nerve discharge including dopamine acetylcholine and the neurokinins (82,83,96,97).

VIII. Effects of Hypoxia on Smooth Muscle and Neurons

Mechanisms of O_2 sensing have also been examined in vascular and neural tissues, but no consensus on the molecules or mechanisms involved has emerged.

Hypoxia inhibits the contraction of systemic vascular smooth muscles (98), but it initiates contraction of pulmonary arterial smooth muscle cells as in the carotid body (67,85–90). O_2 sensing in these tissues may arise from direct effects of hypoxia on ion channels, by an effect on ion channels mediated through some second messenger such as H_2O_2, or by an effect on oxidative metabolism (98–102).

Some experimental observations indicate the difference in behavior of ion channels in the two kinds of smooth muscle. Hypoxia directly suppresses L-type Ca^{2+}-channel activity in systemic smooth muscle cells (99). Also, hypoxia reduces potassium currents in cultured rat pulmonary but not mesenteric arterial myocytes (100–102). The K^+ current depression in pulmonary arterial smooth muscle cells depolarizes them, opening voltage-gated Ca^{2+} channels so that Ca^{2+} enters, causing the smooth muscles to contract (100–102). The same sequence of events does not occur in smooth muscle from systemic arteries.

Wolin and others (67,85–90) believe that the reduction of superoxide and H_2O_2 formation by hypoxia lessens the production of cGMP, inhibits K^+ channels, and causes smooth muscle contraction in pulmonary arteries. Archer et al. (63) have a similar theory.

Ulrich and colleagues showed that high P_{CO} in the dark relaxes ileal smooth muscle by activating guanylate cyclase and cGMP production (103,104). It has been proposed that inhibition of CO binding to a cytochrome P450–dependent monooxygenase reduced hemoprotein is the basis for ductus arteriosis smooth muscle relaxation in the lamb (105). Utz and Ulrich found that the photoreversible CO-dependent relaxation occurred with an action spectrum of 422 nm maximum, which was superimposable on that of guanylate cyclase (103). They could not detect any involvement of the respiratory chain. This group also found that CO inhibited the Ca^{2+} rise induced by depolarization (104). Lin and McGrath found that CO decreased $[Ca^{2+}]$ in systemic vascular smooth muscle, thereby relaxing and dilating the blood vessel (106). This was accompanied by a rise in cGMP.

ATP levels can be modulated by hypoxia in systemic arteries but not in carotid body glomus cells according to Lopez-Barneo et al. (43). Opening of ATP-sensitive K^+ channels in coronary artery cells by several minutes of severe hypoxia causes hyperpolarization, closure of Ca^{2+} channels, and muscle relaxation (100). Glibenclamide, a potent inhibitor of ATP-sensitive K^+ channels, caused vasoconstriction, and cromakalim, an opener of K^+ channels, produced vasodilation (107).

Interference with oxidative metabolism by blockade of the respiratory chain to produce a fall in ATP has also been reported to relax the smooth muscle of the ductus arteriosis. The involvement of the respiratory chain in vascular smooth muscle sensing of oxygen lack has been found by one group (98) but denied by others (103–105).

O_2-sensitive K^+ channels have also been found in central neurons, although the channels do not all have the same properties. Some of them work through alterations of cytosolic factors (ATP, Ca^{2+}, etc.). Other investigators have found a direct mechanism for low P_{O_2} sensing in the membrane, which is not affected by H_2O_2 generation or by P_{CO} (380 torr) (108).

IX. Changes in the Ventilatory Response to Hypoxia with Time

One of the fascinating aspects of the effects of hypoxia are the changes that occur in ventilation and hypoxic sensitivity with the length of hypoxia and the maturity of the responder. Except in a few instances, the cellular basis of these rather complex alterations is poorly understood.

Some of the fluctuations in sensitivity probably occur because of the effects if hypoxia on different tissues occurring with different time courses, i.e., the effects of hypoxia on the brain may alter the arterial chemoreceptor signal (efferent control) or modify the interpretation of that signal by central respiratory neurons.

X. Erythropoietin Production During Hypoxia

Recent studies have considerably advanced our understanding of how hypoxia increases the production of erythropoietin (68,69). The physiological stimulus for erythropoietin production is a decrease in O_2 supply (68,69). In fetal life, erythropoietin is produced in the liver, but after birth erythropoetin is formed in renal peritubular and the renal cortex in interstitial cells and perhaps endothelial cells (69). The hormone is a 30.4 kDa glycoprotein which targets hemopoietic tissue, where it binds with receptors on erythroid progenitor cells. These then differentiate into functioning red blood cells (69).

Studies in the human hepatoma cell line Hep 3B have shown that hypoxia, cobalt, and nickel chloride stimulate erythropoietin production through a common pathway. On the basis of these studies it was proposed that the O_2 sensor for erythropoietin production is a heme protein located outside the mitochondria (109). When the O_2 tension is low, the heme protein is in the deoxy form, which initiates the process leading to erythropoietin gene activation. When O_2 tension is high, the heme protein is in an inactive oxy form. Cobalt or nickel can substitute for the ferrous ion in the porphyrin ring, locking the heme protein in the deoxy form so that erythropoietin formation is stimulated (109).

Red blood cell production also increases in rats inhaling 0.1% CO because the high-affinity CO binding to hemoglobin will inhibit O_2 release from hemoglobin, thus lowering the effective O_2 supply (69). On the other hand, the induction of erythropoietin in Hep 3B cells following exposure to 1% O_2 in vitro is markedly inhibited by the presence of 10% CO presumably because it attaches to the O_2 heme sensor locking in the oxy position (109,110). CO does not diminish the induction of erythropoietin by cobalt- or nickel-substituted heme because CO does not combine with these hemes (69). These observations support the hypothesis for O_2 sensing by a heme protein (which may deoxygenate cytochrome P450) (69,109).

There are similarities in the stimuli that produce erythopoietin and vascular endothelial growth factor. Cobalt also enhances the expression of vascular endothelial growth factor (110). Chronic cobalt administration to rats will give rise to increases in the volume of glomus cells (type I) along with hematocrit.

One study found erythropoietin production to be inhibited by H_2O_2, but glutathione is not involved in erythropoietin formation (112). However, the role of oxygen radicals in erythropoietin production needs further investigation. Studies in Hep 3B cells described later indicate that hypoxia leads to the formation of one or more protein factors, which induce the erythropoietin gene. The activation of the protein factors may depend on the level of reactive O_2 intermediates like peroxides (68).

In humans there is a single copy of the erythropoietin gene located in the q^{11}

to q^{22} region of chromosome 7 (69,110). In some cell types second messenger molecules such as cGMP and cAMP may also participate in gene activation (68). New evidence indicates that O_2 sensing activates erythropoietin genes via a protein factor (HIF-1 or hypoxia inducible factor). The same factor also seems to induce vascular endothelial growth factor genes and seems to be participating in metabolic changes that occur with prolonged hypoxia (112a).

XI. Ventilatory Acclimatization to Hypoxia

Studies of adaptation to chronic hypoxia demonstrate that a complex interaction of a number of different systems is involved. The cellular basis of this interplay of respiratory cardiovascular and neural factors is the focus of many investigators.

The gradual increase in breathing during hypoxia, known as ventilatory acclimatization, has been extensively studied. Mountaineers ascending to high altitudes continue to hyperventilate over and above the acute response (113). Animal experiments have shown that this response is initiated by the carotid body (114) and does not occur in carotid body denervated animals (115) or during systemic hypoxia if the carotid body is vascularly isolated and maintained normoxic (114). It has been demonstrated that continued hypoxia after the initial rise in chemosensory discharge causes a further increase after a delay of 2–3 hours, explaining the increase in Po_2 and decrease in CO_2 that has been observed in intact animals and (116,117). It is highly likely that the acclimatization is due to an increase in humans carotid body activity. The roll-off of the acute ventilatory response to hypoxia that is observed a few minutes after exposure to hypoxia, however, is not caused by any corresponding change in carotid body activity, but rather seems to involve the release of inhibitor neurotransmitters in the brain (118). Thus the changes in ventilation as hypoxia is prolonged seem to involve a rather complicated interplay of responses to hypoxia in different tissues.

The role of various neurotransmitters in ventilatory acclimatization or even the response to acute hypoxia remains controversial, but increases in production of dopamine in PC-12 cells have been used to surrogate for the evaluation of carotid body activity. Acute hypoxia does cause dopamine release in the carotid body and dopamine content to decrease (125–131). But whether dopamine release has an excitatory or inhibitory effect has not been settled (see, e.g., 126, 130). Norepinephrine, another carotid body catecholamine, seems to inhibit chemoreceptor activity (113). Chronic hypoxia stimulates its synthesis, but to a lesser extent than dopamine (127,128). The enlarged carotid bodies in chronic hypoxia contain a high concentration of dopamine. The carotid body is also known to secrete a higher concentration of dopamine as it enlarges (132,133). Despite the uncertainties concerning its effects, a number of studies have focused on dopamine metabolism in the carotid body during hypoxia.

In the chronically hypoxic group, dopamine content and its turnover rate increased 14- to 15-fold (127,128). Barer et al. (129) observed a blunted hypoxic ventilatory response in rats that lived for more than 2 weeks in an environment of severe hypoxia as compared to normoxic rats. According to Barer et al. (129) and Bee and Pallot (130), increased utilization of dopamine would tend to suppress the ventilatory response to hypoxia. Tatsumi et al. found a similar result in the cat exposed to severe hypoxia for 21 days (126). With less severe hypoxia the chemosensory response is augmented, even though dopamine content and turnover rate are also increased (128,129,131). Thus, a correspondence between dopamine secretion and sensory discharge in chronic hypoxia has not yet been established.

XII. Blunted Ventilatory and Carotid Chemosensory Function with Prolonged Hypoxia

Studying adult Peruvian high-altitude natives, Severinghaus and colleagues (119) found that the ventilatory response to hypoxia is blunted at altitude and remains blunted for a long time after the return to sea level (120). Similar results were found in the adult natives in the Himalayas (121,122).

It seems reasonable to believe that Type I cells may function differently when hypoxia response is blunted. Wyatt et al. (56) showed that glomus cells of rats which developed blunted ventilatory response to hypoxia did not depolarize, even though they had K channels that were closed by low Po_2 and K^+ conductance was suppressed. Progress in this work would be illuminating.

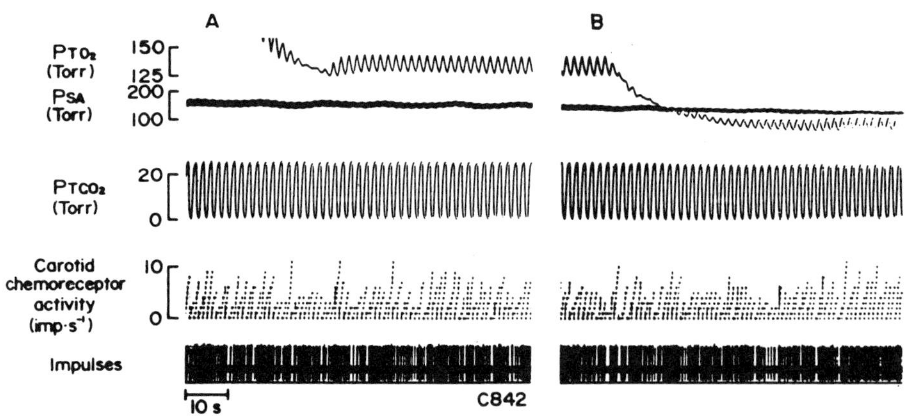

Figure 4 Effects of lowering end-tidal Po_2 on carotid chemoreceptor activity in a cat that was exposed to 100% O_2 for 65 hours. (A) Effect of lowering Po_2 from hyperoxia (100% O_2) to air breathing. (B) Effect of further lowering of Po_2. There was no effect of hypoxia on chemoreceptor activity. Traces (top to bottom) the tracheal Po_2 (P_{TCO_2}); systemic arterial pressure (Psa); tracheal Pco_2 (P_{TCO_2}); carotid chemoreceptor activity (in impulses per second); and raw impulses.

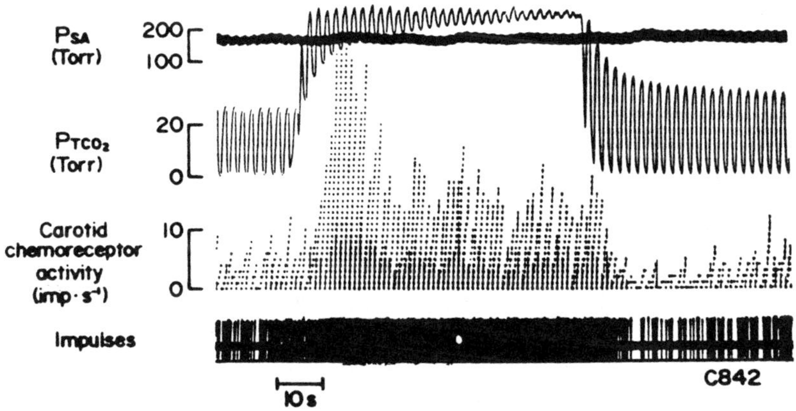

Figure 5 Effects of hypercapnia during hyperoxia: effects of sudden hypercapnia during acute hyperoxia on carotid chemoreceptor activity and ventilation in a control cat. Traces (top to bottom) are systemic arterial pressure (Psa); tracheal P_{CO_2} (P_{TCO_2}); carotid chemoreceptor activity (in impulses/s); and impulses.

In this context, a blunted carotid chemosensory function has also been found in cats exposed to chronic hyperoxia (123,124) (Fig. 4). The chemosensory response to CO_2 was, however, augmented (Fig. 5). This separation of O_2 and CO_2 response has been seen in other studies. These findings are not compatible with the idea that K^+ channel opening and closing of Type I cells occurs in both the response to O_2 and CO_2 by the carotid body (54,55,61).

XIII. Effect of Chronic Hypoxia on the Carotid Body and Carotid Body Cells In Vivo and In Vitro

Exposure to high altitude or normobaric hypoxia at times causes substantial enlargement of the carotid bodies in humans and in other animal species (114,115,130–134). The cells of carotid body blood vessels hypertrophy with chronic hypoxia and the vessels become filled high hematocrit blood (increases from 39% to 70%) (132). Bee and Pallot found evidence for hyperplasia of the carotid body cells by light microscopy (130). Quantitative morphometric analysis using stereological techniques at an ultrastructural level showed that the Type I rather than the Type II cells enlarge. The mean volume of Type I cells increased from 320 μm^3 in controls to 1212 μm^3 in chronically hypoxic rats. Similar observations have been made by Pegingnot et al. (133) and are universally accepted.

Studies in cell culture by Nurse and coworkers have provided data on the effects of chronic hypoxia on the transmembrane movement of ions (135,136). Cellular hypertrophy and modifications of channel function occurs in cultured rat glomus cells exposed to chronic hypoxia (6% O_2 for 1–3 weeks) (135,136). These cells undergo changes in the absence of neural input or input from the blood or cardiovascular system. The mechanisms responsible for initiating and controlling the cellular hypertrophy remain unknown.

Changes in cell size were accompanied by changes in membrane currents and probably on channels as well. While inward Na^+ current increased significantly, because of increased cell size, the K^+ current density decreased, whereas the inward Ca^{2+} current density remained unaffected. It is impossible to know whether these in vitro changes lead to increases or decreases in O_2 sensitivity or chemosensory discharge in vivo. Studies measuring chemosensory discharge show that hypoxia of short-term duration increases (116,118,125,132) and of long-term duration decreases (105) O_2 sensitivity. These changes do not seem to be explained by the changes reported in Type I cell size or number.

A more recent study was done to compare the type I cells isolated from chronically hypoxic neonatal rats which have blunted hypoxic ventilatory responses of cells from normoxic animals (56). Although the density of O_2-sensitive K^+ channels was decreased in the cells from hypoxic compared to cells from normoxic animals, acute hypoxia caused similar reversible inhibition of K^+ currents in type I cells from both groups. Type I cells from both normoxic and chronically hypoxic animals possess O_2-sensitive K^+ currents. Resting membrane potentials were also similar in normoxic and hypoxic rats. However, acute hypoxia failed to depolarize the Type I cells from chronically hypoxic rats. Charybodotoxin (20 nM) also failed to suppress the Ca^{2+}-activated K^+ current in hypoxic cells unlike normoxic cells (58). This suggests that absence of charybdotoxin Ca^{2+}-sensitive K^+ channel, rather than a decrease in O_2-sensitive K^+ channel, is responsible for the lack of O_2 sensitivity in the carotid body of the chronically hypoxic rats.

A major problem in the cultured cell experiments has to do with the range of Po_2 sensitivity of O_2-sensitive K^+ channels (K-O_2 channel). Excised patches of glomus cell membrane have K^+-O_2 channels (49), which are suppressed at levels of Po_2 that do not excite the intact carotid body (13). The maximum inhibition of K^+ current is obtained at a Po_2 above 85 torr, but the carotid body in vivo is maximally excited at much lower levels of Po_2 (49). Thus, there seems to be a discrepancy between the lower level of Po_2 which stimulates chemosensory discharge and that which inhibits K^+-O_2 channel in the membrane patch.

Possible reasons for this difference are as follows: the patch preparations used eliminate the modulation of ionic current by intracellular messengers, which usually occurs in vivo. For example, cAMP is known to increase during acute hypoxia and augment chemosensory response (138); cAMP analogs applied to the whole cell inhibit K^+ oxygen-sensitive currents (49). These effects of cAMP are lost in the membrane patch experiments (49). Therefore, it has been suggested that some of the apparent differences between in vitro and in vivo studies can be explained by taking into account the decrease in the effectiveness of intracellular messengers such as cAMP. Another possibility is that intracellular O_2-sensing mechanisms may be coupled to several different types of K^+ channels (47,55). Confounding the interpretation of these findings are the vast differences between the degree and duration of hypoxic exposure in the studies reported. For example, Nurse et al. incubated cells taken from normoxic rats at a Po_2 of 49 torr for 3 weeks, while Peers used 70 torr for 3 days for cells taken from chronically hypoxic rats (61,136).

XIV. Effects of Hypoxia on Gene Expression

We now know that changes in O_2 tension can initiate changes in the expression of a variety of genes (69). Hepatoma cells (Hep 3B) produce a hypoxia-inducible factor (HIF-1 nucleoprotein) with about the same time course as erythropoietin formation (69). Cobalt chloride, like severe prolonged hypoxia, produces erythropoietin in these cells and also causes HIF-1 formation. HIF-1 formation can be blocked by the transcription inhibitor actinomycin D and by the protein synthesis inhibitor cycloheximide (139). HIF-1 has been shown to bind to the 3′-flanking enhancing regions of the erythropoietin gene (erythropoietin), but this requires phosphorylation to induce gene activity (139–142). HIF-1 has been shown to consist of different subunits, which contact DNA directly. The erythropoietin enhancer has three parts, two of which are essential for induction by hypoxia; the third amplifies the induction signal (140,141). Binding is transient and is quickly and easily reversible with high O_2. In kidney cells other factors besides HIF-1 may be needed to produce erythropoietin (69). The increased levels of erythropoietin produced by hypoxia are only in part due to increased gene transcription (69). Erythropoietin mRNA half-life is also markedly prolonged by hypoxia (69). Interestingly, other studies have indicated that HIF-1 is also produced in cell lines that do not produce erythropoietin, suggesting that HIF-1 might activate other genes besides erythropoietin (139,141). Semenza et al. have shown that HIF-1 has a more general regulatory role and that RNAs encoding glycolytic enzymes such as aldolase, phosphoglycerate kinase, and pyruvate kinase were produced when Hep 3B or Hela cells were exposed to hypoxia (140). The formation of these mRNAs did not occur when cycloheximide was used to prevent HIF-1 formation (139,140). Sequences from these genes containing HIF-1–binding sites were shown to be activated in transient expression assays using 1% O_2 to stimulate HIF-1 formation (140). The studies strongly suggest that HIF-1 regulates several glycolytic genes as well as erythropoietin and that HIF-1 functions as an important mediator of adaptation responses to hypoxia.

Other gene pathways may be affected by hypoxia (Fig. 6). Hypoxia has been shown to increase the formation of platelet-derived growth factor, endothelin, interleukin-1A, and vascular endothelial growth factor (69,143–146). Exposure of rats to 10% hypoxia for even one hour increases mRNA for tyrosine hydroxylase in Type I carotid body cells (133). Recent studies by Milhorn and coworkers and by Prabhakar and coworkers suggest that the increase of tyrosine hydroxylase probably involves the activation of immediate early response genes of the fos-jun family

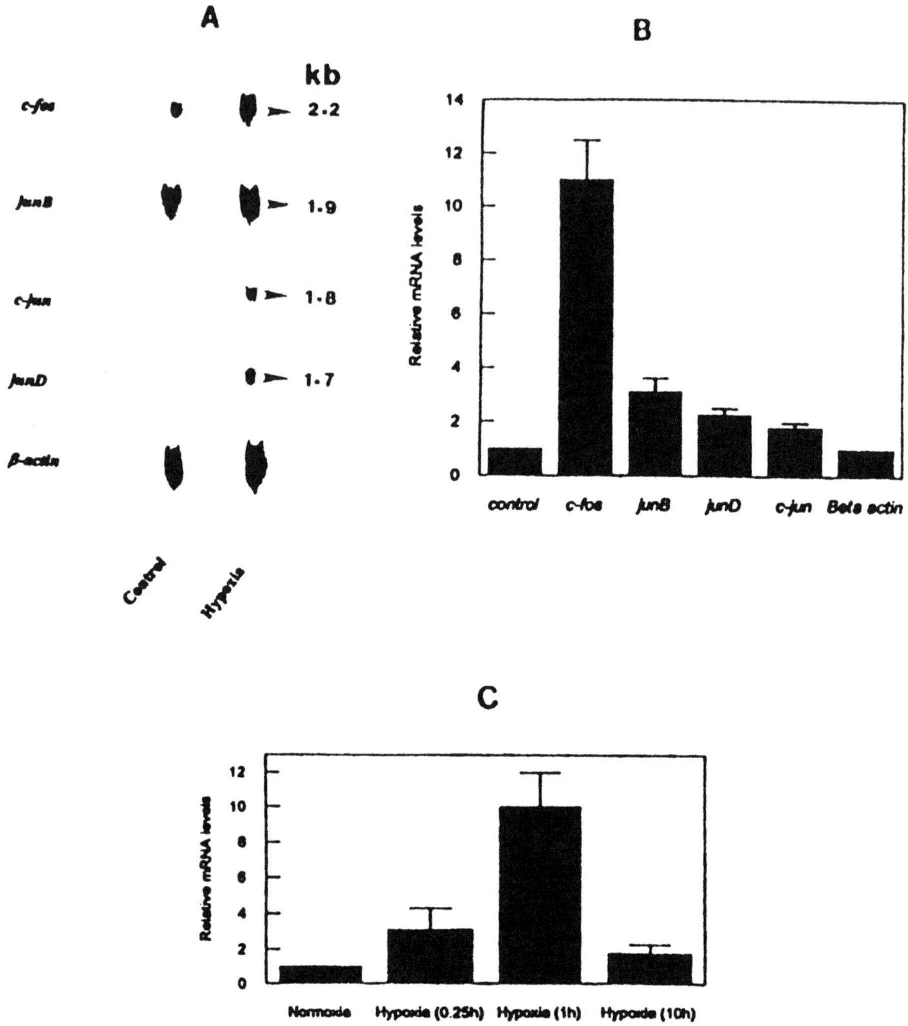

Figure 6 Effects of hypoxia on IERG expression in PC-12 cells. (A) Northern blot assay showing mRNAs encoding c-*fos, jun*B, *jun*D, and β-actin during control (normoxia; 21% O_2) and after 1 hour of hypoxia (5% O_2). Average results are summarized in (B). Effects of different durations of hypoxia in c-*fos* mRNA expression are presented in (C). Results presented are mRNA changes relative to normoxic controls (mRNA are normalized for 28s mRNA). Data presented are mean ± SEM from five experiments in B and three experiments in C.

(144,145). Immediate early response genes (IERG) are comprised of several different groups of which the fos-jun family have been most extensively studied (146). Exposure of a number of different kinds of cells (PC-12, neuroblastoma, etc) to a variety of stimuli produce fos and jun, the protein products of the c-*fos* and c-*jun* genes. These two proteins form heterodimers (AP-1), which then bind to the promoter 3′ region of a number of different genes (146). AP-1 can also be formed by jun-jun homodimers but not by fos-fos homodimers. It has been shown in studies in vitro that AP-1 binding is affected by redox state and is greater under hypoxic conditions, which seems to require the integrity of crucial SH groups present in AP-1 (147).

In vivo in rats and cats moderate hypoxia (10% O_2 for 30 minutes) will induce fos formation in the ventral and dorsal medulla and rather diffusely in other brain regions. A roughly similar distribution is obtained by exposure of the same animals to 15% CO_2 in O_2. In studies that focused on the nucleus tractus solitarii, fos formation by hypoxia could be blocked by NMDA, a glutamate antagonist, suggesting that glutamate may be needed for fos production in vivo (145,148).

Hypoxia produces increased fos and AP-1 binding in PC-12 cells, but not in neuroblastoma cells (145,149), indicating that cells do not respond in a uniform way. Fos, though, can be induced in neuroblastoma cells by other stimuli. Studies of mRNA formation indicate that low levels of O_2 in cultures of PC-12 cells (Po_2 = 40 torr) increase transcription of c-*fos* and c-*jun* and other genes such as *jun*B and *jun*D (149). It is of interest that nitric oxide exposure of neuroblastoma cells augments fos formation.

Both transcription of the tyrosine hydroxylase gene and the stability of tyrosine hydrolase mRNA are augmented by hypoxia (150). The 3′ promoter region of the TH gene contains both AP-1–and HIF-1–binding sites (44). Recent studies show that mutations of the AP-1–binding site drastically reduce the formation of TH mRNA by hypoxia, which suggests the importance of the fos-jun family in the hypoxic induction of tyrosine hydroxylase (44) but does not eliminate HIF-1 as well. It is not clear if the heme protein involved in sensing hypoxia for HIF-1 formation will also activate IERGs. However, Hep3B cells in culture will produce VEGF, a highly specific mitogen for endothelial cells, upon exposure to hypoxia and will increase levels of *jun*B and c-*jun* (110). In addition, cobalt chloride, which leads to increases in both erythropoietin and VEGF, also increases c-*jun* and c-*fos* (109).

Adaptations to chronic hypoxia may frequently involve activation of genes and the formation of new or increased amounts of proteins which alter the phenotypic properties of the cells. Some genes may begin to be activated within 6 minutes of hypoxic exposure. Responses to chronic hypoxia seem to involve many organs but may be coordinated by the activation of a relatively small set of genes, which produce protein products that bind to different target genes in different cell types. These gene-mediated changes may occur in the carotid body itself and may alter signaling cascades and affect the sensitivity of Type I cells to chronic hypoxia. Hypoxia-inducible factor may be a part of a widespread oxygen-sensing mechanisms (32,140,147).

Our understanding at this time is that HIF-1αs are continually degraded under normoxia (151,152). This degradation during normoxia is a remarkable feature that requires iron-dependent von-Hippel-Lindau ligase complexes (153,154). This oxygen-dependent reaction is still unraveling (154,155). Hypoxia reverses this effect on normoxia rapidly within a few minutes so that HIF-1αs then accumulate and combine with the constitutively produced HIF-1β to HIF-1. These are then translocated into the nucleus to set the tone of various genes and mRNA expressions. These reactions may require a few minutes to hours to accomplish (151,152). However, the initial oxygen sensing that involves plasma membrane and cytoplasm may be over within split seconds. Despite all the advances in the area of gene regulation, the molecules that initiate the mechanisms of oxygen sensing still remain elusive.

Acknowledgments

We are grateful to Mary Pili and Stella Rogers for their secretarial assistance. Supported in part by grants HL-43413-11 and HL-50180-06.

References

1. Wilson DF, Mokashi A, Chugh D, Vinogradov S, Osanai S, Lahiri S. The primary oxygen sensor of the cat carotid body is cytochrome a3 of the mitochondrial respiratory chain. FEBS Lett 1994; 351:370–374.
2. Duchen MR, Biscoe TJ. Mitochondrial function in type I cells isolated from rabbit arterial chemoreceptors. J Physiol 1992; 450:13–31.
3. Duchen MR, Biscoe TJ. Relative mitochondrial membrane potential and $[Ca^{2+}]$ I in the type I cells isolated from the rabbit carotid body. J Physiol 1992; 546:33–61.
4. Mulligan E, Lahiri S, Storey BT. Carotid body O_2 chemoreception and mitochondrial oxidative phosphorylation. J Appl Physiol 1981; 51:438–446.
5. Lahiri S, Rumsey WL, Wilson DF, Iturriaga R. Contribution of in vivo microvascular Po_2 in the cat carotid body chemo transduction. J Appl Physiol 1993; 75:1035–1043.
6. Goldberg, MA, Bunning SP, Bunn, HF. Regulation of the erythropoietin gene: evidence that the oxygen sensor is a hemeprotein. Science 1988; 242:1412–1415.
7. Rotta A, Capena A, Hurado A, Velasquez T, Chavez R. Pulmonary circulation at sea level and at high altitudes. J Appl Physiol 1956; 9:328–336.
8. Fay FS, Nair P, Whalen WJ. Mechanisms of oxygen induced contraction of ductus arteriosus. In: Reivich M, Coburn R, Lahiri S, Britton C, eds. Tissue Hypoxia and Ischemia. New York: Plenum Press, 1977:123–134.
9. Acker H, Delpiano M, Degner F. The meaning of the Po_2 field for the chemoreceptive process. In: Acker H, O'Regan RG, eds. Physiology of the Peripheral Arterial Chemoreceptors. Amsterdam: Elsevier, 1983:89–115.
10. Whalen WJ, Nair P. Oxidative metabolism and tissue Po_2 of the carotid body. In: Acker H, O'Regan RG, eds. Physiology of the Peripheral Arterial Chemoreceptors. Amsterdam: Elsevier, 1983:117–132.
11. Buerk DG, Iturggiaga R, Lahiri S. Testing the metabolic hypothesis of O_2 chemoreception in the cat carotid body in vitro. J Appl Physiol 1994; 76:1317–1323.

12. Gonzalez C, Vicario I, Alvarez L, Rigul R. Oxygen sensing in the carotid body. Biol Signals 1995; 4:245–256.
13. Lahiri S. Chemophores in O_2 chemoreception: the carotid body model. NIPs 1994; 9: 161–165.
14. Fay FS. Oxygen consumption of the carotid body. Am J Physiol 1970; 218:518–523.
15. Lahiri S, Buerk DG, Chugh D, Osanai S, Mokashi A. Reciprocal photolabile O_2 consumption and chemoreceptor excitation by carbon monoxide in the cat carotid body: evidence for cytochrome a3 as the primary O_2 sensor. Brain Res 1995; 684:194–200.
16. Lahiri S. Role of arterial O_2 flow in peripheral chemoreceptor excitation. Fed Proc 1980; 39:2648–2652.
16a. Acker H. The involvement of nerve terminals in the NY paraganglionic chemoreceptor system. Ann NY Acad Sci 1987; 519:369–384.
17. Floyd WF, Neil E. Arch Int Pharmocodyn Ther 1952; 91:230–239.
18. Osanai S, Chugh DK, Mokashi A, Lahiri S. Stimulus interaction between CO and CO_2 in the cat carotid body chemoreception. Brain Res 1996; 711:56–63.
19. Hornbein TF, Roos A. Specificity of H ion concentration as a carotid body chemoreceptor stimulus. J Appl Physiol 1963; 18:580–584.
20. Lahiri S, DeLaney RG. Stimulus interaction in the response of carotid body chemoreceptor single afferent fibers. Respir Physiol 1975; 24:249–266.
21. Gray BA. Response of the perfused carotid body to changes in pH and Pco_2. Respir Physiol 1968; 4:229–245.
22. Black AMS, McCloskey DI, Torrance, RW. The responses of carotid body chemoreceptors in the cat to sudden changes of hypercapnic and hypoxic stimuli. Respir Physiol 1971; 13:36–49.
23. Iturriaga R, Mokashi A, Lahiri, S. Dynamics of carotid body responses in vitro in the presence of CO_2-HCO3: role of carbonic anhydrase. J Appl Physiol 1993; 75:1587–1594.
24. Lahiri S, Itturiaga R, Mokashi A, Botre F, Chugh D, Osanai S. Adaptation to hypercapnia vs intracellular pH in cat carotid body response in vitro. J Appl Physiol 1996; 80: 1090–1099.
25. Beidler LM. Taste receptor stimulation. Prog Biophys Mol Biol 1962; 12:109.
26. McDonald DM. Peripheral chemoreceptors. Structure-function relationships of the carotid body. In: Hornbein TF, ed. Regulation of Breathing. Part I. New York: Marcel Dekker, 1981:105–319.
27. Lahiri S, Mokashi A, Nishino T, Mulligan E. Comparison of aortic and carotid chemoreceptor responses to hypercapnia and hypoxia. J Appl Physiol 1981; 51:55–61.
28. Youngson C, Nurse CA, Yeger H, Cutz E. Oxygen sensing in airway chemoreceptors. Nature 1993; 365:153–155.
29. Czyzyk-Krezeska MR, Bayliss DA, Lawson EE, Millhorn DE. Regulation of tyrosine hydroxylase gene expression in the rat carotid by hypoxia. J Neurochem 1992; 58: 1538–1546.
30. Duchen MR, Biscoe TJ. Relative mitochondrial membrane potential and [Ca_2+] I in type I cells isolated from the rabbit carotid body. J Physiol 1992; 450:33–61.
31. Czyzyk-Krezeska MR, Dominiski Z, Kole R, Millhorn DE. Hypoxia stimulates binding of the cytoplasmic protein to a pyrimidine-rich sequence in the 3′-untranslated region of rat tyrosine hydroxylase mRNA. J Biol Chem 1994; 269:9940–9945.
32. Fandrey J. Hypoxia-inducible gene expression. Respir Physiol 1995; 101:1–10.

33. Wang ZZ, Bredt DS, Fidone SJ, Stensaas LJ. Neurons synthesizing nitric oxide innervate the mammalian carotid body. J Comp Neurol 1993; 336:429–432.
34. Prabhakar NR, Kumar GK, Chang CH, Agani FH, Haxhiu MA. Nitric oxide in sensory function of the carotid body. Brain Res 1993; 625:16–22.
35. Chugh DK, Katayama M, Mokashi A, Bebout DE, Ray KK, Lahiri S. Nitric oxide-related inhibition of carotid chemosensory nerve activity in the cat. Respir Physiol 1994; 97:147–156.
36. Grimes PA, Mokashi A, Stone R, Lahiri S. Nitric oxide synthase in autonomic innervation of the cat carotid body. J Auton Nerv Syst 1995; 54:80–88.
37. Lahiri S, Rumsey WL, Wilson DF, Iturriaga R. Contribution of in vivo microvascular Po_2 in the cat carotid body chemotransduction. J Appl Physiol 1993; 75:1035–1043.
38. Rumsey WL, Iturriaga R, Spergel D, Lahiri S, Wilson DF. Optical measurements of the dependence of chemoreception on oxygen pressure in the cat carotid body. Am J Physiol 1991; 261:C614–C622.
39. Itturiaga R, Lahiri S, Mokashi A. Carbonic anhydrase and chemoreception in the cat carotid body. Am J Physiol 1991; 261:C565–C573.
40. Black AMS, McCloskey DI, Torrance RW. The responses of carotid body chemoreceptions in the cat to sudden changes of hypercapnic and hypoxic stimuli. Respir Physiol 1971; 13:36–49.
41. Nurse, CA. Carbonic anhydrase and neuronal enzymes in cultures glomus cells of the carotid body of the rat. Cell Tissue Res 1990; 261:55–71.
42. Peers C. Ionic channels in Type I carotid body cells. In: O'Regan RG, Nolan P, McQueen DS, Patterson D, eds. Arterial Chemoreceptors: Cell to System. New York: Plenum Press, 1994:29–40.
43. Lopez-Barneo J, Lopez-Lopez JR, Urena J, Gonzalez C. Chemotransduction in the carotid body: K^+ currents modulated by Po_2 in type I chemoreceptor cells. Science 1988; 241:580–582.
44. Delpiano MA, Hescheler J. Evidence for a Po_2-sensitive K^+ channel in the type-I cell of the rabbit carotid body. FEBS Lett 1989; 249:195–198.
45. Lopez-Lopez J, Gonzalez D, Urena J, Lopez-Barneo J. Low Po_2 selectively inhibits K channel activity in chemoreceptor cells of the mammalian carotid body. J Gen Physiol 1989; 93:1001–1015.
46. Hescheler J, Delpiano JA. Ionic currents on carotid body type I cells and the effects of hypoxia and NaCN. In: Eyzaquirre C, Fidone SJ, Fitzgerald RS, Lahiri S, McDonald DM, eds. Arterial Chemoreception. New York: Springer-Verlag, 1990:58–62.
47. Lopez-Lopez JR, De Luis DA, Gonzalez C. Properties of a transient K^+ current in chemoreceptor cells of rabbit carotid body. J Physiol (London) 1993; 460:15–32.
48. Ganfornina, MD, Lopez-Barneo J. Potassium channel types in arterial chemoreceptor cells and their selective modication by oxygen. J Gen Physiol 1992; 100:401–426.
49. Ganfornina MD, Lopez-Barneo J. Single K^+ channels in membrane patches of arterial chemoreceptor cells are modulated by O_2 tension. Proc Natl Acad Sci USA 1991; 88: 2927–2930.
50. Lopez-Barneo, J. Oxygen-sensitive ion channels: How ubiquitous are they? TINS 1994; 17:133–135.
51. Gonzalez C, Almaraz L, Plesso A, Rignal R. Oxygen and chemoreception in the carotid body chemoreceptors. TINS 1992; 15:146–153.
52. Acker H. Mechanisms and meaning of cellular oxygen sensing in the organism. Respir Physiol 1994; 95:1–10.

53. Acker H, Xue D. Mechanisms of O_2 sensing in the carotid body in comparison with other O_2-sensing cells. NIPS 1995; 10–24–216.
53a. Shirahata M, Fitzgerald RS. Dependency of hypoxic chemotransduction in cat carotid body on voltage-gated calcium channels. J Appl Physiol 1991; 71:1052–1069.
53b. Roy A, Rozanov C, Iturriaga R, Mokasi A, Lahiri S. Acid-sensing by carotid body is inhibited by blockers of voltage-sensitive Ca^{2+} channels. Brain Res 1997; 769:396–399.
54. Fieber LA, McCloskey EW. J Neurophysiol 1993; 70:1378–1384.
55. Buckler KJ, Vaughan-Jones RD. Role of intracellular pH and {Ca_2+} i in acid chemoreception in type–cells of carotid body. In: O'Regan RG, Nolan P, McQueen DS, Paterson D, eds. Arterial Chemoreceptors: Cell to System. New York: Plenum Press, 1994:41–55.
56. Wyatt CN, Wright C, Bee D, Peers C. O_2-sensitive K+ currents in carotid body chemoreceptor cells from normoxic and chronically hypoxic rats and their roles in hypoxic chemotransduction. Proc Natl Acad Sci 1995; 295–299.
57. Cheng PM, Donnelly DF. Relationship beetween changes of glomus cell current and neural response of rat carotid body. J Neurochem 1995; 74:2077–2086.
58. Osanai S, Buerk DG, Mokashi A, Chugh DK, Lahiri S. Lack of effect of charybdotoxin on cat carotid body chemosensory discharge: a denial of the role of O_2-sensitive K+ Ca currents. Brain Res (in press).
59. Buckler KJ, Vaughan-Jones RD. Effects of hypoxia on membrane potential and intracellular calcium in rat neonatal carotid body type I cells. J Physiol (London) 1994; 476:423–428.
60. Buckler KJ. Role of potassium channels in mediating the hypoxia evoked depolarization of carotid body type-1 cells. In: Zapta P, Eyzaquirre C, Torrance RW, eds. Frontiers in Arterial Chemoreception. New York: Plenum Press, 1966:
60a. Roy A, Rozanov C, BuerkDG, Mokashi A, Lahiri S. Supression of glomus cell K+ conductance by 4-aminopyridine is not related to [Ca^{2+}], dopamine release and chemosensory discharge from carotid body. Brain Res 1998; 785:228–235.
60b. Lahiri S, Roy A, Rpzanov C, Mokashi A. K^+ current modulated by PO2 in Type cells in rat carotid body is not a chemosensor. Brain Res 1998; 794:162–165.
61. Peers C. Ionic channels in type I carotid body cells. In: O'Reagan R, Nolan P, McQueen DS, Patterson D, eds. Arterial Chemoreceptors: Cell to System. New York: Plenum Press, 1994:29–40.
62. Mulligan E, Lahiri S. The separation of carotid chemoreceptor responses to O_2 and CO_2 by oligomycin and antimyocin A. Am J Physiol 1982; 242:C200–C206.
63. Archer SL, Huang J, Henry T, Peterson D, Weir KE. A redox-based O_2 sensing in rat pulmonary vasculature. Cir Res 1993; 73:1100–1112.
64. Wolin MS. Reactive oxygen species and vascular signal transduction mechanisms. Microcirculation 1995; 3:1–17.
65. Franco-Obregon A, Urena J. Lopez-Barnes J. Oxygen-sensitive calcium channels in vascular smooth muscle and their possible role in hypoxic arterial relaxation. Proc Natl Acad Sci 1995; 92:4715–4719.
66. Prabhakar NR, Dinerman JL, Againi FH, Solomon S. Carbon monoxide: a role in carotid body chemoreception. Proc Natl Acad Sci. 1995:92–1997.
67. Burke TM, Wolin MS. Hydrogen peroxide elicits pulmonary arterial relaxation and guanylate cyclase activation. Am J Physiol 1987; 21:H721–H732.
68. Bunn HF, Payton RO. Oxygen sensing and molecular adaptation to hypoxia. Physiol Rev 1996; 76:839–886.

69. Jelkman W. Erythropoietin: Structure control of production and function. Physiol Rev 1992; 72:449–489.
69a. Fandrey J, Frede S, Ehleben W, Ponwol T, Acker H, Jelkmann W. Cobalt chloride and desferrioxamine antagonize the inhibition of erythropoietin production by reactive oxygen species. Kidney Int 1997; 51:492–496.
70. Lopez-Lopez J, Gonzalez C. Time course of K^+ current inhibition by low oxygen in chemoreceptor cells of adult rabbit carotid body. Effect of carbon monoxide. FEBS Lett 1992; 299:251–254.
71. Lahiri S, Ray, DK, Chugh D, Iturriaga R, Mokashi A. CO-binding chromophores in oxygen chemoreception in the carotid body. In: O'Regan RG, Nolan P, McQueen DS, Patterson DJ, eds. Arterial Chemoreceptors: Cell to System, New York: Plenum Press, 1994:149–153.
72. Lahiri S. Iturriaga R, Mokashi A, Ray DK, Chugh D. CO reveals dual mechanisms of O_2 chemoreception in the carotid body. Respir Physiol 1993; 94:227–240.
73. Wilson DF, Mokashi A, Chugh D, Vinogradove S, Osanai, S, Lahiri S. The primary oxygen sensor of the cat carotid body is cytochrome a3 of the mitochondrial respiratory chain. FEBS Lett 1994; 351:370–374.
74. Warburg O, Negelein E. Über die photochemische Dissoziation bei inermitterender Bleichuterung und das absolute Absorpsionsspektrumn des Athmungsferments. Biochem Z 1928; 202:202–228.
75. Melnick JL. The photochemical spectrum of cytochrome oxidase. J Biol Chem 1942; 146:385–390.
76. Castor LN, Chance B. Photochemical action spectra of carbon monoxide-inhibited respiration. J Biol Chem 1955; 217:453–465.
78. Lahiri S. Carotid body O_2 chemoreception: respiratory and non-respiratory aspects. Biol Signals 1995; 4:257–262.
79. Fidone SJ, Gonzalez C, Yoshikazi K. A study of the relationship between dopamine release and chemosensory discharge from rabbit carotid body in vitro: preliminary findings. In: Belmonte C, Pallot DJ, Acker H, Fidone S, eds. Arterial chemoreceptors. Leicester: Leicester University Press, 1981:209–219.
80. Donnelly DF. Electrochemical detection of catecholamine release from rat carotid body in vitro. J Appl Physiol 1993; 75:2330–2337.
81. Buerk DG, Lahiri S, Chugh D, Mokashi A. Electrochemical detection of rapid DA release kinetics during hypoxia in perfused-superfused cat CB. J Appl Physiol 1995; 78:830–837.
82. Urena J, Fernandez-Chcosr R, Benot AR, Alvarez de Toledo G, Lopez-Barnes J. Hypoxia induced voltage-dependent Ca^{2+} entry and quantal dopamine secretion in carotid body glomus cells. Proc Natl Acad Sci 1994; 91:10209–10211.
83. Donnelly DF. Does catecholamine secretion mediate the hypoxia-induced increase in nerve activity? Biol Signals 1995; 4:304–309.
84. Gonzalez C, Almaraz L, Obesso A, Rigual R. Carotid body chemoreceptors: from natural stimuli to sensory discharges. Physiol Rev 1994; 74:829–900.
85. Burke-Wolin TM, Wolin MS. H_2O_2 and cGMP may function as an O_2 sensor in the pulmonary artery. J Appl Physiol 1989; 66:167–170.
86. Burke-Wolin TM, Wolin S. Inhibition of cGMP associate pulmonary arterial relaxation to H_2O_2 and O_2 by ethanol. Am J Physiol 1990; 258:H1267–H1270.
87. Cherry PD, Omas HA, Farrell KA, Stuart JS, Wolin MS. Superoxide anion inhibits

cGMP-associated bovine pulmonary arterial relaxation. Am J Physiol 1990; 239: H1056–H1062.
88. Omar HA, Mohazzab H, Mortelliti MP, Wolin MS. O_2 dependent modulation of calf pulmonary artery tone by lactate: role of H_2O_2 and cGMP. Am J Physiol 1993; 264: L141–L145.
89. Paky A, Michael JR, Burke-Wolin T, Wolin MS, Gurtner GH. Endogenous production of superoxide by rabbit lungs: effects of hypoxia or metabolic inhibitors. J Appl Physiol 1993; 74:2868–2874.
90. Wolin MS, Burke TM. Hydrogen peroxide elicits activation of bovine pulmonary arterial soluble guanylate cyclase by a mechanism associated with its metabolism by catalase. Biochem Biophys Res Commun 1987; 143:20–25.
90a. Lander HM. An essential role for free radicals and derived species in signal transduction. FASEB J 1997; 11:118–124.
91. Lahiri S. Chemical modification of carotid body chemoreception by sulfhydryls. Science 1981; 212:1065–1066.
92. Dinauer MC. A murine model of chronic granulomatous disease generated by targeted disruption of the mouse gphphon gene (abstr.). Blood 1993; 82:328A.
93. Hoth M, Penner R. Calcium release-activated calcium current in rat mast cells. J Physiol 1993; 465:359–386.
94. Putney Jr. JW. A model for receptor-regulated calcium entry. Cell 1986; 7:1–12.
95. Lahiri S, Osanai S, Buerk DG, Mokashi M, Cugh DK. Thepsigargin enhances carotid body chemosensory discharfe in response to hypoxia in zero $[Ca^{2+}]_e$: evidence for intracellular Ca^{2+} release. Brain Res 1996; 709:141–144.
96. Fitzgerald RS, Shirahata M. Acetylcholine and carotid body excitation during hypoxia in the cat. J Appl Physiol 1994; 76:1566–1574.
97. Prabhakar NR, Landis SC, Kumar GK, Mullikin-Kilpatrick D, Cherniack NS. Substance P and neurokinin A in the cat carotid body: Localization, exogenous effects and changes in response to arterial Po_2. Brain Res 1989; 481:205–214.
98. Katayama T, Coburn RF, Fillers WS, Baron CB. Oxygen sensors in vascular smooth muscle. J Appl Physiol 1994; 77:2088–2092.
99. Franco-Obegon A, Urena J, Lopez-Barneo J. Proc Natl Acad Sci USA 1995; 92:4715–4719.
100. Weir EK, Archer SL. The mechanism of acute hypoxic pulmonary vasoconstriction: the tale of two channels. FASEB J 1995; 9:183–189.
101. Post JM, Hume JR, Archer SL, Weir EK. Direct role for potassium channel inhibition in hypoxic pulmonary vasoconstriction. Am J Physiol 19092; 262:C882–C890.
102. Yuan XJ, Goldman WF, Tod ML, Rubin LJ, Blaustein MP. Hypoxia reduces potassium currents in cultured rat pulmonary but not mesenteric arterial monocytes. Am J Physiol 1993; 264:L116–L123.
103. Utz J. Ulrich V. Carbon monoxide relaxes ileal smooth muscle through activation of guanylate cyclase. Biochem Pharmacol 1991; 41:1195–1201.
104. Trischmann U, Klockner U, Isenberg G, Utz J, Ulrich V. Carbonmonoxide inhibits depolarization-induced Ca rise and increases cyclic GMP in visceral smooth muscle cells. Biochem Pharmacol 1991; 41:237–241.
105. Coceani F, Hamilton NC, Labdue J, Olley PM. Cytochrome P450-linked monooxygenase: involvement in the lamb ductus arteriosus. Am J Physiol 1984; 246:H640–H643.

106. Lin H, McGrath JJ. Carbon monoxide effects on calcium levels in vascular smooth muscle. Life Sci 1988; 43:1813–1816.
107. Quast U, Cook NS. In vitro and in vivo comparison of two K+ channel openers, diazoxide and cromakalim, and their inhibition by glibenclamide. J Pharm Exp Ther 1989; 250:261–271.
108. Jiang C, Haddad GG. A direct mechanism for sensing low oxygen levels by central neurons. Proc Natl Acad Sci 1994; 91:7198–7201.
109. Goldberg M, Dunning S, Bunn HF. Regulation of the erythropoietin genome: evidence that the oxygen sensor is a heme protein. Science 1988; 242:1412–1415.
110. Goldberg MH, Schneider TJ. Similarities between the oxygen-sensing mechanisms regulating the expression of vascular endothelial growth factor and erythropoietin. J Biol Chem 1994; 269:4355–4359.
111. DiGiulio C, Data PG, Lahiri S. Chronic cobalt causes hypertrophy of glomus cells in the carotid body. Am J Physiol 1991; 261:C102–C05.
112. Fandrey J, Frede S, Jelkmann W. Role of hydrogen peroxide in hypoxia-induced erythropoietin production. Biochem J 1994; 303:507–510.
112a. Gassman M, Wenger RH. HIF-1, a mediator of the molecular response to hypoxia. N Physiol Sci 1997; 12:214–218.
113. Bisgard GE, Forster HV. Ventilatory responses to acute and chronic hypoxia. In: Fregley MJ, Blatteis C, eds. Handbook of Physiology: Environmental Physiology. New York: Oxford University Press, 1996; 1207–1240.
114. Busch MA, Bisgard C, Rorster HV. Ventilatory acclimatization to hypoxia is not dependent on arterial hypoxemia. J Appl Physiol, 1985; 58:1874–1880.
115. Lahiri S, Edelman NH, Cherniack NS, Fishman AP. Role of carotid chemoreflex in respiratory acclimatization to hypoxemia in goat and sheep. Respir Physiol 1981; 46: 367–382.
116. Nielsen AM, Bisgard GE, Vidruk EH. Carotid chemoreceptor activity during acute and sustained hypoxia in goats. J Appl Physiol 1988; 65:1796–1802.
117. Howard LSGE, Robbins PA. Alterations in respiratory control during eight hours of isocapnic and poikilocapnic hypoxia in humans. J Appl Physiol (in press).
118. Vizek M, Pickett CK, Weil JV. Increased carotid body hypoxic sensitivity during acclimatization to hypobaric hypoxia. J Appl Physiol 1987; 63:2403–2410.
119. Severinghaus JW, Bainton CK, Carcelen A. Respiratory insensitivity to hypoxia in chronically hypoxic man. Res Physiol 1966; 1:308–334.
120. Lahiri S, Brody JS, Velasquez T, Motoyama EK, Simpser M, Dleaney RG, Polgar G. Pulmonary adaptation to high altitude: genetic vs. environment. Nature 1976; 261: 133–135.
121. Milledge JS, Lahiri S. Respiratory control in lowlanders and Sherpa highlanders at altitude. Respir Physiol 1967; 2:310–322.
122. Curan LS, Zhuan J, Droma T, Moore LG. Hypoxic ventilatory response in Tibetan residents of 4400 m compared with 3658 m. Respir Physiol 1995; 100:223–230.
123. Lahiri S, Mulligan E, Andronikou S, Shirahata M, Mokashi M. Carotid body chemosensory function in prolonged normobaric hyperoxia in the cat. J Appl Physiol 1987; 62:1924–1931.
124. Mokashi A, Lahiri S. Aortic and carotid body chemoreception in prolonged hyperoxia in the cat. Respir Physiol 1991; 86:233–243.
125. Hanbauer I, Hellstrom S. The regulation of dopamine and noradrenaline in rat carotid body and its modification by denervation and hypoxia. J Physiol 1978; 282:21–34.

126. Tatsumi K, Pickett CK, Weil JV. Decreased carotid body hypoxic sensitivity in chronic hypoxia: role of dopamine. Respir Physiol 1995; 101:47–57.
127. Haubauer I, Karroum F, Hellstrom S, Lahiri S. Effects of hypoxia lasting up to one month on the catecholamine content in rat carotid body. Neuroscience 1981; 6:81–86.
128. Pequingnot JM, Collet-Emard JM, Dalmaz Y, Peyrin L. Dopamine and norepinephrine dynamics in rats carotid body during long-term hypoxia. J Autonom Nerv Systm 1987; 21:9–14.
129. Barer GR, Edwards CW, Jolly AI. Changes in the carotid body and the ventilatory response to hypoxia in chronically hypoxic rats. Clin Sci 1976; 50:311–313.
130. Bee D, Pallot DJ. Hypoxic ventilatory response. Carotid body cell division and dopamine content during early hypoxic exposure in rats. J Appl Physiol (in press).
131. Olson EB Jr, Dempsey J. Rat as a model for human-like ventilatory adaptation to chronic hypoxia. J Appl Physiol 1978; 44:763–769.
132. McGregor KH, Gil J, Lahiri S. A morphometric study of the carotid body in chronically hypoxic rats. J Appl Physiol 1984; 57:1430–1438.
133. Pequignot JM, Hellstrom S, Johansson C. Intact and sympathectomized carotid bodies of long-term hypoxic rats: a morphometric ultrastructural study. J Neurochem 1984; 13:481–493.
134. Katz DM, markey KA, Goldstein M, Black IB. Expression of catecholaminergic characteristic by primary sensory neurons in the normal adult rat in vivo. Proc Natl Acad Sci 1981; 80:3526–3530.
135. Stea A, Nurse CA. Whole-cell and perforated patch recording from O_2-sensitive rat carotid body cells grown in short- and long-term culture. Pflugers Arch 1991; 418: 93–101.
136. Nurse CA. Carotid body adaptation to hypoxia: cellular and molecular mechanisms in vitro. Biol Signals 1995; 4:286–291.
137. Barnard P, Audronikou S, Pokorski M, Smatresk N, Mokashi A, Lahiri S. Time-dependent effect of hypoxia on carotid body chemosensory function. J Appl Physiol 1987; 63:685–691.
138. Wang WJ, Cheng GF, Yoshizaki K, Dinger B, Fidone S. The role of cyclic AMP in chemoreception in the rabbit carotid body. Brain Res 1991; 540:96–104.
139. Wang IGL, Semenza GL. General involvement of hypoxia-inducible factor 1 in transcriptional response to hypoxia. Proc Natl Acad Sci 1993; 90:4304–4308.
140. Semenza GL, Roth PH, Fang HM, Wang GI. Transcriptional regulation of gene encoding glycolytic enzymes by hypoxia inducible factor-1. J Biol Chem. 1994; 269:23757–23763.
141. Wang GL, Semenza GL. Characterization of hypoxia-inducible factor-1 and regulation of DNA binding activity by hypoxia. J Biol Chem 1993; 268:21513–21518.
142. Wang GI, Semenza GL. Purification and characterization of hypoxia inducible factor-1. J Biol Chem 1995; 270:1230–1237.
143. Millhorn DE, Czyzyk-Krezeska M, Bayliso DA, Lawson EE. Regulation of gene expression by hypoxia. Sleep 1993; 161:S44–48.
144. Norris ML, Millhorn DE. Hypoxia-induced protein binding to O_2 responsive sequences in the tyrosine hydroxylase gene. J Biol Chem 1996; 270:23774–23779.
145. Cherniack NS, Prabhakar N, Haxhiu M. Possible genomic mechanisms involved in modeling and control of ventilation. In: Semple SJC, Adams L, Whipp BJ, eds. New York: Plenum Press, 1995:89–94.

146. Morgan JI, Curran T. Stimulus transcription coupling in the nervous system: involvement of the inducible proto-oncogenes fos and jun. Ann Rev Neurosci 1991; 14:421–451.
147. Xanthoudakis S, Miao G, Wang F, Pon YCE, Curran T. Redox activation of Fos-Jun DNA binding activity is mediated by a DNA repair enzyme. EMBO J 1992; 11:3323–3335.
148. Haxhiu MA, Stronl KP, Cherniack NS. The N-methyl-D-aspartate receptor pathway is involved in hypoxia-induced c-Fos protein expression in the rat nucleus of the solitary tract. J Autonom Nerv Syst 1995; 55:65–68.
149. Czyzky-Krzesak MF, Funair BA, Millhorn DE. Hypoxia increases rate of transcription and stability of tyrosine hydroxylase mRNA in pheochromocytoma PC-12 cells. J Biol Chem 1994; 269:760–764.
150. Prabhakar NR, Shenoy B, Simonson MS, Cherniack NS. Cell selective induction and transcriptional activation of immediate early genes by hypoxia. Brain Res 1995; 67: 266–270.
151. Salceda S, Caro J. Hypoxia-inducible factor 1α(HIF-1α) protein is rapidly degraded by the ubiquitin-proteasome system under normoxic conditions. Its stabilization by hypoxia depends on redox-induced changes. J Biol Chem 1997; 272:22642–22647.
152. Semenza GL. Hypoxia-inducible factor 1: master regulator of O_2 homeostasis. Curr Opin Genet Dev 1998; 8:588–594.
153. Maxwell PH, Wiesener MS, Chang GW, Clifford SC, Vaux EC, Cockman ME, Wykoff CC, Pugh CW, Maher ER, Ratcliffe PJ. The tumour suppressor protein VHL targets hypoxia-inducible factors for oxygen-dependent proteolysis. Nature 1999; 399:271–275.
154. Wenger RH. Mammalian oxygen sensing, signalling and gene regulation. J Exp Biol 2000; 203:1253–1263.
155. Guillemin K, Krasnow MA. The hypoxic response: huffing and hifing. Cell 1997; 89: 9–12.

5

Molecular/Metabolic Defense and Rescue Mechanisms for Surviving Oxygen Lack

From Genes to Pathways*

PETER W. HOCHACHKA

University of British Columbia
Vancouver, British Columbia, Canada

I. Cell Level Paradigms for Hypoxia Tolerance

It is well known in biology that certain vertebrate species have evolved unusual capabilities for surviving prolonged periods without oxygen or with greatly reduced supplies of oxygen. Studies of such species [some being so anoxia tolerant they are referred to as "facultative" anaerobes (1)] have revealed several seemingly universal strategies of hypoxia adaptation. Two of the most significant of these include severe downregulation of energy turnover (2–6) and upregulation of energetic efficiency of ATP-producing pathways (7). The latter involve stoichiometric efficiencies. In hypoxia adaptation, pathways that maximize the yield of ATP per mole oxygen are favored, while in anoxia adaptation, anaerobic pathways that maximize the yield of ATP per mole of H^+ formed in the fermentation are favored (8,9). In hypoxia or anoxia adaptation, the ratio of anaerobic/aerobic metabolic potential may well be upregulated, coincident with upregulation of glycogen (fermentable substrate) and of buffering capacities. Even if all these mechanisms are useful in surviv-

* Only literature up to 1996 was surveyed for this chapter.

ing hypoxia, evaluation of such hypoxia defense strategies having naturally evolved in hypoxia-tolerant animals shows that suppression of energy turnover supplies the greatest protection against, and hence advantage in, hypoxia. The uniquely large advantage of this defense strategy is widely appreciated by many biologists (4,5,10–12). In one of his last personal communications to the authors, the great comparative physiologist Kjell Johansen referred to this strategy as "turning down to the pilot light"; like earlier workers in this field, he was acutely aware of its relative importance. Although recognized as a kind of hallmark of reversible entry into and return from states of severe oxygen deprivation, a number of unexplained problems have remained. First, it has not been clear how cells/tissues "know" when to turn on their hypoxia defense mechanisms. Second, it has not been known which pathways of ATP demand and ATP supply are downregulated or by how much. Additionally, it has been unclear how membrane electrochemical gradients are stabilized and what gene expression and protein expression level adjustments are involved in reorganization of cell structure and function under oxygen limitation. In recent studies of cells and tissues from the aquatic turtle, brain cortical slices were used to probe electrophysiological properties of neurons under anoxia (13–16) while isolated liver hepatocytes were used to probe cell level biochemical responses to anoxia (17–21). In combination with independent lines of research in several other laboratories, these studies supply the raw material for formulating a synthetic cell level model or general hypothesis of hypoxia tolerance, which goes a long way to answering the above unresolved questions.

II. Detecting When Oxygen Becomes Limiting

The traditional views of cell level responses to oxygen limitation are formalized in the concept of the Pasteur effect: as ATP generation by oxygen-based mitochondrial metabolism begins to decline due to oxygen lack, the energetic deficit is made up by activation of anaerobic ATP-generating pathways. In turtle liver cells, this kind of oxygen sensing would require activation of anaerobic glycolysis (since this is the only known mechanism for ATP generation without oxygen in these cells). Two telling observations argue potently against this as the mechanims for detecting hypoxia in cells tolerant to oxygen lack. First, hypoxia-tolerant systems rarely activate anaerobic metabolism to make up energy deficits; they favor a reduced energy turnover state instead. Equally importantly, responses to falling oxygen supplies begin at concentrations much higher than the Km(O_2) for mitochondria (termed oxygen conformity). Indeed, the observation of oxygen conformity first led workers in this area to postulate the occurrence of a mechanism for sensing molecular oxygen as the cell level means for detecting hypoxia (9,10,12,22) (a point to be further discussed below).

III. ATP Demand Pathway = ATP Supply Pathway During Hypoxia

In view of the above, it was not surprising to find that when liver hepatocytes encounter oxygen lack, they suppress energy turnover by a factor of almost 10-fold (17,18). In this condition, a mole of ATP could sustain these cells 10 times longer than under normoxic conditions. Without comparable regulation of ATP demand pathways, this mechanism would quickly deplete ATP supplies during hypoxia; thus it is important to know quantitatively which processes are turned down. To this end, we assessed the main energy demand functions under normoxia (energetically balanced by the oxygen consumption rates under these conditions) and compared these to the energy sinks remaining under anoxia. We found that under normoxic conditions, the main energy demand processes (Table 1) are protein synthesis, protein degradation, Na^+K^+ pumping, urea biosynthesis, and glucose biosynthesis. The ATP demands of these processes account for pretty well all of the ATP production expected from oxygen consumption (17–21).

What happens to the same processes in turtle liver cells under anoxia is most instructive. Under these conditions, the ATP demand of protein turnover drops to less than 10% of normoxic rates; urea biosynthesis drops to essentially zero, as does the biosynthesis of glucose. Although the ATP requirements of the Na^+K^+ ATPase are also drastically reduced, the suppression in percentage terms is less than for overall ATP turnover. As a result, under anoxic conditions, the Na^+ pump becomes the cell's dominant energy sink, accounting for up to 75% of the ATP demand of the cell (17). Although there may well be other minor energy demand processes remaining under anoxic conditions, the quantitative ATP requirements of the energy demand pathways identified during anoxia fully account for the ATP generated by anaerobic glycolysis (Table 1).

Table 1 The Main ATP Demand Pathways During Normoxia and Anoxia in Turtle Hepatocytes

	ATP demand (μmol ATP $\times$ g^{-1} $\times$ h^{-1})		
Pathway	Normoxia	Anoxia	% Suppression
Total	67.0	6.3	94
Na^+ pump	19.1	4.8	75
Protein synthesis	24.4	1.6	93
Protein breakdown	11.1	0.7	94
Urea synthesis	2.0	0.6	70
Gluconeogenesis	11.4	0.0	100

Source: Modified from Refs. 17, 19, 20.

IV. Coupling Metabolism and Membrane Functions During Hypoxia

While in relative terms ion pumping (as assessed by the activity of Na^+K^+ ATPase) is the single largest ATP sink during anoxia in turtle hepatocytes, its absolute ATPase or pumping activity is only a small fraction (about one fourth) of normoxic levels. Despite this, experimental estimates indicate that the electrochemical potential in anoxic liver cells is essentially the same as in normoxia (17). To account for the large-scale drop in absolute Na^+K^+ ATPase activity and the simultaneous maintenance of normal electrochemical gradients would seem to require a similar magnitude decrease in cell membrane permeability [termed generalized ''channel arrest'' in the literature (10)].

In an earlier synthesis of information in this research field (10), a channel arrest component of a hypoxia tolerance theory postulated (1) that hyoxia-tolerant cells would have an inherent low permeability (either low channel densities or low channel activities) and (2) that they would sustain a further suppression of membrane permeability to ions when exposed to oxygen lack (further channel arrest by either suppression of channel densities or channel activities). Turtle liver cells display both of these characteristics (especially when compared to mammalian homologs); i.e., they fit the classical definition of metabolic and channel arrest as two telling signatures of hypoxia tolerance.

In turtle cortical cells, in contrast, only the first criterion is met: a background electrical conductivity that is unusually low and when compared at biological temperatures (15 vs. 37°C) can be as low as 1/25 the conductivity of cell membranes of rat cortical cells (15). When exposed to acute oxygen lack (for up to several hours), there is no further channel arrest and therefore no further decrease in background conductivity of these neuronal cells. This may be one reason why the main energy-saving mechanism in turtle brain (23,24) is downregulation of firing rates or synaptic transmission [termed ''spike arrest'' by Sick et al. (24)] presumably through adenosine-mediated downregulation of excitatory amino acid (especially glutamate) release with concomitant increase in release of inhibitory amino acids (23,25). It may also explain why the metabolic suppression in brain is substantially less than in liver cells, down to about 1/2 rather than 1/10 of normoxic rates (16).

An additional consequence of these results is that the ATP turnover rate of anoxic turtle cortical brain cells is higher than in turtle liver cells. Direct microcalorimetric measurements indicate that heat fluxes in anoxic brain cortical tissue are about threefold higher than in anoxic liver cells (16,18). In vivo, of course, anoxic brain survival depends solely upon anaerobic glycolysis (13,26) fueled by plasma glucose derived ultimately from liver glucose. This appears to indicate that in the turtle brain, hypoxia tolerance equals a more modest metabolic arrest than in liver cells plus the arrest not of ''leakage'' channels but of channel functions associated with synaptic transmission, in order to remain in energy balance.

V. Hypoxic Sensitivity of Protein Synthesis

Protein turnover is the only energy-requiring process in normoxic liver cells that is more costly than ion pumping (17–20); hence evaluating this ATP sink under oxygen limiting conditions is of particular importance. For perspective, it is important to note that one of the first and most dramatic effects of hypoxia on cell metabolic functions is a rapid and large magnitude inhibition of protein synthesis. The decline can be so rapid that its time course is hard to quantify accurately with currently used techniques for measuring protein biosynthesis (27).

A hypoxia-induced block in protein synthesis could occur at the level of gene transcription or translation. In animal models, the mechanism for an initial hypoxia-induced block in protein synthesis is unclear, but in plant systems hypoxic suppression of protein synthesis is mediated by a translational block affecting both initiation and elongation. Inhibition of initiation is not yet adequately analyzed, but inhibition of elongation appears to depend upon an accumulation of EF1Â, an elongation factor which at low pH appears to form nonfunctional complexes with polysome-associated mRNA (28). The main role for this elongation factor is to present amino acyl-tRNA to the A site of ribosomes, so it is easy to visualize how failure to dissociate from polysomes would prevent peptidyl synthase and translocation. In hypoxia-sensitive systems, such as rat hepatocytes (27), hypoxia-induced translational arrest and thus blockade of protein synthetic capacity seems to remain general for at least 2 hours.

The acute or defense phase of hypoxia tolerance is postulated to blend smoothly into a secondary series of processes; in the literature these are variously termed immediate-early gene responses (40) or acclimatory expression adjustments (2,3). As the combined effect of these processes is to reactivate some mRNA translational capacities and probably to consolidate and stabilize the cell at strikingly reduced ATP turnover rates (at Kjell Johansen's ''pilot light''), these combined processes are referred to as a rescue phase for establishing hypoxia tolerance. The rescue phase includes:

1. Heme protein-based, hypoxia-sensing, and signal transduction pathways that activate one-cycle gene expression for the production of key transcription factors (such as HIF1); one-cycle gene regulation may also allow continued production of key elongation factors and so seemingly ''rescue'' the cell translational capacities for specific mRNAs during continued oxygen limitation.
2. Sets of two-cycle, hypoxia-sensitive genes (probably including genes for glycolytic enzymes) whose expression is upregulated during prolonged oxygen limitation (protein products of these genes are presumably involved in stabilizing cell operations at severely suppressed ATP turnover rates during hypoxia; the gene for EF1Â could be regulated by such two-cycle, rather than one-cycle, gene-activation circuits).
3. Sets of two-cycle, hypoxia-regulated genes (possibly including genes for

enzymes in gluconeogenesis and the Krebs cycle) whose expression is downregulated during prolonged oxygen limitation (protein products of these genes presumably are not needed, or are not as critical, for survival during prolonged oxygen lack).

4. Sets of more complex, two- or three-cycle, hypoxia-regulated genes (such as *cfos* and *cjun*) whose products appear as tertiary messengers in the expression regulation of (probably numerous) other housekeeping genes (involved in restructuring, consolidation, and stabilization of cell functions at the severely suppressed ATP turnover rates requisite for surviving prolonged hypoxia). These latter, more complex regulatory phenomena make it easier to appreciate features of the oxygen-limited cell such as suppressed proteolysis, increased protein stability, and induction of chaperone proteins, features again indicative of extensive reprogramming of the normoxic cell to generate the hypoxia-tolerant cell.

From our currently available data base, it appears that the combination of these molecular processes allows cells and tissues of hypoxia-tolerant species to greatly extend the length of time they are able to survive under hypoxic or even anoxic conditions. However, it is worth emphasizing that summaries only supply key highlights and key processes of hypoxia tolerance. Further studies are required for evaluating how useful these studies of good facultative anaerobes are in explaining hypoxia responses observed in humans under mild or severe hypoxia. Some of the processes (such as downregulation of protein synthesis and adjustments in ATP-generating metabolic pathways) may well be similar in the two very different kinds of biological systems; others, such as three-cycle gene-mediated cell reprogramming, may occur to a lesser extent (or not at all) in humans in hypobaric hypoxia. Further studies are also required to fill in the details of such generalized frameworks of the nature of hypoxia tolerance. Finally, and perhaps most daunting of all, is the challenge of assessing whether or not these strategic mechanisms can be transferred from hypoxia-tolerant to hypoxia-sensitive cells. This—a kind of Holy Grail for many workers in the field—supplies young researchers with a good target for the next millennium.

Acknowledgments

This work was supported by an NSERC Research Grant. Special thanks are extended to Dr. Michael Vayda for bringing his work on hypoxia-sensitive protein synthesis to our attention.

References

1. Storey KB, Hochachka PW. J Biol Chem 1974; 249:1423–1429.
2. Hochachka PW, Somero GN. Biochemical Adaptation. Princeton, NJ: Princeton University Press, 1984:1–521.

3. Hochachka PW, Guppy M. Metabolic Arrest and the Control of Biological Time. Cambridge, MA: Harvard University Press, 1987:1–237.
4. Storey KB, Storey JM. Q Rev Biol 1990; 65:145–193.
5. Hand SC. In: Hochachka PW, Lutz PL, Sick T, Rosenthal M, van den Thillart G, eds. Surviving Hypoxia. Boca Raton, FL: CRC Press, 1993:171–185.
6. Guppy M, Fuery CJ, Flanigan JE. Comp Biochem Physiol 1994; 109B:175–189.
7. Hochachka PW. In: Hochachka PW, Lutz PL, Sick T, Rosenthal M, van den Thillart G, eds. Surviving Hypoxia. Boca Raton, FL: CRC Press, 1993:127–135.
8. Hochachka PW. In: Sutton JR, Houston CS, Coates G, eds. Hypoxia and Molecular Medicine. Queen City Printers, Inc., Burlington, VT: 1993:146–155.
9. Hochachka PW. Muscles as Molecular and Metabolic Machines. Boca Raton, FL: CRC Press, 1994:1–157.
10. Hochachka PW. Science 1986; 231:234–238.
11. Hand SC. In: Gilles R, ed. Advances in Comparative and Environmental Physiology. New York: Springer-Verlag, 1991:1–47.
12. Land SC, Bernier NJ. In: Hochachka PW, Mommsen TP, eds. Biochemistry and Molecular Biology of Fishes. Amsterdam: Elsevier Science, 1995:381–405.
13. Doll CJ. In: Hochachka PW, Lutz PL, Sick T, Rosenthal M, van den Thillart G, eds. Surviving Hypoxia. Boca Raton, FL: CRC Press, 1993:389–400.
14. Doll CJ, Hochachka PW, Reiner PB. Am J Physiol 1991; 261:R1321–R1324.
15. Doll CJ, Hochachka PW, Reiner PB. Am J Physiol 1993; 265:R929–R933.
16. Doll CJ, Hochachka PW, Hand SC. J Exp Biol 1994; 191:141–153.
17. Buck LT, Hochachka PW. Am J Physiol 1993; 265:R1020–R1025.
18. Buck LT, Hochachka PW, Schon A, Gnaiger E. Am J Physiol 1993; 265:R1014–R1019.
19. Land SC, Buck LT, Hochachka PW. Am J Physiol 1993; 265:R41–R48.
20. Land SC, Hochachka PW. Am J Physiol 1994; 266:C1028–C1036.
21. Land SC, Hochachka PW. Proc Natl Acad Sci USA 1995; 92:7505–7509.
22. Thurman RG, Nakagawa Y, Matsumura T, Lemasters JJ, Misra UK, Kauffman FC. In: Hochachka PW, et al., eds. Surviving Hypoxia—Mechanisms of Control and Adaptation. Boca Raton, FL: CRC Press, 1993:329–340.
23. Lutz PL. Ann Rev Physiol 1992; 54:619–637.
24. Sick TJ, Perez-Pinon M, Lutz PL, Rosenthal M. In: Hochachka PW, Lutz PL, Sick T, Rosenthal M, van den Thillart G, eds. Surviving Hypoxia. Boca Raton, FL: CRC Press, 1993:351–363.
25. Nilsson G. In: Hochachka PW, Lutz PL, Sick T, Rosenthal M, van den Thillart G, eds. Surviving Hypoxia. Boca Raton, FL: CRC Press, 1993:401–413.
26. Chih CP, Rosenthal M, Lutz PL, Sick TJ. Am J Physiol 257, R854–R859.
27. Buc-Calderon P, Lefebvre V, van Steenbrugge M. In: Hochachka PW, Lutz PL, Sick T, Rosenthal M, van den Thillart G, eds. Surviving Hypoxia. Boca Raton, FL: CRC Press, 1993:271–280.
28. Vayda ME, Shewmaker CK, Morelli JK. Plant Mol Biol 1995; 28:751–757.
29. Eckardt KU. Nephron 1994; 67:7–23.
30. Firth JD, Ebert BL, Pugh CW, Ratcliffe PJ. Proc Natl Acad Sci USA 1994; 91:6496–6500.
31. Firth JD, Ebert BL, Ratcliffe PJ. J Biol Chem 1995; 270:21021–21027.
32. Wang GL, Semenza GL. J Biol Chem 1995; 270:1230–1237.
33. Semenza GL, Roth PH, Fang FM, Wang GL. J Biol Chem 1994; 269:23757–23763.
34. Maxwell PH, Pugh CW, Ratcliffe PJ. Proc Natl Acad Sci USA 1993; 90:2423–2427.

35. Wang GL, Semenza GL. Proc Natl Acad Sci USA 1993; 90:4304–4308.
36. Keitzmann T, Schmidt H, Probst I, Jungermann K. FEBS Lett 1992; 311:251–255.
37. Keitzmann T, Schmidt H, Unthan-Feschner K, Probst I, Jungermann K. Biochem Biophys Res Comm 1993; 195:792–798.
38. Webster KA, Murphy BJ. Can J Zool 1988; 66:1046–1058.
39. Murphy BJ, Robin ED, Tapper DP, Wong RJ, Clayton DA. Science 1984; 223:707–709.
40. Webster KA, Discher DJ, Bishopric NH. Circ Res 1994; 74:679–686.
41. Webster KA, Discher DJ, Bishopric NH. J Biol Chem 1993; 268:16852–16858.
42. Prabhakar NB, Chenoy BC, Cherniak NS. FASEB J 1995; 9(4):3769.
43. Murphy BJ, Laderoute KR, Short SM, Sutherland RM. Br J Cancer 1991; 63:69–73.
44. Benjamine IJ, Horie S, Greenberg ML, Alpern RJ, Williams RS. J Clin Invest 1992; 89:1685–1689.
45. Ransome LJ, Verma IM. Ann Rev Cell Biol 1990; 6:539–557.
46. Schlingensiepen KH, Wollnik F, Kunst M, Schlingensiepen R, Herdegen T, Brysch W. Cell Mol Neurobiol 1994; 14:487–505.
47. Bergelson S, Pinkus R, Daniel V. Cancer Res 1994; 54:36–40.
48. Das DK, Engelman RM, Kimura Y. Cardiovasc Res 1993; 27:578–584.
49. Hermes-Lima M, Storey KB. Am J Physiol 1993; 265:R646–R652.
50. Moffat GJ, McLaren AW, Wolff CR. J Biol Chem 1994; 269:16397–16402.
51. Bedoya FJ, Flodstrom M, Eizirik DL. Biochem Biophys Res Comm 1995; 210:816–821.
52. Rice ME, Lee EJK, Choy Y. J Neurochem 1995; 64:1790–1799.
53. Jackson DC. In: Sutton JR, Houston CS, Coates G, eds. Hypoxia and Molecular Medicine. Burlington, VT: Queen City Printers, Inc., 1993:119–126.
54. Venkatachalam MA, Winberg JM. In: Hochachka PW, Lutz PL, Sick T, Rosenthal M, van den Thillart G, eds. Surviving Hypoxia. Boca Raton, FL: CRC Press, 1993:473–494.

6

Control of Breathing at High Altitude

CURTIS A. SMITH and JEROME A. DEMPSEY

University of Wisconsin
Madison, Wisconsin

THOMAS F. HORNBEIN

University of Washington
School of Medicine
Seattle, Washington

I. Introduction

This chapter will focus on the mechanisms of ventilatory responses to hypoxia from the initial responses of the first seconds to minutes of hypoxic exposure up to many years of residence in hypoxic environments. We will discuss how breathing may influence performance at extreme altitude (see also Chapter 20). Space requires that our approach be a selective one. We will not discuss ventilatory responses to hypoxia comprehensively and/or historically, as a number of excellent reviews of this topic are available (1–6). Rather, we will attempt to highlight what we believe are new, unresolved, or controversial questions in this area that would benefit from further research. We will only touch upon control of breathing during sleep at altitude and genetic/evolutionary considerations of ventilatory control in ethnic groups that have been resident at high altitudes for centuries, as these topics are discussed in Chapters 3 and 21.

II. Background and Definitions

Upon ascent to altitude or exposure to hypoxia, awake mammals attempt to protect tissue oxygenation. Initially, compensation is primarily via hyperventilation and some selective vasodilation. Later, changes in hematocrit, renal mechanisms, and tissues may also occur.

Three distinct phases of the ventilatory response to hypoxia are generally recognized:

1. Acute hyperventilatory responses, those occurring in seconds to minutes, are thought to be mediated primarily by peripheral chemoreceptors in conjunction with secondary inhibitory feedback from central chemoreceptors.
2. Responses occurring over hours to days are usually termed short-term acclimatization or, simply, acclimatization. Deacclimatization is the time-dependent return to eupneic ventilation following a return to normoxic conditions and may be of importance in understanding the acclimatization process. The mechanisms involved in acclimatization and deacclimatization have been long debated and remain controversial to this day but may include both central and peripheral processes.
3. Responses occurring over weeks to years represent long-term acclimatization. These responses are observed in long-term altitude sojourners and those who are native to high altitudes.

There may also be extremely long-term, evolutionary adaptations in populations that have resided at altitude for generations, but this question is beyond the scope of this review (see Chapter 3).

III. Acute Responses to Hypoxia

The acute (seconds to minutes) ventilatory responses to hypoxia are the net result of several known and putative feedback mechanisms. During this interval, most studies in unanesthetized humans, goats, and cats found that ventilation increased to a peak (magnitude was dependent on the severity of the hypoxia). This increase in ventilation results chiefly from hypoxic stimulation of the peripheral chemoreceptors, primarily the carotid bodies. Immediately following this increase, ventilation declines somewhat over the subsequent 10–30 minutes, although remaining above control levels under both poikilocapnic and isocapnic conditions (7–13). However, other studies in the dog and the goat found no evidence of a ventilatory decline during this period (14,15). This decline in ventilation (hypoxic ventilatory decline, or HVD, termed "roll-off" by some workers) is somewhat surprising, and its mechanism is the subject of considerable investigation. While the decline is relatively small and short-lived when contrasted with ventilatory acclimatization (Fig. 1), we have chosen to emphasize this topic more than is customary because we felt that

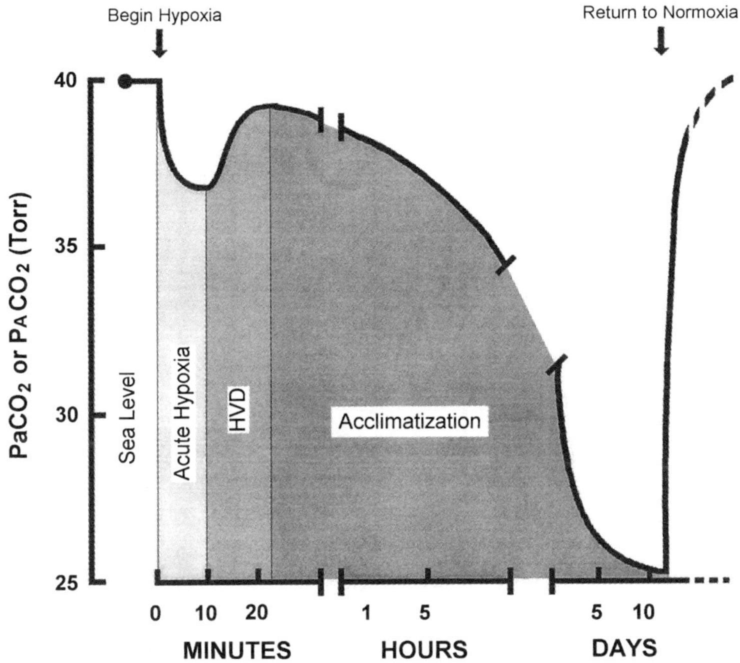

Figure 1 Idealized time course of ventilatory acclimatization (using P_{aCO_2} or P_{ETCO_2} as indices of $\dot{V}A$) compiled from five different studies in humans. SL = Sea level; HVD = hypoxic ventilatory decline. (Adapted from Refs. 10, 41, 120, 122, 124.)

an understanding of the mechanisms underlying HVD might provide useful insights into the broader topic of ventilatory adaptation to hypoxia.

A. Carotid Body Hypoxia

The only well-established stimulatory mechanism operating in the acute time period is the increased neural output from the peripheral chemoreceptors, chiefly the carotid bodies (see also Sec. III.D). Following a few second overshoot (16), carotid chemoreceptor output remains virtually constant (Fig. 2) during the period of ventilatory decline (17–19); therefore, the decline in ventilation is not due to a decline in sensor output.

B. Hypocapnia

As noted above, hypoxic-induced hyperventilation results in hypocapnia, which may itself diminish breathing by several possible mechanisms.

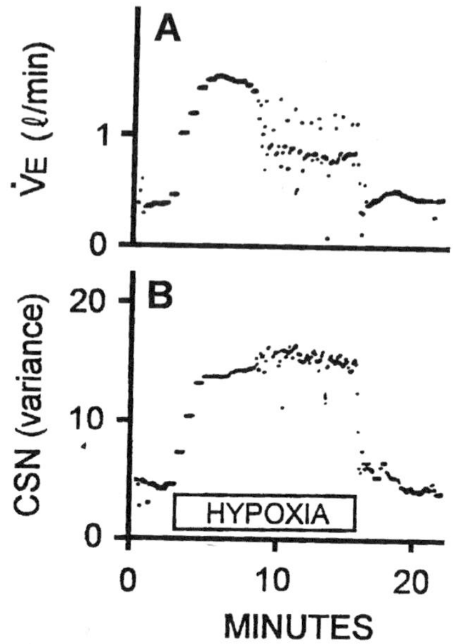

Figure 2 Comparison of ventilation and carotid sinus nerve activity in one anesthetized cat in response to isocapnic hypoxia. Note the essentially constant nerve activity during hypoxic ventilatory decline. (From Ref. 137.)

C. Central Hypocapnia at the Central Chemoreceptors

Central Hypocapnia via Increased Cerebral Blood Flow

In naturally occurring (poikilocapnic) hypoxia, Pa_{O_2} increases and Pa_{CO_2} decreases over time. During the hypoxic ventilatory decline period, this trend will reverse for a time, which should tend to stimulate, rather than inhibit, ventilation. Despite these systemic changes, hypocapnia at the level of the central chemoreceptors, enhanced by hypoxic cerebral vasodilatation, could inhibit ventilation (20). The difficulty here, as Neubauer et al. have pointed out, is that the hypocapnia secondary to the cerebral vasodilation reaches a steady state in about 5 minutes and therefore does not correlate with the time course of hypoxic ventilatory decline (21–24).

Peripheral Hypocapnia

The hyperventilation that results from natural hypoxic exposure causes systemic hypocapnia, and the major inhibitory effect of this hypocapnia has usually been assumed to be mediated by the central chemoreceptors. However, this hypocapnia is also sensed by the carotid body. Studies of hypoxic ventilatory decline using isocapnic conditions have demonstrated that hypocapnia is not obligatory for its development (8–10,25). Indeed, HVD is more readily demonstrated under isocapnic

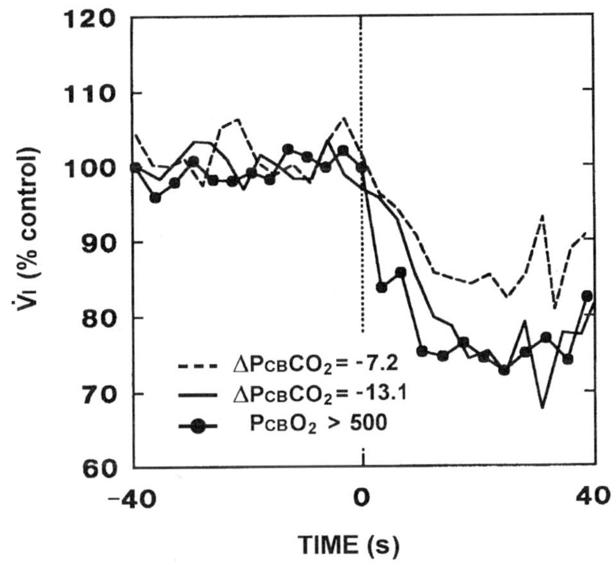

Figure 3 Time course of response of the ventilatory response to specific carotid body hypocapnia or hyperoxia in awake dogs using the vascularly isolated carotid body perfusion technique. Note that the time course and magnitude of the decrease in ventilation is virtually identical whether the inhibition was caused by hypocapnia (about 13 torr less than eupneic value) or hyperoxia. (From Ref. 29.)

conditions. Several studies have produced estimates of 10–50% for the peripheral contribution to the ventilatory response to CO_2 (26–28). Although hypocapnia at the level of the carotid body does inhibit ventilation, even in the presence of hypoxia (Fig. 3) (29,30), carotid body hypocapnia is probably not directly related to hypoxic ventilatory decline because (1) the time course of hypoxic ventilatory decline does not correlate with changes in systemic Pa_{CO_2} and (2) hypoxic ventilatory decline can still be observed under isocapnic conditions.

D. A Role for Hypoxic Depression?

Hypoxic depression of neurons involved in ventilatory control is frequently invoked to explain hypoxic ventilatory decline. Depression by hypoxia is seen consistently in anesthetized preparations, usually with CO-induced hypoxemia and with Pa_{CO_2} and blood pressure maintained constant (e.g., Refs. 31–33); however, the situation in awake animals is far less clear. In contrast to anesthetized preparations, awake carotid body denervated dogs, rats, ponies, sheep, and goats either showed no change in ventilation or actually hyperventilated when exposed to hypoxia (33–37). This ventilatory stimulation has been shown over a wide range of Pa_{O_2} (30–50 torr) and time (minutes to days). Utilizing extracorporeal circulation to isolate the carotid bodies of awake goats, Engwall et al. observed that brain hypoxia caused hyperventilation even when the carotid body was maintained normoxic and normocapnic (Fig. 4) (38). Similar results have been obtained recently by Curren et al. (38a) during non-REM sleep in the dog. In the foregoing studies the hyperventilation was due to increases in both V_T and fb. This response clearly differed from the frank tachypnea described in carotid body denervated cats (39) in which hypoxic depression of higher brain centers was hypothesized to disinhibit the respiratory centers in the medulla. In addition, Engwall et al. (38) showed that both inspiratory and expiratory muscle EMGs increased such that the ventilatory response to brain hypoxia was indistinguishable from the ventilatory response to moderate whole-body hypercapnia. Further, the ventilatory response to hypoxic stimulation of the isolated carotid body was the same regardless of whether the brain was hypoxic or normoxic. This latter finding has also been observed in an anesthetized preparation (31) even when hypoxic depression was present. Finally, Dahan et al. (40) have demonstrated in human subjects that peripheral chemoreceptor input seems to be required for the development of HVD. All these observations lead one to doubt that hypoxic depression of the respiratory controller contributes to HVD.

Indirect measures of central nervous system (CNS) neuronal function suggest

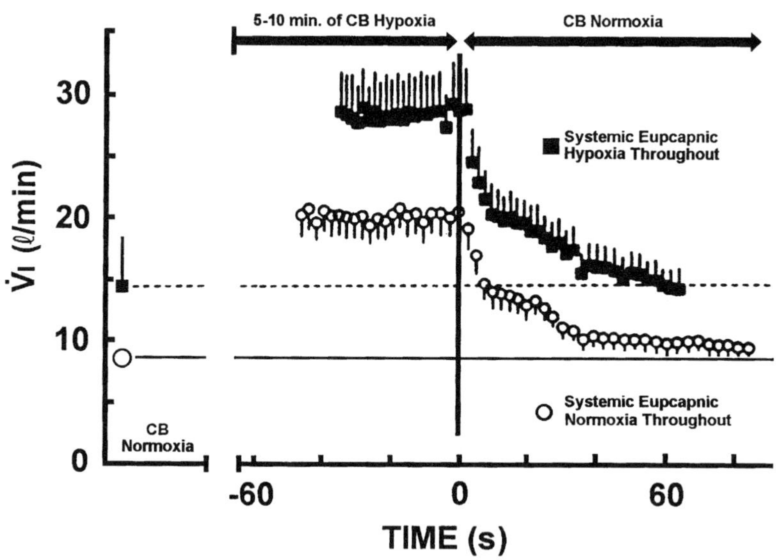

Figure 4 Hyperventilation in an unanesthetized goat in response to brain hypoxia; carotid bodies were maintained normoxic and normocapnic by means of extracorporeal perfusion of vascularly isolated carotid bodies. Open circle = brain normoxia; filled square = brain hypoxia. Note: (1) Increased ventilation in response to brain hypoxia is additive with superimposed carotid body hypoxia. (2) The time-dependent poststimulus hyperpnea indicative of STP was unaffected by brain hypoxia. (Adapted from Ref. 38.)

that ventilatory acclimatization accompanies a general hyperexcitability (41–43). Recent evidence from tissue slice preparations of medulla and hypothalamus shows that most neurons in these regions were excited by hypoxia (44–49). Presently, our ability to extrapolate these results to the whole animal is limited because of the uncertainty of the exact cell types recorded from and the relatively short durations of hypoxia employed.

Another mechanism affected by acute hypoxia is the ventilatory manifestation of short-term potentiation (STP). STP is a property of central neurons in which output increases over time (seconds) in the face of a constant input or stimulus, and this increased output persists for a time (''afterdischarge,'' manifested in ventilation as a continued but waning hyperpnea) when the stimulus is removed (50–52). In the case of ventilation, it is an excitatory influence that tends to counteract transient inhibitory feedback and stabilize breathing. In sleeping humans, if the duration of hypoxic exposure was <1 minute (53), ventilation returned to normal over the next few breaths, i.e., the usual poststimulus hyperpnea indicative of STP was still present. However, in the same subjects, hypoxic exposures of >5 minutes that were either isocapnic or hypocapnic resulted in diminished or absent poststimulus hyperpnea. In contrast, when awake goats were exposed to systemic hypoxia (including brain) for many minutes and square-wave changes in Po_2 (with normocapnia maintained) were imposed at the isolated, perfused carotid body, there was no effect on the poststimulus hyperpnea unless systemic CO_2 was allowed to fall (Fig. 4) (38). Similar observations were made in the anesthetized, normocapnic cat with CO-induced hypoxemia (32). While the goat data cited above are the most direct evidence we are aware of suggesting a lack of effect of CNS hypoxia on STP in unanesthetized animals, species differences cannot be ruled out. If STP is affected by hypoxia, it raises some intriguing questions: (1) Are STP and hypoxic ventilatory decline linked to some common mechanism? (2) If so, will STP return over time as acclimatization proceeds?

In summary, moderate levels of CNS hypoxia probably have a net stimulatory effect on the respiratory control centers in unanesthetized animals. On the cellular level many CNS neurons have been shown to depolarize in the presence of hypoxia, and there is a plausible mechanism to explain this depolarization. It is not yet clear if STP is affected by CNS hypoxia in unanesthetized animals, but the best available data suggest it is not. In sum, it seems unlikely that a direct depressive effect of hypoxia on the CNS can account for hypoxic ventilatory decline (see Sec. III.E).

E. Changes in Central Neuroeffector Levels

A fourth, and perhaps the most likely, mechanism of hypoxic ventilatory decline is the build-up of inhibitory neurotransmitters and/or neuromodulators, and/or the decline of excitatory ones, in the CNS. Recent evidence in support of this idea has been provided by Long et al. (25), who showed that the hypoxic ventilatory decline was immediately apparent when awake cats were reexposed to hypoxia after 5 minutes of normoxia. They suggested that their findings were consistent with a hypoxia-induced change in the level of some neuroeffector that remained altered for some minutes following return to normoxia. Georgopoulos et al. (54) made similar observations and came to similar conclusions in studies of awake humans made hypoxic (Sa_{O_2} = 80%) at three different CO_2 levels, returned to normoxia for a few minutes, and then reexposed to the same level of isocapnic hypoxia.

Several neuroeffector candidates could potentially explain this persistent ventilatory depression with repeated hypoxic exposure.

GABA

Gamma-aminobutyric acid (GABA) is a known inhibitory neurotransmitter in the CNS (55). Central GABA levels have been shown to increase in response to hypoxia (56–59). GABA and GABA analogs have been shown to depress ventilation in anesthetized animals (60–63). GABA antagonists have been shown to reverse partially the ventilatory depression of hypoxia in sedated piglets (64) and anesthetized cats (65). Yamada et al. (66) have shown in anesthetized cats that GABA can exert inhibitory effects on ventilation when applied directly to Schlaefke's area on the ventrolateral medulla. Thus, GABA looks like a good candidate to mediate hypoxic ventilatory decline. However, there are some caveats: rather extreme hypoxia is required to elicit increases in GABA levels. Most studies find a critical F_{IO_2} of about 0.08 (56,57). One study of neonatal rats (58) reported increases in GABA at milder levels of hypoxia (F_{IO_2} = 0.12), but a stimulus-response relationship appeared to be lacking, as a F_{IO_2} of 0.06 elicited no further increase in GABA levels. There are well-known differences in neonatal vs. adult responses to hypoxia, and this may account for the discrepancy. In general, however, the extreme hypoxia that seems to be required in the adult mammal causes one to question the relevance of GABA in nonpathological situations at terrestrial altitudes.

Another problem is that, to our knowledge, no studies of ventilatory responses to GABA agonists or antagonists have been performed in unanesthetized preparations. GABA antagonists such as bicuculline have convulsant properties that make them difficult to use in the unanesthetized preparation. Studies in the unanesthetized animal might help elucidate the role of GABA in hypoxic ventilatory decline.

Adenosine

Adenosine is another neuromodulator that could be involved in hypoxic ventilatory decline. Adenosine has been shown to have an inhibitory effect on CNS corticospinal neurons (67,68). In the fetal sheep brain adenosine concentrations have been shown to increase with a time course that is compatible with the development of hypoxic ventilatory decline when the dam breathed 0.09 F_{IO_2} (69). In anesthetized cats and rats Eldridge et al. (70) and Hedner et al. (71) showed that intracerebroventricular administration of an adenosine agonist inhibited ventilation. Similar observations have been obtained with microdialysis techniques in the piglet (72). In anesthetized preparations, adenosine antagonists partially reduced the hypoxic ventilatory decline (73–75). Similar observations were reported in studies of unanesthetized human subjects (76). However, adenosine had no effect on HVD in awake adult goats (77).

In summary, the bulk of the evidence suggests a role for adenosine in HVD, but additional studies in unanesthetized models would seem to be required before this can be firmly established.

Lactic Acid

Brain acidosis is usually thought of as a potent stimulus to ventilation (also see Sec. IV). Several studies, however, have demonstrated that decreased pH in the extracellular fluid around the ventral medullary surface can inhibit ventilation (22,75) and that pretreatment with dichloroacetate (a blocker of lactic acid production) eliminated ventilatory depression until very low oxygen contents were reached (78). Relating these results directly to hypoxic ventilatory decline is problematic for two reasons: (1) the studies cited here generally required rather severe degrees of hypoxemia to elicit decreases in pH and (2) Pa_{CO_2} was kept normal in these studies. Hypocapnia and hypoxia are synergistic in the production of lactic acid. Musch et al. (79) showed, in rats exposed to moderate-to-severe hypoxia, that brain tissue lactacidosis could be prevented (moderate hypoxia) or reduced (severe hypoxia) if hypocapnia was prevented. Inasmuch as the normal physiological response to hypoxia is hyperventilation, the lactic acid–mediated inhibition of ventilation might occur even sooner under physiological (i.e., hypocapnic) conditions. More recently, Aaron et al. (80) found that dichloroacetate enhanced the hyperventilatory response when it was administered to unanesthetized goats exposed to moderate hypoxia. These observations are consistent with a depressive role for lactic acid.

Severinghaus (81) has put forward a hypothesis to explain the role of lactic acid in HVD. He suggests that the chemosensitive neurons near the ventral medullary surface readily produce lactic acid when exposed to hypoxia (82,83). During the initial hypoxic exposure when ventilation is increased as a result of the hypoxic stimulus at the carotid bodies, the chemosensitive medullary neurons produce lactic acid and thereby increase intracellular $[H^+]$. This increased $[H^+]$ reduces the extracellular fluid (ECF)-to-intracellular fluid (ICF) $[H^+]$ gradient and tends to hyperpolarize the neurons. The reduced chemosensor output causes a reduction in ventilation that results in an increased systemic P_{CO_2}. Due to the poor buffering of ECF relative to ICF, this increased P_{CO_2} tends to restore the ECF-ICF $[H^+]$ gradient and normalize breathing. This is an attractive hypothesis that could account for the pattern of ventilatory stimulation/inhibition/recovery that characterizes HVD. A potential limitation of this hypothesis is that it does not explain why carotid bodies appear to be required for the development of HVD.

Endogenous Opioids

Inhibition of ventilation by endogenous opioids has been clearly demonstrated in the neonate (84,85) and is thought to be a vestigial fetal response to the stress of hypoxia (20). In the unanesthetized adult, however, no role for endogenous opioids has been found (11,13,86,87). Accordingly, we will not elaborate on this topic but refer the reader to two excellent reviews on the subject (22,88).

Endogenous Benzodiazepines

We are aware of only one study dealing with the role of endogenous benzodiazepines in the ventilatory response to hypoxia (89). These authors found that inhibition of benzodiazepine receptors with a specific antagonist had, if anything, a slight inhibitory effect on ventilation during hypoxia. Thus, these receptors do not appear to be a likely candidate for the mediation of hypoxic ventilatory decline.

In summary, there seems little doubt that changes in excitatory and inhibitory neuroeffectors occur. The essential issues are what causes these changes and how significant they are. Possible mechanisms include a direct effect of hypoxia on CNS neurons and/or an effect related to magnitude and duration of input from the peripheral chemoreceptors.

F. Metabolic Rate

A fifth potential cause of hypoxic ventilatory decline is a decrease in metabolic rate secondary to systemic hypoxia. What this means is that ventilation could decrease secondary to a hypoxia-induced fall in metabolic rate without hypoventilation. This effect is known to exist in neonates of most mammalian species and adults of most species smaller than about 5 kg (90,91). The effect decreases progressively as mass increases and specific metabolic rate decreases (91). The decrease in $\dot{V}O_2$ occurred very rapidly (91) and remained stable during the period of hypoxic ventilatory decline. $\dot{V}CO_2$ may change more slowly, however, for reasons that are not clear but may be related to CO_2 stores. In humans exposed to moderate hypoxia metabolic rate remained normal (e.g., Ref. 92). Similar observations have been obtained in ponies (35) and goats (93). Thus, it is difficult to invoke hypometabolism to explain hypoxic ventilatory decline in species that are as large as human adults. An important caveat here is that in neonates and smaller mammals, hypometabolism and the disparity between the time course of changes in $\dot{V}O_2$ and $\dot{V}CO_2$ may affect at least the first few minutes of the response to hypoxia, which could raise significant interpretive problems. In addition, in the rat at least, there is a marked increase in V_D that has reached a plateau by 1 hour of exposure to hypoxia (94). Therefore, some index of $\dot{V}_A$ would appear to be essential in studies of hypoxic ventilatory decline, particularly in small mammals.

In summary, in the absence of a change in $PaCO_2$ or PaO_2, the carotid body chemoreceptors maintain an essentially constant output during the period of hypoxic ventilatory decline, suggesting that, although the carotid bodies are essential for hypoxic ventilatory decline to occur, the decline must depend on CNS processes. Considerable potential for redundancy and interaction of central mechanisms exists; some or all may play a role in hypoxic ventilatory decline, and the mechanisms that are employed may well change depending on the severity of hypoxia. A major unanswered question here is the significance of hypoxic ventilatory decline in the overall process of ventilatory acclimatization.

IV. Short-Term Acclimatization

Short-term acclimatization is the progressive increase in ventilation that ensues over hours to days of hypoxic exposure (Fig. 1). While the time course of this response varies between species, the essential features are the same in all mammals that have been studied. These are (1) a progressive hyperventilation that reaches a plateau in hours to days, (2) persistent alkalosis in the blood and bulk CSF, and (3) continued hyperventilation upon return to normoxia that diminishes slowly over hours to days (see Sec. IV.C). Despite extensive research since the nineteenth century, the mechanism(s) underlying acclimatization remains poorly understood, although important advances have been made in the past decade.

A. Carotid Bodies

It is often assumed that carotid body sensitivity increases over time during hypoxic exposure. This assumption is based on a number of studies of hypoxic ventilatory response in humans before and after several days of hypoxic exposure (95–99). Increased sensitivity could also be inferred simply from the increased ventilation in the face of increasing Pa_{O_2} and decreasing Pa_{CO_2} as acclimatization proceeds. Unfortunately, studies of this type do not tell us if the carotid chemosensor's sensitivity is increasing, if the central integrating mechanism is becoming sensitized, or both. More specific studies are required to elucidate these questions.

Considerable evidence has accumulated to show that the carotid bodies are required for acclimatization to hypoxia. Carotid body–denervated, unanesthetized dogs, goats, ponies, and sheep have been shown to not acclimatize to hypoxia (33,35–37,93). Evidence suggests that the carotid bodies could account completely for ventilatory acclimatization. Using the vascularly isolated, perfused carotid body technique in the unanesthetized goat, Busch et al. and Bisgard et al. separated the circulation to the carotid body from the rest of the systemic circulation (including brain and medullary chemosensitive and integrating structures) (100,101). When they rendered the carotid body hypoxic for several hours while maintaining the rest of the body normoxic (either poikilocapnically or isocapnically), they observed essentially normal acclimatization. From these data they concluded that brain hypoxemia or hypocapnia were not required for acclimatization (see Sec. IV.C). Weizhen et al. (102) used identical techniques to extend these observations by maintaining the carotid bodies normoxic and normocapnic while the brain was maintained hypoxic. They showed that prolonged CNS hypoxia alone did not promote acclimatization, i.e., upon return to normoxia Pa_{CO_2} returned to its original control value (Fig. 5). Normally, of course, Pa_{CO_2} is falling and Pa_{O_2} is increasing over time of hypoxic exposure; possible inhibitory effects of these changing stimuli at the CNS remain to be determined. At the level of the carotid body, recent work in the unanesthetized goat with vascularly isolated and exogenously perfused carotid body (15) has shown that carotid body hypoxic sensitivity to isocapnic hypoxia increases over time of

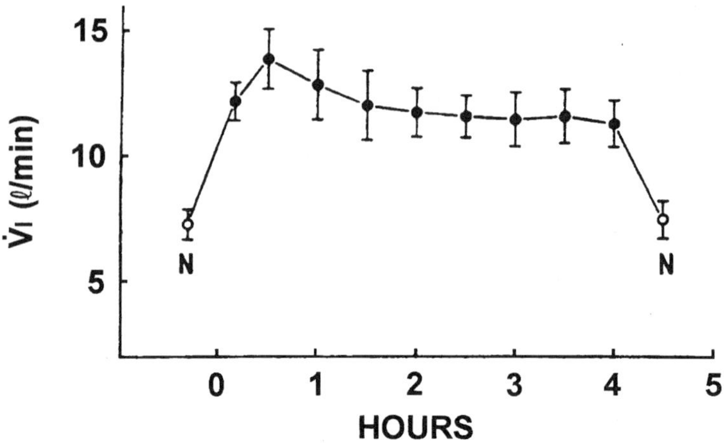

Figure 5 Lack of acclimatization in response to 4 hours of brain hypoxia with carotid bodies held normoxic and normocapnic using the vascularly isolated carotid body perfusion technique in awake goats. Note the stable ventilation over the period indicating no acclimatization. (From Ref. 102.)

hypoxic exposure, and the level of P_{CO_2} at the carotid body seems to have no effect on this increase.

In summary, carotid body hypoxia alone is capable of causing acclimatization. Is this time-dependent increase in ventilation due to changes at the medullary integrator or at the carotid body? It has been shown that carotid sinus nerve activity increases over time when hypoxia is the stimulus (Fig. 6) (103) but not when CO_2 is the stimulus (104). At the cellular and molecular level, chronic hypoxia has been shown to increase excitability in cultured rat glomus cells associated with differential changes in ionic currents (105). In terms of ventilatory response, at the level of the carotid body hypoxia is the only stimulus that will cause acclimatization. It has been demonstrated that long-term hypercapnia at the carotid body (using the isolated carotid body perfusion technique in the unanesthetized goat) stimulates breathing but does not cause acclimatization (106). This observation suggests that the carotid body can differentiate between stimuli and encode this information into afferent carotid sinus nerve traffic. While a discussion of carotid body transduction processes is beyond the scope of this chapter (see Chapter 5 and Ref. 107), we do note that attempts to find a difference in carotid sinus nerve discharge patterns, which might relate to how O_2 versus CO_2 are encoded in the carotid sinus nerve, have so far been unsuccessful (108).

Many neurochemicals could be involved in the carotid body chemoreceptor responses to hypoxia. Space does not permit us to discuss this extensively; an excellent review of this material can be found in Bisgard and Forster (6). A few key points should be noted, however.

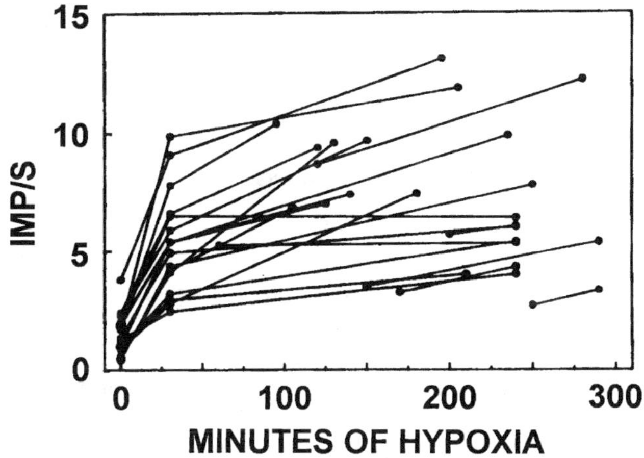

Figure 6 Response of single carotid body chemoreceptor afferents to prolonged isocapnic hypoxia. The control discharge rate is indicated at time 0. Note the progressive increase in activity over time. (From Ref. 136.)

Dopamine and norepinephrine appear to be the predominant neurochemicals in the carotid body and have been the most extensively investigated. Whether or not dopamine or norepinephrine play a role in carotid body chemoreceptor-mediated ventilatory acclimatization to hypoxia is controversial, as there is negative evidence from the goat (summarized in Ref. 6) and positive evidence from the cat (14,109). Regardless of whether there is a role for either of these neurochemicals, there is a clear link between hypoxia and the genetic regulation of catecholamine production. Czyzyk-Krzeska et al. (110) showed that hypoxia regulates the gene that codes for tyrosine hydroxylase, the rate-limiting enzyme in the production of catecholamines, in dopaminergic cells of the mammalian carotid body.

Other neurochemicals present in the carotid body could also be involved in the acclimatization process. These include nitric oxide, substance P, neurokinin A, enkephalins, VIP, CGRP, atrial natruretic peptide, and neuropeptide Y (111–114). To date, little is known about the role of these compounds in the carotid body.

A number of key questions remain unanswered with respect to these carotid body neurochemicals: Is there a role for these neurochemicals in ventilatory acclimatization, especially the noncatecholamines? Are there species differences in response to these neurochemicals? What are their mechanisms of action at the receptor level? Do different neurochemicals interact, and, if so, how?

B. Central Nervous System

The fact that carotid bodies could account for ventilatory acclimatization does not rule out a role for processes in the brain, but it does appear that carotid bodies are mandatory for manifestation of any central acclimatization processes. Central processes may be more important under nonlaboratory conditions where Pa_{O_2} is increasing and Pa_{CO_2} is decreasing over time as acclimatization progresses. Several processes could contribute to ventilatory acclimatization.

Sensitization by Peripheral Afferent Inputs

Afferent input from peripheral chemoreceptors may contribute to increased respiratory motor output over time. Dwinell et al. (15) showed in the goat with a vascularly isolated and perfused carotid body that carotid body hypoxic sensitivity was elevated after 4 hours of hypoxic carotid body perfusion. Millhorn et al. (115), in anesthetized, vagotomized, and paralyzed cats, showed that serotonin antagonists blocked a long-term facilitation of phrenic activity that occurred secondary to carotid sinus nerve stimulation. They suggest that the time course of this facilitation is such that it could contribute to ventilatory acclimatization to hypoxia, thus implicating serotonin in this process (see Sec. IV.C). However, Olson (116) has shown that serotonin depletion by parachlorophenylalanine in unanesthetized rats had no effect on ventilatory acclimatization, thus questioning the conclusions of Millhorn et al. Critical comparisons between these two studies are difficult because (1) serotonin levels change over time in hypoxia, (2) serotonin levels may not reflect turnover at the relevant sites, (3) all serotonin pools may not be equally accessible to pharmacological agents (117,118), and (4) species differences and presence of anesthesia may also play a role. Further experiments to clarify the role of serotonin would seem to be required.

In the context of enhanced central respiratory activity secondary to increased carotid chemoreceptor input, it is interesting to note that recent evidence has shown that nitric oxide generated by increased carotid sinus nerve input to the nucleus tractus solitarius can act as a retrograde messenger in an L-glutamate–releasing positive feedback system that serves to augment ventilation during hypoxia (119). Could this process contribute to some of the time-dependent increase in ventilation observed during acclimatization (i.e., in addition to direct carotid body afferent inputs)? This novel idea deserves further study.

CNS Acid-Base

Central nervous system acid-base changes certainly occur during the period of acclimatization to hypoxia, and it has long been thought that such changes must account for most or all of ventilatory acclimatization. However, whether or not there is in fact a role for central acid-base adjustments in acclimatization, and, if so, the precise mechanisms of action, remain highly controversial.

One theory proposes that acclimatization results from regulation of the bulk CSF $[H^+]$ over time in hypoxia. That is, hypoxia initially results in a carotid body–mediated hyperventilation and respiratory alkalosis. The hypocapnia alkalosis in turn would tend to inhibit ventilation peripherally and, especially, centrally such that ventilatory output would be the net result of ongoing carotid body stimulation and the concomitant respiratory alkalosis. In the face of continued carotid body stimulation, CSF [H^+] would then be regulated progressively back toward normoxic control levels, restoring much of the central ventilatory drive that was lost and thereby increasing ventilation (e.g., Ref. 120). Similarly, deacclimatization (i.e., continued, but reduced, hyperventilation upon return to normoxic conditions; see below) would be accounted for by the respiratory acidosis produced in CSF when the carotid body component of ventilatory drive is abruptly removed by normoxia resulting in reduced ventilation and increased Pa_{CO_2}. This continued hyperventilation would then diminish over time as bulk CSF [H^+] was regulated back to normoxic eupneic values (121).

Others have demonstrated that bulk CSF [H^+] changes in the wrong directions over time to account for acclimatization or deacclimatization. In humans (122–124), dogs (25), and ponies (35,126) exposed to various altitudes (3100–4300 m), ventilation and $[H^+]_{CSF}$ were negatively correlated or uncorrelated at all time points; i.e., ventilation was increasing at a time when CSF [H^+] was falling. Moreover, changes in CSF [H^+] were compensated to the same extent as in arterial plasma. There are also observations in the anesthetized cat showing that [H^+] on the medullary surface did not correlate with ventilation even though it increased during hypoxic exposure (74). When acclimatized humans and ponies were abruptly returned to normoxia and followed for up to 13 hours, $[H^+]_{CSF}$ (lumbar in the humans; cisternal in the ponies) again was negatively correlated with ventilation (127), i.e., ventilation continued to decrease in the face of increasing CSF [H^+]. These findings do not rule out the possibility that the acid-base status of a compartment of the medullary interstitial fluid, which cannot be measured directly, is somehow being regulated such that neurons at some point along a CSF-to-blood [H^+] gradient would be stimulated appropriately (128,129). However, manipulation of this gradient had no effect on acclimatization in human subjects (130). In addition, Musch et al. (79) have shown that, in rats exposed to hypoxia for up to seven days, changes in medullary intracellular pH and [lactate] did not correlate with ventilatory acclimatization.

Jennings (131) has suggested another alternative, namely that the moiety that is sensed by the central chemosensors is the strong ion difference of CSF. The strong ion difference is defined as [Σ strong cations] − [Σ strong anions] (132). The strong ion difference is an especially important variable in CSF because, in an essentially protein-free solution like CSF, the strong ion difference and CO_2 determine [H^+] (132,133). Using this approach, Jennings has shown a correlation between CSF strong ion difference and ventilation for many of the studies cited above. He further suggests that the control of CSF strong ion difference might be mediated by angiotensin II which is known to stimulate ventilation. However, this approach does not appear to hold for the one available study of deacclimatization over periods greater than one hour (127). A re-analysis of these data done by the authors of this chapter shows essentially no correlation between CSF strong ion difference and ventilation during this interval. It should be noted that this does not necessarily invalidate Jennings' approach but rather may point to some unique characteristics of the deacclimatization process (see Section IV.C).

C. Deacclimatization

It was long assumed that deacclimatization, i.e., continued but falling, time-dependent hyperventilation following return to normoxia after acclimatization to hypoxia (Fig. 1), was simply the reverse of the acclimatization process. A considerable amount of evidence has accumulated that suggests that this is not the case. Dempsey et al. (127) pointed out that, in deacclimatizing humans and ponies, the measurable chemical stimuli to breathing were changing in the wrong directions to account for the observed continued hyperventilation (also see Sec. IV.B). More recent evidence has shown that, whether in goats acclimatized secondary to specific carotid body hypoxia or whole-body hypoxia and in the presence of either systemic poikilocapnia or isocapnia, ventilatory acclimatization occurred. However, the poikilocapnic goats showed continued hyperventilation upon return to normoxic conditions, but the isocapnic goats did not, even if the latter had also been exposed to systemic hypoxia (100,101,134,135). Carotid body afferent activity in the goat was not elevated upon return to normoxia after ventilatory acclimatization had occurred despite an apparent increase in carotid body sensitivity (136). Further, in the cat deacclimatization persisted after carotid sinus nerve section if the intact animal had first been allowed to acclimatize (137). In humans 24 hours of mild systemic hypocapnia (produced by voluntary hyperventilation) was followed by a persistent hyperventilation, and this after-effect was amplified when hypoxia accompanied the prolonged hypocapnia (138). Taken together, these studies indicate that acclimatization and deacclimatization may, at least in part, utilize different processes. Acclimatization requires carotid bodies and may be largely mediated by them, whereas deacclimatization appears to be an exclusively central process linked to prolonged CNS hypocapnia. Hypoxia may amplify the ventilatory effects but is not required. The mechanism of this central process remains obscure. Recent work in the anesthetized rat (131) showed that repeated 2-minute periods of electrical carotid sinus nerve stimulation or repeated 5-minute periods of isocapnic hypoxia resulted in long-term facilitation; i.e., increased phrenic motor output persisted for up to 30 minutes following stimulation. This long-term facilitation could be linked to increased serotonin levels, as suggested by Millhorn et al. (115) (see Sec. IV.B). However, more recently, Olson (116) has shown that deacclimatization occurred normally in serotonin-depleted rats. We must conclude that, to date, a clear role for serotonin in ventilatory acclimatization or deacclimatization has not been established.

In summary, regulation of CNS $[H^+]$ as a mechanism of ventilatory acclimatization to hypoxia is an elegant hypothesis with considerable logical appeal. However, it has been difficult to establish any major role for CNS $[H^+]$ in the control of ventilatory acclimatization and deacclimatization, particularly when the entire time course of the process is considered. The apparent dependence of ventilatory deacclimatization on the presence of hypocapnia during the acclimatization period suggests that deacclimatization may have some sort of secondary link to reduced $[H^+]$, perhaps changes in neuromodulators and/or other sorts of plasticity, but additional study is required to resolve this question.

V. Long-Term Acclimatization

In some populations indigenous to or resident for many years at high altitude, the ventilatory response to hypoxia may be much diminished compared to that of the acclimatized lowlander (139,140). This ''blunting'' of the hypoxic response is also evidenced by a resting, eupneic Pa_{CO_2} 1–3 mmHg higher than that of the acclimatized sojourner at the same altitude. This difference is particularly apparent during moderate to heavy exercise at high altitude, as shown by the highlander's near-isocapnic hyperpneic response to mild through moderate exercise in contrast to the sojourner's tachypneic hyperventilation (Fig. 7) (141). While tachypnea and dyspnea are commonly experienced during moderate exercise in the sojourner (Fig. 8) (142), these responses are only rarely experienced in the highlander (141). Furthermore, when the sojourner was suddenly made hyperoxic during exercise, ventilation was reduced precipitously and great relief from dyspnea was experienced, whereas the highlander's ventilation was not reduced by hyperoxia until heavy work loads were reached, and they were often unaware that their hypoxemia had been eliminated (141). Another manifestation of the blunted hypoxic ventilatory response may be the relative absence of periodic breathing during sleep in hypoxia in Sherpa highlanders (see Chapter 21) (140).

Interestingly, blunting appears to be less prevalent in Tibetans indigenous to altitudes >3600 m than in the Andean Quechua and Aymara, suggesting a genetic/evolutionary adaptation to high altitude (see Chapter 3) (143,144). In other highlanders, the blunting of hypoxic ventilatory response seems to result from many years of prolonged postnatal hypoxic exposure (3). In humans, the blunted response to exercise in hypoxia is achieved in native lowlanders who move to high altitude as young children (3,95). Animals native to high altitudes, such as the llama or yak (144,145), or calves born at high altitudes (146) do not exhibit blunted ventilatory responses to acute hypoxia, although these animals, like human highlanders, tend to show less hyperventilation during resting, air-breathing eupnoea. In the rat, exposure to hypoxia in the early (preweaning) neonatal period was required to achieve a blunted hypoxic response (147), but blunting could be produced after 3–4 weeks of exposure to 5500 m altitude in adult cats (148).

What causes this blunted response to hypoxia? Certainly, the carotid body must be implicated. Specific pharmacological stimulation of carotid bodies in highlanders produced a ventilatory response less than in sojourners at the same altitude (149). Additional support for this idea is provided by the observation of increased size and dopamine content of carotid bodies in chronically hypoxic animals (3). Further, cats chronically exposed to severe hypoxia show blunting of both carotid sinus nerve activity and global ventilatory response to acute hypoxia (148). Dopaminergic mechanisms have been implicated (109), but opioids have been ruled out (150) as mediators of this reduced carotid body sensitivity.

There are some other clues to the mechanism causing blunting: (1) ventilatory response to CO_2 is not abnormal in the highlander, (2) the response to acute hyperoxia is hyperventilation in some highlanders, much as it is in a carotid body-

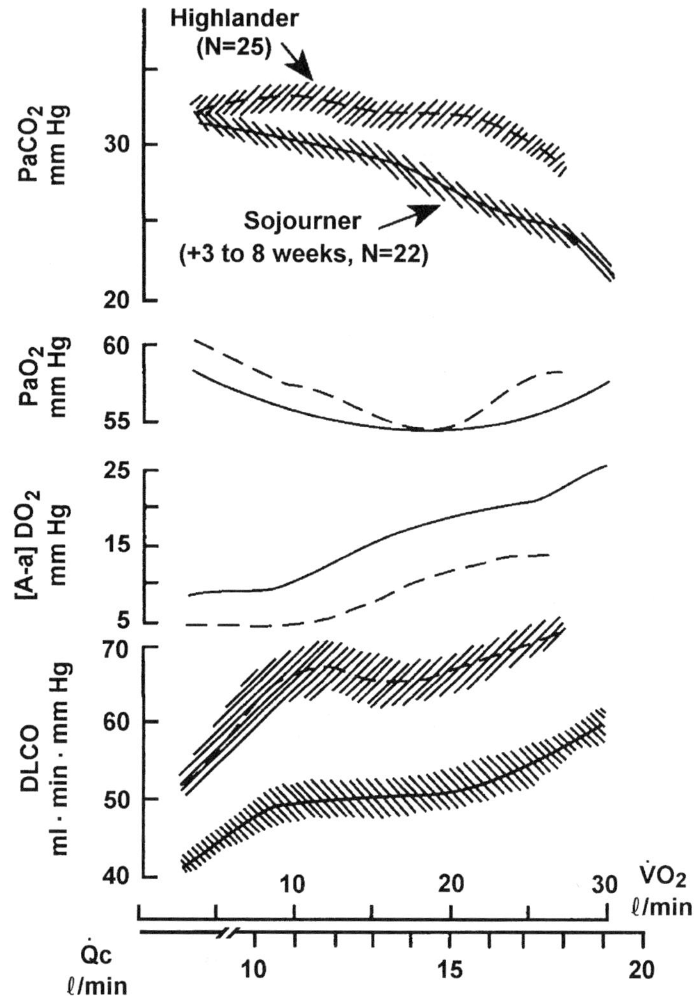

Figure 7 Pulmonary gas transport during progressive exercise in the sojourner to 3100 m (+8 weeks) vs. the second-to-third generation native or long-term resident at 3100 m (highlanders). Note that the level of hyperventilatory response is much less in the highlander, and yet the arterial Po_2 is similar because the alveolar-to-arterial Po_2 difference is much narrower owing to the enhanced diffusion capacity. (Data from Ref. 141.)

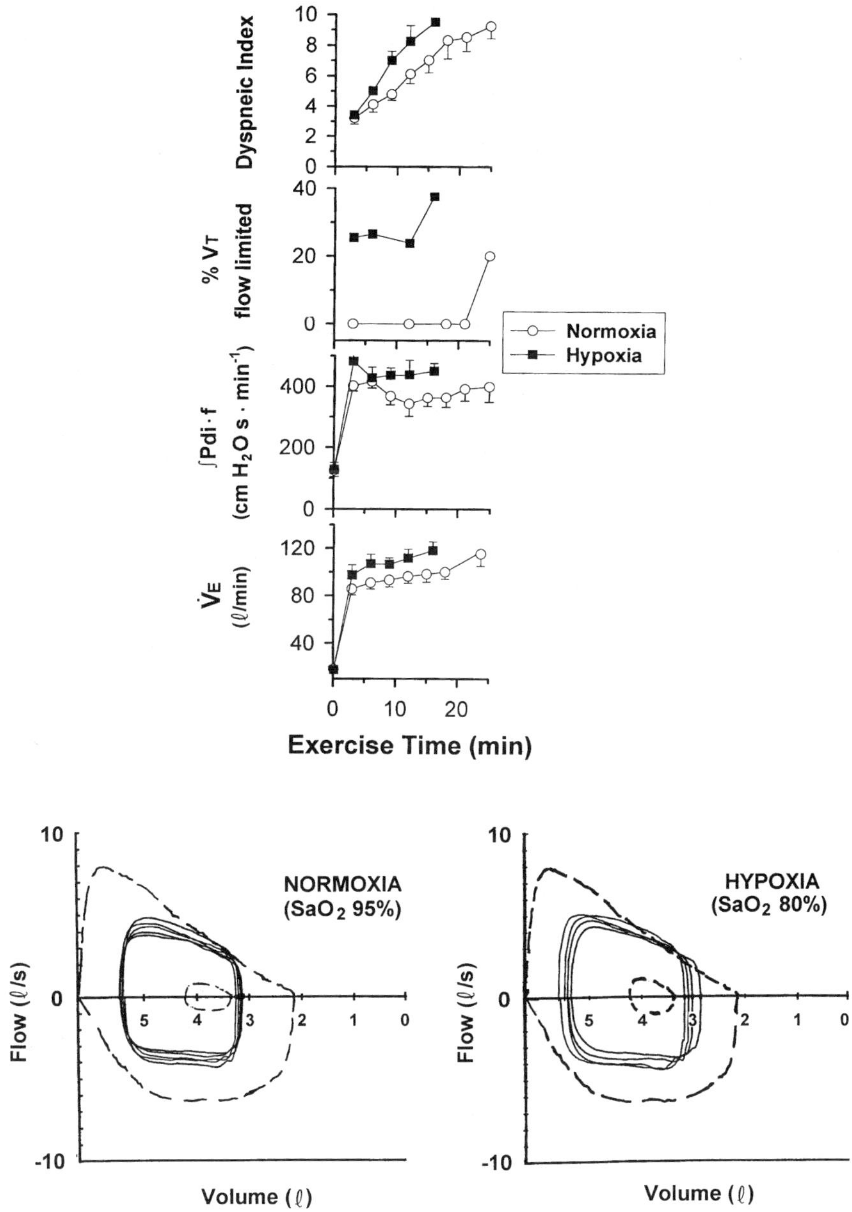

Figure 8 Effects of prolonged heavy exercise (85–90% of $\dot{V}O_2$max) in acute hypoxia (SaO_2 = 80%) and normoxia (SaO_2 = 95%) on ventilation, the percent of the $\dot{V}E$ that is flow limited during expiration, the force output of the diaphragm per minute, and ratings of dyspneic sensations (on a 10-point scale). (Data from Ref. 142.)

denervated animal (37), and (3) midcollicular decerebration in the chronically hypoxic animal alleviates much of the blunted hypoxic response (151). These data, although incomplete, suggest that the highlander's altered regulation of breathing is very specific to the hypoxic condition and attributable to both a reduced chemoresponsiveness of the carotid bodies per se, along with a reduced central integration of sensory afferent inputs.

VI. Exercise Hyperpnea at High Altitude

Any sojourner upon first arriving at high altitude will readily attest to the heightened dyspnea they perceive during exercise. The magnitude of this hyperventilation and the accompanying hypocapnia increases with increasing exercise intensity. Further, after weeks or months at high altitude, the sojourner experiences a substantial tachypnea, hyperventilation, and respiratory alkalosis during sustained exercise of moderately heavy intensity (see Fig. 8). When exercise is combined with hypoxia, the ventilatory response is more than additive (3,152). In fact, hypoxia and other carotid body stimuli, whether physiological (i.e., CO_2) or pharmacological (i.e., Doxapram HCl), commonly show synergistic ventilatory responses when combined. In the case of exercise, we speculate that the enhanced sensory input from the hypoxic carotid bodies is multiplied by simultaneous feed-forward–type input to the medullary respiratory controller from locomotor areas of the higher CNS.

A number of important consequences derive from this magnified ventilatory response to exercise in the sojourner to high altitudes. On the plus side, the hyperventilatory response increases Pa_{O_2} and hence arterial O_2 content during exercise. For example, at 4000 m altitude, the Pa_{O_2} during moderately heavy exercise in the acclimatized sojourner averages about 40 mmHg and Pa_{CO_2} about 17 mmHg (152). Without hyperventilation Pa_{O_2} would have been in the 25–30 mmHg range, roughly equivalent to that found in a climber at the summit of Mount Everest (8848 m). Thus, hyperventilation, especially during exercise, has a potent ''altitude-sparing'' effect.

Unfortunately, this sparing of arterial O_2 content may come at considerable cost. The work of breathing is 40–60% greater than at sea level at any given exercise intensity (152,153). Although air density is reduced significantly at very high altitudes (i.e., low barometric pressures), the resultant reduction in airway resistance is more than offset by the increasing turbulent flow produced by high flow rates. At very heavy work loads in hypoxia, the increased ventilatory requirement will push the sojourner closer and closer to the limits of his maximum flow-volume loop. Significant expiratory flow limitation will cause additional ventilatory work by increasing flow resistance, distorting the chest wall, and causing end-expiratory lung volume to rise—the latter resulting in shortened inspiratory muscles and increased elastic work of breathing. Furthermore, powerful stimulation of breathing frequency could disrupt coordination of the normal synchronization of locomotion and breathing pattern that one normally accomplishes readily at sea level, particularly during running.

Hypoxia probably hastens the fatigue of respiratory muscles. This possibility has been demonstrated in humans by showing that heavy endurance exercise to exhaustion caused a reduction in the maximum pressure generated by the diaphragm in response to supramaximal phrenic nerve stimulation (154). This reduced maximal force output or diaphragmatic "fatigue" was similar following hypoxic and normoxic exercise, even though the exercise time to exhaustion was reduced by one half in hypoxia (142). In exercising rats, glycogen depletion and lactic acid accumulation in the diaphragm were markedly enhanced when exercise was conducted in hypoxia (155). The combination of an increased workload placed on the respiratory muscles plus reduced O_2 transport to these muscles, along with an increased level of circulating metabolites from hypoxic limb locomotor muscles, are factors that might contribute to enhanced diaphragmatic fatigue during hypoxic exercise (142).

$\dot{V}O_2$max and exercise performance are decreased with increasing altitude (see Chapter 20). Does ventilatory work contribute to this limitation? First, despite the increased work of breathing and possible diaphragm fatigue, the sojourner, even at very high altitudes, is clearly capable of a substantial alveolar hyperventilation during heavy exercise. At $\dot{V}O_2$max, minute ventilation at high altitude approximates that at sea level, even though $\dot{V}O_2$max is reduced at high altitudes. Accordingly, ventilatory "failure" in terms of the adequacy of alveolar oxygenation and CO_2 elimination certainly does not occur during heavy exercise at high altitudes. Thus, despite diaphragmatic fatigue and expiratory flow limitation, the arsenal of "accessory" inspiratory and expiratory muscles appears to be sufficient to generate the appropriate magnitude of pleural pressure to meet the increased requirements for ventilation and ventilatory work during exercise at high altitudes.

There are two more likely ways in which the hyperventilation and increased ventilatory work during exercise at high altitudes might contribute to decrements in exercise performance. First, the O_2 requirement of exercise hyperpnea per se, particularly under conditions where flow limitations exist throughout a large portion of the breath, has been shown to require as much as 15% of the $\dot{V}O_2$max (156). These high levels of respiratory muscle work have recently been shown to promote a substantial "steal" of blood flow from limb locomotor muscles during maximum exercise (157) and may, therefore, limit locomotor muscle endurance performance. Second, the perception of dyspnea is extreme during heavy exercise in hypoxia (see Fig. 8). The additional sensory input to the cortex from chemoreceptors and chest wall mechanoreceptors is manifest by sensations that are foreign and highly unpleasant to the exercising sojourner at high altitude. Certainly, this discomfort could contribute to a sense of overall "effort" and whole-body "fatigue," which may curtail exercise performance. In addition, brain hypoxia leading to cerebral dysfunction may also limit locomotor performance (see Sec. VII and Chapter 13). The hyperventilation of heavy exercise might exacerbate brain hypoxia because of the reduction in cerebral blood flow that has been shown to accompany the hypocapnia of exercise at high altitude (152).

Finally, the magnitude of one's ventilatory responsiveness to hypoxia may have a bearing on exercise performance at high altitude; this possibility is discussed in Section VII (see Chapter 20). An enhanced ventilatory response to hypoxia would increase alveolar oxygenation even during heavy exercise (see above). In the absence of other more efficient acclimatization mechanisms, this hyperventilatory response is extremely important to the sojourner. However, an excessive ventilatory responsiveness to exercise in hypoxia will also increase the work of breathing with all of the attendant consequences outlined above, especially the steal of blood flow from exercising limbs. The net effect of the opposing influences may be estimated by recalling that O_2 transport to working muscle is the product of arterial O_2 content and limb blood flow, the former being protected by alveolar hyperventilation and the latter being jeopardized by the attendant increase in respiratory muscle work.

On the other hand, highlanders do not appear to be disadvantaged by a lesser ventilatory response to exercise (see Chapter 3). Because their alveolar to arterial O_2 difference is much narrower, $Pa{O_2}$ and $Sa{O_2}$ remain comparable to that in the hyperventilating sojourner throughout exercise (see Fig. 7). The reason for the smaller A-a$D{O_2}$ in exercising highlanders is a much larger pulmonary diffusion capacity, both at rest and during exercise (Fig. 7) (158,159). This highly efficient pulmonary gas exchange probably requires true structural adaptation as a consequence of lifelong hypoxia.

VII. Breathing and Human Performance at High Altitude

The final question we wish to explore is: Does how we breathe affect how we fare at high altitude? The more vigorous the ventilatory response to hypoxia (HVR), the higher will be the alveolar $P{O_2}$ and hence the arterial $P{O_2}$ and arterial oxygen saturation (although as indicated in the prior section, other factors during exercise may serve to widen the alveolar-arterial oxygen tension difference). Individual variability in HVR between high and low responders may span an order of magnitude (160,161). Interindividual variability in HVR appears to have a genetic underpinning (see also Chapter 3) (162,163). Our purpose in this section will be to review how variability of HVR relates to various aspects of human performance at high altitude. Presumably a higher HVR, by elevating $Pa{O_2}$, should be an asset at high altitude and especially at extreme elevations, i.e., above 7500–8000 m, where, because of the steepness of the oxyhemoglobin dissociation curve, small changes in $Pa{O_2}$ net large increases in arterial oxygen content.

The HVR is often measured by relating changes in moment-to-moment ventilation resulting from breathing progressively more hypoxic inspired gas to concurrent changes in $P{O_2}$, either end-tidal or arterial, or to arterial oxygen saturation. Stimulation of breathing by hypoxia causes the $Pa{CO_2}$ to fall, the hypocapnic alkalosis blunting the hypoxic ventilatory response, a poikilocapnic HVR. To more clearly assess the hypoxic responsiveness of the peripheral chemoreceptors, HVR is measured with $Pa{CO_2}$ held constant at the resting level by adding carbon dioxide to the inspired gas, the isocapnic HVR (12). As we have seen with acclimatization, the resting $Pa{CO_2}$ falls adding an element of uncertainty to tracking changes in HVR over time; methods have been proposed that attempt to standardize for these changes in resting P_{CO_2} that take place over the course of time of hypoxia (99,164). As indicated earlier, both measurements of HVR during acclimatization and the progressive increase in ventilation occurring over days to weeks at altitude, even as the chemical stimuli, P_{O_2} and P_{CO_2}, lessen in potency, indicate that hypoxic gain is increasing (95–99). The magnitude of the HVR in a given individual correlates moderately well with the vigor of exercise hyperpnea and, perhaps to a somewhat lesser extent, with the ventilatory responsiveness to carbon dioxide (HCVR) (165–169).

We can speculate on potential advantages and disadvantages of being gifted with an exceptionally high or low HVR. Endurance athletes commonly possess a low HVR, the benefits of which during performance might well be a lower metabolic cost of ventilation and less dyspnea (160). One can also imagine this same attribute as playing a role in those with chronic obstructive pulmonary disease, contributing to the distinction between those classified as ''pink puffers'' and ''blue bloaters.'' Schoene noted that the HVR of 14 mountaineers who had climbed above 7470 m was greater than that of 10 normal controls and significantly so than that of 10 endurance athletes. The putative advantage from a greater HVR at high altitude, particularly at extremely high altitude, is that thanks to the shape of the oxygen-hemoglobin dissociation curve, even a small increase in Pa_{O_2} will yield a large increase arterial oxygen content. Although controlled trials on a sufficient number of subjects of capacity to perform work are lacking, observations of climbing performance on several Himalayan expeditions suggest that those possessing high HVR tend to perform better than those with low HVR, even though the latter may be highly trained aerobic athletes with high $\dot{V}_{O_2}max$ at sea level (170–172). These conclusions are based upon such things as maximum climbing and sleeping altitude and subjective appraisal of individual performance and well-being. Inevitably, there are exceptions, one of the more noteworthy being that one of the two who first climbed Everest without supplemental oxygen possessed a quite low HVR at sea level (173). Although HVR clearly increases with acclimatization, what is less well established is how it changes in a given individual over the course of time. In particular, might someone with a low HVR at sea level acquire a significantly more vigorous responsiveness with acclimatization to high altitude?

Does HVR influence how quickly and how well lowlanders acclimatize upon ascending to high altitude? Reeves and colleagues noted that a lower resting end-tidal P_{CO_2} in those living at low elevations was associated with a higher sea level HVR, implying first of all that even sea level Pa_{O_2} might influence resting ventilation (174). In addition, both sea level P_{ETCO_2} and HVR predicted the magnitude of increase in breathing, as assessed by decrease in P_{ETCO_2} after one and 19 days at 4300 meters. Associations of sea level HVR with quality of adaptation have been noted in a number of studies. Individuals with lower HVRs may be more susceptible to acute high-altitude illnesses, including acute mountain sickness (AMS) (175–179) and high altitude pulmonary edema (180–183). No similar information exists with regard to an association between HVR and susceptibility to high-altitude cerebral edema. The link between a low HVR and AMS has not been universally observed, which is not surprising considering that many other factors dominate, including rate of ascent, altitude gained, and possibly age, gender, and other considerations (see Chapter 22). Acute mountain sickness has been associated with an increased fluid retention and weight gain (184). The peripheral chemoreceptors likely play a role in modulating sodium and water balance with hypoxia (185). Recently Swenson demonstrated in humans that those exhibiting a greater altitude-induced natriuresis and diuresis also evinced a higher HVR (186). No association with plasma concentrations of known salt and water-regulating hormones, renin, aldosterone, atrial natriuretic peptide, or antidiuretic hormone was found (186). These observations indicate a peripheral chemoreceptor influence on renal regulation of water balance but leave us guessing as to what the mechanistic link between the carotid body and kidney may be.

For lowlanders acclimatizing to high altitude, a brisk HVR seems to confer an advantage, netting both lower susceptibility to acute illnesses of high altitude and an enhanced capacity to perform physical work at extremely high altitudes. This latter gift may come with strings attached. A number of observers have reported evidence of mild, generally transient brain injury in some individuals who have climbed to extremely high altitudes. In looking for a physiological basis for the variability in susceptibility, Hornbein et al. indeed found an association with HVR, but it was the opposite of the one they had anticipated (187), those with higher HVR appeared to suffer more residual neurobehavioral deficits than those with lower HVR. Yet, as cited above, these same individuals appeared capable of better physical performance while at high altitude (172). One explanation proffered for this unexpected association of higher HVR with greater brain injury was that while a higher HVR increased arterial oxygen content, the greater hypocapnia caused more profound cerebral vasoconstriction; while exercising muscle received more oxygen the brain received less. Because arterial Po_2 and Pco_2 so powerfully influence cerebral blood flow, an association of CBF with HVR seems plausible, but thus far no information exists to indicate whether those with high HVR might possess heightened cerebrovascular reactivity to oxygen and CO_2 as well. This association between HVR and brain injury becomes more relevant as increasing numbers attempt to climb Mount Everest, including a small subset who choose to do so without the use of supplemental oxygen. Another possible contributor to brain hypoxia is the greater tendency toward periodic breathing noted in those with higher HVR (140,188). Even though average arterial oxygenation appears to be improved in those with higher HVR, the nadirs of Sao_2 associated with apneic pauses at extreme altitude might result in brief, repeated moments of such severe arterial hypoxemia as to cause brain injury.

Hormonal changes have been associated with HVR. For example, Raff et al. noted that hypoxia, acting by way of the carotid bodies in dogs, can stimulate cortisol secretion (189). In premenopausal women, HVR is increased during the luteal phase of the menstrual cycle (169,190,191). Studies in both experimental animals and humans (192) indicate that progesterone increases carotid chemoreceptor output while estrogen may have an enhancing effect acting centrally. HVR in premenopausal women exceeds that of men when corrected for differences in body size. Pregnancy enhances HVR further. Whether this gender difference confers a functional advantage at high altitude is a question that has been only partially explored. Premenopausal women appear to be less susceptible to high-altitude pulmonary edema, and those living permanently at high altitude are less likely to experience chronic mountain sickness prior to menopause (see Chapters 23 and 24). Whether a higher HVR of premenopausal women might enable better physical performance at extreme altitude is a question that has not been examined.

What happens to HVR in permanent high-altitude populations is discussed elsewhere in this volume (see Chapters 3 and 24). Initial observations that those born at or living for years at high altitude possess a lower HVR than acclimatized lowlanders have been extended by more recent studies indicating that not all high-altitude dwellers experience such blunting. Ethnic differences appear to play a role, with those of Tibetan origin showing greater variability and higher HVRs than the high-altitude Aymara of South America and comparable to the HVRs of acclimatized lowlanders (143).

References

1. Kellogg RH. Altitude acclimatization, a historical introduction emphasizing the regulation of breathing. Physiologist 1968; 11:37–57.
2. Dempsey JA, et al. Mediation of ventilatory adaptations. Physiol Rev 1982; 62:262–346.
3. Weil JV. Ventilatory control at high altitude. In: Cherniack NS, Widdicombe JG, eds. Handbook of Physiology, Section 3, The Respiratory System. Vol. 2. Bethesda: American Physiology Society, 1986:703–727.
4. Dempsey JA, Schoene RB. Pulmonary system adaptations to high altitude. In: Bone RC, ed. Pulmonary and Critical Care Medicine, Respiratory Physiology. Vol. 1, Part B6. St. Louis: Mosby Year Book Publishers, 1993:1–22.
5. Neubauer JA, Bisgard GE. Peripheral and central effects of hypoxia. In: Dempsey JA, Pack AI, eds. Regulation of Breathing. 2d ed. New York: Marcel Dekker, 1994:617–668.
6. Bisgard GE, Forster HV. Ventilatory responses to acute and chronic hypoxia. In: Fregly MJ, Blatteis CM, eds. Handbook of Physiology, Environmental Physiology. New York: Oxford University Press, 1996:1207–1239.
7. Gershan WM, Forster HV, Lowry TF, et al. Effect of metabolic rate on ventilatory roll-off during hypoxia. J Appl Physiol 1994; 76:2310–2314.
8. Long WQ, Giesbrecht GG, Anthonisen NR. Ventilatory response to moderate hypoxia in awake chemodenervated cats. J Appl Physiol 1993; 74:805–810.
9. Weil JV. Ventilatory responses to CO_2 and hypoxia after sustained hypoxia in awake cats. J Appl Physiol 1994; 76:2251–2252.
10. Easton PA, Slykerman LJ, Anthonisen NR. Ventilatory response to sustained hypoxia in normal adults. J Appl Physiol 1986; 61:906–911.

11. Kagawa S, Stafford MJ, Waggener TB, Severinghaus JW. No effect of naloxone and hypoxia-induced ventilatory depression in adults. J Appl Physiol 1982; 52:1030–1034.
12. Weil JV, Zwillich CW. Assessment of ventilatory response to hypoxia: methods and interpretation. Chest 1976; 70:124–128.
13. Freedman A, Scardella AT, Edelman NH, Santiago TV. Hypoxia does not increase CSF or plasma beta-endorphin activity. J Appl Physiol 1988; 64:966–971.
14. Cao K-Y, Zwillich CW, Berthon-Jones M, Sullivan CE. Ventilatory response to sustained eucapnic hypoxia in the adult conscious dog. Respir Physiol 1992; 89:65–73.
15. Dwinell MR, Janssen PL, Bisgard GE. Effects of carotid body hypocapnia during ventilatory acclimatization to hypoxia. J Appl Physiol 1997; 82:118–124.
16. Hornbein TF, Roos A, Griffo AJ. Transient effect of sudden mild hypoxia on respiration. J Appl Physiol 1961; 16:11–14.
17. Barnard P, Andronikou S, Pokorski M, Smatresk N, Mokashi A, Lahiri S. Time-dependent effect of hypoxia on carotid body chemosensory function. J Appl Physiol 1987; 63:685–691.
18. Vizek M, Pickett CK, Weil JV. Increased carotid body hypoxic sensitivity during acclimatization to hypobaric hypoxia. J Appl Physiol 1987; 63:2403–2410.
19. Nye PC. Identification of peripheral chemoreceptor stimuli. Med Sci Sports Exerc 1994; 26:311–318.
20. Neubauer JA, Melton JE, Edelman NH. Modulation of respiration during brain hypoxia. J Appl Physiol 1990; 68:441–451.
21. Brown DL, Lawson EE. Brain stem extracellular fluid pH and respiratory drive during hypoxia in newborn pigs. J Appl Physiol 1988; 68:1055–1059.
22. Neubauer JA, Santiago TV, Posner MA, Edelman NH. Ventral medullary pH and ventilatory responses to hyperperfusion and hypoxia. J Appl Physiol 1985; 58:1659–1668.
23. Suzuki A, Nishimura M, Yamamoto H, Miyamoto K, Kishi F, Kowakami Y. No effect of brain blood flow on ventilatory depression during sustained hypoxia. J Appl Physiol 1989; 66:1674–1678.
24. Yoshida K, Meyer JS, Sakamoto K, Honda J. Autoregulation of cerebral blood flow: electromagnetic flow measurements during acute hypertension in the monkey. Circ Res 1966; 19:726–738.
25. Long W, Lobchuk D, Anthonisen NR. Ventilatory responses to CO_2 and hypoxia after sustained hypoxia in awake cats. J Appl Physiol 1994; 76:2262–2266.
26. Heeringa J, Berkenbosch A, de Goede J, Olievier CN. Relative contribution of central and peripheral chemoreceptors to the ventilatory response to CO_2 during hyperoxia. Respir Physiol 1979; 37:365–379.
27. Berkenbosch A, van Dissel J, Olievier CN, de Goede J, Herringa J. The contribution of the peripheral chemoreceptors to the ventilatory response to CO_2 in anaesthetized cats during hyperoxia. Respir Physiol 1979; 37:381–390.
28. DeGoede J, Berkenbosch A, Ward DS, Bellville JW, Olievier CN. Comparison of chemoreflex gains obtained with two different methods in cats. J Appl Physiol 1985; 59:170–179.
29. Smith CA, Saupe KW, Henderson KS, Dempsey JA. Ventilatory effects of specific carotid body hypocapnia in dogs during wakefulness and sleep. J Appl Physiol 1995; 79:689–699.
30. Smith CA, Henderson KS, Harms CA, Zeng Y-J, Dempsey JA. Ventilatory effects of specific carotid body hypocapnia and hypoxia in awake dogs. J Appl Physiol 1997; 82:791–798.

31. Melton JE, Neubauer JA, Edelman NH. CO_2 sensitivity of cat phrenic neurogram during hypoxic respiratory depression. J Appl Physiol 1988; 65:736–743.
32. Melton JE, Yu QP, Neubauer JA, Edelman NH. Modulation of respiratory responses to carotid sinus nerve stimulation by brain hypoxia. J Appl Physiol 1992; 73:2166–2171.
33. Bouverot P, Bureau M. Ventilatory acclimatization and csf acid-base balance in carotid chemodenervated dogs at 3550 m. Pflugers Arch 1975; 361:17–23.
34. Olson EB, Jr., Vidruk EH, Dempsey JA. Carotid body excision significantly changes ventilatory control in awake rats. J Appl Physiol 1988; 64:666–671.
35. Forster HV, Bisgard GE, Rasmussen B, Orr JA, Buss DD, Manohar M. Ventilatory control in peripheral chemoreceptor-denervated ponies during chronic hypoxemia. J Appl Physiol 1976; 41:878–885.
36. Lahiri S, Edelman NH, Cherniack NS, Fishman AP. Role of carotid chemoreflex in respiratory acclimatization to hypoxemia in goat and sheep. Respir Physiol 1981; 46: 367–382.
37. Smith CA, Bisgard GE, Nielsen AM, Daristotle L, Kressin NA, Forster HV, Dempsey JA. Carotid bodies are required for ventilatory acclimatization to chronic hypoxia. J Appl Physiol 1986; 60:1003–1010.
38. Engwall MJA, Smith CA, Dempsey JA, Bisgard GE. Ventilatory afterdischarge and central respiratory drive interactions in the awake goat. J Appl Physiol 1994; 76:416–423.
38a. Curran AK, Rodman JR, Eastwood PR, Henderson KS, Dempsey JA, Smith CA. Ventilatory responses to specific CNS hypoxia in sleeping dogs. J Appl Physiol 2000; 88: 1840–1852.
39. Miller MJ, Tenney SM. Hypoxia-induced tachypnea in carotid-deafferented cats. Respir Physiol 1975; 23:31–39.
40. Dahan A, Ward D, van den Elsen M, Temp J, Berkenbosch A. Influence of reduced carotid body drive during sustained hypoxia on hypoxic depression of ventilation in humans. J Appl Physiol 1996; 81:565–572.
41. Forster HV, Dempsey JA, Chosy LW. Incomplete compensation of CSF $[H^+]$ in man during acclimatization to high altitude (4,300 m). J Appl Physiol 1975; 38:1067–1072.
42. Kayser B, Bokenkamp R, Binzoni T. Alpha-motoneuron excitability at high altitude. Eur J Appl Physiol Occup Physiol 1993; 66:1–4.
43. O'Reilly JP, Haddad GG. Chronic hypoxia in vivo renders neocortical neurons more vulnerable to subsequent acute hypoxic stress. Brain Res 1996; 711:203–210.
44. Haddad GG, Donnelly DF. O_2 deprivation induces a major depolarization in brain stem neurons in the adult but not in the neonatal rat. J Physiol 1990; 429:411–428.
45. Dillon GH, Waldrop TG. In vitro responses of caudal hypothalamic neurons to hypoxia and hypercapnia. Neuroscience 1992; 51:941–950.
46. Nolan PC, Waldrop TG. In vivo and in vitro responses of neurons in the ventrolateral medulla to hypoxia. Brain Res 1993; 630:101–114.
47. Nolan PC, Waldrop TG. In vitro responses of VLM neurons to hypoxia after normobaric hypoxic acclimatization. Respir Physiol 1996; 105:23–33.
48. Kramer JM, Nolan PC, Waldrop TG. In vitro responses of neurons in the periaqueductal gray to hypoxia and hypercapnia. Brain Res 1999; 835:197–203.
49. Horn EM, Waldrop TG. Oxygen-sensing neurons in the caudal hypothalamus and their role in cardiorespiratory control. Respir Physiol 1997; 110:219–228.

50. Gesell R, Hamilton MA. Reflexogenic components of breathing. Am J Physiol 1941; 133:694–719.
51. Eldridge FL, Gill-Kumar P. Central neural respiratory drive and afterdischarge. Respir Physiol 1980; 40:49–63.
52. Eldridge FL. Central integration of mechanisms in exercise hyperpnea. Med Sci Sports Exerc 1994; 26:319–27.
53. Badr MS, Skatrud JB, Dempsey JA. Determinants of poststimulus potentiation in humans during NREM sleep. J Appl Physiol 1992; 73:1958–1971.
54. Georgopoulos D, Walker S, Anthonisen NR. Increased chemoreceptor output and ventilatory response to sustained hypoxia. J Appl Physiol 1989; 67:1157–1163.
55. Curtis DR, Duggan AW, Felix D, Johnston GAR. GABA, bicuculline and central inhibition. Nature 1970; 226:1222–1224.
56. Wood JD, Watson WJ, Ducker AJ. The effect of hypoxia on brain γ-aminobutyric acid levels. J Neurochem 1968; 15:603–608.
57. Weyne J, Van Leuven F, Leusen I. Brain amino acids in conscious rats in chronic normocapnic acid levels. Respir Physiol 1977; 31:231–239.
58. Iversen K, Hedner T, Lundborg P. GABA concentrations and turnover in neonatal rat brain during asphyxia and recovery. Acta Physiol Scand 1983:91–94.
59. Hagberg H, Lehman A, Sandberg M, Nystrom B, Jacobson I, Hamberger A. Ischemia-induced shift of inhibitory and excitatory amino acids from intra- to extracellular compartments. J Cerebral Blood Flow Metab 1985; 5:413–419.
60. Hedner T, Hedner J, Bergman B, Iversen K, Jonason J. Effects of GABA and some GABA analogues on respiratory regulation in the preterm rabbit. Biol Neonate 1983; 43:134–145.
61. Hedner J, Hedner T, Wessberg P, Jonason J. An analysis of the mechanism by which γ-aminobutyric acid depresses ventilation in the rat. J Appl Physiol 1984; 56:849–856.
62. Kneussel MP, Pappagianopoulos P, Hoop B, Kazemi H. Reversible depression of ventilation and cardiovascular function by ventriculocisternal perfusion with gamma aminobutyric acid in dogs. Am Rev Respir Dis 1986; 13:1024–1028.
63. Taveira da Silva AM, Hartley B, Hamosh P, Quest JA, Gillis RA. Respiratory depressant effects of GABA α- and β-receptor agonists in the cat. J Appl Physiol 1987; 62: 2264–2272.
64. Huang J, Suguihara C, Hehre D, Lin J, Bancalari E. Effects of GABA receptor blockage on the respiratory response to hypoxia in sedated newborn piglets. J Appl Physiol 1994; 77:1006–1010.
65. Melton JE, Neubauer JA, Edelman NH. GABA antagonism reverses hypoxic respiratory depression in the cat. J Appl Physiol 1990; 69:1296–301.
66. Yamada KA, Norman WP, Hamosh P, Gillis RA. Medullary ventral surface GABA receptors affect respiratory and cardiovascular function. Brain Res 1982; 248:71–78.
67. Phillis JW, Kostopoulos GK, Limacher JJ. Depression of corticospinal cells by various purines and pyrimidines. Can J Pharmacol 1974; 52:1226–1229.
68. Phillis JW, Kostopoulos GK, Limacher JJ. A potent depressant action of adenine derivatives on cerebral cortical neurons. Eur J Pharmacol 1975; 30:125–129.
69. Koos BJ, Mason BA, Punla O, Adinolfi AM. Hypoxic inhibition of breathing in fetal sheep: relationship to brain adenosine concentrations. J Appl Physiol 1994; 77:2734–2739.

70. Eldridge FL, Millhorn DE, Kiley JP. Respiratory effects of a long-acting analog of adenosine. Brain Res 1984; 301:273–280.
71. Hedner T, Hedner P, Wessberg P, Jonason J. Regulation of breathing in the rat: indications for a role of central adenosine mechanisms. Neurosci Lett 1982; 33:147–151.
72. Yan S, Laferriere A, Zhang C, Moss IR. Microdialyzed adenosine in nucleus tractus solitarii and ventilatory response to hypoxia in piglets. J Appl Physiol 1995; 79:405–410.
73. Long WA, Lawson EA. Neurotransmitters and biphasic respiratory response to hypoxia. J Appl Physiol 1984; 57:213–222.
74. Millhorn DE, Eldridge FL, Kiley JP, Waldrop TG. Prolonged inhibition of respiration following acute hypoxia in glomectomized cats. Respir Physiol 1984; 57:331–340.
75. Javaheri S, Teppema LJ, Evers JA. Effects of aminophylline on hypoxemia-induced ventilatory depression in the cat. J Appl Physiol 1988; 64:1837–1843.
76. Easton PA, Anthonisen NR. Ventilatory response to sustained hypoxia after pretreatment with aminophylline. J Appl Physiol 1988; 64:1445–1450.
77. Gershan WM, Forster HV, Lowry TF, Garber AK. Effect of theophylline on ventilatory roll-off during hypoxia in goats. Respir Physiol 1996; 103:157–164.
78. Neubauer JA, Simone A, Edelman NH. Role of brain lactic acidosis in hypoxic depression of respiration. J Appl Physiol 1988; 65:1324–1331.
79. Musch TI, Dempsey JA, Smith CA, Mitchell GS, Bateman NT. Metabolic acids and $[H^+]$ regulation in brain tissue during acclimatization to chronic hypoxia. J Appl Physiol 1983; 55:1486–1495.
80. Aaron EA, Forster HV, Lowry TF, Korducki MJ, Ohtake PJ. Effect of dichloroacetate on Pa_{CO_2} responses to hypoxia in awake goats. J Appl Physiol 1996; 80:176–181.
81. Severinghaus JW. Hypoxic ventilatory depression may be due to central chemoreceptor cell hyperpolarization. Adv Exp Med Biol 1995; 393:257–260.
82. Xu FD, Spellman MJ, Jr., Sato M, Baumgartner JE, Ciricillo SF, Severinghaus JW. Anomalous hypoxic acidification of medullary ventral surface. J Appl Physiol 1991; 71:2211–2217.
83. Xu F, Sato M, Spellman MJ, Jr., Mitchell RA, Severinghaus JW. Topography of cat medullary ventral surface hypoxic acidification. J Appl Physiol 1992; 73:2631–2637.
84. DeBoeck C, Van Reempts P, Rigatto H, Chernick V. Naloxone reduces decrease in ventilation induced by hypoxia in newborn infants. J Appl Physiol 1984; 56:1507–1511.
85. Grunstein MM, Hazinski TA, Schleuter MA. Respiratory control during hypoxia in newborn rabbits: implied action of endorphins. J Appl Physiol 1981; 51:122–130.
86. Schaeffer JJ, Haddad GG. Ventilatory response to moderate and severe hypoxia in adult dogs: role of endorphins. J Appl Physiol 1988; 65:1383–1388.
87. Wheeler MD, Farber JP. Naloxone administration and ventilation in awake cats. Brain Res 1983; 258:343–346.
88. Santiago TV, Edelman NH. Opioids and breathing. J Appl Physiol 1985; 59:1675–1685.
89. Pokorski M, Paulev P-E, Szereda-Przestaszewska M. Endogenous benzodiazepine system and regulation of respiration in the cat. Respir Physiol 1994; 87:33–45.
90. Mortola J, Gautier H. Interaction between metabolism and ventilation: Effects of respiratory gases and temperature. In: Dempsey JA, Pack AI, eds. Regulation of Breathing. Vol. 79. 2d ed. New York: Marcel Dekker, 1994:1011–1064.
91. Frappell P, Lanthier C, Baudinette RV, Mortola JP. Metabolism and ventilation in

acute hypoxia: a comparative analysis in small mammalian species. Am J Physiol 1992; 262:R1040–R1046.

92. Dempsey JA, Forster HV, Birnbaum ML, Reddan WG, Thoden J, Grover RF, Rankin J. Control of exercise hyperpnea under varying durations of exposure to moderate hypoxia. Respir Physiol 1972; 16:213–231.
93. Forster HV, Bisgard GE, Klein JP. Effect of peripheral chemoreceptor denervation on acclimatization of goats during hypoxia. J Appl Physiol 1981; 50:392–398.
94. Olson EB, Jr. Physiological dead space increases during initial hours of chronic hypoxemia with or without hypocapnia. J Appl Physiol 1994; 77:1526–1531.
95. Forster HV, Dempsey JA, Birnbaum ML, Reddan WG, Thoden J, Grover RF, Rankin J. Effect of chronic exposure to hypoxia on ventilatory response to CO_2 and hypoxia. J Appl Physiol 1971; 31:586–592.
96. White DP, Gleeson K, Pickett CK, Rannels AM, Cymerman A, Weil JV. Altitude acclimatization: influence on periodic breathing and chemoresponsiveness during sleep. J Appl Physiol 1987; 63:401–412.
97. Schoene RB, Roach RC, Hackett PH, Sutton JR, Cymerman A, Houston CS. Operation Everest II: ventilatory adaptation during gradual decompression to extreme altitude. Med Sci Sports Exerc 1990; 22:804–810.
98. Goldberg SV, Schoene RB, Haynor D, et al. Brain tissue pH and ventilatory acclimatization to high altitude. J Appl Physiol 1992; 72:58–63.
99. Sato M, Severinghaus JW, Powell FL, Xu F-D, Spellman MJ. Augmented hypoxic ventilatory response in men at altitude. J Appl Physiol 1992; 73:101–107.
100. Busch MA, Bisgard GE, Forster HV. Ventilatory acclimatization to hypoxia is not dependent on arterial hypoxemia. J Appl Physiol 1985; 58:1874–1880.
101. Bisgard GE, Busch MA, Forster HV. Ventilatory acclimatization to hypoxia is not dependent on cerebral hypocapnic alkalosis. J Appl Physiol 1986; 60:1011–1015.
102. Weizhen N, Engwall MJA, Daristotle L, Pizarro J, Bisgard GE. Ventilatory effects of prolonged systemic (CNS) hypoxia in awake goats. Respir Physiol 1992; 87:37–48.
103. Nielsen AM, Bisgard GE, Vidruk EH. Carotid chemoreceptor activity during acute and sustained hypoxia in goats. J Appl Physiol 1988; 65:1796–1802.
104. Engwall MJA, Vidruk EH, Nielsen AM, Bisgard GE. Response of the goat carotid body to acute and prolonged hypercapnia. Respir Physiol 1988; 74:335–344.
105. Nurse C. Carotid body adaptation to hypoxia: cellular and molecular mechanisms in vitro. Biol Signals 1995; 4:286–291.
106. Bisgard GE, Busch MA, Daristotle L, Berssenbrugge AD, Forster HV. Carotid body hypercapnia does not elicit ventilatory acclimatization in goats. Respir Physiol 1986; 65:113–125.
107. Lahiri S, Rozanov C, Cherniack NS. Altered structure and function of the carotid body at high altitude and associated chemoreflexes. High Altitude Med Biol 2000; 1:63–74.
108. Niu WZ, Engwall MJA, Bisgard GE. Two discharge patterns of carotid body chemoreceptors in the goat. J Appl Physiol 1990; 69:734–739.
109. Tatsumi K, Pickett CK, Weil JV. Decreased carotid body hypoxic sensitivity in chronic hypoxia: role of dopamine. Respir Physiol 1995; 101:47–57.
110. Czyzyk-Krzeska MF, Bayliss DA, Lawson EE, Milhorn DE. Regulation of tyrosine hydroxlyase gene expression in the rat carotid body by hypoxia. J Neurochem 1992; 58:1538–1546.

111. Wang ZZ, Bredt DS, Fidone SJ, Stensaas LJ. Neurons synthesizing nitric oxide innervate the mammalian carotid body. J Comp Neurol 1993; 336:419–432.
112. Kummer W. Three types of neurochemically defined autonomic fibers innervate the carotid baroreceptor and chemoreceptor regions in the guinea pig. Anat Embryol 1990; 181:477–489.
113. McQueen DS. Pharmacological aspects of putative transmitters in the carotid body. In: Acker J, O'Regan RG, eds. Physiology of the Peripheral Chemoreceptors. Amsterdam: Elsevier, 1983:149–196.
114. Prabhakar NR, Landis SC, Kumar GK, Mullikin-Lilpatrick D, Cherniack NS. Substance P and neurokinin A in the cat carotid body: localization, exogenous effects and changes in content in response to arterial pO_2. Brain Res 1989; 481:205–214.
115. Millhorn DE, Eldridge FL, Waldrop TG. Prolonged stimulation of respiration by endogenous central serotonin. Respir Physiol 1980; 42:171–188.
116. Olson EB, Jr. Ventilatory adaptation to hypoxia occurs in serotonin-depleted rats. Respir Physiol 1987; 69:227–235.
117. Steinman JL, Carlton SM, Haber B, Willis WD. Differential effects of *p*-chlorphenylalanine on indoleamines in brainstem nuclei and spinal cord of rats. Brain Res 1987; 426:297–309.
118. Carlton SM, Steinman JL, Hillman GR, Willis WD. Differential effects of *p*-chlorophenylalanine on indoleamines in brainstem nuclei and spinal cord of rats. II. Identification of immunohistochemically stained structures using computer-assisted image enhancement techniques. Brain Res 1987; 426:310–322.
119. Ogawa H, Mizusawa A, Kikuchi Y, Hida W, Miki H, Shirato K. Nitric oxide as a retrograde messenger in the nucleus tractus solitarii of rats during hypoxia. J Physiol 1995; 486:495–504.
120. Severinghaus JW, Mitchell RA, Richardson BW, Singer MM. A respiratory control at high altitude suggesting active transport regulation of CSF pH. J Appl Physiol 1963; 18:1155–1166.
121. Crawford RD, Severinghaus JW. CSF pH and ventilatory acclimatization to altitude. J Appl Physiol 1978; 45:275–283.
122. Dempsey JA, Forster HV, DoPico GA. Ventilatory acclimatization to moderate hypoxemia in man. The role of spinal fluid (H^+). J Clin Invest 1974; 53:1091–1100.
123. Forster HV, Soto JR, Dempsey JA, Hosko MJ. Effect of sojourn at 4,300 m altitude on electroencephalogram and visual evoked response. J Appl Physiol 1975; 39:109–113.
124. Weiskopf RB, Gabel RA, Fencl V. Alkaline shift in lumbar and intracranial CSF in man after 5 days at high altitude. J Appl Physiol 1976; 41:93–97.
125. Bureau M, Bouverot P. Blood and CSF acid-base changes, and rate of ventilatory acclimatization of awake dogs to 3,550 m. Respir Physiol 1975; 24:203–216.
126. Orr JA, Bisgard GE, Forster HV, Buss DD, Dempsey JA, Will JA. Cerebrospinal fluid alkalosis during high-altitude sojourn in unanesthetized ponies. Respir Physiol 1975; 25:23–37.
127. Dempsey JA, Forster HV, Bisgard GE, Chosy LW, Hanson PG, Kiorpes AL, Pelligrino DA. Role of cerebrospinal fluid [H^+] in ventilatory deacclimatization from chronic hypoxia. J Clin Invest 1979; 64:199–205.
128. Fencl V, Miller TB, Pappenheimer JR. Studies on the respiratory response to disturbances of acid-base balance, with deductions concerning ionic composition of cerebral interstitial fluid. Am J Physiol 1966; 210:459–472.

129. Pappenheimer JR. The ionic composition of cerebral extracellular fluid and its relation to control of breathing. Harvey Lect 1967; 6:71–93.
130. Dempsey JA, Forster HV, Chosy LW, Hanson PG, Reddan WG. Regulation of CSF[HCO_3^-] during long-term hypoxic hypocapnia in man. J Appl Physiol 1978; 44: 175–182.
131. Jennings DB. The physicochemistry of [H+] and respiratory control: roles of P_{CO_2}, strong ions, and their hormonal regulators. Can J Physiol Pharmacol 1994; 72:1499–1512.
132. Stewart PA. How to Understand Acid-Base. A Quantitative Acid-Base Primer for Biology and Medicine. New York: Elsevier, 1981.
133. Nattie EE. Ionic mechanisms of cerebrospinal fluid acid-base regulation. J Appl Physiol 1983; 54:3–12.
134. Engwall MJA, Bisgard GE. Ventilatory responses to chemoreceptor stimulation after hypoxic acclimatization in awake goats. J Appl Physiol 1990; 69:1236–1243.
135. Ryan ML, Hedrick MS, Pizarro J, Bisgard GE. Carotid body noradrenergic sensitivity in ventilatory acclimatization to hypoxia. Respir Physiol 1993; 92:77–90.
136. Bisgard GE. The effect of prolonged stimulation of afferent activity of the goat carotid body. In: Acker H, Trzebski A, O'Regan R, eds. Chemoreceptors and Chemoreceptor Reflexes. New York: Plenum Press, 1990:165–170.
137. Vizek M, Pickett CK, Weil JV. Biphasic ventilatory response of adult cats to sustained hypoxia has central origin. J Appl Physiol 1987; 63:1658–1664.
138. Dempsey JA, Thomson JM, Alexander SC, Forster HV, Chosy LW. Respiratory influences on acid-base status and their effects on O_2 transport during prolonged muscular work. In: Howald H, Poortmans JR, eds. Adaptations to Prolonged Exercise. Basel: Birkhauser Verlag, 1975:56–64.
139. Milledge JS, Lahiri S. Respiratory control in lowlanders and Sherpa highlanders at altitude. Respir Physiol 1967; 2:310–322.
140. Lahiri S, Maret KH, Sherpa MG, Peters RMJ. Sleep and periodic breathing at high altitude: Sherpa natives vs. sojourners. In: West J, Lahiri S, eds. High Altitude and Man. Bethesda: American Physiological Society, 1984:73–90.
141. Dempsey JA, Reddan WG, Birnbaum ML, Forster HV, Thoden JS, Grover RF, Rankin J. Effects of acute through life-long hypoxic exposure on exercise pulmonary gas exchange. Respir Physiol 1971; 13:62–89.
142. Babcock M, Johnson BD, Pegelow DF, Suman OE, Griffin D, Dempsey JA. Hypoxic effects on exercise-induced diaphragmatic fatigue in normal healthy humans. J Appl Physiol 1995; 78:82–92.
143. Beall CM, Strohl KP, Biangero J, Williams-Blangero S, Almasy LA, Decker MJ, Worthman CM, Goldstein MC, Vargas E, Villena M, Soria R, Alarcon AM, Gonzales C. Ventilation and hypoxic ventilatory response of Tibetan and Aymara high altitude natives. Am J Phys Anthropol 1997; 104:427–447.
144. Brooks JGI, Tenney SM. Ventilatory response of llama to hypoxia at sea level and high altitude. Respir Physiol 1968; 5:269–278.
145. Lahiri S. Unattenuated ventilatory hypoxic drive in ovine and bovine species native to high altitude. J Appl Physiol 1972; 32:95–102.
146. Bisgard GE, Ruiz AV, Grover RF, Will JA. Ventilatory acclimatization to 3400 meters altitude in the Hereford calf. Respir Physiol 1974; 21:271–296.
147. Okubo S, Mortola J. Control of ventilation in adult rats hypoxic in the neonatal period. Am J Physiol 1990; 259:R836–R841.

148. Tatsumi K, Pickett CK, Weil JV. Attenuated carotid body hypoxic sensitivity after prolonged hypoxic exposure. J Appl Physiol 1991; 70:748–755.
149. Forster HV, Dempsey JA, Vidruk EH, DoPico G. Evidence of altered regulation of ventilation during exposure to hypoxia. Respir Physiol 1974; 20:379–392.
150. Pokorski M, Lahiri S. Effects of naloxone on carotid body chemoreception and ventilation in the cat. J Appl Physiol 1981; 51:1533–1538.
151. Tenney SM, Ou LC. Hypoxic ventilatory response of cats at high altitude: an interpretation of 'blunting'. Respir Physiol 1977; 30:185–199.
152. Dempsey JA, Gledhill N, Reddan W, Forster HV, Hanson PG, Claremont AD. Pulmonary adaptation to exercise: effects of exercise type and duration, chronic hypoxia and physical training. Ann NY Acad Sci 1977; 301:243–261.
153. Thoden JS, Dempsey JA, Reddan WG, Birnbaum ML, Forster HV, Grover RF, Rankin J. Ventilatory work during steady-state response to exercise. Fed Proc 1969; 28:1316–1321.
154. Johnson BD, Babcock M, Suman O, Dempsey JA. Exercise-induced diaphragmatic fatigue in healthy humans. J Physiol 1993; 460:385–405.
155. Fregosi R, Dempsey JA. The effects of exercise in normoxia and acute hypoxia on respiratory muscle metabolites. J Appl Physiol 1986; 60:11274–11283.
156. Aaron EA, Johnson BD, Seow KC, Dempsey JA. Oxygen cost of exercise hyperpnea: measurement. J Appl Physiol 1992; 72:1810–1817.
157. Harms CA, Babcock MA, McClaran SR, Pegelow DF, Nickele GA, Nelson WB, Dempsey JA. Respiratory muscle work compromises leg blood flow during maximal exercise. J Appl Physiol 1997; 82:1573–1583.
158. Cerny FC, Dempsey JA, Reddan WG. Pulmonary gas exchange in nonnative residents of high altitude. J Clin Invest 1973; 52:2993–2999.
159. Johnson RLJ, Cassidy SC, Grover RF, Schutte JE, Epstein RH. Functional capacities of lungs and thorax in beagles after prolonged residence at 3,100 m. J Appl Physiol 1985; 59:1773–1782.
160. Schoene RB. Control of ventilation in climbers to extreme altitude. J Appl Physiol 1982; 53:886–890.
161. Weil JV, Byrne-Quinn E, Sodal IE, et al. Hypoxic ventilatory drive in normal man. J Clin Invest 1970; 49:1061–1072.
162. Collins DD, Scoggin CH, Zwillich CW, Weil JV. Hereditary aspects of decreased hypoxic response. J Clin Invest 1978; 62:105–110.
163. Kawakami Y, Yoshikawa T, Shida A, Asanuma Y, Murao M. Control of breathing in young twins. J Appl Physiol 1982; 52:537–542.
164. Robbins PA. Hypoxic ventilatory decline: site of action. J Appl Physiol 1995; 79:373–374.
165. Lugliani R, Whipp BJ, Sears C, Wasserman K. Effect of bilateral carotid-body resection on ventilatory control at rest and during exercise in man. N Engl J Med 1971; 285:1105–1111.
166. Martin BJ, Sparks KE, Zwillich CW, Weil JV. Low exercise ventilation in endurance athletes. Med Sci Sport Exerc 1979; 11:181–185.
167. Martin B, Weil JV, Sparks K, McCullough RE, Grover RF. Chemical drives to breathe as determinants of exercise ventilation. Chest 1978; 73:283.
168. Martin BJ, Weil JV, Sparks KE, McCullough RE, Grover RF. Exercise ventilation correlates positively with ventilatory chemoresponsiveness. J Appl Physiol 1978; 45: 557–564.

169. Schoene RB, Robertson HT, Pierson DJ, Peterson AP. Respiratory drives and exercise in menstrual cycles of athletic and nonathletic women. J Appl Physiol 1981; 50:1300–1305.
170. Kojima A, Masuyama A, Hasako K, Kouchiyama S, Shinozaki T, Tatsumi K, Kimura K, Houda Y. Japanese high altitude climbers in Mt. Everest Expedition revealed higher hypoxic ventilatory drive than the others (abstr). In: Sutton JR, Coates G, Remmers JE, eds. Hypoxia: The Adaptations. Philadelphia: BC Dekker, 1990:293.
171. Masuyama S, Kimura H, Sugita T, Kuriyama T, Tatsumi K, Kunitomo F, Okita S, Tojima H, Yuguchi Y, Watanabe S, et al. Control of ventilation in extreme-altitude climbers. J Appl Physiol 1986; 61:500–506.
172. Schoene RB, Lahiri S, Hackett PH, Peters RM, Milledge JS, Pizzo CJ, Sarnquist FH, Boyer SJ, Graber DJ, Maret KH, et al. Relationship of hypoxic ventilatory response to exercise performance on Mount Everest. J Appl Physiol 1984; 56:1478–1483.
173. Schoene RB, Hackett PH, Roach RC. Blunted hypoxic chemosensitivity at altitude and sea level in an elite high altitude climber. In: Sutton JR, Houston CS, Coates G, eds. Hypoxia and Cold. New York: Praeger, 1987:532.
174. Reeves JT, McCullough RE, Moore LG, Cymerman A, Weil JV. Sea-level P_{CO_2} relates to ventilatory acclimatization at 4,300 m. J Appl Physiol 1993; 75:1117–1122.
175. Anholm JD, Houston CS, Hyers TM. The relationship between acute mountain sickness and pulmonary ventilation at 2,835 meters (9,300 feet). Chest 1979; 75:33–36.
176. King AB, Robinson SM. Ventilation response to hypoxia and acute mountain sickness. Aerosp Med 1972; 43:419–421.
177. Mathew L, Gopinathan PM, Purkayastha SS, Sen Gupta J. Chemoreceptor sensitivity and maladaptation to high altitude in man. Eur J Appl Physiol 1983; 51:137–144.
178. Matsuzawa Y, Fujimoto K, Kobayashi K, et al. Blunted hypoxic ventilatory drive in subjects susceptible to high-altitude pulmonary edema. J Appl Physiol 1989; 66:1152–1157.
179. Moore LG, Harrison GL, McCullough RE, McCullough RG, Micco AJ, Tucker A, Weil JV, Reeves JT. Low acute hypoxic ventilatory response and hypoxic depression in acute altitude sickness. J Appl Physiol 1986; 60:1407–1412.
180. Bartsch P, Paul A, McCullough RE, Kucherer H, Hohenhaus E. Hypoxic ventilatory response and hypoxic pulmonary vascular response in HAPE-susceptible subjects. In: Sutton JR, Houston CS, Coates G, eds. Hypoxia and the Brain. Burlington: Queen City Press, 1995:265–270.
181. Hackett PH, Roach RC, Schoene RB, Harrison GL, Mills WJ, Jr. Abnormal control of ventilation in high-altitude pulmonary edema. J Appl Physiol 1988; 64:1268–1272.
182. Hohenhaus E, Paul A, McCullough RE, Kucherer H, Bartsch P. Ventilatory and pulmonary vascular response to hypoxia and susceptibility to high altitude pulmonary oedema. Eur Respir J 1995; 8:1825–1833.
183. Matsuzawa Y, Kobayashi T, Fujimoto K, Kubo K, Sekiguchi M. Hypoxic ventilatory response and pulmonary gas exchange during exposure to high altitude in subjects susceptible to high altitude pulmonary edema (HAPE). In: Sutton JR, Houston CS, Coates G, eds. Hypoxia and the Brain. Burlington: Queen City Press, 1995:331.
184. Hackett PH, Rennie ID. Fluid retention and relative hypoventilation in acute mountain sickness. Respiration 1982; 43:321–329.
185. Honig A. Peripheral arterial chemoreceptors and reflex control of sodium and water homeostasis. Am J Physiol 1989; 257:R1282–R1302.
186. Swenson ER, Duncan TB, Goldberg SV, Ramirez G, Ahmad S, Schoene RB. Diuretic

effect of acute hypoxia in humans: relationship to hypoxic ventilatory responsiveness and renal hormones. J Appl Physiol 1995; 78:377–383.

187. Hornbein TF, Townes BD, Schoene RB, Sutton JR, Houston CS. The cost to the central nervous system of climbing to extremely high altitude. N Engl J Med 1989; 321:1714–1719.

188. Masuyama S, Kohchiyama S, Shinozaki T, Okita S, Kunitomo F, Tojima H, Kimura H, Kuriyama T, Honda Y. Periodic breathing at high altitude and ventilatory responses to O_2 and CO_2. Jpn J Physiol 1989; 52:523–535.

189. Raff H, Tzankoff SP, Fitzgerald RS. Chemoreceptor involvement in cortisol responses to hypoxia in ventilated dogs. J Appl Physiol 1982; 52:1092–1096.

190. Takano N. Changes in ventilation and ventilatory response to hypoxia during the menstrual cycle. Pflugers Arch 1984; 402:312–316.

191. White DP, Douglas NJ, Pickett CK, Weil JV, Zwillich CW. Sexual influence on the control of breathing. J Appl Physiol 1983; 54:874–879.

192. Tatsumi K, Hannhart B, Pickett CK, Weil JV, Moore LG. Influences of gender and sex hormones on hypoxic ventilatory response in cats. J Appl Physiol 1991; 71:1746–1751.

7

Mechanics of Breathing

JOSEPH MILIC-EMILI

McGill University
Montreal, Quebec, Canada

BENGT KAYSER

University of Geneva
Geneva, Switzerland

HENRY GAUTIER

Atelier de Physiologie Respiratoire
Faculté de Médecine Saint-Antoine
Paris, France

I. Introduction

Ventilation is higher at altitude than at sea level (Table 1) (1,2). Reinhold Messner, describing his and Peter Habeler's approach to the summit of Mount Everest, the first to climb Mount Everest without supplementary bottled oxygen, stated: "Breathing becomes such a strenuous business that we scarcely have strength to go on," and upon reaching the summit: "I am nothing more than a single, narrow, gasping lung, floating over the mists and the summits" (3). Expressions like these from many other mountaineers venturing to high altitude suggest the possibility that ventilation might limit exercise at least at extremely high altitude. In this chapter we provide an overview of how respiratory mechanics at high altitude may affect the work of breathing and exercise performance (see also Chapter 20). We will focus mainly on research performed on humans.

Altitude has both a direct effect on respiratory mechanics because of decreased density of ambient air and an indirect effect linked to hypoxic responses. The lower density should lead to a decrease in airway resistance and dynamic work of breathing together with increased maximal inspiratory and expiratory flows. On the other hand, hypoxia may affect the static lung volumes through increased intrathoracic blood

Table 1 Ventilation at Rest and During Exercise at Different Barometric Pressures (PB) in a Decompression Chamber (Operation Everest II)

PB (mmHg)	Altitude (m)	PIO_2 (mmHg)	VE (L/min) Rest[a]	VE (L/min) 120 watts[b]
760	0	149	11	48
429	4,800	80	15	72
347	6,300	63	21	92
282	8,100	49	37	162
252	8,848	42	42	184

PIO_2 = calculated inspired partial pressure of O_2; VE = minute ventilation corrected to BTPS.
[a] From Ref. 1.
[b] From Ref. 2.

volume and pulmonary edema, and pulmonary dynamics through changes in bronchomotor tone. The latter changes depend on (1) time at altitude—acute exposure (<24 hours), short-term acclimatization (days to weeks), and long-term acclimatization (years to generations)—and (2) the population studied—subjects born and living at low altitude (lowlanders, most commonly Caucasians) or subjects born and living at high altitude (highlanders, commonly Indians of South America and Tibetans of the Himalayas).

In lowlanders, the changes in respiratory mechanics with altitude have been assessed in field studies carried out on mountains as well as in simulated altitude experiments in which the effect of low barometric pressure (PB) per se was studied in decompression chambers (hypobaric hypoxia). In order to evaluate the effects of hypoxia per se without possible side effects of reduced PB, gas mixtures low in oxygen have also been administered at sea level (normobaric hypoxia). Similarly, at reduced PB (both on mountains and in decompression chambers), the effects of oxygen administration have also been investigated, but only for relatively short periods of O_2 administration. In a few acute and short-term experiments highlanders were studied at sea level.

II. Lung Volumes

First we will describe the effects of altitude on the static subdivisions of lung volume, and next those on ventilatory capacity.

A. Static Volumes

Acute Exposure (<24 Hours)

Decompression Chamber

According to Bert (4), the first measurements of the vital capacity (VC) during acute exposure to hypoxia were made by von Vivenot in 1868. In a low-pressure chamber at PB of 424 mmHg (corresponding to an altitude of 4650 m), the vital capacity of two subjects was reduced by 9 and 13%. Bert found that his own vital capacity was reduced by 32% at PB of 430 mmHg. In 1932, Schneider (5) reported that 24 subjects exposed to a simulated altitude of 6100 m had an average 8% decrease in VC. He stated that up to about 3000 m of simulated altitude there was no appreciable change in VC, and that at 6100 m the reduction in VC was virtually abolished by pure oxygen breathing. In 1934, Hurtado et al. (6) reported that in three subjects exposed to a simulated altitude of 5,000 m the vital capacity decreased on average by 9%. They also found that the residual volume (RV), measured with the nitrogen washout method, increased by 45%. As a result the total lung capacity (TLC) was essentially unchanged. In all these early studies, the vital capacity values were not corrected to BTPS conditions. Such corrections were made by Rahn and Hammond (7) in a systematic study on 18 subjects exposed to 3050, 4300, and 5500 m. Like Schneider (5), they found no significant change in VC at 3050 m with a small but significant reduction, averaging 2.4 and 3.8% at 4300 and 5500 m, respectively. Without correction for BTPS conditions, the change in VC at 5500 m relative to sea level would be about 5% greater, for a spirometer temperature of 20°C. Rahn and Hammond (7) also acutely exposed four subjects to very low barometric pressures, ranging from 349 to 141 mmHg. Because the subjects were not acclimatized to high altitude, in these experiments the inspired oxygen tension was in all instances kept higher than 94 mmHg. At PB of 349 mmHg, corresponding to 6100 m, the VC was the same as at sea level, in line with the previous observations of Schneider (5). At lower PB, however, the VC was less than at sea level, the reduction averaging 4% at PB of 226 mmHg and 7% at PB of 141 mmHg. In eight normal subjects breathing pure oxygen exposed to PB of 380 and 190 mmHg, Finkelstein et al. (8) found a significant reduction of VC averaging 5 and 15%, respectively. These changes were higher than those found by Rahn and Hammond (7) at similar altitudes and were significant at PB of 380 mmHg.

After 4 hours at simulated altitude of 4900 m, Gray et al. (9) found a 7% decrease in vital capacity in five subjects, whereas after 5 hours at 4300 m Coates et al. (10) did not find any significant change in the VC of four subjects but a 78% increase in RV measured with the helium dilution method. As a result, TLC was increased by 21%. They ascribed the rise in RV to increased gas trapping. Indeed, they found a concomitant increase in closing capacity [(phase IV + RV)/TLC], which was due almost entirely to the increase in residual volume. By contrast, the closing volume (phase IV, % VC) did not change in line with the results of Gray et al. (9), who found no significant changes in either the phase IV volume or the closing volume. The increase in TLC observed by Coates et al. (10) was surprisingly large as compared to the 5% increase reported by Saunders et al. (11) during acute normobaric hypoxia simulating an altitude of about 4300 m (see below).

Normobaric Hypoxia

To nine of the subjects who had previously served in the decompression chamber experiments (see above), Rahn and Hammond (7) administered 14.2, 11.8, and 9.9% oxygen at sea level. Only with 9.9% oxygen did they find a significant reduction in VC (1.4%), which was smaller than the 3.8% found in the same subjects with the same inspired P_{O_2} in the decompression chamber. By contrast, with a Pa_{O_2} reduced to 40–50 mmHg, such as found at an altitude of about 4300 m, and a Pa_{CO_2} maintained at 38–40 mmHg, Saunders et al. (11), using a body plethysmograph, found no change in VC, whereas the TLC, FRC, and RV increased significantly by 5, 9, and 31%, respectively. All of these changes were reversed within 3 minutes of pure O_2 breathing. They suggested that the rise in TLC and FRC could be explained by the concomitant loss of elastic recoil of the lungs. By contrast, when lung volumes were measured by inert gas dilution, with a degree of hypoxia similar to that used by Saunders et al. (11), Goldstein et al. (12) found no significant changes in TLC, VC, closing volume, and closing capacity. They suggested that the changes in TLC reported by Saunders et al. (11) were due to artefacts inherent in plethysmography. It should be noted, however, that by helium dilution method Coates et al. (10) also found a significant increase in TLC at simulated altitude of 4300 m in a decompression chamber. In line with Saunders et al. (11), an average increase in FRC of 14%, also measured with a body plethysmograph, has been also reported by Garfinkel and Fitzgerald (13) in 43 subjects breathing 11% oxygen at sea level.

In short, the scanty literature dealing with the effects of acute simulated altitude exposure on RV, FRC, TLC, and closing capacity has yielded conflicting results, and further systematic studies are needed. In contrast, according to most studies there is a small but significant reduction in VC when humans are acutely exposed to equivalent altitudes higher than about 3000 m ($P_B < 525$ mmHg). At altitudes between 3000 and 5000 m, this reduction in VC is completely reversed by oxygen inhalation, while at higher altitudes the reversal is only partial. The latter phenomenon may be related in part to expansion of the gases below the diaphragm. Indeed, Rahn and Hammond (7) noted that in many of their subjects the forced expiration at very high simulated altitudes elicited abdominal pain. The observation that, at the same P_B, the reduction in VC is greater with hypobaric hypoxia than with hypobaria alone has been attributed to engorgement of pulmonary blood vessels and decreased respiratory muscle force under hypoxic conditions (7).

Short-Term Exposure (Days to Weeks)

Decompression Chamber

Only two studies have been concerned with short-term exposure to altitude in a decompression chamber. In the study carried out during Operation Everest II by

Welsh et al. (14), subjects resided for 40 days with PB decreasing progressively to 240 mmHg, corresponding to an altitude of 8844 m. The VC began to decrease significantly at PB lower than 429 mmHg (equivalent altitude of 4572 m), diminishing by 14% at PB of 240 mmHg (Fig. 1). Unfortunately, in these experiments the effect of oxygen administration on VC was not assessed. However, Ulvedal et al. (15) studied subjects breathing enriched-oxygen mixtures ($P_{IO_2} > 160$ mmHg) at PB of 380, 260, and 190 mmHg for 30, 14, and 17 days, respectively. They found a significant and sustained 3% decrease in VC at PB of 380 and 260 mmHg and 7% at PB of 190 mmHg. These changes were smaller than those reported by Welsh et al. (14).

Field Studies

Mosso in 1897 (16) was the first to study the effects of a short-term sojourn on mountains on VC. On Monte Rosa (4560 m), an average decrease of 10% in VC (not corrected to BTPS) was observed in eight subjects. In 1932 on Pike's Peak (4300 m), Schneider (5) found a similar decrease in VC, which ranged from 7 to 15% (also not corrected to BTPS) on the first day of residence at altitude with a tendency to a return to sea level values in the following days. At the same altitude,

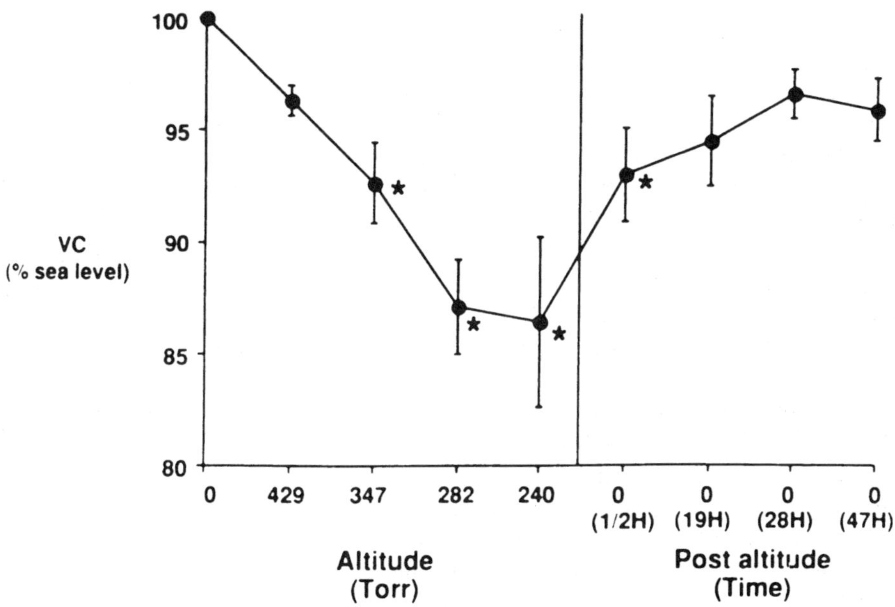

Figure 1 Vital capacity, VC (mean ± SEM), decreased significantly (* $p < 0.05$ from sea level) and progressively with chronic altitude exposure (6096–8844 m). There was rapid partial recovery 30 minute after descent ($n = 7$ at 0, 429, and 347 torr and for all postaltitude measurements; $n = 6$ at 282 torr; $n = 4$ at at 240 torr). (From Ref. 14.)

a transient decrease in VC (3–4% corrected to BTPS) has also been found in several subsequent studies (7,17–20). The decrease was partially reversed when pure oxygen was breathed for 15 minutes (5), whereas when it was administered for only 3 minutes no significant changes were observed (20).

In four subjects studied at 4300 m, Tenney et al. (20), using the N_2 washout method, found a 10% increase of FRC and RV on days 3–7 of altitude sojourn. The corresponding changes were somewhat greater on days 1–3, but there was marked variability in response between the four subjects. Using the same method, after a 3-day sojourn at the same altitude, Cruz (17) found no significant change in FRC in 6 subjects, while Jaeger et al. (18) found a significant 10% increase in RV in 17 individuals. After a 30-day sojourn at 5366 m, Mansell et al. (21), using the helium dilution method, found a significant increase in TLC (18%), reflecting entirely a large increase in RV (78%), while VC did not change significantly.

During short-term sojourns at altitudes lower than 4000 m, the changes in static lung volumes are usually small, if any. Indeed, no significant changes in VC, FRC, and TLC were observed after a 1- to 6-week sojourn at either 3100 m (22,23) or 3660 m (24). Using a body plethysmograph, Gautier et al. (25) made daily measurements of RV, FRC, TLC, and VC in nine subjects during a 6-day sojourn at an altitude of 3460 m. In comparison to sea level values, there was no change in either TLC or RV, while there was a small but significant drop in FRC only on the first day at altitude (Fig. 2). Small but significant decreases in VC were also found on the second and third days at altitude. These changes in VC were reversed by breathing an oxygen-enriched mixture (P_{IO_2} = 150–160 mmHg).

According to the above results, during short-term altitude exposure, a significant decrease in VC is found only above 4500 m, while during acute exposure it may be observed at lower altitudes. The reason for this discrepancy is not clear because the mechanisms that may contribute to the reduction in VC during short-term altitude exposure are essentially the same as those pertaining to acute exposure, as elegantly reviewed by Rahn and Hammond (7). These may include decreased respiratory muscle force as well as an increase in thoracic blood volume and interstitial fluid associated with hypoxia (26). The latter may explain the observation that after return from high to low altitude, it takes several days to restore the VC to its sea level value, as shown in Figure 1.

At altitudes lower than 4000 m, there are no significant sustained changes in RV, FRC, and TLC. At higher altitudes, however, all of these volumes increase. According to Mansell et al. (21), the increase in RV may reflect an increase in closing capacity due to interstitial lung edema and/or loss of lung recoil. By contrast, the increase in FRC and TLC is probably due to the fact that at altitudes higher than 4000 m the effect of increased thoracic blood volume is more than counterbalanced by the concomitant loss in lung elastic recoil (see Sec. III.A).

Long-Term Exposure (Years to Generations)

The first systematic study of subjects born and living at high altitude was made by Hurtado (27), who showed that young Indians born in Morococha (4540 m) had a

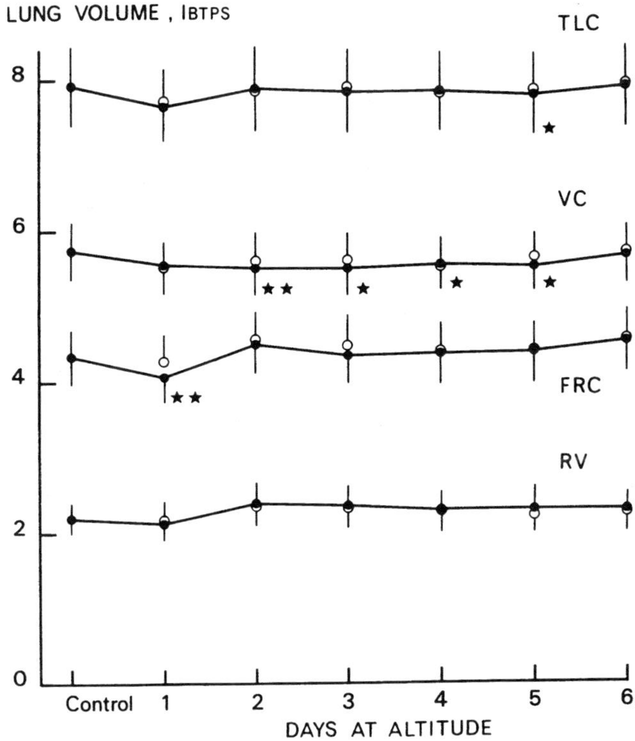

Figure 2 Average values (±SEM) of lung volumes at sea level (control) and during a 6-day sojourn at 3460 m of nine subjects (filled circles). Breathing ambient air (P_{IO_2} = 93 mmHg) (open circles). Breathing a mixture with P_{IO_2} = 150–160 mmHg. TLC, total lung capacity; VC, vital capacity; FRC, functional residual capacity; RV, residual volume. Levels of significance between altitude and sea level values: * $p < 0.05$; ** $p < 0.01$. (From Ref. 25.)

38% larger RV than individuals of the same ethnicity born and living at sea level. The vital capacity was approximately the same in both groups, and, as a result, the TLC was larger in the highland natives. In similar studies carried out at 4350 m, a greater FRC was also found in highland natives (17). Other studies carried out in the Andes (3850 m) have shown that the VC is larger in high-altitude natives than in sea level dwellers of the same ethnic origin (28,29).

In Nepalese Sherpas born and living above 3340 m and studied at 4243 m, the vital capacity was 12% higher than that predicted for sea level dwelling Caucasians (30). At altitudes of 3500–4500 m in eastern Kashmir (31) and at 3660 m in Lhasa (32), the VC of the highland natives was found to be higher than that of lowland natives acclimatized to those altitudes. Similarly, a relatively high VC was also found in Peruvian and Bolivian native children born and living at high altitude (4270 and 3600 m, respectively) (33,34).

From studies carried out in the Andes, it appears that the increased lung volume of highland natives is mainly the result of adaptations to hypoxia occurring during growth. Indeed, after the age of 11 years, the VC of Peruvian boys living at an altitude of 4270 m is higher than in Peruvian boys living at a lower altitude or at sea level (33). Furthermore, Peruvian natives born at sea level but acclimatized to high altitude as adults exhibit smaller VC than Peruvian natives acclimatized during growth or born and living at altitude (35). The above observations are in agreement with results obtained in rats. After prolonged exposure to altitude (3540 m), at a given distending pressure the lungs of young rats were larger than those of control rats kept at sea level. In contrast, under similar experimental conditions adult rats did not show any change in lung volume (36). Interestingly, Sekhon et al. (37) have shown that in growing rats exposed to hypobaric hypoxia, the lung volumes were slightly but significantly larger than in growing rats subjected to normobaric hypoxia. This suggests that decreased P_B per se may have a small growth-promoting effect on lung volume.

The effects of descent of high altitude natives to sea level on lung volumes have been investigated in Andean Indians. In Bolivian highlanders born and living above 3500 m, a 1-week sojourn at low altitude (420 m) did not induce any change in TLC or its components (38), a finding replicated in Peruvian natives of 3500–4500 m who, after 2 and 37 days at sea level, did not change their TLC or its subdivisions, which remained larger than predicted for sea level dwellers (39).

In contrast to non-Caucasian natives, Caucasians living at high altitude do not appear to have an increased VC. In Leadville (3100 m), 126 normal Caucasians, aged 18–61 years, were studied to establish the normal predicted values for this altitude. The results indicate that (1) the regression equation for VC was the same for the subjects who were born at altitude and the subjects who had moved to altitude as adults and (2) the predicted values at Leadville were virtually identical to those predicted from sea level regressions (40). In lifelong Leadville residents the VC was also found to be within the range of the predicted values for lowlanders, while the TLC was significantly larger (23). In children of European ancestry born and living at 3600 m, the VC was also found to be within the normal range of sea level Caucasian children (34).

B. Dynamic Volumes and Maximal Flows

In 1946, Rahn et al. (41) pointed out that the effective area of the maximal static volume-pressure (V-P) diagram, and hence the maximal potential work available per breathing cycle, decreases with reduced barometric pressure because of compressibility of gas (Fig. 3). This, however, is more than compensated for by a decrease in airway resistance resulting from the lower density of air and possibly also from bronchodilatation (see Sec. III.B), so that the maximal voluntary ventilation (MVV) actually increases at altitude.

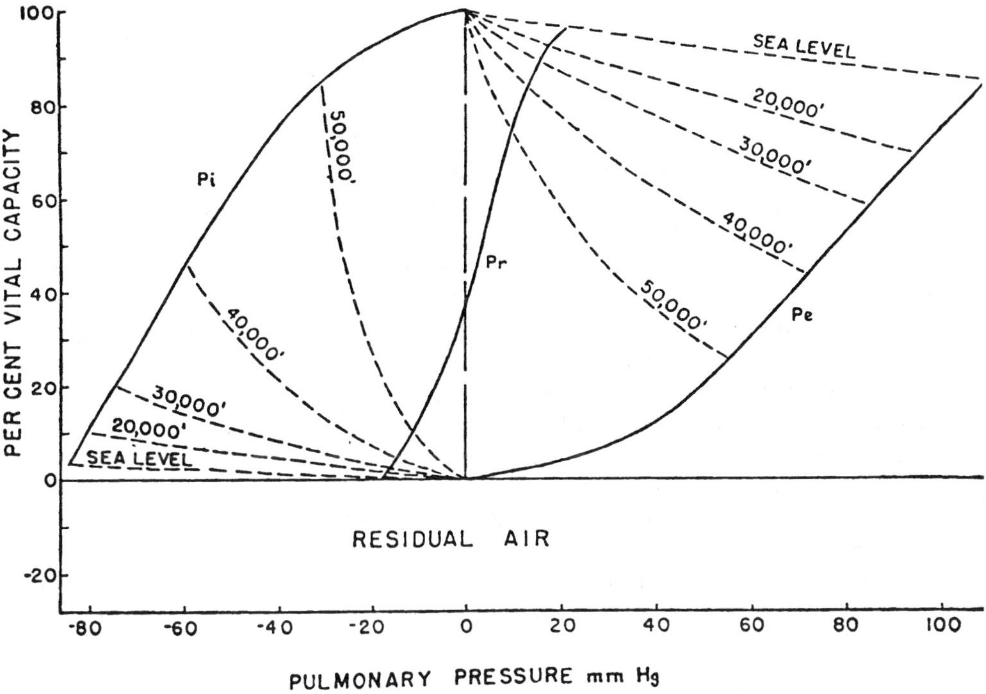

Figure 3 The effect of altitude on the dynamic expiratory and inspiratory pressures that can be exerted starting, respectively, at maximum inspiration and maximum expiration (broken lines). Pi and Pe, maximal static inspiratory and expiratory pressures. Pr, relaxation volume-pressure curve of total respiratory system. (From Ref. 41.)

In 1954, Cotes (42) measured the MVV in a decompression chamber at simulated altitudes of 3050, 5200, and 8200 m (acute exposure) in a group of subjects breathing an oxygen-enriched mixture. Compared to sea level, the MVV increased by 13, 24, and 31%, respectively. In similar acute experiments carried out at equivalent altitudes of 5500 and 10300 m, Finkelstein et al. (8) found an increase in MVV of 15 and 24%, respectively. Ulvedal et al. (15), also in a decompression chamber, measured MVV in subjects breathing oxygen-enriched mixtures during short-term exposure to a simulated altitude of 10,200 m (hypobaric normoxia). Compared to sea level, on days 3 to 17, the MVV rose on average by 37%. The above authors concluded that the main factor influencing the MVV during acute and short-term exposure to altitude is the reduction in air density, not hypoxia per se.

In 1958, Pugh (43) found that, in three subjects breathing ambient air during a sojourn on a high mountain (6400 m), the MVV was about 20% higher than at lower altitude (2100 m). The increase in MVV at altitude was confirmed in several field studies carried out at various altitudes (3700–4350 m) on a larger number of individuals during sojourns ranging from 3 to 65 days (17,19,24,44).

Figure 4 depicts the average increase in MVV, relative to sea level, found at different altitudes in the various field studies as well as in the decompression chamber. There are no consistent differences between the results obtained in the field and decompression chamber studies although hypoxia was present only in the former. Also see in Figure 4 is the substantial scatter of the data. This is not surprising because the changes in MVV with altitude depend on many factors apart from gas density (force and speed of activation of both inspiratory and expiratory muscles, thoracic gas compressibility and mechanical properties of the respiratory system, respiratory frequency, etc). By contrast, the effects of decreased air density on maximal expiratory flows ($\dot{V}$max) can readily be predicted. An increase in maximal expiratory flows at altitude has been reported in several studies in which a body plethysmograph was not used (8,14,19,21,29). However, because of artefacts due to gas compressibility, the use of a body plethysmograph is mandatory for measurements of $\dot{V}$max at altitude, and hence the above results are difficult to interpret (41,45). Figure 5 shows the maximal expiratory flows obtained by Gautier et al. (25) using a body plethysmograph in nine subjects during a 6-day sojourn at an altitude of 3460 m. Peak flow (PF) increased significantly on the first day at altitude and remained at about the same level thereafter. The maximal flows at 50 and 75% VC were signifi-

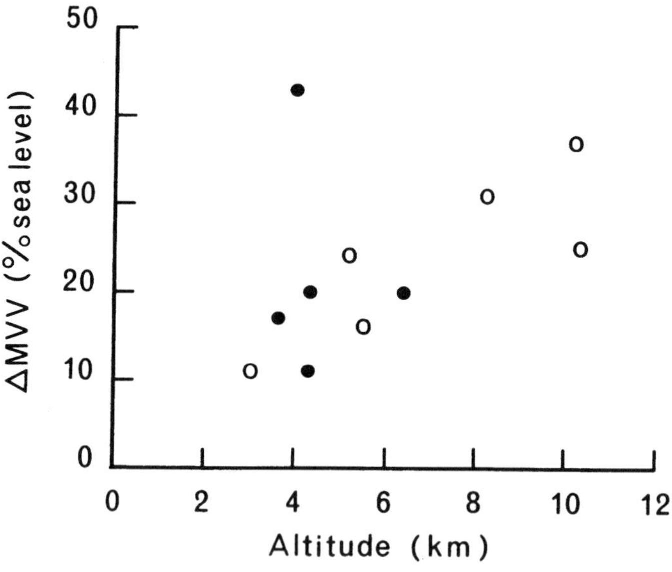

Figure 4 Percent change of maximal voluntary ventilation relative to sea level (ΔMVV, %) found at different altitudes by various authors in field studies (filled circles) and decompression chamber (open circles). (From Refs. 8,15,17,19,24,42–44.)

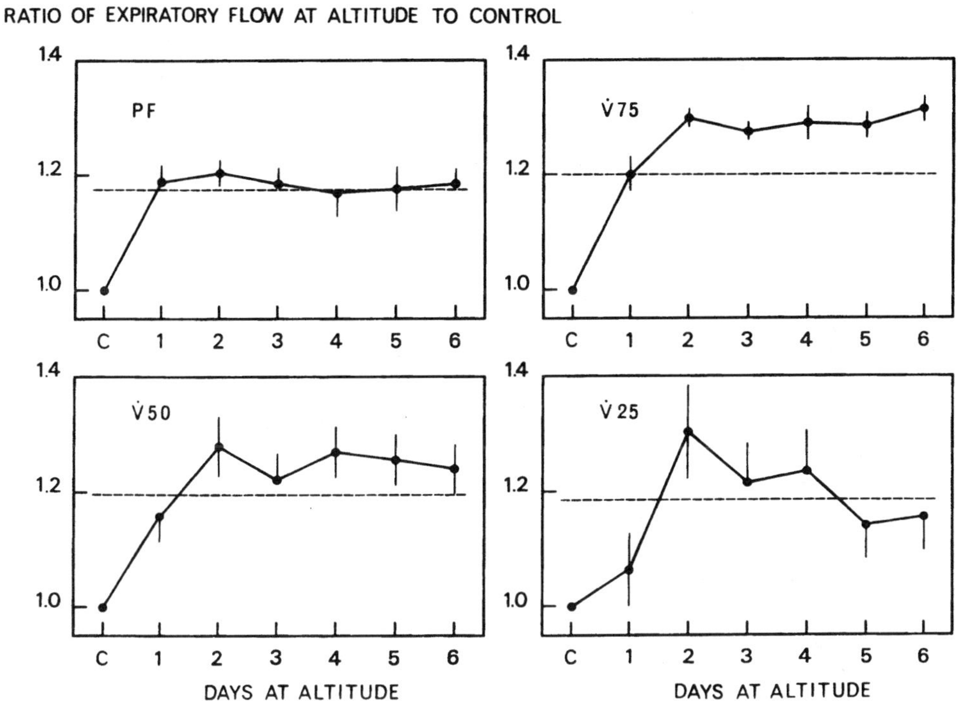

Figure 5 Average values (±SEM) of peak expiratory flow (PF) and maximum expiratory flows ($\dot{V}$) at 75, 50, and 25% of vital capacity at altitude, expressed as a fraction of corresponding values at sea level (broken lines). Predicted values of PF and V at altitude. For further explanation, see text. (From Ref. 25.)

cantly increased from the first day at altitude but reached a plateau only by the second day. Maximum flow at 25% VC increased until day 2 but then progressively decreased in the following days. Also shown in Figure 5 are the values expected solely on the basis of decreased air density. The latter were predicted according to Wood and Bryan (46). Except for $\dot{V}$max at 25% VC, the data for day 1 closely fit the predicted values. Thereafter the flows exceeded the predicted values on all days for $\dot{V}$max at 75 and 50% VC and on 3 days for $\dot{V}$max at 25% VC, suggesting that there was a bronchodilatation during that period. Since after day 1 at altitude there was a loss in elastic recoil of the lungs (see Sec. III.A), the dilatation of the flow-limiting segments of the airways must have been substantial, as per se the loss of lung recoil should have decreased $\dot{V}$max (47). Loss in recoil, however, may explain the progressive decrease in $\dot{V}$max at 25% VC after day 2 at altitude.

To our knowledge, apart from Gautier et al. (25), there are only two other studies of peak expiratory flow made at altitude on lowlanders, namely that of Cruz (17) during the third day at 4350 m and that of Mansell et al. (21) after 9–30 days at 5366 m. In the former study the ratio of peak flow at altitude relative to sea level averaged 1.14 compared with a ratio predicted according to Wood and Bryan (46) of 1.23, while in the case of Mansell et al. (21), the corresponding experimental ratio averaged 1.21 compared with a predicted value of 1.32. The results of both studies are in disagreement with those of Gautier et al. (25), which showed a good agreement between predicted and experimental values of peak flow (Fig. 5). It should be noted, however, that peak flow is effort-dependent (46,48), and hence predictions based on changes in gas density alone may be inherently less precise than in the case of $\dot{V}$max.

In most studies, a slight increase of the forced expired volume in one second (FEV_1) has been observed at altitude (8,14,17,19,21,24,25,29,44). The relative stability of FEV_1 at altitude in the face of the marked increase in $\dot{V}$max probably mainly reflects the fact that as a result of gas compression, the FEV_1 is exhaled at lower thoracic gas volumes (and hence lower maximal flows) than at sea level.

At 3660 m, Lefrançois et al. (24) measured FEV_1 and VC in 18 Bolivian natives born and living at this altitude and in 10 Caucasian lowlanders after a 30-day sojourn at that altitude. No significant differences were observed. However, in 126 residents at 3100 m, Kryger et al. (40) found a slightly but consistently greater FEV_1 than that predicted for sea level Caucasians while the VC was in the range of predicted.

In order to evaluate the role of genetic and environmental factors, Brody et al. (29) measured $\dot{V}$max at 50 and 25% VC in Peruvian native highlanders (3850 m) and lowlanders (800 m), both studied at their respective residence levels. The

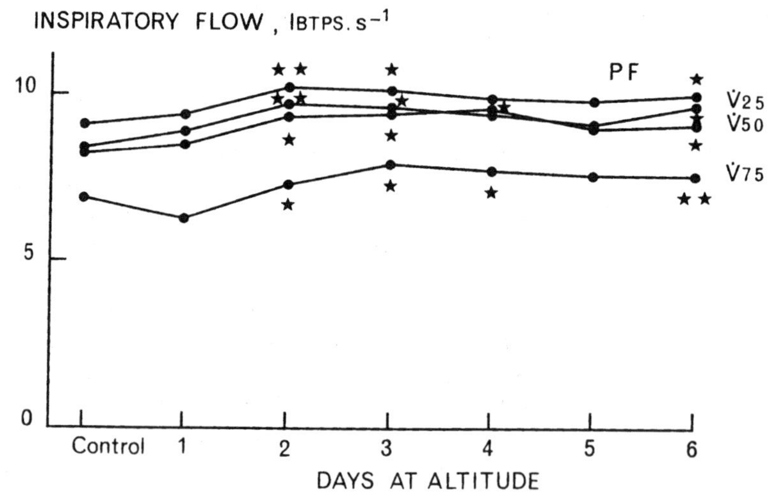

Figure 6 Average values of peak inspiratory flow (PF) and maximum inspiratory flows at 75, 50, and 25% vital capacity. Levels of significance between altitude and sea level values as in Figure 2. (From Ref. 25.)

highlanders had a higher VC than the lowlanders (respectively 116 and 84% of predicted). Because the values of $\dot{V}$max were lower in highlanders than in lowlanders, the authors suggested that the airways, which form in fetal life, do not participate in adaptation to altitude, and that the large lungs of highlanders are due to postnatal environmental hypoxic stimulation of their growth. It should be noted, however, that Brody et al. did not measure volume with a body plethysmograph and, therefore, the values of $\dot{V}$max 50 and 25% VC were obtained at a lower thoracic gas volume (TGV) in the highlanders than in the lowlanders because of the different PB at which the measurements were made. At low ambient PB, the thoracic gas volume during an FVC maneuver is necessarily lower than at higher PB, because the gas compressibility is greater (Fig. 3). As a result, the lower values of $\dot{V}$max found in the highlanders may merely reflect artefacts due to gas compressibility. Clearly, further studies using body plethysmography are needed to assess the iso-TGV changes in $\dot{V}$max with altitude.

Using body plethysmography, Gautier et al. (25) provided the only available values of maximal inspiratory flows at altitude. As shown in Figure 6, the maximal inspiratory flows increased significantly by the second day of sojourn.

III. Mechanical Properties of the Respiratory System

A. Static Volume-Pressure Curves and Compliance

In four subjects after 3 days at 4100 m, Kronenberg et al. (49) measured the static lung compliance (Cst,L) during stepwise lung inflation from FRC to 70–80% TLC and found a significant 20% reduction relative to sea level. Using the same method on four subjects at an altitude of 4300 m, a small (statistically insignificant) increase in Cst,L was found during 2–10 days at altitude (50). In both of these contradictory studies, the lung volume history was similar but different from that used in conventional studies of the elastic properties of the lung (lung deflation from TLC). In 10 subjects after 30 days at 3660 m, Lefrançois et al. (24) measured the static V-P curves of the lung and chest wall during stepwise deflation from TLC. They stated that both curves did not change at altitude. However, inspection of their data reveals a parallel shift to the left of the V-P curve of the lung, reflecting a small loss of lung recoil at all volumes considered. Subsequently, Brody et al. (29) measured the static V-P curve of the lungs in three subjects after 3 days at 3850 m. They stated that lung elastic recoil was unchanged but did not provide the actual data. Mansell et al. (21) measured the static deflation V-P curve of the lungs in three subjects after 9 days and in four subjects after 30 days at an altitude of 5366 m. Except at TLC, at all volumes (% TLC), the elastic recoil of the lung was reduced at altitude, the difference increasing progressively from TLC to FRC. However, Cst,L in the midrange of lung volume (FRC + 1 L) remained the same at altitude as at sea level. Similar results were obtained by Gautier et al. (25) in nine subjects during a 6-day sojourn at 3460 m, as shown in Figure 7. During altitude sojourn, the static V-P curve of lungs was shifted progressively to the left. The decrease in elastic recoil,

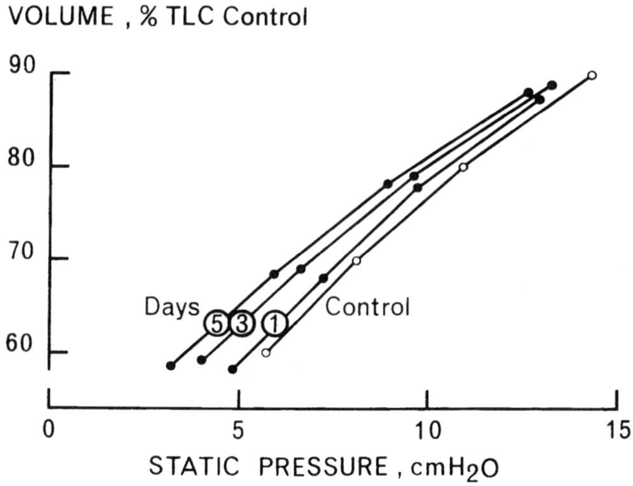

Figure 7 Average static deflation V-P relationships of lung at sea level and for the first, third, and fifth days at altitude of 3460 m. Volumes are expressed as percent of total lung capacity (% TLC). (From Ref. 25.)

which was almost complete by day 4, averaged about 2 cmH_2O and was significant on days 4–6. Cst,L measured from 60 to 70% TLC did not change significantly.

Jaeger et al. (18) measured the quasi-static deflation V-P curve of the lungs on 11 soldiers who participated in a 72-hour field exercise at an altitude of 3000–4300 m. They found a rightward shift of the V-P curve above FRC while below FRC the curve was shifted to the left. In line with Frank et al. (51), they interpreted this change as suggestive of lung congestion and/or interstitial edema. Thus, the discrepancy with the results of Mansell et al. (21) and Gautier et al. (25) may be ascribed to the fact that the soldiers performed strenuous exercise.

The loss of elastic recoil of the lung during short-term exposure to altitude may explain the increase in RV and closing capacity. Indeed, an increase in closing capacity at altitude has been reported by Coates et al. (10) and Jaeger et al. (18). The loss of lung recoil at altitude should also per se cause an increase in the elastic equilibrium volume of the respiratory system (Vr) and hence in FRC (25). However, Gautier et al. (25) found that on days 2–6 at an altitude of 3460 m the FRC was the same as at sea level (Fig. 2) in spite of a significant loss of lung recoil (Fig. 7). They attributed this phenomenon to a rightward shift of the static V-P curve of the chest wall due to increased thoracic blood volume. In fact, the changes in FRC at altitude depend on the balance between the loss in recoil of the lung, which should increase Vr, and the opposing effect of loss in recoil of the chest wall due to increased thoracic blood volume (25). In the subjects of Gautier et al., these two effects were balanced on days 2–6, and hence there was no change in FRC relative to sea level. On day 1, however, there was very little change in elastic recoil, and hence the FRC actually decreased probably as a result of increased thoracic blood volume. The sustained increase in FRC found at altitudes higher than 4000 m (20,21) probably reflects a predominant effect of loss of elastic recoil of the lungs, which may also explain the concomitant increase in TLC.

Brody et al. (29) measured the static deflation characteristics of the lungs in Peruvian native highlanders (3850 m) and lowlanders (800 m), both studied at their respective residence levels. Lung recoil at FRC and TLC, as well as the size-corrected V-P curves, were similar in the two groups, suggesting that in highlanders there is no loss in elastic recoil. This is in contrast to the loss of lung recoil found in lowlanders during short-term exposure to altitude. Brody et al. (29) also found that, at any given static transpulmonary pressure, the lung volumes of Peruvian highlanders, and hence also Cst,L, was higher than in the Peruvian lowlanders, reflecting their larger lungs. A similar conclusion was reached by Mortola et al. (52), who compared the respiratory system compliance of Bolivian infants born at high altitude (3600 m) with that of Bolivian infants of similar ethnic origin born at 400 m. They found that in the highland infants the respiratory system compliance was 33% higher but concluded that this reflected the larger lungs of highlanders due to their hypoxic environment rather than a genetic characteristic. These results are compatible with the concept that fetal hypoxia at high altitude is more marked than at low altitude and that the developing fetus can recognize environmental hypoxia and respond to it.

Several investigators have reported that altitude has no significant effect on dynamic lung compliance (17,21,24,25).

B. Resistance

Using the interrupter technique in six lowlanders exposed to 4350 m for 3 days, Cruz (17) found a 7% decrease in resistance versus 17% expected by modification in air density and proposed this as evidence for reduction in large airway size due to bronchoconstriction. At 5366 m, Mansell et al. (21) found a decrease in pulmonary resistance (R_L) averaging 29% relative to sea level. In the study of Gautier et al. (25), R_L and airway resistance (Raw) decreased by an average of 14 and 17%, respectively, on day 1 at altitude, and a further substantial decrease of R_L was found on day 2 (Fig. 8). Thereafter R_L remained essentially constant. The drop in resistance caused by the decrease in gas density at altitude can be predicted according to the formula proposed by Varène et al. (53):

$$Rx/Rc = 0.57 + 0.44\ P_B$$

where P_B is in atmospheres, and Rx/Rc is the ratio of resistance at the altitude corresponding to P_B to sea level resistance. At the altitude of the study of Gautier et al. (25) (P_B = 0.67 ATA), the Rx/Rc ratio should amount to 0.86. On day 1 at altitude the measured R_L, expressed as a fraction of the corresponding value at sea level, was in agreement with the prediction (0.86), indicating that the change in R_L could be entirely explained by decreased air density. After day 1, however, the

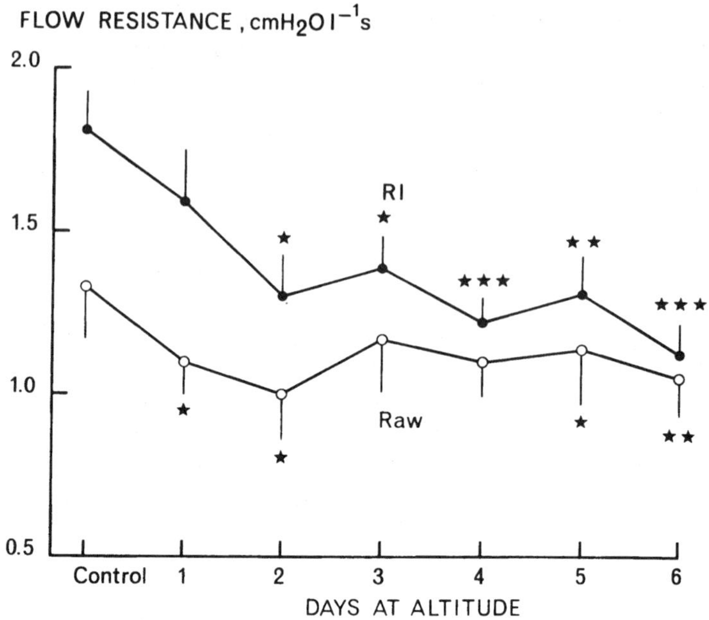

Figure 8 Average values (±SEM) of airway resistances (Raw, open circles) and pulmonary flow resistances (RL, filled circles) at sea level and at altitude. Levels of significance between altitude and sea-level values: $*p < 0.05$; $**p < 0.01$; $***p < 0.001$. (From Ref. 25.)

observed ratios were smaller than predicted, averaging 0.7 over the 2- to 6-day span. This suggests the occurrence of bronchodilatation (see below). Raw was also smaller than predicted during the stay at altitude. These results are consistent with the time course of the changes in maximum inspiratory and expiratory flows observed at altitude (Figs. 5 and 6).

C. Mechanisms Affecting Lung Mechanics at Altitude

It has been suggested that engorgement of the pulmonary vascular bed develops in individuals who ascend to high altitude (18,26,31,49,54,55). The available evidence, however, indicates that acute elevation in pulmonary blood volume has relatively little effect on the static V-P curve of the lungs, except at lung volumes lower than FRC, where Pst,L tends to decrease (51). Accordingly, the substantial loss in Pst,L observed over the lung volume range between 60 and 90% TLC (Fig. 7) cannot be ascribed to changes in pulmonary blood volume. Both the magnitude and direction of the changes in static V-P curve of the lung and in resistance observed after day 1 of sojourn at altitude are similar to those found by De Troyer et al. (56) on normal subjects after administration of a β_2-adrenergic agent (fenoterol). They reported that this drug caused a reduction of Raw associated with a shift to the left of the static deflation V-P curve of the lungs. In addition, the maximum expiratory flows increased. They postulated that increased activity of the β_2-adrenergic system caused a dilatation not only of the bronchial tree, but also a relaxation of smooth muscle in alveolar ducts or other contractile elements in the lung parenchyma leading to loss of elastic recoil of the lungs. Vincent et al. (57) found analogous results after administration of 1.2 mg atropine sulfate (iv) to normal volunteers. Thus, the decrease of resistance found during sojourns at altitude might be explained in part by a change in activity of the β_2-adrenergic and/or cholinergic systems. Increases in levels of catecholamines at high altitude have been reported (58). In this connection it should be noted that both the hypoxia and the hypocapnia present at altitude may be expected to cause increased resistance (59,60). Such an effect has not been reported except by Cruz (17), but it may become important at very high altitudes when the degree of hypocapnia and hypoxemia becomes more severe.

At 4350 m, Cruz (17) measured resistance by the interrupter technique in eight high-altitude natives whose values were similar to those of six low-altitude natives studied after 3 days at the same altitude. Analogous results were found by Lefrançois et al. (24) at 3660 m. However, in the highlanders endowed with large lungs, the non–size-corrected values of resistance would be expected to be lower than in lowlanders.

IV. Work of Breathing

In theory, the maximal mechanical work available for a single breathing cycle is given by the area subtended by the curves relating the maximal voluntary static inspiratory and expiratory pressures to lung volume (Fig. 3). At sea level, for young male adults it amounts to 20–30 cal per breath. However, this potential work is never realized during the actual breathing movements. Agostoni and Fenn (48) demonstrated that the maximal inspiratory work that a subject can achieve decreases with increasing flow, in line with the force-velocity relationship of skeletal muscles. Other mechanisms, however, also contribute to limit the maximal work per breath; these include (1) the speed of activation of the respiratory muscles at the beginning of forced inspirations and expirations and (2) gas compressibility (61). As shown in Fig. 3, the effective area of the maximal static V-P diagram, and hence the maximal potential work available per breathing cycle, decreases with reduced barometric pressure because of compressibility of gas. As a result of decreased air density and bronchodilatation, the work of breathing required from the respiratory muscles for a given ventilation should decrease at altitude, particularly at high ventilations.

Figure 9 shows the relationship between the mechanical power of breathing and ventilation in a normal adult breathing oxygen at different altitudes. These curves were obtained in a decompression chamber at various simulated altitudes (rapid ascent) ranging from 34 to 7500 m (62). Also shown in Figure 9 is a curve for simulated altitude of 8848 m (corresponding to the top of Mt. Everest), which was obtained by extrapolation. Whereas the respiratory power requirements for a

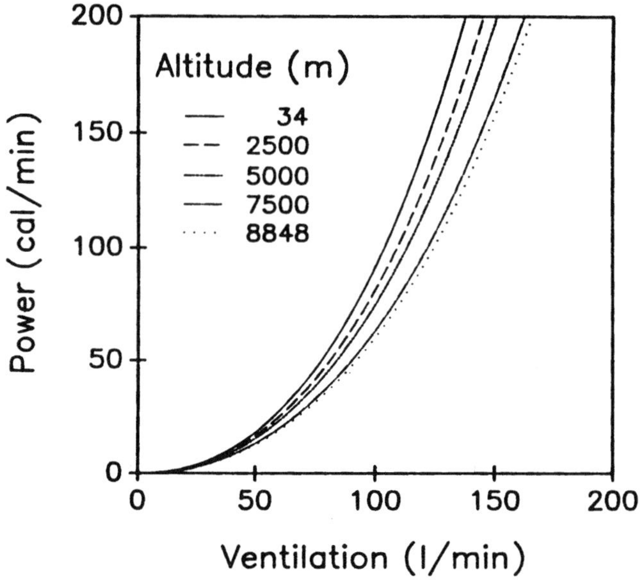

Figure 9 Relationships between respiratory mechanical power (cal/min) and ventilation (L/min) at different simulated altitudes in a normal adult. The curve for 8848 m (corresponding to the top of Mt. Everest) was obtained by extrapolation. (From Ref. 62.)

given ventilation decrease with altitude, the opposite is true in terms of the ventilation required for a given O_2 consumption during muscular exercise (see Chapter 20). In fact, breathing at extreme altitudes is a particularly interesting situation because the maximum oxygen consumption ($\dot{V}O_2max$) is so low whilst the maximum exercise ventilation is so high. For example, during a simulated ascent to Mt. Everest, the steady-state (5–8 min) $\dot{V}O_2max$ in three healthy young men averaged 1.12 L/min, while the corresponding ventilation was 183.5 L/min (2). The respiratory mechanical power required for this ventilation and altitude should amount to 260 cal/min (62). Assuming a mechanical efficiency of 0.2 and a caloric equivalent per milliliter of O_2 of 4.86 cal, the corresponding oxygen cost of breathing would amount to 0.27 L/min, or 24% of $\dot{V}O_2max$ (vs. 7% at sea level) (63). In contrast, a mechanical efficiency of only 0.05 should require a respiratory O_2 cost of 1.07 L/min, or 95% of $\dot{V}O_2max$. In this case, virtually all of the O_2 uptake would be required to meet the demands of the respiratory muscles with almost nothing left for the rest of the body, let alone for external work! Clearly, a low mechanical efficiency of breathing should severely limit the work tolerance at extreme altitudes.

It should be noted that the above analysis was based on results obtained on three individuals. Furthermore, during sojourn at altitude flow resistance may decrease as a result of bronchodilatation (see Sec. III.B). However, even if in the three subjects exercising at simulated altitude corresponding to Mt. Everest (see above) the mechanical power requirements of breathing were reduced to one half of those predicted in Figure 9 for that altitude (i.e., 130 instead of 260 cal/min, at efficiency of 0.05), the net $\dot{V}O_2$ available for the body would amount to only 0.59 L/min. This is barely sufficient to satisfy the resting metabolic needs of the body. In this connection, Petit et al. (62) underestimated the mechanical power output of the respiratory muscles because their measurements did not include the work due to gas compressibility (61) and distortion of the chest wall (64), which at altitude may be substantial.

In contrast to high altitude, breathing in hyperbaric environments increases the work of breathing (65). Exercise in dense environments is primarily limited by expiratory mechanisms, namely expiratory flow limitation (66), which does not appear to be a problem at high altitude.

There are only two reports on changes in mechanical work of breathing from sea level to altitude. In six lowlanders, Cruz (17) reported that at rest the mechanical work on the lung per liter of ventilation was 22% higher at an altitude of 4350 m than at sea level in the same subjects. Surprisingly, this was due to an increase in resistive work. It should be noted, however, that at rest, the work of breathing is very small even at altitude.

Cibella et al. (67) measured Wrs during submaximal exercise on a bicycle ergometer in four lowlanders at sea level and after a 1 month sojourn at 5050 m. In only one subject was Wrs, for a given $\dot{V}E$, consistently lower at altitude than sea level, as expected from decreased air density (62). In contrast, in the other three individuals, the relationship of Wrs to $\dot{V}E$ was the same at altitudes as at sea level. The latter was attributed mainly to bronchoconstriction due to severe hypoxia and hypocapnia. In addition, interstitial pulmonary edema is likely to develop during exercise at very high altitude (68).

At 3660 m, Lefrançois et al. (24) measured the respiratory mechanical power at rest and exercise up to 120 watts in five highlanders and three lowlanders after a 30-day sojourn at that altitude. For any given ventilation, the mechanical power was similar in both groups. However, at any given exercise level, the mechanical power requirements were smaller in the highlanders reflecting their lower ventilation. Thus, the lower ventilatory requirements during exercise of highlanders put them at an advantage relative to lowlanders. In this respect, the highlanders with large lungs should also be at an advantage because of smaller respiratory mechanical power requirement for any given ventilation. Unfortunately, Lefrançois et al. (24) did not measure the respiratory mechanical output during maximal exercise. In this connection, the resistive work of breathing depends on the flow waveform, being minimal with constant flow inflation and deflation (69). During heavy exercise at sea level both inspiratory and expiratory flows approach constant flow (70,71). Whether such an optimal pattern is also adopted during exercise at altitude is not known.

At sea level, the work per breath during maximal exercise amounts to about 20% of the maximal potential work (69). No such information is available for altitude. However, this percentage should be considerably greater at altitude in view of the decrease in maximal potential work.

There is only one report on the oxygen cost of breathing at altitude (44). The oxygen cost of breathing was measured during voluntary hyperventilation. In 11 lowlanders after 16 days at 4000 m, the cost of breathing per liter of ventilation was 36% lower than at sea level, probably reflecting in part lower respiratory power due to decreased density (Fig. 9). However, the oxygen cost of breathing during voluntary hyperventilation is usually substantially higher than during spontaneous breathing (69).

At sea level, exercise performance is usually not limited by ventilation in normal subjects. In contrast, at high altitudes, the respiratory requirement probably plays a more important role in limiting exercise performance. In spite of this, during maximal exercise at a simulated altitude of 8848 m, the dyspnea ratings were similar and not, as expected, higher than those for leg activity (72). Thus, even at that very high altitude, exercise limitation was attributed to both peripheral and respiratory muscle activity. This surprising finding could be explained by the hypothesis that during exercise at very high altitudes there is preferential blood supply to the respiratory muscles, which should benefit the performance of the respiratory muscles but at the expense of the peripheral muscles. Clearly further studies on mechanical power, energy cost of breathing, blood supply to the respiratory muscles, etc., are needed, particularly at the very high ventilations encountered during exercise at high altitude.

In conclusion, while at sea level there is a considerable body of knowledge of respiratory mechanics, as detailed in a recent review (69), for altitude, there is substantial information available only with respect to lung volumes. In contrast, the studies of respiratory mechanics at high altitude are rather scanty. This is particularly true for the work of breathing, which must play an important role in limiting exercise performance at high altitudes.

Acknowledgments

The authors thank Ms. M. Gras for typing this manuscript and Ms. J. Chandellier for artwork. They also wish to thank Drs. W. A. Whitelaw and J. P. Mortola for their helpful comments and suggestions.

References

1. Reeves JT, Welsh CH, Wagner PD. The heart and lungs at extreme altitude. Thorax 1994; 49:631–633.
2. Sutton JR, Reeves JT, Wagner PD, Groves BM, Cymerman A, Malconian MK, Rock PB, Young PM, Walter SD, Houston CS. Operation Everest II: oxygen transport during exercise at extreme simulated altitude. J Appl Physiol 1988; 64:1309–1321.
3. Messner R. Everest: Expedition to the Ultimate. London: Kaye and Ward, 1979.
4. Bert P. La Pression Barométrique. Paris: Masson, 1878.

5. Schneider EC. The vital capacity of the lungs at low barometric pressure. Am J Physiol 1932; 100:426–432.
6. Hurtado A, Kaltreider N, McCann WS. Respiratory adaptation to anoxemia. Am J Physiol 1934; 109:626–637.
7. Rahn H, Hammond D. Vital capacity at reduced barometric pressure. J Appl Physiol 1952; 4:715–724.
8. Finkelstein S, Tomashefski JF, Shillito FH. Pulmonary mechanics at altitude in normal and obstructive lung disease patients. Aerospace Med 1965; 36:880–884.
9. Gray GW, Rennie IDB, Houston CS, Bryan AC. Phase IV of the single-breath nitrogen washout curve on exposure to altitude. J Appl Physiol 1973; 35:227–230.
10. Coates G, Gray G, Mansell A, Nahmias C, Powles A, Sutton J, Webber C. Changes in lung volume, lung density, and distribution of ventilation during hypobaric decompression. J Appl Physiol 1979; 46:752–755.
11. Saunders NA, Betts MF, Pengelly LD, Rebuck AS. Changes in lung mechanisms induced by acute isocapnic hypoxia. J Appl Physiol 1977; 42:413–419.
12. Goldstein RS, Zamel N, Rebuck AS. Absence of effects of hypoxia on small airway function in humans. J Appl Physiol 1979; 47:251–256.
13. Garfinkel F, Fitzgerald RS. The effect of hyperoxia, hypoxia and hypercapnia on FRC and occlusion pressure in human subjects. Respir Physiol 1978; 33:241–250.
14. Welsh CH, Wagner PD, Reeves JT, Lynch D, Cink TM, Armstrong J, Malconian MK, Rock PB, Houston CS. Operation Everest II: Spirometric and radiographic changes in acclimatized humans at simulated high altitudes. Am Rev Respir Dis 1993; 147:1239–1244.
15. Ulvedal F, Morgan TE Jr, Cutler RG, Welch BE. Ventilatory capacity during prolonged exposure to simulated altitude without hypoxia. J Appl Physiol 1963; 18:904–908.
16. Mosso A. Fisiologia dell'uomo sulle Alpi. Milan: Treves, 1897.
17. Cruz JC. Mechanics of breathing in high altitude and sea level subjects. Respir Physiol 1973; 17:146–161.
18. Jaeger JJ, Sylvester JT, Cymerman A, Berberich JJ, Denniston JC, Maher JT. Evidence for increased intrathoracic fluid volume in man at high altitude. J Appl Physiol 1979; 47:670–676.
19. Shields JL, Hannon JP, Harris CW, Platner WS. Effects of altitude acclimatization on pulmonary function in women. J Appl Physiol 1968; 25:606–609.
20. Tenney SM, Rahn H, Stroud RC, Mithoefer JC. Adaptation to high altitude: changes in lung volumes during the first seven days at Mt. Evans, Colorado. J Appl Physiol 1953; 5:607–613.
21. Mansell A, Powles A, Sutton J. Changes in pulmonary PV characteristics of human subjects at an altitude of 5,366 m. J Appl Physiol 1980; 49:79–83.
22. Cerny FC, Dempsey JA, Reddan WG. Pulmonary gas exchange in non-native residents of high altitude. J Clin Invest 1973; 52:2993–2999.
23. Degraff AC Jr, Grover RF, Johnson RL Jr, Hammond JW Jr, Miller JM. Diffusing capacity of the lung in Caucasians native to 3,100 m. J Appl Physiol 1970; 29:71–76.
24. Lefrançois R, Gautier H, Pasquis P. Mécanique ventilatoire chez l'homme à haute altitude. C R Soc Biol 1969; 163:2037–2042.
25. Gautier H, Peslin R, Grassino A, Milic-Emili J, Hannhart B, Powell E, Miserocchi G, Bonora M, Fischer JT. Mechanical properties of the lungs during acclimatization to altitude. J Appl Physiol 1982; 82:1407–1415.

26. Roy SB, Guleria JS, Khanna PK, Talwar JR, Manchanda SC, Pande JN, Kaushik VS, Subba PS, Wood JE. Immediate circulatory response to high altitude hypoxia in man. Nature 1968; 217:1177–1178.
27. Hurtado A. Animals in high altitudes: resident man. In: Dill DB, Adolph EF, Wilber CG, eds. Handbook of Physiology. Adaptation to Environment. Washington, DC: American Physiology Society, 1964:843–860.
28. Lahiri S, Delaney RG, Brody JS, Simpser M, Velasquez T, Motoyama EK, Polgar C. Relative role of environmental and genetic factors in respiratory adaptation to high altitude. Nature 1976; 261:133–135.
29. Brody JS, Lahiri S, Simpser M, Motoyama EK, Velasquez T. Lung elasticity and airway dynamics in Peruvian natives to high altitude. J Appl Physiol 1977; 42:245–251.
30. Hackett PH, Reeves JT, Reeves CD, Grover RF, Rennie D. Control of breathing in Sherpas at low and high altitude. J Appl Physiol 1980; 49:374–379.
31. Kamat SR, Rao TL, Sarma BS, Venkataraman C, Raju VRK. Study of cardiopulmonary function on exposure to high altitude. Am Rev Respir Dis 1972; 106:414–431.
32. Sun SF, Droma TS, Zhang JG, Tao JX, Huang SY, McCullough RG, McCullough RE, Reeves CS, Reeves JT, Moore LG. Greater maximal O_2 uptakes and vital capacities in Tibetan than Han residents of Lhasa. Respir Physiol 1990; 79:151–162.
33. Frisancho AR. Human growth and pulmonary function of a high altitude Peruvian Quechua population. Human Biol 1969; 41:365–379.
34. Greksa LP, Spielvogel H, Paz-Zamora M, Caceres E, Paredes-Fernandez L. Effect of altitude on the lung function of high altitude residents of European ancestry. Amer J Phys Anthropol 1988; 75:77–85.
35. Frisancho AR. Functional adaptation to high altitude hypoxia. Science 1975; 187:313–319.
36. Burri PH, Weibel ER. Morphometric evaluation of changes in lung structure due to high altitude. In: Porter R, Knight J, eds. High Altitude Physiology. Edinburgh: Churchill Livingstone, 1971:15–30.
37. Sekhon HS, Wright JL, Thurlbeck WM. Pulmonary function alterations after 3 wk of exposure to hypobaria and/or hypoxia in growing rats. J Appl Physiol 1995; 78:1787–1792.
38. Paz Zamora M, Coudert J, Ergueta Collao J, Vargas E, Gutierrez N. Respiratory and cardiocirculatory responses of acclimatization of high altitude natives (La Paz, 3500 m) to tropical lowland (Santa Cruz, 420 m). In: Brendel W, Zink RA, eds. High Altitude Physiology and Medicine. New York: Springer-Verlag, 1982:21–27.
39. Jones RL, Man SFP, Matheson GO, Parkhouse WS, Allen PS, McKenzie DC, Hochachka PW. Overall and regional lung function in Andean natives after descent to low altitude. Respir Physiol 1992; 87:11–24.
40. Kryger M, Alrich F, Reeves JT, Grover RF. Diagnosis of airflow obstruction at high altitude. Amer Rev Respir Dis 1978; 117:1055–1058.
41. Rahn H, Otis AB, Chadwick LE, Fenn WO. The pressure-volume diagram of the thorax and lung. Amer J Physiol 1946; 146:161–178.
42. Cotes JE. Ventilatory capacity at altitude and its relation to mask design. Proc R Soc B 1954; 143:32–39.
43. Pugh LGCE. Muscular exercise on Mount Everest. J Physiol Lond 1958; 141:233–261.
44. Mazess RB. The oxygen cost of breathing in man: effects of altitude, training, and race. Am J Phys Anthropol 1968; 29:365–375.

45. Ingram RH Jr, Schilder DP. Effect of gas compression on pulmonary pressure, flow, and volume relationship. J Appl Physiol 1966; 21:1821–1826.
46. Wood LDH, Bryan AC. Effect of increased ambient pressure on flow-volume curve of the lung. J Appl Physiol 1969; 27:4–8.
47. Hyatt RE. Dynamic lung volumes. In: Fenn WO, Rahn H, eds. Handbook of Physiology. Respiration. Washington, DC: American Physiology Society, 1965:1381–1397.
48. Agostoni E, Fenn WO. Velocity of muscle shortening as a limiting factor in respiratory air flow. J Appl Physiol 1960; 15:349–353.
49. Kronenberg RS, Safar P, Lee J, Wright F, Noble W, Wahrenbrock E, Hickey R, Nemoto E, Severinghaus JW. Pulmonary artery pressure and alveolar gas exchange in man during acclimatization to 12,470 ft. J Clin Invest 1971; 50:827–837.
50. Raymond L, Severinghaus JW. Static pulmonary compliance of man during altitude hypoxia. J Appl Physiol 1971; 31:785–787.
51. Frank NR, Radford EP, Whittenberger JL. Static volume-pressure interrelations of the lungs and pulmonary blood vessels in excised cat lungs. J Appl Physiol 1959; 14:167–173.
52. Mortola JP, Rezzonico R, Fisher JT, Villena-Cabrera N, Vargas E, Gonzales R, Pena F. Compliance of the respiratory system in infants born at high altitude. Am Rev Respir Dis 1990; 142:43–48.
53. Varène P, Timbal J, Jacquemin C. Effect of different ambient pressures on airway resistance. J Appl Physiol 1967; 22:699–706.
54. Kleiner JP, Nelson WP. High altitude pulmonary edema. J Am Med Assoc 1975; 234: 491–495.
55. Reeves JT, Halpin J, Cohn JE, Daoud F. Increased alveolar-arterial oxygen difference during simulated high-altitude exposure. J Appl Physiol 1969; 27:658–661.
56. De Troyer A, Yernault JC, Rodenstein D. Influence of beta-2 agonist aerosols on pressure-volume characteristics of the lungs. Am Rev Respir Dis 1978; 118:987–995.
57. Vincent NJ, Knudson R, Leith DE, Macklem PT, Mead J. Factors influencing pulmonary resistance. J Appl Physiol 1970; 29:236–243.
58. Wolfel EE, Selland MA, Mazzeo RS, Reeves JT. Systemic hypertension at 4,300 m is related to sympatho adrenal activity. J Appl Physiol 1994; 76:1643–1650.
59. Libby PM, Briscoe WA, King TKC. Relief of hypoxia-related bronchoconstriction by breathing 30 per cent oxygen. Am Rev Respir Dis 1981; 123:171–175.
60. Newhouse MT, Becklake MR, Macklem PT, McGregor M. Effect of alterations in end-tidal CO_2 tension on flow resistance. J Appl Physiol 1964; 19:745–749.
61. Jaeger MJ, Otis AB. Effects of compressibility of alveolar gas on dynamics and work of breathing. J Appl Physiol 1964; 19:83–91.
62. Petit JM, Milic-Emili G, Troquet J. Travail dynamique pulmonaire et altitude. Rev Med Aeronaut 1963; 2:276–279.
63. Milic-Emili G, Petit JM, Deroanne R. Mechanical work of breathing during exercise in trained and untrained subjects. J Appl Physiol 1962; 17:43–46.
64. Goldman MD, Grimby G, Mead J. Mechanical work of breathing derived from rib cage and abdominal V-P partitioning. J Appl Physiol 1976; 41:752–763.
65. Hesser CM, Linnarsson D, Fragraeus L. Pulmonary mechanics and work of breathing at maximal ventilation and raised air pressure. J Appl Physiol 1981; 50:747–753.
66. Van Liew HD. Mechanical and physical factors in lung function during work in dense environments. Undersea Biomed Res 1983; 10:255–264.

67. Cibella F, Cuttitta G, Romano S, Grassi B, Bonsignore G, Milic Emili J. Respiratory energetics during exercise at high altitude. J Appl Physiol 1999; 86:1785–1792.
68. West JB. Left ventricular filling pressure during exercise. Chest 1998; 113:1695–1697.
69. Milic-Emili J. Work of breathing. In: Crystal RG, West JB, eds. The Lung. New York: Raven Press, 1991:1065–1075.
70. Proctor DF, Hardy JB. Studies of respiratory airflow. Bull Johns Hopkins Hosp 1949; 85:253–280.
71. Lafortuna CL, Minetti AE, Mognoni P. Modelling the airflow pattern in humans. J Appl Physiol 1984; 57:1111–1119.
72. Sutton J, Balcomb A, Killian K, Green HJ, Young PM, Cymerman A, Reeves JT, Houston CS. Breathlessness at altitude. In: Jones NL, Killian KL, eds. Breathlessness. The Campbell Symposium. Hamilton (ON): Boehringer Ingelheim Canada, 1992:143–148.

8

Gas Exchange

PETER D. WAGNER

University of California, San Diego
La Jolla, California

The essence of high altitude is hypoxia, reducing arterial PO_2. This chapter deals mainly with how pulmonary gas exchange is modified by altitude. Both rest and exercise are considered, and O_2 transport is analyzed in terms of (1) effects of reduced inspired PO_2, (2) diffusion limitation of O_2 exchange, and (3) ventilation/perfusion relationships. Acute altitude exposure and prolonged, acclimatized situations are described separately. Hypoxia is of course the result of barometric pressure (P_B) decreasing with altitude. While air at any altitude remains approximately 21% O_2, the virtually exponential fall in P_B with altitude reduces ambient PO_2. When it is further remembered that as soon as air is inhaled into the large airways it becomes warmed to body temperature (37°C usually) and saturated with water vapor [contributing a partial pressure of 47 mmHg (at 37°C) at all altitudes], it is clear that inspired PO_2 in the airways (PIO_2) is rapidly reduced by ascent to altitude. Table 1 indicates pertinent values derived from the U.S. Standard atmosphere relating PO_2 to altitude.

The U.S. Standard atmosphere is only a general guide to effects of altitude on inspired PO_2. Geographical, thermal, seasonal, and meteorological factors will cause variation from the figures in Table 1. These variations may be relatively large and biologically substantial. Perhaps the best example is that the U.S. atmosphere

Table 1 Inspired Po_2 at Altitude

Altitude (m)	Altitude (ft.)	P_B (torr)	P_{IO_2} (saturated, 37°C) (torr)
0	0	760	149
1000	3281	674	131
2000	6562	596	115
3000	9843	526	100
4000	13123	463	87
5000	16404	405	75
6000	19685	354	64
7000	22966	308	55
8000	26247	267	46
9000	29528	231	39

predicts P_B on the Everest summit to be 236 torr while data obtained during the American Medical Research Expedition to Everest in 1981 (1) indicated a P_B of 253 torr, 17 mm higher. While at sea level, a fall in P_B of 17 torr from 760 to 743 will cause P_{IO_2} (at 37°C, saturated) to fall by 3.5 torr (0.2093×17), and thus arterial Po_2 to fall similarly in a normal subject, the arterial saturation will fall by about 0.2%, a trivial amount. The 17 torr difference in P_{IO_2} on the Everest summit translates into the same 3.5 torr drop in P_{IO_2}, but with arterial Po_2 normally at about 30–35 torr, the fall in arterial saturation would be about 6%. This is not considered trivial under conditions of extreme hypoxia. More discussion of barometric pressure and its effects on inspired Po_2 can be found in Ref. 2.

Many excursions to altitude over the years have examined gas exchange, usually focusing on noninvasive measurements (3–5). This has necessarily limited the amount of data collected, especially at extreme altitudes. Two studies in particular (using healthy subjects), however, have resulted in large and complete data sets. One focused on acute exposure to P_B of 429 torr (~15,000 ft) on a single day, while the other, called Operation Everest II (OEII) (7), studied subjects gradually decompressed to $P_B \approx 250$ torr over 42 days. Both of these hypobaric chamber studies sampled arterial and pulmonary arterial blood and used the multiple inert gas elimination technique to characterize pulmonary gas exchange. For these reasons, and to promote internal consistency and to present the most complete picture possible, the chapter will be based mostly on the data from these two projects. The philosophy of this chapter is more to describe the physiological responses to altitude exposure from these studies than to present an exhaustive literature survey of necessarily sparse data. In general, the limited data from many other studies are in agreement with those of the two projects featured. With this brief introduction, the effects of altitude on pulmonary gas exchange are now discussed.

I. Gas Exchange at Rest

A. Consequences of the Fall in Inspired Po_2 with Altitude on Alveolar Po_2

As inspired Po_2 falls, so too must alveolar Po_2 (PAO_2). Hypoxemia ensues such that there is ventilatory stimulation via the carotid chemoreceptors, and this helps to mitigate the obligatory fall in alveolar Po_2 that follows a reduction in PIO_2. Before considering the effects of hyperventilation, it is instructive to calculate the hypothetical effects of altitude on alveolar Po_2 in the absence of any ventilatory stimulation. Such a calculation serves to point out the criticality of the ventilatory response to hypoxia. This can be done using the alveolar gas equation (8):

$$PAO_2 = PIO_2 - PACO_2/R + PACO_2 \cdot FIO_2 \cdot (1\text{-}R)/R \qquad (1)$$

In Eq. (1), $PACO_2$ is alveolar Pco_2, R is respiratory exchange ratio, and FIO_2 inspired O_2 fractional concentration. Typically at rest, alveolar (or arterial, essentially the same) Pco_2 is 40 torr at sea level and R is 0.8, while FIO_2 is 0.2093 independent of altitude.

Figure 1 shows how alveolar Po_2 falls essentially linearly with altitude, using Eq. (1) and data from the two above-mentioned studies (6,7). In Figure 1, the closed circles reflect the calculated values of PAO_2 assuming alveolar Pco_2 remained con-

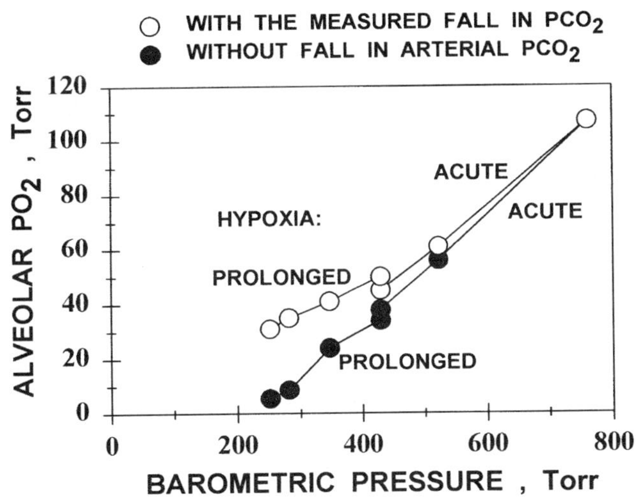

Figure 1 Fall in alveolar Po_2 with increasing altitude. Open circles reflect actual data, and closed circles reflect hypothetical values had progressive compensatory hypocapnia not developed with altitude. The break at $P_B = 430$ torr separates different data sets from acute (6) and prolonged (7) hypoxic exposures. Hyperventilation is of increasing adaptive importance the higher the altitude.

stant at sea level values, and the open circles are actual data from the studies. The highest barometric pressure of 760 torr is that of sea level; the lowest of 253 torr corresponds to the Everest summit.

Thus, the criticality of the ventilatory response is seen when P_B falls below about 400 torr, and increasingly so the higher the altitude. Up to common skiing altitudes (about 10,000 ft., P_B about 500 torr or greater) there would be little reduction in P_{AO_2} from absence of the hypoxic ventilatory response. Due to the shape of the O_2-Hb dissociation curve (essentially flat at P_{O_2} values >60 torr), there is also negligible effect on arterial O_2 saturation. However, at 4000 m (P_B 463 torr), failure to hyperventilate would cause an almost 10 torr fall in P_{AO_2}. At about 350 torr, P_{AO_2} would be some 15 torr lower, at less than 25 torr. At the Everest summit equivalent, alveolar P_{O_2} is still maintained at or above 30 torr (Fig. 1) whereas in the absence of hyperventilation it would have plummeted to 5 torr. There are many accounts of O_2-unaided ascents of Mt. Everest, which must clearly be dependent in large measure on the ability to maintain alveolar and thus arterial P_{O_2} by hyperventilation.

Several studies conducted during mountaineering expeditions measuring alveolar P_{O_2} and P_{CO_2} are summarized in Figure 2. Interest is focused on extreme altitude (P_B < 350 torr) in this figure. There is a progressive, linear fall in both P_{AO_2} and P_{ACO_2} as altitude is gained. Resting P_{AO_2} on the Everest summit or equivalent appears to be between 30 and 35 torr; corresponding P_{ACO_2} is between 8 and 11 torr. There is an apparent discrepancy between the data of AMREE (The American Medical Research Expedition to Everest) and those of OEII (closed and open circles, respectively, Fig. 2). The higher P_{AO_2} and lower P_{ACO_2} of AMREE have been interpreted as reflecting greater ventilatory acclimatization to altitude (2). While the time to reach the summit was only 40 days in OEII (9) and more than 70 in AMREE (10), other explanations of the OEII-AMREE differences are possible. Thus, at about 282 torr, where OEII examined six subjects (11) and AMREE four (12), the data are bunched so closely that there is very little difference in P_{AO_2} or P_{ACO_2} between them (Fig. 2). This is mirrored by recent data from the British Everest Expedition (BEE) (13). The obvious OEII-AMREE difference is thus seen at the summit altitude. Here, the AMREE data reflect a single subject who, even at lower altitudes, had a generally lower P_{ACO_2} and higher hypoxic ventilatory response than most of the AMREE subjects studied (12). The respiratory gas exchange ratio on the summit for this person was 1.49, probably reflecting considerable acute hyperventilation. Further data will be necessary to determine the reason for these differences, but there is similarity in alveolar P_{O_2} and P_{CO_2} average AMREE, BEE, and OEII subjects at the one common altitude (of $P_B \sim 282$ torr) where multiple subjects were available for study.

Figure 3 shows the mean arterial P_{O_2} and P_{CO_2} in resting subjects from the two studies mentioned [reflecting acute (6) and prolonged (7) altitude exposure]. These data agree with those of others estimated indirectly or measured directly (14–16). The top panel shows the effects of the remarkable fourfold increase in resting

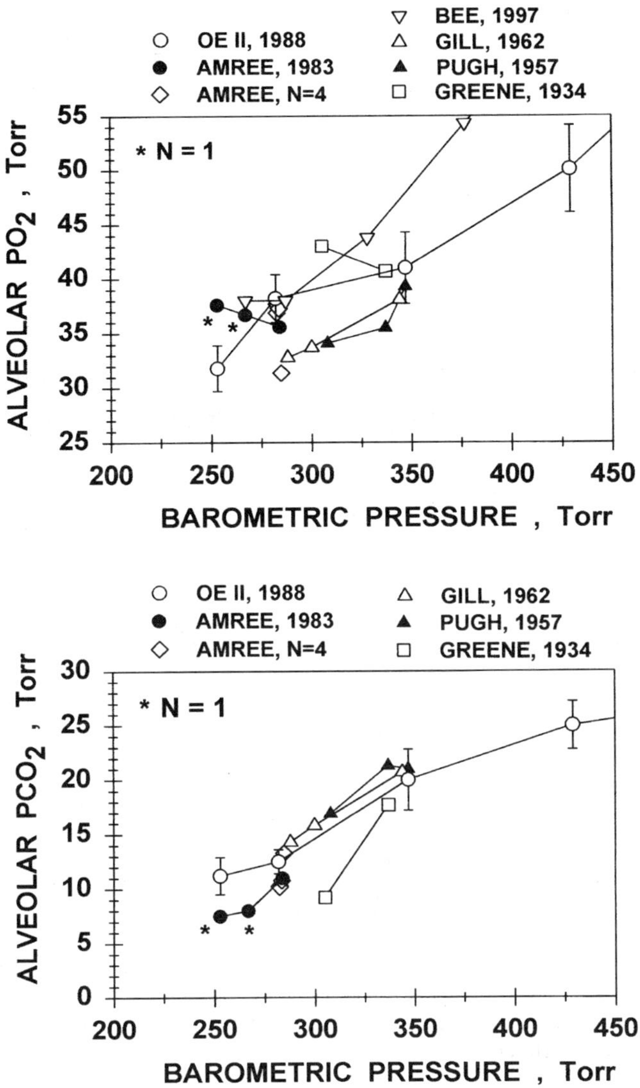

Figure 2 Collection of data from several sources depicting alveolar P_{O_2} and P_{CO_2} as a function of altitude. Given the small numbers of subjects, intersubject variation, and experimental error, there is remarkable agreement at P_B = 275–300 torr among AMREE, BEE, and OEII. Differences appear at the summit, but the environmental and technical differences and small numbers of subjects preclude clear interpretation.

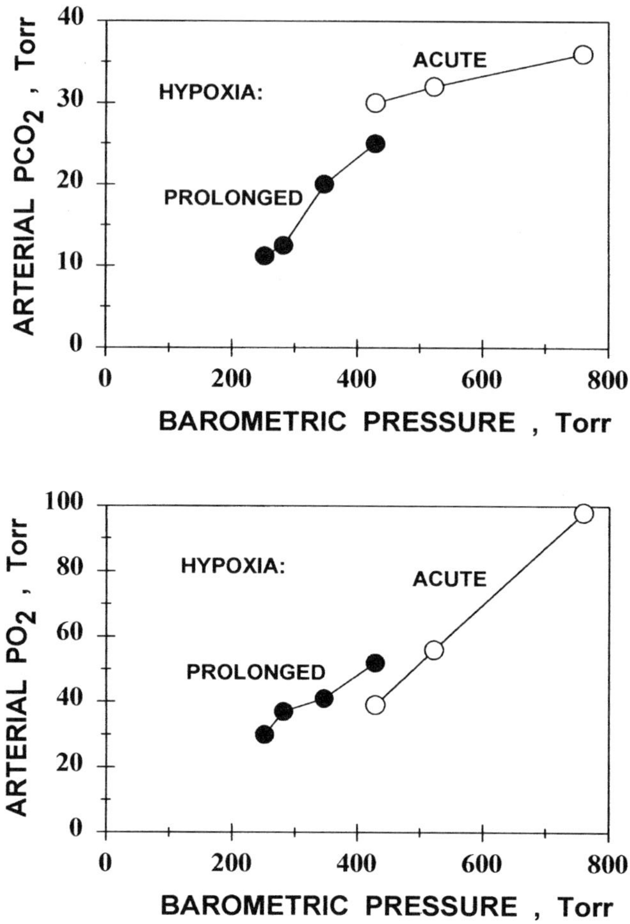

Figure 3 Resting arterial P_{CO_2} (top) and P_{O_2} (bottom) for acute hypoxic exposures to P_B = 429 torr (about 15,000 ft.) and prolonged exposure up to the Everest summit (P_B = 253 torr). These values mirror those in alveolar gas of Figures 1 and 2 and also reflect a lower alveolar-arterial P_{O_2} difference with prolonged exposure.

ventilation from sea level to the Everest summit, which correspondingly causes a fourfold reduction in alveolar and thus arterial P_{CO_2} in these normal subjects. This one-for-one relationship between ventilation and P_{ACO_2} indicates no change in resting metabolic rate with altitude. In the field, additional stresses of anxiety and cold may well impose changes in metabolic rate that would additionally influence the ventilation actually measured at altitude.

Figure 3 also shows a discontinuity between the overlapping data from the two studies at the 430 torr P_B point: arterial P_{CO_2} was lower and P_{O_2} higher in chronic exposure at the same altitude. While this could reflect study population differences, sea level data for the two groups were not different, which suggests that this is not the explanation. The differences at PB 430 torr more likely reflect the well-known further increase in ventilation (at any altitude) comparing prolonged to acute exposure. These differences are generally thought to be due to removing the braking effect on ventilation of alkalosis in blood and CSF, which are present to a greater extent in acute than chronic exposure and to greater hypoxic sensitivity of the carotid chemoreceptors (see Chapter 6 for further details).

B. Pulmonary Gas Exchange Efficiency

There is a small additional difference between acute and chronic altitude exposure that affects arterial PO_2 and PCO_2 and is not explained by differences in ventilation per se. If the alveolar-arterial PO_2 difference ($AaPO_2$) is computed by subtracting the measured arterial PO_2 from the calculated alveolar PO_2 [Eq. (1)], one sees a small but systematic difference: the $AaPO_2$ is higher in acute than prolonged altitude exposure (Fig. 4). Reasons for this are discussed later in this chapter, but the smaller $AaPO_2$ with chronic altitude exposure helps to preserve arterial PO_2 for any given level of ventilation.

The efficiency of pulmonary gas exchange is therefore changed, even at rest, with acute or chronic exposure to altitude. Several factors could contribute to this. Changes in air density could affect intrapulmonary gas distribution; increases in ventilation, by changing gas flow rates, could also alter inspired gas distribution.

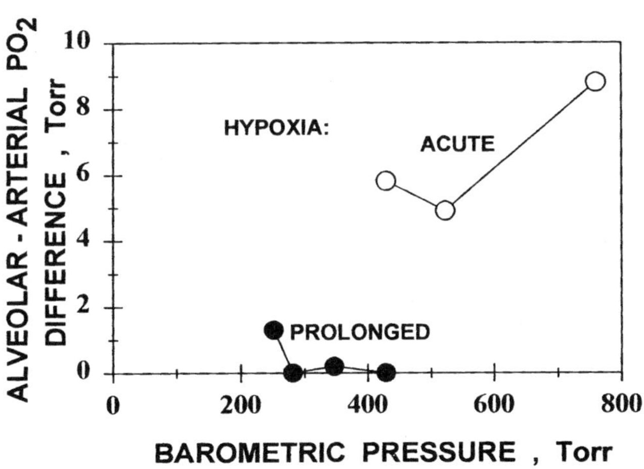

Figure 4 Resting alveolar-arterial PO_2 difference as a function of altitude (6,7). From normal sea level values of <10 torr, the $AaPO_2$ progressively falls with altitude to values close to zero. There appears to be a reduction in $AaPO_2$ with chronic exposure compared to acute exposure at the same altitude. Absolute values of $AaPO_2$ are generally quite small.

The sympathetically mediated tachycardia and increase in cardiac output of acute altitude exposure could lead to blood flow distribution changes, and the development of hypoxic pulmonary vasoconstriction could also redistribute blood flow within the lungs, especially if vasoconstriction occurs unevenly as has been suggested (17). Because alveolar Po_2 is reduced with increasing altitude, vulnerability to alveolar-capillary diffusion limitation increases (18). Finally, to the extent that high-altitude pulmonary edema (HAPE) develops, even subclinically, ventilation and/or blood flow distribution could change and shunts might develop.

Assessing the combined effects of all of these potential phenomena is complex. In particular, for a given (constant) level of ventilation/perfusion ($\dot{V}A/\dot{Q}$) inequality, the $AaPo_2$ falls with increasing altitude as alveolar Po_2 falls onto the steeper part of the O_2-Hb dissociation curve. This is shown in Figure 5 using a computer model having a constant level of $\dot{V}A/\dot{Q}$ mismatching. Log SDQ, an abbreviation for the second moment of the $\dot{V}A/\dot{Q}$ distribution about its mean, on a log scale, is used as an index of the amount of dispersion or heterogeneity of $\dot{V}A/\dot{Q}$ ratios within the lung. Three examples are illustrated: a normal amount of inequality (log SDQ = 0.4) and two moderately increased values of log SDQ (0.7 and 1.0). There are two contributing factors to this reduction in the $AaPo_2$. First, as Po_2 falls onto the steep portion of the O_2-Hb dissociation curve, the range of Po_2 values from low to high $\dot{V}A/\dot{Q}$ ratios is reduced and this will mathematically bring the alveolar and arterial Po_2 values closer together. However, as West showed many years ago (19), a second factor reducing $AaPo_2$ is the increase in overall $\dot{V}A/\dot{Q}$ ratio of the lung due to hyperventilation. This is also a consequence of the shape and slope of the O_2-Hb dissociation curve.

Consequently, the fall in $AaPo_2$ (at rest) with altitude cannot be assumed to reflect an improvement in $\dot{V}A/\dot{Q}$ relationships. Figure 4 is a good example of this dilemma: Does the progressive fall in $AaPo_2$ reflect improving $\dot{V}A/\dot{Q}$ relationships, or is this only a reflection of what is shown in Figure 5? The latter explanation is explored below.

Gas exchange at altitude is further affected by behavior of the $AaPo_2$ when caused by alveolar-capillary diffusion limitation (18). Figure 5 shows how the $AaPo_2$ due to diffusion limitation would *increase* with ascent to altitude even if the pulmonary O_2 diffusing capacity (DL_{O_2}) were to remain constant. These calculations use several values for DL_{O_2}, each held constant with altitude. None is so low as to produce any $AaPo_2$ at rest at sea level as the figure shows.

Examination of Figure 4 and other data (15) in light of the predictions modeled in Figure 5 suggests that because the $AaPo_2$ in fact falls with altitude, diffusion limitation at rest does not occur. Indeed, the fall in measured $AaPo_2$ appears consistent with a constant and normal level of $\dot{V}A/\dot{Q}$ mismatch. Figure 6 replots the data of Figure 4 with the calculations of Figure 5 superimposed (as a dashed line for the case of log SDQ = 0.4, a normal value), and the agreement between the actual data and those computed for log SDQ = 0.4 is generally good. Thus, it is reasonable to postulate that in young normal subjects at altitude there is on average a pattern of gas exchange that simply reflects predictable mathematical effects of reduction in

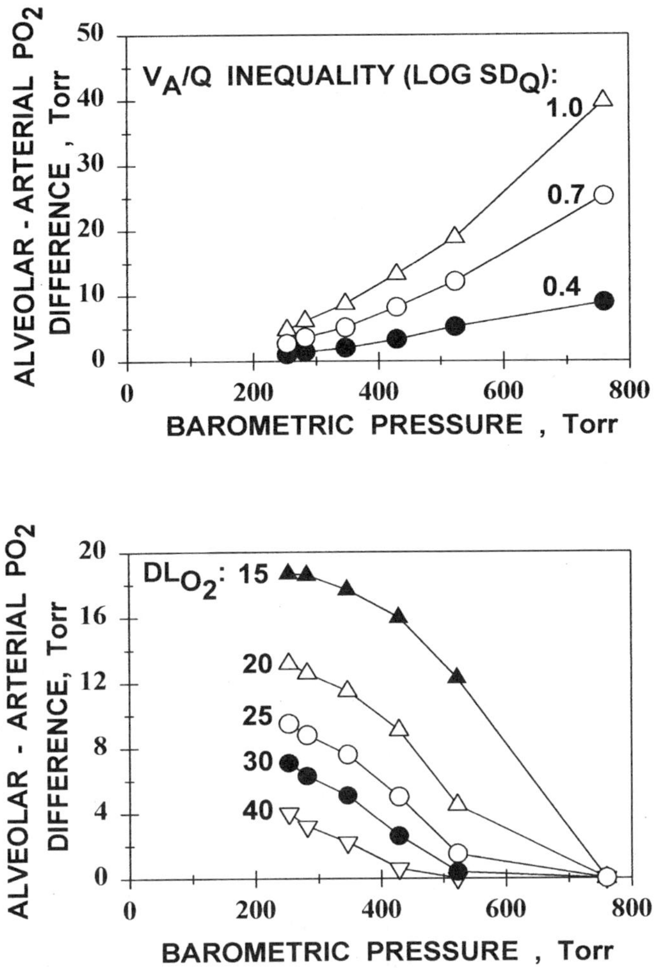

Figure 5 Calculated values for the $AaPo_2$ as a function of altitude in a lung model with a fixed degree of ventilation/perfusion inequality indicated by the dispersion parameter log SDQ (upper panel). The $AaPo_2$ for any value of $\dot{V}A/\dot{Q}$ inequality falls progressively with altitude, with values converging as altitude is gained. The lower panel shows predicted $AaPo_2$ at different altitudes in five lung models, each having a fixed value of oxygen diffusing capacity (DLo_2). Under these resting conditions, $AaPo_2$ is zero at sea level, but rises progressively with altitude according to the value of DLo_2, in sharp contrast to the opposite behavior of the effects of $\dot{V}A/\dot{Q}$ inequality.

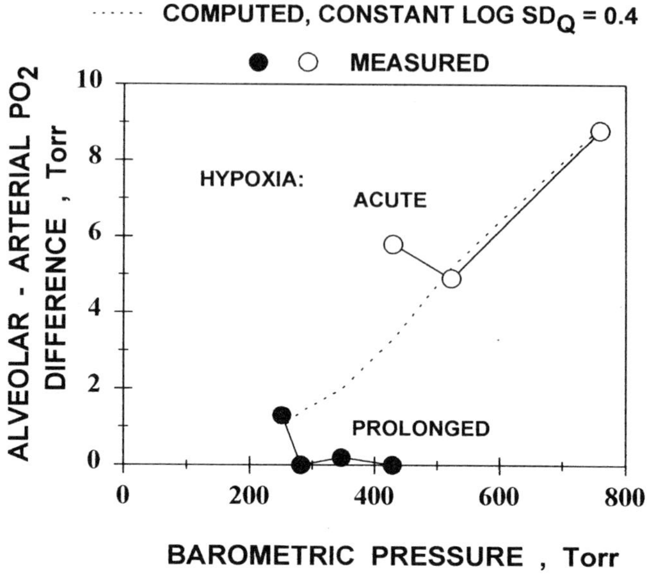

Figure 6 Measured resting alveolar-arterial Po_2 difference in subjects acutely and chronically exposed to hypoxia from sea level to the Everest summit (from Fig. 4). Also plotted from Figure 5 is the expected fall in $AaPo_2$ for a constant, normal degree of $\dot{V}A/\dot{Q}$ mismatch (log SDQ = 0.4), dashed line. Note the scale of the ordinate spanning only 10 torr. No point is more than 2–3 torr from the predicted line. Thus, the measured values are broadly consistent with the expected behavior of a normal lung without diffusion limitation.

PIo_2 without change in $\dot{V}A/\dot{Q}$ relationships or development of diffusion limitation. However, caution is needed in arriving at this conclusion because the $AaPo_2$ is inherently noisy and the curves of Figure 5 tend to converge at low PB, reducing the ability to infer changes in $\dot{V}A/\dot{Q}$ mismatch from the behavior of $AaPo_2$.

To better understand how altitude does (or does not) affect resting pulmonary gas exchange, methodology is required that is not only independent of inspired Po_2, but that can distinguish between $\dot{V}A/\dot{Q}$ mismatch and diffusion limitation. The multi-

Figure 7 Ventilation/perfusion inequality at rest as a function of altitude in acute (○) (6) and prolonged (●) (7) chamber exposures. The top panel shows mean data set against the 95% upper confidence limit for log SDQ of 0.6. The two points marked with an asterisk (*) are in the abnormal range, and as described in the text were observed shortly after periods of rapid decompression. The middle panel shows individual subject data from the same studies, showing variability in individual responses. The bottom panel also shows individual data from the chronically exposed subjects with different symbols according to initial sea level resting ventilation/perfusion mismatch. See text for more details.

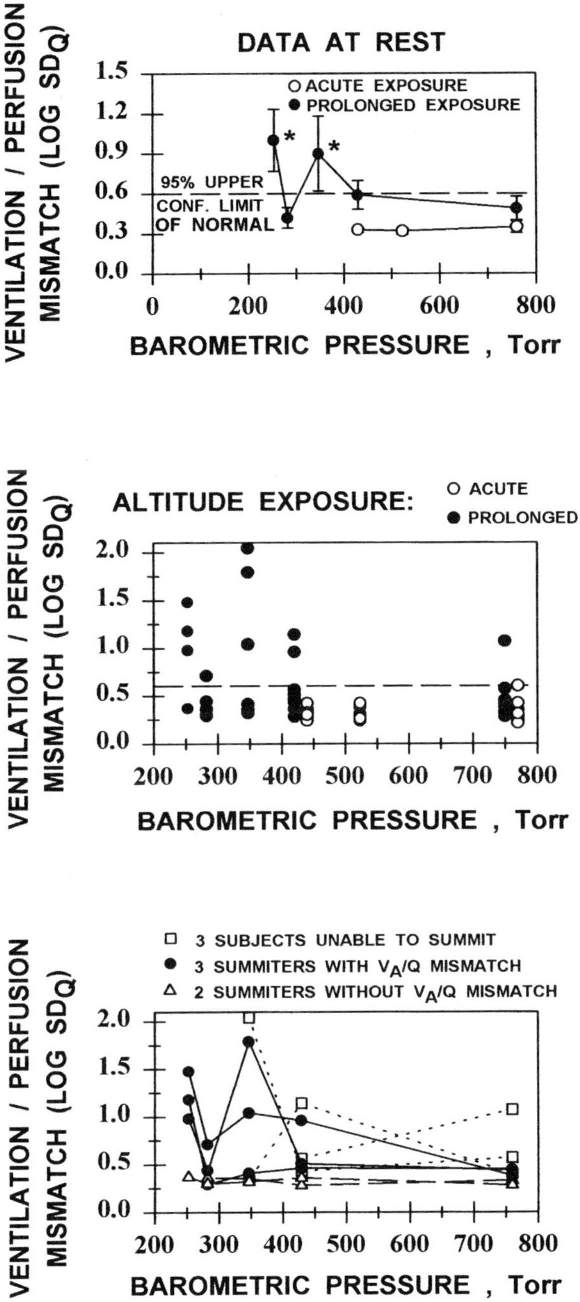
DATA AT REST
O ACUTE EXPOSURE
● PROLONGED EXPOSURE
95% UPPER CONF. LIMIT OF NORMAL
VENTILATION / PERFUSION MISMATCH (LOG SDQ)
BAROMETRIC PRESSURE , Torr
ALTITUDE EXPOSURE:
O ACUTE
● PROLONGED
VENTILATION / PERFUSION MISMATCH (LOG SDQ)
BAROMETRIC PRESSURE , Torr
□ 3 SUBJECTS UNABLE TO SUMMIT
● 3 SUMMITERS WITH VA/Q MISMATCH
Δ 2 SUMMITERS WITHOUT VA/Q MISMATCH
VENTILATION / PERFUSION MISMATCH (LOG SDQ)
BAROMETRIC PRESSURE , Torr

ple inert gas elimination method (20–22) offers both of these possibilities and has been used in chamber simulations of altitude. Figure 7 shows how $\dot{V}_A/\dot{Q}$ mismatch at rest changes with altitude, based on the studies cited earlier (6,7).

Figure 7 is complex. Consider first the data for acute altitude exposure (open circles of top and middle panels). The top panel shows mean data; the middle panel all individual data. In both panels, the amount of $\dot{V}_A/\dot{Q}$ mismatch (i.e., log SDQ) that represents, at sea level, the mean plus 2 SD (i.e., approximate 95% upper confidence limit of normal) is shown by the horizontal dashed line and is 0.6.

All data for acute exposure (Fig. 7, open circles) lie within the 95% confidence limits. This is true at sea level and at both altitudes (corresponding to about 10,000 and 15,000 ft. above sea level). Acute exposure in this study was ascent to the altitude in question over several minutes in a hypobaric chamber. Altitude over this range had no net effect on $\dot{V}_A/\dot{Q}$ inequality. This is in spite of known alterations in ventilatory and circulatory components that acutely accompany hypoxia. In particular, any hypoxia-induced pulmonary hypertension that might more uniformly distribute blood flow up and down the lungs is too subtle to reduce the very small amount of $\dot{V}_A/\dot{Q}$ mismatch present in the normal lungs at sea level (i.e., log SDQ of 0.3–0.4 as the open circles of Fig. 7 show for most subjects). Also, the very small rise in pulmonary artery pressure at rest in hypoxia in the group of subjects in Figure 7 would not likely cause even interstitial pulmonary edema.

Sustained exposure to hypoxia for 2 weeks at an altitude of about 3800 m (at the University of California White Mountain Barcroft Laboratory) revealed no evolution of $\dot{V}_A/\dot{Q}$ changes—again, minimal amounts of $\dot{V}_A/\dot{Q}$ mismatch were noted both before and at the end of the altitude exposure (23), a situation discussed in more detail later in this chapter.

This appears not to be the case when decompression to P_B levels lower than those of normal human habitation is examined. Operation Everest II (OEII) addressed this situation (Fig. 7, all three panels, closed circles). The data up to 15,000 ft. (P_B = 430 torr) are essentially compatible with those of the acute study (open circles). Most subjects were indistinguishable in their $\dot{V}_A/\dot{Q}$ mismatching at sea level and 15,000 ft. as the figure shows. Of potential importance, the subjects who were less altitude-tolerant, i.e., who failed to complete OEII, were those subjects with the greatest degree of resting $\dot{V}_A/\dot{Q}$ mismatch at sea level (open squares, Fig. 7, lower panel). Omitting these subjects makes the two subject groups indistinguishable in $\dot{V}_A/\dot{Q}$ relationships from sea level to 15,000 ft. Statistically, even considering all subjects of both groups, they were not different at the altitudes available for direct comparison (P_B = 760 and 430 torr). On the other hand, at P_B = 347 (~20,000 ft.) and P_B = 253 torr (~Everest summit) there was clear evidence of $\dot{V}_A/\dot{Q}$ mismatch of moderately severe proportions (Fig. 7). There was great variability among subjects, as the data show. Yet at P_B = 282 torr, about 25,000 ft., $\dot{V}_A/\dot{Q}$ relationships were again within normal limits.

How can these inconsistent results be explained? One possibility is the technical irreproducibility of results, but extensive experience has shown that the coefficient of variation of the log SDQ parameter is only about 6–7% when duplicate data are averaged (24). The top panel of Figure 7 shows a factor of 2 separating log SDQ mean values at P_B = 282 torr from those at both 347 and 253 torr. This variability is therefore not explained by technical factors.

A common thread connects the data of P_B = 347 and 253 torr: Rapid decompression immediately proceeding studies at each of these two altitudes (25). On the other hand, data at P_B = 282 torr were collected only after subjects had been at that pressure for several days. Combined with independent clinical assessment of rales in more than one subject (when $\dot{V}_A/\dot{Q}$ mismatch was greater and shunt was present) at P_B = 347 torr (25), a more tenable hypothesis for the abnormal degrees of $\dot{V}_A/\dot{Q}$ mismatch at P_B = 347 and 253 torr is acute high-altitude pulmonary edema (HAPE). Individual log SDQ values of 1.5–2.0 as noted in Figure 7 at these altitudes are as high as seen in the intensive care setting in patients with respiratory distress syndromes, requiring assisted ventilation (26).

An intriguing hypothesis is suggested by the examination of individual resting data (Fig. 7, lower panel). Those subjects with the most sea level $\dot{V}_A/\dot{Q}$ mismatch (open squares) were unable to "summit," becoming intolerant to altitude. Those subjects with the least sea level mismatch (open triangles) had the smoothest clinical experience throughout OEII, never developed abnormal $\dot{V}_A/\dot{Q}$ mismatch, and were able to reach the summit. Those with intermediate initial sea level values of $\dot{V}_A/\dot{Q}$ mismatch (closed circles) summited successfully, but developed occasionally substantial $\dot{V}_A/\dot{Q}$ mismatch during periods of rapid ascent.

While the small number of subjects requires care not to overinterpret the findings, the OEII results do suggest a relationship between normoxic $\dot{V}_A/\dot{Q}$ relationships at sea level and subsequent development of pulmonary gas exchange problems at altitude. Furthermore, only two of the eight OEII subjects escaped without some $\dot{V}_A/\dot{Q}$ mismatch developing, and this is consistent with the idea that altitude-related gas exchange problems, presumably on the basis of some degree of HAPE, may be more prevalent at extreme altitude than current estimates of HAPE incidence would imply. These possibilities require testing with larger numbers of subjects.

II. Gas Exchange During Exercise

A. Overview

Physical activity is so much a part of ascent to altitude that there continues to be great interest in pulmonary gas exchange during exercise at altitude, both in terms of overall ventilatory and cardiovascular responses and with respect to pulmonary exchange efficiency, which is addressed in this section. Overall ventilatory and cardiovascular responses are covered in detail in other chapters of this volume. In acute altitude exposure, both total ventilation and cardiac output increase at a greater rate (with exercise intensity) than at sea level over the same work range (2,27). However, maximal values of both ventilation and cardiac output are similar to sea level values (Fig. 8), reached of course at lower workloads than at sea level. The mechanisms of these heightened responses are discussed in Chapters 6 and 9.

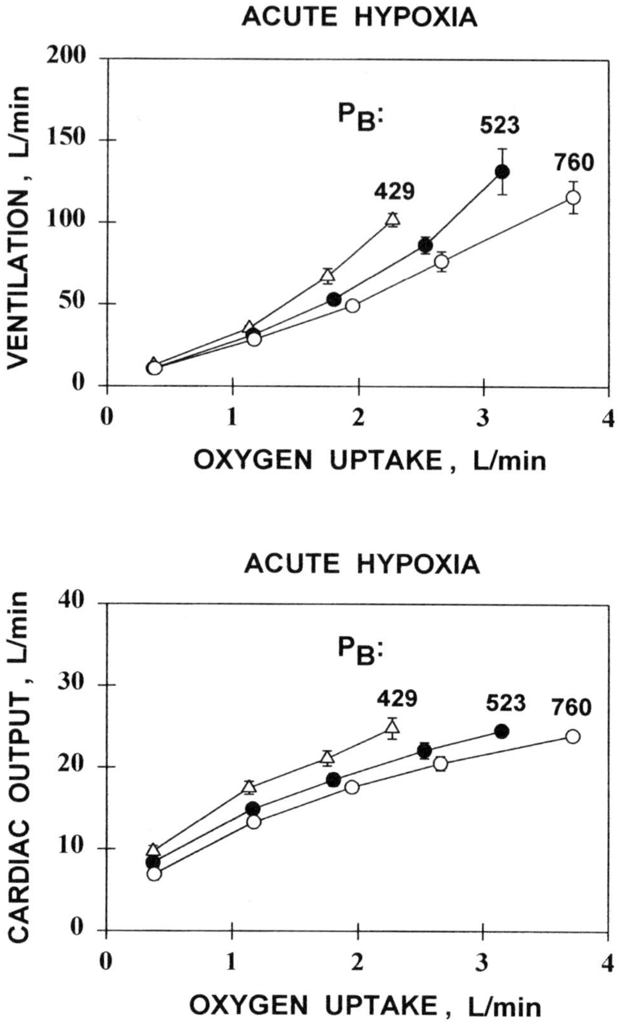

Figure 8 Effects of acute altitude exposure on ventilation and cardiac output at rest and during exercise. Both are higher at altitude than at sea level at a given oxygen uptake, but maximal values are essentially independent of altitude (6).

With chronic altitude exposure, the ventilatory response to exercise is even more exaggerated than with acute exposure, but conversely, the cardiac output response declines and closely follows that for sea level (28,29) (Fig. 9). These differences are likely explained by (1) enhanced ventilatory responses to chronic hypoxia as the blood and CSF alkalosis is gradually compensated by renal bicarbonate excretion and carotid body sensitivity is increased and (2) reduced heart rate response to

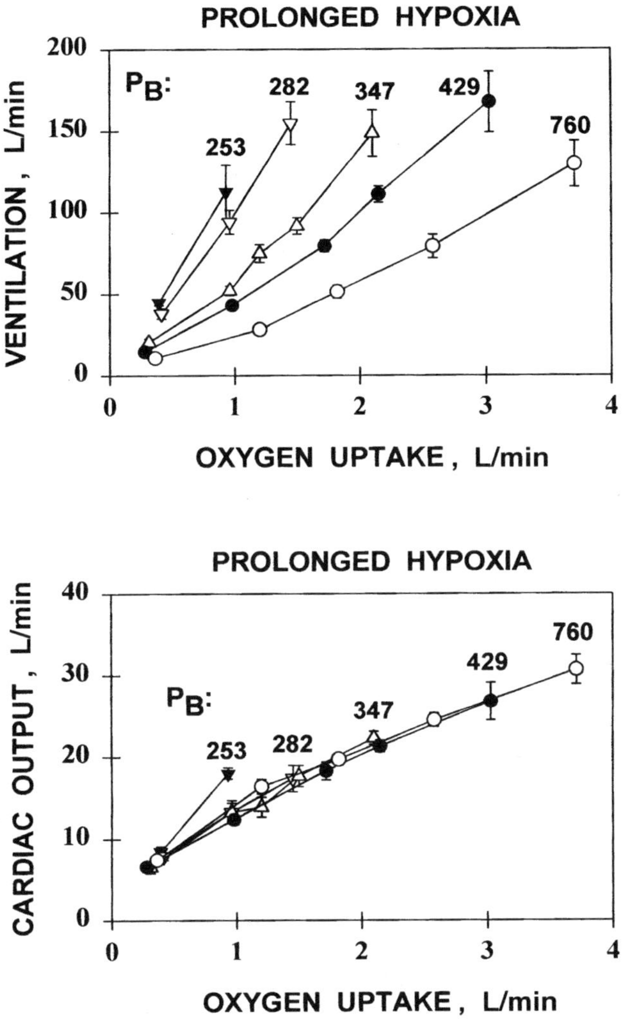

Figure 9 Ventilation and cardiac output at rest and during exercise in prolonged altitude exposure (7). In contrast to acute exposure (Fig. 8), ventilatory responses are even more enhanced than with acute exposure while the cardiac output response essentially reverts to that seen at sea level. Moreover, maximal cardiac output falls progressively with altitude (in contrast to acute exposure).

hypoxia with chronic exposure possibly related to sympathetic receptor downregulation (see Chapters 6 and 9 for much fuller discussion). Stroke volume is also found to be slightly reduced. Such global responses in ventilation and cardiac output affect alveolar P_{O_2} and P_{CO_2} as well as mixed venous P_{O_2} and P_{CO_2}.

Gas exchange efficiency in the lungs suffers during exercise at altitude. Several factors are responsible, but their relative importance changes with altitude. Even at sea level, gas exchange is compromised during exercise with the development of a modest amount of $\dot{V}_A/\dot{Q}$ mismatch at moderate and heavy levels of work. During maximal effort, alveolar-capillary diffusion limitation can be observed at sea level, and there is evidence that the more elite the athlete, the more diffusion limitation can be detected (30–34).

At intermediate altitudes (10,000–15,000 ft.) both $\dot{V}_A/\dot{Q}$ mismatch and diffusion limitation are more in evidence during exercise than at sea level (31,35). At extreme altitudes, there is still more $\dot{V}_A/\dot{Q}$ mismatch and more diffusion limitation as well (7). Diffusion limitation is expected as alveolar P_{O_2} falls onto the steep portion of the O_2-Hb dissociation curve (18) but appears for the most part not to reflect any pathological reduction in diffusing capacity of the lung.

In contrast, ventilation/perfusion mismatch develops as the result of some abnormal process (or processes) created by altitude and exercise in combination. Indirect evidence points to transient pulmonary edema as the basic mechanism (7,36,37).

While both $\dot{V}_A/\dot{Q}$ inequality and diffusion limitation are more apparent the higher the altitude, the implications of Figure 5 must be kept in mind when assessing the effects of these two phenomena on O_2 exchange. Thus, in spite of worsening $\dot{V}_A/\dot{Q}$ relationships with altitude, the effects on arterial P_{O_2} are minimal (Fig. 5), while diffusion limitation increasingly affects arterial P_{O_2} (Fig. 5). On average, $\dot{V}_A/\dot{Q}$ inequality and diffusion limitation are of similar importance to the AaP_{O_2} during sea level exercise, but diffusion limitation becomes much more of a factor at high altitude. Figure 10 shows this relationship for the acutely exposed subjects as well as those of OEII. Despite sometimes major degrees of $\dot{V}_A/\dot{Q}$ mismatch at altitude (see below), diffusion limitation remains the dominant contributor to the AaP_{O_2}, reducing the arterial O_2 concentration by some 4 $mL \cdot dL^{-1}$ at peak exercise on the Everest "summit" as emulated in OEII.

B. Ventilation/Perfusion Relationships During Exercise at Altitude

As mentioned above, $\dot{V}_A/\dot{Q}$ mismatch worsens at any exercise level at altitude. Figure 11 shows this for both acute (6) and chronic (7) altitude exposures, using the same studies as discussed for resting conditions above. Because altitude reduces the component of the AaP_{O_2} due to $\dot{V}_A/\dot{Q}$ mismatch (Fig. 4), the only reliable data on $\dot{V}_A/\dot{Q}$ inequality come from the multiple inert gas elimination technique. The increase in $\dot{V}_A/\dot{Q}$ inequality of Figure 11 is not likely due to hypobaria per se. Sea level studies reducing F_{IO_2} but keeping P_B = 760 torr have produced the same exercise-induced worsening of mismatch as seen in hypobaric exercise breathing room air (35).

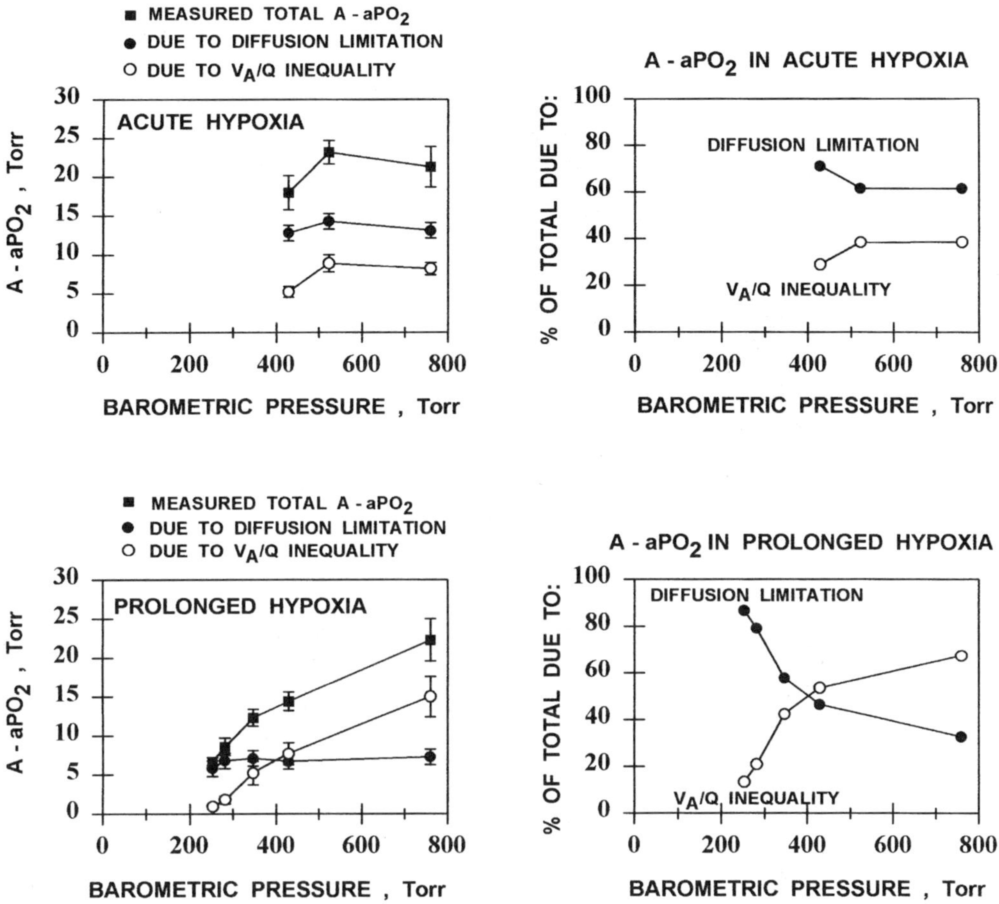

Figure 10 Components of the alveolar-arterial Po_2 difference ($AaPo_2$) as a function of altitude. Data reflect peak exercise levels at each altitude. Top panels reflect acute altitude exposure (6), bottom panels prolonged exposure (7). Total $AaPo_2$ declines progressively with altitude. The fraction due to diffusion limitation increases progressively, while that due to $\dot{V}_A/\dot{Q}$ inequality decreases. Differences between acute and chronic hypoxic groups at sea level presumably reflect differences in pulmonary diffusing capacity and $\dot{V}_A/\dot{Q}$ relationships.

Table 2 summarizes the changes in $\dot{V}_A/\dot{Q}$ mismatch observed to date during exercise in acute hypoxia produced either by hypobaric or normobaric reduction in F_{IO_2}. The changes are all in the same direction, but the magnitude is small and therefore of little importance to O_2 transport.

Chronically hypoxic subjects (Fig. 11) exhibit more variability in response than those studied during acute hypoxia. Three subject responses from OEII make this point (Fig. 12). Subject 3 was essentially insensitive to exercise and altitude in

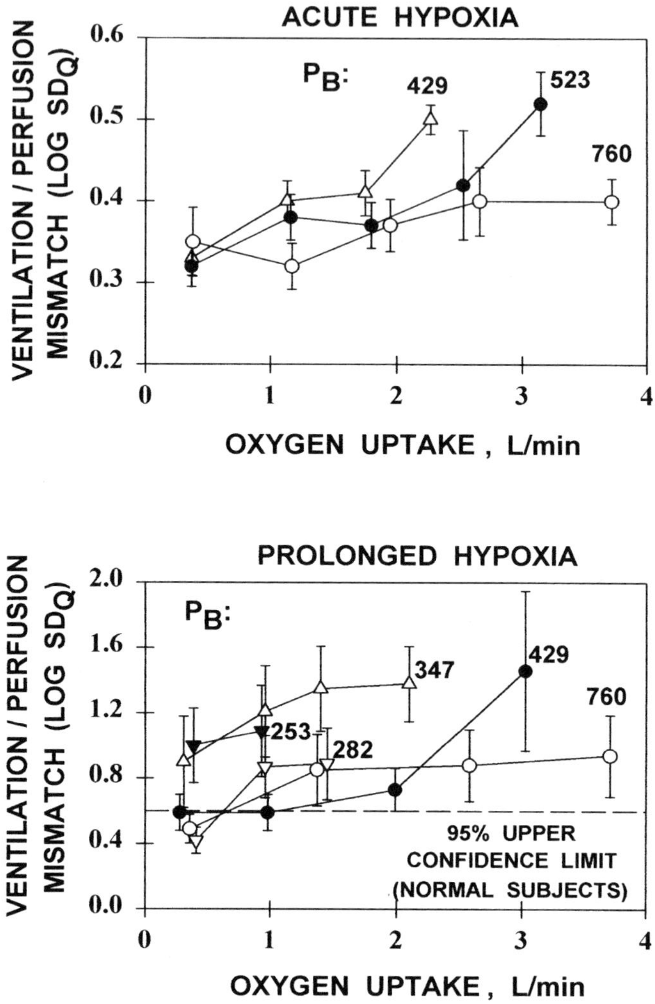

Figure 11 Amount of ventilation/perfusion mismatch (characterized by log SDQ values) in acute (upper panel) and prolonged (lower panel) altitude exposures at rest and during exercise. Note the different ordinate scales. Under both conditions, $\dot{V}_A/\dot{Q}$ inequality increases with exercise and is greater at a given oxygen uptake at altitude than at sea level. There is considerable individual variability, and most subjects show highly abnormal patterns with chronic hypoxia. Log SDQ values above about 1 equate to considerable, clinically significant disturbances of gas exchange.

Table 2 Exercise-Induced $\dot{V}A/\dot{Q}$ Mismatch During Acute Hypoxia: Literature Summary (mean ± SE)

Ref.	PB (torr)	FIO_2	PIO_2 (torr)	Log SDQ at rest	Log SDQ heavy exercise	$\dot{V}O_2$ ($L \cdot min^{-1}$) heavy exercise
6	760	0.21	149	0.35 ± 0.04	0.40 ± 0.03	3.72
6	523	0.21	100	0.32 ± 0.02	0.52 ± 0.04	3.15
6	429	0.21	80	0.33 ± 0.02	0.50 ± 0.02	2.27
35	760	0.21	149	0.38 ± 0.02	0.48 ± 0.02	2.74
35	760	0.12	86	0.41 ± 0.03	0.62 ± 0.09	2.31

terms of developing $\dot{V}A/\dot{Q}$ inequality (Fig. 12, top panel). Only at PB = 347 torr at the heaviest level was there an increase in mismatch to above normal as defined by the 95% upper confidence limit at sea level of the dispersion parameter log SDQ. In sharp contrast, subject 6 developed unusually large degrees of inequality on exercise (log SDQ exceeding 1.5) at several altitudes. Subject 9 showed little or no mismatch at sea level or moderate altitude but developed severe inequality at PB = 347 torr, the lowest pressure at which data could be obtained from him. Recall that normally, log SDQ ranges from about 0.3 to 0.6 at rest at sea level and that the most severely ill patients with respiratory failure in the intensive care setting typically have log SDQ values of 2.0–2.5. Subject 9 clearly became ill at PB = 347 and developed rales on clinical exam consistent with pulmonary edema. Such pathological signs were not seen in subject 6 (middle panel). Why there were marked intersubject differences is not known.

Figures 11 and 12 indicate that altitude increases the degree of exercise-induced $\dot{V}A/\dot{Q}$ mismatch. Figure 13 shows that a given level of exercise causes greater $\dot{V}A/\dot{Q}$ mismatch the higher the altitude, both in acute and chronic altitude exposure.

Subjects with a prior history of high-altitude pulmonary edema (HAPE) display a number of physiological differences from HAPE-free subjects: In prior HAPE victims, the hypoxic pulmonary vasoconstriction response is exaggerated (38), recent evidence suggests a higher wedge pressure during exercise as well (39), implying greater pulmonary capillary pressures than in normal subjects. Exercise-induced $\dot{V}A/\dot{Q}$ mismatch is greater in prior HAPE victims than in normal subjects (40), and in these studies it was also found that $\dot{V}A/\dot{Q}$ mismatch was related closely to pulmonary artery pressure (Fig. 14) but not to total ventilation or cardiac output. Thus, as pulmonary artery pressure rose with increasing exercise, so too did $\dot{V}A/\dot{Q}$ inequality. It is therefore clear that exercise at altitude worsens $\dot{V}A/\dot{Q}$ relationships in normal subjects and more so in prior HAPE victims.

The key question, of course, is what causes $\dot{V}A/\dot{Q}$ relationships to deteriorate during exercise, especially so at altitude? Some possible reasons are discussed in the following paragraphs.

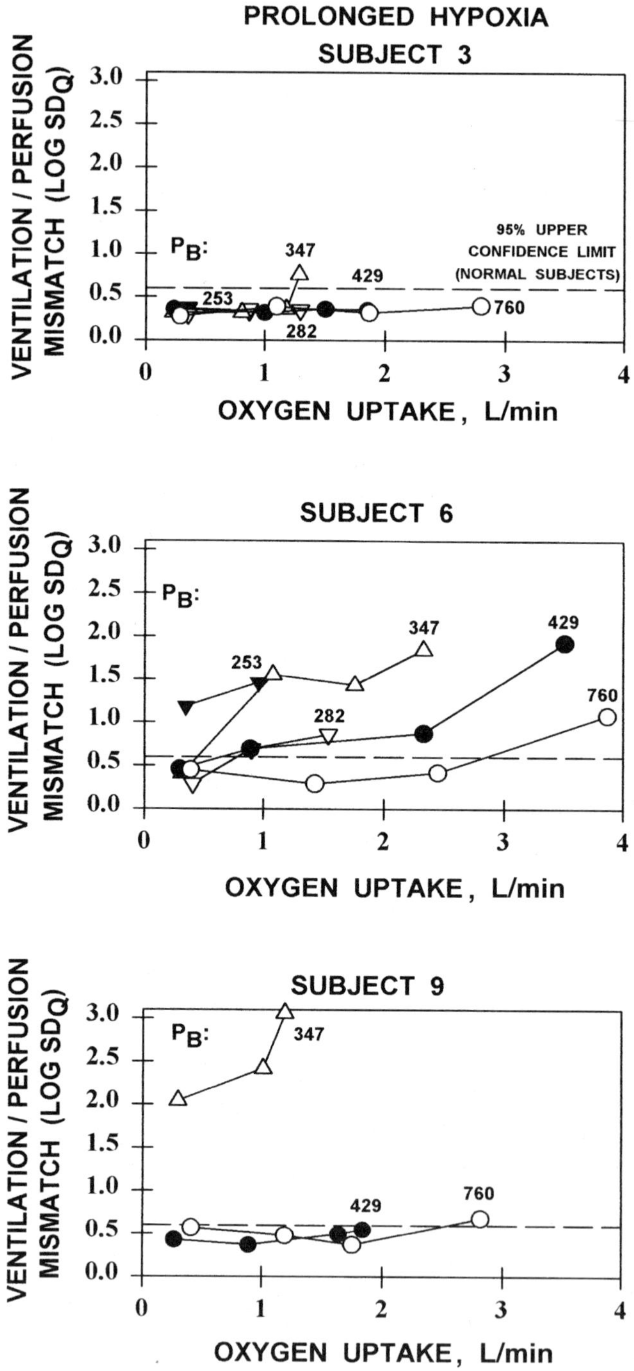
PROLONGED HYPOXIA
SUBJECT 3
VENTILATION / PERFUSION MISMATCH (LOG SDQ)
PB:
347
429
253
282
760
95% UPPER CONFIDENCE LIMIT (NORMAL SUBJECTS)
OXYGEN UPTAKE, L/min
SUBJECT 6
VENTILATION / PERFUSION MISMATCH (LOG SDQ)
PB:
347
429
253
760
282
OXYGEN UPTAKE, L/min
SUBJECT 9
VENTILATION / PERFUSION MISMATCH (LOG SDQ)
PB:
347
760
429
OXYGEN UPTAKE, L/min

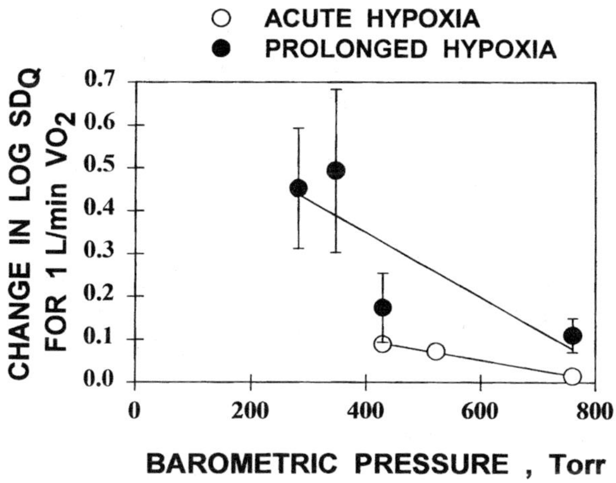

Figure 13 Summary of the effects of altitude on $\dot{V}A/\dot{Q}$ mismatch expressed as the increase in $\dot{V}A/\dot{Q}$ inequality (log SDQ) per liter per minute increase in oxygen uptake during exercise. In both acute and prolonged exposures, the increase in log SDQ per liter $\dot{V}O_2$ is progressively greater at altitude than sea level, while prolonged exposure caused greater changes.

One possibility is alterations in ventilation distribution due to altered gas density, magnified by the high gas flow rates of exercise. Exercise- or cold, dry air–induced bronchoconstriction could be a factor as well. However, despite extreme $\dot{V}A/\dot{Q}$ mismatch observed in OEII, spirometric tests performed immediately after exercise failed to provide any evidence at all of airways obstruction, central or peripheral (41). Moreover, $\dot{V}A/\dot{Q}$ inequality persists beyond the time for essentially complete return of ventilation to resting values (36), suggesting that high rates of airflow do not cause the $\dot{V}A/\dot{Q}$ mismatch of exercise. Finally, prior HAPE victims were found to have more $\dot{V}A/\dot{Q}$ mismatch but lower minute ventilation during exercise than normal subjects (40), supporting this conclusion. For all of these reasons, factors related to airway mechanics and/or total ventilation are thought not to be causative of greater $\dot{V}A/\dot{Q}$ mismatch on exercise, at sea level or at altitude.

Figure 12 Individual responses of ventilation/perfusion relationships to exercise and altitude in three subjects from Operation Everest II (prolonged exposure). Subject 3 (top) was essentially invulnerable to the combined effects of exercise and altitude, log SDQ staying within the normal range for virtually the entire study. Subject 6 shows data that, while normal at rest (except on the summit), increased to above normal values at each altitude as exercise was undertaken. Subject 9 (bottom) showed little response to exercise at altitudes up to 15,000 ft. (PB = 429) but became clinically ill at the next altitude studied (PB = 347 torr). Chest rales indicated acute high altitude pulmonary edema, and log SDQ values reached levels equivalent to those seen in patients in intensive care with severe lung disease.

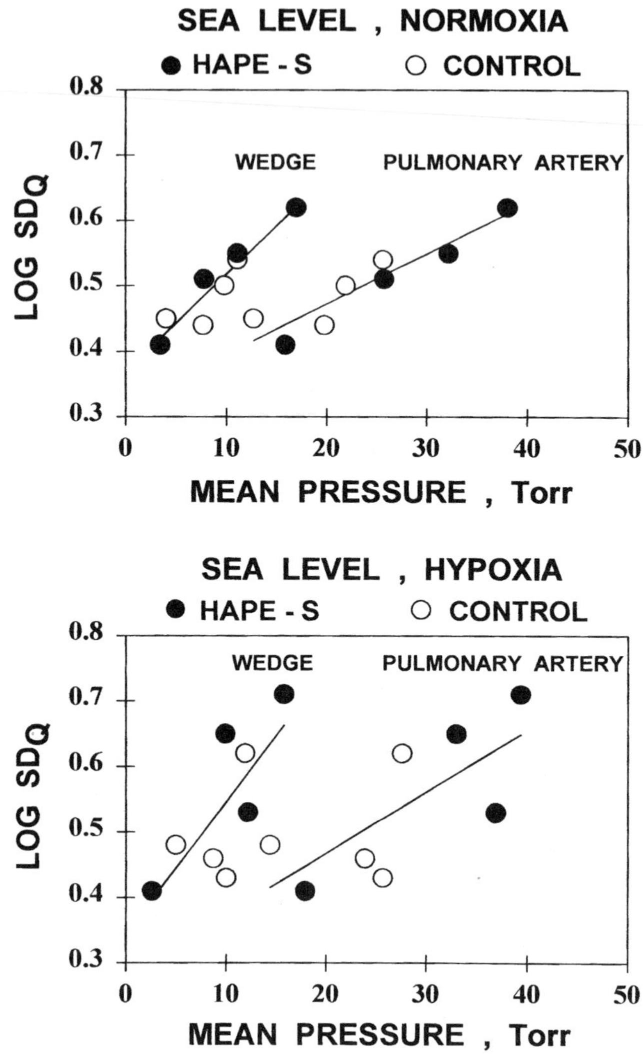

Figure 14 Relationship of $\dot{V}_A/\dot{Q}$ inequality (log SDQ) to pulmonary vascular pressures in normal control subjects (open circles) and subjects with a prior history of high altitude pulmonary edema (HAPE-S, closed circles). Normoxic exercise data are shown in the upper panel, those during acute hypoxia in the lower panel. Each data point is the average for all subjects at a given exercise level in each group. There is a close linear relationship between $\dot{V}_A/\dot{Q}$ inequality and pulmonary vascular pressures (whether gauged by wedge or pulmonary artery values) in both normoxia and hypoxia. Prior HAPE subjects develop both higher pressures and more $\dot{V}_A/\dot{Q}$ inequality but the relationship between log SDQ and pressure is the same for both groups.

A second possibility is that $\dot{V}A/\dot{Q}$ mismatch is the result of transient, mild pulmonary edema. Except for extreme conditions (e.g., in OEII at ≥20,000 ft.), this is probably interstitial rather than alveolar and is fairly mild. It is well known that transcapillary fluid flux is greatly increased during exercise due to increases in intravascular pressure (42), and it seems reasonable to postulate that, if lymph clearance capacity is transiently exceeded during heavy exercise, interstitial edema may develop. This should be related to pulmonary hypertension and should therefore become more apparent with altitude (due to hypoxic vasoconstriction). The available observations support these expectations. The persistence of $\dot{V}A/\dot{Q}$ mismatch beyond the postexercise period required for ventilation and cardiac output to return to resting levels is further consistent with this idea (36), and the elimination of $\dot{V}A/\dot{Q}$ worsening by breathing 100% O_2 during exercise at altitude is further evidence supporting the relationship between hypoxia, pulmonary artery pressure, and $\dot{V}A/\dot{Q}$ mismatch (43). Unfortunately, direct confirmation of this hypothesis in humans is lacking because the tools needed to identify minor degrees of interstitial edema do not exist. Exercise-induced perivascular cuffing has been observed in the pig (37), supporting the hypothesis of $\dot{V}A/\dot{Q}$ inequality caused by edema. Some reports of radiographic techniques support the development of mild edema but remain somewhat equivocal (44).

In OEII in at least one subject, acute pulmonary edema developed during exercise at PB = 347 torr. There were rales (25), gross $\dot{V}A/\dot{Q}$ mismatch (see Fig. 12) and a large shunt of over 30% of the cardiac output, similar to values seen in many intensive care unit (ICU) patients with respiratory failure (26). Whether the more common, milder degrees of exercise-induced mismatch differ only by degree or rather by basic mechanism from the HAPE observed in the above subject cannot be resolved, but Figure 14 suggests the former, since there is a smooth, single relationship between pulmonary artery pressure and $\dot{V}A/\dot{Q}$ mismatch during exercise.

C. Pulmonary Diffusion Limitation of O_2 Exchange During Exercise at Altitude

Diffusion of O_2 in the lungs is usually thought of (a) as responsible for mixing of inspired and alveolar gas and (b) as the process of transfer of O_2 from alveolar gas to capillary blood. Diffusion limitation becomes more and more apparent with increasing altitude at a given workload. The problem is O_2 diffusion between alveolar gas and capillary blood, rather than any diffusive limitation to gas mixing in the alveolar spaces. The latter is barely detectable as a gas exchange defect (45) while the former becomes a major factor limiting exercise at altitude.

Why should diffusion limitation become a problem at altitude? The answer is deduced best by examining Fick's first law of diffusion:

$$\text{FLUX} = A \cdot k \cdot \alpha \cdot [PAO_2 - PCAPO_2]/(\sqrt{MW} \cdot T) \qquad (2)$$

where the diffusive FLUX is a function of the alveolar surface area available (A), the physical properties of the alveolar wall (represented by a diffusion constant k), the solubility (α) of O_2 in the Hb-free water of the blood-gas barrier, the molecular weight (MW) of O_2, the thickness of the alveolar blood-gas barrier (T), and the P_{O_2} difference between alveolar gas (P_{AO_2}) and the pulmonary capillary (P_{CAPO_2}) at any instant. For a given capillary blood volume (Vc), the FLUX of O_2 is equivalent to the rate of change of blood O_2 concentration $[O_2]$. This in turn is equal to the product of the rate of change of blood O_2 partial pressure, P_{CAPO_2}, and the instantaneous slope of the O_2-Hb dissociation curve, β. Thus:

$$\text{FLUX} = d[O_2]/dt \times V_C/100 = \beta \cdot dP_{CAPO_2}/dt \cdot V_C/100 \tag{3}$$

Combining Eqs. (2) and (3) gives:

$$P_{CAPO_2}/dt = (A/(T \cdot V_C)) \cdot (100k) \cdot (\alpha/(\beta\sqrt{MW})) \cdot (P_{AO_2} - P_{CAPO_2}) \tag{4}$$

The key point of Eq. (4) is that the rate of equilibration (i.e., the rate of rise of P_{CAPO_2}) is dependent on the gas-specific ratio $\alpha/(\beta\sqrt{MW})$. Between sea level and altitude, there is likely no major change in pulmonary microvascular structure [i.e., A, T, Vc, or k of Eq. (4)], but the ratio α/β is very dependent on altitude. While α (O_2 water solubility) is a constant independent of altitude, β (the slope of the O_2 dissociation curve) increases with altitude, as shown in Figure 15. Since β is in the denominator of Eq. (4), the rate of diffusion equilibration falls substantially as altitude is gained. During exercise at sea level, β is about 0.2 mL $O_2 \cdot dL^{-1} \cdot torr^{-1}$, while on the Everest summit, OEII data show that β is about 0.7. Even the probably longer capillary transit time associated with a reduced maximal cardiac output cannot overcome this effect of increase in β.

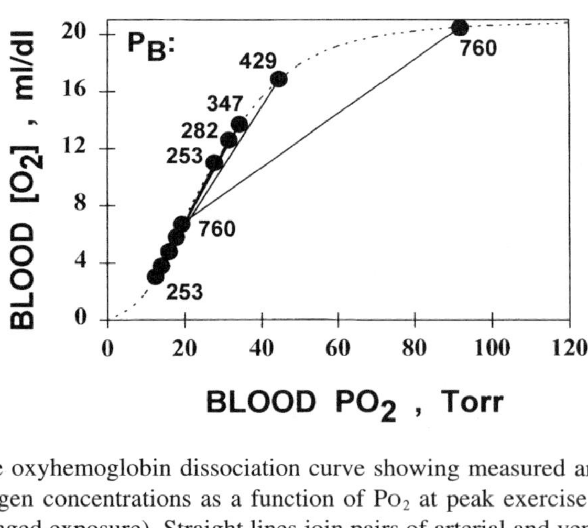

Figure 15 The oxyhemoglobin dissociation curve showing measured arterial and pulmonary arterial oxygen concentrations as a function of P_{O_2} at peak exercise during Operation Everest II (prolonged exposure). Straight lines join pairs of arterial and venous points, showing that the average slope of the oxyhemoglobin dissociation curve increases with altitude. This increase in average slope retards diffusion equilibration in the pulmonary capillaries as explained in the text.

It is this mathematical consequence of the O_2-Hb curve shape that underlies the susceptibility of the lungs to diffusion limitation at altitude. It is not necessary to invoke pathological changes in O_2-diffusing capacity per se. In fact, OEII data indicate the converse—pulmonary O_2 diffusion capacity at a given $\dot{V}O_2$ actually *increases* with altitude (7), probably due to the associated increase in pulmonary artery pressure and/or blood hemoglobin level, both of which would, if anything, increase the effective surface area due to more red cells in more perfused capillaries. The assessment of O_2-diffusing capacity (DLO_2) at altitude is complicated by the development of $\dot{V}A/\dot{Q}$ inequality and its small but not insignificant (Fig. 10) contributions to the total $AaPO_2$. However, using the multiple inert gas elimination technique, DLO_2 can be estimated (46). The portion of the total $AaPO_2$ due to measured $\dot{V}A/\dot{Q}$ inequality is subtracted from the actual $AaPO_2$ and the DLO_2 value necessary to account for the remainder of the $AaPO_2$ calculated. A particular assumption is that throughout the lungs DLO_2 and blood flow ($\dot{Q}$) are correlated such that $DLO_2/\dot{Q}$ ratios are uniform. While this is an untestable hypothesis, it results in the smallest overall value for DLO_2 required to explain the data and is a useful parameter in a comparative sense (to determine if DLO_2 is affected by altitude, for example). Calculations of DLO_2 in this manner at sea level are not always reliable. This is seen when the $AaPO_2$ is due mostly to $\dot{V}A/\dot{Q}$ inequality.

Using this approach, DLO_2 was assessed during exercise in acute altitude exposure to $PB = 523$ and 429 (6) and also during the prolonged exposure of OEII (7). Figure 16 shows the effects of progressive altitude on DLO_2 at peak exercise in these

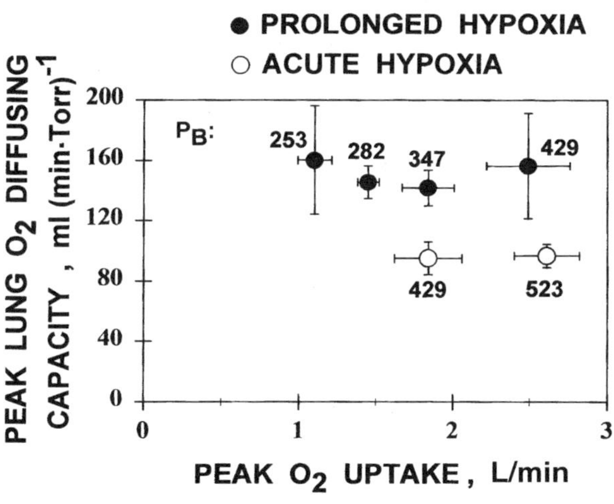

Figure 16 Calculated diffusing capacity of the lungs for oxygen (DLO_2) during peak exercise in both prolonged and acute hypoxia. While peak $\dot{V}O_2$ clearly falls progressively with altitude (indicated by barometric pressure, PB) peak DLO_2 in each group is unaffected by altitude.

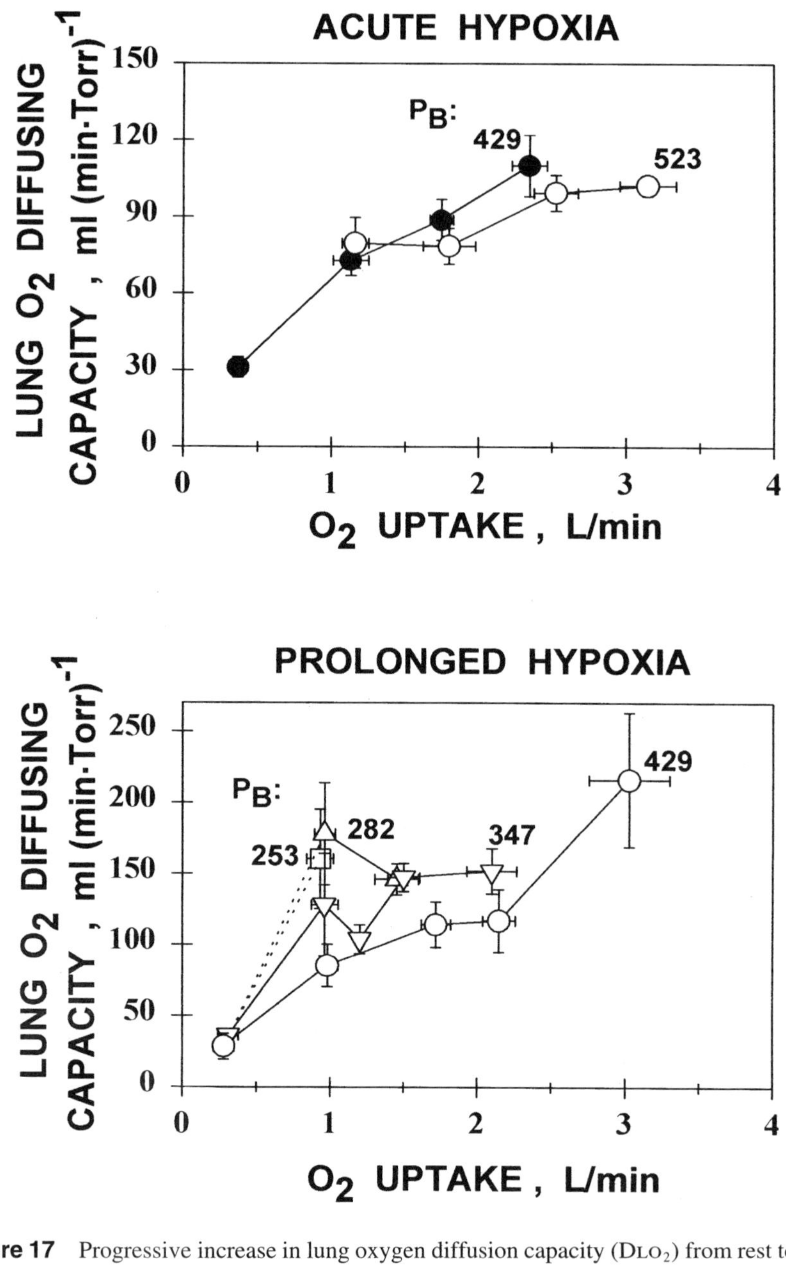

Figure 17 Progressive increase in lung oxygen diffusion capacity (DLO_2) from rest to exercise in acute and in prolonged hypoxic exposures. The relationship is not measurably dependent upon altitude in acute hypoxia (upper panel), but in the lower panel, at an oxygen uptake of 1 L · min^{-1}, DLO_2 increases progressively with altitude.

two scenarios (sea level data not computed). Neither acute nor chronic exposure has any effect on DLO_2 at peak exercise. More data need be collected to be confident of these results.

If one examines how estimated DLO_2 varies with exercise intensity from the same studies, one obtains the results depicted in Figure 17. DLO_2 increases with exercise at any altitude in both acute and chronic exposure, consistent with the older reports using carbon monoxide to measure diffusing capacity (47). In acute hypoxia, the relationship is not measurably different between the altitudes studied (Fig. 17, top), but with prolonged hypoxia DLO_2 clearly increased with altitude at submaximal, constant $\dot{V}O_2$ (e.g.,$\sim$1 L · min^{-1}, Fig. 17, bottom).

Figure 18 relates DLO_2 at 60 watts ($\dot{V}O_2 \sim 1$ L · min^{-1}) to PB using the data of Figure 17 and demonstrates this increase with altitude. One can imagine at least two explanations: (1) greater numbers of perfused capillaries at altitude due to pulmonary hypertension or (2) greater DLO_2 due to the increase in [Hb] with altitude. If one uses standard correction equations (48) for the effect of the rise in [Hb], from about 15 to about 18 g dL^{-1} on diffusing capacity, seen in OEII (Fig. 18, lower panel), there is still a large increase in DLO_2 that is not accounted by increases in [Hb]. Figure 18, middle panel, shows for the OEII subjects a relationship between mean pulmonary artery pressure and DLO_2. The vascular pressure data at 429 torr were determined by interpolation using measured values (at the same work rate) at PB = 760, 347, 282, and 253 torr. The conclusion from Figure 18 is that DLO_2 at constant $\dot{V}O_2$ increases substantially with altitude probably due to greater perfusion of the capillary network. This is speculative for several reasons: it depends on data from small numbers of subjects. Second, [Hb] in the OEII subjects did not increase as much as expected due to very frequent blood sampling for the large number of research projects, so that the importance of [Hb] to DLO_2 may have been underestimated. Third, the moderately extensive degrees of $\dot{V}A/\dot{Q}$ inequality described above at high altitude may have in fact reduced DLO_2 if alveolar or interstitial edema developed, since edema could increase blood-gas barrier thickness. If so, DLO_2 would have been even higher at high altitude had such problems not developed.

Finally, the reader is again referred to Figure 10 for a summary of the relative contributions of diffusion limitation and $\dot{V}A/\dot{Q}$ mismatch to the $AaPO_2$ during both acute and chronic altitude exposure. This figure is consistent with Figure 5, and the anticipated effects of these two processes on gas exchange as altitude is gained.

III. Effects of Acclimatization on Pulmonary Gas Exchange

To this point, this chapter has dealt with either acute hypoxic exposures over the course of minutes or prolonged exposures with steady continuing altitude ascent over several weeks. A totally different issue arises when subjects acutely ascend to a given altitude and then remain at that altitude for several days to weeks, achieving acclimatization. Does acclimatization at a given altitude affect pulmonary gas exchange, either at rest or during exercise? Very little work has explored this question

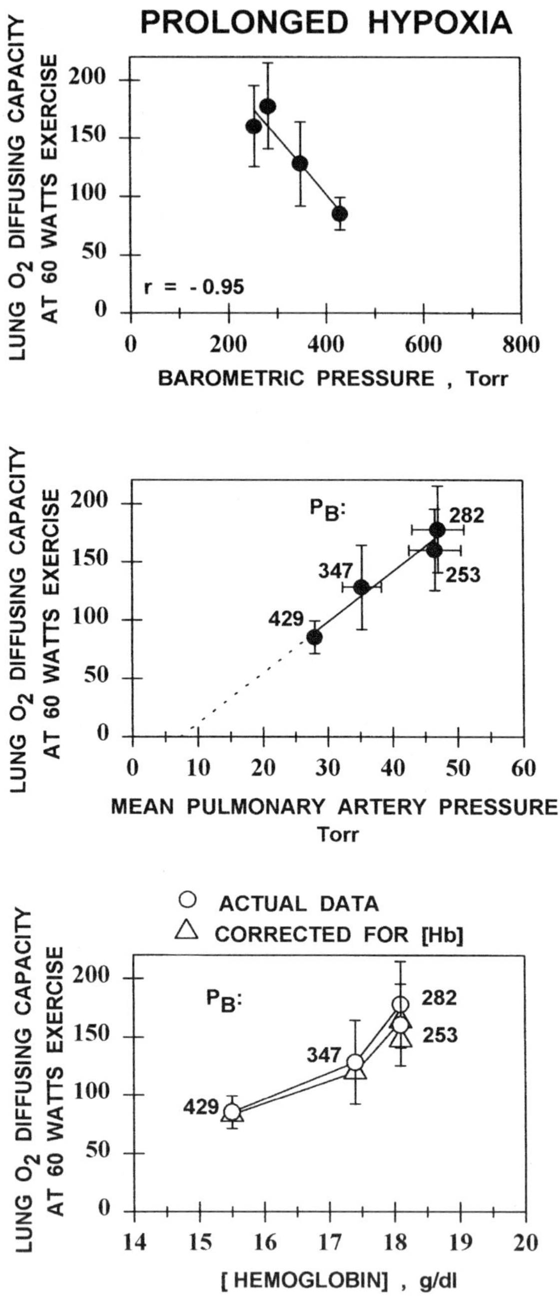

PROLONGED HYPOXIA
LUNG O2 DIFFUSING CAPACITY AT 60 WATTS EXERCISE
r = - 0.95
BAROMETRIC PRESSURE , Torr
PB:
282
253
347
429
MEAN PULMONARY ARTERY PRESSURE Torr
ACTUAL DATA
CORRECTED FOR [Hb]
[HEMOGLOBIN] , g/dl

because field and chamber studies alike have rarely opted to spend several days at any one altitude, let alone permit serial studies at any one altitude.

This issue was addressed at the White Mountain Research Station in California, where pulmonary gas exchange was assessed before and after 2 weeks of residence at 3800 m (P_{IO_2} = 90 torr). Data were collected both at rest and during exercise (23).

Acclimatization led to an increase in ventilation as expected, with concomitant increases in arterial P_{O_2} and decreases in arterial P_{CO_2}. While these changes were anticipated, there was an unexpected small fall in the alveolar-arterial P_{O_2} difference (AaP_{O_2}) at any level of exercise. This was consistent with a prior report of increasing arterial O_2 saturation after acclimatization (49). Using the multiple inert gas elimination technique, it was shown that of the possible reasons for a lower AaP_{O_2} (less $\dot{V}_A/\dot{Q}$ mismatch, higher lung O_2-diffusing capacity, or greater diffusion equilibration due to higher ventilation and lower cardiac output after acclimatization), only the latter was found to occur. In other words, improved gas exchange was simply the consequence of a longer pulmonary capillary red cell transit time due to reduction in cardiac output and higher alveolar P_{O_2} from increased ventilation (increasing the alveolar-capillary P_{O_2} difference and thus the rate of diffusion). It is of interest to note that systemic O_2 delivery (the product of arterial O_2 concentration and cardiac output) was unaffected by acclimatization—the reduced cardiac output was balanced by increased blood $[O_2]$. This observation is consistent with the known lack of effect of such short-term acclimatization on maximal $\dot{V}_{O_2}$ (50).

IV. Muscle Tissue Gas Exchange

This chapter has focused on pulmonary gas exchange, since major changes occur in the lungs and cardiovascular systems as described. However, the processes allowing O_2 to be extracted from muscle microcirculatory blood and to reach the mitochondria are also subject to effects of altitude and may also contribute to exercise limitation.

It has long been known that even in healthy subjects exercising maximally at sea level, venous blood from muscle is never fully O_2-depleted (51). This failure to completely extract O_2 could reflect a transport limitation for O_2 in muscle (due to

---

Figure 18 Analysis of potential causes of the increase in D_{LO_2} (constant work rate, 60 watts) with increasing altitude. Top panel shows the significant increase in D_{LO_2} with altitude. Middle panel relates D_{LO_2} to mean pulmonary artery pressure at these altitudes, with an almost proportional, linear relationship between the two. When D_{LO_2} is plotted against blood hemoglobin concentration (lower panel), there is also a positive correlation, but correction of D_{LO_2} for the increase in hemoglobin concentration according to standard formulas shows that hemoglobin concentration is a minor contributor to the increase in D_{LO_2} with altitude. See text for more details.

blood flow/$\dot{V}O_2$ heterogeneity, shunts, or diffusion limitation), or it could reflect maximal mitochondrial O_2 utilization rates such that the residual venous O_2 reflects unusable O_2 that mitochondria could not burn due to their being in a state of maximal ATP turnover. To distinguish between these two alternatives, one can increase arterial O_2 availability by increasing P_{IO_2}, muscle blood flow, or [Hb] and determine whether maximal $\dot{V}O_2$ is increased or not. Each of these manipulations has been studied, and each increases maximal $\dot{V}O_2$, allowing the conclusion that the residual venous O_2 represents an O_2 transport limitation in the muscle (see Ref. 52 for review). These findings are consistent with the fall in maximal $\dot{V}O_2$ with altitude and the immediate restoration of $\dot{V}O_2$max (completely in acute hypoxia and almost completely in prolonged hypoxia) when the hypoxia is eliminated by breathing 100% O_2.

As normal subjects exercise maximally at increasing altitudes, both arterial and muscle venous O_2 levels fall in parallel with maximal $\dot{V}O_2$ (53). Figure 19 shows how the principal components of the O_2 transport chain respond to maximal exercise in subjects of OEII at each altitude. Mixed venous O_2 levels are still considerably greater than zero, even on the summit. More interestingly, there are correlations between maximal $\dot{V}O_2$ and total O_2 transport, venous P_{O_2} (at $\dot{V}O_2$max), and mean capillary P_{O_2}, as Figure 20 shows from the acute hypoxic exposure (6) and OEII data (11), respectively. Note that in OEII the correlations of $\dot{V}O_2$max to venous and mean capillary P_{O_2} fail below P_B = 347 torr.

The existence of these relationships is postulated to shed light on the role of O_2 transport in muscle in contributing to the reduction in $\dot{V}O_2$max at altitude. The data of Figure 20 (acute hypoxia) are compatible with the notion of a constant conductance for O_2 in muscle between the red cell and the mitochondria as P_B is varied. Conductance is the ratio of maximal $\dot{V}O_2$ to the mean P_{O_2} difference between the capillaries and the mitochondria and is the slope of the line in Figure 20 (lower panel) if mitochondrial P_{O_2} at $\dot{V}O_2$max is negligibly small (54). If mean capillary P_{O_2} falls in proportion to muscle venous P_{O_2} as altitude is gained, this conductance will be proportional to the slope of the plot of $\dot{V}O_2$max to venous P_{O_2} at the various altitudes (Fig. 20, middle). Figure 20 does indicate proportionality and thus a constant conductance as altitude is changed.

This conductance is clearly finite since complete O_2 extraction does not occur, but whether it is limited by diffusive properties of the intramuscular O_2 transport pathway or rather by inhomogeneity of blood flow with respect to metabolic rate or by muscle shunts cannot be deduced from the data presented. Animal studies in which mean capillary P_{O_2}, but not convective O_2 delivery, is changed by altering Hb P_{50} show $\dot{V}O_2$max changing in proportion to mean capillary P_{O_2} at constant O_2 delivery (55). Such an outcome is not compatible with inhomogeneity but is the expected consequence of diffusion limitation. The residual O_2 in venous blood is thus hypothesized to result from diffusion limitation of intramuscular O_2 transport. Yet other studies have pointed to the importance of the amount of capillary structure in muscle as the key determinant of the total muscle conductance (56), and in addition, theoretical calculations suggest that muscle diffusive conductance is a more

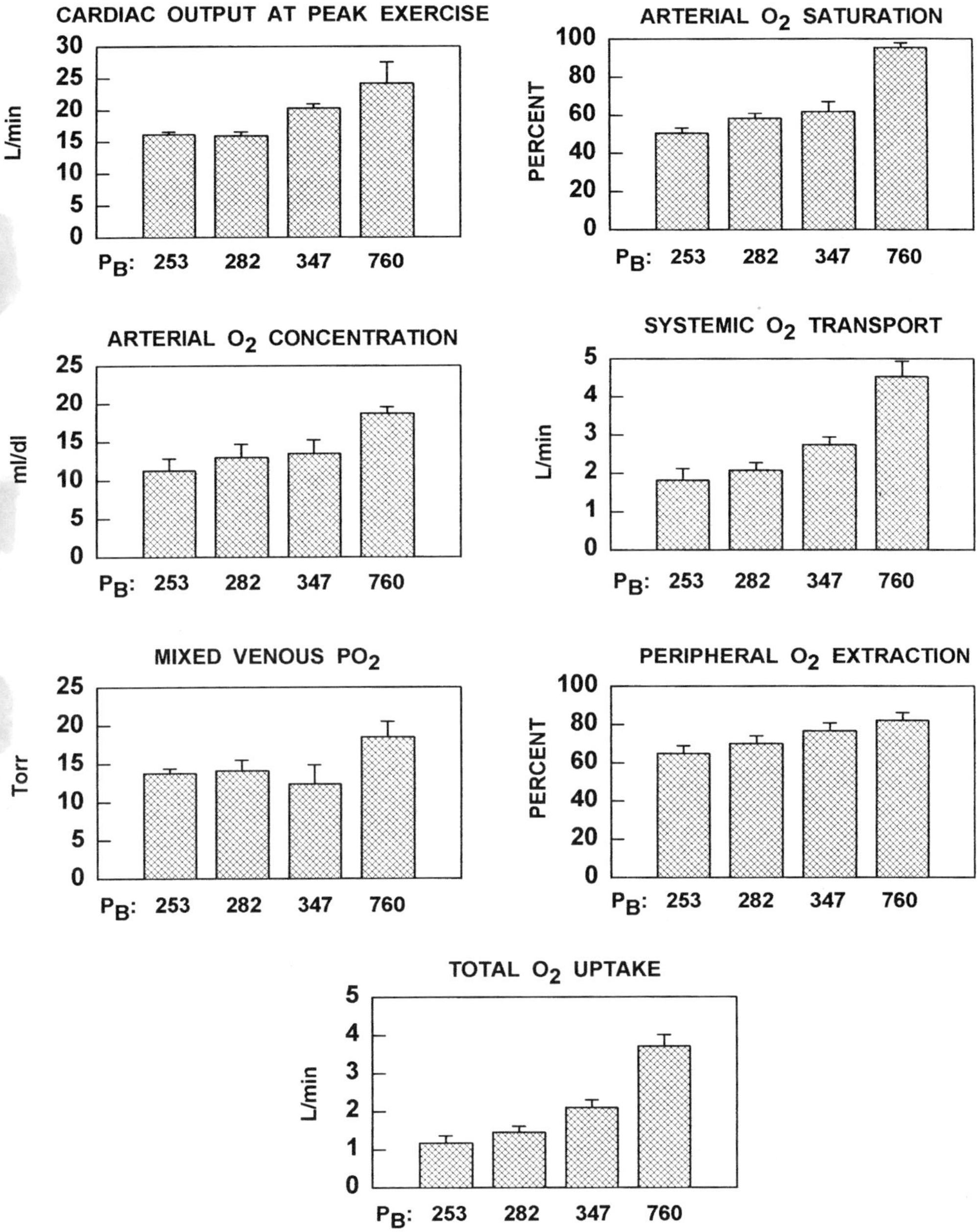

Figure 19 Principal systemic oxygen transport variables in the subjects of Operation Everest II at altitudes from sea level to the Everest summit. Systemic oxygen transport falls progressively with altitude because both cardiac output and arterial oxygen saturation fall with altitude exposure. Peripheral oxygen extraction also falls, which appears to buffer Po_2 of mixed venous blood. For further interpretation, see text.

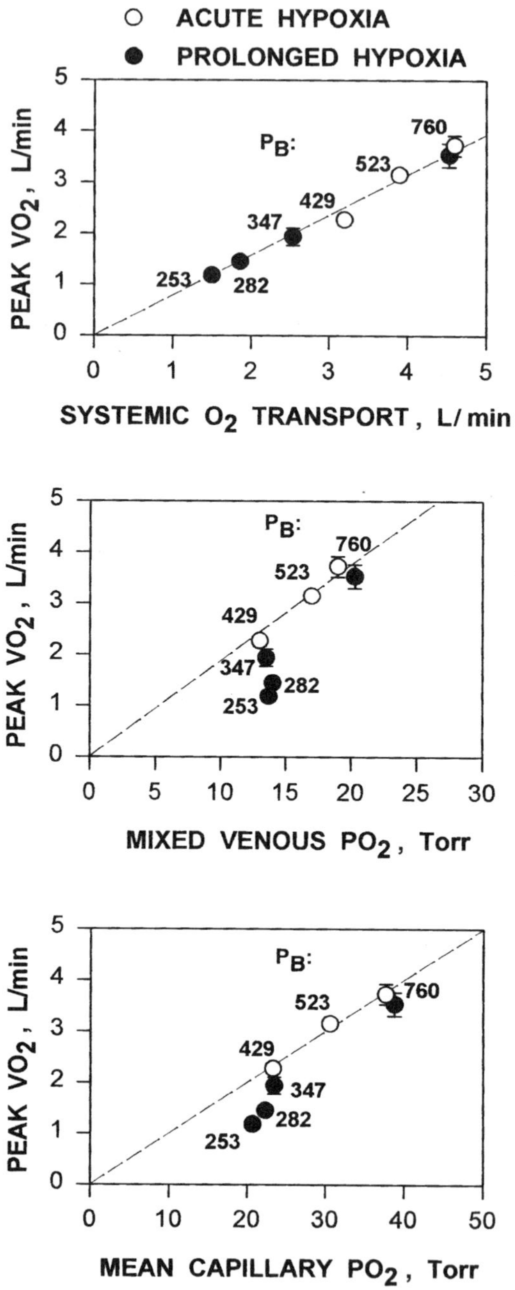
ACUTE HYPOXIA
PROLONGED HYPOXIA
PEAK VO2, L/min
SYSTEMIC O2 TRANSPORT, L/min
PB:
760
523
429
347
282
253
PEAK VO2, L/min
MIXED VENOUS PO2, Torr
PEAK VO2, L/min
MEAN CAPILLARY PO2, Torr

important determinant of mitochondrial O_2 supply at altitude than at sea level (57). The muscle biopsy data of Oelz and coworkers (58) in elite climbers reaching the summit of Everest without supplemental O_2 showed increased muscle capillarity, and it is tempting to speculate that a key element of climbing success is muscle capillarity facilitating diffusive unloading of O_2. Much work needs to be done in this area to better understand the factors that are important to maximal O_2 transport to, and extraction by, exercising muscle at altitude.

The data on prolonged hypoxia (Fig. 20) are not so easy to interpret as those for acute hypoxia (Fig. 20), since while peak $\dot{V}O_2$ remains closely tied to total O_2 transport (top panel), the relationship between peak $\dot{V}O_2$ and mean capillary PO_2 (lower panel) loses proportionality at extreme altitude (P_B = 282 and 253 torr). Between sea level and P_B = 347 torr, the results are compatible with previous data in fit subjects during acute hypoxic exposure. They indicate a muscle O_2 conductive limit that contributes to a predictable, declining ceiling on $\dot{V}O_2max$ at altitude by preventing complete O_2 extraction. At higher altitudes, the apparent conductance has fallen some 20%. (Peak $\dot{V}O_2$ is lower at a given capillary PO_2.) This fall in conductance has occurred despite a reduction in average muscle fiber diameter that reduces average diffusion distances within muscle (59) and which should therefore enhance O_2 conductance. One possibility is that loss of muscle mass at extreme altitude reduces peak O_2 consumption and thus total conductance. Another is that because of the considerable reduction in maximal cardiac output at extreme altitude (Fig. 19), perfusion of muscle capillaries is more nonhomogeneous than at lower altitudes. One possibility has been excluded—loss of muscle capillaries with severe hypoxia: biopsy data suggests that muscle capillarity was not reduced in OEII (59).

However, since the venous blood sampled for the analysis of maximal O_2 transport is mixed venous and not muscle venous in origin, another possibility is increasing contamination of muscle venous blood with venous blood from nonexercising tissues (kidney, brain, gut). This contamination would become evident as a higher than anticipated mixed venous PO_2 since muscle venous PO_2 is less than mixed venous (53). This could play a greater role at extreme altitude compared to sea level because of the steadily falling maximal cardiac output, which implies a greater relative contribution to mixed venous blood by nonexercising tissue blood

Figure 20 Relationship between peak oxygen uptake and total oxygen transport (upper panel), mixed venous PO_2 (middle panel) and calculated mean capillary PO_2 (lower panel) in acute (○) and prolonged (●) hypoxic exposures. In all three panels there is an essentially linear and proportional relationship for acute hypoxia, indicating a finite peripheral muscle oxygen conductance for oxygen that interacts with systemic oxygen transport to limit peak $\dot{V}O_2$. In prolonged hypoxia, there is a closely proportional relationship between $\dot{V}O_2$ and systemic oxygen transport, but that between $\dot{V}O_2$ and venous or mean capillary PO_2 is proportional only to modest altitude. At extreme altitudes, peak $\dot{V}O_2$ is less than expected for calculated mean capillary PO_2, suggesting a fall in muscle oxygen conductance. See text for further interpretation.

at high altitude. Direct femoral venous blood sampling offers a better chance of choosing among these various explanations, especially because muscle blood flow rather than cardiac output can be measured using the same catheters.

References

1. West JB, Lahiri S, Maret KH, et al. Barometric pressures at extreme altitudes on Mt. Everest: physiological significance. J Appl Physiol 1983; 54:1188.
2. Ward MP, Milledge JS, West JB. In: Ward MP, Milledge JS, West JB, eds. High Altitude Medicine and Physiology. London: Chapman and Hall Ltd., 1989:27–44.
3. Greene R. Observations on the composition of alveolar air on Everest, 1933. J Physiol Lond 1934; 32:481.
4. Pugh LGCE. Resting ventilation and alveolar air on Mount Everest: with remarks on the relation of barometric pressure to altitude in mountains. J Physiol (Lond) 1957; 135: 590.
5. Gill MB, Milledge JS, Pugh LGCE, et al. Alveolar gas composition at 21,000 to 25,000 feet (8,400–7,830 m). J Physiol (Lond) 1962; 163:373.
6. Wagner PD, Gale GE, Moon RE, et al. Pulmonary gas exchange in humans exercising at sea level and simulated altitude. J Appl Physiol 1986; 61(1):260.
7. Wagner PD, Sutton JR, Reeves JT, et al. Operation Everest II: pulmonary gas exchange during a simulated ascent of Mt. Everest. J Appl Physiol 1987; 63(6):2348.
8. Rahn H, Fenn WO. A Graphical Analysis of the Respiratory Gas Exchange. Washington, DC: American Physiological Society, 1955.
9. Houston CS. Acclimatization to hypoxia: Operations Everest I and II. Ann Sports Med 1988; 4(4):171.
10. West JB. Everest—the Testing Place. New York: McGraw-Hill Book Co., 1985.
11. Sutton JR, Reeves JT, Wagner PD, et al. Operation Everest II: oxygen transport during exercise at extreme simulated altitude. J Appl Physiol 1988; 64:1309.
12. West JB, Hackett PH, Maret KH, et al. Pulmonary gas exchange on the summit of Mount Everest. J Appl Physiol 1983; 55:678.
13. Peacock AJ, Jones PL. Gas exchange at extreme altitude: results from the British 40th Anniversary Everest Expedition. Eur Respir J 1997;
14. West JB, Lahiri S, Gill MB, et al. Arterial oxygen saturation during exercise at high altitude. J Appl Physiol 1962; 17:617.
15. Hurtado A. Animals in high altitudes: resident man. In: Dill ed. Adaptation to the Environment. Washington, DC: American Physiological Society, 1964:843.
16. Kreuzer F, Tenney SM, Mithoeffer JC, et al. Alveolar-arterial oxygen gradient in Andean natives at high altitude. J Appl Physiol 1964; 19:13.
17. Hultgren HN. High altitude pulmonary edema. In: Hegnauer AH, ed. Biomedicine Problems of High Terrestrial Altitude. New York: Springer-Verlag, 1969:131.
18. West JB, Wagner PD. Predicted gas exchange on the summit of Mt. Everest. Respir Physiol 1980; 42:1.
19. West JB. Ventilation/perfusion inequality and overall gas exchange in computer models of the lung. Respir Physiol 1969; 7:88.
20. Evans JW, Wagner PD. Limits on $\dot{V}A/\dot{Q}$ distributions from analysis of experimental inert gas elimination. J Appl Physiol 1977; 42(6):889.

21. Wagner PD, Naumann PF, Laravuso RB. Simultaneous measurement of eight foreign gases in blood by gas chromatography. J Appl Physiol 1974; 36(5):600.
22. Wagner PD, Laravuso RB, Uhl RR, et al. Continuous distributions of ventilation-perfusion ratios in normal subjects breathing air and 100% O_2. J Clin Invest 1974; 54(1):54.
23. Bebout DE, Story D, Roca J, et al. Effects of altitude acclimatization on pulmonary gas exchange during exercise. J Appl Physiol 1989; 67:2286.
24. Wagner PD, Hedenstierna G, Bylin G, et al. Reproducibility of the multiple inert gas elimination technique. J Appl Physiol 1987; 62:1740.
25. Houston CS, Sutton JR, Cymerman A, et al. Operation Everest II: man at extreme altitude. J Appl Physiol 1987; 63(2):877.
26. Dantzker DR, Brook CJ, DeHart P, et al. Ventilation/perfusion distributions in the adult respiratory distress syndrome. Am Rev Respir Dis 1979; 120:1039.
27. Cerretelli P. Gas exchange at high altitude. In: West JB, ed. Organism and Environment. New York: Academic Press, 1980:97.
28. Pugh LGCE, Gill MB, Lahiri S, et al. Muscular exercise at great altitudes. J Appl Physiol 1964; 19:431.
29. Pugh LGCE. Cardiac output in muscular exercise at 5,800 m (19,000 ft). J Appl Physiol 1964; 19:441.
30. Gledhill N, Froese AB, Dempsey JA. Ventilation to perfusion distribution during exercise in health. In: Dempsey JA, Reed CE, eds. Muscular Exercise and the Lung. Madison: University of Wisconsin Press, 1977:325.
31. Torre-Bueno JR, Wagner PD, Saltzman HA, et al. Diffusion limitation in normal humans during exercise at sea level and simulated altitude. J Appl Physiol 1985; 58(3):989.
32. Dempsey JA, Reddan WG, Birnbaum ML, et al. Effects of acute through life-long hypoxic exposure on exercise pulmonary gas exchange. Respir Physiol 1971; 13:62.
33. Dempsey JA, Hanson PG, Henderson KS. Exercise-induced arterial hypoxemia in healthy human subjects at sea level. J Physiol (Lond) 1984; 355:161.
34. Powers SK, Lawler J, Dempsey J, et al. Effects of incomplete pulmonary gas exchange on $\dot{V}O_2max$. J Appl Physiol 1989; 66:2491.
35. Hammond MD, Gale GE, Kapitan KS, et al. Pulmonary gas exchange in humans during normobaric hypoxic exercise. J Appl Physiol 1986; 61(5):1749.
36. Schaffartzik W, Poole DC, Derion T, et al. $\dot{V}A/\dot{Q}$ distribution during heavy exercise and recovery in humans: implications for pulmonary edema. J Appl Physiol 1992; 72(5):1657.
37. Schaffartzik W, Arcos J, Tsukimoto K, et al. Pulmonary interstitial edema in the pig after heavy exercise. J Appl Physiol 1993; 75(6):2535.
38. Fasules JW, Wiggins JW, Wolfe RR. Increased lung vasoreactivity in children from Leadville, Colorado, after recovery from high-altitude pulmonary edema. Circulation 1985; 72(5):957.
39. Eldridge MW, Podolsky A, Richardson RS, et al. Pulmonary hemodynamics response to exercise in subjects with prior high-altitude pulmonary edema. J Appl Physiol 1996; 81(2):911.
40. Podolsky A, Eldridge MW, Richardson RS, et al. Exercise-induced $\dot{V}A/\dot{Q}$ inequality in subjects with prior high altitude pulmonary edema. J Appl Physiol 1996; 81(2):922.

41. Welsh CH, Wagner PD, Reeves JT, et al. Operation Everest II: spirometric and radiographic changes in acclimatized humans as simulated high altitudes. Am Rev Respir Dis 1993; 147:1239.
42. Coates G, O'Brodovich H, Jefferies AL, et al. Effects of exercise on lung lymph flow in sheep and goats during normoxia and hypoxia. J Clin Invest 1984; 74:133.
43. Gale GE, Torre-Bueno JR, Moon RE, et al. Ventilation/perfusion inequality in normal humans during exercise at sea level and simulated altitude. J Appl Physiol 1985; 58(3): 978.
44. Anholm JD, Bourne JC, Pai RG, et al. Hypoxic pulmonary vasoconstriction is not necessary for development of radiographic evidence of pulmonary edema following exercise at moderate altitude. Hypoxia Brain 1995:312.
45. Hlastala MP, Scheid P, Piiper J. Interpretation of inert gas retention and excretion in the presence of stratified inhomogeneity. Respir Physiol 1981; 46:247.
46. Hammond MD, Hempleman SC. Oxygen diffusing capacity estimates derived from measured $\dot{V}_A/\dot{Q}$ distributions in man. Respir Physiol 1987; 69:129.
47. Johnson RL Jr, Spicer WS, Bishop JM, et al. Pulmonary capillary blood volume, flow and diffusing capacity during exercise. J Appl Physiol 1960; 15:893.
48. Cotes JE, Dabbs JM, Elwood PC, et al. Iron-deficiency anemia: its effect on transfer factor for the lung, diffusing capacity and ventilation and cardiac frequency during submaximal exercise. Clin Sci 1972; 42:325.
49. Bender PB, McCullough RE, McCullough RG, et al. Increased exercise Sa_{O_2} independent of ventilatory acclimatization at 4300 m. J Appl Physiol 1989; 66:2733.
50. Bender PR, Groves BM, McCullough RE, et al. Oxygen transport in exercising leg in chronic hypoxia. J Appl Physiol 1988; 65:2592.
51. Pirnay F, Lamy M, Dujardin J, et al. Analysis of femoral venous blood during maximum exercise. J Appl Physiol 1972; 33:289.
52. Wagner PD. Determinants of maximal oxygen transport and utilization. In: Massaro D, ed. Annual Reviews of Physiology. vol. 58. Palo Alto, CA: Annual Reviews, 1996:21.
53. Roca J, Hogan MC, Story D, et al. Evidence for tissue diffusion limitation of $\dot{V}O_2$max in normal humans. J Appl Physiol 1989; 67:291.
54. Honig CR, Gayeski TEJ, Federspiel WJ, et al. Muscle O_2 gradients from hemoglobin to cytochrome: new concepts, new complexities. Adv Exp Med Biol 1984; 169:23.
55. Hogan MC, Bebout DE, Wagner PD. Effect of increased Hb-O_2 affinity on $\dot{V}O_2$max at constant O_2 delivery in dog muscle in situ. Med Sci Sports Exerc 1991; 23S36.
56. Mathieu-Costello O. Comparative aspects of muscle capillary supply. Ann Rev Physiol 1993; 55:503.
57. Wagner PD. Algebraic analysis of the determinants of $\dot{V}O_2$max. Respir Physiol 1993; 93:221.
58. Oelz O, Howald H, di Prampero PE, et al. Physiological profile of world-class high-altitude climbers. J Appl Physiol 1986; 60:1734.
59. Green HJ, Sutton JR, Cymerman A, et al. Operation Everest II: adaptations in human skeletal muscle. J Appl Physiol 1989; 66(5):2454.

9

The Cardiovascular System at High Altitude

Heart and Systemic Circulation

EUGENE E. WOLFEL

University of Colorado Health Sciences Center
Denver, Colorado

BENJAMIN D. LEVINE

Institute for Exercise and Environmental Medicine
University of Texas Southwestern Medical Center
Dallas, Texas

I. Introduction

> We climbed a hundred meters, mouths open, trying to suck in air, resting after a few steps, then again a few more.
>
> —Reinhold Messner, May 1978

The cardiovascular system has been a major focus of interest in high-altitude physiology and medicine since it was first observed that symptoms in normal individuals at altitude were similar to those experienced by patients with circulatory insufficiency at sea level (1). Breathlessness, excessive fatigue, and tachycardia in healthy climbers during exertion or in patients with high-altitude illness at rest suggested a potential cardiovascular abnormality. Some of the key historical figures in hypoxia research—Douglas and Haldane on Pikes Peak; Barcroft in the Peruvian Andes—created an early controversy by making directionally opposite observations regarding an increase (2) or decrease (3) in cardiac output after acclimatization to altitude. Although the magnitude and temporal progression of this adaptation has been more precisely worked out over the past 50 years (4), controversy still exists regarding its mechanisms and physiological consequences.

The major role of the cardiovascular system is to transport oxygen and sub-

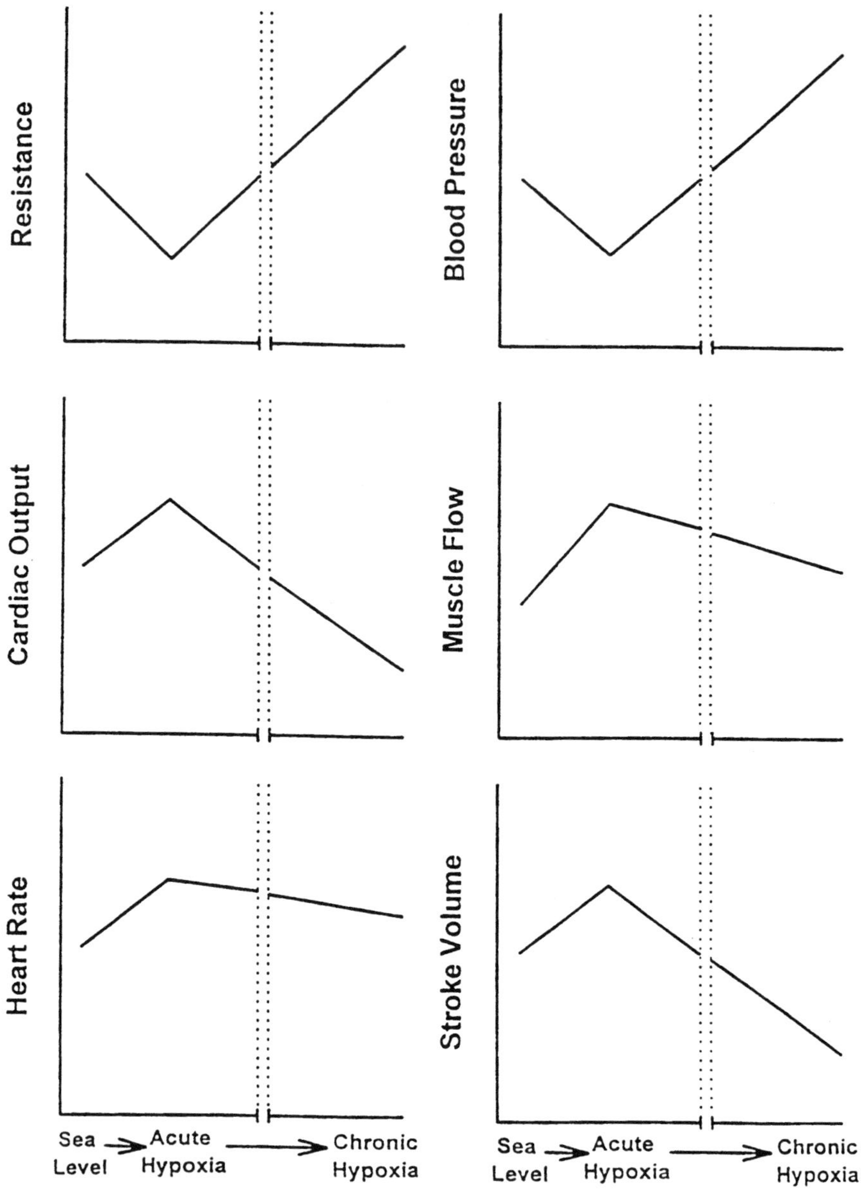

Figure 1 Schematic of the cardiovascular hemodynamic changes that occur with acute and chronic hypoxia compared to sea level. Heart rate, cardiac output, muscle flow, and stroke volume initially rise with acute hypoxia but decrease to levels at or below sea level with sustained exposure. Blood pressure and vascular resistance initially fall with acute hypoxia but rise progressive with prolonged hypoxia.

strate to support the metabolic demands of the body. Moreover, during increases in physical activity, the cardiovascular system must not only maintain oxygen transport to the vital organs, but also augment peripheral oxygen delivery to skeletal muscle by as much as 10-fold during maximal exercise. Both the systemic (cardiac output) and regional mechanisms by which the cardiovascular system matches oxygen supply with oxygen demand are challenged by a hypoxic environment. The response appears to be a central activation via the sympathetic nervous system coupled with local vascular control mechanisms directing regional distribution of flow. Thus gross changes in the pump function of the heart (affecting heart rate and stroke volume) are ''fine-tuned'' at a local level to ensure that organ-specific metabolic demands are met. In addition, the heart is not only a pump but a neurohumoral organ as well, both initiating and responding to changes in autonomic (sympathetic and parasympathetic) tone and circulating neurohormones. Moreover, recent insights regarding the peripheral vasculature as a sensor and modulator of circulatory control via both endothelium-dependent and -independent mechanisms have improved understanding of the precision by which the cardiovascular system operates to serve its primary function of supporting oxygen transport. Finally, acute responses are modified by chronic adaptations that restore circulatory function towards normoxic levels over time periods that may range from a few days or weeks for the sea level sojourner to years for the high-altitude native. A qualitative summary of the acute and chronic effects of hypoxia on the cardiovascular system is presented in Figure 1. The data supporting this summary will be reviewed, and areas of controversy, with specific attention to the most important issues for future research, will be highlighted.

II. Acute Hypoxia

Acute hypoxia decreases Pa_{O_2} *and arterial oxygen content, which activates the sympathetic nervous system primarily via peripheral chemoreceptors. The carotid bodies appear to respond to partial pressure of oxygen, at least with respect to ventilatory signals, while local vascular beds seem to be regulated by arterial oxygen content. This activation is the key acute adaptive response of the cardiovascular system. However, peripheral vasoconstriction does not occur due to release of local vasodilatory substances and ''functional sympatholysis.'' Thus, heart rate and cardiac output increase, but systemic vascular resistance and blood pressure decrease transiently.*

With acute exposure to normobaric or hypobaric hypoxia, the partial pressure of oxygen in the air, alveoli, and blood is reduced, and oxyhemoglobin desaturates, reducing arterial oxygen content. Although the specific stimulus to the cardiovascular system is surprisingly unclear, the ultimate circulatory response is an increase in systemic blood flow in order to maintain tissue oxygen delivery. Classic studies by Grollman on Pikes Peak (4300 m) demonstrated that there is an approximately 40% increase in resting cardiac output within the first few days of ascent to high altitude (4). Similar observations have been made under more rigorously controlled conditions in the laboratory, with equivalent degrees of normobaric (5,6) or hypobaric (7–9) hypoxia. The magnitude of the increase in cardiac output appears proportional to the hypoxia, with no apparent increase at altitudes below 700 m (7), reaching as much as 75% increase at altitudes approaching 5000 m (7–9). This response appears to be independent of gender (7,10) or age (11).

A. Components of Cardiac Output—Acute Hypoxia

Heart Rate

The increase in cardiac output at rest with acute high-altitude exposure is for the most part due to an elevation in heart rate (7–9,12). Resting heart rate is a function of both the intrinsic heart rate and the balance of sympathetic and parasympathetic neural activity, which may vary widely among individuals. Intrinsic heart rate does not change with acute hypoxia [i.e., no change in the face of combined adrenergic and vagal blockade (13)]; therefore, autonomic mechanisms must be invoked to explain the increase in heart rate.

The evidence for sympathetic activation is compelling (see Chapter 13). Indirect measures such as plasma and urinary catecholamines (14–16) have shown a consistent elevation in both norepinephrine and epinephrine concentrations with acute altitude exposure, with epinephrine increasing to a greater degree (15). However, such data must be interpreted cautiously as norepinephrine clearance also increases prominently with acute hypoxia (17). Frequency analysis of heart rate variability has also been used as a means of analyzing changes in autonomic function, with high-frequency variability (0.15–0.40 H) reflecting modulation of parasympathetic activity and low-frequency variability (0.04–0.15 H) modulated by both sympathetic and parasympathetic activity. However, this growing body of data attempting to use dynamic analysis of heart rate variability as an index of autonomic function at altitude (18–26) suffers from the frequent failure to quantify many of the variables that markedly influence heart rate variability, such as ventilatory volume and respiratory rate (27), cardiac volume and mechanical function (28), arterial stiffness and pulse amplification, arterial baroreflex function (29), and sinus node responsiveness (30), all of which are altered by altitude exposure. Thus, it may be difficult to unravel true changes in sympathetic or parasympathetic function at altitude (or any other condition) from analysis of heart rate variability alone (31).

Fortunately, recent direct measurements of efferent postganglionic sympathetic activity to skeletal muscle have confirmed a proportional relationship between the magnitude of hypoxia (32) or high altitude (33–35) and sympathetic activation. It is this increase in sympathetic activity that is likely responsible for the global stimulation of the circulatory system in the face of acute hypoxia.

However, sympathetic activation by itself cannot explain all of the increase in heart rate or cardiac output. For example, beta-blockade alone does not completely abolish the heart rate response to acute hypoxia (12,13). In one study, blockade reduced the hypoxia-induced increase in cardiac output and heart rate by only 50% (12), suggesting an important component of vagal withdrawal in addition to sympathetic activation. Further evidence for vagal withdrawal during hypoxia was presented by Eckberg et al., who noted the rapid onset of shortening of the R-R interval, indicative of vagal withdrawal, after brief periods of hypoxia deemed to be too transient to fully activate the sympathetic nervous system (36). This vagal contribution was confirmed in a study by Koller et al., in which vagal blockade with atropine in combination with beta-blockade prevented any further increase in resting heart rate with hypoxia (13). These observations are at odds with the traditional concept that stimulation of peripheral chemoreceptors by hypoxia leads to an increase rather than a decrease in efferent vagal activity (37). However, cardiac vagal motoneurons appear to be inhibited by central inspiratory neuronal activity as well as by afferents from thoracic stretch receptors (38). Thus, bradycardia, rather than tachycardia, is present during apnea (37) when these inhibitory influences are silent. Moreover, R-R shortening was noted almost exclusively during inspiration in the study by Eckberg et al. cited above (36). These findings support the role of vagal withdrawal in the tachycardia associated with acute hypoxia.

In summary, it appears that hypoxia causes sympathetic activation directly via stimulation of peripheral chemoreceptors (39) and vagal withdrawal indirectly via increases in ventilation, resulting ultimately in tachycardia and an increase in cardiac output. The recent observation that afferent fibers from both baroreflex and chemoreflex receptor populations synapse very closely together in the nucleus tractus solitarius (40) raises the intriguing possibility that regulation of the cardiovascular and ventilatory responses to acute hypoxia may be neurally connected. Preliminary observations by Asano et al. (41) demonstrating a relationship between urinary norepinephrine excretion and ventilation with altitude acclimatization support this notion and suggest an important area for future research.

Stroke Volume

Changes in stroke volume play a minor role in the acute cardiovascular response to hypoxia. Although myocardial contraction velocity is augmented, consistent with sympathetic activation (42), and end-systolic volume is slightly reduced, this increase in contractility may be offset by a reduction in end-diastolic volume resulting in only a small increase or no change in stroke volume (4,7–9), particularly when the measurements are performed within 1–2 hours after exposure. Most importantly, even severe acute hypoxia is not associated with depression of myocardial contractile function in normal hearts, although sustained hypoxia equivalent to 10,000 m altitude will result in heart failure in dogs if adrenergic compensatory mechanisms are inhibited (43). Changes in stroke volume, however, become much more important during acclimatization and will be discussed in more detail below.

Changes in Peripheral Vascular Resistance

Activation of the sympathetic nervous system under a variety of conditions results in peripheral vasoconstriction. However, with the sympathetic activation of acute hypoxia, this vasoconstriction is absent. In fact, in virtually every study published, acute hypoxia results in vasodilation in all vascular beds except the lung (7,44–47) directly in proportion to the reduction in arterial oxygen content. An example of the dramatic ability of hypoxia to override sympathetically mediated vasoconstriction at the local level is shown in Figure 2 (44). In this study, lower body negative pressure was used to pool blood in the lower extremities, thus reducing central blood volume and unloading cardiopulmonary and arterial baroreceptors (27,48). This technique, when applied at relatively low levels of suction in healthy subjects under normoxic conditions, always leads to a prominent increase in sympathetic nerve activity and increase in limb vascular resistance, with maintenance of arterial pressure. In contrast, when hypoxia, induced by breathing 10% oxygen, is superimposed on lower body negative pressure, forearm vascular resistance fails to rise, and the ability to maintain blood pressure is compromised (see Fig. 2). This compromise in blood pressure control during orthostatic stress may explain the increased incidence of orthostatic hypotension and syncope recently reported in some otherwise healthy individuals early after ascent to high altitude (49).

The capacity of local, metabolic control mechanisms to override sympathetic vasoconstriction has been well described during exercise and has been termed "functional sympatholysis" (50,51). The combination of global activation of the sympathetic nervous system with a resultant increase in heart rate and cardiac output, accompanied by simultaneous local vasodilation resulting in distribution of flow to vascular beds with the greatest metabolic demand, thus is a physiological strategy

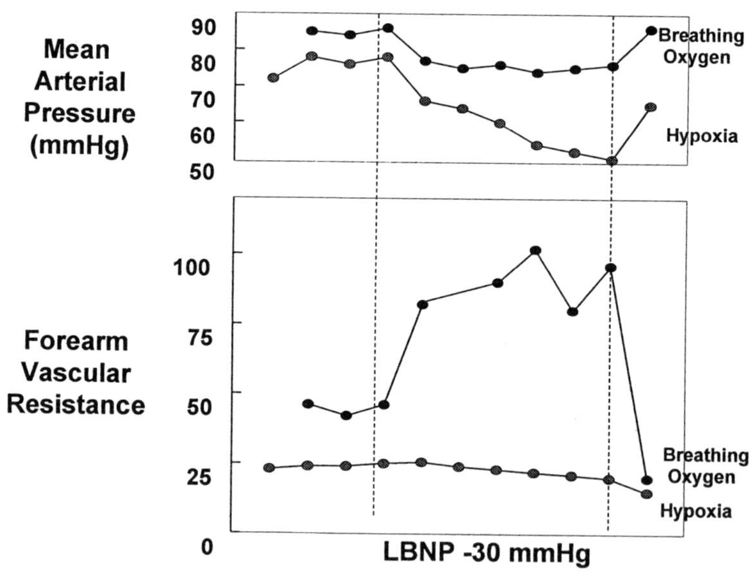

Figure 2 Hypoxia prevents the rise in forearm vascular resistance with lower body negative pressure (LBNP) and the ability to maintain blood pressure is compromised. (Data from Ref. 44.)

common to both exercise and hypoxia. During exercise, the mechanism of functional sympatholysis appears to be muscular contraction–induced opening of ATP-sensitive potassium channels by endothelium-derived hyperpolarizing factor (52) mediated, at least in part, by nitric oxide (53).

Whether a similar mechanism is operative during hypoxia is not known. In the coronary arteries, acute hypoxia causes neurally mediated cholinergic vasodilation (54). However, there is no evidence that such a mechanism exists in the peripheral circulation. Furthermore, beta-adrenergic blockade in the forearm does not alter the vasodilator response to acute hypoxia (12), providing evidence against direct adrenergic vasodilation. The increment in leg blood flow associated with hypoxia, both at rest as well as during exercise, precisely matches the decrease in arterial oxygen content, keeping O_2 delivery constant (45,46). Any putative mechanism must explain this remarkably tight regulation of peripheral blood flow to oxygen availability and metabolic demand.

Recently an elegant study by Ellsworth et al. suggested that the red cell itself might be a potential regulator of the vascular resistance, thus providing a unifying hypothesis for the mechanism of ''sensing'' of arterial oxygen content (55). They demonstrated that the release of ATP from red blood cells, as a direct function of hypoxemia and acidosis, could bind to P2y receptors on the vascular endothelium, resulting in a propagated vasodilation both proximally and distally. Whether these ATP-sensitive receptors are the same ones mediating exercise-induced local vasodilation (52) remains to be determined.

Whether sympathetic vasoconstriction or hypoxic vasodilation prevails ultimately determines what happens to arterial blood pressure; the outcome for any given individual probably depends on the absolute altitude achieved, the rapidity of early acclimatization responses, which may quickly restore arterial oxygen content despite persistent hypoxia, and the timing of the blood pressure measurement. When blood pressure measurements are made very early—within the first hour of exposure—most studies show a decrease in total peripheral resistance and a small but measurable reduction in blood pressure (7,11,44,56). However, when measurements are made even a few hours after exposure, the acute increase in hemoglobin concentration from reduction in plasma volume may tip the balance toward sympathetic vasoconstriction, and modest hypertension has been observed under chronic conditions as discussed below (57). With the increase in hemoglobin concentration there is a local increase in arterial oxygen content, which may activate some peripheral sensor that decreases flow by vasoconstriction in order to match oxygen supply with demand. A key question arises: What is the hypoxic stimulus to both peripheral chemoreceptors and peripheral vasculature—partial pressure of oxygen or arterial oxygen content? It is clear that the receptors in the carotid body that mediate ventilation respond directly to Pa_{O_2} (58). Whether the receptors that mediate sympathetic activation respond similarly is unknown, but likely. However, as discussed above, the peripheral circulation, as well as other organs such as the kidney, respond to arterial oxygen content such that anemia results in vasodilation or synthesis and release of erythropoietin even with a normal Pa_{O_2}. This distinction may be critically important in understanding the chronic response to hypoxia and is an important area of future research.

In summary, acute hypoxia stimulates peripheral chemoreceptors to increase sympathetic nerve activity, resulting in an increase in heart rate and systemic blood flow (cardiac output). Local vasodilating mechanisms, responding to reductions in arterial oxygen content result in a reduction in peripheral resistance and a modest transient reduction in arterial pressure.

Key Unanswered Questions—Acute Hypoxia

What are the biological signals derived from oxygen that are being sensed by the cardiovascular system, and what are the pathway and feedback loops which regulate them?

Are the cardiovascular and ventilatory systems linked together via neural connections in the nucleus tractus solitarius (NTS) facilitating matching of ventilation and oxygen transport during hypoxia?

Is the sympathetic system necessary for successful acclimatization, or can it be potentially harmful both to healthy individuals and to those with high-altitude illness or underlying disease?

Is endothelial function the important factor in the modulation of functional sympatholysis?

III. Sustained Hypoxia

Acclimatization occurs predominantly in other organ systems, with the cardiovascular system responding to, rather than initiating changes in oxygen availability. Thus, as Cao_2 *increases from first hemoconcentration and ventilatory acclimatization, and later increased red cell mass, the hyperdynamic state of the circulation diminishes and both cardiac output and peripheral blood flow return toward normal. This decrease in blood flow may be adaptive by allowing more diffusion time for extraction of oxygen, both in the lung as well as in skeletal muscle and other organs. Sympathetic activation persists, particularly at higher altitudes, and thus, resting heart rate remains elevated. This chronic sympathetic hyperactivity leads to downregulation of beta receptors, which contributes to a reduction in rest and exercise heart rate. Blood pressure gradually rises as the peripheral mechanisms responsible for sympatholysis abate, but sympathetic activation actually increases, possibly as a function of increased chemosensitivity. Changes in stroke volume play a much greater role in the response to sustained hypoxia, with a consistently reported decrease at all altitudes studied. However, ventricular contractile function remains normal, and the reduction in stroke volume is entirely a consequence of ventricular filling and the Starling mechanism.*

In contrast to the acute response to hypoxia, in which the cardiovascular system plays a primary role in restoring oxygen transport toward normal, the response of the cardiovascular system to sustained hypoxia depends more on adaptation by other organ systems. There is an ongoing interplay between the ventilatory, metabolic, and hematological adaptations, which improve O_2 delivery and result in secondary changes in heart rate, stroke volume, cardiac output, blood pressure, and peripheral vascular resistance over time at high altitude. Other factors, such as severity and duration of hypoxia, level of activity, nutritional status, and fluid balance, may also play a role.

Ventilatory acclimatization, the key early adaptive response, is characterized by increases in alveolar ventilation, which maximize P_{AO_2} and shift the oxyhemoglobin dissociation curve to the left. Moreover, within the first few hours to days of altitude exposure, there is a prominent decrease in plasma volume that increases the hemoglobin concentration. Together, these adaptations substantially increase the Ca_{O_2} within the first few days of sustained altitude exposure (see Chapters 6 and 15).

A. Cardiac Adaptations—Sustained Hypoxia

Heart Rate

In the context of increased chronic sympathetic activation with sustained hypoxia, resting heart rate has been shown to be consistently elevated at altitudes greater than 3000 m. The degree of elevation in resting heart rate is dependent on the duration and severity of sustained hypoxia (Fig. 3). The greater the altitude, the higher the resting heart rate, with little evidence for a decrease toward sea level values over 2–3 weeks. Studies examining the effects of progressively increasing altitude (i.e.,

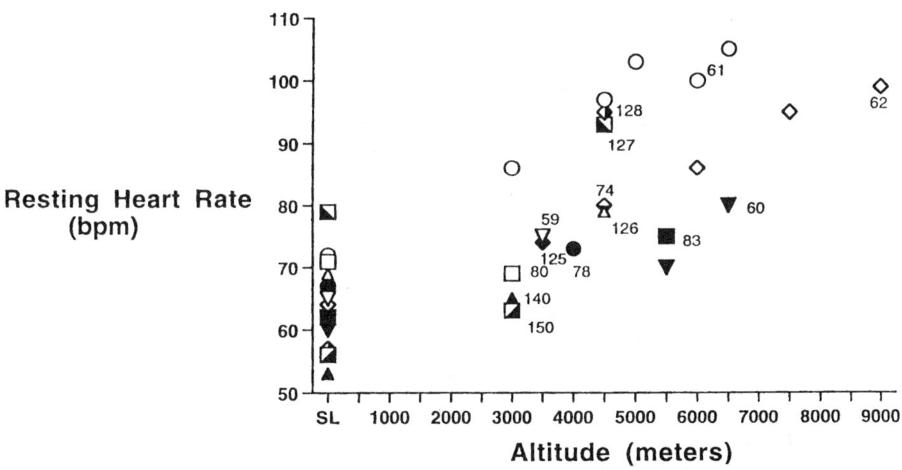

Figure 3 Mean resting heart rates with varying degrees of exposure to high altitude. Numbers identify each individual study listed in the references. There appears to be a linear relationship between progressive altitude and resting heart rate.

superimposing a more severe hypoxic stimulus upon prior acclimatization) have shown further increases in resting heart rate with each increment in altitude. For example, in the American Medical Research Expedition to Mount Everest, resting heart rate increased 20 bpm from sea level to 6300 m (60). Similar data have been obtained in a hypobaric chamber study, confirming that hypobaric hypoxia per se, rather than the stresses of mountaineering, is responsible for this tachycardia (61,62). Thus, even in the setting of adequate acclimatization, the ability to respond to a further increase in hypoxia with cardioacceleration remains present.

Although beta blockade prior to ascent prevents most of the increase in resting heart rate with sustained altitude exposure (63,64), confirming the essential role of hypoxia-induced sympathetic activation in this response, when beta blockade (65) or supplemental oxygen (62,66) has been administered acutely after acclimatization has occurred, there is a persistent elevation of heart rate above sea level baseline values. The mechanism for this difference is uncertain, but it may well be that more prolonged administration of supplemental oxygen or sustained beta blockade at altitude would restore heart rate to the sea level value. Indeed, this explanation is likely to be true, as after a few days at sea level heart rate does return to normal (60,67). The decrease in plasma volume during the first few weeks of altitude exposure may unload both cardiopulmonary and arterial baroreceptors such that sympathetic activity remains above baseline even in the absence of chemoreflex activation. Moreover, a heightened chemoreflex associated with acclimatization could well yield an increased basal firing rate even under normoxic conditions. Finally, such hypovolemia would also be expected to cause vagal withdrawal so that the tachycardia with sustained hypoxia may not be entirely dependent on sympathetic activation.

Indeed, chronic sympathetic hyperactivity usually leads to a downregulation of cardiac beta receptors (68), reducing the tachycardia response to catecholamines. Such a downregulation has been well described in animals and humans after acclimatization to high altitude (69–73). Thus, the heart rate response to acute altitude exposure may be dominated by sympathetic influences, while with more prolonged hypoxia exposure, some attenuation of this response occurs. Effects of the parasympathetic nervous system, along with other reflex interactions, appear to have an increasingly important influence on the resting heart rate.

Despite this chronic increase in resting heart rate, cardiac output is consistently depressed after acclimatization to high altitude, often to below baseline sea level values. Resting hemodynamic data from a large number of studies are summarized in Table 1. Although these results are consistent in studies performed at moderate high altitude with the subjects remaining at a given altitude over the course of acclimatization (Table 1), important differences are observed if hypoxia is incremental and more extreme, as seen in a typical climbing expedition. For example, in the studies performed on Pikes Peak (Table 2), although arterial oxygen content increased over the course of acclimatization, cardiac output actually continued to decrease so that conductive O_2 transport was reduced compared to sea level and did not change with acclimatization (63,74). In contrast, in Operation Everest II (Table 3) (62,75), resting cardiac output actually increased above 7000 m but remained

Table 1 Effects of Sustained Hypoxia on Resting Cardiac Hemodynamics

Altitude (m)	*n*	Age (yr)	Duration (days)	Method	Responses (%)[a] CO	SV	HR	Ref.
3658	50	21–34	10	imped.	−26*	−36*	+18*	59
3800	5	20–73	21–28	CO_2 rb	−4	−11*	+9*	78
4300	8	18–24	21	dye dilut.	−8	−20*	+18*	127
4350	4	20–22	10	dye dilut.	−21*	−22*	+15*	126
3100	8	23–32	10	Fick	−7	nc	nc	80
4300	4	20–21	14	dye dilut.	+28*	−23*	+31*	128
4300	12	21–28	21	dye dilut	−17*	−30*	+11*	63,74

[a] % changes in responses from sea level.
rb = Rebreathing; dye dilut. = indocyanine green dye method; imped. = impedance; nc = no change; CO = cardiac output, SV = stroke volume, HR = heart rate.
* $p < 0.05$ vs sea level.

Table 2 Resting Hemodynamics in Healthy Men at 4300 m—Pikes Peak, Colorado

(n = 12)	Sea level[a]	Acute hypoxia[b]	Sustained hypoxia[c]
Heart rate (bpm)	72 ± 5	80 ± 4*	80 ± 4*
Cardiac output (L/min)	6.6 ± 0.6	6.7 ± 0.6	5.5 ± 0.6*
Stroke volume (mL)	91 ± 6	84 ± 6	70 ± 7*†
Vo_2 (mL/min)	290 ± 17	316 ± 14	376 ± 14*
a-v o_2 (vol%)	4.4 ± 0.5	4.7 ± 0.8	6.8 ± 0.8*†
CaO_2 (vol%)	18.4 ± 0.3	15.3 ± 0.3*	18.3 ± 0.4†
O_2 delivery (L O_2/min)	1230 ± 120	1010 ± 90*	1015 ± 96*
MAP (mmHg)	88 ± 3	86 ± 3	107 ± 3*†
SVR (dynes-sec-cm^{-5})	1176 ± 116	1162 ± 126	1772 ± 204*†

[a] Sea level studies performed in Palo Alto, California.
[b] Acute hypoxia data obtained within 4 hours of arrival at 4300 m, Pikes Peak.
[c] Prolonged hypoxia studies performed at 21 days on Pikes Peak.
MAP = mean arterial pressure; SVR = systemic vascular resistance.
* $p < 0.05$ versus sea level; † $p < 0.05$ versus acute hypoxia. The increase in a-v o_2 with sustained hypoxia was related to the decrease in cardiac output (80%) and the increase in Vo_2 (20%).
Source: Refs. 63, 74.

constant for a given oxygen uptake during exercise. However, conductive O_2 delivery was nevertheless decreased, similar to the Pikes Peak studies due to the progressive increase in altitude and superimposition of acute on chronic hypoxia. In both groups of studies, metabolic demand was defended by an increase in peripheral oxygen extraction, which appears to have resulted in a relative ''sparing'' effect on cardiac output and systemic blood flow. When diffusion gradients for O_2 transfer

Table 3 Resting Hemodynamic Measurements with Progressive High Altitude—Operation Everest II

	Sea level	6100 m	7260 m	8848 m
Heart rate (bpm)	64 ± 4	86 ± 6*	95 ± 6*	99 ± 6*
Cardiac output (L/min)	6.3 ± 1.2	5.0 ± 1.1	7.3 ± 1.5	8.6 ± 0.8
Stroke volume (mL)	107 ± 8	72 ± 6*	69 ± 10*	81 ± 11*
CaO_2 (vol%)	17.9 ± 1.2	15.7 ± 2.0*	13.6 ± 1.7*	11.8 ± 1.9*
a-v o_2 (vol%)	5.7 ± 0.5	6.4 ± 1.1	5.7 ± 1.0	4.6 ± 0.2
O_2 delivery (mL O_2/min)	1125 ± 144	786 ± 220*	992 ± 155	1018 ± 152
Vo_2 (mL/min)	360 ± 20	306 ± 25	406 ± 16	386 ± 17
MAP (mmHg)	96 ± 3	96 ± 3	90 ± 2	96 ± 9
SVR (dynes-sec-cm^{-5})	1219 ± 200	1536 ± 48*	986 ± 107*	893 ± 300*

Simulated altitude where PIo_2 = 63 torr (6100 m); 49 torr (7620 m); 43 torr (8840 m).
* $p < 0.05$ versus sea level.
Source: Refs. 62, 106, 138.

are reduced at very low Po_2, a more prolonged capillary transit time would allow more time for oxygen transfer. Thus, a higher cardiac output might trade off a higher conductive O_2 delivery for a reduced diffusional oxygen transfer such that actual net O_2 flux could actually be decreased. In this context, Wagner presented a model suggesting that even at maximal exercise after acclimatization to high altitude, when maximal cardiac output is markedly reduced compared to sea level (see below), if one could further increase cardiac output, it would not lead to an increase in oxygen uptake (76). At present, a mechanism for the sensing and effecting of such a matching of diffusion time to diffusion gradients is unknown.

Stroke Volume

Since heart rate is consistently increased with sustained altitude exposure, the reduction in cardiac output must be mediated by a decrease in stroke volume, and such a decrease has been uniformly observed. With sustained hypoxic exposure, reductions in stroke volume have been reported at rest (Table 1), with upright tilt (77), and at all levels of submaximal and at maximal exercise (Tables 4,5). Reductions in stroke volume can be seen as early as a few hours after exposure to hypobaric hypoxia during submaximal exercise (8). However, consistent reductions at rest and at all levels of exercise including maximal exercise are usually seen only after 2 days at high altitude (78) with some continued decline over the first week, after which stroke volume stabilizes for a given altitude (79). With progressively increasing altitude there appears to be further reductions in stroke volume (62), although differences in techniques make precise determination of the effect of progressive prolonged hypoxia on stroke volume difficult. Since stroke volume is end-diastolic volume minus end-systolic volume, it will be reduced only if end-*systolic* volume is *increased* due to impaired contractile function, or if end-*diastolic* volume is *reduced* due to impaired left ventricular filling. Each of these possibilities will be considered in turn.

Cardiac Contractile (Systolic) Function

Early studies evaluating the hemodynamic responses to sustained hypoxia suggested that decreased cardiac contractile function contributed to the reductions in stroke volume (80–82). However, the measures used in these studies were extremely load dependent and misleading in the setting of sustained hypoxia associated with a prominent reduction in plasma volume and LV end-diastolic volume. More recently, Fowles and Hultgren normalized LV ejection indices to left ventricular end-diastolic dimensions and confirmed that contractile function was actually enhanced rather than impaired for a given preload (83), similar to observations made after diuretic treatment (84).

The most convincing evidence that cardiac function is well preserved during sustained hypoxia comes from the Operation Everest II project. In this study, two-dimensional echocardiograms were performed at progressively greater simulated altitudes up to and including that equivalent to the summit of Mount Everest, both

Table 4 Sustained Hypoxia—Submaximal Exercise Hemodynamic Changes

Altitude (m)	n	Duration	Method	Oxygen uptake at altitude, L/min (mL/kg/min)	Responses (%)[a] CO	SV	HR	Ref.
3800	5	3–4 wks	CO_2 rb	1.88 (23.3)	−7*	−14*	+9*	78
4300	8	3 wks	dye dilut.	1.70 (23.0)	nc	−16*	+6*	127
3100	8	10 days	Fick	1.60 (20.5)	−15*	−11*	+4	80
4300	4	2 wks	dye dilut.	1.57 (20.3)	nc	−15*	+8*	127
4350	4	10 days	dye dilut.	1.43 (19.6)	−23*	−32*	+24*	126
3100	6	2–3 wks	N_2O	1.62 (21.4)	nc	−8	+13*	139,140
4300	12	3 wks	dye dilut.	1.77 (24.7)	−19*	−26*	+8*	63,74
5800	4	2–3 mos	C_2H_2 rb	1.48 (23.2)	nc	−26*	+14*	132
6100	6	3–4 wks	thermo/Fick	1.49 (19.8)	−13	−28*	+17*	62
7620	6	1–2 wks	thermo/Fick	1.45 (19.1)	−7	−30*	+16*	62
8840	6	20 min	thermo/Fick	1.18 (15.5)	+8	−26*	+25*	62

[a] % changes in responses from sea level; changes relate to same absolute oxygen uptake at sea level and altitude.

rb = Rebreathing; dye dilut. = indocyanine green dye method; C_2H_2 = acetylene; thermo = thermodilution; nc = no change; CO = cardiac output, SV = stroke volume, HR = heart rate.

* $p < 0.05$ vs. sea level.

Table 5 Sustained Hypoxia—Maximal Exercise Hemodynamic Changes

Altitude (m)	n	Duration	Method	Peak oxygen uptake at altitude, L/min (mL/kg/min)	Responses (%)[a] CO	SV	HR	Ref.
4300	8	3 wks	dye dilut.[b]	2.42 (32.7)	+4	+14*	−4	127
4300	4	2 wks	dye dilut.	2.61 (33.9)	−22*	−19*	−3	128
4350	4	10 days	dye dilut.	2.41 (33.0)	−29*	−24*	−7*	126
5000	6	10 wks	CO_2 rb[c]	not available	−23*	−24*	nc	141
5800	4	2–3 mos	C_2H_2 rb	2.15 (33.1)	−29*	−5	−25*	132
6100	6	3–4 wks	thermo/Fick	2.10 (27.8)	−16*	−4	−14*	62
7620	6	1–2 wks	thermo/Fick	1.45 (19.2)	−31*	−30*	−23*	62
8848[d]	6	20 min	thermo/Fick	1.12 (14.8)	−30*	−14†	−26*	62

[a] % changes from sea level.
[b] Potential concerns about methodology of dye dilution (see Ref. 128).
[c] Studies done immediately after descent under more normoxic conditions.
[d] Only 3 subjects studied at simulated 8848 m altitude.
rb = Rebreathing, dye dilut. = indocyanine green dye method; thermo = thermodilution; C_2H_2 = acetylene; nc = no change; CO = cardiac output; SV = stroke volume; HR = heart rate.
* $p < 0.05$ vs sea level.

at rest and during submaximal exercise (85). As noted above and in Table 3, stroke volume was decreased with chronic altitude exposure, even at 8848 m. However, at each altitude, end-systolic volume was smaller than at sea level. Virtually all indices of left ventricular systolic function, including left ventricular ejection fraction, mean normalized systolic ejection rate, and peak systolic pressure/end-systolic volume, were either unchanged or improved even at the most extreme altitudes. Invasive measurements of cardiac filling pressure allowed the construction of Starling ventricular function curves and demonstrated that they were superimposable among sea level and altitude studies (62). In other words, stroke volume was always appropriate for the LV filling pressure, which was reduced at altitude. There was no increase in stroke volume with acute administration of 100% O_2, arguing against any hypoxic depression of contractile function. These data confirm that in normal individuals LV contractile function is not diminished during sustained exposure to high altitude, and thus an increase in LV end-systolic volume cannot explain the reduction in stroke volume observed at altitude, even under the most extreme hypoxic conditions.

Although it is now clear that myocardial contractile function is preserved in the setting of prolonged hypoxia, the mechanisms responsible for myocardial protection from severe hypoxia are not well understood. Various candidates for the development of a ''hypoxia-tolerant'' myocardium include heightened sympathetic activity (86), a shift in myocardial metabolism to glucose and lactate, with increased ATP yield per molecule of oxygen (87), transcriptional changes in mitochondrial and nuclear genes (88,89), and the induction of various cellular regulatory factors that preserve mitochondrial function such as hypoxia-inducible factor (89), nuclear factor (NF)-κB (90) and heat shock proteins (91,92), which appear to be activated by sustained hypoxia in various animal and tissue culture models. In addition, sustained hypoxia exposure may result in downregulation of some cellular functions, such as the inducible nitric oxide synthase (iNOS) transcriptional response to circulating cytokines, which could provide protection against myocardial and vascular endothelial damage (90). Although none of these mechanisms have been definitely shown to occur in humans at high altitude, these studies present compelling data that changes on the cellular level do occur with prolonged exposure to hypoxia. A combination of these molecular and cellular adaptations could result in a more energy-efficient, normally functioning contractile state of the myocardium in the setting of prolonged hypoxia. Some of these genetic adaptations may also explain the tolerance of high-altitude populations as demonstrated by nearly normoxic cardiovascular function despite life-long exposure to significant hypoxia.

The reduction in stroke volume with sustained altitude exposure therefore must be due to a reduction in LV end-diastolic volume (LVEDV), which has been universally observed in echocardiographic studies at altitude (79,83,85,93). For any given level of left ventricular contractility or afterload, stroke volume is determined to a large extent by left ventricular filling as a function of the Frank-Starling mechanism (94). Left ventricular end-diastolic pressure (LVEDP) is determined by

LVEDV, and this relationship is influenced by cardiac chamber stiffness. Because of the curvilinear nature of the left ventricular pressure-volume relationship, cardiac stiffness is dynamic with the instantaneous stiffness, defined as dP/dV, dependent on the specific value of left ventricular volume (95). Cardiac stiffness, LVEDV and consequently SV may therefore be altered by either (1) shifting up (hydrated state) or down (dehydrated state) on any individual pressure-volume curve or (2) changing the underlying pressure-volume relationship. These changes may occur with either an alteration in intravascular volume or a sudden change in extracardiac influences (pericardial or pulmonary mechanical restraint) or via a specific cardiac adaptation such as bedrest deconditioning (96). These concepts are illustrated in Figure 4.

One obvious mechanism for the early reduction in both filling pressure (62) and stroke volume with sustained altitude exposure is a reduction in plasma volume. Numerous studies have documented this adaptation, which ranges from 20 to 30% (97,98) (see also Chapter 15). Plasma volume is reduced within the first few hours

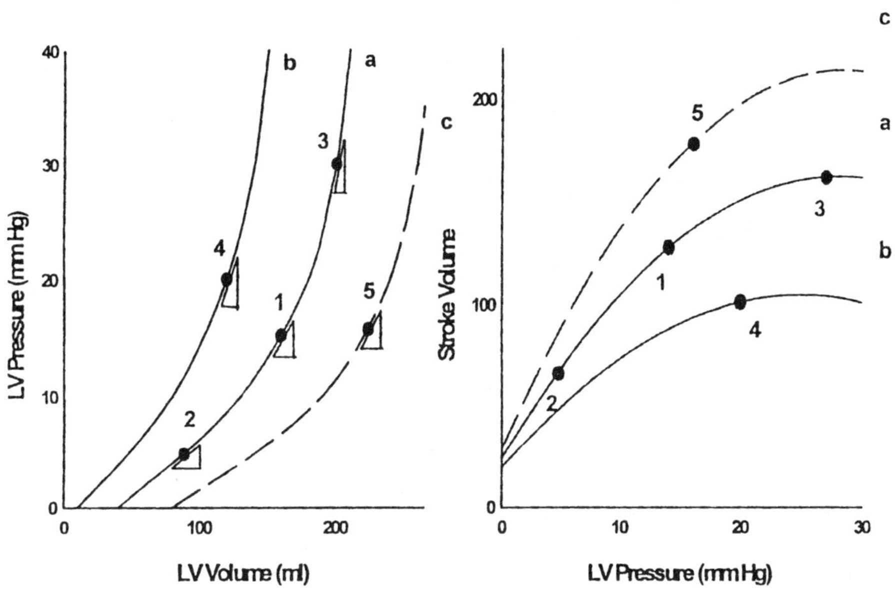

Figure 4 Left ventricular pressure-volume relationships. Point 1 on curve represents the normal supine position on the pressure-volume (left) and Starling (right) curves. Acute dehydration would cause a shift to point 2, to a less stiff region (smaller slope) of the P-V curve and a large fall in stroke volume (SV) during orthostasis. In contrast, hyperhydration would shift to point 3, where changes in SV would be modest despite large changes in LVEDP. For a given left ventricular volume, changes in the underlying P-V relationship could result in an increase in cardiac compliance (curve c) or decrease in compliance (curve b) due to intrinsic or extrinsic factors.

to days at altitude and is responsible for acute hemoconcentration (99). When this reduction in plasma volume is prevented with CO_2 breathing (100), the reduction in stroke volume is also prevented. Over time, a gradual increase in red cell mass occurs, such that blood volume gradually increases. For example, in most of the studies from Pikes Peak there was approximately a 15–20% decrease in plasma volume throughout the period of residence and a smaller (2–5%) decrease in blood volume after 3 weeks at 4300 m due to the progressive increase in red cell mass (74,101). Figure 5 shows the relationship between the changes in blood and plasma volume and the reductions in stroke volume and cardiac output with sustained hypoxia at 4300 m. In these studies, the changes in blood volume accounted for some of the decrease in stroke volume after acclimatization, particularly during submaximal exercise.

However, other mechanisms are probably also involved since the reduction in stroke volume at altitude is seen early in the acclimatization period and does not appear to be totally ameliorated by the gradual increase in blood volume over time. Alexander et al. infused 500 mL of dextran into two subjects acclimatized to 3100 m (80). In these two subjects, pulmonary capillary wedge pressure increased modestly, suggesting increased cardiac filling, and stroke volume was increased but not to sea level values. Unfortunately, the small numbers of subjects studied and the rather small volume increases and attendant pressure changes preclude definitive conclusions from these data. Moreover, as depicted in Figure 4, a rise in filling pressure does not necessarily equate with a rise in LVEDV, particularly if ventricular interdependence and pericardial constraints are operative (102). However, a recent study performed in young men in a hypobaric chamber (Operation COMEX) demonstrated that intravenous volume infusion improved maximal oxygen consumption at simulated altitudes of 7000 m, especially in those subjects with the lowest plasma volume from prolonged hypoxia exposure (103). Although no determinations of stroke volume were obtained with these infusions, the increase in exercise capacity was assumed to be related to increases in cardiac output from an increased intravascular volume, suggesting that both convection and diffusion limitations may be operative in the reduced physical performance at extreme altitudes.

A second possible explanation for the reduction in stroke volume may be that it occurs as a consequence of the increase in heart rate. Chronic increases in heart rate, for example, in models using chronic pacing, are always associated with decreases in stroke volume, even before LV function deteriorates (104). This result may be due to redistribution of blood volume to the venous capacitance associated

Figure 5 Simple linear regression of changes in blood volume and plasma volume with changes in cardiac output and stroke volume during submaximal exercise between sea level and 18–20 days at 4300 m. Data taken from combined studies in 12 men on Pikes Peak in 1988 and 1991 (74,101). Changes in both cardiac output and stroke volume between sea level and 4300 m are correlated with changes in both plasma volume and total blood volume.

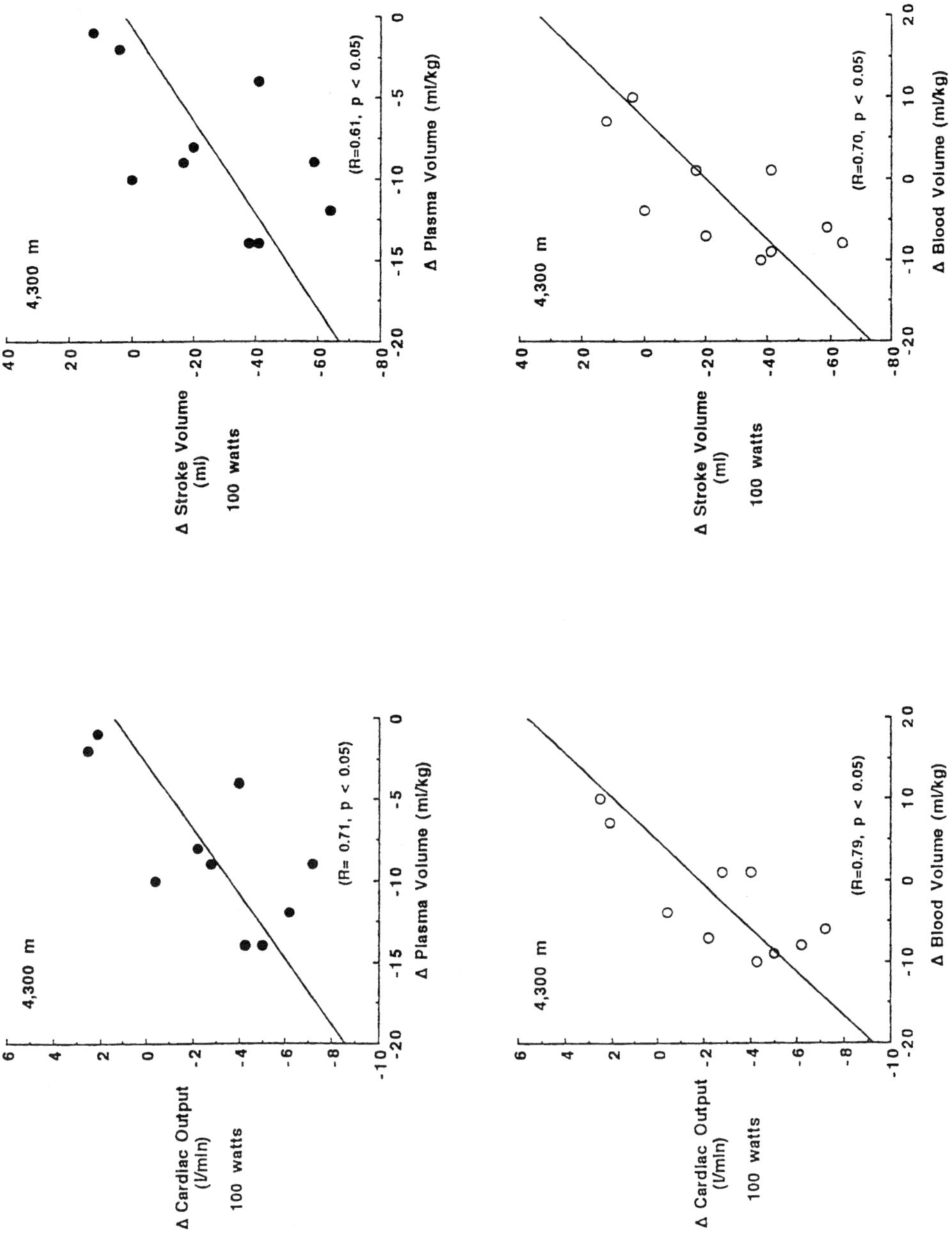
4,300 m
Δ Stroke Volume (ml)
100 watts
(R=0.61, p < 0.05)
Δ Plasma Volume (ml/kg)
4,300 m
Δ Stroke Volume (ml)
100 watts
(R=0.70, p < 0.05)
Δ Blood Volume (ml/kg)
4,300 m
Δ Cardiac Output (l/min)
100 watts
(R= 0.71, p < 0.05)
Δ Plasma Volume (ml/kg)
4,300 m
Δ Cardiac Output (l/min)
100 watts
(R=0.79, p < 0.05)
Δ Blood Volume (ml/kg)

with transiently increased cardiac output or as a function of increased peripheral vascular resistance. Supportive evidence for this concept comes from the observation that when subjects are beta-blocked prior to ascent to altitude (63), the increase in heart rate is prevented and the decrease in stroke volume is substantially reduced. However, this explanation is not sufficient, as stroke volume is reduced with sustained altitude exposure even at maximal exercise, when maximal heart rate is lower than at sea level.

Finally, the reduced *left* ventricular filling at altitude may be due to impaired *right* ventricular function. Because the right heart is not well suited to geometric modeling, volume data similar to that of the LV are not available for the right ventricle at altitude. The role of the right ventricle may be particularly important because hypoxia causes pulmonary vasoconstriction and sustained pulmonary hypertension at altitude (105,106) (see also Chapter 11), resulting in numerous electrocardiographic reports of right ventricular strain or overload (60,67,107). Because the right ventricle does not have the contractile reserve available to the left ventricle, it may not be able to perform optimally against acute or subacute increases in afterload (108). Anecdotal reports of acute right ventricular dilatation have been made in climbers with high-altitude pulmonary edema, and RV dilatation has been documented by echocardiography even in well subjects at extreme altitude in OE II (85) and more recently in a group of Japanese climbers acclimatized in the field to altitudes above 6000 m (93). More dramatic echo evidence of RV failure has been obtained in some ultra-endurance athletes competing in a high-altitude endurance race (109). In fact, right ventricular function is probably particularly important at maximal exercise, when ventricular filling time is shortest. In a recent study (110), patients born without a right ventricle who had undergone the Fontan operation (passive conduit directly from the systemic veins into the pulmonary artery) had completely normal LV stroke volume and cardiac output response at 3100 m, both at rest and submaximal exercise, but had a depression in stroke volume at maximal exercise, indicating that right ventricular dysfunction could contribute to the reductions in stroke volume during exercise with sustained hypoxia.

The potential significance of the right ventricle for supporting LV filling has recently been highlighted in some patients with congestive heart failure (111). In this study, patients with heart failure and depressed LV stroke volume were subjected to lower body negative pressure with pooling of blood in the lower part of the body. Contrary to what occurs in normal individuals, these patients had an increase rather than a decrease in cardiac output, associated with an increase in LVEDV and SV. Thus, in some patients RV dilation may impair LV filling through ventricular interdependence and pericardial constraint with shift of the interventricular septum from right to left. Although such an effect is theoretically possible at altitude, it is unlikely to play a major role in most individuals because LV filling pressure when measured has been uniformly low, providing evidence against pericardial constraint. However, much more work remains to be done to examine right ventricular function at altitude.

Diastolic Function

Although LV systolic function is normal with sustained hypoxia, abnormalities of LV diastolic function could also contribute to the impaired ventricular filling. The filling of the heart during diastole is a dynamic, complex process involving several distinct hemodynamic phases and interactions among a number of variables. These include left ventricular chamber stiffness, dependent on both intrinsic myocardial and extrinsic factors such as pericardial and pulmonary mechanical constraint, but also active myocardial relaxation, atrioventricular coupling, the dynamic establishment of atrioventricular and intraventricular pressure gradients, and extracardiac factors. Abnormalities in intracellular calcium transport, secondary to hypoxia, could result in abnormal left ventricular relaxation or compliance. With brief exposure to varying degrees of acute hypoxia induced by breathing hypoxic gas mixtures, 10 young men were found to have abnormalities in left and right ventricular diastolic function on cardiac echo-Doppler studies (112). However, traditional Doppler measures of diastolic function are extremely load dependent (113), and it is difficult to sort out frank changes in relaxation and/or compliance from the well-described hemodynamics of acute altitude exposure. Probably the strongest evidence against abnormalities of diastolic function at altitude come from the low pulmonary capillary wedge pressures measured in OEII (62). However, it is possible that abnormalities of ventricular compliance or distensibility may be masked by the reduced ventricular volumes observed in subjects during sustained hypoxia (compare point 2 with point 5 in Fig. 4).

At least two plausible mechanisms exist that could lead to reduced distensibility after sustained altitude exposure. The first is a reduction in physical activity with associated cardiac atrophy (96). Two weeks of bedrest deconditioning in normal healthy subjects results in both atrophy of the heart and reduced LV distensibility, associated with a reduction in LVEDV and SV at any given filling pressure. Although complete pressure/volume curves have not been obtained after sustained altitude exposure, examination of the reported PCW pressure and LVEDV at rest from the data in OEII (62,85) suggests a shift upward and to the left, i.e., an increase in pressure for any given volume, consistent with decreased ventricular distensibility. Certainly the confinement associated with chamber or mountain hut studies, as well as the hypoxia-induced decrease in aerobic power, could conceivably lead to cardiovascular deconditioning, particularly if the subjects were very fit and active prior to the studies.

A second mechanism that could explain this decreased distensibility is increased pulmonary mechanical constraint associated with expansion of lung volumes. Data in dogs after pneumonectomy show dramatic increases in cardiac distensibility, and recent data from space flight suggest that removal of pulmonary mechanical constraint can also lead to a dramatic increase in distensibility in humans (114,115). Although there are no data available on the role of enhanced lung volume on diminishing cardiac distensibility, this mechanism could play some role because of the typically large ventilatory adaptations seen with acclimatization to high altitude.

B. Systemic Circulatory Changes—Sustained Hypoxia

General Systemic Vascular Responses

The systemic circulation involves the entire circulatory system, including the heart and its interaction with regional vascular beds, which determines overall circulatory function. Although the local peripheral circulation appears to be precisely regulated by arterial oxygen content, the systemic circulation seems to behave differently. One of the key questions regarding the cardiovascular response to sustained hypoxia has been whether sympathetic activation persists or diminishes over the course of prolonged exposure to high altitude. Indirect measurements of blood and urinary catecholamines over 3 weeks on Pikes Peak suggest that circulating norepinephrine actually increases rather than decreases over time at altitude (15,16,116). Direct measurements of muscle sympathetic nerve activity (MSNA) have confirmed that with short-term (2–4 days) exposure to altitudes up to 4500 m, MSNA is slightly greater than with acute hypoxic exposure in healthy subjects (34). In contrast, MSNA is markedly augmented in climbers who are susceptible to high-altitude pulmonary edema, both acutely and even more so with sustained altitude exposure, forming a potential mechanistic link between the magnitude of sympathetic activation and the illnesses of high altitude (34). Most recently, MSNA was recorded from a relatively large number of well-acclimatized subjects who had been at altitudes above 5200 m for more than one month as part of the 1998 Danish Medical Research Expedition to the Andes. In contrast to the opinion that sympathetic activity would decrease over time, MSNA in these well-acclimatized subjects was dramatically elevated (117). These data suggest that the well-described increase in peripheral chemosensitivity, which has been documented for the ventilatory response to hypoxia, may lead not only to a progressive increase in ventilation during acclimatization (see Chapter 7) but also to a similar progressive increase in sympathetic nerve activity despite improvements in arterial oxygen content. This hypothesis needs to be tested by further experiments in subjects during sustained hypoxia.

Regional Vascular Responses

Since oxygen delivery (blood flow $\times$ Cao_2) seems to be precisely regulated at a local level with acute hypoxia, it is reasonable to assume that the changes in peripheral blood flow with sustained hypoxia would also mirror the changes in Cao_2, and in fact that is the case.

The largest body of work examining the effect of sustained hypoxia on peripheral blood flow has been performed at 4300 m on the summit of Pikes Peak. Leg blood flow was measured using the thermodilution technique under both acute and more than 3 weeks at rest and during exercise (63,74,118). In general, these studies used a common paradigm: baseline measurements at sea level, followed by acute exposure to 4300 m, and then repeat measurements after 2–3 weeks at 4300 m. Based on these experiments, the following conclusions can be drawn: (1) leg blood flow was consistently elevated with acute hypoxia at the same metabolic demand (same oxygen uptake) compared to sea level, thereby maintaining oxygen delivery; (2) with acclimatization, the increase in arterial oxygen content was associated with a decrease in leg blood flow and a concomitant increase in oxygen extraction at the same metabolic demand; and (3) the addition of supplemental oxygen to acclimatized subjects resulted in a further reduction in leg blood flow, confirming the tight coupling of leg blood flow to oxygen content for any given oxygen requirement.

Similar findings have been reported for the nonexercising forearm of men at the same high altitude (119), i.e., a progressive increase in forearm vascular resistance over 7 days of acclimatization to 4300 m (Fig. 6). Moreover, regardless of whether the upper or lower extremity was studied, and independent of metabolic activity, the observed peripheral vasoconstriction during acclimatization was related to elevated urinary and arterial norepinephrine levels, indicating that enhanced sympathetic stimulation was likely playing an important role (16,119). Thus, functional sympatholysis diminished with sustained hypoxia as acclimatization restored arterial O_2 content and sympathetic activity appeared to increase. Intriguingly, hypocapnia might be playing an important role in mediating changes in limb blood flow and vascular resistance, as forearm flow and resistance were unaltered at altitude under conditions of normocapnic hypoxia (120). These data provide support for a potentially significant interaction between ventilatory and cardiovascular chemoreflexes.

Finally, upon return to sea level after a sojourn at altitude, climbers who had been exposed to altitudes ranging from 3500 to 8400 m were found to have a 26–34% reduction in muscle blood flow, determined by ^{133}Xe clearance, during submaximal exercise (121), associated with persistent elevations in hemoglobin concentration and arterial oxygen content. Thus, even with restoration of normal Pa_{O_2}, the surfeit of oxygen availability continues to regulate oxygen delivery via alterations in limb blood flow.

Thus, the peripheral circulation develops adaptive responses with sustained hypoxia that maintain tissue oxygen uptake and substrate delivery. Both at rest and during exercise, in a small (forearm) or a large (legs) muscle bed, vasoconstriction appears to be a common mechanism during acclimatization as arterial oxygen content increases and a persistent sympathetic activation is unmasked by withdrawal of functional sympatholysis.

Coronary Circulation

The coronary circulation has a number of features that distinguish it from the peripheral circulation and may influence the adaptation to hypoxia. First of all, coronary blood flow is regulated by myocardial oxygen demand, which in turn is determined by three main factors: heart rate, contractility, and wall stress. Wall stress itself is determined both immediately prior to the onset of contraction (preload), and after the onset of contraction (afterload). Whether myocardial oxygen demand is increased,

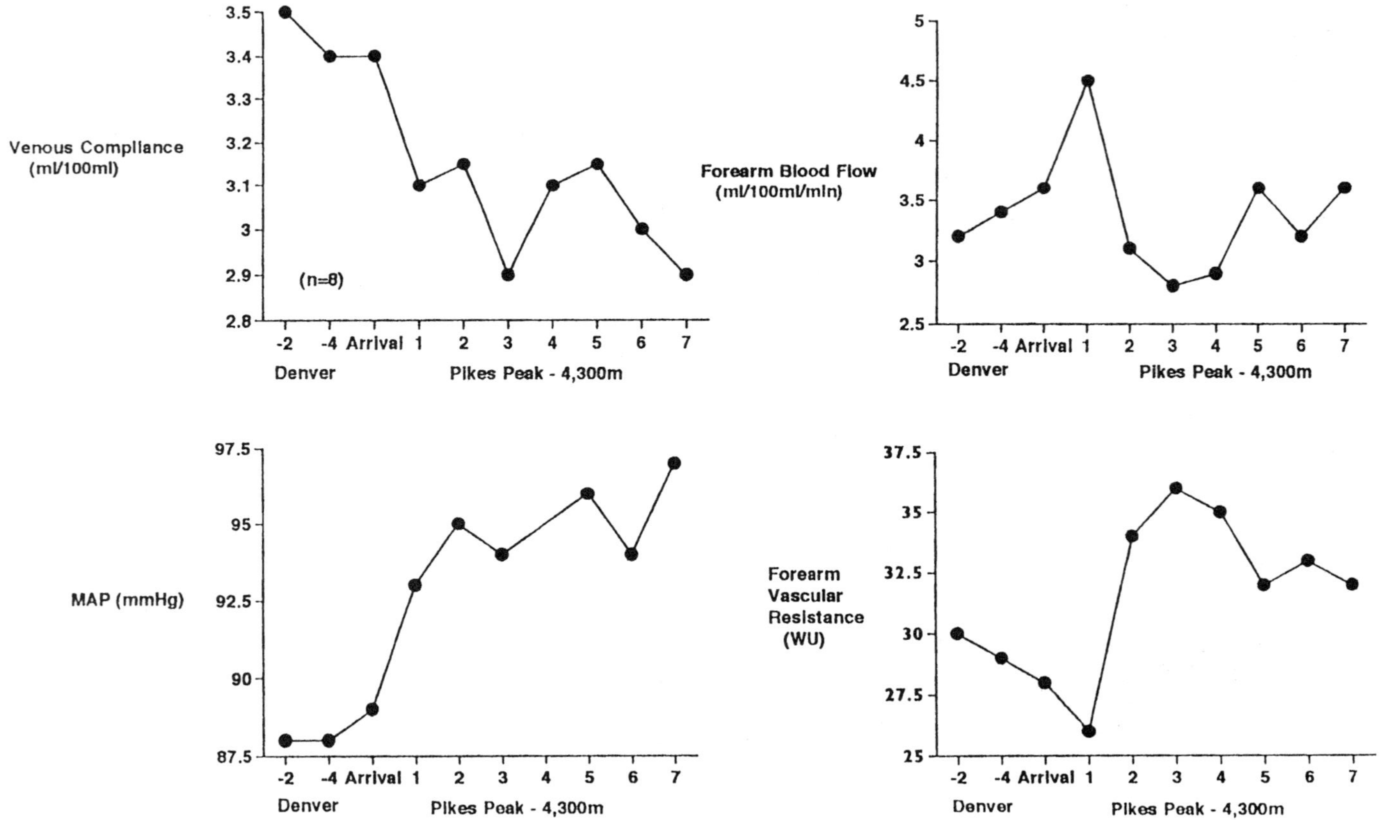

Figure 6 Sustained hypoxia at 4300 m results in a reduction in forearm compliance (upper left) and blood flow (upper right) along with increases in mean arterial pressure (MAP) (lower left) and forearm vascular resistance (lower right). Subjects were residents of 1600 m (Denver, CO). (Figures redrawn from original data in Ref. 119.)

decreased, or unchanged during sustained hypoxia is not clear and depends on a complex interplay among these factors, some of which are increased (heart rate, contractility, blood pressure, afterload) while others are maintained at a lower level (preload). In addition, oxygen extraction is extremely high in the myocardium even at rest, and thus most of the adaptive range of myocardial metabolic activity must be met by alterations in coronary blood flow. For normal young individuals without coronary artery disease, this vasodilator reserve is adequate to meet myocardial oxygen demands despite extreme hypoxia, even under conditions of maximal exercise. Thus in both Operation Everest II (67) and the American Medical Research Expedition to Mount Everest (AMREE) (60), there were no symptoms or ECG evidence of ischemia in subjects exercising maximally (Vo_2 = 1.17 L/min) at 8848 m despite extraordinary metabolic derangements including an Sao_2 as low as 49% and pH of 7.52 (67).

Similar to the peripheral circulation, with acute hypoxia coronary blood flow increases in proportion to the reduction in arterial oxygen content (122). However, with sustained hypoxia the situation is quite different. Studies performed in both high-altitude residents and recently acclimatized lowlanders suggest that coronary blood flow is decreased compared to sea level. For example, high-altitude Andean natives demonstrated a progressive decline in coronary blood flow at rest with increasing altitude, compared to sea level natives, reaching 30% at the highest altitudes (123). These findings were related to a lower heart rate, which contributed to a decreased myocardial O_2 demand as well as to an increased red cell mass, which allowed a relatively greater coronary arteriovenous O_2 difference. However, similar reductions in coronary blood flow have also been observed in newly acclimatized young men after only 10 days at 3100 m (Leadville, CO) (124). Like high-altitude natives, these recently acclimatized lowlanders also had a 30% decrease in coronary blood flow compared to sea level values, though without a concomitant reduction in heart rate and myocardial O_2 consumption. Although coronary venous Po_2 was appropriately low (though not less at 3100 m that at sea level), coronary O_2 extraction was increased as reflected by a decrease in coronary sinus O_2 content and saturation. Thus, it appeared that coronary blood flow was regulated to maintain a constant myocardial O_2 tension at high altitude, similar to the findings in the peripheral circulation during submaximal exercise when skeletal muscle metabolic rate is similarly high.

Systemic Blood Pressure

There has been some controversy as to how the myriad of cardiovascular responses to sustained hypoxia affects systemic blood pressure. Most of the controversy, however, probably relates to the timing of blood pressure measurements. With acute exposure to altitude, as described above, blood pressure usually falls due to a decrease in systemic vascular resistance. However, with acclimatization stroke volume and cardiac output fall, even in the face of persistent tachycardia, and peripheral vascular resistance gradually rises as ventilatory and hematological acclimatization restore arterial oxygen content towards normal. This acclimatization response may occur relatively rapidly, particularly at moderate altitudes of 3000–4000 m. Thus, persistent sympathetic activation due to hypoxia is no longer offset by functional sympatholysis and locally mediated vasodilation, and blood pressure rises over time (42). Increases in systemic arterial pressure and vascular resistance have been noted in numerous studies of high-altitude acclimatization. For example, native lowlanders who have spent 2–4 weeks at altitudes varying from 3000–5000 m have been shown consistently to have elevations in arterial pressure at rest and during submaximal and maximal exercise when compared to sea level values (63,65,74,78,118,125–128). Conversely, high-altitude residents at 3100 m (Leadville, CO) have been shown to have lower systolic and diastolic blood pressures at rest and during submaximal exercise after 10 days at sea level (82). Because cardiac output is decreased rather than increased with sustained altitude exposure, studies using β-blockers prior to ascent have not observed a reduction in the systemic blood pressure elevation associated with chronic hypoxia (57,63), thus raising the importance of α-adrenergic vasoconstriction in this process. In support of this concept, healthy young women treated with the α-adrenergic blocker prazosin were found to have a decline in both diastolic and mean arterial pressure during exercise after 48 hours of hypoxia in a hypobaric chamber simulating an altitude of 4300 m when compared to the unblocked state (129). α-Adrenergic blockade also blunted the systemic pressor response to the inhalation of hypoxic gas (130). However, administration of prazosin to women prior to a 12-day sojourn at 4300 m blunted but did not abolish the rise in systemic blood pressure observed with ambulatory monitoring (131). Thus, sympathetic blockade with either α- or β-blockers was not sufficient to prevent the systemic pressor response at moderate high altitude. Moreover, few data exist regarding the relative importance of other neurohormones such as angiotensin, aldosterone, or vasopressin, all of which may be involved in some degree with raising blood pressure during sustained high-altitude exposure. These mediators may be particularly important in sustaining the pressor response to sustained hypoxia, as acute administration of 35% O_2 after acclimatization to 4300 m did not return blood pressure to sea level values (118).

Although several studies at altitudes between 3000 and 5000 m have reported increases in systemic arterial pressure, no increases in blood pressure or systemic vascular resistance have been observed at more extreme altitudes (106,132). At these extreme altitudes, the degree of hypoxia is profound and the acclimatization process cannot restore arterial oxygen content to normal as it can at lower altitudes due to the profound stimulus to peripheral vasodilation with reduction in peripheral resistance.

Few data exist regarding blood pressure on recovery from altitude. A study of 47 men who were taken from sea level to 3658 m for a 10-day sojourn reported a progressive rise in resting diastolic blood pressure that was associated with an increase in total urinary catecholamines (125). On return to sea level, diastolic blood pressure remained elevated for several days while urinary catecholamines returned to normoxic levels within the first 24 hours. Insufficient data were available to provide a satisfying explanation and more research is needed in this area.

Finally, elevations in systemic arterial pressure do not seem to persist in lowlanders who remain at altitude for several months or longer (133). Sympathetic activity may diminish with continued residence at the same altitude. For example, urinary catecholamine levels have been reported to return to sea level values in subjects who remained at altitudes greater than 3000 m for more than 90 days (134). However, no direct data regarding sympathetic nerve activity are available in sea level natives over such a prolonged period of time. In addition, highlanders who were native to this altitude had similar urinary catecholamines levels as lowlanders at sea level. Moreover, studies in high-altitude populations have shown that systemic hypertension is uncommon and blood pressures in highlanders tend to be lower than in sea level natives (135). Thus, elevations in systemic arterial pressure appear to be part of the early (i.e., days to weeks) acclimatization process to high altitude associated with heightened sympathetic stimulation but may gradually abate over time as the sea level native becomes more like a high-altitude native.

In summary, although the mechanisms are not entirely clear, there is general agreement that over a period of days to weeks, the cardiovascular adaptations to sustained hypoxia include persistent and probably augmented sympathetic activation, resulting in tachycardia with a reduced stroke volume that together is not sufficient to maintain cardiac output at sea level values. Coronary and peripheral oxygen transport is maintained via increases in Cao_2, and regional blood flow is precisely regulated to support delivery of oxygen and substrate to the tissues. With acclimatization, a gradual reduction in coronary and peripheral blood flow is associated with sympathetic activation that is now unopposed by regional vasodilation, leading to increases in total peripheral resistance and systemic hypertension.

Key Unanswered Questions—Sustained Hypoxia

Despite all the studies cited, the mechanism and feedback loops for the reduction in cardiac output between acute and sustained hypoxia have not been elucidated. Is it all a consequence of changes in preload, or are other factors involved, such as the central nervous system? Are there cellular mechanisms that govern blood flow and what variable is sensed?

What is the physiological purpose for the apparent matching of tissue oxygen supply to demand?

Which comes first—is heart rate increased from chemoreceptor mechanisms and stroke volume passively reduced, or is stroke volume reduced via neurohumoral mechanisms and heart rate increased reflexively?

Is the systemic vasoconstriction with sustained hypoxia an adaptive or maladaptive response?

Is the elevation in systemic blood pressure with sustained hypoxia of short duration completely governed by enhanced sympathetic activity, or are there local vascular factors that play a major role?

IV. Cardiovascular Function During Exercise

As compared to sea level responses, submaximal exercise during acute hypoxia is associated with an increased heart rate and no decrease in stroke volume and therefore an increased cardiac output for any given absolute value of oxygen uptake. The same response has been shown with leg blood flow during acute hypoxia. However, with sustained hypoxia, submaximal exercise is associated with an increased heart rate but a decrease in stroke volume for any given absolute level of oxygen uptake. The usual sea level relationship between cardiac output and oxygen uptake is preserved in some but not all studies at more moderate altitudes of 2500–4500 m, while other studies report a reduced cardiac output for a given oxygen uptake. With more extreme hypoxia, altitudes greater than 5000 m, the cardiac output and oxygen uptake relationship is similar to sea level.

Maximal exercise capacity is reduced to a similar degree with both acute or sustained hypoxia, although the mechanisms for the reduction differ depending on the duration of hypoxia. With acute hypoxia of short duration, heart rate, stroke volume, and cardiac output are generally maintained at sea level values with reductions in arterial oxygen content being responsible for the decrease in exercise capacity. With sustained hypoxia, reduction in maximal exercise capacity is associated with a decreased maximal heart rate and cardiac output that is as yet not clearly explained. Whether the reduction in cardiac output at maximal exercise with acclimatization is responsible for the decrease in maximal oxygen uptake or is a consequence of it is one of the most intriguing questions in regard to human adaptation to high altitude.

One of the key features of the cardiovascular system that ultimately determines work capacity of the organism is its remarkable range of responsiveness: oxygen uptake can increase 10- to 20-fold from rest to maximal exercise depending on the state of physical conditioning. Cardiac output must increase by nearly an order of magnitude in order to supply skeletal muscle with substrate to support this metabolic demand. Adaptations that are adequate to ensure substrate delivery at rest may be inadequate during high levels of exercise. Many investigators have studied the acute and sustained effects of hypoxia on submaximal and maximal exercise performance, and a detailed review of exercise physiology in hypoxia is presented in Chapter 20. However, a number of important points related to cardiovascular function during exercise deserve emphasis and comment.

V. Acute Hypoxia—Submaximal Exercise

When compared to normoxic exercise, with acute hypoxia, submaximal exercise is associated with an increase in muscle blood flow, heart rate, and cardiac output for any given absolute work rate, thus enabling the maintenance of oxygen uptake at sea level values in the face of reduced arterial oxygen content (60,62,63,74,118) (Figs. 7,8). Exposure to hypoxia of short duration (hours) is not associated with any

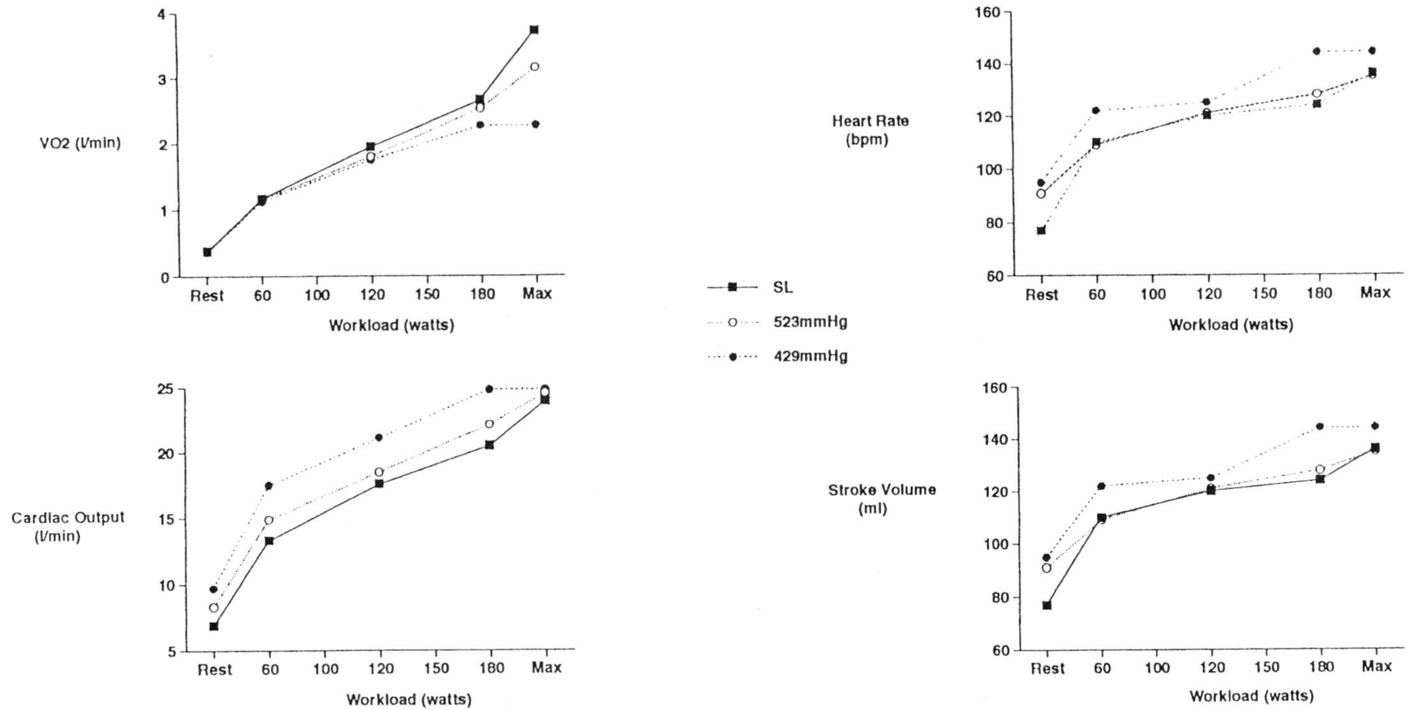

Figure 7 Cardiac hemodynamic responses to acute hypoxia at simulated altitudes of 3048 m (523 torr) and 4572 m (429 torr) (9). The Vo_2-workload relationship is maintained as at sea level, while both heart rate and cardiac output increases in a progressive fashion with greater hypoxia at a given workload. Stroke volume remains at or above sea level values. Although maximal Vo_2 is lower than at sea level; both maximal heart rate and cardiac output are unchanged compared to sea level with less than one hour of hypoxia exposure.

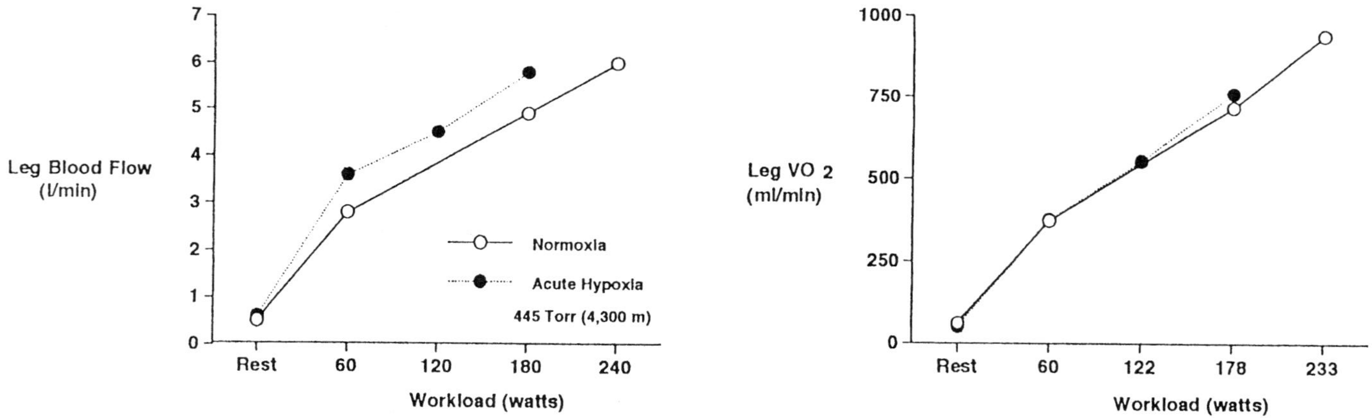

Figure 8 Leg blood flow is greater for a given workload with acute hypobaric hypoxia compared to normoxia. During acute hypoxia, the relationship between leg Vo_2 and workload is similar to that at sea level. (Figures drawn from data from Ref. 118.)

reduction in stroke volume. The magnitude of the increase in heart rate is proportional to the severity of hypoxia, and the changes in heart rate during submaximal exercise are similar to those of resting heart rate. One simple explanation for this increase in heart rate during submaximal exercise is that the same *absolute* work rate in hypoxia represents a greater *relative* work rate (greater percentage of maximal oxygen uptake) compared with normoxia, and relative oxygen uptake appears to govern the cardiovascular responses to exercise (136). Yet when workload is adjusted for % maximal oxygen uptake, heart rate is still somewhat greater in hypoxia than in normoxia (8). Increases in sympathetic stimulation, especially the elevation in arterial epinephrine, may be partly responsible for this increase in heart rate (15). However, studies in both humans and animals with β-adrenergic blockade during acute hypoxia continue to show some rise in cardiac output and heart rate with no effect on the vasodilator effect of hypoxia (12). Thus, peripheral factors involved in "functional sympatholysis" may also play a major role in the augmentation of cardiac output during acute hypoxia by reducing peripheral resistance and afterload.

Despite the increases in cardiac output, conductive oxygen delivery is still somewhat reduced compared to sea level (8,9), notwithstanding the existence of substantial systemic and regional blood flow reserve during submaximal exercise. Therefore, oxygen uptake is preserved additionally by a decrease in venous Po_2, thereby increasing peripheral oxygen extraction (76). Intracardiac filling pressures during submaximal exercise in the setting of acute hypoxia are no different than with normoxia except for the elevation in pulmonary arterial pressures (9), thus excluding systolic or diastolic dysfunction as a cause of this relative reduction in oxygen delivery. Thus, for reasons that are not entirely clear, activation of the cardiovascular system during submaximal exercise is restrained during acute hypoxia, and oxygen utilization is preserved by a combination of both increased blood flow and increased peripheral extraction.

VI. Acute Hypoxia—Maximal Exercise

All studies performed within an hour after exposure to hypoxia have shown that maximal heart rate and cardiac output are unchanged from sea level, although this concept has been challenged recently, at least with respect to maximal heart rate (137). Maximal oxygen uptake is decreased because of an obligatory reduction in maximal convective oxygen transport (8,9). Similar to the situation during submaximal exercise, oxygen delivery is decreased compared to sea level. However, at maximal exercise tissue oxygen extraction is unable to compensate for the reduced convective oxygen transport and maximal oxygen uptake falls (8,9). Intracardiac filling pressures remain at or below sea level values with a hypoxia-related rise in pulmonary arterial pressures (9).

VII. Sustained Hypoxia—Submaximal Exercise

The key difference between acute and sustained hypoxia during exercise is the apparent ''cardiac sparing'' (reduced heart rate, stroke volume, and cardiac output) that occurs due to adaptations in other organ systems such as increased ventilation, enhanced tissue oxygen extraction, and increases in hemoglobin concentration. Unlike with acute hypoxia, there is less cardiac output and limb blood flow at a given absolute work rate, and conductive oxygen delivery is supported to a greater extent by the increase in oxygen content in arterial blood. However, the degree of hypobaric hypoxia (P_B), the duration of exposure, and the type of exercise, i.e., submaximal or maximal, influence the specific physiological changes in cardiovascular function during exercise over time at high altitude.

As opposed to acute hypoxia, where cardiac output is elevated for a given oxygen uptake, with sustained hypoxia the cardiac output decreases so that over time at altitudes >5000 m, the relationship between cardiac output and oxygen uptake is similar to that at sea level (61,62,66,106,132,138).

In contrast, at more moderate altitudes (3100–4500 m), where there is both less direct hypoxic vasodilation and less sympathetic activation, the decrease in cardiac output with acclimatization may result in an output that is actually less than that at sea level for the same oxygen uptake (63,74,78,80,126) (Fig. 9), although the magnitude of reported decreases varies widely (127,128,139,140) (see Table 4). However, studies on climbers who have spent an extended period of time at high altitude have demonstrated consistent reductions in both stroke volume and cardiac output at several submaximal workloads even upon return to sea level, indicating that there is an alteration in the cardiac output–oxygen uptake relationship from prolonged exposure to hypoxia (141,142).

Although heart rate remains elevated at the same absolute or relative submaximal work rate during sustained hypoxia exposure (63,75,143), stroke volume is universally lower than at sea level (Fig. 10) and is primarily responsible for the reduction in cardiac output during acclimatization. Similar to observations made at rest, pulmonary capillary wedge pressure is also decreased during submaximal exercise, indicating that reductions in cardiac filling or preload and not decreased contractility are responsible for the decrease in stroke volume. Thus, even with activation of the muscle pump during exercise, central blood volume remains reduced during sustained hypoxia resulting in a decreased stroke volume.

One of the intriguing regulatory questions regarding the cardiovascular response to exercise after acclimatization to hypoxia is how the balance between systemic and local blood flow and oxygen extraction is determined. Although at least some of the reduction in cardiac output may be offset (if not induced) by an increased oxygen-carrying capacity of the blood, total oxygen delivery virtually always is reduced compared to sea level, and there is a greater dependence on peripheral extraction (66,132,138) (Fig. 11). This response occurs in spite of substantial heart rate and blood flow reserve, which could, if more substantially engaged, restore oxygen delivery to normal levels (63,74).

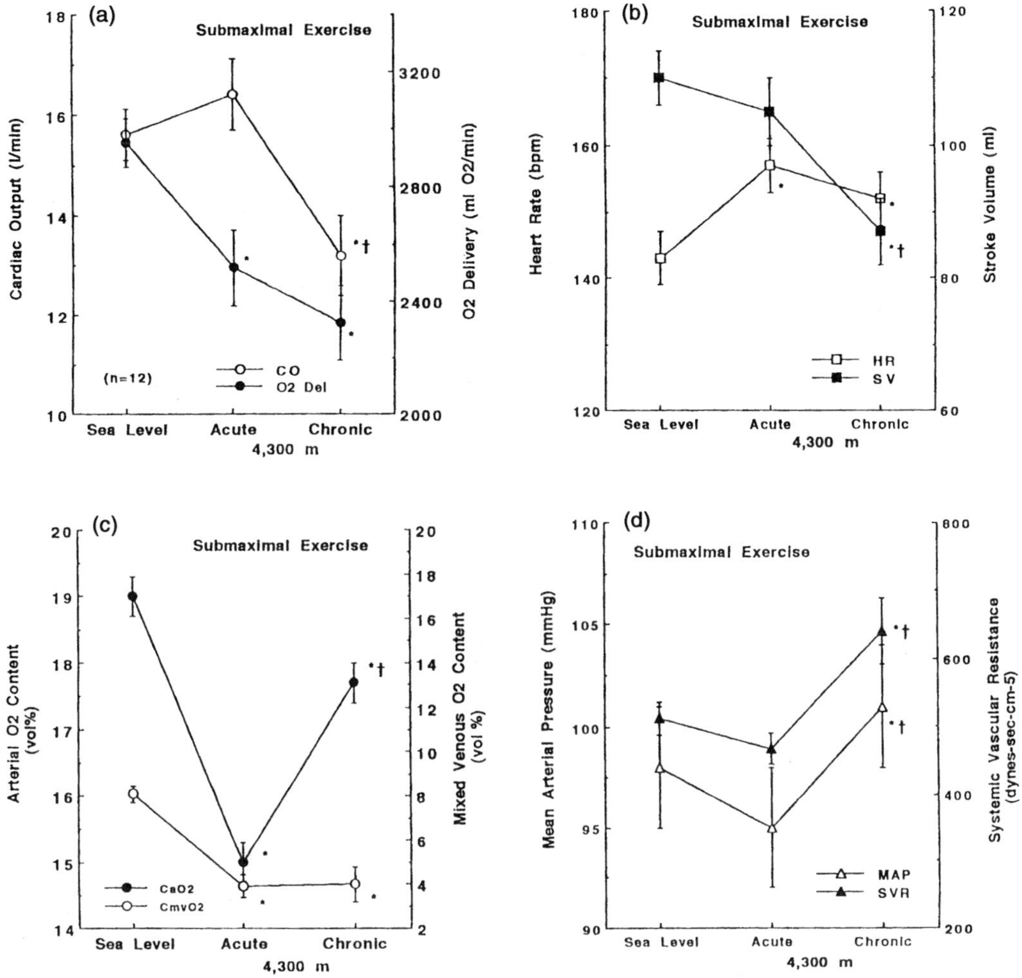

Figure 9 Figures drawn from combined data of 12 male subjects in studies in 1988 and 1991 at 4300 m (63,74). Submaximal exercise was at the same absolute workload (100 W) which was 50% Vo_2max at sea level and 65% Vo_2max at both acute and chronic altitude. (a) Cardiac output and oxygen delivery decreases with prolonged hypoxia. (b) Heart rate increases while stroke volume falls. (c) Arterial oxygen content falls dramatically with acute hypoxia but recovers with acclimatization, while mixed venous oxygen content decreases with acute hypoxia and remains at the same level with sustained hypoxia. (d) Mean arterial pressure and vascular resistance decrease with acute hypoxia but both rise dramatically with prolonged hypoxia.

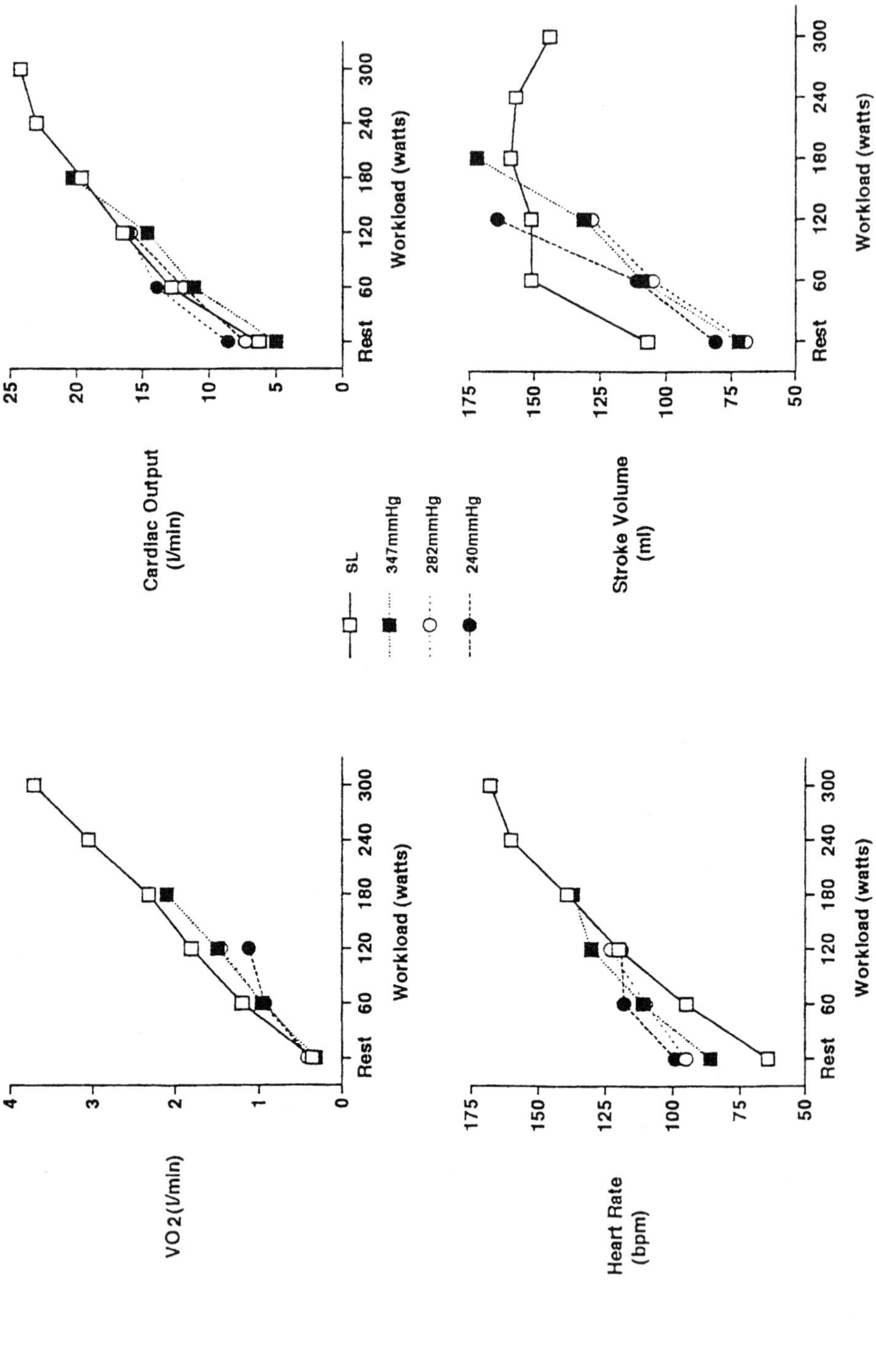
VO2(l/min)
Heart Rate
(bpm)
Cardiac Output
(l/min)
Stroke Volume
(ml)
Workload (watts)
Rest 60 120 180 240 300
SL
347mmHg
282mmHg
240mmHg

In fact, with prolonged acclimatization at the same altitude, there appears to be a reduction over time in heart rate at the same level of submaximal exercise and submaximal $\dot{V}O_2$, which contributes to the reduction in cardiac output (63,74, 128,143). Administration of supplemental O_2 during submaximal exercise in acclimatized subjects residing at high altitude has resulted in a further reduction in heart rate, suggesting that oxygen content of the blood is an important determinant of the heart rate response (62,66). Augmented vagal tone does not appear to be responsible for this reduction in exercise heart rate during acclimatization as heart rate increases with atropine to the same degree, both at sea level and after acclimatization to 4300 m (99). However, acute β blockade while at altitude does reduce submaximal heart rate to near sea level values (63,65), emphasizing the importance of hypoxia-induced sympathetic activation in this adaptation. The reductions in heart rate with acclimatization are therefore probably related to a decreased responsiveness to sympathetic stimulation as a result of downregulation of cardiac β-adrenergic receptors (69,70).

The mechanism by which arterial oxygen content regulates both central and peripheral blood flow during exercise remains unknown. Studies in animals suggest that it is the active oxygen-carrying capacity of red blood cells and not the hematocrit that influences changes in blood flow and vascular resistance (144–146). More investigation is required to determine the sensing mechanism and factors responsible for this regulation of blood flow during submaximal exercise with prolonged hypoxia.

VIII. Sustained Hypoxia—Maximal Exercise

Virtually all studies of maximal exercise after acclimatization to sustained hypoxia show a reduction in maximal cardiac output (see Table 5). In contrast to the hemodynamic changes during submaximal exercise, reductions in both heart rate and stroke volume are responsible for the reduced cardiac output at maximal exercise to varying degrees depending on the altitude in question. For example, at moderate altitudes of 3000–4000 m, reductions in stroke volume are predominant (126,127,128,143) and are similar in magnitude to those observed during submaximal exercise. However, at altitudes of ≥4000 m, reductions in maximal heart rate appear to play an increasingly important role (62,66,147,148). This reduction in maximal cardiac output, compounded by decreased arterial oxygen content, results in a decrease in peak

Figure 10 Cardiac hemodynamic responses during exercise with progressive hypoxia exposure in the Operation Everest II study (62). The $\dot{V}O_2$-workload (upper left) and the cardiac output–workload relationships (upper right) are maintained during submaximal exercise at all altitudes. With progressive hypoxia, heart rate increases (lower left) and stroke volume decreases (lower right). The response of stroke volume to progressive exercise differs between sea level and extreme hypoxia. Contrast these findings with the hemodynamic responses with acute hypoxia in Figure 7.

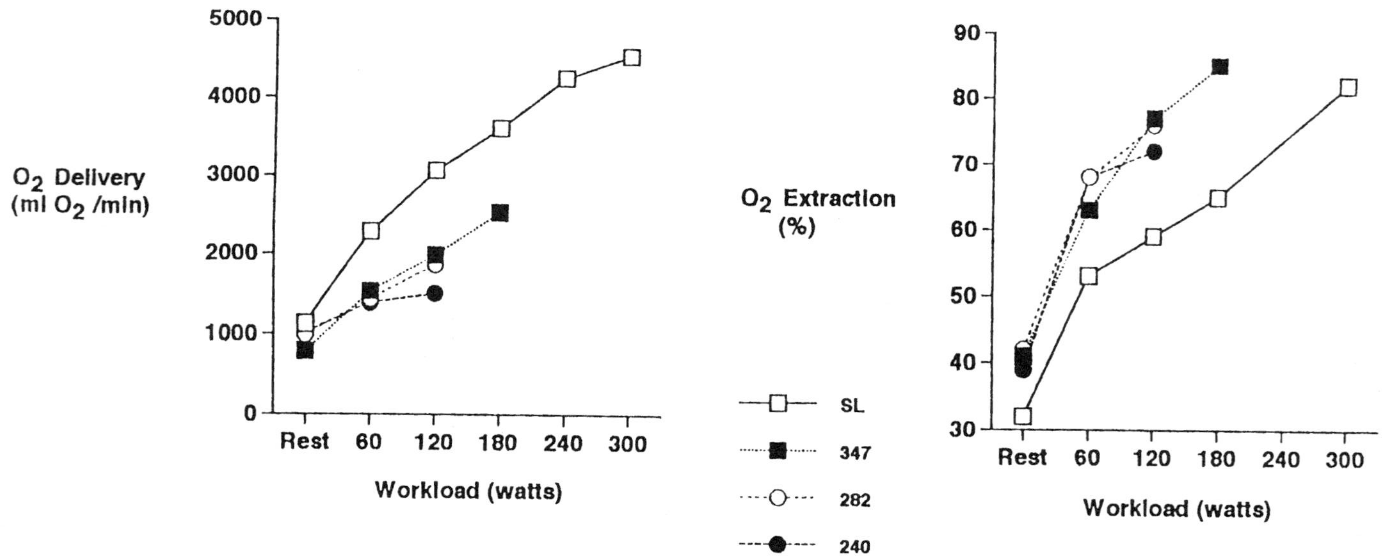

Figure 11 With progressive hypoxia exposure, oxygen uptake is maintained by an increase in oxygen extraction (a-v o_2/CaO_2) as oxygen delivery is decreased compared to sea level. (From Ref. 138.)

O_2 delivery. Thus, despite appropriate acclimatization, maximal oxygen uptake remains reduced compared to sea level, and similar to the reduction observed with acute hypoxia in all studies at >3000 m, though the reductions in conductive oxygen delivery are offset to some extent by increased tissue O_2 extraction. Still unclear is whether the reduction in cardiac output at maximal exercise is a primary cause of or secondary to the reduction in maximal oxygen uptake at high altitude (149).

In contrast to the elevated heart rate at rest and during submaximal exercise, there is a consistent reduction in maximal heart rate at virtually all altitudes studied (Fig. 12). The degree of reduction in heart rate is dependent on the duration and severity of sustained hypoxia exposure with the greatest reductions at the most extreme altitudes. Many studies have reported reductions in maximal heart rate after 2–3 weeks residence at 4300 m (63,74,126–128,150). At altitudes of >5800 m there is a 25% reduction in maximal heart rate associated with a >60% reduction in maximal oxygen uptake (62,66,147,148). The mechanism for this reduction in maximal heart rate could be related to a direct depressant effect of hypoxia itself on either the sinus node or central cortical irradiation (central command), a secondary effect of reduced work capacity, and therefore reduced afferent feedback from exercising skeletal muscle, or specific alterations in autonomic nervous system activity associated with sustained hypoxia. In practice, these different mechanisms have been very difficult to differentiate.

Studies with inhalation of supplemental oxygen to simulate normoxic conditions after prolonged exposure to hypoxia have yielded mostly expected results. For example, at an altitude of 4300 m, abrupt restoration of normoxic conditions in

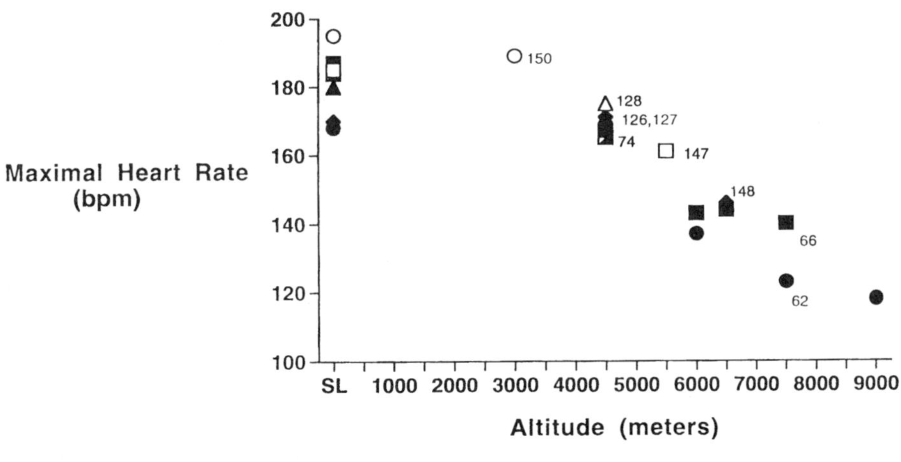

Figure 12 Maximal heart rate falls with sustained hypoxia exposure. The average duration of high-altitude exposure in each study was at least 2 weeks. Maximal heart rate began to decrease above altitudes of 4000 m. Numbers identify each specific study in the reference list.

recently acclimatized subjects led to an increase in heart rate and cardiac output with a definite increase in maximal workrate and oxygen uptake (128). However, cardiac output was still less than at sea level. Similar findings were seen at an altitude of 5800 m with improvement in heart rate and oxygen uptake at maximal exercise with oxygen inhalation (66). In contrast, with more prolonged, progressive exposure to more extreme altitude, there was little benefit in heart rate and work capacity with supplemental oxygen (62). It is likely, however, that prominent cardiovascular deconditioning played a substantial role in reducing work capacity in these studies. The degree of restoration of maximal heart rate with acute oxygen administration in acclimatized subjects may depend on the degree of attenuation of sympathetic activity (65)—if sympathetic activity is already near maximal, then supplemental oxygen results in relatively minimal additional tachycardia. Thus, autonomic nervous system activity likely plays either a primary or a facilitative role in reducing maximal heart rate with prolonged hypoxic exposure.

In addition to the known blunting of the heart rate response to enhanced sympathetic activity with prolonged hypoxia, early studies suggested an additional enhancement of parasympathetic activity with chronic hypoxia, as maximal heart rate was restored to sea level values with atropine in recently acclimatized subjects to 4600 m (151). Moreover, in lifelong high-altitude residents, atropine produces a greater increase in exercise heart rate compared to recently acclimatized subjects (152), suggesting an enhancement of vagal influences. However, studies with both β-adrenergic blockade and parasympathetic blockade in climbers residing at altitudes of >5000 m suggest that the degree of parasympathetic modulation of the heart is intimately influenced by the extent of cardiac sympathetic activity (65). In an elegant study by Savard et al., climbers who had the greatest reduction in adrenergic responsiveness also had the greatest increase in maximal heart rate with atropine. Conversely, if sympathetic responsiveness was minimally reduced, then the response to atropine was less prominent. Thus, the interplay between sympathetic and parasympathetic regulation of heart rate during maximal exercise makes a simple determination of a single, responsible neural pathway somewhat tenuous. Interestingly, in all the studies where maximal heart rate was increased with atropine, there was no improvement in maximal oxygen uptake, suggesting that the decreased maximal heart rate with acclimatization is unlikely to be a primary cause of the reduction in peak work capacity.

Finally, a key unanswered question is the degree to which the reduction in maximal cardiac output at high altitude contributes to the reduction in maximal oxygen uptake. At present, most evidence points to the converse: that the reduction in oxygen availability/utilization is the primary determinant of systemic and regional blood flow (149). Lines of evidence that support this hypothesis include (1) the relationship between cardiac output and oxygen uptake remains similar to that at sea level, even at maximal exercise at extreme altitude; (2) addition of supplemental oxygen results in an immediate increase in work rate, oxygen uptake, heart rate, and cardiac output; (3) increases (65,151) or decreases (143) in heart rate by pharmacological intervention fail to alter maximal oxygen uptake; and (4) even autologous blood transfusion at altitude, which increases blood volume and red cell mass, fails to increase maximal oxygen uptake at altitude (153), arguing for a primary role of reduced diffusion gradients for oxygen as the cause of reduced oxygen utilization. Issues related to distribution of cardiac output and degree of regional autonomic activation during maximal exercise with prolonged hypoxia remain to be elucidated before this hypothesis can be universally accepted (154).

In summary, during exercise at high altitude, the cardiovascular system functions to provide the delivery of oxygen in relation to the degree of oxygen utilization. During acute hypoxia, blood flow is increased to levels greater than expected for a given oxygen uptake at sea level. With prolonged hypoxia, there appears to be a ''cardiac-sparing'' effect on cardiac output and limb blood flow in relation to the degree of improvement in arterial oxygenation by the acclimatization process. There is a complex interaction between oxygen delivery, diffusion, and extraction to maintain oxygen uptake during submaximal exercise. At maximal exercise, oxygen uptake is reduced but the relationship between cardiac output and oxygen uptake remains similar to sea level and the factors responsible for the reduction in cardiac output remain to be determined. There clearly is an interplay between the autonomic nervous system and the degree of hypoxia which influences the cardiovascular responses to exercise with prolonged hypoxia.

A. Key Unanswered Questions—Exercise During Acute and Sustained Hypoxia

What is the mechanism for the increased blood flow responses during acute hypoxia? Is it all sympathetic stimulation?

What factors regulate the degree of peripheral oxygen extraction during both acute and sustained hypoxia? Peripheral venous oxygen tension remains the same with acute and sustained hypoxia. Why?

What are the relative roles of diminished cardiac response to sympathetic stimulation and enhanced parasympathetic stimulation with sustained hypoxia?

Do the changes in cardiac output and peripheral blood flow play a primary or secondary role in the reduced exercise capacity during sustained hypoxia?

IX. High-Altitude Residents and Populations

Exposure to high altitude over many years leads to alterations in hemodynamic, autonomic, metabolic, and coronary circulatory changes that may enhance tolerance to chronic hypoxia. High-altitude natives taken to sea level have a persistent reduction in stroke volume and cardiac output despite the introduction of normoxic conditions. Thus, vascular remodeling and other structural changes may occur with long-term residence at altitude that are not readily reversible with normoxia. An enhanced utilization of glucose by the heart in high-altitude natives results in greater oxygen efficiency (more high-energy phosphate production per molecule of oxygen) and may be an important factor in the preservation of cardiac function with prolonged hyp-

oxia. Autonomic balance favors enhanced parasympathetic and reduced sympathetic activity in high-altitude natives, though definitive studies have not yet been done. The significance of these findings for cardiovascular function has not been determined. There may be favorable changes in the coronary circulation of high-altitude natives that enhance myocardial blood flow and aid in myocardial protection. The combination of reduced coronary blood flow and enhanced vascularity may be a favorable adaptation to chronic hypoxia. The mechanisms for these coronary responses remain to be determined. Finally, the distinction between lifelong acclimatization to hypoxia versus population-based, genetic adaptations in the phenotypic cardiovascular changes of the high-altitude native have yet to be determined.

The study of high-altitude natives provides an opportunity to examine more prolonged exposure to chronic hypoxia and to differentiate the cardiovascular adaptations both quantitatively and qualitatively from the short-term (days to weeks) acclimatization responses discussed above. In particular, the responses to normoxia (sea level) in such individuals can provide information on the reversibility of the cardiovascular adaptations to chronic hypoxia. These populations are usually limited to elevations of 3000–4500 m. Interpretation of information from high-altitude residents is influenced by many factors including the altitude of residence, the physical activity patterns of the subjects, and the possible role of genetic changes that may have resulted in advantageous physiological adaptations to a low oxygen environment. Men of European ancestry whose parents moved from low altitude to Leadville, Colorado, differ genetically from natives of the Peruvian Andes and inhabitants of the Tibetan plateau whose ancestors have spent several generations at high altitude, though it is not clear whether these differences are a direct result of high-altitude residence.

Early reports on the cardiovascular function of the Sherpas of Nepal suggested that this high altitude population might have unique physiological characteristics that explained their superior exercise capacity under conditions of extreme hypoxia (6,132). These responses include a greater heart rate and cardiac output at maximal exercise along with lower minute ventilation, a higher PCO_2, and a normal blood pH compared to recently acclimatized lowlanders. Intriguingly, many of these adaptations of Asian highlanders are directionally opposite to those obtained in lowlanders who have acclimated well to prolonged hypoxia. Whether inhabitants of high regions elsewhere in the world behave similarly is a question for future study.

In general, hemodynamic changes with prolonged hypoxia in acclimatized lowlanders are similar in high altitude residents. One of the key differences between high altitude natives and recently acclimatized lowlanders is the substantially greater red cell mass in the former (see Chapter 3). This difference appears to result in a more prominent reduction in cardiac volumes. For example, the higher the hematocrit with progressive altitude residence, the lower the cardiac output secondary to decreases in stroke volume (155). These changes appear to persist with lifelong exposure to high altitude as Tibetans at 3,658 m have lower stroke volumes at rest and during exercise compared to sea level natives examined in the Operation Everest II study, even when corrected for differences in body size (62,156) (Fig. 13). Cardiac

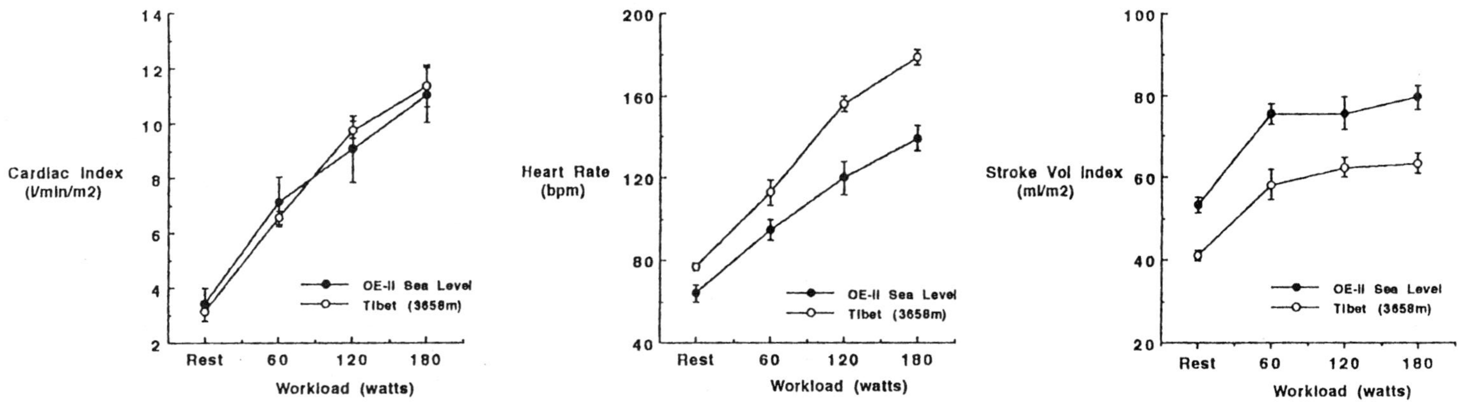

Figure 13 Native Tibetans have higher heart rates and lower stroke volume indices at rest and during exercise compared to North American men. Cardiac index, however, is similar between the groups. Vo_2 values were similar at identical workloads in both subject groups. Figures drawn from data on sea level subjects from Operation Everest II (62) and from lifelong residents of Tibet at 3658 m (156).

index, however, was similar for a given oxygen uptake in the high-altitude residents compared to acclimatized sea level natives, as a greater heart rate offset the smaller stroke volume. This observation confirms the similarity in regulatory mechanisms between the two populations and provides additional evidence supporting the reliability of the finding of smaller stroke volume.

The differences in stroke volume between these two diverse populations might also be explained by differences in fitness between the North American subjects and the Tibetans. For example, the Operation Everest II subjects were generally very fit and included a number of endurance-trained athletes, resulting in a mean peak oxygen uptake of 51.2 ± 9.0 (SD) mL/kg/min at sea level. In contrast, the Tibetan subjects were drawn from the general population in Lhasa and had a peak oxygen uptake more typical of untrained subjects of 43.4 ± 4.0 mL/kg/min.

Comparing the stroke volumes between these highlanders and lowlanders also provides indirect evidence against the hypothesis that impaired right ventricular function could be partially responsible for the reduction in left ventricular stroke volume. For example, when compared to recently acclimatized subjects or high-altitude residents in both North and South America, the pulmonary artery pressures of Tibetan natives were significantly lower at similar O_2 tensions (156). Thus, if pulmonary hypertension was responsible for impairing left ventricular filling, the Tibetans would be expected to have higher, rather than lower stroke volumes.

When high-altitude residents are taken to sea level, their cardiac outputs and stroke volumes are lower than those of sea level natives (80,82,157). With short-term residence at sea level, cardiac output and stroke volume increase but they remain lower than in sea level natives, raising the possibility of a structural remodeling. However, most of the change in stroke volume could be attributed to changes in plasma volume (9), and differences in fitness, diet, blood pressure, and other factors between the two populations make definitive conclusions from short-term deacclimatization studies uncertain. Interestingly, the breathing of a high-oxygen mixture to simulate sea level conditions while the subjects remained at altitude did not result in any significant hemodynamic changes (82). However, when cardiac hemodynamics were studied in high-altitude natives after 2 years of residence at sea level, the findings were similar to sea level natives, indicating that the majority of the cardiovascular changes with chronic hypoxia are reversible with prolonged exposure to normoxia. This observation argues against any permanent structural changes or fundamental genetic differences (158).

One of the key questions regarding long-term exposure to chronic hypoxia is whether the dramatic sympathetic activation recently reported with short-term acclimatization persists or gradually abates (see discussion above). This response may be very different in individuals born in a hypoxic environment, compared with those who were born at sea level but have migrated to high altitudes. Studies with β-adrenergic blockade and atropine in Tibetans compared to newly acclimatized Han Chinese suggest greater parasympathetic activity in the high-altitude natives (152), manifested by greater increases in heart rate with atropine. Heart rate variability studies have also shown a greater respiratory sinus arrhythmia (20), though such differences simply may be a function of differences in respiratory rate and tidal volume. This is an important area for future research.

One of the most intriguing findings in high-altitude natives relates to alterations in myocardial metabolism. Early studies in native high-altitude Peruvians suggested a greater reliance on carbohydrate metabolism for energy sources by the resting myocardium (159). This difference was suggested by a greater coronary sinus respiratory quotient (0.81 in lowlanders vs. 0.91 in high-altitude residents) as well as greater myocardial extraction of lactate and glucose compared to free fatty acids. More recent studies on Sherpa men using ^{31}P-magnetic resonance spectroscopy also support this theory: lower ratios of phosphocreatine to ATP found in these natives are consistent with a threefold increase in free ADP concentration in the myocardium (160). These metabolic conditions would accommodate the higher enzyme kinetic constants (K_M) of the enzymes phosphoglycerate kinase and pyruvate kinase, indicating enhanced capacity for carbohydrate metabolism. Similar findings were also reported in both Quechua and Sherpa subjects in whom an enhanced glucose uptake compared to sea level natives was observed with positron emission tomography (161). It remains to be determined whether this alteration in myocardial substrate utilization is a phenotypic expression of an altered genetic myocardial adaptation in high-altitude natives or a common acclimatization response seen with more short-term exposure to high altitude. The persistence of myocardial preference for glucose after 3–4 weeks of residence at sea level in these high-altitude natives would support a more permanent adaptation (160,161). On the other hand, with short-term (3 weeks) exposure to a moderate altitude of 4300 m there is a definite increase in glucose and lactate utilization by skeletal muscle (162–164). Whether there is also an early shift in myocardial metabolism with short-term hypoxia exposure is unknown. The functional significance of these observations is unclear, however, since myocardial function has been demonstrated to be entirely normal in sea level natives performing maximal exercise under extreme hypoxia.

Limited studies available in high-altitude natives also suggest that there are changes in coronary anatomy and physiology that may be protective in the setting of chronic hypoxia. Resting coronary blood flow has been shown to decline progressively with increasing altitude (123). However, myocardial oxygen consumption was also lower at rest in these high-altitude natives, a finding that differs from studies of recently acclimatized newcomers to altitude (124). Differences in methodology and the possible inclusion of patients with chronic mountain sickness make these data difficult to interpret. Pathological studies on high-altitude natives also demonstrate that lifelong exposure to hypoxia is associated with a more abundant coronary vascular bed with a greater density of peripheral branching of smaller coronary vessels (165). The significance of these findings remains unclear and is hard to separate from population differences in diet, physical activity, and genetic susceptibility to coronary artery disease. However, it is possible that coronary adaptations to chronic hypoxia, similar to the increased collateralization seen with chronic myocardial ischemia (166), could result in more effective myocardial vascularization with greater surface area of oxygen diffusion. Such adaptations could be responsible for a possible protective effect of long-term residence at high altitude against coronary artery disease (167).

In summary, there are various hemodynamic, autonomic, metabolic, and possibly coronary structural adaptations that occur in long-term residents at high altitude that may contribute to preservation of cardiovascular function. The time course and the factors responsible for many of these changes have not been determined. Despite the known increase in pulmonary artery pressure with sustained hypoxia, certain high-altitude populations appear to have a blunted hypoxia-pressor response in the pulmonary vascular bed. The heterogeneous responses to sustained high-altitude exposure may relate to different genetic susceptibility as well as to alterations in controlling mechanisms of cardiovascular function.

A. Key Unanswered Questions—High-Altitude Residents and Populations

What is the time course for the various cardiovascular adaptations observed in high-altitude natives?

Are there actual genotypic changes that result from generations of chronic hypoxia exposure, or are there changes reversible with prolonged normoxic exposure?

How do alterations in autonomic activity influence the cardiovascular adaptations to chronic hypoxia in high-altitude residents?

X. Clinical Correlation

Although the cardiovascular system appears to function normally even during exercise at extreme altitude, the effect of acute and sustained hypoxia on cardiac function in patients with underlying cardiovascular disease has not been well defined. The role of hypoxia in the induction of myocardial ischemia, cardiac arrhythmia, and heart failure in these patients is based on limited available clinical data; however, the application of theoretical concepts will provide a clearer understanding of potential pathophysiological responses (see Chapter 25).

Although the cardiovascular system appears to function well at high altitude in healthy individuals, even during high-intensity exercise at extreme altitudes, exposure to acute and prolonged hypoxia in patients with cardiovascular disease could result in the new onset or accentuation of symptoms when compared to living at sea level. Clinical issues of particular concern include the provocation of myocardial ischemia, the occurrence of unstable coronary syndromes including acute myocardial infarction, susceptibility to cardiac arrhythmias, and the occurrence or worsening of heart failure. Exposure to high altitude presents several potentially deleterious stresses that may exacerbate the clinical manifestations of underlying cardiac disease. These include the degree of hypoxia itself, alkalosis, heightened sympathetic activity, elevated systemic blood pressure and heart rate, and increased blood viscosity. Exercise at altitude also may be an important factor, as any submaximal workload at altitude is at a higher percentage of maximal oxygen uptake than at sea level and thereby requires increased cardiac work. However, the "cardiac-sparing" effects of acclimatization (reductions in heart rate and cardiac output) as well as the known decreases in heart size with prolonged hypoxia exposure may diminish any potential cardiac risks.

A. Myocardial Ischemia and Infarction

In patients with known coronary disease, exposure to the hypoxia of high altitude may result in myocardial ischemia, especially during exercise. There is limited information available to evaluate the risk of ischemia in patients with coronary disease. Extreme hypoxia alone does not appear to cause any clinical or ECG evidence of ischemia in the absence of atherosclerotic disease (60,67). However, the vasomotor response of coronary arteries, especially via endothelial-dependent mechanisms, has been shown to be abnormal in patients with atherosclerotic coronary disease. For example, intracoronary infusion of acetylcholine, an endothelial-dependent vasodilator in normal vessels, causes vasoconstriction in diseased coronary vessels (168). A similar observation has been noted during exercise at sea level in coronary patients where vasoconstriction, rather than vasodilation, can occur in diseased coronary segments (169). A potential mechanism for this abnormal coronary vasoconstriction is enhanced sympathetic activity (170), such as occurs during exercise. There clearly is enhanced sympathetic activity on arrival at high altitude and further activation with prolonged exposure. Whether this increased sympathetic activity influences coronary vasomotor tone at high altitude is not known; however, the combination of exercise and hypoxia could theoretically result in enhanced coronary vasoconstriction in patients with CAD. The role of other vasoconstrictors such as endothelin and angiotensin II in coronary vasomotor control during exercise with hypoxia has not been investigated, but they do appear to be important in other hyperadrenergic states such as heart failure. Despite this condition of heightened sympathetic activity during hypoxia, counterregulatory mechanisms may serve to blunt or prevent adverse coronary vascular tone. Acute exposure of endothelial cells to hypoxia leads to production of nitric oxide (NO), an endothelial-dependent vasodilator (171,172). NO production may serve to protect the coronary tree from vasoconstriction early during hypoxia, but there are no long-term data on NO production by endothelial cells during chronic hypoxia.

The pathogenesis of acute coronary syndromes including myocardial infarction are related to atherosclerotic plaque disruption in a coronary vessel. Local factors that may lead to these events include a lipid-rich plaque, vasoconstriction, residual thrombus in the area of the plaque, and the degree of stenosis (173). Systemic factors are also important and include increased catecholamine levels, an abnormal coagulation and fibrinolysis profile, abnormal metabolic states (diabetes, homocysteinemia), and shear forces in the vessels as can occur with hypertension. Although there have not been any reported coagulation abnormalities with hypoxia (174), other factors such as hypertension, heightened sympathetic activity, and possibly coronary vasoconstriction could lead to plaque disruption in the susceptible patient with lipid-laden plaques while at altitude. Exercise may be a critical factor as it has been found to be an important cause of serious plaque rupture at sea level (175).

Despite these theoretical reasons for increased cardiac risk at high altitude, the limited available data suggest that altitude exposure is generally well tolerated by patients with coronary disease. Coronary patients who were residents at 1600 m developed objective evidence of myocardial ischemia at a lower exercise workload upon initial arrival at 3100 m (176). However, after 5 days of acclimatization to the higher altitude, their ischemic pattern returned to their baseline level (177). In these partially altitude-acclimatized individuals, there was no change in the ischemic threshold (heart rate–systolic blood pressure product at the onset of ischemia) with acute or prolonged exposure to hypoxia. In contrast, sea level natives with coronary artery disease studied under conditions of acute hypoxia in a hypobaric chamber at a simulated altitude of 2500 m did develop myocardial ischemia at a lower hemodynamic threshold compared to normoxic conditions, raising the possibility of coronary vasoconstriction during hypoxia (11). However, these patients when studied after 5 days of residence at 2500 m only developed ischemia at their sea level threshold. Thus, it appeared that the initial adverse response to acute hypoxia was reversible with more prolonged exposure. Finally, in 90 male subjects over the age of 40 years, who underwent ECG-telemetry monitoring during alpine skiing at 2500 m, the frequency of silent ST-segment depression was 5.6% (178). There were no adverse clinical events in these men who were exercising at >80% of their maximal sea level heart rate. Unfortunately, no sea level exercise ECG information was available in these men, preventing a direct comparison.

B. Cardiac Arrhythmias

There is limited information on the occurrence of cardiac arrhythmias in heart-diseased patients at high altitude. Other than occasional premature ventricular and atrial beats (67), there does not appear to be an increased risk of arrhythmias with altitude exposure in normal subjects despite significant hypoxia and alkalosis. In elderly patients with either an increased risk for or known coronary disease, there were no significant arrhythmias noted by either short-term ECG monitoring at rest or exercise electrocardiography with acute or more prolonged exposure to 2500 m (11). In these same subjects, there were no hypoxia-related abnormalities in resting signal–averaged ECG recordings, a sensitive marker for the presence of a myocardial electrophysiological substrate conducive for arrhythmia production. No occurrence of significant cardiac arrhythmias has been reported in any study of heart disease patients at high altitude; however, most of these studies did not include patients with significant left ventricular systolic dysfunction, a group more prone to develop arrhythmias. In one small study of patients with a reduced mean left ventricular ejection fraction (LVEF) of 39%, no arrhythmias were reported during exercise at 2500 m (179). Further work at similar and higher altitudes is required in a larger number of patients with a greater clinical likelihood of cardiac arrhythmias before any definitive statement can be made about the risks of arrhymogenesis with altitude exposure.

C. Heart Failure

Although the myocardium functions normally under conditions of acute and sustained hypoxia, it is unclear how a heart with reduced ventricular systolic function would tolerate the increased demands of acute and prolonged hypoxia. The heightened sympathetic activity, demand for increased cardiac output with acute hypoxia, and the systemic vasoconstriction that occurs with more prolonged exposure may result in further deterioration of ventricular function and may produce symptoms of decompensated heart failure. Although the usual cardiac response to prolonged hypoxia is a reduction in end-diastolic volume with a maintained or increased ejection fraction (85), echocardiographic studies in coronary subjects living in Denver, CO (1600 m), who spent 5 days at 3100 m demonstrated an increase in end-diastolic and end-systolic dimensions with a fall in LVEF from 51% at 1600 m to 37% at 3100 m (180). These results have not been confirmed in another study at 2500 m in which sea level residents with known coronary disease studied after 5 days at altitude had no echocardiographic abnormalities with sustained moderate hypoxia (11). Patients with known left ventricular dysfunction who were exercised at 2500 m were found to have the same decrement in exercise capacity as normal age-matched subjects (179). No adverse cardiovascular events were noted after 2 days at moderate altitude in these patients.

Recently, sea level residents with chronic heart failure were studied while exercising at different degrees of acute normobaric hypoxia equivalent to altitudes varying from 1000 to 3000 m (181). These patients, with reduced exercise capacities when compared to normal subjects, had a similar response to progressive hypoxia in regards to arterial oxygen saturation, exercise ventilation, heart rate, and blood pressure. In normal subjects there was a 3% decrement in maximal exercise capacity for each 1000 m in elevation. Heart failure patients with a sea level maximal uptake greater than 15 mL/kg/min had a similar decrement of 5% per each 1000 m. However, patients with maximal oxygen uptake less than 15 mL/kg/min at sea level had a more marked reduction in exercise capacity, with an 11% reduction for each 1000 m of elevation. This reduction in exercise capacity probably reflected the inability to increase cardiac output in response to acute reductions in arterial oxygenation. No data are available in more prolonged hypoxia exposure in these heart failure patients. In addition, all these patients were on active heart failure medication that may have modified the cardiovascular responses to hypoxia.

D. Clinical Guidelines

Based on the above information, patients with cardiac disease should be evaluated at sea level prior to ascent to altitude to insure a stable disease state (182) (see also Chapter 25). In addition, patients with known coronary disease should limit their physical activity during the first few days at altitude to allow the favorable effects of acclimatization to occur. Guidelines for patients with left ventricular dysfunction are more difficult to determine, but patients with mild to moderate ventricular dysfunction should be able to tolerate moderate altitude without serious adverse consequences. In questionable cases, nocturnal oxygen supplementation should be considered and medications used at sea level should be continued at altitude. Patients with more significant cardiac impairment at sea level, especially with significant exercise limitation, should not travel to high altitude in the absence of supplemental oxygen to simulate normoxic conditions.

In summary, there is limited information on altitude tolerance in patients with cardiovascular disease. Hypoxia-induced coronary vasoconstriction is a particularly intriguing concept that needs to be explored. The limited data would suggest that moderate altitude exposure (2500–3100 m) is well tolerated by stable coronary patients with normal or moderately depressed ventricular function. Although difficult to obtain, more information is needed in this important area of high-altitude medicine.

E. Key Unanswered Questions—Clinical Correlations

Can hypoxia alone lead to instability of coronary vascular lesions and vasomotor tone with the precipitation of unstable coronary syndromes and possibly acute myocardial infarction?

Does the cardiac response to exercise at sea level in coronary disease patients predict the response during either acute or sustained hypoxia?

What are the implications of altitude exposure on patients with reduced ventricular function at sea level? Are they at increased risk for arrhythmias, cardiogenic pulmonary edema, pulmonary hypertension, and right heart failure?

Does exposure to acute or sustained hypoxia increase the likelihood of arrhythmias in patients with cardiovascular disease? What is the electrophysiological mechanisms of cardiac conduction during hypoxia?

What guidelines are reliable to assess cardiac risk at high altitude?

Acknowledgments

This chapter is dedicated to the memory of Herbert Hultgren, M.D., who was a pioneer in the investigation of cardiovascular responses to high altitude in a variety of settings. His work in lowlanders exposed to high altitude, in high-altitude natives, and in patients with high-altitude pulmonary edema has provided important information that has contributed to our understanding of the functioning of the cardiovascular system and systemic circulation at altitude. His mentoring and constant guidance in the pursuit of high-altitude research as well as his genuine concern for his colleagues will be greatly missed. We also wish to acknowledge the investigators and subjects of the various Pikes Peak studies, Operation Everest II, and the Tenth Mountain Division Study. Their efforts have greatly contributed to the advancement of knowledge in this important area of high-altitude physiology and medicine.

References

1. Bert P. La Pression Barometrique. Masson, Paris, 1878. English translation by Hitchcock MA and Hitchcock FA. Columbus, Ohio: College Book Co., 1943:343–346.
2. Douglas CG, Haldane JA, Henderson Y, Schneider EC. Physiological observations made on Pikes Peak, Colorado, with special reference to adaptation to low barometric pressures. Phil Trans R Soc London Ser B 1913; 203:185–381.
3. Barcroft J, Binger CA, Bock AV, Doggart JH, Forbes HS, Harrop G, Meakins JC, Redfield AC. Observations upon the effect of high altitude on the physiological processes of the human body, carried out in the Peruvian Andes, chiefly at Cerro de Pasco. Phil Trans R Soc London Ser B 1923; 211:351–480.
4. Grollman A. Physiological variations of the cardiac output of man. VII. The effect of high altitude on the cardiac output and its related functions: an account of experiments conducted on the summit of Pikes Peak, Colorado. Am J Physiol 1930; 93:19–40.
5. Doyle JT, Wilson JS, Warren JV. The pulmonary vascular responses to short-term hypoxia in human subjects. Circulation 1952; 5:263–270.
6. Wescott RN, Fowler NO, Scott RC, Hauenstein VD, McGuire J. Anoxia and human pulmonary vascular resistance. J Clin Invest 1951; 19:957–970.
7. Vogel JA, Harris CW. Cardiopulmonary responses of resting man during early exposure to high altitude. J Appl Physiol 1967; 22:1124–1128.
8. Stenberg J, Ekblom B, Messin R. Hemodynamic response to work at simulated altitude, 4,000 m. J Appl Physiol 1966; 21:1589–1594.
9. Wagner PD, Gale GE, Moon RE, Torre-Bueno JR, Stolp BW, Saltzman HA. Pulmonary gas exchange in humans exercising at sea level and simulated altitude. J Appl Physiol 1986; 61:260–270.
10. Wagner JA, Miles DS, Horvath SM. Physiological adjustments of women to prolonged work during acute hypoxia. J Appl Physiol 1980; 49:367–373.
11. Levine BD, Zuckerman JH, deFilippi CR. Effect of high-altitude exposure in the elderly: The Tenth Mountain Division Study. Circulation 1997; 96:1224–1232.
12. Richardson DW, Kontos HA, Raper AJ, Patterson JL Jr. Modification by beta-adrenergic blockade of the circulatory responses to acute hypoxia in man. J Clin Invest 1967; 46:77–85.
13. Koller EA, Drechsel S, Hess T, Macherel P, Boutellier U. Effects of atropine and propranolol on the respiratory, circulatory, and ECG responses to high altitude in man. Eur J Appl Physiol 1988; 57:163–172.
14. Cunningham WL, Becker EJ, Kreuzer F. Catecholamines in plasma and urine at high altitude. J Appl Physiol 1965; 20:607–610.
15. Mazzeo RS, Bender PR, Brooks GA, Butterfield GE, Groves BM, Sutton JR, EE Wolfel, Reeves JT. Arterial catecholamine responses during exercise with acute and chronic high-altitude exposure. Am J Physiol 1991; 261:E419–E424.
16. Mazzeo RS, Wolfel EE, Butterfield GE, Reeves JT. Sympathetic response during 21 days at high altitude (4,300 m) as determined by urinary and plasma catecholamines. Metabolism 1994; 43:1226–1232.

17. Leuenberger U, Gleeson K, Wroblewski K, Prophet S, Zelis R, Zwillich C, Sinoway L. Norepinephrine clearance is increased during acute hypoxemia in humans. Am J Physiol 1991; 261:H1659–H1664.
18. Hughson RL, Yamamoto Y, McCullough RE, Sutton JR, Reeves JT. Sympathetic and parasympathetic indicators of heart rate control at altitude studied by spectral analysis. J Appl Physiol 1994; 77:2537–2542.
19. Yamamoto Y, Hoshikawa Y, Miyashita M. Effects of acute exposure to simulated altitude on heart rate variability during exercise. J Appl Physiol 1996; 81:1223–1229.
20. Passino C, Bernardi L, Spadacini G, Calciati A, Robergs R, Anand I, Greene R, Martignoni E, Appenzeller O. Autonomic regulation of heart rate and peripheral circulation: comparison of high altitude and sea level residents. Clin Sci 1996; 91(suppl):81–83.
21. Ponchia A, Noventa D, Bertaglia M, Carreta R, Zaccaria M, Miraglia G, Pascotto P, Buja G. Cardiovascular neural regulation during and after prolonged high altitude exposure. Eur Heart J 1994; 15:1463–1469.
22. Farnelli CC, Kayser B, Binzoni T, Cerretelli P, Girardier L. Autonomic nervous control of heart rate at altitude (5,050 m). Eur J Appl Physiol 1994; 69:502–507.
23. Perini R, Milesi S, Biancardi L Veicsterinas A. Effects of high altitude acclimatization on heart rate variability in resting humans. Eur J Appl Physiol 1996; 73:521–528.
24. Yamamoto Y, Hughson RL, Sutton JR, Houston CS, Cymerman A, Fallen EL, Kamath MV. Operation Everest II: an indication of deterministic chaos in human heart rate variability at simulated high altitude. Biol Cybern 1993; 69:205–212.
25. Lipsitz LA, Hsahimoto F, Lubowsky LP, Mietus J, Noody GB, Appenzeller O, Goldberger AL. Heart rate and respiratory rhythm dynamics on ascent to high altitude. Br Heart J 1995; 74:390–396.
26. White DB. Autonomic cardiac regulation during acute and chronic hypoxia assessed by spectral analysis of R-R interval variability. In: Houston CS, Coates G, eds. Hypoxia: Women at Altitude. Burlington, VT: Queen City Printers Inc, 1997:124–133.
27. Brown TE, Beightol LA, Koh J, Eckberg DL. Important influence of respiration on human R-R interval power spectra is largely ignored. J Appl Physiol 1993; 75:2310–2317.
28. Levine BD, Zhang R, Zuckerman JH, Morrow M. The effect of cardiac mechanics on heart rate variability. Circulation 1996; 94(8):I–368.
29. Piepoli M, Sleight P, Leuzzi S, Valle F, Spadacini G, Passino C, Johnston J, Bernardi L. Origin of respiratory sinus arrhythmia in conscious humans: an important role for arterial carotid baroreceptors. Circulation 1997; 95:1813–1821.
30. Grassman E, Blomqvist CG. Absence of respiratory sinus arrhythmia: a manifestation of the sick sinus syndrome. Clin Cardiol 1983; 6:151–154.
31. Eckberg DL. Sympathovagal balance: a critical appraisal. Circulation 1997; 96:3224–3232.
32. Somers VK, Mark AL, Abboud FM. Interaction of baroreceptor and chemoreceptor reflex control of sympathetic nerve activity in normal humans. J Clin Invest 1991; 87: 1953–1957.
33. Saito M, Mano T, Iwase S, Koga K, Abe H, Yamazaki Y. Responses in muscle sympathetic activity to acute hypoxia in humans. J Appl Physiol 1988; 65:1548–1552.
34. Duplain H, Vollenweider L, Delabays A, Nicod P, Bartsch P, Scherrer U. Augmented sympathetic activation during short-term hypoxia and high altitude exposure in subjects susceptible to high altitude pulmonary edema. Circulation 1999; 99:1713–1718.

35. Roach R, Vissing SF. Unpublished results, personal communication, 1998.
36. Eckberg DL, Bastiw H III, Scruby AE. Modulation of human sinus node function by systemic hypoxia. J Appl Physiol 1982; 52:570–577.
37. Sommers VK, Abboud FM. Chemoreflex control of cardiac vagal activity. In: Levy MN, Schwartz PJ, eds. Vagal Control of the Heart. Armonk, NY: Futura Publishing Co, 1994:403–416.
38. Spyer KM. Neural organisation and control of the baroreceptor reflex. Rev Physiol Biochem Pharmacol 1981; 88:23–123.
39. Kahler RL, Goldblatt A, Braunwald E. The effects of acute hypoxia on the systemic venous and arterial systems and on myocardial contractile force. J Clin Invest 1962; 41:1553–1563.
40. Mifflin SW, Felder RB. Synaptic mechanisms regulating cardiovascular afferent inputs to solitary tract nucleus. Am J Physiol 1990; 259:H653–H661.
41. Asano A, Mazzeo RS, McCullough RE, Wolfel EE, Reeves JT. Relation of sympathetic activation to ventilation in man at 4,300 m altitude. Aviat Space Environ Med 1997; 68:104–110.
42. Hultgren H. The systemic circulation. In: High Altitude Medicine. Stanford, CA: Hultgren Publications, 1997:33–63.
43. Pool PE, Covell JW, Chidsey CA, Braunwald E. Myocardial high energy phosphate stores in acutely induced hypoxic heart failure. Circ Res 1966; 19:221–229.
44. Heistad DD, Abboud FM. Circulatory adjustments to hypoxia. Circulation 1980; 61: 463–470.
45. Rowell LB, Saltin B, Kiens B, Christensen NJ. Is peak quadriceps blood flow in humans even higher during exercise during hypoxia? Am J Physiol 1986; 251:H1038–H1044.
46. Knight DR, Schaffartzik W, Poole DC, Hogan MC, Bebout DE, Wagner PD. Effects of hyperoxia on maximal leg O_2 supply and utilization in men. J Appl Physiol 1993; 75:2586–2594.
47. Hartley LH, Vogel JA, Landowne M. Central, femoral, and brachial circulation during exercise in hypoxia. J Appl Physiol 1973; 34:87–90.
48. Mancia G, Mark AL. Arterial baroreflexes in humans. In: Shepherd JT, Aboud FM, eds. Handbook of Physiology, Section 2: The Cardiovascular System, Vol. 3. Bethesda, MD: American Physiological Society, 1983:795–813.
49. Nicholas R, O'Meara P, Calonge N. Is syncope related to moderate altitude exposure? JAMA 1992; 268:904–906.
50. Remensnyder JP, Mitchell JH, Sarnoff SJ. Functional sympatholysis during muscular activity. Circ Res 1962; 11:370–380.
51. Hansen J, Thomas GD, Harris SA, Parsons WJ, Victor RG. Differential sympathetic neural control of oxygenation in resting and exercising human skeletal muscle. J Clin Invest 1996; 98:584–596.
52. Thomas GD, Hansen J, Victor RG. ATP-sensitive potassium channels mediate contraction-induced attenuation of sympathetic vasoconstriction in rat skeletal muscle. J Clin Invest 1997; 99:2602–2609.
53. Thomas GD, Victor RG. Nitric oxide mediates contraction-induced attenuation of sympathetic vasoconstriction in rat skeletal muscle. J Physiol 1997; 506:817–826.
54. Hackett JG, FM Abboud, AL Mark, Schmid PG, Heistad DD. Coronary vascular responses to stimulation of chemoreceptors and baroreceptors: evidence for reflex activation of vagal cholinergic innervation. Circ Res 1972; 31:8–17.

55. Ellsworth ML, Forrester T, Ellis CG, Dietrich HH. The erythrocyte as a regulator of vascular tone. Am J Physiol 1995; 269:H2155–H2161.
56. Hultgren H. Circulatory responses to acute hypoxia in normal subjects at sea level. In: High Altitude Medicine. Stanford, CA: Hultgren Publications, 1997:34.
57. Wolfel EE, Selland MA, Mazzeo RS, Reeves JT. Systemic hypertension at 4,300 m is related to sympathoadrenal activity. J Appl Physiol 1994; 76:1643–1650.
58. Nielsen AM, Bisgard GE, Vidruk EH. Carotid chemoreceptor activity during acute and sustained hypoxia in goats. J Appl Physiol 1988; 65:1796–1802.
59. Hoon RS, Balasubramanian V, Mathew OP, Tiwari SC, Sharma SC, Chadha KS. Effect of high-altitude exposure for 10 days on stroke volume and cardiac output. J Appl Physiol 1977; 42:722–727.
60. Karliner JS, Sarnquist FF, Graber DJ, Peters RM Jr, West JB. The electrocardiogram at extreme altitude: experience on Mount Everest. Am Heart J 1985; 109:505–513.
61. Houston CS, Riley RL. Respiratory and circulatory changes during acclimatization to high altitude. Am J Physiol 1947; 140:565–588.
62. Reeves JT, Groves BM, Sutton JR, Wagner PD, Cymerman A, Malconian MK, Rock PB, Young PM, Houston CS. Operation Everest II: preservation of cardiac function at extreme altitude. J Appl Physiol 1987; 63:531–539.
63. Wolfel EE, Selland MA, Cymerman A, Brooks GA, Butterfield GE, Mazzeo RS, Grover RF, Reeves JT. O_2 extraction maintains O_2 uptake during submaximal exercise with β-adrenergic blockade at 4,300 m. J Appl Physiol 1998; 85:1092–1102.
64. Moore LG, Cymerman A, Huang SY, McCullough RE, McCullough RG, Rock PB, Young A, Young P, Weil JV, Reeves JT. Propranolol blocks metabolic rate increase but not ventilatory acclimatization to 4,300 m. Respir Physiol 1987; 70:195–204.
65. Savard GK, Areskog NH, Saltin B. Cardiovascular response to exercise in humans following acclimatization to extreme altitude. Acta Physiol Scand 1995; 154:499–509.
66. Pugh LGCE, Gill MB, Lahiri S, Milledge JS, Ward MP, West JB. Muscular exercise at great altitudes. J Appl Physiol 1964; 19:431–440.
67. Malconian M, Rock P, Hultgren H, Donner H, Cymerman A, Groves BM, Reeves JT, Alexander J, Sutton JR, Nitta M, Houston CS. The electrocardiogram at rest and exercise during a simulated ascent of Mount Everest (Operation Everest II). Am J Cardiol 1990; 65:1475–1480.
68. Bristow MR, Ginsburg R, Minobe W, Cubiccioti RS, Sageman WS, Lurie K, Billingham ME, Harrison DC, Stinson EB. Decreased catecholamine sensitivity and β-adrenergic receptor density in failing human hearts. N Engl J Med 1982; 307:205–211.
69. Voelkel NF, Hegstrand L, Reeves JT, McMurtry IF, Molinoff PB. Effects of hypoxia on density of β-adrenergic receptors. J Appl Physiol 1981; 50:363–366.
70. Kacimi R, Richalet JP, Corsin A, Abousahl I, Crozatier B. Hypoxia-induced downregulation of β-adrenergic receptors in rat heart. J Appl Physiol 1992; 73:1377–1382.
71. Richalet JP, Larmignat P, Rathat C, Keromes A, Baud P, Lhoste F. Decreased cardiac response to isoproterenol infusion in acute and chronic hypoxia. J Appl Physiol 1988; 65:1957–1961.
72. Antezana AM, Kacimi R, Le Trong JL, Marchal M, Abousahl I, Dubray C, Richalet JP. Adrenergic status of humans during prolonged exposure to the altitude of 6,542 m. J Appl Physiol 1994; 76:1055–1059.

73. Maher JT, Manchanda C, Cymerman A, Wolfe DL, Hartley LH. Cardiovascular responsiveness to β-adrenergic stimulation and blockade in chronic hypoxia. Am J Physiol 1975; 228:477–481.
74. Wolfel EE, Groves BM, Brooks GA, Butterfield GE, Mazzeo RS, Moore LG, Sutton JR, Bender PR, Dahms TE, McCullough RE, McCullough RG, Huang SY, Sun SF, Grover RF, Hultgren HN, Reeves JT. Oxygen transport during steady-state submaximal exercise in chronic hypoxia. J Appl Physiol 1991; 70:1129–1136.
75. Houston CS, Sutton JR, Cymerman A, Reeves JT. Operation Everest II: man at extreme altitude. J Appl Physiol 1987; 63:877–882.
76. Wagner PD. A theoretical analysis of factors determining VO_2 max at sea level and altitude. Respir Physiol 1996; 196:329–343.
77. Fulco CS, Cymerman A, Rock PB, Farese G. Hemodynamic responses to upright tilt at sea level and high altitude. Aviat Space Environ Med 1985; 56:1172–1176.
78. Klausen K. Cardiac output in man at rest and work during and after acclimatization to 3,800 m. J Appl Physiol 1966; 21:609–616.
79. Alexander JK, Grover RK. Mechanism of reduced cardiac stroke volume at high altitude. Clin Cardiol 1983; 6:301–303.
80. Alexander JK, Hartley LH, Modelski M, Grover RF. Reduction in stroke volume during exercise in man following ascent to 3,100 m altitude. J Appl Physiol 1967; 23: 849–858.
81. Balasubramanian V, Behl A, Das GS, Waldhwa AK, Mathew OP, Hoon RS. Effect of digoxin and diuretics on high altitude left ventricular dysfunction. Circulation 1978; 57:1180–1185.
82. Hartley LH, Alexander JK, Modelski M, Grover RF. Subnormal cardiac output at rest and during exercise in residents at 3,100 m altitude. J Appl Physiol 1967; 23:839–848.
83. Fowles RE, Hultgren HN. Left ventricular function at high altitude examined by systolic time intervals and M-mode echocardiography. Am J Cardiol 1983; 52:862–866.
84. Buch J, Egeblad H, Hansen P, Kjaergard H, Waldorf S, Steiness E. Correlation between changes in systolic time intervals and left ventricular end-diastolic diameter after preload reduction. Br Heart J 1980; 44:668–671.
85. Suarez J, Alexander JK, Houston CS. Enhanced left ventricular systolic performance at high altitude during Operation Everest II. Am J Cardiol 1987; 60:137–142.
86. Webster KA, Bishopric NH. Molecular regulation of cardiac myocyte adaptation to chronic hypoxia. J Mol Cell Cardiol 1992; 24:741–752.
87. Mazer CD, Stanley WC, Hicker RF, Neese RA, Cason BA, Demas KA, Wisneski JA, Gertz EW. Myocardial metabolism during hypoxia: Maintained lactate oxidation during increasing glycolysis. Metabolism 1990; 39:913–918.
88. Wiesner RJ. Adaptation of mitochondrial gene expression to changing cellular energy demands. News Physiol Sci 1997; 12:178–184.
89. Bunn HF, Poyton RO. Oxygen sensing and molecular adaptation to hypoxia. Physiol Rev 1996; 76:839–887.
90. Kacimi R, Long CS, Karliner JS. Chronic hypoxia modulates the interleukin-1β stimulated inducible nitric oxide synthase pathway in cardiac myocytes. Circulation 1997; 96:1937–1943.
91. Lau S, Patnaik N, Sayen R, Mestril R. Simultaneous over-expression of two stress proteins in rat cardiomyocytes and myogenic cells confers protection against ischemia-induced injury. Circulation 1997; 96:2287–2294.

92. Zimmerman LH, Levine RA, Farber HW. Hypoxia induces a set of stress proteins in cultured endothelial cells. J Clin Invest 1991; 87:908–914.
93. Hirata K, Ban T, Jinnouchi Y, Kubo S. Echocardiographic assessment of left ventricular function and wall motion at high altitude in normal subjects. Am J Cardiol 1991; 68:1692–1697.
94. Sagawa K, Maughan L, Suga H, Sunagawa K. Physiologic determinants of the ventricular pressure-volume relationship. In: Sagawa K, Maughan L, Suga H, Sunagawa K, eds. Cardiac Contraction and the Pressure-Volume Relationship. New York: Oxford University Press, 1988:110–170.
95. Mirsky I. Assessment of diastolic function: suggested methods and future considerations. Circulation 1984; 69:836–841.
96. Levine BD, Zuckerman JH, Pawelczyk JA. Cardiac atrophy after bed-rest deconditioning: A non-neural mechanism for orthostatic intolerance. Circulation 1997; 96:517–525.
97. Jung RC, Dill DB, Horton R, Horvath SM. Effect of age on plasma aldosterone levels and hemoconcentration at altitude. J Appl Physiol 1971; 31:593–597.
98. Surks MI, Chinn KSK, Matoush LO. Alterations in body composition in man after acute exposure to high altitude. J Appl Physiol 1966; 21:1741–1746.
99. Grover RF, Weil JV, Reeves JT. Cardiovascular adaptations to exercise at high altitude. Exer Sports Sci Rev 1986; 14:269–302.
100. Grover RF, Reeves JT, Maher JT, McCullough RE, Cruz JC, Denniston JC, Cymerman A. Maintained stroke volume but impaired arterial oxygenation in man at high altitude with supplemental CO_2. Circ Res 1976; 38:391–396.
101. Grover RF, Selland MA, McCullough RG, Dahms TE, Wolfel EE, Butterfield GE, Reeves JT, Greenleaf JE. β-Adrenergic blockade does not prevent polycythemia or decrease in plasma volume in men at 4,300 m altitude. Eur J Appl Physiol 1998; 77: 264–270.
102. Tyberg JV. Coupled or uncoupled pericardial constraint: Effects on cardiac chamber interactions. J Appl Physiol 1996; 81:1797–1798.
103. Richalet JP. Operation COMEX. In: Hypoxia: Into the Next Millennium, Proceedings of the 11th International Hypoxia Symposium, Jasper, Alberta, Canada, 1999.
104. Ross J Jr, Linhart JW, Braunwald E. Effect of changing heart rate in man by electrical stimulation of the right atrium: studies at rest, during exercise, and with isoproterenol. Circulation 1965; 32:549–556.
105. Hultgren H, Kelly J, Miller H. Pulmonary circulation in acclimatized man at high altitude. J Appl Physiol 1965; 20:233–238.
106. Groves BM, Reeves JT, Sutton JR, Wagner PD, Cymerman A, Malconian MK, Rock PB, Young PM, Houston CS. Operation Everest II: elevated high-altitude pulmonary resistance unresponsive to oxygen. J Appl Physiol 1987; 63:521–530.
107. Das BK, Tewari SC, Parashar SK, Akhtar M, Grover DN, Ohri VC, Dutta SK, Chatterjee JC. Electrocardiographic changes at high altitude. Indian Heart J 1983; 35:30–37.
108. Abel FL, Waldhausen JA. Effects of alterations in pulmonary vascular resistance on right ventricular function. J Thorac Cardiovasc Surg 1967; 54:886–894.
109. Davila-Roman VG, Guest TM, Tuteur PG, Rowe WJ, Ladenson JH, Jaffe AS. Transient right but not left ventricular dysfunction after strenuous exercise at high altitude. J Am Coll Cardiol 1997; 30:468–473.
110. Garcia JA, McMinn SB, Zuckerman JH, Fixler DE, Levine BD. The role of the right

ventricle during hypobaric hypoxic exercise: insights from patients after the Fontan operation. Med Sci Sports Exerc 1999; 31:269–276.
111. Atherton JJ, Moore TD, Lele SS, Thomson HL, Galbraith AJ, Belenkie I, Tyberg JV, Frenneaux MP. Diastolic ventricular interaction in chronic heart failure. Lancet 1997; 349:1720–1724.
112. Cargill RI, Kiely DG, Lipworth BJ. Adverse effects of hypoxaemia on diastolic filling in humans. Clin Sci 1995; 89:165–169.
113. Garcia MJ, JD Thomas, Klein AL. New Doppler echocardiographic applications for the study of diastolic function. J Am Coll Cardiol 1998; 32:865–875.
114. Buckey JC, Gaffney FA, Lane LD, Levine BD, Watenpaugh DE, Wright SJ, Yancy CW Jr, Blomqvist CG. Central venous pressure in space. J Appl Physiol 1996; 81: 19–25.
115. White RJ, Blomqvist CG. Central venous pressure and cardiac function during spaceflight. J Appl Physiol 1998; 85:738–746.
116. Mazzeo RS, Brooks GA, Butterfield GE, Podolin DA, Wolfel EE, Reeves JT. Acclimatization to high altitude increases muscle sympathetic activity both at rest and during exercise. Am J Physiol 1995; 269:R201–207.
117. Sander M, Hansen J, Saltin B. Unpublished observations.
118. Bender PR, Groves BM, McCullough RE, McCullough RG, Huang SY, Hamilton AJ, Wagner PD, Cymerman A, Reeves JT. Oxygen transport to exercising leg in chronic hypoxia. J Appl Physiol 1988; 65:2592–2597.
119. Weil JV, Byrne-Quinn E, Battock DJ, Grover RF, Chidsey CA. Forearm circulation in man at high altitude. Clin Sci 1971; 40:235–246.
120. Cruz JC, Grover RF, Reeves JT, Maher JT, Cymerman A, Denniston JC. Sustained venoconstriction in man supplemented with CO_2 at high altitude. J Appl Physiol 1976; 40:96–100.
121. Cerretelli P, Marconi C, Deriaz O, Giezendanner D. After effects of chronic hypoxia on cardiac output and muscle blood flow at rest and exercise. Eur J Appl Physiol 1984; 53:92–96.
122. Hellems HK, Ord JW, Talmers FN, Christensen RC. Effects of hypoxia on coronary blood flow and myocardial metabolism in normal human subjects. Circulation 1963; 16:893–898.
123. Moret PR. Coronary blood flow and myocardial metabolism in man at high altitude. In: Porter R, Knight J, eds. High Altitude Physiology: Cardiac and Respiratory Aspects. Edinburgh: Churchill Livingstone, 1971:131–148.
124. Grover RF, Lufschanowski R, Alexander JK. Alterations in the coronary circulation of man following ascent to 3,100 m altitude. J Appl Physiol 1976; 41:832–838.
125. Sharma SC, Balasubramanian V, Mathew OP, Hoon RS. Serial studies of heart rate, blood pressure, and urinary catecholamine excretion on acute induction to high altitude (3,658 m). Indian J Dis Chest 1977; 19:16–20.
126. Vogel JA, Hartley LH, Cruz JC, Hogan RP. Cardiac output during exercise in sea-level residents at sea level and high altitude. J Appl Physiol 1974; 36:169–172.
127. Vogel JA, Hansen JE, Harris CW. Cardiovascular responses in man during exhaustive work at sea level and high altitude. J Appl Physiol 1967; 23:531–539.
128. Saltin B, Grover RF, Blomqvist CG, Hartley LH, Johnson RL Jr. Maximal oxygen uptake and cardiac output after 2 weeks at 4,300 m. J Appl Physiol 1968; 25:400–409.
129. Wolfel EE, Reeves JT, Lindenfeld J, Rock PB, Fulco C, Muza S, Butterfield GE, Braun B, Zamudio S, Moore LG. Women at altitude: Reduction in peak exercise systemic

blood pressure with alpha-adrenergic blockade upon exposure to simulated altitude. FASEB J 1998; 12:A56.

130. Muza SR, Rock PB, Butterfield GE, Wolfel EE, Moore LG. Women at Altitude: Effect of alpha-blockade on the ventilatory and systemic arterial pressure responses to hypoxia and hypercapnia. 11th International Hypoxia Symposium Proceedings, Jasper, Alberta, Canada, 1999.
131. Wolfel EE, Tokeshi J, Meertens M, McCullough RG, Rock PB, Butterfield GE, Moore LG. Alpha-adrenergic blockade limits the increase in ambulatory blood pressure observed in young women after short-term residence at 4,300 m altitude. FASEB J 1999; 13:A784.
132. Pugh LGCE. Cardiac output in muscular exercise at 5,800 m (19,000 ft). J Appl Physiol 1964; 19:441–447.
133. Schneider EC, Hedblom CA. Blood pressure with special reference to high altitudes. Am J Physiol 1908; 23:90–104.
134. Sharma SC, Hoon RS, Balasubramanian V, Chadha KS. Urinary catecholamine excretion in temporary residents of high altitude. J Appl Physiol 1978; 44:725–727.
135. Marticorena E, Ruiz L, Severino J, Galvez J, Penaloza D. Systemic blood pressure in white men born at sea level: changes after long residence at high altitudes. Am J Cardiol 1969; 23:364–368.
136. Lewis SF, Taylor WF, Graham RM, Pettinger WA, Schutte JE, Blomqvist CG. Cardiovascular responses to exercise as functions of absolute and relative work load. J Appl Physiol 1983; 54:1314–1323.
137. Roach R. Exercise at high altitude. Symposium on Mountain Medicine. Med Sci Sports Exer 1999; 31(suppl):S–39.
138. Sutton JR, Reeves JT, Wagner PD, Groves BM, Cymerman A, Malconian MK, Rock PB, Young PM, Walter SD, Houston CS. Operation Everest II: oxygen transport at extreme simulated altitude. J Appl Physiol 1988; 64:1309–1321.
139. Thomson JM, Dempsey JA, Chosy LW, Shahidi NT, Reddan WG. Oxygen transport and oxyhemoglobin dissociation during prolonged muscular work. J Appl Physiol 1974; 37:658–664.
140. Dempsey JA, Thomson JM, Forster HV, Cemy FC, Chosy LW. HBO_2 dissociation in man during prolonged work in chronic hypoxia. J Appl Physiol 1975; 38:1022–1029.
141. Ferretti G, Boutellier U, Pendergast DR, Moia C, Minetti AE, Howald H, di Prampero PE. Oxygen transport system before and after exposure to chronic hypoxia. Int J Sports Med 1990; 11(suppl 1):S15–S20.
142. Steinacker JM, Liu Y, Boning D, Halder A, Maassen N, Thomas A, Stauch M. Lung diffusion capacity, oxygen uptake, cardiac output and oxygen transport during exercise before and after an Himalayan expedition. Eur J Appl Physiol 1996; 74:187–193.
143. Moore LG, Cymerman A, Shao-Yung H, McCullough RE, McCullough RG, Rock PB, Young A, Young PM, Bloedow D, Weil JV, Reeves JT. Propranolol does not impair exercise oxygen uptake in normal men at high altitude. J Appl Physiol 1986; 61:1935–1941.
144. Fan F, Chen RYZ, Schuessler GB, Chien S. Effects of hematocrit variations on regional hemodynamics and oxygen transport in the dog. Am J Physiol 1980; 238:H545–H552.
145. Lindenfeld J, Weil JV, Travis VL, Horwitz LD. Hemodynamic responses to normovolemic polycythemia at rest and during exercise in dogs. Circ Res 1985; 56:793–800.
146. Lindenfeld J, Weil JV, Horwitz LD. Oxygen content, not viscosity regulates cardiac output during induced polycythemia in dogs. Clin Res 1983; 31:201A.

147. Cerretelli P. Limiting factors to oxygen transport on Mount Everest. J Appl Physiol 1976; 40:658–667.
148. West JB, Boyer SJ, Graber DJ, Hackett PH, Maret KH, Milledge JS, Peters RM Jr, Pizzo CJ, Samaja M, Sarnquist FH, Schoene RB, Winslow RM. Maximal exercise at extreme altitudes on Mount Everest. J Appl Physiol 1983; 55:688–698.
149. Wagner PD. Why does the maximal exercise cardiac output fall during altitude residence and is it important? Proceedings of the 11th International Hypoxia Symposium, Jasper, Alberta, Canada, 1999.
150. Reeves JT, Grover RF, Cohn JE. Regulation of ventilation during exercise at 10,200 feet in athletes born at low altitude. J Appl Physiol 1967; 22:546–554.
151. Hartley LH, Vogel JA, Cruz JC. Reduction of maximal exercise heart rate at altitude and its reversal with atropine. J Appl Physiol 1974; 36:362–365.
152. Zhuang J, Droma T, Sutton JR, McCullough RE, McCullough RG, Groves BM, Rapmund G, Janes C, Sun SF, Moore LG. Autonomic regulation of heart rate response to exercise in Tibetan and Han residents of Lhasa (3,658 m). J Appl Physiol 1993; 75:1968–1973.
153. Young AJ, Sawka MN, Muza SR, Boushel R, Lyons T, Rock PB, Freund BJ, Waters R, Cymerman A, Pandolf KB, Valeri CR. Effects of erythrocyte infusion on VO_2 max at high altitude. J Appl Physiol 1996; 81:252–259.
154. Rowell LB. Why does the maximal exercise cardiac output fall during altitude residence and is it important? Proceedings of the 11th International Hypoxia Symposium, Jasper, Alberta, Canada, 1999.
155. Moret P, Covarrubias E, Coudert J, Duchosal F. Cardiocirculatory adaptations to chronic hypoxia. Acta Cardiol 1972; 27:285–305.
156. Groves BM, Droma T, Sutton JR, McCullough RG, McCullough RE, Zhuang J, Rapmund G, Sun S, Janes C, Moore LG. Minimal hypoxic pulmonary hypertension in normal Tibetans at 3,658 m. J Appl Physiol 1993; 74:312–318.
157. Vogel JA, Hartley LH, Cruz JC. Cardiac output during exercise in altitude natives at sea level and high altitude. J Appl Physiol 1974; 36:173–176.
158. Sime F, Penaloza D, Ruiz L. Bradycardia, increased cardiac output, and reversal of pulmonary hypertension in altitude natives living at sea level. Br Heart J 1971; 33: 647–657.
159. Moret P. Myocardial metabolism: acute and chronic adaptation to hypoxia. Med Sport Sci 1985; 19:48–63.
160. Hochachka PW, Clark CM, Holden JE, Stanley C, Ugurbil K, Mennon RS. ^{31}P magnetic resonance spectroscopy of the Sherpa heart: A phosphocreatine/adenosine triphosphate signature of metabolic defense against hypoxia. Proc Natl Acad Sci USA 1996; 93:1215–1220.
161. Holden JE, Stone CK, Clark CM, Brown WD, Nickles RJ, Stanley C, Hochachka PW. Enhanced cardiac metabolism of glucose in high-altitude natives: adaptation against chronic hypoxia. J Appl Physiol 1995; 79:222–228.
162. Brooks GA, Butterfield GE, Wolfe RR, Groves BM, Mazzeo RS, Sutton JR, Wolfel EE, Reeves JT. Increased dependence on blood glucose after acclimatization to 4,300 m. J Appl Physiol 1991; 70:919–927.
163. Brooks GA, Butterfield GE, Wolfe RR, Groves BM, Mazzeo RS, Sutton JR, Wolfel EE, Reeves JT. Decreased reliance on lactate during exercise after acclimatization to 4,300 m. J Appl Physiol 1991; 71:333–341.
164. Brooks GA, Wolfel EE, Groves BM, Bender PR, Butterfield GE, Cymerman A, Maz-

zeo RS, Sutton JR, Wolfe RR, Reeves JT. Muscle accounts for glucose disposal but not lactate appearance during exercise after acclimatization to 4,300 m. J Appl Physiol 1992; 72:2435–2445.

165. Arias-Stella J, Topilsky M. Anatomy of the coronary circulation at high altitude. In: Porter R, Knight J, eds. High Altitude Physiology: Cardiac and Respiratory Aspects. Edinburgh: Churchill Livingstone, 1971:149–157.
166. Fuster V, Frye RL, Kennedy MA, Connolly DC, Makin HT. The role of collateral circulation in the various coronary syndromes. Circulation 1979; 59:1137–1144.
167. Mortimer EA Jr, Monson RR, MacMahon B. Reduction in mortality from coronary heart disease in men residing at high altitude. N Engl J Med 1977; 296:581–585.
168. Ludmer PL, Selwyn AP, Shook TL, Wayne RR, Mudge GH, Alexander RW, Ganz P. Paradoxical vasoconstriction induced by acetylcholine in atherosclerotic coronary arteries. N Engl J Med 1986; 315:1046–1051.
169. Gage JE, Hess OM, Murakami T, Ritter M, Grimm J, Krayenbuehl HP. Vasoconstriction of stenotic coronary arteries during dynamic exercise in patients with classic angina pectoris: reversibility by nitroglycerin. Circulation 1986; 73:865–876.
170. Vita JA, Treasure CB, Yeung AC, Vekshtein VI, Fantasia GM, Fish RD, Ganz P, Selwyn AP. Patients with evidence of coronary endothelial dysfunction as assessed by acetylcholine infusion demonstrate marked increase in sensitivity to constrictor effects of catecholamines. Circulation 1992; 85:1390–1397.
171. Pohl U, Busse R. Hypoxia stimulates the release of endothelium-derived relaxant factor (EDRF). Am J Physiol 1989; 256:H1595–H1600.
172. Busse R, Pohl U, Kellner, Klemm U. Endothelial cells are involved in the vasodilator response to hypoxia. Pflugers Arch 1983; 397:78–80.
173. Fuster V. Mechanisms leading to myocardial infarction: Insights from studies of vascular biology. Circulation 1994; 90:2126–2146.
174. Bartsch P, Waber U, Haeberli A, et al. Enhanced fibrin formation in high altitude pulmonary edema. J Appl Physiol 1987; 63:752–757.
175. Burke AP, Farb A, Malcom GT, Liang Y, Smialek JE, Virmani R. Plaque rupture and sudden death related to exertion in men with coronary artery disease. JAMA 1999; 281:921–926.
176. Morgan BJ, Alexander JK, Nicoli SA, Brammell HL. The patient with coronary disease at altitude: observations during acute exposure to 3,100 m. J Wilderness Med 1990; 1:147–153.
177. Alexander JK. Personal communication. Unpublished results.
178. Grover RF, Tucker CE, Robinson-McGroarty S, Travis RR. The coronary stress of skiing at high altitude. Arch Intern Med 1990; 150:1205–1208.
179. Erdmann J, Sun KT, Masar P, Niederhauser H. Effects of exposure to altitude on men with coronary artery disease and impaired left ventricular function. Am J Cardiol 1998; 81:266–270.
180. Alexander JK, Abinader EG, Sharif DS, Morgan BJ, Brammell HL. Left ventricular function in coronary heart disease at high altitude. Circulation 1988; 78(suppl II):II-6.
181. Agostoni P, Cattadori G, Guazzi M, Bussotti M, Conca C, Lomanto M, Marenzi G, Guazzi MD. Exercise capacity in heart failure patients during an hypoxia-altitude simulation test. J Am Coll Cardiol 1999; 33(suppl A):195A.
182. Hultgren HN. Coronary heart disease and trekking. J Wilderness Med 1990; 1:154–161.

10

The Pulmonary Circulation at High Altitude

JOHN T. REEVES and KURT R. STENMARK

University of Colorado Health Sciences Center
Denver, Colorado

I. Introduction

The pulmonary circulation changes profoundly at high altitude as pulmonary hypertension appears. Modest pulmonary hypertension at high altitude may be beneficial in some circumstances, but severe pulmonary hypertension is nearly always detrimental. On arrival at altitude, acute hypoxia raises pulmonary arterial pressure, and with the chronic hypoxia of continued residence, pressure rises even more. But the relationship between the acute and the chronic hypoxic responses is far from clear. We hypothesize that the key to the magnitude of the pulmonary hypertension at altitude is how the pulmonary arteriolar walls thicken in the transition from arrival to chronic residence. Understanding this transition is difficult because both acute hypoxic vasoconstriction and chronic thickening of the arteriolar walls are extremely variable within and between species, and the mechanisms for either are only dimly seen.

Our approach in the following review is to use available data to describe the lung circulation before, during, and after the transition from arrival at altitude to chronic residence and to examine possible mechanisms involved. We will use human data where possible as well as animal data where necessary. The susceptibility of the lung circulation to pulmonary hypertension at altitude seems to be greatest in the newborn because the mechanisms which have maintained high lung pressures and resistance in fetal life are powerfully retained in the neonate (69). We will therefore examine in detail the altitude effects in the very young. Most of the relevant human data are from young adults, from whom vascular reactivity and the responses to exercise provide clues to altered lung circulatory control at altitude, and these also will be examined in some detail. The aged may be susceptible to the pulmonary hypertensive effects of altitude when they suffer the ravages of heart and lung diseases as presented elsewhere (95). We refer the reader to a recent review by Grover (27) and other excellent reviews (23,30,69,92,93,108,115,116,123,124).

Perhaps to appreciate where we are and how far we have come, it is important to look specifically at the history of the lung circulation at high altitude. In doing so, we are struck by two facts: first, the history is quite brief, and second, the important foundations of our knowledge were laid by South American investigators working in the Andes.

II. History

The general history of humans at high altitude going back for several centuries is reported elsewhere in this volume. The body of pulmonary circulatory research conducted in humans between the years 1928 and 1962, particularly by investigators based in Lima, Peru, is remarkable and has been amply confirmed by subsequent workers. In order to follow the development of ideas, we have attempted to highlight the salient features of these studies and put them into context with other concepts simultaneously developed at low and high altitudes around the world.

Observations related to the lung circulation in humans at high altitude began about 1928 (64–67) when Carlos Monge (Lima) described heart failure and an increase in the second heart tone in a subset of men with chronic polycythemia (often called Monge's disease), thus raising the possibility of increased pulmonary arterial pressure. Hurtado, a colleague of Monge, noted in a 1942 review (48) that no direct measurement had been made of the pulmonary arterial pressure at high altitudes, but its possible elevation may be inferred from the permanent congestive condition (46,47) and the greater volume of blood contained in the lungs (47) and from the high frequency of right ventricular preponderance in the electrocardiograms taken in apparently healthy natives (12). High altitude residence was considered the culprit, because all signs and symptoms resolved when the polycythemic patient went to live at low altitude (46).

Hurtado had referred to the large Argentinean expedition of 1936 to the Bolivian Altiplano led by Capdehourat (12) (Fig. 1). Electrocardiograms had been obtained at approximately 4000 m in young men who were asymptomatic and who had passed the preemployment physical examination to work in the mine at Catavi. The electrocardiograms suggested right ventricular hypertrophy because they showed a mean QRS electrical axis in the frontal plane (standard leads I, II, and

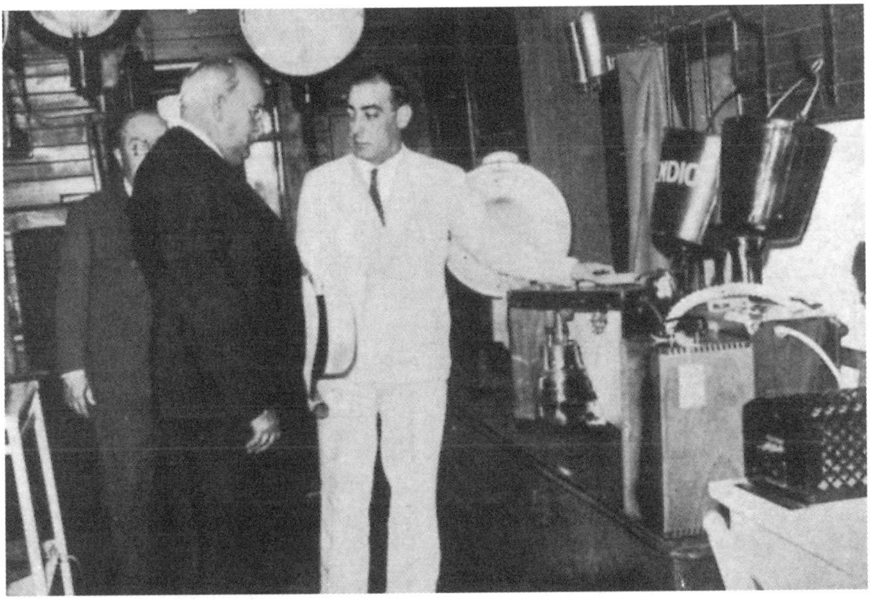

Figure 1 Photograph of E. L. Capdehourat, right, showing the President of Argentina some of the equipment for the expedition to the Bolivian Altiplano (12).

III) exceeding +90° in more than 20% of the 78 subjects examined, four times the percentage expected in sea level dwellers (7). Histograms of the electrical axis (Fig. 2) show the skewing to the right in the Bolivian high-altitude natives. Subsequent electrocardiographic studies in children and adults of Leadville, Colorado (3100 m), showed findings similar to those obtained at 4000 m in Peru (84). To our knowledge, the electrocardiograms from the Altiplano were the first quantitative data ever reported that suggested right ventricular hypertrophy (and hence pulmonary hypertension) in normal human residents at high altitude.

In 1938 Rotta reported that chest radiographs showed larger hearts in normal residents living at high altitude in Peru than in residents at sea level (Lima), but he was unable to specify which chambers were enlarged. However, at the 1946 Congress of Cardiology in Mexico City he presented a comprehensive study (100) of young men native to 4540 m in Morococha, Peru, compared with those living in Lima at 200 m. The distribution of transverse cardiac diameters indicated radiographic cardiac enlargement at altitude (Fig. 3A). Further, using cardiac fluoroscopy and electrocardiography, he specified that the right ventricle was enlarged. Also, he obtained phonocardiograms, which suggested the presence of pulmonary hypertension, and he noted an elevated venous pressure, which indicated increased resistance to filling of the right ventricle. The two decades of studies from the high altitudes of Bolivia and Peru were prescient because they predicted (48) elevation of pulmonary

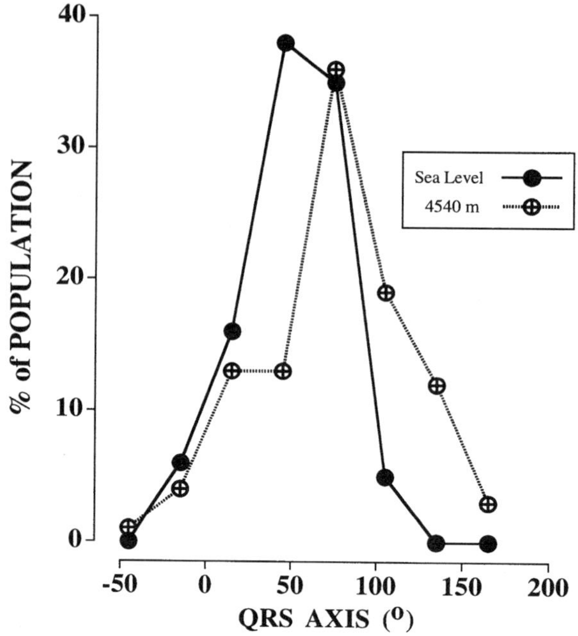

Figure 2 Histogram of the electrocardiographic frontal mean QRS axis as reported from normal adult residents of the Bolivian Altiplano (12) (crossed circles), and sea level (New Orleans, 7) (filled circles). An axis of ≥90° is considered right axis deviation and is consistent with a large right ventricular muscle mass. Seen is that the high-altitude populations are skewed toward greater right axis deviations. QRS axis from 3100 m (not shown) (84) is more to right than at sea level but less rightward than from the Bolivian Altiplano. (Adapted from Refs. 7,12.)

arterial pressure before hypoxia was known to cause acute pulmonary vasoconstriction and before cardiac catheterization was available to measure pulmonary arterial pressure.

However, there were rapid developments elsewhere. In 1946 von Euler and Liljestrand demonstrated that acute hypoxia raised the pulmonary arterial pressures of cats (119). About the same time Cournand reported measuring pulmonary arterial pressure in humans during cardiac catheterization (15), and in 1947 his laboratory found acute hypoxia raised human pulmonary arterial pressure (for reviews, see Refs. 23, 72). Maybe acute hypoxia on arrival at high altitude increased pulmonary arterial pressure and continued residence increased it further. Rotta and colleagues lost little time in setting about to make the measurements, and in 1949 (99) and 1952 (98) they reported the first pulmonary arterial pressures ever obtained in any species at high altitude. The full reports in English (80,97) were a *tour de force* in that pulmonary arterial pressures and blood flows were measured (1) in native residents of Lima, (2) in native high-altitude residents of Morococha at 4540 m, (3) in newcomers living for 1 year at Morococha, (4) in residents of Morococha with chronic mountain sickness (Monge's disease), and (5) in some of the high-altitude residents during the breathing of 35% oxygen to simulate sea level Po_2. Their measurements established the co-existence of arterial oxygen desaturation and pulmonary hypertension (Fig. 3B). Thus residents at high altitude with low oxygen saturations also had elevated pulmonary arterial pressure, supporting the concept that chronic hypoxia caused chronic pulmonary hypertension.

In addition, the data showed that newcomers to Morococha had less hypoxemia and less pulmonary hypertension than did healthy people born and living there (Fig. 3B). Further, two subjects who had lived in Morococha for more than 25 years and who had excessive polycythemia (chronic mountain sickness) had the most severe hypoxemia and the highest pulmonary arterial pressures. Taken together (Fig. 3B), the results indicated that the elevated pressure was related to the severity of the hypoxia and that contributing factors were likely to be the duration of residence at altitude and the presence of polycythemia.

Rotta et al. must have expected these findings based on prior clinical, laboratory, and literature findings. However, they could not have expected the poor pulmonary vasodilator responses when their subjects breathed a high oxygen mixture (97). While 35% oxygen tended to lower pulmonary arterial pressure in the various groups, the cardiac output fell more than the pressure in each instance. Thus, supplemental oxygen tended to raise, not lower, total pulmonary resistance! How were they to explain, then, that acute reversal of the chronic hypoxia failed to reduce the elevated vascular resistance? The suggestion by Rotta et al. (97) showed considerable insight: "anoxia cannot be considered as a possible isolated factor in the pulmonary hypertension." Clearly these authors realized that the pulmonary hypertension of chronic hypoxia was something more than the maintenance of acute hypoxic vasoconstriction. One possibility they considered was that 15–25 minutes of high oxygen might not be sufficient to reverse the elevated resistance, and, indeed, in a subsequent study (80), Sime et al. showed normalization of pressures and resistances in normal or polycythemic men taken for 2 years to sea level (104). As discussed below, we continue to debate the sequence of events and mechanisms by which chronic pulmonary hypertension at high altitude replaces acute hypoxic vasoconstriction.

So far, the measurements and observations in the Andes had been largely confined to adult men, but events unfolding simultaneously in England pointed to the importance of studying children at high altitude, especially in the immediate newborn period. In Cambridge, Barcroft, who had begun to study the fetal circulation in the late 1920s, and knowing of the low Po_2 in fetal blood, referred to "Mt. Everest in utero" (9). Geoffrey Dawes, who continued the Barcroft tradition, began his classic studies in the lamb in 1948 at the Nuffield Institute in Oxford. Research in his laboratory showed in the near-term fetus that inflating the lung with air or oxygen at birth caused pressure to fall in the pulmonary artery, but not in the aorta (17). Within a few days after a normal birth of lambs, pulmonary pressure had fallen

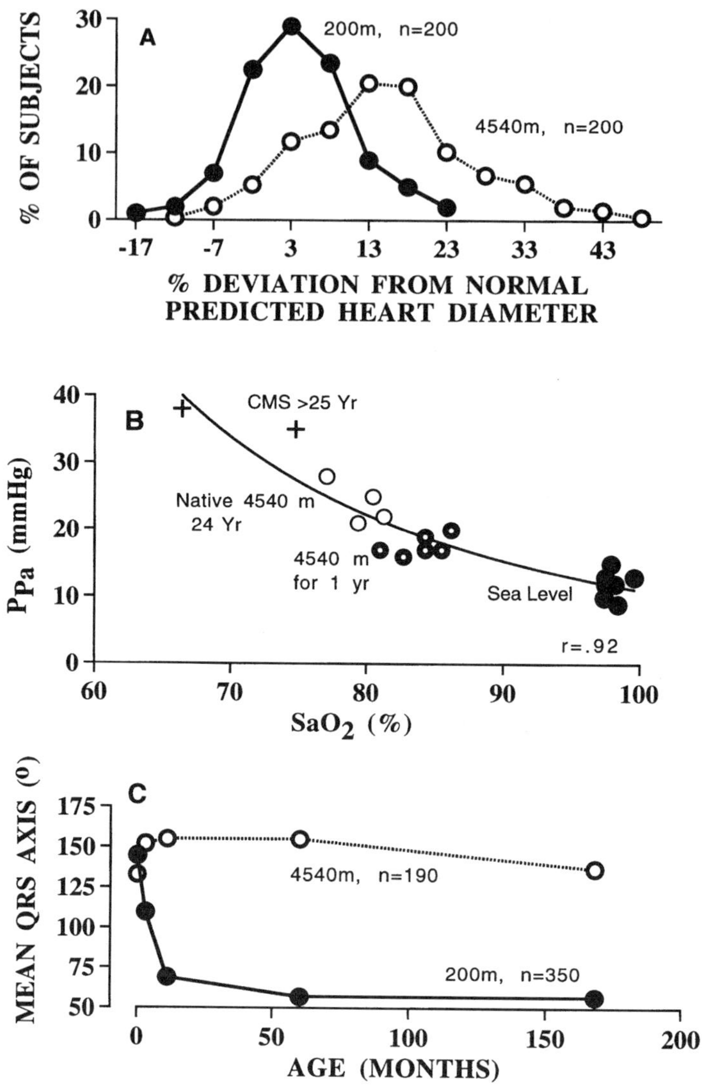

Figure 3 Early measurements from Peru relating to the lung circulation. (A) Histograms of the transverse diameter of the heart from frontal chest roentgenograms in 200 native residents of Lima at 200 m elevation (filled circles) and in 200 native residents of Morococha at 4540 m elevation (unfilled circles). (Adapted from Ref. 100.) Subjects at high altitude, but not at sea level, had hearts that were larger than predicted normal. (B) Mean pulmonary arterial pressures as related to systemic arterial oxygen saturations from individual adult native residents of Lima (filled circles), from Lima residents who had lived at Morococha for one year (partially filled circles), from 24-year-old healthy residents of Morococha (unfilled circles), and from two Morococha residents older than 25 years with chronic mountain sickness (CMS, +). Shown is the curvilinear regression (unbroken line) through all data. Seen is that

to the stable nadir that would be maintained throughout life (17). Similar findings were obtained in human newborns (1). One potentially important mechanism for survival of mammalian species might be hypoxic vasoconstriction in the fetal lung to limit blood flow while the organ was relatively dormant and fluid filled. However after birth, high lung blood flow becomes essential for oxygenation of the newborn, and reversal of hypoxic pulmonary vasoconstriction is one of several mechanisms promoting the increased lung blood flow. But what if the neonate breathed an hypoxic gas after birth? In 1957 Rowe showed (101) that even brief inhalation of low-oxygen mixtures caused large pulmonary hypoxic vasopressor responses. Thus, children born at high altitude might not enjoy a rapid relief of their fetal pulmonary hypertension because their alveoli would be filled with air at a low Po_2. The newborns at altitude might be denied the full vasodilating effect of the high Po_2 available to newborns at sea level.

It was important, then, that the Peruvian investigators look for clues relative to the persistence of pulmonary hypertension following birth at high altitude. Their strategy was first to use noninvasive measurements to compare, for varying postnatal ages, children born at sea level with those born at altitude. Naturally they turned to the children in Lima and Morococha, respectively (77,78). Using the electrocardiogram to provide an index of right ventricular muscle mass, they found that at birth there was comparable right axis deviation at birth in both Lima and Morococha. These findings were consistent with high pulmonary artery pressures and thick right ventricles (Fig. 3C), known to exist in the fetus. However, findings from the two populations quickly diverged. Of 350 children examined in Lima, the right axis deviation rapidly regressed after birth to normal sea level values (Fig. 3C), but the axes of the 190 children in Morococha did not. Rather, in Morococha the right axis deviation persisted near the newborn levels. The data implied a rapid atrophy of the right ventricle at sea level, but not at 4540 m. Even in 14-year-olds in Morococha, the electrocardiographic axis was not different from that of the newborn. The inspired Po_2 in Morococha is of the order of 80 mmHg, and the alveolar Po_2 would therefore be expected to be between approximately 40–50 mmHg, which apparently is low enough to delay or even prevent the normal postnatal regression of the right ventricle.

Because the electrocardiogram is an insensitive indicator of ventricular hypertrophy, Arias-Stella and Recavarren (6) sought direct evidence by weighing the right and left ventricles from children of Lima and Morococha who died from accidents or acute illness. At birth, the ratios of right ventricular to total ventricular weights

---

pressures appeared to increase with hypoxemia, were higher in natives than in newcomers, and were highest in CMS patients. (Adapted from Ref. 97.) (C) Mean frontal electrocardiographic axis of the QRS complex at various ages in healthy residents of Lima (filled circles) and Morococha (unfilled circles). Seen is that the right axis deviation persisted after birth at high altitude but not at sea level. (Adapted from Ref. 77.)

showed no difference between high- and low-altitude children (Fig. 4A), supporting the electrocardiographic findings. Also supporting the electrocardiogram were the decreasing ratios of right to total ventricular weight as children grew older at sea level and the much smaller decreases in the weight ratios with age in Morococha (Fig. 4A). The results were consistent with the persistence of right ventricular hypertrophy up to 10 years, the age of the oldest children from whom hearts were obtained.

Thus, the electrocardiogram and the postmortem examination of the right ventricle pointed to a persistence of pulmonary hypertension following birth, but there had been no histological examinations of the pulmonary vascular tree at different ages after birth at high altitude, and the time was ripe for such an examination. In 1951 in Minnesota, Civin and Edwards (13) had shown for humans that fetal pulmonary arterioles had thick media but that the media became thin after birth. Wagenvoort (120), then in Edwards's laboratory, published quantitative measurements showing the time course of the postnatal thinning to adult levels by 6–12 weeks after birth. Because chronic hypoxia after birth at high altitude might sustain lung arteriolar thickening, Arias-Stella and Saldaña (5) examined the lungs from persons dying at low and high altitudes. Their findings were that arteriolar medial thickness decreased, as expected, after birth at sea level, but at high altitude thickened media persisted for many months and, to some extent, throughout life (Fig. 4B).

Taken together, all of these data suggested the persistence of fetal pulmonary hypertension after birth at high altitude, but there had been no pressure measurements in healthy infants following birth at high altitude. The measurements available from sea level showed a prompt fall in pressure within minutes to hours after birth (Fig. 5A) (1,102,103). To obtain high-altitude data, cardiac catheterization measurements were made in healthy children in Morococha by Sime et al. (105) and reported initially by Penalosa et al. in 1962 (80). They showed that the pulmonary arterial pressures were sustained near the fetal levels for weeks and months after birth (Fig. 5A). These early studies in humans suggested that infants might be particularly susceptible to chronic hypoxic pulmonary hypertension. Cardiac catheterization studies conducted nearly simultaneously at 3100 m in Leadville, Colorado (117), showed that elevated pulmonary arterial pressures persisted for years in adolescents and young adults (Fig. 5A).

Thus by June of 1962, South American investigators, together with those from Colorado, had set in place principles relating to the lung circulation at high altitude:

1. Residence at high altitude raised resting pulmonary arterial pressure in newcomers and in the native born.
2. Certain life-long residents who had hypoxemia and excessive polycythemia (Monge's disease) also had excessive pulmonary hypertension.
3. The pulmonary hypertension in normal or polycythemic individuals could be reversed by a period of residence near sea level.
4. In children born at high altitude, there were failures of the normal regression in right ventricular hypertrophy, the regression of pulmonary arterio-

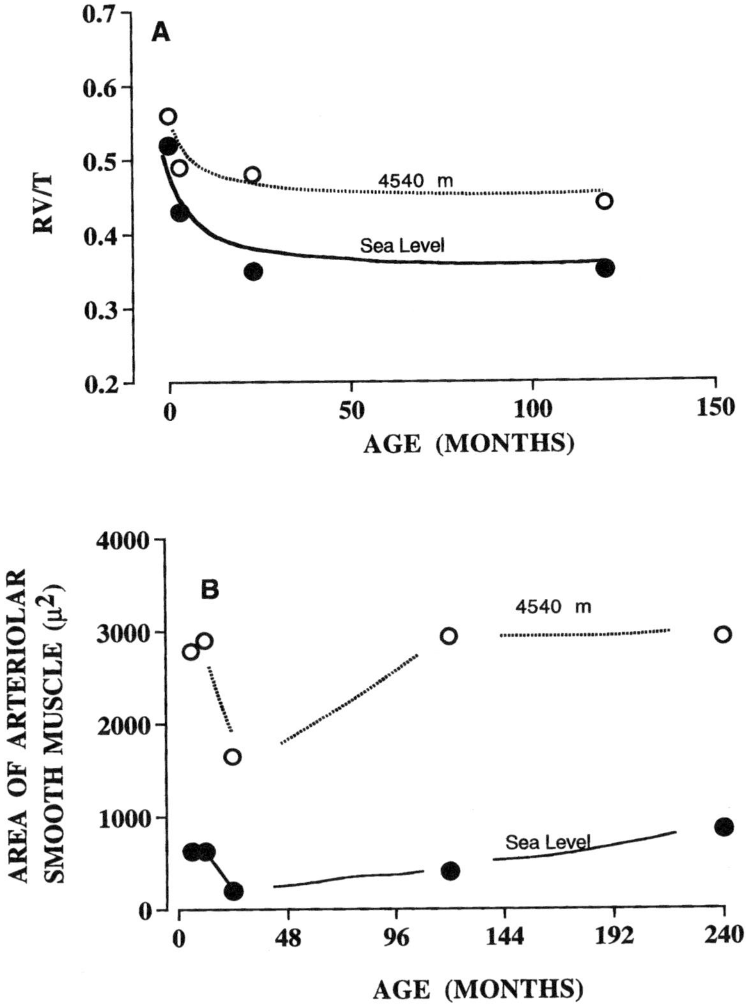

Figure 4 Early measurements from Peru relating to the right ventricle and hypertrophy of lung arterioles. (A) Ratios of the weight of the right ventricle to total ventricular weight from persons of various ages dying acutely and thought to be previously healthy as obtained in Lima (sea level, filled circles) and Morococha (4540 m, unfilled circles). Curvilinear regression lines are drawn through each data set. Seen are that high right ventricular weight ratios persist at high altitude but not at sea level. (Data recalculated from Ref. 6.) (B) Pulmonary arteriolar medial smooth muscle in residents of various ages dying acutely in Lima (filled circles) or Morococha (unfilled circles). Seen is that the media thickness is increased at high altitude. (Data recalculated from Ref. 5.)

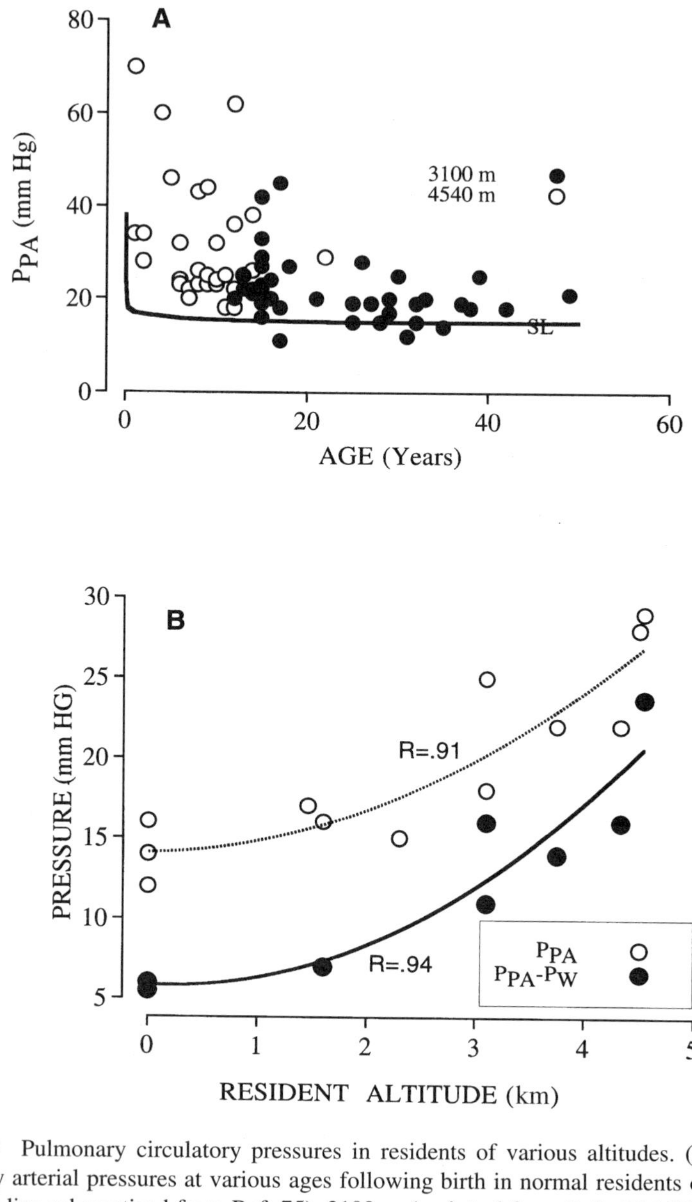

Figure 5 Pulmonary circulatory pressures in residents of various altitudes. (A) Resting pulmonary arterial pressures at various ages following birth in normal residents of sea level (unbroken line schematized from Ref. 75), 3100 m (replotted from Refs. 23,100), and 4540 m (replotted from Ref. 88). Compared to sea level, pulmonary hypertension persists after birth at both high altitudes shown. (B) Resting pulmonary arterial pressures (open circles) and pulmonary arterial minus wedge (filled circles) in native residents of various altitudes. (Data replotted from Refs. 8,36,44,97,117.)

lar vascular smooth muscle, and the fall in pulmonary arterial pressure and resistance.

5. As discussed below, exercise, even in healthy residents, was accompanied by greater pulmonary hypertension than at sea level and a blunting of an exercise-related fall in vascular resistance.

Subsequent investigators who have measured pulmonary hemodynamics have shown the altitude-related pulmonary hypertension in terms of pulmonary arterial pressure as well as the pressure gradient from pulmonary artery to wedge (Fig. 5B). However, the early investigations from the Andes set the challenge for the search for the mechanisms by which high altitude altered the pulmonary circulation. The mechanisms must take account of differences between acute hypoxic pulmonary vasoconstriction and chronic altitude-induced hypertrophy/hyperplasia of the pulmonary arterioles, both of which seem particularly to affect the newborn. The current chapter focuses on these mechanisms.

III. Mechanisms

A. Acute Hypoxic Pulmonary Vasoconstriction

Direct Effects on Smooth Muscle

The question of the mechanism of hypoxic pulmonary vasoconstriction has been the subject of intense investigation around the world since von Euler's report (119). Hypoxic vasoconstriction is intrinsic, and probably unique, to the lung circulation. Early investigations by Duke and Killick (18) and subsequently by others, established that lungs removed from the body and artificially perfused, with or without blood (58), retained their capacity for hypoxic pulmonary vasoconstriction. In the isolated lung, the pressor response develops progressively over 1–5 minutes during exposure to alveolar hypoxia and tends to regress rather more rapidly on return to normoxia (Fig. 6A). Further, isolated arterioles (<300 mm diameter) from lungs, but not from systemic organs, constricted in response to hypoxia (54). At issue was whether the constriction was mediated by a direct effect of hypoxia on the vascular smooth muscle cells or indirectly by vasoconstrictor substances from other cells. In evaluating this issue, Archer suggested criteria (3,4) that need to be satisfied.

1. The response is intrinsic to the lungs and lung arterioles.
2. The threshold for the response is a Po_2 of approximately 55 mmHg in isolated lungs (an alveolar Po_2 of 65 mmHg in intact man).
3. There is a rapid onset and offset of the constriction.
4. Calcium moves from outside to inside the smooth muscle cell.
5. There is depolarization of the smooth muscle membrane (Fig. 6B).

While naturally occurring vasoactive substances, including catecholamines, histamine, angiotensin II, 5-hydroxytryptamine, prostaglandins, and leukotrienes, all

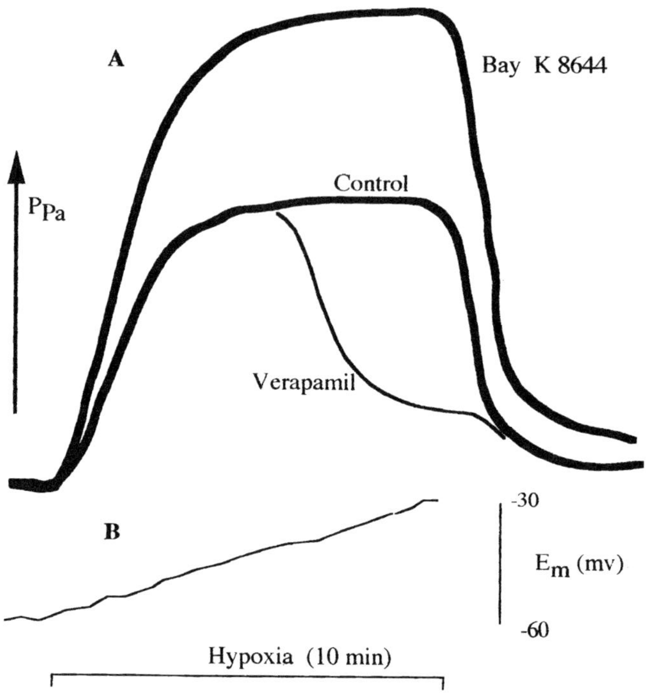

Figure 6 (A) Schematized representation of hypoxic pulmonary pressor responses in blood perfused isolated rat lungs before (control) and after the administration of the calcium channel blocker, verapamil, and a compound facilitating calcium entry in to cells, Bay K8644. (Adapted from Refs. 60,124.) (B) Schematized decrease in the transmembrane potential of isolated pulmonary vascular smooth muscle cells during exposure to hypoxia. (Adapted from Ref. 124.)

have been found to influence the magnitude of the vasoconstrictor response, none of them have been shown to satisfy the above criteria (3,4,116,124).

The possibility that the mechanism was not mediated through vasoactive substances, but more directly via the vascular smooth muscle was suggested by the requirement for extracellular calcium in the vasoconstrictor response to hypoxia (59,60). Blockade of calcium channels by verapamil inhibited the hypoxic pulmonary vasoconstriction in isolated lungs, while facilitation of calcium entry by a calcium channel agonist, such as the compound Bay K8466, augmented the response (Fig. 6A). Hypoxia appeared to depolarize the pulmonary vascular smooth muscle cell (Fig. 6B), allowing the entry of calcium, which then initiated the contraction (33,54,55). For the carotid body type I cells (i.e., the O_2-sensing cells), the possibility that the calcium entered the cell via a voltage-dependent K^+ channel was indicated by the experiments of Lopez-Barneo (53). If, as suggested by Torrance (111), there is a mechanism for O_2 sensing that is common to several different organs, then the pulmonary arterioles might be controlled in a way analogous to the carotid body. In smooth muscle from small (<600 mm) pulmonary arteries (82,83,124), and even from isolated pulmonary arterial smooth muscle cells (55,82), evidence has been accumulated supporting the role for K^+ channels in the hypoxic vasoconstrictor response. It seems likely that hypoxia inhibits an outward potassium current, resulting in depolarization of the pulmonary vascular smooth muscle cell membrane, which then allows calcium entry through voltage-dependent calcium channels followed by contraction (for review, see Ref. 124). By contrast, in systemic arteries hypoxia increases potassium outflow through ATP-dependent potassium channels resulting in membrane hyperpolarization and smooth muscle relaxation. These studies seem to be pointing to differences in K^+ channel populations and functions in large versus small lung arteries and in systemic versus pulmonary arterioles (124). We are moving toward a greater understanding of the mechanisms of hypoxic pulmonary vasoconstriction, where the emphasis is now on the membrane of the smooth muscle cell itself, and a modulating role for vasoactive substances (116). The questions now are more focused, relating to the type and structure of the K^+ channels involved, their mechanism of activation, how they transduce the Ca^+ entry, how hypoxia is specifically involved in the processes, and the influences of various modulating mechanisms.

Modulation of Hypoxic Pulmonary Vasoconstriction

Hypoxic pulmonary vasoconstriction is widely variable among healthy persons (90). It also varies markedly with age of an individual and among the different mammalian species. The variability is probably related to the multiple factors that influence the magnitude of the response. The K^+ and Ca^{2+} ion channels themselves are under multiple influences such as vascular tone, cytochrome P-450 enzymes, and a variety of eicosanoids (32,33). In addition, neighboring cells influence smooth muscle. The capacity of the endothelium to produce nitric oxide (NO) or prostacyclin with increasing shear stress, the release of constrictor or dilator substances from mast cells, or the presence of inflammatory mediators will all affect the tone of the lung arterioles and their response to acute hypoxia. The amount of the vascular smooth muscle and even matrix proteins are determinants of the response. Review of these and other factors can be found elsewhere (3,4,115,123).

B. Transition from Acute Hypoxic Vasoconstriction to Chronic High-Altitude Pulmonary Hypertension

Acute hypoxic vasoconstriction and its modulation by chemical substances, as discussed above, are mechanisms likely set in motion when individuals first arrive at high altitude. But if chronic high-altitude pulmonary hypertension is not simply the persistence of acute vasoconstriction, then we must look to other mechanisms. As the focus of this review is related to the transition from acute to chronic hypertension, we must consider how the transition is controlled and whether it is reversible. Finally, there is the philosophical question of whether high-altitude pulmonary hypertension serves a useful purpose. As is often the case with questions, they are easier to pose than to answer. However, the starting point must be a description of the time course of the pulmonary arterial pressure changes at high altitude in living individuals.

Pulmonary Arterial Pressure Changes over Time at Altitude

Changing pulmonary arterial pressure over time at high altitude would be compatible with the concept that chronic pulmonary hypertension is not just the maintenance of acute hypoxic vasoconstriction. If, compared to arrival at altitude, for example, pressure subsequently increased more than could be accounted for by an increase in flow, then either functional or structural changes in the lung vascular bed would be expected. Selected and representative data from three species are examined below.

Humans

In young men having cardiac catheters in place at 3800 m, pulmonary arterial pressure doubled from the sea level value in the first 24 hours and then remained elevated at this level for 3 days (Fig. 7) (50). Cardiac output rose on arrival and subsequently fell, resulting in a progressive increase over 3 days at altitude in the calculated total pulmonary resistance. Figure 7 also compares the measurements in these recent arrivals with life-long altitude residents of Leadville, Colorado (117). Even though the Leadville residents were at only 3100 m, they had higher pulmonary arterial pressures and higher calculated total pulmonary resistances than observed in subjects residing over 3 days at the higher altitude of 3800 m (50). Taken together, the data suggest that the pulmonary vascular resistance in humans increases over the first days at altitude, but with life-long residence, pressure and resistance increase further.

Cattle

Because of their large size, cattle can have repeated catheterizations with relative ease, and as in humans, measurements in cattle have been made in the actual high-altitude environment. Thus, for example, hemodynamic data are available in 10 yearling steers brought from near sea level to reside for up to 9 weeks at 3900 m on Mt. Evans, Colorado (24). Pulmonary arterial pressure progressively rose (Fig. 7). In three animals, one of which developed right heart failure, pressures reached mean values of approximately 100 mmHg. Cardiac output was unchanged or decreased at altitude, and therefore total pulmonary resistance progressively rose (Fig. 7). The average increase in pressure was 47 mmHg at 3900 m, i.e., greater than the 9 mmHg increase caused by acute hypoxia at low altitude. A prior study (128) in cattle taken from sea level to 3050 m also showed large increases in pulmonary pressure and resistance.

Rats

When placed in altitude or hypoxic chambers, rats' pulmonary arterial pressure can be measured daily, but the catheterization procedure is usually done when the rats

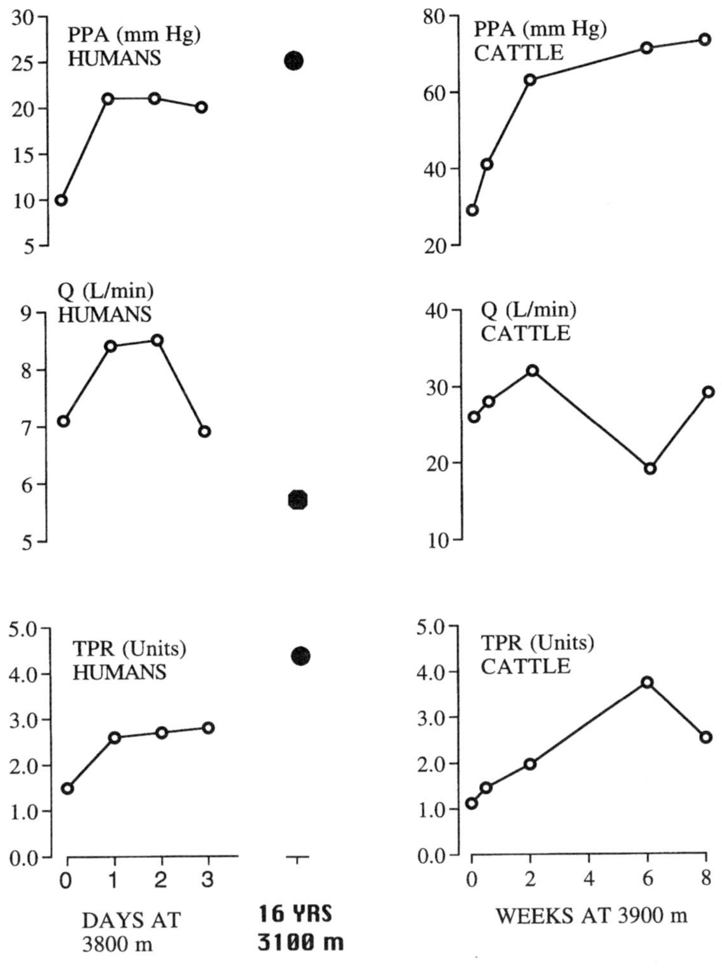

Figure 7 Left panels: Pulmonary arterial pressure (PPA, top), blood flow (Q, middle), and total pulmonary resistance (TPR, bottom) in humans during 3 days at 3800 m. (Adapted from Ref. 41.) Also shown for comparison, in filled circles, are pressure, flow, and resistance group means from young adults living an average of 16 years in Leadville, Colorado, at 3100 m elevation. (Adapted from Ref. 117.) Right panels: Pulmonary arterial pressures (PPA, top), blood flow (Q, middle), and total pulmonary resistance (TPR, bottom) in cattle during 8-week residence at 3900 m. (Adapted from Ref. 24.)

have been removed to room air. In reported studies, whether or not the rats breathed hypoxic gas at the time of measurement (to simulate the hypoxia at which they had been kept), pulmonary arterial pressure progressively rose (Fig. 8). For example, in one study (87), after approximately one month at half atmosphere, mean pressures approached 60 mmHg, a value higher than seen during acute hypoxia in sea level rats.

Thus, measurements in humans and other species point to the same general conclusion, namely, that pulmonary hypertension is progressive over time at altitude. Further, during chronic high-altitude residence the pressures are higher than can be accounted for by acute hypoxic vasoconstriction at low altitude. Clearly, factors not present during acute hypoxia operate during the development of chronic high-altitude pulmonary hypertension.

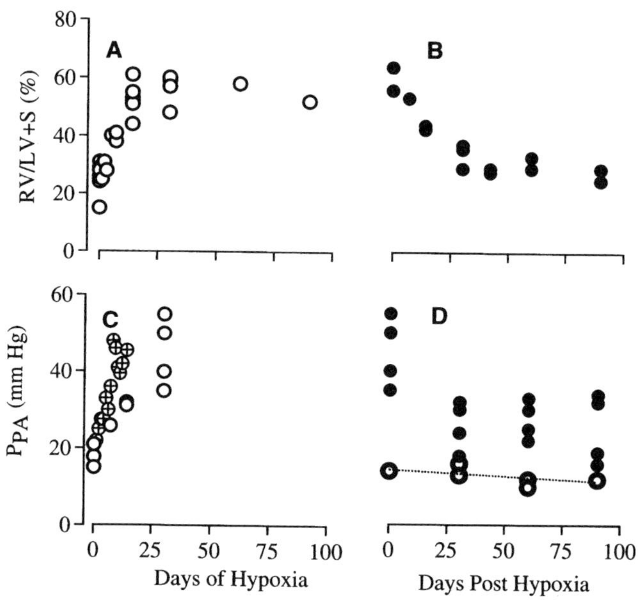

Figure 8 Measurements in rats during development and regression of hypoxic pulmonary hypertension. (A & B) Percent right ventricular of total ventricular weight during the development (unfilled circles, panel A, left) of hypoxic pulmonary hypertension and its regression after return to normoxia (filled circles, panel B, right) as collected from several studies (39,42,68,70,87,88,106,114). (C & D) Mean pulmonary arterial pressure during the development (unfilled or cross-filled circles, panel C, left) of hypoxic pulmonary hypertension and its regression after return to normoxia (filled circles, panel D, right) in rats as collected from several studies (68,70,87,88,106,114). (Cross-filled circles are data from Ref. 88, where hypoxia was maintained during measurement.)

Factors (Excluding Wall Thickening) Contributing to Chronic Hypoxic Pulmonary Hypertension

Polycythemia

Increasing polycythemia increases the viscosity of blood and could increase pulmonary arterial pressure. One wonders whether polycythemia that occurs over time in humans and most other mammals has a major role in the development of altitude-related pulmonary hypertension. In rabbits, a species with a brisk erythropoietic response to chronic hypoxia, there was a parallel increase in right ventricular systolic pressure and hematocrit (changes of 65 and 62%, respectively) during several weeks at 4300 m, suggesting that blood viscosity played an important role in the pulmonary hypertensive response (91). Polycythemia and pulmonary hypertension co-existed as central features in chronic mountain sickness (Fig. 3B). However, cattle with pulmonary arterial pressures increasing by more than 300% during high-altitude residence had no increase in hematocrit (24,128). Hultgren reported no relationship of pulmonary arterial pressure or resistance to hematocrit in 30 healthy residents at 3750 m in Peru (45). Thus, an increase in viscosity may contribute to, but is not necessary for, pulmonary hypertension at high altitude.

Augmented Hypoxic Pulmonary Vasoconstriction

Another possibility to account for the high pulmonary pressures at altitude is that hypoxic pulmonary vasoconstriction increases with continued residence in an hypoxic environment. If so, either the hypoxic stimulus or the response to it increases. For normal subjects, we can dismiss an increase in the severity of the hypoxic stimulus, because from the time of arrival at altitude to a residence of many years duration, hypoxemia does not increase. The question of an increased response to a given hypoxic stimulus (stronger vasoconstriction) is more difficult to answer for humans, because we do not have hypoxic pressor responses at sea level and at altitude in the same subjects. However, resistance changes are available from Leadville, Colorado, at 3100 m (100) in 15 subjects, who breathed 13% O_2 ($P_{I_{O_2}}$ = 59 mmHg) for a few minutes. As a group, total pulmonary resistance increased by 28% (Fig. 9A). Eight of the subjects did not increase pressure with hypoxia, and for them an acute hypoxic pulmonary pressor response was absent (Fig. 9B). The other 7 subjects had increases of less than 100%. The little human data available in normal altitude residents suggest the acute hypoxic pulmonary vasoconstrictor response is not stronger than in sea level residents. However, there may be subsets of the population with strong responses, as in Leadville children, who develop high-altitude pulmonary edema when they get upper respiratory infections (21).

For chronically hypoxic rats, the answer is clearer than in humans. When rats that have been at high altitude for 4–6 weeks have their lungs removed and artificially perfused with blood, the acute hypoxic pressor responses are blunted (59,114) (Fig. 10). Thus, for hypoxic P_{O_2} values in the physiological range, pressor reactivity is reduced, not increased. However, pressor responses are increased to other vasoconstrictor agonists such as angiotensin II (Fig. 10), norepinephrine, or prostaglandin

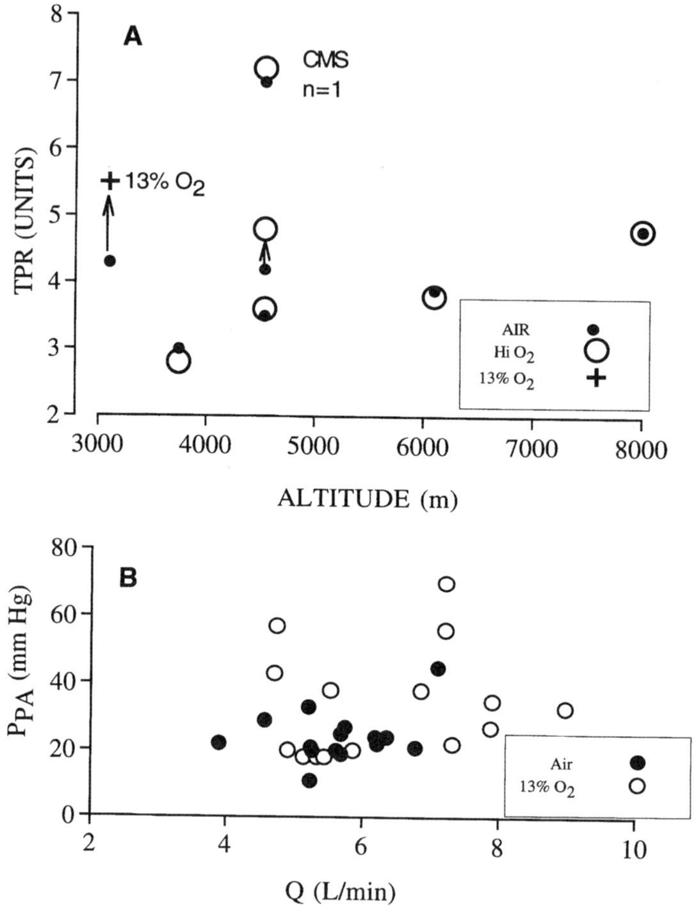

Figure 9 Effect of acute hypoxia or normoxia in persons exposed chronically to high altitude. (A) Measurements of total pulmonary resistance at various altitudes in persons breathing air (filled circles), 13% oxygen (cross), or high oxygen mixtures (unfilled circles). (From Refs. 29,44,97,117.) (B) Individual pulmonary arterial pressures and flows in 15 young adults breathing air (filled circles) and then 13% oxygen (unfilled circles) in Leadville, Colorado, at 3100 m. (From Ref. 100.)

F_{2a}. These results are paradoxical. The increased arteriolar muscularity at altitude (96) allows greater responses to vasoactive substances, yet the response to hypoxia is reduced. Further, the mystery is enhanced by the remarkable finding that when rats that had been at altitude for weeks were returned to low altitude for several hours, pressor responses to hypoxia increased well above those of the low-altitude controls (Fig. 10). The reasons for the blunting by hypoxia and the posthypoxic enhancement are not clear. However, the results for many normal humans and for

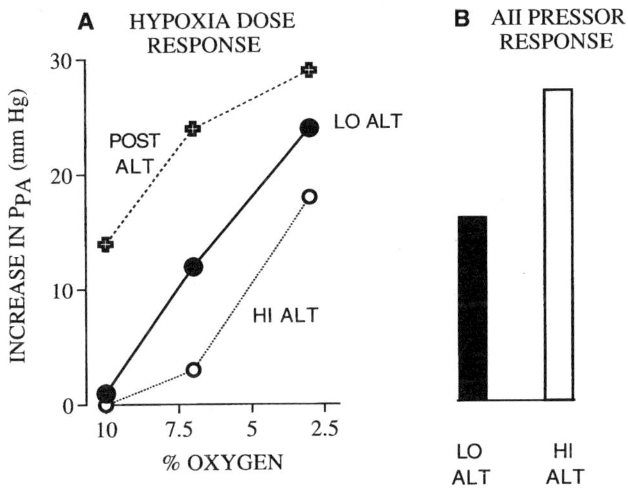

Figure 10 Pulmonary pressor responses to hypoxia or angiotensin II in isolated lungs from low-altitude rats (filled symbols), rats living for 4 weeks at high altitude (unfilled symbols), or rats that have been removed for 3–5 days from the hypoxic environment (crosses). (Adapted from Ref. 59.)

rats are compatible with the concept that during high-altitude residence the pressor response to a given hypoxic stimulus is not increased and, further, that the greater pulmonary hypertension with chronic than acute hypoxia is not explained by a stronger acute hypoxic vasoconstriction.

Blunted Hyperoxic Vasodilation

The question of hypoxic reactivity can also be addressed by examining the "other side of the coin," namely the ability of high oxygen to reverse acutely chronic hypoxic pulmonary hypertension (97). Several studies suggest that while human residents of high altitude eliminate hypoxia by breathing O_2-enriched mixtures for 5–25 minutes, they fail to attain normal sea level pulmonary hemodynamics (29,44). Furthermore, the results from the collected studies suggest that the relief of hypoxia usually lowers the pulmonary blood flow along with the pressure so that resistance is not changed (Fig. 9A). Similar findings are reported for chronically hypoxic neonatal calves with severe pulmonary hypertension, namely, 5–10 minutes of 100% oxygen fail to lower pulmonary vascular resistance (107).

These data from both human and animal studies with both increasing and decreasing hypoxic stimuli lead to the concept that chronic pulmonary hypertension cannot be explained fully by increasing vasoreactivity. That is, an increasing contractile response to a given hypoxic stimulus may not account for altitude-related pulmonary hypertension. In fact, the results often point in the opposite direction, namely that chronic high-altitude exposure blunts specifically and reversibly the ability of acute hypoxia to raise pulmonary arterial pressure. The change from an hypoxic-responsive to an hypo-responsive lung circulation may develop in a few days. In the newborn calf the relatively fixed component of hypoxic pulmonary hypertension increases within 2 weeks (Fig. 11). Although acute vasomotor reactivity to hypoxia persists at high altitude to a variable extent among subjects and among species, altitude-related pulmonary hypertension appears to be more than just sustained acute hypoxic pulmonary vasoconstriction.

Is Acute Hypoxic Vasoconstriction Required for the Development of Pulmonary Hypertension at High Altitude?

In humans, there is a very large variation in the pressor response to acute hypoxia (Fig. 12A) and also in the magnitude of chronic high altitude-related pulmonary hypertension (Fig. 12B), but the measurements are in different populations, and the relation of the acute to the chronic response cannot be established. However, a clue that the acute response may be dissociated from the chronic response is suggested from data in high-altitude residents. No correlation existed for the relation of pulmonary pressor response to 13% or 14% O_2 to the resting pulmonary arterial pressure in 21 healthy residents of Leadville, Colorado (117) (Fig. 12C) or in 5 residents of Lhasa, Tibet, respectively (28). These data neither established nor denied that the acute hypoxic response determines the chronic response to altitude. However, they have indicated that the acute response was variable whether or not high altitude–related pulmonary hypertension was present, and suggested that in the high altitude

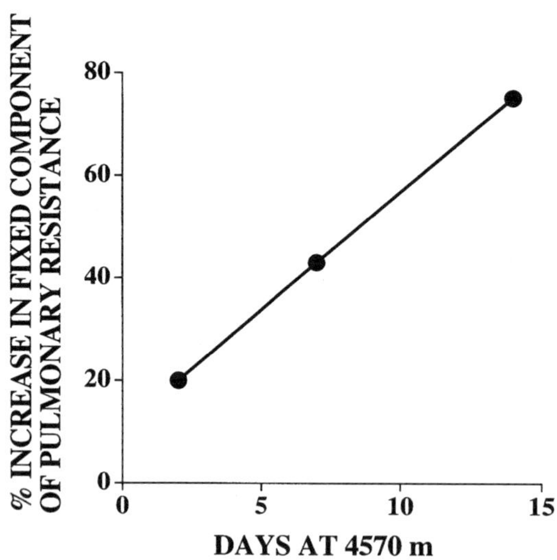

Figure 11 Development during 2 weeks of altitude exposure of an increased fixed component (unresponsive to 1 hour of normoxia and/or normoxia plus acetylcholine infusion) in newborn calves with severe pulmonary hypertension. (Data redrawn from Ref. 20.)

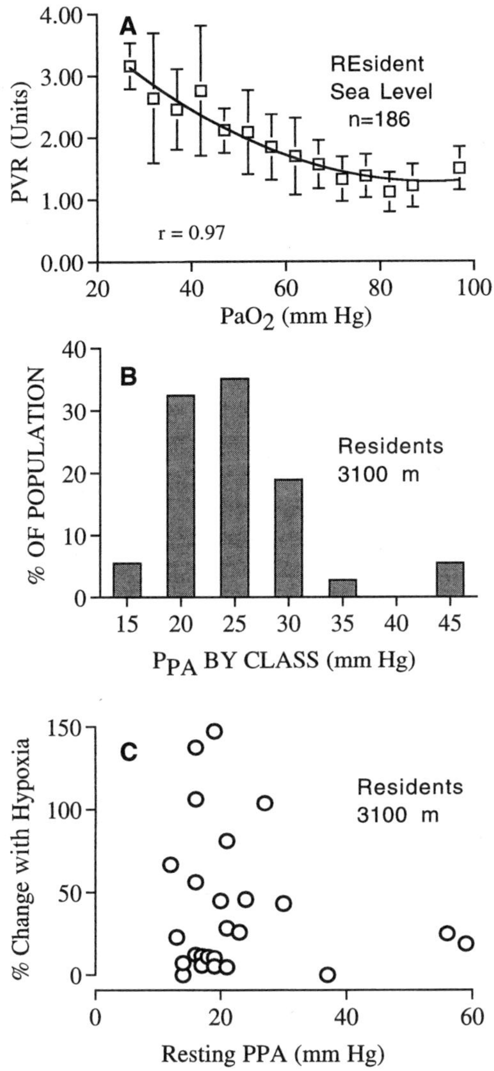

Figure 12 Human variability in pulmonary hypertension with hypoxia. (A) Measurements (n = 186) of pulmonary vascular resistance (PVR) in 50 persons near sea level and without lung disease, breathing air and a series of hypoxic gases. Shown for 5 mmHg classes of pulmonary arterial pressure are mean and one standard deviation. (Adapted from Ref. 90.) (B) Distribution of pulmonary arterial pressure in 37 residents of Leadville, Colorado, illustrating the variation in chronic hypoxic pulmonary hypertension in this population. (Adapted from Ref. 117.) (C) Percent change in pulmonary arterial pressure in 25 of the persons shown in B, above, during breathing of acute hypoxia, 13% O_2. The pressor response was variable, but most of the cohort increased pressures less than 50%.

resident factors other than the strength of the acute response might determine the chronic response.

In cattle, by contrast, a strong relationship was seen between the acute hypoxic response at low altitude to the severity of chronic pulmonary hypertension observed when the individuals were taken to altitude (127). Thus, steers with small acute hypoxic pressor responses developed less pulmonary hypertension at altitude than those with large acute responses. However, the significance of the observation is not clear because the cattle came from two distinct strains bred specifically to be either susceptible or resistant to hypoxic pulmonary hypertension. Further, the susceptible breed had from an early age thicker pulmonary arterioles than the resistant breed (113). Thus, factors other than the strength of the acute hypoxic response could have determined the responses both to acute and to chronic hypoxia in the two herds.

There are interesting findings in the rat, where banding the left main pulmonary artery to reduce its lumen markedly reduces the flow to the left lung and the arterial pressure distal to the band. When such rats were exposed to chronic hypoxia, the arteriolar walls in the unbanded right lung were greatly thickened, as expected (86). However, distal to the band, arteriolar walls in the left lung were not thicker than those in the left lung in comparably banded rats at sea level. Presumably, reducing pressure and flow to one lung ''protected'' the arterioles from the chronic hypoxia–related wall thickening. If acute hypoxic vasoconstriction occurred and was sustained in both lungs, but only one lung developed remodeled artrioles, then acute vasoconstriction would not account for all of the chronic remodeling. Presumably high pressure (or flow) is required. A possible scenario is that acute hypoxic vasoconstriction raises pressure, which initiates a series of events, which then lead to wall thickening. However, many factors contribute to acute hypoxic vasoconstriction and to the chronic pulmonary hypertension at altitude, and in each individual and species the factors may differ. Therefore, we must be cautious in assigning a role for the acute response in determining altitude-related pulmonary hypertension.

Thickening of the Pulmonary Arteriolar Wall in Adolescents and Adults

It is generally accepted that chronic hypoxia increases the thickness of the pulmonary arterioles walls. Thickened walls encroach upon and narrow the lumen, with the result that pulmonary arterial pressure and resistance increase. Evidence for such a scenario includes the following:

1. The thickness of the medial layer in the arteries, but much less in veins, increases over time at high altitude in humans and animals, and the increase is relatively greater in the smaller than in the larger arterioles (34,39,62,121). Thus, the greatest smooth muscle response is at the level of the arteriolar resistance vessels, where lumenal narrowing would be most effective in raising resistance.
2. Perhaps even more important, precapillary arterioles that are poorly muscularized or even devoid of smooth muscle at low altitude become muscu-

larized at high altitude and could be particularly effective in obstructing the lung circulation (34,42,62,63,106,121).

3. The development of a thick smooth muscle layer lags some hours or days behind the increase in pressure, consistent with an initial vasoconstriction, followed by a muscularization process (61,87,106).
4. When animals with altitude-induced pulmonary hypertension return to low altitude, both pressure and medial thickness decrease, compatible with a causative role for vascular smooth muscle in the hypertension (42,88).
5. With relief of hypoxia the decrease in pressure and medial thickness occur over many weeks, a time course consistent with smooth muscle atrophy, and not just relief of vasoconstriction (87,88,106).
6. Administration of agents such as heparin and collagen synthesis inhibitors, which are not vasodilators but which inhibit the increase of smooth muscle, also blunt the rise in pressure at altitude (30,31,37,49).
7. Because the perivascular space is limited, increasing wall thickness of some magnitude must encroach upon the lumen (108).
8. All layers, including the adventitia (85,108,109), participate in the increasing wall thickness at high altitude, and where measurements have been made lumenal narrowing has been observed (108,109).
9. Electron microscopy has demonstrated endothelial enlargement (38), muscular evaginations (68), and longitudinal muscle in the intima (121), all of which would act to narrow the lumena of small arteries.

With so many pieces of evidence from so many laboratories, it is perhaps surprising that the case is not airtight for structural remodeling of the vascular wall as the principal mechanism for chronic hypoxic pulmonary hypertension. However, some uncertainty remains. Whether there is (42,62) or is not (68) permanent loss of small arteries with chronic hypertension is hotly debated. Also to be resolved is the question of chronic hypoxic vasoconstriction, where there could be an hypoxia-related increased tonus, which requires more than a few minutes to be relieved by high oxygen. Finally, we are not aware of studies that describe, for various sized vessels, the extent of hypoxic wall thickening that will result in narrowed lumena, nor are there data establishing the relationship (for various sized arterioles) of lumenal narrowing to increased resistance in chronic hypoxia. Thus, increased wall thickening occurs in chronic hypoxia, it may be of great magnitude, and it is strategically located to obstruct the lung circulation, but relevant studies still need to be done.

Regression of Altitude-Related Chronic Pulmonary Hypertension and Vascular Wall Thickening

Measurements in Humans

At issue is the difficult question of whether the pulmonary circulation becomes entirely normal when an individual comes to live at low altitude following a period of residence at high altitude. For 11 healthy male natives of 4540 m going for 2 years to sea level (80), the answer was that the group mean pulmonary arterial pressure fell from 24 to 12 mmHg and calculated pulmonary vascular resistance became normal. Rats (39) and calves (109) with chronic hypoxic pulmonary hypertension brought to low altitude have shown normal resting hemodynamics and normal vasculature by light microscopy.

However, there are clues that following relief of chronic hypoxia the lung circulation may retain abnormal features, depending on the severity of the pulmonary hypertension in hypoxia, the duration of the residence in normoxia, the kind of examination performed, and the subsequent stresses to which the lung circulation is subjected. For example, when a resident at 4540 m with chronic mountain sickness went for 2 months at sea level, he lowered his pulmonary arterial pressure from 62 to 24 mmHg, but not to normal levels. More remarkable is the case of an apparently healthy 15-year-old native of Leadville, Colorado, who by chance was discovered to have mean resting and exercising pulmonary arterial pressures of 44 and 109 mmHg at 3100 m (25) (Fig. 13A). Eleven months after she had moved to sea level, her pressure had fallen to 17 mmHg. Although this pressure at sea level was in the upper normal range, there were two clues that her lung circulation was not entirely normal. First, during exercise her pulmonary arterial pressure rose to 38 mmHg, about double that predicted for her estimated workload. Second, 6 months following her return to Leadville, her resting and exercising pressures at 3100 m were 49 and 72 mmHg. The pressures on return to Leadville in relation to O_2 uptakes (Fig. 13A) indicated that her lung circulation had returned to that originally observed at 3100 m. The findings suggested that 11 months at sea level were not sufficient to restore a completely normal lung circulation as judged by her response to exercise and the challenge of returning to high altitude.

Measurements in Experimental Animals

Particularly in young rats exposed to chronic hypoxia, pulmonary arterial pressures may not become normal even after several months of normoxia (88). Similar but less impressive results have been seen in older rats (88). Light microscopic studies also suggest that chronic high-altitude rats retain a larger than normal number of muscularized arterioles after return to sea level (106). Further, when young rats exposed for only a few days to altitude are returned to sea level for many months and then reexposed to altitude, they develop pulmonary hypertension of greater magnitude and with greater rapidity than do rats that have never had altitude exposure (40,41). These data in rats are reminiscent of a report of Grover et al. (25). Even though the species, the relative ages, and the exposure times are different, the results taken together raise the possibility that once the lung circulation has been exposed to chronic hypoxia, some individuals or species retain the capacity for abnormal responses to subsequent hypoxic stress.

One mechanism for the development of smooth muscle in precapillary vessels was reported by Sobin et al. (106) in a remarkable electron micrographic study in the rat. They found that by 24 hours of either normobaric or hypobaric hypoxia,

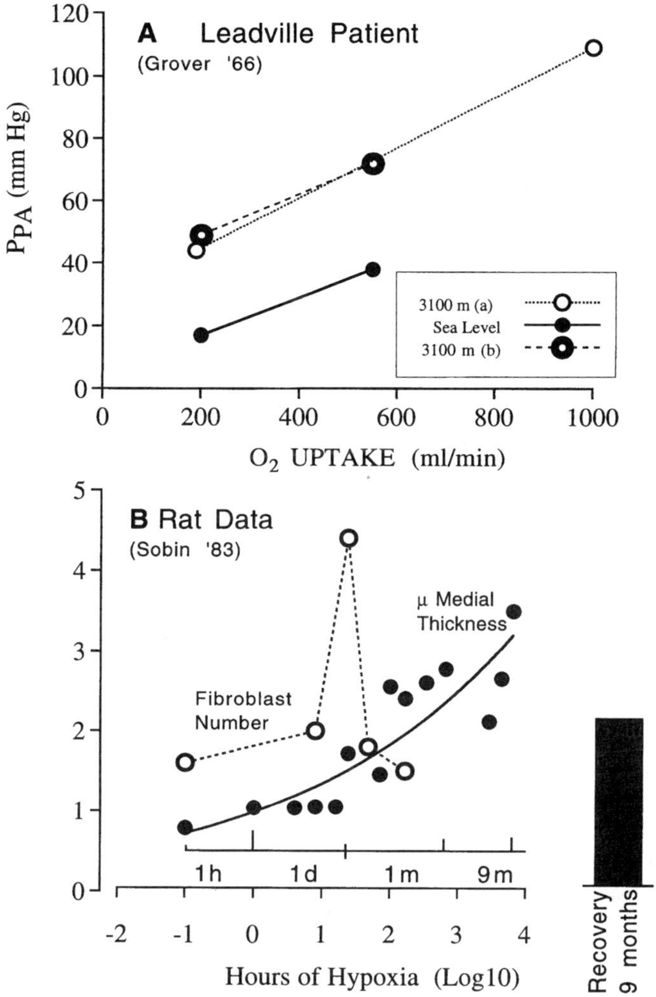

Figure 13 Regression of hypoxic pulmonary vascular changes on removal to low altitude. (A) Pulmonary arterial pressures and blood flows at rest and during exercise in a 15-year-old girl from Leadville, at discovery in Leadville (open circles), following 11 months residence near sea level (filled circles), and 6 months after return to Leadville (partially filled circles). (Adapted from Ref. 25.) (B) Measurements on a logarithmic time scale showing (filled symbols) the development of medial thickness of normally nonmuscular pulmonary arterioles over 9 months at high altitude, and the partial regression of the thickness after 9 months recovery at low altitude. Open circles show the pattern of increase of interstitial fibroblasts neighboring these arterioles for the first part of the high-altitude exposure. (Adapted from Ref. 106.)

precapillary vessels of the diameter (~25 m), which normally had no smooth muscle, began to show smooth muscle within their walls, a consequence of migration and transformation of adventitial fibroblasts. As the smooth muscle continued to increase over 9 months of altitude exposure, so did wall thickness (Fig. 13B). But when some of the rats were returned to sea level for 9 months, smooth muscle was still found in normally nonmuscular arteries (Fig. 13B). The findings were in general agreement with those of Rabinovitch et al. (87), who reported by light microscopy that by day 2 of simulated altitude, smooth muscle began to appear in some arterioles that were normally nonmuscular. They also found that the proportion of these muscularized arterioles increased along with the increase in pulmonary arterial pressure. A subsequent report (88) indicated that following recovery in normoxia, smooth muscle was found in normally nonmuscular arterioles. Taken together, these findings raise the possibility that once very distal, normally nonmuscular arterioles become muscularized at altitude, they may retain some smooth muscle for a very long time after return to sea level.

Origin of the Arteriolar Smooth Muscle

How does the smooth muscle get into arterioles that usually contain none? Two routes, which are not mutually exclusive, have been proposed. The first is that there is direct peripheral extension from normally muscular arteries to progressively smaller and smaller arterioles (40). The other proposed route is that cells that resemble fibroblasts (pericytes, transitional cells, intermediate cells) exist in the in the wall or interstitium surrounding the nonmuscular arteries. These cells increase in number with hypoxia, invade the arteriolar wall, and become transformed into smooth muscle cells (61–63,106). Evidence favoring the scenario are electronmicrographs in rats (106) showing a threefold increase in fibroblast-like cells in the interstitium within 24 hours of hypoxic exposure (Fig. 13B). The decrease in fibroblast number after 24 hours (Fig. 13B) is accompanied by an increase in muscularity within the arterioles, as though there had been migration of fibroblasts from the interstitium to the cell wall, where they were transformed into smooth muscle. Migration and transformation are dynamic processes that cannot be seen in fixed tissues and therefore must be inferred. However, fibroblasts, including those in the pulmonary arterial adventitia, are known to proliferate and migrate in response to several vasoconstrictors and growth factors (76). Fibroblasts are also potentially pleomorphic and are considered sometimes to be precursors to smooth muscle cells (109). If further research confirms that hypoxia causes fibroblast migration and transformation to medial smooth muscle cells, then one would wonder if the process is poorly reversible to account for an occasional slow and incomplete resolution of pressure and muscularization on return to normoxia.

Hypertrophy/Hyperplasia of the Newly Formed Muscle

No matter whether the first smooth muscle cells arrive at the previously nonmuscular arterioles by direct extension along the vascular tubes or by migration of other cells, it seems likely that smooth muscle cells will increase in number and/or size. At issue are the factors that promote the increasing muscularization. Stretch of walls of conduit arteries from rats will rapidly induce DNA synthesis of the pulmonary arterial smooth muscle in 1–3 days (125). In young calves with hypoxic pulmonary hypertension, the ratio of diploid to tetraploid cells in 1-mm-diameter arteries was increased, suggesting that hypertrophy (75) and not only hyperplasia had occurred. Stretch of pulmonary arteries also induces increased production of adventitial matrix within hours (112). Although neither flow cytometry studies to assess ploidy nor in vitro experiments of connective tissue synthesis are reported for the smallest arterioles, available studies raise the possibility that physical factors can influence wall thickness in pulmonary arteries.

Chemical factors are certainly important, as recent work has shown for angiotensin II (AII) (70,71). Earlier literature was clear that inhibiting angiotensin II formation by administering angiotensin-converting enzyme (ACE) inhibitors reduced hypoxic pulmonary hypertension (130). However, equally clear was the fact that rat lung angiotensin content and ACE activity decreased with chronic hypoxia (74). It was paradoxical that angiotensin was important in producing hypoxic pulmonary hypertension, and yet its concentration decreased in hypoxic lung. The resolution of the paradox was that loss of angiotensin II from the parenchyma decreased its content in whole lung, yet there was an impressive increase in angiotensin II activity and expression in the arterioles (70) in parallel with the developing chronic pulmonary hypertension. Of particular interest was that the finding was limited to the smallest arterioles as determined by immunohistochemistry and in situ hybridization. Further, the development of new muscle was substantially blocked by ACE or by AII receptor inhibition. AII is only one participant in a cascade whereby paracrine and autocrine growth factors (22) enhance lung arteriolar muscularity.

Thickening of the Pulmonary Arterial Wall in the Newborn

A Unique Circulation in Structure and Function

Abundant evidence exists that the newborn lung circulation differs from that in the adult in having, e.g., (1) cuboidal rather than flattened endothelial cells (38), (2) increased medial arteriolar muscularity, (3) increased arteriolar adventitial thickness, (4) reduced cross-sectional area of the microvascular bed, (5) an augmented vasoconstrictor response to hypoxia and other vasoactive agents, and (6) an augmented capacity to develop chronic pulmonary hypertension in response to a variety of stimuli (88,107,114). In addition, the newborn at sea level demonstrates its restricted lung vascular bed when it stands and begins to walk only a few minutes after birth. Pulmonary arterial pressures increase from 47 mmHg at rest to 66 mmHg during walking (89). Mild exercise in older calves (89), and as discussed below in adult humans, causes little increase in pressure. Also, newborn cattle develop more quickly than do adults a blunted response to pulmonary vasodilators at high-altitude exposure (19). The newborn lung circulation clearly differs from that of the adult.

A Unique Capacity for Smooth Muscle Replication

Vascular smooth muscle replication rates are extremely high (up to 80% in 24 hours) in the fetus (rat aorta) and decrease toward birth (14). Further, early in gestation the smooth muscle replicates in an autocrine fashion, independent of stimulation by known growth factors. As the fetus matures and replication slows, the cells become dependent on growth factors from other cells. At birth, for both the rat and the calf, smooth muscle replication is still higher than in the adult (14,108). When the newborn calf is placed in an hypoxic environment, smooth muscle replication increases. Of particular interest, at least for the large pulmonary arteries, is that not all medial cells appear to increase their replication rate, but mitosis occurs in rather selective populations within the media (126). These observations raise the possibility that particular smooth muscle phenotypes, which might persist from fetal life, are retained in the newborn and are stimulated by the hypoxic environment to an active replicative state. Preliminary data from cell cultures indicate a remarkable developmentally related variation in capacity for pulmonary arteriolar smooth muscle replication where the replication rates are high in the term fetus, which exceed the normoxic newborn, which exceed the adult (129). Perhaps most remarkable of all is that the hypoxic newborn has even higher rates than those of the term fetus. The preliminary data further suggest that a protein kinase C (PKC) pathway, which has been stimulated by chronic hypoxia in the newborn calf pulmonary arteries, is uniquely related to the capacity for smooth muscle cell replication (129).

Replication of Fibroblasts and Matrix Production in the Neonate

As noted above, fibroblasts lie in the adventitia of the pulmonary arterioles in close relation to the smooth muscle cells. Fibroblasts too are stimulated in chronic hypoxia (16,85). Further as in the case of the smooth muscle, specific subpopulations of fibroblasts proliferate, where the rates of proliferation are greatly enhanced in the hypoxic newborn as compared to the adult (16). Both fibroblasts and the smooth muscle cells show increased production of matrix in the chronically hypoxic calf, and there may well be interaction of the fibroblast and matrix with the smooth muscle cell. In a sense, then, the pulmonary circulation at birth is a ''loaded gun'' with a large amount of arteriolar smooth muscle poised to constrict and to replicate. For the neonate, chronic hypoxia may, in fact, pull the trigger.

IV. Hemodynamics in Exercising Adult Humans

The foregoing discussion has been largely limited to rest, but life requires physical exertion. Based on the preceding sections, the following discussion assumes that high altitude residence increases resistance to blood flow through the lungs, whether the increase might be from hypoxic tone, loss of vasodilating capacity, blood viscosity, and/or vascular remodeling. Exercise is a stress that not only increases pulmonary blood flow, but also introduces stimuli (decreased mixed venous Po_2, increased neural activity, augmented release of vasoactive substances), which could affect the lung circulation (93). The question is, therefore, whether altitude-related changes in the lung circulation alter its function during exercise, and if so, what are the consequences? To approach this question we must first look at the lung circulation in normal persons exercising at sea level.

A. Exercise in Normal Subjects at Sea Level

A recent extensive review (93) has suggested that at sea level the lung circulation is markedly affected by rising downstream pressures in the left heart. Review of published data for men and women performing supine and sitting cycle exercise of varying intensities indicated:

1. The pulmonary wedge (left atrial) pressure rose with increasing oxygen uptake and pulmonary blood flow.
2. Pulmonary arterial pressure correlated so strongly with wedge pressure that nearly 90% of the variation in the pulmonary arterial pressure could be accounted for by the variation in wedge pressure.
3. The pulmonary gas-exchanging surface, the arterioles, capillaries, and venules, are all thin walled and distensible in humans and other mammals.
4. The increasing microvascular pressures with increasing exercise distend and recruit the microcirculation, which results in increasing diffusion surface area and improved pulmonary oxygen transport.

Inquiry into the mechanism of the increased wedge pressure (in supine and sitting subjects) suggested that it arose from the left heart, possibly largely from the inherent limitations to ventricular filling by the normal pericardium (93). The evidence that implicated pericardial limitation was:

1. The wedge pressures correlated strongly with the left ventricular end diastolic pressure at rest and during exercise, suggesting the wedge pressure was a reliable indicator of left ventricular filling pressure.
2. With progressively heavier exercises, the left ventricular end-diastolic pressure rose progressively, but the end-diastolic volume did not. This suggested that left ventricular filling was on the steep portion of the pressure-volume curve.
3. Stroke volume related to end-diastolic volume and not to end-diastolic pressure, confirming that during moderate and heavier exercises filling of the left ventricle was limited.
4. Pericardial pressure-volume curves from animal experiments were compatible with a close application of a poorly compliant pericardium to the ventricles.
5. In animals, removal of the pericardium led to increased ventricular volumes and, during exercise, increased stroke volumes and oxygen uptakes.

Taken together, the data suggested that the left heart and particularly the pericardium limited exercise-related ventricular filling, thereby raising filling pressures, which increased pulmonary wedge pressure and pulmonary arterial pressure, dilated the thin walled pulmonary microcirculation and contributed importantly to the increased pulmonary O_2 diffusing capacity (93).

B. The Pulmonary Circulation During Exercise on Acute Exposure to High Altitude

For persons arriving at high altitude, the question is whether acute hypoxia alters the pulmonary circulatory response to exercise. On the one hand, hypoxic pulmonary vasoconstriction would be expected to raise resting resistance. On the other hand, if wedge and pulmonary arterial pressures increased during exercise, the increasing intralumenal pressures could oppose vasoconstriction of delicate thin-walled arterioles (10,57). In dog lobes, as the left atrial pressure was progressively increased, the acute hypoxic pulmonary pressor response diminished, and became extinguished at about 25 mmHg (10), compatible with the concept that an increased microvascular pressure had opposed the hypoxic pulmonary vasoconstriction.

Human data (73) (Fig. 14) indicate that for a given pulmonary flow, during exercise in subjects acutely breathing hypoxic gas mixtures, the pulmonary arterial pressures (P_{PA}), wedge pressures (P_W), and driving pressures ($P_{PA} - P_W$) were not different from those during exercise in normoxia. The administration of almitrine, a pharmaceutical agent that augments the ventilatory response to hypoxia and may augment acute pulmonary vasoconstriction, had no effect on the exercise pressures (Fig. 14). Also, data reported by Wagner et al. (122) indicated that during an acute exposure to high altitude in a chamber, exercise pulmonary arterial pressures and wedge pressures, as related to flow, were similar to those observed in the same subjects at sea level (Fig. 14). The concept arises that acute hypoxia, by altering neither the rise in wedge pressure during exercise nor the relation of wedge to pulmonary arterial pressure, has little influence on the control of the exercise pulmonary circulation in normal humans. If so, one might speculate that a microcirculatory pressure rise during exercise opposes and offsets hypoxic vasoconstriction in the normal, thin-walled, lung arterioles.

C. Exercise and the Pulmonary Circulation During Altitude Acclimatization

As discussed above, the transition from acute hypoxic pulmonary vasoconstriction to structurally altered lung arterioles begins during the first few days at high altitude and results in higher resting pulmonary arterial pressure and resistance. Thus, one might expect that remodeling of the arterioles could affect the pressure-flow relationships in the lung during exercise. Human pulmonary arterial pressures during exercise through the first 3 days at altitudes of 3900 and 4300 m are available from Kronenberg et al. (50) and Vogel (118), respectively. These measurements show that for a given flow, the altitude exposure is associated with a higher pulmonary arterial pressure during exercise (Fig. 15). There was a parallel upward shift in pressure-flow diagram, which, by analogy with the Starling resistor (for review, see

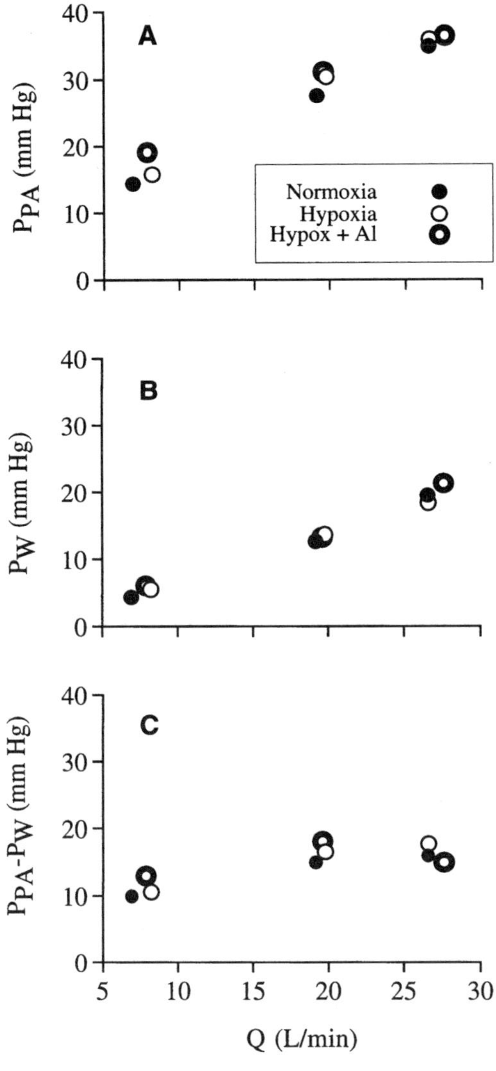

Figure 14 Measurements against blood flow in humans at rest and during exercise in six humans breathing air (normoxia, filled circles), during acute altitude exposure (unfilled circles), and during altitude exposure following the administration of an agent (almitrine, hypox + al, partially filled circles), which putatively potentiates hypoxic pulmonary vasoconstriction. (A) Mean pulmonary arterial pressure. (B) Pulmonary wedge pressure. (C) Pressure gradient from artery to wedge. (Data from Refs. 73,122.)

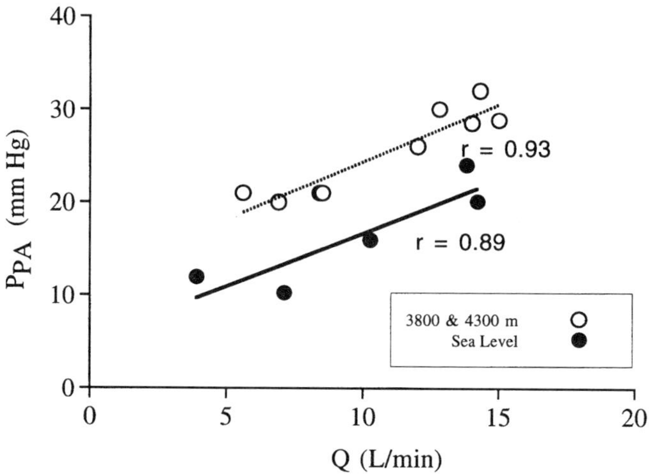

Figure 15 Pulmonary arterial pressures and flows in healthy men at rest and during exercise at low altitude (filled circles) and after 1–3 days at high altitude. Correlation coefficients are shown. (Data from Refs. 50,118.)

Ref. 81) may be interpreted as an increase in capillary resistance. Unfortunately, wedge pressures were not reported in these studies, but one expects the site of the increased resistance to be in the microcirculation.

D. Lung Circulation in Newcomers to Extreme Altitude, Operation Everest II

Important insight into how chronic hypoxia increased pulmonary vascular resistance and affected the lung circulation was obtained by making sequential measurements in healthy individuals over time as they were taken to extreme, simulated high altitude in Operation Everest II (OE II) (29,43,94). Over nearly 7 weeks, healthy men were progressively decompressed in an altitude chamber to the equivalent elevation of Mt. Everest. Pulmonary hemodynamics were measured at sea level, again at a barometric pressure of 347 mmHg (~6100 m), at 282 mmHg (~7620 m), and at 253 mmHg (8848 m, not shown). At rest and during exercise, pulmonary arterial pressures increased with increasing altitude (Fig. 16A). Mean resting and exercise pulmonary arterial pressures for the subjects having measurements made at 282 mmHg reached 34 and 54 mmHg, respectively. At comparable pulmonary blood flows, resting and exercise pulmonary arterial pressures at sea level during normoxia and acute altitude exposure were only approximately 15 and 20 mmHg, respectively. Thus, normal subjects exposed chronically to extreme altitude developed moderately severe pulmonary arterial hypertension.

Wedge pressures measured during exercise at high altitude were less than at sea level (Fig. 16B). With increasing exertion, the blunted rise in wedge pressure

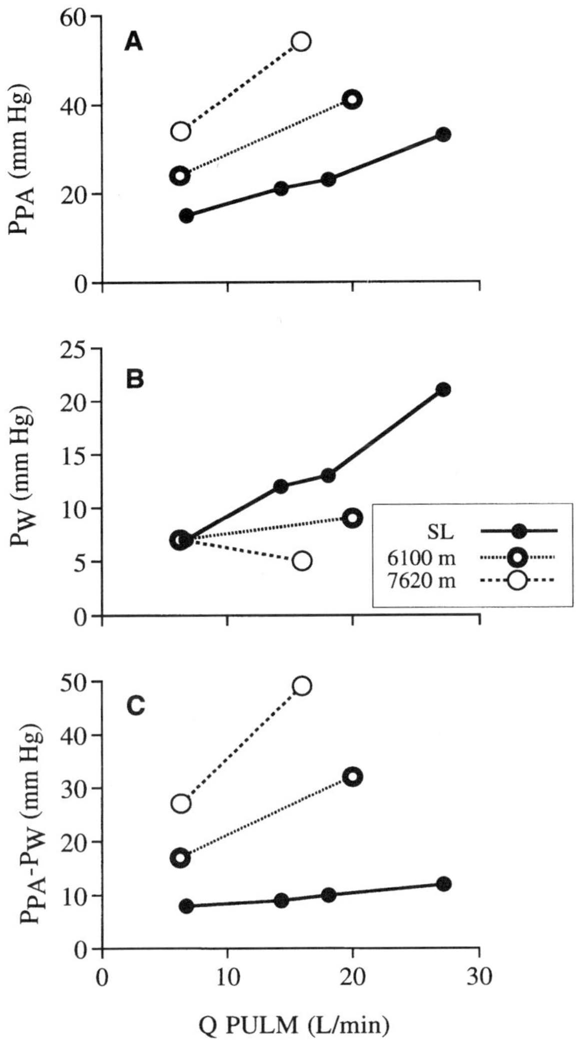

Figure 16 Pressure measurements against flow from Operation Everest II in normal residents of sea level (filled circles), after 3–4 weeks at 6100 m (partially filled circles), and after 6–7 weeks at approximately 7620 m (open circles). (A) Pulmonary arterial pressure. (B) Wedge pressure. (C) Pressure gradient from pulmonary artery to wedge. (Data from Refs. 29,92,94.)

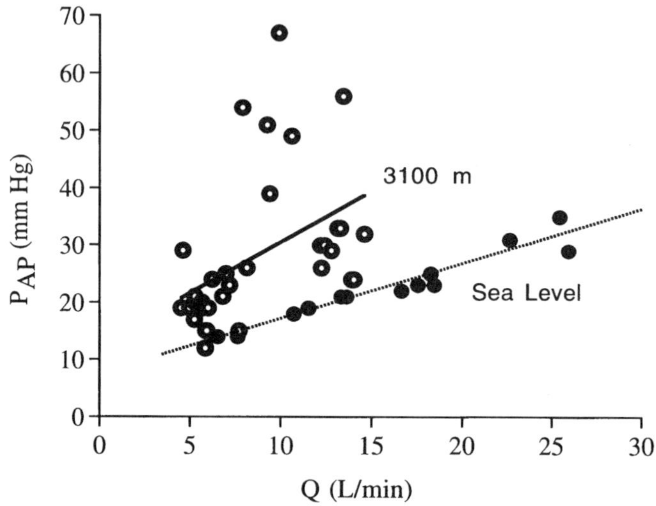

Figure 17 Variability in pulmonary arterial pressure and flow at rest and during exercise in young adult residents of Leadville (partially filled circles), compared to group mean values of normal persons at sea level (filled circles). (Data from Refs. 92,117.)

contributed strongly to the increase in the pulmonary artery to wedge pressure gradient ($P_{PA} - P_W$) at altitude. The low wedge pressures during exercise at altitude differed from the increasing wedge pressure with exercise in normoxic subjects at sea level as well as in those exposed to acute hypoxia or acute altitude (compare Fig. 14 and 16B). These observations raised the possibility that time at altitude allowed the development of some factor that inhibited the rise in wedge pressure with altitude.

At sea level the pressure gradient driving flow through the lung ($P_{PA} - P_W$) increased slightly from rest to exercise as the blood flow tripled (Fig. 16C). Therefore, pulmonary vascular resistance fell by more than 50% (29). However, at the altitudes shown the higher exercise-induced pulmonary arterial pressures and the lower wedge pressures resulted in large increases in driving pressure. The $P_{PA} - P_W$ pressure gradient during exercise increased nearly fivefold from sea level to the 7620 m altitude (Fig. 16C). The slope of the pressure-flow line increased, compatible with an increasing precapillary resistance to blood flow (81). Oxygen inhalation for 5 minutes had little effect on pulmonary vascular resistance. Taken together, the findings are compatible with structural changes in the walls of the lung arterioles at these very high altitudes (29).

E. Pulmonary Circulation During Exercise in High-Altitude Natives

Reports from Peru in 1962 and 1963 (79,80) indicated that more severe pulmonary hypertension was induced by relatively mild exercise in high altitude (4540 m) than

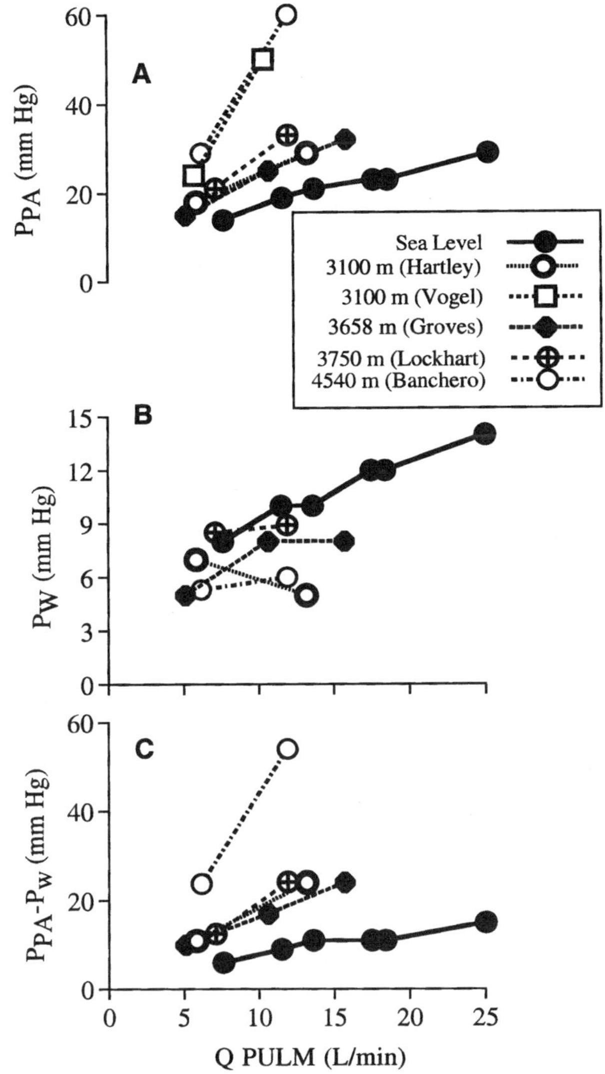

Figure 18 Group mean resting and exercise pressures and flows from normal persons native to various altitudes. (A) Pulmonary arterial pressure. (B) Pulmonary wedge pressure. (C) Pressure gradient from artery to wedge. (Adapted from Refs. 8,44,80,92,117.)

in sea level residents. A more detailed study from the same group was published in 1966 (8). Lockhart et al. (52) reported data from La Paz, Bolivia (3750 m). Exercise pressures in North American residents of 3100 m were reported by Vogel (117) and Hartley (36). Data from Tibetan residents of Lhasa (3600 m) were reported by Groves et al. (28). The exercise pulmonary arterial pressures from all altitudes were clearly elevated above those at sea level (Fig. 17A). With increasing altitude of residence, the exercise pulmonary arterial pressures tended to increase, with the highest pressures being seen at the greatest altitude of 4540 m. In addition, for the highest pressures, the pressure-flow lines (Fig. 18A) were not parallel to that at sea level, but the slopes were greater, suggesting increased precapillary resistance (81). The collected data also show great variability in exercise pulmonary arterial pressures. For example, in Leadville at 3100 m, the high school students studied by Vogel et al. (117) had much higher pressures than the 29-year-old men studied by Hartley et al. (36). Even when age was minimized as a factor by just considering the high school students (117), interindividual variability was very great (117), where for example one student had an exercise pressure of 23 mmHg, while for a comparable exercise intensity in another student, the pressure reached 68 mmHg (Fig. 17). In this cohort of students, much of the variation in the exercise pressure could be attributed to the variation in the resting pressure and resistance, where the higher the resting pressures or resistances, the greater the exercise pressures and resistances (respectively, $r = 0.6$ and 0.7, $p < 0.01$). Another factor related to the observed variability in exercise pressures was the population being studied. For example, young Tibetan men at 2658 m had relatively low exercise pulmonary arterial pressures (Fig. 18A) (28). The authors proposed that populations long resident at high altitude developed relatively low resistance lung circulations as an adaptation to hypoxia (see Sec. VI).

An important observation from the reviewed data was that the exercise wedge pressures at sea level tended to be higher than those at altitude (Fig. 18B). (Factors relating to the changes in wedge pressure during exercise at sea level versus those at altitude are considered in Sec. VI. F.) The lower exercise wedge pressures during exercise at altitude than at sea level contributed to the wider pressure gradient from pulmonary artery to wedge ($P_{PA} - P_W$) observed at altitude (Fig. 18C). Thus, the higher exercise-related pressure gradient at high altitude than at sea level resulted both from higher pulmonary arterial pressures and lower wedge pressures. At 4540 m the calculated vascular resistance did not fall with exercise and tended to increase (8). These data suggest that the regulation of the lung circulation is likely different for residents at altitude than for those at sea level (see Sec. IV.F).

If so, the altered regulation of the lung circulation during exercise in altitude residents may be due to their increased pulmonary vascular smooth muscle, as suggested by several pieces of evidence. For example, breathing high oxygen acutely lowered the exercise resistance in Leadville natives (Fig. 19), even though it had little effect at rest. Apparently during exercise there developed an increased vascular tone, which could be reversed by high oxygen. Further insight into the mechanism

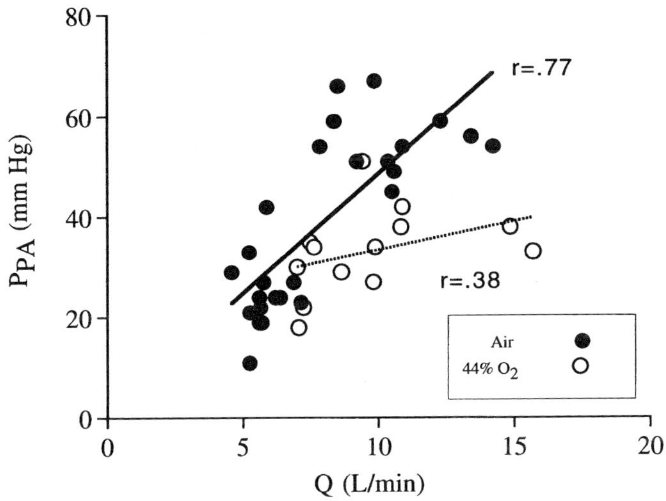

Figure 19 Decrease during breathing 44% O_2 in exercise pulmonary arterial pressure in normal young adult residents of Leadville. (Data from Ref. 117.)

of altered control was provided by Lockhart et al. (51,52) in a remarkable study in natives of La Paz (3750 m). Exercise increased both pressure and resistance, and assuming blood flow was equally divided between the two lungs, they could estimate resistance to flow through one lung. They then approached the question of whether increased pressure per se could dilate the lung circulation in high altitude natives. To increase pressure they forced the entire cardiac output through one lung using a balloon catheter (Fig. 20). During exercise with all flow diverted to one lung, higher pressures were achieved and the lung vascular resistance fell, even though there was more severe hypoxemia with occlusion. The results were compatible with a pressure-induced dilation of the arteriolar vessels. In a second group of subjects, high oxygen breathing also reduced pulmonary vascular resistance during exercise, compatible with relief of hypoxic vasoconstriction (Fig. 20). From consideration of all the data available to them, the authors concluded that both alveolar and mixed-venous oxygen tensions contributed to the pulmonary vascular tone during exercise. They also concluded that raising the pressure could dilate the lung circulation of the high-altitude native. Of interest, the fall in exercise resistance with balloon occlusion of one main artery was comparable to that observed during oxygen breathing without balloon occlusion (Fig. 20). Apparently either oxygen breathing or high pressure during air breathing could lower exercise resistance, though in neither case to the sea level normal values. The results are compatible with the concept that increased arteriolar smooth muscle could account for some of the differences between sea level and altitude residents in lung vascular function.

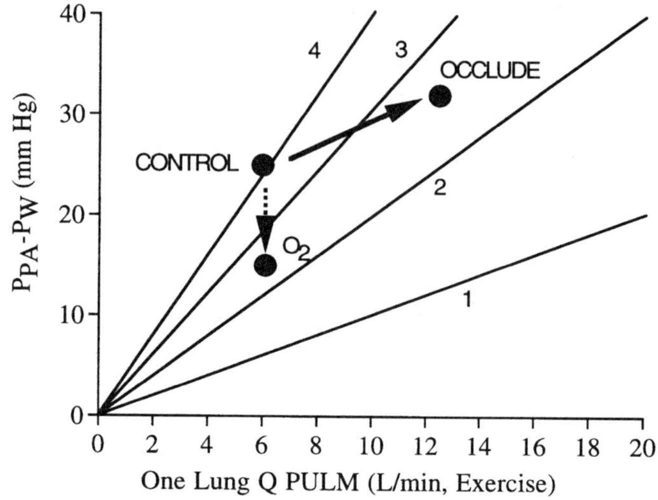

Figure 20 Pressure gradient during exercise from pulmonary artery to wedge against the flow through one lung in normal residents of 3750 m in La Paz. Shown passing through the origin are isopleths of resistance from 1 through 4 units, as labeled. Estimated exercise flow through one lung did not change when subjects changed from air (control) to high O_2 (O_2), but pressure and resistance fell. Exercise with unilateral occlusion of one main pulmonary artery (occlusion) raised flow and pressure to the contralateral lung, but resistance fell. (Data from Refs. 51,52.)

F. Factors Related to the Lung Circulation During Exercise: Normoxia, Acute Hypoxia, and Chronic Hypoxia

For a given pulmonary blood flow during exercise, pulmonary hemodynamics in normal sea level residents are markedly different from from the hemodynamics in subjects exposed to chronic hypoxia (altitude residents and subjects in OE II), summarized in Table 1. As indicated in the previous sections, hypoxia alone does not account for the difference, because subjects exposed to acute hypoxia, i.e., breathing hypoxic gases, or exposed briefly to hypobaria in a chamber have exercise hemodynamics no different from sea level residents (Table 1). At issue are the factors that cause pulmonary hemodynamics to become altered over time at altitude.

A key issue is the rise in wedge pressure with exercise at sea level but not at altitude (see Secs. IV.D,E). A review of sea level exercise wedge pressure measurements (93) found them to relate well to left ventricular end diastolic pressures, indicating that the elevations with exercise were not artifacts, but represented increased ventricular filling pressures. Other factors proposed (93) to raise wedge and left atrial diastolic pressure during exercise at sea level included (1) increased left ventricular filling volume and increased afterload pressure, both of which would in-

Table 1 Changes with Exercise in Hemodynamic Variables That Affect the Lung Circulation[a]

	Normoxia	Hypoxia, Acute	Hypoxia, Chronic
Wedge pressure (PW)	+++	+++	±
Pulmonary arterial pressure (PPA)	+++	+++	+++++
Pressure gradient (PPA-PW)	±	±	+++++
Stroke volume	+++	+++	+

[a] Summary for normal subjects living at sea level (normoxia), for subjects breathing hypoxic mixtures or having brief exposure to hypobaria in an altitude chamber (hypoxia, acute), and for residents at high altitude or subjects at extreme altitude in Operation Everest II (hypoxia, chronic).
Shown are approximate changes with exercise relative to normal resting sea level values. Residents at high altitude do not include subjects from the Tibetan population (see Sec. VI).

crease ventricular filling pressure via the Starling mechanism, (2) pericardial limitation of left ventricular stretch during the filling phase, and (3) the presence of a pressure gradient across the mitral valve with very high exercise blood flows. On going from rest to very mild exercise, both stroke volume and left ventricular filling pressure rose, consistent with operation of the Starling mechanism. On going from mild to moderate and heavy exercise, wedge and left ventricular end-diastolic pressures rose, but stroke volume did not increase further. For example, in Operation Everest II at sea level, stroke volumes reached a plateau at about 150 mL, considered to represent pericardial limitation of ventricular filling. The literature review suggested that in many (but not all) subjects at sea level, pericardial limitation to left ventricular filling may have accounted for much of the rise in left atrial pressure during exercise at moderate intensities. If so, an increase in stroke volume with exercise would be necessary for the full increase in wedge pressure to occur.

Key issues for the present chapter are the factors that cause the wedge pressures to be less with chronic hypoxia than with normoxia or acute hypoxia. Considering the measurements from OE II, we probably can rule out left ventricular dysfunction at altitude as causing the blunted increase in wedge pressure with exercise, because cardiac function was preserved in these subjects (94,110) and, further, left heart failure should have raised, not lowered, wedge pressure. End-systolic, end-diastolic, and stroke volumes were smaller in these subjects during chronic hypoxia than at sea level. Heart rates for a given cardiac output were higher than at sea level, and stroke volumes were smaller. At 7620 m altitude, stroke volumes of less than 150 mL were not associated with elevations of wedge pressure. Possibly these smaller stroke volumes [and smaller end-diastolic volumes (110)] represented ventricular volumes not at the limits of pericardial compliance. Based on data from Operation Everest II, a wedge pressure versus stroke volume curve may be drawn to approximate for these subjects the compliance characteristics of their pericardia (Fig. 21). An hypothesis to be tested, therefore, is that the wedge pressure will not

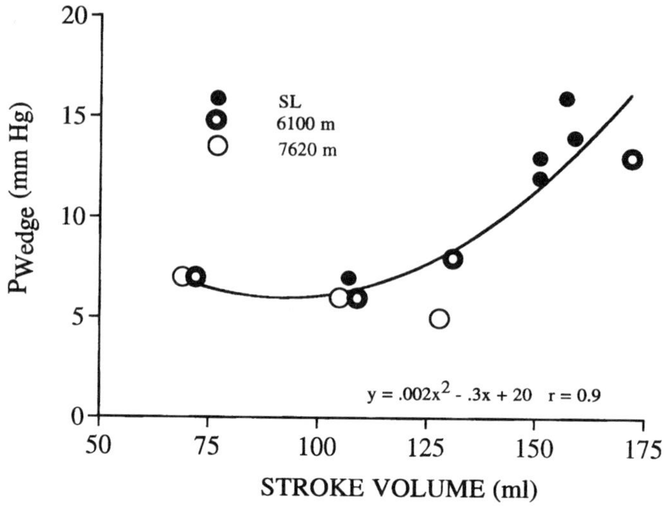

Figure 21 Wedge pressure against stroke volume from Operation Everest II. A regression line through all the data and the regression equation are shown. (Data from Refs. 29,94.)

rise in altitude-exposed subjects until left ventricular filling becomes restricted by the pericardium.

One wonders why the stroke volumes during exercise are less in altitude than in sea level residents. Exercise stroke volumes may become reduced within a few days of exposure to moderate altitude (26). The reduction in stroke volume at altitude accompanies a reduction in plasma volume and has been prevented by preventing plasma volume reduction (26). Whether or not the increasing pulmonary vascular resistance contributes to the reduction in exercise stroke volume during chronic altitude exposure is not known. However, one might speculate that contributing to the reduced stroke volume over time at altitude are several factors, including tachycardia from increased sympathetic tone, reduced plasma volume, and possibly increased pulmonary vascular resistance.

V. Schema and Unanswered Questions

A. Schema

As indicated from the above considerations, the behavior of the lung circulation at altitude, either at rest or during exercise, depends on the vascular remodeling which occurs over time. The remodeling results in both structural and functional alteration of the lung circulation. That alteration occurs primarily in lung arterioles and may be said to involve at least four general steps (131):

1. Detection by the vessel wall of altered physical and hemodynamic forces
2. The relay of the signal to the endothelial, smooth muscle, and fibroblast cells involved in the process of remodeling
3. Synthesis of substances that may initiate and promote cell division and hypertrophy and other substances that oppose these changes
4. Alteration and thickening of the composition of vessel wall cells and matrix

These general steps may be incorporated into a tentative and rather simplified schema (Fig. 22). The schema, though incomplete, indicates that all cell types within the vascular wall participate in the altitude-related changes, that there are factors such as hypertrophy, proliferation, matrix synthesis, and chemotaxis, which contribute to thickening of the cell wall, and other factors that oppose thickening (11). Further, there are general influences that contribute to the interindividual and inerspecies variability of the vascular remodeling. The most important of these influences seem to be the postnatal age of the individual, the duration of the hypoxic exposure, and genetic susceptibility. Examples of chemical mediators are indicated.

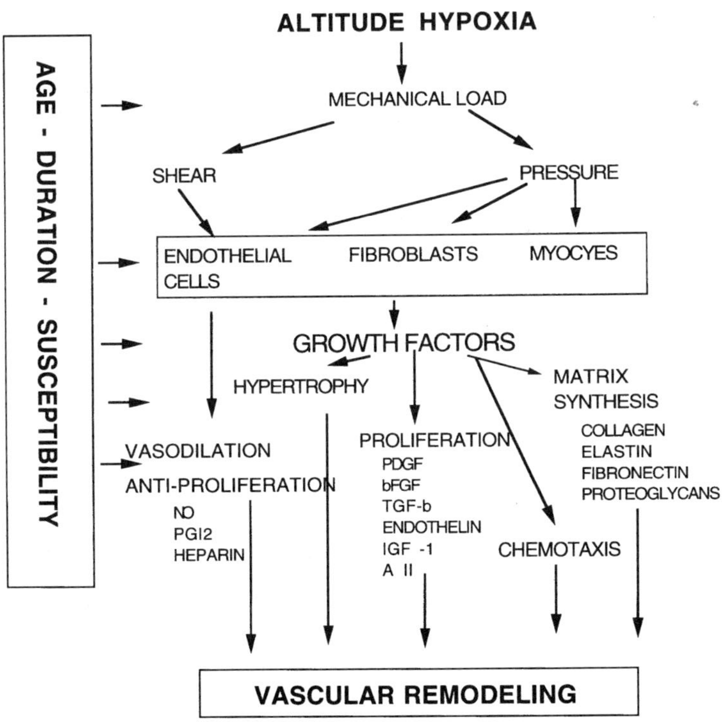

Figure 22 Hypothetical schema for steps by which chronic hypoxia at altitude leads to remodeling of the pulmonary arterioles. (Adapted from Ref. 11.)

B. Unanswered Questions

The schematic overview in Figure 22 simply illustrates likely pathways for vascular remodeling but does not portray the mechanisms involved in these pathways. It is fair to say that for each of the arrows shown, the mechanisms are not understood or are, at best, poorly understood. The mechanisms that control alterations in the lung circulation from chronic hypoxia remain important physiological and clinical questions.

Questions also remain as to how the remodeled lung circulation behaves in humans at altitude. Does the increased pulmonary hypertension in altitude residents distend the thickened arterioles, or is the pressure effect offset by hypoxic vasoconstriction? At sea level each mmHg intralumenal pressure rise increases the diameter of the microcirculatory vessels by approximately 2% (for review, see Ref. 93). The result is dilation and recruitment, which contribute to an increased diffusing capacity. We need to know to what extent increasing intralumenal pressure increases diameter in lung vessels at altitude. If the microcirculation is not well distended during exercise in altitude residents, is diffusing capacity impaired thereby? Further, a role specifically during exercise has been proposed (51,52) for pulmonary vasoconstriction from hypoxic alveoli and hypoxemic mixed venous blood. One wonders if exercise alters the vasoconstriction/vasodilation ratio toward constriction. Clarification is needed for roles of structural changes in the arteriolar walls, the capacity for passive stretch, and induced vasomotor tone. Finally, does the increased peripheral lung resistance constitute a bottleneck for cardiac output in the high-altitude dweller, and is this a factor in the failure of wedge pressure to rise with exercise? These and other questions continue to emphasize the need for further clinical and basic research.

VI. Teleology

> [I]f you take a just little piece of lung and make it hypoxic, it will constrict (its blood vessels) very strongly—maybe 80% of the blood flow will be diverted away from that region of lung. But it makes hardly any difference to pressure because it's such a small piece. If you climb a mountain, however, and have the whole lung hypoxic . . . you can't divert any blood flow from one region of the lung to another. All you can do is increase pressure (56).

One wonders what useful purpose is served by the increased pulmonary arterial pressure that, with few exceptions, is found in people and animals living at high altitude. Of course, we are not sure of the answer to this question. When there is a relatively large vertical distance from the bottom to the top of the lung, as in resting upright adult humans or in cattle, the high arterial pressure may make the lung blood flow more even by increasing flow to the top of the lung, possibly improving gas exchange (27). But one can hardly imagine that such an advantage would accrue to the rat, mouse, or guinea pig, which have very small lungs. With the resting pulmonary arterial pressure already elevated at altitude, one wonders what further advantage would result from a large increase during exercise.

Perhaps large increases in pulmonary arterial pressure are not advantageous and may even be hurtful. Most populations of mammals, including humans, which have been at high altitude for thousands of years have relatively little pulmonary hypertension. For example, Tibetans, a population with perhaps the longest residence at altitude, have relatively little exercise pulmonary hypertension (28). The yak, a member of the bovine species in the Himalayas, has low resting pulmonary artery pressure and thin-walled pulmonary arterioles (20,35). A related species, the llama, which is native to the Andes, has a small but significant rise in pressure on going from low to high altitude (27).

That harmful pulmonary hypertension occurs at high altitude is most clearly seen in infants and young adults, particularly from poorly adapted populations. For example, fatal pulmonary hypertension on the Tibetan plateau was observed in infants born to Chinese migrants of Han ancestry and but rarely in infants of Tibetan parentage (40). Chronic pulmonary hypertension leading to right heart failure was seen in young Indian soldiers stationed high in the Himalayas (2). These findings in humans and those discussed above in young rats and newborn calves reinforce the susceptibility of the young to high-altitude pulmonary hypertension as originally reported from the Andes. The concept is that in the perinatal period there is potential for rapid proliferation of vascular cells and increased production of matrix (109). There are in the newborn muscular arterioles well-developed mechanisms for hypoxic pulmonary vasoconstriction—mechanisms likely needed for a healthy fetus which are still present at birth. Thus, the pulmonary circulation of the newborn is unfortunately positioned to be susceptible to the hypoxia of high altitude. One can imagine that high-altitude species may need enough chronic hypoxic pulmonary hypertension to get them through the fetal and newborn periods, but not enough to cause right heart failure in infancy and adolescence.

References

1. Adams FH, Lind J. Physiologic studies on the cardiovascular status of normal newborn infants (with special reference to the ductus arteriosus). Pediatrics 1957; 19:431–437.
2. Anand IS, Chandrashekhar Y. Subacute mountain sickness syndromes: role of pulmonary hypertension. In: Sutton JR, Coates G, Houston CS, eds. Hypoxia and Mountain Medicine, eds. Burlington, VT: Queen City Printers, 1992:241–251.
3. Archer SL, Weir EK. Mechanisms in hypoxic pulmonary hypertension. In: Wagenvoort CA, Denolin H, eds. Pulmonary Circulation: Advances and Controversies. Amsterdam: Elsevier, 1989:87–113.
4. Archer SL, McMurtry IF, Weir EK. Mechanisms of acute hypoxic and hyperoxic changes in pulmonary vascular reactivity. In: Weir EK, Reeves JT, eds. Pulmonary Vascular Physiology and Pathophysiology, New York: Marcel Dekker, 1989:241–290.

5. Arias-Stella J, Saldana M. The muscular pulmonary arteries in people native to high altitude. Conference on the Normal and Abnormal Pulmonary Circulation June 13–16, 1962. Med Thorac 1962; 19:292–301.
6. Arias-Stella J, Recavarren S. Right ventricular hypertrophy in native children living at high altitude. Am J Pathol 1962; 41:55–64.
7. Ashman R, Byer E. The normal human ventricular gradient I. Factors which affect its direction and its relation to the mean QRS axis. Am Heart J 1943; 25:16–35.
8. Banchero N, Sime F, Peñalosa D, Cruz J, Gamboa R, Marticorena E. Pulmonary pressure, cardiac output, and arterial saturation during exercise at high altitude and at sea level. Circulation 1966; 33:249–262.
9. Barcroft J. Researches on Prenatal Life. Oxford: Blackwell Scientific Publications, 1946.
10. Benumof JL, Wahrenbrock EA. Blunted hypoxic pulmonary vasoconstriction by increased lung vascular pressures. J Appl Physiol 1975; 38:846–850.
11. Bishop JE, Reeves JT, Laurent GJ. Pulmonary Vascular Remodelling. London: Portland Press, 1995.
12. Capedehourat EL, Morera V, Nanclares A, Cotino LE, Coppola A, Martinez A. Estudio electrocardiografico. In: Estudios sobre la "Biologia del Hombre de Altitud." Buenos Aires: Ministerio de Justica e Instrucción Pública, 1937:235–263.
13. Civin WH, Edwards JE. Postnatal structural changes in intrapulmonary arteries and arterioles. Arch 1951; 51:192.
14. Cook CL, Weiser MCM, Schwartz PE, Jones CL, Majack RA. Developmentally timed expression of an embryonic growth phenotype in vascular smooth muscle cells. Circ Res 1994; 74:189–196.
15. Cournand A. Measurement of the cardiac output in man using right heart catheterization, description of the technique, discussion of the validity and of place in the study of the circulation. Fed Proc 1945; 4:207.
16. Das M, Dempsey EC, Stenmark KR. Unique growth properties of pulmonary arterial adventitial fibroblasts from neonatal calves with pulmonary hypertension. Am J Res Crit Care Med 1995; 151:A733.
17. Dawes GS, Mott JC, Widdicombe JG, Wyatt DG. Changes in the lungs of the newborn lamb. J Physiol 1953; 121:141–162.
18. Duke H, Killick EM. Pulmonary vascular responses of isolated perfused cat lungs to anoxia. J Physiol 1952; 117:303–316.
19. Durmowicz AG, Orton EC, Stenmark KR. Progressive loss of vasodilator responsive component of pulmonary hypertension in neonatal calves exposed to 4570 m. Am J Physiol 1993; 265:H2175–2183.
20. Durmowicz AG, Hofmeister S, Kadyraliev TK, Aldashev AA, Stenmark KR. Functional and structural adaptation of the yak pulmonary circulation to residence at high altitude. J Appl Physiol 1993; 74:2276–2285.
21. Fasules JW, Wiggins JW, Wolfe RR. Increased lung vasoreactivity in children from Leadville, Colorado after recovery from high altitude pulmonary edema. Circulation 1985; 72:957–962.
22. Fishel RS, Thourani V, Eisenberg SJ, Shai SY, Corson MA, Nabel EG, Bernstein KE, Berk BC. Fibroblast growth factor stimulates angiotensin converting enzyme expression in vascular smooth muscle cells. J Clin Invest 1995; 95:377–387.
23. Fishman AP. A physician-scientist's tale. In: Wagner WW Jr, Weir EK, eds. The Pul-

monary Circulation and Gas Exchange. Armonk, NY: Futura Publishing Co., 1994: 381–387.
24. Grover RF, Reeves JT, Will DH, Blount SG Jr. Pulmonary vasoconstriction in steers at high altitude. J Appl Physiol 1963; 18:567–574.
25. Grover RF, Vogel JHK, Voigt GC, Blount SG Jr. Reversal of high altitude pulmonary hypertension. Am J Cardiol 1966; 18:928–932.
26. Grover RF, Reeves JT, Maher JT, McCullough RE, Denniston JC, Cymerman A. Maintained stroke volume but impaired arterial oxygenation in man at high altitude with supplemental CO_2. Circ Res 1976; 38:391–396.
27. Grover RF. Pulmonary hypertension: The price of high living. In: Wagner WW Jr, Weir EK, eds. The Pulmonary Circulation and Gas Exchange. Armonk, NY: Futura Publishing Co, 1994:317–342.
28. Groves BM, Droma T, Sutton JR, McCullough RG, McCullough RE, Zhuang J, Rapmund G, Sun S-F, Janes C, Moore LG. Minimal hypoxic pulmonary hypertension in normal Tibetans at 3658 m. J Appl Physiol 1993; 74:312–318.
29. Groves BM, Reeves JT, Sutton JR, Wagner PD, Cymerman A, Malconian MK, Rock PB, Young PM, Houston CS. Operation Everest II: elevated high-altitude pulmonary resistance unresponsive to oxygen. J Appl Physiol 1987; 63:521–530.
30. Hales CA. Pulmonary hypertension: vasoconstriction and vascular remodeling. Semin Resp Med 1985; 7:136–140.
31. Hales CA, Kradin RL, Brandsetter RD, Zhu Y. Impairment of hypoxic pulmonary artery remodeling by heparin in mice. Am Rev Respir Dis 1983; 128:747–751.
32. Harder DR, Gebremedhin D, Roman RJ. Influence of physical and environmental factors on ion channels in arterial smooth muscle. In: Weir EK, Hume JR, Reeves JT, eds. Ion Flux in Pulmonary Vascular Control. New York: Plenum Press, 1993:105–110.
33. Harder DR, Madden JA, Dawson CA. Hypoxic induction of Ca^{2+}-dependent action potentials in small pulmonary arteries of the cat. J Appl Physiol 1985; 59:1389–1393.
34. Harris P, Heath D. The Human Pulmonary Circulation. 2d ed. Edinburgh: Churchill Livingstone, 1977.
35. Harris P. The pulmonary circulation of some domestic animals at high altitude: the real story. In: Wagner WW Jr, Weir EK, eds. The Pulmonary Circulation and Gas Exchange. Armonk, NY: Futura Publishing Co, 1994:283–315.
36. Hartley LH, Alexander JK, Modelski M, Grover RF. Subnormal cardiac output at rest and during exercise in residents at 3100 m altitude. J Appl Physiol 1967; 23:839–848.
37. Hassoun PM, Thompson BT, Hales CA. Partial reversal of hypoxic pulmonary hypertension by heparin. Am Rev Respir Dis 1992; 145:193–196.
38. Haworth SG. Pulmonary vascular remodeling in neonatal pulmonary hypertension. Chest 1988; 93:1335–1385.
39. Heath D, Edwards C, Winston M, Smith P. Effects on the right ventricle, pulmonary vasculature, and carotid bodies of the rat exposed to, and on recovery from, simulated high altitude. Thorax 1973; 28:24–28.
40. Heath D. Pulmonary vascular disease in Sheffield, the Andes, Tibet, and Tanzania. In: Wagner WW Jr, Weir EK, eds. The Pulmonary Circulation and Gas Exchange. Armonk, NY: Futura Publishing Co, 1994:265–282.
41. Herget J, Hample V. Perinatal hypoxia increases hypoxic pulmonary vasoconstriction

in adult rats recovering from chronic exposure to hypoxia. Am Rev Respir Dis 1990; 142:619–624.
42. Hislop A, Reid L. Changes in the pulmonary arteries of the rat during recovery from hypoxia-induced pulmonary hypertension. Br J Exp Path 1977; 58:653–662.
43. Houston CS, Sutton JR, Cymerman A, Reeves JT. Operation Everest II: man at extreme altitude. J Appl Physiol 1987; 63:877–883.
44. Hultgren HN, Kelly J, Miller H. Effect of oxygen upon pulmonary circulation in acclimatized man at high altitude. J Appl Physiol 1965; 20:239–243.
45. Hultgren HN, Kelly J, Miller H. Pulmonary circulation in acclimatized man at high altitude. J Appl Physiol 1965; 20:233–238.
46. Hurtado A. Studies at high altitude. Am J Physiol 1932; 100:487–505.
47. Hurtado A. Aspectos Fisiologicos e Pathologicos de las Vida en la Altura. Editorial Rimac, Lima, 1937.
48. Hurtado A. Chronic mountain sickness. J Am Med Assoc 1942; 120:1278–1282.
49. Kerr JS, Ruppert CL, Tozzi CA. Reduction of chronic hypoxic pulmonary hypertension in the rat by an inhibitor of collagen production. Am Rev Respir Dis 1987; 135:300–306.
50. Kronenberg RS, Safar P, Lee J, Wright F, Noble W, Wahrenbrock E, Hickley R, Nemoto E, Severinghaus JW. Pulmonary artery pressure and alveolar gas exchange in man during acclimatization to 12,470 ft. J Clin Invest 1971; 50:828–837.
51. Lockhart A, Saiag B. Altitude and human pulmonary circulation. Clin Sci London 1981; 60:599–605.
52. Lockhart A, Zelter M, Mensch-Dechene J, Antezana G, Paz-Zamora M, Vargas E, Coudert J. Pressure-flow relationships in pulmonary circulation of normal highlanders. J Appl Physiol 1976; 41:449–456.
53. Lopez-Barneo J, Lopez-Lopez JR, Urena J, Gonzales C. Chemotransduction in the carotid body: K^+ current modulated by PO_2 in type I chemoreceptor cells. Science 1988; 242:580–582.
54. Madden JA, Dawson CA, Harder DR. Hypoxia-induced activation in small isolated pulmonary arteries from the cat. J Appl Physiol 1985; 59:113–118.
55. Madden JA, Vadula MS, Kurup VP. Effects of hypoxia and other vasoactive agents on pulmonary and cerebral artery smooth muscle cells. Am J Physiol 1992; 7:L384–L393.
56. Davis D. At the head of the table. Penn Med 1995; (fall):12–21.
57. Marshall C, Marshall B. Site and sensitivity of hypoxic pulmonary vasoconstriction. J Appl Physiol 1983; 55:711–716.
58. McMurtry IF, Davidson AB, Reeves JT, Grover RF. Inhibition of hypoxic pulmonary vasoconstriction by calcium antagonists in isolated rat lungs. Circ Res 1976; 38:99–104.
59. McMurtry IF, Petrun MD, Reeves JT. Lungs from chronically hypoxic rats have decreased pressor responses to acute hypoxia. Am J Physiol 1978; 235:H104–H109.
60. McMurtry IF. BAY K 8644 potentiates and A23187 inhibits hypoxic vasoconstriction in rat lungs. Am J Physiol 1985; 18:H741–H746.
61. Meyrick B, Reid L. Hypoxia and incorporation of 3H-thymidine by cells of the rat pulmonary arteries and alveolar wall. Am J Pathol 1979; 96:51–70.
62. Meyrick B, Reid L. The effect of continued hypoxia on rat pulmonary arterial circulation. Lab Invest 1978; 38:188–200.

63. Meyrick BO, Perkett EA. The sequence of cellular and hemodynamic changes of chronic pulmonary hypertension induced by hypoxia and other stimuli. Am Rev Respir Dis 1989; 140:1486–1489.
64. Monge C, Heraud C, Encinas E, Hurtado A. Estudios fisiologicos sobre el hombre de los Andes. Facultad de Medicina, 1928.
65. Monge C. La enfermedad de los Andes (sindromes eritremicos). Med Per Lima 1929; 2:7.
66. Monge C. La enfermidad de los Andes (sindromes eritremicos). Facultad de Medicina, Lima, 1928.
67. Monge C. Les erythremies de l'altitude. Paris Cie, 1929.
68. Mooi W. Muscular evaginations and disappearing arteries in hypoxia. In: Heath D, ed. Aspects of Hypoxia. Liverpool: Liverpool University Press, 1986:1–18.
69. Morin FC, Stenmark KR. Persistent pulmonary hypertension of the newborn: state-of-the-art. Am J Respir Crit Care Med 1995; 151:2010–2032.
70. Morrell NW, Atochina EN, Morris KG, Danilov SM, Stenmark KR. Angiotensin converting enzyme expression is increased in small pulmonary arteries of rats with hypoxia-induced pulmonary hypertension. J Clin Invest 1995; 96:1823–1833.
71. Morrell NW, Morris KG, Stenmark KR. Role of angiotensin converting enzyme and angiotensin II in development of hypoxic pulmonary hypertension. Am J Physiol 1995; 269:H1186–H1194.
72. Motley HL, Cournand A, Werko L, Himelstein A, Dresdale D. The influence of short periods of induced acute anoxia upon pulmonary artery pressures in man. Am J Physiol 1947; 150:315–320.
73. Naeije R, Mélot C, Georges Niset, Delecroix M, Wagner PD. Mechanisms of improved arterial oxygenation after peripheral chemoreceptor stimulation during hypoxic exercise. J Appl Physiol 1993; 74:1666–1671.
74. Oparil S, Narkates AJ, Jackson RM, Ann HS. Altered angiotensin converting enzyme in lung and extrapulmonary tissues of hypoxia adapted rats. J Appl Physiol 1988; 65: 218–227.
75. Orton EC, LaRue SM, Ensley B, Stenmark K. Bromodeoxyuridine labeling and DNA content of pulmonary arterial medial cells from hypoxia-exposed and nonexposed healthy calves. Am J Vet Res 1992; 53:1925–1930.
76. Peacock AJ, Dawes KE, Shock A, Gray AJ, Reeves JT, Laurent GJ. Endothelin-1 and endothelin-3 induce chemotaxis and replication of pulmonary artery fibroblasts. Am J Respir Cell Mol Biol 1992; 7:492–499.
77. Penalosa D, Gamboa R, Dyer J, Echevarria M, Marticorena E. The influence of high altitudes on the electrical activity of the heart I. Electrocardiographic and vectorcardiographic observations in the newborn, infants, and children. Am Heart J 1960; 59:111–128.
78. Penalosa D, Gamboa R, Dyer J, Echevarria M, Marticorena E. The influence of high altitudes on the electrical activity of the heart II. Electrocardiographic and vectorcardiographic observations in adolescence and adulthood. Am Heart J 1961; 61:101–128.
79. Penaloza D, Sime F, Banchero N, Gamboa R, Cruz J, Marticorena E. Pulmonary hypertension in healthy man born and living at high altitudes. Am J Cardiol 1963; 11:150.
80. Penaloza D, Sime F, Banchero N, Gamboa R. Pulmonary hypertension in healthy man born and living at high altitudes, Fifth Annual Conference on Research in Emphysema, Aspen, CO, June 13–16, 1962. Med Thorac 1962; 19:449–460.
81. Permutt S. Pulmonary mechanics and the pulmonary blood vessels. In: Wagner W,

Weir EK, eds. The Pulmonary Circulation and Gas Exchange. Armonk, NY: Futura Publishing Co., 1994:147–165.

82. Post JM, Gelband CH, Hume JR. $[Ca^{2+}]_i$ inhibition of K^+ channels in canine pulmonary artery. Circ Res 1995; 77:131–139.
83. Post JM, Hume JR, Archer SL, Weir EK. Direct role for potassium channel inhibition in hypoxic pulmonary vasoconstriction. Am J Physiol 1992; 262:C882–C890.
84. Pryor R, Weaver WF, Blount SG Jr. Electrocardiographic observation of 493 residents living at high altitude (10150 feet). Am J Cardiol 1965; 16:494–499.
85. Rabinovitch M. Elastase, remodeling of the extracellular matrix, and pulmonary hypertension. Semin Respir Crit Care Med 1994; 15:199–206.
86. Rabinovitch M, Konstam MA, Gamble WJ, Papanicolau N, Aronovitz MJ, Treves S, Reid L. Changes in blood flow affect vascular response to chronic hypoxia in rats. Circ Res 1983; 52:432–441.
87. Rabinovitch M, Gamble WJ, Nadas AS, Miettinen OS, Reid L. The rat pulmonary circulation after chronic hypoxia. Hemodynamic and structural features. Am J Physiol 1979; 236:H818–H827.
88. Rabinovitch M, Gamble WJ, Miettinen OS, Reid L. Age and sex influence on pulmonary hypertension of chronic hypoxia and on recovery. Am J Physiol 1981; 240:H62–H72.
89. Reeves JT, Leathers JE. Pulmonary hypertension with muscular exercise in the newborn calf. J Appl Physiol 1965; 20:249–252.
90. Reeves JT, Groves BM. Approach to the patient with pulmonary hypertension. In: Weir EK, Reeves JT, eds. Pulmonary Hypertension. Mt. Kisko, NY: Futura, 1984:1–44.
91. Reeves JT, Grover EB, Grover RF. Circulatory responses to high altitude in the cat and rabbit. J Appl Physiol 1963; 18:575–579.
92. Reeves JT, Dempsey JA, Grover RF. Pulmonary circulation during exercise. In: Weir EK, Reeves JT, eds. Pulmonary Vascular Physiology and Pathophysiology. New York: Marcel Dekker, 1989:107–133.
93. Reeves JT, Taylor AE. Pulmonary hemodynamics and fluid exchange in the lungs during exercise. In: Rowell LB, Shepherd JT, eds. Handbook of Physiology, Section 12, Exercise: Regulation and Integration of Multiple Systems. New York: American Physiological Society, Oxford University Press, 1996:585–613.
94. Reeves JT, Groves BM, Sutton JR, Wagner PD, Cymerman A, Malconian MK, Rock PB, Young PM, Houston CS. Operation Everest II: preservation of cardiac function at extreme altitude. J Appl Physiol 1987; 63:531–539.
95. Regensteiner JG, Moore LG. Migration of the elderly from high altitude in Colorado. J Am Med Assoc 1985; 253:3124–3128.
96. Reid LM. Structural remodeling of the pulmonary vasculature by environmental change and disease. In: Wagner WW Jr, Weir EK, eds. The Pulmonary Circulation and Gas Exchange. Armonk, NY: Futura Publishing Co, 1994:77–110.
97. Rotta A, Canepa A, Hurtado A, Velasquez T, Chavez R. Pulmonary circulation at sea level and at high altitudes. J Appl Physiol 1956; 9:328–336.
98. Rotta A, Canepa A, Velasquez T, Hurtado A, Aste Salazar H, Chavez R. Presión de la arteria pulmonar en el hombre que vive a 4500 metros de altitud. Rev Argentina Cardiol 1952; 19:374–375.
99. Rotta A, Miranda A, Chavez R. Heart catheterization at high altitudes. Symposium International de Biología de Altitud, 1949.

100. Rotta A. Physiologic condition of the heart in the natives of high altitudes. Am Heart J 1947; 33:669–676.
101. Rowe RD, James LS. The pattern of response of pulmonary and systemic arterial pressures in newborn and older children to short periods of hypoxia. AMA J Dis Child 1957; 93:13.
102. Rudolph AM, Auld PAM, Gulinko RJ, Paul MH. Pulmonary vascular adjustments in the neonatal period. Pediatrics 1961; 28:28–34.
103. Saling E. Neue Untersuchungsergebnisse über den Kreislauf des Kindes unmittelbar nach der Geburt. Arch Gynäk 1961; 194:287–306.
104. Sime F, Penalosa D, Ruiz L. Bradycardia, increased cardiac output, and reversal of pulmonary hypertension in altitude natives living at sea level. Br Heart J 1971; 33: 647–657.
105. Sime F, Banchero N, Penaloza D, Gamboa R, Cruz J, Marticorena E. Pulmonary hypertension in children born and living at high altitudes. Am J Cardiol 1963; 11:143–149.
106. Sobin S, Tremer HM, Hardy JD, Chiodi HP. Changes in arteriole in acute and chronic hypoxic pulmonary hypertension and recovery in rat. J Appl Physiol 1983; 55:1445–1455.
107. Stenmark KR, Fasules J, Hyde DM, Voelkel NF, Henson J, Tucker A, Reeves JT. Severe pulmonary hypertension and arterial adventitial changes in newborn calves at 4300 m. J Appl Physiol 1987; 62:821–830.
108. Stenmark KR, Dempsey EC, Badesch DB, Frid M, Mecham RP, Parks WC. Regulation of pulmonary vascular wall cell growth: developmental and site specific heterogenicity. Eur Respir Rev 1993; 16:629–637.
109. Stenmark KR, Mecham RP, Durmowicz AG, Parks WC. Persistence of the fetal pattern of tropoelastin gene expression in severe neonatal pulmonary hypertension. J Clin Invest 1994; 93:1234–1242.
110. Suarez A, Alexander JK, Houston CS. Enhanced ventricular systolic performance at high altitude during Operation Everest II. Am J Cardiol 1987; 60:137–142.
111. Torrance RW. The idea of a chemoreceptor. In: Fishman AP, Hecht HH, eds. The Pulmonary Circulation and Interstitial Space. Chicago: University of Chicago Press, 1969:223–239.
112. Tozzi CA, Poiani GJ, Harangozo AM, Boyd CD, Riley DJ. Pressure-induced connective tissue synthesis in pulmonary artery segments is dependent on intact endothelium. J Clin Invest 1989; 84:1005–1012.
113. Tucker A, McMurtry IF, Reeves JT, Alexander AF, Will DH, Grover R. Lung vascular smooth muscle as a determinant of pulmonary hypertension at high altitude. Am J Physiol 1975; 228:762–767.
114. Tucker A, Migally N, Wright ML, Greenlees KJ. Pulmonary vascular changes in young and aging rats exposed to 5486 m altitude. Respiration 1984; 46:246–257.
115. Voelkel NF. Mechanisms of hypoxic pulmonary hypertension. Am Rev Respir Dis 1986; 133:1186–1195.
116. Voelkel NF. Pulmonary hypertension: From phenomenology to a molecular understanding of disease mechanisms. In: Weir EK, Hume JR, Reeves JT, eds. Ion Flux in Pulmonary Vascular Control. New York: Plenum Press, 1993:1–14.
117. Vogel JHK, Weaver WF, Rose RL, Blount SG, Grover RF. Pulmonary hypertension on exertion in normal man living at 10,150 feet (Leadville, Colorado). Med Thorac 1962; 19:461–477.

118. Vogel HK, Goss JE, Mori M, Brammel HL. Pulmonary circulation in normal man with acute exposure to high altitude (14,260 feet). Circulation 1966; 34:III-233.
119. von Euler US, Liljestrand G. Observations on the pulmonary arterial blood pressure in the cat. Acta Physiol Scand 1946; 12:301–320.
120. Wagenvoort CA, Neufeld HN, Edwards JE. The structure of the pulmonary arterial tree in fetal and early postnatal life. Lab Invest 1961; 10:751.
121. Wagenvoort CA, Wagenvoort N. Diseases of the pulmonary circulation. In: Silver MD, ed. Cardiovascular Pathology. New York: Churchill Livingstone, 1983.
122. Wagner PD, Gale GE, Moon RE, Torre-Bueno J, Stolp BW, Saltzman H. Pulmonary gas exchange in humans exercising at sea level and simulated altitude. J Appl Physiol 1986; 61:260–270.
123. Wagner W. Pulmonary circulatory control through hypoxic vasoconstriction. Semin Respir Med 1985; 7:124–135.
124. Weir EK. The mechanism of acute hypoxic pulmonary vasoconstriction: the tale of two channels. FASEB J 1995; 9:183–189.
125. Weiser MCM, Majack RA, Tucker A, Orton EC. Static tension is associated with increased smooth muscle cell DNA synthesis in rat pulmonary arteries. Am J Physiol 1995; H1133–H1138.
126. Whorley JD, Frid MG, Orton EC, Belknap JK, Stenmark KR. Hypoxia selectively induces proliferation in a specific subpopulation of smooth muscle cells in bovine neonatal pulmonary arterial media. J Clin Invest 1995; 96:273–281.
127. Will DH, Hicks JL, Card CS, Reeves JT, Alexander AF. Correlation of acute with chronic hypoxic pulmonary hypertension in cattle. J Appl Physiol 1975; 38:495–498.
128. Will DH, Alexander AF, Reeves JT, Grover RF. High altitude induced pulmonary hypertension in normal cattle. Circ Res 1962; 10:172–177.
129. Xu Y, Stenmark KR, Walchak SJ, Ruff LJ, Dempsey EC. Pulmonary artery smooth muscle cells from hypoxia-exposed neonatal calves retain and/or re-express fetal-like growth properties. Submitted.
130. Zakheim RM, Mattioli L, Molteni A, Mullis KB, Bartley J. Prevention of pulmonary vascular changes of chronic alveolar hypoxia by inhibition of angiotensin I-converting enzyme in the rat. Lab Invest 1975; 33:57–61.
131. Zhao L, Winter RJD. Pulmonary vascular remodeling. In: Pulmonary Circulation. London: Chapman & Hall Medical, 1996.

11

Cerebral Circulation at High Altitude

JOHN W. SEVERINGHAUS

University of California
San Francisco, California

I. Introduction

The effect of high altitude upon CBF, appearing simple at first, has proved complex. With acute ascent to about 4000 m altitude, cerebral blood flow (CBF) rises by 30–60%. Over the next few days to weeks it falls, and after months to years it is not different from sea level normal values, despite chronic severe hypoxemia. This fall is not due to insensitivity to hypoxia because in natives of altitude, CBF falls in response to acute hyperoxia. There is no evidence of decreased cerebral metabolic rate for oxygen ($CMRo_2$), which remains normal even in hypoxia severe enough to obtund consciousness and increase brain lactic acid. In this chapter we examine the physiological responses of cerebral circulation to short- and long-term hypoxia.

Cerebral vessels receive conflicting signals in acute hypoxia: While hypoxia is a vasodilator, the hypoxic hyperventilation lowers Pco_2, causing cerebral vasoconstriction. The magnitude of change of CBF depends upon the relative strengths of four reflex mechanisms: (1) hypoxic ventilatory response (HVR), (2) hypercapnic ventilatory response (HCVR), (3) hypoxic cerebral vasodilation, and (4) hypocapnic cerebral vasoconstriction. In addition, the ventilatory responses to O_2 and CO_2 interact. For example, hypercapnia greatly increases the hypoxic ventilatory response, and hypocapnia diminishes it. The variations in these four sensitivities between individuals may determine individual susceptibility to altitude sickness and explain why some subjects have failed to show increased CBF in acute hypoxia.

The task of this chapter is to examine the evidence about the several mechanisms and their interactions, consider the possible roles played by them in the pathological cerebral signs and symptoms one encounters in climbing and living on mountains, and identify questions about these phenomena begging further investigation. How does hypoxia cause vasodilation? To what extent can hypocapneic vasoconstriction limit hypoxic vasodilation, and can this cause injury? Does hypoxia directly dilate arterioles or require ECF (extracellular fluid) messengers such as H^+, K^+, adenosine, or NO? Does high CBF cause or contribute to acute mountain sickness (AMS), high-altitude cerebral edema (HACE), or brain injury? Do some of the signs or symptoms of AMS or HACE depend on the CBF response to altitude? Does acetazolamide help in acclimatization by increasing CBF at altitude? Are there other unrecognized mechanisms downregulating CBF in chronic hypoxia? Does brain decrease its inherent metabolic rate in response to chronic hypoxia?

II. Brain Oxidative Biochemistry and Its Effect on Flow

$CMRo_2$ is maintained constant despite hypoxia by the Pasteur effect by which increased ADP stimulates glycolysis. Isolated mitochondria are able to maintain a constant O_2 consumption until solution Po_2 falls below 0.5 torr (1). Average cortex tissue Po_2 is about 9 mmHg as determined with recessed, calibrated gold-plated microelectrodes (2). Neuronal mitochondrial cytochrome is normally not fully saturated with O_2, and thus the redox state is not fully oxygenated, as shown by the rise in NAD with vasodilation (3). Thus, even in normoxia the brain exhibits some "anaerobic metabolism," detected by measuring the difference between arterial and cerebral venous lactate. Lactic acid crosses the capillary endothelial blood-brain barrier (BBB) using a stereospecific, facilitated, saturable, passive transporter (4).

III. Normal Cerebral Blood Flow Regulation

CBF regulation involves responses in both large arteries and precapillary sphincters. Smooth muscles of arteries larger than about 50 μm constrict or dilate in response to variation of intra-arterial pressure, exhibiting the curious stretch response, which actually reduces lumen diameter when arterial pressure rises. This response is independent of the sympathetic innervation (5,6). These arteries regulate pressure supplied to the microcirculation but do not control flow. Rather, terminal arterioles and precapillary sphincters control flow, moment to moment, in response to local chemical and neural mediators (7). At rest, red cells travel through the capillary network in 100–300 ms along 150–500 μm paths (8). Blood flow in individual capillaries in brain is cyclic, resulting in oscillations in tissue Po_2 (4–10 min^{-1}) (9–11). Vasodilators increase the velocity of the red cells (8), the number of capillaries perfused at any moment, and the length of each open period (12), suggesting this open-shut system to be a major component of cerebral autoregulation by which flow is independent of pressure between approximately 60 and 150 mmHg (13), but rapidly responsive to metabolic demands partly mediated via locally released NO (14). During and following hypoxia, autoregulation may be disrupted (15,16).

CBF increases in response to both decreased O_2 supply and increased O_2 demand. With neuronal activation, O_2 consumption rises within milliseconds, and local CBF rises within 1–2 s (14). The hyperemic response to step hypoxia is almost equally rapid (17). ECF adenosine concentration also rises within a minute, but the distance between neurons and arterioles is too great for diffusion of neuronally released adenosine or lactic acid H^+ to arterioles to account for the rapid flow increase. K^+ is also a possible mediator (see Sec. III.C).

A. Hypoxia

During acute hypoxia, red blood cell (RBC) velocity increases in all brain capillaries (using intravital microscopy) (8). CBF rises similarly in proportion to the severity of isocapnic hypoxia in all mammals. However, when $P\text{CO}_2$ is not controlled during hypoxia, there is considerable variability between individuals and species because of the wide variability of the HVR, which determines how far $P\text{CO}_2$ falls. A strong HVR and major fall of $P\text{CO}_2$ in experimental subjects led to an early impression that hypoxic vasodilation was a threshold phenomenon, scarcely beginning until $Pa\text{O}_2$ had fallen to about 30 mmHg (18).

In their original study in young male volunteers using their N_2O method (19), Kety and Schmidt found a 35% increase in CBF while breathing 10% O_2 (resulting in $Sa\text{O}_2$ = 65%), which caused $Pa\text{CO}_2$ to fall from 40 to 36 mmHg. Reported responses in humans at altitude have varied from no immediate increase to rises of the order of 30–60% during the initial hours or days. In awake sheep after 30 minutes of hypoxia to $Pa\text{O}_2$ = 40 mmHg, CBF increased by as much as 250% in spite of a 6 mmHg fall of $Pa\text{CO}_2$ (20). CBF rose more than 300% in rats at $Pa\text{O}_2$ = 24 mmHg (21). Some effects on CBF of acute hypoxia may persist after reoxygenation. Trojan and Kapitola (22) found residual high blood flow through the medulla, cerebellum, subcortical regions, and cerebral cortex in adult rats 20 hours after completing an 8-hour exposure to 7000 m altitude (in a chamber).

The direct vasodilating effect of hypoxia on CBF may be separated from the negative feedback change of $Pa\text{CO}_2$ that usually accompanies hypoxia by adding CO_2 to the inspired air to keep end-tidal $P\text{CO}_2$ constant, a procedure known as isocapnic hypoxia. Using isocapnia in nine healthy male volunteers, Cohen et al. (23) showed that a step reduction of $Pa\text{O}_2$ to 34.6 ± 1.6 mmHg (SE) (≈66% $Sa\text{O}_2$) increased CBF about 70% (from 0.45 to 0.77 mL g^{-1} min^{-1}) accompanied by a 27% rise in glucose consumption (CMR_{glu}) and a fourfold rise in cerebral lactate production (CMR_{lac}), with no significant fall of $CMR\text{O}_2$. This isocapnic rise of CBF was twice that reported at the same $Sa\text{O}_2$ by Kety and Schmidt (19) when $Pa\text{CO}_2$ was allowed to fall. In a direct comparison of the effect of P_{CO_2} on hypoxic vasodilation in sheep, Yang et al. (24) reported that step reduction of Pa_{O_2} to 40 mmHg increased CBF 156% in isocapnia but only 68% when P_{CO_2} was allowed to fall freely (from 36 to 28 mmHg). These several studies provide a simple guideline that the effect of isocapnia is to double the hypoxic increase of CBF compared to poikilocapnia.

The relationship of CBF to Pa_{O_2} is hyperbolic, rather like the hypoxic ventilatory response. Since the ventilatory response proved to be a linear function of arterial Sa_{O_2}, several studies have sought similar linearity for the plot of CBF vs. Sa_{O_2}. Ashwal et al. (25) found that, in fetal lambs made hypoxic by acute maternal isocapnic hypoxemia, CBF was an approximately linear function of fetal Sa_{O_2} down to nearly zero, at which point flow was increased to about 250% of control. However, Jensen et al. (26) found the relationship between flow and Sa_{O_2} to be hyperbolic in humans. In six normal males at sea level and during 5 days at 3810 m altitude using 5-minute steps to four levels of isocapnic hypoxia, they measured the relationship of CBFv (v for velocity, measured by transcranial doppler, TCD) to Sa_{O_2} (Fig. 1). Acute hypoxia to 70% Sa_{O_2} increased CBFv by 35% at sea level and by 46% after 5 days at altitude, indicating a 34% increase in the sensitivity of cerebral vessels to acute hypoxia induced by this 5-day acclimatization period. At sea level the acute

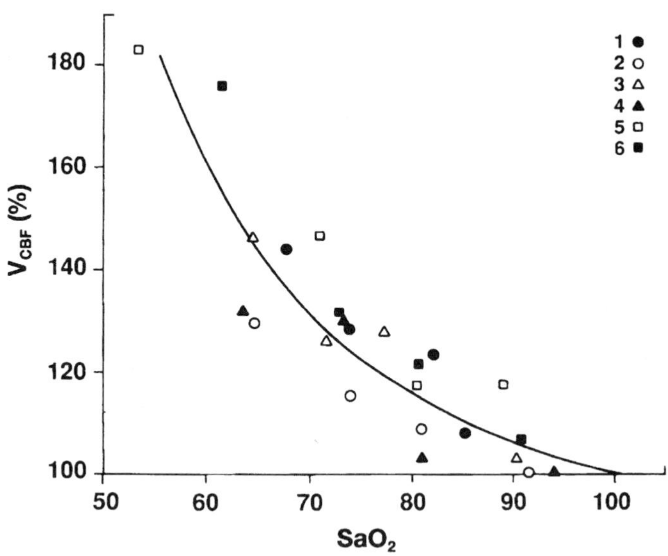

Figure 1 CBFv% (% of control CBF velocity by TCD) in response to 5 minute steps to four levels of isocapnic step hypoxia in normal subjects after 3–4 days at 3810 m altitude. The hyperbolic empirical relationship is: CBFv% $100(1 + X[(60/(S_aO_2\text{-}40)) - 1])$. Factor X was found to average 0.35 ± 0.11 at sea level and 0.46 ± 0.08 after 5 days at altitude. X may be interpreted as the fractional increase in CBFv induced by acute isocapnic hypoxia at 70% Sa_{O_2}. (From Ref. 26.)

increase of CBF was about two-thirds of the sensitivity reported by Cohen et al. 30 years earlier.

Hypoxia had surprisingly little effect on the high energy state of cortex, cerebellum, or brain stem in lightly anesthetized rats at 35, 29, and 23 mmHg, as determined by a rapid (<5 s) in situ freezing method (27). At the lowest Po_2 (23 mmHg), ADP showed small increases, but no change in ATP or AMP was detected. Lactate rose from <2 to 13 mM kg^{-1}, pyruvate levels tripled, and regional CBF rose to five to seven times control in all three regions. Autoradiographic methods were used to show that the vasodilatory response to both hypercapnia and hypoxia was greatest among brain stem gray matter structures, intermediate among cortical and diencephalic gray matter structures, and least in white matter (28). In subcortical white matter, hypoxia increased glucose consumption fivefold but flow less than twofold, suggesting localized anaerobic glycolysis and acidosis. In highly active regions of gray matter, in comparison, glucose consumption rose less than 50% while flow rose 200–500%. This observation may fit with other evidence that white matter is especially vulnerable to hypoxic damage (29,30).

B. Anemia

Isovolemic hemodilution increases RBC velocity three- to fourfold and increases RBC flux to a moderate degree, with a relatively small decrease in capillary hematocrit, under normal and compromised arterial blood supply (8). Hemodilution increases cerebral blood flow in both normal and polycythemic patients (31–33). In newborn lambs this effect was extremely variable. Reduction of hematocrit from 55% to 10% by means of exchange transfusions increased CBF sevenfold in a few animals, but scarcely doubled it in others (34). Both the lowered viscosity and lowered O_2 content are presumed to contribute, but because of the strong effect of O_2 content, it has been difficult to demonstrate a separate effect of viscosity. The role of viscosity was reported to be insignificant in a study of the influence of severe acute normovolemic anemia on CBF and $CMRo_2$ in normocapnic rats under nitrous oxide anesthesia (35,36). The arterial hemoglobin content was reduced to values of about 12, 9, 6, and 3 g dL^{-1} by replacing blood with equal volumes of plasma. The CBF increase was proportional to the reduction in hemoglobin content, rising to five to six times normal in extreme anemia, but $CMRo_2$ remained unchanged. Cerebral venous Po_2 and Sao_2 did not decrease below normal values, indicating that tissue hypoxia did not develop. At 3 g dL^{-1} tissue, lactate content increased moderately, and associated changes in other carbohydrate metabolites and in amino acids suggested that tissue hypoxia was present. However, since the concentrations of phosphocreatine and adenine nucleotides remained constant, this hypoxia must have been slight. Since the increase in CBF at hemoglobin concentrations of below 9 g dL^{-1} was far in excess of that expected from the decrease in viscosity, hyperemia was thought to be due primarily to the reduction in arterial O_2 content.

More recently, MetHb has been used to separate the effects of viscosity from O_2 content. Red cells containing metHb cannot deliver oxygen and thus lower O_2

content without affecting viscosity (37). Two groups of awake anemic lambs were transfused, one with normal blood, the other with metHb-inactivated red cells. In both groups, Hct was raised from about 20 to 40%. CBF fell 50% using normal O_2-carrying erythrocytes but also fell 28% with MetHb cells, suggesting that about half of the effect must be due only to viscosity. The effect of viscosity was quite variable, being greatest in those lambs with the highest CBF. The role of viscosity in reducing CBF during polycythemia has not been reported but would be of interest for those with high-altitude polycythemia in view of the report that CBF in altitude natives has the same inverse linear relationship of Hct as in sea level natives (38).

A compilation of data from various studies (39) of anemic, normal, and polycythemic subjects revealed a hyperbolic fourfold variation of (A-V)Δ_{O2} with hematocrit (Hct) (Fig. 2). From these data, CBF values may be estimated to vary with Hct from 1.0 to 0.29 mL g^{-1} min^{-1} by the following process: normal global CBF in humans at sea level is 0.45 mL g^{-1} min^{-1} at normal Hct = 45%. $CMRo_2$ is independent of oxygen supply down to a critical threshold. Because (A-V)o_2 difference (vol%) is a hyperbolic function of Hct, CBF is inversely proportional to Hct (38).

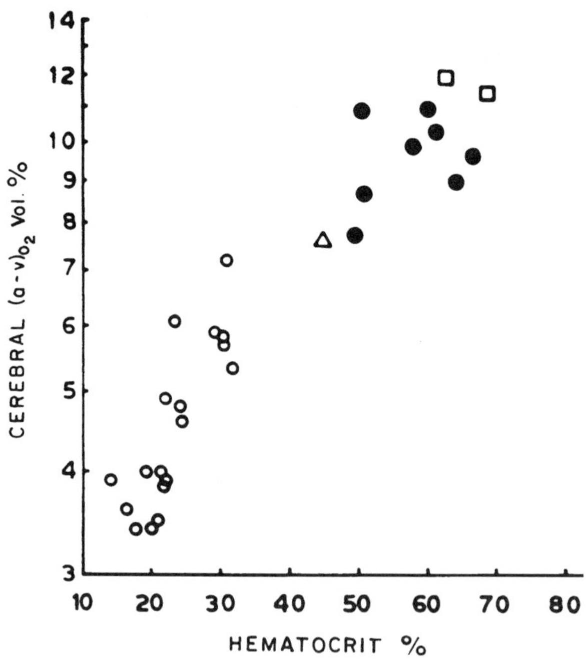

Figure 2 The (A-V)ΔO_2 content during acute normoxia in various groups including natives of the Bolivian altiplano as a function of Hct. The normal relationship of CBF to Hct derived from these data is given in the text. ○: anemia; △: normal mean; ●: chronic polycythemia (altitude); □: chronic polycythemia (other). (From Ref. 39.)

By assuming $(A\text{-}V)\Delta O_2 = CMRo_2/CBF$, where $CMRo_2$ is invariant, the relationship of normoxic global CBF to Hct from these data may be approximated by:

$$CBF = [0.675 - 0.005 \text{ Hct}] \ (\text{mL} \cdot \text{g}^{-1} \cdot \text{min}^{-1})$$

In order to use this relationship to factor out Hct in studies at varying times and in varying groups with other normal control CBF values (Q_{45}), the relationship may be written:

$$Q_{45} = 90\text{CBF}/[135\text{-Hct}]$$

C. Carbon Monoxide and Its Effect on the O_2 Dissociation Curve

CO hypoxia is a more potent vasodilator than either hypoxia or anemia, actually increasing O_2 delivery ($CaO_2 \times CBF$) above normal. Koehler et al. (40) compared CBF responses to hypoxia induced by low alveolar Po_2 or by carbon monoxide in awake adult sheep. The arterial O_2 content (CaO_2) was equally reduced stepwise to 50–60% of the control value. CBF and O_2 delivery increased more with CO hypoxia than with alveolar hypoxia. This CBF difference was proportional to the magnitude of the leftward shift of the oxygen dissociation curve (ODC) in those given CO. Despite the higher O_2 delivery with CO hypoxia, $CMRo_2$ fell by 16% in adults but was maintained in newborns, while $CMRo_2$ was unchanged during alveolar hypoxia in both age groups. CO lowers capillary Po_2 by inducing a leftward shift of ODC (oxygen dissociation curve), especially at low saturation, because it eliminates the S shape at the bottom. For example, 20% COHb reduces the Po_2 of blood with 24% O_2Hb from 17 (normal) to 12 mmHg. In normal blood, $Po_2 = 12$ mmHg at 13% O_2Hb. The critical Po_2 is approximately 12 mmHg, below which $CMRo_2$ begins to fall. Thus, COHb severely limits unloading of O_2 from Hb. This is a considerably greater effect than other causes of left-shifted ODC, such a low 2,3-DPG.

The importance of the position of the ODC as a determinant of cerebral blood flow supports the presence of a highly sensitive, tissue Po_2–dependent mechanism regulating the cerebral circulation. To test whether the CO effect was simply due to the shift of the ODC or to additional cellular effects, Koehler et al. (40,41) performed isovolemic exchange transfusions on unanesthetized newborn lambs, replacing their high–O_2-affinity hemoglobin with low-affinity adult sheep donor blood. Exchange transfusion resulted in an average increase in P_{50} of 10 mmHg and in a 14% decrease of regional cerebral blood flow and cerebral O_2 delivery. Induction of CO hypoxia (20–40% COHb) after the exchange transfusion returned P_{50} to the control level, and restored both cerebral O_2 delivery and fractional O_2 extraction to the pretransfusion values. Thus it is the fall in P_{50}, rather than a direct tissue effect of CO, which is responsible for the relative cerebral overperfusion during CO hypoxia. They concluded that capillary Po_2, not capillary O_2 content, determined CBF, an interpretation similar to that of Woodson using hypocapnic left shifts of the ODC (see Sec. III.D) (42).

D. Mediators of Acute Hypoxic Cerebral Vasodilation

How does reduced capillary Po_2 dilate arterioles? Attention is now focused on those glial cells whose feet are wrapped around arterioles and whose heads are in tight contact with local neurons (13). Flow is increased by local neuronal activity (43–45), increasing ECF $[K^+]$ (46), adenosine (45,47), and hypercapnia (48). Other vasoactive messengers might include intravascular NO generated in endothelia (44,49) and a variety of autocoids and cytokines (13). While hypoxia probably operates through several of these mechanisms, the evidence to be presented suggests that it may have other still unknown direct vasodilating mechanisms.

Potassium

When a neuron depolarizes, K^+ ions diffuse out from the neurons through the ECF space into adjacent glial cells whose cellular membranes are freely permeable to K^+. Experimental elevation of ECF K^+ up to about 15 mmol L^{-1} causes cerebral vasodilation, while higher levels cause vasoconstriction (46,50). This is one putative mechanism by which neuronal activity signals the glial cell to cause vasodilation. The effector at ''glial feet'' may be either K^+ or adenosine (46,51). In cerebral blood vessels, four different potassium channels have been described, one responding to ATP, one to calcium, and two types of rectifier channels, the ATP channels being most clearly linked to the hypoxic response (51,52). However, I have found no evidence that hypoxia causes a rise in ECF K^+ in brain.

Lactic Acid

As noted in Sec. III.A, normally CBF limits brain tissue Po_2 such that ECF and CSF lactate average 1.5 mM, twice the arterial level. Increasing CBF by hyperoxic hypercapnia lowers CSF lactate to arterial levels. One might expect altitude hypoxia to increase brain lactate production. Early studies in dogs led to the conclusion that lactic acidosis was the main hypoxic vasodilator (15), but studies at high altitude have not supported this possibility. Cain and Dunn (53) showed that acute exposure of dogs to 21,000 ft altitude (chamber) produced only transient increases in brain and cerebral venous blood lactate, the small rises having returned to control levels within 8 hours.

Silver recorded rat brain ECF Po_2, pH, $[K^+]$, $[Cl^-]$, $[Ca^{2+}]$, lactate, and blood flow using ion-specific microelectrodes (17). Increased blood flow in response to close arterial injection of hypoxic blood occurred within 1–2 seconds of the fall in tissue Po_2 before changes in either pH or potassium. The immediate effect of step hypoxia was a rise of local CBF and ECF pH, excluding lactic acid as the mediator of vasodilation. The same conclusion was reached by Siesjo et al. using different methods (rapid freezing and brain chemical analysis) (43,50).

Adenosine

Hypoxia to $Pao_2 = 45$ mmHg does not significantly reduce ATP, CP, or the cerebral metabolic rate for O_2 ($CMRo_2$) (54). It is therefore somewhat surprising to find that the final breakdown product of ATP, adenosine, is a potent cerebral endogenous vasodilator (47). In both brief (30 s) and longer (5 min) hypoxia, brain adenosine increases rapidly from 0.34 ± 0.08 (SE) to 1.65 ± 0.33 mM kg^{-1}, paralleling temporally the changes in cerebral blood flow (47). By using extremely rapid freezing (within 1 s) this group found that adenosine concentrations remained stable between Pao_2 200 and 100 mmHg, doubled at Pao_2 = 50 mmHg, and increased sevenfold when Pao_2 reached 30 mmHg. In prior studies using a slower in situ freezing technique, no increases in adenosine or its metabolites had been detected during comparable hypoxia. Three other groups have now documented an adenosine-hypoxic vasodilator link (55–57). Furthermore, pial arterial dilation by hypoxia measured through an implanted cranial window in rats was enhanced by inhibiting adenosine uptake and blocked by adenosine blockade with theophylline (57). Thus, adenosine plays an important role in hypoxic cerebral vasodilation, probably being expressed by glial cells at their capillary feet.

Nitric Oxide

Nitric oxide, formerly called endothelial-derived relaxing factor (EDRF), is generated by endothelial cells and acts locally on the arteriolar smooth muscle. While NO is intimately involved in the CBF response to hypercapnia and hypocapnia (58), it seems to play a less easily defined role in hypoxic vasodilation (8,33,49). NO inhibition by N-nitro-L-arginine methyl ester (L-NAME) did not alter the increases in CBF elicited by ventilation with 8% oxygen for 25 s (59), but the selective nNOS inhibitor 7-nitroindazole completely blocked hypoxic hyperemia in other studies (33,60).

Other Speculative Mechanisms

It now seems probable that cerebral vasodilation in hypoxia is signaled through glial cells which ''connect'' neurons to the nearest arteriolar smooth muscle cells. Adenosine, H^+, K^+, and NO may serve as vasodilator messengers from glia to smooth muscle, but of these only adenosine and NO are confirmed transmitters in hypoxia. It remains unknown whether oxygen supply to some critical metabolic process must be impaired to cause vasodilation, or whether some non–oxygen-consuming way of sensing Po_2 is at work. There are preliminary reports that oxygen directly affects the opening of potassium channels, but that mechanism is still unexplained since there is no known chromophore or metal reaction site on K^+ channels.

E. Effects of Hypocapnia on CBF

The Pco_2–pH_{ECF}—CBF Relationship

In normoxia, CBF falls about 3–4% per mmHg decrease of $Paco_2$ from normal (48). ECF H^+ is assumed to mediate the response to Pco_2 (61–63). Local vasodilation occurs with topically applied weak fixed acids (13). A fall of Pco_2 to 20 mmHg reduces CBF by half, but further hypocapnia has little effect, vasoconstriction being limited by tissue hypoxia. Hypercapnia increases CBF to a maximum two to three times normal. Brain P_{CO_2} is about 6 mmHg higher than Pa_{CO_2}. At normal Pa_{CO_2} of 40 mmHg, a fall of 1 mmHg causes a rise of 0.01 pH in CSF and ECF. The blood-brain barrier consists of endothelial cells with tight junctions, which effectively block arterial blood nonrespiratory acid-base changes from both the arteriolar smooth muscle cells and from most of the brain ECF (62,64,65). Even in fetal lambs, where the blood-brain barrier is generally thought to be incompletely formed, and thus more permeable, variation of pHa from 6.56 to 7.59 had no detectable influence on CBF at constant Pa_{CO_2} (25).

The relationship between brain ECF pH and CBF at altitude was difficult to unravel, in part because investigators have studied lumbar CSF as a proxy for brain ECF, and partly because of the potential for error in measuring pH in poorly buffered CSF and its variation with P_{CO_2} before and during sampling. The evidence now suggests that as long as the carotid chemoreceptor drive remains elevated by hypoxia, the balance of respiratory drives is tilted, the increased ventilation causing medullary respiratory center and cerebral ECF pH to remain above normal. This alkalosis presumably continues to exert a vasoconstricting effect, in opposition to hypoxic vasodilation (66,67). Acclimatization gradually increases Pa_{O_2} as ventilation rises. In addition, capillary P_{O_2} may rise if blood pH falls, reducing the alkaline Bohr shift. P_{50} (measured at pH = 7.4) was found to be elevated 4 mmHg during the early days of acclimatization according to Lenfant et al. (68) but 2 mmHg or less in a similar test by Weiskopf et al. (69).

Site of Action of CO_2: Neural Cells or Endothelium?

In order to answer this controversial question. Severinghaus and Lassen (70) determined the time constant of the CBF response to step hypocapnia. Their rationale was that the washout of CO_2 from the endothelium would be very rapid compared to its predicted 1–2 minutes 63% washout from bulk brain tissue. In gray matter, CBF ~1 mL g^{-1} min^{-1}, which predicts a CO_2 washout time constant of 1 minute. White matter CBF is about one third that of gray. CBF was estimated from the jugular vein to arterial Sa_{O_2} difference, assuming constant CMR_{O_2}. Following sudden hyperventilation to a constant end-tidal P_{CO_2} of 20 mmHg, the CBF time constant was 20–25 seconds (70). This suggested that the flow regulation responds directly to arteriolar smooth muscle ECF pH, not to gray matter P_{CO_2}.

Effect of pH on Blood O_2 Affinity

Blood O_2 affinity is quantified as P_{50}, the P_{O_2} resulting in 50% O_2 saturation of Hb. Under standard conditions, at pH = 7.4, T = 37°C, P_{50} in adults at sea level averages 26.6 mmHg. Affinity changes induced by pH, temperature, and 2,3-DPG affect P_{O_2} proportionally at all levels of O_2 saturation, so a 10% rise of P_{50} means multiplying by 1.10 all P_{O_2} values of the ODC. Acute respiratory alkalosis not only decreases cerebral blood flow, but also increases the affinity of hemoglobin for oxygen (Bohr effect), which thereby lowers capillary and brain tissue P_{O_2} and increases brain and blood lactate concentrations (68,71). For example, P_{50} falls 20% to 21.3 mmHg at pH = 7.57 and BE = −10 mM L^{-1} (the values obtained at Pb = 250 mmHg in volunteers acclimatized in a chamber simulating ascent of Everest) (72). An acute left shift in the ODC when combined with hemorrhagic shock diminished shock tolerance with evidence of increased sympathetic outflow and increased mortality (42). In 10 newborn lambs Rosenberg (73) showed that as $PaCO_2$ was reduced from 46 to 12 mmHg, $CMRo_2$ was unchanged, while brain lactate production increased. Exchange transfusion of normal rats with low P_{50} blood resulted in a major increase in cerebral flow. At high altitude the standard P_{50} (at pH = 7.4) rises slightly, but because of hyperventilation alkalosis, the net effect is a left shift in most subjects. At altitude, a left shift increases O_2 loading in the lung about enough to compensate for its impairment of unloading in tissues.

Interaction of Hypocapnic Vasoconstriction and Bohr Effect on O_2 Delivery

The effect on O_2 delivery of the conflicting responses of hypoxic vasodilation, hypocapnic vasoconstriction, and alkaline Bohr shift depends on the relative strengths of these three regulatory mechanisms. Each is subject to interindividual variability, and this may be important in functioning at high altitude. There may be pharmacological ways of limiting the vasoconstriction or Bohr shift that would ease life at extreme altitude. A comparative physiological approach has been used to identify patterns of the balance of these three effects in species that tolerate altitude well. Some birds, especially bar-headed geese, fly over Everest at altitudes as high as 9000 m. Grubb et al. (74) reported that artificially ventilated Pekin ducks have weaker hypocapnic cerebral vasoconstriction than mammals but a mammal-like hypoxic vasodilation (75).

Faraci et al. (76) compared the effects of hypoxia in awake bar-headed geese (*Anser indicus*) and Pekin ducks (*Anas platyrhynchos*). While $Paco_2$ decreased to about 7 mmHg at 28 mmHg Pao_2 in both, arterial O_2 content (CaO_2) in geese (10.4 mL dL^{-1}) was significantly higher than in ducks (4.1 mL dL^{-1}) due to a left-shifted ODC in the geese. O_2 delivery to the brain of geese was the same as or higher than that of ducks, even though cerebral blood flow increased more than fivefold in ducks but less than threefold in geese. Faraci and Fedde then confirmed Grubb's report, finding that in artificially ventilated, normoxic geese, hypocapnia to as low as $Paco_2$ = 7 mmHg did not significantly reduce CBF below normal (77), although hypercapnia ($Paco_2$ = 60 mmHg) tripled CBF. These geese also were found to have a greater cerebral capillary density and higher diffusing capacity, permitting them to consume O_2 at the lower capillary Po_2 resulting from their strongly left-shifted ODC.

Bickler and Julian (11) studied the effect of extreme hypocapnia on cortical blood flow and tissue Po_2 (using O_2 and H_2 microelectrodes) in anesthetized normoxic geese (*Anser domesticus*). CBF decreased as $Paco_2$ fell to 20 mmHg but was not affected by further Pco_2 reduction, similar to the findings of Faraci and Fedde. This threshold for maximum vasoconstriction has been reported in mammals, including humans, and has been attributed to reduced O_2 delivery since CBF will fall farther with 100% inspired O_2. Despite the plateau in CBF during severe hypocapnia, tissue P_{O_2} continued to decline as Pa_{CO_2} fell but may have reached a plateau below 15 mmHg (Fig. 3). They noted that, at very low P_{CO_2}, reduction of Pa_{O_2} to 30–40 mmHg had no significant effect on brain tissue P_{O_2}, which may be interpreted as suggesting that CBF increased (not measured), or that more lactic acid entered the capillary blood, maintaining P_{O_2} by the Bohr effect, as it does in exercising muscle.

Thus, in high-flying geese, CBF is less reduced by hypocapnia than in mammals. During severe hypocapnic hypoxia, the left-shifted ODC delivers a higher Sa_{O_2} to brain at very low Pa_{O_2}. The brain apparently tolerates the low capillary P_{O_2} well, flow being less than in ducks at similar Pa_{CO_2} and Pa_{O_2}. Apparently, geese utilize higher O_2 content despite a lower P_{O_2} because diffusing capacity and capillary density are greater. Unfortunately, as far as I could determine, none of the studies of awake bar-headed geese have measured CBF during the combination of severe hypocapnia and hypoxia expected at 9000 m.

Role of Nitric Oxide in Hypercapnia

NO is involved in the CBF increase caused by hypercapnia (13,58,78). NO affects both major cerebral arteries and arteriolar sphincters (79). In rats an inhibitor of NO

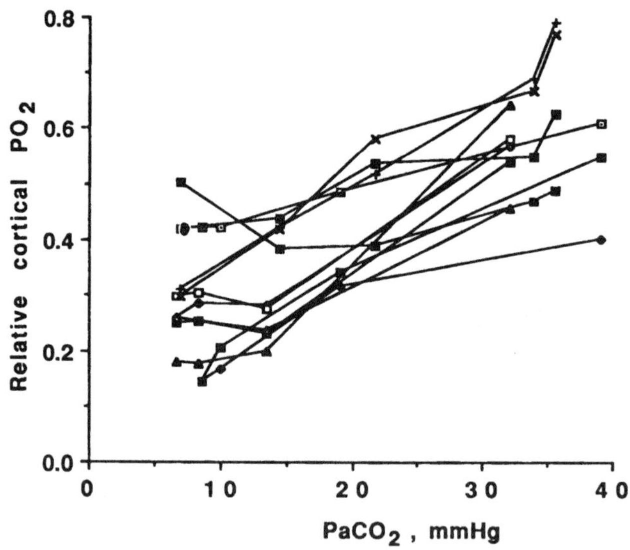

Figure 3 Effect of normoxic hypocapnia on duck brain tissue P_{O_2} (implanted polarographic electrodes). Although CBF reduction reached a plateau at about 20 mmHg, as in humans, brain tissue P_{O_2} continued to fall in some birds with P_{CO_2} reduction, due largely to the left-shift of the ODC curve (Bohr effect of high pH). Symbols represent different birds. (From Ref. 11.)

synthase decreased normocapnic CBF by 40% despite an increase of arterial blood pressure and attenuated CO_2 reactivity (49,80). However, the vasoconstriction of hypocapnia is unimpaired by blockade of NO synthesis.

Acetazolamide and CBF

Acetazolamide (Diamox®) is widely used to reduce AMS during acclimatization and improve sleep at altitude. Its mechanism of action on several aspects of blood gas transport and acid-base balance are complex and have led to controversy about how it helps the sojourner at altitude. Acetazolamide blocks carbonic anhydrase (81), primarily in red blood cells and renal tubules, but also in brain cells. Unloading of CO_2 from HCO_3^- in red cells is delayed in the lung, so blood CO_2 content is increased but alveolar P_{CO_2} is reduced. $PaCO_2$ rises slowly after blood leaves the lung so its value in tissues depends on the time from lung to tissue. Respiration is stimulated (82,83) by a rise of P_{CO_2} in the medullary respiratory chemoreceptors, as in all tissues, because metabolic CO_2 cannot rapidly be transported and converted in red cells into HCO_3^- (84). The ventilatory rise improves oxygenation during acute hypoxia (85) without an increase in lactate (86,87).

Acetazolamide increases CBF in dogs (88). In 10 adult patients given 1 g acetazolamide, CBF rose by 66% within 30 minute (89). Oral administration of 1 g of acetazolamide to eight normoxic subjects caused a 38% increase in CBF (90, 91). However, at sea level continued administration does not sustain high CBF in normoxic volunteers (90,92).

Acetazolamide increases brain bicarbonate concentration (93) and raises tissue P_{CO_2} about 6 mmHg, while Pa_{CO_2} is significantly lower when blood reaches capillaries having left the lung at the low alveolar P_{CO_2}. Acetazolamide increases cortex surface P_{O_2} 16–20 mmHg, eliminating a proposed explanation of the vasodilation as due to Bohr shift hypoxia (94). Acetazolamide causes brain carbonic acidosis (95), reducing brain surface pH by 0.1 pH within 3 minutes (96) (Fig. 4), while intracellular pH (MRS using ^{31}P) does not fall (limit of detection ± 0.06) (97).

Several authors have suggested that little of the benefit of acetazolamide at altitude is due to increased CBF, but that the beneficial effects are more due to increased ventilation raising Pa_{O_2}, affording a significant increase of the arterial oxygen content (98). However, the definitive studies at altitude remain to be conducted, since one must consider the effect of the drug on ventilation, P_{CO_2}, P_{O_2}, Sa_{O_2}, acid-base balance, and CBF together. My impression is that CBF is higher with acetazolamide at altitude at the lower P_{CO_2} it causes than it would have been at that P_{CO_2} without the drug.

IV. CBF Changes During Acclimatization

A. Time Course of CBF Change at Altitude

The effect of altitude on CBF depends in part on the changes of P_{O_2} and P_{CO_2}. With rapid ascent to high altitude, ventilation immediately increases slightly reducing

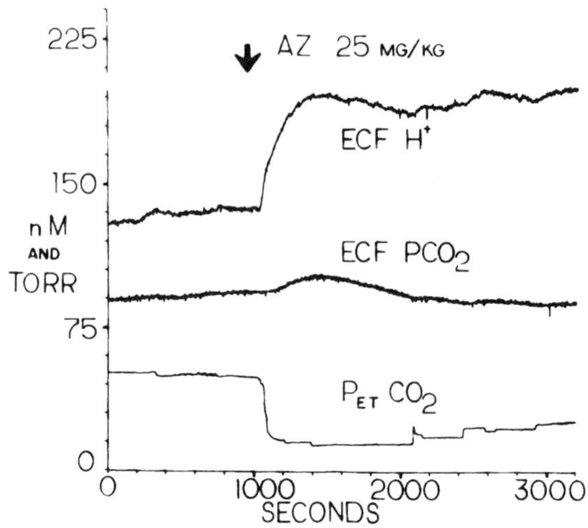

Figure 4 Acetazolamide (25 mg/kg IV in a dog) causes a rapid acidification of brain ECF (surface flat pH electrode) even when tissue P_{CO_2} (flat surface electrode) is held constant (by increased ventilation). This effect cannot be explained by the effect on red cell carbonic anhydrase and indicates a rise of tissue carbonic acid due to delay of dehydration of metabolically produced H^+ and HCO_3^- to CO_2 within the brain. (From Ref. 96.)

Pa_{CO_2}. During the next hours, days, and weeks the process of acclimatization results in further increase in ventilation, rise of P_{O_2} and fall of P_{CO_2}, affecting CBF. While the literature implies considerable variability of this CBF response, in general, in humans at moderate altitudes, CBF usually rises initially and then falls slowly over the first week at altitude. Do the concurrent blood gas changes account for the observed CBF time course?

The first measurements of CBF during acclimatization to high altitude were obtained by Severinghaus et al. in seven normal men using the N_2O method of Kety and Schmidt (19,99). After 6–12 hours at 3810 m, CBF was increased 24% from the sea level value with $Pa_{CO_2} = 35.0$, $Pa_{O_2} = 43.5$ mmHg (Fig. 5). Acute normoxia caused CBF to return to sea level values, although Pa_{CO_2} remained low (35.1 mmHg). After 3–5 days at this altitude, CBF had decreased but was still 13% above sea level control values ($Pa_{O_2} = 51.2$, $Pa_{CO_2} = 29.7$ mmHg). Acute normoxia again caused CBF to fall to sea level control, while Pa_{CO_2} rose only to 30.9 mmHg.

The continued hyperventilation with acute normoxia demonstrates a resetting of the medullary ventilatory CO_2 chemoreceptor ''set point.'' When 4% CO_2 was added to the normoxic gas, Pa_{CO_2} rose from 30.9 to 35.2 mmHg. CBF increased 34% above sea level control. We may estimate that at sea level, to increase CBF 34% would require Pa_{CO_2} to be raised about 7 mmHg. Thus the set point of the relationship of CBF to Pa_{CO_2} had been reset downward by about 10 mmHg, a resetting similar to that of ventilatory control.

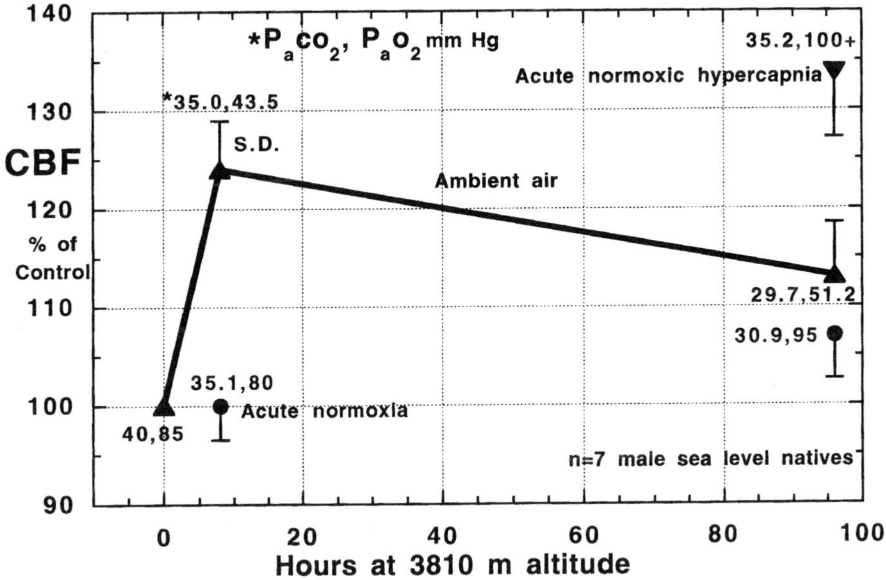

Figure 5 CBF in five newcomers after 6–12 hours and 3–5 days at 3810 m altitude. The acute responses to normoxia and normoxic hypercapnia show the resetting of hypercapnic response to a lower P_{CO_2}. (Replotted data from Ref. 99.)

This time course of prompt rise and gradual decline of CBF with hypoxia has been observed by others. Jensen et al. reported that in 12 lowlanders ascending to 3475 m, CBF was increased 24% at 24 hours and was still 4% above sea level values after 6 days (100). In nine others acclimatized to 3200 m, ascent to 4785–5430 m increased CBF 53% above estimated sea level values. Similarly, in 13 soldiers transported to 3700 m altitude, Roy et al. (101) found that CBF was 40% above control at 12–36 hours of hypoxia, and diminished to 4% above control after 4 days. Gradual fall of CBF with time in chronic hypoxia has also been reported by others (102–105).

The ventilatory and blood gas changes in acute hypoxia suggest that the rise of CBF should occur within minutes and then decrease over days. However, this dominance of initial vasodilation has not always been evident. Huang et al. (102) used TCD to measure the CBFv in the internal carotid and vertebral arteries of six healthy men (Fig. 6). After 2–3 days a CBF increase of 20% was observed. Subsequently (days 4–12) CBFv declined to values close to those observed originally at sea level. However, 2–4 hours after arrival at the summit of Pikes Peak (4300 m), they reported (but excluded from the plot of Fig. 6) that CBFv in both arteries was only slightly increased above sea level values, even though the lowest arterial O_2 saturations (Sa_{O_2}) were observed on arrival. This lack of prompt increase of CBF

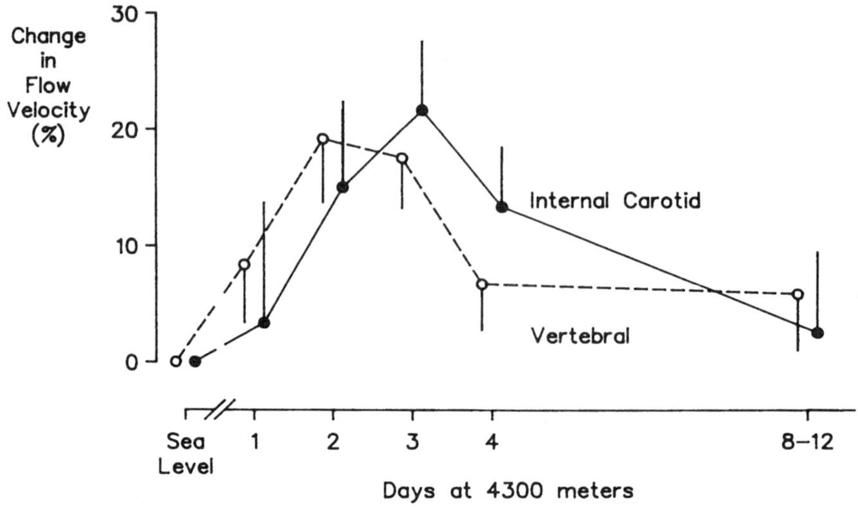

Figure 6 CBF measured by TCD in men during acclimatization on Pikes Peak was only slightly elevated on day 1, but increased on days 2 and 3, followed by a reduction on days 4–12. The failure to promptly rise at altitude was unexpected and unexplained, but was presumed to indicate a vasoconstrictive effect of the initial hypocapnia. (From Ref. 102.)

was thought to be probably due to an initial vigorous hypocapnic vasoconstriction.

Attempts to assess the effect of exercise on CBF have been performed on lowlanders on Pikes Peak and comparing Han to Tibetans in Lhasa (106,107). These studies are insufficient to conclude that increases of CBF exceed those that might be expected from the increased regional brain neural activity of motor cortex involved in work. Serial studies using TCD face problems of probe replacement, changing vessel diameter, and altered peak-to-mean Doppler pulse flow signal ratio when heart rate and arterial pressure rise (108).

In nine healthy subjects, Koppenhagen (109) investigated the effect on CBF of ascent in a chamber over 25–30 minutes. They found no change in CBF after 2 hours at 4000 m and a 10% decrease after 2 hours at 7000 m. Why CBF did not increase acutely was not explained, but again might be because in these subjects hypocapnic cerebral vasoconstriction dominated acutely. P_{CO_2} was not reported.

Furthermore, in animal studies not all investigators have observed the tendency toward normalization of CBF with time, especially with more extreme hypoxia. LaManna et al. (110) noted that in rats, despite an increased Hct and thus increased oxygen-carrying capacity, regional cerebral blood flow remained elevated after 4 weeks in a chamber at 0.5 atm, was still elevated after 4 hours of normoxia despite increased Hct, and only returned to normal after 24 hours of normoxia. Aritake et al. reported that, in cats, CBF was elevated even after 4.5 months of hypoxia, with a final inspired O_2 concentration of 8% and Hct = 56% (111), and again flow was still elevated when these animals were made acutely normoxic.

Finally, as we describe below, CBF is not elevated in either natives or long-term residents of high altitude despite low Sa_{O_2} and Pa_{O_2}. Can we account for the slow reduction of CBF in terms of the known physiological factors regulating cerebral arteriolar tone?

B. Mechanisms Normalizing CBF with Time

The initial rise of CBF in hypoxia and its variability are consistent with what is known about acute responses to combinations of low P_{O_2} and low P_{CO_2}. Nevertheless, Pa_{O_2} remains normal below sea level, and acute hyperoxia does slow CBF in altitude natives (39). Why then does CBF not remain elevated? A major question is whether there are other factors than changing blood gases that cause CBF to decline toward sea level values during long-term hypoxia. I cite eight variables that may be involved:

Rising Pa_{O_2} and Falling Pa_{CO_2}

At a constant altitude, the acclimatization process results in a rise of ventilation, which increases Pa_{O_2} and reduces Pa_{CO_2}. In some subjects this may be aided by resolution of acute pulmonary dysfunction or HAPE. Qualitatively at least both changes would be expected to decrease CBF. Of the two, Krasney et al. suggest that Pa_{CO_2} may be the more important (112). In sheep during 4 days of constant arterial hypoxia, there was no fall of CBF if Pa_{CO_2} was not permitted to fall. When P_{CO_2} was allowed to fall while arterial P_{O_2} was held constant, CBF did fall with time (Fig. 7).

The Resetting of CSF Acid-Base Balance

The effect of Pa_{CO_2} on both CBF and ventilatory control have been shown to be mediated by ECF and CSF pH (99). Peripheral carotid body hypoxic stimulation at altitude causes an immediate fall of Pa_{CO_2}, which elevates ECF pH. This alkalinity decreases central ventilatory drive and constricts cerebral arterioles, counteracting the hypoxic vasodilation. The subsequent acclimatization process involves a resetting of the relationship of pH to Pa_{CO_2} by a change of ECF and CSF HCO_3^-.

CSF HCO_3^- falls, typically from 24 to 18 mM/L at 4000 m in about 5 days, mostly during the first day. Pa_{CO_2} falls in proportion to the fall of CSF HCO_3^-, while CSF pH remains alkaline. This may be regarded as a metabolic acidosis compensating for the respiratory alkalosis of brain ECF, but it is not initially accompanied by an equivalent blood metabolic acidosis.

With continued hypoxia, neural output from carotid body hypoxic chemosensitivity is upregulated. Over about 2 weeks this may double the hypoxic ventilatory response (113). This further reduces Pa_{CO_2}, which increases CSF alkalinity and in

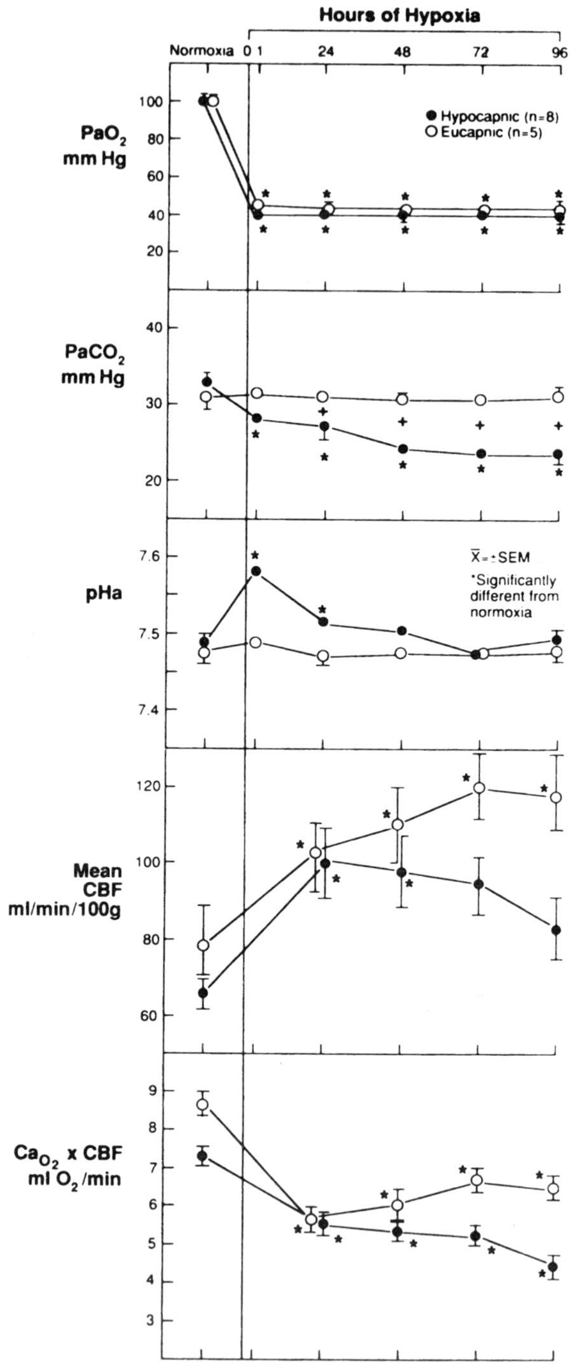
Hours of Hypoxia
Normoxia 0 1 24 48 72 96
PaO2 mm Hg
Hypocapnic (n=8)
Eucapnic (n=5)
PaCO2 mm Hg
pHa
X̄=±SEM
*Significantly different from normoxia
Mean CBF ml/min/100g
CaO2 x CBF ml O2/min

turn presumably results in additional fall of CSF HCO_3^-. The rise of Pa_{O_2} with time slightly offsets the net effect of this increased gain in sensitivity.

The balance of these two processes determines whether CSF pH falls, stays constant, or rises. The consensus is that CSF pH remains alkaline or even rises during the first few weeks, because the second process, carotid body upregulation, although slower, usually is stronger than the first, the compensatory metabolic acidosis. The resulting alkalosis is probably responsible for the falling CBF.

When a subject acclimatized to altitude is made acutely normoxic, the removal of the hypoxic drive to peripheral chemoreceptors raises Pa_{CO_2} enough to cause CBF and CSF pH to fall to normal (114). Because CSF HCO_3^- is low and can only change slowly, this residual metabolic acidosis of CSF and ECF cause ventilation to remain somewhat elevated. Calculations suggest that the pH decrement at ventral medullary chemoreceptors needed to account for the continued hyperventilation and hypocapnia is about 0.005 units (114), too small to affect CBF measurably.

Rising P_{50}

An increase of P_{50} facilitates unloading of O_2 to tissue, but the overall benefit depends on both pulmonary loading and tissue unloading, and this is dependent upon the altitude involved. At moderate elevations (3500–5000 m), a right shift of the ODC should raise tissue P_{O_2} because uptake in the lung is less affected, $P_{A_{O2}}$ being on a still relatively flat part of the ODC. At higher altitude and lower alveolar P_{O_2}, Sa_{O_2} should be adversely affected by a right-shifted ODC, offsetting the advantage to unloading of oxygen at the tissues. At about 4000 m, a P_{50} increase of 2–4 torr (69,115) might contribute slightly to tissue oxygenation, thereby enabling decreasing CBF with time.

Rising Hct

CBF is inversely proportional to Hct at altitude as at sea level (see Sec. III.B), but the increase of Hct is too slow and modest at the moderate altitudes of these studies to play a significant role in this short-term adaptation of CBF.

Autonomic Nervous System

Based on studies of sheep at simulated high altitude, Curran-Everett et al. (116) reported that increased sympathetic tone might increase cerebrovascular resistance, causing decreased CBF over time. However, no evidence exists that sympathetic tone increases with time at altitude, and in fact Ueno et al. found decreased adrener-

---

Figure 7 The CBF response to a step to steady hypoxia depends upon whether Pa_{CO_2} is held constant or not. With isocapnia, CBF continues to rise while with poikilocapnia, it slowly falls. (From Ref. 104.)

gic receptor density in major cerebral arteries in pregnant sheep and their fetuses after 110 days at 3810 m (117).

Capillary Morphology

A hypothetical remodeling of the microcirculation resulting in shorter critical diffusion paths between capillaries and mitochondrial cytochrome might enhance oxygen flux comparable to that shown in muscle with training (118). This possibility is supported by evidence for angiogenesis with prolonged hypoxia (119–121). After 49 days exposed to 7% O_2, rat cerebral cortex, striatum, and hippocampus showed a significant increase of vessel density, but the changes were small despite the prolonged and severe hypoxia. One would not predict much improvement in diffusing capacity in humans acclimatizing to altitude, though a change in vascular density may influence CBF in natives of altitude.

Biochemical Changes

In sheep, 4 days of constant hypoxia at constant normal P_{CO_2} resulted in sustained elevation of CBF (112). I have found no evidence of changes in intermediary metabolic pathways coupling hypoxia and CBF. If any adaption to hypoxia had occurred, one would anticipate reduction of cerebral vasodilation in response to an acute hypoxic challenge. But the contrary was found—CBF sensitivity actually *increased* in response to acute hypoxic (Fig. 1) after 5 days at 3810 m (26,122).

Brain Metabolic Rate

If CMR_{O_2} were to decrease with prolonged hypoxia, CBF could also fall. In Quechua natives of the Peruvian altiplano, Hochachka et al. (123) found regional glucose metabolic rates reduced by as much as 20% by positron emission tomography (PET). The greatest differences from sea level dwellers were found in areas associated with higher cortical functions. The authors speculate that low cerebral metabolic rate could protect against hypoxia as is presumed to be the case in cold-blooded amphibians and reptiles. Acute hypoxia does not decrease CMR_{O_2}, nor does CMR_{O_2} appear to be decreased with acclimatization of lowlanders. No direct measurements of CMR_{O_2} in highlanders have been reported.

To summarize, the reasons for gradual fall of CBF toward sea level values with time at altitude are likely multifactorial. Major contributors are the gradually falling P_{CO_2} with associated rises of Pa_{O_2} and possibly CSF pH. Increases of P_{50} and Hct may also play a lesser role. However, those factors that have been measured fail to explain why CBF is normal despite severe hypoxia in natives of high altitude. Whether individuals with low HVR who are more hypoxic and less hypocapnic at altitude have higher CBF than those with more vigorous HVR is not known. The cerebral vasoconstrictive effect of a strong HVR on CBF in climbers at extreme altitudes has been proposed as possibly contributing to subsequent residual neuropsychological deficits reported in some mountaineers (124) (see Sec. VII).

V. Cerebral Circulation with Prolonged Hypoxia

Since CBF falls to normal with time at altitude in newcomers, it is not surprising that it has usually been found normal or below normal in natives of high altitude. The challenge is to determine whether the low flow is adequately explained by known factors.

CBF is inversely related to Hct both at sea level (31) and in natives at high altitude (Fig. 2). At normal sea level Hct values, CBF was the same in altitude natives as the accepted normal in sea level natives. Marc-Vergnes et al. (125) reported a 20% lower CBF in 16 natives of the Bolivian altiplano compared with sea level natives. They made no correction for increased Hct. In five of their group, cisternal CSF pH averaged 7.357, no different from sea level values. They noted that the effects on CBF of increased and decreased alveolar O_2 were greater than in sea level natives, a contrast to the diminished change of ventilation with change of Po_2 in these lifelong residents of high altitude. CBF was reduced despite decreased ventilation and higher Pco_2 in those with a blunted HVR.

Brain oxygen delivery is affected by hemoglobin oxygen affinity. Acute hypoxia elevates P_{50} 2–4 mmHg in newcomers, which tends to raise tissue Po_2. Winslow et al. reported P_{50} values averaging 31.2 ± 1.9 mmHg in natives of the Andean altiplano at pH = 7.4, 2.0 mmHg higher that group's sea level average (115). This could help explain a chronically low CBF.

In eight normal adult natives of Cerro de Pasco, Peru (4300 m), with a mean Hct = 58%, Milledge and Sørensen (39) found that breathing 100% O_2 increased the arterial-internal jugular O_2 content difference from 7.89 ± 1.01 to 9.58 ± 1.17 mL dL^{-1}. This was not caused by hyperventilation induced by hyperoxia, since these subjects, on average, showed no change in $Paco_2$ with 100% O_2. Their mean (a-v) O_2 content difference while breathing air at 4300 m was greater than that of normals at sea level, indicating a subnormal CBF. Assuming no change in $CMRo_2$, they showed an 18% decrease of CBF with hyperoxia. Thus, vasodilation by hypoxia persists after a lifetime of exposure to high altitude.

Because newcomers develop increasing carotid chemoreceptor drive, and therefore a lower $Paco_2$ than natives, their CSF pH remains somewhat alkaline for weeks or months. In some natives and long-term residents at altitude, a gradual loss of peripheral hypoxic chemosensitivity reduces the ventilatory drive and slowly normalizes arterial and CSF pH. Natives of the Peruvian altiplano were found to have normal arterial and CSF pH (126) but elevated pHa (7.44) by Winslow (115). The subjects of Sørensen's study (38) showed no abnormalities in brain glucose or lactate metabolism compared with sea level values. The lactate/glucose index of 3.7 was unchanged by either one hour of breathing 100% O_2 or in some subjects by short periods of reduced inspired O_2 at altitude (LaPaz, 3800 m).

In calves, CBF is normal in chronic hypoxia and is insensitive to acute Po_2 changes, while in cats and rats CBF may remain elevated in chronic hypoxia and fail to fall with acute normoxia. Human studies suggest that after long residence, or birth, at high altitude, CBF in chronic hypoxia may be normal or subnormal. This low flow is not explained by P_{CO_2}, CSF pH, or Hct. But in such persons CBF remains sensitive to acute changes of P_{O_2} in either direction. We are left to attribute the unexpectedly low flow to the slightly elevated P_{50}, biochemical changes, increased diffusing capacity with increased capillarity, and/or reduced CMR_{O_2}. Determining the roles of these factors in high altitude natives is a reasonable research target.

VI. CBF in AMS and HACE: Cause or Effect?

The possible relationships of CBF to altitude illnesses have been much discussed and studied to answer such questions as: Does the high CBF at altitude cause or exacerbate AMS and HACE (122,127,128)? Do subjects with AMS have a poor ventilatory response to hypoxia, a lower Sa_{O_2} and a higher P_{CO_2}, which should result in higher CBF (129)?

A. Does CBF Differ in Those with AMS or HACE Versus Those Without?

Reeves et al. (130) found no correlation of CBF with headache in newcomers at altitude. Jensen et al. reported that, in nine normal subjects, ascending from 3200 to 4785–5430 m, CBF increased 53% above estimated sea level values. Increases in CBF were similar in subjects with or without AMS (100). Baumgartner et al. (131) assessed CBFv of both middle cerebral arteries by TCD in 23 subjects at 490 m and over 3 days after a rapid ascent to the Capana Margarita laboratory on Monta Rosa (4559 m). After ascent, CBFv increased 48% in the subjects who developed AMS and 27% in subjects who did not. The rise of CBFv correlated inversely with arterial P_{O_2} but was unrelated to blood pressure, Pa_{CO_2}, or [Hb]. They concluded that subjects with AMS had a higher CBFv than healthy subjects due to a lower arterial P_{O_2}, presumably representing a poorer HVR.

It is unlikely that studies of this kind can answer the question of etiology of AMS and HACE because those conditions may impede CBF by the increased intracranial pressure.

B. Attempts to Treat AMS with CO_2

For many years (since Mosso proposed acapnia to be the cause of AMS in 1898) (132), there has been a lingering suspicion that hypocapnia contributed to the illness. Attempts to test this have been made by adding CO_2 to inspired air during hypoxic exposure.

To determine the role of CO_2, Yang et al. (24) exposed chronically instrumented ewes to 96 hours of hypoxia (Pa_{O_2} = 40 mmHg) in an environmental chamber. One group was permitted to become hypocapnic (Pa_{CO_2} = 27 mmHg), while the other was kept eucapnic (Pa_{CO_2} = 37 mmHg). AMS, estimated from food and water intakes and behavior, occurred in 9 of 12 with hypocapnia and 9 of 9 with normocapnia ($p < 0.05$). Intracranial pressure and the ICP-sagittal sinus pressure gradient increased in AMS sheep in both groups. CBF was greater in the normocapnic animals. Cerebral O_2 and glucose uptakes were normal in both groups. Brain edema, reflected by elevated wet-to-dry tissue weight ratios, occurred only in AMS sheep. They concluded that AMS is associated with cerebral edema despite normal brain aerobic metabolism and that CO_2 supplementation at constant Pa_{O_2} did not reduce AMS, intracranial pressure, or brain tissue edema.

Maher et al. (133) tested whether prevention of hypocapnia and alkalosis would ameliorate the symptoms of acute mountain sickness (AMS). Five subjects were exposed to simulated high altitude for 4 days but kept isocapnic by adding 3.8% CO_2 to the chamber. Four other subjects were allowed to become hypocapnic. Alveolar oxygen tensions (55–60 mmHg) were held equal (by adjusting barometric pressure). They reported that severity of symptoms was ''clearly greater'' in normocapnic than in hypocapnic subjects (no symptom scoring was done).

Harvey et al. (134) added 3% CO_2 to inhaled air in six subjects with AMS during a medical expedition to 5400 m. However, they made no effort to keep alveolar or arterial P_{O_2} constant, so the resulting higher alveolar P_{CO_2} and rise of ventilation increased alveolar oxygen tension by 24–40%. Symptoms of AMS were rapidly relieved. In three subjects cerebral blood flow increased by 17–39%, so that oxygen delivery to the brain would have been considerably improved. They pronounced their findings a ''rediscovery'' despite the preceding paper by Maher et al. using constant alveolar P_{O_2} showing an opposite result.

The claim of beneficial effect of addition of 3% CO_2 to air was soon challenged by Bärtsch et al. (135), who randomly allocated 20 mountaineers with AMS at 4559 m to three treatment groups: (1) with 33% O_2, (2) with 3% CO_2 in air, and (3) an air control. Thirty-three percent O_2 significantly relieved symptoms of AMS and reduced CBF, but CO_2 addition did not significantly ameliorate AMS, despite the rise of P_{CO_2}, ventilation, and alveolar P_{O_2}.

The differing results in these studies appear to be explained by differences in P_{O_2} and Sa_{O_2} caused by experimental design. When CO_2 is added to inspired air, it only slightly reduces $P_{I_{O2}}$ but raises $P_{A_{O2}}$ by inducing hyperventilation. Such studies find an improvement in or less AMS. When Pa_{O_2} is constant at differing Pa_{CO_2} levels, AMS is found to be more severe with the higher Pa_{CO_2}. This is explained in part by the effect in the lung of the higher Pa_{CO_2}, which reduces Sa_{O_2} (due to the Bohr effect) at constant Pa_{O_2}. While this pulmonary loss is partly offset by higher capillary P_{O_2}, facilitating unloading, the net effect at the altitude used in these studies is to reduce capillary P_{O_2}. At constant Pa_{O_2} and Sa_{O_2}, there is no evidence of a beneficial effect of isocapnia, despite their (assumed) higher CBF. It now seems clear that low P_{CO_2} is beneficial rather than contributing to AMS.

C. The Proposed Role of High CBF in Causing AMS

The relationships of hypoxia and CO_2 to intracranial dynamics and CBF was investigated in sheep and goats by Yang et al. (78). In six unanesthetized goats subjected to hypobaric hypoxia at a simulated 4000 m altitude for 2 hours CBF increased from 1.46 ± 0.11 to 1.85 ± 0.08 mL g^{-1} min^{-1}. ICP increased from 15.4 ± 1.8 to 27.4 ± 8.8 cm H_2O. With both 2 and 72 hours of hypoxia exposure, cerebral water content increased and intracranial compliance fell (136).

In a review of this work, Krasney (128) postulated that HACE is in some way due to hypoxic cerebral vasodilation and elevated cerebral capillary hydrostatic pressure, which results in reduced brain compliance with compression of intracranial structures in the absence of altered global brain metabolism. He postulated that these primary intracranial events elevate peripheral sympathetic activity, which acts neurologically in the lung, possibly in concert with pulmonary capillary stress failure, to cause HAPE, and in the kidney to promote salt and water retention. He proposed that the adrenergic responses are likely modulated by striking increases of aldosterone, vasopressin, and atrial natriuretic peptide, compounded by increase in sympathetic neural activity.

D. Evidence That High CBF Without Hypoxia Fails to Cause AMS

To test whether high CBF alone could cause cerebral edema, Yang and Krasney (137) kept sheep for 4 days in elevated CO_2 (52–55 mmHg). CBF remained about twice normal, and $CMRo_2$ was increased both during exposure and in the posthypercapnic period. They observed no symptoms like those of AMS or HACE, although brain water rose slightly from 79.8 ± 0.24 to 80.3 ± 0.2% ($p < 0.05$). High CBF is unlikely to be the cause of HACE, and other factors must be sought.

E. Possible Role of VEGF and Angiogenesis in HACE

Retinal petechial hemorrhages found in many climbers at extreme altitude suggest a pathological process involving cerebral circulation, which may mirror changes throughout the brain. These hemorrhages may be a result of the first step in a process called angiogenesis (138). Tissue hypoxia is the initiating stimulus of angiogenesis, initiating growth of new capillaries into hypoxic tissue. Hypoxia causes expression in rat brain of vascular endothelial growth factor (VEGF), formerly termed vascular permeability factor (VPF) (139). VEGF attacks and dissolves capillary basement membranes, permitting plasma and red cell leakage. Harik et al. found increased microvascular protein at one week and increased DNA in adult rats after 3 weeks at 0.5 A (119), and Mironov et al. reported increased capillary segment length (121). Dexamethasone inhibits angiogenesis. Twelve reports show that it is effective in preventing and treating AMS and HACE (138).

VII. Brain Injuries at Extreme Altitude

A major unanswered question is whether hypocapnic cerebral vasoconstriction plus the Bohr shift of the ODC contribute to AMS, neuropsychological functional limitation, or brain cellular injury. There is evidence that the brain is injured in some climbers at extreme altitudes, but it is not clear whether low blood flow may be involved (124).

Song et al. (140) reported cerebral thrombi in several climbers who had gone higher than 5000 m for longer than 3 weeks. They speculated that the cause was hemoconcentration resulting from secondary polycythemia and dehydration at altitude. In the fetus and infant, hypoxemia, whether from high altitude or other causes, is associated with increased cerebrovascular morbidity. Longo et al. (141) compared cerebral arteries obtained from normoxic and chronically hypoxic sheep adults and fetuses. Long-term hypoxemia was associated with generalized increase in base-soluble protein (5–50%), a depression of the maximum potassium-induced tensions (16–49%), and a depression of the relaxation responses to S-nitroso-*N*-acetylpenicillamine (1–11%), which releases nitric oxide into solution upon hydration. They concluded that chronic hypoxemia depresses cerebral vascular smooth muscle and endothelial hypoxic response to a greater extent in the fetus than in adults.

VIII. Personal Comment Regarding Hypoxic Distress

During the past decade my colleagues and I have tested pulse oximeter accuracy by multiple brief (2–3 min) steps to Sa_{O_2} values as low as 50% in several hundred subjects in order to calibrate instruments. I have noted that acute hypoxia uniformly evokes greater anxiety and hyperventilation on the first trial. Step hypoxia, even with voluntary hyperventilation and low Pa_{CO_2}, may be less well tolerated at 75% on the first run than at 55% on later runs. Since we allow only 1–4 minutes at high P_{O_2} between runs, we presume this subsequent relief of distress might be related to the phenomenon called hypoxic ventilatory decline in (HVD) in which ventilatory drive fades over 5–20 minutes to half or less at constant P_{O_2} and P_{CO_2}. As a subject of hypoxic ventilatory or CBF experimentation, I have noted extreme anxiety on some hypoxic tests at levels of saturation that I found easily tolerable at other times. Hypoxia stimulates the sympathetic nervous system and catechol secretion. I note this observation because HVD may play a role in the variability of the rise in CBF in acute hypoxia, even at altitude, primarily through its effect on ventilation and P_{CO_2} since most evidence suggests that catecholamines and sympathetic nerves have no direct effect on CBF.

IX. Summary

Cerebral vasodilation by arterial hypoxemia occurs within 1 second and is mediated primarily by adenosine. Its site of expression is unknown but must be very close to the arteriolar smooth muscle in view of the rapid response, which tends to implicate the glial feet that invest all arterioles. H^+ from lactic acidosis may also contribute later. K^+ and NO, known cerebral vasodilators, are not proved to be involved in hypoxic response. K^+ dilates with neural activation, and NO with hypercapnia. During the first hours at altitude, CBF rises 30–60% and then over days declines toward normal sea level values. The magnitude of the initial rise depends on the altitude and four factors: the individual cerebral vascular and ventilatory sensitivities to both O_2 and CO_2. The reason for the subsequent reduction of flow remains poorly understood, but it is at least partly explained by the concurrent rise of Pa_{O_2} and fall of Pa_{CO_2}, though not by polycythemia or reduced sensitivity of brain or cerebral vessels to hypoxia. Later effects of hypoxia with effects on CBF include reduced CSF HCO_3^-, increasing and then decreasing peripheral chemoreceptor sensitivity, and rising Hct. None of the known factors modulating CBF (including polycythemia) can fully account for the evidence that in several studies, natives of high altitude and acclimatized lowlanders have been found to have normal or subnormal CBF (compared with sea level normals). It seems probable that some remodeling of the capillaries increases local O_2-diffusing capacity. Reduced brain CMR_{O_2} has been reported in humans native to high altitude, which could help explain the low CBF. AMS and HACE may be accompanied by high CBF, because Pa_{O_2} is lower, but high CBF (if caused by hypercapnia) fails to cause AMS or HACE. Hypocapnic vasoconstriction may contribute to the subtle brain injury seen in mountaineers after return from extremely high altitudes.

References

1. Li Y, Naqui A, Frey TG, Chance B. A new procedure for the purification of monodisperse highly active cytochrome c oxidase from bovine heart. Biochem J 1987; 242: 417–423.
2. Fennema M, Wessel JN, Faithful NS, Erdmann W. Tissue oxygen tension in the cerebral cortex of the rabbit. Adv Exp Med Biol 1989; 248:451–460.
3. Bickler PE, Litt L, Severinghaus JW. Effects of acetazolamide on cerebrocortical NADH and blood volume. J Appl Physiol 1988; 65:428–433.
4. Nemoto EM, Severinghaus JW. Stereospecific permeability of rat blood-brain barrier to lactic acid. Stroke 1974; 5:81–84.
5. Harder DR. Pressure-dependent membrane depolarization in cat middle cerebral artery. Circ Res 1984; 55:197–202.
6. Faraci FM, Heistad DD. Regulation of large cerebral arteries and cerebral microvascular pressure. Circ Res 1990; 66:8–17.
7. Ngai AC, Winn HR. Modulation of cerebral arteriolar diameter by intraluminal flow and pressure. Circ Res 1995; 77:832–840.
8. Hudetz AG. Blood flow in the cerebral capillary network: a review emphasizing observations with intravital microscopy. Microcirculation 1997; 4:233–252.
9. Clark LC, Misrahy G, Fox RP. Chronically implanted polarographic electrodes. J Appl Physiol 1958; 13:85–91.
10. Ozanne GM, Vilnis V, Severinghaus JW. Implications of O_2 wave shapes and synchrony for regulation of cerebral microcirculation. In: Harper M, Jennett WB, Miller JD, Rowan JO, eds. Blood Flow and Metabolism in the Brain. Edinburgh: Churchill-Livingston, 1975:9.3–9.7.
11. Bickler PE, Julian D. Regional cerebral blood flow and tissue oxygenation during hypocarbia in geese. Am J Physiol 1992; 263:R221–225.

12. Frankel HM, Garcia E, Malik F, Weiss JK, Weiss HR. Effect of acetazolamide on cerebral blood flow and capillary patency. J Appl Physiol 1992; 73:1756–1761.
13. Wahl M, Schilling L. Regulation of cerebral blood flow: a brief review. Acta Neurochirurg 1993; 59 (suppl):3–10.
14. Cholet N, Seylaz J, Lacombe P, Bonvento G. Local uncoupling of the cerebrovascular and metabolic responses to somatosensory stimulation after neuronal nitric oxide synthase inhibition. J Cereb Blood Flow Metab 1997; 17:1191–1201.
15. Kogure K, Scheinberg P, Reinmuth OM, Fujishima M, Busto R. Mechanisms of cerebral vasodilatation in hypoxia. J Appl Physiol 1970; 29:223–229.
16. Freeman J, Ingvar DH. Elimination by hypoxia cerebral blood flow autoregulation and EEG relationship. Exp Brain Res 1968; 5:61–71.
17. Silver IA. Cellular microenvironment in relation to local blood flow. Ciba Found Symp 1978; 56:49–67.
18. Shimojyo S, Scheinberg P, Kogure K, Reinmuth OM. The effects of graded hypoxia upon transient cerebral blood flow and oxygen consumption. Neurology 1968; 18:127–133.
19. Kety SS, Schmidt CF. The effects of altered arterial tensions of carbon dioxide and oxygen on cerebral blood flow and cerebral oxygen consumption of normal young men. J Clin Invest 1948; 27:484–492.
20. Iwamoto J, Curran-Everett DC, Krasney E, Krasney JA. Cerebral metabolic and pressure-flow responses during sustained hypoxia in awake sheep. J Appl Physiol 1991; 71:1447–1453.
21. Johannsson H, Siesjo BK. Blood flow and oxygen consumption of the rat brain in profound hypoxia. Acta Physiol Scand 1974; 90:281–282.
22. Trojan S, Kapitola J. [Reactive hyperemia in the brain of rats after high altitude hypoxia]. Sb Lek 1990; 92:97–102.
23. Cohen PJ, Alexander SC, Smith TC, Reivich M, Wollman H. Effects of hypoxia and normocarbia on cerebral blood flow and metabolism in conscious man. J Appl Physiol 1967; 23:183–189.
24. Yang SP, Bergo GW, Krasney E, Krasney JA. Cerebral pressure-flow and metabolic responses to sustained hypoxia: effect of CO_2. J Appl Physiol 1994; 76:303–313.
25. Ashwal S, Dale PS, Longo LD. Regional cerebral blood flow: studies in the fetal lamb during hypoxia, hypercapnia, acidosis, and hypotension. Pediatr Res 1984; 18:1309–1316.
26. Jensen JB, Sperling B, Severinghaus JW, Lassen NA. Augmented hypoxic cerebral vasodilation in man during five days at 3810 M altitude. J Appl Physiol 1996; 80: 1214–1218.
27. Macmillan V, Salford LG, Siesjo BK. Metabolic state and blood flow in rat cerebral cortex, cerebellum and brainstem in hypoxic hypoxia. Acta Physiol Scand 1974; 92: 103–113.
28. Cavazzuti M, Duffy TE. Regulation of local cerebral blood flow in normal and hypoxic newborn dogs. Ann Neurol 1982; 11:247–257.
29. Hamilton AJ, Cymmerman A, Black PM. High altitude cerebral edema. Neurosurgery 1986; 19:841–849.
30. Myers RE. Four patterns of perinatal brain damage and their conditions of occurrence in primates. Adv Neurol 1975; 10:223–234.
31. Harrison MJ. Influence of haematocrit in the cerebral circulation. Cerebrovasc Brain Metab Rev 1989; 1:55–67.

32. Chapler CK, Cain SM. The physiologic reserve in oxygen carrying capacity: studies in experimental hemodilution. Can J Physiol Pharmacol 1986; 64:7–12.
33. Todd MM, Farrell S, Wu B. Cerebral blood flow during hypoxemia and hemodilution in rabbits: different roles for nitric oxide? J Cereb Blood Flow Metab 1997; 17:1319–1325.
34. Holzman IR, Tabata B, Edelstone DI. Blood flow and oxygen delivery to the organs of the neonatal lamb as a function of hematocrit. Pediatr Res 1986; 20:1274–1279.
35. Borgstrom L, Johannsson H, Siesjo BK. The influence of acute normovolemic anemia on cerebral blood flow and oxygen consumption of anesthetized rats. Acta Physiol Scand 1975; 93:505–514.
36. Johannson H, Siesjo BK. Brain energy metabolism in anesthetized rats in acute anemia. Acta Physiol Scand 1975; 93:515–524.
37. Hudak ML, Koehler RC, Rosenberg AA, Traystman RJ, Jones MDJ. Effect of hematocrit on cerebral blood flow. Am J Physiol 1986; 251:H63–H70.
38. Sørensen SC, Lassen NA, Severinghaus JW, Coudert J, Zamora MP. Cerebral glucose metabolism and cerebral blood flow in high-altitude residents. J Appl Physiol 1974; 87:305–310.
39. Milledge JS, Sørensen SC. Cerebral arteriovenous oxygen difference in man native to high altitude. J Appl Physiol 1972; 32:687–689.
40. Koehler RC, Traystman RJ, Zeger S, Rogers MC, Jones MDJ. Comparison of cerebrovascular response to hypoxic and carbon monoxide hypoxia in newborn and adult sheep. J Cereb Blood Flow Metab 1984; 4:115–122.
41. Koehler RC, Traystman RJ, Rosenberg AA, Hudak ML, Jones MDJ. Role of O_2-hemoglobin affinity on cerebrovascular response to carbon monoxide hypoxia. Am J Physiol 1983; 245:H1019–1023.
42. Woodson RD. Functional consequences of altered blood oxygen affinity. Acta Biol Med Ger 1981; 40:733–736.
43. Siesjo BK, Johannsson H, Norberg K, Salford I. Brain function metabolism and blood flow in moderate and severe arterial hypoxia. In: Ingvar DH, Lassen NA, eds. Brain Work. The Coupling of Function, Metabolism and Blood Flow in the Brain. Copenhagen: Munksgaard, 1975:101–125.
44. Faraci FM, Brian JEJ. Nitric oxide and the cerebral circulation. Stroke 1994; 25:692–703.
45. Janigro D, Wender R, Ransom G, Tinklepaugh DL, Winn HR. Adenosine-induced release of nitric oxide from cortical astrocytes. Neuroreport 1996; 7:1640–1644.
46. Kuschinsky M, Wahl M, Bosse O, Thurau K. Perivascular potassium and pH as determinants of local pial arterial diameter in cats. A microapplication study. Circ Res 1972; 31:240–247.
47. Winn HR, Rubio R, Berne RM. Brain adenosine concentration during hypoxia in rats. Am J Physiol 1981; 241:H235–242.
48. Reivich M. Arterial P_{CO_2} and cerebral hemodynamics. Am J Physiol 1964; 206:25–34.
49. Buchanan JE, Phillis JW. The role of nitric oxide in the regulation of cerebral blood flow. Brain Res 1993; 610:248–255.
50. Astrup J, Heuser D, Lassen NA, Nilsson BB, Norberg K, Siesjo BK. Evidence against H^+ and K^+ as main factors for the control of cerebral blood flow: a microelectrode study. Ciba Foundation Symp 1978; 56:313–337.

51. Taguchi H, Heistad DD, Kitazono T, Faraci FM. ATP-sensitive K^+ channels mediate dilatation of cerebral arterioles during hypoxia. Cir Res 1994; 74:1005–1008.
52. Kitazono T, Faraci FM, Taguchi H, Heistad DD. Role of potassium channels in cerebral blood vessels. Stroke 1995; 26:1713–1723.
53. Cain SM, Dunn JE. Transient arterial lactic acid changes in unanesthetized dogs at 21,000 feet. Am J Physiol 1964; 206:1437–1440.
54. Hamer J, Wiedemann K, Berlet H, Weinhardt F, Hoyer S. Cerebral glucose and energy metabolism, cerebral oxygen consumption, and blood flow in arterial hypoxaemia. Acta Neurochir (Wien) 1978; 44:151–160.
55. Morii S, Ngai AC, Ko KR, Winn HR. Role of adenosine in regulation of cerebral blood flow: effects of theophylline during normoxia and hypoxia. Am J Physiol 1987; 253:H165–175.
56. Laudignon N, Farri E, Beharry K, Rex J, Aranda IV. Influence of adenosine on cerebral blood flow during hypoxic hypoxia in the newborn piglet. J Appl Physiol 1990; 68: 1534–1541.
57. Meno JR, Ngai AC, Winn HR. Changes in pial arteriolar diameter and CSF adenosine concentrations during hypoxia. J Cereb Blood Flow Metab 1993; 13:214–220.
58. Underwood MD, Iadecola C, Sved A, Reis DJ. Stimulation of Cl area neurons globally increases regional cerebral blood flow but not metabolism. J Cereb Blood Flow Metab 1992; 12:844–855.
59. McPherson RW, Koehler RC, Traystman RJ. Hypoxia, alpha-2-adrenergic, and nitric oxide dependent interactions on canine cerebral blood flow. Am J Physiol 1994; 266: H476–482.
60. Hudetz AG, Shen H, Kampine JP. Nitric oxide from neuronal NOS plays critical role in cerebral capillary flow response to hypoxia. Am J Physiol 1998; 274:H982–989.
61. Betz E, Kozak R, Heredia A. Die Bedeutung des pH der Gehrinrinde für die Regulation der corticalen Durchblutung. Pflugers Arch Ges Physiol 1966; 289:R3.
62. Harper AM, Bell RA. Effect of metabolic acidosis and alkalosis on the blood flow through the cerebral cortex. J Neurol Neurosurg Psychiatry 1963; 26:341–352.
63. Lambertsen C, Semple SJG, Smyth MG, Gelfand R. H^+ and P_{CO_2} as chemical factors in respiratory and cerebral circulatory control. J Appl Physiol 1961; 16:473–478.
64. Mitchell RA, Carman CT, Severinghaus JW, Richardson BW, Singer MM, Shnider S. Stability of cerebrospinal fluid pH in chronic acid base disturbances in blood. J Appl Physiol 1965; 20:27–30.
65. Severinghaus JW. Regulation of respiration and CSF acid base balance. In: Loeppke JA, Riedesel ML, eds. Oxygen Transport to Human Tissues. New York: Elsevier North Holland, Inc, 1982; 231–241.
66. Weiskopf RB, Gabel RA, Fencl V. Alkaline shift in lumbar and intracranial CSF in man after 5 days at high altitude. J Appl Physiol 1976; 41:93–97.
67. Forster HV, Dempsey JA, Chosy LW. Incomplete compensation of CSF H^+ in man during acclimatization to high altitude (4300 M). J Appl Physiol 1975; 38:1067–1072.
68. Lenfant C, Torrence JD, Reynafarje C. The shift of the O_2 Hb dissociation curve at altitude: mechanism and effect. J Appl Physiol 1971; 30:625–631.
69. Weiskopf RB, Severinghaus JW. Lack of effect of high altitude on hemoglobin oxygen affinity. J Appl Physiol 1972; 33:276–277.
70. Severinghaus JW, Lassen NA. Step hypocapnia to separate arterial from tissue P_{CO_2} in the regulation of cerebral blood flow. Circ Res 1967; 20:272–278.

71. Kennealy JA, McLennan JE, Loudon RG, McLaurin RL. Hyperventilation-induced cerebral hypoxia. Am Rev Respir Dis 1980; 22:407–412.
72. Houston CS, Sutton JR, Cymerman A, Reeves JT. Operation Everest II. Man at extreme altitude. J Appl Physiol 1987; 63:877–882.
73. Rosenberg AA. Response of the cerebral circulation to profound hypocarbia in neonatal lambs. Stroke 1988; 19:1365–1370.
74. Grubb B, Mills CD, Colacino JM, Schmidt-Nielsen K. Effect of arterial carbon dioxide on cerebral blood flow in ducks. Am J Physiol 1977; 232:H596–601.
75. Grubb B, Jones JH, Schmidt-Nielsen K. Avian cerebral blood flow: influence of the Bohr effect on oxygen supply. Am J Physiol 1979; 236:H744–749.
76. Faraci FM, Kilgore DLJ, Fedde MR. Oxygen delivery to the heart and brain during hypoxia: Pekin duck vs. bar-headed goose. Am J Physiol 1984; 247:R69–75.
77. Faraci FM, Fedde MR. Regional circulatory responses to hypocapnia and hypercapnia in bar-headed geese. Am J Physiol 1986; 250:R499–504.
78. Yang YB, Sun B, Yang Z, Wang J, Pong Y. Effects of acute hypoxia on intracranial dynamics in unanesthetized goats. J Appl Physiol 1993; 74:2067–2071.
79. Faraci FM. Role of endothelium-derived relaxing factor in cerebral circulation: large arteries vs. microcirculation. Am J Physiol 1991; 261:H1038–H1042.
80. Wang Q, Paulson OB, Lassen NA. Effect of nitric oxide blockade by NG-nitro-L-arginine on cerebral blood flow response to changes in carbon dioxide tension. J Cereb Blood Flow Metab 1992; 12:947–953.
81. Maren TH, Mayer E, Wadsworth BC. Carbonic anhydrase inhibition. I. The pharmacology of Diamox, 2-acetylamino-1,3,4-thiadiazole-5-sulfonamide. Bull Johns Hopkins Hosp 1954; 95:199–255.
82. Tomashefski JF, Chinn HI, Clark RT. Effect of carbonic anhydrase inhibition on respiration. Am J Physiol 1954; 177:451–454.
83. Mithoefer JC. Inhibition of carbonic anhydrase: its effect on carbon dioxide elimination by the lungs. J Appl Physiol 1959; 14:109–115.
84. Brzezinski J, Kjallquist A, Siesjo BK. Mean carbon dioxide tension in the brain after carbonic anhydrase inhibition. J Physiol 1967; 188:13–24.
85. Carter ET, Clark RT. Effects of carbonic anhydrase inhibition during acute hypoxia. J Appl Physiol 1958; 13:47–52.
86. Cain SM, Dunn JE. Increase of arterial oxygen tension at altitude by carbonic anhydrase inhibition. J Appl Physiol 1965; 20:882–884.
87. Cain SM, Dunn JE. Low doses of acetazolamide to aid accommodation of men to altitude. J Appl Physiol 1966; 21:1195–1200.
88. Posner JB, Plum F. Toxic effects of carbon dioxide and acetazolamide in hepatic encephalopathy. J Clin Invest 1960; 39:1246–1258.
89. Ehrenreich DL, Gurns RA, Alman RW, Fazekas JF. Influence of acetazolamide on cerebral blood flow. Arch Neurol 1961; 5:125–134.
90. Friberg L, Kastrup J, Rizzi D, Jensen JB, Lassen NA. Cerebral blood flow and end-tidal P_{CO_2} during prolonged acetazolamide treatment in humans. Am J Physiol 1990; 258:H954–959.
91. Lassen NA, Friberg L, Kastrup J, Rizzi D, Jensen JJ. Effects of acetazolamide on cerebral blood flow and brain tissue oxygenation. Postgrad Med J 1987; 63:185–187.
92. Huang SY, McCullough RE, McCullough RG, Micco AJ, Manco-Johnson M, Weil JV, Reeves JT. Usual clinical dose of acetazolamide does not alter cerebral blood flow velocity. Respir Physiol 1988; 72:315–326.

93. Kjallquist A, Siesjo BK. Increase in the intracellular bicarbonate concentration in the brain after acetazolamide. Acta Physiol Scand 1966; 68:255–267.
94. Cotev S, Lee J, Severinghaus JW. The effects of acetazolamide on cerebral blood flow and cerebral tissue Po_2, Anesthesiology 1968; 29:471–477.
95. Severinghaus JW, Hamilton RN, Cotev S. Carbonic acid production and the role of carbonic anhydrase in decarboxylation in brain. Biochem J 1969; 114:703–705.
96. Bickler PE, Litt L, Banville DL, Severinghaus W. Effects of acetazolamide on cerebral acid-base balance. J Appl Physiol 1988; 65:422–427.
97. Vorstrup S, Jensen KE, Thomsen C, Henricksen O, Lassen N, Paulson OB. Neuronal pH regulation: constant normal intracellular pH is maintained in brain during low extracellular pH induced by acetazolamide—31P NMR study. J Cerebral Blood Flow Metab 1989; 9:417–421.
98. LaManna JC, McCracken KA. Carbonic anhydrase inhibition and cerebral cortical oxygenation in the rat. Adv Exp Med Biol 1990; 277:335–343.
99. Severinghaus JW, Chiodi H, Eger Eld, Brandstater B, Hornbein TF. Cerebral blood flow in man at high altitude. Role of cerebrospinal fluid pH in normalization of flow in chronic hypocapnia. Circ Res 1966; 19:274–282.
100. Jensen JB, Wright AD, Lassen NA, Harvey TC, Winterborn MH, Raichle ME, Bradwell AR. Cerebral blood flow in acute mountain sickness. J Appl Physiol 1990; 69: 430–433.
101. Roy SB, Guleria JS, Khanna PK, Talwar JR, Manchanda SC, Pande JN, Kaushik VS, Subba PS, Wood JE. Immediate circulatory response to high altitude hypoxia in man. Nature 1968; 217:1177–1178.
102. Huang SY, Moore LG, McCullough RE, McCullough RG, Micco AJ, Fulco C, Cymerman A, Manco-Johnson M, Weil JV, Reeves JT. Internal carotid and vertebral arterial flow velocity in men at high altitude. J Appl Physiol 1987; 63:395–400.
103. Bocking AD, Gagnon R, White SE, Homan J, Milne KM, Richardson BS. Circulatory responses to prolonged hypoxemia in fetal sheep. Am J Obstet Gynecol 1988; 159: 1418–1424.
104. Krasney JA, Jensen JB, Lassen NA. Cerebral blood flow does not adapt to sustained hypoxia. J Cereb Blood Flow Metab 1990; 10:759–764.
105. Manohar M, Parks CM, Busch M, Bisgard GE. Bovine regional brain blood flow during sojourn at a simulated altitude of 3500 m. Respir Physiol 1984; 58:111–122.
106. Huang SY, Tawney KW, Bender PR, Groves BM, McCullough RE, McCullough RG, Micco AJ, Manco-Johnson M, Cymerman A, Greene ER, et al. Internal carotid flow velocity with exercise before and after acclimatization to 4,300 m. J Appl Physiol 1991; 71:1469–1476.
107. Huang SY, Sun S, Droma T, Zhuang J, Tao JX. McCullough RG, McCullough RE, Micco AJ, Reeves JT, Moore LG. Internal carotid arterial flow velocity during exercise in Tibetan and Han residents of Lhasa (3,658 m). J Appl Physiol 1992; 73:2638–2642.
108. Poulin MJ, Syed RJ, Robbins PA. The use of transcranial doppler to assess cerebral blood flow during exercise in humans. FASEB J 1998; 12:A1116.
109. Koppenhagen K. Cerebral blood flow under hypobaric conditions: effects of pentoxifylline ('Trental' 400). Pharmatherapeutica 1984; 4:1–5.
110. LaManna JC, McCracken KA, Strohl KP. Changes in regional cerebral blood flow and sucrose space after 3–4 weeks of hypobaric hypoxia (0.5 ATM). Adv Exp Med Biol 1989; 248:471–477.

111. Aritake K, Mayer HM, Fritschka E, Cervos-Navarro J, Takakura K. [Cerebral hemodynamics in chronic hypoxic hypoxia]. No To Shinkei 1986; 38:363–369.
112. Krasney JA, Hajduczok G, Miki K, Matalon S. Peripheral circulatory response to 96 hours of eucapnic hypoxia in conscious sheep. Respir Physiol 1985; 59:197–211.
113. Sato M, Severinghaus JW, Powell FL, Xu FD, Spellman MJ. Augmented hypoxic ventilatory response in men at altitude. J Appl Physiol 1992; 73:101–107.
114. Crawford RD, Severinghaus JW. CSF pH and ventilatory acclimatization to altitude. J Appl Physiol 1978; 45:275–283.
115. Winslow RM, Monge CC, Statham NJ, Gibson CG, Charache S, Whittembury J, Moran O, Berger RL. Variability of oxygen affinity of blood: human subjects native to high altitude. J Appl Physiol 1981; 51:1411–1416.
116. Curran-Everett DC, Meredith MP, Krasney JA. Acclimatization to hypoxia alters cerebral convective and diffusive O_2 delivery. Respir Physiol 1992; 88:355–371.
117. Ueno N, Zhao Y, Zhang L, Longo LD. High altitude-induced changes in α_1-adrenergic receptors and Ins(1,4,5)P3 responses in cerebral arteries. Am J Physiol 1997; 272: R669–R674.
118. Knight DR, Schaffartzik W, Poole DC, Hogan MC, Bebout DE, Wagner PW. Effects of hyperoxia on maximal leg O_2 supply and utilization in man. J Appl Physiol 1993; 75:2586–2594.
119. Harik SI, Hritz MA, LaManna JC. Hypoxia-induced brain angiogenesis in the adult rat. J Physiol 1995; 485:525–530.
120. Patt S, Sampaolo S, Theallier-Janko A, Tschairkin I, Cervos-Navarro J. Cerebral angiogenesis triggered by severe chronic hypoxia displays regional differences. J Cereb Blood Flow Metab 1997; 17:801–806.
121. Mironov V, Hritz MA, LaManna JC, Hudetz AG, Harik SI. Architectural alterations in rat cerebral microvessels after hypobaric hypoxia. Brain Res 1994; 660:73–80.
122. Lassen NA. Increase of cerebral blood flow at high altitude: its possible relation to AMS. Int J Sports Med 1992; 13 (suppl 1):S47–48.
123. Hochachka PW, Clark CM, Brown WD, Stanley C, Stone CK, Nickles RJ, Zhu GG, Allen PS, Holden JE. The brain at high altitude: hypometabolism as a defense against chronic hypoxia? J Cerebral Blood Flow Metab 1994; 14:671–679.
124. Hornbein TF, Townes BD, Schoene RB, Sutton JR, Houston CS. The cost to the central nervous system of climbing to extremely high altitude. N Engl J Med 1989; 321:1714–1719.
125. Marc-Vergnes JP, Antezana G, Coudert J, Gourdin D, Durand J. [Cerebral blood flow and energy metabolism, and acid-base equilibrium in cerebrospinal fluid in high altitude residents]. J Physiol (Paris) 1974; 68:633–654.
126. Severinghaus JW, Carcelen AB. Cerebrospinal fluid in man native to high altitude. J Appl Physiol 1964; 19:319–321.
127. Foulke GE. Altitude-related illness. Am J Emerg Med 1985; 3:217–226.
128. Krasney JA. A neurogenic basis for acute altitude illness. Med Sci Sports Exerc 1994; 26:195–208.
129. Otis SM, Rossman ME, Schneider PA, Rush MP, Ringelstein EB. Relationship of cerebral blood flow regulation to acute mountain sickness. J Ultrasound Med 1989; 8:143–148.
130. Reeves JT, Moore LG, McCullough RE. Headache at high altitude is not related to internal carotid arterial blood velocity. J Appl Physiol 1985; 59:909–913.

131. Baumgartner RW, Bärtsch P, Maggiorini M, Waber U, Oelz O. Enhanced cerebral blood flow in acute mountain sickness. Aviat Space Environ Med 1994; 65:726–729.
132. Mosso A. Life of Man on the High Alps. London: T. Fisher Unwin, 1898.
133. Maher JT, Cymerman A, Reeves JT, Cruz JC, Denniston JC, Grover RF. Acute mountain sickness: increased severity in eucapnic hypoxia. Aviat Space Environ Med 1975; 46:826–829.
134. Harvey TC, Raichle ME, Winterborn MH, Jensen J, Lassen NA, Richardson NV, Bradwell AR. Effect of carbon dioxide in acute mountain sickness: a rediscovery. Lancet 1988; 2:639–641.
135. Bärtsch P, Baumgartner RW, Waber U, Maggorini M, Oelz O. Comparison of carbon-dioxide-enriched, oxygen-enriched, and normal air in treatment of acute mountain sickness. Lancet 1990; 336:772–775.
136. Curran-Everett DC, Iwamoto J, Meredith MP, Krasney JA. Intracranial pressures and O_2 extraction in conscious sheep during 72 h of hypoxia. Am J Physiol 1991; 261: H103–H109.
137. Yang SP, Krasney JA. Cerebral blood flow and metabolic responses to sustained hypercapnia in awake sheep. J Cerebral Blood Flow Metabolism 1995; 15:115–123.
138. Severinghaus JW. Hypothetical roles of angiogenesis, osmotic swelling, and ischemia in high-altitude cerebral edema. J Appl Physiol 1995; 79:375–379.
139. Xu FP, Severinghaus JW. Rat brain VEGF expression in alveolar hypoxia: possible role in high-altitude cerebral edema. J Appl Physiol 1998; 85:53–57.
140. Song SY, Asaji T, Tanizaki Y, Fujimaki T, Matsutani M, Okeda R. Cerebral thrombosis at altitude: its pathogenesis and the problems of prevention and treatment. Aviat Space Environ Med 1986; 57:71–76.
141. Longo LD, Hull AD, Long DM, Pearce WJ. Cerebrovascular adaptations to high-altitude hypoxemia in fetal and adult sheep. Am J Physiol 1993; 264:R65–72.

12

The High-Altitude Brain

MARCUS E. RAICHLE

Washington University Medical Center
St. Louis, Missouri

THOMAS F. HORNBEIN

University of Washington School of Medicine
Seattle, Washington

I. Introduction

While a sojourn at altitude can be an enjoyable and at times thrilling experience, brief visits to even modest altitudes quickly remind most of us of our limitations. Most obvious is the breathlessness we experience on climbing a short flight of stairs or navigating a modest incline in a trial. More sinister, though, because it is usually not apparent to us, is the decline in some of our most sophisticated cognitive skills due to the effect of altitude on the function of our brain. In many instances safety in these spectacular yet hazardous environments is dependent upon proper and timely decisions about what courses of action to take and what to avoid. Such decisions depend upon a functioning brain. Some of the tragedies visited upon humans at altitude, such as the events on Everest in 1996, may have resulted from a critical loss of judgment as the result of brain hypoxia (1). That the mortality during the descent from the summit of Everest is three times greater for those who did not use oxygen than in those who did lends credence to this possibility (2).

This chapter focuses on the effect of high altitude on human brain *function*. What we address in this review are not the high metabolic demands of the resting brain but the additional metabolic and circulatory requirements associated with perception, thought, and action. Lack of sufficient oxygen is certainly a plausible explanation for the changes in perception, thought, and action observed, given the brain's large appetite for oxygen and the reduced availability of oxygen at altitude. As will become apparent, the exact mechanism or mechanisms underlying the neurobehavioral abnormalities associated with high altitude remain largely unknown. A variety of factors undoubtedly contribute. We will enumerate these factors in the hope that future research can focus on promising aspects of the problem.

We begin this review with a look at the behavioral effects of altitude on brain function as well as several sea level conditions that mimic some of the effects of altitude. We then consider the metabolic requirements for normal brain function and, finally, the manner in which the function of the brain might be disrupted during and after exposure to high altitude.

II. Effect of Acute Hypoxia on Behavior

Nothing caught the attention of the public regarding the potential dangers of an acute exposure to high altitude more than did the exploits of early balloonists. The flavor of these escapades is captured in the commentary by Glaisher on the experiences of himself and his colleague Coxwell in 1862 as they ascended to heights thought to be in excess of the summit of Mt. Everest (3):

> In consequence, however, of the rotatory motion of the balloon, which had continued without ceasing since leaving the earth, his valve-line had become entangled, and he (Coxwell) had to leave the car and mount into the ring to readjust it. I then looked at the barometer, and found its reading to be 9 and 3/4 inches, still decreasing fast, implying a height exceeding 29,000 feet. Shortly after I laid my arm upon the table, possessed of its full vigour, but on being desirous of using it I found it powerless—it must have lost its power momentarily; trying to move the other arm, I found it powerless also. Then I tried to shake myself, and succeeded, but I seemed to have no limbs. In looking at the barometer my head fell over my left shoulder; I struggled and shook my body again, but could not move my arms. Getting my head upright for an instant only, it fell on my right shoulder; then I fell backwards, my back resting against the side of the car and my head on its edge. In this position my eyes were directed to Mr. Coxwell in the ring. When I shook my body I seemed to have full power over the muscles of the back, and considerably so over those of the neck, but none over either my arms or my legs. As in the case of the arms, so all muscular power was lost in an instant from my back and neck. I dimly saw Mr. Coxwell, and endeavored to speak, but could not. In an instant intense darkness overcame me, so that the optic nerve lost power suddenly, but I was still conscious, with as active a brain as at the present moment whilst writing this. I thought I had been seized with asphyxia, and believed I should experience nothing more, as death would come unless we speedily descended: other thoughts were entering my mind when I suddenly became unconscious as on going to sleep.

Thirteen years after Glaisher and Coxwell's flight, the Zenith took off from Paris carrying Tissandier, Croce-Spinelli, and Sivel and bags containing oxygen provided by Paul Bert to enable a high ascent. But hypoxia dulled the mind and the three forgot to use the oxygen. When the balloon settled back to earth only Tissandier was alive to describe their journey toward the stars, his two companions dead from a surfeit of too little oxygen.

While the experience of Glaisher and Coxwell tends to suggest that the major impairments short of unconsciousness were, first, motoric and then visual, the experience of Tissandler, Croce-Spinelli, and Sivel (4) more truly capture much of the subsequent literature. This literature suggests that a critical impairment in higher cognitive functions is the earliest and most insidious consequence of exposure to high altitude. We can suspect that Glaisher experienced the same difficulties, one of the signs of which is failure to recognize one's own limitations. Triggered by these early balloon ascents to explore the troposphere and compelled particularly by concerns about pilot performance in aircraft of World War II, the literature on real-time effects of acute hypoxia is venerable, largely descriptive, and limited in scope. Much early work is reviewed by Stickney and Van Liere (5) and Tune (6). Seminal among these early studies is the account published by McFarland (7) in 1937 as the first in a series of papers on the effect of hypoxia on mental performance at high altitude. Utilizing a variety of observations and tests of sensory, motor, and cognitive function, McFarland observed that individuals taken rapidly (hours) to 15,000–16,500 feet exhibited impairment in both simple and complex psychological performance. Motor functions, such as handwriting, were also impaired but to a lesser extent, and sensory modalities were affected little if at all. We will focus upon a sample of more recent studies that fundamentally reinforce these earlier observations.

Because of the well-recognized sensitivity of dark adaptation to acute hypoxia and concern about the consequence to night vision of pilots, increasingly sophisticated inquiries into visual deficits of graded acute hypoxia have accumulated in the literature. Kobrick and coworkers have examined the effects on peripheral as well as central vision (8). The time taken to respond to peripheral visual stimuli increases with increasing altitude between sea level and 17,000 feet. The slowing is less when the individual is concentrating on a central visual target, perhaps because the central task increases alertness. The authors concluded that the hypoxic-induced delay is a consequence of a perceptual deficit rather than an effect on retinal receptors per se. This theme of perceptual deficit recurs, as we shall see.

Other consequences of acute hypoxia have been reviewed by Ernsting (9) and others. Response or reaction times are a usual component of such studies. In general, reaction times to a variety of visual or auditory stimuli are increased with progressive hypoxia, with the effect being greater the more demanding the cognitive task involved in the response. An intriguing approach evaluating acute hypoxia to arterial saturations equivalent to up to 15,000 feet while keeping end-tidal carbon dioxide tension constant demonstrated a hypoxic dose–related decrement in finger-tapping but with no significant effect on digit symbol and a number of other tests noted to be abnormal during or after sustained hypoxic exposure (10). Unfortunately, no normally hypocapnic group was studied to determine the extent to which a higher cerebral blood flow with isocapnia might have mitigated the magnitude of the changes. Even as low as 5000 feet (simulated), response to a visual positioning test administered during light work was slowed and more variable, although with learning these differences disappeared (11). Other described decrements are in short-term memory storage and recall. Fowler et al. (12), in a study designed to assess storage and retrieval from short-term memory, concluded that acute hypoxia did not directly impair either of these processes but rather slowed "the central executive or working memory, thereby reducing its capacity to process information." They also posed the intriguing question whether this impairment of "working memory" is a direct effect of hypoxia on what they referred to as the central executive or a consequence of depression of an earlier, preprocessing stage in the reception of information. Relating this study with Kobrick's observation that retinal receptors seemed unaffected by acute hypoxia suggests that the decrements in function might be occurring at higher levels of processing of information. Yet a "preprocessing" failure fits with the observations of Crow and Kelman (13), who noted that at 12,000 feet the error rate of recall of a short-term memory test was the same after a 32-second delay as after a 2-second delay, suggesting that "the defect may be mainly a failure of registration (failure to record a single digit, perhaps failure to perceive?) rather than a uniform loss of information in the interval before recall." This possible explanation may apply as well to observations of effects of more sustained hypoxia on short-term memory (see below) and conceivably also to residual deficits noted after the hypoxia has been relieved.

While the generic response to hypoxia, either acute or chronic, is one of slowed or impaired performance, not all behavioral tests are altered, even with rather extreme hypoxia. Indeed some tests actually show improvement. For example, response times on simple reaction time paradigms are shortened. Usually such improvement is ascribed to learning effects, but a number of observations, extending over many decades, indicate better performance with acute hypoxia even when learning effects are unlikely or are adequately controlled for. In a recent study, an acute, mild hypoxic challenge (3450 m equivalent altitude) resulted in shortened time to recognize briefly presented letters (14), replicating earlier observations (15). As Kelman and Crow pointed out subsequently, this absence of deterioration was observed only for simpler tests requiring little learning or complex cognition (16). At this stage one can only speculate about where and how acute hypoxia may be acting to excite or inhibit pathways to account for at least a transitory improvement in visual perception.

With improvements in aircraft design, including cabin pressurization, rapid ascent to extremely high altitudes became increasingly possible after World War II. This capability caused researchers to worry about the possibility of sudden cabin depressurization and its effects on pilot performance. How much useful time does an individual have before loss of consciousness prevents the institution of life-saving measures? Luft and colleagues explored this issue by studying rapid decompression to 55,000 feet from 30,000 feet in healthy individuals breathing 100% oxygen (17). From their work and a review of the work of others they concluded that there was a minimum latency or "free interval" prior to the loss of consciousness, regardless of the altitude achieved. During the time to unconsciousness (TUC), no discernible impairments in behavior (they had their subjects performing a writing task) could be detected, and the subjects reported no symptoms. In their entire series of experiments on decompression to 55,000 feet (17), unconsciousness occurred regularly after a latent period of 15–17 seconds provided that the exposure time exceeded 6 seconds. Reasoning from estimates of the systemic circulation time and the desaturation time of hemoglobin, they were led to conclude that the brain continues to function unimpaired for several seconds after the oxygen supply in the surrounding capillaries has reached its lowest level. Obviously, this is such a short period of time that detailed analysis of behavior would be difficult. Nevertheless, such timing information is important in eventually understanding how sudden increases in metabolic demands might be met by a vasculature that tends to respond in seconds rather than milliseconds (18–21). Another important aspect of this work on the TUC was the observation that humans born and living at altitude have significantly longer TUC than do their sea level counterparts (22). Chronic adaptations to altitude will become important when we consider the neurobiology of high-altitude behavioral symptoms later in this chapter.

III. Effect of More Sustained Hypoxia on Behavior

A. Experiences at Altitude: Awake

The history of Mt. Everest climbs is replete with anecdotal reports of cognitive impairments (23,24). Greene, the physician for the 1933 expedition, quoted Hingston's observation on Everest in 1924 (23): "Though the mind is clear, yet there was a disinclination for effort. It was far more pleasant to sit about than to do a job of work that required thought. . . . Though mental work is a burden at high altitudes, yet with an effort it can be done. . . . The main effect of altitude is a mental laziness that determination can overcome."

Hallucinations, particularly the sensed presence of an extra companion, surface periodically in this high-altitude literature, though to be sure, altitude is not essential to such wanderings of the mind (23):

> (Frank Smythe) was convinced when climbing alone, and of course without oxygen equipment, at about 28,000 feet (8530 metres), that he had with him a friendly companion, to whom at one point he offered a sandwich. He also had visual hallucinations, seeing above the ridge of the mountains some curious dark bodies which came to be known as "Frank's pulsating teapots." Yet his judgment was otherwise unimpaired. At 28,100 feet (8565 metres), with the summit less than 1000 feet (305 metres) above him, he was able soberly to consider the pros and cons of continuing, weighing his feeling of physical fitness and the early hour against the bad condition of the snow, and coming to the unwelcome conclusion that he must return. Smythe never suffered from the irritability that occasionally appeared in all the rest of us who went high. It seemed that chronic anoxia increased his placidity. At low altitudes a little inclined to take offense, at great ones he emanated an infectious happiness.

Smythe, acclimatized, was climbing at an altitude almost identical to that which rendered Glaisher unconscious some 70 years earlier. And Glaisher's companion, Coxwell, was still functioning even as Glaisher was fading away. These contrasting observations indicate that adaptations to high altitude can take place given an appropriate period of acclimatization, and, additionally, that individual susceptibility to hypoxia can vary (see also Ref. 22). And Ryn (25) reported a variety of bizarre behaviors, including confusion and psychoses, among a group of Polish mountaineers even at relatively moderate altitudes, although many of these individuals had prior abnormalities that may have contributed importantly to the observations made.

Once again Ross McFarland and the International High Altitude Expedition to Chile have contributed much to our early appreciation of these phenomena (26,27). McFarland described the sensory, motor, emotional, and psychosomatic responses during the course of acclimatization over 3 months to increasing altitudes up to 20,140 feet. While variability in response to sensory and motor testing was increased at 15,000 feet, changes in tests of auditory, visual, and eye-hand coordination responses did not become statistically significantly different from sea level until the station at Aucanquilcha (17,500 feet) was reached. For those best acclimatized, changes even at 20,140 feet were modest, presumably in part because of the slow rate of ascent. McFarland noted, as have many before and since, that concentration was diminished and that impairments were perhaps more in the time taken to perform a task than in the quality of the performance (i.e., the number of errors committed, a speed vs. accuracy tradeoff). He describes the observations of Hingston of performance of simple arithmetical tests by colleagues on the 1924 Mount Everest expedition (27): "By increased effort the problems could be solved quite easily with no loss of accuracy. Nearly everyone experienced mental lassitude, however, and although the mind was clear, there was a disinclination toward effort, especially if the task required thought." Behaviorally, in addition to this disinclination to carry out a task and associated mental laziness, individuals commonly exhibited more critical attitudes concerning the behavior of others, increased irritability, slow thought, repetitiousness, and difficulty remembering.

More explicit testing of task performance during acclimatization to high altitude has been carried out on many subsequent mountaineering expeditions and also in simulated ascents to high altitude. As with acute hypoxia, performance on many but not all tests deteriorated with increasing altitudes and with increasing time at high altitude. Comparing speed of card-sorting of three individuals an average of 2 and 11 weeks after establishing residence at 19,000 feet, Gill et al. found an increase in the number of responses delayed beyond 2.0 seconds; comparison to near sea level performance was not presented but frequency of errors was unaffected (28). Using a more complex card-sorting paradigm over 48 hours after rapid simulated ascent to 15,000 feet (PB 429 mmHg), Cahoon noted that tasks demanding greater cognition showed greater decrements in speed and accuracy than those predominantly motor in character; and the more complex the cognitive task was, the more impaired was the performance (29). Speed was sacrificed to maintain accuracy. Perhaps most interesting of all, the greatest impairment was seen at the initial 3-hour testing, long before symptoms of acute mountain sickness reached maximum severity. Learning may have played a major role in explaining the improvement over time, but no sea level control group was evaluated to assess this possibility. Examining cognitive performance during a simulated 40-day ascent to the equivalent of the summit of Mt. Everest (Operation Everest II), Kennedy et al. came to similar conclusions, namely that tests of cognition (grammatical reasoning and pattern recognition) were more consistently and intensely affected at 25,000 feet (7620 m, Pb 282 mmHg) compared to performance at 15,000 feet (4572 m, Pb 429 mm Hg) than was a pure motor function (finger-tapping) (30). Code substitution and the Strenberg test, a presumed measure of short-term memory [see Fowler's interpretation above regarding interpretation of this test (12)] were significantly impaired but with appreciably greater interindividual variability. The lack of change in finger-tapping capability contrasts with impairment of this function observed in some individuals subsequent to severe hypoxic exposure. These authors also reinforce the observation that speed is sacrificed to accuracy. A recent simulated ascent of Everest (Everest-Comex 97) noted that psychomotor performance and mental efficiency deteriorated progressively above 5500–6500 m, but the changes did not become statistically significant in the eight subjects until an equivalent altitude of 8000 m and above was reached (31).

Adding to the impression that complex cognitive functions are more likely to be impaired than primary sensorimotor functions with acclimatization are a number of other observations. Lieberman et al. (32) observed speech, motor performance, and cognition in members of a 1993 Everest expedition. They found that the time required to respond (i.e., comprehend sentences) increased with increasing altitude from 5300 to 7150 m; additionally, the speech-motor ability to communicate difference between the "voiced stop" consonants b, d, and g and their "unvoiced" counterparts p, t, and k deteriorated with ascent.

Regard et al. reported cognitive changes during the 48 hours following rapid ascent to the Monte Rosa laboratory at 4559 m (33). Those who developed symptoms of acute mountain sickness showed mild impairment of short-term memory but performed better on conceptual tasks than did their asymptomatic counterparts, these differences being apparent even before maximum symptomatic discomfort occurred. As noted above, while improvement in performance on some tests has been observed with both acute and more sustained exposure to high altitude (14,34,35), the observation that those faring less well on most tests may actually perform better on some, if verifiable, poses the intriguing question—why? To these observations Shukitt-Hale et al. added quantification of associated symptom and mood after 4–7 hours at nearly comparable altitude, relating these measures to severity of AMS score (36). Symptoms correlated best with AMS score, then mood, and lastly measures of performance (37).

Not always is significant impairment apparent with a stay at extremely high altitudes (34). Petiet et al. found in eight women ascending to 20,500 feet in the Himalaya that cognitive function was intact with the exception of decrements of complex abstract reasoning and word-finding ability (38). But psychosocial and physiological perturbations were more apparent. Tolerance toward companions diminished while level of anxiety rose. These latter changes are as likely attributable to the risks inherent in the undertaking and the social dynamics of the involved community as to an effect of hypoxia per se. In contrast to the learning of new information, similar robustness of accuracy and latency of retrieval of stored information was reported by Nelson et al. (39); no average decrements were observed between sea level and greater than 6400 m on Mount Everest. But at the same time, individuals' assessments of their own performance, a ''feeling of knowing,'' an element of what is referred to as metacognition, was diminished both on the mountain and after return to low altitude in contrast to the intactness of information retrieval. While this sense of uncertainty, perhaps with associated anxiety, has often been described anecdotally in the mountaineering literature, many other reports marvel at the capacity of summit-motivated climbers to ignore or fail to experience what would seem at times to be objective reasons for doubt (1).

Sharma et al. studied 25 healthy soldiers during a stay of up to 2 years at 4000 m (40). Overall efficiency (speed + accuracy) of a simple test of eye–hand coordination declined progressively for 10 months; only after 13 months could the benefits of acclimatization on performance begin to be seen, and even after 24 months performance at altitude was still less than that measured initially at sea level A possible reason that this test failed to show the oft-reported sacrifice of speed to preserve accuracy is that the main demand of this particular test was not cognitive but one of eye-hand motor coordination. Nevertheless, the improvement after a year at altitude poses interesting questions as to why.

Objective correlates of these behavioral alterations, such as in electrical activity or imaged appearance, have been sought only to a limited extent thus far. Two studies have reported residual changes in event-related potentials in subjects exposed to high altitude. One of the earliest examinations of evoked responses was by Forster et al. published in 1975, of seven healthy lowlanders during 12 days at 4300 m (41). Increased EEG frequency and decreased amplitude of both EEG and visual event–related potentials seemed compatible with central nervous system arousal, which the authors speculated might somehow relate to the behavioral and physiological manifestations of this brief high-altitude exposure. Auditory event–related potentials following rapid ascent to the same altitude were observed by Wesenstein et al. in 10 healthy males (42). Electrical changes were correlated with performance, namely reaction time and counting errors. While amplitude and latency of earlier waves were unaffected relative to sea level, after 8 hours of exposure the amplitude of the P300 component decreased while its latency increased. Reaction time also increased. The association of a temporal correlation between auditory response latency and reaction time was noted, but the extent to which the two are linked is far from clear.

B. Experiences at Altitude: Sleep

Significant alterations in the pattern of sleep have been noted at altitude for many years (e.g., see Ref. 43). More recent studies have provided a much more detailed view of these disturbances, including sleep staging with electroencephalography at altitude (44) and detailed documentation of disturbances in respiration during sleep (44,45). Because of the known deleterious effect of sleep deprivation on waking brain function (see Ref. 46 and Chapter 21 for review), these observations become important in considering the pathophysiology of brain dysfunction observed at altitude. Disturbances in respiration during sleep leading to increased hypoxia may serve to compound the effect of sleep disturbances on brain function at altitude.

IV. Residual Behavioral Effects of Exposure to Hypoxia

Before Mount Everest was first climbed in 1953, many wondered whether acclimatized humans could even survive at its summit, much less climb there. Hillary and Tenzing proved it could be done while breathing supplemental oxygen during the climb and that one could stand quietly about up there without losing consciousness breathing only ambient air. A quarter century later, Peter Habeler (47) and Reinhold Messner (48) showed that some could even climb there without supplemental oxygen. Although breaking of that physiological barrier may have come as no great surprise by 1978, long before the doubts had taken a somewhat different shape: at what cost? Noting that the brain, an aerobic organ that demands its substrates in real time, is most vulnerable to a brief lack of oxygen, the question arose as to whether exposure to such extremes of terrestrial hypoxia could result in injury to the brain.

This question of residual brain damage after climbs to high altitude was fed by intriguing and generally unsubstantiated anecdotes about Mt. Everest climbers whose employment had to be altered after return from being high. A typical anecdote concerned the after-effects in a climber who spent 5 nights at 25,000 feet. This climber was an A.D.C. at Government House in Calcutta, and as part of his duties he had to introduce guests to one another at social functions. Before going to Mt. Everest he could remember and instantly recall the names of some 100 guests; after returning he found that his ability to perform this function was impaired, and he had to use a list (24).

So far as we have been able to determine, the earliest nonanecdotal report of residual behavior abnormalities after exposure to high altitude was the report about Himalayan climbers published by Ryn in his Ph.D. thesis in 1970 and a year later in the literature (25). He recounted an astonishingly high incidence of severe behavioral, neurological, and electroencephalographic abnormalities in a group of Polish mountaineers climbing to about 5300 m in the Himalayas. Unfortunately, this study lacked behavioral measurements on the subjects before going to altitude. The subjects seemed a curious bunch in regard to personality in the first place. A number of the climbers displayed unusual personality attributes even before going to higher altitudes, and the stresses they faced were not limited to hypoxia. They suffered exposure to cold and lack of food and water, which might have affected their post-expedition behavior. Not being able to replicate the incidence or severity of brain dysfunction reported by Ryn, many subsequent investigators have tended to minimize the role of hypoxia per se in the results he reported. Nevertheless, this seminal report certainly alerted people to the possibility that climbing to high altitude could have permanent behavioral residua, stimulating much subsequent research.

In 1983, Clark et al. published the results of careful behavioral evaluations of 22 Himalayan climbers on several expeditions, all going above 5300 m and some as high as 7700 m before a few used supplemental oxygen to reach the summit of Mt. Everest (49). Although a few of their subjects were convinced that their mental function was poorer after altitude exposure, on average sophisticated behavioral testing failed to bear this out. No significant impairments were found, and indeed performance on some tests improved, presumably due to practice.

One other study, that by Jason et al. in 1989 evaluating the behavioral performance of members of a Canadian Everest Expedition, also came up with a negative result (34). None of the 12 climbers studied exhibited any change from their preexpedition performance. Nor did they discern any measurable deficits in function even while at high altitude. Perhaps Canadians are tougher specimens than others, or alternatively, these authors suggest that their rate of ascent and time living at great heights amounted to a lesser dose of hypoxia than that endured in other studies. We now know that individuals vary greatly in how they fare, and this too could contribute to the absence of an overall impairment.

But all other studies, which are still not many, report detectable behavioral changes following return from high altitude. The observations reported in 1984 by Townes et al. on 21 members of the American Medical Research Expedition to Mount Everest (AMREE) and a subsequent larger report in 1989 of observations on 35 mountaineers to Everest and Tirich Mir, including the AMREE subset and participants in OEII, describe statistically significant deficits measured relatively early, i.e., within a month and mostly with one week, after return to low altitude (50,51). In addition similar measurements were performed upon six members of OEII before and within 24 hours of returning to sea level. These various groups yielded rather similar results: declines in visual short and long-term memory, but verbal long-term memory was impaired only in the OEII group; increased errors on the aphasic screening test; and impaired finger-tapping speed in the mountaineers and unilaterally only in the six subjects of OEII. In the AMREE group, evaluated again a year later, while memory and aphasic deficits appeared to have recovered, finger-tapping speed remained impaired in 13 of 16 tested.

These findings were nicely complemented by similar evaluations performed by Regard et al. on eight "world class" mountaineers, all of whom had climbed above 8500 m one or more times without the use of supplemental oxygen (52). Measurements were performed between 2 and 10.5 months (an average of 7 months) after return from high altitude. These climbers did not serve as their own controls but were compared to a matched group of elite rock climbers who had never been high. All the high-altitude climbers showed some performance decrements that were more than two standard deviations from the mean for the control group. Five subjects, in particular, seemed to be more affected, exhibiting mild impairment of concentration and short-term memory and cognitive flexibility (i.e., the ability to shift concepts and control errors). A finger test of motor function was diminished in two but not statistically so for the group as a whole. Three of the most behaviorally impaired also displayed EEG abnormalities. Time above 8000 m ranged from 48 to 358 hours and above 8500 m from 3 to 19 hours. Of particular interest was that the extent of impairment bore no apparent relationship to the number of hours of exposure to extreme altitude nor to the time elapsed between exposure and testing. The great individual variability in tolerance is noteworthy, and we shall return to this later.

Observations on two separate expeditions by Cavaletti et al. of climbers spending time above 5200, m and up to 7400 m in a few instances, add to the hypothesis that exposure to high altitude can have lasting detrimental effects on behavioral performance (53,54). Studied 75 days after descent, a total of 21 subjects displayed memory and concentration deficits, again with much interindividual variability. Petiet and colleagues reported decrements in visual confrontational naming and abstract reasoning in women on a Himalayan expedition (38).

In addition to the EEG abnormalities noted by Ryn (25) and Regard and colleagues (52), Garrido et al. reported MRI abnormalities in 12 of 26 climbers tested from 26 days to 3 years after exposure to >7000 m without supplemental oxygen (55). All had normal neurological exams, but 15 were stated to have residual behavioral impairment upon return to sea level, though it was not made clear whether preascent control studies were obtained or what actual tests were performed, and no data on behavioral outcomes were presented in the paper. Also, preascent controls for the imaging studies were not performed, rather the climbers were compared with a matched control group. MRI changes were characteristic of cortical atrophy in five, periventricular hyperintensity lesions in T_2-weighted images in another five, and both types of lesions in a final two individuals. These changes did not correlate with a variety of variables including age, gender, clinical symptoms, behavioral impairment, or hypoxic dose. Their significance, therefore, is unclear. A study using MRI in seven ''elite'' Sherpas, performed by this same group, again without baseline control measurements, compared them to 21 high-altitude mountaineers who had ascended above 8000 m (56). No behavioral assessment information was provided for these Sherpas. Only one had evidence of MRI abnormalities, and that one was the only Sherpa who experienced neurological difficulties while at altitude. In apparent contrast, all the climbers had at least some symptoms while at altitude, which included headache, insomnia, irritability, and other rather nonspecific complaints. The authors cautiously concluded that this MRI difference between Sherpas and acclimatized mountaineers implied that highlanders' brains might be better capable of withstanding the hypoxic stress of great heights than lowlanders. While the hypothesis is intriguing, the mode of subject selection is distinctly nonrandom. These seven Sherpas were selected because of their track record of exceptional high-altitude performance. Yet this study brings back memories from long ago of a study published by Tulio Velasquez, who observed the acute hypoxic tolerance of high-altitude Peruvian natives (22). He found impressively long periods of consciousness upon exposure to hypoxia equivalent to 10,000–13,000 m altitude. He commented on the immensely greater tolerance to acute hypoxia than that of the sea level inhabitant, but whether the acute hypoxia tolerance of highlanders exceeds that of *acclimatized* lowlanders has not been evaluated, as far as we know.

V. Sea Level Forms of Chronic Hypoxemia

In addition to the above detailed experience with residual behavioral deficits in humans following exposure to high altitude, there are several other clinical conditions in which a period of sustained hypoxia is associated with varying degrees of residual behavioral deficits. These conditions include chronic obstructive pulmonary disease (COPD), sleep apnea, and cyanotic congenital heart disease.

A. Chronic Obstructive Pulmonary Disease

Those with COPD represent a sea level model of some aspects of life at high altitude (although many of the physical stresses of altitude are, of course, not duplicated in such patients). These patients are afflicted with a chronic impairment of gas exchange as a consequence of pathological changes in the lung, commonly associated with a long history of smoking. A major difference is that these individuals with arterial hypoxemia are not the same, healthy physical specimens that one tends to see climbing at high altitude, though at the extremes, some of them become as dependent on supplemental oxygen as is the average Everest climber. In 1973 Krop and colleagues were the first to perform neurobehavioral testing on individuals with chronic hypoxemia from COPD (57). They found that those with a Pa_{O_2} below 55 mmHg had greater deficits on tests of simple motor and perceptual motor abilities than did those with a Pa_{O_2} above 55 mmHg. Of more relevance to our discussion of residual injury, a month of continuous oxygen therapy resulted in improvement in performance while breathing supplemental oxygen. One could conclude either that continuous oxygen breathing reversed brain injury or that the benefit seen was a consequence of the increased F_{IO_2} per se and might not have been apparent if tests had been performed breathing air at the end of the month of continuous oxygen therapy.

This study set the stage for what followed nearly a decade later. First, Grant et al. noted the correlation between severity of arterial hypoxemia and neurobehavioral impairment (58). Then, under the auspices of the National Institutes of Health (NIH), trials to assess the efficacy of Intermittent Positive Pressure Breathing (IPPB Trial) and continuous versus nocturnal oxygen therapy (NOTT Trial) were performed on a large number of patients with COPD (59). These studies, when their populations were combined, showed clearly that neurobehavioral impairment was greater as hypoxemia became more severe. But of particular relevance to the question of residual deficits with improved oxygenation were the results of the NOTT Trial. After 6 months of continuous versus only nocturnal oxygen breathing, neurobehavioral improvement was similar in both groups, but by the end of 12 months the neurobehavioral performance of the nocturnal oxygen group had begun to worsen while those on continuous oxygen therapy showed further improvement. This favorable response to continuous oxygen therapy implies that the initial deficits breathing ambient air were a consequence of reversible hypoxic injury to the brain, analogous to the recovery observed in memory and aphasia deficits in members of AMREE a year later.

B. Sleep Apnea

One other clinical model of hypoxemia in which neurobehavioral disturbances have been documented is sleep disordered breathing or sleep apnea syndrome. Individuals with these sleep problems can experience hypoxemia during sleep, leading to the same neurobehavioral deficits seen in high-altitude climbers or those hypoxemic with COPD. But they also suffer from significant sleep deprivation, which can result in similar performance deficits. The challenge of researchers has been to separate the contributions of nocturnal hypoxemia from sleep deficits in accounting for the observed impairment of performance. Although not all studies have succeeded in identifying a hypoxic contribution to the overall neurobehavioral impairment in these individuals, several studies suggest it does play a role. Bédard et al. reported that increased severity of obstructive sleep apnea correlated with increased cognitive deficits (60). They felt that reductions seen in general intellectual performance as well as in executive and psychomotor tasks could be attributed to hypoxemia while attention and memory decrements were a reflection of diminished vigilance due to sleep deprivation. Conversely, Roehrs et al., comparing individuals with sleep-disordered breathing to those with COPD, found the former fared less well on tests requiring vigilance while those with COPD exhibited the kind of psychomotor decrements associated with hypoxemia (61). Thus, while it may be possible to separate effects of sleep deprivation from hypoxemia, the magnitude of the contribution from the latter, though possible, is not yet clearly defined.

C. Cyanotic Congenital Heart Disease

A final clinical model of hypoxemia with neurobehavioral consequences occurs in young children with cyanotic congenital heart disease. In a study of this condition Newburger and colleagues (62) examined the neurobehavioral consequences of the timing of corrective surgery. Children in their study group underwent corrective surgery anywhere from 6 months to 6 years of age. When controlled for social index, age at repair was inversely correlated with cognitive function. Interestingly, there was no discernible relation between the severity of the insult, as measured by systemic arterial oxygen saturation, and later cognitive deficits. In contrast, age at repair correlated poorly with cognitive function in children with ventricular septal defect, an acyanotic congenital heart defect.

VI. Normal Brain Function

A. Requirements at Rest

In setting the stage for a consideration of changes in brain function, which might underlie the behavioral abnormalities associated with or resulting from an exposure to high altitude (hypoxia), it seems appropriate to review briefly the metabolic requirements of the normal adult human brain. Readers interested in a more in depth coverage of the general topic of brain metabolism may wish to consult the splendid review by Siesjo (63). A review of the relationship between brain function and brain circulation and metabolism can be found in Refs. (64,65).

The adult human brain normally accounts for approximately 15% of the cardiac output and 20% of the total oxygen consumption of the body. The oxygen consumed by the brain is used almost entirely for the oxidation of glucose to carbon dioxide and water. Theoretically, oxidation of one mole of glucose yields 36–38 moles of ATP, which is used as an energy source for the maintenance of membrane ionic gradients and various biosynthetic processes (63). Measurements of whole brain metabolism, beginning with the earliest quantitative measurements of Kety and Schmidt (66), have indicated considerable constancy in *whole brain* metabolism despite a range of behavioral states. However, measurements of whole brain metabolism obscure local moment-to-moment adjustments in blood flow and metabolism that are associated with the constantly changing activity of neurons and astrocytes (67).

B. Behaviorally Induced Circulatory Changes

While studies of behaviorally induced changes in regional brain metabolism and blood flow have been under investigation since the 1950s (67,68), the advent of imaging techniques, especially positron emission tomography (PET) and more recently functional magnetic resonance imaging (fMRI), coupled with behavioral paradigms from cognitive psychology have refined the approach immensely. The past 10 years have seen an exponential growth of research in this area, making possible a more accurate assessment of metabolic and circulatory changes occurring in human brain function in relation to behavior (69,70). In order to appreciate subsequent discussions, we will introduce a representative example of this work.

The current functional imaging strategy compares images of blood flow (PET) or blood flow–related changes (fMRI; see below) taken in a control state to those obtained when the brain is engaged in a task of interest. Great care is taken to choose the control state and the task state so as to isolate as well as possible a limited number of mental operations when subtracting the two states. Subtracting or otherwise comparing blood-flow-dependent measurements made in the control state from those made in the task state should isolate those parts of the brain uniquely responsible for the performance of the task. The details of the image manipulations necessary for the accomplishment of such comparisons and the subsequent statistical analysis of the resulting data are beyond the scope of this review.

A typical example of this type of work, and one that will have many useful features when considering the effect of high altitude on cognition, is taken from our earlier work (71–73). We studied the manner in which the normal brain organizes the act of reading a word aloud. In the experimental results depicted in Figure 1, normal young adults were scanned with PET in a series of hierarchically organized tasks designed to look at word reading. Because reading aloud in the experiment involved opening the eyes and maintaining visual fixation on a television monitor while lying with the head immobilized in a PET camera, the first task depicted is maintaining visual fixation on a television monitor. This is compared to lying quietly in the camera with eyes closed (i.e., the images obtained with eyes closed are subtracted from those obtained during visual fixation with the resulting difference images shown in the top row of Fig. 1). Not surprisingly, eye opening with visual fixation results in significant activation of areas of the visual cortex compared to resting quietly with eyes closed.

With this level of baseline brain activation in mind, it is possible to appreciate the response of the brain when subjects are presented with common English nouns that appeared just below the fixation point on the television monitor at the rate of 40–60 words per minute. Their only instruction was to maintain fixation. The images in the second row of Figure 1 represent those *additional* areas of the brain activated as the result of this passive perception of words. Note that these images (row 2, Fig. 1) represent the *difference* between looking at a blank television monitor and looking at the monitor while words are presented. It is clear that even with this rather blurred representation of foci of activity in the brain, multiple areas within the human visual system are activated by the passive perception of words. An active area of investigation has been to relate these patterns of brain activation to specific levels of the analysis of words (i.e., visual features, letters, orthographic regularities, and meaning) (74).

Next subjects were asked to read aloud the same words as they appeared on the television monitor while another image of brain blood flow was obtained. Subtracting the image obtained during passive word viewing (row 2, Fig. 1) from this image resulted in the difference image in row 3 of Figure 1. This image should reflect the areas of the brain concerned with the motoric aspects of reading words aloud. Not surprisingly, activity was observed in the primary motor cortices bilaterally and the paramedian cerebellum. In addition, there was prominent activity in the Sylvian-insular cortices bilaterally ($Z = 20$, row 3, Fig. 1). Not shown in the selected images in Figure 1 was prominent activity in the supplementary motor areas ($Z = 30$ along the anterior midline).

To ensure that subjects consciously thought about the meaning of the words they were reading aloud, they were asked to say aloud an appropriate use or verb for the nouns visually presented (e.g., see ''knife,'' say ''cut''). It should be pointed out that in addition to the explicit addition of lexical semantics to the task of reading aloud, verb generation also introduces the concepts of *inhibition* and novelty (75). These are important concepts in considering the behavioral changes at altitude and the manner in which the brain implements strategies to deal with them.

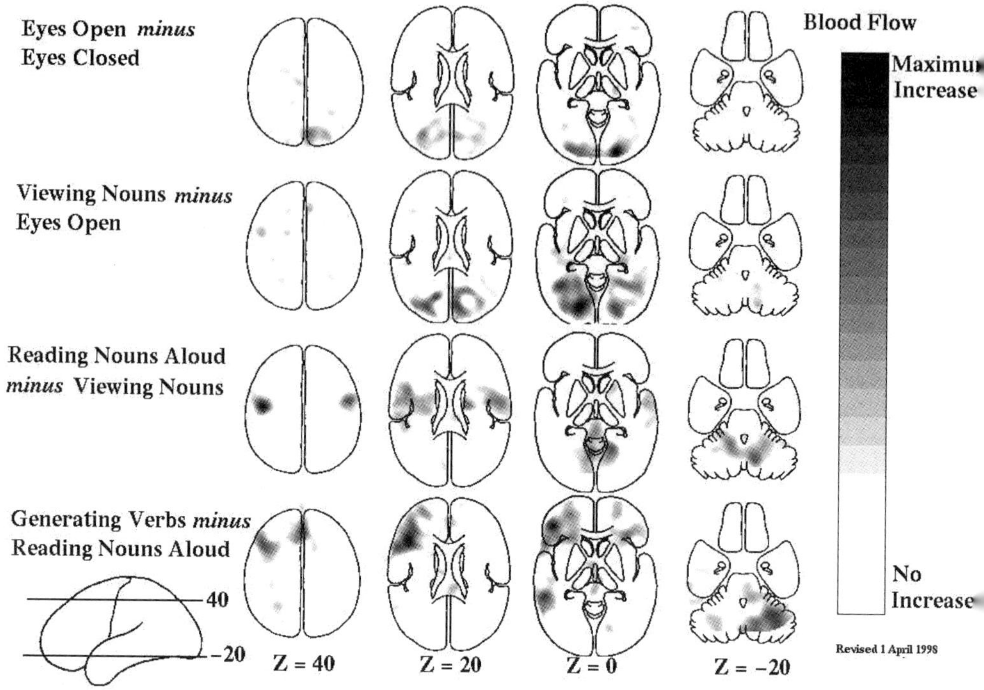

Figure 1 Four different task states are represented in these mean blood flow difference images obtained with positron emission tomography (PET) from a group of normal subjects performing a hierarchically designed set of language tasks. Each row represents the mean difference images between the designated task state and a control state. The labels on the left indicate the task state and the control state for each row of blood flow difference images. The markings "Z = 40" indicate millimeters above and below a horizontal plane through the brain marked "Z = 0." These horizontal images are oriented with the front of the brain on the top and the left side to the readers left. The top row indicates the difference between a control state of eyes closed and a "task" state of eyes open and fixing on a small crosshair in the middle of a television monitor. Note the changes in the visual cortex associated with eye opening and visual fixation. The second row indicates the differences between eyes open and fixing on the television monitor and viewing common English nouns appearing one at a time below the fixation point. Note the increased blood flow over and above that observed with opening the eyes (row 1) in the back of the brain in areas known to be concerned with visual perception. The third row indicates the differences between passively viewing nouns (row 2) and reading them aloud as they appear on the monitor. Note the changes in primary motor cortex bilaterally (Z = 40), sylvian-insular cortex bilaterally (Z = 20), and cerebellum (Z = −20) associated with speaking aloud the words seen on the television monitor. The fourth row indicates the differences between the control state of reading aloud (row 3) and saying aloud an appropriate verb for each noun as it appears on the screen. Note the extensive changes in the left frontal and temporal cortices and the right cerebellar hemisphere. Not shown are increases in the anterior cingulate gyrus and decreases in the Sylvian-insular cortices in areas seen active in reading nouns aloud (row 3, Z = 20). Taken together, these blood flow difference images, arranged in this hierarchical fashion, are designed to demonstrate the spatially distributed nature of the processing by task elements going on in the normal brain during a complex cognitive task. (From Refs. 71, 72.)

Subtracting the PET blood flow images obtained during reading nouns aloud (row 3, Fig. 1) from generating verbs aloud for the same common English nouns revealed activation in the anterior cingulate cortex (Z = 40, row 4, Fig. 1), the left prefrontal cortex and the left frontal operculum (Z = 40 through 0, row 4, Fig. 1), the left posterior temporal cortex (Z = 0, row 4, Fig. 1), and the right lateral cerebellar hemisphere (Z = −20, row 4, Fig. 1).

Several general observations are apparent from the data presented in Figure 1. First, reading aloud, as an example of a complex cognitive task involving verbal output, involves activity in a complex network of areas spatially distributed from the cerebellum to the frontal cortex. Although the traditional language areas of the left hemisphere (i.e., so-called Broca's and Wernicke's areas) are involved, many additional areas of the brain, not necessarily confined to the left hemisphere, are involved as well. Second, involvement of the areas is task-specific. For example, Broca's and Wernicke's areas, shown active during verb generation (row 4, Fig. 1), do not appear to be active during the simple, learned act of reading aloud common English nouns, whereas they become active during verb generation and other more complex cognitive tasks. Third, the cerebellum appears to contribute significantly not only to the motoric aspects of speech production with activity in paramedian areas (see row 3, Fig. 1), but also to the cognitive aspects of word generation, with striking activity in the right cerebellar hemisphere (compare Z = −20 in rows 3 and 4, Fig. 1).

Although the first of these three general observations is not particularly surprising, it is certainly important to emphasize (i.e., that language and other complex tasks are supported by an extensive, spatially distributed network of brain areas). The other two observations regarding the task-specific nature of Broca's and Wernicke's areas and the role of the cerebellum deserve some additional comment.

Appreciating that Broca's and Wernicke's areas were not active in all language-production tasks and the more general implications of this observation with regard to the execution of novel versus learned tasks emerged from a number of subsequent observations. First, it was noted that areas of the Sylvian-insular cortex bilaterally active during reading nouns aloud (row 3, Z = 20, Fig. 1) were not active during verb generation (73). Second, practice of the verb generation task not only led to improvement in reaction time (i.e., the time necessary to speak the appropriate verb aloud), but also to the disappearance of activity in the areas shown in row 4 of Figure 1 and the emergence of activity in the Sylvian-insular cortex bilaterally. In essence, the brain circuitry for the performance of the verb-generation task had shifted dramatically during the course of learning the task. Performance of the verb-generation task in the learned state could not be distinguished from simply reading words aloud.

This discovery led to the conclusion that two distinct routes or brain circuits exist between word perception in the visual cortices and speech production from motor cortices (73). Furthermore, the use of these two routes reflects the degree to which a given response is novel or practiced. Novelty would appear to require a unique neural circuitry to inhibit habitual responses and instantiate new modes of responding. Obviously, the ability of the mature nervous system to benefit from the convenience of smoothly functioning habits yet accommodate changing contingencies is provided for in such an organizational scheme. The circuitry depicted in Figure 1 (row 4) for the execution of novel mental operations involves areas of the cerebral cortex, especially the frontal lobes, which mature late in the course of nervous system development and are vulnerable to normal aging and degenerative diseases such as Alzheimer's disease. This later maturation may explain the more reflexive nature of behaviors in young children as well as the elderly, especially those with Alzheimer's disease (for further development of this theme, see Refs. 75,76). *Of importance to this review is the possibility that the special vulnerability of these frontal areas may be made manifest at high altitude in otherwise healthy, young adults.* As a result, reflexive behaviors may dominate, especially under stressful critical circumstances. Thus, individuals without well-honed mountaineering skills (i.e., reflexes or habits) could be at a distinct disadvantage in the presence of a life-threatening emergency.

That cerebellum in the human is important for the cognitive as well as motoric aspects of our behavior is dramatically demonstrated in Figure 1. The role of the cerebellum in the motoric aspect of speech is clearly shown during reading aloud ($Z = -20$, row 3, Fig. 1). These areas of the cerebellum can readily be distinguished from the large area of the right cerebellar hemisphere that becomes active when reading aloud is subtracted from verb generation ($Z = -20$, row 4, Fig. 1). This latter subtraction provides a direct demonstration of an internal mental operation, the motoric aspects of which have been subtracted away. Although others had speculated that the cerebellum might play an important role in human cognition (e.g., for review see Ref. 77), the evidence for this assertion was limited. Functional brain imaging in normal humans has provided that evidence (72).

Armed with this new information about a specific role of the *normal* cerebellum in human cognition, studies of patients with specific injuries to the cerebellum commenced with a completely new set of questions (e.g., see Ref. 78). Patient evaluation was now based on an understanding of the normal functional brain anatomy. It was quickly discovered (78) that patients with lesions of the cerebellum had, in addition to their traditional motoric deficits, specific cognitive abnormalities that could be clearly demonstrated when tasks were designed around an understanding of the circumstances that utilized that part of the brain. In particular, such patients exhibit an inability to learn new rule-based behaviors. A particularly striking element of their performance was a seeming unawareness of performance errors. These performance deficits occurred despite normal intelligence, language function, and the seeming ability to comprehend the rules of novel tasks when explained to them (78).

Some of the *behavioral* abnormalities experienced at extreme altitude may result from abnormalities of the cerebellum, a part of the brain known to be vulnerable to hypoxia and ischemia. There is already an intimation that the cerebellum might be affected at altitude, given the frequency with which finger-tapping abnormalities are noted in climbers both during and after high-altitude exposure. Future considerations of the possible role of the cerebellum in altitude-induced behavioral abnormalities should take advantage of its now recognized role in cognition. One would hypothesize that the acquisition of skills, both motoric and cognitive, would be impaired (78–80).

VII. Metabolic Requirements of Cognition

While many had assumed that these behaviorally induced changes in local blood flow (Fig. 1) would be reflected in local changes in oxidative metabolism (e.g., see Ref. 63), evidence from brain imaging studies with positron emission tomography (PET) (81,82) indicate otherwise (Fig. 2). In these studies we demonstrated that in normal, awake adult humans, stimulation of the visual or somatosensory cortices resulted in dramatic increases in blood flow but minimal changes in oxygen consumption. Increases in glucose utilization occurred in parallel with blood flow (82),

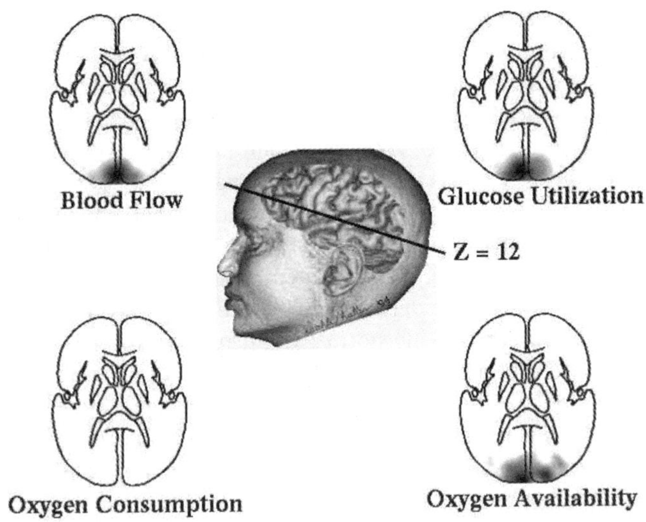

Figure 2 The circulatory and metabolic changes occurring in the visual cortex of normal human subjects during the presentation of a simple visual stimulus (i.e., a reversing annular checkerboard). The orientation of these horizontal slices is shown on the three-dimensional MRI reconstruction of the head and brain in the lower half of the figure. The horizontal slices shown in the upper row are oriented such that the left side of the brain is to the reader's left and anterior is on top. Note the increases in blood flow (top left), glucose utilization (top center), and oxygen availability occurring in visual cortices during stimulation. The increase in oxygen availability in the area of activation (top right) is due to the fact that blood flow increases out of proportion to the increase in oxygen consumption (i.e., supply exceeds demand for oxygen). Quantitative analysis of these data indicates that blood flow and glucose utilization increase proportionately, whereas oxygen consumption does not. (From Ref. 82.)

an observation fully anticipated by the work of others (83,84). That glucose utilization increases much in excess of the changes in oxygen consumption is contrary to most popularly held notions of brain energy metabolism (63). These results suggested that the additional metabolic requirements associated with increased neuronal activity were being supplied largely through glycolysis alone (Fig. 2).

Complementing these observations were detailed metabolic studies of a phenomenon known as cerebellar diaschisis (85,86). In this condition, there is a reduction in blood flow and metabolism in the hemisphere of the cerebellum contralateral to an injury, usually a stroke. These changes in the cerebellum are thought to reflect a reduction in neuronal activity within the cerebellum due to reduced input from the cerebral cortex. Blood flow is reduced significantly more than oxygen consumption. Taken together with the observations from functional brain imaging with PET (82,87), these findings suggest that blood flow changes more than oxygen consumption in the face of both increases as well as decreases in local neuronal activity.

These results were not entirely unanticipated. Studies of epilepsy in well-oxygenated, passively ventilated experimental animals (88) had indicated that blood flow increased in excess of the oxygen requirements of the tissue during the increased neuronal activity of a seizure discharge leading to an increase in the brain venous oxygen content. Because of the increase in blood pressure associated with the seizure discharge, the fact that blood flow exceeded the oxygen requirements of the tissue was attributed to a loss of cerebral autoregulation (88). However, experiments by Cooper and colleagues refuted this explanation (90,91). They demonstrated that oxygen availability locally in the cerebral cortex of patients who remain awake during surgery for the treatment of intractable epilepsy increased during changes in behavioral activity (e.g., looking at pictures, manual dexterity, reading). These changes in oxygen availability occurred in the absence of any changes in blood pressure and were observed during normal brain function.

Surprisingly, these observations were largely ignored until the work of Fox and colleagues called attention to the phenomenon in normal subjects by providing quantitative measurements not only of blood flow but also of oxygen consumption and glucose utilization (82,87). Their work conclusively established that during increases in local brain activity associated with perception, thought, and action, blood flow and glucose utilization in the involved brain areas increase much more than does oxygen consumption.

That blood flow and glucose utilization change more than oxygen consumption during changes in neuronal activity (Fig. 3) has been used to advantage in the development of functional brain imaging with MRI (92–94). Because deoxyhemoglobin is paramagnetic (92,95), any change in the amount of deoxyhemoglobin in venous blood is reflected in a local change in the MRI signal intensity (i.e., T2*) (92). When blood flow changes more than oxygen consumption in either direction, there is a reciprocal change in the amount of deoxyhemoglobin present locally in the tissue. Laboratories throughout the world take advantage of this phenomenon [usually referred to as the blood oxygen level dependent or BOLD signal after Ogawa (92)] using functional MRI or fMRI, as it is often called, to map functional circuitry in

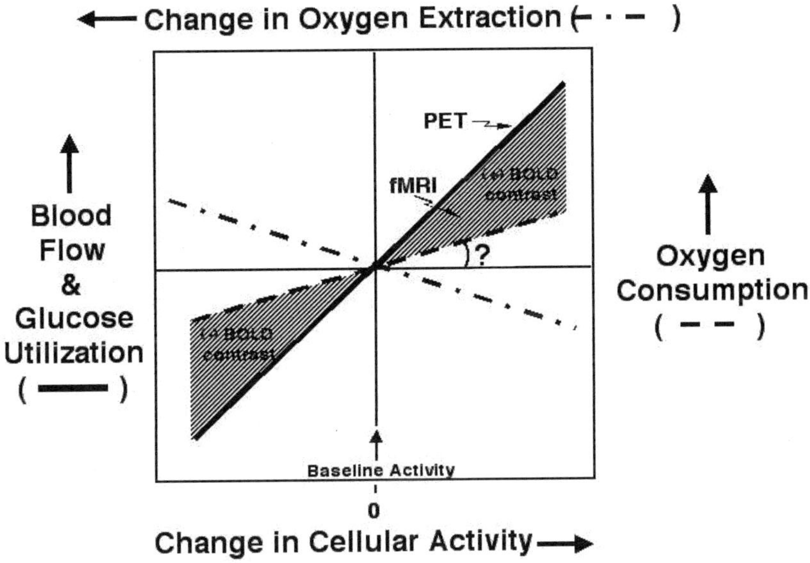

Figure 3 A summary of currently available data on the relationship of blood flow, glucose utilization, and oxygen consumption to the cellular activity of the brain during changes in functional activity is shown in this figure. The changes occurring in blood flow and glucose utilization exceed changes in oxygen consumption. The degree to which oxygen consumption actually changes, if at all, remains to be determined. Positron emission tomography (PET) measures the changes in blood flow. Functional magnetic resonance imaging (fMRI) measures a blood oxygen level–dependent [BOLD (92)] signal or contrast that arises when changes in blood flow exceeds changes in tissue oxygen consumption.

the normal human brain (89,93,94). This technique now provides daily demonstrations that blood flow changes more than oxygen consumption in the normal brain during both increases and decreases in neuronal activity (for a recent summary of work in this area see Ref. 70).

Physiological interpretation of these observations is presently controversial. One hypothesis is most eloquently articulated by Pierre Magistretti and colleagues (for review see Refs. 96,97). The basis for this theory is that glutamate, the major excitatory neurotransmitter in the brain, once released from nerve terminals, is removed from the synapse by uptake into adjacent astrocytes. Within the astrocytes the glutamate is converted to glutamine. The energy needed to convert glutamate to glutamine within the astrocyte is provided by glycolysis. No oxygen is required. Processing glutamate in this manner has the advantage of not requiring an increase in blood flow to provide the additional oxygen required. Increases in blood flow to the brain do not begin for several seconds following the onset of neuronal activity and do not reach their maximum for several more seconds (for example, see Refs. 18,19). Thus, requiring that oxygen delivery to the brain accommodate change in neuronal activity that can occur on a millisecond time scale would be most inefficient and potentially disruptive to normal functional activity.

Additional support for this hypothesis comes from in vivo observations that increases in neuronal activity are associated with glycogenolysis in astrocytes (98), a convenient source of readily available energy for such a process, located in a cell uniquely equipped enzymatically for the process (97,99). Finally, some measurements of tissue lactate with magnetic resonance spectroscopy (MRS) in humans (100–102) have shown an increase. The changes in the expected by-product of glycolysis, however, have always been small and have not been universally found. A more direct and sensitive technique is substrate-induced bioluminescence in laboratory animals (103). This technique has shown localized increases in tissue lactate during physiologically induced increases in neuronal activity.

Alternative hypotheses have emerged to explain the observed discrepancy between changes in blood flow, glucose utilization, and oxygen consumption. One popular hypothesis is based on optical imaging work of physiologically stimulated visual cortex by Grinvald and associates (20,104), who measured changes in reflected light from the surface of visual cortex in anesthetized cats. Using wavelengths of light sensitive to deoxyhemoglobin and oxyhemoglobin, they noted an almost immediate increase in deoxyhemoglobin concentration with stimulation followed after a brief interval by an increase in oxyhemoglobin, which is greater in magnitude and extends over a much larger area of the cortex than do the changes in deoxyhemoglobin. They interpret these results to mean that increases in neuronal activity are associated with highly localized increases in oxygen consumption that stimulate a vascular response, delayed by several seconds, that is large in relation to both the magnitude of the increase in oxygen consumption and the area of cerebral cortex that is actually active. In other words, by their theory increases in neuronal activity in the cerebral cortex are associated with a reactive hyperemia triggered by a lack of oxygen that then delivers to the active area of cortex and its surroundings oxygen in excess of the metabolic needs. Recall, however, that such a process takes many seconds to accomplish (18,105).

Support for the hypothesis of Grinvald et al. (20,104) comes from theoretical work of Buxton and Frank (106), who their modeling work showed that in an idealized capillary tissue cylinder in the brain, an increase in blood flow in excess of the increased oxygen metabolic demands of the tissue is needed in order to maintain proper oxygenation of the tissue. In their model this increase in blood flow results from the poor diffusivity and solubility of oxygen in brain tissue. According to their theory, blood flow remains coupled to oxidative metabolism but in a nonlinear fashion designed to overcome the diffusion and solubility limitations of oxygen in brain tissue in order to maintain adequate tissue oxygenation. It is of interest that Leniger-Follert and Lubbers anticipated this theoretical possibility in 1976 (107).

While the hypothesis that reactive hyperemia is a normal consequence of increased neuronal activity merits careful consideration, several observations remain unexplained. First, reactive hyperemia does not account for the increased glucose utilization that parallels the change in blood flow observed in normal humans (82) and laboratory animals (21,103). Second, it does not agree with the observations of Woolsey and associates, who have done detailed studies of the vasculature and its response to changes in neuronal activity in the rodent whisker-barrel system (21). In their studies utilizing several techniques to characterize the functional and anatomic relationships between a whisker-barrel and its vasculature, Woolsey et al. demonstrated a remarkably tight spatial relationship between changes in neuronal activity within a single barrel and the response of the vascular supply to that barrel. There is little evidence in these studies for spatially diffuse reactive hyperemia surrounding the stimulated area of cortex. Additionally, Woolsey and associates (21) provide within the same article previously unpublished data from Jones, Hand, and Greenberg showing a close spatial correspondence between increased glucose metabolism in the rodent whisker-barrel measured with the deoxyglucose method (108) and blood flow measured with iodoantipyrine (109). These results complement similar observations in rats (103) and humans (82). Also, the hypothesis put forth by Buxton and Frank (106) does not provide an explanation for the observation that blood flow and oxygen utilization *at rest* do not change in the face of increasing hypoxia until a Pao_2 of approximately 30 mmHg is reached (e.g., see Ref. 110). Finally, recent data (111) demonstrate that acutely reducing the Pao_2 from 100.7 to 45.0 in nine normal young adults at rest had no influence on resting blood flow as well as on the blood flow response in visual cortex to visual stimulation. These data provide strong evidence that oxygen delivery to the brain is not the primary reason for the blood flow response to increased functional activity in the human brain.

So, what are we to conclude from these observations? At the moment the data collectively support the idea that local increases as well as decreases in neuronal and astrocyte activity in the brain are accompanied by blood flow changes in excess of any accompanying change in the oxygen consumption. While paired data on glucose metabolism and blood flow are limited at the moment, those that we have (21,82,103) suggest that blood flow changes are accompanied by changes in glucose metabolism of approximately equal magnitude and spatial extent. If this is true, impairment of brain function at altitude and the resulting behavioral abnormalities cannot simply be attributed to a lack of sufficient oxygen because a significant component of the added energy requirements of increased neuronal activity associated with cognitive activities may well be supplied largely by glycolysis and depend on normal astrocyte as well as neuronal function. It will be from this perspective that we shall approach our discussion of brain abnormalities at altitude that follows.

One final caveat should be mentioned. Because blood flow and glucose utilization increase together and more than oxygen utilization during increases in neuronal activity, we might assume that the increase in blood flow serves to deliver the needed glucose. Recent data from Powers and colleagues (112) suggest otherwise. They noted no change in the magnitude of the regional blood flow response to physiological stimulation of the human brain during stepped hypoglycemia, concluding, therefore, that the increase in blood flow associated with physiological brain activation was not regulated by a mechanism matching local cerebral glucose supply to local cerebral glucose demand (112). One of the important unknowns that will emerge from this review is that we presently do not know why blood flow changes so dramatically or reliably during changes in neuronal activity!

Before turning to a discussion of the effect of hypoxia and altitude on the brain, it is important to review briefly what is known about the brain during sleep because sleep is known to be disturbed at altitude (see Chapter 21) and because sleep disturbances are know to have significant behavioral consequences.

VIII. Metabolic Requirements of Sleep

Adequate sleep is important to normal, waking brain function (113,114). That sleep is disturbed at altitude (44,115–117) must be considered when searching for the causes of hypoxia-induced behavioral abnormalities. Measurements of brain circulation and metabolism during sleep is a most challenging task, largely because of the difficulty obtaining normal sleep in subjects in a laboratory setting equipped to make such measurements coupled with the time needed to make measurements of oxygen consumption and glucose utilization (approximately 30–45 minutes) (118–120). Nevertheless, some recent data on blood flow changes measured with PET do provide insight into the dynamic events occurring within the brain.

Seymour Kety and colleagues performed the early studies of brain blood flow and metabolism during sleep shortly after their development of the first quantitative means of making such measurements in humans (121). At the time of their research, rapid eye movement (REM) sleep had not yet been recognized: thus their measurements relate only to so-called slow wave sleep (SWS). During SWS they reported an increase in blood flow but no increase in oxygen consumption and no changes in arterial blood gases. Some doubt about these results has subsequently arisen because the average increase in blood flow for all subjects was primarily the result of large changes in only one subject (see Ref. 122 for critical review). Subsequent reports of blood flow during SWS (123,124) have shown a decrease. Whether this blood flow decrease is accompanied by a decrease in oxygen consumption has yet to be determined.

Detailed regional blood flow measurements in humans during SWS have only recently been reported in humans (125–127) and reveal striking regional changes. Using PET, marked reductions in blood flow were observed in central core structures such as the dorsal pons and mesencephalon, thalamus, and basal ganglia. At the level of the cerebral cortex the most striking reductions of blood flow during SWS appeared to occur in medial, orbital, prefrontal cortex and along the dorsal midline in the region of the anterior cingulate, the posterior cingulate, and adjacent precuneus.

Recently at least two groups have been able to capture the changes in brain blood flow during REM sleep and dreaming (127–129). Their studies reveal significant increases in blood flow in the brainstem, thalamus, bilaterally in the amygdala, the anterior cingulate cortex, and the right parietal operculum. Reductions in blood flow were noted across a wide area of the dorsolateral prefrontal cortex bilaterally, in parietal cortex, and in the posterior cingulate and in the precuneus [recall that these latter two areas also showed reductions in SWS (125,127)]. The exact purpose of the events occurring in the brain during REM sleep remains largely unknown. On the other hand, a recently presented hypothesis by Benington and Heller concerning the purpose of SWS (130) is supported by considerable empirical data and merits our consideration.

Benington and Heller (130) argue that the primary purpose of SWS is to restore brain glycogen stores. Recall from our earlier discussion that glycogen is contained within the astrocyte. Glycogen is used by the astrocyte as a source of energy during transient increases in functional activity (98) because of the rapidity with which glycolytic energy can be provided from this source. The small additional amounts of energy required under these circumstances (131) are accommodated in a much more timely fashion than through a change in blood flow, which requires several seconds to commence (18,19). The increased energy requirements of the astrocytes under these circumstances involve the conversion of the excitatory amino acid transmitter glutamate (96,132) taken up from the adjacent synapse. This conversion is the primary means by which the build-up of toxic levels of glutamate at excitatory synapses is prevented. The energy needed for this process is produced by glycolysis alone, circumventing the need for an instantaneous increase in blood flow that would be necessary to supply the oxygen needed for the complete conversion of glucose to carbon dioxide and water.

The events surrounding SWS may well be important for an understanding of the cognitive difficulties experienced at altitude and even some of the long-term sequelae. Sleep deprivation is common at high altitude (133). If the proposal by Benington and Heller (130) is correct (see above), reduced or absent SWS will lead to a depletion of brain (astrocyte) glycogen, thus compromising the ability of astrocytes to remove and detoxify the glutamate from excitatory synapses. Not only does sleep deprivation lead to disturbances in behavior (113), possibly as a consequence of metabolic disturbances associated with glycogen depletion at excitatory synapses, but it may also predispose the brain to damage through the accumulation of excess amounts of glutamate. While such a formulation is entirely speculative at the moment, it does provide the basis for future experimentation that is clearly needed.

IX. Neurobiology of the High-Altitude Brain

A. Real-Time Effects

By real-time effects we mean those occurring as a consequence of hypoxia while it is present. In a later section we will explore the residual consequences from the hypoxic event after return to normal levels of oxygenation.

Acute Exposure to Altitude

With sudden and brief (minutes to a very few hours) high-altitude exposure, hypoxia may cause CNS perturbations as a consequence of the hypoxia per se or secondary to hypoxically mediated hypocapnia or alterations in neurotransmitter or hormonal levels.

Hypoxia

Even a cursory review of the literature on the neurobiology of hypoxia reveals frequent confusion between hypoxia and ischemia. Reports often suggest that brain hypoxia was studied when in fact the experiments involved brain ischemia. The usual reason for this confusion is that investigators overlooked the importance of adequate brain perfusion when attempting to study the effects of hypoxia alone. When blood pressure falls, as it frequently does in such experiments, or an artery is ligated, hypoxia is converted to ischemia (134). And, as we will show, this dramatically changes the effect on the brain.

The literature on the effect of acute hypoxia on *overall* brain metabolism extends over many years and, in some respects, is remarkable for its consistency. Representative of this literature are studies in laboratory animals (135,136) as well as in humans (137,138) (110) in which brain energy metabolism was measured under the stress of acute hypoxia both at sea level (110) and at altitude (137). Several observations emerge. First, whole brain oxygen consumption in normal awake human subjects is unaffected even with severe degrees of hypoxemia. Even when arterial oxygen tensions fall well below 30 mmHg, an apparent threshold for the initiation of increases in blood flow (110,138) (see also Chapter 11), oxygen consumption is unaffected.

Second, despite the fact that oxygen consumption does not change even with significant degrees of hypoxemia, brain glucose consumption and lactate production increase in humans (138), a finding consistent with observations in laboratory animals suggesting an increase in glycolytic flux (63,135,136). While it might be attractive to attribute this increase in glycolysis to inadequate energy production, an alternative explanation appears more likely (139). In spontaneously breathing subjects, hypoxia-induced hyperventilation results in decreased blood and tissue carbon dioxide. Systemically, this alkalosis is compensated for by excretion of bicarbonate. In the brain, adaptation to prolonged mild hypoxia produces near normal pH_i. The decrease in tissue carbon dioxide and resulting tissue alkalosis is balanced by an increase in glycolysis and an increased turnover in glycolytically produced ATP (139). ATP and PCr are at control levels. This observation implies that glycolytic increases in mild hypoxia are driven by acid-base considerations rather than in response to inadequate energy production (141). LaManna has speculated (141) that the mechanism underlying this alkalinization-induced glycolysis reflects activation of normal processes of metabolic coupling that are exaggerated under hypoxic conditions.

Third, the energy state of brain tissue remains remarkably normal even during significant degrees of hypoxia. Levels of PCr, ATP, ADP, and AMP are unaffected by arterial oxygen tensions as low as 40 mmHg. ATP, ADP, and AMP can remain normal even to arterial oxygen tensions as low as 20 mmHg when tissue lactate has begun to rise and PCr is falling. Estimates of the arterial oxygen tensions in man on the summit of Mt. Everest are in the range of 28 to 30 mmHg (142), which are well within the range of values necessary to preserve oxygen-dependent energy stores in brain according to these data. These observations, made many years ago, remain unchallenged and indicate that brain tissue energy stores are remarkably resistant to hypoxia provided the brain is adequately perfused by the maintenance of normal blood pressure. On the basis of these observations, one might conclude that the brain should function normally despite a significant reduction in available oxygen, yet we know this is not true. As a result we must dig deeper for an explanation of the behavioral disturbances.

Hypocapnia

Respiratory alkalosis is a normal initial accompaniment of exposure to high altitude. An alkalotic blood pH is generally viewed as the normal consequence of a hypoxia-induced increase in ventilation (see Chapter 6). The notion that respiratory alkalosis per se might have deleterious effects on brain function has received spotty attention with conflicting research results over the years. Given the uncertainty explaining the behavioral effects of high altitude, it is appropriate to review the colorful history of what is known about the possible effects of respiratory alkalosis on brain function.

The possibility that carbon dioxide might play a significant role in behavioral symptoms at altitude probably comes from early work on the control of ventilation. It was believed that hypocapnia impeded breathing (143). Following this logic, Angelo Mosso suggested that carbon dioxide played a significant role in the pathophysiology of acute mountain sickness (AMS). In his classic 1898 monograph *Life of Man on the High Alps* (144), he coined the term "acapnia" and strongly suggested that the cause of AMS was a reduction in body carbon dioxide due to the hyperventilation occurring at altitude. He based his ideas, in part, on his own observation in an early altitude chamber that tolerance to hypoxia was clearly enhanced when carbon dioxide rose in the inhaled gas mixture.

The observation that carbon dioxide was useful in the treatment of carbon monoxide poisoning stimulated further work during the early 1900s on its effect on respiration and oxygen delivery to the body. This work engaged the efforts of many prominent respiratory physiologists of the time including Haldane, Henderson, and Douglas. In a very thoughtful summary of this work, Haldane and Priestley (145) commented that the effectiveness of carbon dioxide administration in cases of carbon monoxide poisoning as well as altitude-induced hypoxia was due to a combination of factors: increased ventilation with consequent improvement in arterial oxygenation, facilitated release of oxygen to the tissue due to the Bohr effect, and possibly, improved tissue blood flow. All of these factors remain potential explanations in those instances in which the acute administration of carbon dioxide has appeared efficacious. In 1943 Gibbs and colleagues published the most dramatic demonstration in the literature to date of the value of carbon dioxide in counteracting the effects of acute hypoxia (146). Their study was done on eight healthy young men breathing gas mixtures of 10, 8, 6, and 4% oxygen in nitrogen and 10, 8, 6, 4, and 2% oxygen plus 5% carbon dioxide in nitrogen. Breathing 6% oxygen without carbon dioxide produced confusion and a markedly slowed electroencephalogram; breathing 4% oxygen produced unconsciousness. Adding 5% carbon dioxide to the gas mixtures produced a normal EEG and no alteration in consciousness down to an oxygen level as low as 2% in inspired air. Despite this observation, interest in the role of carbon dioxide waned when oxygen was recognized as the more efficient means of relieving the symptoms of hypoxia! Finally, attempts to use more prolonged carbon dioxide administration to prevent symptoms of AMS actually resulted in more symptoms (147).

More recent work by Harvey and colleagues (148) demonstrating an almost immediate amelioration of symptoms of AMS upon administration of 3% carbon dioxide in air served to rekindle interest in the role of carbon dioxide in the behavioral abnormalities associated with high altitude. Two other groups (149,150), prompted by this report, reexamined the use of carbon dioxide in the treatment of acute mountain sickness and found little or no effect as compared to the use of oxygen.

There is an additional aspect to the possible role of carbon dioxide, acting though brain pH, in influencing brain function at altitude. This relates to what some have called the "pH paradox" (141,151). According to this theory mild tissue acidosis, long thought to be damaging to neurons, is actually protective in the time period immediately following an episode of ischemia. We will return to the implications of the "pH paradox" for residual brain damage following high-altitude exposure in a later section of this chapter. Suffice it to say here that consideration should be given to this phenomenon when trying to understand the seemingly paradoxical observation that while a vigorous ventilatory response to hypoxia appears to contribute to climbing success (152), it results in more residual behavioral impairment (51).

Neurotransmitters

In spite of the preserved state of brain energy stores during hypoxia, the turnover of several neurotransmitters appears to be altered. Frequently cited and studied examples in a literature that is largely over a decade old include dopamine, norepinephrine, and serotonin (153–156), whose rate-limiting synthetic enzymes tyrosine and tryptophan hydroxylase require oxygen, as well as acetylcholine (155). The extant data indicate that hypoxia has a somewhat variable effect depending upon the duration of exposure (153,156) and the age of the subject (157). Nevertheless, sufficient information exists to at least suggest hypotheses concerning the role of selected neurotransmitters in the pathophysiology of altitude-induced behavioral abnormalities. We offer two examples, beginning with dopamine.

In reviewing the cognitive abnormalities that characterize the behavioral changes experienced at high altitude, impairment in short-term or working memory [i.e., the ability to hold in memory for a brief period of time a limited number of items; for example, a telephone number (158)] stands out as a recurring theme. Work by Fuster (159) and Goldman-Rakic (160) has contributed much to our understanding of the underlying neurobiology of this important type of memory. Studying the activity of neurons in the dorsolateral prefrontal cortex of primates, investigators

(161,162) have observed sustained firing of individual neurons during delay periods between the presentation of an instruction to perform a task and the signal to commence its performance. This sustained firing, which is often quite specific to the nature of the task and the neuron under observation, is thought to represent a means by which the brain retains information on line in the absence of external cues. Failure to maintain this firing is almost invariably associated with errors in behavioral performance (162). Of importance here is the fact that the ability of neurons to perform in this manner is very dependent upon the concentration of dopamine in their environment (163,164).

By combining iontophoretic analysis of dopamine receptors, using selective dopamine D1 agonists and antagonists, with single-cell recording during behavior in an awake primate, Williams and Goldman-Rakic (164) found that D1 receptor activation was critical to the response properties of prefrontal cortical neurons. Their work revealed an inverted U-shaped function for the role of the dopamine D1 receptor in prefrontal cortex mechanisms for working memory. When either prefrontal dopamine levels or D1 activity are below the optimal range, as may occur with aging or Parkinson's disease, or above the optimal range, as may occur in stress and amphetamine psychosis, working memory performance is impaired. Of related interest is a recent description of cognitive deficits in members of a 1993 Mt. Everest expedition that were characterized as "similar to that noted for Parkinson's disease" (32)!

Another important role for dopamine concerns the processes associated with the motivational control of goal-directed behaviors (165,166). Projections from dopamine neurons in the ventral tegmental area to the basal ganglia and regions of the orbital and medial frontal cortices provide the means by which the release of dopamine can signal the expectation of a reward and, hence, significantly influence behavior. Hypoxia-induced impairment in dopamine synthesis might well be considered among the potential causes of impaired decision making exhibited in subjects at extreme altitudes. The particular impairments to be considered are likely reflected in their most extreme form in patients with lesions of the orbital and medical frontal cortex (167–169). Future research might well consider the experimental paradigms employed in the evaluation of such patients.

A number of studies indicate a role for norepinephrine in the cognitive function of a variety of species (see Ref. 170 for an introduction to this literature). One of the critical features of the energy requirements of cognition is the availability of glucose from glycogen within astrocytes. Evidence indicates that norepinephrine may be important to that process in facilitating the breaking down of glycogen (171–174). In particular, Harik and colleagues (171,172) have demonstrated that following stimulation of the norepinephrine-depleted cortex, the breakdown of glycogen, primarily residing in astrocytes, is slowed along with the recovery of cortical ATP. It would not be difficult to imagine that this would have a deleterious effect on the energy requirements associated with cognition (see above). Consistent with the role of norepinephrine in cognitive function have been demonstrations that adrenergic agonists can improve memory performance under certain circumstances (170). What is lacking presently is a study examining the role of norepinephrine in memory or other cognitive functions in the setting of hypoxia.

Hormones

Few would dispute that climbing to extreme altitudes is stressful. Glucocorticoids are released during stress (175). Considerable interest has been stimulated by observations that elevated levels of glucocorticoids produced by chronic stress can damage cells of the hippocampus (176,177). Stimulated by these observations, Newcomer and colleagues (178) investigated the effect of dexamethasone treatment on declarative memory performance (i.e., paragraph recall) in normal adult human subjects over a 4-day period. They observed a progressive decline in memory performance over the 4-day period of dexamethasone administration that was completely reversed upon cessation of the medication. The observed memory impairment did not seem to be associated with generalized deficits in attention and arousal. The dose of dexamethasone used raised circulating glucocorticoids levels only to the upper boundary of the normal range of values. Thus, excessive circulating levels of glucocorticoids are not required to produce reversible effects on memory.

The mechanism by which glucocorticoids might impair declarative memory performance and ultimately damage hippocampal cells has been under active investigation. A primary focus of this work has been the manner in which glucocorticoids impair cellular energy metabolism (179). Evidence would suggest that glucocorticoids can impair the uptake of glucose into cell groups including some brain regions such as the hippocampus (180). This impaired uptake occurs in concert with hepatic gluconeogenesis in what appears to be a strategy of energy redistribution favoring muscle tissue, the tissue most likely to have increased energy demands during a variety of stressors and whose glucose uptake is unaffected by glucocorticoids (181). Unfortunately, it would appear that declarative memory function becomes the victim of such a strategy.

In light of our discussions on the metabolic requirements of cognition and, in particular, the role of glycolysis and glycogenolysis in the astrocyte, earlier information on the cellular regulation of glucose transport in such cells is of interest. Working in cultured cells of glial and neuronal origin, Keller and colleagues (182) and Lange and associates (183) concluded some years ago that glucose uptake in cells of neuronal origin seems to be regulated at the level of phosphorylating enzymes, whereas in cells of glial origin the control point appears to be in the cell membrane. If this is indeed true, one might hypothesize that the metabolic events within the astrocyte are particularly vulnerable to the actions of glucocorticoids. Additionally, in light of the ameliorative effects of glucose supplementation on the hippocampal cellular toxicity of glucocorticoids (179), we wonder whether carbohydrate intake by climbers going to extreme altitudes might affect cognitive function.

One final comment with regard to the actions of glucocorticoids seems appropriate. The use of dexamethasone in the prevention and treatment of acute mountain sickness (AMS) has been advocated (184). While the data indicate that symptoms of AMS are fewer in subjects treated with dexamethasone, no attempts have been made to evaluate the effect of this drug on cognitive performance. In particular, it would be valuable in future studies to determine whether any impairment in declarative memory was evident in those treated with dexamethasone despite improvements in AMS.

Prolonged Exposure to Altitude

Comparing the response of early balloonists to hypoxia with that of somewhat acclimatized climbers of Mt. Everest tells us that duration of hypoxia exposure matters. Additionally, to be born at altitude, especially into a family long resident there, appears to confers some special benefits (22,185). The question is, how are those benefits conferred? Studies of both humans and animals exposed for prolonged periods to hypoxic environments suggest that adaptations within the central nervous system may play a role.

While short-term exposure of experimental animals to hypoxia leads to little or no capillary recruitment in the brain, as would be manifest by a change in the permeability–surface area product (186), chronic exposure may produce an absolute increase in the number of perfused capillaries (187,188). Further, recent evidence would suggest that this microvascular proliferation in response to hypoxia is induced and controlled by hypoxia-induced vascular endothelial growth factor (VEGF) (189). VEGF itself is probably regulated by a protein transcription factor known as ARNT (190), which may be critical to an organism's response to both hypoxia and hypoglycemia. ARNT is of interest in that genetic mutants lacking this protein show an abnormal response to hypoxia (190). Such information hints at a direction in which research seeking to explain individual differences in susceptibility to hypoxia should proceed.

Accompanying the increase in capillary density stimulated by hypoxia is an increase in the glucose transporter at the blood-brain barrier (191). Using rats exposed to a hypoxic environment, Harik and associates (191) detected an increase in glucose transporter protein in isolates of brain microvessel after one week of hypobaric hypoxia. After 3 weeks in the hypobaric, hypoxic environment there was a demonstrable increase in the unidirectional flux of glucose across the blood-brain barrier, indicating an increase in the capacity of brain capillaries to transport glucose from blood to brain.

Studies of cerebral glucose metabolism in chronic hypoxia are conflicting. Working in an animal model of chronic hypoxia, Harik and colleagues (192) found an increase in the cerebral metabolic rate for glucose of up to 40%. Their animals had been in a hypobaric hypoxic environment for 3 weeks. These results are to be contrasted with studies on individuals who had resided at high altitude all of their lives. In an earlier study Sorensen and colleagues (193) found no evidence of an increase in brain glucose consumption or lactate production. This study employed brain arteriovenous measurements for glucose and oxygen as well as various metabolites; blood flow was measured using a modification of the classic Kety-Schmidt technique (see Ref. 64 for details of this technique).

Recently, Hochachka and colleagues (194) measured brain glucose consumption in high-altitude natives using the deoxyglucose technique (108) modified for use in humans with PET (119). Metabolic rates for glucose quantified in 26 regions of the brains of these high-altitude natives indicated systematically *lower* region-by-region metabolic rates than in a control group of lowlanders. The reductions were most pronounced in regions of the frontal cortices but were seen throughout the brain. Three weeks of deacclimatization at sea level resulted in further reductions in glucose metabolism. The authors of this study (194) suggest that brain energy production has become more dependent upon glycolysis in the face of a lifelong shortage of oxygen in the environment.

In support of this hypothesis they point to the remarkable tolerance of a number of species including the turtle (195), the harbor seal (196), and the bar-headed goose (197,198). Among the metabolically defining characteristics of these species as well as other with exceptional hypoxia tolerance (see Ref. 199 for review) is an increased reliance on glycolysis as a source for brain energy. This possibility is nicely supported by the much earlier work of McDougal and colleagues (200), who studied the effects of anoxia upon energy sources and selected metabolic intermediates in the brain of fish, frogs, and turtles and reported much higher resting levels of glycogen and much slower declines in glycolytic intermediates following decapitation.

Acceptance of the results of Hochachka and colleagues (194) should be done with caution. The deoxyglucose technique used in their measurements make important assumptions about the relationship of deoxyglucose (the tracer) to glucose (the substance being traced) that are embodied in what is known as the ''lumped constant.'' The lumped constant used in the calculation of the local brain metabolic rate for glucose from the behavior of labeled deoxyglucose accounts for the differences in brain uptake and metabolism of deoxyglucose (108). Any change in the uptake or metabolism of either glucose or deoxyglucose as the result of hypoxia (e.g., Ref. 191) could lead to a change in the lumped constant and, hence, the computed brain metabolic rate for glucose. Before we can accept their results as definitive it will be necessary to have measurements of the lumped constant in high-altitude natives, preferably those on whom they made their measurements. This will be extremely important as the implications of a significantly reduced metabolic rate for glucose on the functioning of a presumably normal human brain are likely to be crucial. In this connection it would be important to know whether there were any cognitive differences between the subjects studied by Hochachka and colleagues (194)—volunteer Quechua natives—and the lowlander controls.

Brain capillary density appears to be increased, especially in high-altitude natives. Attempts should be made through the acquisition of autopsy materials to confirm this observation. As a result of capillary proliferation, and possibly in addition to it, there is a facilitated entry of glucose into the brain. Does the brain rely more heavily on glucose through glycolysis alone as an energy substrate in these people? Presently we do not know. It seems unlikely that overall energy demands of the brains of these people are less and, as a result, met through a greater reliance on glycolysis, present data notwithstanding (194). Are people born at high altitude cognitively different from lowlanders? We don't know.

B. Residual Effects

As reviewed earlier, it is clear that residual behavioral deficits can result from a visit to high altitude. This functional impairment implies brain damage. The questions remain: What is the nature of the insult that causes such damage? And what compensatory changes have occurred that either accentuate or hide the behavioral effects of this damage?

Hypoxia

Simon has recently addressed the role of hypoxia in a most convincing manner (201). Reviewing both the classical literature on the subject of hypoxic brain damage and some of his own recent work, Simon concludes that hypoxia alone cannot explain the residual cognitive deficits experienced by high-altitude climbers. Data presented in his review clearly show, as have others before, that tissue oxygen tensions can be made to fall to extremely low levels without evidence of damage, as assessed by today's most sophisticated cellular techniques, as long as blood flow to brain tissue is maintained. On the other hand, ischemia of a similar duration, while not producing tissue oxygen tensions as low as those seen with hypoxia, nevertheless produces significant residual tissue injury. This injury is preceded by a marked rise in the tissue glutamate concentration that is not seen during hypoxia. One is left to wonder whether hypoxia alone can produce permanent brain injury or whether ischemia must be a component of the insult.

Ischemia

If not hypoxia but ischemia, then how does ischemia arise in the climber at high altitudes? While blood pressure is not routinely measured, there is no clinical evidence that the average healthy climber is hypotensive to the point of symptoms and ischemia. Several other possibilities are worth considering.

Beginning with a report by Frayser and colleagues (202), high-altitude climbers who have had their retinas examined at altitude have frequently been noted to have small hemorrhages (203). The nature of the vascular pathology underlying this abnormality is unknown. It does raise the question of whether similar vascular lesions are occurring in the brain and, if so, are they contributing to small areas of focal ischemia? At the present time we have no answers for such questions, but they should be borne in mind when looking for a possible cause of focal brain ischemia.

On a more global scale one cannot help but be impressed by the association of a vigorous hypoxic ventilatory drive and residual behavioral impairment (51,152). The simplest interpretation of this observation might be that hypoxia coupled with intense hypocapnic cerebral vasoconstriction is sufficient to render some areas of the brain, possibly those more vulnerable in the first place (76), to ischemic damage. To carry this interpretation a bit further, one might also imagine that because a vigorous hypoxic ventilatory drive confers added ability to climb, such individuals achieve greater heights and, hence, greater degrees of hypoxia/ischemia. There is, of course, an additional element to be considered—the effect of tissue alkalosis produced by the hypoxic ventilatory drive. As reviewed earlier, tissue acidosis paradoxically protects brain tissue from ischemic damage, whereas alkalosis does not. Thus, the combination of severe hypoxia, intense cerebral vasoconstriction, and perhaps tissue alkalosis may conspire to produce tissue ischemia in high-altitude climbers. This, however, is a hypothesis for which we have no direct evidence at the present time. It is of interest that one of the explanations for the remarkable high-altitude flying ability of birds such as the bar-headed goose (204) is that their cerebral vasculature does not constrict in response to hypocapnia even though retaining a normal responsivity to hypercapnia.

We would need much additional evidence to determine whether ischemia is an essential contributor to the residual brain damage in climbers to high altitude. If it could be established that ischemia did occur at altitude in otherwise normotensive, healthy adults, then the mechanisms of cell death under intense investigation in laboratories throughout the world could be brought to bear on the problem (for a recent review, see Ref. 205). These mechanisms include excitotoxic cell death with the excitatory neurotransmitter glutamate as the leading culprit; programmed cell death or apoptosis that may be set in motion by an ischemic insult but plays out its deadly effects over a much longer period of time (a recent series of articles summarizes our understanding of apoptosis; see Ref. 206 for introduction); and oxidative stress or the generation of excess oxygen radicals (207). This chapter is not the place to consider these or other mechanisms of brain cell death in detail. An important first step is to determine whether, in fact, ischemia is essential to damage in the first place.

Hormonal Damage

Finally, we should return to the matter of hormonal contribution to brain injury and, in particular, the example we used. The literature we reviewed indicates that prolonging the exposure of the brain to elevated levels of glucocorticoids leads to cell loss, especially in the hippocampus where it has been sought most often (176,177,208,209). How long is long enough? We have no definitive answer to that question. Also, how high must glucocorticoids rise in order to cause damage? We do know that reversible changes in memory can occur even when glucocorticoids levels rise to the high normal range (178). Do such levels, when sustained, cause permanent damage? We do not know. Finally, what factors might reduce the brain's resistance to cell death initiated in this manner? Again, we do not now have answers, but the high-altitude environment about which we have been speaking is rich in factors that potentially threaten the life of brain cells.

Compensatory Changes and the Assessment of Injury

The brain does not remain static after injury. Compensatory changes can and do take place, which may, in fact, obscure the full extent of the injury (210). Behavioral assessment of climbers returning from high altitude is not sufficient to determine the extent of damage if any is present. What is needed is careful assessment of brain morphology and function using the best available imaging techniques. With this type of information in hand, the brain no longer remains a black box. If particular types of information processing use different circuits after an exposure to high altitude than before, we have relatively good circumstantial evidence that damage has occurred in the system normally assigned to the task. Such a question can be answered with today's functional imaging techniques, such as fMRI, where data from single subjects are of such high quality as to permit interpretation without group averaging. Such imaging would be a particularly powerful approach if control subjects not exposed to high altitude or other stresses that might injure the brain were also studied on more than one occasion on the same task.

Behavioral assessment of brain function is only as good as the questions you ask. Suspecting damage to a particular area or circuit and framing the questions you ask in light of that information may be particularly powerful in revealing hidden but serious cognitive problems. The example of the cerebellum in cognition that we used earlier (see Ref. 78) is a particularly useful demonstration of this point.

X. Suggestions for Future Research

Below we suggest a number of avenues for future research that might help elucidate the effect of high altitude on brain function and, more importantly, lead to safer exposure to this environment.

1. *Resting, quantitative, functional imaging studies of brain circulation and metabolism.* As revealed in this chapter, we have little detailed information on the regional changes in brain circulation and metabolism occurring under conditions of acute hypoxia. Are specific brain regions more vulnerable than others? Why? Are such changes related to focal resting metabolic rates?
2. *Controlled study of brain morphometry before and after exposure to high altitude.* High-quality MRI can now detect even small changes in brain anatomy. To date there has been no study in which subjects were imaged with state-of-the-art MRI techniques before and after going to altitude, with these observations being compared to detailed behavioral evaluations of the subjects.
3. *Evaluation of the therapeutic effects of glucose infusions at altitude.* If one of the causes of altitude-induced behavioral abnormalities is the result of problems in glucose metabolism due to stress-induced elevations of

glucocorticoids, then one might anticipate an improvement in performance at altitude and less residual deficits with glucose supplementation. This suggestion is motivated by the observation that glucose supplementation reverses glucocorticoid-induced hippocampal damage in experimental animals (179) and memory impairment in patients with senile dementia of the Alzheimer type (211–213). It may be more than a coincidence that mountaineers have celebrated the achievement of new heights with Wilson's Celebrated Kendal Mint Cake! Maybe this should be viewed as more of a necessity than an enjoyable tradition.

4. *Evaluation of the behavioral effects of dexamethasone in the treatment of AMS*. While dexamethasone is efficacious in the prevention and treatment of AMS, the possibility that its use might impair memory at high altitude deserves to be evaluated. These effects might even be accentuated when the drug is used at altitude.
5. *Cognitive evaluation of high-altitude natives compared with lowlander controls*. Probably such a study would be a difficult undertaking because of cultural differences among other things. However, the results would be most interesting considering the suggestions that brain metabolism is significantly different in high-altitude natives and also in view of the permanent psychometric changes observed in sea level dwellers with hypoxia as a consequence of COPD.
6. *Determine whether there is actually greater capillary density in high-altitude natives*. Animal studies suggest that capillaries proliferate in a hypoxic environment. Do high-altitude natives exhibit such increased vascularity? If so, does this result from environmental pressures that have selected those with a genetic predisposition to greater capillary density?

References

1. Krakauer J. Into Thin Air. New York: Villard, 1997.
2. Huey RB, Eguskitza X. Supplemental oxygen and mountaineer death rates on Everest and K2. JAMA 2000; 284:181.
3. Glaisher J, Flammarion C, De Fonvielle E, Tissandier G. Ascent from Wolverhampton. In: West JB, ed. High Altitude Physiology. Stroudsberg, PA: Hutchinson Ross, 1981: 104–110.
4. Bert P. Barometric Pressure. Bethesda, MD: Undersea Medical Society, 1978.
5. Stickney JC, Van Liere EJ. Acclimatization to low oxygen tension. Physiol Rev 1953; 33:13–34.
6. Tune GS. Psychological effects of hypoxia: review of certain literature from the period 1950 to 1963. Percept Mot Skills 1964; 19:551–562.
7. McFarland RA. Psycho-physiological studies at high altitude in the Andes. I. The effects of rapid ascents by aeroplane and train. J Comp Psychol 1937; 23:191–225.
8. Kobrick JL. Effects of hypoxia on peripheral visual response to dim stimuli. Percept Mot Skills 1975; 41:467–474.

9. Ernsting J. Prevention of hypoxia—acceptable compromises. Aviat Space Environ Med 1978; 49(3):495–502.
10. Berry DT, et al. Isocapnic hypoxemia and neuropsychological functioning. J Clin Exp Neuropsychol 1989; 11(2):241–251.
11. Denison DM, Ledwith F, Poulton EC. Complex reaction times at simulated cabin altitudes of 5,000 feet and 8,000 feet. Aerospace Med 1966; 37:1010–1013.
12. Fowler B, Pralic H, Brabant M. Acute hypoxia fails to influence two aspects of short-term memory: implications for the source of cognitive deficits. Aviat Space Environ Med 1994; 65:641–645.
13. Crow TJ, Kelman GR. Effect of mild acute hypoxia on human short-term memory. Br J Anaesth 1971; 43:548–552.
14. Schlaepfer TE, Bartsch P, Fisch HU. Paradoxical effects of mild hypoxia and moderate altitude on human visual perception. Clin Sci (Colch) 1992; 83(5):633–636.
15. Kelman GR, Crow TJ, Bursill AE. Effect of mild hypoxia on mental performance assessed by a test of selective addition. Aerospace Med 1969; 40:301–303.
16. Kelman GR, Crow TJ. Impairment of mental performance at a simulated altitude of 8,000 feet. Aerospace Med 1969; 40:981–982.
17. Luft UC, Clamann HG, Opitz E. The latency of hypoxia on exposure to altitude above 50,000 feet. J Aviat Med 1951; 22:117–122.
18. Boynton GM, Engel SA, Glover GH, Heeger DJ. Linear systems analysis of functional magnetic resonance imaging in human VI. J Ncurosci 1996; 16(13):4207–4221.
19. Buckner RL, Bandettini PA, O'Craven KM, Savoy RL, Petersen SE, Raichle ME, Rosen BR. Detection of cortical activation during averaged single trials of a cognitive task using functional magnetic resonance imaging. Proc Natl Acad Sci USA 1996; 93(25):14878–14883.
20. Malonek D, Grinvald A. Interactions between electrical activity and cortical microcirculation revealed by imaging spectroscopy: implications for functional brain mapping. Science 1996; 272:551–554.
21. Woolsey TA, Rovainen CM, Cox SB, Henegar MH, Liang GE, Liu D, Moskalenko YE, Sui J, Wei L. Neuronal units linked to microvascular modules in cerebral cortex: response elements for imaging the brain. Cerebral Cortex 1996; 6(5):647–660.
22. Velasquez T. Tolerance to acute anoxia in high altitude natives. J Appl Physiol 1959; 14:357–362.
23. Greene R. Mental performance in chronic anoxia. Br Med J 1957; 1:1028–1031.
24. Pugh LGCE. The effect of acute and chronic exposure to low oxygen supply on consciousness. In: Schaeffer KE, ed. Environment. New York: Macmillan Press; 1962: 106–116.
25. Ryn Z. Psychopathology in alpinism. Acta Med Pol 1971; 12(3):453–467.
26. McFarland RA. Psycho-physiological studies at high altitude in the Andes. II. Sensory and motor responses during acclimatization. J Comp Psychol 1937; 23:227–258.
27. McFarland RA. Psycho-physiological studies at high altitude in the Andes. III. Mental and psychosomatic responses during gradual adaptation. J Comp Psychol 1937; 24: 147–188.
28. Gill MB, Poulton EC, Carpenter A. Falling efficiency at sorting cards during acclimatization at 19000 ft. Nature 1964; 203:436.
29. Cahoon RL. Simple decision making at high altitude. Ergonomics 1972; 2:157–164.
30. Kennedy RS, Dunlap WP, Banderet LE, Smith MG, Houston CS. Cognitive perfor-

mance deficits in a simulated climb of Mount Everest: Operation Everest II. Aviat Space Environ Med 1989; 60:99–104.

31. Abraini J, et al. Cognitive performance during a simulated climb of mount everest: implications for brain function and central adaptive processes under chronic hypoxia stress. Pflugers Archiv 1998; 436:553–559.
32. Lieberman P, Protopapas A, Kanki BG. Speech production and cognitive deficits on Mt. Everest. Aviat Space Environ Med 1995; 66:857–864.
33. Regard M, Landis T, Casey J, Maggiorini M, Bärtsch P, Oelz O. Cognitive changes at high altitude in healthy climbers and in climbers developing acute mountain sickness. Aviat Space Environ Med 1991; 62:291–295.
34. Jason GW, Pajurkova EM, Lee RG. High-altitude mountaineering and brain function: Neuropsychological testing of members of a Mount Everest expedition. Aviat Space Environ Med 1989; 60:170–173.
35. Mackinstosh JH, Thomas DJ, Olive JE, Chesner IM, Knight RJ. The effect of altitude on tests of reaction time and alertness. Aviat Space Environ Med 1988; 59:246–248.
36. Shukitt-Hale B, Banderet LE, Lieberman HR. Relationships between symptoms, moods, performance, and acute mountain sickness at 4,700 meters. Aviat Space Environ Med 1991; 62:865–869.
37. Shukitt BL, Banderet LE. Mood states at 1600 and 4300 meters terrestrial altitude. Aviat Space Environ Med 1988; 59:530–532.
38. Petiet CA, Townes BD, Brooks RJ, Kramer JH. Neurobehavorial and psychological functioning of women exposed to high altitude mountaineering. Percept Mot Skills 1988; 67:443–452.
39. Nelson TO, et al. Cognition and metacognition at extreme altutides on Mount Everest. J Exp Psychol Gen 1990; 119(4):367–374.
40. Sharma VM, Malhotra MS, Baskaran AS. Variations in psychomotor efficiency during prolonged stay at high altitude. Ergonomics 1975; 18:511–516.
41. Forster HV, Soto RJ, Dempsey JA, Hosko MJ. Effect of sojourn at 4300m altitude on electroencephalogram and visual evoked response. J Appl Physiol 1975; 39:109–113.
42. Wesenstein N, Crowley J, Balkan T, Kamimori G, Iwanyk E, Pearson N, Devine J, Belenky G, Cyerman A. Effects of simulated high altitude exposure on long-latency event related brain potentials and performance. Aviat Space Environ Med 1993; 64: 30–36.
43. Ravenhill TH. Some experiences of mountain sickness in the Andes. J Trop Med Hygiene 1913; 1620:313–320.
44. Nicholson AN, Smith PA, Stone BM, Bradwell AR, Coote JH. Altitude insomnia: studies during an expedition to the Himalayas. Sleep 1988; 11:354–361.
45. Lahiri S, Maret KH, Sherpa MG, Peters RM. Sleep and periodic breathing at high altitude: Sherpa natives versus sojourners. In: West JB, Lahiri S, eds. High Altitude and Man. Bethesda, MD: American Physiological Society, 1984: 73–91.
46. Hobson JA. The Dreaming Brain. New York: Basic Books, 1988.
47. Habeler P. The Lonely Victory. Mt. Everest'78 New York: Simon and Schuster, 1978.
48. Messner R. Everest: Expedition to the Ultimate. New York: Oxford University Press, 1979.
49. Clarke CF, Heaton RK, Wiens AN. Neuropsychological functioning after prolonged high altitude exposure in mountaineering. Aviat Space Environ Med 1983; 54:202–207.
50. Townes BD, Hornbein TF, Schoene RB, Sarnquist FH, Grant I. Human cerebral func-

tion at extreme altitude. In: West JB, Lahiri S, eds. High Altitude and Man. Bethesda, MD: American Physiological Society, 1984:31–36.

51. Hornbein TF, Townes BD, Schoene RB, Sutton JR, Houston CS. The cost to the central nervous system of climbing to extremely high altitude. N Engl J Med 1989; 321:1714–1719.
52. Regard M, Oelz O, Brugger P, Biol D, Landis T. Persistent cognitive impairment in climbers after repeated exposure to extreme altitude. Neurology 1989; 39:210–213.
53. Cavaletti G, Garavaglia P, Arrigoni G, Tredici G. Persistent memory impairment after high altitude climbing. Int J Sports Med 1990; 11:176–178.
54. Cavaletti G, Tredici G. Long-lasting neuropsychological changes after a single high altitude climb. Acta Neurol Scand 1993; 87:103–105.
55. Garrido E, Castello A, Ventura JL. Cortical atrophy and other brain magnetic resonance imaging (MRI) changes after extremely high-altitude climbs without oxygen. Intl J Sport Med 1993; 14:232–234.
56. Garrido E, Segura R, Capdevila A, Pujol J, Javierre C, Ventura JL. Are Himalayan Sherpas better protected against brain damage associated with high altitude climbs? Clin Sci 1996; 90:81–85.
57. Krop HD, Block AJ, Cohen E. Neuropsychologic effects of continuous oxygen therapy in chronic obstructive pulmonary disease. Chest 1973; 64(3):317–322.
58. Grant I, Heaton RK, MeSweeney AJ, Adams KM, Timms RM. Neurophysiologic findings in hypoxemic chronic obstructive pulmonary disease. Arch Intern Med 1982; 142: 1470–1476.
59. Grant I, Heaton RK. Neuropsychiatric abnormalities in advanced COPD. In: Petty TL, ed. Chronic Obstructive Pulmonary Disease. New York: Marcel Dekker, 1985:355–373.
60. Bédard M-A, et al. Obstructive sleep apnea syndrome: pathogenesis of neuropsychological deficits. J Clin Exp Neuropsychol 1991; 13:950–964.
61. Roehrs T, et al. Neuropsychological function in obstructive sleep apnea syndrome (OSAS) compared to chronic obstructive pulmonary disease (COPD). Sleep 1995; 18(5):382–388.
62. Newburger JW, Silbert AR, Buckley LP, Fyler DC. Cognitive function and age at repair of transposition of the great arteries in children. N Engl J Med 1984; 310:1495–1499.
63. Siesjo BK. Brain Energy Metabolism. New York: John Wiley & Sons, 1978.
64. Raichle ME. Circulatory and metabolic correlates of brain function in normal humans. In: Plum F, ed. Handbook of Physiology: The Nervous System V. Higher Functions of the Brain. Bethesda, MD: American Physiological Society, 1987:643–674.
65. Raichle ME. Behind the scenes of functional brain imaging: a historical and physiological perspective. Proc Natl Acad Sci USA 1998; 95:765–772.
66. Kety SS, Schmidt CF. The nitrous oxide method for the quantitative determination of cerebral blood flow in man: theory, procedure, and normal values. J Clin Invest 1948; 27:107–119.
67. Raichle ME. Behind the scenes of function brain imaging: A historical and physiological perspective. Proc Natl Acad Sci 1998; 95:765–772.
68. Raichle ME. A brief history of human functional brain mapping. In: Toga AW, Mazziotta JC, eds. Brain Mapping. The Systems. San Diego: Academic Press, 2000:33–75.
69. Posner MI, Raichle ME. Images of Mind. New York: W. H. Freeman & Company, 1994.

70. Posner MI, Raichle ME. National Academy of Sciences Colloquium: Neuroimaging of Human Brain Function. Proc Natl Acad Sci 1998; 95(3):763–898.
71. Petersen SE, Fox PT, Posner MI, Mintum M, Raichle ME. Positron emission tomographic studies of the cortical anatomy of single-word processing. Nature 1988; 831: 585–589.
72. Petersen SE, Fox PT, Posner MI, Mintun MA, Raichle ME. Positron emission tomographic studies of the processing of single words. J Cogn Neurosci 1989; 1:153–170.
73. Raichle ME, Fiez JA, Videen TO, Macleod AK, Pardo JV, Fox PT, Petersen SE. Practice-related changes in human brain functional anatomy during nonmotor learning. Cerebral Cortex 1994; 4:8–26.
74. Fiez JA, Petersen SE. Neuroimaging studies of word reading. Proc Natl Acad Sci 1998; 95:914–921.
75. Dempster FN. The rise and fall of inhibitory mechanisms: toward a unified theory of cognitive development and aging. Dev Rev 1992; 12:45–75.
76. Braak H, Braak E. Aspects of cortical destruction in Alzheimer's disease. In: Hyman BT, Duykaerts C, Christen Y, eds. Connections, Cognition and Alzheimer's Disease. Berlin: Springer-Verlag, 1997:1–16.
77. Leiner HC, Leiner AL, Dow RS. Does the cerebellum contribute to mental skills? Behav Neurosci 1986; 100:443–454.
78. Fiez JA, Petersen SE, Cheney MK, Raichle ME. Impaired non-motor learning and error detection associated with cerebellar damage. A single case study. Brain 1992; 115(Pt 1):155–178.
79. Raichle ME, Fiez JA, Videen TO, MacLeod AM, Pardo JV, Fox PT, Petersen SE. Practice-related changes in human brain functional anatomy during nonmotor learning. Cerebral Cortex 1994; 4(1):8–26.
80. Van Mier H, Tempel LW, Perlmutter JS, Raichle ME, Petersen SE. Changes in brain activity during motor learning measured with PET: effects of hand of performance and practice. J Neurophysiol 1998; 80:2177–2199.
81. Fox PT, Raichle ME. Focal physiological uncoupling of cerebral blood flow and oxidative metabolism during somatosensory stimulation in human subjects. Proc Natl Acad Sci USA 1986; 83:1140–1144.
82. Fox PT, Raichle ME, Mintun MA, Dence C. Nonoxidative glucose consumption during focal physiologic neural activity. Science 1988; 241:462–464.
83. Sokoloff L. Relationships among local functional activity, energy metabolism, and blood flow in the central nervous system. Fed Proc 1981; 40:2311–2316.
84. Yarowsky P, Kadekaro M, Sokoloff L. Frequency-dependent activation of glucose utilization in the superior cervical ganglion by electrical stimulation of cervical sympathetic trunk. PNAS 1983; 80:4179–4183.
85. Martin WR, Raichle ME. Cerebellar blood flow and metabolism in cerebral hemisphere infarction. Ann Neurol 1983; 14(2):168–176.
86. Yamauchi H, Fukuyama H, Kimura J. Hemodynamic and metabolic changes in crossed cerebellar hypoperfusion. Stroke 1992; 23:855–860.
87. Fox PT, Raichle ME. Focal physiological uncoupling of cerebral blood flow and oxidative metabolism during somatosensory simulation in human subjects. PNAS 1986; 83: 1140–1144.
88. Plum F, Posner JB, Troy B. Cerebral metabolic and circulatory responses to induced convulsions in animals. Arch Neurol 1968; 18:1–13.

89. Bandettini PA, Wong EC, Hinks RS, Tlkofsky RS, Hyde JS. Time course EPI of human brain function during task activation. Magnet Reson Med 1992; 25:390–397.
90. Cooper R, Crow HJ, Walter WG, Winter AL. Regional control of cerebral vascular reactivity and oxygen supply in man. Brain Res 1966; 3:174–191.
91. Cooper R, Papakostopoulos D, Crow HJ. Rapid changes of cortical oxygen associated with motor and cognitive function in man. In: Harper AM, Jennett WB, Miller JD, Rowan JO, eds. Blood Flow and Metabolism in the Brain. New York: Churchill Livingstone, 1975: 14.8–14.9.
92. Ogawa S, Lee TM, Kay AR, Tank DW. Brain magnetic resonance imaging with contrast dependent on blood oxygenation. Proc Natl Acad Sci 1990; 87:9868–9872.
93. Ogawa S, Tank DW, Menon R, Ellermann JM, Kim S-G, Merkle H, Ugurbil K. Intrinsic signal changes accompanying sensory stimulation: functional brain mapping with magnetic resonance imaging Proc Natl Acad Sci 1992; 89:5951–5955.
94. Kwong KK, Belliveau JW, Chesler DA, Goldberg IE, Weiskoff RM, Poncelet BP, Kennedy DN, Hoppel BE, Cohen MS, Turner R, Cheng HM, Brady TJ, Rosen BR. Dynamic magnetic resonance imaging of human brain activity during primary sensory stimulation. Proc Natl Acad Sci USA 1992; 89:5675–5679.
95. Pauling L, Coryell CD. The magnetic properties and structure of hemoglobin, oxyghemoglobin and caronmonoxyhemoglobin. PNAS 1936; 22:210–216.
96. Tsacopoulos M, Magistretti PJ. Metabolic coupling between glia and neurons. J Neurosci 1996; 16:877–885.
97. Bittar PG, Charnay Y, Pelleria L, Bouras C, Magistretti P. Selective distribution of lactate dehydrogenase isoenzymes in neurons and astrocytes of human brain. J Cerebral Blood Flow Metab 1996; 16:1079–1089.
98. Swanson RA, Morton MM, Sagar SM, Sharp FR. Sensory stimulation induces local cerebral glycogenolysis: demonstration by autoradiography. Neuroscience 1992; 51: 451–461.
99. Harley CA, Bielajew CH. A comparison of glycogen phosphorylase α and cytochrome oxidase histochemical staining in rat brain. J Compar Neurol 1992; 322:377–389.
100. Prichard JW, Rothman DL, Novotny E, Hanstock CC, Shulman RG. Lactate rise detected by H NMR in human visual cortex during physiological stimulation. Proc Natl Acad Sci 1991; 88:5829–5831.
101. Sappey-Marinier D, Calabrese G, Fein G, Hugg JW, Biggins C, Weiner MW. Effect of photic stimulation on human visual cortex lactate and phosphates using ^{1}H and ^{31}P magnetic resonance spectorscopy. J Cerebral Blood Flow Metab 1992; 12:584–592.
102. Frahm J, Kruger G, Merboldt K-D, Kleinschmidt A. Dynamic uncoupling and recoupling of perfusion and exidative metabolism during focal brain activation in man. Magnet Reson Med 1996; 35:143–148.
103. Ueki M, Linn F, Hossmann K-A. Functional activation of cerebral blood flow and metabolism before and after global ischemia of rat brain. J Cerebral Blood Flow Metab 1988; 8:486–494.
104. Malonek D, Dirnagl U, Lindauer U, Yamada K, Kanno I, Grinvald A. Vascular imprints of neuronal activity: relationships between the dynamics of cortical blood flow, oxygenation, and volume changes following sensory stimulation. Proc Natl Acad Sci 1997; 94:14826–14831.
105. Buckner RL, Goodman J, Burock M, Rotte M, Koutstaal W, Schacter D, Rosen B, Dale AM. Functional-anatomic correlates of object priming in humans revealed by rapid presentation event-related fMRI. Neuron 1998; 20:285–296.

106. Buxton RB, Frank LR. A model for the coupling between cerebral blood flow and oxygen metabolism during neural stimulation. J Cerebral Blood Flow Metab 1997; 17:64–72.
107. Leniger-Follert E, Lubbers DW. Behavior of microflow and local PO_2 of the brain cortex during and after direct electrical stimulation. Pfluegers Arch 1976; 366:39–44.
108. Sokoloff L, Reivich M, Kennedy C, Des Rosiers MH, Patlak CS, Pettigrew KD, Sakurada O, Shinohara M. The [^{14}C]deoxyglucose method for the measurement of local glucose utilization: theory, procedure and normal values in the conscious and anesthetized albino rat. J Neurochem 1977; 28:897–916.
109. Sakaruda O, Kennedy C, Jehle J, Brown JD, Carbin GL, Sokoloff L. Measurement of local cerebral blood flow with iodo [^{14}C]antipyrine. Am J Physiol 1978; 234:H59–H66.
110. Shimojyo S, Scheinberg P, Kogure K, Reinmuth OM. The effect of graded hypoxia ypon transient cerebral blood flow and oxygen consumption. Neurology 1968; 18:127–133.
111. Mintun MA, Lundstrom BN, Snyder AZ, Vlassenko AG, Shulman GL, Raichle ME. Blood flow and oxygen delivery to human brain during functional activity: theoretical modeling and experimental data. PNAS, in Press.
112. Powers WJ, Hirsch IB, Cryer PE. Effect of stepped hypoglycemia on regional cerebral blood flow response to physiological brain activation. Am J Physiol 1996; 270 (Heart Circ Physiology 39):H554–H559.
113. Horne JA. Sleep function with particular reference to sleep deprivation. Ann Clin Res 1985; 17:199–208.
114. Hobson JA. Sleep and dreaming. J Neurosci 1990; 10:371–382.
115. Joern AT, Shurley JT, Brooks RE, Buenter CA, Pierce CM. Short-term changes in sleep patterns on arrival at the south polar plateau. Arch Intern Med 1970; 125:649–654.
116. Guenter CA, Joern AT, Shurley JT, Pierce CM. Cardiorespiratory and metabolic effects in men on the south polar plateau. Arch Intern Med 1970; 125:630–637.
117. Natani K, Shurley JT, Pierce CM, Brooks RE. Long-term changes in sleep patterns in men on the south polar plateau. Arch Intern Med 1970; 125:655–659.
118. Mintun MA, Raichle ME, Martin WR, Herscovitch P. Brain oxygen utilization measured with O-15 radiotracers and positron emission tomography. J Nuclear Med 1984; 25(2):177–187.
119. Reivich M, Kuhl D, Wolf A, Greenberg J, Phelps M, Ido T, Casella V, Hoffman E, Alavi A, Sokoloff L. The [18F] fluorodeoxyglucose method for the measurement of local cerebral glucose utilization in man. Circ Res 1979; 14:127–137.
120. Phelps ME, Huang SC, Hoffman EJ, Selin C, Sokoloff L, Kuhl DE. Tomographic measurement of local cerebral glucose metabolic rate in humans with (F-18)2-fluoro-2-deoxy-D-glucose: validation of method. Ann Neurol 1979; 6:371–388.
121. Mangold RL, Sokoloff L, Conner E, Kleinerman J, Therman POG, Kety S. The effects of sleep and lack of sleep on the cerebral circulation and metabolism of normal young men. J Clin Invest 1955; 84:1092–1100.
122. Kennedy C, Gillin JC, Mendelson W, Suda S, Miyaoka M, Ito M, Nakamura RK, Storch W, Pettigrew J, Mishkin M, Sokoloff L. Local cerebral glucose utilization in non-rapid eye movement sleep, Nature London 1982; 297:325–327.
123. Madsen PL, Schmidt JF, Wildschiodtz G, Friberg L, Holm S, Vorstrup S, Lassen NA. Cerebral O2 metabolism and cerebral blood flow in humans during deep and rapid-eye-movement sleep. J Appl Physiol 1991; 70(6):2597–2601.

124. Townsend RE, Prinz PN, Obrist WD. Human cerebral blood flow during sleep and waking. J Appl Physiol 1973; 35(620–625).
125. Maquet P, Degueldre C, Delfiore G, Aerts J, Peters J-M, Luxen A, Franck G. Functional anatomy of human slow wave sleep. J Neurosci 1997; 17(8):2807–2812.
126. Hofle N, Paus T, Reutens D, Fiset P, Gotman J, Evans AC, Jones BE. Regional cerebral blood flow changes as a function of delta and spindle activity during slow wave sleep in humans. J Neurosci 1997; 17(12):4800–4808.
127. Braun AR, Balkin TJ, Wesensten NJ, Carson RE, Varga M, Baldwin P, Selbie S, Gelenky G, Herscovitch P. Regional cerebral blood flow throughout the sleep-wake cycle. Brain 1997; 120:1173–1197.
128. Maquet P, Peters JM, Aerts J, Delfiore G, Degueldre C, Luxen A, Franck G. Functional neuroanatomy of human rapid-eye-movement sleep and dreaming. Nature 1996; 383: 163–166.
129. Braun AR, Balkin TJ, Wesensten NJ, Gwadry F, Carson RE, Varga M, Baldwin P, Belenky G, Herscovitch P. Dissociated pattern of activity in visual cortices and their projections during human rapid eye movement sleep. Science 1998; 279:91–95.
130. Benington JH, Heller HC. Restoration of brain energy metabolism as the function of sleep. Prog Neurobiol 1995; 45:347–360.
131. Creutzfeldt OD. Neurophysiological correlates of different functional states of the brain. In: Ingvar DH, Lassen NA, eds. Brain Work: The Coupling of Function, Metabolism and Blood Flow in the Brain. Copenhagen: Munksgaard, 1975:22–47.
132. Magistretti PJ, Pellerin L, Rothman DL, Shulman RG. Energy on demand. Science 1999; 283:496–497.
133. Nicholson AN, Smith PA, Stone BM, Bradwell AR, Coote JH. Effects of acetazolamide and temazepam on sleep at high altitude (abstr). Postgrad Med J 1987; 63:191.
134. Pulsinelli WA, Duffy TE. Local cerebral glucose metabolism during controlled hypoxemia in rats. Science 1979; 204:626–629.
135. Duffy TE, Nelson SR, Lowry OH. Cerebral carbohydrate metabolism during acute hypoxia and recovery. J Neurochem 1972; 19:959–977.
136. MacMillan V, Siesjo BK. Brain energy metabolism in hypoxemia. Scand J Clin Lab Invest 1972; 30:127–136.
137. Severinghaus JW, Chiodi H, Eger EI, Brandstater B, Hornbein TF. Cerebral blood flow in man at high altitude. Role of cerebrospinal fluid pH in normalization of flow in chronic hypoxia. Circ Res 1966; 19:274–282.
138. Cohen PJ, Alexander SC, Smith TC, Reivich M, Wollman H. Effects of hypoxia and normocarbia on cerebral blood flow and metabolism in conscious man. J Appl Physiol 1967; 23:183–189.
139. Siesjo BK. Metabolic control of intracellular pH. Scand J Clin Lab Invest 1973; 32: 97–104.
140. Lauro K, LaManna J. Cerebral oxygen and metabolic consumption model of the compensatory adaptations in chronic hypobaric hypoxia in the rat. FASEB J 1994; 8: A1047.
141. LaManna J. Hypoxia and the pH paradox. In: Ince C, Kesecioglu J, Akpir K, Telci L, eds. Oxygen Transport to Tissue. New York: Plenum Press, 1996:283–292.
142. West JB. Man on the summit of Mount Everest. In: West JB, Lahiri S, eds. High Altitude and Man. Bethesda, MD: American Physiological Society, 1984:5–18.
143. Henderson Y. Adventures in Respiration. Modes of Asphyxiation and Methods of Resuscitation. Baltimore, MD: Williams and Wilkins, 1938.

144. Mosso A. Life of Man in the High Alps. 2d ed. London: T Fisher Unwin, 1898.
145. Haldane JS, Priestley JG. Respiration. New Haven, CT: Yale University Press, 1935.
146. Gibbs FA, Gibbs EL, Lennox WG, Nims LF. The value of carbon dioxide in counteracting the effects of low oxygen. J Aviat Med 1943; 14:250–261.
147. Maher JT, Cymerman A, Reeves JT, Cruz JC, Denniston JC, Grover RF. Acute mountain sickness: Increased severity in eucapnic hypoxia. Aviat Space Environ Med 1975; 46:826–829.
148. Harvey TC, Raichle ME, Winterborn MH, Jensen J, Lassen NA, Richardson NV, Bradwell AR. Effect of carbon dioxide in acute mountain sickness: a rediscovery. [Published erratum appears in Lancet 1988; 2(8614):808.] Lancet 1988; 2(8612):639–641.
149. Hackett PH. Carbon dioxide breathing and acute mountain sickness (letter). Lancet 1989; 1:272.
150. Bartsch P, Baumgartner RW, Waber U, Maggiorini M, Oelz O. Comparison of carbon-dioxide-enriched, oxygen-enriched, and normal air in treatment of acute mountain sickness. Lancet 1990; 2:772–775.
151. Tombaugh GC. Mild acidosis delays hypoxic spreading depression and improves neuronal recovery in hippocampal slices. J Neurosci 1994; 14:5635–5643.
152. Schoene RB, Saxon R, Pierson DJ. Control of ventilation in climbers of extreme altitude. In: Sutton JR, Jones NL, Houston CS, eds. Hypoxia: Man at Altitude. New York: Thieme-Stratton, 1982:195.
153. Koob GF, Annau Z. Behavioral and neurochemical alterations induced by hypoxia in rats. Am J Physiol 1974; 227:73–78.
154. Davis JN. The role of oxygen in the regulation of brain tyrosine hydoxylation. In: Almgren O, Carlsson A, Engel J, eds. Chemical Tools Catecholamine Research. New York: North-Holland, 1975:31–36.
155. Gibson GE, Duffy TE. Impaired synthesis of acetylcholine by mild hypoxic hypoxia or nitrous oxide. J Neurochem 1981; 36:28–33.
156. Olson EB, Vidruk EH, McCrimmon DR, Dempsey JA. Monoamine neurotransmitter metabolism during acclimatization to hypoxia in rats. Respir Physiol 1983; 54:79–96.
157. McNamara MC, Gingras-leatherman JL, Lawson EE. Effect of hypoxia on brainstem concentration of biogenic amines in postnatal rabbits. Dev Brain Res 1986; 25:253–258.
158. Baddeley A. Working memory. Science 1992; 255:556–559.
159. Fuster JM. The Prefrontal Cortex; Anatomy, Physiology, and Neuropsychology of the Frontal Lobe. 3rd ed. Philadelphia: Lippincott-Raven, 1997.
160. Goldman-Rakic PS. Circuitry of primate prefrontal cortex and regulation of behavior by representation memory. In: Plum F, ed. The Nervous System. Bethesda, MD: American Physiological Society, 1987:373–417.
161. Zipser D, Kehoe B, Littlewort G, Fuster J. A spiking network model of short-term memory. J Neurosci 1993; 13:3406–3420.
162. Funahashi S, Chafee MV, Goldman-Rakic PS. Prefrontal neuronal activity in rhesus monkeys performing a delayed anti-saccade task. Nature 1993; 365:753–756.
163. Sawaguchi T, Goldman-Rakic PS. D1 dopamine receptors in prefrontal cortex: involvement in working memory. Science 1991; 251:947–950.
164. Williams GV, Goldman-Rakic PS. Modulation of memory fields by dopamine D1 receptors in prefrontal cortex. Nature 1995; 376:572–575.
165. Schultz W, Dayan P, Montague PR. A neural substrate of prediction and reward. Science 1997; 275:1593–1599.

166. Schultz W, Tremblay L, Hollerman JR. Reward prediction in primate basal ganglia and frontal cortex. Neuropharmacology 1998; 37:421–429.
167. Rolls ET, Hornak J, Wade D, McGrath J. Emotion-related learning in patients with social and emotional changes associated with frontal lobe damage. J Neurol Neurosurg Psychiatry 1994; 57:1518–1524.
168. Eslinger PJ, Damasio AR. Severe disturbance of higher cognition after bilateral frontal lobe ablation: patient EVR. Neurology 1985; 35:1731–1741.
169. Bechara A, Damasio AR, Damasio H, Anderson SW. Insensitivity to future consequences following damage to human prefrontal cortex. Cognition 1994; 50:7–15.
170. Arnsten AFT, Cai JX, Goldman-Rakic PS. The alpha-2 agonist guanfacine improves memory in aged monkeys without sedative or hypotensive side effects: evidence for alpha-2 receptor subtypes. J Neurosci 1988; 8:4287–4298.
171. Harik SI, Busto R, Martinez E. Norepinephrine regulation of cerebral glycogen utilization during seizures and ischemia. J Neurosci 1982; 2:409–414.
172. Harik SI, LaManna JC, Light AI, Rosenthal M. Cerebral norepinephrine: influence on cortical oxidative metabolism in situ. Science 1979; 206:69–71.
173. Kuschinsky W, Suda S, Bunger R, Yaffe S, Sokoloff L. The effects of intravenous norepinephrine on the local coupling between glucose utilization and blood flow in the rat brain. Pflugers Archiv 1983; 398:134–138.
174. Magistretti PJ, Schorderet M. Norepinephrine and histamine potentiate the increases in cyclic adenosine 3′ :5′-monophosphate elicited by vasoactive intestinal polypeptide in mouse cerebral cortical slices: mediation by alpha-1-adrenergic and H1-histaminergic receptors. J Neurosci 1985; 5:362–368.
175. Munck A, Guyre P, Holbrook N. Physiological functions of glucocorticoids in stress and their relations to pharmacological actions. Endocrinol Rev 1984; 5:25–44.
176. Sapolsky RM, Krey LC, McEwen BS. Prolonged glucocorticoid exposure reduces hippocampal neuron number: implications for aging. J Neurosci 1985; 5:1222–1227.
177. Stein-Behrens B, Mattson MP, Chang I, Yeh M, Sapolsky R. Stress exacerbates neuron loss and cytoskeletal pathology in the hippocampus. J Neurosci 1994; 14:5373–5380.
178. Newcomer JW, Craft S, Hershey T, Askins K, Bardgett ME. Glucocorticoid-induced impairment in declarative memory performance in adult humans. J Neurosci 1994; 14: 2047–2053.
179. Sapolsky RM. Glucocorticoid toxicity in the hippocampus: reversal by supplementation with brain fuels. J Neurosci 1986; 6:2240–2244.
180. Horner HC, Packan DR, Sapolsky RM. Glucocorticoids inhibit glucose transport in cultured hippocampal neurons and glia. Neuroendocrinology 1990; 52:57–64.
181. Munck A. Glucocorticoid inhibition of blucose uptake by peripheral tissues: old and new evidence, molecular mechanisms and physiological significance. Perspect Biol Med 1971; 14:265–289.
182. Keller K, Lange K, Noske W. D-glucose transport in cultured cells of neural origin: the membrane as possible control point of glucose utilization. J Neurochem 1981; 36: 1012–1017.
183. Lange K, Keller K, Ludwig WD, Monden I, Reinsch I, Brandt U. Rapid cellular regulation of D-glucose transport in cultured neural cells. J Neurochem 1982; 39:1594–1600.
184. Johnson TS, Rock PB, Fulco CS, Trad LA, Spark RF, Maher JT. Prevention of acute mountain sickness by dexamethasone. N Engl J Med 1984; 310:683–686.
185. Niermeyer S, Yang P, Shanmina, Drolkar, Zhuang J, Moore LG. Arterial oxygen satu-

ration in Tibetan and Han infants born in Lhasa, Tibet. N Engl J Med 1995; 333:1248–1252.

186. Chen J-L, Wei L, Bereczki D, Hans F-J, Otsuka T, Acuff V, Richardson G, Patlak C, Fenstermacher J. Virtually unaltered permeability-surface area products imply little capillary recruitment in brain with hypoxia. Microcirculation 1994; 1:35–47.
187. Mironov V, Hritz MA, LaManna JC, Hudetz AG, Harik SI. Architectural alterations in rat cerebral microvessels after hypobaric hypoxia. Brain Res 1994; 660:73–80.
188. Harik SI, Hritz MA, LaManna JC. Hypoxia-induced brain angiogenesis in the adult rat. J Physiol (London) 1995; 485:525–530.
189. Stone M, Itin A, Alon T, Pe'er J, Gnessin H, Chan-Ling T, Keshet E. Development of retinal vasculature is mediated by hypoxia-induced vascular endothelial growth factor (VEGF) expression by neuroglia. J Neurosci 1995; 15:4738–4747.
190. Maltepe E, Schmidt JV, Baunoch D, Bradfield CA, Simon MC. Abnormal angiogenesis and responses to glucose and oxygen deprivation in mice lacking the protein ARNT. Nature 1997; 386:403–407.
191. Harik SI, Behmand RA, LaManna JC. Hypoxia increases glucose transport at blood-brain barrier. J Appl Physiol 1994; 77:896–901.
192. Harik SI, Lust WD, Jones SC, Lauro KL, Pundik S, LaManna JC. Brain glucose metabolism in hypobaric hypoxia. J Appl Physiol 1995; 79:136–140.
193. Sorensen SC, Lassen NA, Severinghaus JW, Coudert J, Paz Zamora M. Cerebral glucose metabolism and cerebral blood flow in high-altitude residents. J Appl Physiol 1974; 37:305–310.
194. Hochachka PW, Clark CM, Brown WD, Stanley C, Stone CK, Nickles RJ, Zhu GG, Allen PS, Holden JE. The brain at high altitude: hypometabolism as a defence against chronic hypoxia. J Cereb Blood Flow Metab 1994; 14:671–679.
195. Perez-Pinzon MA, Chan CY, Rosenthal M, Sick TJ. Membrain and synaptic activity during anoxia in the isolated turtle cerebellum. Am J Physiol 1992; 263:R1057–R1063.
196. Kerem D, Elsner R. Cerebral tolerance to asphyxial hypoxia in the harbor seal. Respir Physiol 1973; 19:188–200.
197. Faraci FM. Circulation during hypoxia in birds. Comp Biochem Physiol 1986; 85A: 613–620.
198. Faraci FM, Feede MR. Regional circulatory responses to hypocapnia and hypercapnia in bar-headed geese. Am J Physiol 1986; 250:R499–R504.
199. Lutz PL. Mechanisms for anoxic survival in the vertebrate brain. Ann Rev Physiol 1992; 54:601–618.
200. McDougal DB, Jr., Holowach J, Howe MC, Jones EM, Thomas CA. The effects of anoxia upon energy sources and selected metabolic intermediates in the brains of fish, frog and turtle. J Neurochem 1968; 15:577–588.
201. Simon RP. CNS response to hypoxia. In: Sutton JR, Houston CS, Coates G, eds. Hypoxia and the Brain. Proceedings of the 9th International Hypoxia Symposium. Burlington, VT: Queen City Printers, 1995:1–7.
202. Frayser R, Houston CS, Bryan AC, Rennie ID, Gray GW. Retinal hemorrhage at high altitude. N Engl J Med 1970; 282:1183–1184.
203. McFadden DM, Houston CS, Sutton JR, Powles AP, Gray GW, Roberts R. High altitude retinopathy. J Am Med Assoc 1981; 245:581–586.
204. Faraci FM, Kilgore DL, Fedde MR. Oxygen delivery to the heart and brain during hypoxia: Pekin duck vs. bar-headed goose. Am J Physiol 1984; 16:R69–R75.
205. Hossmann K-A. The Hypoxic Brain. Insights from Ischemia Research. In: Roach RC,

Wagner PD, Hackett PH, eds. Hypoxia, into the Next Millennium. New York: Kluver Academic/Plenum Publishers, 1999:155–169.

206. Miller LJ, Marx J. Apoptosis. Science 1998; 281:1301.
207. Coyle JT, Puttfarcken P. Oxidative stress, glutamate and neurodegenerative disorders. Science 1993; 262:689–695.
208. Lupien S, Lecours AR, Lussier I, Schwartz G, Nair NPV, Meaney MJ. Basal cortisol levels and cognitive deficits in human aging. J Neurosci 1994; 14:2893–2903.
209. Sheline YL, Wang PW, Gado MH, Cernansky JG. Hippocampal atrophy in recurrent major depression. PNAS 1996; 93:3908–2913.
210. Buckner RL, Corbetta M, Schatz J, Raichle ME, Petersen SE. Preserved speech abilities and compensation following prefrontal damage. PNAS 1996; 93:1249–1253.
211. Craft S, Zalten G, Baker L. Glucose and memory in mild senile dementia of the Alzheimer type. J Clin Exp Neuropsychol 1992; 14:253–267.
212. Craft S, Dagogo-Jack SE, Wiethop B, Murphy CM, Nevins R, Fleischman S, Rice V, Newcomer J, Cryer PE. The effects of hyperglucemia on memory and hormone levels i dementia of the Alzheimer type: a longitudinal study. Behav Neurosci 1993; 107: 926–940.
213. Manning C, Ragozzino M, Gold PE. Glucose enhancement of memory in patients with probable senile dementia of the Alzheimer type. Neurobiol Aging 1993; 14:523–528.

13

Autonomic Nervous System

DOUGLAS R. SEALS and PAMELA PARKER JONES

University of Colorado
Boulder, Colorado

KEVIN P. DAVY

Colorado State University
Fort Collins, Colorado

The autonomic nervous system plays a critical role in mediating many of the physiological adjustments to hypoxemia (1–4). The focus of this chapter will be on the autonomic nervous system responses to hypoxemia in humans, with emphasis, where data are available, on high altitude. Because the duration of exposure to altitude appears to influence the nature of the autonomic nervous system adjustments, the responses to both acute (minutes to hours) and sustained (days and longer) hypoxemia will be considered. Experimental data from sea level studies of humans (and animals) breathing low inspired O_2 or in a hypobaric chamber to simulated altitude and field studies conducted at high altitude provide the substrate for this discussion. Because humans must perform the activities of daily life while at altitude, the influence of hypoxemia during exercise and orthostatic stress will also be discussed. Where information is available, we also will examine the mechanisms responsible for mediating autonomic nervous system adjustments to hypoxemia as well as their known or possible physiological significance. Finally, the potential modulatory effects of factors such as age, gender, and chronic physical activity status will be mentioned, although few data are available at present.

In light of the relatively large amount of information that does exist as well as its substantial importance in circulatory control, our primary emphasis will be on the sympathetic nervous system (SNS) and adrenal medullary release of epinephrine. We will consider data on both average ''whole-body'' measures of SNS activity and, because of the marked regional specificity of the SNS, also SNS outflow to selective organs. Our discussion of the parasympathetic nervous system will be confined to the possible role of cardiac vagal activity in the regulation of heart rate during hypoxemia. Throughout we will emphasize that which is not currently known or well understood as it pertains to the need for further research. For other views of this and related topics, the reader is referred to several previous reviews (1–9).

I. Sympathoadrenal System

A. Acute Hypoxemia

Resting Conditions

Whole-Body SNS Activity

Studies conducted at high altitude as well as several laboratory investigations have used urinary and/or plasma concentrations of norepinephrine (NE), the primary neurotransmitter released from sympathetic nerve endings, as measures of total systemic (i.e., whole-body) SNS activity. Both of these approaches have serious limitations due in part to effects of hypoxemia on NE clearance (10–12); thus, the values obtained can fail to reflect, in magnitude or even direction, true changes in SNS activity during hypoxemia (13,14).

With this caveat in mind, urinary total catecholamine (norepinephrine + epinephrine + dopamine) concentrations have been reported to be either increased (15) or unchanged (16) in humans upon initial exposure to moderately high altitude (~3600–4660 m) compared to sea level. Mazzeo and colleagues (17) found increased urinary NE concentrations in young adult men after one day at Pikes Peak (4300 m) compared to their sea level values; however, arterial plasma NE levels were not elevated (17,18). Venous plasma NE concentrations consistently have been reported to be unchanged with acute hypoxemia induced either by exposure to high altitude or low O_2 breathing (3,6,9,14,19–21). These observations have been interpreted as indicating that SNS activation does not occur with acute hypoxemia (5,9,17). However, acute hypoxemia evokes increases in heart rate and cardiac output (systemic arterial flow) and regional blood flow (1), which affect both NE spillover into (i.e., appearance rate) and clearance from the plasma compartment (12,22). Thus, SNS activity could be increased in response to acute hypoxemia without a corresponding increase in plasma NE concentrations.

In support of the possibility that acute hypoxemia influences plasma NE kinetics in humans, Leuenberger and colleagues (13) reported that low O_2 breathing sustained for 25–30 minutes increased whole-body plasma NE spillover rate by 46%, consistent with an increase in SNS activity. Because of an increase in plasma NE clearance, plasma NE concentrations increased only about 20%. Inspection of their data indicate, however, that plasma NE clearance increased significantly in only two of six subjects. Therefore, it is unclear from these results whether acute hypoxemia uniformly affects plasma NE clearance in humans.

These data on plasma NE kinetics suggest that acute hypoxemia *may* evoke physiologically significant net whole-body SNS activation in humans, even though plasma NE concentrations may not reflect this activation because of an increase in plasma NE clearance. Additional evaluation of plasma NE kinetics, however, is needed to confirm these laboratory observations and to determine whether acute exposure to high altitude really does increase whole-body SNS activity in humans.

Skeletal Muscle SNS Activity

Acute systemic hypoxemia excites most muscle sympathetic vasoconstrictor neurons isolated in anesthetized (vagally intact, artificially ventilated) cats (23). SNS activity to skeletal muscle (MSNA) also can be measured directly (i.e., intraneurally) in conscious humans using the microneurographic procedure (24). In addition, limb arterial-venous plasma NE differences and limb plasma NE spillover measurements can provide insight, albeit indirect, into MSNA. Over the last decade several studies have examined the effects of acute hypoxemia on MSNA in resting humans (3,13,14,25–28).

Laboratory studies in humans either breathing low O_2 gas mixtures or exposed to simulated altitude in a chamber indicate that moderate to severe acute hypoxemia is required to evoke increases in microneurographically recorded MSNA (13,14,25,27). Severe acute hypoxemia produces as much as a 300% increase in MSNA, without affecting antecubital venous plasma NE concentrations (3,14). Therefore, plasma NE concentrations do not reflect SNS activation to skeletal muscle. The findings of Leuenberger et al. (13) that forearm plasma NE spillover increased ~100% while microneurographically measured total minute MSNA increased ~180% in subjects breathing low O_2 provides additional support for the conclusion that acute hypoxemia stimulates an increase in SNS outflow to inactive skeletal muscle in humans.

In studies using low inspired O_2, the MSNA response has a latency of up to several minutes, with the rate and magnitude of increase proportional to the degree of hypoxemia imposed (3,14). The latency of the response, the level of hypoxemia at which MSNA increases, and the magnitude of increase in MSNA all vary widely among individual subjects (3,14). Upon return to normoxia, MSNA can remain elevated for one hour or longer (3,25). No consistent modulatory influence of ventilatory pattern or Pa_{CO_2} on the MSNA response has been observed (3,14).

Recent observations at high altitude support these findings. Mazzeo et al. (17) reported a shift from net uptake of NE from the leg at sea level toward net NE release from the leg after 4 hours of exposure to Pikes Peak in young adult male subjects. These experimental findings demonstrate that acute hypoxemia can evoke a marked increase in sympathetic outflow to skeletal muscle under resting conditions.

The mechanism(s) responsible for hypoxemia-induced muscle sympathetic activation have not been determined. Although some investigators have assumed that arterial chemoreceptors mediate the response (27,28), several observations, including the long latency of the response (3), suggest that other mechanisms may be involved. A direct influence of hypoxemia on the central nervous system also has been hypothesized (3). Thus, the nature of the mechanism(s) underlying hypoxemia-induced stimulation of MSNA in humans is a key question for future investigations.

SNS Activity to the Heart

In cats, acute systemic hypoxemia evokes a marked increase in cardiac SNS activity (29). The increase occurs despite a corresponding elevation in arterial blood pressure, which normally results in baroreflex-mediated sympathoinhibition.

Although cardiac SNS activity can be estimated in humans by measuring coronary sinus plasma NE kinetics (10,30,31), no one has yet used this technique to study the effects of acute hypoxemia. Instead, the behavior of cardiac sympathetic nerves has been inferred from the effects of β-adrenergic receptor blockade on the increase in heart rate evoked by acute hypoxemia. β-Blockade attenuates, but does not abolish, the tachycardia attendant to acute hypoxemia in humans (20,32) and in dogs (33). For example, Richardson et al. (20) reported that about half of the normal 30 bt/min rise in heart rate with low O_2 breathing was abolished in men pretreated with propranolol. Plasma concentrations of NE and epinephrine did not increase in their subjects in response to acute hypoxemia (see also below), suggesting that increased cardiac sympathetic nerve release of NE, rather than the β-adrenergic stimulatory effects of elevations in circulating catecholamines, was responsible for the tachycardia. Some, but not all (19), data from high altitude support this view in that the increase in heart rate during the initial day of exposure to 4300 m is not observed after β-blockade (17,34).

Collectively, these results suggest that acute hypoxemia may stimulate cardiac SNS activity, at least in neurons innervating the sino-atrial node. The mechanisms underlying this response, at least in humans, are unknown. In cats, the increase in cardiac SNS activity with acute systemic hypoxemia is observed after chemoreceptor denervation, and direct stimulation of the carotid body does not result in an increase in cardiac SNS activity (29). Thus, the increase in cardiac SNS activity does not appear to be mediated via carotid chemoreceptors. Instead, a direct stimulatory effect on the CNS has been proposed (20,29), as suggested for MSNA by Seals and Rowell (3).

Again, more direct measures of cardiac SNS activity (e.g., cardiac NE spillover) are needed to determine the influence of acute hypoxemia on SNS outflow to the heart in humans. Such technically challenging experimental procedures, however, may be difficult to perform at high altitude.

During Concurrent Physiological Stress

Some information is available concerning the modulatory effect of acute hypoxemia on the SNS response to physical stress such as dynamic exercise, hyperthermia, and orthostasis.

Whole-Body and Regional SNS Activity

The increases in arterial and venous plasma NE concentrations normally seen with submaximal exercise performed under normoxic conditions are increased with acute hypoxemia (6,9,17,18). This augmentation likely occurs because the plasma NE response to dynamic exercise is influenced by the relative work rate (i.e., percent of maximal capacity) (3,37). Because acute hypoxemia decreases maximum work capacity, a particular absolute submaximal work rate becomes a higher relative work rate.

An increased release of NE from active muscle appears to contribute significantly to this augmented whole-body plasma NE response (35). Microneurographic measurements of MSNA along with venous plasma NE concentrations during small-muscle dynamic exercise indicate that augmented sympathetic outflow to inactive as well as active skeletal muscle contributes to the SNS response to hypoxemic exercise (36). Augmented plasma NE responses even at equivalent relative work rates during hypoxemic versus normoxic large-muscle dynamic exercise (37) suggest an effect of acute hypoxemia on sympathetic regulation that is independent of relative work rate, although not all have found this additional effect (6,9). Thus, currently no consensus exists as to whether acute hypoxemia exerts a primary influence on sympathetic control during exercise or whether the effect is secondary to the associated increase in the relative work rate.

As with exercise, acute hypoxemia appears to augment the SNS response to both hyperthermia and orthostasis in humans. Rowell et al. (38) demonstrated that the increases in venous plasma NE concentrations with progressive hyperthermia were greater during low O_2 versus room air breathing. In addition, the increments in microneurographically measured MSNA and venous plasma NE concentrations in response to progressive hypovolemia produced by graded lower body negative pressure are augmented during acute hypoxemia (26). Under these conditions, SNS regulation of arterial blood pressure appears to be well preserved in most humans. However, in a minority of subjects during combined hypoxemic-orthostatic stress, epinephrine release from the adrenal medulla is stimulated and plasma levels rise (3,4,26). If the rise is rapid and marked, it is associated with a vasovagal response, which includes sympathoinhibition, vasodilation, bradycardia, and sudden arterial hypotension (3,4,26). The neural mechanism initiating this sequence of events is unknown but may involve the activation of a cardiac depressor reflex (3,4,26).

In summary, the whole-body SNS response to several types of physical stress appears to be augmented during acute hypoxemia. Increased activation of sympathetic nerves to skeletal muscle is likely to be an important contributor to this response.

Adrenal Medullary Release and Circulating Levels of Epinephrine

The effect of acute hypoxemia on the release of epinephrine from the adrenal medulla at rest and during physical stress has been deduced from changes in plasma epinephrine concentrations. Interpretation based on this measure is subject to the same limitations as those for plasma NE concentrations. Recent measurements of plasma epinephrine kinetics in humans clearly demonstrate that plasma concentrations do not necessarily reflect adrenal medullary release of epinephrine under certain conditions (39). However, we need to understand the influence of acute hypoxemia on the plasma concentrations of epinephrine per se because of the substantial cardiovascular and metabolic effects of circulating epinephrine.

Twenty-four hour urinary concentrations of epinephrine have been reported to increase during the initial day of exposure to both 4560 m of simulated altitude (40) and at Pikes Pike (4300 m) (17). Increases in arterial, though not venous, concentrations of epinephrine at rest have been observed after only 4 hours of exposure to 4300 m (9,17,18). However, the available experimental data on the effects of acute hypoxemia on urinary and plasma epinephrine concentrations in humans are inconsistent, with studies reporting no change, small increases, and large increases compared to normoxic values (5,6,9). Factors contributing to these discrepant findings may include the nature of the hypoxemic stimulus (normobaric vs. hypobaric hypoxemia), the degree of hypoxemia, the duration of hypoxia, attainment of true resting conditions, the site of blood sampling, and differing group characteristics (e.g., physical training status) (9).

The effects of acute hypoxemia on the plasma epinephrine responses to submaximal exercise appear to be similar to those described for plasma NE. Increases in plasma epinephrine in response to a given absolute work rate are augmented by acute hypoxemia, whereas both similar and greater plasma epinephrine responses have been reported at the same relative work rate (6,9,18,37). Thus, as with SNS activity, it presently is unclear whether the influence of hypoxemia on epinephrine release from the adrenal medulla or on circulating plasma levels of epinephrine during standardized submaximal exercise is a primary influence or simply a secondary effect of reduced maximal work capacity (increased relative work rate).

As with plasma NE concentrations, the increases in plasma epinephrine levels in response to progressive hyperthermia in humans are greater under conditions of acute hypoxemia (38). In contrast, in most humans plasma epinephrine concentrations do not increase during combined orthostatic stress plus acute hypoxemia (3,4,26). However, in some humans, this combined stress will evoke either slow and progressive or rapid and marked increases in plasma epinephrine levels; as described above, the latter is associated with frank bradycardia and acute hypotension (3,4,26). This "susceptible" subset of individuals do not differ from their "resistant" counterparts in any obvious manner with regard to subject characteristics, etc.

A better understanding of the direct effects of acute hypoxemia on epinephrine release from the adrenal medulla at rest and during concurrent stress in humans will require studies of plasma epinephrine kinetics (39).

Physiological Significance of SNS Activation by Acute Hypoxemia

What might be the physiological significance (advantage) of SNS activation at rest and during physical stress in response to acute hypoxemia?

In regard to circulatory control, at least two possibilities seem reasonable. First, the reduction in arterial O_2 content (O_2-carrying capacity) associated with acute hypoxemia necessitates an immediate increase in heart rate and cardiac output in order to maintain systemic O_2 delivery. Thus, the activation of cardiac sympathetic nerves and increased adrenal medullary release of epinephrine, together, produce a strong β-adrenergic stimulus at the sino-atrial node.

Second, acute hypoxemia is known to increase the concentrations of a number of substances with vasodilator properties and to produce relaxation in vascular smooth muscle (1,3,4). Elevated arterial plasma concentrations of epinephrine such as those observed upon acute exposure to high altitude (17,18) are known to dilate vascular smooth muscle, particularly in skeletal muscle (1,4). Despite these strong locally and humorally mediated vasodilatory influences, only modest vasodilation is observed in skeletal muscle and splanchnic organs in hypoxemic humans (1,4). Perhaps the increase in sympathetic vasoconstrictor outflow to skeletal muscle is a baroreflex adjustment intended to counteract the vasodilatory effects of acute hypoxemia. Skeletal muscle is a large vascular bed with substantial systemic hemodynamic impact. Thus, sympathetic vasoconstriction to this and possibly other regions might be evoked in order to maintain arterial blood pressure and ensure proper perfusion of vital organs in the face of hypoxemia-induced vasodilation. However, it is not presently clear whether the SNS vasoconstriction is secondary to hypoxemic vasodilation (i.e., baroreflex-mediated) or is a primary effect of hypoxemia.

Finally, altitude sickness associated with acute exposure to high altitude may be mediated, at least in part, by the sympathoadrenal system (5). This concept is supported by the observations that urinary excretion of catecholamines is greater in subjects with acute mountain sickness compared to healthy subjects (41) and that symptoms of acute altitude illness are less severe in subjects pretreated with propranolol compared to placebo (42).

B. Sustained Hypoxemia

At Rest and During Concurrent Physiological Stress

Whole-Body and Peripheral SNS Activity

Over the initial 7–12 days of exposure to high-altitude 24-hour urinary NE concentrations increase progressively and then plateau (5,9,15–17). At rest, the changes in both venous and arterial plasma NE concentrations are consistent with these findings (9,17–19,21). Moreover, net NE uptake in the resting leg shifts to release over the course of 21 days at 4300 m (17,35), presumably contributing to the elevations in urinary and plasma concentrations of NE with sustained/continued exposure to high altitude.

The influence of continued hypoxemia on SNS activity during exercise has been inferred from plasma NE responses. At a fixed absolute submaximal work rate, plasma NE levels mirror the increases observed under resting conditions (9,18,19,35). In addition, the net release of NE from the exercising leg increases over 21 days of exposure to high altitude (35), suggesting that active skeletal muscle contributes importantly to the augmented whole-body plasma NE response to submaximal exercise. The magnitude of the elevations in plasma NE at rest and during submaximal exercise with sustained exposure to high altitude are greater at 6542 m than 4350–4800 m, indicating that the intensity of the SNS response parallels the degree of hypoxemia (43). In contrast to submaximal exercise, plasma NE concentrations during maximal exercise appear to be reduced with sustained exposure to high altitude (44,45). The mechanisms responsible for this phenomenon have not been determined.

We might reasonably conclude that the elevations in urinary and plasma NE concentrations indicate a progressive stimulation of the SNS over the initial 7–10 days of exposure to high altitude, followed by a sustained elevation thereafter (5,9,17). However, alternative explanations are possible. For example, cardiac output (systemic blood flow) after an initial increase subsequently falls at the same time that plasma NE concentrations are rising (2), and leg blood flow also is reduced (46). Therefore, the increases in plasma NE concentrations with sustained exposure to high altitude may not accurately reflect elevations in SNS activity because of simultaneous changes in plasma NE clearance. Microneurographic recordings of MSNA together with whole-body and regional measurements of plasma NE kinetics are needed to document the pattern and magnitude of SNS behavior in response to chronic hypoxemia.

Mechanisms and Significance

The increases in urinary and plasma NE levels over time at high altitude correlate poorly with the hypoxemic stimulus, but much better with decreasing plasma volumes (9). If the increases in NE concentrations do reflect increases in SNS activity, the relation with reductions in plasma volume may be due to decreases in cardiac filling (2) and cardiopulmonary baroreceptor unloading with consequent reflex sympathoexcitation (9). The increases in plasma NE concentrations over 21 days at high altitude also have been reported to be inversely correlated with changes in $P_{ET_{CO_2}}$ (9), suggesting the possibility of an association with the present state of hyperventilation.

What might be the physiological or pathophysiological significance of an increase in plasma NE concentration over time at high altitude? Grover and colleagues (2) speculate that the β-adrenergic stimulation produced by elevated circulating NE levels supports left ventricular contractility (i.e., in the face of a decline in filling pressure), stimulates substrate mobilization (particularly fat metabolism), plays a role in the elevation of resting metabolic rate, and increases red blood cell production.

The increases in urinary and plasma NE concentrations over 21 days at Pike's Peak correlate with increases in 24-hour levels of arterial blood pressure and with elevations in systemic vascular resistance during standardized submaximal exercise (7,46,47), suggesting that elevations in plasma NE levels may have a physiological effect. How an increase in systemic vascular resistance and arterial blood pressure above sea level values helps to maintain homeostasis is not obvious. On the other hand, these elevations in arterial pressure observed with short-term exposure to high altitude appear to be transient because prolonged exposure (years) is not consistently associated with elevated blood pressure (2).

Pathophysiologically, the increases in SNS activity, systemic vascular resistance, and arterial pressure may be important for patients with essential hypertension who travel to and reside at high altitude (2,47). Moreover, the elderly and patients with heart disease are known to be at elevated risk of life-threatening ventricular arrhythmias, particularly so under conditions of high adrenergic stimulation (30). Thus, the elevated plasma NE levels associated with the initial 1–2 weeks of stay at high altitude might have important implications for older adults and patients with chronic cardiovascular disease. Little or no information exists, however, as to whether these concerns are realized upon sojourn to high altitude.

SNS-Adrenergic Regulation of Heart Rate

In rats, sustained hypoxemia results in an increase in NE turnover in the heart (48), suggesting an increase in cardiac SNS activity. As with acute hypoxemia, however, no data are available as to the effects of sustained hypoxemia on cardiac plasma NE kinetics, the current state-of-the-art measure of SNS activity to the heart in humans (30). As such, the effects of chronic hypoxemia on SNS-adrenergic regulation of cardiac function have been investigated primarily using pharmacological approaches to study the control of heart rate and the properties of cardiac adrenergic receptors.

The primary effect of sustained hypoxemia is to blunt the heart rate response to changes in β-adrenergic stimulation. Sustained hypoxemia attenuates the tachycardia in response to isoproterenol and also diminishes the slowing with β-receptor blockade at rest and during exercise (6,8,21,49,50). Consistent with these observations, the increase in heart rate with increases in plasma NE concentrations from rest to exercise is smaller with prolonged exposure to high altitude (6,8,21,43). This attenuation of β-adrenergic control of cardiac chronotropic function can be observed within the first few days at altitude (6,8), and it appears to be related to the duration of exposure (8). This blunting of chronotropic responsiveness may be due to a downregulation of β-adrenergic function with sustained exposure to high levels of NE (6,21,43). Reductions in β-adrenergic receptor density as well as postreceptor changes appear to contribute to the attenuated response (6,8,43,49–51). Perhaps the reduced β-adrenergic responsiveness associated with chronic hypoxemia has a cardioprotective effect, limiting the magnitude of tachycardia-mediated increases in myocardial O_2 demand in the face of sustained systemic hypoxemia (6,8,51).

Epinephrine

The effects of chronic exposure to high altitude on urinary and plasma concentrations of epinephrine are less consistent than those for NE. Over time at altitude, urinary 24-hour concentrations of epinephrine demonstrate no consistent pattern of change (6,9), nor do venous plasma concentrations of epinephrine show consistent change at rest or during exercise at a fixed absolute submaximal work rate (6,9). On Pikes Peak, Mazzeo and colleagues (17,18) found that although arterial plasma epinephrine concentrations at rest and during standardized submaximal dynamic exercise are elevated on the initial day at altitude, after 21 days plasma epinephrine levels are significantly lower under both conditions, albeit still higher than sea level values. The mean changes in arterial plasma epinephrine concentrations were inversely related to changes in Sa_{O_2} (18), i.e. as Sa_{O_2} increased with acclimatization, plasma epinephrine levels declined. Young and colleagues (19) found that the venous plasma epinephrine response to the same relative work rate of submaximal dynamic exercise was increased by about twofold after 3 days at the summit of Pikes Peak, but subsequently (8 and 20 days) plasma epinephrine concentrations during exercise declined to sea level values.

Thus, the influence of chronic exposure to high altitude on adrenal medullary release and circulating levels of epinephrine in humans at rest and during concurrent physical stress such as exercise is not well understood at present. A consistent finding is that the plasma epinephrine response to submaximal exercise is augmented during the initial few days of exposure to high altitude and thereafter returns toward, if not to, sea level values. Perhaps as the arterial oxygen content increases, the hypoxic stimulus for adrenal medullary release of epinephrine decreases (5,9,18). Better measures of epinephrine release and plasma kinetics will be necessary before a more definitive understanding is possible.

The initial increases and subsequent declines in plasma epinephrine concentrations observed over days at high altitude could have a number of effects. First, along with an acute increase in MSNA, epinephrine could contribute to the early elevation and subsequent decline in resting metabolic rate observed under these conditions (5). Second, the reduction in circulating epinephrine could play a role in the early elevation and subsequent decline in heart rate observed over time at high altitude (2,17). Third, the decline in plasma epinephrine concentrations over time during submaximal exercise has been postulated to play a role in the marked reduction in plasma lactate concentrations, presumably by reducing β-adrenergic stimulation of muscle glycolysis (5,9,19). Finally, the decline in plasma epinephrine concentrations with sustained exposure could be involved in the reduction of symptoms associated with altitude illness (5).

II. Parasympathetic Nervous System

The primary cardiac effect of arterial chemoreceptor stimulation is an increase in efferent vagal nerve activity and bradycardia (1,4,29). However, pulmonary afferents strongly oppose these effects in intact animals (1,4,29). Thus, as stated earlier, in humans and in many other animal species, systemic hypoxemia evokes tachycardia, not bradycardia (1,4,29). In this section we will address the possible role of cardiac vagal activity in mediating the tachycardia associated with acute hypoxemia, as well as the seemingly paradoxical observation that maximal exercise heart rate is reduced with sustained exposure to high altitude. A major constraint to this discussion is the paucity of experimental data concerning cardiac vagal function in hypoxemic humans.

A. Tachycardia with Acute Hypoxemia

At least four lines of experimental evidence support the concept that a portion of the tachycardia evoked by acute hypoxemia is mediated by a reduction in cardiac vagal nerve activity.

First, as noted above, only about half of the tachycardia in response to acute hypoxemia appears to be mediated by an increase in cardiac SNS activity and consequent β-adrenergic stimulation of the sino-atrial node (20). Because acute changes in heart rate are strictly under ANS control, this observation alone suggests that cardiac vagal withdrawal accounts for the remaining response.

Second, in dogs (33) ~40% of the tachycardia evoked by acute systemic hypoxemia is eliminated either by vagal blockade with atropine, or by chemical sympathectomy (adrenalectomy plus 6-hydroxydopamine), whereas the combination of atropine and sympathectomy completely abolishes the increase in heart rate.

Third, in some humans acutely exposed either to normobaric (32) or hypobaric (19) hypoxia, pretreatment with propranolol fails to prevent an increase in heart rate. However, combined ANS blockade with atropine and propranolol abolishes the heart rate response to acute exposure to simulated altitude (32).

Fourth, in dogs (52) acute hyperoxia produces a bradycardia that is not influenced by β-blockade but which is completely prevented by pretreatment with atropine. This observation indicates that hyperoxia-evoked bradycardia is mediated by an increase in cardiac vagal activity.

Considered together these observations support the view that the increase in heart rate evoked by acute hypoxemia is mediated by a combination of cardiac sympathetic neural activation and a reduction in cardiac vagal nerve activity, with each of these two mechanisms contributing roughly equally to the tachycardic response.

B. Cardiac Vagal Tone with Sustained Hypoxemia

Little is known about the influence of chronic hypoxemia on cardiac parasympathetic regulation of heart rate. In three young adult males exposed to 4300 m altitude, Hughson and colleagues (53) reported that heart rate was increased at day 4 compared to sea level and remained elevated at this level throughout days 11 and 12 of altitude. Time and frequency domain analysis of heart rate variability revealed that two conventional indices of cardiac vagal modulation of heart rate, i.e., the standard deviation of the R-R intervals and the high-frequency power, were reduced to the same extent at these two time points. Thus, in these subjects, 12 days of exposure to high altitude resulted in a sustained tachycardia that was associated with reductions in these two heart rate variability–derived estimates of cardiac vagal modulatory tone.

Muscarinic receptor blockade with atropine also has been used to gain insight into changes in cardiac vagal tone during sustained hypoxemia in humans. A greater increase in heart rate upon administration of atropine would be indicative of a greater level of cardiac vagal tone and vice versa. In humans, Grover and colleagues (2) found the same increase in heart rate at rest and during submaximal exercise with administration of atropine in male subjects at sea level and after 7 days at 4350 m. With regard to life-long exposure to chronic hypoxemia, the magnitude of the increase in heart rate at rest with acute administration of atropine in native Tibetan residents of high altitude has been reported to be the same as that demonstrated by acclimatized (Han) lowlanders (54). The results of these two investigations imply that in humans cardiac vagal tone is not altered by either short- or long-term acclimatization to high altitude.

In goats (49), heart rate after compared with before atropine was not statistically different in animals studied initially under normoxic conditions (102 ± 9 vs. 151 ± 10 bt/min) compared to responses after 10 days in a decompression chamber at ~4300 m (117 ± 7 vs. 147 ± 6); however, the *increase* in heart rate in response to atropine was ~40% smaller in the acclimatized animals (30 vs. 49 bt/min). Thus, in contrast to the conclusion of Grover et al. (2), cardiac vagal tone, at least in goats, might be lower after 10 days at simulated high altitude. This observation would be consistent with the Hughson et al. (53), finding that two of their measures of cardiac vagal modulation of heart rate in humans were lower than sea level control values after 11–12 days of exposure to Pikes Peak.

These observations suggest that: (1) a reduction in cardiac vagal tone may contribute to the tachycardia observed over the initial one to two weeks of exposure to high altitude in humans; (2) a return of cardiac vagal tone toward sea level values may contribute to the decline in heart rate with more prolonged exposure to high altitude in humans; and (3) there are no obvious differences in vagal tone between life-long residents of high altitude and acclimatized lowlanders under resting conditions. However, given the inconsistencies in the available data, further investigation is needed to better define the influence of chronic hypoxemia on cardiac vagal modulation of heart rate in humans.

C. Cardiac Vagal Tone and Reduced Maximal Heart Rate

An ongoing controversy involves the mechanism(s) underlying the reduction in maximal heart rate observed at altitudes greater than ~4000 m (2,6). Part of the reduction may be due to the blunted responsiveness of the sino-atrial node to β-adrenergic stimulation. However, could increased cardiac vagal tone contribute to this phenomenon (2,55)?

The experimental approach used has been to measure heart rate at maximal work rates under control (placebo) conditions and after administration of atropine. Atropine does not affect maximal heart rate measured at sea level (2). Thus, during exercise at high altitude, a higher maximal heart rate after atropine compared to the control condition would support the claim that elevated cardiac vagal tone is playing an inhibitory role.

The available information on this issue yields equivocal results. An average 11 bt/min increase in maximal heart rate was observed following pretreatment with atropine in subjects acclimatized to 4600 m for approximately one week (55), whereas maximal heart rate was 5 bt/min higher after atropine in native Tibetan residents of high altitude (54). In contrast, lowlanders (Han) acclimatized for 1–2 years at ~3600 m actually demonstrated a decrease in maximal heart rate of 6 bt/min after pretreatment with atropine versus placebo (54).

These findings leave us uncertain whether the cardiac vagus plays a role in limiting maximal heart rate in individuals acclimatized to high altitude (6). Importantly, the apparent paradox associated with such a "vagal hypothesis" would require physiological clarification. That is, how could cardiac vagal tone be reduced or unchanged at rest and during submaximal exercise under acute and sustained hypoxemic conditions, respectively, yet increased vagal tone be responsible for the lower heart rate at maximal exercise? Obviously, additional experimental work will be required to provide more definitive insight into this issue.

III. Influence of Age, Gender, and Physical Activity/Fitness Status

Little is known about the influence of age, gender, and physical activity/fitness status on autonomic nervous system function during hypoxemia. All studies of acute and sustained hypoxia have involved young and early middle-aged adult males. Similarly, no data are available on ethnic differences even though studies at high altitude often have employed native residents.

A. Age and Gender

We recently conducted a study to determine the influence of age (56) and gender on the MSNA responses to acute hypoxemia in humans. Groups of older men, premenopausal women, and young adult male controls were studied at rest under conditions of room air and isocapnic normobaric hypoxemia. We had previously demonstrated that MSNA at rest increases with age in both males and females but is lower in premenopausal females than in males at a particular age (57). In addition, we also noted recently that the MSNA responses to other forms of acute stress (e.g., isometric exercise, mental stress, cold stimulation, and mild orthostatic stress) are not consistently related to age or gender (58,59).

Aging

Because ventilatory and cardiac chronotropic responsiveness to acute hypoxemia has been reported to be reduced in older adults (60), we hypothesized a similarly diminished MSNA response. Our findings, however, only partially support this postulate. We found no difference in the MSNA response to a 15-minute period of breathing 12% inspired O_2 in the older and young adult males. With 15 minutes of 10% inspired O_2, the absolute increase in MSNA also was similar in the two groups, although the percentage increase was less in the older men because of their higher normoxic baseline levels of MSNA. The increases in heart rate were smaller in the older men in both conditions, but the regulation of arterial blood pressure was similar in the two groups. Thus, there is no obvious effect of age on the absolute SNS adjustments to systemic hypoxemia in healthy adult humans.

Gender

We found no gender-related differences in the peak ΔMSNA/ΔSaO_2 responses to the 12 and 10% inspired O_2 levels in healthy young adult subjects (unpublished results). However, the premenopausal females demonstrated a shorter latency in their increase in MSNA during the hypoxemia stimuli and a more rapid recovery to normoxic baseline levels after returning to room air breathing compared with male controls. The females also demonstrated a greater increase in heart rate and a modest elevation in diastolic arterial blood pressure, whereas the ventilatory responses were identical in the two groups. These preliminary observations suggest that although the peak increase in MSNA to moderate hypoxemia is not obviously related to gender, healthy young adult females demonstrate more rapid increases and reductions in MSNA in response to and following the hypoxemic stimulus, respectively. As such, gender may contribute to the interindividual variability in the SNS adjustments to acute systemic hypoxemia in humans.

B. Physical Activity

Grover and colleagues (2) have suggested an intriguing analogy between the physiological adaptations to sustained exposure to hypoxia and endurance exercise training. According to their concept, exercise and altitude may produce similar physiological adaptations in the O_2 transport system. Therefore, the autonomic nervous system of physically active humans may respond differently to hypoxemia than that of sedentary individuals. Unfortunately, little information exists addressing this possibility. We previously found that MSNA at rest and in response to several types of acute stress was not different in endurance-trained compared to untrained young adults under normoxic conditions (61). In addition, in our studies on the effect of acute hypoxemia on MSNA and circulatory control during supine rest and orthostatic stress (3,14,26), we did not note any differences that could be related to fitness, but these studies were not designed specifically to assess this potential influence.

Kjaer et al. (62,63) addressed the effects of habitual exercise status on the sympathoadrenal response to hypoxemia. They found that plasma NE concentrations at rest and during exercise at the same relative work rate during acute hypoxemia (breathing low O_2 gas) were similar in endurance trained versus sedentary males. However, acute hypoxemia evoked increases in plasma epinephrine concentrations at rest and during exercise only in the trained men. These findings were consistent with previously published data suggesting that endurance training is associated with a greater capacity for adrenal medullary secretion of epinephrine.

Thus, fitness level may modulate the autonomic nervous system response to acute hypoxemia in humans. What happens with sustained exposure to high altitude is yet to be determined.

IV. Concluding Remarks

Hypoxemia produces changes in autonomic nervous system function in the human. In part because of a greater overall amount of data, more easily controlled experimental conditions, and the ability to employ more sophisticated and precise experimental procedures, at present more is known about the autonomic nervous system adjustments to acute than chronic hypoxemia. Thus, extensive additional research, particularly on the influence of chronic hypoxemia, is needed. In order to produce definitive (i.e., clearly interpretable) information, future studies of the effects of chronic hypoxemia may need to employ simulated altitude in a well-controlled, fully instrumented laboratory (chamber) environment if technical limitations exist in facilities at high altitude. Finally, more attention needs to be focused on the mechanisms by which acute and chronic hypoxemia evoke changes in autonomic nervous system function in humans, as well as the physiological and pathophysiological significance of these changes.

References

1. Rowell L, Blackmon J. Human cardiovascular adjustments to acute hypoxemia. Clin Physiol 1987; 7:349–376.
2. Grover R, Weil J, Reeves J. Cardiovascular adaptation to exercise at high altitude. In: Pandolf K, ed. Exercise and Sport Sciences Reviews. New York: Macmillan, 1986: 269–302.
3. Seals DR, Rowell LB. Influence of acute hypoxemia on muscle sympathetic nerve discharge at rest and during stress in healthy humans. In: Sutton JR, Houston CS, Coates G, eds. Hypoxia and Molecular Medicine. Burlington: Queen City Printers, 1993:30–52.
4. Rowell LB. Human Cardiovascular Control. New York: Oxford University Press, 1993.
5. Reeves J, Moore L, Wolfel E, Mazzeo R, Cymerman A, Young A. Activation of the sympatho-adrenal system at high altitude. In: Ueda G, et al., eds. High Altitude Medicine. Matsumoto: Shinshu University Press, 1992:10–23.
6. Richalet J. Heart and adrenergic system in hypoxia. In: Sutton J, Coates G, Remmers J, eds. Hypoxia: The Adaptations. Toronto: PC/Dekker Inc, 1990:231–240.
7. Wolfel E. Sympatho-adrenal and cardiovascular adaptation to hypoxia. In: Sutton J, Houston C, Coates G, eds. Hypoxia and Molecular Medicine. Burlington: Queen City, 1993:62–80.
8. Savard G. Autonomic regulation during exercise in chronic hypoxia. In: Sutton J, Hous-

ton C, Coates G, eds. Hypoxia and Molecular Medicine. Burlington: Queen City, 1993: 18–29.
9. Mazzeo R. Pattern of sympathoadrenal activation at altitude. In: Sutton J, Houston C, Coates G, eds. Hypoxia and Molecular Medicine. Burlington: Queen City, 1993:53–61.
10. Esler M, Jennings G, Korner P, et al. Assessment of human sympathetic nervous system activity from measurements of norepinephrine turnover. Hypertension 1988; 11:3–20.
11. Goldstein DS, McCarty R, Polinsky RJ, Kopin IJ. Relationship between plasma norepinephrine and sympathetic neural activity. Hypertension 1983; 5:552–559.
12. Dimsdale J, Ziegler M. What do plasma and urinary measures of catecholamines tell us about human response to stress? Circulation 1991; 83 (suppl II):II-36–II-42.
13. Leuenberger U, Gleeson K, Wroblewski K, et al. Norepinephrine clearance is increased during acute hypoxemia in humans. Am J Physiol 1991; 261:H1659–H1664.
14. Rowell LB, Johnson DG, Chase PB, Comess KA, Seals DR. Hypoxemia raises muscle sympathetic activity but not norepinephrine in resting humans. J Appl Physiol 1989; 66:1736–1743.
15. Cunningham W, Becker E, Kreuzer F. Catecholamines in plasma and urine at high altitude. J Appl Physiol 1965; 20:607–610.
16. Sharma S, Balasubramanian V, Mathew O, Hoon R. Serial studies of heart rate, blood pressure, and urinary catecholamine excretion on acute induction to high altitude (3,658 m). Indian J Dis Chest 1977; 19:16–20.
17. Mazzeo R, Wolfel E, Butterfield G, Reeves J. Sympathetic responses during 21 days at high altitude (4,300 m) as determined by urinary and arterial catecholamines. Metabolism 1994; 43:1226–1232.
18. Mazzeo R, Bender P, Brooks G, et al. Arterial catecholamine responses during exercise with acute and chronic high-altitude exposure. Am J Physiol 1991; 261:E419–E424.
19. Young A, Young P, McCullough R, Moore L, Cymerman A, Reeves J. Effect of beta-adrenergic blockade on plasma lactate concentration during exercise at high altitude. Eur J Appl Physiol 1991; 63:315–322.
20. Richardson D, Kontos H, Raper A, Patterson J. Modification by beta-adrenergic blockade of the circulatory responses to acute hypoxia in man. J Clin Invest 1967; 46:77–85.
21. Richalet J-P, Larmignat P, Rathat C, Keromes A, Baud P, Lhoste F. Decreased cardiac response to isoproterenol infusion in acute and chronic hypoxia. J Appl Physiol 1988; 65:1957–1961.
22. Chang P, Kriek E, Krogt Jvd, Brummelen Pv. Does regional norepinephrine spillover represent local sympathetic activity? Hypertension 1991; 18:56–66.
23. Gregor M, Janig W. Effects of systemic hypoxia and hypercapnia on cutaneous and muscle vasoconstrictor neurones to the cat's hindlimb. Pflugers Arch 1977; 368:71–81.
24. Wallin BG, Fagius J. Peripheral sympathetic neural activity in conscious humans. Ann Rev Physiol 1988; 50:565–576.
25. Morgan B, Crabtree D, Palta M, Skatrud J. Combined hypoxia and hypercapnia evokes long-lasting sympathetic activation in humans. J Appl Physiol 1995; 79:205–213.
26. Rowell LB, Seals DR. Sympathetic activity during graded central hypovolemia in hypoxemic humans. Am J Physiol 1990; 259:1197–1206.
27. Saito M, Mano T, Iwase S, Koga K, Abe H, Yamazaki Y. Responses in muscle sympathetic nerve activity to acute hypoxia in humans. J Appl Physiol 1988; 65:1548–1552.

28. Somers V, Mark A, Zavala D, Abboud F. Contrasting effects of hypoxia and hypercapnia on ventilation and sympathetic activity in humans. J Appl Physiol 1989; 67:2101–2106.
29. Downing S, Siegel J. Baroreceptor and chemoreceptor influences on sympathetic discharge to the heart. Am J Physiol 1963; 204:471–479.
30. Esler MD, Thompson JM, Turner AG, et al. Effects of aging on the responsiveness of the human cardiac sympathetic nerves to stressors. Circulation 1995; 91:351–358.
31. Esler M, Jennings G, Korner P, Blomberg P, Sacharias N, Leonard P. Measurement of total and organ-specific norepinephrine kinetics in humans. Am J Physiol 1984; 247: E21–E28.
32. Koller E, Drechsel S, Hess T, Macherel P, Boutellier U. Effects of atropine and propranolol on the respiratory, circulatory, and ECG responses to high altitude in man. Eur J Appl Physiol 1988; 57:163–172.
33. Hammill S, Jr WW, Latham L, Frost W, JVWeil. Autonomic cardiovascular control during hypoxia in the dog. Circ Res 1979; 44:569–575.
34. Moore L, Cymerman A, Huang S-Y, et al. Propranolol does not impair exercise oxygen uptake in normal men at high altitude. J Appl Physiol 1986; 61:1935–1941.
35. Mazzeo R, Brooks G, Butterfield G, Podolin D, Wolfel E, Reeves J. Acclimatization to high altitude increases muscle sympathetic activity both at rest and during exercise. Am J Physiol 1995; 269:R201–R207.
36. Seals DR, Johnson DG, Fregosi RF. Hypoxia potentiates exercise-induced sympathoexcitation in humans. J Appl Physiol 1991; 71:1032–1040.
37. Escourrou P, Johnson D, Rowell L. Hypoxemia increases plasma catecholamine concentrations in exercising humans. J Appl Physiol 1984; 57:1507–1511.
38. Rowell L, Brengelmann G, Savage M, Freund P. Does acute hypoxemia blunt sympathetic activity in hyperthermia? J Appl Physiol 1989; 66:28–33.
39. Esler M, Kaye D, Thompson J, et al. Effects of aging on epinephrine secretion, and on regional release of epinephrine from the human heart. J Clin Endocrinol Metab 1995; 80:435–442.
40. Becker E, Kreuzer F. Sympathoadrenal response to hypoxia. Pflugers Arch 1968; 304: 1–10.
41. Hoon R, Sharma S, Balasubramanian B, Chadha K, Matheew O. Urinary catecholamine excretion on acute induction to high altitude (3658m). J Appl Physiol 1976; 41:631–633.
42. Fulco C, Rock P, Reeves J, Trad L, Young P, Cymerman A. Effects of propranolol on acute mountain sickness (AMS) and well-being at 4300 meters of altitude. Aviat Space Eviron Med 1989; 60:679–683.
43. Antezana A-M, Kacimi R, Trong J-LL, et al. Adrenergic status of humans during prolonged exposure to the altitude of 6,542 m. J Appl Physiol 1994; 76:1055–1059.
44. Sutton J, Green H, Young P, Rock P. Plasma vasopressin, catecholamines and lactate during exhaustive exercise at extreme simulated altitude. Can J Appl Sport Sci 1986; 11:43P.
45. Bouissou P, Peronnet F, Brisson G, Helie R, Ledoux M. Metabolic and endocrine response to graded exercise under acute hypoxia. Eur J Appl Physiol 1986; 55:290–294.
46. Wolfel E, Groves B, Brooks G, et al. Oxygen transport during steady-state submaximal exercise in chronic hypoxia. J Appl Physiol 1991; 70:1129–1136.
47. Wolfel E, Selland M, Mazzeo R, Reeves J. Systemic hypertension at 4,300 m is related to sympathoadrenal activity. J Appl Physiol 1994; 76:1643–1650.

48. Johnson T, Young J, Landsberg L. Sympathoadrenal responses to acute and chronic hypoxia in the rat. J Clin Invest 1983; 71:1263–1272.
49. Maher J, Denniston J, Wolfe D, Cymerman A. Mechanism of the attenuated cardiac response to beta-adrenergic stimulation in chronic hypoxia. J Appl Physiol 1978; 44: 647–651.
50. Voelkel N, Hegstrand L, Reeves J, McMurty I, Molinoff B. Effects of hypoxia on density of beta-adrenergic receptors. J Appl Physiol 1981; 50:363–366.
51. Richalet J-P, Kacimi R, Antezana A-M. The control of cardiac chronotropic function in hypobaric hypoxia. Int J Sports Med 1992; 13(suppl 1):S22–S24.
52. Lodato RF, Jubran A. Response time, autonomic mediation, and reversibility of hyperoxic bradycardia in conscious dogs. J Appl Physiol 1993; 74:634–642.
53. Hughson R, Yamamoto Y, McCullough R, Sutton J, Reeves J. Sympathetic and parasympathetic indicators of heart rate control at altitude studied by spectral analysis. J Appl Physiol 1994; 77:2537–2542.
54. Zhuang J, Droma T, Sutton J, et al. Autonomic regulation of heart rate response to exercise in Tibetan and Han residents of Lhasa (3,658 m). J Appl Physiol 1993; 75: 1968–1973.
55. Hartley L, Vogel J, Cruz J. Reduction of maximal exercise heart rate at altitude and its reversal with atropine. J Appl Physiol 1974; 36:362–365.
56. Davy KP, Jones PP, Seals DR. Influence of age on the sympathetic neural adjustments to alterations in systemic oxygen levels in humans. Am J Physiol 1997; 273:R690–R695.
57. Ng AV, Callister R, Johnson DG, Seals DR. Age and gender influence muscle sympathetic nerve activity at rest in healthy humans. Hypertension 1993; 21:498–503.
58. Jones P, Spraul M, Matt K, Seals D, Skinner J, Ravussin E. Gender does not influence sympathetic neural reactivity to stress in healthy humans. Am J Physiol 1996; 270: H350–H357.
59. Ng AV, Callister R, Johnson DG, Seals DR. Sympathetic neural reactivity to stress does not increase with age in healthy humans. Am J Physiol 1994; 267:H344–H353.
60. Kronenberg R, Drage CW. Attenuation of the ventilatory and heart rate responses to hypoxia and hypercapnia with aging in normal men. J Clin Invest 1973; 52:1812–1819.
61. Seals DR. Sympathetic neural adjustments to stress in physically trained and untrained humans. Hypertension 1991; 17:36–43.
62. Kjaer M, Galbo H. Effect of physical training on the capacity to secrete epinephrine. J Appl Physiol 1988; 64:11–16.
63. Kjaer M, Bangsbo J, Lortie G, Galbo H. Hormonal response to exercise in humans: influence of hypoxia and physical training. Am J Physiol 1988; 254:R197–R203.

14

The Effects of Altitude on Skeletal Muscle

HOWARD J. GREEN

University of Waterloo
Waterloo, Ontario, Canada

JOHN R. SUTTON[†]

Exercise Research Centre
University of Sydney
Sydney, Australia

I. Altitude and Skeletal Muscle

A. An Overview

Skeletal muscle is one of the most malleable tissues in mammals, capable of extensive reorganization and remodeling when stressed. This phenotypic plasticity has been demonstrated using a variety of models such as alterations in hormonal milieu, gravitational force, and contractile activity (1). Chronic low intensity activity, for example, can induce a fundamental change in both the type and extent of a variety of cellular proteins (2). In general, these adaptations appear directed at insuring the functional integrity of the excitation and contraction processes in the cell and providing for the continuous supply of energy derived from the aerobic metabolism during sustained activity (2).

The muscle cell may also be subjected to a variety of other provocative agents. Hypoxia, for instance, can potentially impair oxygen availability, perturbing mitochondrial respiration and energy state, with concomitant disruption in a wide range of cellular functions, including the processes involved in translating neural motor commands into force generation. How does the muscle cell adjust to acute hypoxia, and what adaptations occur if the hypoxic stimuli is sustained over days, weeks,

† Deceased.

months, years, and even generations? The answers to these questions, which represent the primary focus of our review, are of fundamental importance if we are to understand how changes occur in performance, a property that represents the primary function of muscle. Our approach has been to provide a comprehensive analysis of the structural and compositional adaptations of the muscle cell to chronic hypoxia and to examine the significance of these adaptations on the metabolic alterations that result both during rest and exercise. Since hypoxia rarely occurs in isolation from other events such as activity, which also influence muscle character, we have also provided a brief review of the training response and attempted, where possible, to emphasize the importance of activity level on the adaptive responses that occur with residence at altitude. For the reader not familiar with the organization of the muscle cell and the concept of muscle fiber types, introductory information is provided to make the review more readable. In Chapter 20, the significance of the cellular adaptations are examined with regard to the impact on mechanical behavior and work performance (3). The extent to which adaptations in the muscle cell can explain the prodigious increases in work tolerance that occur with acclimatization remains an intriguing question.

B. The Muscle Cell—Adaptable Elements

Adaptations in the muscle cell that affect the contractile response are of two types, namely those involved in excitation and contraction and those involved in the provision of the fuel and energy requirements for the excitation and contraction processes (Fig. 1). Since muscle cells are specialized in their ability to convert a neural command into a discrete motor or mechanical response, many unique structures or organelles are required for transmitting the excitation signal to the interior of the cell and for converting it into the generation of force. The neural impulses are communicated across the neuromuscular junction via a chemical transmitter, acetylcholine (Ach), where they trigger an action potential in a specialized region of the surface membrane of the cell, the motor end plate. The action potential then spreads over the plasma membrane or sarcolemma where at selected points the signal is conducted to the interior of the fiber by an invagination of the sarcolemma, the t-tubule system. Once inside the cell, the excitation signal must be conducted to another specialized organelle, the sarcoplasmic reticulum (SR), which has near exclusive control over the cytosolic free calcium levels (Ca_f^{2+}), by virtue of its ability to store, release, and sequester Ca^{2+}.

The myofibrillar apparatus, which consists of the regulatory proteins, troponin and tropomyosin, and the contractile proteins, actin and myosin, is responsible for sensing the elevation of Ca_f^{2+} that occurs following the transmission of the excitation signal from the t-tubule to the Ca^{2+} release channel of the SR and removing the inhibitory constraints on force development. To initiate contraction, Ca^{2+} is initially bound to a specialized subunit of troponin (Troponin C), which controls a series of changes in two other troponin subunits (Troponin T and Troponin I) and tropomyosin, allowing actin and myosin to go from a dissociated or weak binding state to a

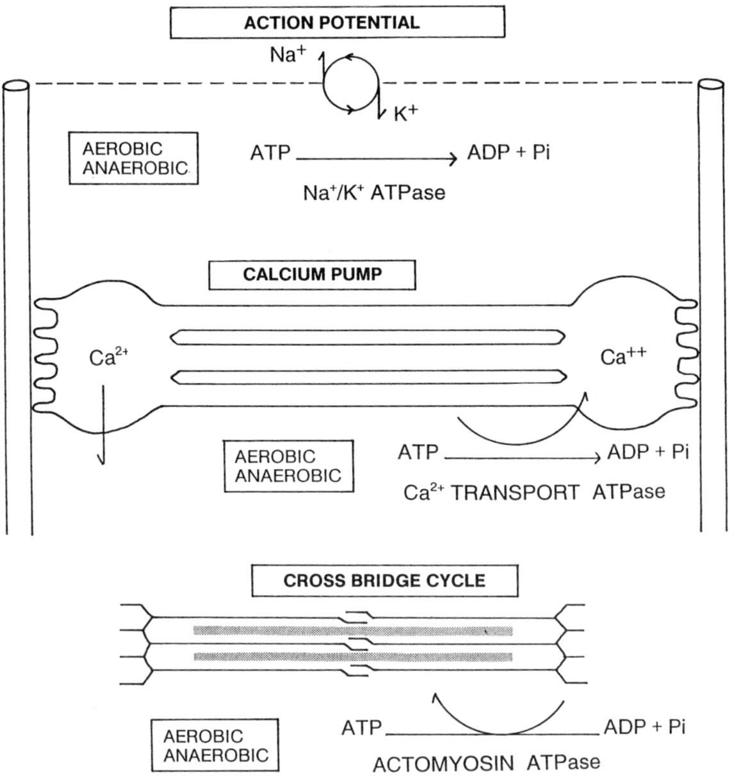

Figure 1 Excitation contraction processes, ATPase enzymes, and energetic processes in the muscle cell. Three ATPases are represented: the sarcolemma Na^+-K^+ ATPase, which functions to pump Na^+ back out of the cell and K^+ in following an action potential; the sarcoplasmic reticulum Ca^{2+}-transport ATPase, which pumps Ca^{2+} back into the sarcoplasmic reticulum; and the actomyosin ATPase, which allows for weak to strong binding of actin and myosin and force development. Aerobic and anaerobic metabolic pathways regenerate ATP at each of the sites.

strong binding, force-generating state (2). Interestingly, the proteins involved in all of these processes are capable of extensive modification if the stimulus controlling gene expression is sufficient.

Since excitation and contraction are energy-dependent processes, provision must be made both for the conversion of ATP to free energy and for the continual regeneration of ATP. The conversion of the chemically bound energy contained in ATP is mediated by specialized ATPase enzymes localized in different regions of the cell. One ATPase, the sarcolemma Na^+-K^+ ATPase, provides the energy for pumping the Na^+ out of the cell and the K^+ back in following an action potential. Reestablishment of resting membrane potential is necessary if the sarcolemma and t-tubule system are to conduct repetitive action potentials, a condition necessary for the tetanic force production. An additional ATPase involved in excitation is located primarily in the longitudinal portion of the SR. This ATPase, referred to as a Ca^{2+}-ATPase, provides the energy necessary for sequestering Ca^{2+} back into the SR during tetany and during relaxation following cessation of the neural stimulus. Actomyosin cycling accounts for the greatest energy consumption is the contracting muscle cell. The specialized ATPase involved here is located on the myosin molecule and is referred to as the actomyosin ATPase. As with the other proteins involved in excitation and contraction, the ATPase enzymes are also capable of extensive alterations.

Highly organized metabolic pathways and segments attempt to maintain the energy potential (phosphorylation potential) of the cell by providing for the regeneration of ATP and preparation of substrates. These specialized pathways consist of high-energy phosphate transfer reactions, oxidative phosphorylation, and glycolysis. High-energy phosphate reactions, of which there are two, given their equilibrium nature and high activity, are extremely responsive to ATP levels and are capable of rapid synthesis of ATP. However, a primary substrate (phosphocreatine) is extremely limited. Glycolysis, which consists of the catabolism of carbohydrates (glucose, glycogen) to lactate, is also capable of producing large amounts of ATP per unit time, but the terminal product produced in this pathway is lactic acid, an excess accumulation of which could have dire consequences to cellular function and contractility. Oxidative phosphorylation involves the aerobic combustion of fats or carbohydrates in the mitochondria for the production of ATP. In the mitochondria, two metabolic pathways are intimately coordinated for aerobic production of ATP. The tricarboxylic acid cycle (TCA) generates reduced coenzymes, while the electron transport chain uses the reduced coenzymes for ATP production. Also located in the mitochondria are the enzymes needed for the membrane transport, activation, and conversion of fatty acids to acetyl CoA, the common intermediate used in mitochondrial respiration, and an enzyme complex, pyruvate dehydrogenase, used to catalyze the conversion of pyruvate to acetyl CoA.

Flux rates in these metabolic pathways and consequently the rate at which ATP can be generated are critically dependent on pathway development (i.e. maximal activity of enzymes) and on the availability of specific substrates. Substrates may be provided by sources outside the cell or from storage depots within the cell. Glycogen and triglycerides represent important fuels, which can be stored within the cell but require special enzymes to mediate their synthesis or to provide for their hydrolysis and delivery to the sites of combustion. Glucose and fatty acids are supplied by extramuscular sources, namely the liver and adipose tissue, respectively. Intracellular availability depends on specialized enzymes (i.e., hexokinase) or membrane transporters (i.e., glucose transporters, GLUT-4, and fatty acid–binding proteins, FABP). Metabolic patterning studies have attempted to examine the organization of each of the pathways and segments within the cell both with regards to the enzyme interrelationships within pathways and between pathways. These studies have provided important insights into the unique way that the muscle cell is designed to meet specific threats to its energy potential.

Both the processes involved in excitation and contraction and the processes involved in energy metabolism display a unique set of properties with coordination evident both within and between the processes. Distinct fiber types appear uniquely designed for specific types of contractile tasks. Two major fiber types are widely recognized: slow-twitch or type I fibers and fast-twitch or type II fibers. Slow-twitch fibers are primarily specialized for sustained, low-velocity, low-intensity work. In this fiber type, the rapidity with which the neural signal can be conducted to the interior of the fiber and the rate at which force can be generated are not important properties. In contrast, fast-twitch fibers are capable of rapid transmission of the excitation signal and rapid rates of force development and high velocities. The differences between the two fiber types are determined by both the type (isoforms) and the amount of the various proteins involved in excitation and contraction. Fast-twitch fibers, for example, possess a highly developed SR capable of rapid calcium cycling and a unique myosin composition designed to allow rapid hydrolysis of ATP and actomyosin cycling. These fibers also possess great potential for ATP resynthesis via high-energy phosphate transfer and glycolysis. In contrast, ATP production in slow-twitch fibers is primarily aerobic-based with the pathways for high-energy phosphate transfer and glycolysis relatively poorly developed. A range of fiber subtypes also exist, particularly for the fast-twitch fibers. As with the major fiber types, fiber subtypes (types IIA, IIB, IIX, IIC) can be routinely identified with histochemical techniques based on the myosin heavy-chain isoform composition. The histochemical procedure exploits the pH lability of the ATPase enzyme, which differs with different myosin heavy chain isoforms (5). Although differences do exist in the proteins constituting the organelles involved in excitation and contraction, the most conspicuous difference between the fiber subtypes is in mitochondrial potential and the processes supporting aerobic metabolism. Other fiber type classification procedures also exist. Type II fibers may also be subclassified as fast-twitch oxidative-glycolytic (FOG) or fast-twitch glycolytic (FG). Although there is some correspondence with the myosin based procedures, differences do exist (5).

In summary, the elements reviewed represent the primary structures that have been examined in a variety of models investigating muscle cell plasticity. Some of these models (hormonal, chronic activity) have been studied much more extensively than the hypoxic (altitude) model, and consequently considerably more is known about both the nature and time course of adaptation. Studies on muscle adaptation to hypoxia have focused largely on the metabolic machinery and supporting processes.

C. Potential Strategies and Implications

Insight into how the muscle cell adapts to oxygen deficient environments can be obtained by examining the possible strategies used to maintain ATP homeostasis during acute hypoxic exposure. These strategies, which may involve alterations in

ATP synthesis pathways or in ATP utilization processes, are best revealed by using contractile activity to increase ATP turnover rates. Accommodation responses that occur rapidly may involve events outside the muscle cell, within the muscle cell, or an interaction between the two. In the absence of compensatory adjustments, acute hypoxia results in reductions in arterial oxygen tension ($Pa{O_2}$), saturation ($Sa{O_2}$), oxygen content ($Ca{O_2}$), and the amount of oxygen available to the skeletal muscle. Two strategies provide for the energy requirements of the muscle cell when confronted with a reduction in oxygen supply. On the one hand, the muscle cell can accept the reduction in oxygen and consequently in the level of oxidative phosphorylation and increase the activation of other energy-supplying metabolic pathways such as high-energy phosphate transfer and glycolysis. However, such a strategy provokes other challenges that the cell must cope with, namely the accumulation of a range of metabolic by-products, which may, in themselves, inhibit vital cellular processes (6) including mitochondrial respiration (7). Under such conditions, the cell may be incapable of using the available oxygen, which could further disrupt cellular homeostasis.

The second strategy may be to protect ATP synthesis rates by oxidative phosphorylation in spite of reduction in substrate oxygen. Such a strategy might be implemented by attempting to insure an uninterrupted supply of oxygen to the mitochondria and/or by enabling the mitochondria to sustain respiration rates at a reduced $p{O_2}$. Adjustments aimed at minimizing the reduction in oxygen availability appear to be a fundamental priority, with changes in ventilatory and circulatory function attempting to protect oxygen delivery to the muscle cell (8). However, compensation in other systems also increases the demand for oxygen and energy. For example, increasing ventilation and cardiac output require increases in ATP. Long-term viability of these tissues would also appear to be fundamentally dependent on adequate and protected levels of aerobic metabolism. Competition between muscles for access to the available oxygen is unavoidable.

Cellular adaptations resulting from long-term exposure to hypoxia aimed at providing the conditions necessary to maintain oxidative phosphorylation at subnormal $P{O_2}$ are also not without cost. Such adaptations would need to involve processes associated with cellular diffusion of oxygen and changes in enzymatic organization and metabolic control (9). Increases in myoglobin and mitochondrial potential, as examples, construed as important alterations to protect cellular respiration are energetically expensive, causing a further drain on cellular O_2 resources. Evidence is accumulating that, at least under some conditions, cellular reorganization may be accomplished by catabolic processes. Decreases in fiber size, which has been reported in several acclimatization studies (10), may shorten intracellular diffusion distances and provide O_2 to the mitochondrial at smaller pressure gradients. However, decreases in muscle mass would be expected to decrease the maximal force-generating capability of the muscle.

Recently, a third potential strategy designed to help cope with hypoxia has been postulated (11). This strategy is based on improving efficiency, that is, allowing the work-related processes in the cell to be performed at lower energy cost. Metabolic efficiency (12) involves both the type of metabolism being used for ATP supply and the type of substrate selected. Such a strategy could also involve modification of one or more excitation and contraction processes itself, whereby functional integrity of the process is protected but at lower energy requirements (13). In this regard, hypoxia has been found, at least in some organisms, to suppress cellular function resulting in a sustained downregulation but at the same time maintaining viability and the capability for normal function once normoxia has been restored (14).

Attempting to isolate the time-dependent specific effects of hypoxia itself on muscle structure, composition, and function is difficult. Even with acute exposure, potential differences in response may exist when the hypoxic stimulus is applied in the laboratory at sea level with other environmental conditions normalized versus soon after arrival at altitude. When the stimulus persists chronically, differences in the response between the laboratory (acclimation) and altitude (acclimatization) may become even more exaggerated. When hypoxia becomes a life-long characteristic, experienced by many who have resided at altitude throughout their life and for generations before, attempting to unravel the adaptations that are specific to hypoxia and not influenced by other environmental and demographic aspects such as cold and diet is problematic at best. Extrapolation of the responses to the high-altitude native from short-term acclimatization studies remains tenuous and always complicated not only by the time of exposure but by species, gender activity level, and race. Adaptations must be viewed as a dynamic process constantly modified by the time of exposure (15,16) (Table 1). The general failure to control for activity level is one

Table 1 Environmental Physiology—Some Useful Terms

Term	Definition
Adaptation[a]	
Genetic	Physiological, morphological, and other changes that have occurred over many generations in a species or strain of animals
Phenotype	Physiological, morphological, and other changes that occur over the lifetime of an individual
Accommodation (habituation)	Changes that occur quickly and are of a temporary nature (can occur within milliseconds and last for several hours)
Acclimatization	Functional compensation extending over a period of days to weeks in response to a complex of environmental factors and in seasonal or climatic changes
Acclimation	Functional compensation over a period of days to weeks in response to a single environmental factor

[a] Obtained from Refs. 15 and 16.

of the more serious limitations in attempting to unravel the isolated effects of chronic hypoxia on muscle structure and function.

Adaptations of muscle to hypoxia remain poorly described compared to ventilatory and circulatory phenomena, primarily because invasive procedures have been required for observation of muscle tissue. Not only have such studies been difficult to conduct, but until recently only few muscles and few properties could be studied, particularly in the human. The development of noninvasive capabilities such as ^{31}P-NMR and the refinement of a spectrum of analytical techniques capable of allowing measurement of a wide range of cellular properties with only small samples of tissue has marked a new era in the study of the adaptation of the human skeletal muscle cell to hypoxia.

To date, investigators have been almost totally preoccupied with examining selected features in the muscle cell related to hypoxia-induced adaptation in energy metabolic potential and related properties. Such studies have concentrated on describing changes in key enzymes in various metabolic pathways and segments. These studies have employed biochemical procedures based on the measurements of maximal activity of the enzymes. Other studies have allowed for ultrastructural and morphometric measurements, which permit the semi-quantitative description of the size and location of key elements in the hypoxia-adapted muscle cell. Frequently these measurements are also complemented by measures of capillarization and cell size, allowing for estimation of the diffusion potential. More recent studies have also addressed the changes in muscle metabolic behavior resulting from altitude exposure and the changes in substrate selection, particularly during the working state. These approaches from the core of the following review.

In any emerging area, initial studies tend to be exploratory in nature, frequently lacking in appropriate controls and sample size. Statistical inference is often problematic, with the ever-persistent risk of misinterpretation of the real or experimental effects. Such is the case with the effects of hypoxia on the muscle cell. Although many studies provide potential building blocks for models of adaptation, the lack of scientific rigor diminishes their current importance. Our approach has been to cite most studies and to address the limitations of specific studies where appropriate. For additional information and perspective on skeletal muscle and altitude adaptation, the reader is referred to several comprehensive articles (12,17–19).

II. Adaptations in the Muscle Cell

A. Physical Activity and Normoxia

Because activity itself is a major determinant of muscle character and composition, we have provided a brief description of the most salient adaptations (20–22). With this information, the reader may be able to compare and contrast the effects of habitual physical activity performed at sea level versus the effects of chronic hypoxia per se and both with and without habitual physical activity.

The original landmark studies of Holloszy and colleagues examining the effects of training, employed 2 hours of running five times a week over 12 weeks using Wistar rats (23). This training resulted in a twofold elevation in mitochondrial potential, as indicated by the maximal activities of a variety of enzymes involved in the citric acid cycle, the electron transport chain, and β-oxidation. Only minor changes were observed in the maximal activities of a range of enzymes involved in glycogenolysis, glycolysis, and high-energy phosphate transfer (23). These results demonstrate a marked and proportional enhancement of the enzymes in mitochondrial function at relatively constant glycolytic potential. Although the magnitude of the change differed by fiber type composition of the muscles examined, the relative changes in the different metabolic pathways within a muscle were qualitatively similar (23).

Other adaptations appear to contribute to the delivery of oxygen to the cell. Training also resulted in pronounced elevations in capillarization (24), and since fiber size in rats does not appear to change with even the most severe training programs, the capillary–to–fiber area ratio was substantially enhanced (24).

Perhaps as a result of species differences, cellular adaptation in humans to similar types of training programs is much more modest (20–22). With the exception of changes in fiber types, restricted mainly to the transformation between fast fiber subtypes (type IIB to type IIA) (20), adaptations appear to be most pronounced in the energy metabolic pathways and in cellular capillarization (20,21). Increases in mitochondrial potential approximating 50% with sustained training appear to be unaccompanied by adaptations in the enzymes of high-energy phosphate transfer and glycolysis (20). The muscle, in effect, becomes more poised for aerobic production of ATP, with free fatty acids serving as a preferred substrate. Even the most challenging experimental programs of prolonged exercise training appear not to result in a diminution of fiber cross-sectional area (20). In fact, the contrary is typically observed, namely an increase in fiber size that extends to the major fiber types and subtypes (25). Increases in capillarization outstrip the increase in fiber size, and increases in capillary diffusion potential can also be observed with training (25–27).

Myoglobin represents another potentially important adaptation since it can act as a store of oxygen in the cell and facilitate the diffusion of oxygen to the mitochondria (28). Changes in muscle myoglobin with exercise training appear to depend on both the species and fiber type composition of the muscle. In the rat, the species most often investigated, large increases in concentration (up to twofold) have been demonstrated with extended training programs consisting of prolonged running (29). The greatest increases have been observed in muscles composed primarily of fast oxidative fibers (FOG, red vastus lateralis), although significant elevations have also been demonstrated in muscles composed predominately of slow fibers (SO, soleus) (30,31). No changes in myoglobin have been reported in the white vastus lateralis (FG fibers), even with training programs varying in intensity (31) and frequency of training sessions (32).

Studies on humans, however, have not been able to demonstrate differences in myoglobin concentration in the vastus lateralis, either between untrained and competitive cyclists (33) or prior to and following prolonged exercise training (34), in sedentary volunteers or in response to supplementary high-intensity training in previously trained cyclists (35). Increases in mitochondrial potential that are typically associated with increased patterns of activity appears to be dissociated from the changes in myoglobin in both rats (30,31) and humans (34). That no changes were found in the vastus lateralis muscle of humans, which is composed of a mixture of all fiber types, suggests that no fiber type responds to training with increases in myoglobin. In fact, in humans, one leg training for 30 minutes a day three to four times per week over 4 weeks resulted in a small but significant decrease in myoglobin (36).

Recently studies have been published examining the effects of different training regimes on proteins and protein isoforms involved in ATP utilization, such as those participating in excitation processes in the muscle cell (37). The sarcolemma Na^+-K^+ ATPase appears to be very sensitive to the level of regular activity with increases reported within the first few days of training. In contrast, the SR Ca^{2+}-ATPase is more resistant to the effects of training. Reductions in SR Ca^{2+}-ATPase can be induced, but only with extreme patterns of submaximal prolonged exercise training. Interestingly, changes in the proteins of the SR appear to be expressed coordinately with myosin and, in particular, the myosin fast to slow heavy chain isoforms. Fiber type changes measured histochemically reflect myosin isoforms shifts.

The cellular adaptations observed following voluntary exercise programs do not represent the limits of adaptability to habitual activity. Extreme patterns of chronic activity induced by electrical stimulation to specific muscles, most commonly in the rat or rabbit, can elicit a complete fast to slow fiber transformation at all levels of organization (5,38). To achieve complete transformation, periods of stimulation have extended for up to 24 hours per day and for in excess of 60 days (2). A comprehensive review is available that provides a complete description of the adaptations that occur with chronic, low-frequency electrical stimulation (2).

In the absence of hypoxia, these are muscular adaptations that remain possible with sustained increases in daily activity. However, the adaptive changes in muscle at altitude could be qualitatively different depending on the nature of the activity. Different types of activity at sea level can result in modifications of the adaptations typically observed with prolonged, aerobic-based activity. At one extreme is the adaptation to high resistance training. This type of training program typically results in large increases in fiber cross-sectional area in the absence of alterations in the metabolic potential, either mitochondrial or glycolytic (22,39). However, since the relatively large increase in fiber size does not appear to be accompanied by changes in the number of capillaries per fiber, capillary density or the number of capillaries per cross-sectional area may be decreased (20,22). This type of adaptation contrasts with that observed with prolonged exercise training (20) and that which apparently occurs as a result of chronic hypoxia (40,41).

Interestingly, it might be expected that in residents involved in trekking or mountaineering, the ascent phase of the climb would be expected to require repeated generations of high force concentric or shortening contractions. Descent, in contrast, would involve lengthening or eccentric-based activity, also characterized by high force outputs. Unlike concentric resistance programs, eccentric training, following the initial period of cellular damage, does not result in muscle hypertrophy and may, in fact, increase aerobic potential (42). In large part the adaptations that do occur would appear to involve processes that allow the muscle protection against the damage that occurs with eccentric activity (43,44).

Recent experiments also demonstrate that if the exercise (i.e., sprint) promotes large increases in high-energy phosphate transfer and glycolytic flux rates, adaptations to these pathways become most emphasized (45). In such circumstances, an upscaling of the glycolytic potential relative to the mitochondrial potential can occur (45) and the muscle cell becomes more poised to meet the energy requirements of the activity by anaerobic combustion of substrates, mainly muscle glycogen. Other elements of the cell also appear to adapt. Sarcolemmal Na^{-}/K^{-} ATPase is increased, potentially providing for an increased ability of the sarcolemmal to conduct high-frequency stimulation characteristic of intense activity (46). To what extent adaptations to this type of training extend to other specialized proteins of the cell has not been defined.

In summary, activity in itself can create a spectrum of adaptive responses in the muscle cell, the type and magnitude being to some degree dependent on the specific strains imposed on the different cellular processes and induced by the character of the activity. The effect of physical activity in modifying the muscular response to chronic hypoxia may well depend on the regimen of activities pursued.

B. Acclimatization, Adaptation, and Altitude

Fiber Type Composition

An intriguing and highly pertinent issue given the differences that exist in performance and efficiency is whether sustained residence at high altitude promotes selection and adaptation based on a preference for specific muscle fiber types (Table 2). The only reported attempt at fiber type classification in high altitude natives was on three Andean Quechuas resident to 3300–4500 m (47). The percent cross-sectional area occupied by type I, IIA, and IIB fibers on tissue samples obtained from the vastus lateralis averaged approximately 68, 16, and 16%, respectively. Fiber type distributions independent of area were not reported. The type I and type IIA fiber percentages are considerably higher and lower, respectively, than one might expect in a sea level group (22,48). However, the fiber type cross-sectional area may give a misleading approximation of the percentage of fibers areas in altitude residents than sea level residents because of fiber size differences (22). At present it is unclear what the effect long-term residence has on fiber type distribution in muscle.

Table 2 Altitude, Adaptation, Acclimatization, and Muscle Fiber Type Distribution

	Fiber types (%)		
	I	IIA	IIB
High-altitude natives,			
Quechuas (n = 3)[a]	67.9 ± 5.4	16.4 ± 3.9	15.6 ± 3.3
Acclimatization			
Operation Everest II (n = 5)[b]			
Pre	48.7 ± 4.0		
Post	50.9 ± 3.5		
Pikes Peak (n = 5)[c]			
Pre	54.9 ± 7.9	35.6 ± 5.6	9.5 ± 2.6
Post	43.0 ± 7.9	43.7 ± 6.3	13.5 ± 1.7

Values are $\bar{x}$ ±SE.

[a] Data obtained from Rosser and Hochachka (47), representing relative contribution of each fiber type to total cross-sectional area.

[b] Data obtained from Green et al. (40).

[c] Data obtained from Green et al. (50).

Brief residence of between 21 to 40 days at moderate to severe hypoxia does not appear to cause alterations in the major fiber types (49,50), nor does brief residence at moderate altitude alter the subtypes (50). Given the general resistance for fiber transformation at the level of the major fiber types even to severe regular exercise (22), it would appear that the high percent of type I fibers observed in experienced climbers (51) represents an inherited property.

Fiber Size and Capillarization

Capillarity measurements in conjunction with fiber size alterations provide important insights into the potential of chronic hypoxia to alter cellular perfusion characteristics. Increasing capillarization during chronic hypoxia could enable reductions in arterial O_2 content to be compensated for by increasing cellular blood flow, lowering velocity and reducing capillary to mitochondria O_2 diffusion distance, thereby encouraging sustained O_2 transfer to mitochondria.

Indeed increases in capillarity were accepted as a characteristic response to chronic hypoxia (8). However, recent reports strongly suggest that although increases in perfusion potential do occur, they are not a consequence of increased capillarization but result from reduction in fiber size. Almost invariably the reductions in fiber size accompany reductions in whole body weight (10).

We are aware of only one study that has examined capillary–fiber size interrelationships in high-altitude natives (52) (Table 3). This study, using muscle from five Sherpas, was part of a larger investigation examining other ultrastructural features, including mitochondrial density and distribution in cells obtained from the vastus

Table 3 Muscle Fiber Area and Capillarization in High-Altitude Natives, Mountaineers, and Sea Level Controls

	Caps/Fiber (n)	Caps/Area (n)	Fiber area (μm^2)
Sherpas ($n = 5$)	1.48	467	3186
Elite mountaineers ($n = 5$)	1.67	542	3108
Swiss expeditions			
Pre ($n = 14$)	1.99*	483	4170
Post ($n = 12$)	1.79	538	3,360*
Sedentary ($n = 5$)	1.39	387*	3630

Note: Biopsies from vastus lateralis. Caps/Fiber, number of capillaries surrounding each fiber; Caps/Area, number of capillaries per unit fiber area; Fiber area, area of fiber.
* Significantly different from Sherpas ($p < 0.05$).
Source: Refs. 41, 52.

lateralis muscle. When compared to sea level, untrained controls, no significant differences were found for the average number of capillaries around each fiber, nor was the average area per fiber of Sherpas significantly different. However, the capillary density (defined as the number of capillaries per unit tissue area) was greater in the high-altitude natives. This adaptation, which would provide a shortened diffusion path for O_2 in the fiber, was a consequence primarily of the reductions (although not significant) in fiber cross-sectional area. However, care must be taken in the interpretation of these results given the larger body size and more trained state of the control group compared to the altitude group. Using a more appropriate control group, identified in a later study, muscle fiber sizes, capillary number, and capillary density were nearly identical to the Sherpas. The significance of regular activity on cell capillarization while resident at altitude also appears important (53). Natives resident to 3700 m who were active displayed a higher number of capillaries per fiber, although sizes were not different between those who were active and those who were not.

Several recent studies have examined fiber morphology and ultrastructure during acclimatization in individuals who were either volunteers in carefully controlled experiments (40,50) or who were part of mountaineering expeditions (52,54,55).

Three weeks at 4300 m (Pikes Peak) had no effect, regardless of fiber type and subtype (I, IIA, IIB), on the number of capillaries per fiber, fiber area, or consequently, the number of capillaries per unit fiber area (50). Possibly because of the dietary control implemented, no reductions in body weight were observed during the period of residence at Pikes Peak.

Operation Everest II, an acclimation study conducted in a hypobaric chamber simulated the ascent to the summit of Mt. Everest (8848 m) over a 40-day period (56) (Fig. 2). Muscle morphometric analyses performed on tissue from vastus lateralis before and after acclimation indicated reductions in fiber size of 20–25% in

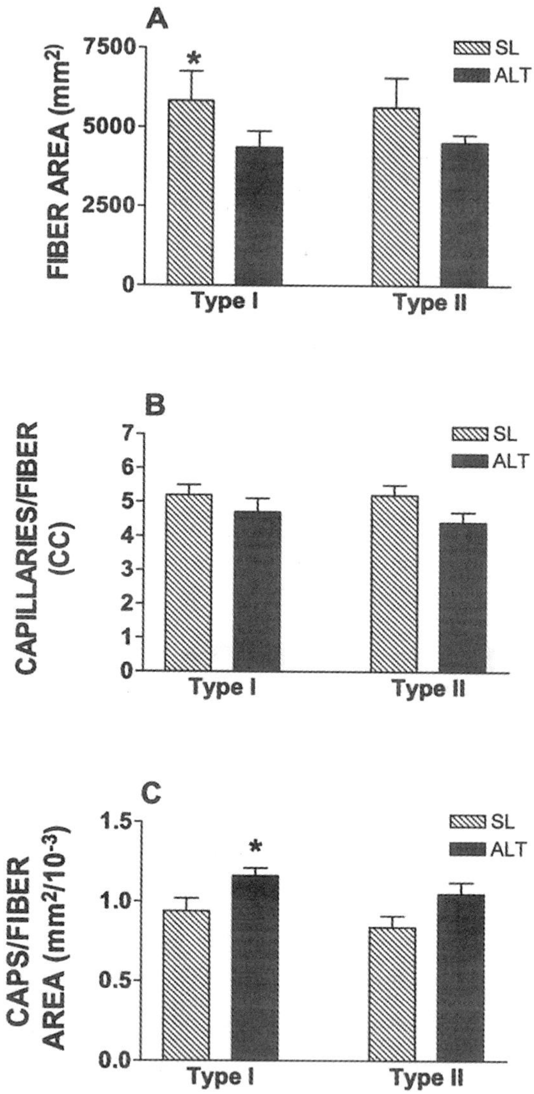

Figure 2 Fiber area (A), capillaries per fiber (B), and capillaries (C) per fiber area in Type I and Type II fibers during acclimatization to simulated altitude (Operation Everest II). (Data from Ref. 40.)

both type I and type II fibers (although only the change in type 1 was significant) and no change in the average number of capillaries in contact with each fiber (40). As with high-altitude natives (52), a pronounced increase, comparable in magnitude for each fiber type (only type 1 fibers significant), was found in the number of capillaries per unit fiber area. In spite of careful dietary control, the extreme hypoxic stress resulted in significant loss of body weight (40). Muscle changes do not appear to be specific to the vastus lateralis, since total muscle volume decreased in both upper arm and thigh (55). Similarly actual residence at altitude (3200 m) for 32 weeks appears to result in the same effect, namely a reduced fiber area with consequent increased capillary density (53).

Mountaineers to Everest (8848 m), all participating in a daily schedule of climbing activity during 55–70 days of chronic altitude exposure, also showed the same effect, namely an increase in the capillary density, which was mediated entirely by reductions in fiber size (41,54). In another study involving climbers to Mt. Lhotse Shar (8398 m), the effects of the expedition on muscle histology were not as clear (41,54). Following the expedition, significant reductions in the number of capillaries per fiber were observed, and since only small and insignificant reductions in fiber size occurred, the capillary density was not significantly changed. It is not evident why the results of the two studies were inconsistent, although the fact that the body weight loss was more severe in the Everest expedition may be significant given the relationship between body weight and muscle mass (57). Whether the apparently greater weight loss was due to dietary deficiencies or a more rigorous activity schedule was not indicated.

It is also possible that the capillarization and fiber size characteristics may to some extent be an inherent characteristic specific to elite climbers. Oelz et al. (51) found that in six world class high-altitude mountaineers, fiber areas were smaller and the number of capillaries per unit area of tissue greater in spite of a similar number of capillaries per fiber than sea level controls. These differences were found 2–12 months after the last of a number of high-altitude climbs, all of which were to peaks in excess of 8500 m without supplementary O_2.

Earlier studies on acclimatization in animals favored increased muscle fiber capillarization as an adaptable property (58). However, not all studies supported this conclusion (59). It has now become clear, at least for moderate to high simulated altitudes (up to 5100 m) and for periods of exposure extending to 14 weeks, that chronic hypoxia does not stimulate muscle angiogenesis (59–61). Moreover, when the effect of changes in body weight is controlled for, no reduction in fiber size occurs during acclimatization and consequently no changes in capillary density can be detected (59,62,63). These results, which were consistent for both rats and guinea pigs and for a number of muscles (soleus, gastrocnemius, tibialis anterior), have recently been confirmed by Bigard et al. (57), who found no evidence of increases in the number of capillaries per fiber in either soleus (SOL), extensor digitorum longus (EDL), or plantaris (PL) muscles in sedentary rats acclimatized for 4 weeks to a simulated altitude of 4000 m. Increases in capillary density were reported in the diaphragm, a change that was mediated in part by small increases in the number of capillaries per fiber (57). The diaphragm, like the heart, has been identified as exhibiting different response patterns to chronic hypoxia than the locomotor muscles (19,64), perhaps because of the increased ventilation and contractile activity that occurs during acclimatization.

Other factors may help explain the variable results noted in earlier studies regarding the effects of chronic hypoxia on capillarization. Cold alone (63) and with chronic hypoxia (65) resulted in angiogenesis. Technical problems associated with identifying capillaries and fiber shrinkage during fixation may also account for the differences observed, particularly in the earlier studies (24,58). Capillary tortuosity, also suggested as a potential analytical problem, does not appear to change at least in rat soleus and gastrocnemius muscles following 5 months at 3800 m (60).

Our present understanding of the effect of training in a hypoxic environment on skeletal muscle is based largely on the new studies performed by Bigard et al. on rats (57,66,67). Training while living in hypobaric hypoxia appears to exert its greatest effect on muscle size and capillarization. Swim training in rats chronically exposed to hypobaric hypoxia for 14 weeks resulted in an increased capillary density in SOL, EDL, and PL (both superficial and deep) (57). In no case was the response to training plus hypoxia potentiated over that obtained when training at sea level. The increase in capillary density was mediated by both an increase in the number of capillaries per fiber and decreases in fiber area. The increase in the number of capillaries per fiber was a generalized response to training independent of the effect of acclimatization. As a result of reductions in body weight with chronic hypoxia, pronounced reductions in fiber area were also observed in all muscles examined. In contrast to endurance training at sea level, where a general decrease in fiber area was observed, training at altitude did not decrease fiber size more than that observed with acclimatization alone. Although training increased the number of capillaries, the major factor promoting enhanced capillary density at altitude was fiber size reduction as a consequence of chronic hypoxia alone.

Mizuno et al. (68) attempted to determine if altitude residence can potentiate the adaptive response in muscle to training in highly trained athletes. In this study, the athletes participated in a controlled training program while residing at 2100 m and then continued training at 2700 m for 2 weeks. Training while living at 2100 m increased muscle cross-sectional area (Type IIA) and the number of capillaries per fiber (type I and type II) in triceps brachii but not in the gastrocnemius muscle (68). No change was observed in capillary supply per cross-sectional area in either muscle during the 2-week training exposure to 2700 m in the trained athletes. Unfortunately, it could not be determined if the change that occurred at 2100 m was due to the increased hypoxic stimulus per se or to an alteration in the training that occurred at the higher altitude.

The effect of hypoxic training on the active but untrained has been compared with training in normoxia (69). The hypoxic condition designed to stimulate a progressive exposure from 4100 to 5700 m over 3 weeks of training was matched to a normoxic condition conducted at the same power output. In a second normoxic condition, training was conducted at a higher absolute intensity to produce the same metabolic effects in the working muscle as less severe hypoxic exercise. When training is conducted at sea level with the hypoxic stimulus only applied during the training, elevations in both fiber size and the number of capillaries per fiber but not in capillary per fiber area density have been reported (70). Significant increases in these parameters were not found for training conducted in normoxia either at the same absolute workload or at a higher absolute work load designed to produce the same relative percentage of $\dot{V}O_2max$.

Enzymatic Organization

Attempts to characterize the enzymatic changes in skeletal muscle in response to hypoxia have a long history. Despite more than 30 years of research on this area, much confusion remains regarding the effect of residence at altitude on the changes in the maximal activities of selected enzymes representative of the various energy metabolic pathways and segments. Ideally, the approach for such studies should be comprehensive, measuring rate-limiting enzymes for the full range of metabolic function; oxidative phosphorylation, β-oxidation, glycogenolysis, glycolysis, glucose phosphorylation, lactate oxidation, and high-energy phosphate transfer. With such an approach not only could the potential of the various metabolic pathways be measured, but the specialization of one pathway relative to another could be described (71). Unfortunately, few enzymes have been measured, and, in some cases, only those of noncritical significance.

A series of papers published in the early 1960s had considerable impact on promoting the proposition that high-altitude native display an upregulation in metabolic potential and, in particular, mitochondrial potential. Tappan et al. (72) found indications (not significant) of elevations in cytochrome *c* activity in a variety of muscles (psoas, biceps, femoris, diaphragm, heart) in guinea pigs native to 4400 m compared to sea level controls. In a follow-up study, Reynafarje (73), using the same species, examined the multienzyme complex of the respiratory chain by measuring the oxidation of reduced pyridine nucleotides in the presence of molecular oxygen. In whole homogenates of the rectus femoris muscle, he found that the reduced forms of dihydropyridine nucleotide oxidase (DPNH-oxidase) was elevated, but trihydropyridine nucleotide (TPNH-oxidase) was not. He attributed these differences to the development of more pigmented, "red" regions of the muscle since the enzyme activities in the dominant red and white areas did not change. In contrast to skeletal muscle, elevations in both DPNH-oxidase and TPNH-oxidase were found in the heart. In both heart and rectus femoris, elevations were observed in pyridine nucleotide transhydrogenase, regarded as a key regulator in biological oxidations. These studies were pivotal since they reinforced the long-held notion (74) of an increased capacity for oxidative phosphorylation in species native to high altitude. Subsequent studies on animals have been contradictory and, in general, have failed to support these initial findings.

Few studies have examined metabolic patterning in the skeletal muscle cell of humans native to high altitude. In one of the earliest studies (73) it was reported that tissue extracted from the sartorius muscle of healthy Andean natives (4400 m) contained higher levels of DPNH-oxidase in homogenates and higher levels of both TPNH-cytochrome *c* reductase and trans-hydrogenase in the mitochondrial fractions as compared to residents of Lima (50 m). Mitochondrial DPNH-cytochrome *c* reductase was not significantly altered, nor was the activity of lactate dehydrogenase (LDH). These results were interpreted to indicate an increased respiratory capacity with no change in glycolytic potential as a consequence of chronic hypoxia.

Two additional studies examining enzymatic differentiation in high-altitude natives were published some 30 years later. In one (75,76), the maximal activities of a number of enzymes were measured in homogenates prepared from the vastus lateralis muscle of Sherpas and Quechuas, both lifetime residents to altitudes of 3300–4500. These results, obtained from six natives from each region, are difficult to interpret because the sea level control consisted of only one individual, a trained marathon runner. However, assuming that the runner is representative for marathons with a mitochondrial potential approximately twofold higher and with a glycolytic potential similar to untrained controls (22), significant differences do not appear to exist between the high-altitude native and the lowlander in the potential for oxidative phosphorylation, β-oxidation, or glycolysis. Some differences in metabolic potential may exist between the two high-altitude native groups, however. Although minimal differences were found in the potential for oxidative phosphorylation and β-oxidation between the Sherpas and Quechuas, a lower glycolytic capacity is clearly indicated in the Sherpas, who displayed values of only 65% and 83% of the Quechuas for PK and LDH, respectively.

When the metabolic potential at the level of the specific fiber types (I, IIA, IIB) was examined in a subsample of the Quechuas ($n = 3$), it was found that MDH activity, a measure of mitochondrial potential, was 20% lower only in type I fibers when compared to the average of two sea level, untrained controls and a marathoner (47); mitochondrial potential in type I fibers of the marathoner was apparently not different from the untrained. The glycolytic potential in type I fibers, as indicated by LDH, was higher (28%) in the Quechuas, as was adenylate kinase (20%), a measure of the potential for high energy phosphate transfer (47), than the controls. The type II fiber types showed no difference in the activities of the three enzymes examined. With regard to the type I fibers, this study suggests that Quechuas have an upregulated glycolytic capacity relative to the mitochondrial capacity as compared to the lowlanders. Care must be exercised in the interpretation of these results, not only because of the very small numbers of altitude natives and lowlanders involved in the study and the lack of appropriate sea level controls, but because of the probable smaller fiber size in altitude natives (52,57,77).

The second study employed electron microscopy to compare the ultrastructure of muscle of high-altitude natives (Sherpas) with that of sedentary, sea level residents (52) (Table 4). This study demonstrated a lower mitochondrial potential in the Sherpas, as indicated by a lower percent volume of mitochondria in the muscle cell. Moreover, the lower mitochondrial volume appeared to be due to a lower interfibrillar density and not a lower subsarcolemmal aggregation. However, the control

Table 4 Intracellular Mitochondria and Lipids in High-Altitude Natives, Mountaineers, and Sea Level Controls

	Interfibrillar mitochondria (%)	Subsarcolemma mitochondria (%)	Total mitochondria (n)	Intracellular lipids (μm^2)
Sherpas ($n = 5$)	3.52	0.43	3.96	0.45
Elite mountaineers ($n = 5$)	4.21*	0.74	4.95*	0.83
Swiss expeditions				
Pre ($n = 14$)	4.88*	0.97*	5.85*	0.81
Post ($n = 12$)	4.23*	0.52	4.74*	0.92
Sedentary ($n = 5$)	4.25*	0.48	4.74*	0.68

Note: Biopsies from vastus lateralis. Values represent volume densities.
* Significantly different from Sherpas ($p < 0.05$).
Source: Ref. 52.

group selected was not only semi-trained, as indicated by their $\dot{V}$max values (61 mL/kg/min) but had a decidedly heavier body mass (+14 kg) (78). Given that training increases mitochondrial potential, the smaller, less trained Sherpas could have a lower mitochondrial density unrelated to an altitude effect. Saltin et al. (53), using citrate synthase (CS) as a mitochondrial marker, found that the levels of CS may be lower in men born and living continuously at 3700 m but the value was dependent on the activity level. Only inactive residents displayed lower CS values than the sea level sedentary group. The results of these studies emphasize the need to control activity level when examining the effect of residence at altitude on enzymatic properties.

A study by Young et al. (79) highlights modern initiatives aimed at describing the effects of altitude acclimatization on human muscle enzymes. Measurements were performed on tissue obtained from the vastus lateralis muscle of males prior to and following 18 days of residence on Pikes Peak (4300 m). Of the enzymes examined—phosphorylase (PHOSPH), hexokinase (HK), MDH, and LDH—none were significantly altered. A similar study also conducted at Pikes Peak for a similar period of acclimatization and using the same muscle also reported no change in mitochondrial potential using succinic dehydrogenase (SDH) as the representative enzyme or in the glycogenolytic potential (PHOSPH) (50). β-Oxidative potential (HOAD) was also unaltered. However, two enzymes were found to be altered. Hexokinase, a measure of glucose phosphorylation potential, was increased, while phosphofructokinase (PFK), acknowledged as the rate-limiting enzyme in glycolysis, was decreased. The PFK reduction appeared not to be due to acclimatization per se since it occurred within hours after arrival at Pikes Peak (50). The lack of a response in mitochondrial potential was also confirmed at the level of the individual cell, where microphotometric determinations of SDH activity failed to demonstrate change with acclimatization. To our knowledge these are the only two studies that have examined the effects of short-term acclimatization on muscle where physical activity at altitude was intentionally minimized. Differences in activity levels of the subjects in the two studies were evident before ascent to altitude, however. Subjects in the Young et al. study (79) were regularly active soldiers in physical training; Green et al. (50) used sedentary individuals. Thus some differences between the two studies (e.g., HK) could have been due to the initial training state. Hexokinase is known to be rapidly increased with training (80) (Table 5).

In Operation Everest II, the 40-day period of progressive acclimation induced reductions in the potential for oxidative phosphorylation (CS, SDH) and glucose phosphorylation (HEX), with no significant change indicated for β-oxidation (HOAD), glycogenolysis (PHOSPH), glycolysis (PFK, PK, LDH), or high-energy phosphate transfer (creatine phosphokinase, CPK) in vastus lateralis. Microphotometric determinations of oxidative potential in single fiber types using NADH-tetrazoleum reductase (NADH-TR), also a measure of oxidative potential, indicated that the reduction was specific to type I fibers. Mitochondrial density, as assessed by electron microscopy in single cells, did not change with acclimatization (55).

Because measurements were made at intervals during Operation Everest II, changes in maximal enzymatic activity at different stages could be determined (40). Regardless of the metabolic pathway or segment, no significant alterations could be found in any metabolic pathway during the first 32 days of exposure to hypobaric

Table 5 Percent Changes in Metabolic Potential with Acclimatization as Indicated by the Maximal Activities of Representative Enzymes

		% Change		
Pathway	Enzyme	Lhotse and Everest[a]	Operation Everest II[b]	Pikes Peak[c]
Citric acid cycle	CS	−23.0*	−29.1*	7.3
	MDH	−19.3*		
	SDH		−28.2*	
Respiratory chain	CYTOX	−23.3*		
β-Oxidation	HADH	−26.5*	−0.06	9.4
Ketone body utilization	HBDH	−26.5		
Glucose phosphorylation	HEX	−7.8	−38.7*	+16.1*
Glycogenolysis	PHOSPH		−19.5	−6.7
Glycolysis	PFK	8.9	−27	−13.5
	GAPDH	−5.5	−12.6	
	LDH	0.6	−13.8	
	PD		17.8	
High-energy phosphate transfer	CPK	−9.2	−20	

[a] Data from Ref. 81.
[b] Data from Ref. 40.
[c] Data from Ref. 50.

hypoxia up to an altitude of 7620 m. However, certain nonsignificant trends were evident. The enzyme defining β-oxidation, HOAD, appeared to increase, and the enzyme recognized as the rate-limiting enzyme in glycolysis, PFK, appeared to decrease. Transient increases in HEX, the enzyme mediating glucose phosphorylation, were also suggested.

Particularly conspicuous are the changes that occurred during the final week of simulated altitude, when the barometric pressure was the lowest and the volunteers the most lethargic. During this period a reduction in all enzymes, regardless of the metabolic pathway, is suggested. This trend could be explained either by the pronounced reduction in physical activity or the severe hypoxia experienced during this time period. The activity level may be particularly important because several of the volunteers had been particularly active until this stage. In general, the significant findings from Operation Everest II, at least during the initial periods of acclimatization, are consistent with previous acclimatization studies (50,79), namely that short-term periods of acclimatization are not accompanied by changes in either mitochondrial potential or glycolytic potential. However, small increases in glucose phosphorylation potential appear to occur.

Two mountaineering expeditions have also extended our understanding of the effects of acclimatization on the metabolic organization in skeletal muscle (54,81) (Tables 3 and 4). In the Mount Everest study, tissue samples were collected at sea level from the vastus lateralis prior to the expedition and again following 6 weeks of residence at 5300 m with sojourns to altitudes in excess of 8000 m (81). As with the findings from Operation Everest II, a study with a comparable altitude ascent profile, reductions in mitochondrial potential (CS, MDH, cytochrome oxidase) were found in the absence of changes in glycolytic potential (PFK, GAPDH, LDH) or high-energy phosphate transfer (CPK). A nonsignificant trend towards a reduction in glucose phosphorylation potential (HEX) was also observed. As with Operation Everest II, the pronounced drop in mitochondrial potential that was observed in the absence of change in the glycolytic potential indicates that the capacity for glycolytic flux relative to oxidative phosphorylation is more emphasized, in conflict with the conclusions reached on high-altitude residents (76). In the high-altitude residents the reduction in the mitochondrial enzymes examined was approximately the same as the reduction in the glycolytic enzymes so that the metabolic pathways remained proportional.

In a complementary study of the members of the Lhotse expedition in addition to the Everest mountaineers, tissue obtained from the same muscle was used to examine ultrastructural and morphometric features (54). Reductions in the percent fiber volume occupied by mitochondria, indicative of a reduction in mitochondrial potential, were clearly evident, supporting the biochemical findings (81). The reductions in mitochondrial density appeared to occur in both the subsarcolemmal and interfibrillar locations, with the loss in the subsarcolemmal region being the most pronounced, particularly near the periphery of the fiber. As with a previous study examining ultrastructural features in high altitude residents by this group (52), no change in intracellular lipid droplet volume density was detected. In this investigation, the delay in extracting the tissue samples following return to sea level, representing some 10–15 days, remains a concern. The reduction in mitochondrial potential attributed to acclimatization may be much greater, given the expected normalization of the oxidative potential of the muscle with withdrawal of the hypoxic stimulus.

In one other study of mountaineers, tissue from the vastus lateralis was examined 2–12 months after the last climb to altitudes above 8500 m and compared with that of sedentary sea level residents (51). No differences were detected in the cellular densities of interfibrillar, subsarcolemmal, or total mitochondria. It is possible that the extended period at normoxia following the expeditions could have provided for recovery of metabolic potential.

Collectively the acclimatization studies on humans indicate a reduction in the potential for oxidative phosphorylation in muscle in the absence of changes in glycolytic potential and the potential for high-energy phosphate transfer. However, the influence of the degree of hypoxia and the time at residence have on the response are not clear. Based on the observations of mountaineers engaged in climbing and those undergoing acclimatization without activity, the same fundamental changes, at least qualitatively, appear to occur. Indeed, it is possible that the effects of chronic hypoxia may be exaggerated when sustained activity is included. This possibility was emphasized some years earlier by Saltin et al. (53), who found that in sea level residents acclimatized to 3700 m for 32 weeks, activity level was critical in the change observed in citrate synthase, a mitochondrial enzyme. Altitude itself does not appear to stimulate increases in mitochondria. In fact, there are indications that if the hypoxic stress is severe, mitochondrial potential may decrease (40).

Acclimatization studies in animals yield conflicting results regarding the alteration in mitochondrial potential of skeletal muscle in response to chronic hypoxia. Tappan and Reynafarje (19), for example, found that cytochrome *c* activity was unchanged in psoas, biceps femoris, and diaphragm of guinea pigs resident at 4500 m for an average of 44 days. Subsequent studies have generally supported the notion that enzymes representative of the citric acid cycle and respiratory chain are not altered by chronic hypoxia.

Recently, Bigard et al. (57,66) examined muscle enzymatic changes in adult Wistar rats following exposure to a simulated altitude of 4000 m for 60 days. Simulated altitude did not alter either the oxidative potential (CS) or the potential for β-oxidation (HOAD) in any of the muscles examined (SOL, EDL, PL) but did increase the capability for glucose phosphorylation (HEX), but only in the SOL and PL. In addition, glycolytic potential (LDH) was modestly but significantly increased in the EDL and PL, although no alteration in the % LDH-I isoform occurred in any of the muscles examined. The chronic hypobaric stimulus also resulted in pronounced losses in both whole body and muscle weights.

This study is important from several perspectives. First, animals remained sedentary under both normobaric and hypobaric conditions. Second, the study demonstrated that enzymatic adaptations can occur with chronic hypoxia, but only to selected metabolic pathways and only in certain muscles. Finally, the study illustrates that pronounced decreases in fiber size can occur without alteration in oxidative potential.

Bigard et al. (57,66,67) also examined the additional effects of training. These studies involved 14 weeks of hypobaric hypoxia (4000 m) plus daily swim training. The enhancement of the adaptations over those observed at sea level depended on both the muscle and metabolic pathway examined. For the EDL and PL, training during chronic hypoxic potentiated the increase in both CS and HOAD, markers of oxidative phosphorylation and β-oxidative potential, respectively. Similarly, in the EDL and PL but not in the SOL, training at altitude blunted the increase in LDH observed with acclimatization alone. Although training did increase the ratio of LDH1 to total LDH, the magnitude of the change was not altered by training at altitude. Hexokinase was only elevated with sea level training and only in the EDL and PL. Chronic hypobaric hypoxia by itself was generally ineffective in altering enzymatic activity. The one exception was HEX, which was elevated in the soleus with acclimation.

These studies raise concerns regarding the form of training employed and the magnitude of the adaptive response elicited with sea level training. Previous studies have shown swim training not to be very effective in increasing metabolic potential (23). A more appropriate strategy might have been to select an established running program, previously shown to elicit large adaptations in the metabolic machinery in a variety of muscles of primary different fiber composition (23). Even so, these studies do support previous research (82) demonstrating an enhancement of the citrical acid cycle and β-oxidation potential during training at 1750 m compared to sea level training.

Enzymatic adaptations in the diaphragm, however, to altitude training appear to be somewhat unique (67). Although CS was increased with training, HOAD was not regardless of whether the training was conducted at sea level or during acclimatization to 4000 m. As with sea level training, HEX and LDH were increased and decreased, respectively, and to similar magnitudes with chronic hypoxia and training. No additional effect of altitude was found in the ratio of the heart-specific enzyme (LDH1) to total LDH.

The length of time spent at altitude may also be important in the enzymatic changes that result, at least in the heart. Pickett (63) subjected rats to an equivalent altitude of 6583 m in a hypobaric chamber for 27 days and found a reduction in heart weight during the first several days of exposure followed by hypertrophy. Succinic dehydrogenase (SDH) activity, used as a measure of mitochondrial potential, increased during the early period and then regressed to prealtitude levels. These results suggest that the earlier increase in SDH was mediated by reductions in size and total protein content of the muscle. Studies on the adaptations in the heart also have implications for skeletal muscle, particularly for skeletal muscles that are persistently active, such as the diaphragm.

Mizuno et al. (68) using trained athletes subjected to additional training at 2700 m while living at 2100 m for 2 weeks found that in the gastrocnemius but not the triceps brachii, significant reductions occurred in CS and HOAD, both measures of the capacity for oxidative phosphorylation and β-oxidation, respectively. No change was observed in PHOSPH or LDH, measures of glycogenolysis and glycolysis. Unfortunately, these investigators could not rule out changes in the training program and a different involvement of the extremity muscles as a potential explanation for the enzymatic changes.

When the hypoxic stimulus is only applied during the training itself, greater increases in oxidative potential have been reported than when the training conducted at the same absolute intensity and duration is performed during normoxia (36). The increase in CS was approximately twofold higher following hypoxic (equivalent to 2300 m) one-leg training in previously untrained males. Although no change in PFK was observed with the training regardless of the environmental condition, LDH was observed to decrease but only with the hypoxic training.

More recently, studies have employed local graded ischemia in an attempt to produce hypoxia in the working muscle while studying training effects. In one study (84), enzymatic changes were examined in the vastus lateralis to one-leg training, using either ischemia or normal blood flow. In trained volunteers, training with ischemia caused a marked elevation in CS in contrast to nonischemic training. Ischemic training also lowered LDH and increased PFK. The work performed during each type of one-leg training was approximately equal. In additional work using untrained subjects and ultrastructural measurements, training in normoxia and hypoxia at the same relative percent of $\dot{V}O_2$ max was suggestive of greater increases in intermyofibrillar density of mitochondria than training in normoxia at a lower absolute work load. Lipid density is also higher after this type of training.

Myoglobin

Presumably because of the observation that muscle cells in animals habituated to altitude appeared redder in color (85), the earliest studies on cellular adaptation concentrated on the myoglobin content of the cell. The elucidation of the role of myoglobin in the transport and facilitated diffusion of oxygen in the muscle cell (28) suggested that increases in myoglobin could represent an important adaptive response. Current information based on studies of animals native to high altitude, most of which were performed more than three decades ago, support this possibility (8).

The first exploration of this question was by Hurtado et al. (85), who found higher myoglobin content of skeletal muscles of dogs native to high altitude than sea level counterparts. Subsequent studies employing rats (64,86), guinea pigs (19), and even humans (73) have confirmed this landmark observation. However, differences in adaptability seem to exist between different muscles. At least in the guinea pig and rat, the hindlimb and forelimb muscles with the exception of the psoas do not appear to be as adaptable as the masseter (19,64). The diaphragm, a muscle that, like the masseter, is continuously active, does not display significantly higher myoglobin levels when compared to the sea level counterparts (19,64). In contrast, the hearts of both guinea pigs (19) and rats (19) reared at high altitude demonstrate a most pronounced increase in myoglobin level.

Only one study has compared myoglobin levels in sea level and high-altitude natives (73). In the only muscle examined, the sartorius, residents of Peru (4400 m) displayed a content approximately 16% higher than residents of Lima (50 m).

Increases in myoglobin in skeletal muscle have been reported following a relatively brief exposure (75 days) to altitude (64). In some muscles (19) these increases are at least as great as the elevations observed in long-term residents. In guinea pigs, myoglobin increases in sartorius, psoas, rectus femoris, and biceps femoris appear to be pronounced with acclimatization. Heart muscle also exhibits large increases in myoglobin content to short-term residence but in the diaphragm shows either nonexistent or relatively modest changes (19,64).

The preponderance of evidence in animals indicates increases in myoglobin as an adaptation to high altitude with both acclimatization and life-long residence. The failure of earlier studies to report increases (87,88) has been attributed to an inability to achieve the acclimatized state because of the conditions of the exposure (64). In one study that reported decreases in myoglobin in gastrocnemius and soleus muscle of the rat, only intermittent severe daily exposure to a simulated altitude of 7620 m was used (88).

Other factors may also affect the magnitude of the response. Regular physical activity, as an example, has been shown to induce large elevations in myoglobin in rats trained at sea level (29), and limited evidence suggests that a similar response might occur with training at altitude (64,89). Clarke et al. (89) found that in hamsters exposed for 6 weeks to 6096 m and kept inactive, no changes in myoglobin occurred in the heart, gastrocnemius, or diaphragm muscles. Since humans apparently do not demonstrate increases in myoglobin at sea level with training (34), increases comparable to those observed in other species following chronic hypoxia and training may be unlikely.

A 6-week period of training at simulated altitude (20,000 m) induced large elevations in myoglobin concentration in heart (200%), diaphragm (50%), and gastrocnemius (50%) muscles of adult hamsters compared to sedentary, acclimatized controls, who had no increase in myoglobin in any of the muscles examined (89). These findings support earlier acclimatization studies where activity was needed to elicit increases at altitude (8). To our knowledge, no comparable studies have been performed in humans. The myoglobin response has been examined, however, to both one-leg (36) and two-leg (90) training during hypoxia. Two-leg supplementary training in already trained cyclists was without effect (same as normoxia), whereas one-leg hypoxic training in the nontrained induced a small but significant elevation in the vastus lateralis. These results suggest that initial training state may also modify the myoglobin response.

Muscle fiber type may also contribute to the adaptive response. Slow-twitch (ST) or fast-twitch oxidative-glycolytic (FOG) fibers, which already possess high myoglobin levels, may not be as responsive as fast-twitch glycolytic (FG) fibers, which have a low myoglobin level (91). Anthony et al. (86) has proffered another explanation for the contradictory results observed in earlier studies. Following 2–10 weeks of exposure to 5486 m, the myoglobin concentration in a group of muscles (heart, diaphragm, gastrocnemius, and thigh muscles) of a group of Wistar rats increased. However, total myoglobin content determined using the muscle weights did not change. These investigators have postulated that the increased myoglobin concentration reported in some studies may be due to a loss of water content. The exception was the heart, which as in a previous study (88) was found to hypertrophy and to increase myoglobin concentration (86). But more recent studies (73) using nitrogen content rather than muscle weights as a measure of muscle protein still reported increases in myoglobin concentration in some nonhuman muscles.

Summary

Physical activity is a major confounding variable when attempting to ascertain the independent effect of chronic hypoxia on muscle adaptability. For example, while the high-altitude dweller may lead a physically active life, volunteers recruited for study are often more sedentary.

Acclimatization studies, although providing for a longitudinal assessment of changes in muscle to varying periods of chronic hypoxia, are also not immune to the limitations cited. The level of physical activity engaged in prior to entry into the hypoxia environment could have profound implications on the changes observed in muscle during the period of chronic exposure. Well-trained individuals, for example, may attempt to sustain the activity level or lower it during residence at altitude. In this group, members of which may have reached the limit or near limit of adaptability, the hypoxic stimulus may be without effect or, indeed, may provoke changes as a result of excessive strain not consistent with what might be recognized as beneficial. On the other hand, if pronounced reductions in activity state occur during the period of altitude residence, this may either counter the effect of hypoxia or exaggerate the effect. The net result may depend on whether chronic hypoxia by itself has a stimulatory or inhibitory effect on the synthesis of specific cellular proteins. If, as was proposed previously, hypoxia upregulates mitochondrial potential (18), reductions in the activity level would be expected to counter the changes. On the other hand, if, as currently hypothesized, sustained hypoxia downregulates mitochondrial potential (40,81), reductions in activity would potentiate the response.

Inactive volunteers or volunteers with no immediate history of regular activity prior to entry into hypoxic environments also present special problems. At least in normoxia, the initiation and continuation of even modest levels of activity can alter muscle composition and function (22). Consequently, if the level of habitual activity changes during the period of acclimatization, characterization of the specific effects hypoxia would be problematic, particularly since the interactive effects of hypoxia and exercise have not been delineated.

In the acclimatization studies completed to date, at least in humans, these considerations seriously cloud the interpretation of the changes in muscle attributed to chronic hypoxia. Acclimatization studies have incorporated untrained (50,92), semi-trained (40,79), and what would appear to be well-trained mountaineers (41). Moreover, activity levels reported in the various studies during the period at altitude have ranged from inactive (50), to the seriously active, as in mountaineers engaged in an expedition (41), to volunteers who altered activity levels during the duration of the period of acclimatization (40). More recently, investigators have also recognized another concern, namely the effects of hypoxia induced only during the exercise itself versus the effects of full-time residence in hypoxia or altitude. It is possible that effects of hypoxia plus training may be obfuscated if the hypoxic stimulus persists by remaining at altitude rather that returning to normoxia after the exercise training session (69,70).

III. Hypoxia and Muscle Metabolism

A key question is how the changes that occur in the composition and ultrastructure of muscle with high-altitude residence affect muscle metabolism. Can the adaptations that occur in muscle composition and structure explain the apparently improved energy homeostasis that occurs with hypoxia? Adaptations in metabolic behavior have been studied both at rest and during activity.

A. Rest and Exercise

The energy charge or phosphorylation potential of the muscle cell provides a measure of whether or not ATP homeostasis has been disturbed by hypoxia. Maintenance of the phosphorylation potential would indicate that either central or peripheral adjustments or both have compensated for reductions in arterial O_2 content consequent to the reduced in PaO_2.

In humans at rest, no alterations in either ATP or PCr concentration in the vastus lateralis muscle have been reported following acute hypobaric hypoxia simulating an altitude of 4300 m (93) or within 4 hours after arrival at 4300 m (50). Further, inosine monophosphate (IMP), a more sensitive measure of changes in energy state given its low concentration and stoichiometric relationship with ATP (94), also is undisturbed (50). Normal ATP concentrations could occur because an increase in glycolytically generated ATP offsets any reduction in oxidative phosphorylation. That glycolysis appears to increase is indicated by the increase in lactate flux during acute exposure to 4300 m, where both lactate appearance and disappearance rates, measured with a stable isotope, were increased approximately fourfold (95). Increase in lactate turnover was not accompanied by significant change in the concentration of lactate in the resting vastus lateralis muscle (50). At constant ATP utilization, increased glycolytic flux, if primarily directed toward ATP generation, should be accompanied by decreases in oxidative phosphorylation of the resting muscle. However, $\dot{V}O_2$ in the resting leg was unchanged (92,96). One popular theory is that increases in glycolytic flux during hypoxia help sustain mitochondrial respiratory rates by providing more reducing equivalents (9).

As with acute ascent, the reduction in glycolytic flux with acclimatization occurred in the absence of changes in leg $\dot{V}O_2$ (92,96). Curiously, however, resting whole body $\dot{V}O_2$ appears to be increased with acclimatization (92), although the increase has not been identified in an earlier study (96). One factor that could potentially contribute to differences between studies is whether the acute measures were obtained at sea level using a brief exposure to hypobaric hypoxia or following ascent to altitude. Sudden ascent to altitude is reported to result in a higher level of blood catecholamines, in particular epinephrine, than hypobaric hypoxia (97,98).

The ability of the resting muscle to protect energy state by maximizing utilization of available O_2 is also indicated by direct measures. Phosphorylation potential, at least in the dog gastrocnemius, remains undisturbed over a wide range of hypoxic conditions (99,100).

The greatest challenge to oxidative phosphorylation during hypoxia occurs during activity, in particular large muscle mass activity, when ATP requirements can be elevated manyfold. To assess the effects of acute hypoxia or ascent to altitude, several different exercise protocols have been employed. These approaches have emphasized changes during submaximal exercise in the nonsteady state and during the steady state and during progressive exercise to voluntary fatigue. Each approach provides a different insight into how working muscle responds to changes in cellular oxygen availability.

Progressive exercise, by gradually increasing ATP requirements, taxes the capabilities of all ATP regenerating metabolic pathways, including oxidative phosphorylation. Although assessment of muscle metabolism at different intensities of exercise is of interest, such measures have more commonly been restricted to a single endpoint, namely fatigue.

As might be expected, acute hypoxia causes a reduction in peak $\dot{V}O_2$ and consequently in the peak level of oxidative phosphorylation in the working muscle, which is directly related to the degree of hypoxia (for review see Ref. 101). Using normobaric hypoxia (14%, O_2 equivalent to 4000 m) and a sensitive breath by breath gas exchange system, Hughson et al. (98) found that peak $\dot{V}O_2$ is depressed at maximal exercise compared to normoxia and that the reductions are observable at approximately 60% of the normoxic $\dot{V}O_2$ peak. Beyond this intensity, the increase in blood lactate is much more rapid than observed in normoxia for the same workload, suggesting a greater rate of glycolysis in response to limitations in ATP generated from mitochondrial respiration. Interestingly, greater increases in blood lactate occur with acute hypoxia than with normoxia even at similar levels of $\dot{V}O_2$ (98,102). At maximal exercise in hypoxia, the concentration of lactate in muscle and blood is at least equal to that during normoxia (98,103). Work intensity is, of course, lower with hypoxia. These observations support the notion of O_2-limited ATP regeneration relatively early in exercise during acute hypoxia with increases in glycolytic rate attempting to compensate. The apparently higher glycolytic rate, as indicated by the blood lactate profile in acute hypoxia, at similar $\dot{V}O_2$ levels to normoxia suggests, as at rest, of an intracellular strategy to compensate for the low pO_2, namely increasing the supply of reducing equivalents to the mitochondria (9). The similarity of the muscle lactate concentration at fatigue between normoxia and hypoxia implies that glycolysis was activated to similar extent between the two conditions.

These interpretations, however, are subject to question for a number of reasons. Without measuring of lactate turnover, which is difficult if not impossible to obtain during non–steady-state exercise, assumptions regarding glycolytic flux based on concentration measures only are tenuous since clearance could have been modified in the hypoxic condition. Clearance rates may depend on the exercise model and the change in muscle blood flow that occurs with hypoxia (104). Moreover, progressive exercise, even with normoxia, results in the recruitment of additional muscles, which is probably intensity and fatigue dependent (92,105). With hypoxia, motor unit recruitment profiles both within and between muscles may change, resulting in force and ATP production being distributed to other muscles. It is possible that the metabolic response at the level of a single muscle or muscle fiber to hypoxia may be substantially different when a given force level is generated nonvoluntarily by induced techniques rather than by voluntary recruitment.

Steady-state submaximal exercise has been the most common model used to examine muscle metabolism in response to acute and chronic hypoxia. Most often the exercise is performed at the same absolute exercise intensity during both normoxia and hypoxia. These studies consistently demonstrate that at low submaximal exercise intensities whole body $\dot{V}O_2$ is unaltered by acute hypoxia up to 4000 m under either simulated (96) or altitude (92) conditions. At higher submaximal intensities, $\dot{V}O_2$ is clearly compromised (96). Quite intriguing are the results of a study that reported reductions in single-leg $\dot{V}O_2$ but not whole body $\dot{V}O_2$ after acute ascent to 4300 m (92), implying an O_2 delivery limitation of mitochondrial metabolism even at intensities of approximately 50% of sea level and 65% of altitude whole body $\dot{V}O_2$ max. The depression in single-leg $\dot{V}O_2$ with acute ascent was unexpected given previous reports of no change in leg $\dot{V}O_2$ at similar exercise intensities and similar degrees of induced hypoxia during sea level experiments (96). These differences may serve to emphasize the different response patterns that may occur between the different acute hypoxic conditions. Acute ascent to altitude, as an example, resulted in a pronounced loss of plasma volume and an increase in hematocrit (92). Although muscle blood flow was not affected by acute hypoxia, extraction of O_2 by the working muscles apparently was, since $\dot{V}O_2$ was depressed (92). Whether the impaired extraction was mechanistically related to the altered blood viscosity that would appear to have occurred is unknown. The increased discrepancy between whole body $\dot{V}O_2$ and leg $\dot{V}O_2$ observed during acute hypoxia as compared to normoxia was attributed to increased work of the respiratory muscles (92). Conceivably, an increase in respiratory $\dot{V}O_2$ in the absence of change in whole body $\dot{V}O_2$ with acute hypoxia could deprive the working leg muscles of $\dot{V}O_2$.

The depression in $\dot{V}O_2$ by the working muscles observed soon after arrival at altitude may also explain the alteration in muscle metabolism in the vastus lateralis and the increase in lactate flux that occurred. Both muscle lactate concentration (50) and lactate appearance rates measured with a stable isotope (95) were increased during prolonged exercise performed soon after arrival at altitude. Although significant effects were not observed, there was a strong trend for the phosphorylation potential in the vastus lateralis to be reduced as well, as evidenced by the lower PCr and the higher IMP concentrations (50).

An additional model that can be used to examine imbalances in ATP production and utilization during acute hypoxia is non–steady-state exercise. Unlike steady-state exercise, which provides for adjustment in metabolic processes, non–steady-state exercise challenges the kinetics of ATP-dependent O_2 utilization. Studies employing different degrees of hypoxia and different exercise intensities to investigate $\dot{V}O_2$ kinetics are rare and, in the case of the muscle metabolic response, nonexistent. Of the few studies examining $\dot{V}O_2$ kinetics with acute hypoxia, it appears that $\dot{V}O_2$ is reduced (106). These results confirm earlier studies using a much less sensitive system for determination of non–steady-state changes in $\dot{V}O_2$ (93,107). Whether the reductions in $\dot{V}O_2$ alters the $\dot{V}O_2$ kinetics across the working muscle is unclear given the increase in $\dot{V}O_2$ utilization that would be expected by the respiratory muscles (92,108). However, if $\dot{V}O_2$ is compromised during the non–steady-state period with acute hypoxia, significant consequences to muscle metabolic behavior might be expected (80), and in fact this appears to be the case. The changes in vastus lateralis, ATP, and PCr at the end of brief cycle exercise do not appear to be different between normoxia and hypobaric hypoxia (equivalent to approximately 4000 m), but lactate concentration with hypoxia is elevated (93,107). At higher submaximal workloads, greater high-energy phosphate depletion is evident with acute hypoxia than with normoxia (107). At maximal exercise there is no difference for either ATP, PCr, or lactate (93). Interestingly, at the lower exercise intensities, although steady-state $\dot{V}O_2$ was not compromised, $\dot{V}O_2$ kinetics were slower (93,107). It is not clear whether initial metabolic perturbations apparently produced during the nonsteady state can be normalized as the exercise continues.

Experiments on animals are typically performed with anesthetization using different hypoxic gas mixtures and electrical stimulation of a localized muscle group (99,109,110). Using this model, blood flow, arterial O_2 delivery, and O_2 extraction can be measured across the working muscle, and these measures can be supplemented either with direct measures of metabolism using non invasive techniques (^{31}P-NMR) (100) or by biochemical analysis obtained from small tissue samples (99). These investigations demonstrate both a circulatory compensation with increases in blood flow to protect submaximal cellular respiration rates (110,111) and apparent compensations at the intracellular level as well (99,100,110). At a given level of $\dot{V}O_2$, hypoxia compared to normoxia results in elevated levels of the putative stimulators of mitochondrial respiration such as ADP and Pi and an elevation in the cytosolic redox potential (99,100). As previously emphasized, these changes appear necessary to enable a given level of respiration to be maintained given the lower cellular pO_2 levels induced by hypoxia (9,112). At higher levels of contractile activity, eliciting peak normoxic muscle $\dot{V}O_2$, compensatory adjustments are limited and declines in mitochondrial respiratory rate are readily observable with hypoxia (110,113). At these intensities, maximal extraction of O_2 by the contracting tissue appears to be compromised (4,113). This finding has led to the hypothesis that under these conditions tissue diffusion may be limiting (4,113).

With acclimatization, adaptations in muscle metabolism in humans during prolonged steady-state exercise are clearly evident. Residence at 4300 m for 3 weeks results in a more protected energy state as compared to acute hypoxia, as indicated by the higher ATP and PCr and lower IMP levels (50). Acclimatization is also accompanied by a reduction in muscle lactate concentration during submaximal exercise (50) (Fig. 3) and by a reduction in both lactate appearance and disappearance rates (95) and net lactate release (92,114). Previous studies using an approximately similar period of acclimatization and similar altitude have reported smaller reductions in muscle pH and lower blood lactates (79). These results, which suggest a lower glycolysis and increased muscle buffering capacity, should be interpreted with caution, since at altitude even after acclimatization exercise time at the same power output as used at sea level was significantly shortened.

Teleologically a reduction in glycogen depletion might be expected to accompany a reduction in glycolysis during exercise following acclimatization, but the results are inconclusive. Of the two studies (50,115) that have examined this issue, both following 18–21 days at 4300 m and at the same absolute work load, one study reported no difference in glycogen depletion rates from acute ascent to altitude (50). However, differences in preexercise glycogen concentrations observed between acute and chronic hypoxia may have masked the expected effect.

Short-term acclimatization does not appear to alter total body $\dot{V}O_2$ during submaximal exercise at a given absolute power output (92,96,115) or peak $\dot{V}O_2$ during progressive exercise (92,96,115). The depression in $\dot{V}O_2$ in the exercising leg observed during acute ascent is gradually normalized toward sea level values (92). Increase in leg $\dot{V}O_2$ with exercise following acclimatization are not always found. Bender et al. (96), using three intensities of two-leg steady-state exercise, were unable to demonstrate an enhancement of leg $\dot{V}O_2$ compared to the same exercise performed prior to acclimatization. An increase is suggested, particularly at the two higher intensities, but the differences were not significant. Acclimatization did significantly depress muscle lactate release, an effect that was evident at all exercise intensities (114).

Alterations in muscle metabolism during acclimatization have also been examined under more severe conditions, namely during Operation Everest II, where progressive hypobaric hypoxia to an altitude equivalent to 8840 m was induced over a 40-day period (Fig. 4). Using a progressive cycle test to fatigue, reductions in both peak blood (116) and muscle lactate (49) concentration were found with acclimatization. This phenomenon, which was first observed more than a half century ago (117) and repeatedly since then, based on measures of blood lactates (for review see Ref. 118) has been labeled the "lactate paradox" and suggests decreased glycolysis and lactate production. The apparent blunting of the glycolytic rate that occurs at peak exercise is also accompanied by a smaller reduction in muscle energy state and tighter metabolic control (49). Given the fact that peak glycolytic rates appear not to be impaired with progressive exercise in acute hypoxia (119), the blunting in "glycolysis" appears due to acclimatization itself (118). However, this issue remains unresolved since most studies have only used an acute hypoxic challenge

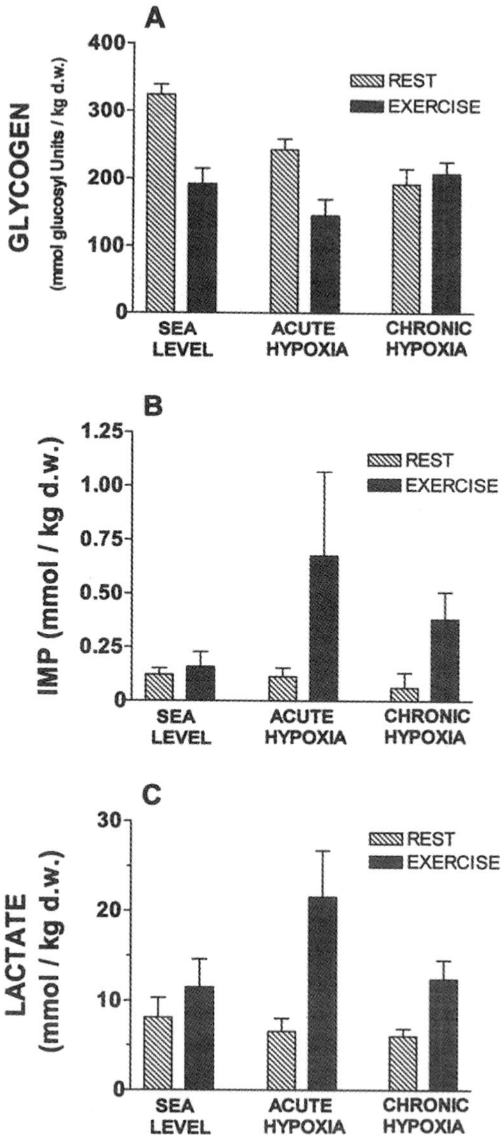

Figure 3 Changes in muscle glycogen (A), lactate (B), and IMP (C) to a standardized submaximal task during acclimatization. Values represent ± SE. Open columns, preexercise; hatched columns, postexercise. (Data from Ref. 50.)

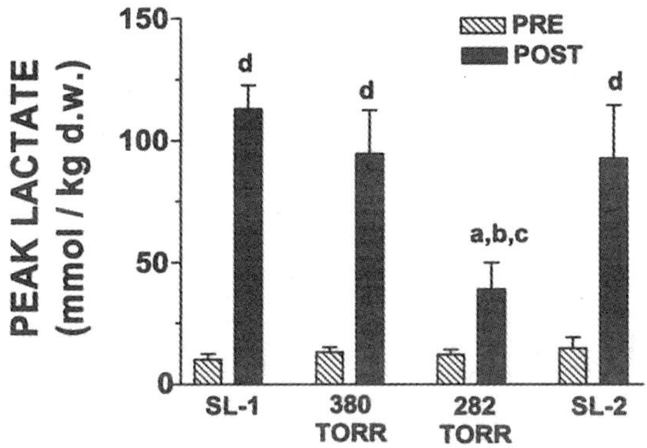

Figure 4 Changes in muscle lactate following progressive cycle exercise to fatigue during acclimatization. Values represent ± SE. SL-1, initial sea level testing; 380 torr and 282 torr, simulated altitude; SL-2, return to sea level. Pre, preexercise; Post, postexercise. a = significantly different from SL-1 postexercise ($p < 0.05$); b = significantly different from SL-1 postexercise ($p < 0.05$); c = significantly different from 380 torr postexercise ($p < 0.05$); d = significantly different from Pre ($p < 0.05$).

equivalent to 4000 m because of safety concerns. It is possible that with more severe hypoxic conditions, a blunting of the peak lactate response acutely similar to that observed with acclimatization may occur.

The apparent inability to activate glycolysis to comparable levels following acclimatization does not appear to be due to substrate availability since muscle glycogen levels are well preserved (49). Moreover, given that work during short-duration, intense activity in which glycolysis represents a major source of ATP production is not compromised with acclimatization (120), it would appear that the "lactate paradox" is specific to large muscle group activity, such as cycling. That muscle lactate concentration can reach near normal sea level concentrations during maximal exercise soon after return to normoxic environments would suggest that central neural drive or muscle excitation failure and not alterations in muscle metabolic organization is, at least partly, responsible (49).

Muscle metabolic adaptations in natives resident to high altitude have unfortunately been limited to studies performed on six Andean Quechua Indians living at between 3700 and 4500 m. These studies were performed near sea level conditions using ^{31}P nuclear magnetic resonance spectroscopy (^{31}P-NMR) localized to the gastrocnemius for noninvasive determinations of muscle metabolic changes (11,77). Using a progressive plantar flexion task to fatigue, it was found that the depression in pH and the increase in the Pi/PCr ratio was much less pronounced than was observed in untrained sea level residents during both normoxia and hypobaric hypoxia (77). Moreover, when the responses were compared to a group of trained distance runners possessing considerably higher $\dot{V}O_2$ max values, the pH and Pi/PCr changes where essentially the same (77). These measures, which are closely related to lactic acid concentration and phosphorylation potential, respectively (121), suggest a much tighter control of oxidative phosphorylation, concomitant with a lower glycolytic rate in the natives compared to the untrained but similar in response to the aerobically trained. The results are similar to what has been observed during Operation Everest II, where at maximal exercise following acclimatization, muscle phosphorylation potential was better protected and was accompanied by a lower lactate concentration (49). In contrast to the local muscle group task employed for the high-altitude Quechuas, the task employed for the acclimatized lowlanders in Operation Everest II involved whole body cycling exercise. In lowlanders, small muscle tasks appear not to demonstrate the effects of acclimatization (120). Whether the high-altitude native display the tighter metabolic control and lactate paradox observed with acclimatization during the cycling exercise, which involves a much greater muscle mass, is unknown.

Metabolic adaptations induced by chronic residence at altitude are likely to occur even for large muscle group activity. In this high-altitude group, blood lactate concentrations are lower than untrained seal level subjects during both submaximal and maximal cycling (11). The lower blood lactate concentration, suggestive of reduced glycogenolysis and reduced glycolysis, need to be confirmed. Needed are direct muscle metabolic measurements, lactate exchange across the working muscles, and isotopic determinations of flux rates. Both for the acclimatized lowlander (114) and the native highlander (19), the "blunted lactate" also occurs during exercise in normoxia. This observation suggests that the adaptation to chronic hypoxia is not only expressed during exercise in hypoxia but can also be transferred to normoxic environments when the challenge to muscle O_2 availability has been normalized. Based on the findings of Operation Everest II, the blunted lactate response at maximal exercise is lost within the first few days of return to sea level (49).

One of the more difficult challenges in high-altitude physiology is unraveling the mechanistic basis for the metabolic adaptations that occur during exercise following chronic hypoxia and whether or not the metabolic adaptations can be related to adaptations in the composition and structure of the muscle cell and surrounding capillary network. Probing for a mechanistic link has been frustrated because of a lack of consensus, particularly in the human, as to what specific adaptations occur in metabolic function, composition, and structure of muscle with chronic hypoxia. It has long been hypothesized that alterations in metabolic function at constant $\dot{V}O_2$ are intimately related to increases in capillary density and mitochondrial potential (22). However, recent evidence has seriously challenged this hypothesis not only with regards to the effects of chronic hypoxia per se but for exercise training while at altitude as well.

Compared to acute hypoxia, residence at 4300 m for up to 3 weeks results in less of of an imbalance between ATP supply and ATP regeneration (50), a lower glycolytic rate (95), less lactate release (92,114), and an increased dependence on blood glucose (97) during prolonged submaximal exercise at constant $\dot{V}O_2$. These metabolic and substrate adaptations are not accompanied by changes in the potential for oxidative phosphorylation, β-oxidation, glycogenolysis, or glycolysis as determined by the maximal activities of representative enzymes (50). Moreover, the metabolic adaptations do not appear to be related to alterations in capillary density since neither fiber size nor the number of capillaries per fiber in any of the fiber types appear altered during the period of acclimatization (50). It thus appears that the hypothesis long invoked to explain the altered metabolic response does not apply at least to relatively short periods of acclimatization. As with acclimatized lowlanders, native highlanders do not appear to display cellular adaptations in composition that can be readily associated with the altered metabolic responses.

Other studies, most notably those involving mountaineering expeditions (54) and severe simulated chronic hypoxia (40), have reported increases in capillary density mediated by pronounced decreases in fiber area. Moreover, such studies demonstrate reductions in mitochondrial potential in the absence of changes in the maximal activities of a number of enzymes representative of glycolytic pathway (40,81).

The muscle metabolic responses to exercise at constant $\dot{V}O_2$ observed in the acclimatized lowlander are suggestive of a tighter control of mitochondrial respiration (i.e., lower ADP, Pi) and downregulation of glycolysis (49) similar to that observed for native highlanders (11). However, the metabolic response to exercise using similar submaximal protocols and mitochondrial respiratory rates needs to be completed before definitive conclusions can be drawn regarding differences between highlanders and acclimatized lowlanders. It would appear that mechanisms other than increased mitochondrial potential are involved.

It is tempting to ascribe some of the metabolic responses observed following acclimatization to extreme altitude to the reduction in fiber size and an increased diffusion potential (122), but such interpretation may be simplistic. To generate a given amount of force following acclimatization with reductions in cross-sectional area, some adjustments would appear necessary, since the force generated per fiber cross-sectional area would increase. Adjustment may occur by increasing the number of fibers recruited either within an active muscle or between synergistic muscles. Given the orderly recruitment of motor units from slow to fast (123), and the fact that the major loss in area appears to be restricted to the type 1 fibers (40), adjustments would appear to be particularly important during activity of low intensity. Regardless of whether compensatory adjustments are made, a greater metabolic disturbance might be expected due simply to the loss of muscle mass. Consequently, it is not clear what the physiological significance is of an increase in capillary density has on exercise metabolism when mediated by a decrease in fiber cross-sectional area.

The general daily activity level during acclimatization compared to that at sea level is also of potential importance, and it is unclear whether the reduction in mitochondrial potential observed during acclimatization to severe hypoxia is mediated by an abrupt reduction in activity level or due to excessive activity. Abrupt reductions in activity result in rapid reductions in mitochondrial potential (20,22), at least under sea level conditions. Intense acute activity has also been shown to reduce maximal mitochondrial respiration in muscle and possibly as a result of reductions in the rate-limiting enzymes involved in mitochondria respiration (7). Moreover, in vascular occlusive diseases, it appears that modest activity may stimulate increases in mitochondria, whereas extreme patterns of activity not only reduce mitochondrial potential but also appear to cause considerable fiber degeneration (124). Regular activity performed while resident at extreme altitude may have similar consequences.

Alternatively, and perhaps more plausibly, the reduction in fiber size may not be related to the pattern of habitual activity but simply represent the product of body weight loss commonly observed during severe chronic hypoxia that occurs because of a loss of appetite or an inability to absorb ingested foods (56). Whether the reduction in fiber size in such situations also results in a disproportionate loss of mitochondria is unclear.

At present it is not possible to ascribe the metabolic adaptations observed with altitude residence to any known muscular adaptation, such as an increase in mitochondrial potential. The same dilemma occurs with short-term training at sea level. Studies have demonstrated, at least qualitatively, the same type of metabolic adaptations during submaximal exercise at sea level as occurs with acclimatization (122). Interestingly, the tighter coupling between ATP synthesis and utilization rates and the apparent reduction in glycolysis occur in the untrained at constant $\dot{V}O_2$ within the first few days of regular exercise and before changes in representative enzymes of metabolic pathways involved in oxidative phosphorylation and glycolysis occur (125). These changes also occur in the absence of any alterations in muscle capillarity and fiber size (125).

Since the metabolic adaptations appear to occur at least in part during the non–steady-state phase of exercise (126), the acceleration in $\dot{V}O_2$ kinetics that result following short-term training may be responsible (127). Oxygen consumption at steady state remains unaffected (127) or indeed may be reduced (128). The increase in $\dot{V}O_2$ kinetics may involve a more rapid adjustment in blood flow. Although adaptations in blood flow during two-leg cycling have not been measured, such appears to be the case with a one-leg, knee extension model (129). Given that there are differences in the blood flow responses between one-leg and two-leg exercise in acute hypoxia (96,104), this potential mechanism needs to be verified with the appropriate exercise model.

Increases in $\dot{V}O_2$ kinetics with short-term training provide an obvious explanation for the more protected energy state and the lower glycolytic rate that has been observed with cycle exercise. With increases in ATP regenerated from oxidative phosphorylation, ATP provided from glycolysis becomes less pronounced. Moreover, a more rapid mitochondrial respiration at the onset of exercise reduces the dependency on PCr hydrolysis as a means of defending ATP homeostasis (9). In effect, less of an imbalance between ATP recruitment and ATP hydrolysis occurs, promoting a high phosphorylation potential in the muscle cell.

Increases in $\dot{V}O_2$ kinetics may also explain the changes in metabolism that have been observed following altitude acclimatization (50). Unfortunately, $\dot{V}O_2$ kinetics have not been examined and muscle tissue samples have not been harvested for analyses during the non–steady-state phase of submaximal exercise. Given that $\dot{V}O_2$ kinetics appear to be significantly slowed with hypobaric hypoxia (106), changes during this period following altitude residence could be meaningful.

The increased blood flow hypothesis, while an inviting mechanism to explain the metabolic adaptations associated with acclimatization (and in native highlanders), is not the only possibility. An additional possibility that has been suggested, based on the exercise response observed in high-altitude natives, is increased metabolic efficiency. Increased metabolic efficiency, defined as the amount of oxidative phosphorylation needed to perform a given amount of mechanical work, has been reported to decrease (11). This conclusion was derived from the persistently lower $\dot{V}O_2$ observed during progressive cycle exercise in high-altitude natives compared to unacclimatized sea level controls. However, at least some of this difference may be explained by the increased work that had to be performed by the lowlanders due to the approximately 19 kg greater body mass. The work performed during cycling is a product not only of the external forces that must be generated to overcome the resistance settings but also the internal component or the work needed to lift the legs during the cycling motion. At 60 revolutions per minute, this component could represent a significant decrease in net mechanical efficiency (130). It would be expected that the substantially greater body mass in the lowlanders would be accompanied by a greater lower limb mass. Yet thus far differences in $\dot{V}O_2$ at the same absolute power output have generally not been observed in previous studies with acclimatization (for review, see Ref. 17).

Although the metabolic adaptations observed with short-term training have potential parallels with acclimatization, there are differences. Short-term training, unlike acclimatization (49) and sustained lifelong residence (77), does not depress the peak blood lactate and supposedly the maximal glycolytic rate observed during progressive exercise. The reasons for the difference remain unclear since with both short-term training and acclimatization there are no obvious peripheral adaptations. Interestingly, with acclimatization the ''lactate paradox'' essentially disappears within the first 48 hours following return to sea level (49). This does not appear to be the case with high-altitude natives since the ''lactate paradox'' remains relatively unperturbed even after 6 weeks at sea level (11). What is particularly curious is why the acclimatized lowlander and the high-altitude native are not able to activate the glycolytic system to a greater extent during progressive work to fatigue, particularly in view of the fact that metabolic by-product accumulation resulting from high-energy phosphate hydrolysis and glycolysis is considerably blunted (49). Previous work has demonstrated rather convincingly that several of the metabolic by-products, in particular, inorganic phosphate (Pi) and hydrogen (H^+), significantly disturb both sarcoplasmic reticulum function (6) and actomyosin cycling (131). In effect, under normal conditions fatigue appears to occur because of both a reduction in activator calcium levels and the ability of the myofibrillar apparatus to translate the activation signal into force. It is interesting that, at least for the high-altitude native, the amount of muscle mass does not seem to influence the expression of the ''lactate'' paradox at maximal exercise since it can be demonstrated in both two-legged cycling and single-leg plantar flexion (77).

B. Maximal Aerobic Power and Acclimatization

An important issue is whether or not the scope of oxidative phosphorylation ($\dot{V}O_2$ peak) during maximal sustainable exercise is extended with altitude acclimatization in lowlanders and in natives resident to high altitude. $\dot{V}O_2$ peak ($\dot{V}O_2$max) is classically measured using an activity involving large muscle groups, most commonly cycling with progressive increments in power output to fatigue. It has been reported that $\dot{V}O_2$ peak with acute exposure to altitude or simulated altitude is decreased at least twice as much in the unacclimatized as in the high altitude native (11). However, atypically low maximum heart rates suggest that a maximal exercise response might not have been achieved in many studies (115). When higher heart rates are obtained in unacclimatized individuals in both normoxia and acute hypoxia (4000 m), the reduction in $\dot{V}O_2$ peak is only 12–13% (98) and within the range reported in more recent studies of high-altitude natives (11,132,133). Short-term acclimatization does not appear to increase $\dot{V}O_2$ peak (92,96,115,134) in spite of the diminished blood lactate concentrations. Another consideration that appears to be important is the training state of the unacclimatized lowlander. Individuals who display high normoxic $\dot{V}O_2$ peak and who experience some degree of arterial desaturation at maximal exercise appear to exhibit the greatest decrement in $\dot{V}O_2$ peak with acute hypoxia (135–137). This observation might also help explain the approximate twofold greater deficit in $\dot{V}O_2$ peak in lowlanders compared to highlanders observed by Hochachka et al. (11), since the lowlanders used in the study were trained athletes with a high $\dot{V}O_2$ peak. In contrast, the $\dot{V}O_2$ peak of the highlanders measured under normoxic conditions was typical of the untrained and occasionally active (98). Based on existing evidence, albeit incomplete, high-altitude natives do not seem to display adaptations that enhance the peak amount of energy that can be obtained from aerobic processes (11).

C. Acclimatization, Adaptation, and Metabolic Control

The final issue to address is whether the metabolic adaptations observed during residence at altitude can be rationalized with the current theories of metabolic control. With acute exposure to hypoxia, submaximal cycling performed at the same power output as at sea level can be accomplished without an impairment in the level of whole body oxidative phosphorylation (92,96). However, reductions in phosphorylation potential and increases in glycolysis appear to occur (50). This response would appear to conform to the model of adaptive cell hypoxia proposed by Connett et al. (9). According to this model and assuming Michaelis-Menten control of enzyme activity, the increased metabolic by products, namely free ADP $(ADP)_f$ and Pi or some combination thereof, in conjunction with an increased mitochondrial redox potential, enables mitochondrial recruitment to be stabilized at normoxic levels despite a lower cellular po_2 (9). In this hypothesis the apparent increase in glycolytic flux observed during exercise with acute hypoxia serves to increase the mitochondrial redox potential by increasing the availability of reducing equivalents and allowing for an undisturbed rate of oxygen reduction by cytochrome oxidase (9,138).

With acclimatization, submaximal $\dot{V}o_2$ for a given power output remains unchanged, but the mitochondrial recruitment and ATP flux rate is achieved with less of a decrease in phosphorylation potential and less of an increase in glycolysis (50). The metabolic adjustment now approaches the sea level or normoxic response pattern. The simplest way to explain the alteration in metabolic control is via increases in the cellular po_2 level, which would, in effect, obviate the need for a reduction in phosphorylation potential and an increase in glycolysis to facilitate a given level of mitochondrial recruitment (138). Measurements of blood flow and O_2 extraction across the working limbs do indicate a shift in the strategy for preserving $\dot{V}o_2$ following acclimatization, namely a reduced dependence on blood flow and an increased extraction (92,96), but it is unclear whether cellular po_2 is affected. If, as proposed for short-term training (129), a more responsive blood flow adjustment occurs during the non–steady-state period, then the tighter metabolic control and depression in glycolysis could easily be explained by less of an imbalance between ATP utilization and ATP synthesis, as a result of increases in oxidative phosphorylation during the rest to work transition (94). Based on current information, the metabolic adaptations during acclimatization cannot be explained by structural or compositional changes at the cell level since capillary density and mitochondrial potential do not increase (50). What effects a reduction in mitochondrial potential (81) and an increase in capillary density, when induced by reductions in cell size (40,41) and observed under more extreme conditions, have on metabolic control and the adaptations observed with acclimatization is unclear.

A reduction of aerobic activity as measured by the $\dot{V}o_2$ peak and obtained by progressive whole body exercise to fatigue is clearly a manifestation of acute hypoxia. The reduction in $\dot{V}o_2$ peak in muscle could occur because of O_2 limited cytochrome turnover (9). Even with moderate levels of exercise and hypoxia (4000 m), $\dot{V}o_2$ is compromised as compared to normoxia (98). At these submaximal exercise levels but not at fatigue, glycolysis appears to compensate for the impairment in oxidative phosphorylation, allowing total ATP synthetic rates to be comparable to normoxia. However, at fatigue, a potentiation of the glycolytic response apparently does not occur since blood (98) and muscle lactates (119) between normoxia and acute hypoxia are not different and mechanical work output is blunted (98). At present, it is unclear whether fatigue, probably induced by the inhibitory effects of increases in metabolic by-products on muscle excitation-contraction sites (6), prevents further recruitment of glycolysis or whether the glycolytic rate is, in itself, inhibited (139).

Acclimatization does not appear to enhance peak oxidative phosphorylation, at least as measured by the $\dot{V}o_2$ peak values obtained during maximal exercise (79,92). Indeed, there is little evidence even in the high-altitude resident to suggest that the range of aerobic activity has been extended over the well-motivated lowlander acutely exposed to hypoxia. The peak $\dot{V}O_2$ however, appears to be achieved with less of a reduction in phosphorylation state and glycolysis (49). This adaptation is peculiar since it implies that even at maximal rates of oxidative phosphorylation in the hypoxic environment, the drive to recruit mitochondrial respiration via elevations in the effectors such as ADP and Pi and increases in mitochondrial redox potential do not appear to prevail as might be expected (9,112). Unlike the unacclimatized state, there is an inability to recruit glycolysis to the same degree to either support mitochondrial respiration (112) or to assist in increasing ATP synthesis (9). To date, no discernible adaptation at the cellular level has been described to explain this behavior. Since metabolic by-product accumulation is blunted at fatigue with acclimatization (49) and ostensibly in the high-altitude native (11), some other mechanism would appear to be involved in causing the neuromuscular failure.

D. Metabolic Adaptation to Altitude—Potential Mechanisms

In the absence of observable cellular level adaptations in the altitude-acclimatized or native resident, such as increases in capillary density or mitochondrial potential, typically used to explain similar exercise metabolic adaptations to long-term training (23), other cellular strategies need to be examined. Several possibilities exist, many of which are highly speculative. These adaptations may involve improvement in cellular diffusion of O_2 to the mitochondria or alterations at the level of the mitochondria and other enzymatic pathways involved in energy production. Reductions in diffusion distance, for example, might be mediated by a better capillary to mitochondria interface, such as has been suggested with training (140) not by simply increasing total mitochondrial density or capillary density but by redistribution. Moreover, rearranging the spatial distribution of mitochondria in the cytoskeleton and reducing the amount of clustering, could significantly reduce the resistance to O_2 transport (141). Reductions in cellular size, if in fact a verifiable adaptation to chronic hypoxia per se, could also mediate increases in O_2 transfer as could increases in myoglobin (28,142). However, the evidence is not compelling, at least in the human, that such adaptations occur.

Alternatively, adaptations might occur to the mitochondrial membrane itself which could facilitate the transport of ATP, ADP, and Pi and reducing equivalents. Changes in adenine nucleotide tanslocase and in the systems for shuttling reducing equivalents could, in theory, facilitate transport and allow for less of an alteration in phosphorylation potential and glycolysis and still maintain mitochondrial respiration during cellular hypoxia (9,138).

Although increases in mitochondrial potential do not appear to occur with chronic hypoxia or residence at altitude, changes in select components of the respiratory chain might be induced. Reynafarje (73) has previously identified elevations in the cytochrome system independent of changes in the citric acid cycle in high-altitude natives compared to sea level controls, and he hypothesized that adaptations at this level may alter the molecular efficiency by which the conformational state of cytochrome *c* oxidase is modulated by the mitochondrial redox state (143). At sea level, evidence has been presented that peak $\dot{V}O_2$ of rat muscle is limited by the level of cytochrome oxidase activity (144). Increases in the potential of the cytochrome oxidase system could also potentially compensate for reductions in pO_2 by allowing a given level of mitochondrial respiration to be sustained at a lower mitochondrial redox potential (112,138). Such an effect might also be realized by a potentiated activation of the mitochondrial dehydrogenates, such as pyruvate dehydrogenase, NAD^+, isocitrate dehydrogenase, and 2-oxoglutarate dehydrogenase by calcium (138,145).

Another exciting possibility that could result in a pronounced alteration in the metabolic response to both acute and chronic hypoxia is via covalent modification of selected enzymes by phosphorylation or by the associated-disassociated level of enzymes to particulate fractions (146,147). In general, current theories of metabolic control are based on simple Michaelis-Menten kinetics with selected by-products of high-energy phosphate transfer reactions assuming primary regulation in the activation of key enzymes (9,112). However, the significance of this form of control during exercise where large increases in ATP regeneration are required has been challenged (94). It has been postulated that changes in metabolite concentration observed in exercise can only hope to provide small changes in flux and not gross activation as required (94). Covalent modification or enzyme association-dissociation could effectively allow large changes in enzyme recruitment as a necessary condition for ATP homeostasis during rest-to-work transitions (94). In some mammalian tissues, such as the heart, hypoxia has been shown to effect alterations in enzyme activities via these mechanisms (147). Regulation of PFK, for example, by covalent modification or the state of enzyme association could explain the glycolytic response to both acute and chronic hypoxia. Moreover, additional control might also occur via changes in other modulators such as fructose-2,6-bisphosphate (F-2,6-P) (147).

Additional adaptations are also possible at the tissue level. Alterations in recruitment strategy, for example, with acclimatization could effectively result in less force generation per fiber by increasing the pool of motor units recruited both within a primary muscle used in the activity or within synergistic muscles (105). This strategy could result in a lower respiratory rate per mitochondrion and lower the levels of ADP and Pi and mitochondria redox needed during cellular hypoxia. Lower levels of the adenine nucleotides could also depress glycolysis at the level of PFK (9). In this respect, the effect could be the same as increasing mitochondrial density and oxidative potential as observed with training (23). Problems in flow heterogeneity within and between muscles have been postulated as a cause of anaerobic glycolysis and increases in lactate concentration observed at relatively low exercise intensities (148). An acclimatization mediated adaptation could potentially reduce flow heterogeneity and improve $\dot{V}O_2$ kinetics, particularly during the nonsteady state.

In summary, numerous possibilities exist that could be mechanistically related to the altered metabolic response observed during exercise following acclimatization and long-term residence at altitude. Since classic cellular adaptations similar to those seen with training do not appear to occur with chronic hypoxia, other explanations appear necessary.

In conclusion, and particularly with regard to the human, major uncertainties remain regarding both the precise nature of the metabolic response to oxygen-deficit environments and the underlying adaptations. Hypoxia-tolerant species potentially can display three fundamental metabolic mechanisms for successful residence at altitude (11). One mechanism, which consists of a decrease in the contribution of anaerobic pathways relative to aerobic metabolism for sustaining ATP synthesis, appears to be the best documented. However, some qualification is warranted. At least in submaximal exercise, following acclimatization reductions in glycolysis do appear to occur (95), but these reductions are not accompanied by improvements in whole body $\dot{V}O_2$ (92). It appears that the elevation in glycolysis observed during acute hypoxic exposure serves not to elevate ATP regeneration but to facilitate a sustained, normoxic level of mitochondrial recruitment (9). With maximal levels of exercise, altitude adaptation also appears to depress glycolysis (11,49), but evidence of a reciprocal increase in peak aerobic metabolism is not compelling.

It has also been proposed that a priority in the hypoxic-tolerant species is to increase the oxidation of carbohydrate over the oxidation of fat because of the increased ATP yield per mole of O_2 consumed (11). Although this response appears logical, it continues to have little experimental support. In the one study that investigated this hypothesis following altitude acclimatization, the contribution of carbohydrate, particularly glucose, during exercise was increased compared to acute hypoxia (97). However, at least in the voluntary exercising human, the concept itself may be challenged, since $\dot{V}O_2$ does not appear to be elevated when fats represent a major fuel source (149,150).

Increased mechanical efficiency, defined as an increase in the mechanical power output to the metabolic (ATP) costs of the activity, has also been suggested as a meaningful adaptation (11). However, few studies have found differences in $\dot{V}O_2$ expenditure during submaximal steady-state work at constant mechanical output following acclimatization (92,96) or between altitude residents and lowlanders equated on the basis of body mass (see Ref. 17). Previous indications of improved mechanical efficiency in high-altitude natives, at least in cycling, might reasonably be explained by differences in internal work as a consequence of the lower body mass and supposedly the lower leg mass in high-altitude residents (130).

In summary, although these hypotheses remain distinctly plausible and worthy of continued investigation, many more studies using a variety of protocols, with regard to both exercise and hypoxic environments, will be needed to examine their tenability. Moreover, additional studies employing many more high-altitude residents at various activity levels must be included and carefully matched to appropriate sea level controls. In this respect, altitude research, particularly with regard to metabolism and substrate selection and associated cellular adaptations, is still in its infancy.

References

1. McComas AJ. Skeletal Muscle. Form and Function. Champaign, IL: Human Kinetics, 1996.
2. Pette D, Staron RS. Mammalian skeletal muscle fiber type transitions. Int Rev Cytol 1997; 170:143–223.
3. Kayser B, Sutton J. Exercise: Performance limits and training at altitude. In: Lung Biology in Health and Disease. 2001.
4. Cain SM. Tolerable limits of hypoxia in muscle. In: Sutton JR, Houston CS, Coats G, eds. Hypoxia: The Tolerable Limits. Indianapolis, IN: Benchmark Press, 1996:79–91.
5. Pette D, Staron RS. Cellular and molecular diversities of mammalian skeletal muscle fibers. Rev Physiol Biochem and Pharmacol 1990; 116:1–75.
6. Korge P, Campbell KB. The importance of the ATPase microenvironment in muscle fatigue: a hypothesis. Int J Sports Med 1995; 16:172–179.
7. Gollnick PD, Bertorci LA, Kelso TB, Witt EH, Hodgson DR. The effect of high intensity exercise on the respiratory capacity of skeletal muscle. Pflügers Arch 1990; 415: 405–413.
8. Monge C, Léon-Velarde F. Physiological adaptation to high altitude: oxygen transport in mammals and birds. Physiol Rev 1991; 71:1135–1172.
9. Connett RJ, Honig CR, Gayeski TEJ, Brooks GA. Defining hypoxia: a systems view of VO_2, glycolysis, energetics and intracellular PO_2. J Appl Physiol 1990; 68:833–842.
10. Green HJ. Muscular adaptations at extreme altitude. Metabolic implications during exercise. Int J Sports Med 1992; 13(suppl 1):S163–S165.
11. Hochachka PW, Stanley C, Mathewson GO, McKenzie DC, Allen PS, Parkhouse WS. Metabolic and work efficiencies during exercise in Andean natives. J Appl Physiol 1991; 70:1720–1730.
12. Hochachka PW. Adaptability of metabolic efficiencies under chronic hypoxia in man. In: Hochachka PW, ed. Surviving Hypoxia. Boca Raton, FL: CRC Press, 1996.
13. Hochachka PW, Dunn JF. The most effective means of protecting tissues against hypoxia. In: Sutton J, Houston C, Jones N, eds. Hypoxia, Exercise and Altitude. Proceedings of the Third Banff International Hypoxia Symposium. New York: Alan R. Liss, Inc., 1981:297–309.
14. Hochachka PW. Metabolic suppression and oxygen availability. Can J Zool 1988; 66: 152–158.
15. Folk GE, Jr. Textbook of Environmental Physiology. 2d ed. Philadelphia: Lea & Febiger, 1974:9–15.
16. Fregly MJ. Adaptations: some general characteristics. In: Fregley MJ, Blatters CM, eds. Handbook of Physiology: Environmental Physiology. New York: Oxford Press, 1996:3–15.
17. Cerretelli P, Hoppeler H. Morphologic and metabolic response to chronic hypoxia: the muscle system. In: Lahiri S, ed. Handbook of Physiology. Adaptation to the Environment. Bethesda, MD: American Physiologic Society, 1996:1155–1181.
18. Hochachka PW. Exercise limitations at high altitude: The metabolic problem and search for its solution. In: Gilles R, ed. Circulation, Respiration and Metabolism. Berlin: Springer-Verlag, 1985:240–249.

19. Tappan DV, Reynafarje R. Tissue pigment manifestations of adaptation to high altitudes. Am J Physiol 1957; 190:99–103.
20. Faulkner JA, Green HJ, White TP. Skeletal muscle response to acute and chronic activity. In: McPherson B, et al., eds. Exercise, Fitness and Health. Champaign, IL: Human Kinetics, 1994.
21. Howald H. Training-induced morphological and functional changes in skeletal muscle. Int J Sports Med 1982; 3:1–12.
22. Saltin B, Gollnick PD. Skeletal muscle adaptability: significance for metabolism and performance. In: Peachy LD, Adrian RH, Geiger SR, eds. Handbook of Physiology. Skeletal Muscle. Baltimore: Williams and Wilkins, 1983:551–631.
23. Holloszy JO, Coyle EF. Adaptations of skeletal muscle to endurance exercise and their metabolic consequences. J Appl Physiol 1984; 56:831–838.
24. Hudlicka O, Brown M, Egginton S. Angiogenesis in skeletal and cardiac muscle. Physiol Rev 1992; 72:369–417.
25. Anderson P, Henriksson J. Capillary supply of the quadriceps femoris muscle of man. Adaptive response to exercise. J Physiol 1977; 270:677–690.
26. Ingjer F. Effects of endurance training on muscle fiber ATPase activity, capillary supply and mitochondrial content in man. J Physiol 1979; 294:419–432.
27. Ingjer F. Capillary supply and mitochondrial content of different skeletal muscle fiber types in untrained and endurance trained men. A histochemical and ultrastructural study. Eur J Appl Physiol 1979; 40:197–209.
28. Wittenberg BA, Wittenberg JB. Transport of oxygen in muscle. Annu Rev Physiol 1989; 51:857–878.
29. Pattengale PK, Holloszy JO. Augmentation of skeletal muscle myoglobin by a program of treadmill running. Am J Physiol 1967; 213:783–785.
30. Hickson RC. Skeletal muscle cytochrome c and myoglobin, endurance and frequency of training. J Appl Physiol 1981; 51:746–749.
31. Harms SJ, Hickson RC. Skeletal muscle mitochondria and myoglobin, endurance and intensity of training. J Appl Physiol 1983; 54:798–802.
32. Gimenez M, Sanderson RJ, Reiss OK, Banchero N. Effect of altitude on myoglobin and mitochondrial protein in canine skeletal muscle. Respiration 1977; 34:171–176.
33. Jansson E, Sylvén C, Nordevang E. Myoglobin in quadriceps femoris muscle of competitive cyclists and untrained man. Acta Physiol Scand 1982; 114:627–629.
34. Svedenhag J, Hendriksson J, Sylvén C. Dissociation of training effects on skeletal muscle mitochondrial enzymes and myoglobin in man. Acta Physiol Scand 1983; 117:213–218.
35. Terrados N, Melichna J, Sylvén C, Jansson E. Decrease in skeletal muscle myoglobin with intensive training in man. Acta Physiol Scand 1986; 128:651–652.
36. Terrados N, Jansson E, Sylvén C, Kaijser L. Is hypoxia a stimulus for synthesis of oxidative enzymes and myoglobin? J Appl Physiol 1990; 68:2369–2372.
37. Green HJ. Cation pumps in skeletal muscle: potential role in muscle fatigue. Acta Physiol Scand 1998; 162:201–213.
38. Pette D, Düsterhöft S. Altered gene expression in fast-twitch muscle induced by chronic low-frequency stimulation. Am J Physiol 1992; 262:R333–R338.
39. Kraemer WJ, Fleck SJ, Evans WJ. Strength and power training: physiological mechanisms of adaptation. In: Holloszy JO, ed. Exercise and Sport Science Reviews. Baltimore: Williams and Wilkins, 1996:363–397.

40. Green HJ, Sutton JR, Cymerman A, Young PM, Houston CS. Operation Everest II: adaptations in human skeletal muscle. J Appl Physiol 1989; 56:2454–2461.
41. Hoppeler H, Kleinert E, Schlegel C, Claassen H, Howald H, Kayar SR, Cerretelli P. Muscular exercise at high altitude: II. Morphologic adaptations of human skeletal muscle to chronic hypoxia. Int J Sports Med 1990; 11(suppl 1):S3–S9.
42. Fridén J, Seger J, Sjostrom M, Ekblom E. Adaptive response in human skeletal muscle to prolonged eccentric training. Int J Sports Med 1983; 4:177–183.
43. Fridén J. Changes in human skeletal muscle induced by long term eccentric exercise. Cell Tissue Res 1984; 236:365–372.
44. Armstrong RB, Warren GL, Warren JA. Mechanisms of exercise-induced muscle fibre injury. Sports Med 1991; 12:184–207.
45. Cadefau J, Casademonat J, Grau JM, Fernándex J, Belaguer A, Vernet M, Cussó R, Urbano-Marguex A. Biochemical and histochemical adaptation to sprint training in young athletes. Acta Physiol Scand 1990; 140:341–351.
46. McKenna MJ, Schmidt TA, Hargreaves M, Cameron L, Skinner SL, Kjeldsen K. Sprint training increases human skeletal muscle Na^+-K^+ ATPase concentration and improves K^+ regulation. J Appl Physiol 1993; 75:173–180.
47. Rosser BWC, Hochachka PW. Metabolic capacity of muscle fibers from high-altitude natives. Eur J Appl Physiol 1993; 67:513–517.
48. Simoneau JA, Bouchard C. Human variation in skeletal muscle fiber-type proportion and enzyme activities. Am J Physiol 1989; 257:E567–E572.
49. Green HJ, Sutton J, Young P, Cymerman A, Houston CS. Operation Everest II: muscle energetics during maximal exhaustive exercise. J Appl Physiol 1989; 66:142–150.
50. Green HJ, Sutton JR, Wolfel EE, Reeves JT, Butterfield GE, Brooks GA. Altitude acclimatization and energy metabolic adaptations in skeletal muscle during submaximal exercise. J Appl Physiol 1992; 73:2701–2708.
51. Oelz O, Howald H, Di Prampero P, Hoppeler H, Claassen H, Jenni R, Bühlmann A, Ferretti G, Brückner J, Veicsteinas A, Gussoni M, Cerretelli P. Physiological profile of world-class high altitude climbers. J Appl Physiol 1986; 60:1734–1742.
52. Kayser B, Hoppeler H, Classen H, Ceretelli P. Muscle structure and performance capacity of Himalayan Sherpas. J Appl Physiol 1991; 70:1938–1942.
53. Saltin B, Nygaard E, Rasmussen B. Skeletal muscle adaptation in man following prolonged exposure to high altitude. Acta Physiol Scand 1980; 109:31A.
54. Hoppeler H, Desplanches D. Muscle structural modifications in hypoxia. Int J Sports Med 1992; 13(suppl 1):S166–S168.
55. MacDougall JD, Green HJ, Sutton JR, Coates G, Cymerman A, Young P, Houston CS. Operation Everest II: structural adaptations in skeletal muscle in response to extreme simulated altitude. Acta Physiol Scand 1991; 142:421–427.
56. Houston CS, Sutton JR, Cymerman A, Reeves JT. Operation Everest II: man at extreme altitude. J Appl Physiol 1987; 63:877–882.
57. Bigard AX, Brunet A, Guezennec CY, Monod H. Effects of chronic hypoxia and endurance training on muscle capillarity in rats. Pflügers Arch 1991; 419:225–229.
58. Banchero N. Cardiovascular responses to chronic hypoxia. Annu Rev Physiol 1987; 49:465–476.
59. Snyder GK, Wilcox EE, Burnham EW. Effects of hypoxia on muscle capillarity in rats. Respir Physiol 1985; 52:135–140.
60. Poole DC, Mathieu-Costello O. Skeletal muscle capillary geometry: adaptation to chronic hypoxia. Respir Physiol 1989; 77:21–30.

61. Yamashita H, et al. Chronic exposure to simulated altitude does not increase angiogenic activity in skeletal muscle of rats. Tohoku J Exp Med 1994; 172:375–379.
62. Sillau AH, Banchero N. Effects of hypoxia on capillary density and fiber composition in rat skeletal muscle. Pflügers Arch 1977; 370:227–232.
63. Sillau AH, Acquin L, Bui MV, Banchero N. Chronic hypoxia does not affect guinea pig skeletal muscle capillarity. Pflügers Arch 1980; 386:39–45.
64. Vaughan BE, Pace N. Changes in myoglobin content of the high altitude acclimatized rat. Am J Physiol 1956; 185:549–556.
65. Banchero N, Kayar SR, Lechner AJ. Increased capillarity in skeletal muscle of growing pigs acclimated to cold and hypoxia. Respir Physiol 1985; 62:245–255.
66. Bigard AX, Brunet A, Guezennec CY, Monod H. Skeletal muscle changes after endurance training at altitude. J Appl Physiol 1991; 71:2114–2121.
67. Bigard AX, Brunet A, Serrurier B, Guezennec CY, Monod H. Effects of endurance training at high altitude on diaphragm muscle properties. Pflügers Arch 1992; 422: 239–244.
68. Mizuno M, Juel C, Bro-Rasmussen T, Mygind E, Schibye B, Rasmussen B, Saltin B. Limb skeletal muscle adaptation in athletes after training at altitude. J Appl Physiol 1990; 68:496–502.
69. Desplanches D, Hoppeler H, Linossier MT, Dennis C, Claasen H, Dormois D, Lacour JR, Geyssant A. Effects of training in normoxia and normobaric hypoxia on human muscle ultrastructure. Pflügers Arch 1993; 425:263–267.
70. Desplanches D, Hoppeler H, Tüscher L, Mayet MH, Spielvogel H, Ferretti G, Kayser B, Levenberger M, Gründenfelder A, Favier R. Muscle tissue adaptations of highlanders to training in chronic hypoxia or acute normoxia. J Appl Physiol 1996; 81:1946–1951.
71. Pette D, Dölken G. Some aspects of regulation of enzyme levels in muscle energy-supplying metabolism. Adv Enz Reg 1975; 13:355–378.
72. Tappen DV, Reynafarje BD, Potter VR, Hurtado A. Alterations in enzymes and metabolites resulting from adaptations to low oxygen tensions. Am J Physiol 1957; 190:93–98.
73. Reynafarje B. Myoglobin content and enzymatic activity of muscle and altitude adaptation. J Appl Physiol 1962; 17:301–305.
74. Hurtado A. Animals in high altitude: resident man. In: Dill DB, Adolph EF, Wilber CG, eds. Handbook of Physiology: Section 4. Adaptation to the Environment. Washington, DC: American Physiologic Society, 1964:843–860.
75. Hochachka PW. Muscle enzymatic composition and metabolic regulation in high altitude adapted natives. Int J Sports Med 1992; 13:S89–S91.
76. Hochachka PW, Stanley C, McKenzie DC, Villena A, Monge C. Enzyme mechanisms for pyruvate-to-lactate flux attentuation: A study of Sherpas, Quechuas and Hummingbirds. Int J Sports Med 1992; 13:S119–S122.
77. Matheson GO, Allen PS, Ellinger DC, Hanstock CC, Gheorghiu D, McKenzie DC, Stanley C, Parkhouse WS, Hochachka PW. Skeletal muscle metabolism and work capacity—a ^{31}P-NMR study of Andean natives and lowlanders. J Appl Physiol 1991; 70:1963–1976.
78. Hoppeler H, Lüthi P, Claasen H, Weibel ER, Howald H. The ultrastructure of the normal human skeletal muscle. Pflügers Arch 1973; 344:217–232.
79. Young AJ, Evans WJ, Fisher EC, Sharp RL, Costill DL, Maher JT. Skeletal muscle metabolism of sea level natives following short-term high altitude residence. Eur J Appl Physiol 1984; 52:463–466.

80. Phillips SM, Green HJ, Tarnopolsky MA, Grant SM. Decreased glucose turnover following short term training is unaccompanied by changes in muscle oxidation potential. Am J Physiol 1995; 269:E222–E230.
81. Howald H, Pette D, Simoneau JA, Uber A, Hoppeler H, Cerretelli P. Muscular exercise at high altitude: III. Effect of chronic hypoxia on muscle enzyme activities. Int J Sports Med 1990; 11(suppl 1):510–514.
82. Terblanche SE, Groenewald JV, van der Linde A, Wolfe-Winskel JM, Jooste PL, Elofsen W. A comparative study of the effect of training at altitude and at sea level on endurance and certain biochemical variables. Comp Biochem Physiol 1984; 78:21–26.
83. Pickett CB, Cascarano J, Wilson MA. Acute and chronic hypoxia in rats. Effect on organismic respiration, mitochondrial protein mass in liver and succinic dehydrogenase activity in liver, kidney and heart. Exp Zool 1979; 210:49–58.
84. Kaijser L, Sunberg CL, Eiken O, Nygren A, Esboemson M, Sylvén C, Jansson E. Muscle oxidative capacity and work performance after training under local leg ischemia. J Appl Physiol 1990; 69:785–787.
85. Hurtado A, Rota A, Merino C, Pons J. Studies of myohemoglobin at high altitude. Am J Med Sci 1937; 194:708–713.
86. Anthony A, Ackerman E, Strother CK. Effects of altitude acclimatization on rat myoglobin. Changes in myoglobin content of skeletal and cardiac muscle. Am J Physiol 1959; 196:512–516.
87. Bowen WJ, Poel WE. The effects of anoxia upon myoglobin concentration. Fed Proc 1948; 7:11.
88. Poel WE. Effect of anoxic anoxia on myoglobin concentration in striated muscle. Am J Physiol 1949; 156:44–51.
89. Clark RT, Jr., Criscuolo D, Coulson CK. Effects at 20,000 ft simulated altitude on myoglobin content of animals with and without exercise. Fed Proc 1948; 11:25.
90. Terrados N, McLichna J, Sylvén C, Jansson E, Kaijser L. Effect of training at simulated altitude on performance and metabolic capacity in competitive road cyclists. Eur J Appl Physiol 1988; 57:203–209.
91. Peter JB, Barnard RJ, Edgerton VR, Gillespie CA, Stemple KE. Metabolic profiles of three types of skeletal muscle in guinea pigs and rabbits. Biochem 1972; 4:2627–2633.
92. Wolfel EE, Groves BM, Brooks GA, Butterfield GE, Mazzeo RS, Moore LG, Sutton JR, Bender PR, Dahms TE, McCullough RE, Huang S-Y, Sun SF, Grover RF, Hultgren HN, Reeves JT. Oxygen transport during steady state, submaximal exercise in chronic hypoxia. J Appl Physiol 1991; 70:1129–1136.
93. Linnarsson D, Karlsson J, Fagraeus L, Saltin B. Muscle metabolites and oxygen deficit with exercise in hypoxia and hyperoxia. J Appl Physiol 1974; 36:399–402.
94. Hochachka PW, Matheson GO. Regulating ATP turnover over broad dynamic work ranges in skeletal muscles. J Appl Physiol 1992; 73:1697–1703.
95. Brooks GA, Butterfield GE, Wolfe RR, Groves BM, Mazzeo RS, Sutton JR, Wolfel EE, Reeves JT. Decreased reliance on lactate during exercise after acclimatization to 4,300 m exercise. J Appl Physiol 1991; 71:333–341.
96. Bender PR, Groves BM, McCullough RE, McCullough RG, Huang S-Y, Hamilton AJ, Wagner PD, Cymerman A, Reeves JT. Oxygen transport to exercising leg in chronic hypoxia. J Appl Physiol 1988; 65:2592–2597.
97. Brooks GA, Butterfield GE, Wolfe RR, Groves BM, Mazzeo RS, Sutton JR, Wolfel EE, Reeves JT. Increased dependence on blood glucose after acclimatization to 4,300 m. J Appl Physiol 1991; 70:919–927.

98. Hughson RL, Green HJ, Sharratt MT. Gas exchange, blood lactate and plasma catecholamines during incremental exercise in hypoxia and normoxia. J Appl Physiol 1995; 79:1134–1141.
99. Hogan MC, Arthur PG, Bebout DE, Hochachka PW, Wagner PD. Role of O_2 in regulating tissue respiration in working dog muscle in situ. J Appl Physiol 1992; 73:728–736.
100. Hogan MC, Nioka S, Brechue WF, Chance B. A ^{31}P-NMR study of tissue respiration in working dog muscle during reduced O_2 delivery conditions. J Appl Physiol 1992; 73:1662–1679.
101. Ferretti G, Boutellier U, Pendergast DR, Moia C, Minetti AE, Howald H, Di Prampero, IV. Oxygen transport system before and after exposure to chronic hypoxia. Int J Sports Med 1990; 11(suppl):S15–S20.
102. Hogan MC, Cox RH, Welch HG. Lactate accumulation during incremental exercise with varied inspired oxygen fractions. J Appl Physiol 1983; 55:1134–1140.
103. Green HJ. Muscle metabolism in chronic hypoxia. In: Sutton JR, Houston CS, Coats G, eds. Hypoxia: The Tolerable Limits. Indianapolis, IN: Benchmark Press, 1988:101–120.
104. Rowell LB, Saltin B, Kiens B, Christensen NJ. Is peak quadriceps blood flow in humans even higher during exercise with hypoxia? Am J Physiol 1986; 251:H1038–1044.
105. Green HJ, Patla AE. Maximal aerobic power: neuromuscular and metabolic considerations. Med Sci Sports Exerc 1992; 24:38–46.
106. Murphy PC, Cuervo LA, Hughson RL. Comparison of ramp and step exercise protocols during hypoxic exercise in man. Cardiovasc Res 1989; 23:825–832.
107. Knuttgen HG, Saltin B. Oxygen uptake, muscle high energy phosphates and lactate in exercise under acute hypoxic conditions in man. Acta Physiol Scand 1973; 87:368–376.
108. Wagner PD, Sutton JR, Reeves JT, Cymerman A, Groves BM, Malconian MK. Operation Everest II: pulmonary gas exchange during a simulated ascent of Mt. Everest. J Appl Physiol 1987; 63:2348–2359.
109. Stainsby WN, Otis AB. Blood flow, oxygen tension, oxygen uptake and oxygen transport in skeletal muscle. Am J Physiol 1964; 206:858–866.
110. Stainsby WN, Brechue WF, O'Drobinak DM, Barclay JF. Effects of ischemic and hypoxic hypoxia on $\dot{V}O_2$ and lactic acid output during tetanic contractions. J Appl Physiol 1990; 68:574–579.
111. Hogan MC, Welch HG. Effect of altered O_2 tensions on muscle metabolism in dog skeletal muscle during fatiguing work. Am J Physiol 1986; 251:C216–C222.
112. Wilson DF. Factors affecting the rate and energetics of mitocondrial oxidative phosphorylation. Med Sci Sports Exerc 1994; 26:37–43.
113. Hogan MC, Roca J, Wagner PD, West JB. Limitation of maximal O_2 uptake and performance by acute hypoxia in dog muscle in situ. J Appl Physiol 1988; 65:815–821.
114. Bender PR, Groves BM, McCullough RE, McCullough RG, Trad L, Young AJ, Cymerman A, Reeves JT. Decreased exercise muscle lactate release after high altitude acclimatization. J Appl Physiol 1989; 67:1456–1462.
115. Young AJ, Evans WJ, Cymerman A, Pandolf KB, Knapik JJ, Maher JJ. Sparing effect of chronic high altitude exposure on muscle glycogen utilization during exercise. J Appl Physiol 1982; 52:857–862.
116. Sutton JR, Reeves JT, Wagner PD, Groves BM, Cymerman A, Malconian MK, Rock

PR, Young PM, Walter SD, Houston CS. Operation Everest II: oxygen transport during exercise at extreme simulated altitude. J Appl Physiol 1988; 64:1309–1321.

117. Edwards HT. Lactate acid at rest and work at high altitude. Am J Physiol 1936; 116: 367–375.
118. Hochachka PW. The lactate paradox. Analysis of underlying mechanisms. Ann Sports Med 1988; 4:184–188.
119. Green HJ. Muscle metabolism during incremental exercise in hypoxia and normoxia. J Appl Physiol 2001; Submitted.
120. Kayser B, Narici M, Binzoni T, Grassi B, Ceretelli P. Fatigue and exhaustion in chronic hypobaric hypoxia: influence of exercising muscle mass. J Appl Physiol 1994; 76: 634–640.
121. Chance B. Metabolic control principles and ^{31}P NMR. Fed Proc 1986; 45:2915–2920.
122. Green HJ. What is the physiological significance of training-induced adaptations in muscle mitochondrial capacity? In: Maughan RL, Shirreffs SM, eds. Biochemistry of Exercise IX. Champaign, IL: Human Kinetic Publishers, 1995:345–359.
123. Cope TC, Pinter MJ. The size principle: Still working after all of these years. News Physiol Sci 1995; 10:280–286.
124. Hoppeler H, Hudlicka O, Uhlmann E, Claassen H. Skeletal muscle adaptations to ischemia and severe exercise. Clin J Sports Med 1992; 2:43–51.
125. Green HJ, Jones S, Ball-Burnett ME, Smith D, Livesey J, Farrance BW. Early muscular and metabolic adaptations to prolonged exercise training in man. J Appl Physiol 1991; 70:2032–2038.
126. Green HJ, Cadefau J, Cussó R, Ball-Burnett M, Jamieson G. Metabolic adaptations to short term training are expressed early in exercise. Can J Physiol Pharmacol 1995; 73:474–482.
127. Phillips SM, Green HJ, MacDonald MJ, Hughson RL. Progressive effect of endurance training on $\dot{V}O_2$ kinetics at the onset of submaximal exercise. J Appl Physiol 1995; 79:1914–1920.
128. Cadefau J, Green HJ, Cussó R, Ball-Burnett M, Jamieson G. Coupling of muscle phosphorylation potential to glycolysis during submaximal exercise of varying intensity following short term training. J Appl Physiol 1994; 76:2586–2593.
129. Shoemaker JK, Phillips SM, Green HJ, Hughson RL. Faster femoral artery blood velocity kinetics at the onset of exercise following short term training. Cardiovasc Res 1996; 31:278–286.
130. Widrick JJ, Freedson PS, Hamill J. Effect of internal work on the calculation of optimal pedalling rates. Med Sci Sports Exerc 1992; 24:376–382.
131. Cooke R, Pate E. The inhibition of muscle contraction by the by-products of ATP hydrolysis. In: Taylor B, ed. Biochemistry of Exercise VII. Champaign, IL: Human Kinetics, 1990:59–72.
132. Favier R, Spielvogel H, Desplanches D, Ferretti G, Kayser B, Gründenfelder A, Leuenberger M, Tüscher L, Caceres E, Hoppeler H. Training in hypoxia vs. training in normoxia in high-altitude natives. J Appl Physiol 1995; 78:2286–2293.
133. Favier R, Spielvogel H, Desplanches D, Ferretti G, Kayser B, Hoppeler H. Maximal exercise performance in chronic hypoxia and acute normoxia in high altitude natives. J Appl Physiol 1995; 78:1867–1874.
134. Niu W, Wu Y, Li B, Chen N, Song S. Effects of long term acclimatization in lowlanders migrating to high altitude: comparison with high altitude residents. Eur J Appl Physiol 1995; 71:543–548.

135. Lawler JS, Powers SK, Thompson D. Linear relationship between VO_2max decrement during exposure to acute hypoxia. J Appl Physiol 1988; 64:1486–1492.
136. Martin D, O'Kroy J. Effects of acute hypoxia on the VO_2max of trained and untrained subjects. J Sports Sci 1993; 11:37–42.
137. Young AJ, Cymerman A, Burse RL. The influence of cardiorespiratory fitness on the decrement in maximal aerobic power of high altitude. Eur J Appl Physiol 1985; 54: 12–15.
138. Wilson DF. Contribution of diffusion to the oxygen dependence on energy metabolism in cells. Experentia 1990; 46:1160–1162.
139. Spriet LL. Phosphofructokinase activity and acidosis during short-term tetanic contractions. Can J Physiol Pharmacol 1991; 69:298–304.
140. Crenshaw AG, Fridén J, Thornell L, Hargens AR. Extreme endurance training: evidence for capillary and mitochondrial compartmentalization in human skeletal muscle. Eur J Appl Physiol 1991; 63:173–178.
141. Jones DP, Aw TY, Sillau AH. Defining resistance to oxygen transfer in tissue hypoxia. Experentia 1990; 46:1180–1185.
142. Green HJ. Muscle energetics, hypoxia and acclimatization. In: Sutton JR, Houston CS, Coates G, eds. Hypoxia: The Tolerable Limits. Indianapolis: Benchmark Press Inc. 1988:101–116.
143. Reynafarje B. Biochemical adaptation to chronic hypoxia of high altitude. Mol Physiol 1985; 8:463–471.
144. McAllister RM, Ogilvie RW, Terjung RL. Impact of reduced cytochrome oxidase activity on peak oxygen consumption of muscle. J Appl Physiol 1990; 69:384–389.
145. McCormack JG, Hallestrap AP, Denton RM. Role of calcium ions in regulation of mammalian intramitochondrial metabolism. Physiol Rev 1990; 70:391–425.
146. Parkhouse WS. Regulation of skeletal muscle metabolism by enzyme binding. Can J Physiol Pharmacol 1992; 70:150–156.
147. Storey KB. A re-evaluation of the Posterior effect: new mechanisms in anaerobic metabolism. Mol Physiol 1985; 8:439–461.
148. Whipp BJ, Lamarra N, Ward SA. Obligatory anaerobiosis from oxygen uptake to blood flow ratio dispersion in skeletal muscle: a model. Eur J Appl Physiol 1995; 71:147–152.
149. Hargreaves M, Kiens K, Richter EA. Effect of increased plasma free fatty acids concentrations on muscle metabolism in exercising man. J Appl Physiol 1991; 70:194–201.
150. Graham TE, Spriet LL. Performance and metabolic responses to high caffeine dose during prolonged exercise. J Appl Physiol 1991; 71:2292–2298.

15

Blood

ROBERT F. GROVER

Arroyo Grande, California

PETER BÄRTSCH

University Hospital
Heidelberg, Germany

Blood, with its many components, plays manifold roles in our responses to high altitude, including changes in the immune system (see Chapter 19). In this chapter we shall limit ourselves to only two functions, oxygen transport and hemostasis. We shall describe both the acute alterations in the volume and composition of the blood following ascent to high altitude as well as the long-term increase in red cell mass (polycythemia), and how these changes serve to enhance oxygen transport from lung to tissue. Variability among individuals in both the regulation of blood composition as well as the patterns of adaptation to hypoxic stress will be emphasized. Further, we shall discuss how hypoxia alters blood clotting and clot lysis.

I. Blood Oxygen Transport

Life based on oxidative metabolism depends upon a continuous supply of O_2 for survival. For most species, this oxygen supply is accomplished by a transport system in which blood is the vehicle, O_2 being bound reversibly to hemoglobin (Hb) within the red blood cells. Multiple factors are involved in this process. Total body O_2 transport, i.e., the amount of O_2 made available to the body per minute, depends upon the quantity of O_2 loaded into each unit of arterial blood (arterial O_2 content) and the volume of blood pumped by the heart each minute (cardiac output). In turn, arterial O_2 content is a function of Hb concentration and HbO_2 saturation. Saturation is determined by arterial O_2 tension and the affinity of Hb for O_2 influenced by the concentration of 2,3-diphosphoglycerate (2,3-DPG) within the red cell, blood temperature, and blood pH. Oxygen delivered to the tissues of the body is a function of the distribution of cardiac output and the quantity of O_2 unloaded from each unit of blood (arterio-venous difference in blood O_2 content). The unloading of O_2 depends upon the affinity of Hb for O_2, i.e., the position of the HbO_2 dissociation curve, and tissue O_2 tension. Marked variability exists among individuals in the way these multiple factors are woven together to maintain O_2 supply. However, for a given individual, the pattern of O_2 transport varies minimally. Each component of the system is under precise regulation at a level that is optimal for that individual.

Obviously the composition of blood is only one component of this complex system. Variability among individuals is clearly evident for Hct. At sea level among normal adults in the United States, males have an Hct between 40 and 49% (5th and 95th percentile, respectively), whereas the normal limits for females are 35 and 44%, with a slight tendency to increase following menopause; for young children the normal limits are 33 and 40% (Fig. 1) (1). These are fairly broad limits, the

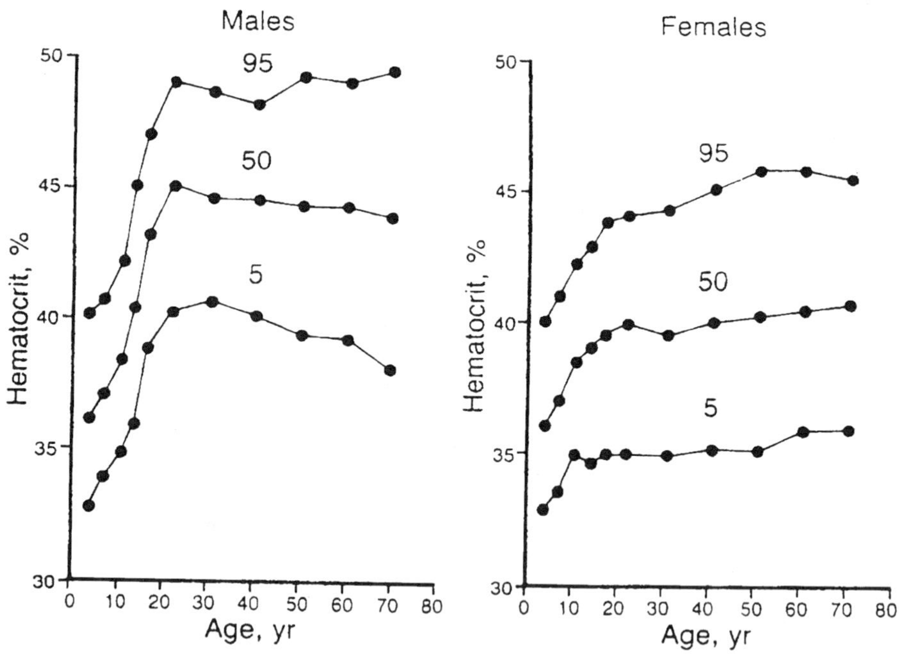

Figure 1 Normal Hct values at sea level as a function of age for 7426 normal males and 7704 normal females in the United States. The 95th, 50th, and 5th percentiles are indicated. (From Ref. 1.)

range being roughly 20% of the mean. However, for a given individual, Hct will vary little over time.

A. Acute Hypoxia: Hypovolemia

Hemoconcentration

One of the first physiological responses observed upon ascent to high altitude was a rapid rise in Hct or Hb concentration (Fig. 2) (2). From a survival standpoint, this provides an early increase in the oxygen-carrying capacity of the blood to counteract the fall in arterial saturation and preserve blood oxygenation. The rise in Hct reflects primarily a shift of water out of the vascular system, i.e., a decrease in plasma volume (PV), a true hemoconcentration, since the concentration of plasma proteins also increases (3). In healthy young men and women, this process appears within the first hours of arrival at high altitude, continues over several days, and frequently decreases PV by as much 20% (Fig. 3). Conversely, when men acclimatized to 3100 m altitude descend to sea level, the prompt fall in the concentration of both Hb and plasma proteins reflects an expansion of PV by an average of 20% over 12 days (4).

The observed decrease in PV appears to be one manifestation of a generalized redistribution of fluid from extracellular to intracellular with no net loss in total body water (5). Hemoconcentration is not a manifestation of hypohydration since it occurs whether or not there is a decrease in body weight (6–8). The mechanism responsible for the decrease in PV is unclear. One might expect activation of the β-adrenergic system to be involved. The autonomic system is important in vascular

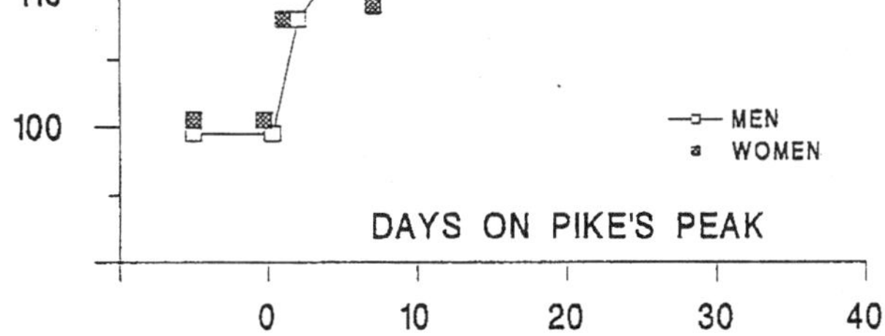

Figure 2 Hemoglobin concentration expressed as a percent of sea level values over the first month following ascent to 4300 m (Pikes Peak). Average values for 3 men and 8 women. (From Refs. 2 and 26.)

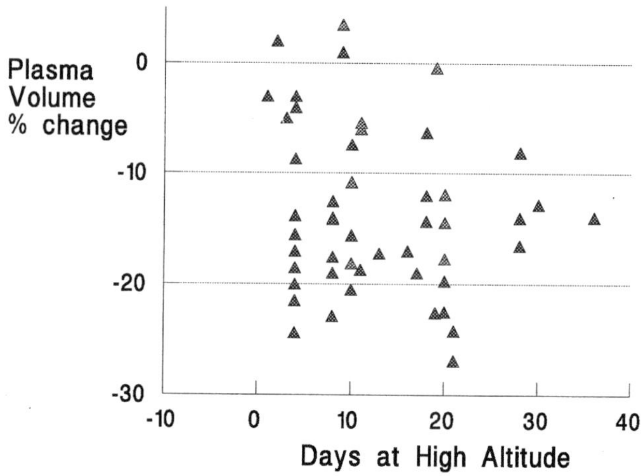

Figure 3 Plasma volume expressed as percent change from sea level following ascent to altitudes of 3800–4500 m. Individual measurements. (From Refs. 3, 6, 8, 16–18, 119.)

volume regulation (9), the large activation of the sympatho-adrenal systems on exposure to high altitude (10), and the complex interactive changes of hormones and peptides (11) that control fluid balance. However, dense β-adrenergic receptor blockade with popranolol was found not to modify the decrease in PV (8). Likewise, α-adrenergic blockade with prazosin has no clear influence on PV reduction (L. G. Moore, personal communication). Changes in PV are also known to be under hormonal control. For example, acute hypoxemia is a potent stimulator for atrial natriuretic peptide (12), which regulates transvascular fluid balance and appears to account for the PV contraction during acute hypoxia (13,14). Perhaps this is the mechanism underlying the rise in Hct on arrival at high altitude.

Blood Volume

Examination of group mean data following ascent to high altitude indicates that a true increase in red cell volume (RCV) requires several weeks to appear, whereas a decrease in PV begins upon arrival, resulting in an initial decrease in total blood volume (BV). This pattern of change was described as early as 1952 by Lawrence et al. (15), who measured RCV and BV using autologous red cells tagged with radioactive phosphorous. Over the subsequent decades, multiple investigators using various techniques including Evans Blue and carbon monoxide have confirmed those early results (see Fig. 6). However, group mean data obscure the very important variability among individuals in these responses. Detailed analysis of published reports by multiple authors that contain individual data (3,6,8,16–18) reveals many individuals in whom RCV apparently decreases by as much as 10% during the first 10 days at high altitude, while an equal number of individuals show increases in

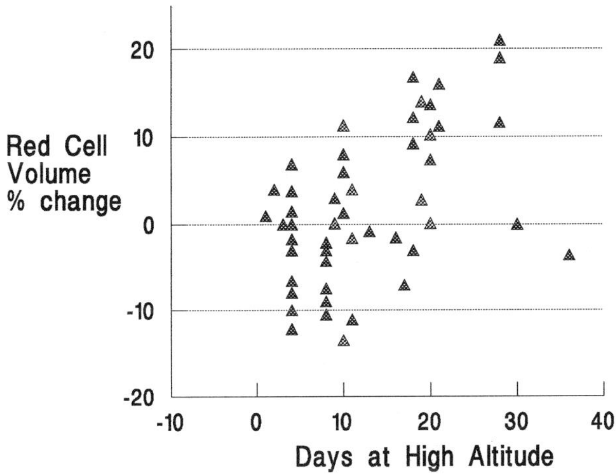

Figure 4 Red cell volume expressed as percent change from sea level following ascent to altitudes of 3800–4500 m. Individual measurements. (From Refs. 3, 6, 8, 16–18, 119.)

RCV of up to 10% during that time period (Fig. 4). For each individual in whom serial measurements were made during the initial 10 days, the increase or decrease in RCV was consistent, suggesting the observations were not the result of methodological inaccuracy. Changes in PV also vary from minimal to decreases in excess of 20% (Fig. 3). Consequently, individual changes in BV range from +3% to −18% (Fig. 5). Decreases in RCV during the first days at high altitude in some individuals are certainly unexpected, since red cell life span has been observed to be normal in newcomers to 4540 m in the Andes (19,20). Whether or not these reported decreases in RCV are more apparent than real is a subject for future investigation using techniques to measure the rate of red cell destruction.

During the first week or two at high altitude, the reductions in BV result in a decrease in ventricular filling pressure and a smaller stroke volume (21). Heart rate fails to compensate and consequently cardiac output falls (22). To maintain oxygen transport, the unloading of oxygen from the blood must increase by lowering mixed venous oxygen saturation and content (widening of total body arterio-venous difference in blood oxygen content); direct measurements confirm this (22).

B. Sustained Hypoxia: Polycythemia

Time Course in Men and Women

True polycythemia develops when residence at high altitude extends over months to years. While the hypoxic stimulus to erythropoiesis begins on arrival at altitude (23), the onset of an increase in RCV varies greatly among individuals (Fig. 4). On average, the expansion of RCV appears to be relatively slow, requiring about a

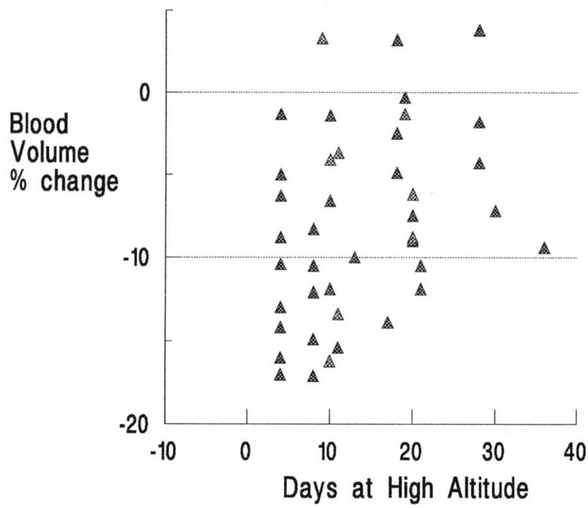

Figure 5 Blood volume expressed as percent change from sea level following ascent to altitudes of 3800–4500 m. Individual measurements. (From Refs. 3, 6, 8, 16–18, 119.)

month to be significant (Fig. 6). Very few data exist on the erythropoietic response to altitude exposures in excess of one month. In the classic investigation conducted by Reynafarje et al. (19,24) 40 years ago, 10 healthy young men in the military were assigned to the mining town of Morococha, Peru, at 5450 m for one year. As seen in Figure 6, the increase in RCV that begins late in the first month at high altitude (mean values for individual data in Fig. 4) continues progressively for about 8 months and then plateaus.

In 1964 Hannon et al. (25) began systematic studies of the hematological, fluid volume, and electrolyte responses of women exposed to high altitude. In that pilot study, they noticed that the rise in Hct and Hb concentration were slower and more modest than in men and thought this might be due to inadequate iron stores. The following year they conducted their first definitive (and now classic) investigation of altitude acclimatization in eight healthy young women taking supplemental iron during a 9-week sojourn at 4300 m on Pikes Peak, Colorado (26). In this study, the rise in Hb concentration and Hct were comparable to that observed in men (Fig. 2). Calculated group mean RCV decreased about 10% soon after ascent. During the first month, there was no increase in RCV, implying that the initial rise in Hct was due primarily to PV contraction. Over the next 5 weeks at 4300 m, RCV increased only 5% in spite of continued iron supplementation. With a sustained reduction in PV of more than 20%, BV after 2 months was still 16% less than at sea level (Fig. 7). Although Hannon et al. published these calculated decreases in RCV and BV in their preliminary report (25), they considered a true reduction of 10% in RCV "highly unlikely" and made no mention of RCV in their subsequent full publication

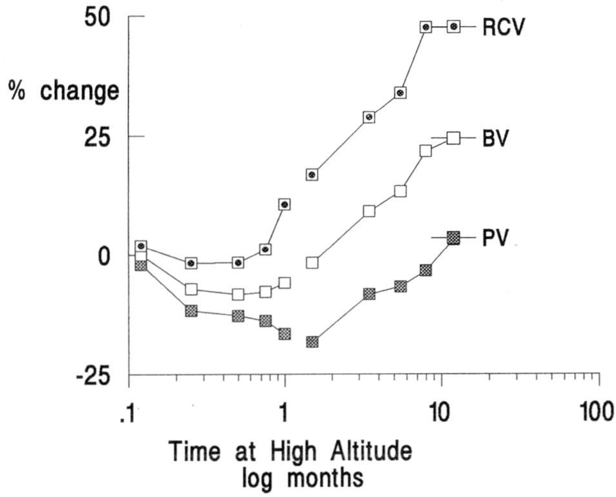

Figure 6 Percent change from sea level in red cell (RCV), plasma (PV), and blood volumes (BV) over one year following ascent to altitudes of 3800–4500 m. For clarity, a logarithmic scale of time at high altitude has been used. Mean values over the first month are derived from the individual data in Figs. 3, 4, and 5. (Data from 1.5 through 12 months are mean values from 10 individuals from Ref. 19.)

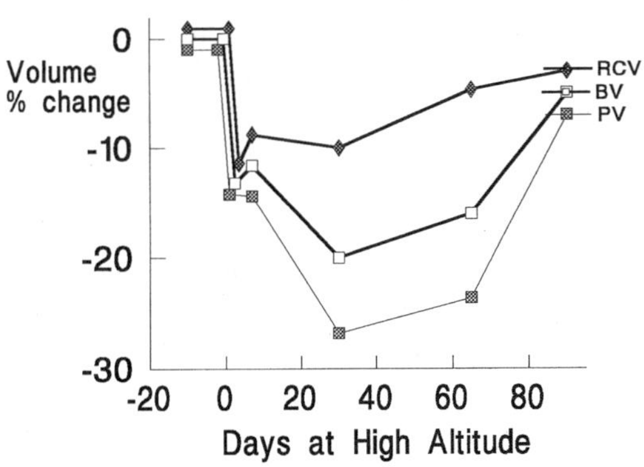

Figure 7 Red cell (RCV), plasma (PV), and blood volumes (BV) expressed as percent change from sea level. Mean values for 8 young women during a 2-month sojourn at 4300 m on Pikes Peak and 1 month following return to sea level. (From Ref. 26.)

(26). However, in a recent study of acclimatization to high altitude in men employing simultaneous measurements of PV using Evans Blue dye and BV using carbon monoxide (8), no systematic errors in PV at altitude were detected, lending credibility to the data collected by Hannon et al. (26). Furthermore, there is little basis for questioning the serial measurements made during the sojourn at high altitude, indicating a polycythemic response even slower than observed in humans. Thus, it appears that in the absence of vigorous iron supplementation (27), most individuals sojourning at high altitude will have an extended delay in realizing their potential increase in RCV.

Polycythemia and Hypoxemia

In the long-term resident at high altitude, there is a sustained augmentation of the total RCV resulting from hypoxic stimulation of erythropoiesis. To ascertain the relationship between the chronic hypoxic stimulus and the long-term response of RCV, Weil et al. (28) examined normal adult male residents of communities at three different altitudes in the United States; Los Angeles, California (sea level), Denver, Colorado (1600 m), and Leadville, Colorado (3100 m). Red cell volumes were measured using autologous chromium-51–labeled red cells. The three populations do in fact constitute a continuous spectrum of hematological and blood gas values. Arterial oxygen saturations ranged from 97.3% to 83.4%, arterial oxygen tensions from 96.0 to 46.5 torr, and RCVs from 22.4 to 41.8 mL/kg.

Data from these 73 individuals revealed a simple linear increase in RCV with decreasing arterial saturation (Fig. 8). However, the relationship between RCV and arterial oxygen tension was more complex, RCV clearly increasing only when oxygen tension fell below 60–70 torr (Fig. 9). Due to the sigmoid shape of the oxyhemoglobin dissociation curve, arterial saturation begins to fall rapidly when oxygen tension reaches this same range. Therefore, the linear correlation of RCV with saturation may be fortuitous and may not imply that arterial desaturation is the physiological stimulus to erythropoiesis, since there is no known biological sensor of saturation per se. At high altitude where ambient oxygen tension is reduced, arterial saturation is highly dependent on arterial oxygen tension, that in turn follows variations in ventilation, as during sleep. Hence, a single determination of arterial saturation may well not be a precise measure of the hypoxemic stimulus over 24 hours.

As with Hct, there is a broad range of normal values for RCV at all levels of arterial oxygenation (Fig. 8). In fact, in the same individuals, the range in RCV will be three times as great as the range in Hct since the latter is a ratio of RCV to the sum of RCV and PV.

Erythropoietin

Erythropoietin (EPO), originating primarily from the kidney, acts by retarding DNA breakdown characteristic of apoptosis and thereby promotes survival in

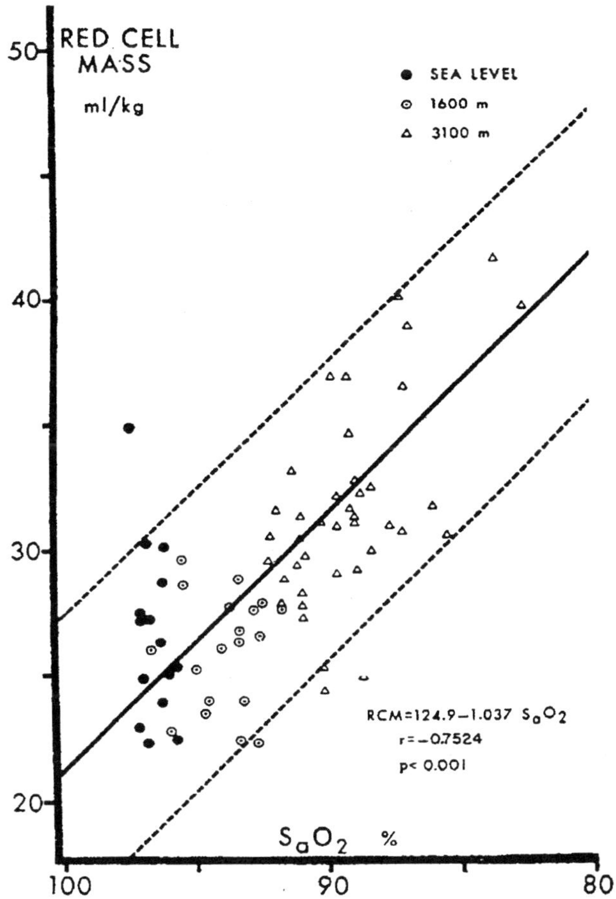

Figure 8 Linear increase in red cell mass with decreasing arterial oxygen saturation ($Sa\text{o}_2$) in adult males living at various altitudes in the United States. (From Ref. 25.)

erythroid progenitor cells in bone marrow (29) leading to greater red cell production. EPO production increases when oxygen tension in the renal cortex is reduced (30), as with exposure to atmospheric hypoxia. Following arrival at high altitude, the serum concentration of EPO rises progressively, reaching a peak on the third day, followed by a decline to near preascent levels within the second week (Fig. 10) (31).

The transient nature of the increase in serum EPO concentration is difficult to explain. Serum EPO concentration reflects the balance between EPO production by the kidney and EPO consumption by the bone marrow. On arrival at high altitude, EPO production exceeds consumption, and EPO concentration rises. However, after

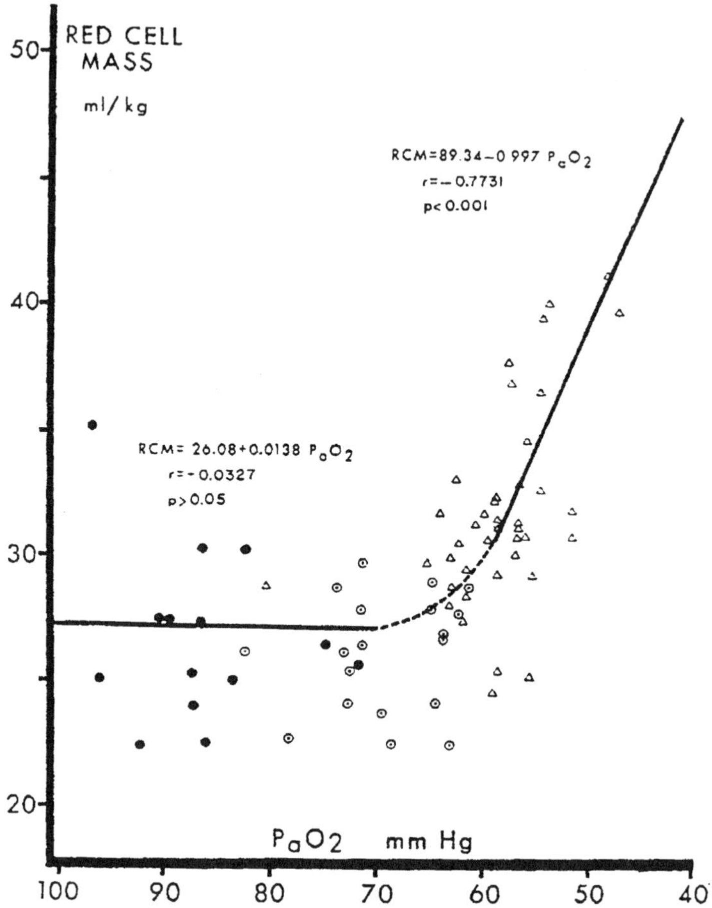

Figure 9 Increase in red cell mass only when arterial oxygen tension (Pao2) falls below about 65 mmHg in adult males living at various altitudes in the United States. (From Ref. 25.)

a latent period, increased erythroid activity leads to greater EPO consumption, resulting in a fall in EPO concentration. A new dynamic equilibrium is then established, manifest by an elevated rate of EPO turnover. Obviously measurements of serum EPO concentration alone will not reflect this sustained increase in EPO production, which maintains the enhanced level of erythropoiesis.

Among individuals ascending to high altitude, variability in the hyperpneic response is well recognized. As a consequence, the intensity of the hypoxic stimulus for EPO production must also be variable, and this may also contribute to the observed variation in the magnitude and rate of change in RCV (Fig. 4).

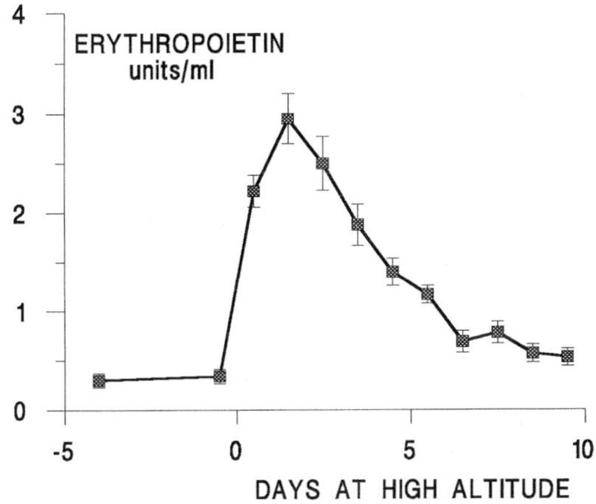

Figure 10 Transient increase in serum erythropoietin concentration during the first week at high altitude. (From Ref. 28.)

Hypoxic stimulation of the autonomic nervous system on arrival at high altitude is well recognized, and this may also be an added stimulus to EPO production. Plasma levels of epinephrine and norepinephrine increase progressively (32). Tachycardia is a familiar manifestation of sympathetic β-receptor stimulation. However, the specific role of β-receptor stimulation in augmenting EPO production is unclear. In the rabbit, β-receptor blockade greatly reduces the initial increase in serum EPO concentration during exposure to acute hypoxia (33). However, in humans, dense β-receptor blockade with propranolol had no apparent effect on the erythropoietic response to a sojourn at high altitude (8).

Iron supply is a major factor determining whether an increase in red cell production can occur in response to a rise in EPO concentration. The elevated red cell protoporphyrin levels observed at 4540 m altitude (24) provide evidence that iron supply is inadequate for the degree of EPO stimulation. At more moderate altitudes, vigorous oral iron supplementation may improve iron supply. When endurance athletes from sea level were taken to 2500 m for 4 weeks, those who were iron deficient (low plasma ferritin) failed to increase RCV. In contrast, athletes who received 200–400 mg elemental iron per day had a significant increase in RCV (27). In view of these findings, the more modest iron supplementation utilized by Hannon et al. (26) may have been inadequate for the menstruating young women who spent 9 weeks at 4300 m and who had only a minimal RCV response.

Oxygen-Hemoglobin Affinity

Because the HbO_2 dissociation curve is relatively flat at oxygen tensions above 65 torr, changes in Pa_{O_2} in this range have little influence on oxygen loading. For example, ascent to 3000 m altitude with lowering of Pao_2 from 95 to 65 torr produces a relatively small decrease in saturation to about 89%. Further ascent (further lowering of Pao_2) causes saturation to fall more rapidly because of the increasing steepness of the dissociation curve.

As saturation falls, so does arterial oxygen content, but this can be offset by increasing Hb concentration. For example, when saturation falls from 95% to 85%, if Hb concentration increases from 15.0 to 16.8 g/dL there will be no reduction in arterial oxygen content (19.4 vol%). Polycythemia accomplishes this.

2,3-Diphosphoglycerate

Various factors modify the affinity of hemoglobin for oxygen, including temperature and pH. Another factor is the red cell content of 2,3-diphospho glycerate (2,3 DPG). On exposure to high altitude, hypoxia stimulates ventilation and Pco_2 falls. This early respiratory alkalosis increases HbO_2 affinity, but also stimulates formation of red cell 2,3 DPG (34), which has the opposite effect. The result is a net decrease in HbO_2 affinity, i.e., the HbO_2 dissociation curve shifts to the right.

Oxygen Delivery

There has been much debate as to the physiological significance of increasing 2,3 DPG at altitude, and it is generally thought to be minimal (35). However, the effect on the coronary circulation may be noteworthy. Unlike most other tissues, the coronary circulation appears to be regulated to maintain constant myocardial oxygen tension under a wide variety of conditions (36). Operation of this homeostatic mechanism is reflected by a constant oxygen tension of 18 torr in coronary sinus (venous) blood. At sea level, the corresponding saturation is 34%. With the shift to the right of the dissociation curve at altitude caused by the increase in red cell 2,3 DPG, if saturation remained 34%, oxygen tension would rise. Regulation to prevent such a rise in tension requires lowering saturation to 22% (Fig. 11). At moderate altitude, where a right shift in the dissociation curve has minimal effect on oxygen loading, the increase in red cell 2,3 DPG would enhance oxygen extraction (widening of the coronary arterio-venous difference in oxygen content), permitting coronary blood flow to decrease. (22). However, at higher altitudes, where increased red cell 2,3 DPG would impair O_2 loading, this potential advantage would be lost. Direct measurements in healthy young adults confirmed a 25–30% decrease in coronary blood flow both at rest and during exercise at 3100 m altitude (37). The same mechanism prevents a fall in mixed venous oxygen tension (37) as cardiac output decreases with adaptation to moderate altitude.

Theoretically, increasing oxygen extraction without lowering venous oxygen tension by decreasing the affinity of hemoglobin for oxygen (raising P_{50}) could enhance oxygen delivery and actually elevate maximal oxygen consumption (Vo_2-max). In an animal model, such enhancement has been demonstrated with a large increase in P_{50} (38). However, in a model of total body oxygen transport, Vo_2max proved to be insensitive to shifts in P_{50} (35). Since the results of modeling and those from in vivo experiments and from intact humans do not agree, the question of

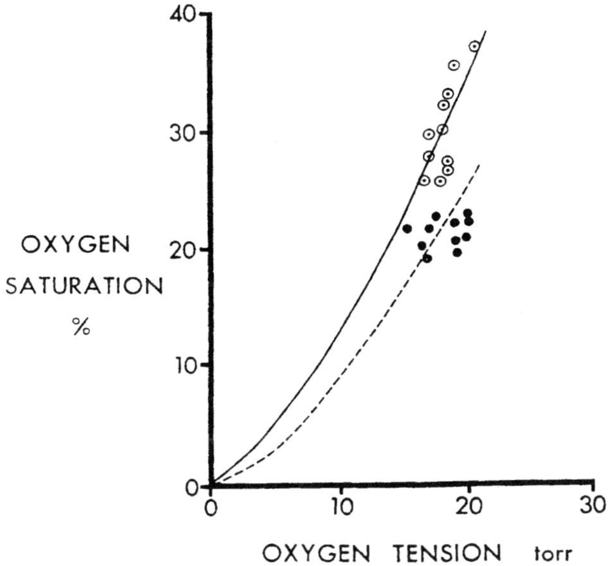

Figure 11 Lower end of the in vivo oxygen-hemoglobin dissociation curve. Simultaneous determinations of coronary sinus blood oxygen saturation and oxygen tension in three normal young men at sea level and after 10 days at 3100 m. The right shift at high altitude permits saturation to fall, i.e., extraction to increase, as oxygen tension is regulated at a constant value of about 18 torr. (From Ref. 34.)

significance in red cell 2,3 DPG and shifts in P_{50} remain unresolved. If the increases do prove to be significant in enhancing oxygen delivery, they could have important beneficial implications for patients with coronary artery disease visiting mountainous regions at moderate altitudes. This could also provide a rationale for Marticorena's taking patients to 3600 m altitude in the Andes for rehabilitation following recovery from myocardial infarct (39).

High-Altitude Residents

Since hypoxia is the basis for the erythropoietic response to high altitude, a correlation between the elevation above sea level and the magnitude of the long-term polycythemic response might be anticipated. This hypothesis has been examined in the Andes of Peru and Bolivia located near the equator, where climate permits long-term residence over a wide range of altitudes from sea level to 5000 m. A reasonable correlation does indeed exist between elevation above sea level (atmospheric hypoxia) and the mean Hb concentration of male residents of a number of Andean communities (Fig. 12) (40).

This relationship between altitude and Hb concentration may not apply equally to all human populations. Published data indicate that for a given altitude, i.e., an

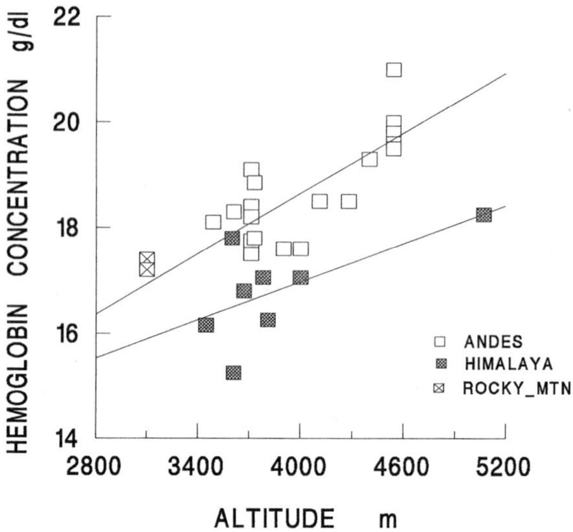

Figure 12 Correlation between the severity of the atmospheric hypoxic stimulus, i.e., altitude above sea level, and the polycythemic response expressed as mean hemoglobin concentration of adult males living in various high altitude communities. Twenty-one Andean samples, 8 Himalayan samples, and 2 samples from the Rocky Mountains. Simple regression lines are indicated. (From Refs. 37–39.)

equivalent hypoxic stimulus, Hb concentrations are lower in Himalayan populations than in Andean populations (Fig. 12). While this population difference may be genetic, such comparisons are difficult to interpret due to the broad range of normal (Fig. 1). Large population samples are required if they are to be representative. Further, factors other than altitude also influence blood composition, e.g., nutrition, parasitic infection, urban air pollution, even age, not to mention pulmonary disease or other causes of hypoventilation. Even the technique of blood sampling will modify the data, e.g., both upright posture and prior exertion elevate Hb concentration. Failure to consider these variables could result in population differences that are more apparent than real.

The hematological response of populations of European extraction to altitude residence is also of interest. We established normal values for blood composition in the population of Leadville, Colorado, situated at 3100 m in the Rocky Mountains. Residents were largely of Caucasian origin, although some were Hispanic. They had lived at this altitude from several years (nonnative) to three to four generations (native). One sample (41) consisted of school children 10–18 years of age, while our adult sample (42) ranged in age from 20 to 78 years. In all, data were obtained from a total of 456 individuals.

Figure 13 summarizes these data. In the adolescent population, Hct increased with age, with no sex difference prior to the onset of puberty. Thereafter, the rise

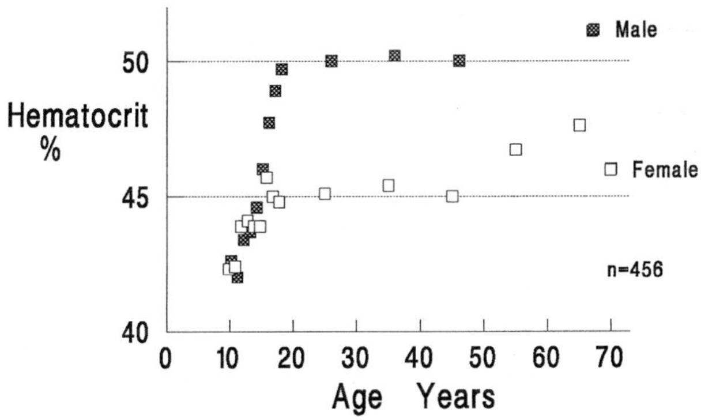

Figure 13 Comparison of mean Hct values between males and females at various ages living in Leadville, Colorado, at 3100 m. (From Refs. 38, 39.)

in Hct was accelerated in the males, reaching adult values by age 18. Among the adults, Hct remained stable; a slight increase beyond age 50 in both men and women was not statistically significant. This population did not exhibit the age-related increase in Hct observed at higher altitude in the Andes (43). Like at sea level, a broad range of normal was observed at all ages and in both sexes, but at every age mean values were clearly higher compared with sea level populations (Fig. 14) and also higher than Himalayan but not Andean populations at the same altitude (Fig. 12).

Excessive Polycythemia

Excessive polycythemia of high altitude, as in chronic mountain sickness (CMS, Monge's disease), appears to be an extension of this relationship between hypoxe-

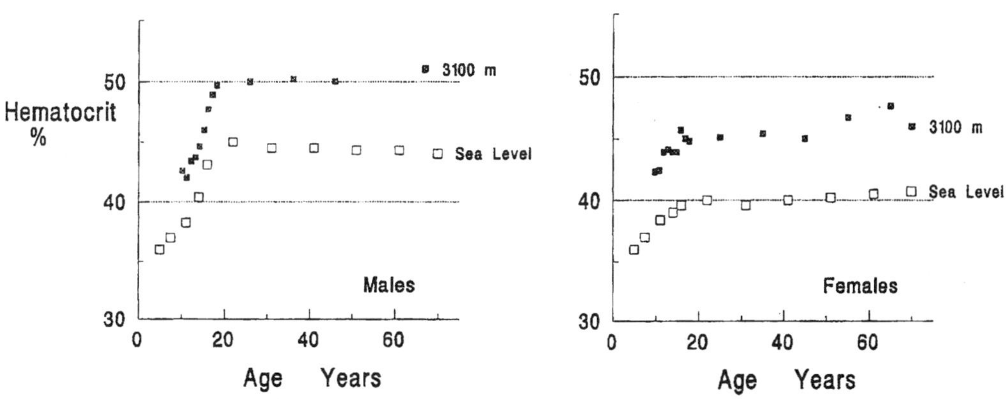

Figure 14 Comparison of mean Hct values for males and females living at sea level and 3100 m altitude in the United States. (From Refs. 1, 38, 39.)

mia and increased RCV (Fig. 8) (and Hct) (see Chapter 4). Among high-altitude residents, those with abnormally high Hcts usually have a greater degree of arterial desaturation. This hypoxemia may be occult, with daytime saturations being in the normal range. However, saturation may fall dramatically during sleep (44). Such nocturnal hypoventilation is not caused by loss of arterial hypoxic chemosensitivity, since many high-altitude residents exhibit a loss of hypoxic ventilatory response (HVR), but only a few hypoventilate (45). Rather, a low HVR permits hypoxemia to persist when it is caused by some other factor, such as obstructive sleep apnea. Chronic obstructive pulmonary disease (COPD) may also increase hypoxemia, but in our experience COPD is not the cause of hypoventilation in CMS (45). Consequently, men with CMS usually respond well to respiratory stimulation, as with medroxyprogesterone (Provera). In a study spanning several years, Kryger et al. (44) found that daily administration of Provera lowered arterial P_{CO_2}, raised arterial P_{O_2} and saturation, relieved symptoms, and produced a sustained reduction in RCV and Hct. If men with CMS are able to relocate to a lower altitude, this too relieves the excessive hypoxemic stimulus and Hct falls to normal levels (46).

C. Speculation

Ascent to high altitude lowers arterial oxygen saturation. Since increasing the oxygen-carrying capacity of the blood could offset this desaturation, it would seem to be beneficial in the overall process of acclimatization (see Chapter 9). Admittedly, this is teleological thinking since physiological processes do not require a purpose, only a mechanism. Walter B. Cannon spoke of the "wisdom of the body," observing that many physiological processes seem to operate to the overall benefit of the organism. However, the polycythemic response to chronic hypoxia may not be all that wise. At the heart of the problem is the fact that as Hct rises, blood viscosity also increases. While an increased Hb concentration may favor oxygen transport, the associated higher blood viscosity may negate that potential benefit by impairing blood flow (47).

At sea level, artificially raising Hb concentration and Hct increases maximum work capacity (V_{O_2}max). In young male track athletes, Buick et al. (48) employed reinfusion of previously stored autologous blood to raise Hb concentration from 15.1 to 16.5 mL/dL with an associated increase in Hct from 44% to 48%. This increase in the oxygen-carrying capacity of the blood produced an increase in V_{O_2}-max from 79.5 to 83.6 mL/kg/min. In terms of oxygen transport, then, raising Hct within the normal range is beneficial. In fact, because the effect is so significant, such "blood doping" for enhancing athletic performance is prohibited.

During acclimatization to high altitude, the increase in Hb concentration and Hct does seem to convey some benefit. When healthy young men spent 2 weeks on the summit of Pikes Peak at 4300 m altitude (49), Hct increased from 46% to 53%. Isovolemic reduction of Hct to 48% reduced arterial blood oxygen content by 10%

and lowered Vo_2max from 39.2 to 36.0 mL/kg/min Clearly, then, the higher Hct of 53% was beneficial in that aerobic working capacity was significantly impaired when Hct was lowered. However, still higher Hcts may not confer additional benefit. Sarnquist et al. (50) studied four mountaineers at 5400 m whose Hct had risen to 58%. By means of isovolemic hemodilution, Hct was lowered to 50%, but in spite of a 13% reduction in blood oxygen-carrying capacity, Vo_2max was not reduced significantly. This implies that at the higher Hct, blood flow was impaired, and presumably lowering Hct improved blood flow consequent to lower blood viscosity, thereby maintaining oxygen transport. Hence, the optimal Hct appears to be about 50–53%.

II. Blood Coagulation and Fibrinolysis

A. Introduction

A number of observations suggest that blood coagulation and fibrinolysis might be altered at high altitude, contributing to the pathophysiology of illnesses associated with acute or chronic exposure. Autopsies after death from high-altitude pulmonary edema (HAPE) or high-altitude cerebral edema show intravascular thrombi, hemorrhages, and extravascular fibrin deposition. Thromboembolism has been reported in healthy individuals at high altitude. In animals, induction of microembolism in the lungs causes a pulmonary edema similar to HAPE in distribution and composition of edema fluid. In addition, activation of blood coagulation is part of a response pattern of endothelial cells to severe hypoxia resulting in suppression of endothelial barrier function (51,52). Thus, these findings suggest that hypoxia may enhance blood clotting and thereby contribute to HAPE and possibly also to HACE by obstructing blood flow and increasing vascular permeability. Blood clotting can result from enhanced prothrombotic activity through activation of platelets or plasmatic coagulation (extrinsic or intrinsic pathway) or from impaired fibrinolysis.

In interpreting studies on hemostasis in vitro measures, such as platelet count, in vitro platelet aggregation or adhesion, coagulation times, plasma concentrations of coagulation factors or inhibitors of coagulation, and in vitro tests of fibrinolysis do not allow conclusions about coagulation and fibrinolysis in vivo. Over the last 10 years, a variety of very sensitive immunoassays have been developed (Fig. 15), which detect changes of the in vivo activity of coagulation and fibrinolysis even when these occur in the physiological range. These assays measure peptides that are released from activated platelets (β-thromboglobulin, BTG) or cleaved from coagulation factors when they are activated (prothrombin fragment 1 + 2, PTF 1 + 2, fibrinopeptide A, FPA) or neo-antigens that are exposed after proteolytic cleavage (fibrin degradation products, FBDP) or formed by binding of a factor to its inhibitor (thrombin-antithrombin III (TAT) complexes or plasmin-α2-antiplasmin (PAP) complexes). Unfortunately, a few studies performed at high altitude or in hypoxia have measured in vivo markers of hemostatic activity.

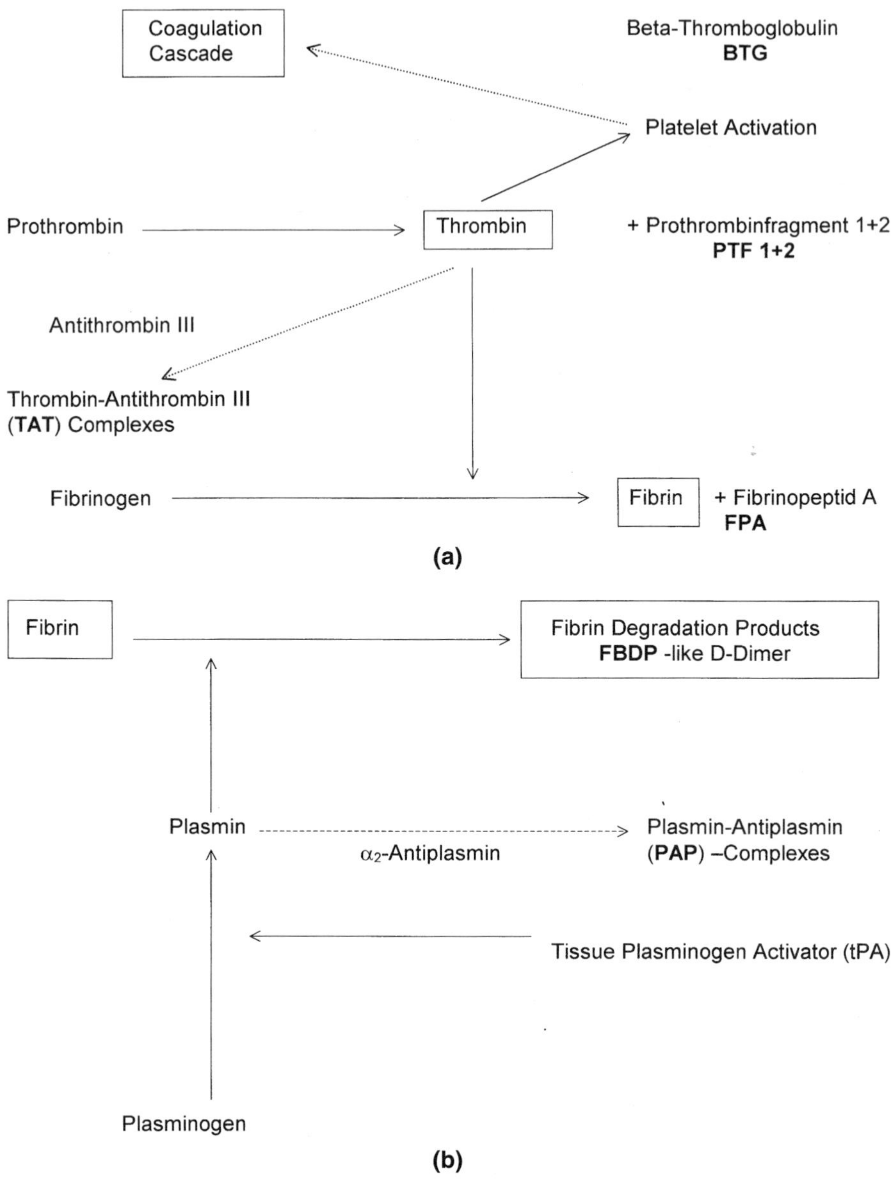

Figure 15 (a) Thrombin activation by prothrombinase complex after coagulation activation via intrinsic or extrinsic pathway. After enzymatic splitting of prothrombin by prothrombinase complex, 1 mol of thrombin and 1 mol of PTF 1+2 are formed. The main thrombin effect is conversion of fibrinogen to fibrin, leading to generation of FPA. Thrombin may activate blood platelets as well, which thereafter release BTG. Thrombin is neutralized by antithrombin III in vivo, leading to formation of TAT. (b) Plasminogen activators, which can be neutralized by plasminogen activator inhibitors, convert plasminogen to plasmin. Circulating plasmin is rapidly inactivated by α-2-antiplasmin, forming plasmin-antiplasmin (PAP) complexes. Plasmin bound to fibrin is inactivated at a several times slower rate by α-2-antiplasmin and will cleave fibrin(ogen) molecules into fibrin(ogen) degradation products.

B. Acute Exposure

Platelets

With two exceptions (53,54), most studies in humans showed no significant change of platelet counts on acute exposure to real or simulated altitudes as high as 4500 m for 12 hours to 4 days (55–59). Some of these studies included subjects susceptible to high-altitude pulmonary edema (HAPE) (56,57) and with acute mountain sickness (AMS) and early HAPE (56); none showed changes in platelet count.

In vitro tests of platelet function (53,54,57,60) and measurements of BTG (56), a marker in vivo platelet activation (Fig. 1), were unchanged compared to baseline after acute exposure to altitudes between 3700 and 4500 m in healthy individuals.

Brief exposure to a simulated altitude of 7000–8500 m (12–26 min, mean $Sa_{O_2} = 62\%$) caused slight significant immediate increase in BTG, which was sustained over 45 minutes after recompression (61). Reduced platelet count and platelet survival time lasting several days were reported after a 2-hour decompression to 6100 m, with hypobaria alone, without hypoxia (62). Interestingly, 3 hours of normoxic decompression from sea level to 380 torr (5500 m) of platelet-rich plasma in vitro increased platelet aggregation (63). Though of questionable relevance to human function at high altitude, these latter two studies imply an effect of acute barometric pressure change per se on platelet activity.

In summary, gradual ascent to altitudes of 4500 m has no significant effect on platelet count on most in vitro assays of platelet activity and markers of in vivo platelet activity, while rapid decompression and/or more severe hypoxia and advanced high-altitude pulmonary edema may enhance platelet activation or aggregation.

Coagulation

Healthy Individuals

Rapid exposure in hypobaric chambers from near sea level to altitudes of 4400 m (58), 4900 m (59), and 7000–8500 m (61) lead consistently to a shortening of activated partial thromboplastin time (aPTT), which can be attributed to a rise in factor VIII. These changes are caused by hypoxia, not hypobaria (59). Plasma levels of fibrinopeptide A (FPA) and of fibrin monomers, both markers of in vivo fibrin formation, did not change compared to baseline. A minor acute altitude challenge from 1600 m (level of residency) to 4150 m or a gradual exposure over 1 month to an altitude of 7620 m during Operation Everest II (OE II) did not result in changes in blood coagulation.

Examinations assessing markers of in vivo activity in individuals transported by railroad to an altitude of 3457 m (64) and in mountaineers without significant symptoms of AMS at 4559 m (56,65,66) revealed no activation of blood coagulation. Calves at 4300 m (67) and steers at 3850 m (68), but not lambs (69), had decreased clotting times as indirect evidence of hypercoagulable states, though markers of in vivo thrombin or fibrin formation were not determined. Calves also had decreased turnover of fibrinogen and platelets, but fibrin and fibrinogen degradation products were not increased (67). Steers and calves, but not lambs, developed excessive pulmonary hypertension (brisket disease) at high altitude.

Exercise per se can induce acceleration of coagulation times, increased plasma levels of factor VIII, and of markers of in vivo activity of thrombin and fibrinolysis (70–72). The exercise-induced increase of FVIII:C appears to be unaltered during gradual exposure up to altitudes of 7620 m (55). Nevertheless, minor but significant increases of FPA (64) occurred after short-term exhaustive exercise at 3457 m and of PTF1 + 2 (65) after 4 hours of moderate exercise at altitudes between 3600 m and 4500 m. Exercise-induced thrombin and fibrin formation might be enhanced by altitude, since comparable exercises at low altitude is not associated with increase of FPA (73) and PTF1 + 2 (72,74).

In summary, these data obtained in chamber and field study indicate that blood coagulation after more or less gradual exposure (over one to several days) to high altitude is not altered at rest. Sudden exposure in a hypobaric chamber leads to a minor increase of FVIII:C and a shortening of aPTT, which, however, do not give rise to increased thrombin or fibrin formation. The question whether exercise at high altitude can cause greater thrombin or fibrin generation than at low altitude has not been sufficiently explored.

AMS and HAPE

Do changes in hemostasis play a role in the evolution of AMS or HAPE? In cross-sectional studies performed at 3692 m increased platelet adhesiveness, plasma levels of fibrinogen, factor VIII, platelet factor 3, and decreased plasma levels of factor XII were found in subjects with HAPE of undetermined duration compared to healthy controls (75–77). As pointed out before, these measures do not allow firm conclusions about in vivo changes in coagulation. Three prospective studies focussing on in vivo activation during the first 3 days after rapid ascent on foot to 4559 m found no activation of coagulation either prior to or during the development of AMS and early HAPE (56,65,66). Exercise-induced changes of coagulation were also comparable between HAPE or AMS susceptibles and controls (65). Advanced cases of HAPE as documented by blood gas analysis and chest radiographs, however, did show in vivo platelet activation and thrombin, and fibrin formation (65,78), which normalized within 12–24 hours after return to low altitude (65). Prospective studies (56,66) imply that platelet activation and fibrin formation may be a consequence rather than a cause of HAPE. The findings with regard to coagulation obtained in advanced cases of HAPE are in accordance with the data reported from autopsies of HAPE victims, who showed thrombi in pulmonary vessels and intra-alveolar fibrin deposition (79,80). In summary, enhanced coagulation or impaired fibrinolysis do not accompany the development of AMS or HAPE, while advanced HAPE gives rise to increased fibrin formation.

Fibrinolysis

Healthy Individuals

Short exposure to normobaric hypoxia [F_{IO_2} = 0.13 for 30 min (81) or F_{IO_2} = 0.115 for 115 min (82) at rest] does not increase blood fibrinolytic activity. Exercise per se enhances fibrinolysis at sea level (70–72). An F_{IO_2} = 0.13 has no additional effect on exercise-induced activation of fibrinolysis (81,83). More severe normobaric hypoxia (F_{IO_2} = 0.10 for 15 min) (84) and prolonged exposure to a simulated altitude of 4150 m for 6–12 hours (57) was associated with an increased in vitro fibrinolytic activity as measured by a shortened euglobulin clot lysis time (ECLT), while exposure to simulated altitude of 7000–8500 m for 20 minutes increased degradation products of fibrinogen or fibrin (61). Rats also showed increased fibrinolytic activity in the aorta or caval vein after exposure to F_{IO_2} = 0.08 for 24 hours or F_{IO_2} = 0.10 for 6 weeks (85).

After passive ascent to 3692 m in vitro fibrinolytic activity was decreased in Himalayan soldiers (77,86) compared to low-altitude controls. A prospective investigation at 3450 m (64), however, found in vitro fibrinolytic activity unchanged compared to sea level. In two studies performed in mountaineers at 4559 m (56,78), parameters of in vivo and in vitro fibrinolytic activity were not significantly different from sea level baseline measurements when AMS and HAPE were absent. Two reports of measurements of plasma levels of D-dimer, a fibrin degradation product of cross-linked fibrin that is an indicator of in vivo fibrinolytic activity, yielded differing results. At an altitude of 3300 m, D-dimer levels were unchanged (87), while they were increased by 100–200% above baseline after 1–3 weeks at an altitude of 6452 m (88).

In summary, the data from laboratory and field studies indicate that acute exposure up to altitudes of 4500 m or to an F_{LO_2} = 0.12 has no significant effect on fibrinolysis, while fibrinolytic activity may be enhanced after exposure to elevations above 5000–6000 m or to an F_{IO_2} of 0.10 or less. In addition, exposure in a chamber may enhance in vitro fibrinolytic activity even at rather low altitudes (57) due to psychological stress, which is known to enhance ECLT (89).

AMS and HAPE

Although in vitro fibrinolytic activity assessed by ECLT is somewhat prolonged, indicating decreased fibrinolytic activity in mountaineers with AMS and HAPE compared to asymptomatic controls, markers of in vivo fibrinolytic activity are not different between groups (56). In vitro fibrinolytic activity was found to be decreased in subjects with HAPE at an altitude of 3692 m (77,86,90). A prospective investigation at an altitude of 4559 m revealed that decreased in vitro fibrinolytic activity, which could be attributed to increased plasma levels of plasminogen activator inhibitor (PAI-1), is a consequence of edema formation. It does not, however, necessarily indicate decreased fibrinolysis in vivo, as fibrin degradation products were greatly increased (56). Furthermore, the fibrinolytic potential assessed after 10 minutes of venous occlusion, which leads to release of tissue plasminogen activator and should reduce in vitro fibrinolysis (ECLT) by 50–75%, was normal in early HAPE. In summary, these data indicate that in vivo fibrinolytic activity is normal in individuals developing AMS or HAPE.

Mechanism of Activation of Coagulation

Increased plasma levels of vWF, which binds FVIII:C and tPA as well as shortening of aPTT and ECLT, are consistently reported to occur with physical (71) and mental stress (89). Most likely the rise of factor VIII accounts for the shortening of aPTT and the rise of tPA for the shortening of the ECLT. Release of vWF and tPA from endothelial cells are most likely a consequence of sympathoadrenergic activation, since it can be induced by infusion of epinephrine (91,92) and partly or completely blocked by propranolol (93,94). Conceivably the increased plasma levels of FVIII: C, of tPA antigen and the shortening of aPTT and ECLT with acute exposure to simulated altitude of 4000 m as well as similar changes in subjects developing AMS and HAPE at 4559 m, can be attributed to nonspecific stresses associated with chamber exposures or high-altitude illness. Indeed, plasma levels of epinephrine and norepinephrine (95), as well as urinary excretion of catecholamines (96), are increased in subjects with AMS compared to asymptomatic controls. Thus, increased in vitro activity of blood coagulation and fibrinolysis in AMS and beginning HAPE can be considered a nonspecific stress response that does not lead to increased fibrin formation and degradation.

Increased in vivo activity of coagulation with enhanced fibrin formation found in five advanced cases of HAPE could be explained by the effects of severe hypoxia and inflammation (78). Mean Pa_{O_2} in these patients was 20 mmHg with a range of 20–27 mmHg, and bronchoalveolar lavage in other individuals with advanced HAPE demonstrated an inflammatory reaction with accumulation of macrophages in the lung (97). Both severe hypoxemia and inflammation may lead to an activation of coagulation, as demonstrated in mice exposed to an $F_{I_{O_2}}$ of 0.06 (corresponding to an altitude of 9400 m). These animals developed thrombosis in the pulmonary vasculature and fibrin deposition in the lung (98,99). Mononuclear phagocytes appear to play a crucial role in this model as thrombosis and fibrin deposition could be abolished by depletion of these cells.

Thrombosis at High Altitude

While anecdotal clinical reports of complications compatible with thrombotic occlusions are not rare in the high-altitude literature, no epidemiological data are available addressing the prevalence of thromboembolic disease after acute or subacute exposure to high altitude. There are case reports of transient ischemic attacks (100), cerebral infarction (101), cerebral venous thrombosis (102,103), and pulmonary embolism (104,105). Other cases reported in the alpine literature have been summarized by Ward (106). In addition, thrombi in the brain and lung are common findings at autopsy in victims or pulmonary and cerebral edema (79,80,107).

From what we know of hemostasis at altitudes as high as 5000 m, hypoxia per se does not seem to present a special risk for thrombosis. The thrombosis seen in HAPE, and most likely also in HACE, at these altitudes is likely a late consequence of severe illness. Coagulation and fibrinolysis between 5000 and 8000 m have not been examined sufficiently to answer the question whether hypoxia at these altitudes, alone or in combination with exercise, increases the risk of thrombosis. The limited data from OEII speak against this possibility (55). Increased levels of D-dimer after several weeks at 6500 m are compatible with increased fibrin formation or increased fibrinolysis or both (88). Rats exposed to an $F_{I_{O_2}}$, corresponding to an altitude of 9100 m, clearly showed thrombosis and fibrin deposition in the lungs (108). Investigations on hemostasis in animal models and humans at relevant altitudes are needed to answer the question whether hypoxia and exercise increase the risk for thrombosis at very high altitudes.

In addition to a possible risk for thromboembolism by severe hypoxia special circumstances associated with exposure to high altitudes, such as dehydration due to increased fluid loss during exercise and insufficient fluid intake, inactivity during bad weather, and polycythemia, may be conducive to thromboembolic events in presumably healthy individuals without other risk factors. Individuals with insufficient inhibitors of coagulation such as protein C deficiency or resistance to activated protein C (e.g., factor V Leiden), which has a prevalence up to 2–5% (109), may be at increased risk for thromboembolism at high altitude.

C. Chronic Exposure

Almost no studies dealing with blood coagulation and fibrinolysis in high-altitude natives or residents are to be found in the literature, and there are none focusing on markers of in vivo activity assessing changes in hemostasis. However, in 1932 Hurtado (110) did report a shortening of clotting time in natives of the Peruvian Andes. The increase of hematocrit with altitude may enhance coagulation, and thus in vivo fibrin formation, since the risk for thrombosis in polycythemia is proportional to the increase of hematocrit (111). No data are available on the prevalence of venous thrombosis at different altitudes or in patients with chronic mountain sickness.

Indeed, thrombo-embolic events associated with atherosclerosis appear to be significantly less frequent in residents or natives at high altitude (112,113). This observation may be attributed to differences in lifestyle, diet, and genetics resulting in a low-risk profile for atherosclerosis. In contrast to these epidemiological findings, platelet adhesiveness in vitro was increased in permanent residents above 3000 m (114) and also during the first half year in temporary residents at 3658 m (115). Platelet count and platelet aggregation were also increased and in vitro fibrinolysis decreased (116) compared to reference groups living at low altitude, suggesting changes that would favor atherosclerosis. We have to consider, however, that in vitro tests do not necessarily reflect in vivo activity.

Singh and Chohan (86) suggested that the development of excessive pulmonary hypertension at altitudes of 3660–5490 m was due to occlusive pulmonary thrombosis resulting from increased platelet adhesiveness and elevated plasma levels of platelet factor 3 and coagulation factors V and VIII. Interestingly, administration of heparin was recently found to impair hypoxia-induced hyperplasia of smooth muscle cells of pulmonary arteries. This impairment effect could, however, be attributed to antiproliferative effects of heparin (117) and not to its anticoagulant properties (118). At present, there is no convincing evidence that increased coagulation or decreased fibrinolysis is associated with the development of abnormal pulmonary artery hypertension during prolonged stay at high altitude.

D. Conclusions and Recommendations

Rapid exposures to altitudes of 4000 m or higher in decompression chambers lead to an in vitro activation of platelets, plasmatic coagulation, and fibrinolysis, which does not, however, result in enhanced in vivo thrombin or fibrin formation. This activation is most likely a nonspecific stress reaction, because such changes can also be evoked by mental stress alone.

Gradual ascent up to altitudes of 4500 m in the mountains and 7620 m in a decompression chamber does not result in significant changes of platelet activation, plasmatic coagulation, or fibrinolysis. This is also true for individuals developing AMS or HAPE at 4559 m. Advanced cases of HAPE, however, show activated platelets and coagulation with enhanced in vivo fibrin formation and fibrin degradation. It is not clear whether this activation is caused by severe hypoxemia, inflammation, structural damage (stress failure), or a combination of all these factors, which may be associated with advanced HAPE.

Hemostasis at altitudes above 5000 m has not been studied sufficiently. Enhanced fibrin formation cannot be excluded after prolonged stay above 6000 m. Heavy prolonged exercise at low altitude leads to enhanced fibrin formation (and also increased fibrinolysis). Thus, the question relevant for those going to extreme altitudes is whether extreme altitude and vigorous exercise act synergistically on hemostasis. Another question of clinical importance concerns individuals with a deficiency of inhibitors of coagulation like protein-C, protein-S those with factor V Leiden, or those who are on drugs that may enhance coagulation, like oral contraceptives. Are some of these individuals at an increased risk for thrombosis at high altitude, and if so, at which altitude does the risk become relevant? Furthermore, our knowledge should be improved about the effects of chronic altitude exposure on hemostasis at rest and exercise in healthy individuals and in those with diseases like pulmonary hypertension and chronic mountain sickness.

Further investigations that address such questions should focus on markers that reflect in vivo activity when careful attention is paid to minimize activation of platelets and coagulation with blood sampling and sample handling. Clean venipunctures, prechilled tubes, refrigerated centrifuges, and liquid nitrogen or dry ice are prerequisites for such examinations. This may explain why data from extreme environments are sparse and will probably remain so.

References

1. Fulwood R, Johnson CL, Bryner JD, Gunter EW, McGrath CR. Hematological and nutritional biochemistry reference data for persons 6 months-74 years of age: United States, 1976–80. DHHS Publication No. (PHS) 83-1682. Bethesda, MD: USPHS, DHHS, National Center for Health Statistics, 1982.
2. Douglas CG, Haldane JS, Henderson Y, Schneider EC. Physiological observations made on Pike's Peak, Colorado, with special reference to adaptation to low barometric pressure. Phil Trans R Soc London, Ser B 1913; 203:185–318.
3. Dill DB, Horvath SM, Dahms TE, Parker RE, Lynch JR. Hemoconcentration at altitude. J Appl Physiol 1969; 27:514–518.
4. Johnson RL, Jr. Rate of red cell and hemoglobin destruction after descent from high altitude. Final Report Contract AF 41609-68-C-0032. Brooks Air Force Base, Texas: USAF School of Aerospace Medicine, 1968.
5. Hannon JP, Chinn KSK, Shields JL. Effects of acute high-altitude exposure on body fluids. Fed Proc 1969; 28:1178–1184.
6. Wolfel EE, Groves BM, Brooks GA, Butterfield GE, Mazzeo RS, Moore LG, Sutton JR, Bender PR, Dahms TE, McCullough RE, McCullough RG, Huang S-Y, Sun S-F, Grover RF, Hultgren HN, Reeves JT. Oxygen transport during steady-state submaximal exercise in chronic hypoxia. J Appl Physiol 1991; 70:1129–1136.
7. Butterfield GE, Gates J, Fleming S, Brooks GA, Sutton JR, Reeves JT. Increased energy intake minimizes weight loss in men at high altitude. J Appl Physiol 1992; 72: 1741–1748.
8. Grover RF, Selland MA, McCullough RG, Dahms TA, Wolfel EE, Butterfield GE, Reeves JT, Greenleaf JE. Beta-adrenergic blockade does not prevent polycythemia or decrease in plasma volume in men to 4300 m altitude. Eur J Appl Physiol 1998; 77: 264–270.
9. Hoyt RW, Honig A. Body fluid and energy metabolism at high altitude. In: Fregly MJ, Blattes CM, eds. Handbook of Physiology: Section 4: Environmental Physiology. V. The Terrestrial Altitude Environment. New York: Oxford University Press, 1996: 1277–1289.
10. Wolfel EE, Selland MA, Mazzeo RS, Reeves JT. Systemic hypertension at 4300 m is related to sympathoadrenal activity. J Appl Physiol 1994; 76:1643–1650.
11. Claybaugh JR, Brooks DP, Cymernan A. Hormonal control of fluid and electrolyte balance at high altitude in normal subjects. In: Sutton JR, Coates G, Houston CS, eds. Hypoxia and Mountain Medicine. Oxford: Pergamon Press, 1992:61–72.
12. Lawrence DL, Skatrud JB, Shenker Y. Effect of hypoxia on atrial natriuretic factor and aldosterone regulation in humans. Am J Physiol (Endocrinol Metab 21) 1990; 258: E243–E248.
13. Albert TSE, Tucker VL, Renkin EM. Acute alveolar hypoxia increases blood-to-tissue albumin transport: role of atrial natriuretic peptide. J Appl Physiol 1997; 82:111–117.
14. Albert TSE, Tucker VL, Renkin EM. Atrial natriuretic peptide levels and plasma volume contraction in acute alveolar hypoxia. J Appl Physiol 1997; 82:102–110.
15. Lawrence JH, Huff RL, Siri W, Wasserman LR, Hennessey TG. A physiological study in the Peruvian Andes. Acta Med Scand 1952; 142:117–131.

16. Faura J, Reynafarje C. Uso de le metandrostenolona para acelerar la eritropoyesis en la exposicion a la altura. Arch Inst Biol Andina (Lima, Peru) 1970; 3:94–98.
17. Surks MI, Chinn KK, Matoush LO. Alterations in body composition in man after acute exposure to high altitude. J Appl Physiol 1966; 21:1741–1746.
18. Myhre LG, Dill DB, Hall FG, Brown DK. Blood volume changes during three-week residence at high altitude. Clin Chem 1970; 16:7–14.
19. Reynafarje C The influence of high altitude on erythropoietic activity. Brookhaven Symp Biol 1957; 10:132–146.
20. Reynafarje C, Berlin NI, Lawrence JH. Red cell life span in acclimatization to high altitude. Proc Soc Exp Biol Med 1954; 87:101–104.
21. Alexander JK, Grover RF. Mechanism of reduced cardiac stroke volume at high altitude. Clin Cardiol 1983; 6:301–303.
22. Alexander JK, Hartley LH, Modelski M, Grover RF. Reduction of stroke volume during exercise in man following ascent to 3,100 m altitude. J Appl Physiol 1967; 23: 849–858.
23. Faura J, Ramos J, Reynafarje C, English E, Finne P, Finch CA. Effect of altitude on erythropoiesis. Blood 1969; 33:668–676.
24. Reynafarje C, Lozano R, Valdivieso J. The polycythemia of high altitude: iron metabolism and related aspects. Blood 1959; 14:433–455.
25. Hannon JP, Shields JL, Harris CW. High altitude acclimatization in women. In: Goddard RF, ed. The Effects of Altitude on Physical Performance. Albuquerque, NM: The Athletic Institute, Lovelace Foundation, 1967:37–44.
26. Hannon JP, Shields JL, Harris CW. Effects of altitude acclimatization on blood composition of women. J Appl Physiol 1969; 26:540–547.
27. Stray-Gundersen J, Levine BD. Iron requirements during altitude training in endurance athletes. Physiologist 1996; 39(5):A4.
28. Weil JV, Jamieson G, Brown DW, Grover RF. The red cell mass—arterial oxygen relationship in normal man. J Clin Invest 1968; 47:1627–1639.
29. Silva M, Grillot D, Benito A, Richard C, Nunez G, Fernandez-Luna JL. Erythropoietin can promote erythroid progenator survival by repressing apoptosis through Bcl-XL and Bcl-2. Blood 1996; 88:1576–1582.
30. Jones RM, Terhaard C, Zullo J, Tenney SM. Mechanism of reduced water intake in rats at high altitude. Am J Physiol 1981; 240:R187–R191.
31. Abbrecht PH, Littell JK. Plasma erythropoietin in men and mice during acclimatization to different altitudes. J Appl Physiol 1972; 32:54–58.
32. Mazzeo RS, Wolfel EE, Butterfield GE, Reeves JT. Sympathetic response during 21 days at high altitude (4,300 m) as determined by urinary and arterial catecholamines. Metabolism 1994; 43:1226–1232.
33. Fink GD, Paulo LG, Fisher JW. Effects of beta adrenergic blocking agents on erythropoietin production in rabbits exposed to hypoxia. J Pharmacol Exp Ther 1975; 193: 176–181.
34. Lenfant C, Torrance JD, Reynafarje C. Shift of the O_2-dissociation curve at altitude: mechanism and effect. J Appl Physiol 1971; 30:625–631.
35. Wagner PD. Insensitivity of VO_{2max} to hemoglobin-P50 at sea level and altitude. Respiration Physiol 1997; 107:205–212.
36. Duvelleroy MA, Mehmel H, Laver MB. Hemoglobin-oxygen equilibrium and coronary blood flow; an analog model. J Appl Physiol 1973; 35:480–484.

37. Grover RF, Lufschanowski R, Alexander JK. Alterations in the coronary circulation of man following ascent to 3,100 m altitude. J Appl Physiol 1976; 41:832–838.
38. Richardson RS, Tagore K, Haseler LJ, Jordan M, Wagner PD. Increased VO_{2max} with right-shifted Hb-O_2 dissociation curve at constant O_2 delivery in dog muscle in situ. J Appl Physiol 1998; 84:995–1002.
39. Marticorena EA, Marticorena JM. Prevencion y rehabilitacion coronaria utilizando las grandes alturas/Coronary prevention and rehabilitation utilizing high altitudes. Lima, Peru: Universidad Nacional Mayor de San Marcos, 1990.
40. Beall CM, Brittenham GM, Macuaga F, Barragan M. Variation in hemoglobin concentration among samples of high-altitude natives in the Andes and the Himalayas. Am J Human Biol 1990; 2:639–651.
41. Treger A, Shaw DB, Grover RF. Secondary polycythemia in adolescents at high altitude. J Lab Clin Med 1965; 66:304–314.
42. Okin JT, Treger A, Overy HR, Weil JV, Grover RF. Hematologic response to medium altitude. Rocky Mtn Med J 1966; 63:44–47.
43. Whittembury J, Monge-C C. High altitude, age, and hematocrit. Nature (London) 1972; 238:278–279.
44. Kryger M, Glas R, Jackson D, McCullough RE, Scoggin C, Grover RF, Weil JV. Impaired oxygenation during sleep in excessive polycythemia of high altitude: improvement with respiratory stimulation. Sleep 1978; 1:3–17.
45. Kryger M, McCullough R, Doekel R, Collins D, Weil JV, Grover RF. Excessive polycythemia of high altitude: role of ventilatory drive and lung disease. Am Rev Respir Dis 1978; 118:659–666.
46. Grover RF, Kryger MH. Polycythemia of chronic mountain sickness. In: Chamberlayne EC, Condliffe PG, eds. Adjustment to High Altitude. NIH Publication No. 83-2496. Bethesda, MD: USDHHS, PHS, NIH, 1983:37–41.
47. Thomas DJ, DuBoulay GH, Marshall J, et al. Effect of haematocrit on cerebral bloodflow in man. Lancet 1977; ii:941–943.
48. Buick FJ, Gledhill N, Froese AB, Spriet L, Meyers EC. Effect of induced erythrocythemia on aerobic work capacity. J Appl Physiol 1980; 48:636–642.
49. Horstman D, Weiskopf R, Jackson RE. Work capacity during 3-wk sojourn at 4,300 m: effects of relative polycythemia. J Appl Physiol 1980; 49:311–318.
50. Sarnquist FH, RB Schoene, PH Hackett, BD Townes. Hemodilution of polycythemic mountaineers: effects on exercise and mental function. Aviat Space Environ Med 1986; 57:313–317.
51. Pinsky DJ, Yan SF, Lawson C, Naka Y, Chen JX, Connolly ES Jr, Stern DM. Hypoxia and modification of the endothelium: implications for regulation of vascular homeostatic properties. Semin Cell Biol 1995; 6:283–294.
52. Yan SF, Ogawa S, Stern DM, Pinsky DJ. Hypoxia-induced modulation of endothelial cell properties: regulation of barrier function and expression of interleukin-6. Kidney Int 1997; 51:419–425.
53. Chatterji JC, Ohri VC, Das BK, Chadha KS, Akhtar M, Bhatacharji P, Tewari SC, Behl A. Platelet count, platelet aggregation and fibrinogen levels following acute induction to high altitude (3200 and 3771 metres). Thromb Res 1982; 26:177–182.
54. Sharma SC. Platelet count on acute induction to high altitude. Thromb Haemost 1980; 43:24.
55. Andrew M, O'Brodovich H, Sutton J. Operation Everest II: coagulation system during prolonged decompression to 282 torr. J Appl Physiol 1987; 63:1262–1267.

56. Bärtsch P, Haeberli A, Franciolli M, Kruithof EKO, Straub PW. Coagulation and fibrinolysis in acute mountain sickness and beginning pulmonary edema. J Appl Physiol 1989; 66:2136–2144.
57. Hyers TM, Scoggin CH, Will DH, Grover RF, Reeves JT. Accentuated hypoxemia at high altitude in subjects susceptible to high-altitude pulmonary edema. J Appl Physiol 1979; 46:41–46.
58. Maher JT, Levine PH, Cymerman A. Human coagulation abnormalities during acute exposure to hypobaric hypoxia. J Appl Physiol 1976; 41:702–707.
59. O'Brodovich HM, Andrew M, Gray GW, Coates G. Hypoxia alters blood coagulation during acute decompression in humans. J Appl Physiol 1984; 56:666–670.
60. Sharma SC, Singh Hoon R. Platelet adhesiveness on acute induction to high altitude. Thromb Res 1978; 13:725–732.
61. Bärtsch P, Haeberli A, Hauser K, Gubser A, Straub PW. Fibrinogenolysis in the absence of fibrin formation in severe hypobaric hypoxia. Aviat Space Environ Med 1988; 59:428–432.
62. Gray GW, Bryan AC, Freedman MH, Houston CS, Lewis WF, McFadden DM, Newell G. Effect of altitude exposure on platelets. J Appl Physiol 1975; 39:648–652.
63. Murayama M. Ex vivo human platelet aggregation induced by decompression during reduced barometric pressure, hydrostatic, and hydrodynamic (Bernoulli) effect. Thromb Res 1984; 83:477–485.
64. Bärtsch P, Schmidt EK, Straub PW. Fibrinopeptide A after strenuous physical exercise at high altitude. J Appl Physiol 1982; 53:40–43.
65. Bärtsch P, Haeberli A, Nanzer A, et al. High altitude pulmonary edema: blood coagulation. In: Sutton JR, Houston CS, Coates G, eds. Hypoxia and Molecular Medicine. Burlington: Queen City Printers Inc, 1993:252–258.
66. Bärtsch P, Lämmle B, Huber I, Haeberli A, Vock P, Oelz O, Straub PW. Contact phase of blood coagulation is not activated in edema of high altitude. J Appl Physiol 1989; 67:1336–1340.
67. Genton E, Ross AM, Takeda YA, Vogel JHK. Alterations in blood coagulation at high altitude. Adv Cardiol 1970; 5:32–40.
68. Grover RF, Reeves JT, Will DH, Blount SGJ. Pulmonary vasoconstriction in steers at high altitude. J Appl Physiol 1963; 18:567–574.
69. Reeves JT, Grover EB, Grover RF. Pulmonary circulation and oxygen transport in lambs at high altitude. J Appl Physiol 1963; 18:560–566.
70. Bärtsch P, Welsch B, Albert M, Friedmann B, Levi M, Kruithof EKO. Balanced activation of coagulation and fibrinolysis after a 2-h triathlon. Med Sci Sports Exerc 1995; 27:1465–1470.
71. Bourey RE, Santoro SA. Interactions of exercise, coagulation, platelets, and fibrinolysis—a brief review. Med Sci Sports Exerc 1988; 20:439–446.
72. Weiss C, Seitel G, Bärtsch P. Coagulation and fibrinolysis after moderate and very heavy exercise in healthy male subjects. Med Sci Sports Exerc 1998; 30:246–251.
73. Vogt A, Hofmann V, Straub PW. Lack of fibrin formation in exercise-induced activation of coagulation. Am J Physiol 1979; 236:H577–H579.
74. Herren T, Bärtsch P, Haeberli A, Straub PW. Increased thrombin—antithrombin III complex after 1 h of physical exercise. J Appl Physiol 1992; 73:2499–2504.
75. Singh I, Chohan IS. Abnormalities of blood coagulation at high altitude. Int J Biometeorol 1972; 16:283–297.

76. Singh I, Chohan IS. Reversal of abnormal fibrinolytic activity, blood coagulation factors and platelet function in high-altitude pulmonary oedema with frusemide. Int J Biometeorol 1973; 17:73–81.
77. Singh I, Chohan IS. Adverse changes in fibrinolysis, blood coagulation and platelet function in high altitude pulmonary oedema and their role in its pathogenesis. Int J Biometeorol 1974; 18:33–45.
78. Bärtsch P, Waber U, Haeberli A, Maggiorini M, Kriemler S, Oelz O, Straub WP. Enhanced fibrin formation in high-altitude pulmonary edema. J Appl Physiol 1987; 63:752–757.
79. Dickinson J, Heath D, Gosney J, Williams D. Altitude-related deaths in seven trekkers in the Himalayas. Thorax 1983; 88:646–656.
80. Nayak NC, Roy S, Narayanan TK. Pathologic features of altitude sickness. Am J Pathol 1964; 45:381–391.
81. Stegnar M, Peternel P, Chen JP. Acute hypoxemia does not increase blood fibrinolytic activity in man. Thromb Res 1987; 45:333–343.
82. Cunningham GM, Boyd G, Windebank J, Moran F, McNicol GP. The effect of oxygen on the fibrinolytic enzyme system in vivo. J Clin Pathol 1971; 24:705–707.
83. Stegnar M, Dolzan N, Hojnik M, Peternel P. The effect of hypoxia during exercise on release of tissue-type plasminogen activator (T-Pa). Fibrinolysis 1990; 4 (suppl 2): 108–109.
84. Mangum M, Venable RH, Boatwright JD, Cocke TB. Hypoxia: A stimulus for tissue plasminogen activator release in humans? Aviat Space Environ Med 1987; 58:1093–1096.
85. Risberg B, Stenberg B. Modulation of tissue fibrinolysis from hypoxia and hyperoxia. Thromb Res 1985; 38:129–136.
86. Singh I, Chohan IS. Blood coagulation changes at high altitude predisposing to pulmonary hypertension. Br Heart J 1972; 34:611–617.
87. Bärtsch P, Shaw S, Franciolli M, Gnädinger MP, Weidmann P. Atrial natriuretic peptide in acute mountain sickness. J Appl Physiol 1988; 65:1929–1937.
88. Roeggla G, Roeggla M, Binder M, Wagner A. Does altitude or exercise induce fibrin degradation in mountaineers? J R Soc Med 1995; 889:239.
89. Le Roux G, Larmignat P, Marchal M, Richalet J-P. Haemostasis at high altitude. Int J Sports Med 1992; 13:S49–S51.
90. Jern C, Eriksson E, Tengborn L, Risberg B, Wadenvik H, Jern S. Changes of plasma coagulation and fibrinolysis in response to mental stress. Thromb Haemost 1989; 62: 767–771.
91. Chohan IS, Singh I, Balakrihsnan K. Fibrinolytic activity at high altitude and sodium acetate buffer. Thrombos Diathes Haemorrh 1974; 32:65–70.
92. Biggs R, Macfarlane RG, Pilling J. Observations on fibrinolysis. Experimental activity produced by exercise or adrenaline. Lancet 1947; 402–405.
93. Cash JD, Woodfield DG, Allan AGE. Adrenergic mechanisms in the systemic plasminogen activator response to adrenaline in man. Br J Haematol 1970; 18:487–494.
94. Ingram GIC, Vaughan Jones R. The rise in clotting factor VIII induced in man by adrenaline: effect of alpha- and beta-blockers. J Physiol 1966; 187:447–454.
95. Cohen RJ, Epstein SE, Cohen LS, Dennis LH. Alterations of fibrinolysis and blood coagulation induced by exercise, and the role of beta-adrenergic-receptor stimulation. Lancet 1968; 11:1264–1266.
96. Hoon RS, Sharma SC, Balasubramanian V, Chadha KS, Mathew OP. Urinary catechol-

amine excretion on acute induction to high altitude (3,658 m). J Appl Physiol 1976; 41:631–633.
97. Schoene RB, Swenson ER, Pizzo CJ, Hackett PH, Roach RC, Mills WJ, Henderson WR, Martin TR. The lung at high altitude: bronchoalveolar lavage in acute mountain sickness and pulmonary edema. J Appl Physiol 1988; 64:2605–2613.
98. Lawson CA, Yan SD, Yan SF, Liao H, Zhou YS, Sobel J, Kisiel W, Stern DM, Pinsky DJ. Monocytes and tissue factor promote thrombosis in a murine model of oxygen deprivation. J Clin Invest 1997; 99:1729–1738.
99. Yan SF, Lawson CA, Stern DM, Pinsky DJ. Hypoxia-mediated modulation of vascular function—Implications for organ preservation and thrombogenesis: Roger S. Mitchell Lecture. Chest 1998; 114:S46–S50.
100. Wohns RN. W. Transient ischemic atacks at high altitude. In: Sutton JR, Houston CS, Coates G, eds. Hypoxia and Cold. New York: Praeger, 1987:536.
101. Clarke CRA. Cerebral infarction at extreme altitude. In: Sutton JR, Houston CS, Jones NL, eds. Hypoxia, Exercise and Altitude. New York: Alan R. Liss, Inc., 1983:453.
102. Fujimaki T, Matsutani M, Asai A, Kohno T, Koike M. Cerebral venous thrombosis due to high-altitude polycythemia. J Neurosurg 1986; 64:148–150.
103. Song S-Y, Asaji T, Tanizaki Y, Fujimaki T, Matsutani M, Okeda R. Cerebral thrombosis at altitude: Its pathogenesis and the problems of prevention and treatment. Aviat Space Environ Med 1986; 57:71–76.
104. Shlim DR, Papenfus K. Pulmonary embolism presenting as high-altitude pulmonary edema. Wild Environ Med 1995; 6:220–224.
105. Wickwire J. Pulmonary embolus and/or pneumonia on K-2. In: Sutton JR, Jones NL, Houston CS, eds. Hypoxia: Man at Altitude. New York: Thieme-Stratton, 1982:173–176.
106. Ward M. Mountain Medicine—A Clinical Study of Cold and High Altitude. London: Crosby Lockwood Staples, 1975.
107. Houston CS, Dickinson J. Cerebral form of high-altitude illness. Lancet 1975; ii:758–761.
108. Viswanathan R, Jain SK, Subramanian S, Puri BK. Pulmonary edema of high altitude. I. Production of pulmonary edema in animals under conditions of simulated high altitude. Am Rev Respir Dis 1969; 100:327–333.
109. Koster T, Rosendaal FR, de Ronde H, Briët E, Vandenbroucke JP, Bertina RM. Venous thrombosis due to poor anticoagulant response to activated protein C: Leiden Thrombophilia Study. Lancet 1993; 342:1503–1506.
110. Hurtado A. Studies at high altitude. Blood observations on Indian natives of the Peruvian Andes. Am J Physiol 1932; 100:487–505.
111. Pearson TC, Wetherley-Mein G. Vascular occlusive episodes and venous haematocrit in primary proliferative polycythaemia. Lancet 1978; ii:1219–1222.
112. Arias-Stella J, Topilsky M. Anatomy of the coronary circulation at high altitude. In: Porter R, Knight J, eds. High Altitude Physiology: Cardiac and Respiratory Aspects. Edinburgh: Churchill Livingstone, 1971:149.
113. Mortimer EA Jr, Monson RR, MacMahon B. Reduction in mortality from coronary heart disease in men residing at high altitude. N Engl J Med 1977; 296:581–585.
114. Sharma SC, Balasubramanian V, Chadha KS. Platelet adhesiveness in permanent residents of high altitude. Thromb Haemost 1979; 42:1508–1512.
115. Sharma SC. Platelet adhesiveness on temporary residents of high altitude. Thromb Res 1981; 21:685–687.

116. Caen JP, Drouet L, Bellanger R, Michel H, Hénon P. Thrombosis, platelet behaviour, fibrinolytic activity, and diet on the Andes plateau. Haemostasis 1973; 2:13–20.
117. Thompson BT, Spence CR, Janssens SP, Joseph PM, Hales CA. Inhibition of hypoxic pulmonary hypertension by heparins of differing in vitro antiproliferative potency. Am J Respir Crit Care Med 1994; 149:1512–1517.
118. Hassoun PM, Thompson BT, Steigman D, Hales CA. Effect of heparin and warfarin on chronic hypoxic pulmonary hypertension and vascular remodeling in the guinea pig. Am Rev Respir Dis 1989; 139:763–768.
119. Jung RC, Dill DB, Horton R, Horvath SM. Effects of age on plasma aldosterone levels and hemoconcentration at altitude. J Appl Physiol 1971; 81:593–597.

16

Renal Function and Fluid Homeostasis

ERIK R. SWENSON

University of Washington School of Medicine
Seattle, Washington

I. Introduction

For many decades, scientific opinion and alpine folklore have claimed that those who acclimatized to high altitude had increased urination (*Höhendiurese*), whereas those who fared poorly exhibited oliguria and edema, suggesting a critical role of the kidneys and fluid balance in transition to an hypoxic environment. Yet even 70 years since the first reports on fluid balance at high altitude, the effects of hypoxia on salt and water homeostasis remain poorly elucidated. Hypoxia can directly affect the kidneys, but, more importantly, its effects on systemic acid-base status, ventilation, neuroendocrine reflexes, and hemodynamics all play a far greater part in altering renal function and fluid balance. Other features of high altitude separate from hypoxia, either alone or in combination, including hypobaria, exercise, and cold, may also significantly perturb renal function.

The aim of this chapter is to describe the alterations in fluid balance at high altitude, focusing primarily on the renal and hormonal responses to acute and chronic hypoxia. The mechanisms and sites at which hypoxia evoke changes in fluid balance and renal function will be discussed. Acute responses are defined as those that occur in the first hours to few days, and chronic responses are those evolving over many days to years. These responses will be contrasted to those developing with more severe hypoxia, although the reader will find greater discussion of these details in the chapters relating to high-altitude illness (Chapters 22–24).

II. Effects of Hypoxia on Salt and Water Balance

A. Renal Function

Urinary Output

A blurred dose-response relationship is found in humans between urine output and sodium excretion with acute hypoxia (Fig. 1). Diuresis and natriuresis occur with mild hypoxia, reaching a maximum between F_{IO_2} of 0.12–0.14. In the range of 0.10–0.12, urine volume and sodium excretion revert to normoxic levels, and below 0.08–0.10 virtually all studies demonstrate antidiuresis and sodium retention. The variability between studies done at equivalent inspired O_2 may be due to differences in rate of change to the final hypoxic level, duration of hypoxia, barometric pressure, amount of exercise, water and salt intake, level of fitness, ventilatory response to hypoxia, and development of acute mountain sickness (AMS).

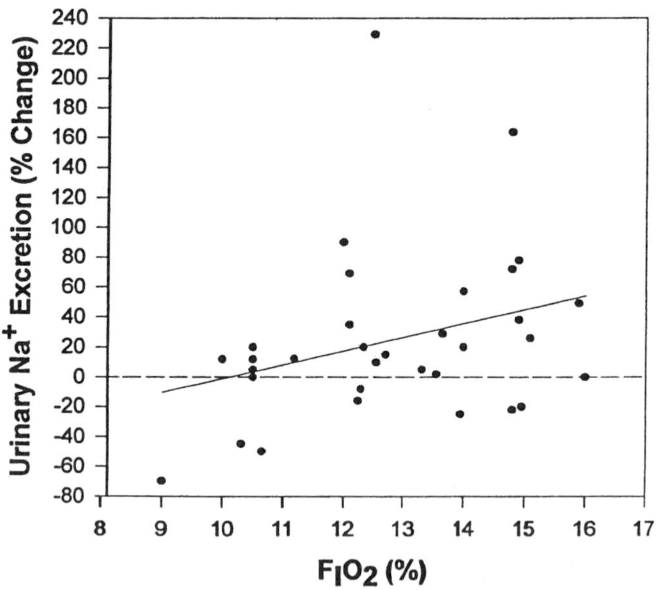

Figure 1 Summary of 32 high-altitude and normobaric hypoxic studies in humans in which sodium excretion was measured between 1 and 24 hours of indicated inspired O_2. Sodium excretion with hypoxia is given as the percent change above or below the preceding normoxic baseline period.

Those with diuresis have increased sodium, potassium, and bicarbonate excretion accompanied by an increase in urinary pH and fall in ammonia and titratable acid (180). Since total volume and solute output rise in parallel, urinary osmolality is not greatly altered (328). The diuretic and natriuretic response occur within the first hour and can persist for 1–2 days. Diuresis and natriuresis can be elicited with subsequent new and greater hypoxic challenges after a period of acclimatization (197). With antidiuresis, particularly associated with ADH release, there is the expected urinary concentration (6,122) with or without depressed solute excretion.

With chronic hypoxia, urinary composition and output in humans and animals generally return to a new steady state dictated by dietary intake and extrarenal fluid and salt losses (180,237,246). In general, adaptation to chronic hypoxia results in persistent net negative fluid balance relative to sea level, but with severe hypoxia fluid retention may persist as subacute and chronic mountain sickness.

Renal Hemodynamics

Glomerular filtration rate (GFR) and renal blood flow (RBF) are critical factors in any renal response since they determine the total water and solute entering the nephron. With systemic hypoxia of mild to moderate intensity ($F_{IO_2} \sim 0.09$–0.14) and of a short duration (hours), humans show little GFR change (8,34,245,328), but there is significant interindividual variability. At high altitude (3450 m), Pauli et al. (247) showed a sustained 5–10% increase in GFR for several days similar to data reported by Olsen et al. (243,244) in subjects studied 2 days after rapid ascent to 4350 m. In contrast, Jankoski et al. (157) found that GFR was reduced almost 50% at 24 hours in a group of subjects at 4300 m. The discrepancy between these studies highlights the critical influence of mountain sickness since none of the subjects in Olsen et al. (243,244) or Pauli et al. (247) reported illness as opposed to a high rate of illness reported in the group studied by Jankoski et al. (157). RBF rises slightly (8–20%) with acute hypoxia in well-acclimatizing humans (8,11,34,245), but this increase is not sustained after several days (243,244,247).

Other mammals differ somewhat in the renal hemodynamic effects of acute hypoxia, but in general, as in humans, 0.13–0.14 F_{IO_2} appears to be a threshold above which GFR and RBF are preserved even if blood pressure drops slightly (119,304,324,353). Between 10 and 13% O_2, RBF and GFR (32,53,57,210,235,338), and below 10% reductions can be as great as 20–40% (57,119,235).

Animal studies point up the importance of hypotension and the negative impact of anesthesia. In rats, O_2 between 0.10 and 0.13 reduces GFR by 40% (235). This is associated with a 25% fall in blood pressure and a reduction in urine output and sodium excretion; however, GFR and urinary output decrements are prevented with blood pressure maintenance by partial occlusion of the infrarenal aorta. Selkurt (299) used venous blood perfusion to study the effect of isolated renal hypoxia and found in contrast that renal blood flow increased 25%, although GFR fell 5–10%. In this situation, urinary volume and sodium excretion rose 25 and 50%, respectively. Rose et al. (278) and Walker (337) found blood pressure and cardiac output rose along with Na excretion in conscious dogs breathing 7–8% without changes in RBF or GFR, suggesting a pressure natriuresis, but Krebs et al. (181) found equal natriuresis with 10% O_2 without the rise in arterial blood pressure or cardiac output.

Becker et al. (29), and Winslow and Monge (347) found even in markedly polycythemic (Hct > 60%) high-altitude Andean natives (4000–5000 m) that RBF is only marginally reduced but renal plasma flow is depressed by 30–40%. Despite a reduction in plasma flow, GFR is well maintained because the filtration fraction rises almost proportionally to the drop in renal plasma flow, and renal oxygen delivery, a-v O_2 content gradients, and renal oxygen consumption are equal to sea level residents (272). However, with greater hypoxia, renal blood flow ultimately is depressed by about 30% in healthy subjects newly residing at 6000 m (3) and by roughly the same magnitude in high-altitude Andean natives with chronic mountain sickness (199). Findings are similar in other mammals except that RBF with long-term adaptation (>1 month) actually increases despite severe polycythemia (29,246) at altitudes corresponding to >5500 m. The mechanism of preserved or augmented renal blood flow in the face of blood hyperviscosity is not known. It may involve a degree of structural vascular remodeling since hypoxia-adapted rats continue to show high renal blood flow with acute correction of their hypoxia (246), or an upregulation of endothelial NO synthesis (287a).

In conclusion, moderate acute systemic or local renal hypoxia reduce renal vascular resistance and cause GFR and RBF to rise slightly or remain unchanged. The mechanism of direct hypoxic renal vasodilation is not established, but like many systemic vascular beds it may involve local release of adenosine (84,236), nitric oxide (50) and vasodilating prostaglandins (61). Hypoxia evoked neuroendocrine influences may act, however, to attenuate direct hypoxic local vasodilation. Furthermore, hypocapnia in spontaneously breathing subjects may accentuate renal vasodilation or limit the vasoconstrictor input via the renal sympathetic nerves (209,338). In contrast, it is well established that if the systemic response to hypoxia causes hypotension, GFR and renal blood flow fall with attendant antidiuresis and antinatriuresis.

Tubular Function

Hypoxia may also influence renal salt and water transport at the level of the tubule. With its high blood flow and low overall oxygen extraction (mean renal Svo_2 is 80–90%), the kidney is conventionally thought to be resistant to hypoxia (11,272). Yet, this overall picture neglects the complexities of the intrarenal vasculature and tubular geometry, metabolism and function that generate a marked heterogeneity of local Po_2 within the cortex and even more prominently in the medulla (51,91).

The cortex and outer medulla is richly perfused to accommodate high rates of glomerular filtration and active salt reabsorption by the proximal tubule. Owing to the presence of a preglomerular oxygen diffusion shunt (297) local Po_2 in the cortex can vary between 20 and 100 mmHg (190). The series arrangement of glomerular perfusion and cortical tubular blood flow dictates that a reduction of RBF generally reduces GFR, which in turn decreases the metabolic work of salt reabsorption and limits the fall in renal tissue Po_2. Interestingly, reduced O_2 delivery to the kidney by anemia or arterial hypoxemia has much less effect on GFR than reduced RBF, with the result that local Po_2 falls in the vicinity of the peritubular erythropoietin-producing cells. This unique perfusion to metabolism (i.e., filtration) relationship makes the kidney an ideal site for erythropoietin production in response to anemia or hypoxemia but appropriately unresponsive to changes in its own blood flow (94).

The inner medulla, in contrast, receives about 10% of RBF. Its relative hypoperfusion is a result of the countercurrent arrangement of the medullary vasculature necessary for urinary concentration and dilution (61,220). The low perfusion in the vasa recta leaves this region surprisingly hypoxic (Po_2 8–10 mmHg) (17,26,190,191). These Po_2 values are close to those in vitro that cause biochemical dysfunction and anatomical derangements in isolated tubules (287,320). The importance and presence of a low perfusion-to-metabolism ratio in the medulla is demonstrated by the increase in renal tissue Po_2 and mitochondrial cytochrome reduction with diuretics, which reduce O_2 consumption by blocking ion transport (10,49,89,254).

Measures of functional tubular integrity such as the Tm for glucose and organic acids appear to be maintained over a wide range of hypoxemia in vivo (11,52) since these functions are located in the better oxygenated renal cortex and because renal oxygen consumption is dominated largely by the metabolic cost of Na, Cl, and HCO_3 reabsorption (68). The medulla would logically be the first region even in the absence of blood flow changes to demonstrate depressed tubular function in hypoxia, such as polyuria and salt wasting.

Although there is no evidence that the diuresis and natriuresis of mild to moderate hypoxia arises from tubular ion pump dysfunction due to ATP depletion, a Fio_2 of 0.12–0.14 causes a downregulation in vivo of several membrane Na^+-transporting proteins in the kidney (H. Mairbäurl, personal communication) similar to that in isolated alveolar epithelial cells (202). Suppression or withdrawal of membrane transporters in hypoxia may be considered a protective response to reduce O_2 consumption in the face of decreasing O_2 availability. In studies of urinary output with global renal oxygenation assessment, natriuresis and diuresis with hypoxia occur before there is any measurable reduction in total oxygen consumption, renal blood flow, O_2 delivery, and GFR. Indeed, antinatriuresis and antidiuresis develop only when there is severe in vivo arterial hypoxemia ($Sao_2 < 50\%$) or very low perfusate Po_2 in the isolated saline perfused kidney (<20 mmHg). Concurrently there are reductions in systemic blood pressure, global RBF, renal Vo_2, renal O_2 delivery, and GFR (57,103,118).

Explanations for the surprising resistance of the kidney and the medulla, in particular, from acute arterial hypoxemia, include the oxygen-sparing metabolic effect of diuresis and natriuresis (68) and a possible selective increase in medullary blood flow with hypoxia mediated by endothelin (vide infra). When GFR falls with more extreme hypoxemia, O_2 consumption falls because of the decreased reabsorptive work with reduced glomerular filtration.

In chronic hypoxia, the findings are of normal or augmented tubular function. High-altitude Andean natives (347) have normal renal acid-base regulatory ability, and respond normally to angiotensin infusion. Delgado (82) showed that the maximal urinary concentration in response to water deprivation and vasopressin and maximal free water reabsorption with a mannitol diuresis were no different between Peruvian sea level dwellers and high-altitude natives. In the case of maximal salt excretion, high-altitude Andean natives can excrete an infused salt load faster than lowlanders (267,269). Gonzalez and colleagues demonstrated enhanced renal compensation to acute metabolic and respiratory acidosis in chronically hypoxic rats, which was not the case with acute respiratory alkalosis (114,116,117,343).

Taken together these studies on tubular function demonstrate the dominating influence of renal blood flow, systemic hemodynamics, and hormonal effects rather than arterial Po_2 per se in affecting tubular function at high altitude. It would appear that levels of acute arterial hypoxemia not associated with reductions in renal blood flow are remarkably well tolerated and do not lead to significant impairment of renal function despite the significant intrarenal Po_2 heterogeneity, which even at sea level predicts that certain regions may verge on critical hypoxia.

B. Nonrenal Salt and Water Losses

The respiratory tract is not a site of significant water loss at high altitude despite sustained high ventilation with mouth breathing of cold dry air that bypasses the heat and water conserving nasal passages. Although Pugh (255) calculated a water loss of ~3 mL per 100 L ventilation with moderate exercise at 5500 m, which if sustained would cause a loss of 200 mL per hour, Hoyt et al. (148) found only a 200 mL/day greater respiratory evaporative water loss in subjects working at 4300 m than that measured at sea level. The expected greater losses with ventilation at altitude earlier were not found because of the much lower temperatures of exhaled air and thus water content measured under environmental conditions.

Water losses through sweating and insensible evaporation can be quite high under conditions of heavy work at high altitude and the low relative humidity. Although gastrointestinal malabsorption of fat and sugars occur with hypoxia, salt and water malabsorption is usually not a consequence. Studies in dogs (329,330) showed that small intestinal salt and water absorption is reduced by only 5–10% with inspired O_2 as low as 8%. Noninfectious diarrhea is rarely a problem at high altitude, and any GI water and salt losses are insignificant when compared to renal and cutaneous losses.

C. Salt and Water Intake

In the first 1 or 2 days of ascent above 7000 feet, water intake in humans may fall by a liter a day even in healthy subjects without symptoms of AMS (73,98,318) but ultimately returns toward normal. Part of the reduced water intake reflects reduced food ingestion, but as in rats, the hypodipsic effect of hypoxia appears greater than the hypophagic effect (99,130,161). Thus, hypoxia has independent effects on thirst as well as appetite. Decreased salt intake not surprisingly follows the decline in caloric intake and can be depressed 30% in the first days of acute hypoxia. In addition to decreased food intake, a reduced salt appetite has been demonstrated in rats with acute hypoxia (30). The hypoxic suppression of thirst does not appear to be mediated by changes in ADH since both normal rats and those with hereditary diabetes insipidus show an equal suppression in water intake (161,162).

D. Total Body Water and Volume Distribution

With adaptation to hypoxia there is a 2 to 3 L loss of total body water in humans over several days (65,155,160,182). The water loss is roughly shared by the plasma, extracellular, and intracellular spaces (155,156,160,182,342), although in goats and sheep there appears to be an increase in the extracellular fluid (ECF) and in swine an increase in the interstitial fluid volume despite the loss of total body water (85,147). The loss is a result of decreased intake, but also of increased respiratory, cutaneous, and renal losses since forced drinking does not entirely prevent the loss of total body water (325). Upon return to normoxia or lower altitudes, the water loss is quickly corrected (131,281).

In chronic hypoxia, hypohydration persists (85,131a,156). The magnitude of hypohydration is greatest in populations resident at high altitude. Ramirez et al. (268) found that total body water was only 43% of total body mass in high-altitude natives of the Colombian Andes compared to 53% in sea level residents, a finding that may be related to a relative insensitivity to ADH and diminished excretion of aquaporin 2 water channels in the urine.

E. Electrolytes and Acid-Base Status

The net result in plasma and extracellular fluid composition is generally one of no change in serum osmolality and sodium concentration acutely (64,98,157). However, with very high altitude (>5400 m), Blume et al. (38) measured a rise in serum sodium and osmolality of 5 mM and 10 mOsm, respectively, in the face of a normal serum ADH. The same blunted ADH vs. osmolality relationship is observed in chronically hypoxic rats (162). These findings can be interpreted as a failure of normal osmoregulation and/or insensitivity of osmoresponsive cells in the CNS. Ramirez et al. (267) showed that this hyporesponsiveness in humans may develop as early as 3 days.

The acute hypocapnia of hypoxia can lead to a slight reduction of serum potassium, both from intracellular uptake and a kaliuresis (107,110,136,206), but not all studies in humans are consistent in this regard (70,305,306,318). The changes in plasma chloride and bicarbonate are those expected of acute and chronic hyperventilation, and in humans do not differ between equally hypocapnic normoxic or hypoxic hyperventilation (110). In general, the compensatory metabolic acidosis of hypoxic hyperventilation reflects the severity of the hypoxia and the balance between the dominating stimulus of enhanced hypoxic ventilatory drive and the smaller but opposing hypocapnic inhibition of the hypoxic response. In humans, the net acid-base effect of hypoxia is a mild respiratory alkalemia with incomplete renal compensation (180), but in other mammals (108,114,116) systemic pH may return to the normal range. Intracellular acid-base status in acute and chronic hypoxia has received little attention. Chronic hypobaric hypoxia in rats leads to enhanced intracellular pH defense capacity (114,116) beyond that provided by the renal generation of buffer equivalents and the greater nonbicarbonate buffering capacity of polycythemic blood.

III. Mechanisms of Changes in Salt and Water Balance at High Altitude

A. Sites of Hypoxic Sensation

The sensation of hypoxia relevant to renal function and fluid balance at high altitude may occur at several sites, including peripheral central chemoreceptors, brain, baroreceptors, heart, lungs, and kidneys. The ultimate response will be a summation of differing hypoxic sensitivities of each site and their neurohumoral response. This complex interaction of several sensors may account for the rather variable and sometimes conflicting responses to hypoxia described throughout this chapter.

Peripheral Chemoreceptors

In addition to their dominant role in the ventilatory response to hypoxia, the peripheral chemoreceptors have considerable influence on renal function and fluid balance, particularly on systemic cardiovascular regulation, the renal vasculature, and sympathetic innervation (208). With the pioneering work of Honig (144) and Behm (30–32), one must now add an important contribution to salt and water balance in hypoxia beyond their influence on renal hemodynamics.

The arterial chemoreceptors are the principal sensor initiating the diuresis and natriuresis of hypoxia. In normoxic animals, peripheral chemoreceptor stimulation increases absolute and fractional excretion of water and sodium, whether by isolated perfusion of the carotid bodies with hypoxemic blood (144,166) or by administration of the peripheral chemoreceptor stimulant, almitrine (18,32). If the peripheral chemoreceptors are denervated, the response is ablated and sodium and water excretion decrease (32,144). When peripheral chemoreceptors are left normoxic and only the brain and kidneys are rendered hypoxic, the response is one of antidiuresis and antinatriuresis (113).

The role of the peripheral chemoreceptors is not as clearly established in man, but a similar function is likely. Almitrine is a diuretic in normoxic subjects (46,179,187) as it is in animals. Swenson et al. (319) took a different approach and showed in subjects with a 10-fold range of hypoxic ventilatory responsiveness (HVR) that a strong correlation exists between a subject's urinary sodium and water excretion over 6 hours with 14% O_2 breathing and HVR (a marker of hypoxic peripheral chemosensitivity and afferent output). These data are shown in Figure 2. However, Hildebrandt et al. (137) could not demonstrate a correlation in subjects studied in the first hour of acute hypoxia, in which a diuretic but not a natriuretic response occurred, suggesting multiple factors promoting urine output with different time courses. Another explanation for the difference is also possible. The subjects studied by Hildebrandt et al. (137) were water loaded to increase urine volume, and the slight hyponatremia that this induces may generate a sodium-conserving response that overrides any natriuretic stimulus from the peripheral chemoreceptors. A similar lack of natriuresis was found by Brauer et al. (46) after 3 hours of 12% O_2 in normal water-loaded subjects.

How hypoxic or pharmacological stimulation of peripheral chemoreceptors provokes diuresis and natriuresis and at what nephron site(s) these occur remain uncertain. Lacking definitive tubular micropuncture studies, the only data are in water-loaded humans, using lithium clearances to detect changes in proximal tubular fluid and Na^+ reabsorption. Christiansen et al. (62) found an increase in lithium clearance in hypocapnic hypoxia, but no change in isocapnic hypoxia, while Brauer et al. (46) studied only hypocapnic hypoxia and found a slightly decreased lithium clearance. Both studies point to a more distal site of suppressed sodium reabsorption with hypoxia. Although respiratory and circulatory responses to hypoxia have been studied in persons who have had bilateral carotid body resection (143), unfortunately the renal responses to hypoxia in this unique group have never been studied.

These renal responses to hypoxia do not depend upon signals carried via the renal nerves because kidney denervation does not eliminate the diuresis or natriuresis. Although other consequences of peripheral chemoreceptor stimulation also might be responsible, the response is still quite strong in anesthetized animals, when changes in $P\text{CO}_2$, acid-base status, minute ventilation, and intrathoracic pressure variations are prevented. Peripheral chemoreceptor mediated natriuresis and diuresis, however, depend upon a normal circulatory volume and low firing rate of cardiovascular stretch and pressure receptors. Karim and Al-Obaidi (166) showed that hypotension from blood loss combined with hypoxia is a very potent stimulant of renal sympathetic nerve activity (RSNA). Under circumstances of combined chemoreceptor stimulation and baroreceptor unloading, natriuretic and diuretic responses are overwhelmed. Salt intake is important since there is an attenuation of hypoxic diuresis with salt restriction (107,120) and a greater response with more liberal intake (110,180,319). The specific mechanism by which peripheral chemoreceptors mediate increased output of sodium and water with hypoxia, by exclusion, thus points to an endogenous humoral factor(s) or intrarenal blood flow redistribution, to be discussed in the following section on mediators of renal salt and water transport.

Peripheral chemoreceptors also are involved in suppression of salt intake, since the response is lost with carotid body denervation, but this is not the case for thirst and water intake suppression (30). Other effects of hypoxic peripheral chemoreceptor stimulation with direct or indirect impact on renal function include generalized activation of the sympathetic nervous system including RSNA (208).

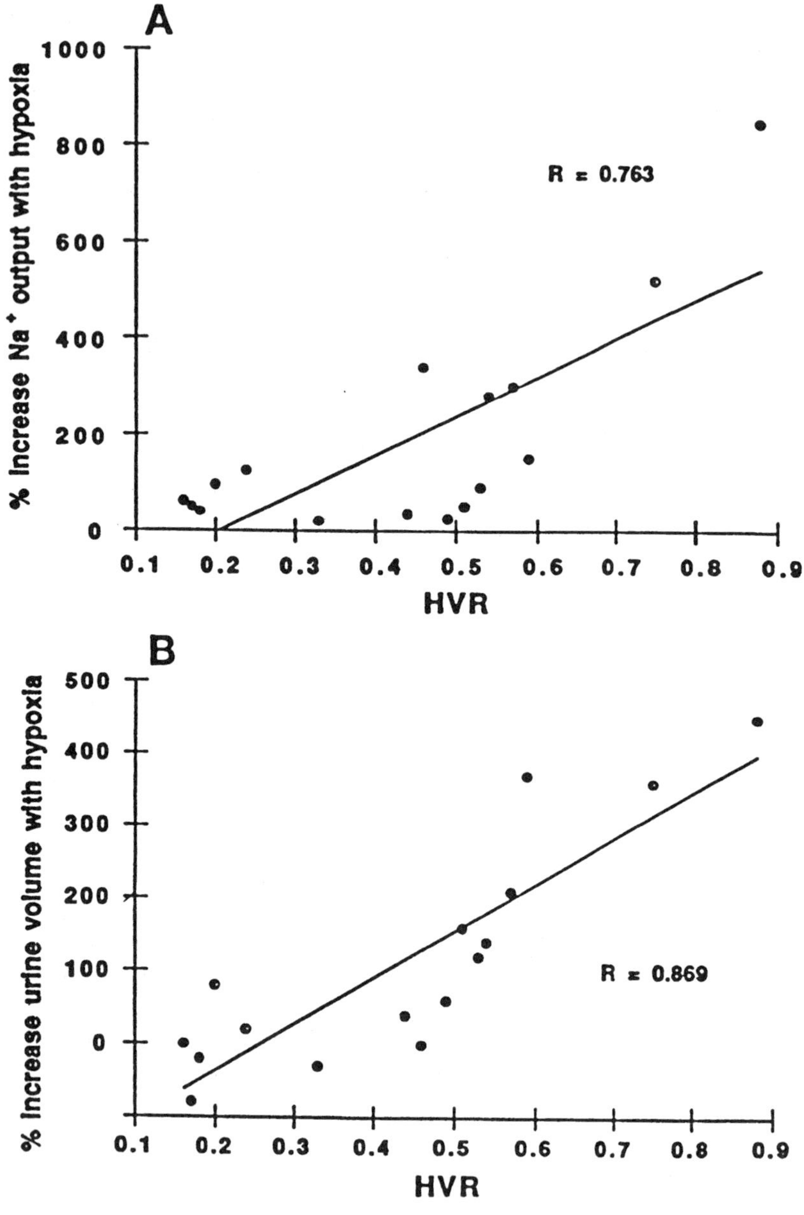

Figure 2 Correlation between isocapnic hypoxic ventilatory response (HVR) and increase in urinary sodium excretion (A) and urinary volume (B) with 6 hours of hypoxia (O_2 = 14%) in 16 healthy human subjects. Correlation coefficients (R) are given.

With some species differences, this is manifested as increased venous tone (75,127,283) and hypertension (144,208). What is intriguing is that hypoxic diuresis and natriuresis occur despite increased RSNA, which otherwise causes salt and water retention (83).

Brain and Sympathetic Nervous System

Severe CNS hypoxia excites vasomotor neurons and sympathetic tone, independent of peripheral chemoreceptor input (102,141,317), and this itself will decrease sodium and water excretion. However, the impact of lesser degrees of hypoxia in the brain on the CNS mediation of renal vascular and tubular responses to hypoxia is heavily dependent upon input from the peripheral chemo- and baroreceptors (102,141,208) with some modulation by the resulting changes in acid-base status and ventilation. When the carotid sinus nerves are sectioned in the rat, the increased sympathetic activation of mild to moderate hypoxia is virtually abolished (102). However, with normal peripheral input (i.e., normoxia at the peripheral chemoreceptors), an equivalent level of hypoxia isolated to the CNS by separate perfusion of the brain with hypoxic blood causes potent renal vasoconstriction, decreased renal blood flow, GFR, and urinary sodium and water excretion (113), suggesting intense renal sympathetic nerve activity (141,301). Thus the net effect of hypoxic sympathetic activation on the kidneys is a complex summation of direct effects on hypoxia in the cardiovascular regulatory areas and peripheral chemoreceptor and baroreceptor afferent information. The baroreceptor response will depend on the overall blood pressure, dictated by a balance of sympathetic-mediated vasoconstriction in some vessel beds (splanchnic and renal) and vasodilation in others (skin) as well as direct local hypoxic vasodilation (151,208).

Kidneys

The erythropoietin-producing cells along the proximal tubule are an important oxygen sensor in the hematopoietic response to hypoxia but are not accorded any role in the monitoring of hypoxia relevant to fluid balance and renal function. Although erythropoietin has no known effects on tubular function (94), it is vasoconstricting in the renal vasculature (134). Thus, its release into the renal interstitium with hypoxia could alter intrarenal hemodynamics and blood flow distribution and secondarily alter renal salt and water handling. Renal "chemoreceptors" have been described in the rat kidney; this name was chosen to denote that they are activated with the chemical milieu of renal ischemia and are different from the better characterized pressure-sensitive mechanoreceptors (270). Whether these "chemoreceptors" or the erythropoietin-producing cells relay neural input to the brain or feedback on efferent RSNA with the hypoxic stress of high altitude is not known, but it is clear that renal nerve afferents may be important in the response of the kidney to other changes in cardiovascular and volume status (315). Since denervation of the kidneys markedly augments the diuretic and natriuretic response to hypoxia (32,144), it is conceivable that afferent information from the kidney may diminish the final response.

Heart and Lungs

It is doubtful whether any chemosensitive sites exist within the heart or lungs of mammals, but changes in ventilation and hemodynamics accompanying the hypoxic state can modulate brain and peripheral chemoreceptor outputs involving renal function. These will be discussed in the following section.

B. Mechanisms of Hypoxic Salt and Water Regulation

Many mediators of altered salt and water regulation in hypoxia have been proposed, including a myriad of hormones, the sympathetic nervous system, intrathoracic pressure and volume receptors, respiratory alkalosis, and neural or hormonal activation arising from stimulation of the peripheral chemoreceptors.

Humoral Factors

Many hormones with salt- and water-regulating properties have been investigated in hypoxia to determine whether their plasma concentrations correlate with the urinary responses to hypoxia. The data remain inconclusive, despite the number of changes in salt- and water-regulating hormones reported to occur with hypoxia (vide infra). Definitive assignment of a hormone's role should include consistent findings across a number of studies and loss or blunting of effect with a specific antagonist or, genetic manipulation of a hormone or its receptor. The lack of consistent findings in classic salt and water hormones studied to date warrants pursuing other newly reported compounds with diuretic and natriuretic action as possible mediators.

ADH

Hypoxic exposure has variable effects on antidiuretic hormone release and action. In conscious humans or animals, ADH is either mildly suppressed (63) or not affected by mild to moderate ($F_{IO_2} > 0.10$) normobaric hypoxia (8,27,76,86,136,319). Hypobaric exposure for several days also does not alter ADH in subjects without illness (64,124,132,176,253). Only with severe hypoxia ($F_{IO_2} < 0.10$) in humans (often associated with the nausea of AMS) and in animal experiments with severe hypoxia and anesthesia do ADH levels rise (6,64,124,129,263,340).

Although isolated peripheral chemoreceptor stimulation may cause ADH release (300), why ADH is suppressed with mild hypoxic stress in vivo is not known. It may involve baroreceptor and chemoreceptor input as well as cortisol release. Hyperventilation, hypocapnia, and increased intrapleural pressure swings (causing distension of low pressure atrial stretch receptors) may inhibit ADH release (18,176,177,328). Hypocapnic suppression of ADH release in hypoxia is supported by studies showing ADH release with maintenance of eucapnia in humans (64) or hypercapnic acidosis in animals (263,340). Without these associated peripheral chemoreceptor-mediated responses, hypoxia under anesthesia or isolated perfusion of carotid chemoreceptors with desaturated blood causes ADH to rise and water excretion to fall (251,300). Raff et al. (262) have shown elevations in circulating cortisol suppress ADH release to a variety of stimuli including hypoxia.

That a diuresis can occur without suppression of ADH, as with almitrine (179), implies that either other factors such as changes in intrarenal osmolality promote water loss (washout of the hypertonic medullary interstitium with changes in medullary blood flow) or that acute hypoxia induces a functional state of renal ADH unresponsiveness (nephrogenic diabetes insipidus). The latter is unlikely because exogenous ADH given to acutely hypoxic dogs and humans is antidiuretic (6,35). Whether suppression of ADH secretion accounts for the diuresis of mild to moderate hypoxia and increased secretion accounts for the antidiuresis of severe hypoxia might be answered with ADH antagonists, which have no effect on hypoxic renovascular response (338).

Renin-Angiotensin-Aldosterone System

Hypoxia may, like many other stresses, cause sympathetically mediated release of renin from the renal juxtaglomerular cells. Renin cleaves its circulating substrate, angiotensinogen, to angiotensin I, which is then further metabolized to its active form, angiotensin II, by angiotensin-converting enzyme (ACE) in the lung vascular endothelium. Angiotensin II is a potent vasoconstrictor and stimulates secretion by the adrenal glands of the antinatriuretic mineralocorticoid, aldosterone, both functions serving to defend circulating volume.

Data on renin activity at high altitude and with hypoxia are conflicting. Some studies found an increase (98,305,306,327) but others no change or a decline (140, 171,201,221,318,319). The differences may rest on whether the subjects had exercised (which raises renin) and whether there was a critical fall in effective circulating volume and thus renal perfusion pressure. Supporting the importance of renal perfusion is the finding of no increased renin secretion in the isolated perfused kidney when perfusion pressure was constant but the Pa_{O_2} of the perfusing blood was lowered to 6 mmHg in contrast to the increased generation of renin when perfusion pressure was lowered (312).

Surprisingly there are few measurements of angiotensin I in hypoxia to ascertain whether generation of angiotensin I is altered. In humans, Vonmoos et al. (334) found that angiotensin I was not altered by one hour of 12% O_2 in the face of a constant plasma renin activity. A possible inhibition of conversion of angiotensin I to angiotensin II has been reported in hypoxic animals (153,192,259) and in humans (334), but the changes are quite small and likely of little consequence. Angiotensin II levels are not elevated with hypoxia (221,334). Although earlier work suggested that hypoxia suppressed serum ACE activity and thus by extension lung ACE activity, this has not been borne out in subsequent studies as a cause for reduced angiotensin II formation and aldosterone release (8,42,69,226).

A remarkable and common finding is a suppression of serum and urinary aldosterone with hypoxia in resting subjects (98,140,171,186,305,306,318,327), particularly on a low-salt diet and with high prehypoxic aldosterone levels (70). Aldosterone suppression occurs even with hypoxic exercise despite a rise in renin (40, 185,201,227,302). Stimulation of the peripheral chemoreceptors does not appear responsible because renin and aldosterone are unaltered in normoxic rats given almitrine (146). Thus it not surprising that other studies find that aldosterone is only minimally suppressed or not all with hypoxia (8,22,266,267,319). With more chronic hypoxia, aldosterone levels remain normal (224,225) and can rise in those with overt illness and fluid retention (3).

Several mechanisms are responsible for hypoxic suppression of aldosterone. Hypoxia directly suppresses ACTH and angiotensin II–stimulated formation of aldosterone by isolated adrenocortical cells (257). Changes in plasma potassium also regulate aldosterone levels (350), thus with acute hypoxic hypocapnia, hypokalemia or the transmembrane shifts of potassium in vivo may suppress aldosterone formation or release (76). Although the effect of hypocapnia on aldosterone synthesis in hypoxia has not been studied, Raff et al. (258) found that acute hypercapnia increased aldosterone synthesis in vitro but did not affect the hypoxic suppression.

In vivo, Colice and Ramirez (70) found that hypoxia did not alter the aldosterone response to angiotensin II in humans but did suppress the response to ACTH (266). Although increased hepatic catabolism of aldosterone in hypoxia is conceivable, this seems unlikely given stability of hepatic blood flow with hypoxia (69). Hypoxic enhancement of dopamine-mediated suppression of aldosterone release was not borne out in studies of humans taking metoclopramide, a dopamine antagonist (334). Since not all studies, in fact, find hypoxic aldosterone suppression, one explanation for natriuresis without aldosterone suppression may be downregulation of aldosterone receptors as shown by Jenq et al. (159) in hypoxic cultured renal cells.

Natriuretic Peptides

Very soon after the discovery of atrial natriuretic peptide (ANP), acute hypoxia was found to stimulate its release (2,13,15,71,86,185,186,266), although an equal number of reports demonstrate that it is not a uniform finding (24,42,66,170,198,266, 267,269,319). Possible factors involved in hypoxia-mediated ANF release include right atrial and ventricular stretch (111) and a direct effect of low Po_2 on atrial tissue (13). Although isolated hypoxic perfusion of the peripheral chemoreceptors causes ANF release (2), this does not occur in humans in response to peripheral chemoreceptor stimulation with almitrine (176,177). ANP release from the lung with hypoxia may also occur, since vagal denervation of the lungs blunts the normal increase in hypoxic pigs (14). Whether hyperventilation and hypocapnia with hypoxia reduce ANP release is not known, but it may be possible given the significant increase in ANP when hypoxia and hypercapnia are combined (280).

The role of ANF as a critical factor in the hypoxic natriuretic response remains uncertain. In some cases, there is diuresis and natriuresis with elevation in ANP and in others water and sodium retention accompanying elevated ANP (24,169). These contradictory findings suggest one of two possibilities: either ANP is not an important contributor to sodium excretion, as has been shown in several nonhypoxic states

(111), or there is a balance of the direct natriuretic effects of ANP and its other systemic actions, such as generalized vasodilation and increased capillary permeability leading to extravascular escape of plasma and decreased circulating blood volume and GFR (87,291,346). The natriuretic actions of ANP include its suppression of angiotensin II–mediated aldosterone synthesis and release (4,45,219), direct inhibition of tubular sodium reabsorption (106), increased GFR (111), and inhibition of ADH release (150,189).

ANP is only one member of several natriuretic peptides that may be important in the hypoxic response. Brain natriuretic peptide (BNP), now known to be synthesized in the cardiac ventricles, has many of the same actions as ANP (183) and is released from the isolated perfused ventricles in response to local hypoxia (323) and in vivo (138).

Urodilatin is another natriuretic peptide with sequence homology to ANP and BNP that is produced only in the kidney (111) and appears to be much better correlated with urinary sodium excretion than ANP in a number of different conditions (112). The evidence is limited, but urinary urodilatin excretion is increased sixfold in the hypoxic isolated perfused kidney and in humans ascending to an equivalent of 6000 m in a hypobaric chamber (178).

The endothelins are a group of vasoactive peptides with many effects beyond blood pressure control (285) and are of considerable potential significance in hypoxia. Acute normobaric and hypobaric hypoxia cause circulating endothelin concentrations to rise about 1.5- to 2-fold (58,230,248,333). In rats, 10% O_2 raises intrarenal and urinary endothelin concentrations (238). In tissue culture rat inner medullary collecting duct cells increase their endothelin secretion two- to threefold in response to 0–3% O_2 (222). Endothelins are potent systemic vasoconstrictors and can depress GFR and sodium excretion when given in high concentrations. However there is evidence that endogenous circulating or intrarenal levels of endothelin may be vasodilating in the kidney, since endothelin receptor blockade increases renal vascular resistance and depresses sodium excretion. Furthermore, when endothelins are infused in low doses in normoxic animals (104,172,249,290) or in association with the mild elevations induced by hypoxic exposure (238), there is diuresis and natriuresis. In these situations, endothelin blocks proximal and distal sodium reabsorption possibly by inhibiting tubular Na^+/K^+-ATPase (352), inhibits ADH action in the renal collecting duct to promote water excretion (293), and inhibits renin secretion (295). The vascular effects of small elevations in endothelin concentrations with hypoxia may promote a redistribution of blood flow to the medulla (286) and help to explain the tolerance of the poorly perfused medulla to arterial hypoxemia and hypoxic diuresis and natriuresis.

Cortisol and Other Steroid Hormones

Acute hypoxia in humans causes 2- to 3-fold increases in circulating ACTH and cortisol within 1–2 hours, with loss of the normal diurnal variation suggesting a stress response (43,98,149,206,267,273,318), although elevated cortisol is not always observed especially in normobaric hypoxic studies (69). In animals, cortisol and ACTH rise with hypoxia (47,146,239,261–263), which requires intact carotid sinus innervation (207). The peripheral chemoreceptors appear to be responsible, since cortisol and ACTH rise in normoxic rats given almitrine and the response is absent with carotid body denervation (146). With more chronic hypoxia, there is a return to normal concentrations (269) except in subjects remaining above 6000 m, who develop chronic mountain sickness (3). Elevations in cortisol may be partly responsible for increased free water clearance at high altitude, independent of its ability to suppress ADH secretion (264) or for the lack of antidiuresis in some cases of ADH elevations (136). Interesting in this regard is increased glucocorticoid receptor expression in cultured hypoxic rat renal cortical epithelial cells (159).

Other adrenal-like steroids have been implicated because diuresis and natriuresis are attenuated with adrenalectomy despite cortisol replacement (194). One candidate could be an endogenous cardiac glycoside, which has Na^+/K^+-ATPase–inhibiting effects (349). These steroid compounds, which are secreted by the adrenal glands and hypothalamus in response to hypoxia (80), raise blood pressure and promote natriuresis (37), making them attractive candidates in the peripheral chemoreceptor-mediated natriuretic response to hypoxia since there are afferent projections from the peripheral chemoreceptors to both these sites. Almost 50 years ago Rein (271) suggested that hypoxia elicited a cardiac glycoside-like substance. DeAngelis et al. (79) reported an elevated endogenous digoxin-like immunoreactive substance (DLIS) in healthy subjects trekking in the Himalayan mountains, which appeared to correlate with urinary sodium excretion, and Bagrov et al. (16) found that a cardiac glycoside-like factor rose with voluntary hypoventilation. However, Swenson et al. (319) did not find that acute normobaric hypoxia elevated plasma DLIS in subjects with diuresis and natriuresis. As yet the exact chemical structure of DLIS has not been elucidated, nor have inhibitors of this agent been tested in hypoxia.

Circulating Catecholamines

Although chronic hypoxic activation of the sympathetic nervous system leads to raised circulating norepinephrine concentrations and increased urinary excretion, norepinephrine levels are generally not elevated with acute hypoxic exposure (42,200,201,215,239,267,282,351). Muscle sympathetic nerve recordings, however, show increased activity. These findings suggest either a heterogeneity of response in the sympathetic nervous system or increased plasma clearance of catecholamines with acute hypoxia. Elevated plasma epinephrine, in contrast, is observed only transiently with acute hypoxia but returns to normal or below normal levels with chronic hypoxia (351). Any consequence of elevated circulating catecholamines as well as increased renal sympathetic nerve activity in hypoxia would be antidiuretic and antinatriuretic by both renal vasoconstriction and direct stimulation of renal tubular sodium and water reabsorption (33,67,341).

Hypoxia increases blood flow to the adrenal cortex by 80% and medulla by

200% even if blood pressure is held constant (48,239). A direct Po_2-mediated vasodilatation appears to account for the increase in cortical blood flow, but the rise in medullary blood flow may be driven more by neural input to the gland (47). Whether these changes in blood flow simply reflect the metabolic demands of increased adrenal corticoid and mineralocorticoid secretion is not known, but an increase in cortical blood flow with the resulting higher tissue Po_2 may be a partial explanation for the suppression of aldosterone secretion with acute hypoxia (257).

Other Natriuretic Factors

All of the established classic salt- and water-regulating hormones known to cause or be associated with fluid retention such as ANP, renin, aldosterone, vasopressin, and angiotensin have been reported to be increased by hypoxia in those with salt and water retention. However, no single hormone or set of hormone changes predictive of natriuresis and diuresis have been consistently associated with hypoxic diuresis and natriuresis (144,319). However, the possibility of small appropriate changes in several hormones that may combine in an additive or synergistic manner cannot be ruled out given the insufficient number of subjects in any one study to attempt such an analysis. More compelling, however, is the possibility that other substances with natriuretic effect will prove more important.

Adrenomedullin is a newly identified peptide synthesized by many organs including the adrenal medulla, kidneys, lung, and heart. It is a potent vasodilator (174) and has natriuretic properties (163,203). Its plasma levels are elevated in renal failure, hypertension, and congestive heart failure (152) and in acute hypoxia of 6 to 48 hours (139a,322a). Similar to endothelin, the net effect of adrenomedullin may be a balance between its antinatriuretic endocrine vasodilator effects on renal perfusion and its natriuretic effects within the kidney (163). Adrenomedullin suppresses angiotensin II–mediated aldosterone production by isolated rat adrenal gland zona glomerulosa cells (216), providing another possible control element in the phenomenon of aldosterone reduction.

Nitric oxide (NO) is generated by a variety of cells in the kidney and has multiple vascular and tubular actions. Its release is stimulated directly by polycythemia (287a), arterial hypoxemia (247a) and local renal hypoxia (136a) as well as secondarily to changes in renal vascular hemodynamics (25), neural tone (322), or other hormones (218). Although reduction of NO production by NO synthase inhibition causes renal vasoconstriction, NOS inhibitors do not cause any further renal vasoconstriction in hypoxia over that observed with hypoxia alone (248). NO is vasodilatory (265), blocks sodium reabsorption at multiple sites along the nephron (277,316), decreases renal oxygen consumption (186a) and alters renin secretion from the juxtaglomerular apparatus (133). Any of these effects may play an important role in hypoxic renal function.

Medullipin is another substance of possible interest in the renal response to hypoxia. It is secreted by renomedullary cells and then metabolized by the liver to yield a vasodilator (231) with natriuretic activity. Medullipin may act as a regulatory hormone to counter the actions of the renin-angiotensin system and may be an integral component in the pressure-natriuresis response discussed below (74). If hypoxia-meditated natriuresis is a pressure-natriuresis-like response, then it would be predicted that acute hypoxic exposure should raise plasma and urinary medullipin levels, but this has not been tested.

A complete listing of other hormones and peptides with natriuretic effects such as epidermal growth factor (121), α-melanocyte–stimulating hormone (214), neuropeptide Y (36), and uroguanylin (97) is beyond the scope of this review. They are mentioned because the mechanism of hypoxic natriuresis and diuresis still remains unexplained and these and other natriuretic substances warrant investigation as potential mediators.

Systemic and Intrarenal Hemodynamics

When hypoxia reduces renal blood flow, with or without systemic hypotension, there is almost invariably sodium and water retention (3,199), in part due to the fall in GFR. When renal blood flow is maintained and systemic blood pressure rises, increased sodium and water excretion might be explained in part by a ''pressure natriuresis and diuresis.''

The mechanism of pressure natriuresis and diuresis arises from the lack of renal medullary blood autoregulation (275). As medullary blood flow increases with increasing blood pressure, there is a steep pressure-natriuresis relationship (~15% increase in sodium excretion per mmHg rise in renal perfusion pressure). The basis for this response is a rising interstitial pressure (105) that suppresses sodium reabsorption at multiple sites by decreasing the hydrostatic pressure differences across the tubule. Furthermore, medullary solute washout from increased blood flow reduces the osmotic and electrochemical gradients favoring sodium and water exit from the tubule (276). A number of factors modify the pressure-natriuresis response, including volume status, renin-angiotensin-ANP concentrations, and the level of RSNA modified by either renal denervation (44) or carotid sinus denervation (90).

Whether pressure natriuresis and diuresis plays a role in the diuresis and natriuresis of acute hypoxia has not been studied adequately. Certainly in animals such as dogs (119,337) that can become hypertensive with hypoxia, it may be important. However, diuresis and natriuresis with hypoxia can still occur in dogs, rabbits, and rats (57,181) and humans (34,110,136,178,211,247,319) who do not become hypertensive. Thus it is obvious that systemic hypertension with hypoxia is not necessary to increase salt and water excretion. Hypoxia appears to lead to a selective increase in medullary blood flow without changing systemic blood pressure (190a), as is seen with low-dose endothelin infusion (286). Thus an extreme sensitivity of the renal perfusion pressure-natriuresis response in the medulla could account for hypoxic natriuresis and diuresis. Results with almitrine in normoxic rats suggest a unique effect of hypoxia since peripheral arterial chemoreceptor stimulation itself does not lead to an increase in medullary blood flow (187a).

Renal Innervation and Hypoxic Responses

The response of the autonomic system to hypoxia is that of generalized sympathetic and parasympathetic activation. The implications for the kidney are increased RSNA and elevated circulating catecholamines, which when unopposed act as antidiuretic and antinatriuretic influences. The antidiuretic and antinatriuretic influence of peripheral chemoreceptor mediated RSNA is revealed by the 30–50% greater diuresis and natriuresis observed in animals whose kidneys are denervated (166,167,187). Although there is recent evidence that renal sympathetic nerves release neuropeptide Y (36) and renal nerves can generate NO (196,213), both of which are natriuretic when given exogenously, it appears that the catecholamine effects of sympathetic nerve stimulation are dominant.

Although the kidney is not thought to have any important nonsympathetic innervation, there is some biochemical and pharmacological evidence for opposing renal cholinergic innervation in the dog (233,252), analogous to the hypoxic stimulation of parasympathetic activity to the heart (208). However, an important parasympathetic mediated diuresis and natriuresis would not be compatible with the greater hypoxic salt and water excretion following renal denervation, and the diuresis and natriuresis with acetylcholine infusion (233) could also be explained by generation of intrarenal NO, which is both vasodilatory and natriuretic.

Cardiopulmonary Innervation and Hypoxic Responses

The ventilatory response to hypoxia is increased ventilation with increased tidal volume and intrathoracic pressure swings, which alone stimulate intrathoracic stretch receptors in the venous, atrial, pulmonary vasculature (173), and lung parenchyma (168). In humans, voluntary normoxic isocapnic hyperventilation at rest (>30 L/min) causes diuresis and natriuresis (77). However, the importance of this finding is unclear since hypoxic hyperventilation in normal resting subjects does not usually exceed 10–15 L/min at least above 10% O_2 (319), and diuresis and natriuresis occur in constantly ventilated animals (144). Furthermore, pulmonary denervation in spontaneously breathing rabbits does not alter the increase in RSNA with inspired hypoxic gas (241), suggesting that pulmonary stretch receptors do not attenuate RSNA in hypoxia.

The hemodynamic response to hypoxia includes an increase in cardiac output, pulmonary artery pressure, and increased venous return. These processes shift blood volume into the central circulation and stimulate intrathoracic low pressure baroreceptors, which increase urine output (88,164,165,188,195,288) and dampen RSNA with hypoxia (208,288). However, data on intrathoracic volume and pressure changes in the vena cavae and atria are not in agreement, and in many cases no elevations in central venous pressures or blood volume with acute hypoxia are found (101,154,166,176,177,232,289,307). Bilateral cervical vagotomy studies would be useful to assess the role that these intrathoracic baroreceptors play in hypoxic diuresis as has been shown with the diuresis following volume expansion (332).

Hypocapnia

Hypocapnia depresses proximal tubular bicarbonate reabsorption (108), and the resulting bicarbonaturia is accompanied by water, Na^+, and K^+ (110). Although hypocapnia alone might account for the urinary changes with hypoxia, normoxic hypocapnia results in more potassium loss and less sodium loss than hypoxic hypocapnia or isocapnic hypoxia (77,107,110,180,298,314). Hypocapnic-mediated sodium bicarbonate excretion does not appear essential because natriuresis occurs with hypoxia in mechanically ventilated animals whose ventilation and $Paco_2$ are fixed (144,166). In spontaneously breathing humans, Swenson et al. (319) found only a very weak correlation between sodium and bicarbonate excretion with acute hypoxia, and Krapf et al. (180) and Gledhill et al. (110) found that hypoxic natriuresis exceeded bicarbonate output two- and fivefold. Further evidence that hypoxic natriuresis in man is not simply a hypocapnic natriuresis is the finding that hypocapnic normoxia and hypocapnic hypoxia, but not isocapnic hypoxia, reduce proximal tubular fluid reabsorption (62).

Hypocapnia may also alter the renal response to hypoxia apart from its effects on bicarbonate reabsorption and acid excretion. Respiratory alkalosis blunts the sympathetic responses to hypoxia in the CNS (102,128) and in the peripheral chemoreceptors (208). A decrease in RSNA (102) may partly explain the higher renal blood flow observed when spontaneously breathing rats were permitted to become hypocapnic in response to hypoxia than in those given CO_2 to prevent hypocapnia (209). Hypocapnia may also alter the balance of salt- and water-regulating hormones and blunt renal sodium avidity, since hypercapnia activates the renin-angiotensin-aldosterone system (5,260,278) and is a potent stimulant of RSNA (203a,303). These findings could explain the adverse effects on AMS symptoms of preventing or failing to achieve a degree of hypocapnia when hypoxic (20,75,126,139,200). For example, Grover et al. (123) showed in humans exposed to hypobaric hypoxia (440 mmHg) that prevention of hypocapnia (breathing 3.7% CO_2) also prevented the negative fluid balance that occurred in control subjects.

These studies reveal a complex interaction of CO_2 and O_2 on peripheral and central chemoreceptors on ventilation (229), hemodynamics (208,250), and renal hormones, but in general it appears that hypocapnia has moderating effects on those neurohumoral factors promoting salt and water retention in hypoxia.

Hypobaria

Whether the reduced barometric pressure of high altitude itself alters renal function and fluid balance is not clear. Exposure to hypobaric normoxia (Pb = 400 mmHg, Pio_2 = 150 mmHg) of water-loaded humans caused no differences in blood pressure and urine output, urinary Na, K, or solute excretion or urinary osmolality, or any significant changes in circulating ADH, cortisol, plasma renin activity, or prolactin over one hour (136), unlike normobaric and hypobaric hypoxia. Shibamato et al. (301) found that hypobaric normoxia (PB = 326 mmHg; Pio_2 = 150 mmHg) alone had no effect on blood pressure, heart rate, central venous pressure, and renal sympathetic nerve activity (RSNA) in anesthetized rabbits in contrast to the equal hypertension and increased RSNA of normobaric and hypobaric hypoxia. Tucker et al. (325) studied water-loaded human subjects and found that they developed an antidiuresis with rapid decompression from 630 to 430 mmHg but not with equivalent normobaric hypoxia. Interestingly, they found that if the subjects breathed 100% oxygen 90 minutes before decompression, there was no antidiuresis.

The effects of chronic hypobaria (Pb = 258 mmHg) were studied by Epstein and Saruta (92,93). Over 9 days normoxic (P_{IO_2} = 150 mmHg) and hyperoxic (P_{IO_2} = 258 mmHg) hypobaria had differing effects on the sudden restriction of dietary sodium (200 mEq/day to 10 mEq/day). They found that hyperoxic hypobaria caused no departures from pattern observed in normobaric subjects. In contrast, the same sodium restriction in hypobaric normoxia was associated with greater sodium and water retention, less weight loss, and a 40% fall in GFR. A smaller fall in GFR was noted by Glatte and Giannetta (109) in a 56-day sojourn at 258 mmHg (P_{IO_2} = 175 mmHg), suggesting that, analogous to the acute studies, there may be some mitigating effect of hyperoxia on the renal responses to hypobaria.

Hypobaria thus appears to possibly enhance a sodium- and water-retaining influence of hypoxia. The reasons for this deserve further study. One explanation for the differences between hypobaric and normobaric hypoxia is that the resulting Pa_{O_2} may be less in hypobaric hypoxia than equivalent normobaric hypoxia due to a deleterious effect of hypobaria on pulmonary capillary permeability and gas exchange efficiency (193). Another possibility is greater activation of the sympathetic nervous system. In humans, hypobaric hypoxia, but not always normobaric hypoxia, is associated with increased circulating norepinephrine (55,215, 267,284). It is interesting to speculate that hyperoxic exposure prior to hypobaric hypoxia may dampen the sympathetic nervous system as it does when given before exercise (135), possibly via a mechanism involving the peripheral chemoreceptors.

Cold

Exposure to cold will itself influence fluid balance. As part of the normal thermoregulatory response, peripheral vasoconstriction shifts blood into the central circulation, causing release of ANP and reducing baroreceptor input, which lead to salt and water elimination by the kidney (cold diuresis). As core temperature falls, further dehydration develops as thirst is suppressed and diuresis continues.

Cold exposure may alter the response to hypoxia, but there are few data. Purshottam et al. (256) showed that severe acute normobaric hypoxia over 6–18 hours (equivalent to 4300 m) caused a mild antidiuresis in young men, but when combined with environmental cold (8°C), the subjects had a diuresis. In rats, combined cold exposure (5°C) and milder hypoxia (12% O_2) caused equally increased urine output for any given water intake, as did cold or hypoxia separately (100). The effect of cold at high altitude on fluid balance may be very difficult to predict in any particular situation due to factors such as the severity of the hypoxia, duration of cold exposure, and whether hypothermia develops. However, in general, most aspects of the acute cold response promote renal salt and water losses that will either augment the diuresis with mild to moderate hypoxia or counteract the antidiuresis of severe hypoxia.

Exercise

Ascent to high altitude generally involves exertion. Given the profound effects of exercise on hemodynamics and renal humoral mediators, it is not surprising that climbing may play a critical role in fluid balance at high altitude. Acute normoxic exercise, especially performed in the upright position (59), causes renal blood flow to decrease in proportion with exercise intensity, although GFR falls only by 25% with maximal exercise. Urinary water and sodium outputs decrease during exercise (308) even in the absence of changes in GFR, elevations in ANP, and an apparent insensitivity to ADH. The antidiuretic and antinatriuretic states persist for 30–60 minutes after cessation of exertion. With normoxic exercise there are increases in ANP, aldosterone, renin, serum potassium, catecholamines, and ACTH (205, 292, 309). ADH rises with short-term heavy exercise but rises less or not at all in well-trained individuals doing lower-level sustained work (336). Interestingly, there appears to be no suppression of aldosterone by ANP, which is observed during resting hypoxemia and in many edematous states including hypoxemic lung disease (205). Exercise also increases RSNA in proportion to the exercise intensity (212,240), whose abolition by renal nerve denervation blunts much of the renal hemodynamic effects and changes in urine output (59).

Exercise in hypoxia produces qualitatively many of the same neurohumoral responses as does normoxic exercise (185,198,335), but a few interesting differences occur depending upon the length of exercise, intensity of the hypoxia, degree of exercise-induced O_2 desaturation, and time of acclimatization. In all studies there is a small suppression of the normal aldosterone increase of normoxic exercise (200,201,227,305). Bouissou et al. (40–42) and Olsen et al. (244) showed that some of these same features are exhibited by exercising subjects at 4350 m after 2–5 days of adaptation, although the rise in plasma renin and ANP with exercise was reduced from that measured in normoxic exercise. The differences between these studies and those performed with acute hypoxic exposures suggest that 2–5 days of acclimatization may blunt certain of the humoral responses to exercise that are considered to promote the fluid retention and acute illness of high altitude (22).

Milledge and colleagues (223–225,228,344,348) have very elegantly explored the role that less vigorous but sustained exercise exerts on fluid balance whether done at low or high altitudes. The virtue of these ''hill-walking'' studies (day-long hiking >20 km at altitudes between 1100 and 3600 m for 5–7 days) is the approximation of the exertional stress of mountaineering on fluid balance. Their studies of subjects maintaining a fixed salt intake while climbing clearly document marked sodium and water retention. Calculations of body fluid compartment changes show an expansion of the extracellular and plasma volumes at the expense of intracellular space. Associated with these changes were sustained elevations in aldosterone, plasma renin, and ANP, but Williams et al. (344) showed no changes in ADH in accord with the studies of Wade (336).

Thus, exercise promotes salt and water avidity and frank retention if the stimulus is maintained. The additional effect of hypoxia on exercise appears small since the studies of Milledge and colleagues show very little difference in salt and water retention between normoxic and hypoxic climbing and hiking. Field studies comparing differing exercise regimens with and without the attendant hypoxemia of that altitude are needed to separate any significant role hypobaria adds to the high-altitude response.

In contrast to upright exercise, there is diuresis and natriuresis with recumbancy (321). Changes in posture influence renal function and may have relevance to fluid balance studies at high altitude. The importance of body position in modulating the renal response was studied by Loeppky et al. (197) in subjects moved from 5400 feet to 10,700 feet either maintained at constant head-down bedrest for 8 days or permitted normal diurnal sedentary activities. They found that head-down bedrest promoted greater urine output and sodium excretion, lower ADH and aldosterone levels, and greater ANP levels than in the normal activity group. There were no significant differences in weight, plasma catecholamine levels, or GFR between groups. Interestingly, despite more negative fluid balance and lack of exercise in recumbent subjects, AMS symptoms were slightly greater.

IV. Conclusions

The potentially advantageous response of negative fluid balance on going to high altitude reflects the dominance of hypoxic stimulation of the peripheral chemoreceptors overriding the negative (or fluid retaining) consequences of cerebral hypoxia and hypoxic systemic vasodilation and hypotension. Without this opposition, especially when hypobaria and exercise are involved, there is intense activation of the sympathetic nervous system and efferent RSNA as well as marked increases in ADH, aldosterone, and ANP. These opposing influences are shown in Figure 3. The complexity of the picture is further made evident by the possibility of other nonperipheral chemoreceptor-mediated mechanisms of natriuresis, such as direct hypoxic downregulation of renal tubular membrane Na transport proteins.

The several ways in which peripheral chemoreceptor afferent activity may oppose hypoxic fluid retention are shown in Figure 3 and include the direct stimulation of ventilation, possible release of a natriuretic substance(s), and sufficient (but not excessive) sympathetic nervous system activation to counteract the direct vasodilating effect of hypoxia. Increased ventilation limits the ultimate magnitude of arterial hypoxemia and generates an acute respiratory alkalosis and hyperpnea. Hypocapnia promotes sodium excretion with the attendant renal loss of bicarbonate and acts to limit RSNA, as does hyperpnea via stimulation of intrathoracic stretch receptors. A natriuretic humoral factor found to be consistently produced with hypoxic stimulation of the peripheral chemoreceptors will be of considerable importance.

A strong teleological argument can be made that diuresis and negative sodium balance may be appropriate preemptive responses to prevent or limit the edema

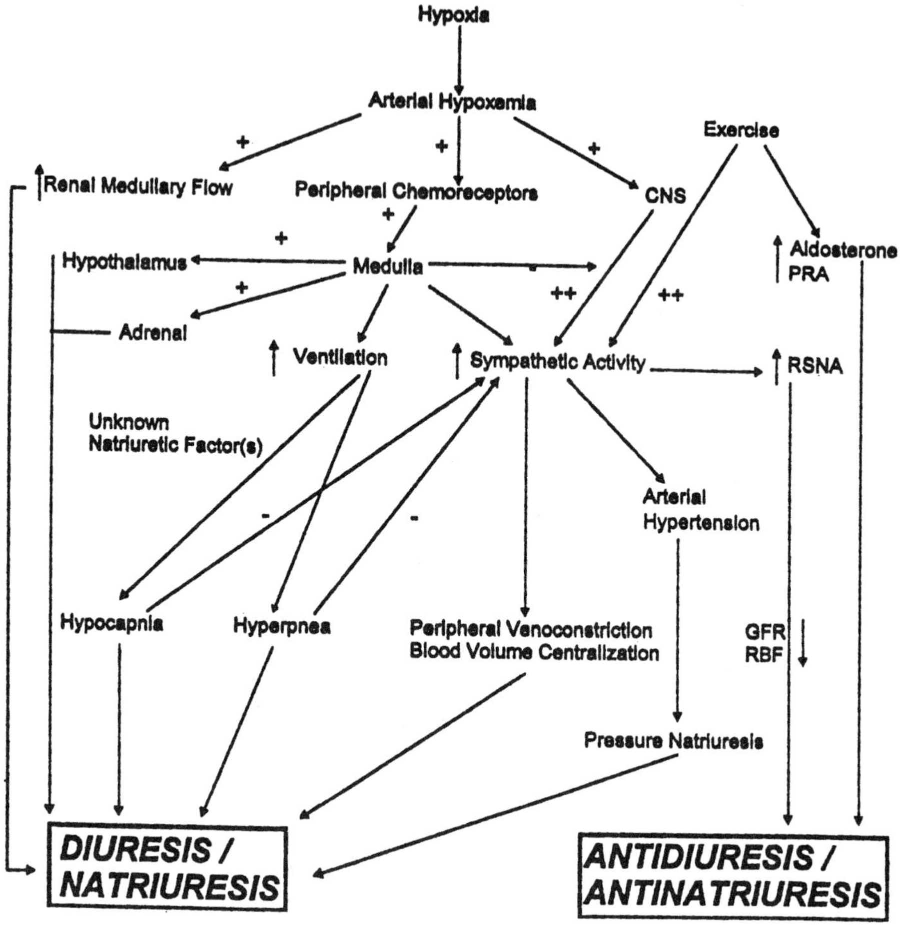

Figure 3 The potential multiple mechanisms by which environmental hypoxia can elicit either diuresis and natriuresis or antidiuresis and antinatriuresis. (+ or ++) reflects stimulation and (−) indicates inhibition of the indicated pathway or response.

states associated with high-altitude illness, but whether there exists a causal link of AMS with fluid retention has not been convincingly established despite considerable investigative interest and controversy (see Chapter 22). Hackett and coworkers (125,126) have clearly shown that subjects ill at high altitude have relative hypoventilation and a preceding interval of salt and water retention, but these results raise the question of cause and effect. One of the earliest field studies, that of Stampfli and Eberle (313), found fluid retention followed rather than preceded the development of AMS. In a hypobaric chamber study, Koller et al. (178) showed that residents of high altitude (1650 m) tolerated a rapid ascent to 6000 m better than lowland residents despite the fact that the latter group had a much greater natriuresis and diuresis than did the highlanders. Furthermore, it is not clear that peripheral chemoreceptor sensitivity plays an important role in fluid balance of climbers, since Bärtsch and colleagues found that urine output in mountaineers with or without AMS and in subjects airlifted to high altitude did not correlate with the hypoxic ventilatory response and that fluid retention was only evident after subjects became ill (Bärtsch et al., 1998; Biollaz et al., personal communication). These aforementioned studies do not rule out the possibility that internal fluid balance differences may be important in the pathogenesis of AMS, as suggested by recent results of Roach et al. (274), who found that subjects developing AMS appeared to differ from those not becoming ill by experiencing an increase in PV that was correlated with AMS. Additional prospective well-controlled field studies emphasizing rigorous renal and compartmental fluid balance measurements are clearly needed.

The physiology of fluid balance at high altitude reflects an integration of many organ systems with (sometimes opposing) inputs and responses. Future investigation in humans should explore further the peripheral chemoreceptor role in hypoxic normo- and hypobaria and in resting vs. exercise conditions. Specific hormonal antagonists will be extremely useful to rule in or out the importance of a particular hormonal change with hypoxia. In animals, anesthesia should be avoided whenever possible. Use of transgenic animals with knockout and/or overexpression of various hormonal mediators and their receptors should be pursued, coupling them with use of humoral antagonists and manipulation of the peripheral chemoreceptors. Greater focus needs to be directed at intrarenal hemodynamics, blood flow distribution and innervation, as well as measurements of newer humoral mediators affecting salt and water transport. Lastly, the effects of hypoxia in the kidney on hormone receptors and ion transporters, their number, binding characteristics, and membrane localization as well as gene expression, will likely yield important new insights into renal physiology in general and hypoxia adaptation in particular.

References

1. Ahmad S, Kenny SM, Scribner BH. Hypertension and a digoxin-like substance in the plasma of dialysis patients: possible marker for a natriuretic hormone. Clin Physiol Biochem 1986; 4:210–216.
2. Al-Obaidi M, Whitaker EM, Karim F. The effect of discrete stimulation of the carotid body chemoreceptors on atrial natriuretic peptide in anaesthetized dogs. J Physiol 1991; 443:529–531.
3. Anand IS, Chandrashenkar Y, Rao SK, Malhotra RM, Ferrari R, Chandana J, Ramesh B, Shetty KJ, Boparai MS. Body fluid compartments, renal blood flow, and hormones at 6000 m in normal subjects. J Appl Physiol 1993; 74:1234–1239.
4. Anderson JV, Struthers AD, Payne NN. Atrial natriuretic peptide inhibits the aldosterone response to angiotensin II in man. Clin Sci 1986; 70:507–512.
5. Anderson RJ, Henrich W, Gross PA, Dillingham MA. Role of the renal nerves, angio-

tensin II and prostaglandins in the antinatriuretic response to acute hypercapnic acidosis in the dog. Circ Res 1982; 50:294–300.

6. Anderson RJ, Pluss RG, Berns AS, Jackson JT, Arnold PA, Schrier RW, McDonald KM. Mechanism of effect of hypoxia on renal water excretion. J Clin Invest 1978; 62:769–777.
7. Armstrong HG. Principles and Practice of Aviation Medicine. Baltimore: Williams & Wilkins, 1939:261–293.
8. Ashack R, Farber MO, Weinberger MH, Robertson GL, Fineberg NS, Manfredi F. Renal and hormonal responses to acute hypoxia in normal individuals. J Lab Clin Med 1985; 106:12–16.
9. Asmussen E, Nielsen M. Studies on the initial increase in O_2-capacity of the blood at low O_2 pressure. Acta Physiol Scand 1945; 9:75–87.
10. Aw TY, Wilson E, Hagen TM, Jones DP. Determinants of mitochondrial O_2 dependence in kidney. Am J Physiol 1987; 253:F440–F447.
11. Axelrod DR, Pitts RF. Effects of hypoxia on renal tubular function. J Appl Physiol 1952; 4:593–601.
12. Ayres PJ, Hurter RC, Williams ES. Aldosterone excretion and potassium retention in subjects living at high altitude. Nature 1961; 191:78–80.
13. Baertschi AJ, Adams JM, Sullivan MP. Acute hypoxemia stimulates atrial natriuretic factor secretion in vivo. Am J Physiol 1988; 255:H295–H300.
14. Baertschi AJ, Jiao J-H, Carlson DE, Campbell RW, Teague WG, Willson D, Gann DS. Neural control of ANF release in hypoxia and pulmonary hypertension. Am J Physiol 1990; 259:H735–H744.
15. Baertschi AJ, Teague WG. Alveolar hypoxia is a powerful stimulus for ANF release in conscious lambs. Am J Physiol 1989; 256:H990–H998.
16. Bagrov AY, Fedorova OV, Austin-Lane JL, Dmitrieva RI, Anderson DE. Endogenous marinobufagenein-like immunoreactive factor and Na+/K+ ATPase inhibition during voluntary hypoventilation. Hypertension 1995; 26:781–788.
17. Baines AD, Adamson G, Wojciechowski P, Pliura D, Kluger R. Effect of modifying O_2 diffusivity and delivery on glomerular and tubular function in hypoxic perfused kidney. Am J Physiol 1998; 274:F744–F752.
18. Bardsley PA, Johnson BF, Barer GR. Natriuresis secondary to carotid chemoreceptor stimulation with almitrine bismethylate in the rat: the effect on kidney function and the response to renal denervation and deficiency of antidiuretic hormone. Biomed Biochim Acta 1991; 50:175–182.
19. Barker ES, Singer RB, Elkinton JR, Clark JK. The renal response in man to acute experimental respiratory acidosis and alkalosis. J Clin Invest 1957; 46:919–923.
20. Bärtsch P, Baumgartner RW, Waber U, Maggiorini M, Oelz O. Comparison of carbon dioxide enriched, oxygen enriched, and normal air in treatment of acute mountain sickness. Lancet 1990; 336:772–775.
21. Bärtsch P, Jülg B, Hohenhaus E. Urine volume in acute mountain sickness is not related to hypoxic ventilatory response. Eur Respir J 1995; 8:625.
22. Bärtsch P, Maggiorini M, Schobersberger W, Shaw S, Rascher W, Girard J, Weidman P, Oelz O. Enhanced exercise-induced rise of aldosterone and vasopressin preceding mountain sickness. J Appl Physiol 1991; 71:136–141.
23. Bärtsch P, Pfluger N, Audetat M, Shaw S, Weidman P, Vock P, Vetter W, Rennie D, Oelz O. Effects of slow ascent to 4559 M on fluid homeostasis. Aviat Space Environ Med 1991; 62:105–110.

24. Bärtsch P, Shaw S, Franciolli M, Gnadinger MP, Weidman P. Atrial natriuretic peptide in acute mountain sickness. J Appl Physiol 1988; 65:1929–1937.
25. Baumann JE, Persson PB, Ehmke H, Nafz B, Kirchheim HR. Role of endothelium-derived relaxing factor in renal autoregulation in conscious dogs. Am J Physiol 1992; 263:F208–F213.
26. Baumgartl H, Leichtweiss H-P, Lubbers DW, Weiss C, Huland H. The oxygen supply of the dog kidney; measurements of intrarenal PO_2. Microvasc Res 1972; 4:247–257.
27. Baylis PH, Stockley RA, Heath DA. Effect of acute hypoxaemia on plasma vasopressin in conscious man. Clin Sci Mol Med 1977; 53:401–404.
28. Becker EL, Schilling JA, Harvey RB. Renal function in man acclimatized to high altitude. J Appl Physiol 1957; 10:79–80.
29. Becker EL. Renal function in polycythemic dogs. J Appl Physiol 1957; 10:75–78.
30. Behm R, Honig A, Griethe M, Schmidt M, Schneider P. Sustained suppression of voluntary sodium intake of spontaneously hypertensive rats (SHR) in hypobaric hypoxia. Biomed Biochim Acta 1984; 43:975–985.
31. Behm R, Gerber B, Habeck J-O, Huckstorf C, Ruckborn K. Effect of hypobaric hypoxia and almitrine on voluntary salt and water intake in carotid body-denervated spontaneously hypertensive rats. Biomed Biochim Acta 1989; 48:689–695.
32. Behm R, Mewes H, Keizer WHD, Unger T, Rettig R. Cardiovascular and renal effects of hypoxia in conscious carotid-body denervated rats. J Appl Physiol 1993; 74:2795–2800.
33. Bello-Reuss E. Effect of catecholamines on fluid reabsorption by the isolated proximal convoluted tubule. Am J Physiol 1980; 238:F347–F352.
34. Berger EY, Gladston M, Horwitz SA. The effect of anoxic anoxia on the human kidney. J Clin Invest 1949; 28:648–652.
35. Bestle MH, Olsen NV, Roach RC, Anderson LJ, Bie P. Renal sensitivity to vasopressin in man in acute hypoxia. FASEB J 1998; 12:A722.
36. Bischoff A, Erdbrügger W, Smits J, Michel M. Neuropeptide Y-mediated diuresis and natriuresis in anaesthetized rats is independent of renal blood flow reduction. J Physiol 1996; 495:525–534.
37. Blaustein MP. Physiological effects of endogenous ouabain: control of intracellular Ca stores and cell responsiveness. Am J Physiol 1993; 264:C1367–C1387.
38. Blume FD, Boyer SJ, Braverman LE, Cohen A, Dirkse J, Mordes JP. Impaired osmoregulation at high altitude: studies on Mt Everest. J Am Med Assoc 1984; 252:524–526.
39. Bouissou P, Guezennec CY, Defer G, Galen F-X, Fiet J, Pesquies PC. Dissociated response of aldosterone from plasma renin activity during prolonged exercise under hypoxia. Horm Metab Res 1988; 20:517–521.
40. Bouissou P, Perronet F, Brisson G, Helie R, Ledoux M. Fluid-electrolyte shift and renin-aldosterone responses to exercise under hypoxia. Horm Metab Res 1987; 19: 331–334.
41. Bouissou P, Perronet F, Brisson G, Helie R, Ledoux M. Metabolic and endocrine responses to graded exercise under hypoxic conditions. Eur J Appl Physiol 1986; 555: 290–294.
42. Bouissou P, Richalet J-P, Galen FX, Lartigue M, Larmignat P, Devaux F, Dubray C, Keromes A. Effect of beta adrenergic blockade on renin-aldosterone and ANF during exercise at altitude. J Appl Physiol 1989; 67:141–146.
43. Brahmachari HD, Malhotra MS, Ramachandran K, Radhakrishnan U. Progressive

changes in plasma cortisol, antidiuretic hormone and urinary volume of normal low-landers during short stay at high altitude. Ind J Exp Biol 1973; 11:454–455.
44. Brand PH, Coyne KB, Kostrzewski KA, Shier D, Metting PJ, Britton SL. Pressure natriuresis and autonomic function in conscious dogs. Am J Physiol 1991; 261:R802–R810.
45. Brands MW, Freeman RH. Aldosterone and renin inhibition by physiological levels of atrial natriuretic factor. Am J Physiol 1988; 254:1011–R1016.
46. Brauer H, Gens H, Lederhos C, Sanchez R, Schuster R, Quies W, Honig A. Cardiorespiratory and renal responses to arterial chemoreceptor stimulation by hypoxia or almitrine in men. Clin Exp Pharmacol Physiol 1996; 23:1021–1027.
47. Breslow MJ, Ball TD, Miller CF, Raff H, Traystman RJ. Adrenal blood flow and secretory relationships during hypoxia in anesthetized dogs. Am J Physiol 1989; 257: H1458–H1465.
48. Breslow MJ. Regulation of adrenal medullary and cortical blood flow. Am J Physiol 1992; 262:H1317–H1330.
49. Brezis M, Agmon Y, Epstein FH. Determinants of intrarenal oxygenation. Effects of diuretics. Am J Physiol 1994; 267:F1059–F1062.
50. Brezis M, Heyman SN, Dinour D, Epstein FH, Rosen S. Role of nitric oxide in renal medullary oxygenation; studies in isolated and intact rat kidneys. J Clin Invest 1991; 88:390–395.
51. Brezis M, Rosen S. Hypoxia of the renal medulla—its implications for disease. N Engl J Med 1995; 332:647–655.
52. Brull L, Divry H. Metabolic and secretory activity of the kidney under anoxemia. Arch Int Physiol 1951; 58:415–423.
53. Bruns FJ. Decrease in renal perfusion pressure and sodium excretion by hypoxia in the dog. Proc Soc Exp Biol Med 1978; 159:48–472.
54. Bryan AH, Ricketts HT. The effect in man of chronic intermittent anoxia on glucose tolerance and the urinary excretion of sodium, potassium, chloride, phosphorus, nitrogen, and 17-ketosteroids. J Clin Endocrinol 1944; 4:450–464.
55. Bubb WJ, Howley ET, Cox RH. Effect of various levels of hypoxia on plasma catecholamines at rest and during exercise. Aviat Space Environ Med 1983; 54:637–640.
56. Burrill MW, Freeman S, Ivy AC. Sodium, potassium and chloride excretion of human subjects exposed to a simulated altitude of eighteen thousand feet. J Biol Chem 1945; 157:297–302.
57. Bursaux E, Poyart C, Bohn B. Renal hemodynamics and renal O_2 uptake during hypoxia in the anesthetized rabbit. Pflugers Arch 1976; 365:213–220.
58. Cargill RI, Kiely DG, Clark RA, Lipworth BJ. Hypoxaemia and release of endothelin-1. Thorax 1995; 50:1308–1310.
59. Castenfors J. Renal function during exercise. Acta Physiol Scand 1967; 70 (suppl 293): 1–40.
60. Chatterji JC, Ohri VC, Chadra KS, Das BK, Akhtar M, Tewari SC, Bhattacharji P, Wadwha A. Serum and urinary cation changes on acute induction to high altitude (3200 and 3771 metres). Aviat Space Environ Med 1982; 53:576–579.
61. Chou SY, Porush JG, Faubert PF. Renal medullary circulation; hormonal control. Kidney Int 1990; 37:1–13.
62. Christiansen H, Hansen JM, Klausen T, Fogh-Andersen N, Kanstrupp IL, Olsen NV. Effects of isocapnic/hypocapnic hypoxemia and hyperventilation on renal function in

normal water-loaded volunteers. In: Sutton JR, Houston CS, Coates G, eds. Hypoxia and the Brain. Burlington, VT: Queen City Printers, 1995:317.

63. Claybaugh JR, Hansen JE, Wozniak DB. Response of antidiuretic hormone to acute exposure to mild and severe hypoxia in man. J Endocrinol 1978; 77:157–160.
64. Claybaugh JR, Wade CE, Sato AK, Cucinell SA, Lane JC, Maher JT. Antidiuretic hormone responses to eucapnic and hypocapnic hypoxia in humans. J Appl Physiol 1982; 53:815–823.
65. Claybaugh JR, Wade CE, Cucinelli SA. Fluid and electrolyte balance and hormonal response to the hypoxic environment. In: Claybaugh JR, Wade CE, eds. Hormonal Regulation of Fluid and Electrolytes. New York: Plenum, 1989:187–214.
66. Clozel J-P, Saunier C, Hartemann D, Allam M, Fischli W. Effects of hypoxia and hypercapnia on atrial natriuretic factor and plasma renin activity in conscious dogs. Clin Sci 1989; 76:249–254.
67. Cogan MG. Neurogenic regulation of proximal bicarbonate and chloride reabsorption. Am J Physiol 1986; 250:F22–F26.
68. Cohen JJ. Relationship between energy requirements for Na+ reabsorption and other renal functions. Kidney Int 1986; 29:32–40.
69. Colice G, Ramirez G. Effect of hypoxemia on the renin-angiotensin-aldosterone system in humans. J Appl Physiol 1985; 58:724–730.
70. Colice G, Ramirez G. Aldosterone response to angiotensin II during hypoxemia. J Appl Physiol 1986; 61:150–154.
71. Colice G, Yen S, Ramirez G, Dietz J, Ou L-C. Acute hypoxia-induced diuresis in rats. Aviat Space Environ Med 1991; 62:551–554.
72. Consolazio CF, Johnson HL, Krzywicki HJ, Daws TA. Metabolic aspects of acute altitude exposure (4300 meters) in adequately nourished humans. Am J Clin Nutr 1972; 25:23–29.
73. Consolazio CF, Matoush LO, Johnson HL, Daws TA. Protein and water balances of young adults during prolonged exposure to high altitude (4300 meters). Am J Clin Nutr 1968; 21:154–161.
74. Cowley AJ. A tribute to Eric Muirhead. Evolution of the medullipin concept of blood pressure control. Nephrol Dial Transplant 1995; 10:1137–1141.
75. Cruz JC, Grover RF, Reeves JT, Maher JT, Cymerman A, Denniston JC. Sustained venoconstriction in man supplemented with CO_2 at high altitude. J Appl Physiol 1976; 40:96–100.
76. Curran-Everett DC, Claybaugh JR, Miki K, Hong SK, Krasney JA. Hormonal and electrolyte responses of conscious sheep to 96 h of hypoxia. Am J Physiol 1988; 255: R274–R283.
77. Currie JCM, Ullmann E. Polyuria during experimental modification of breathing. J Physiol 1961; 155:438–455.
78. Danielson BG, Gransjo G, Persson E, Uhlfendahl HR, Wolgast M. Renal function in the dog in acute disturbances of the acid-base balance. Acta Anaesth Scand 1973; 17: 88–102.
79. DeAngelis C, Farrace S, Urbani L, Porcu S, Ferri C, D'Amelio R, Santucci A, Balsano F. Effects of high altitude exposure on plasma and urinary digoxin-like immunoreactive substance. Am J Hypertension 1992; 5:600–607.
80. DeAngelis C, Haupert GT. Hypoxia triggers release of an endogenous inhibitor of Na^+-K^+-ATPase from midbrain and adrenal. Am J Physiol 1998; 274:F182–F188.
81. DeAngelo SA. Urinary output and phosphorus excretion in human subjects during

prolonged exposures at low simulated altitudes. Proc Soc Exp Biol Med 1946; 44:13–17.

82. Delgado C. Funcion de concentration renal en al habitante de la altura. MD thesis. Lima, Peru: Cayetano Heredia University, 1969.
83. DiBona GF, Kopp EC. Neural control of renal function. Physiol Rev 1997; 77:75–197.
84. Dinour D, Brezis M. Effects of adenosine on intrarenal oxygenation. Am J Physiol 1991; 261:F787–F791.
85. Durkot MJ, Hoyt RW, Darigrand A, Hubbard LJ, Kamimori GH, Cymerman A. Chronic hypobaric hypoxia decreases intracellular and total body water in microswine. Comp Biochem Physiol 1996; 114A:117–121.
86. du Souich P, Saunier C, Hartemann D, Sautegau A, Ong H, Larose P, Babini R. Effect of moderate hypoxia on atrial natriuretic factor and arginine vasopressin in normal man. Biochem Biophys Res Comm 1987; 148:906–912.
87. Ebert TJ, Groban L, Muzi M, Hanson M, Cowley AW. ANP-mediated volume depletion attenuates renal responses in humans. Am J Physiol 1992; 263:R1303–R1308.
88. Echtenkamp SF, Gilmore JP. Intravascular mechano-receptor modulation of renal sympathetic nerve activity in the cat. Am J Physiol 1980; 238:H801–H808.
89. Eckhardt KU, Kurtz A, Bauer C. Erythropoietin formation is related to proximal tubular function. Am J Physiol 1989; 256:F942–F947.
90. Ehmke H, Persson PB, Seyfarth M, Kirchheim HR. Neurogenic control of pressure natriuresis in conscious dogs. Am J Physiol 1989; 256:F63–F70.
91. Epstein FH, Balaban RS, Ross BD. Redox state of cytochrome aa3 in isolated perfused rat kidney. Am J Physiol 1982; 243:F356–F363.
92. Epstein M, Saruta T. Effects of an hyperoxic hypobaric environment on renin-aldosterone in normal man. J Appl Physiol 1973; 34:49–52.
93. Epstein M, Saruta T. Effects of simulated high altitude on renin-aldosterone and Na homeostasis in normal man. J Appl Physiol 1972; 33:204–210.
94. Erslev AJ, Caro J, Besarab A. Why the kidney? Nephron 1985; 41:213–216.
95. Foldi M, Solti F, Rev J, Koltay E. Wirkung der Hypoxie auf die Nierenfunktion beim Menschen. Acad Sci Hung 1956; 10:43–47.
96. Fontaneles MC, Villar-Palasi C, Fong G, Larner J, Guerrant L. Partial characterization of an ANF/urodilatin like substance released from the perfused rat kidney under hypoxia. Braz J Med Biol Res 1993; 26:75–79.
97. Forte LR, Fan X, Hamra FK. Salt and water homeostasis: uroguanylin is a circulating peptide hormone with natriuretic activity. Am J Kidney Dis 1996; 28:296–304.
98. Frayser R, Rennie ID, Gray GW, Houston CS. Hormonal and electrolyte response to exposure to 17,500 feet. J Appl Physiol 1975; 38:636–642.
99. Fregly MJ. Thirst immediately following removal of rats from graded levels of hypoxia. Proc Soc Exp Biol Med 1971; 138:448–453.
100. Fregly MJ, Nelson EL, Tyler PE. Water exchange in rats exposed to cold, hypoxia, and both combined. Aviat Space Environ Med 1976; 47:600–607.
101. Fritts HW, Odell JE, Harris P, Braunwald EW, Fishman AP. Effects of acute hypoxia on the volume of blood in the thorax. Circulation 1960; 22:216–219.
102. Fukoda Y, Sato A, Suzuki A, Trzebski A. Autonomic nerve and cardiovascular responses to changing oxygen and carbon dioxide levels in the rat. J Autonomic Nerv Syst 1989; 28:61–74.

103. Galat JA, Robinson AV, Rhodes RS. Effect of hypoxia on renal flow. J Trauma 1988; 28:955–961.
104. Garcia R, Lachance D, Thibault G. Positive inotropic action, natriuresis and atrial natriuretic factor release induced by endothelin in the conscious rat. J Hypertens 1990; 8:725–731.
105. Garcia-Estan J, Roman RJ. Role of renal interstitial hydrostatic pressure in the pressure diuresis response. Am J Physiol 1989; 256:F63–F70.
106. Garvin JL. ANF inhibits norepinephrine-stimulated fluid absorption in rat proximal straight tubules. Am J Physiol 1992; 263:F581–F585.
107. Gennari FJ, Cohen JJ. Role of the kidney in potassium homeostasis. Lessons from acid-base disturbances. Kidney Int 1975; 8:1–5.
108. Gennari FJ, Goldstein MB, Cohen JJ. The nature of the renal adaptation to chronic hypercapnia. J Clin Invest 1972; 51:1722–1730.
109. Glatte HV, Giannetta CL. Study of man during a 56 day exposure to an oxygen-helium atmosphere at 258 mmHg total pressure: renal response. Aerospace Med 1966; 37: 559–562.
110. Gledhill N, Beirne GJ, Dempsey JA. Renal response to short-term hypocapnia in man. Kidney Int 1975; 8:376–386.
111. Goetz KL. Renal natriuretic peptide (urodilatin) and atriopeptin: evolving concepts. Am J Physiol 1991; 261:F921–F932.
112. Goetz K, Drummer C, Ahu JL, Leadley R, Fiedler F, Gerzer R. Evidence that urodilatin, rather than ANP, regulates renal sodium excretion. J Am Soc Nephrol 1990; 1: 867–874.
113. Gomori P, Kovacs AGB, Takacs L, Foldi M, Szabo G, Nagy Z, Wiltner W. The control of the renal circulation in hypoxia. Acta Med Hung 1960; 16:43–60.
114. Gonzalez NC, Clancy RL. Acid-base regulation in prolonged hypoxia: effect of increased PCO_2. Respir Physiol 1986; 64:213–237.
115. Gonzalez NC, Clancy RL. Acid-base regulation in prolonged hypoxia: effect of increased PCO_2. Respir Physiol 1986; 64:213–237.
116. Gonzalez NC, Clancy RL. Intracellular pH regulation during prolonged hypoxia in rats. Respir Physiol 1986; 65:331–339.
117. Gonzalez NC, Albrecht T, Sullivan LP, Clancy RL. Compensation of respiratory alkalosis after acclimatization to simulated altitude. J Appl Physiol 1990; 69:1380–1386.
118. Gotshall RW, Miles DS, Sexson WR. Renal oxygen consumption during progressive hypoxemia in the anesthetized dog. Proc Soc Exp Biol Med 1983; 174:363–367.
119. Gotshall RW, Miles DS, Sexson WR. The combined effects of hypoxemia and mechanical ventilation on renal function. Aviat Space Environ Med 1986; 57:782–786.
120. Gougoux A, Kaehny WD, Cohen JJ. Renal adaptation to chronic hypocapnia: dietary restraints in achieving H+ retention. Am J Physiol 1975; 229:1330–1337.
121. Gow CB, Phillips PA. Epidermal growth factor as a diuretic in sheep. J Physiol 1994; 477:27–33.
122. Granberg P-O. Effect of acute hypoxia on renal hemodynamics and water diuresis in man. Scand J Clin Lab Invest 1962; 14(suppl 63):1–61.
123. Grover RF, Reeves JT, Maher JT, McCullough RE, Cruz JC, Denniston JC, Cymerman A. Maintained stroke volume but impaired arterial oxygenation in man at high altitude with supplemental CO_2. Circ Res 1976; 38:391–396.
124. Hackett PH, Forsling ML, Milledge J, Rennie D. Release of vasopressin in man at altitude. Horm Metab Res 1978; 10:571.

125. Hackett PH, Rennie D, Grover RF, Reeves JT. Acute mountain sickness and the edemas of high altitude: A common pathogenesis? Respir Physiol 1981; 46:383–390.
126. Hackett PH, Rennie D, Hofmeister SE, Grover RF, Grover ET, Reeves JT. Fluid retention and relative hypoventilation in acute mountain sickness. Respir 1982; 43:321–329.
127. Hainsworth R, Karim F, McGregor KH, Wood LN. Responses of abdominal vascular resistance and capacitance to stimulation of the carotid chemoreceptors in anesthetized dogs. J Physiol 1983; 334:409–419.
128. Hainsworth R, Rankin JA, Soladoye AO. Effect of cephalic carbon dioxide tension on the cardiac inotropic response to carotid chemoreceptor stimulation in dogs. J Physiol 1985; 358:405–416.
129. Hanley DF. Wilson DA, Feldman MA, Traystman RJ. Peripheral chemoreceptor control of neurohypophysial blood flow. Am J Physiol 1988; 254:H742–H750.
130. Hannon JP. Nutrition at high altitude. In: Horvath SM, Yousef MK, eds. Environmental Physiology: Aging, Heat and Altitude. New York: Elsevier/North Holland, 1981:335–350.
131. Hannon JP, Chinn KS, Shields JL. Alterations in serum and extracellular electrolytes during high-altitude exposure. J Appl Physiol 1971; 31:266–273.
131a. Hannon JP, Roger BG. Body composition of mice following exposure to 4300 and 6100 meters. Aviat Space Environ Med 1975; 46:1232–1235.
132. Harber MJ, Williams JD, Morton JJ. Antidiuretic hormone excretion at high altitude. Aviat Space Environ Med 1981; 52:38–40.
133. He X-R, Greenberg SG, Briggs JP, Schnermann J. Effect of nitric oxide on renin secretion II: studies in the perfused juxtaglomerular apparatus. Am J Physiol 1995; 268: F953–F959.
134. Heidenreich S, Rahn KH, Zideck W. Direct vasopressor effect of erythropietin on renal resistance vessels. Kidney Int 1991; 39:259–265.
135. Hesse B, Kanstrup IL, Christiansen NJ, Ingemann-Hansen T, Hansen JF, Halkjaer-Kristensen J, Petersen FB. Reduced norepinephrine response to dynamic exercise in human subjects during oxygen breathing. J Appl Physiol 1981; 51:176–178.
136. Heyes MP, Farber MO, Manfredi F, Robertshaw D, Weinberger M, Fineberg N, Robertson G. Acute effects of hypoxia on renal and endocrine function in normal humans. Am J Physiol 1982; 243:R265–R270.
136a. Heyman SN, Goldfarb M, Darmon D, Brezis M. Tissue oxygenation modifies nitric oxide bioavailability. Microcirc 1999; 6:199–203.
137. Hildebrandt W, Offenbacher A, Schuster M, Swenson ER, Bärtsch P. Diuretic effect of hypoxia, hypocapnia, and hyperpnea in humans. Relation to hormones and O_2 chemosensitivity. J Appl Physiol 2000; 88:599–610.
138. Hill NS, Klinger JR, Warburton RR, Pietras L, Wrenn DS. Brain natriuretic peptide: possible role in the modulation of hypoxic pulmonary hypertension. Am J Physiol 1994; 266:L308–L315.
139. Hodes R, Larrabee MG. The relation between alveolar carbon dioxide tension and susceptibility to decompression sickness. Am J Physiol 1946; 147:603–615.
139a. Hofbauer K-H, Jensen BL, Kurtz A, Sandner P. Tissue hypoxygenation activities of the adrenomedullin system in vivo. Am J Physiol 2000; 278:R513–R519.
140. Hogan RP, Kotchen TA, Boyd AE, Hartley LH. Effect of altitude on renin-aldosterone system and metabolism of water and electrolytes. J Appl Physiol 1973; 35:385–390.
141. Hohenhaus E, Paul A, McCullough RE, Kucherer H, Bärtsch P. Ventilatory and pulmo-

nary vascular response to hypoxia and susceptibility to high altitude pulmonary edema. Eur Respir J 1995; 8:1825–1833.

142. Hoka S, Bosnjak, ZJ, Arimura H, Kampine JP. Regional venous outflow, blood volume, and sympathetic nerve activity during severe hypoxia. Am J Physiol 1989; 256: H162–H170.
143. Honda Y. Respiratory and circulatory activities in carotid body-resected humans. J Appl Physiol 1992; 73:1–8.
144. Honig A. Peripheral arterial chemoreceptors and reflex control of sodium and water homeostasis. Am J Physiol 1989; 257:R1282–R1302.
145. Honig A, Landgraf R, Ledderhos C, Quies W. Plasma vasopressin levels in healthy young men in response to stimulation of the peripheral arterial chemoreceptors by almitrine bismethylate. Biomed Biochem Acta 1987; 46:1043–1049.
146. Honig A, Wedler B, Opperman H, Gruska S, Schmidt M. Effect of arterial chemoreceptor stimulation with almitrine bismethylate on plasma renin activity, aldosterone, ACTH and cortisol in anaesthetized, artificially ventilated cats. Clin Exp Pharmacol Physiol 1996; 23:106–110.
147. Hoyt RW, Durkot MJ, Forte VA, Hubbard LJ, Trad LA, Cymerman A. Hypobaric hypoxia (380 torr) decreases intracellular and total body water in goats. J Appl Physiol 1991; 71:509–513.
148. Hoyt RW, Jones TE, Baker-Fulco CJ, Schoeller DA, Schoene RB, Schwartz RA, Askew EW, Cymerman A. Doubly labeled water measurement of human energy expenditure during exercise at high altitude. Am J Physiol 1994; 266:R966–R971.
149. Humpeler E, Skrabal F, Bartsch G. Influence of exposure to moderate altitude on the plasma concentration of cortisol, aldosterone, renin, testosterone and gonadotropins. Eur J Appl Physiol 1980; 45:167–180.
150. Iitake K, Share L, Crofton JT, Brooks DP, Ouchi Y, Blaine EH. Central atrial natriuretic factor reduces vasopressin secretion in rats. Endocrinology 1986; 119:438–440.
151. Iriki M, Pleschka K, Walther O-E, Simon E. Hypoxia and hypercapnia in asphyxic differentiation of regional sympathetic activity in the anaesthetized rabbit. Pflugers Arch 1971; 328:91–102.
152. Ishimitsu T, Nishikimi T, Saito Y, Kitamura K, Eto T, Kangawa K, Matsuo H, Omae T, Matsuoka H. Plasma levels of adrenomedullin, a newly identified hypotensive peptide, in patients with hypertension and renal failure. J Clin 1994; 94:2158–2161.
153. Jackson RM, Narkates AJ, Oparil S. Impaired pulmonary conversion of angiotensin I to angiotensin II in rats exposed to chronic hypoxia. J Appl Physiol 1986; 60:1121–1127.
154. Jaeger JJ, Sylvester JT, Cymerman A, Berberich JJ, Denniston JC, Maher JT. Evidence for increased intrathoracic fluid volume in man at high altitude. J Appl Physiol 1979; 47:670–676.
155. Jain SC, Bardhan J, Swamy YV, Krishna B, Nayar HS. Body fluid compartments in humans during acute high-altitude exposure. Aviat Space Environ Med 1980; 51:234–236.
156. Jain SC, Bardhan J, Swamy YV, Grover A, Nayar HS. Body water metabolism in high altitude natives during and after a stay at sea level. Int J Biometeorol 1981; 25:47–52.
157. Jankoski AH, Whitten BK, Shields JL, Hannon JP. Electrolyte patterns and regulation in man during acute exposure to high altitude. Fed Proc 1969; 28:1185–1189.

158. Jezek V, Ourednik A, Daum S, Krouzkova L. Effect of short term hypoxemia on the potassium and sodium metabolism and on cardiac contraction. Acta Med Scand 1965; 177:175–180.

159. Jenq W, Rabb H, Wahe M, Ramirez G. Hypoxic effects on the expression of mineralocorticoid and glucocorticoid receptors in human renal cortex epithelial cells. Biochem Biophys Res Comm 1996; 218:444–448.

160. Johnson HL, Consolazio CJ, Matoush LO, Krzywicki HJ. Nitrogen and mineral metabolism at altitude. Fed Proc 1969; 28:1195–1198.

161. Jones RM, LaRochelle FT, Tenney SM. Role of arginine vasopressin on fluid and electrolyte balance in rats exposed at high altitude. Am J Physiol 1981; 240:R182–186.

162. Jones RM, Terhard C, Zullo J, Tenney SM. Mechanisms of reduced water intake in rats at high altitude. Am J Physiol 1981; 240:R187–191.

163. Jougasaki M, Wei C-H, Aarhus LL, Heublin DM, Sandberg SM, Burnett JC. Renal localization and actions of adrenomedullin: a natriuretic peptide. Am J Physiol 1995; 268:F657–F663.

164. Kaczmarczyk G, Drake GA, Eisele R, Mohnhaupt R, Noble MI, Simgen B, Stubbs J, Reinhardt HW. Role of the cardiac nerves in the regulation of sodium excretion in conscious dogs. Pfluger Arch 1981; 390:125–130.

165. Kappagoda CT, Karim F, MacKay D. Effects of combined carotid chemoreceptor and atrial receptor stimulation on renal blood flow in anaesthetized dogs. J Physiol 1983; 336:91–100.

166. Karim F, Al-Obaidi M. Modification of carotid chemoreceptor-induced changes in renal haemodynamics and function by carotid baroreflex in dogs. J Physiol 1993; 466: 599–610.

167. Karim F, Poucher SM, Summerill RA. The effects of stimulating carotid chemoreceptors on the renal hemodynamics and function in dogs. J Physiol 1987; 392:451–462.

168. Kaufman MP, Cassidy SS. Reflex effects of lung inflation and other stimuli on the heart and circulation. In: Scharf SM, Cassidy SS, eds. Heart-Lung Interactions in Health and Disease. New York: Marcel Dekker, 1989:339–363.

169. Kawashima A, Kubo K, Matsuzawa Y, Kibayashi T, Sekiguchi M. Hypoxia-induced ANP secretion in subjects susceptible to high altitude pulmonary edema. Respir Physiol 1992; 89:309–317.

170. Kawashima A, Kubo K, Harai K, Yoshikawa S, Matsuzawa Y, Kibayshi T. Plasma levels of atrial natriuretic peptide under hypoxic conditions in normal subjects. Respir Physiol 1989; 76:79–92.

171. Keynes RJ, Smith GW, Slater JDH, Brown MM, Brown SE, Payne NN, Jowett TP, Monge CC. Renin and aldosterone at high altitude in man. J Endocr 1982; 92:131–140.

172. King AJ, Brenner BM, Anderson S. Endothelin: a potent renal and systemic vasoconstrictor. Am J Physiol 1989; 256:F1051–F1058.

173. Kirchheim HR. Systemic arterial baroreceptor reflexes. Physiol Rev 1976; 56:100–176.

174. Kitamura K, Kangawa K, Kawamoto M, Ichiki Y, Nakamura S, Matsuo H, Ito T. Adrenomedullin; a novel hypotensive peptide isolated from human pheochromocytoma. Biochem Biophys Res Commun 1993; 192:553–560.

175. Kjeldsen K, Damgaard F. Influence of prolonged carbon monoxide exposure and high

altitude on the composition of blood and urine in man. Scand J Clin Lab Invest 1968; 22(suppl 103):20–25.

176. Koller EA, Bischoff M, Buhrer A, Felder L, Schopen M. Respiratory, circulatory, and neuropsychological responses to acute hypoxia in acclimatized and non acclimatized subjects. Eur J Appl Physiol 1991; 62:67–72.

177. Koller EA, Buhrer A, Felder L, Schopen M, Valloton MB. Altitude diuresis, endocrine and renal responses to acute hypoxia of acclimatized and non acclimatized subjects. Eur J Appl Physiol 1991; 62:228–234.

178. Koller EA, Lesniewska B, Buhrer A, Bub B, Kohl J. The effects of acute altitude exposure in Swiss highlanders and lowlanders. Eur J Appl Physiol 1993; 66:146–154.

179. Koller EA, Schopen M, Keller M, Lang RE, and Valloton MB. Ventilatory, circulatory, endocrine, and renal effects of almitrine infusion in man. A contribution to high altitude physiology. Eur J Appl Physiol 1989; 58:419–425.

180. Krapf R, Beeler I, Hertner D, Hulter H. Chronic respiratory alkalosis: the effect of sustained hyperventilation on renal regulation of acid-base regulation. N Engl J Med 1991; 324:1394–1401.

181. Krebs MO, Naydowski M, Jorres D, Reinhardt HW, G Kaczmarczyk. Zunahme der Natriumexkretion wacher Hunde bei akuter hypokapnischer Hypoxie. Anaesthesist 1992; 92(suppl):S198.

182. Krzywicki HJ, Consolazio CJ, Johnson HL, Nielsen WC, Barnhart RA. Water metabolism in humans during acute high-altitude exposure (4300 m). J Appl Physiol 1971; 30:806–809.

183. La Villa G, Stefani L, Zurli C, Lazzeri C, Guerra CT, Barletta G, Bandinelli R, Strazzula G, Franchi F. Acute effects of physiological increments of brain natriuretic peptide in humans. Hypertension 1995; 26:628–633.

184. Lamperi S, Fumagalli G. L'effeto de l'ipossia sulla funzione renale del soggetto normale e del nefropatico. Minerva Nefrol 1959; 6:65–74.

185. Lawrence DL, Shenker Y. Effect of hypoxic exercise on atrial natriuretic factor and aldosterone regulation. Am J Hypert 1991; 4:341–347.

186. Lawrence DL, Skatrud JB, Shenker Y. Effect of hypoxia on the atrial natriuretic factor and aldosterone regulation in humans. Am J Physiol 1990; 258:E243–E248.

186a. Laycock SK, Vogel T, Forfia PR, Tuzman J, Xiaobin X, Ochoa M, Thompson CI, Hintze TH. Role of nitric oxide in the control of renal oxygen consumption and the regulation of chemical work in the kidney. Circ Res 1998; 82:1263–1271.

187. Ledderhos C, Quies W, Schuster R, Peters R. Renal hemodynamics and excretory function of healthy young men during stimulation of their peripheral arterial chemoreceptors by almitrine bismethylate. Biomed Biochim Acta 1987; 46:1035–1042.

187a. Ledderhos C, Gross V, Cowley AW. Pharmacological stimulation of arterial chemoreceptors in conscious rats produces different responses in renal cortical and medullary blood flow. Clin Exp Pharmacol Physiol 1998; 25:536–540.

188. Ledsome JR, Kan W-O. Reflex changes in hindlimb and renal vascular resistance in response to distension of the isolated pulmonary arteries of the dog. Circ Res 1977; 40:64–72.

189. Lee J, Malvin RL, Claybaugh JR, Huang BS. Atrial natriuretic factor inhibits vasopressin secretion in conscious sheep. Proc Soc Exp Biol Med 1987; 185:272–276.

190. Leichtweiss H-P, Lubbers DW, Weiss C, Baumgartl H, Reschke W. The oxygen supply of the rat kidney: measurement of intrarenal PO_2. Pflugers Arch 1969; 309:328–349.

190a. Leonard BG, Malpas SC, Denton KM, Madden AC, Evans RG. Differential control of ultrarenal blood flow during reflex measures in sympathetic nerve activity. Am J Physiol 2001; 280:R62–R68.

191. Leonhardt KO, Landes RR. Oxygen tension of the urine and renal structures. N Engl J Med 1963; 269:115–121.

192. Leuenberger PJ, Stalcup SA, Mellins RB, Greenbaum LM, Turino GM. Decrease in angiotensin I conversion by acute hypoxia in dogs. Proc Soc Exp Biol Med 1978; 158: 586–589.

193. Levine ZBD, Kubo K, Kobayashi T, Fukushima M, Shibamoto T, Ueda G. Role of barometric pressure in pulmonary fluid balance and oxygen transport. J Appl Physiol 1988; 64:419–428.

194. Lewis RA, Thorn GW, Koept GF, Dorrance SS. The role of the adrenal cortex in acute anoxia. J Clin Invest 1942; 21:33–46.

195. Linden RJ, Mary ASG, Weatherill D. The responses in renal nerves to stimulation of atrial receptors, carotid sinus baroreceptors, and carotid chemoreceptors. Q J Exp Physiol 1981; 66:179–191.

196. Liu GL, Liu L, Batrajas L. Development of NOS-containing neuronal somata in the rat kidney. J Autonomic Nerv Sys 1996; 58:81–88.

197. Loeppky JA, Roach RC, Selland MA, Scotto P, Luft FC, Luft UC. Body fluid alterations during head-down bed rest in men at moderate altitude. Aviat Space Environ Med 1993; 64:265–274.

198. Lordick F, Hauck RW, Senekowitsch R, Emslander HP. Atrial natriuretic peptide in acute hypoxia-exposed healthy subjects and in hypoxaemic patients. Eur Respir J 1995; 8:216–221.

199. Lozano R, Monge C. Renal function in high altitude natives and in high altitude natives with chronic mountain sickness. J Appl Physiol 1965; 20:1026–1027.

200. Maher JT, Cymerman A, Reeves JT, Cruz JC, Denniston JC, Grover RF. Acute mountain sickness: increased severity in eucapnic hypoxia. Aviat Space Environ Med 1975; 46:826–829.

201. Maher JT, Jones LG, Hartley LH, Williams GH, Rose LI. Aldosterone dynamics during graded exercise at sea level and high altitude. J Appl Physiol 1975; 39:18–22.

202. Mairbäurl H, Wodopia R, Eckes S, Schulz S, Bärtsch P. Impairment of cation transport in A549 cells and rat alveolar epithelial cells by hypoxia. Am J Physiol 1997; 273: L797–L7806.

203. Majid DSA, Kadowitz PJ, Coy DH, Navar G. Renal responses to intra-arterial administration of adrenomedullin in dogs. Am J Physiol 1996; 270:F200–F205.

203a. Malpas SC, Shweta A, Anderson WP, Head GA. Functional response to graded increases in renal nerve activity during hypoxia in conscious rabbits. Am J Physiol 1996; 270:R1489–R1499.

204. Malhotra MS, Brahmachari HD, Sridiharan K, Purshottam T, Ramachandran K, Radhakrishnan U. Electrolyte changes at 3500 m in males with and without high altitude pulmonary edema. Aviat Space Environ Med 1975; 46:409–412.

205. Mannix ET, Palange P, Aronoff GR, Manfredi F, Farber MO. Atrial natriuretic peptide and the renin-aldosterone axis during exercise in man. Med Sci Sports Exer 1990; 22: 785–789.

206. Maresh CM, Noble BJ, Robertshaw KL, Harvey JS. Aldosterone, cortisol and electrolyte responses to hypobaric hypoxia in moderate-altitude natives. Aviat Space Environ Med 1985; 56:1078–1084.

207. Marotta SF. Roles of aortic and carotid chemoreceptors in activating the hypothalamic-hypophysial-adrenocortical system during hypoxia. Proc Soc Exp Biol Med 1972; 141: 915–927.
208. Marshall JM. Peripheral chemoreceptors and cardiovascular regulation. Physiol Rev 1994; 74:543–594.
209. Marshall JM, Metcalfe JD. Influences of the cardiovascular response to graded levels of systemic hypoxia of the accompanying hypocapnia in the rat. J Physiol 1989; 410: 381–394.
210. Marshall JM, Metcalfe JD. Effects of systemic hypoxia on the distribution of cardiac output in the rat. J Physiol 1990; 426:335–353.
211. Masuda A, Sakakibara Y. Shn B, Toma S, Nakajima Y, Honda Y. Interrelationship between cardiopulmonary responses and sympathetic nerve activities during sustained hypoxia in man. Jpn J Physiol 1990; 40(suppl):S57.
212. Matsukawa K, Mitchell JH, Wall TP, Wilson LB. The effect of static exercise on renal sympathetic nerve activity in conscious cats. J Physiol (London) 1991; 434:453–467.
213. Mattson DL, Bellehumeur TG. Neural nitric oxide synthase in the renal medulla and blood pressure regulation. Hypertension 1996; 28:297–303.
214. Mazbar SA, Wiedemann E, Humphreys MH. Mechanism of the natriuretic effect of unilateral carotid artery traction in the rat. J Am Soc Nephrol 1990; 1:266–271.
215. Mazzeo RS, Wolfel EE, Butterfield GE, Reeves JT. Sympathetic response during 21 days at high altitude (4300 m) as determined by urinary and arterial catecholamines. Metabolism 1994; 43:1226–1232.
216. Mazzocchi G, Rebuffat P, Gottardo G, Nussdorfer GC. Adrenomedullin and calcitonin gene-related peptide inhibit aldosterone secretion in rats, acting via a common receptor. Life Sci 1996; 58:839–844.
217. McKenna OC, Angelakos ET. Acetylcholinesterase-containing nerve fibers in the canine kidney. Circ Res 1968; 23:645–651.
218. McLay JS, Chatterje PK, Mistry SK, Weerakody RP, Jardine AG, McKay NG, Hawksworth GM. Atrial natriuretic factor and angiotensin II stimulate nitric oxide release from human proximal tubular cells. Clin Sci 1995; 89:527–531.
219. Metzler CH, Ramsey DJ. Physiological doses of atrial peptide inhibit angiotensin II stimulated aldosterone secretion. Am J Physiol 1989; 256:R1155–R1159.
220. Michel CC. Renal medullary microcirculation: architecture and exchange. Microcirculation 1995; 2:125–139.
221. Millar EA, Angus RM, Nally JE, Clayton R, Thomson NC. Effect of hypoxia and beta agonists on the activity of the renin-angiotensin system in normal subjects. Clin Sci 1995; 89:273–276.
222. Miller RL, Kohan DE. Hypoxia regulates endothelin-1 production by the inner medullary collecting duct. J Lab Clin Med 1998; 131:45–48.
223. Milledge JS, Bryson EI, Catley DM. Sodium balance, fluid homeostasis and the renin-aldosterone system during the prolonged exercise of hill walking. Clin Sci 1982; 62: 595–604.
224. Milledge JS, Catley DM, Ward MP. Renin-aldosterone and angiotensin converting enzyme during prolonged altitude exposure. J Appl Physiol 1983; 55:699–702.
225. Milledge JS, Catley DM, Williams ES, Withey WR, Minty BD. Effect of prolonged exercise at altitude on the renin-aldosterone system. J Appl Physiol 1983; 55:413–418.
226. Milledge JS, Catley DM. Angiotensin converting enzyme activity and hypoxia. Clin Sci 1987; 72:149.

227. Milledge JS, Catley DM. Renin, aldosterone and converting enzyme during exercise and acute hypoxia in humans. J Appl Physiol 1982; 52:320–323.
228. Milledge JS, McArthur S, Morice A, Luff N, Abrahams R, Thomas PS. Atrial natriuretic peptide and exercise-induced fluid retention in man. J Wild Med 1991; 2:94–101.
229. Moore LG, Huang SY, McCullough RE, Sampson JB, Maher JT, Weil JV, Grover RF, Alexander JK, Reeves JT. Variable inhibition by falling CO_2 of hypoxic ventilatory response in humans. J Appl Physiol 1984; 56:207–210.
230. Morganti A, Giussani M, Sala C, Gazzano G, Marana I, Pierini A, Savoia MT, Ghio F, Cogo A, Zanchetti A. Effects of exposure to high altitude on plasma endothelin-1 levels in normal subjects. J Hypertens 1995; 13:859–865.
231. Muirhead EE. Medullipin system of blood pressure control. News Physiol Sci 1990; 5:241–244.
232. Nahas GG, Visscher MB, Matther GW, Haddy FJ, Warner HR. Influence of hypoxia on the pulmonary circulation of nonnarcotized dogs. J Appl Physiol 1954; 6:467–474.
233. Nahmod VE, Lanari A. Abolition of autoregulation of renal blood flow by acetylcholine. Am J Physiol 1964; 207:123–127.
234. Nesarajah MS, Matalon S, Krasney JA, Farhi LE. Cardiac output and regional oxygen transport in the acutely hypoxic conscious sheep. Respir Physiol 1983; 53:161–172.
235. Neylon M, Marshall JM, Johns EJ. The effects of systemic hypoxia on renal function in the anaesthetized rat. J Physiol 1995; 487:497–511.
236. Neylon M, Marshall JM. The role of adenosine in the respiratory and cardiovascular response to systemic hypoxia in the rat. J Physiol 1991; 440:529–545.
237. Neylon M, Marshall JM, Johns EJ. The effects of chronic hypoxia on renal function in the rat. J Physiol 1997; 501:243–250.
238. Nir A, Clavelli AL, Heublein D, Aarhus LL, Burnett JC. Acute hypoxia and endogenous renal endothelin. J Am Soc Nephrol 1994; 4:1920–1924.
239. Nishijima MK, Breslow MJ, Raff H, Traystman RJ. Regional adrenal blood flow during hypoxia in anesthetized, ventilated dogs. Am J Physiol 1989; 256:H94–H100.
240. O'Hagan KP, Bell LB, Mittelstadt SW, Clifford PS. Effect of dynamic exercise on renal sympathetic activity in conscious rabbits. J Appl Physiol 1993; 74:2099–2104.
241. O'Hagan KP, Bell LB, Clifford PS. Effects of pulmonary denervation on renal sympathetic and heart rate responses to hypoxia. Am J Physiol 1995; 269:R923–R929.
242. Ochwaldt B. Über Bikarbonatausscheidung und Kohlensäuresystem im Harn während akuter Hypoxia. Pflugers Arch Ges Physiol 1947; 249:452–469.
243. Olsen NV, Hansen JM, Kanstrup I-L, Richalet J-P, Leyssac PP. Renal hemodynamics, tubular function, and response to low dose dopamine during acute hypoxia in humans. J Appl Physiol 1993; 74:2166–2173.
244. Olsen NV, Kanstrup I-L, Richalet J-P, Hansen JM, Plazen G, Galen F-X. Effects of acute hypoxia on renal and endocrine function at rest and during graded exercise in hydrated humans. J Appl Physiol 1992; 73:2036–2043.
245. Olsen NV, Christiansen H, Klausen T, Fogh-Anderson N, Plum I, Kanstrup I-L, Hansen JM. Effects of hyperventilation and hypocapnic/normocapnic hypoxemia on renal function and lithium clearance in humans. Anesthesiology 1998; 89:1389–1400.
246. Ou LC, Silverstein J, Edwards RB. Renal function in rats chronically exposed to high altitude. Am J Physiol 1984; 247:F45–49.
247. Pauli HG, Truniger B, Larsen JK, Mulhausen RO. Renal function during prolonged

exposure to hypoxia and carbon monoxide. Scand J Lab Clin Invest 1968; 22(suppl 103):55–60.

247a. Pearson PJ, Evora PR, Discigil B, Schaff HV. Hypoxia increases vasodilator release from internal mammary artery and saphenous vein grafts. Ann Thorac Surg 1998; 65: 1220–1225.

248. Perella MA, Edell ES, Krowka MJ, Cortese DA, Burnett JC. Endothelium-derived relaxing factor in pulmonary and renal circulations during hypoxia. Am J Physiol 1992; 263:R45–R50.

249. Perico N, Cornejo R, Benigni P, Malachini B, Ladny JR, Remuzzi G. Endothelin induces diuresis and natriuresis in the rat by acting on proximal tubular cells through a mechanism mediated by lipoxygenase products. J Am Soc Nephrol 1991; 2:57–69.

250. Persson PB. Modulation of cardiovascular control mechanisms and their interaction. Physiol Rev 1996; 76:193–344.

251. Philbin DM, Baratz RA, Patterson RW. The effect of carbon dioxide on plasma antidiuretic hormone levels during intermittent positive pressure breathing. Anesthesiology 1970; 33:345–349.

252. Pirola JP, Alvarez AL, Balda MS, Finkielman S, Nahmod VE. Evidence for cholinergic innervation in dog renal tissue. Am J Physiol 1989; 257:F746–F754.

253. Porchet M, Contat H, Waeber B, Nussberger J, Brunner HR. Response of plasma arginine vasopressin levels to rapid changes in altitude. Clin Physiol 1984; 4:435–438.

254. Prasad P Edelman RR, Epstein FH. Noninvasive evaluation of intrarenal oxygenation with MRI. Circulation 1996; 94:3271–3275.

255. Pugh LGCE. Animals in high altitude: man above 5000 meters-mountain exploration. In: Handbook of Physiology. Adaptation to the Environment. Washington, DC: Am Physiol Soc 1964:861–868.

256. Purshottam T, Pahwa ML, Brahmachari HD. Effects of six hours hypoxic and cold exposure on urinary electrolyte and catecholamine excretion. Aviat Space Environ Med 1978; 49:62–65.

257. Raff H, Ball DL, Goodfriend TL. Low oxygen selectively inhibits aldosterone secretion from bovine adrenocortical cells in vitro. Am J Physiol 1989; 256:E640–E644.

258. Raff H, Jankowska B. Effect of CO_2/pH on the aldosterone response to hypoxia in bovine adrenal cells in vitro. Am J Physiol 1993; 265:R820–R825.

259. Raff H, Maselli J, Reid IA. Correlation of plasma angiotensin II concentration and plasma renin activity during acute hypoxia in dogs. Clin Exp Pharmacol Physiol 1985; 12:91–94.

260. Raff H, Roarty TP. Renin, ACTH, and aldosterone during acute hypercapnia and hypoxia in conscious rats. Am J Physiol 1988; 254:R431–R435.

261. Raff H, Shinsako J, Dallman MF. Renin and ACTH responses to hypercapnia and hypoxia after chronic carotid chemodenervation. Am J Physiol 1984; 247:R412–R417.

262. Raff H, Shinsako J, Keil LC, Dallman MF. Feedback inhibition of adrenocorticotropin and vasopressin responses to hypoxia by physiological increases in endogenous corticosteroids in dogs. Endocrinology 1984; 114:1245–1249.

263. Raff H, Shinsako J, Keil LC, Dallman MF. Vasopressin, ACTH, and corticosteroids during hypercapnia and graded hypoxia in dogs. Am J Physiol 1983; 244:E453–E458.

264. Raff H. Glucocorticoid inhibition of hypophysial vasopressin secretion. Am J Physiol 1987; 252:R635–R644.

265. Raij L, Baylis C. Glomerular actions of nitric oxide. Kidney Int 1995; 48:20–32.

266. Ramirez G, Bittle PA, Hammond M, Ayers CW, Dietz JR, Colice GL. Regulation of aldosterone secretion during hypoxemia at sea level and moderately high altitude. J Clin Endocrinol Metab 1988; 67:1162–1165.
267. Ramirez G, Hammond M, Bittle PA, Dietz JR, Colice GL. Sodium excretion and hormonal changes during salt loading at moderately high altitude and acute hypoxemia at sea level. Aviat Space Environ Med 1992; 63:891–898.
268. Ramirez G, Pineda DO, Bittle PA, Rabb HA, Rosen R, Vesely D, Sesaki S. Partial renal resistance to arginine vasopressin as an adaptation to high altitude living. Aviat Space Environ Med 1998; 69:58–65.
269. Ramirez G, Pineda DO, Bittle PA, Agosti SJ, Rabb HA, Dietz JR. Salt excretory capacity in natives adapted to moderate high altitude living after acute mobilization to sea level. Aviat Space Environ Med 1995; 66:1063–1070.
270. Recordati GM, Moss NG, Waselkov L. Renal chemoreceptors in the rat. Circ Res 1978; 43:534–543.
271. Rein H. Über ein Regulationsystem 'Milz-Leber' für den oxidativen Stoffwechsel der Korpwebe und besonders des Herzens. Naturwissenschaften 1949; 8:233–239.
272. Rennie D, Lozano R, Monge C, Sime F, Whittembury J. Renal oxygenation in male Peruvian natives living permanently at high altitude. J Appl Physiol 1971; 30:450–456.
273. Richalet J-P, Rutgers V, Bouchet P. Diurnal variation of acute mountain sickness, color vision and plasma cortisol and ACTH at high altitude. Aviat Space Environ Med 1989; 60:105–111.
274. Roach RC, Maes D, Riboni K, Icenogle M, Loeppky J. Increased plasma volume at simulated altitude and the onset of acute mountain sickness. FASEB J 1998; 12:A57.
275. Roman RJ, Cowley AJ, Garcia-Estan J, Lombard H. Pressure-diuresis in volume-expanded rats: cortical and medullary hemodynamics. Hypertension 1988; 12:168–172.
276. Roman RJ, Zou A-P. Influence of the medullary circulation on the control of sodium excretion. Am J Physiol 1993; 265:R963–R973.
277. Romero JC, Lahera V, Salom MG, Biondi ML. Role of the endothelium-dependent relaxing factor nitric oxide on renal function. Am J Soc Nephrol 1992; 2:1371–1387.
278. Rose CE, Kimmel DP, Godine RL, Kaiser DL, Carey RM. Synergistic effects of acute hypoxemia and hypercapnic acidosis in conscious dogs. Renal dysfunction and activation of the renin-angiotensin system. Circ Res 1983; 53:202–213.
279. Rose CE, Kimmel DP, Godine RL, Kaiser DL, Carey RM. Synergistic effects of acute hypoxemia and hypercapnic acidosis in conscious dogs. Renal dysfunction and activation of the renin-angiotensin system. Circ Res 1983; 53:202–213.
280. Rose CE, Ragsdale NV, Carey RM. Combined acute hypoxemia and hypercapnic acidosis increases atrial natriuretic peptide in conscious dogs. Miner Electrolyte Metab 1992; 18:24–34.
281. Rose MS, Houston CS, Fulco CS, Coates G, Sutton JR, Cymerman A. Operation Everest II: nutrition and body composition. J Appl Physiol 1988; 65:2545–2551.
282. Rostrup M. Catecholamines, hypoxia and high altitude. Acta Physiol Scand 1998; 162: 389–399.
283. Rothe CF, Maass-Moreno R, Flanagan AD. Effects of hypercapnia and hypoxia on the cardiovascular system; vascular capacitance and aortic chemoreceptors. Am J Physiol 1990; 259:H932–H939.

284. Rowell LB, Blackmon JR. Lack of sympathetic vasoconstriction in hypoxemic humans at rest. Am J Physiol 1986; 251:H562–H570.
285. Rubanyi GM, Polokoff MA. Endothelins: molecular biology, biochemistry, pharmacology, physiology and pathophysiology. Pharmacol Rev 1994; 46:325–415.
286. Rubinstein I, Gurbanov K, Hoffman A, Better OS, Winaver J. Differential effect of endothelin-1 on renal regional blood flow: role of nitric oxide. J Cardiovasc Pharmacol 1995; 26(suppl 3):S208–S210.
287. Ruegg CE, Mandel LJ. Bulk isolation of renal PCT and PST: differential responses to anoxia and hypoxia. Am J Physiol 1990; 259:F176–F185.
287a. Ruszchitzka FT, Wenger RH, Stallach T, Quascling T, de Wit C, Wagner K, Labugger R, Kelm M, Nell G, Rulicke T, Shaw S, Lindberg RL, Rodenwaldt B, Lutz H, Bauer C, Luscher TF, Gassmann M. Nitric oxide prevents cardiovascular disease and determines survival in polyglobulic mice over expressing erythropoietin. Proc Natl Acad Sci 2000; 97:11609–11613.
288. Rutherford JD, Vatner SF. Integrated carotid chemoreceptor and pulmonary inflation reflex control of peripheral vasoactivity in conscious dogs. Circ Res 1978; 43:200–208.
289. Saltz SB, Beller GA, Giamber SR, Alpert JS. Circulatory response to acute hypobaric hypoxia in conscious dogs. Aviat Space Environ Med 1976; 47:129–132.
290. Sandgaard NCF, Bie F. Natriuretic effect of non-pressor doses of endothelin-1 in conscious dogs. J Physiol 1996; 494:809–818.
291. Sarelius IH, Huxley VH. A direct effect of atrial peptide on arterioles of the terminal microvasculature. Am J Physiol 1990; 258:R1224–R1229.
292. Schmidt W, Brabant G, Kroger G. Atrial natriuretic peptide during and after maximal and submaximal exercise under normoxic and hypoxic conditions. Eur J Appl Physiol 1990; 61:398–407.
293. Schnermann J, Lorena JN, Briggs JP, Keiser JA. Induction of water diuresis by endothelin in rats. Am J Physiol 1992; 263:F516–F526.
294. Schirlo C, Bub A, Reize C, Bühre A, Kohl J, Koller EA. Volume changes in the forearm and lower limbs during 2 h of acute hypobaric hypoxia in non-acclimatized subjects. Eur J Appl Physiol 1997; 75:124–131.
295. Scholz H, Kramer BK, Hamann M, Gotz K-H, Kurtz A. Effects of endothelins on renin secretion from rat kidneys. Acta Physiol Scand 1995; 155:173–182.
296. Schonholzer G, Gross F, Marthaler F. Untersuchungen über die Säure-, Elektrolyt- und Wasserausscheidung in Hochgebirge unter besondere Berucksichtigung der Tag- und Nachtrhythmen. Helv Physiol Pharmacol Acta 1948; 6:713–730.
297. Schurek HJ, Jost U, Baumgartl H, Bertram H, Heckmann U. Evidence for a glomerular oxygen diffusion limitation shunt in rat renal cortex. Am J Physiol 1990; 259:F910–F915.
298. Sehy JT, Roseman K, Arruda JAL, Kurtzman NA. Characterization of distal hydrogen ion secretion in acute respiratory alkalosis. Am J Physiol 1978; 235:F203–F208.
299. Selkurt EE. Influence of hypoxia on renal circulation and on excretion of electrolytes and water. Am J Physiol 1953; 172:700–708.
300. Share L, Levy MN. Effect of carotid chemoreceptor stimulation on plasma antidiuretic hormone titer. Am J Physiol 1966; 210:157–161.
301. Shibamato T, Uematsu H, Matsuda Y, Fujita T, Sawano F, Saeki Y, Hayashi T, Koyama S. Acute effect of hypobaria and hypoxia on renal nerve activity in anaesthetized rabbits. Acta Physiol Scand 1992; 144:47–53.

302. Shigeoka JW, Colice GL, Ramirez G. Effect of normoxic and hypoxic exercise on renin and aldosterone. J Appl Physiol 1985; 59:142–148.
303. Shirahata M, Nishino T, Honda Y, Itoh K, Yonezawa T. Effects of hypercapnia on renal nerve activity. Jpn J Physiol 1985; 35:391–399.
304. Silvette H. Some effects of low barometric pressure on kidney function in the white rat. Am J Physiol 1944; 140:374–386.
305. Slater JDH, Tuffley RE, Williams ES, Beresford CH, Sonksen PH, Edwards HT, Ekins RP, McLaughlin M. Control of aldosterone secretion during acclimatization to hypoxia in man. Clin Sci 1969; 37:327–341.
306. Slater JDH, Williams ES, Edwards RHT, Ekins RP, Sonksen PH, Beresford CH, McLaughlin M. Potassium retention during the respiratory alkalosis of mild hypoxia in man; its relationship to aldosterone secretion and other metabolic changes. Clin Sci 1969; 37:311–326.
307. Smith EE, Crowell JW. Influence of hypoxia on mean circulatory pressure and cardiac output. Am J Physiol 1967; 212:1067–1069.
308. Smith HW. The Kidney. Structure and Function in Health and Disease. New York: Oxford University Press, 1951.
309. Somers VK, Anderson JV, Conway J. Atrial natriuretic peptide is released by dynamic exercise in man. Horm Metab Res 1986; 18:871–872.
310. Somers VK, Mark AL, Zavala DC, Abboud FM. Influence of ventilation and hypocapnia on sympathetic nerve responses to hypoxia in normal humans. J Appl Physiol 1989; 67:2095–2100.
311. Sorensen SS, Madsen JK, Pedersen EB. Systemic and renal effect of intravenous infusion of endothelin in healthy human volunteers. Am J Physiol 1994; 266:F411–F418.
312. Spath JA, Daugherty RM, Scott JB, Haddy FJ. Effect of local hypoxia on renin activity in renal venous plasma. Proc Soc Exp Biol Med 1971; 137:484–488.
313. Stampfli VR, Eberle A. Menge, spezifisches Gewicht und Leitfähigkeit des menschlichen Harns im Hochgebirge. Helv Physiol Acta 1944 (Suppl III):221–232.
314. Stanbury SW, Thompson AE. The renal response to respiratory alkalosis. Clin Sci 1952; 11:357–374.
315. Stella A, Zanchetti A. Functional role of renal afferents. Physiol Rev 1991; 71:659–699.
316. Stoos BA, Garcia NH, Garvin JL. Nitric oxide inhibits sodium reabsorption in the isolated perfused cortical collecting duct. Am J Soc Nephrol 1995; 6:89–94.
317. Sun M-K, Reis DJ. Hypoxia selectively excites vasomotor neurons of rostral ventrolateral medulla in rats. Am J Physiol 1994; 266:R245–R256.
318. Sutton JR, Viol GW, Gray GW, McFadden MD, Keane PM. Renin, aldosterone, electrolyte, and cortisol responses to hypoxic decompression. J Appl Physiol 1977; 43: 421–424.
319. Swenson ER, Duncan TB, Goldberg SV, Ramirez G, Ahmad S, Schoene RB. Diuretic effect of acute hypoxia in humans: relationship to hypoxic ventilatory responsiveness and renal hormones. J Appl Physiol 1995; 78:377–383.
320. Takano T, Soltoff SP, Murdaugh S, Mandel LJ. Intracellular respiratory dysfunction and cell injury in short-term anoxia of the rabbit proximal tubule. J Clin Invest 1985; 76:2377–2384.
321. Thomas S. Some effects of change if posture on water and electrolyte excretion by the human kidney. J Physiol 1957; 139:337–352.

322. Thomson SC, Vallon V. Alpha 2 adrenoreceptors determine the renal response to nitric oxide in the rat glomerulus and proximal tubule. J Am Soc Nephrol 1995; 6:1482–1490.
322a. Toepfer M, Hartman G, Schlosshauer M, Hautmann H, Tschop M, Huber RM. Adrenomedullin: a player at high altitude? Chest 1998; 113:1428.
323. Toth M, Vourinen KH, Voulteenaho O, Hassinen IE, Uusimaa PA, Leppaluoto J, Ruskohao H. Hypoxia stimulates release of ANP and BNP from perfused rat ventricular myocardium. Am J Physiol 1992; 266:H1572–H1580.
324. Tucker A, Horvath SM. Regional blood flow responses to hypoxia and exercise in altitude-adapted rats. Europ J Appl Physiol 1974; 33:139–150.
325. Tucker A, Reeves JT, Robertshaw D, Grover RF. Cardiopulmonary response to acute altitude exposure: water loading and denitrogenation. Respir Physiol 1983; 54:363–380.
327. Tuffley RE, Rubenstein D, Slater JDH, Williams ES. Serum renin activity during exposure to hypoxia. J Endocrinol 1970; 48:497–510.
328. Ullmann E. Acute anoxia and the excretion of water and electrolyte. J Physiol 1961; 155:417–437.
329. Van Liere EJ, David NA, Lough DH. Absorption of water from the small intestine at various degrees of anoxemia. Am J Physiol 1941; 115:239–244.
330. Van Liere EJ, Sleeth CK. Absorption of sodium chloride from the small intestine at various degrees of anoxemia. Am J Physiol 1936; 90:309–312.
331. Vander AJ. Effects of acetylcholine, atropine and physostigmine on renal function in the dog. Am J Physiol 1966; 206:492–498.
332. Vatner SF, Manders WT, Knight DR. Vagally mediated regulation of renal function in conscious primates. Am J Physiol 1986; 250:H546–H549.
333. Vollenweider L, Savcic M, Loffler B-M, Nicod P, Bärtsch P, Scherrer U. Exaggerated endothelin-1 release in high altitude pulmonary edema susceptible subjects. Eur Respir J 1995; 8:322S.
334. Vonmoos S, Nussberger J, Waber B, Biollaz J, Brunner HR, Leuenberger P. Effect of metoclopramide on angiotensin, aldosterone and atrial peptide during hypoxia. J Appl Physiol 1990; 69:2072–2077.
335. Voulteenajo O, Koistinen P, Martikkala V, Takala T, Leppaluoto J. Effect of physical exercise in hypobaric conditions on atrial natriuretic peptide secretion. Am J Physiol 1992; 263:R647–R652.
336. Wade CE. Response, regulation and actions of vasopressin during exercise: a review. Med Sci Sports Exer 1984; 16:506–511.
337. Walker BR. Diuretic response to acute hypoxia in the conscious dog. Am J Physiol 1982; 253:F440–F446.
338. Walker BR, Brizzee BL. Renal vascular response to combined hypoxia and hypercapnia in conscious rats. Am J Physiol 1988; 254:R552–R558.
339. Walker BR, Brizzee BL. Cardiovascular responses to hypoxia and hypercapnia in barodenervated rats. J Appl Physiol 1990; 68:678–686.
340. Wang BD, Sundet WD, Goetz KL. Vasopressin in plasma and cerebrospinal fluid of dogs during hypoxia or acidosis. Am J Physiol 1984; 247:E449–E455.
341. Wang T, Chan YI. Neural control of distal tubular bicarbonate and fluid transport. Am J Physiol 1989; 257:F72–F76.
342. Westerterp KR, Robach P, Wouters L, Richalet JP. Water balance and acute mountain sickness before and after arrival at high altitude of 4350 m. J Appl Physiol 1996; 80:1968–1972.

343. Widener G, Sullivan LP, Clancy RL, Gonzalez NC. Renal compensation to hypercapnia in prolonged hypoxia. Respir Physiol 1986; 65:341–350.
344. Williams ES, Ward MP, Milledge JS. Effect of exercise of seven consecutive days hill-walking on fluid homeostasis. Clin Sci 1979; 56:305–314.
345. Williams ES. Electrolyte regulation during the adaptation of humans at high altitude. Proc R Soc Lond B Biol Sci 1966; 165:266–280.
346. Williamson JR, Holmberg SW, Chang K, Marvel J, Sutera SP, Needleman P. Mechanisms underlying atriopeptin-induced increases in hematocrit and vascular permeation in rats. Circ Res 1989; 64:890–899.
347. Winslow RM, Monge C. Renal Function in high-altitude polycythemia. In: Winslow RM, Monge C, eds. Hypoxia, Polycythemia, and Chronic Mountain Sickness. Baltimore: Johns Hopkins University Press, 1987:119–141.
348. Withey WR, Milledge JS, Williams ES, Minty BD, Bryson EI, Luff NP, Older WJ, Beeley JM. Fluid and electrolyte homeostasis during prolonged exercise at altitude. J Appl Physiol 1983; 55:409–412.
349. Woolfson RG, Posten I, de Wardener HE. Digoxin-like inhibitors of active sodium transport and blood pressure; current status. Kidney Int 1994; 46:297–309.
350. Young D. Analysis of long term potassium regulation. Endocrinol Rev 1985; 6:24–45.
351. Young PM, Rose MS, Sutton JR. Operation Everest II: plasma lipid and hormonal responses during a simulated ascent of Mt Everest. J Appl Physiol 1989; 66:1430–1435.
352. Ziedel ML, Brady HR, Kone BC, Gullans SR, Brenner BR. Endothelin, a peptide inhibitor of Na^+-K^+-ATPase in intact renal tubular epithelial cells. Am J Physiol 1989; 257:C1101–C1107.
353. Zillig B, Schuler G, Truniger B. Renal function and intrarenal hemodynamics in acutely hypoxic and hypercapnic rats. Kidney Int 1978; 14:58–67.

17

Metabolic Response of Lowlanders to High-Altitude Exposure

Malnutrition Versus the Effect of Hypoxia

GEORGE A. BROOKS

University of California
Berkeley, California

GAIL E. BUTTERFIELD[†]

Palo Alto VA Medical Center
Palo Alto, California

I. Introduction

The acute physiological response to hypobaric hypoxia encountered at elevations above 10,000 feet has significant nutrition implications which have generated a body of ''conventional wisdom'' about the long-term effects of such exposure on nutrient need and energy substrate utilization, a conventional wisdom that needs to be reconsidered. Acute altitude exposure is associated with a cachexia due in part to loss of appetite. Under such circumstances metabolism may be biased toward catabolism of body protein and fat stores, resulting in body wasting and use of less oxygen-efficient fuels. Accordingly many of the ''chronic'' physiological responses in energy substrate utilization attributed to altitude exposure may really be a consequence of malnutrition. Recognizing the confound presented by this situation to the understanding of true physiological response to hypoxia in our experiments, we have imposed nutritional control sufficient to maintain body weight and protein and fat stores. Thus, cachexia and malnutrition are differentiated from the other metabolic responses to hypoxia. Under these circumstances, altitude exposure shifts energy substrate utilization away from fat and protein and toward carbohydrate dependence.

---

† Deceased.

Thus, the study of true metabolic fuel response to hypobaric hypoxia may be in its infancy.

II. Acute Altitude Exposure

Acute exposure (0–24 hours) to altitudes greater than 10,000 feet carries with it reports of anorexia, diuresis, increased metabolic energy need, and shifts in circulating levels of fuel metabolites suggesting changes in energy substrate utilization from normal sea level metabolism.

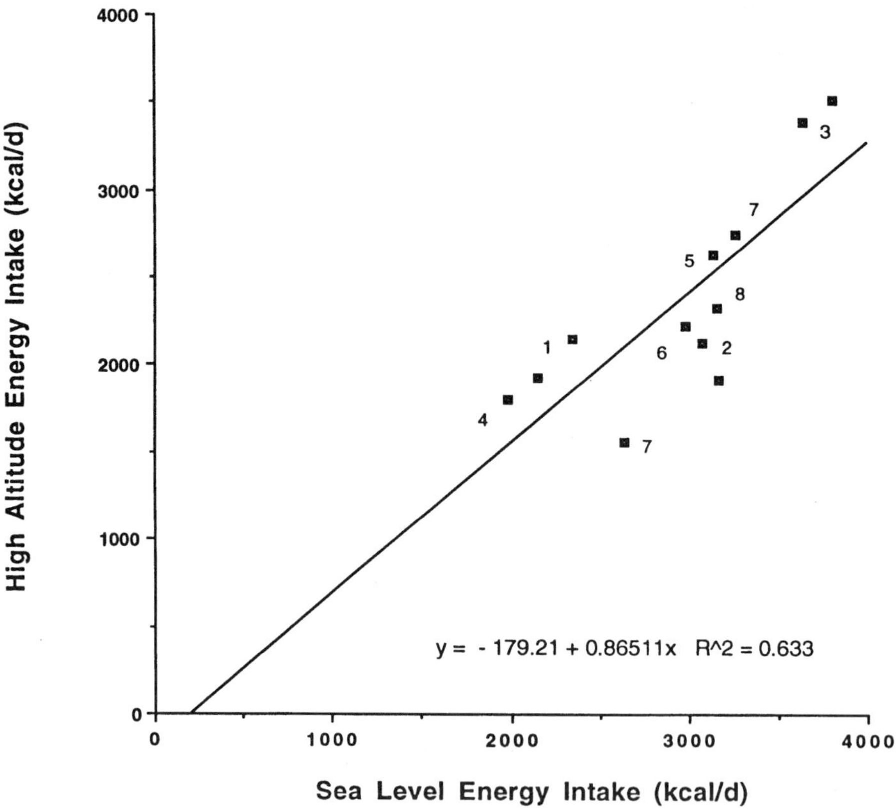

Figure 1 Relationship between sea level energy intake (kcal/d) and high-altitude energy intake (kcal/d) compiled from studies involving ad libitum food intake and providing accurate measures of both parameters. Numbers next to points on figure indicate reference used. 1: From Ref. 1. 2: From Ref. 72. 3: From Ref. 2. 4: From Ref. 3. 5: From Ref. 29. 6: From Ref. 4. 7: From Ref. 5. 8: From Ref. 6.

Reports of anorexia upon acute exposure to altitude are frequent (1–7). Accompanying this anorexia may be acute mountain sickness (AMS), including nausea and vomiting. The resultant decrease in energy intake may, over a prolonged period, lead to weight loss. Careful assessment of studies in which food intake has been documented both at sea level and at altitude suggest a fairly consistent decrement in energy intake during altitude exposure of approximately 180 kcal/day (Fig. 1). Such a deficiency in energy intake carries with it a deficient intake of most other essential nutrients, complicating the interpretation of research on the need for those nutrients in response to hypobaric hypoxia as well.

Accompanying anorexia, diuresis is frequently reported (4,5,8). This diuresis serves to concentrate hemoglobin and thus improve the peripheral delivery of oxygen during the initial hours at altitude (9). Diuresis may be a mandatory part of the initial successful acclimatization to altitude exposure (10); individuals who are unable to diurese suffer significant symptoms of AMS, whereas those who do diurese adjust to acute altitude exposure more comfortably. The loss of body water inevitably contributes to the weight loss seen at altitude, and more water loss than is necessary for appropriate adaptation to acute exposure may adversely affect performance.

These responses, which occur within hours of exposure to hypobaric hypoxia, may include a severe increase in basal energy needs, which continues over the first 2–3 days of altitude exposure. The initial elevation in basal energy requirement has been documented to be as high as 30% above sea level values (11,12) but declines after 2–3 days to less than 10% above sea level values.

Associated with the decrease in energy intake and the increased basal energy needs appears to be a shift in energy substrate metabolism. Blood glucose levels upon acute exposure to altitude have been shown to be decreased over those at sea level (13–15), glycogen stores are spared (16), and circulating glycerol (16,17), triglycerides (16), and free fatty acids (17,18) are elevated over sea level values. These data have previously been interpreted to suggest that fuel substrate choice upon acute exposure to altitude has shifted away from carbohydrate and toward fat (16) as the predominant energy source.

III. Acclimatization

Over the long run, the failure to ingest even sea level quantities of food at altitude, with the accompanying increase in basal energy requirement, leads to significant weight loss in individuals upon chronic exposure to altitude (Table 1). The composition of the weight loss is controversial because monitoring of body composition at altitude is complicated by diuresis, which changes the composition of body compartments (8). Thus, the usual equations used for computing body composition, which are based on assumptions of consistency of composition of lean and fat tissue, are not operative (19,20). Studies have shown weight to be lost as water (8), lean and water (21), fat, lean and water (22) and fat and lean (4,6,23), or just fat (24). Studies that include measures of nitrogen balance (1,11) frequently show a negative balance,

Table 1 Rate and Composition of Weight Loss at Altitude

Ref.	Altitude (m)	Time at altitude (d)	Total weight loss (kg)	Rate of weight loss (g/d)	Composition of weight loss	
					kg lean	kg fat
1	4300	28	2.66	95		
			3.80	136		
2	4300	6	0.88	147		
			1.13	188		
3[a]	4300	7	1.00	143		
4	<5400	23	1.90	83	0.56	1.34
	>5400	26	4.00	154	2.80	1.20
5	4800–6000	20	3.95	198		
6	8846	38	7.40	196	5.05	2.51
7	2400–4300	31	1.90	61	0.90	2.80
12	4300	21	2.20	104		
23	7000	10	2.20	220	0.80	1.40
71	4300	9	4.27	474		
72	4300	12	3.54	295	2.24	1.29
			3.96	330	2.51	1.46
73	4846	16	4.50	281		

[a] Study on women.

suggesting a significant loss of lean tissue. Interesting to note, studies in which strenuous training (2,7) have been part of the regimen, show maintenance of lean tissue even in the face of negative energy balance, as has been shown by others at sea level (25). The issue of body composition changes at altitude has been reviewed recently elsewhere (26,27).

Attempts to ensure adequate food intake to counter the anorexia and elevated need have had mixed results. In studies where food is unpalatable (1), intakes are low; in studies where food intake is encouraged, intakes may match sea level values (7). It is only in studies where food intake is enforced either in military troops (28,29) or through experimental design (12) that adequate intakes have been attained to meet altitude needs. This issue has been reviewed recently elsewhere (30).

The elevation in basal energy requirements seen upon acute exposure declines over time, returning to sea level values, by most reports, in 2–3 weeks (11,31,32). Interestingly, in the one study done using female subjects (11), the decline in metabolic rate was more rapid than that seen in most studies on men, the women achieving both sea level appetite and metabolic rate within 1 week of exposure. The magnitude of the elevation of basal metabolic rate increases total energy requirements during exposure to altitude by 300 kcal/day above sea level values. Thus, the total energy deficit experienced by an individual allowed to eat *ad libitum* at altitude (and thus consuming an average of 180 kcal/day less at altitude than at sea level; see Fig. 1) is approximately 480 kcal/day. Reports of lethargy at altitude are also reported (12).

The diuresis so positively associated with acute hypoxic exposure appears to continue through the first 7 days of acclimatization, and *ad libitum* water consumption is unable to keep pace with losses (1). In addition, insensible water losses through respiration and skin evaporation in the dry air at altitude further contributes to a potential water imbalance. When reported, altitude exposure is associated with consistent negative water balances (1).

With chronic exposure to altitude, the energy substrate shifts seen upon acute exposure (decrease in circulating glucose, increase in circulating free fatty acids) remain, and a paradoxical decline in lactate accumulation in response to exercise has been noted (33). Thus, the assumption has been made that although carbohydrate feeding is shown to decrease symptoms of AMS (34), fat may be the ''preferred'' fuel at altitude. Several extra- and intracellular factors are known or suspected to affect the balance of substrate utilization during exercise and at altitude. Overall, the regulation is complex and poorly understood (35).

Further exacerbating the energy imbalance created by anorexia and elevated energy requirement, significant malabsorption may occur with chronic exposure to altitude. Early studies by Pugh (36) describe ''greasy stools'' being produced by individuals living at 17,000 feet, suggesting that malabsorption of fat is particularly prevalent. However, others (12,28) have not been able to confirm such fat malabsorption, and work using stable isotopes of palmitate given to sojourners at 5500 m shows no significant malabsorption of fat under such circumstances (37). Possible malabsorption of carbohydrate and protein have also been ruled out (12) as sources of energy loss at least at 4300 m. Recent work by Westerterp et al. (23) evaluates protein and energy absorption at 6542 m and indicates that inadequate intake, rather than malabsorption, is the primary contributor to energy deficit with acclimatization to these higher altitudes.

In summary, past reports of acute and chronic exposure to hypobaric hypoxia include a decrease in energy intake accompanied by a decline in nitrogen balance resulting in a decrease in lean body mass, a decrease in total body water and total body weight, and a decrease in basal metabolic rate (BMR), with a shift in metabolism toward fat as a major fuel. In addition, lethargy and malabsorption have been noted.

IV. Malnutrition

This description of the consequences of acclimatization looks remarkably like the classic description of the response to starvation (38). Specifically, reports of the response to inadequate energy intakes include negative nitrogen balance, weight loss, a decline in basal energy needs, significant diuresis, and a shift toward utilization of fat as a primary energy source.

In fact, most diet programs, whether they are severe or not, indicate initial weight loss to be primarily in the form of fluid (39). Over time, the composition of weight loss shifts so that the fluid component is less and the component of weight loss presented by lean body mass and fat mass is increased. The proportion of weight loss as lean body mass or fat mass may be altered by the level by exercise performed (40); the greater the exercise, the better the maintenance of lean body mass. Such an effect of exercise may account for some of the discrepancy among studies in the composition of tissue lost at altitude, as some studies were conducted using actively training military recruits and other sedentary subjects.

This decline in lean body mass leads to a decrease in basal energy needs (38), lean body mass being considered the primary determinant of basal metabolic rate. This response leads one to speculate about the reason for the decline in basal metabolic rate seen in most studies of individuals chronically exposed to altitude. This decline may mirror the decline in metabolically active tissue.

In addition, under circumstances of negative energy balance, blood levels of circulating free fatty acids are increased, and mobilization of fat stores represent the major source of energy (41).

Finally, a decrease in usual energy intake has been shown to result in a decrease in strenuous activities (38,42) and a decrement in appetite (38), the latter suggesting a downward spiral with deficits in energy balance, which may lead under circumstances of altitude exposure to the continued decrements in energy intake.

This similarity between the response of an individual to exposure to hypobaric hypoxia and the response to inadequate energy intake and starvation (2,26) led us to question the paradigm previously developed from altitude studies in which individuals were allowed to eat *ad libitum* that anorexia and weight loss were almost inevitable consequences of altitude exposure and that the apparent shift toward fat metabolism represents a normal response to altitude. The question became whether the responses seen in fuel use were a consequence of altitude or of the semistarvation state created by the acute responses to altitude. Thus, we undertook a series of studies in which we attempted to accommodate the acute response by maintaining energy balance, minimizing weight loss and diuresis, and preventing malnutrition. The resultant studies allowed us to study the chronic response to altitude separately from the response to malnutrition.

V. Studies Uncomplicated by Malnutrition

The primary difference between our work and that of others is that we enforced a dietary regimen designed to initially mimic energy intakes required to maintain body weight at sea level and, subsequently, to adjust to altitude requirements for increased energy. Thus, energy intake at altitude was *greater* by 200–400 kcal/day than it was at sea level. A reasonable fluid intake (2 L/d as food and drink and at least an additional 1 L/d as water) was also enforced in an attempt to cover water losses

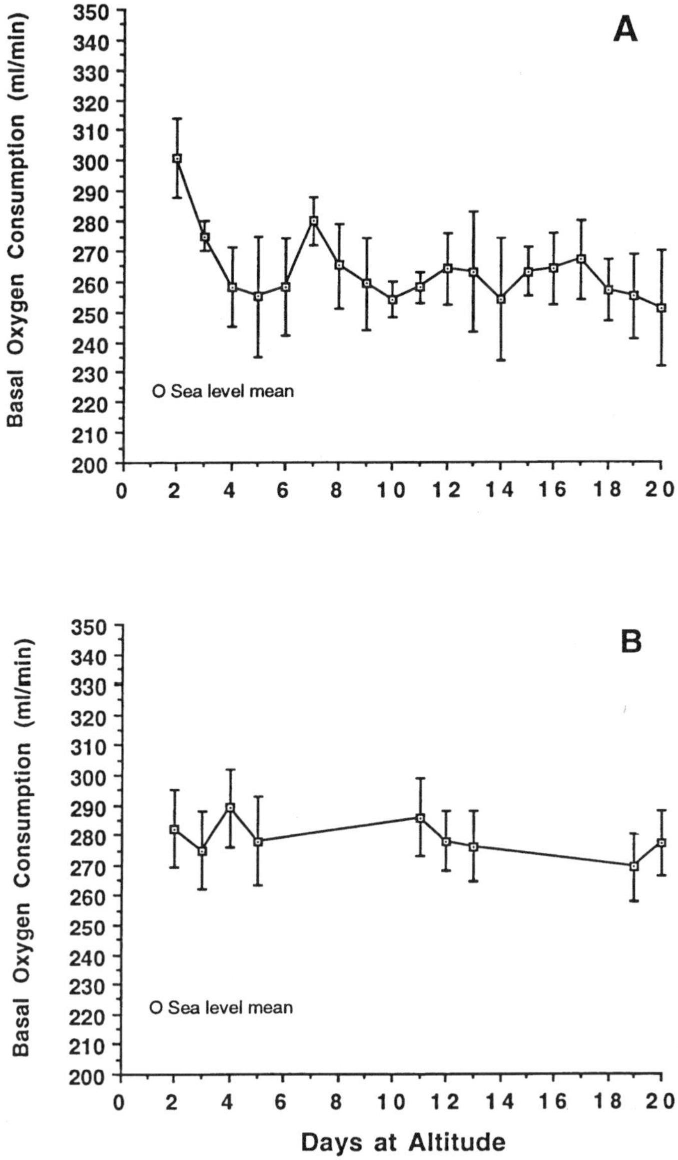

Figure 2 Basal oxygen consumption in men exposed to 3 weeks of hypobaric hypoxia at 4300 m. (A) Represents 7 individuals in whom energy intake was matched to changes in basal energy expenditure after the first week of exposure. (B) Represents 11 individuals in whom energy intake was matched to changes in basal energy expenditure from the first day of altitude exposure. All values are mean ± SEM.

due to diuresis and insensible losses. Activity patterns were kept constant between sea level and altitude so that detraining would not occur. Finally, metabolic fuel stores were standardized by providing a meal 12 hours before all exercise testing. Below we emphasize the results that our collaborations have contributed. We are most comfortable in interpreting metabolic data obtained with these nutrition controls. Further, we note that the results obtained to date have been derived from male subjects. Consequently, variations due to gender and the menstrual cycle remain to be determined.

A. Energy Requirements

We, too, found basal energy requirements to be elevated upon acute exposure to altitude, especially when energy intake was not matched to need over the first week. Basal energy needs were as high as 40% above sea level values during the first 2 days of exposure and declined over a period of 3–4 days to stabilize at about 17% above sea level values (Fig. 2A). The spike in basal energy consumption seen upon

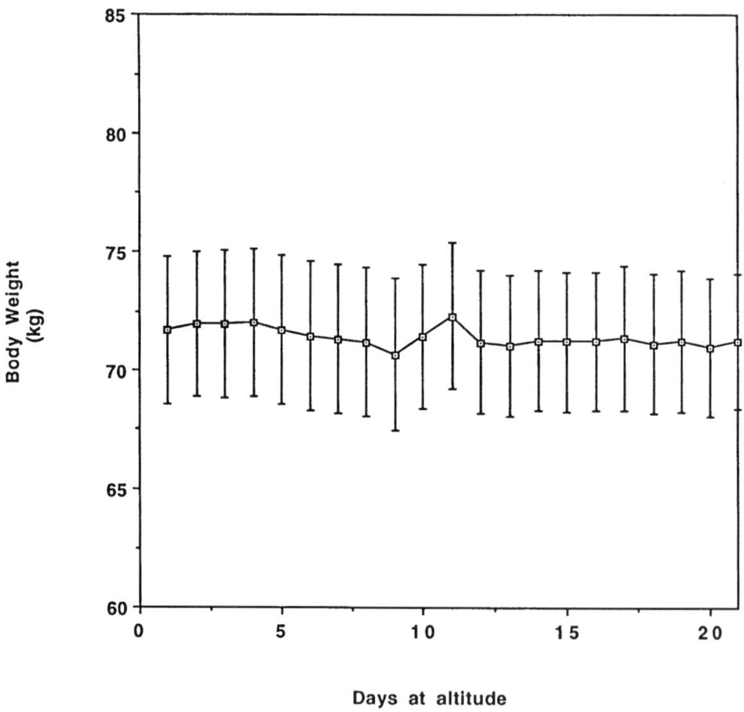

Figure 3 Body weight in five men consuming energy sufficient to cover energy expenditure while residing at 4300 m for 21 days. All values are mean ± SEM.

acute exposure was eliminated when energy intake was matched to energy requirement from the first day of exposure (Fig. 2B).

The mechanism by which basal energy need is elevated is unknown, although the sympathetic nervous system has been implicated. Moore et al. (32) showed β-blockade to eliminate the peak in basal energy requirement in individuals chronically exposed to altitude; however, the subjects in that experiment were not in energy balance. In a similar attempt to block epinephrine involvement in the metabolic response to altitude, we conducted an experiment in which we gave propranolol to 6 of 11 subjects, but we found no difference in basal metabolic rate between the two groups of subjects, and all showed elevated basal energy needs throughout the 3 weeks of exposure to altitude (Fig. 2B). Thus, it may be that the spike in basal energy needs seen during the first days at altitude may be a response to the lack of energy intake, whereas the continued elevation in basal metabolic rate (BMR) may

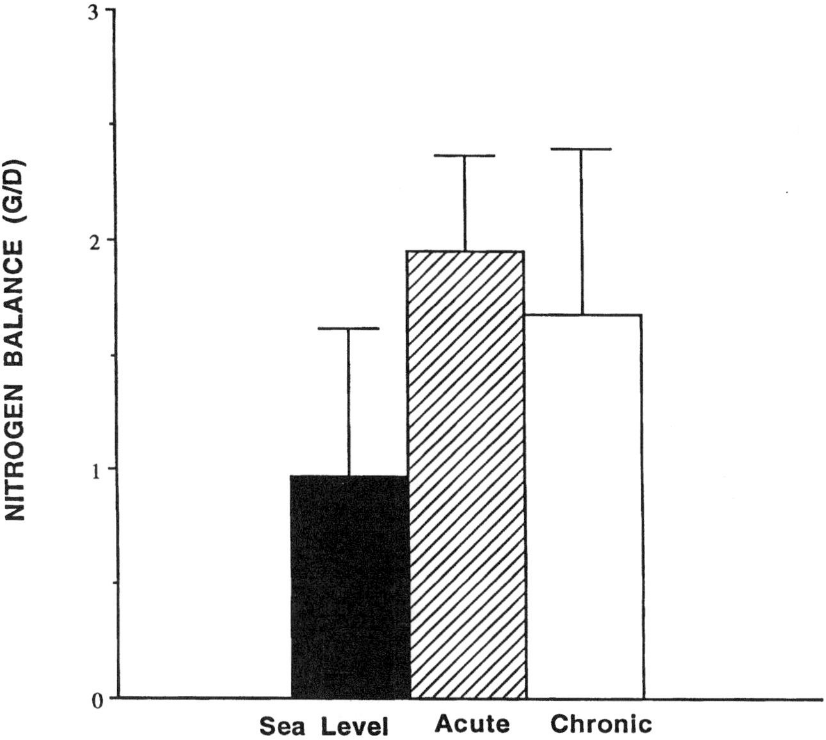

Figure 4 Nitrogen balance in five men fed adequate energy and 1.2 g protein/kg body weight/day while residing at sea level for 3 days and at 4300 m for 21 days. "Acute" indicates values calculated for first 6 days at altitude. "Chronic" indicates values calculated for days 17–19 at altitude. All values are mean ± SEM.

result from the contrived work of breathing at altitude coupled with the increased heat loss through respiratory heat and water losses.

The basal oxygen consumption after acclimatization in individuals in energy balance (260 mL/min or 221 mL O_2/kg/h) matched closely that seen in high altitude natives [247 mL O_2/kg/h (43)]. Total energy requirement for maintenance of body weight (and nitrogen balance, see below) in these relatively inactive individuals was 3583 ± 351 kcal/day, approximately 400 kcal above that required to maintain body weight at sea level (3280 ± 299 kcal/d). Total energy requirement would be further elevated if activity were increased. Measurement of energy expenditure using doubly labeled water in recruits training at 3700 m (7) suggest energy intake must be between 3500 and 4000 kcal/day in such circumstances. Work in climbers at higher

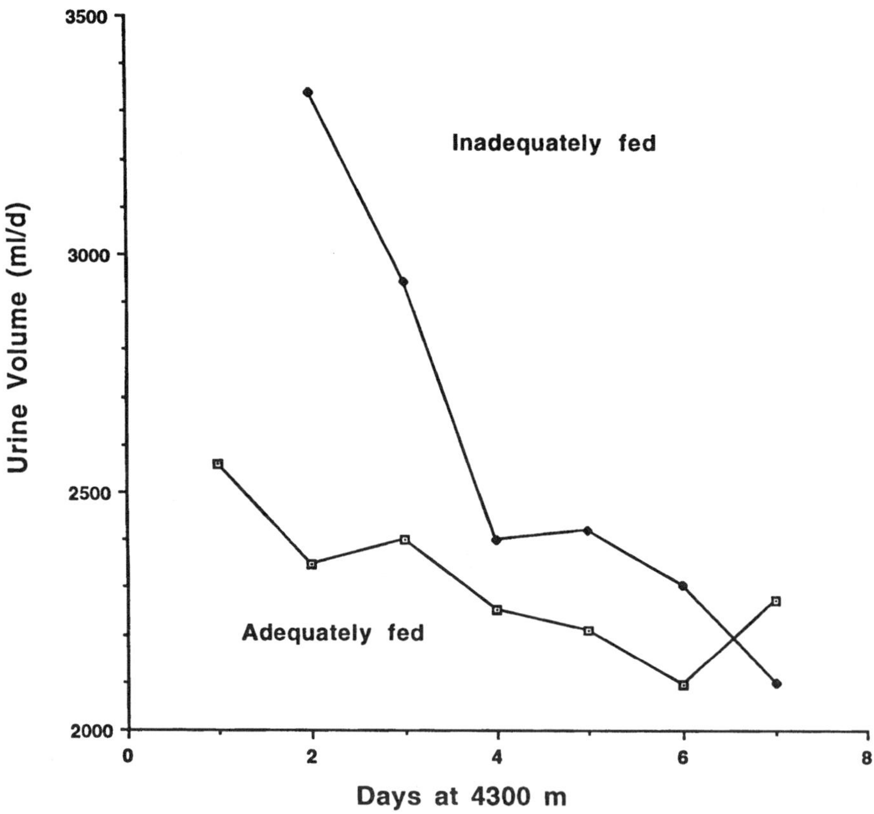

Figure 5 Total daily urine volume produced by men acutely exposed to 4300 m (14,110 ft). Inadequately fed group ($n = 7$) were required to drink 4 L of fluid from food and water each day and consumed sea level energy intake as food and formula for first 7 days at altitude. Adequately fed group ($n = 11$) had same fluid requirements, but energy intake was matched to energy need from the first day at altitude. Diet composition was the same in both groups.

elevations suggest similar total energy requirements, although their activity pattern between climbing days is more sedentary than that of the troops (23).

B. Body Composition

Maintenance of energy balance in subjects resulted in maintenance of body weight (Fig. 3) and nitrogen balance (Fig. 4). Because the measure of body composition at altitude is complicated by shifts in body water, we used nitrogen balance as a monitor of lean body mass (44); if it is negative, tissue is lost. This measure was consistently maintained at equilibrium or above in the individuals given energy to cover increased energy need due to elevated basal metabolic rate, suggesting maintenance of lean tissue in the circumstance of constant body weight.

C. Diuresis

The issue of fluid homeostasis at altitude is more completely treated elsewhere in this volume. At issue here is how much of the diuresis seen during acclimatization to high altitudes is the inevitable consequence of appropriate adaptations, and how much is the result of fluid loss accompanying loss of lean body mass. Figure 5 depicts urine volumes for individuals on standardized fluid and food intakes during the first week of acclimatization to 4300 m. The higher curve was generated by individuals consuming sea level energy intakes, not compensated during the first week at altitude to increases in basal energy needs; the lower curve was generated by individuals consuming the same fluid intake, but energy intake was compensated during the first week to address increased basal energy needs, and neither body weight nor nitrogen balance changed significantly from that seen at sea level. Thus, these data suggest that at least part of the diuresis seen with acclimatization may be consequent to weight loss, and thus avoidable.

VI. Metabolic Consequences of Altitude Exposure Without Malnutrition

Within the context of appropriate nutritional controls, we have utilized isotope tracers and mass balance (arterial-venous difference) measurements to determine the effects of acute and chronic hypoxia on the balance of substrate utilization at altitude. Our results obtained on men showing a shift toward carbohydrate (glucose and lactate) and away from lipid (glycerol and free fatty acids) are strikingly similar to those reported recently by Zinker et al. (45) on dogs made to breathe hypoxic gas mixtures. These results indicate a Pasteur-like effect of hypoxia on the balance of substrate utilization. Thus, in contrast to ''conventional wisdom,'' which interprets hypoglycemia and hyperlipidemia to reflect decreased glucose and increased lipid metabolism, we interpret these results to be due, in large part, to increased glucose and decreased lipid clearance.

A. Glucose

Arterial Glucose Concentration and Delivery

In response to hypoxemia, cardiac output and tissue perfusion increase. As blood flow increases to maintain arterial O_2 delivery, it also increases arterial delivery of all substrates, including glucose, during rest and exercise. Thus, within the first several hours of altitude exposure arterial glucose level is unaffected and decreases only slightly (5–7%) after chronic exposure (Fig. 6) (13,14,46).

In our experiments we have measured limb blood flow and (a-v) differences during rest and a standard exercise task (leg ergometer cycling at 100 W, which elicited approximately 50% of sea level $\dot{V}O_2max$ and 65% of altitude $\dot{V}O_2max$). This power output was selected because it affords steady-state conditions eliciting constant arterial metabolite concentrations and isotopic enrichments and allows whole-body as well as working limb (leg) oxygen consumption (VO_2) to be maintained equivalent to values at sea level. Although other work (35) predicts that carbohydrate utilization should increase at altitude due to the decline in maximal oxygen consumption (and thus the relative increase in work load at the same absolute power output), our experimental design allows for comparison of resting fuel utilization at sea level and altitude, where intensity of exercise is not the issue and in addition allows comparison of acute and chronic responses to the same absolute (and relative) exercise power output at altitude, with significant results.

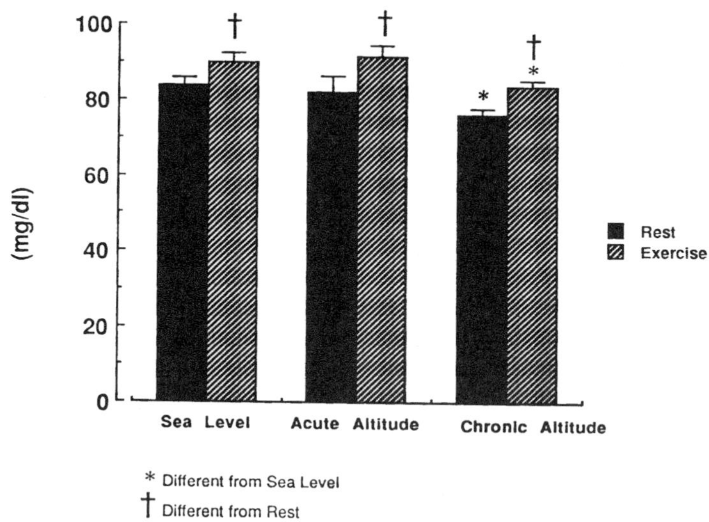

Figure 6 Mean arterial glucose concentration (±SEM) in 6 men during rest and exercise (50% VO_2max) studied at sea level, upon acute exposure to 4300 m, and after a 3-week (chronic) residency at 4300 m altitude. (Data from Ref. 13.)

From limb blood flow and a-v measurements, net glucose uptake can be calculated as:

$$\dot{G} = 2\ \dot{Q}\ (a\text{-}v)_G$$

where: Q is the blood flow in one limb (leg) and $(a\text{-}v)_G$ is the arterial-venous glucose concentration difference (47).

Our results (46,47) indicate that limb net glucose uptake is increased during rest and exercise at altitude. Upon acute altitude exposure, limb blood flow increases over sea level, and it is not possible to detect any immediate change in the a-v difference for glucose, which is small at rest and decreases during exercise as the relative gain in muscle blood flow exceeds the gain in glucose uptake. Therefore, in response to the stresses of exercise and altitude, tissue substrate delivery is maintained or increased, and muscle glucose uptake is not limited by delivery during rest or exercise.

Because muscle glucose uptake is increased at altitude and not limited by delivery (13,46,47), glucose uptake must be controlled by intracellular processes such as muscle cell membrane (sarcolemmal) transport, phosphorylation of glucose to the glycolytic intermediate glucose-6-phosphate, or the glycolytic pathway, which has several controls (48). In most contexts, contemporary investigators conclude that glucose uptake is transport (meaning plasma to cytosol) limited (49–51). Also in contemporary thinking, muscle glucose transport across the sarcolemma is thought to be limited by activity of glucose transport proteins. Because in most cases, glucose-6-phosphate levels in muscle are low and do not change in response to insulin or other means of stimulated glucose uptake, and because we (52) have observed neither glucose-1- nor fructose-6-phosphate levels to increase in muscle as the result of altitude exposure, the normative conclusion for the present must be that consequent to acute altitude exposure, transport, rather than phosphorylation, limits entry of plasma glucose into the glycolytic pathway in skeletal muscle at altitude.

Hepatic Glucose Production and Its Control

Because it is extremely difficult to measure hepatic glucose production, even if hepatic vein catheters are inserted, hepatic glucose production rate in humans is usually estimated from the rate of appearance (Ra) of glucose using an "irreversible," hydrogen-labeled isotopic tracer, such as [6,6-^{2}H]glucose. In the dynamic steady state,

$$R_a = R_d = \frac{\text{Tracer infusion rate}}{\text{Blood isotope enrichment}}$$

From venous tracer infusion and arterial blood sampling, we know that on acute exposure to altitude, arterial glucose concentration is not measurably different from that at sea level, but total body glucose production (appearance, Ra) and uptake (disappearance, Rd) rates are slightly increased (Fig. 7). Moreover, our measurements of muscle glucose uptake during exercise agree well with tracer-calculated rates of glucose disappearance (Fig. 8). Thus, it is apparent that the subtle elevation in muscle and extramuscular tissue glucose disposal at altitude is matched by a similar augmentation in the rate of hepatic glucose production so that only minor alterations in arterial glucose concentration result (Fig. 6).

Based on the information available, it is difficult to comprehend how endocrine signaling accomplishes the matching of glucose production (Ra) to disposal (Rd) at altitude. Upon acute altitude exposure, circulating insulin is unaffected at rest and falls during exercise as at sea level (13,46). Similarly, in two investigations we have been unable to detect an effect of chronic altitude exposure on insulin secretion. Thus, it is difficult to ascribe to insulin a role in changing glucose production and utilization at altitude.

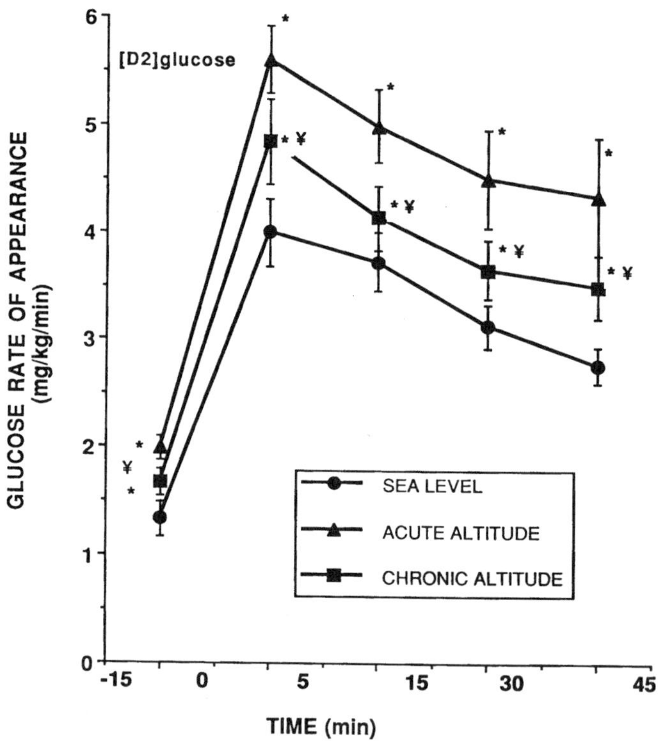

Figure 7 Mean (±SEM) arterial glucose rate of appearance as determined by primed-continuous infusion of [6,6-^{2}H]glucose (D_2-glucose) in five men during rest and exercise (50% $\dot{V}O_2$max) studied at sea level, upon acute exposure to 4300 m, and after a 3-week (chronic) residency at 4300 m altitude. (Data from Ref. 46.)

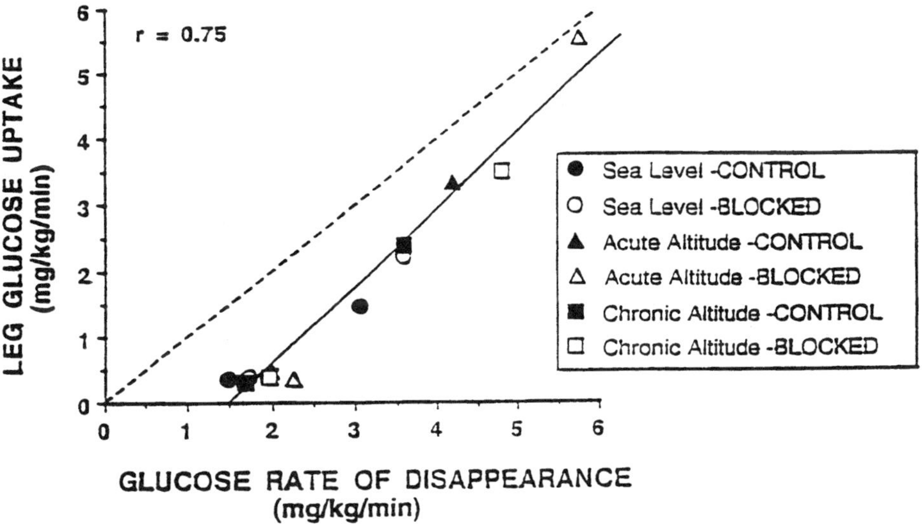

Figure 8 Correlation between leg net glucose uptake [as determined from the (a-v) for glucose and leg blood flow] and systemic glucose rate of disappearance [as determined from primed-continuous infusion of D_2-glucose] in six β-blocked and five unblocked control men during rest and exercise (50% Vo_2max) studied at sea level, upon acute exposure to 4300 m, and after a 3-week (chronic) residency at 4300 m altitude. (Data from Ref. 46.)

Upon acute altitude exposure, circulating epinephrine is elevated (53,55), and epinephrine is known to effect hepatic glucose production (56,57). However, we have observed a similar or augmented rise in glucose Ra at altitude even in the face of dense β-adrenergic blockade by propranolol (46,54). We can understand that β-blockade would limit muscle glycogenolysis as well as lipolysis in adipose and other fat depots and by these means indirectly increase glucose utilization during exercise. However, in postabsorptive, but otherwise well-nourished subjects resting at sea level, muscle glycogenolysis is nil, and glucose oxidation accounts for total body carbohydrate oxidation (49). Therefore, it is not likely that the observed increase in glucose utilization in resting but β-blocked subjects acutely exposed to altitude can be due to inhibition of muscle glycogenolysis. Similarly, as will be discussed below, the bias towards increased glucose use in resting, acutely altitude-exposed males cannot be ascribed to diminished vascular delivery of free fatty acids. Consequently, for the present we must conclude that acute hypoxia uncomplicated by energy deficit provokes intracellular signals to increase glucose and decrease lipid use. Further, given that the increase in glucose use upon acute exposure is matched by a similar increase in glucose production, even in the face of dense β-blockade, epinephrine cannot be an indispensable signal to increased glucose and decreased lipid use at altitude.

Chronic altitude exposure and physical exercise result in significant alterations in circulating norepinephrine, another catecholamine known to have powerful metabolic as well as circulatory effects (57). However, at present the metabolic consequences of alterations in sympathetic activity and levels of circulating norepinephrine are unclear. Acutely, circulating epinephrine is elevated during rest upon altitude exposure, but norepinephrine is not (Fig. 9). In contrast, after residence at altitude, resting norepinephrine is elevated while epinephrine subsides towards sea level values (53,55). Thus, we have difficulty in correlating catecholamine levels with glucose appearance rates during rest and exercise at high altitude, because our most recent results (Fig. 7) indicate that glucose production is greatest during rest and exercise upon acute exposure, but the greatest norepinephrine response is ob-

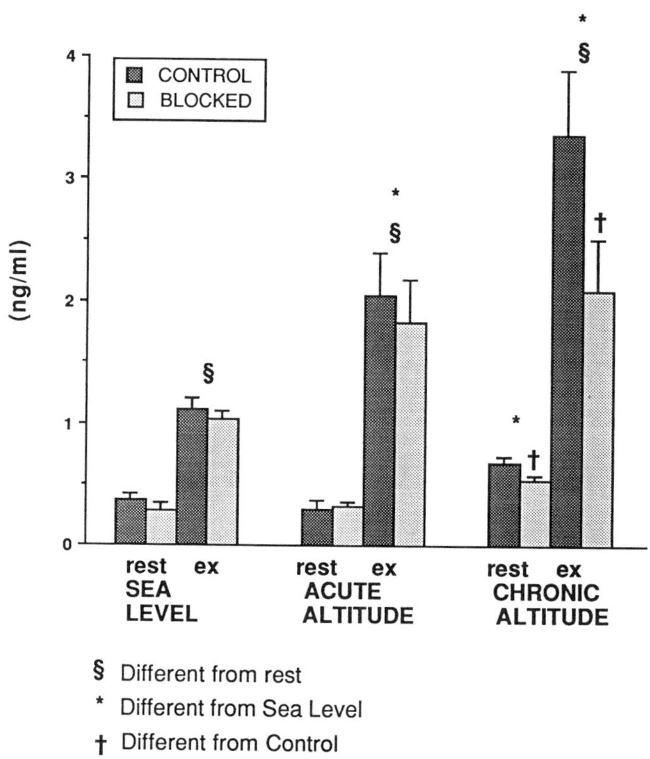

(A)

Figure 9 Mean (±SEM) arterial norepinephrine (A) and epinephrine (B) concentrations in six β-blocked and five unblocked control men during rest and exercise (50% $\dot{V}O_2max$) studied at sea level, upon acute exposure to 4300 m, and after a 3-week (chronic) residency at 4300 m altitude. * Different from sea level ($p < 0.05$); § different from chronic altitude ($p < 0.05$); † different from acute altitude ($p < 0.05$). (Data from Ref. 55.)

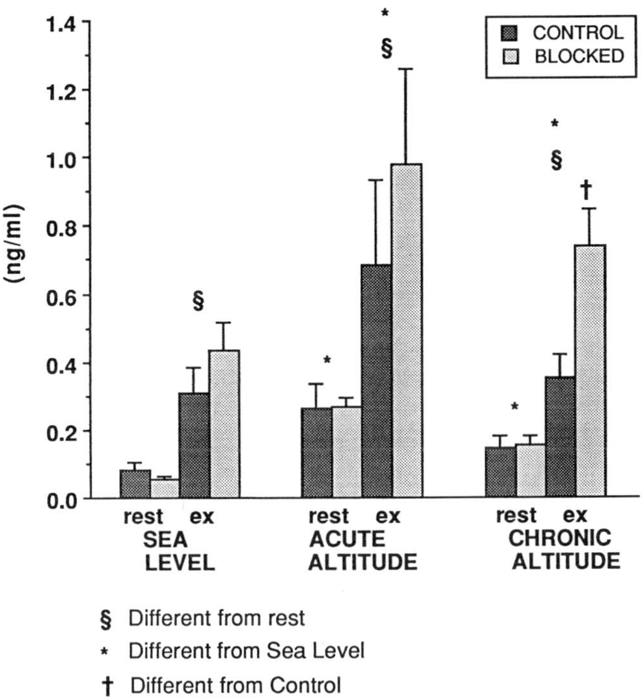

(B)

Figure 9 Continued

served after chronic altitude exposure (Fig. 9A). Moreover, glucose Ra is greatest when epinephrine is greatest (Fig. 9B), but β-blockade increases, does not decrease, glucose appearance.

In the past, we (53) attributed the initial elevation in circulating epinephrine and subsequent diminution of that parameter, with accompanying increase in circulating norepinephrine, to differences in rates of catecholamine secretion. Most recently, from a-v difference and blood flow measurements across active and inactive limbs at sea level and altitude, we (55) have observed increased mean net norepinephrine release from the leg (''spill'') during rest and exercise after chronic, but not acute altitude exposure. Because terminal sympathetic nerves release norepinephrine and because the rate of release can exceed local capacities for catabolism, norepinephrine ''spill'' is taken as a measure of sympathetic nervous system activity. Thus, from limb norepinephrine release measurements, we can deduce that chronic exposure stimulates sympathetic nervous system activity, and α-adrenergic stimulation of the liver through its sympathetic innervation as well as via circulating norepinephrine may, in part, be responsible for the augmented hepatic glucose production after chronic exposure to high altitude. Moreover, because our attempts to blunt with propranolol the rise in glucose use at altitude failed, it is tempting to ascribe a more definitive role to α- as opposed to β-adrenergic mechanisms in the regulation of hepatic glucose production at altitude. However, in the temporary absence of studies employing α-adrenergic blockade at altitude, it is appropriate to note only that both norepinephrine and glucose flux increase at altitude and that we lack definitive studies to postulate a causal relationship between the two.

Like norepinephrine, circulating glucagon is increased with chronic altitude exposure (Fig. 10). Unlike insulin, arterial glucagon rises slightly during exercise, but though statistically significant, the changes in circulating glucagon we have ob-

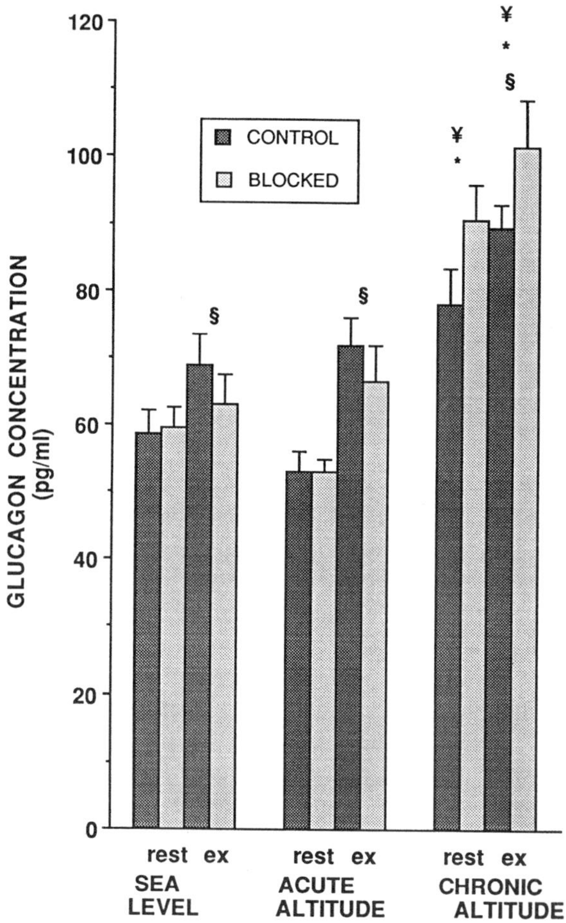

Figure 10 Mean (±SEM) arterial glucagon concentrations in six β-blocked and five unblocked control men during rest and exercise (50% Vo_2max) studied at sea level, upon acute exposure to 4300 m, and after a 3-week (chronic) residency at 4300 m altitude. (Data from Ref. 46.)

served during exercise are small. And, as with norepinephrine, the changes in glucagon are difficult to reconcile with the patterns in hepatic glucose production during rest and exercise. For instance, in our most recent study (46), acute exposure increased glucose Ra and Rd during exercise. However, the glucagon response on acute exposure was the same as at sea level. Possibly, glucagon and norepinephrine interact to play key roles in maintaining HGP during chronic altitude exposure.

Fractional Glucose Oxidation, the Control of Glycolysis, and the Lactate Paradox

At altitude glucose Ra and Rd as well as muscle net uptake are elevated and highly correlated (Fig. 8). Similarly, glucose oxidation is elevated at altitude, which can be understood in terms of mass action or upregulation at the beginning of the pathway (49), as well as regulation at the end of the pathway. As shown in Figure 11, altitude, exercise, and β-blockade all increase the percentage of glucose flux oxidized. Thus, altitude and the other variables studied likely also result in stimulation of pyruvate dehydrogenase (PDH), thus directing a greater portion of the glycolytic flux to oxidation and less to lactate. An acclimatization-induced activation of PDH could be

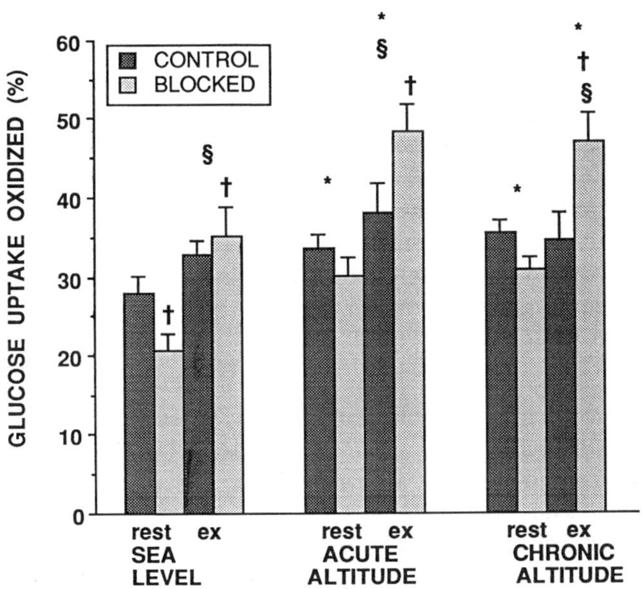

Figure 11 Mean (±SEM) percentage of glucose uptake oxidized (determined from primed-continuous infusion of [1-^{13}C]glucose and measurement of $^{13}CO_2$ and total CO_2 excretion in breath) in six β-blocked and five unblocked control men during rest and exercise (50% $\dot{V}O_2max$) studied at sea level, upon acute exposure to 4300 m, and after a 3-week (chronic) residency at 4300 m altitude. (Data from Ref. 46.)

responsible for diverting more of the glycolytic flux to oxidation, thus explaining the so-called "lactate paradox" (33).

B. Lactate

Arterial Lactate Concentration and Delivery

Our understanding of the subject of lactate responses to altitude benefits from the perspective of the "lactate shuttle" hypothesis (59). This hypothesis considers that lactate is a metabolic intermediate that possesses the characteristics of high rates of intercellular exchange and rapid vascular conductance. In fact, in subjects exercising at sea level (60) and altitude (13), we have observed that lactate flux through the blood exceeds that of glucose. Thus, lactate exchange between tissue sites of production and removal represents an important means for delivery of oxidizable substrate as well as gluconeogenic precursor during exercise (47,48,59). From our experiments conducted on subjects exercising at high altitude, we have developed interpretations much in contrast to traditional views that elevations in blood lactate during exercise at altitude signal tissue O_2 lack. Our results indicate that an elevation in circulating lactate is one of the compensatory adjustments to the stresses of exercise and environment. And, though seemingly at odds with previous views on the causes and consequences of lactacidemia at altitude, our results and conclusions are similar to those of Gutierrez et al. (61), who showed hypoxia to increase lactate uptake by rabbit muscle in situ.

Figure 12 shows the arterial lactate in men during exercise at the power output that elicits 50% of Vo_2max at sea level, upon acute exposure to 4300 m, and after a 3-week acclimatization period. As previously mentioned, the exercise task was selected so that whole-body and working leg rates of O_2 consumption could be maintained, thus minimizing, or eliminating, the possibility of tissue O_2 lack (62). Our results (48) are similar to those observed many times previously, i.e., an elevated circulating lactate level upon acute exposure to altitude and a blunted, though still elevated lactate response after acclimatization (63,64).

Lactate Production

We used ^{13}C-lactate tracer to evaluate the influences of altitude and exercise on lactate production in men. Although we were unable to establish that working muscle was singularly responsible for blood lactate appearance during exercise, it is clear that the elevated circulating lactate level during exercise at constant power and Vo_2 was attributable to increased lactate production (Fig. 13) (47,48).

Muscle Lactate Exchange (Uptake and Release)

During the continuous submaximal exercises we have studied, net lactate release from working muscle is responsible for the initial rise in circulating lactate at exercise onset. However, with continued work at sea level the v-a difference for lactate across muscle changes from positive (release) to zero (no release) (Fig. 14). During

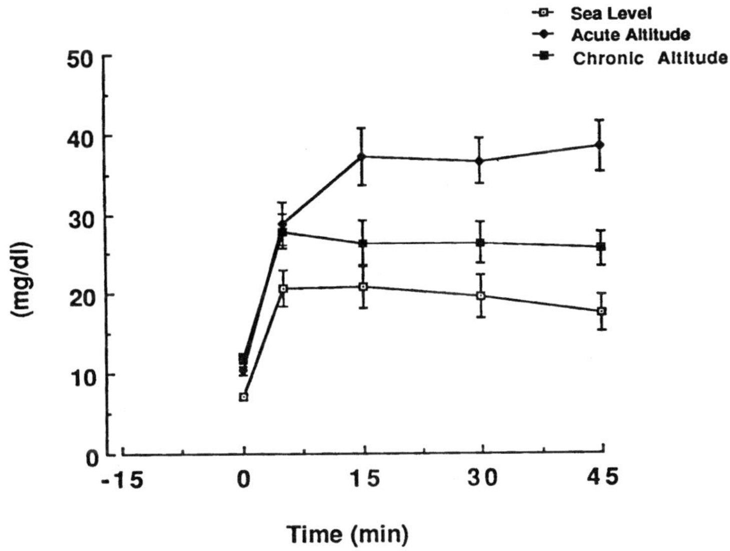

Figure 12 Mean arterial lactate concentration (±SEM) in six men studied during rest and exercise (50% Vo_2max) studied at sea level, at the same power output upon acute exposure to 4300 m, and after a 3-week continuous (chronic) exposure to 4300 m altitude. (From Ref. 13.)

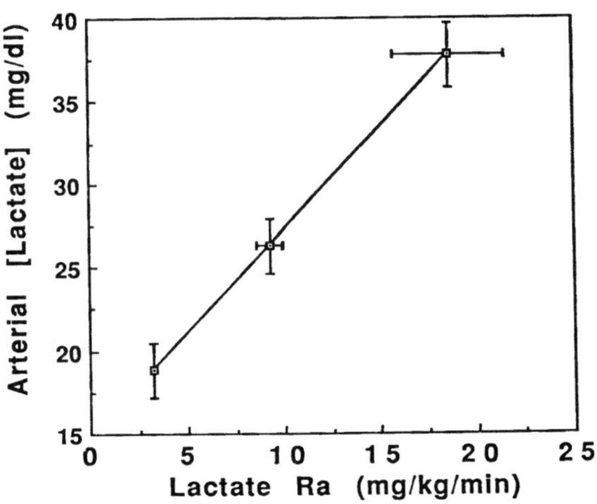

Figure 13 Relationship between arterial lactate concentration and lactate rate of appearance (Ra) in six men studied during rest and exercise (50% Vo_2max) studied at sea level, at the same power output upon acute exposure to 4300 m, and after a 3-week residency at 4300 m altitude. (Data from Ref. 13.)

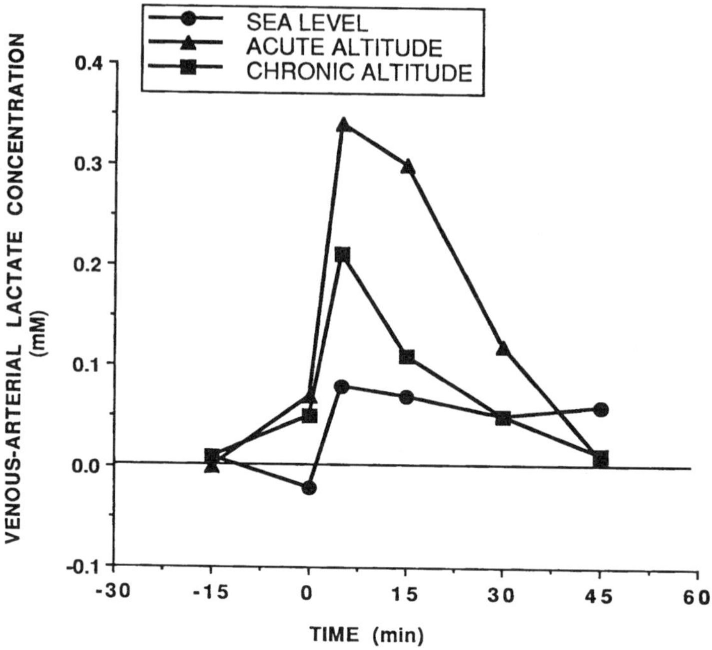

Figure 14 Mean femoral venous-arterial lactate concentration differences (±SEM) in five men studied during rest and exercise (50% $\dot{V}O_2$max) studied at sea level, at the same power output upon acute exposure to 4300 m, and after a 3-week residency at 4300 m altitude. (Data from Ref. 13.)

exercise at altitude, net lactate release from working muscle is increased over sea level, but as at sea level, the larger lactate release from muscle working at altitude is attributable mostly to exercise onset. Therefore, the maintenance of stable and elevated circulating lactate (Fig. 12) at altitude cannot be attributed solely to lactate production in working muscle. The extramuscular source of lactate is unidentified and likely will not be identified with current methodologies unless the liver or some other major tissue site is involved, as the v-a difference for lactate across other low-flow tissues such as skin and adipose is likely small and flow difficult to determine.

Above we emphasized the role of vascular conductance [= (flow) (arterial concentration)] in the delivery and utilization of substrates during exercise and at altitude. In the 1988 Pikes Peak experiment in which we (47) utilized [3-^{13}C]lactate tracer, we showed (Fig. 15) that lactate uptake in working muscle is directly related to arterial lactate concentration. Further, oxidation accounts for essentially all the lactate taken up by muscle working at altitude. We interpret these results to mean

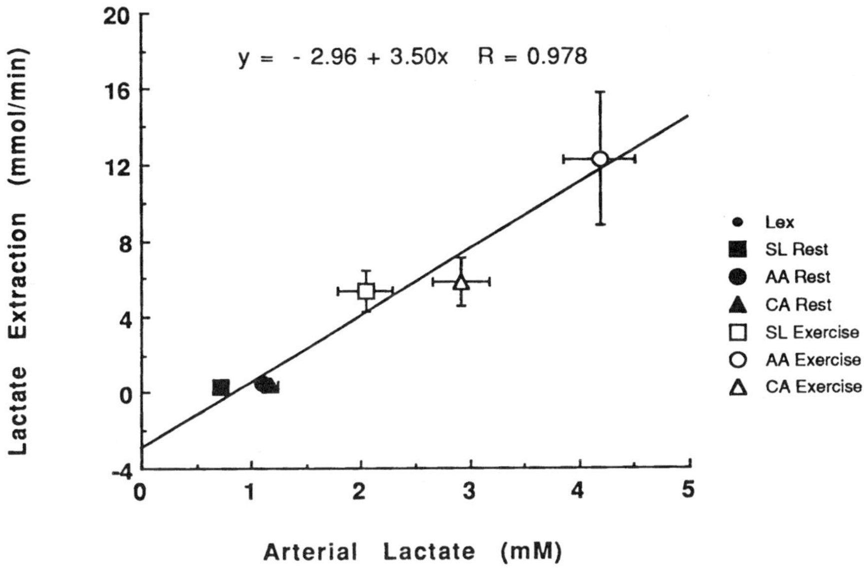

Figure 15 Relationship between muscle lactate extraction and arterial lactate concentration in six men studied during rest and exercise (50% Vo_2max) studied at sea level, at the same power output upon acute exposure to 4300 m, and after a 3-week residency at 4300 m altitude. (From Ref. 60.)

that, contrary to previous thought, during sustained exercise upon acute altitude exposure, when arterial lactate concentration is greatest, significant consumption and oxidation of lactate occur in active skeletal muscle. Thus, the use of lactate as a fuel source at altitude depends on availability, a result consistent with presence of a sarcolemmal lactate transporter (65).

C. Glucose-Lactate Interactions

Dual isotope technology has allowed simultaneous measurements of glucose and lactate fluxes. At rest, lactate appearance (Ra) approximates 30–50% of glucose disappearance (Rd). However, during even mild exercise [e.g., 40–50% Vo_2max at sea level (59,60)] lactate Ra equals or exceeds glucose Rd. As previously noted, altitude exposure increases glucose flux, but the increase in lactate flux during exercise is far greater than the glucose flux (47,58). Upon acute exposure to high altitude, lactate flux exceeds the glucose flux four- to fivefold. These data indicate a role for glycogen in supplying substrate during exercise upon acute altitude exposure in adequately nourished subjects. As with the increase in glucose metabolism, the increase in rate of lactate utilization supports the conclusion that at altitude, there is increased use of carbohydrate energy sources.

D. Lipid

Fatty Acid Utilization

Upon initial exposure to altitude, resting circulating free fatty acid (FFA) and glycerol levels are similar to those observed at sea level. However, after residency at 4300 m, circulating FFA and glycerol levels rise (17) (Fig. 16). The rise in circulating FFAs with chronic exposure has been interpreted as a switch to increased lipid utilization. Despite the elevation in circulating lipids, after a 3-week sojourn at 4300 m we observed minimal a-v differences for glycerol and FFA (Fig. 17) during rest and exercise. Therefore, an important distinction between muscle consumption of FFAs and glucose and their respective circulating levels is realized. Elevated FFA and glycerol levels after acclimatization are largely due to reduced tissue uptake, an extraordinary finding, given the usual assumption that glycerol and FFA levels are predictive of the rate of lipid mobilization and oxidation.

Our understanding of the uptake of FFAs by muscle is limited. Recently, Turcotte et al. (66) identified the presence of a sarcolemmal fatty acid–binding protein (FABP). However, neither they nor others have as yet determined whether the sarcolemmal FABP is mobilized from intracellular sites during exercise or if altitude acclimatization affects downregulation or internalization of existing sarcolemmal fatty acid–binding proteins.

Given the data available indicating suppression of glycerol release and FFA uptake from working limbs at altitude, it is probably appropriate to conclude that intramuscular lipolysis as well as FFA uptake is suppressed by acclimatization. The basis for this conclusion is the absence of glycerol release (17) (Fig. 15). Because skeletal muscle and adipose tissues lack glycerol kinase (the enzyme necessary to recycle to triacylglycerol, the glycerol released as the result of lipolysis), we can interpret the absence of glycerol release from muscle after acclimatization to mean suppression of lipolysis within muscle.

The result of suppressed glycerol release also assists in the interpretation of the acclimatization-imposed limits of FFA uptake. Glycerol release in absence of significant FFA uptake could be interpreted to indicate utilization of intramuscular lipids and an acclimatization-imposed limitation in muscle FFA uptake such as could be imposed by downregulation of the sarcolemmal FABP. However, because after acclimatization limb muscle glycerol release is nil, for the present we can conclude that feed-forward regulation of glycolysis results in mitochondrial acetyl-CoA formation and downregulation of mitochondrial FFA uptake. For working cardiac muscle it has been demonstrated that increased glycolytic flux results in activation of acetyl-CoA carboxylase (ACC) and formation of malonyl-CoA, a potent inhibitor of mitochondrial carnitine palmityl carboxylase 1 (CPT 1) (67,68). Thus, we interpret our results to indicate that chronic altitude exposure both decreases intramuscu-

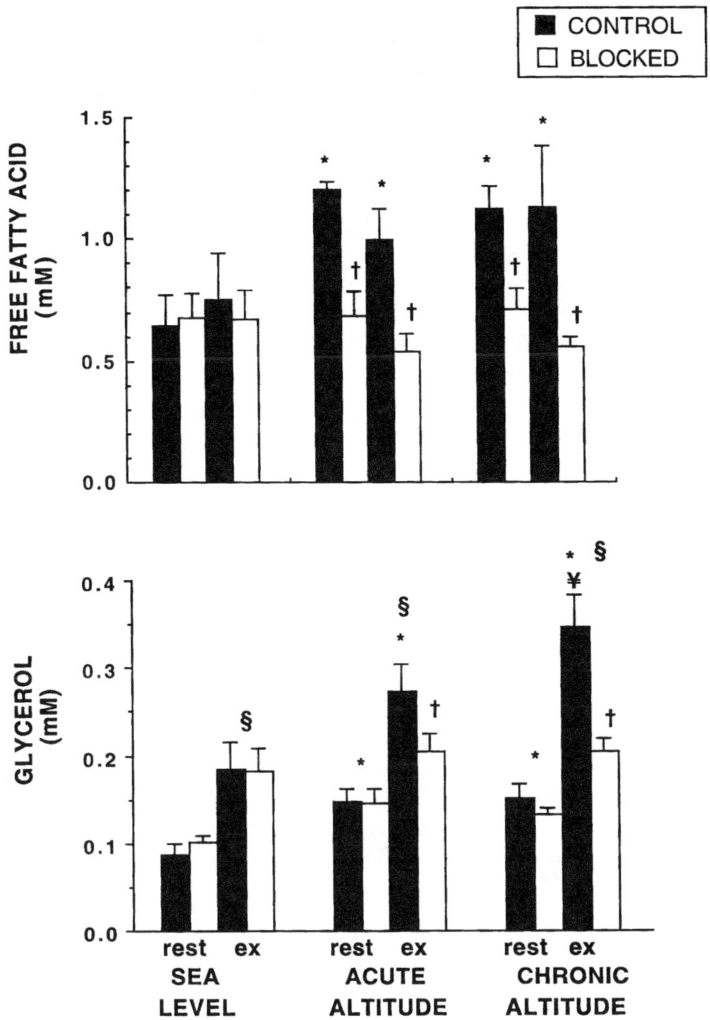

Figure 16 Mean arterial free fatty acid and glycerol concentrations (±SEM) in five control and six β-blocked male subjects at rest and during exercise at sea level, upon acute altitude exposure, and after 3 weeks at 4300 m. § Different from rest ($p < 0.05$); * different from sea level ($p < 0.05$); ¥ different from acute altitude ($p < 0.05$); † different from control ($p < 0.05$). (Data from Ref. 17.)

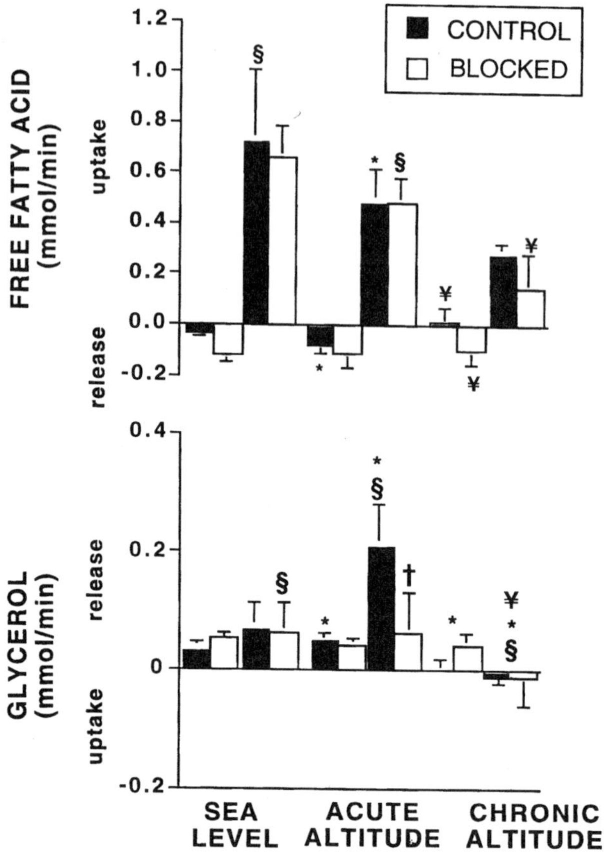

Figure 17 Mean free fatty acid and glycerol exchanges across the legs (±SEM) in five control and six β-blocked male subjects at rest and during exercise at sea level, upon acute altitude exposure, and after 3 weeks at 4300 m. § Different from rest ($p < 0.05$); * different from sea level ($p < 0.05$); ¥ different from acute altitude ($p < 0.05$); † different from control ($p < 0.05$). (From Ref. 17.)

lar lipolysis and suppresses mitochondrial uptake of activated free fatty acids in adequately nourished subjects.

Absolute or Relative Exercise Effects on Substrate Utilization at Altitude

Altitude exposure and increments in exercise intensity both have the effect of shifting the balance of substrate toward carbohydrates and away from lipids (35). In our experiments on Pikes Peak we controlled the dietary status of subjects, and we required them to perform the same absolute exercise tasks at sea level and altitude.

We believe these protocols demonstrate the effects of hypoxia, independent of cachexia, on the balance of substrate utilization.

For the sojourner at altitude, it remains that altitude exposure may depress appetite, and food availability or palatability may exacerbate this situation, leading to significant energy deficit. Therefore, the ill-prepared and ill-informed mountaineer with inadequate dietary energy and carbohydrate can be expected to have difficulty with prolonged exercise tasks. Each task is relatively more difficult at altitude, but the normal shift toward muscle glycogen and blood glucose with intense exercise at sea level could be limited by availability of fuel sources at altitude. The forced catabolism of lipid and lean tissue not only weakens the individual but yields fuel sources with lesser enthalpies per unit O_2 consumed and, in any case, less preferred by working muscle. Therefore, whether one's perspective is from that of nutrition or exercise physiology, whether one considers the perceived effort or the actual effect of hypoxia on muscle metabolism, the primary concern should be for providing carbohydrate-rich foods for the altitude sojourner.

VII. Nutrient Recommendations for High Altitude

Based on the observations above, we recommend the following:

1. *Energy*: Increase intake by 500 kcal/day above appetite (30), emphasizing carbohydrate-rich foods.
2. *Water*: Increase intake to greater than 3 L/day to cover diuresis and insensible losses.
3. *Other nutrients*: No adequate information is available regarding need at altitude for other nutrients in adequately nourished individuals. A recent review commissioned by the Committee on Military Nutrition Research (69) suggests there is no specific advantage from augmented intake of any vitamins or minerals, despite implications of some research that vitamin E (70) may be needed in increased amounts at altitude. These latter studies were based on providing supplements to inadequately nourished individuals, making it hard to differentiate the effect of malnutrition from the effect of hypoxia. Future work in the field of micronutrient nutrition and metabolism at altitude must consider prior status as well.

References

1. Consolazio CF, Matoush LO, Johnson HL, Daws TA. Protein and water balances of young adults during prolonged exposure to high altitude (4300m). Am J Clin Nutr 1968; 21:154–161.
2. Consolazio CF, Johnson HL, Krzywicki HJ, Daws TA. Metabolic aspects of acute altitude exposure (4300m) in adequately nourished humans. Am J Clin Nutr 1972; 25:23–29.

3. Hannon JP, Klain GJ, Sudman DM, Sullivan FJ. Nutritional aspects of high-altitude exposure in women. Am J Clin Nutr 1976; 29:604–613.
4. Boyer SJ, Blume D. Weight loss and changes in body composition at high altitude. J Appl Physiol 1984; 57:1580–1585.
5. Guilland JC, Klepping J. Nutritional alterations at high altitude in man. Eur J Appl Physiol 1985; 54:517–523.
6. Rose MS, Houston CS, Fulco CS, Coates G, Sutton JR, Cymerman A. Operation Everest II: nutrition and body composition. J Appl Physiol 1988; 65:2545–2551.
7. Worme JD, Lickteig JA, Reynolds RD, Deuster PA. Consumption of a dehydrated ration for 31 days at moderate altitudes: energy intake and physical performance. J Am Dietetic Assoc 1991; 91:1543–1549.
8. Krzywicki HJ, Consolazio CF, Johnson HL, Nielsen WC, Barnhart RA. Water metabolism in humans during acute high-altitude exposure (4300 m). J Appl Physiol 1971; 30:806–809.
9. Honig A. Peripheral arterial chemoreceptors and reflex control of sodium and water homeostasis. Am J Physiol 1989; 257:R1282–R1302.
10. Hackett PH, Rennie D, Grover RF, Reeves JT. Acute mountain sickenss and edemas of high altitude: A common pathogenesis? Respir Physiol 1981; 46:383–390.
11. Hannon JP, Sudman DM. Basal metabolic and cardiovascular function of women during altitude acclimatization. J Appl Physiol 1973; 34:471–477.
12. Butterfield GE, Gates J, Brooks GA, Groves BM, Mazzeo RS, Sutton JR, Reeves JT. Energy balance and weight loss during three weeks at 4,300m. J Appl Physiol 1992; 72:1741–1748.
13. Brooks GA, Butterfield GE, Wolfe RR, Groves BM, Mazzeo RS, Sutton JR, Wolfel EE, Reeves JT. Increased dependence on blood glucose after acclimatization to 4,300m. J Appl Physiol 1991; 70:919–927.
14. Houston CS. Going Higher—The Story of Man at Altitude. Boston: Little, Brown, 1987.
15. Johnson JL, Consolazio CF, Burk RF, Davis TA. Glucose ^{14}C-UL metabolism in man after abrupt altitude exposure (4300 m). Aerosp Med 1974; 45:849–854.
16. Young AJ, Evans WJ, Cymerman A, Pandolf KB, Knapik JJ, Maher JT. Sparing effect of chronic high-altitude exposure on muscle glycogen utilization. J Appl Physiol 1982; 52:857–862.
17. Roberts AC, Butterfield GE, Reeves JT, Wolfel EE, Brooks GA. Acclimatization to 4,300 m altitude decreases reliance on fat as substrate. J Appl Physiol 1996; 81:1762—1771.
18. Young PM, Sutton JR, Green HJ, Reeves JT, Rock PB, Houston CS, Cymerman A. Operation Everest II: metabolic and hormonal responses to incremental exercise to exhaustion. J Appl Physiol 1992; 73:2574–2579.
19. Fulco CS, Cymerman A, Pimental NA, Young AJ, Maher JT. Antrhopometric changes at high altitude. Aviat Space Environ Med 1985; 56:220–224.
20. Fulco CS, Hoyt RW, Baker-Fulco CJ, Gonzalez J, Cymerman A. Use of bioelectrical impedance to assess body composition changes at high altitude. J Appl Physiol 1992; 72:2181–2187.
21. Withey WR, Milledge JS, Williams ES, Minty BD, Bryson EI, Luff NP, Older MWJ, Beeley JM. Fluid and electrolyte homeostasis during prolonged exercise at altitude. J Appl Physiol 1983; 55:409–412.
22. Cymerman A, Pandolf KB, Young AJ, Maher JT. Energy expenditure during load carriage at high altitudes. J Appl Physiol 1981; 51:14–18.

23. Westerterp KR, Kayser B, Wouters L, LeTrong JL, Richalet JP. Energy balance at high altitude of 6,542 m. J Appl Physiol 1994; 77:862–866.
24. Surks MI, Chin KSK, Matoush LO. Alterations in body composition in man after acute exposure to high altitude. J Appl Physiol 1966; 21:1741–1746.
25. Todd K, Butterfield GE, Calloway DH. Protein utilization in young men under two conditions of energy balance and three levels of work. J Nutr 1984; 114:2107–2118.
26. Brouns F. Nutritional aspects of health and performance at lowland and altitude. Int J Sports Med 1992; 13:S100–S106.
27. Butterfield GE. Elements of energy balance at altitude. In: Sutton JR, Coates G, Remmers JE, eds. Hypoxia, and Adaptations. Philadelphia: B.C. Decker, Inc., 1990: 88–93.
28. Rai RM, Malhotra MS, Dimri GP, Sampathkumar T. Utilization of different quantities of fat at high altitude. Am J Clin Nutr 1975; 28:242–245.
29. Sridharan K, Malhotra MS, Upadhayay TN, Grover SK, Dua GL. Changes in gastrointestinal function in humans at altitude of 3,500m. Eur J Appl Physiol 1982; 50:148–154.
30. Butterfield GE. Maintenance of body weight at altitude: in search of 500 kcal/day. In: Nutritional Needs in Cold and High Altitude Environments. Washington, DC: National Academy Press, 1996.
31. Stock MJ, Norton NG, Ferro-Luzzi A, Evans E. Effect of altitude on dietary-induced thermogenesis at restand during light exercise in man. J Appl Physiol 1978; 45:345–349.
32. Moore L, Cymerman GA, Huang SY, McCullough RE, McCullough RG, Rock PB, Young A, Young PM, Weil JV, Reeves JT. Propranalol blocks metabolic rate increase but not ventilatory acclimatization to 4300 m. Respir Physiol 1987; 70:195–205.
33. Reeves JT, Wolfel EE, Green HJ, Mazzeo RS, Young AJ, Sutton JR, Brooks GA. Oxygen transport during exercise at high altitude and the lactate paradox: lessons from Operation Everest II and Pikes Peak. Exercise and Sport Sciences Reviews, Vol. 20. Baltimore: Williams & Wilkins, 1992:275–296.
34. Consolazio CF, Matoush LO, Johnson HL, Krzywicki HL, Daws TA, Issac GJ. Effects of high-carbohydrate diets on performance and clinical symptomatology after rapid ascent to high altitude. Fed Proc 1969; 28:937–943.
35. Brooks GA, Mercier J. The balance of carbohydrate and lipid utilization during exercise: The 'crossover' concept, brief review. J Appl Physiol 1994; 76:2253–2261.
36. Pugh LGCE. Physiological and medical aspects of the Himalayan Scientific and Mountaineering Expedition, 1960, 1961. Br Med J 1962; 2:621–627.
37. Imray CHE, Chesner I, Wright A, Neoptolemeusand JP, Bradwell AR. Fat absorption at altitude: a reapprasial. Int J Sports Med 1992; 13:87.
38. Keys A, Brozek J, Henshel A, Mickelson O, Taylor HL. The Biology of Human Starvation. Minneapolis: The University of Minnesota Press, 1950.
39. Fisler JS, Brenick EJ. Starvation and semistarvation diets in the management of obesity. Ann Rev Nutr 1987; 7:465–484.
40. Belko AZ, Obarzanek E, Roach R, Rotter M, Urban G, Weinberg S, Roe DA. Effects of aerobic exercise and weight loss on riboflavin requirements of moderately obese, marginally deficient young women. Am J Clin Nutr 1984; 40:553–561.
41. Saudek CD, Felig P. The metabolic events of starvation. Am J Med 1976; 60:117–126.
42. Gorsky RD, Calloway DH. Activity pattern changes with decreases in food intake. Human Bio 1983; 55:577–586.

43. Picon-Reategui E. Basal metabolic rate and body composition at high altitudes. J Appl Physiol 1961; 16:431–434.
44. Calloway DH. Nitrogen balance in men with marginal intake of protein and energy. J Nutr 1975; 105:914–923.
45. Zinker BA, Namdaran K, Wilson R, Lacy DB, Wasserman DH. Acute adaptation of carbohydrate metabolism to decreased arterial PO2. Am J Physiol 1994; 266:E921–929.
46. Roberts AC, Reeves JT, Butterfield GE, Mazzeo RS, Sutton JR, Wolfel EE, Brooks GA. Altitude and β-blockade augment glucose utilization during exercise. J Appl Physiol 2000; 80:605–615.
47. Brooks GA, Wolfel EE, Groves BM, Bender PR, Butterfield GE, Cymerman A, Mazzeo RS, Sutton JR, Wolfe RR, Reeves JT. Muscle accounts for glucose disposal but not lactate release during exercise after acclimatization to 4,300 m. J Appl Physiol 1992; 72:2435–2445.
48. Connett RJ, Honig CR, Gayeski TEJ, Brooks GA. Defining hypoxia: a systems view of VO_2, glycolysis, energetics and intracellular PO_2. J Appl Physiol 1990; 68:833–842.
49. Brooks GA, Fahey T, White T. Exercise Physiology: Human Bioenergetics and Its Application, 2nd ed. Mountain View, CA: Mayfield Publishing Company, 1996.
50. Cushman SW, Wardzala LJ. Potential mechanism of insulin action on glucose transport in the isolated adipose cell. Apparent translocation of intracellular transport systems to the plasma membrane. J Biol Chem 1980; 255:4558–4762.
51. Suzuki K, Kono T. Evidence that insulin causes translocation of glucose transport activity to the plasma membrane from an intracellular storage site. Proc Natl Acad Sci USA 1980; 77:2542–2545.
52. Green JH, Sutton JR, Wolfel EE, Reeves JT, Butterfield GE, Brooks GA. Altitude acclimatization and energy metabolic adaptations in skeletal muscle during exercise. J Appl Physiol 1992; 73:2701–2708.
53. Mazzeo RS, Bender PR, Brooks GA, Butterfield GE, Groves BM, Sutton JR, Wolfel EE, Reeves JT. Arterial catecholamine responses during exercise with acute and chronic high-altitude exposure. Am J Physiol 1991; E419–E424.
54. Mazzeo RS, Brooks GA, Butterfield GE, Cymerman A, Roberts AC, Selland M, Wolfel EE, Reeves JT. β-Adrenergic blockade does not prevent lactate response to exercise after acclimatization to high altitude. J Appl Physiol 1994; 76:610–615.
55. Mazzeo RS, Brooks GA, Butterfield GE, Podolin DA, Wolfel EE, Reeves JT. Acclimatization to high altitude increases muscle sympathetic activity both at rest and during exercise. Am J Physiol 269 1995; R201–R207.
56. Kjaer M, Farrell PA, Chistensen NJ, Galbo H. Increased epinephrine response and inaccurate glucoregulation in exercising athletes. J Appl Physiol 1986; 61:1693–1700.
57. Winder WW. Role of cyclic AMP in regulation of hepatic glucose production during exercise. Med Sci Sports Exerc 1988; 20:551–560.
58. Brooks GA. Lactate: glycolytic end product and oxidative substrate during sustained exercise in mammals—the ''lactate shuttle.'' In: Gilles R, ed. Comparative Physiology and Biochemistry—Current Topics and Trends, Vol. A, Respiration–Metabolism–Circulation. Berlin: Springer-Verlag, 1984:208–218.
59. Stanley WC, Wisneski JA, Gertz EW, Neese RA, Brooks GA. Glucose and lactate interrelations during moderate intensity exercise in man. Metabolism 1988; 37:850–858.
60. Brooks GA, Butterfield GE, Wolfe RR, Groves BM, Mazzeo RS, Sutton JR, Wolfel

EE, Reeves JT. Decreased reliance on lactate during exercise after acclimatization to 4,300m. J Appl Physiol 1991; 71:333–341.

61. Gutierrez G, Fernandez E, Hurtado FJ, Kiiski R, Chakravathy S, Ronco JJ, Arbelaz G. Hydroxymalonate inhibits lactate uptake by rabbit hindlimb. J Appl Physiol 1994; 76: 2735–2741.
62. Wolfel EE, Bender PR, Brooks GA, Butterfield GE, Groves BM, Mazzeo RS, Sutton JR, Reeves JT. Oxygen transport during steady state, submaximal exercise in chronic hypoxia. J Appl Physiol 1991; 70:1129–1136.
63. Dill DB, Edwards HT, Folling A, Oberg SA, Pappenheimer AM, Talbot JH. Adaptations of the organism to changes on oxygen pressure. J Physiol 1931; 71:47–63.
64. Edwards HT. Lactic acid at rest and work at high altitudes. Am J Physiol 1936; 116: 367–375.
65. Roth DA, Brooks GA. Lactate transport is mediated by a membrane-borne carrier in rat skeletal muscle sarcolemmal vesicles. Arch Biochem Biophys 1990; 279:377–385.
66. Turcotte LP, Kiens B, Richter EA. Saturation kinetics of palmitate uptake in perfused skeletal muscle. FEBS Lett 1991; 279:327–329.
67. Lopaschak GD, Belke DD, Gamble J, Ztoi T, Schonekess BO. Regulation of fatty acid oxidation in mammalian heart in health and disease. Biochem Biophys Acta 1994; 1213: 263–276.
68. Saddik M, Gamble J, Witters LA, Lopaschuk GD. Acetyl-CoA carboxylase regulation of fatty acid oxidation in the heart. J Biol Chem 1993; 268:25836–25845.
69. Committee on Military Nutrition Research. Conclusions and Recommendations. In: Nutritional Needs in Cold and High Altitude Environments. Washington, DC: National Academy Press, 1996.
70. Simon-Schnass M. Nutrition at high altitude. J Nutr 1992; 122:778–781.
71. Winton BK, Janoski AH. Effects of high altitude and diet on lipid components of human serum. Fed Proc 1969; 28:983–984.
72. Krzywicki HJ, Consolazio CF, Matoush LO, Johnson HL, Barnhart RA. Body composition changes during exposure to altitude. Fed Proc 1969; 28:1190–1194.
73. Bradwell AR, Foote JH, Milles JJ, Dykes PW, Forster PJE, Chesner I, Richardson NV. Effect of acetozolamide on exercise performance and muscle mass at high altitude. Lancet 1986;1001–1005.

18

The Endocrine System

JEAN-PAUL RICHALET

Université Paris 13, Bobigny, France

I. Introduction

Adaptation to a new environment needs an information system that first informs the body of the characteristics of the environment and second triggers biological responses, which may be more or less "adaptive" to new environmental conditions. The nervous system and the endocrine (or neuroendocrine) system are where the perception and the adaptive processes occur. This chapter will review what we know to date about the effects of high-altitude hypoxia on the endocrine system.

We need to distinguish, at our state of knowledge, the mechanisms of action of some endocrine pathways involved in regulation loops with negative feedback control systems and the effects of hypoxia on the metabolism of some hormones that do not appear, based on present knowledge, in a regulation loop intending to restore a normal cellular oxygenation (1). The only endocrine or neuroendocrine systems clearly involved in a purposely adaptive regulation loop are erythropoiesis and blood oxygen transport and adrenergic activation and increase in blood flow. Other endocrine responses are not clearly involved in an adaptive process, such as the inhibition of the renin-angiotensin-aldosterone system or an increase in cortisol or thyroid hormone release.

The effect of acute and sustained hypoxia on some endocrine systems has been extensively studied, e.g., hormones involved in water and electrolyte balance or the adrenergic system. Much less is known about others, e.g., thyroid or sex hormones. For more extensive information about some aspects of the endocrine response at high altitude, see also two other reviews (2,3).

II. Hypoxia and Signal Processing in Hormonal Systems

Hormonal receptors are classified as membrane receptors (cathecolamines, angiotensin, ADH, growth hormone, parathormone, prolactine) and intracellular receptors, nuclear or cytoplasmic (cortisol, aldosterone, estrogens, progesterone, T3, T4). Little is known about the effect of hypoxia on nuclear receptors. However, the expression of these receptors can be modulated by hypoxia (4). More is known about some membrane receptors, e.g., G-protein–mediated receptors such as β-adrenoceptors, muscarinic receptors, or adenosinergic receptors. Chronic hypoxia induces a downregulation of β-adrenoceptors and adenosine receptors and an upregulation of muscarinic receptors in rat hearts (5). These processes of desensitization (and sensitization) occur commonly in hormone receptors in response to various stressors. In the case of rat hearts, the mechanisms involved are probably an increased expression of the gene for Gi protein and a decreased functional activity of Gs (6).

Extensive studies have been done on the erythropoietin (EPO) system and especially on the factors involved in the synthesis and release of EPO. Hypoxia-inducible factors (HIF-1) may be released in hypoxia and modulate the expression of various genes coding not only for EPO but also for other factors such as the vascular endothelium growth factor (VEGF) (7).

Exposure to hypoxia has been shown to produce both down- and upregulation. For example, release of cortisol is activated and release of aldosterone is blunted by hypoxia, although these two hormones are secreted by adjoining zones in the same organ. Some endocrine or neuroendocrine pathways are activated in response to the hypoxic stressor; other pathways are blunted. A resistance appears to certain stimuli, limiting the effects of the activated pathways. Finally, an optimal adaptation to the stressful environment would depend on an adequate balance between activation and resistance (Fig. 1). One can speculate about a common pathway involving HIFs in the modulation (activation or inhibition) of the secretion of all hormones in hypoxic conditions. Extensive studies are in progress on the mechanisms of oxygen sensing at the cellular and molecular levels (see Chapter 4).

A. Stress Hormones

Stress is a generic term for any circumstance that tends to disturb an individual from its "normal" resting equilibrium. In that sense, hypoxia is a stress that induces an endocrine reaction from adrenal glands. Whether the physiological response of these

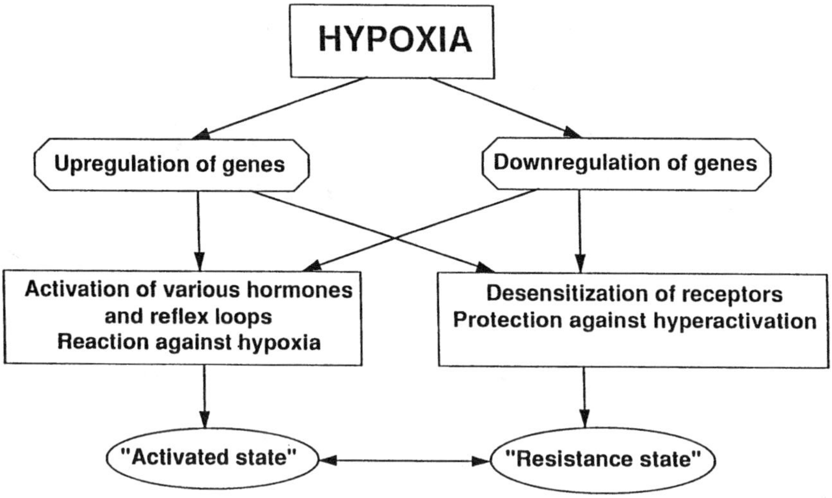

Figure 1 An optimal adaptation to hypoxia would depend on an adequate balance between upregulation and downregulation processes.

glands to hypoxia is specific or similar to the response to other stressors remains to be clarified.

Cortisol

Both acute and sustained altitude hypoxia stimulate the adrenal cortex (Fig. 2). Plasma concentrations of ACTH and cortisol are increased, often transiently (7–15). However, not all studies report a change in plasma cortisol (16–19).

In one of the earliest studies, published in 1962, performed on 10 men exposed for 2–3 weeks at and above 6500 m, no variation was found in urinary 17-OH-corticosteroids. However, inability to store samples at low temperature might have altered the hormones and compromised the analysis (20). In 10 subjects exposed for 2 weeks at 4300 m, urinary 17-OH-corticosteroids were transiently (1–7 days) increased but had returned to normal after 14 days (8). In men studied while climbing at and above 5450 m during an Everest expedition, urine corticosteroids were essentially normal (21). Brief and moderate hypoxia does not significantly increase cortisol (16), but increases ACTH during exercise (17,22). Exercise increased plasma cortisol similarly in normoxia and in acute hypobaric hypoxia (3000 m altitude) (19).

Acute Mountain Sickness

Clinical status may influence the release of corticosteroids. Acute mountain sickness (AMS) was associated with antidiuresis and elevated 17-OH-corticosteroids in urine

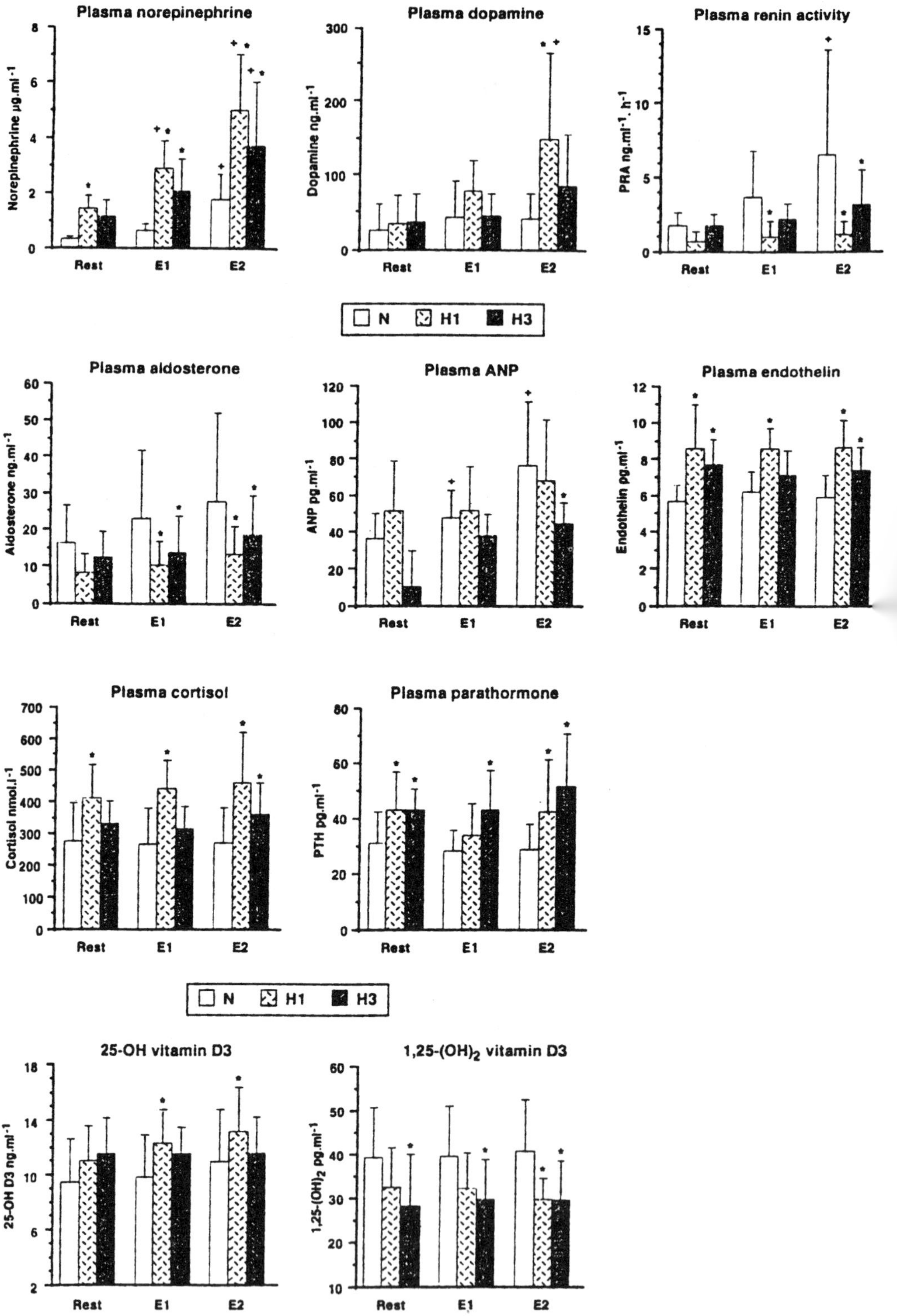
Plasma norepinephrine
Norepinephrine µg.ml⁻¹
Plasma dopamine
Dopamine ng.ml⁻¹
Plasma renin activity
PRA ng.ml⁻¹.h⁻¹
N
H1
H3
Plasma aldosterone
Aldosterone ng.ml⁻¹
Plasma ANP
ANP pg.ml⁻¹
Plasma endothelin
Endothelin pg.ml⁻¹
Plasma cortisol
Cortisol nmol.l⁻¹
Plasma parathormone
PTH pg.ml⁻¹
N
H1
H3
25-OH vitamin D3
25-OH D3 ng.ml⁻¹
1,25-(OH)2 vitamin D3
1,25-(OH)2 pg.ml⁻¹
Rest
E1
E2

(23). Of 10 subjects susceptible to high-altitude pulmonary edema (HAPE), when flown to 3500 m, 4 developed HAPE. Cortisol increased in all subjects but declined more rapidly in the 4 contracting HAPE. The authors suggest that the failure of the adrenocortical response to altitude stress in susceptible subjects might contribute to HAPE (24). In 4 men exposed to a simulated altitude of 4760 m for 2 days, a rise in plasma cortisol preceded onset of AMS symptoms (12). Acute normobaric (10.5% O_2) or hypobaric (5100 m) hypoxia induced an increase in cortisol associated with hypotension and hypokalemia (10). In sea level dwellers exposed to a simulated altitude of 4270 m, cortisol increased, while in subjects residing at 1830–2200 m exposed to 4270 m, there was no increase in cortisol and no AMS symptoms (25). In subjects taken to 4559 m for 6–42 hours, cortisol rose in those with severe AMS but not in well-acclimatized subjects (26) (Fig. 3). Mean diurnal cortisol and ACTH concentrations were found increased in 8 subjects exposed for 79 hours to 4350 m. The diurnal rhythm of cortisol was maintained at high altitude and was accompanied by a parallel variation in AMS score, with maximal values at 8 a.m. (27). At one week at 6542 m plasma cortisol was elevated, decreasing with subsequent acclimatization, further supporting an association with AMS (28,29).

Hypoxia Associated with Other Stressors

When altitude exposure is associated with another stress such as exercise or danger, increase in cortisol levels is more likely. In 8 subjects exposed to 4550 m simulated altitude for 2 hours, cortisol did not increase at rest but was increased during and after exercise (12). In 5 climbers studied at 5400, 6600, and 2 at 7000 m, cortisol excretion was normal; however, in one subject exposed to great climbing stress, cortisol and aldosterone excretion increased (30). In a group of soldiers exposed for more than 10 weeks at altitudes above 6000 m, cold (−30°C), danger, and strenuous exercise, cortisol was elevated (14). A case report of a man on cortisone therapy (25 mg daily) who developed clinical signs of adrenocortical insufficiency while trekking in Nepal raises the question whether persons on corticoid replacement therapy might need to increase their dosage at high altitude (31).

What triggers cortisol release at high altitude? Cortisol secretion rate was studied in dogs exposed to a low F_{IO_2} or carbon monoxide with spontaneous or controlled hyperventilation. ACTH was not the sole controller of cortisol secretion during hypoxic stress. Other factors can also interfere, such as adrenal blood flow (higher in spontaneous ventilation), stimulation of pulmonary stretch receptors (which can inhibit ACTH response), and low adrenal tissue P_{O_2} (32). Cortisol response to acute hypoxia (11% O_2 for 10 min) could be mediated by a reflex action from the carotid

Figure 2 Variations of plasma hormones concentration at rest and with exercise (E1 = 40% of V_{O_2}max, E2 = 70% of V_{O_2}max) at sea level (N) and after one (H1) and 3 (H3) weeks at 6542 m. Hypoxia vs. normoxia. $^*p < 0.05$ or better. Exercise vs. rest: $^+p < 0.05$ or better. (From Ref. 29.)

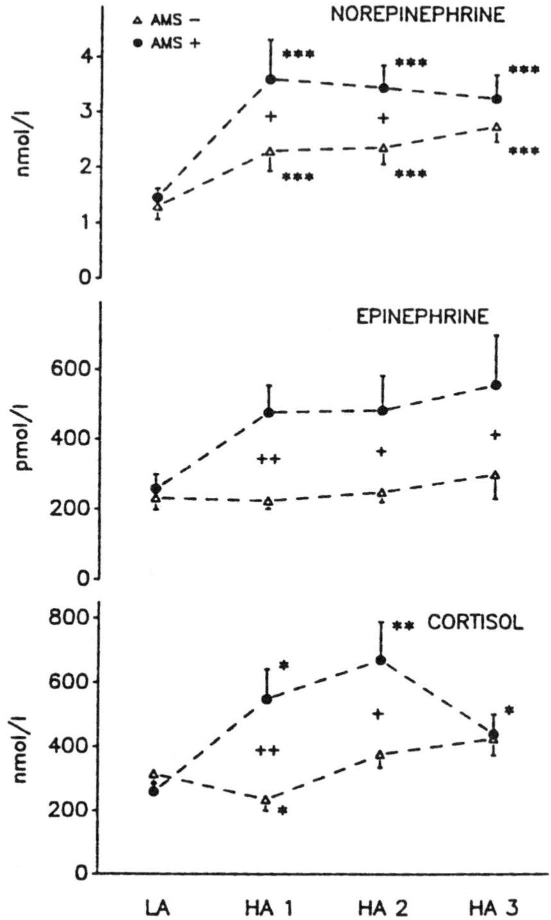

Figure 3 Norepinephrine, epinephrine, and cortisol (mean ± SE) in plasma of 14 subjects without AMS (AMS-, open triangle) and 11 subjects with AMS (AMS-, closed circle) at 550 m (LA) and 6 h (HA 1), 18 h (HA 2), and 42 h (HA 3) after arrival at 4559 m. Significant differences within a group compared with corresponding baseline values: $*p < 0.05$, $**p < 0.01$, $***p < 0.001$. Significant differences between groups ($^{+}p < 0.05$ and $^{++}p < 0.01$) refer to single investigations. (From Ref. 26.)

bodies and also by nonchemoreceptor mechanisms. A possible relationship between hypoxic ventilatory response (HVR) and plasma cortisol is evoked (33,34).

Whatever the mechanism, the stimulation of adrenal cortex seems to be a common feature at high altitude. Whether it is a specific response to hypoxia or a nonspecific response to stress remains to be established.

B. Adrenergic System

Altitude hypoxia activates the adrenergic system: plasma and urine catecholamine concentrations increase as does neuronal activity in adrenergic fibers. With sustained exposure to hypoxia, hyperplasia of the adrenal medulla occurs (see Chapter 13).

At high altitude, increase in plasma norepinephrine concentration is observed both at rest and during exercise (35–37). Consequently, heart rate increases at rest and during submaximal exercise. However, moderate or brief hypoxia does not always increase catecholamine levels (22,38). In general, adrenergic activity at high altitude shows an initial increase followed by a further elevation in the next few days or weeks, after which a decrease is seen. However, plasma norepinephrine concentration always remains higher than at sea level. Dopamine has also been shown to increase in hypoxia (39,40). Permanent inhabitants of high altitude, however, have plasma catecholamines levels similar to those of sea level natives (41).

A major characteristic of the adrenergic system in response to prolonged hypoxia is the desensitization of beta-adrenergic receptors. A number of studies of the heart of humans, rats, and guinea pigs show:

- A decrease in maximal heart rate during exercise (42)
- A lower heart rate response to adrenergic activation (exercise or isoproterenol infusion) (43–45)
- A downregulation of beta-receptors (36,45–47)
- A modulation of G protein gene expression and function (6)

This blunting of beta-adrenergic response is balanced by the upregulation of muscarinic receptors, the downregulation of receptors to adenosine, and the upregulation of alpha-adrenergic receptors (5,48). Teleologically, we might view the decreased maximal heart rate in chronic hypoxia as preventing myocardial oxygen need from exceeding supply and protecting the myocardium against the risk of tissue anoxia. This blunting is a clear mechanism of autoregulation of cardiac function and oxygen balance within the myocardium (42,49).

Adrenergic receptors are found not only in the heart and blood vessels but in many other organs. The adrenergic system is involved in the control of numerous functions and endocrine systems, e.g., control of blood glucose and substrate utilization, renin release by the kidney, thyroid hormones. Thus, activation of the adrenergic system and subsequent receptor desensitization induced by hypoxia may influence hormonal responses.

III. Hormones of Fluid Homeostasis

Three main hormonal systems are involved in the regulation of blood volume and electrolyte concentrations: the renin-angiotensin-aldosterone system, antidiuretic hormone, and atrial natriuretic peptide. Each system responds to hypoxia differently, but the net result is an adequate fluid balance. How these systems interact to influence fluid homeostasis is dependent on many factors, including severity of hypoxia, duration of acclimatization, and whether the individual suffers from AMS (see Chapter 16).

A. Renin-Angiotensin-Aldosterone System

The first indirect information concerning aldosterone and altitude was obtained from 4 subjects during an expedition to the Himalayas in 1956: salivary Na^+ was elevated and $[K^+]$ lowered, suggesting a decrease in aldosterone (50). Confirmation was obtained at the Vallot observatory in 1960: a decrease in K^+ and aldosterone excretion in urine was found in 7 subjects exposed for 24 days at 4350 m. The authors postulated that the aldosterone decrease was due to an increase in thoracic intravascular volume (right atrium), associated with hyperventilation (7,51).

Since these first observations, numerous papers have described the hypoxia-induced modifications of the renin-aldosterone system in normally acclimatizing subjects. Although almost all studies describe low plasma aldosterone concentration (PAC), they report variable responses of plasma renin activity (PRA) in hypoxia. No clear explanation is given for the inhibition of the renin-aldosterone system in hypoxia. In some studies, subjects were not studied in a steady state, e.g., a few minutes after stopping exercise or without control or even measurement of fluid and mineral intake. Moreover, methods used for the measurement of PRA frequently have a greater intra-assay variability than those used for measurement of PAC, leading to a decreased likelihood that hypoxia-induced variations of PRA will reach statistical significance. If we exclude the immediate effects of acute hypoxia (often associated with stress) and consider only studies performed in a reasonable number of humans with prolonged (>1 day) altitude exposure, resting PRA was found to be increased (9,52), unchanged (12,14,53,54), or decreased (11,18,28,55–58). PAC or aldosterone urinary excretion was more consistently decreased (Fig. 2) (12,18,28,51,54–56,58,60,61), although some report an unchanged (53) or even increased (9,11,14) aldosterone response, possibly resulting from reduced Na^+ intake. If we examine only studies on humans after prolonged hypoxia (more than one or two days), a smaller variation of PRA than of PAC, so called "dissociation" between PRA and PAC, was found by some authors (12,52,54,57,62) but not by others (9,28,56,58,59,63).

Factors resulting in decreased renin production are (1) increase in renal perfusion pressure (RPP) via renal baroreceptors, (2) decrease in β-adrenergic activity, (3) increase in distal sodium load (macula densa), and (4) release of various humoral factors such as endothelin or prostaglandins.

Independent of renin variations, the decrease in aldosterone secretion could be by a (1) decreased activity of angiotensin-converting enzyme (ACE), (2) increased release of atrial natriuretic peptide (ANP), (3) variation in the K^+ intracellular/extracellular balance, (4) inhibition of aldosterone synthesis via adrenal hypoxia and reduced sensitivity of adrenal glands to angiotensin II, (5) increased dopamine release, and (6) hypoxemia-induced stimulation of the carotid bodies (64).

Angiotensin-Converting Enzyme

Although some studies gave evidence of a hypoxia-induced inhibition of ACE in dogs (65), rats (66), and humans (67), most authors seem now to agree that a decrease in ACE activity does not explain low aldosterone release (68–71). Hypoxia-mediated change in pulmonary ACE occurring as a consequence of reduced pulmonary vascular transit time does not explain aldosterone suppression, since normoxic exercise (which reduces transit time) is associated with elevated aldosterone (69).

Atrial Natriuretic Peptide

ANP has been shown to inhibit both renin and aldosterone release, as well as aldosterone action on the distal tubule (72). ANP has been evoked as a possible mediator for the hypoxia-induced inhibition of the renin-aldosterone system (26,69,71,73). Small increments of plasma ANP (30 pmol/L) suppressed PRA and inhibited the aldosterone response to angiotensin II in 6 normal men (74). The aldosterone but not the cortisol response to ACTH is inhibited by acute or chronic hypoxia (3000 m) in humans. The authors speculate that ANP may have mediated this inhibitory effect (75).

Potassium

The decrease in aldosterone secretion could also be linked to a slight hypokalemia related to the hyperventilation-induced alkalosis. In most studies, changes in resting plasma [K^+] are insignificant; however, small changes in the extra/intracellular gradient of K^+ in the adrenal zona glomerulosa could influence aldosterone production, even in the absence of clear changes in plasma [K^+]. In sheep exposed to 96 hours of normobaric hypoxia, plasma [K^+] is preserved by movement of K^+ from the intracellular compartment; this change in K^+ distribution could inhibit aldosterone secretion during hypoxia (76). K^+ efflux from the intracellular compartment and renal K^+ retention preserve plasma [K^+] despite negative whole body K^+ balance induced by anorexia and the alkalosis-induced cellular K^+ efflux. The net effect on plasma [K^+] may be variable and depend on the ventilatory response to hypoxia. Finally, the inhibition of aldosterone helps preserve resting membrane potential and the extra-intracellular gradient of K^+ (76). From a teleologic point of view, respiratory alkalosis induces a decrease in plasma [K^+]. To maintain plasma [K^+], aldosterone decreases and urinary K^+ excretion decreases; moreover, low aldosterone increases Na^+ excretion, which reduces the risk for edema (71).

Adrenal Cortex

The size of the glomerula zona in the adrenal cortex, where aldosterone is produced, is reduced in dogs, rabbits, and rats exposed to 1520–4260 m for 5 months (77). A direct effect of hypoxia on the zona glomerulosa has been proposed by Raff et al. (33). A decrease in aldosterone response to angiotensin II was found in cultured cells. However, a direct effect of O_2 is questionable. The hypothesis has been put forward of a hypoxia-induced inhibition of an O_2-dependent enzyme (cytochrome P450 and 18-hydroxylase) within the adrenal zona glomerulosa (71,78,79). Evidence also exists against the hypothesis of reduced sensitivity of the adrenal gland to angiotensin II (69,80). The high plasma cortisol and low plasma aldosterone levels observed by many authors at high altitude are compatible with an inhibition of the late metabolic pathway between corticosterone and aldosterone, with the cortisol pathway itself being unaffected by low O_2 availability. In a suspension of adrenal cells, low Po_2 selectively and reversibly inhibited aldosterone formation and response of aldosterone to angiotensine II, cAMP, and ACTH stimulation. From studies on the conversion of corticosterone to aldosterone in dispersed bovine adrenal glomerulosa cells, Brickner et al. suggest that the sensitivity of 18-hydroxylase to O_2 could account for the dissociation of PRA and aldosterone during hypoxia in vivo (79). When these cells were exposed to CO and/or hypoxia, the conversion of corticosterone to aldosterone was not affected by CO, suggesting that the site of O_2 sensitivity of the cytochrome P450 enzyme involved in the inhibition of 18-hydroxylase is not the heme CO-binding site (81). Exposure of rats to 3 days of hypoxia led to a decrease in aldosterone formation by decreasing the expression of the gene for P-450c11AS, the late-pathway enzyme leading to aldosterone synthesis (82). The expression of genes coding for the steroid receptors could also be modulated by hypoxia: in human renal cortex epithelial cells, hypoxia downregulates mineralocorticoid receptors and upregulates glucocorticoid receptors (4).

Dopamine

Dopamine plays an important role in water and electrolyte handling in humans through modulation of the renin-angiotensin-aldosterone system. Dopamine inhibits aldosterone secretion in humans at low and at high Na^+ intake (83). Dopamine inhibits angiotensin II–stimulated aldosterone secretion during Na^+ deficiency (10 mEq Na^+/d), but not in Na^+-repleted normal subjects (84). Na^+ deprivation inhibits dopamine and increases renin, both actions leading to increased aldosterone (85). Six hours of hypoxia (7% O_2) in cats induced an increase in urine dopamine, which correlated with increased excreted sodium (39). During 1 hour of exposure to 12% O_2, the metoclopramide-induced increase in aldosterone was blunted by hypoxia (86). Basal plasma dopamine was higher in hypoxia in subjects exposed for 2 days to 4350 m (40). In the same study, dopamine-induced diuresis was inhibited in hypoxia, by an effect on distal tubule (possibly by increasing ADH), while dopamine-induced increase in glomerular filtration rate was inhibited probably because of high adrenergic basal tone (40). Since dopamine has been shown to inhibit the aldosterone response to angiotensin II or orthostatism (85), it may have contributed to the depressed aldosterone response to exercise in hypoxia (29,40).

Chemoreceptors

The stimulation of arterial chemoreceptors of normoxic cats by almitrine bismesylate induced hypocapnia, alkalosis, hypokalemia, increase in PRA and aldosterone, with a decrease in the ratio of aldosterone to PRA. Thus, hyperventilation produced the opposite effect of hypoxia on the renin-angiotensin-aldosterone system, although the aldosterone:PRA ratio was similarly reduced (87). In chronically hypoxic rats, natriuretic response to almitrine is blunted. There is an interaction between arterial chemoreceptor and cardiovascular stretch receptor effects: hypovolemia attenuates the hypoxia-induced diuresis (34). In a diet-controlled study in 10 subjects exposed for 3 days to the simulated altitude of 3450 m, supine and standing PRA was decreased despite increased plasma catecholamines. Stimulation of intrathoracic receptors could inhibit PRA, as during immersion (55). Acute (1 day) or prolonged (1 week) exposure to 4300 m induced a decrease in forearm venous compliance. This response is primarily triggered by hypocapnia and potentiated by hypoxia, probably mediated by increased sympathetic activity (88). This phenomenon may induce an increase in central blood volume and explain the decrease in aldosterone through the activation of intrathoracic receptors. In cats, the suppression of plasma aldosterone in acute hypoxia could be due to a reflex mechanism involving carotid chemoreceptors via an increase in nerve activity to the adrenal cortex (64). This pathway may involve dopamine (74) or serotonin (89).

Many other factors may affect the renin-angiotensin-aldosterone response to hypoxia.

Gender

The blunting of the renin-aldosterone system was found to be more pronounced in women than in men (29). We have no explanation for this observation, and in general, information concerning women's responses to high altitude is quite limited. Gender differences could also affect the levels of hormones at high altitude as it does at sea level. For example, at sea level aldosterone is elevated during the midluteal phase, with a greater increase with exercise, probably because of high progesterone levels in this phase (90).

Acute Mountain Sickness

A link between the inhibition of the renin-aldosterone system and the efficiency of acclimatization has been proposed (Fig. 4) (26). The plasma concentration of aldosterone would not decrease at high altitude in subjects suffering from severe AMS or HAPE, which could contribute to increased distal tubule reabsorption and water retention. However, in some studies no correlation is apparent between AMS score and modifications in the hormonal profile (29,91). The severity of AMS symptoms was even related to the decrease in aldosterone secretion (56). However, this is difficult to explain, since spironolactone reduces AMS symptoms. Exercise-induced increase in aldosterone was higher in subjects with AMS (63). Increases in NE, ACTH, and cortisol were also greater in the AMS group. Vomiting can reduce salt intake and favor aldosterone increase in those with AMS (63). However, resting PRA and aldosterone did not vary with altitude and were not significantly different in those with or without AMS when ascent is slow (91). High altitude without exer-

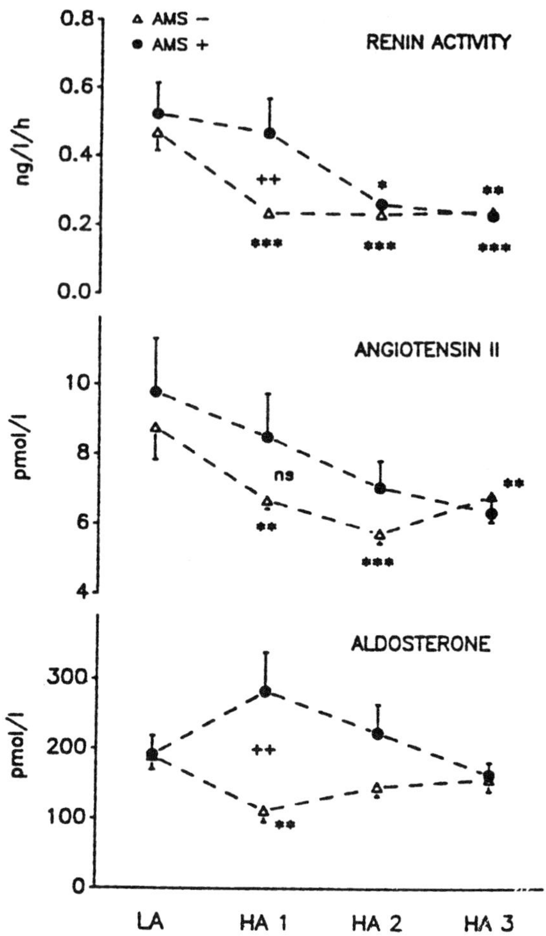

Figure 4 Renin activity, angiotensin II, and aldosterone (mean ± SE) in plasma of 14 subjects without AMS (AMS-, open triangle) and 11 subjects with AMS (AMS-, closed circle) at 550 m (LA) and 6 h (HA 1), 18 h (HA 2), and 42 h (HA 3) after arrival at 4559 m. Significant differences within a group compared with corresponding baseline values: $*p < 0.05$, $**p < 0.01$, $***p < 0.001$. Significant differences between groups ($^{+}p < 0.05$ and $^{++}p < 0.01$) refer to single investigations. (From Ref. 26.)

cise or AMS induces an increase in urine volume and Na^+ excretion and a decrease in aldosterone, renin being variable. At altitude, exercise or AMS decreases urine volume and Na^+ excretion and increases plasma volume and aldosterone (92). In seated subjects exposed for 5 hours to 4115 m in a hypobaric chamber, plasma $[K^+]$ decreased, PRA tended to decrease; aldosterone, cortisol, and ACTH increased. ANP increased slightly but did not inhibit the aldosterone response to hypoxia. Aldosterone is probably responsible for the sodium retention in subjects suffering from AMS, and this effect overrides the natriuretic effect of ANP (93).

High-Altitude Natives

During exercise, the rate of increase of PRA and aldosterone was less in high-altitude natives at 2660 m than in sea level natives at sea level (94). In high-altitude natives studied at La Paz (3600 m), plasma concentration of active renin at rest was high and aldosterone response to renin during exercise was attenuated, particularly in polycythemic subjects (41). The renin response to adrenergic activation during exercise was blunted in sea level natives exposed to 4350 m but not in high-altitude natives at 3600 m (41). In subjects residing at 1830–2200 m exposed to 4270 m, there was no clear decrease in aldosterone and no AMS symptoms (25,60).

Diet and Hydration

In 6 subjects, low Na intake (10 mEq/d) increased and high Na intake (200 mEq) decreased aldosterone secretion. Conversely, a low-K diet (40 mEq/d) decreased aldosterone and high K (200 mEq/d) increased aldosterone. [K^+], independent of Na^+ and angiotensin II, can regulate basal aldosterone secretion and sensitize the adrenal response to ACTH, although Na^+ restriction has a more potent effect (95). The level of hydration and salt intake also determines the secretion of these hormones. A modest level of hypohydration could cause an important increase in PRA and aldosterone (96).

Severe Hypoxia and Associated Stress

Other associated stressors can contribute to the hormonal responses to hypoxia. Increased PRA in severe hypoxic stress is mediated through stimulation of intrarenal β-adrenergic receptors. Rats with adrenalectomy or β-blockers do not survive acute exposure to 7925 m (97). Increase in PRA, aldosterone, and cortisol may be associated with the stress of rapid transport to 5300 m in subjects supine for 2 hours (9). Acute exposure to 6000 m induced an elevation in ADH, PRA, and aldosterone and a reduction in urine flow (98). This pattern of response is more related to psychological stress and AMS than to a physiological response to prolonged hypoxia. Similarly, in a group of soldiers exposed for more than 10 weeks to altitudes above 6000 m, cold (−30°C), danger (military post near a border), and strenuous exercise, aldosterone was elevated while PRA was normal (14). Exposure to extreme altitude (above 5000 m) induces clear-cut physiological responses that are more difficult to explore at moderate altitude, allowing clearer explanations of control mechanisms. For example, data obtained on the Sajama expedition (10 subjects for 3 weeks at 6542 m), clearly disallow the dissociation hypothesis in chronic conditions, requiring new explanations for the hypoxia-induced blunting of the renin-aldosterone system (28,29). In the Sajama study, mean blood pressure (MBP) was elevated and variations of resting PRA could be partly accounted for by the variations in blood pressure. Adrenergic activity was increased in hypoxia when PRA was decreased. A blunted response of renal β-receptors could be evoked by analogy to the downregulation observed in cardiac and lymphocyte β-receptors (36,46). The PRA response to exercise, probably mediated by adrenergic stimulation, was abolished in the Sajama study (29). This inhibition could be due to a local or general release of endothelin by the endothelium. Endogenous endothelin has a renal vasoconstriction effect (99), and the isoproterenol-induced increase in renin release is inhibited by endothelin (100). Norepinephrine would induce a vasoconstriction of afferent arterioles and endothelin a (greater) vasoconstriction of efferent arterioles, thus resulting in increased perfusion pressure and blunting of PRA release. The decrease in plasma aldosterone observed in the Sajama study was fully explained by the depressed PRA. In our opinion, the aldosterone secretion is decreased in prolonged hypoxia mainly as a direct result of lower renin release. The renin-aldosterone system appears to be suppressed, without dissociation between renin and aldosterone. It is probably a multifactorial phenomenon where increase in renal perfusion pressure, altered K^+ balance, blunted adrenergic responsiveness, increased dopamine release, or local adrenal hypoxia could interact in response to hypoxia and acid-base disturbances due to hyperventilation.

Moderate Hypoxia

In studies where the hypoxic stress was moderate or of short duration, changes in PRA and aldosterone are less clear-cut. In 10 subjects exposed to 2000 m with uncontrolled diet, aldosterone increased while PRA tended to decrease. Plasma $[K^+]$, elevated at the end of the stay (perhaps due to high muscular activity), could have stimulated aldosterone secretion (11). Acute hypoxia (17% O_2) for 10 minutes at rest followed by 3 bouts of 5- to 7-minute exercise induced a nonsignificant decrease in PRA and a significant decrease in aldosterone (101). With short (1 hour) exposure to 14.5% O_2, osmolality increased less during exercise, PRA increased less, aldosterone increased much less when compared to normoxia (38). Mild exercise and short exposure to hypoxia (2 hours at 13.3% O_2) did not affect plasma PRA and aldosterone (102). In 12 subjects exposed for 30 minutes to 16% O_2, exercise aldosterone was lower and PRA insignificantly lower in hypoxia (73).

Exercise

PRA increases with exercise (101,103) as a function of relative work load (38,103,104); thus any study on the renin-angiotensin-aldosterone system done during or just after exercise should control the level of work load. Exercise stimulates the secretion of renin, while hypoxia induces a decrease in renin (105). Exercise-induced increase in PRA and aldosterone was found to be blunted in hypoxia (Fig. 5) (29,58,59). In a number of studies performed by Milledge and coworkers, the exercise-induced increase in aldosterone was smaller than the increase in PRA. The authors found this dissociation between PRA and aldosterone at 3100 m in the Alps (106), during a climbing expedition to Mt. Kongur (>4500 m) (57), and after 2–4 weeks above 6300 m on Everest (62). However, the authors could not control hydra-

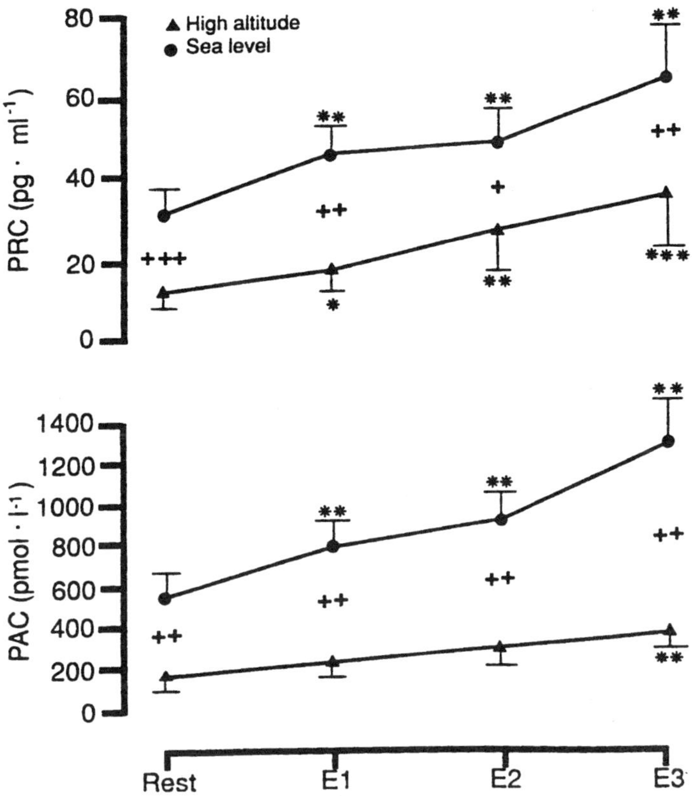

Figure 5 Plasma active renin concentration (PRC) and plasma aldosterone concentration (PAC) at rest and during exercise with increasing work loads (periods E1, E2, and E3) at sea level and high altitude (4350 m). Data are means ± SE. $^{*}p < 0.05$, $^{**}p < 0.01$, $^{***}p < 0.001$, exercise compared with rest; $^{+}p < 0.05$ and $^{++}p < 0.01$, $^{+++}p < 0.001$, high altitude vs. sea level. (From Ref. 58.)

tion or exercise level. When hydration was maintained, PRA and aldosterone decreased with mild and prolonged (from 10:00 a.m. to 4:00 p.m.) exercise and were not influenced by hypoxia (107). The lower level of hormones at 4:00 p.m. probably reflects the nadir of diurnal variation rather than the effect of exercise.

Hemorrhage

In hypoxic rats, basal PRA was unchanged and aldosterone decreased; in response to hemorrhage, ACTH and cortisol increased probably due to a decrease in $[K^+]$. The aldosterone response to ACTH was blunted with acute hypoxia and returned to normal with acclimatization (108). Hypoxia attenuated the renin response to hemorrhage in goats exposed for 2 hours to $F_{IO_2} = 0.10$. The development of hypotension occurred faster in hypoxia. The onset of increasing ADH occurred earlier in hypoxia, possibly inhibiting the PRA increase, while ANP was slightly higher in hypoxia (109).

In summary, normal acclimatization to high altitude is associated with a clear suppression of the renin-aldosterone system. With the development of AMS, this blunting may be less intense, thereby contributing to salt and water retention. The precise mechanisms by which the release of renin and/or aldosterone is inhibited by hypoxia remain to be clarified. Intrarenal hemodynamics and release of local humoral factors may play an important role. Future studies may focus on the endothelial response to hypoxia within the kidney.

B. Antidiuretic Hormone

The major action of ADH is to facilitate the reabsorption of free water from the glomerular filtrate. Water deprivation stimulates and water load decreases ADH secretion. At high altitude ADH has been less extensively studied than the renin aldosterone system. Additionally, this hormone is released into the blood in intermittent bursts, resulting in varying blood concentrations and rendering isolated blood sampling inadequate to assess physiological changes.

In dogs, both plasma and cerebrospinal fluid [ADH] increase in acute, severe hypoxia (1 hour, 10% O_2) resulting in antidiuresis, but no change is seen with more moderate hypoxia (12.5% or 15% O_2) (110,111). Cortisol inhibits this ADH release (112,113). In conscious dogs exposed to 10% O_2 for 40 minutes, hypoxia-induced ADH release was blocked by meclofenamate (cyclooxygenase inhibitor), suggesting a possible role of peripherally produced prostaglandins (114).

In humans, ADH secretion could be affected by various reflexes originating in peripheral receptors. ADH release decreases as the result of arterial chemoreceptor stimulation by almitrine. Interaction with low pressure cardiovascular receptors or osmoreceptors is likely to occur (87). Hyperventilation-induced increase in thoracic blood volume would be offset by a decrease in ADH and an increase in ANP (115). Breathing 10% O_2 for 4 hours has no effect on blood ADH in humans (116). Subjective and objective episodes of distress (nausea, malaise, vasovagal syndrome) may elicit ADH secretion, coinciding with inadequate short-term adjustment to altitude. Acute severe hypoxia stimulates ADH release possibly by way of the carotid bodies (117). Acute normobaric (10.5% O_2) or hypobaric (5100 m) hypoxia induced an increase in plasma [ADH] and a decrease in urine output, associated with hypotension and hypokalemia in humans (10). ADH secretion increases over the first 24 hours of exposure to 465 torr, then decreases to normoxic levels; CO_2 supplementation prevents this transient increase (118). Sustained exposure of 13 subjects to 5400 (7–9 days) and 6300 m (7–17 days) resulted in an increase in serum osmolality with no increase in ADH, suggesting a decreased sensitivity of ADH to the osmotic stimulus and therefore possible hypothalamic posterior pituitary dysfunction (119) (Fig. 6). Acute exposure to 6 hours of normobaric hypoxia (12% O_2) induced a slight increase in ADH in subjects with a fixed sodium intake (120). Most studies

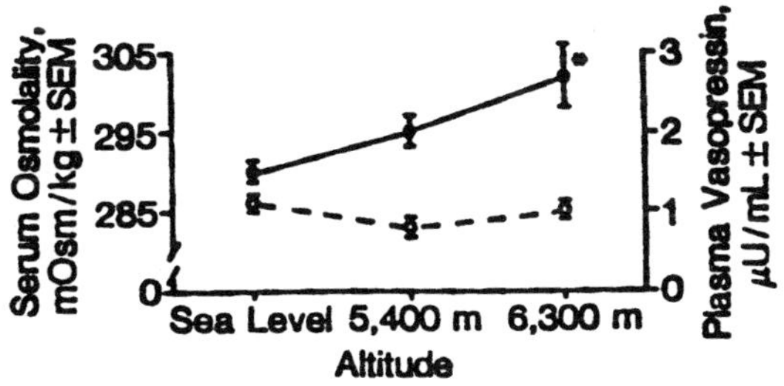

Figure 6 Serum osmolality (solid line) and plasma arginine-vasopressin concentrations (dashed line) at sea level, 5400 m, and 6300 m. Each value represents mean ± SEM of 13 subjects. Osmolality at 6300 m (indicated by asterisk) is greater than at 5400 m ($p < 0.05$) and sea level ($p < 0.01$). Osmolality at 5400 m is also greater than at sea level ($p < 0.05$). Normal range for serum osmolality is 280–295 mOsm/kg. Mean arginine-vasopressin level after overnight dehydration is reported to be 7.2 μU/mL using this assay. (From Ref. 119.)

indicate that moderate hypoxia decreases or has no effect on ADH (121), while severe or poorly tolerated hypoxia increases ADH.

In rats results are contradictory. Hypobaric hypoxia (359 mmHg, 4 hours) induced an increase in urine output in normal rats, possibly due to inhibition of ADH release. In rats with diabetes insipidus, hypoxia induced oliguria (122). Both normal rats and rats with congenital diabetes insipidus exposed to 0.5 atm for 4 days reduced water intake and were oliguric. While the normal rats increased body weight and lung water, the oliguria in the diabetes insipidus group cannot be attributed to increased ADH, but rather to decreased water intake. Thus, rats with diabetes insipidus seem protected against HAPE because they are more dehydrated than normal rats (123).

The relationship between ADH and AMS is unclear. ADH was found to be elevated in HAPE (23,24,124) but not in subjects with benign AMS only or in normally acclimatized subjects (124). Exercise-induced elevation of ADH was found in subjects prior to (63) or while experiencing (118) AMS. Release of ADH in the central nervous system could increase brain capillary permeability and favor AMS and HACE.

In conclusion, ADH is released in excess in severe hypoxia and probably in subjects suffering from AMS or HAPE, while moderate altitude does not induce important changes in ADH release. Whether ADH is responsible for water retention and development of these diseases or the increase in [ADH] is in response to these stresses remains to be established.

C. Atrial Natriuretic Peptide

Atrial natriuretic peptide (ANP) is released by the myocytes of the heart mainly in response to stretching of atrial wall. ANP affects primarily the cardiovascular and renal systems and renal function. It decreases cardiac output and blood pressure by decreasing efferent adrenergic activity to the heart and intravascular volume and venous return. ANP relaxes vascular smooth muscle by inhibiting angiotensin II. ANP increases Na^+ and water excretion, inhibits renin and aldosterone secretion, and inhibits the action of aldosterone on the distal tubule. In vitro and in vivo in rats, ANP inhibits the response of aldosterone to AII, ACTH, and K^+ (125). As part of the overall system of water and electrolyte control with aldosterone and ADH, ANP could play a role in the fluid shifts observed at high altitude.

Evaluating the effect of altitude hypoxia per se on ANP release is complicated by altitude-induced changes in vascular volumes and pressures as well as other factors such as body position and state of hydration. Even so, the preponderance of evidence suggests that ANP does not play a major role in water and electrolyte regulation at high altitude.

In 8 hydrated seated subjects exposed for 2 hours to 10% O_2, ANP increase correlated with increase in diastolic arterial pressure (126). Brief hypoxia (10 minutes, 10% O_2) increased pulmonary arterial pressure and plasma ANP without increase in right atrial or wedge pressure (127). In subjects exposed for 1 week to 5400 m, studied supine in the morning, ANP, diuresis, and natriuresis increased (18). As a whole, in humans, ANP was found unchanged or moderately elevated in hypoxia (18,26,40,58,61,91,104), perhaps by a direct effect of hypoxia on myocytes or mediated by an increased atrial stretch due to pulmonary vasoconstriction (128) and/or increased central blood volume (129).

Factors that might trigger an increase in ANP in hypoxia are pulmonary hypertension, neural pathways, increased heart rate, and increased atrial stretch.

Hypoxia is known to stimulate ANP release in rat atria (130) and in vitro, independent of atrial pressure or volume (131). In rats exposed to 10% O_2 for 3 weeks, there was a reversible induction of right ventricle ANP synthesis due to hypoxia and not to hypertrophy (132). Myocyte stretch-induced increase in myocardial O_2 consumption and tissue hypoxia may represent the underlying mechanisms of ANP release. Acute hypoxia (10.5% O_2, 40 min), which induced an increase in diuresis and natriuresis in rats, was associated with elevated plasma ANP and a lowered systemic arterial pressure (133). Acute normobaric hypoxemia stimulated ANP secretion in rabbits, basal ANP being correlated with decreasing Pa_{O_2} and increasing basal central venous pressure (134). In anesthetized pigs, a neural reflex, probably of pulmonary arterial origin (stretch receptors), may mediate about 50% of the ANP response to hypoxia (128).

Acute normobaric hypoxia (PI_{O_2} = 92.3 mmHg) had no effect on ANP at rest and reduced the exercise (60% of V_{O_2}max) response. There was a high correlation between changes in ANP and osmolarity during exercise (135). There was a lesser increase in ANP response to maximal exercise in hypobaric hypoxia (70 min at 3000 m) than in normoxia, but the response of the precursor of ANP was not modified: hypoxia could stimulate the degradation or the receptor binding of ANP (19). Exercise-induced increase in ANP was absent after 16 days at high altitude: acclimatization blunted the ANP response to exercise, probably by decreasing stroke volume and cardiac output (53). Since at the same time renin and aldosterone levels were not modified, the authors concluded that ANP has little effect on the renin-angiotensin-aldosterone system at high altitude.

In an expedition of 22 subjects to 4300 m with uncontrolled diet but controlled position (supine, seated), ANP increased on the first 2 days at altitude. Na^+ excretion at high altitude and, unexpectedly, ANP levels at low altitude were inversely related to AMS score. The authors suggest that in less AMS-susceptible subjects, ANP would be higher, inhibiting aldosterone and favoring diuresis (61). While ascending to 4559 m, a group of subjects with severe AMS showed less diuresis and natriuresis and, conversely to the former study, a greater increase in plasma ANP than subjects without AMS. ANP may contribute to edema formation by its effect on transcapillary fluid exchange (26). With exposure to high altitude, an initial shift of fluid out of the vascular space into the interstitial and/or intracellular space, diuresis, and decreased total body water would occur in normally acclimatizing subjects. In individuals with AMS, fluid would be retained and result in edema. Subjects who suffered from severe AMS in spite of a slow ascent (3 days to 4559 m) showed weight gain and elevated ANP with right atrial enlargement at echography, probably as a consequence of high pulmonary artery pressure (91). In a study performed at 4300 m with diet-controlled, resting ANP did not change and was not related to AMS (53). The plasma ANP rose significantly in HAPE-susceptible subjects in response to 10% O_2 inhalation (four times greater than the increase in control subjects). The increase in aldosterone was not statistically different between the two groups, but a positive correlation was found between increase in ANP and aldosterone when all subjects were pooled. In HAPE subjects PAP increased from 16 to 33 mmHg, while in control subjects the increase in PAP (from 14 to 26) was insignificant (136). Hypoxia-induced hyperventilation and tachycardia increase venous return and central blood volume, which tends to inhibit ADH and stimulate ANP, in order to increase diuresis and reduce central blood volume (129). ANP infused in anesthetized pigs reduces the hypoxia-induced pulmonary hypertension. ANP may act as an endogenous modulator of the pulmonary pressor response to hypoxia (137). In 5 patients with HAPE, ANP was elevated. ANP can be beneficial by favoring a pulmonary vasodilation but can also aggravate hypoxia by opening pulmonary arteriovenous shunts (138). Endogenous ANP may modulate the subacute and chronic (>6 hours) phases of hypoxic pulmonary hypertension (139). Lower plasma ANP at rest and exercise was found in HA natives at 3600 m despite higher pulmonary arterial pressures, suggesting an adaptation to chronic distension of atrial stretch receptors involved in central blood volume regulation (41).

In male subjects exposed to cold (10°C) for 2 hours, mean skin temperature decreased from 31.2 to 22.6°C, while ANP slightly increased (140). This diuretic effect of cold, mediated through ANP, may contribute to the diuresis frequently observed at high altitude when cold stress is associated with hypoxia. However, in a group of soldiers exposed for more than 10 weeks to altitudes above 6000 m, cold (−30°C), danger, and strenuous exercise, ANP was normal (14).

In conclusion, ANP release could be important in hypoxia in decreasing pulmonary artery pressure, thus decreasing right ventricular afterload and pulmonary capillary fluid filtration. ANP may also increase fluid shift out of the circulation, thus increasing hematocrit and facilitating the action of EPO on the profileration of red cell progenitors (141). However, once again a distinction should be made between those acclimating normally and those who have difficulty: subjects with HAPE who develop pulmonary hypertension have a higher plasma ANP, probably because of greater distension of the right atrium. Normal subjects have little or no modification of plasma ANP at high altitude.

IV. Thyroid Hormones

Thyroid hormones play an important role in regulating the overall rate of body metabolism and are critical for normal growth and development. Thyroid hormones increase oxygen consumption and heat production. They also increase oxygen availability by increasing ventilation and cardiac output as well as red blood cell mass. In that sense, they could play a role in long-term adaptation to altitude hypoxia.

The hypoxia of high altitude stimulates thyroid hormone release. An increase in protein-bound iodine, first reported in 1966 (142), has been confirmed by observations of increased uptake and release of iodine from the thyroid gland (143,144). Most authors agree that high altitude induces an elevation in plasma concentration of both free and total T4 fractions of thyroxin (142,145–155). T3 also increases (145,146,151–153), although to a lesser extent than T4. Furthermore, a decreased concentration of T3 in erythrocytes has been found, combined with elevated levels of T3 in plasma (152). A few authors report no increase in T3, but rather an increase in reverse T3 (rT3) (147,149,156), perhaps because of an inhibition of the conversion of T4 to T3 due to hypoxia, with a concomitant rise in rT3 concentration and an elevated T4:T3 ratio (149,156). Such changes are also seen in other stressful states and are possibly caused by an increased release of corticosteroids. T3 and T4 concentrations rose mostly after a couple of days at high altitude, reached a maximum after 1–2 weeks, and then declined to (151) or toward (145,152) sea level values. Exercise-induced elevation of thyroxin was potentiated at altitude (153). Exhaustion may play some role, for depressed levels of thyroid hormones were observed after an expedition to 4360–6194 m under strenuous and difficult conditions (156). A concomitant elevation of basal metabolic rate and thyroid hormones at high altitude suggests that these changes are causally related (154). A higher urinary excretion of iodine along with increased thyroidal iodine uptake was observed in 15 men after 4 days at 4300 m, which could be due to the elevated plasma levels (143). All thyroid parameters are normalized within a few days after return to sea level (148,151,152).

Most studies show unaltered baseline thyroid-stimulating hormone (TSH) concentration at high altitude despite elevated levels of T3 and T4 (146–148,150–153). This unaltered level of TSH is remarkable because this hormone is considered to be the main regulator of thyroid activity, and high levels of thyroid hormones either suppress this pituitary stimulus or reflect the increased TSH concentration. Neither an elevation preceding the rise in T3 and T4 nor a decrease due to an expected feedback response was found in TSH levels. In a study where TSH was depressed by administration of L-thyroxin, a clear increase of T3 and T4 was still observed on exposure to simulated altitude (152). The few studies done at higher altitudes are inconsistent. In 1983 Mordes et al. observed a slight increase of TSH besides the elevated thyroid hormones, suggesting an altered feedback regulation after exposure to 5400 and 6300 m (149). In contrast, a slight decrease in TSH with high T3 and T4 was found after about 6 months at 5080 and 6300 m (145). At sea level, about 4 hours after returning from an expedition to 4360–6194 m, levels of TSH and T3 were significantly lowered, while T4 was unchanged (156).

Several explanations for the TSH-independent thyroxin rise have been proposed. Every change in plasma hormone levels must be caused by either a modified secretion rate, an altered clearance, or hemoconcentration due to vascular shift. In 1966 Surks investigated the elevated concentrations of total plasma proteins compared to the free thyroxin increase and concluded that the T4 rise was far greater and could not be explained simply by dehydration and hemoconcentration (157). The thyroxin degradation rate increased during the first 3 days at altitude and thereafter remained slightly elevated (154), thus refuting decreased clearance as a possible cause of T4 elevation. Studies of thyroxin-binding capacity yield variable results. In men trekking at altitudes of 3505–3658 m, thyroxin-binding globulin (TBG) levels were found sufficiently elevated to explain the T4 rise (146,155). However, little or no change in TBG and thyroxin-binding prealbumin were found by others (142,149,151,152,157), suggesting that a binding alteration is not likely to account for the T4 rise. Unchanged (148) or increased (145) T3 uptake has been found at high altitude, suggesting a normal binding capacity, supported by the observation that free hormone levels increase in parallel with total levels (145,148,150). A shift from extravascular to the intravascular compartment (154), shrinkage of distribution pools, or a decreased plasma clearance could also account for the observed T4 rise (151).

The T3 and T4 increase at high altitude is often accompanied by an increase in noradrenaline plasma concentrations, reflecting a well-documented rise in sympathetic activity (154). The thyroid gland is innervated by sympathetic branches and adrenaline can stimulate hormone secretion via β-adrenergic receptors. Thyroid hormones, particularly T4, rise during intense exercise, possibly due to an adrenergic influence, since TSH does not change (158). β-Blockade has been used for many years to relieve symptoms of hyperthyroidism. Investigations have shown a direct inhibition of the gland by β-blockers. The augmented levels of catecholamines may be the main cause of the increased level of thyroid hormones at high altitude. Further studies, especially with β-blockade, need to be carried out to investigate this possible explanation.

The unchanged level of TSH in the presence of increased levels of T3 and T4 could reflect an altered feedback regulation or disturbed hypophyseal function. But most studies do not show any evidence of a hypophyseal malfunction as a completely normal TSH response is seen after TRH administration (147,150–153). Indeed, during an expedition to 5400–6300 m, an increased TSH response to TRH was found (149).

High levels of thyroid hormones in prolonged and severe hypoxia suggest hyperthyroidism as a possible protective adaptation to this extreme environment (145). In fact, increased thyroid hormones release would be beneficial in hypoxia since these hormones increase the levels of 2,3-diphosphoglycerate in erythrocytes, facilitating oxygen release to the tissues, by shifting the oxyhemoglobin dissociation curve to the right (see Chapter 16) (159). In contrast, no difference in thyroid response was observed between subjects susceptible to AMS and resistant subjects (155), suggesting that thyroid hormones play a minor role in the early phase of acclimatization.

Cold, a common concomitant factor at high altitude, may also influence thyroid hormone levels. T3, likely to be the most important thyroid hormone for cold habituation, decreases with cold exposure, while T4 and TSH remain unaltered (160,161). The low T3 values found by Hackney et al. (156) may be explained by this cold influence, but not their observation of decreased TSH and unchanged T4. T3 and T4 were elevated on exposure to simulated hypoxia at +22 to +24°C, demonstrating that thyroid hormones increase independent of cold exposure (152).

In summary, plasma concentrations of thyroid hormones, in particular T4, increase with acute and sustained hypoxia. This increase is not accompanied by any consistent change in TSH level but is probably due to the increased adrenergic activity. In many studies done in the mountains, cold exposure may have contributed to the stimulation of the thyroid gland.

V. Hormones of Calcium and Phosphate Balance

A complex interaction of three hormones (parathormone, calcitonin, and vitamin D) on various tissues (intestine, bone, and kidney) is responsible for maintenance of normal concentrations of calcium and phosphate in body fluids. Few studies have been performed on the effect of high altitude on hormones controlling calcium and phosphate metabolism.

High altitude seems to stimulate parathormone (PTH) release but may lead to a progressive desensitization of PTH receptors. Overall calcium regulation is not altered.

In four men exposed for 6 days to 3450 m, plasma ionized calcium decreased, excretion of nephrogenous cAMP decreased, but serum intact parathormone (PTH) and 1,25-dihydrocholecalciferol [$1,25(OH)_2D_3$] did not change, suggesting an impaired renal responsiveness to PTH. The failure of PTH to increase in response to hypocalcemia may reflect impaired PTH secretion (162).

In 10 lowlanders living for 3 weeks at 6542 m during the Sajama expedition, total serum calcium and phosphate were unchanged, while PTH and $25(OH)D_3$ were elevated and $1,25(OH)_2D_3$ was decreased (Fig. 2). These observations could be explained either by a hypoxia-induced inhibition of renal 1α-hydroxylase, resulting in a tendency to hypocalcemia and a secondary hyperparathyroidism, or by a resistance to PTH (28,29). The conversion of $25(OH)D_3$ into $1,25(OH)_2D_3$ within the kidney by the enzyme 1α-hydroxylase might be O_2-dependent (163). In response to this decreased production of active vitamin D, PTH would be released in excess, maintaining a normal level of plasma calcium in spite of a low calcium intake. A low-calcium diet could stimulate PTH release without evoking significant change in plasma ionized calcium (Ca^{2+}). In the Sajama study, high PTH and low $1,25(OH)_2D_3$ levels could be due to a desensitization of PTH receptors (resistance to PTH). In fact, recent observations of 10 subjects at Observatoire Vallot (4350 m) showing a blunted urinary cyclic AMP response to PTH favor the resistance to PTH hypothesis (164). A slight deficiency in $25(OH)D_3$ could have stimulated $1,25(OH)_2D_3$ production in normoxia; then exposure to intense solar radiation during the expedition normalized $25(OH)D_3$ and $1,25(OH)_2D_3$ values. However, a deficiency in $25(OH)D_3$ is not sufficient to explain low $1,25(OH)_2D_3$ associated with low-calcium diet; hypoxia-induced inhibition of 1α-hydroxylase might thus play an important role. It is well known that respiratory alkalosis induces hypocalcemia, hypophosphatemia, and hypophosphaturia (165). In rats, hyperventilation-induced hypocapnia and respiratory alkalosis blunted the phosphaturic effect of PTH. Content of cAMP increased in the kidney (166). Renal failure is accompanied by altered vitamin D metabolism in the kidney, which induces an increase in PTH and a normal plasma Ca^{2+}, determining a "secondary hyperparathyroidism." A similar pattern of vitamin D and PTH profile was found in the Sajama study, suggesting that a modification of the set point for calcium and phosphate related hormones might have been induced by renal hypoxia (29).

In cattle, exercise associated with high altitude (3500 m) resulted in increased plasma PTH. This could be due to a direct effect of sympathetic activation on β-adrenoceptors located on parathyroid cells. Dopamine, which stimulates PTH release in vitro, GH, and cortisol could also have a stimulatory effect on PTH secretion (167).

In summary, altitude seems to stimulate PTH release and inhibit active vitamin D, resulting in an unaltered calcium and phosphate balance. Further research should focus on other factors that might interfere with bone resorption and formation. Moreover, the desensitization of PTH receptors, which are G protein–coupled receptors like β-adrenoceptors, should be examined as a protective mechanism similar to what occurs within the heart with desensitization of adrenergic system.

VI. Hormones Controlling Blood Glucose

Changes in insulin and glucagon have been little studied at high altitude. Resting blood glucose has been reported to be reduced (153,168), unchanged (169), or transiently increased (13) at high altitude. Acute (1 hour) and, in a more marked manner, prolonged (3 weeks) exposure to 4300 m increased utilization of blood glucose during exercise (170). Altitude and β-blockade augment glucose utilization during submaximal exercise (171). Resting insulin has been found to be increased at high altitude (13), but this could be a nonspecific response secondary to the increase in cortisol. Plasma insulin increases after prolonged exposure to high altitude such as in Operation Everest II (172) and the Sajama expedition (173). Serum insulin and C-peptide were found increased in acute hypoxia during intense exercise by sedentary subjects (174). Tolerance to glucose decreases with acute hypoxia, suggesting that the decrease in peripheral use of glucose during hypoxic exercise could be an additional factor accounting for hypoxia-induced hyperglycemia (175,176). In vitro studies on Langerhans cells have shown that severe hypoxia could directly inhibit insulin secretion (177,178). Hypoxia-induced activation of the adrenergic system could suppress insulin release (177). This mechanism could facilitate hepatic glycogenolysis and increase blood glucose. Absence of increased blood glucose in spite of elevated cortisol, catecholamines, and growth hormone suggests a partial blunting of the action of these hormones in hypoxia or might result from the counteracting increased activity of insulin. Fasting glucagon was found unchanged at high altitude (172).

A better knowledge of altitude-induced changes in hormones controlling blood glucose will arise from studies of insulin receptors and their possible upregulation in hypoxia. Diabetics often ask for information about a potential risk at high altitude. Until now, no controlled study is available on diabetes and altitude. However, it is reasonable to think that the stresses of high altitude (hypoxia, cold, intense exercise, limited available food, dehydration) are not favorable factors for an efficient control of blood glucose in insulin-dependent diabetics.

VII. Other Hormones

A. Prolactin

Prolactin release is stimulated by TRH and inhibited by dopamine increase in response to various stressors, including exercise, hyperthermia, fasting, surgery. The reports on prolactin response to high altitude are few and inconclusive.

Studies on acute hypoxic exposure have mostly been performed in male lowlanders, while chronic hypoxia has been more frequently studied in female high-altitude natives. Prolactin concentration in women born and living at 4340 m, compared to that of women at sea level, is diminished in chronic hypoxia (179,180). Increased fertility and decreased frequency of menstrual disturbances have also been observed among high-altitude women compared with women living at sea level (179,180). Dopamine, an inhibitor of prolactin, has been suggested as a cause of this alteration because dopamine concentration increases at high altitude (see above). The exercise-induced increase in prolactin is well documented (181–185), though the mechanism is unclear. During exercise in normobaric hypoxia, plasma prolactin was found to be increased (Fig. 7) (185), unchanged (184), or decreased (182), when compared to normoxia. A blunted prolactin response was observed during exercise in normoxia with concomitant face-cooling, suggesting that cold exposure may di-

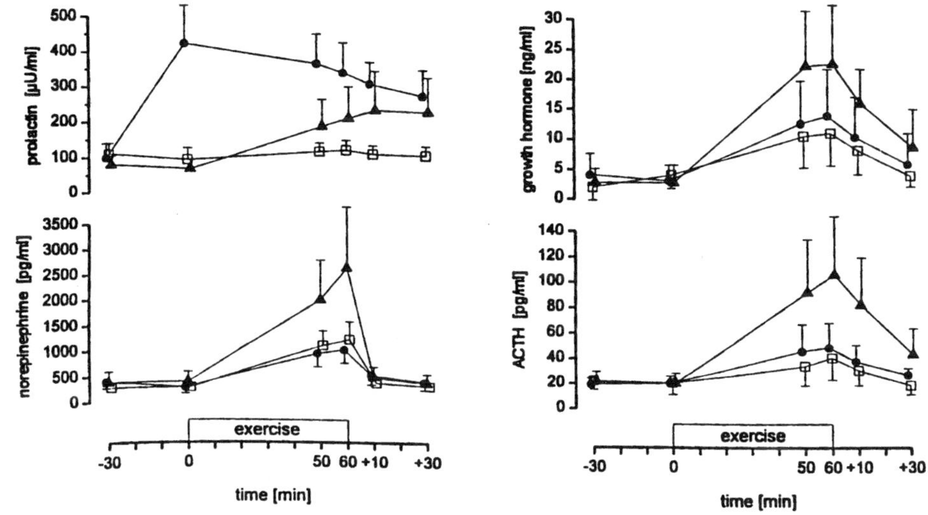

Figure 7 Plasma prolactin, norepinephrine, growth hormone, and ACTH concentrations at different time points before, during, and after exercise in hyperoxia (100% O_2, closed circles), hypoxia (14% O_2, closed triangles), and normoxia (open squares). Gas mixtures were inhaled during exercise as well as 30 minutes before and after. Data are means ± SD. (From Ref. 185.)

minish the exercise-induced prolactin release (181). Elevated resting levels of prolactin have been described in acute hypoxia, similar to the changes found in other situations of stress. In 8 men exposed for 18 days at 3500 m without any heavy exercise, basal prolactin increased and returned to normal within a week on descent to sea level (186). In the same report, intermittent exposure of 6 subjects to simulated altitude of 3500 m induced a prolactin increase, while intermittent cold stress in a cold chamber did not provoke any prolactin response. In contrast, decreased resting concentrations of prolactin were found at simulated altitudes of 2438, 3658, and 4572 m (184). In 10 subjects exposed at 495 mmHg, prolactin increased within 2 hours and rapidly came back to normal (118).

Prolactin secretion is under the influence of numerous regulators, which could be altered by hypoxia. Dopamine and possibly noradrenaline inhibit prolactin secretion, and since both increase with hypoxia, either could provoke prolactin depression at altitude. A possible regulatory effect of opioids on prolactin has been evaluated: no alteration of the prolactin response to exercise was seen after a treatment with an opioid antagonist (183). However, a possible effect of endorphins on prolactin regulation at high altitude has not been investigated. High work intensity, blood lactate level, a fat-rich diet, along with certain hormonal peptides induce a prolactin increase and could explain the varying results at altitude. Whatever the mechanism that alters prolactin baseline levels, it seems to be overridden by hypothalamic influence since the prolactin response to TRH administration is not altered in hypoxia (149,150).

In conclusion, reports on prolactin concentration in acute hypoxia are contradictory, but prolactin secretion seems to be altered only on acute exposure as a nonspecific stress. In women exposed to chronic hypoxia, prolactin levels are decreased when compared to women at sea level.

B. Endothelin and Endothelial Function

The endothelium has potent effects on hemostasis, vasomotor tone, and vascular permeability. The endothelial cell produces vasodilators, PGI2 (with cAMP as second messenger) and endothelium-derived relaxing factor (EDRF), now identified as nitrogen monoxide (NO) (with cGMP as second messenger). Shear stress, among other factors, stimulates NO production through the activation of a K^+ channel. The endothelial cell produces also vasoconstrictor agents such as endothelin, endothelium-derived growth factor (EDGF), and endothelium-derived contracting factor (EDCF2). Endothelin increases catecholamine, renin, and ADH action and, like EDGF, modulates smooth muscle growth. EDCF2, stimulated by arachidonic acid, acetylcholine, and ionophore A231 87, inhibits NO. Endothelin is a potent stimulus for ANP secretion by cultured rat myocytes, through activation of voltage-dependent Ca^{2+} channels (blocked by nicardipine) (187). Endothelin increases the synthesis and secretion of ANP in cultured myocardial cells of neonate rats and in the isolated perfused heart (188). Endothelin inhibits renin release from isolated rat glomeruli via a calcium entry mechanism; the isoproterenol-induced increase in renin release is inhibited by endothelin (100). However, in most studies, arterial administration of endothelin induced an increase in renin and aldosterone (189).

It was discovered for the first time in 1985 that hypoxia releases a diffusible vasoconstrictor substance from the canine vascular endothelium (190). Thirty minutes of hypoxia (5% O_2/95% N_2) induced a 71% increase in the release of endothelin from rat mesenteric arteries in vivo (191). Hypoxia induces endothelin gene expression and secretion in cultured human endothelium (umbilical vein culture, exposed to 1 hour at Po_2 = 30 torr) (192). Endothelin-1 and endothelin receptor gene expression is enhanced in rat lungs exposed to chronic hypoxia (193). Endothelial cells are sensitive O_2 sensors and release platelet-derived growth factor (PDGF-B) in human umbilical vein endothelial cells: there is a transcriptional upregulation of PDGF-B by hypoxia (194). Hypoxia increases endothelial permeability of cultured cells and shifts the balance of cell surface coagulant properties, activating Factor X and suppressing thrombomodulin (195).

In humans there are few studies on the effect of hypoxia on endothelin release. No change in endothelin response to maximal exercise in acute hypobaric hypoxia (70 min at 3000 m) was noted when compared to normoxia (19). In 22 subjects acutely exposed to 4559 m, plasma endothelin increased twofold, correlated with pulmonary arterial pressure; nifedipine lowered pulmonary pressure but had no effect on plasma endothelin (196). Acute exposure to cold increases plasma endothelin-1 (140) and may have additive effects with hypoxia at high altitude. The Sajama expedition showed the first data on plasma endothelin during exercise and prolonged hypoxia (Fig. 2). Although plasma endothelin may not reflect local concentrations of released endothelin, the observation that plasma endothelin did increase with hypoxic exposure, even at rest, but not with severe exercise suggested that shear stress is not a major stimulus for endothelin release at high altitude and that metabolic changes in the endothelial cell, induced by hypoxia, may play a central role (29,197).

Prostaglandins are released by a number of tissues, particularly by the endothelium of various vascular beds. Considered as intrarenal hormones, they play an important role in kidney function, modulating the release of renin and ADH. Prostaglandins induce natriuresis when infused, and blocking the synthesis of prostaglandins by indomethacin provokes antinatriuresis. Mostly for technical reasons, few measurements of prostaglandins in plasma and urine have been performed at high altitude. PGE2 and 6-keto-PGF1α increase in subjects exposed to 4350 m. The increase reaches a maximum after 1–3 days, paralleling clinical signs of AMS, while concentrations correlate with severity and time course of AMS (198). High prostaglandin concentrations have been found in the liquid from bronchoalveolar lavage in subjects with HAPE (199). The origin of these mediators is not known, though they may come from the degradation of endothelial cell membrane phospholipids by activation of phospholipase A_2 and cyclooxygenase. They may play an important role in the inflammatory response related to hypoxic stress, particularly during the development of pulmonary edema (197).

C. Growth Hormone

Growth hormone (GH) is synthesized and released from the hypophysis under the hypothalamic control of GH-releasing hormone (stimulating) and somatostatin (inhibiting). Many other factors also modulate GH synthesis, release, and cellular responsiveness. GH has an important anabolic effect on various tissues. The pulsatile pattern of GH secretion complicates ascertaining changes in baseline levels.

In high-altitude residents, several studies have documented elevated levels of GH (13,200,201) with maintained circadian rhythm (13). Sea level residents acutely exposed to high altitude, on the other hand, did not show any change in the resting level of GH (200,201). However, Anand et al. found greatly increased values of GH in men after about 10 weeks of exposure to 6000 m, possibly due to exertion of the subjects, the long duration in a stressful and cold environment, since exercise and stress, as well as chronic hypoxia, are all considered to induce a GH increase (14).

During intense exercise in normoxia, GH concentration increases (202–204) and rapidly falls when exercise is stopped. One of the mediators of GH action, IGF-1 (insulin-like growth factor), also increases during heavy exercise, though the increase is small, brief, and shows a GH-independent pattern (205). The exercise-induced GH increase is strongly potentiated in hypoxia (Fig. 7), producing a much earlier and faster rise than in normoxia (12,185,200,202,206). However, it has been suggested that the influence of hypoxia on GH levels during exercise is mainly due to changes in relative workload, rather than to an effect of hypoxia per se (207).

The GH response to hypothalamic stimulation (by GH-releasing hormone) was potentiated in 12 men at 2600 m compared to 10 men living at sea level (150), although, in contrast to previous studies (200,201), basal GH levels in highlanders and lowlanders did not differ. The concentration of IGF-1 was elevated in high-altitude subjects, while the binding protein IGFBP-3 was similar in the two groups (150). However, in subjects exposed for 3 weeks at 6542 m, IGF-1 was normal (173), as it was in marathon runners at 4000 m (208).

The altitude-induced GH increase could play a role in modifying the metabolism to satisfy increased needs at altitude, both during short-term exercise and during long-term exposure (201). However, the mechanisms responsible for GH increase at high altitude are unclear. Lactate has been proposed to stimulate GH during exercise, but studies have failed to prove any correlation between these parameters; furthermore, exercise lactate is decreased at high altitude (200,209). Hypoglycemia induces rapid GH secretion, but increased plasma levels of glucose are seen in chronic hypoxia and during exercise in acute hypoxia, i.e., situations when GH levels are elevated, making hypoglycemia an unlikely cause (12,201). A more plausible explanation would be that GH release causes the increased plasma insulin for a given level of plasma glucose.

Thyroid hormones are important for normal secretion of GH, and hypothyroidism is associated with a GH deficiency, but too high levels of thyroid hormones have the opposite effect, inhibiting GH (210). Thus, the increased level of thyroid hormones at high altitude could contribute to GH elevation. The biphasic effects of thyroid hormones, depending on the absolute thyroid hormone concentration, could explain the variability of individual responses. The time courses for the hormonal changes differ, however, in that thyroid hormones increase quickly with acute hypoxia, while GH baseline level is unaltered initially, increasing only with sustained hypoxia. GH release, induced by exercise or falling levels of metabolites, is reduced by α-adrenergic blockade and potentiated by β-adrenergic blockade (211,212), demonstrating the influence of catecholamines on the regulation of GH release. The elevation of these transmitters in hypoxia might also contribute to the variable GH responses at high altitude. The feedback mechanism of GH release is complex and involves GH itself besides its mediators, the somatomedins, and finally GH-releasing hormone. The hypothalamic factor in small doses will decrease rather than increase GH secretion, perhaps via a somatostatin interaction. Interindividual threshold variation might result in either an increased or decreased response at altitude via this mechanism.

In summary, resting GH is not affected by acute hypoxia, whereas the exercise-induced GH response is potentiated by acute hypoxic exposure. With sustained hypoxia and in high-altitude residents, GH concentration is generally increased. Mechanisms accounting for those effects remain speculative.

D. Sex Hormones

In this chapter we focus only on male hormones, leaving the female hormones to Chapter 3, where the whole physiology of women at high altitude is addressed. The male sex hormones, including testosterone, luteinizing hormone (LH), and follicle-stimulating hormone (FSH), are integrated in a number of regulatory systems with complex feedback mechanisms. Hypothalamic gonadotropin-releasing hormone (GnRH) stimulates the hypophysis to secrete LH, an important stimulus for testosterone production, and FSH, influencing testosterone production indirectly by sensitizing LH receptors. Testosterone increases quickly and transiently during short-term exercise (15,22,203,204,213); in contrast, prolonged exercise decreases the testosterone level (15,204).

Acute normobaric hypoxia caused no alteration of the exercise-induced testosterone response (22). Acute exposure to altitudes ranging from 1650 to 3048 m has been reported to augment resting levels of testosterone within a few days (11,214), an effect possibly related to increased adrenal activity. Others have recorded a decrease in testosterone at 3048–3500 m (186,215,216). An influence from a concomitant high prolactinemia (86), as well as an increased level of estradiol (216), have been proposed as possible explanations for a decrease in testosterone. In 1978, Vander et al. observed no significant testosterone change, but the two subjects with the most pronounced symptoms of AMS manifested large decreases, perhaps due to their general malaise (217). Guerra-Garcia in 1971 found a decreased urinary excretion of testosterone on acute exposure to 4250 m but normal levels in high-altitude natives (218). Consistently, chronic hypoxia does not induce any change in testosterone levels of adult high-altitude (3400–4300 m) residents (219), although an earlier increase in testosterone and onset of puberty have been recorded in young high-altitude males (220). In elderly high-altitude men, high levels of testosterone were associated with increased hemoglobin and more hypoxemia during sleep, potentially compromising adaptation to high altitude (221).

Exercise in acute normobaric hypoxia provoked no changes in LH level (22). During intense, anaerobic exercise the LH-testosterone relationship is modified, i.e., testosterone increases without any significant elevation of LH (213). At high altitude, on the other hand (without intense exercise), LH and FSH have a tendency to decrease. Several studies have shown diminished levels of one or both hormones on acute hypoxic exposure (11,186,216,222). Interestingly, as with intense exercise, the LH change did not parallel that of testosterone, perhaps reflecting influences of other mediators, e.g., catecholamines (186). In a group intermittently (4 hours per day for 6 days) exposed to a simulated altitude of 3500 m, LH increased transiently the first day and then returned to normal levels (186). Intermittent moderate cold exposure in another group significantly decreased LH levels, suggesting that cold has an important influence on LH levels (186). Results from high-altitude expeditions may be influenced by this cold effect, possibly accounting for the observed inconsistency. No change in either LH, FSH, or testosterone was found during exposure to 4300 m for 6 days (217). High-altitude and sea level residents had similar

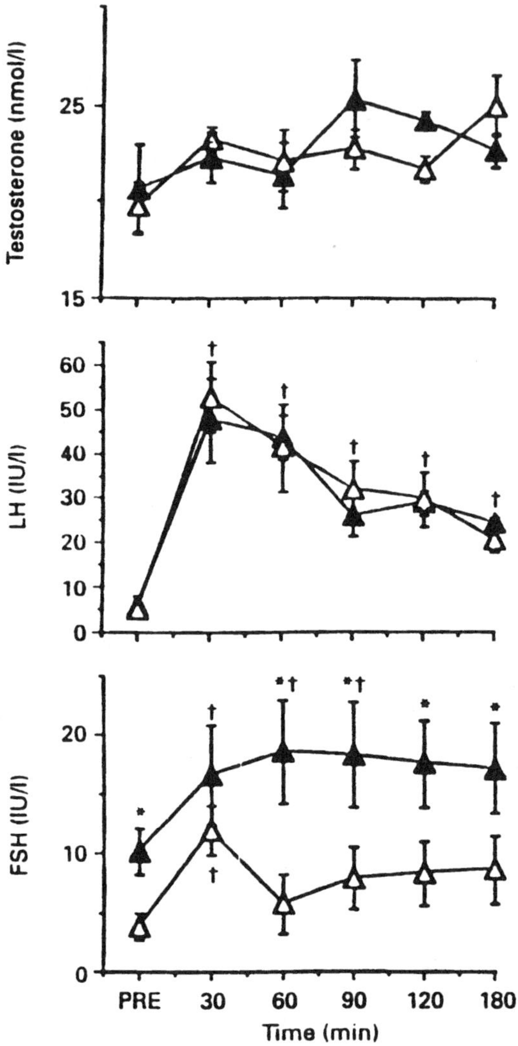

Figure 8 Plasma testosterone, LH, and FSH concentrations after exogenous administration of gonadotropin-releasing factor (GnRH) in 10 men at sea level (closed triangles) and 12 men living at high altitude, 2600 m (open triangles). $*p < 0.05$ between groups; $^{+}p < 0.05$ from baseline. (From Ref. 150.)

levels of LH and FSH (219). LH and testosterone responses to clomiphene, a gonadotropin stimulus, were similar in men living at high altitude and sea level, while FSH response to this stimulus was slightly increased (219). The hormonal response to TRH in 12 male residents at 2600 m was variable: LH response was similar to that in sea level natives, while FSH response was blunted in high-altitude residents (Fig. 8) (150).

A study made in rats suggested that male sex hormones do not have an important role in the development of chronic mountain sickness, whereas female sex hormones suppress both the polycythemic and cardiopulmonary responses during chronic hypoxic exposure (223). Another study in sheep evidenced a gender difference in the hypoxic pulmonary vasoconstrictor response. An attenuation of this response arises in the female at the time of puberty and may be mediated by estradiol (224). Progesterone, on the other hand, is likely to increase the hypoxic ventilatory response in polycythemic high-altitude residents (225). However, no evidence has been found that progesterone might mediate the increased HVR during pregnancy (226).

In summary, LH and FSH decrease with acute hypoxia, while testosterone response is variably described, increasing at moderate altitudes and decreasing at higher altitudes. The hypophyseal axis seems disturbed, as LH and testosterone show

Table 1 Direction of Hormonal Changes in Response to Hypoxia, at Rest, with Exercise, and in the Case of AMS or HAPE

	Normal acclimatization at rest	Normal acclimatization with exercise	AMS or HAPE
Norepinephrine	↗	↗	↗
Epinephrine	→	↗	
Dopamine	↗	↗	
Cortisol	↗	↗	↗ ↗
Renin	↘→	↘→	→↗
Aldosterone	↘	↘	→↗
ADH	→	→	↗
ANP	→↗	→↗	↗
Prostaglandins	↗		↗ ↗
T3, T4	↗	↗	
Parathormone	↗		
Insulin	↗	↗	
Glucagon	→		
Prolactin	↘		
Endothelin	↗	↗	
Growth hormone	↗	↗ ↗	
Testosterone	↗ ↘		
LH, FSH	↘		

different patterns. However, high-altitude natives have normal hormone levels, and the modifications are probably only transient, normalizing on chronic exposure.

VIII. Summary and Future Directions

The direction of hormonal changes in response to acute and chronic hypoxia of high altitude, to the extent it is known, is summarized in Table 1. Among future research challenges in studying responses of human hormones to high altitude is to design clinical outcome studies, which can distinguish the effect of hypoxia from that of many other stressors altering hormone levels. Such studies might better be done under simulated conditions rather than in the field. The perspectives of fundamental research in the domain of endocrines at high altitude are clearly directed toward the study of receptors, gene expression in hormonal systems, and the possible role of hypoxia-inducible factors.

References

1. Richalet JP. Oxygen sensors in the organism. Examples of regulation under altitude hypoxia in mammals. Comp Biochem Physiol 1997; 118A:9–14.
2. Hoyt R, Honig A. Body fluid and energy metabolism at high altitude. In: Handbook of Physiology. New York: Oxford University Press, 1996:1277–1289.
3. Raff H. Endocrine adaptation to hypoxia. In: Handbook of Physiology. New York: Oxford University Press, 1996:1259–1275.
4. Jenq W, Rabb H, Wahe M, Ramirez G. Hypoxic effects on the expression of mineralocorticoid and glucocorticoid receptors in human renal cortex epithelial cells. Biochem Biophys Res Commun 1996; 218:444–448.
5. Kacimi R, Richalet JP, Crozatier B. Hypoxia-induced differential modulation of adenosinergic and muscarinic receptors in rat heart. J Appl Physiol 1993; 75:1123–1128.
6. Kacimi R, Moalic JM, Aldashev A, Vatner DE, Richalet JP, Crozatier B. Differential regulation of G protein expression in rat hearts exposed to chronic hypoxia. Am J Physiol 1995; 38:H1865–1873.
7. Ayres PJ, Hurter WG, Williams ES. Aldosterone excretion and potassium retention in subjects living at high altitude. Nature 1961; 191:78–80.
8. Moncloa F, Donayre J, Sobrevilla LA, Guerra-Garcia R. Endocrine studies at high altitude. II. Adrenal cortical function in sea-level natives exposed to high altitudes (4300 meters) for two weeks. J Clin Endocrin 1965; 25:1640–1642.
9. Frayser R, Rennie ID, Gray GW, Houston CS. Hormonal and electrolyte response to exposure to 17,500 ft. J Appl Physiol 1975; 38:636–642.
10. Heyes MP, Farber MO, Manfredi F, Robertshaw D, Weinberger M. Acute effects of hypoxia on renal and endocrine function in normal humans. Am J Physiol 1982; 243 (Reg Int Comp Physiol 12):R265–R270.
11. Humpeler E, Skrabal F, Bartsch G. Influence of exposure to moderate altitude on the plasma concentration of cortisol, aldosterone, renin, testosterone, and gonadotropins. Eur J Appl Physiol 1980; 45:167–176.
12. Sutton JR, Viol GW, Gray GW, McFadden MD, Keane PM. Renin, aldosterone, elec-

trolyte, and cortisol responses to hypoxic decompression. J Appl Physiol 1977; 43: 421–424.

13. Sawhney RC, Malhotra AS, Singh T. Glucoregulatory hormones in man at high altitude. Eur J Appl Physiol 1991; 62:286–291.
14. Anand IS, Chandrashekhar Y, Rao SK, Malhotra RM, Ferrari R, Chandana J, Ramesch B, Shetty KJ, Boparai MS. Body fluid compartments, renal blood flow, and hormones at 6,000m in normal subjects. J Appl Physiol 1993; 74:1234–1239.
15. Marinelli M, Roi GS, Giacometti M, Bonini P, Banfi G. Cortisol, testosterone and free testosterone in athletes performing a marathon at 4000m altitude. Horm Res 1994; 41: 225–229.
16. Colice GL, Ramirez G. Effect of hypoxemia on the renin-angiotensin-aldosterone system in humans. J Appl Physiol 1985; 58:724–730.
17. Bouissou P, Fiet J, Guezennec CY, Pesquies PC. Plasma adrenocorticotrophin and cortisol responses to acute hypoxia at rest and during exercise. Eur J Appl Physiol 1988; 57:110–113.
18. Tunny TJ, van Gelder J, Gordon RD, Klemm SA, Hamlet SM, Finn WL, Carney GM, Maher C. Effects of altitude on atrial natriuretic peptide: the bicentennial Mount Everest Expedition. Clin Exp Pharmacol Physiol 1989; 16:287–291.
19. Vuolteenaho O, Koistinen P, Martikkala V, Takala T, Leppaluoto J. Effect of physical exercise in hypobaric conditions on atrial natriuretic peptide secretion. Am J Physiol 1992; 263 (Reg Int Comp Physiol 32): 647–652.
20. Hornbein TF. Adrenal cortical response to chronic hypoxia. J Appl Physiol 1962; 17: 246–248.
21. Siri WE, Cleveland AS, Blanche P. Adrenal gland activity in Mt Everest climbers. Fed Proc 1969; 28:1251–1256.
22. Bouissou P, Péronnet F, Brisson G, Hélie R, Ledoux M. Metabolic and endocrine responses to graded exercise under acute hypoxia. Eur J Appl Physiol 1986; 55:290–294.
23. Singh I, Khanna PK, Srivastava MC, Lal M, Roy SB, Subramanyam. Acute mountain sickness. N Engl J Med 1969; 280:175–218.
24. Singh I, Malhotre MS, Khanna PK, Nanda RB, Purshottam T, Upadhyay TN, Radhakrishnan U, Brahmachari HD. Changes in plasma cortisol, blood antidiuretic hormone and urinary catecholamines in high-altitude pulmonary oedema. Int J Biometeor 1974; 18:211–221.
25. Maresh CM, Noble BJ, Robertson KL, Harvey JS. Aldosterone, cortisol and electrolyte responses to hypobaric hypoxia in moderate-altitude natives. Aviat Space Environ Med 1985; 56:1978–1984.
26. Bärtsch P, Shaw S, Francioli M, Gnadinger MP, Weidmann P. Atrial natriuretic peptide in acute mountain sickness. J Appl Physiol 1988; 65:1929–1937.
27. Richalet JP, Rutgers V, Bouchet P, Rymer JC, Keromes A, Duval-Arnould G, Rathat C. Diurnal variations of acute mountain sickness, colour vision, and plasma cortisol and ACTH at high altitude. Aviat Space Environ Med 1989; 60:105–111.
28. Richalet JP, Antezana AM, Bienvenu A, Marchal M, Souberbielle JC, Cauchy E, Le Trong JL, Déchaux M, Kacimi R, Bonaldi V, Westerterp K, Kayser B, Dubray C. Physiological factors in survival at extreme altitude. In: Sutton JR, Houston CS, Coates G, eds. Hypoxia and Molecular Medicine. Burlington: Queen City Printers, 1993:235–251.
29. Richalet JP, Déchaux M, Bienvenu A, Souberbielle JC, Antezana AM, Cauchy E.

Erythropoiesis and renal function at the altitude of 6,542 m. Jpn J Mount Med 1995; 15:135–150.
30. Pines A, Slater JD, Jowett TP. The kidney and aldosterone in acclimatization at altitude. Br J Dis Chest 1977; 71:203–207.
31. Westendorp RGJ, Frölich M, Meinders AE. What to tell steroid-substituted patients about the effects of high altitude? Lancet 1993; 342:310–311.
32. Raff H, Tzankoff SP, Fitzgerald RS. ACTH and cortisol responses to hypoxia in dogs. J Appl Physiol 1981; 51:1257–1260.
33. Raff H, Tzankoff SP, Fitzgerald RS. Chemoreceptor involvement in cortisol responses to hypoxia in ventilated dogs. J Appl Physiol 1982; 52:1092–1096.
34. Honig A. Salt and water metabolism in acute high-altitude hypoxia: role of peripheral arterial chemoreceptors. NIPS 1989; 4:109–111.
35. Richalet JP, Mehdioui H, Rathat C, Vignon P, Kéromès A, Herry JP, Sabatier C, Tanche M, Lhoste F. Acute hypoxia decreases cardiac response to catecholamines in exercising humans. Int J Sports Med 1988; 9:157–162.
36. Antezana AM, Kacimi R, Le Trong JL, Marchal M, Abousahl I, Dubray C, Richalet JP. Adrenergic status of humans during prolonged exposure to the altitude of 6542 m. J Appl Physiol 1994; 76:1055–1059.
37. Mazzeo RS, Wolfel EE, Butterfield GE, Reeves JT. Sympathetic response during 21 days at high altitude (4,300 m) as determined by urinary and arterial catecholamines. Metab Clin Exp 1994; 43:1226–1232.
38. Bouissou P, Péronnet F, Brisson G, Helié R, Ledoux M. Fluid-electrolyte shift and renin-aldosterone responses to exercise under hypoxia. Horm Metab Res 1987; 19: 331–334.
39. Claustre J, Peyrin I. Free and conjugated catecholamines and metabolites in cat urine after hypoxia. J Appl Physiol 1982; 52:304–308.
40. Olsen NV, Hansen JM, Kanstrup IL, Richalet JP, Leyssac PP. Renal hemodynamics, tubular function, and the response to low-dose dopamine during acute hypoxia in humans. J Appl Physiol 1993; 74: 2166–2173.
41. Antezana AM, Richalet JP, Noriega I, Galarza M, Antezana G. Hormonal changes in normal and polycythemic high altitude natives. J Appl Physiol 1995; 79:795–800.
42. Richalet JP. The heart and adrenergic system in hypoxia. In: JR Sutton, G Coates, JE Remmers, eds. Hypoxia: The Adaptations. Toronto: B.C. Decker Inc., 1990: 231–240.
43. Richalet JP, Larmignat P, Rathat C, Kéromès A, Baud P, Lhoste F. Decreased human cardiac response to isoproterenol infusion in acute and chronic hypoxia. J Appl Physiol 1988; 65:1957–1961.
44. Richalet JP, Le Trong JL, Rathat C, Merlet P, Bouissou P, Keromes A, Veyrac P. Reversal of hypoxia-induced decrease in human cardiac response to isoproterenol infusion. J Appl Physiol 1989; 67:523–527.
45. León-Velarde F, Richalet JP, Chavez JC, Kacimi R, Rivera-Chira M, Palacios JA, Clark D. Hypoxia- and normoxia-induced reversibility of autonomic control in Andean guinea pig heart. J Appl Physiol 1996; 81:2229–2234.
46. Kacimi R, Richalet JP, Corsin A, Abousahl I, Crozatier B. Hypoxia-induced downregulation of β-adrenergic receptors in rat heart. J Appl Physiol 1992; 73:1377–1382.
47. Voelkel NF, Hegstrand L, Reeves JT, McMurty IF, Molinoff PB. Effects of hypoxia on density of β-adrenergic receptors. J Appl Physiol 1981; 50:363–366.
48. León-Velarde F, Richalet JP, Molinatti G, Crozatier B. Up-regulation of cardiac

a-adrenergic receptors (a-AR) in rats exposed to prolonged hypoxia. Tenth Hypoxia Symposium. Lake Louise, Canada, Feb. 18–22, 1997.
49. Richalet JP, Kacimi R, Antezana AM. The control of chronotropic function in hypobaric hypoxia. Int J Sports Med 1992; 13:S22–S24.
50. Williams ES. Salivary electrolyte composition at high altitude. Clin Sci 1961; 21:37–42.
51. Williams ES. Electrolyte regulation during the adaptation of humans to life at high altitude. Proc R Soc Lond Ser B 1966; 165:266–280.
52. Ramirez G, Hammond M, Agosti SJ, Bittle PA, Dietz JR, Colice GL. Effects of hypoxemia at sea level and high altitude on sodium excretion and hormonal levels. Aviat Space Environ Med 1992; 63:891–898.
53. Rock PB, Kraemer WJ, Fulco CS, Trad LA, Malconian MK, Rose MS, Young PM, Cymerman A. Effects of altitude acclimatization on fluid regulatory hormone response to submaximal exercise. J Appl Physiol 1993; 75:1208–1215.
54. Slater JD, Tuffley RE, Williams ES, Beresford CH, Sonksen PH, Edwards RH, Ekins RP, McLaughlin M. Control of aldosterone secretion during acclimatization to hypoxia in man. Clin Sci 1969; 37:327–341.
55. Kotchen TA, Hogan RP, Boyd AE, Li TK, Sing HC, Mason JW. Renin, noradrenaline and adrenaline responses to simulated altitude. Clin Sci 1973; 44:243–251.
56. Hogan RP, Kotchen TA, Boyd AE, Hartley LH. Effect of altitude on renin-aldosterone system and metabolism of water and electrolytes. J Appl Physiol 1973; 35:385–390.
57. Milledge JS, Catley DM, Ward MP, Williams ES, Clarke CR. Renin-aldosterone and angiotensin-converting enzyme during prolonged altitude exposure. J Appl Physiol 1983; 55:699–702.
58. Olsen NV, Kanstrup IL, Richalet JP, Hansen JM, Plazen G, Galen FX. Effects of acute hypoxia on renal and endocrine function at rest and during graded exercise in hydrated subjects. J Appl Physiol 1992; 73:2036–2043.
59. Maher JT, Jones LG, Hartley LH, Williams GH, Rose LI. Aldosterone dynamics during graded exercise at sea level and high altitude. J Appl Physiol 1975; 39:18–22.
60. Maresh CM, Noble BJ, Robertson KL, Seip RL. Adrenocortical responses to maximal exercise in moderate-altitude natives at 447 Torr. J Appl Physiol 1984; 56:482–488.
61. Milledge JS, Beeley JM, McArthur S, Morice AH. Atrial natriuretic peptide, altitude and acute mountain sickness. Clin Sci 1989; 77:509–514.
62. Milledge JS, Catley DM, Blume FD, West JB. Renin, angiotensin converting enzyme and aldosterone in humans on Mount Everest. J Appl Physiol 1983; 55:1109–1113.
63. Bärtsch P, Maggiorini M, Schobersberger W, Shaw S, Rascher W, Girard J, Weidmann P, Oelz O. Enhanced exercise-indiced rise of aldosterone and vasopressin preceding mountain sickness. J Appl Physiol 1991; 71:136–143.
64. Schmidt W, Wedler V, Zingler C, Ledderhops C, Honig A. Kidney function during arterial chemoreceptor stimulation. II. Suppression of plasma aldosterone concentration due to hypoxic-hypercapnic perfusion of the carotid bodies in anesthtized cats. Biomed Biochim Acta 1985; 44:711–722.
65. Stalcup SA, Lipset JS, Legant PM, Leuenberger PJ, Mellins RB. Inhibition of converting enzyme activity by acute hypoxia in dogs. J Appl Physiol 1979; 46:227–234.
66. Oparil S, Narkates AJ, Jackson RM, Soo HS. Altered angiotensin-converting enzyme in lung and extrapulmonary tissues of hypoxia-adapted rats. J Appl Physiol 1988; 65:218–227.

67. Milledge JS, Catley DM. Angiotensin converting enzyme response to hypoxia in man: its role in altitude acclimatization. Clin Sci 1984; 67:453–456.
68. Milledge JS, Catley DM. Angiotensin converting enzyme activity and hypoxia. Clin Sci 1987; 72:149.
69. Colice GL, Ramirez G. Aldosterone response to angiotensin II during hypoxemia. J Appl Physiol 1986; 61:150–154.
70. Claybaugh JR, Wade CE, Cucinell SA. Fluid and electrolyte balance and hormonal response to the hypoxic environment. In: Claybaugh JR, Wade CE, eds. Hormonal Regulation of Fluid and Electrolytes. New York: Plenum, 1989:187–214.
71. Raff H. The renin-angiotensin-aldosterone system during hypoxia. In: Lahiri S, Cherniack NS, Fitzerland RS, eds. Response and Adaptation to Hypoxia. Organ to Organelle. New York: Oxford University Press, 1991:211–222.
72. Weidmann PB, Hellmueller DE, Uehlinger RE, Lang MP, Gnaedinger L, Hasler S, Shaw S, Bachmann C. Plasma levels and cardiovascular, endocrine, and excretory effects of atrial natriuretic peptide during different sodium intakes in man. J Clin Endocrinol Metab 1986; 62:1027–1036.
73. Lawrence DL, Shenker Y. Effect of hypoxic exercise on atrial natriuretic factor and aldosterone regulation. Am J Hypertens 1991; 4:341–347.
74. Cuneo RC, Espiner EA, Nicolls MG, Yandle TG, Livessey JH. Effect of physiological levels of atrial natriuretic peptide on hormone secretion: inhibition of angiotensin-induced aldosterone secretion and renin release in normal man. J Clin Endocrinol Metab 1987; 65:765–772.
75. Ramirez G, Bittle PA, Hammond M, Ayers CW, Dietz JR, Colice GL. Regulation of aldosterone secretion during hypoxemia at sea level and moderately high altitude. J Clin Endocrinol Metab 1988; 67:1162–1165.
76. Curran Everett DC, Claybaugh JR, Miki K, Hong SK, Krasney JA. Hormonal and electrolyte responses of conscious sheep to 96h of hypoxia. Am J Physiol 1988; 255: R274–R283.
77. Hartroft PM, Bischoff MB, Bucci TJ. effects of chronic exposure to high altitude on the juxtaglomerular complex and adrenal cortex of dogs, rabbits and rats. Fed Proc 1969; 28:1234–1237.
78. Raff H, Ball DL, Goodfriend TL. Low oxygen selectively inhibits aldosterone secretion from bovine adrenocortical cells in vitro. Am J Physiol 1989; 256:E640–E644.
79. Brickner RC, Jankowski B, Raff H. The conversion of corticosterone to aldosterone is the site of the oxygen sensitivity of the bovine adrenal zona glomerulosa. Endocrinology 1992; 130:88–92.
80. Keynes RJ, Smith GW, Slater JDH, Brown MM, Brown SE, Payne NN, Jowett TP, Monger CC. Renin and aldosterone at high altitude in man. J Endocrinol 1982; 92: 131–140.
81. Raff H, Jankowski B. Inhibition of aldosterone release by hypoxia in vitro: interaction with carbon monoxide. J Appl Physiol 1994; 76:689–693.
82. Raff H, Jankowski BM, Engeland WC, Oaks MK. Hypoxia in vivo inhibits aldosterone synthesis and aldosterone synthase mRNA in rats. J Appl Physiol 1996; 81:604–610.
83. Carey RM, Thorner MO, Ortt EM. Dopaminergic inhibition of Metoclopramide-induced aldosterone secretion in man. J Clin Invest 1980; 66:10–18.
84. Drake CR, Ragsdale NV, Kaiser DL, Carey RM. Dopaminergic suppression of angiotensin II-induced aldosterone secretion in man: differential responses during sodium loading and depletion. Metabolism 1984; 33:696–702.

85. Carey RM, Sen S. Recent progress in the control of aldosterone secretion. In: Recent Progress in Hormone Research. New York: Academic Press, 1986:251–289.
86. Vonmoos S, Nussberger J, Waeber B, Biollaz J, Brunner HR, Leuenberger P. Effect of metoclopramide on angiotensins, aldosterone, and atrial peptide during hypoxia. J Appl Physiol 1990; 69:2072–2077.
87. Honig A, Wedler B, Schmidt M, Gruska S, Twal A. Suppression of the plasma aldosterone to renin activity ratio in anesthetized cats after pharmacological stimulation of the peripheral arterial chemoreceptors with almitrine bismesylate. Biomed Biochim Acta 1987; 12:1055–1059.
88. Weil JV, Byrne-Quinn E, Battock DJ, Grover RF, Chidsey CA. Forearm circulation in man at high altitude. Clin Sci 1971; 40:235–246.
89. Maestri E, Camellini L, Rossi G, Rossi GP, Dotti C, Marchesi M, Gnudi A. Serotonin regulation of aldosterone secretion. Horm Metabol Res 1988; 20:457–459.
90. De Souza MJ, Maresh CM, Maguire MS, Kraemer WJ, Flora-Ginter G, Goetz KL. Menstrual status and plasma vasopressin, renin activity, and aldosterone exercise responses. J Appl Physiol 1989; 67:736–743.
91. Bärtsch P, Pfluger N, Audetat M, Shaw S, Weidmann P, Vock P, Vetter W, Rennie D, Oelz O. Effects of slow ascent to 4559m on fluid homeostasis. Aviat Space Environ Med 1991; 62:105–110.
92. Milledge JS. Salt and water control at altitude. Int J Sports Med 1992; 13:S61–S62.
93. Westendorp RGJ, Roos AN, Simons M, Wertheim W, Bosch FH, Frölich M, Meinders AE. effects of hypoxia and atrial natriuretic peptide on aldosterone secretion in healthy subjects. J Appl Physiol 1993; 75:534–539.
94. Colice GL, Lawrason J, Munsef A, Bittle P, Dietz J, Ramirez G. Hormonal response to exercise in high altitude natives and COPD patients. Aviat Space Environ Med 1993; 64:512–516.
95. Williams GH, Dluhy RG, Underwood RH. The relationship of dietary potassium intake to the aldosterone stimulating properties of ACTH. Clin Sci 1970; 39:489–496.
96. Francesconi RP. Endocronological responses to exercise in stressful environments. Exer Sport Sci Rev 1988; 16:255–284.
97. Rahman H, Siddiqui HH, Kumar R. Simulated altitude and plasma renin activity in rats. Ind J Med Res 1980; 72:445–448.
98. Okazaki S, Tamura Y, Hatano T, Matsui N. Hormonal disturbances of fluid-electrolyte metabolism under altitude exposure in man. Aviat Space Environ Med 1984; 55:200–205.
99. Marsen TA, Schramek H, Dunn MJ. Renal actions of endothelin: Linking cellular signaling pathways to kidney disease. Kidney Int 1994; 45:336–344.
100. Rakugi H, Nakamaru M, Saito H. Endothelin inhibits renin release from isolated rat glomeruli. Biochem Biophys Res Commun 1988; 155:1244–1247.
101. Shigeoka JW, Colice GL, Ramirez G. Effect of normoxemic and hypoxemic exercise on renin and aldosterone. J Appl Physiol 1985; 59:142–148.
102. Story DA, Miller BR, Shield CM, Bowes G. Atrial natriuretic factor during hypoxia and mild exercise. Aviat Space Environ Med 1991;62:287–290.
103. Convertino VA, Veil LC, Bernauer EM, Greenleaf JE. Plasma volume, osmolality, vasopressin, and renin activity during graded exercise in man. J Appl Physiol 1981; 50:123–128.
104. Bouissou P, Richalet JP, Galen FX, Lartigue M, Larmignat P, Devaux F, Dubray C,

Keromes A. Effect of beta-adrenoceptor blockade on renin-aldosterone and alpha-ANF during exercise at altitude. J Appl Physiol 1989; 67:141–146.
105. Richalet JP, Galen FX, Bouissou P, Lartigue M, Rutgers V, Kéromès A. Secretion of active and inactive renin fractions at exercise in high altitude hypoxia. Biochemistry of exercise, 7th Int. Conference, London, Ontario 1–4 June, 1988.
106. Milledge JS, Catley DM, Williams ES, Withey WR, Minty BD. Effect of prolonged exercise at altitude on the renin-aldosterone system. J Appl Physiol Respirat Environ Exercise Physiol 1983; 55:413–418.
107. Meehan RT. Renin, aldosterone and vasopressin response to hypoxia during 6 hours of mild exercise. Aviat Space Environ Med 1986; 57:960–965.
108. Raff H, Sandri RB, Segerson TP. Renin, ACTH, and adrenocortical function during hypoxia and hemorrhage in conscious rats. Am J Physiol 1986; 250:R240–R244.
109. Eichinger MR, Claybaugh JR. Hypoxia attenuates the renin response to hemorrhage. Am J Physiol 1992; 263:664–669.
110. Wang BC, Sundet WD, Goetz KL. Vasopressin in plasma and cerebrospinal fluid of dogs during hypoxia or acidosis. Am J Physiol 1984; 247:E449–E455.
111. Rose CE, Anderson RJ, Carey RM. Antidiuresis and vasopressin release with hypoxemia and hypercapnia in conscious dogs. Am J Physiol 1984; 247:127–134.
112. Raff H, Shinsako J, Keil LC, Dallman MF. Feedback inhibition of adrenocorticotropin and vasopressin responses to hypoxia by physiological increases in endogenous plasma corticosteroids in dogs. Endocrinology 1984; 114:1245–1249.
113. Raff H. Glucocorticoid inhibition of neurohypophysial vasopressin secretion. Am J Physiol 1987; 252:R635–R644.
114. Walker BR. Inhibition of hypoxia-induced ADH release by meclofenamate in the conscious dog. J Appl Physiol 1983; 54:1624–1629.
115. Koller EA, Buhrer A, Felder L, Schopen M, Vallotton MB. Altitude diuresis: endocrine and renal responses to acute hypoxia of acclimatized and non-acclimatized subjects. Eur J Appl Physiol 1991; 62:228–234.
116. Forsling ML, Milledge JS. Effect of hypoxia on vasopressin release in man. J Physiol (London) 1977; 207:52–53.
117. Share L, Levy MN. Effect of carotid chemoreceptor stimulation on plasma antidiuretic hormone titer. Am J Physiol 1966; 210:157–161.
118. Claybaugh JR, Wade CE, Sato AK, Cucinell SA, Lane JC. Antidiuretic hormone responses to eucapnic and hypocapnic hypoxia in humans. J Appl Physiol 1982; 53: 815–823.
119. Blume FD, Boyer SJ, Braverman LE, Cohen A, Dirkse J, Mordes JP. Impaired osmoregulation at high altitude (studies on Mount Everest). JAMA 1984; 252:524–526.
120. Swenson ER, Duncan TB, Goldberg SV, Ramirez G, Ahmad S, Schoene RB. Diuretic effect of acute hypoxia in humans: relationship to hypoxic ventilatory responsiveness and renal hormones. J Appl Physiol 1995; 78:377–383.
121. Claybaugh JR, Hansen JE, Wozniak DB. Response of antidiuretic hormone to acute exposure to mild and severe hypoxia in man. J Endocrinol 1978; 77:157–160.
122. Guiol C, Montastruc P, Prévost MC. Renal effect of acute hypobaric pressure breathing in normal and diabetes insipidus rats. J Physiol (Paris) 1986; 81:41–44.
123. Tenney SM, Jones RM. Water balance and lung fluids in rats at high altitude. Respir Physiol 1992; 87:397–406.
124. Hackett PH, Forsling ML, Milledge JS, Rennie ID. Release of vasopressin in man at altitude. Horm Metab Res 1978; 10:571.

125. Chartier L, Schiffrin E, Thibault G, Garcia R. Atrial natriuretic factor inhibits the stimulation of aldosterone angiotensin II-induced steroidogenesis in vivo. Endocrinology 1984; 115:2026–2028.
126. Du Souich P, Saunier C, Hartemann D, Sautegeau A, Ong H, Larose P, Babini R. Effect of moderate hypoxemia on atrial natruretic factor and arginine vasopressin in normal man. Biochem Biophys Res Comm 1987; 148:906–912.
127. Kawashima A, Kubo K, Hirai K, Yoshikawa S, Matsuzawa Y, Kobayashi T. Plasma levels of atrial natriuretic peptide under acute hypoxia in normal subjects. Respir Physiol 1989; 76:79–92.
128. Baertschi AJ, Jiao JH, Carlson DE, Campbell RW, Teague WG. Neural control of ANF release in hypoxia and pulmonary hypertension. Am J Physiol 1990; 259 (Heart Circ Physiol 28):735–744.
129. Koller EA, Schopen M, Keller M, Lang RE, Vallotton MB. Ventilatory, circulatory, endocrine, and renal effects of almitrine infusion in man: a contribution to high altitude physiology. Eur J Appl Physiol 1989; 58:419–425.
130. McKenzie JC, Tanaka I, Inagami T, Misono KS, Klein RM. Alterations in atrial and plasma atrial natriuretic factor content during development of hypoxia-induced pulmonary hypertension in the rat. Proc Soc Exp Biol Med 1986; 181:459–463.
131. Baertschi AJ, Hausmaninjer C, Walsh RS, Mentzer RM, Pence RA. Hypoxia induced release of atrial natiuretic factor (ANF) from the isoled rat and rabbit heart. Biochem Biophys Res Comm 1986; 410:427–433.
132. Stockmann PT, Will DH, Sides SD, Brunnert SR, Wilner GD, Leanhy KM, Wiegand RC, Needleman P. Reversible induction of right ventricular atriopeptin synthesis in hypertrophy due to hypoxia. Circ Res 1988; 63:207–213.
133. Colice G, Yen S, Ramirez G, Dietz J, Ou LC. Acute hypoxia-induced diuresis in rats. Aviat Space Environ Med 1991; 62:551–554.
134. Baertschi AJ, Milton Adams J, Sullivan MP. Acute hypoxemia stimulates atrial natriuretic factor secretion in vivo. Am J Physiol 1988; 255:H295–H300.
135. Schmidt W, Brabant G, Kröger C, Strauch S, Hilgendorf A. Atrial natriuretic peptide during and after maximal and submaximal exercise under normoxic and hypoxic conditions. Eur J Appl Physiol 1990; 61:398–407.
136. Kawashima A, Kubo K, Matsuzawa Y, Kobayashi T, Sekiguchi M. Hypoxia-induced ANP secretion in subjects susceptible to high-altitude pulmonary edema. Respir Physiol 1992; 89:309–317.
137. Adnot S, Chabrier PE, Brun-Buisson C, Viossat I, Braquet P. Atrial natriuretic factor attenuates the pulmonary pressor response to hypoxia. J Appl Physiol 1988; 65:1975–1983.
138. Cosby RL, Sophocles AM, Durr JA, Peerinjaquet CL, Tee B, Schrier RW. Elevated plasma atrial natruretic factor and vasopressin in high-altitude pulmonary edema. Ann Intern Med 1988; 109:796–799.
139. Jin H, Yang RH, Chenb YF, Jackson RM, Itoh H, Mukoyama M, Nakao K, Imura H, Oparil S. Atrial natriuretic peptide in acute hypoxia-induced pulmonary hypertension in rats. J Appl Physiol 1991; 71:807–814.
140. Hassi J, Rintamaki H, Ruskoaho H, Leppaluoto J, Vuolteenaho O. Plasma levels of endothelin-1 and atrial natriuretic peptide in men during a 2-hour stay in a cold room. Acta Physiol Scand 1991; 142:481–485.
141. Baertschi A, Teague WG. Alveolar hypoxia is a powerful stimulus for ANF release in conscious lambs. Am J Physiol 1989; 256:H990–H998.

142. Surks MI. Elevated PBI, free thyroxine and plasma protein concentration in man at high altitude. J Appl Physiol 1966; 21:1185–1190.
143. Moncloa F, Guerra-Garcia R, Subauste C, Sobrevilla LA, Donayre J. Endocrine studies at high altitude I. Thyroid function in sea level natives exposed for two weeks to an altitude of 4300 meters. J Clin Endocrin 1966; 26:1237–1239.
144. Rawal SB, Singh MV, Tyagi AK, Chaudhuri BN. Thyroidal handling of radioiodine in sea level residents exposed to hypobaric hypoxia. Eur J Nucl Med 1993; 20:16–19.
145. Basu M, Pal K, Malhotra AS, Prasad R, Sawhney RC. Free and total thyroid hormones in humans at extreme altitude. Int J Biometeorol 1995; 39:17–21.
146. Chakraborty S, Samaddar J, Batabyal SK. Thyroid status of humans at high altitude. Clin Chim Acta 1987; 166:111–113.
147. Férézou J, Richalet JP, Sérougue C, Coste T, Wirquin E, Mathé D. Reduction of postprondial lipemia after acute exposure to high altitude hypoxia. Int J Sports Med 1993; 14:78–85.
148. Kotchen TA, Mougey EH, Hogan RP, Bovd AE, Pennigton LL, Mason JW. Thyroid responses to simulated altitude. J Appl Physiol 1973; 34:165–168.
149. Mordes JP, Blume FD, Boyer S, Zheng MR, Braverman LE. High-altitude pituitary-thyroid dysfunction on Mount Everest. N Engl J Med 1983; 308:1135–1138.
150. Ramirez G, Herrera R, Pineda D, Bittle PA, Rabb HA, Bercus BB. The effects of high altitude on hypothalamic-pituitary secretory dynamics in men. Clin Endocrinol 1995; 43:11–18.
151. Rastogi GK, Malhotra MS, Srivastava MC, Sawhney RC, Dua GL, Sridharan K, Hoon RS, Singh I. Study of the pituitary-thyroid functions at high altitude in man. J Clin Endocrin Metab 1977; 44:447–452.
152. Sawhney RC, Malhotra AS. Thyroid function in sojourners and acclimatised low landers at high altitude in man. Horm Metab Res 1991; 23:81–84.
153. Stock MJ, Chapman C, Stirling JL, Campbell IT. Effects of exercise, altitude and food on blood hormone and metabolite levels. J Appl Physiol Respirat Environ Physiol 1978; 45:350–354.
154. Surks MI, Beckwitt HJ, Chidsey CA. Changes in plasma thyroxine concentration and metabolism, catecholamine excretion and basal oxygen consumption in man during acute exposure to high altitude. J Clin Endocrinol 1967; 27:789–799.
155. Wright AD. Birmingham Medical Research Expeditionary Society 1977 Expedition: thyroid function and acute mountain sickness. Postgrad Med J 1979; 55:483–486.
156. Hackney AC, Feith S, Pozos, Seale J. Effects of altitude and cold exposure on resting thyroid hormone concentrations. Aviat Space Environ Med 1995; 66:325–329.
157. Surks MI. Metabolism of human serum albumin in man during acute exposure to high altitude (14,100 feet). J Clin Invest 1966; 45:1442–1451.
158. Hackney AC, Gulledge T. Thyroid hormone responses during an 8-hour period following aerobic and anaerobic exercise. Physiol Res 1994; 43:1–5.
159. Snyder LM, Reddy WJ. Mechanism of action of thyroid hormones on erythrocyte 2,3-diphosphoglyceric acid synthesis. J Clin Invest 1970; 49:1993–1998.
160. Hesslink RL, D'Alesandro MM, Armstrong DW, Reed HL. Human cold air habituation is independent of thyroxine and thyrotropin. J Appl Physiol 1992; 72:2134–2139.
161. Savourney G, Caravel JP, Barnavol B, Bittel JHM. Thyroid hormone changes in a cold air environment after local cold acclimation. J Appl Physiol 1994; 76:1963–1967.
162. Krapf R, Jaeger P, Hulter HN, Fehlman C, Takkinen R. Chronic respiratory alkalosis

induces renal PTH-resistance, hyperphosphatemia and hypocalcemia in humans. Kidney Int 1992; 42:727–734.

163. Kawashima H, Kurokawa K. Metabolism and sites of action of vitamin D in the kidney. Kidney Int 1986; 29:98–107.
164. Souberbielle JC, Richalet JP, Garabedian M, Sachs C, Bourdeau A, Déchaux M. Effect of high altitude hypoxia on calcium metabolism and bone markers. XII Int Conf on Calcium Regulating Hormones. Bone 1995; 16:210S.
165. Mostellar M, Tuttle E. Effects of alkalosis on plasma concentration and urinary excretion of inorganic phosphate in man. J Clin Invest 1964; 43:138–149.
166. Hoppe A, Melter M, Berndt TJ, Knox FG, Angielski S. Effect of respiratory alkalosis on renal phosphate excretion. Am J Physiol 1982; 243:F471–F475.
167. Blum JW, Bianca W, Naf F, Kunz P, Fischer JA, Daprada M. Plasma catecholamine and parathyroid hormone responses in cattle during treadmill exercise at simulated high altitude. Horm Metab Res 1979; 11:246–251.
168. Blume FD, Pace N. Effect of translocation to 3800m altitude on glycolysis in mice. J Appl Physiol 1967; 23:75–79.
169. Sawhney RC, Malhotra AS, Singh T, Rai RM, Sinha KC. Insulin secretion at high altitude in man. Int J Biometeor 1986; 30:231–238.
170. Brooks GA, Butterfield GE, Wolfe RR, Groves BM, Mazzeo RS, Sutton JR, Wolfel EE, Reeves JT. Increased dependence on blood glucose after acclimatization to 4,300 m. J Appl Physiol 1991; 70:919–927.
171. Roberts AC, Reeves JT, Butterfield GE, Mazzeo RS, Sutton JR, Wolfel EE, Brooks GA. Altitude and β-blockade augment glucose utilization during submaximal exercise. J Appl Physiol 1996; 80:605–615.
172. Young PM, Rose MS, Sutton JR, Green H. Operation Everest II: plasma lipid and hormonal responses during a simulated ascent of Mount Everest. J Appl Physiol 1989; 66:1430–1435.
173. Richalet JP, Souberbielle JC, Antezana AM, Déchaux M, Le Trong JL, Bienvenu A, Daniel F, Blanchot C, Zittoun J. Control of erythropoiesis in humans during prolonged exposure to the altitude of 6542 m. Am J Physiol 1994; 266 (Reg Int Comp Phys 35): R756–R764.
174. Kullmer T, Gabriel H, Jungmann E, Haak T, Morbitzer D, Usadel KH, Kindermann W. Increase of serum insulin and stable C-peptide concentrations with exhaustive incremental graded exercise during acute hypoxia in sedentary subjects. Exp Clin Endocrinol 1995; 103:156–161.
175. Janoski A, Anderson JW. Glucose tolerance and insulin secretion in man during acute exposure to high altitude. Clin Res 1968; 16:269.
176. Sutton JR. Effects of acute hypoxia on the hormonal response to exercise. J Appl Physiol 1977; 42:587–592.
177. Baum D. Stress hyperglycemia and the adrenergic regulation of pancreatic hormones in hypoxia. Metabolism 1980; 29:1176–1185.
178. Dionne KE, Colton CK, Yarmush ML. Effect of hypoxia on insulin secretion by isolated rat and canine islets of langerhans. Diabetes 1993; 42:12–21.
179. Gonzales GF, Carrillo CE. Low serum prolactin levels in native women at high altitude. Int J Gynecol Obstet 1993; 43:169–175.
180. Gonzales GF, Villena A. Biomedical factors associated to high fertility at high altitude. Acta Andina 1995; 4:95–101.
181. Brisson GR, Boisvert P, Péronnet F, Quirion A, Senécal L. Face cooling-induced re-

duction of plasma prolactin response to exercise as part of an integrated response to thermal stress. Eur J Appl Physiol 1989; 58:816–820.

182. Bouissou P, Brisson GR, Péronnet F, Hélie R, Ledoux M. Inhibition of exercise-induced blood prolactin response by acute hypoxia. Can J Spt Sci 1987; 12:49–50.
183. Farrell PA, Gustafson AB, Garthwaite TL, Kalkhoff RK, Cowley AW, Morgan WP, Jr. Influence of endogenous opioids on the response of selected hormones to exercise in humans. J Appl Physiol 1986; 61:1051–1057.
184. Knudtzon J, Bogsnes A, Norman N. Changes in prolactin and growth hormone levels during hypoxia and exercise. Horm Metabol Res 1989; 21:453–454.
185. Strüder HK, Hollmann W, Platen P. Increased prolactin response to hyperoxia at rest and during endurance exercise. Int J Sports Med 1996; 17:390–392.
186. Sawhney RC, Chhabra PC, Malhotra AS, Singh T, Riar SS, Rai RM. Hormone profiles at high altitude in man. Andrologia 1985; 17:178–184.
187. Fukuda Y, Hirata Y, Yoshimi H, Kojima T, Kobayashi Y, Yanagisawa M, Masaki T. Endothelin is a potent secretagogue for atrial natriuretic peptide in cultured rat atrial myocytes. Biochem Biophys Res Commun 1988; 155:167–172.
188. Gardner DG, Newman ED, Nakamura KK, Nguyen PT. Endothelin increases the synthesis and secretion of atrial natriuretic peptide in neonatal rat cardiocytes. Am J Physiol 1991; 261:E177–E182.
189. Moe O, Tejedor A, Campbell WB, Alpern RJ, Henrich WL. Effects of endothelin on in vitro renin secretion. Am J Physiol 1991; 260:E521–E525.
190. Rubanyi GM, Vanhoutte PM. Hypoxia releases a vasoconstrictor substance from the canine vascular endothelium. J Physiol 1985; 364:45–46.
191. Rakugi H, Tabuchi Y, Nakamaru M, Nagano M, Higashimori K, Mikami H, Ogihara T, Suzuki N. Evidence for endothelin-1 release from resistance vessels of rats in response to hypoxia. Biochem Biophys Res Commun 1990; 169:973–977.
192. Kourembanas S, Marsden PA, McQuilan LP, Faller DV. Hypoxia induces endothelin gene expression and secretion in cultured human endothelium. J Clin Invest 1991; 88:1054–1057.
193. Li H, Chen SJ, Chen YF, Menq QC, Durand J, Oparil S, Elton TS. Enhanced endothelin-1 and endothelin receptor gene expression in chronic hypoxia. J Appl Physiol 1995; 77:1451–1459.
194. Kourembanas S, Hannan RL, Faller DV. Oxygen tension regulates the expression of the platelet-derived growth factor-B chain gene in human endothelial cells. J Clin Invest 1990; 86:670–674.
195. Ogawa S, Gerlach H, Esposito C, Pasagian-Macaulay A, Brett J, Stern D. Hypoxia modulates the barrier and coagulant function of cultured bovine endothelium. J Clin Invest 1990; 85:1090–1098.
196. Goerre S, Wenk M, Bärtsch P, Luscher TF, Niroomand F, Hohenhaus E, Oelz O, Reinhart WH. Endothelin-1 in pulmonary hypertension associated with high-altitude exposure. Circulation 1995; 91:359–364.
197. Richalet JP. High-altitude pulmonary oedema. Still a place for controversy? Thorax 1995; 50:923–929.
198. Richalet JP, Hornych A, Rathat C, Aumont J, Larmignat P, Rémy P. Plasma prostaglandins, leukotrienes and thromboxane in acute high altitude hypoxia. Respir Physiol 1991; 85:205–215.
199. Schoene RB, Hackett PH, Henderson WR, Sage EH, Chou M, Roach RC, Mills Jr

WJ, Martin TR. High altitude pulmonary edema. Characteristics of lung lavage fluid. JAMA 1986; 256:63–69.

200. Raynaud J, Drouet L, Martineaud JP, Bordachar J, Coudert J, Durand J. Time course of plasma growth hormone duing exercise in humans at altitude. J Appl Physiol Respir Environ Exercise Physiol 1981; 50:229–233.
201. Sawhney RC, Malhotra AS. Circadian rhytmicity of growth hormone at high altitude in man. Ind J Physiol Pharmacol 1991; 35:55–57.
202. Sutton JR. The hormonal responses to exercise at sea level and at altitude. Hypoxia, exercise and altitude. In: Sutton JR, Houston CS, Jones NL, eds. Hypoxia, Exercise, and Altitude. Burlington: Queen City Publishers, 1983:325–338.
203. Sutton JR, Coleman MJ, Casey J, Lazarus L. Androgen responses during physical exercise. Br Med J 1973;1:520–522.
204. Viru A. Plasma hormones and physical exercise. A review. Int J Sports Med 1992; 13:201–209.
205. Cappon J, Brasel JA, Mohan S, Cooper DM. Effects of brief exercise on circulating insulin-like growth factor I. J Appl Physiol 1994; 76:2490–2496.
206. Schmidt W, Doré S, Hilgendorf A, Strauch S, Gareau R, Brisson GR. Effects of exercise during normoxia and hypoxia on the growth hormone-insulin-like growth factor I axis. Eur J Appl Physiol 1995; 71:424–430.
207. Kjær M, Banhsbo J, Lortie G, Galbo H. Hormonal response to exercise in humans: influence of hypoxia and physical training. Am J Physiol 1988; 254:R197–R203.
208. Banfi G, Marinelli M, Roi GS, Colombini A, Pontillo M, Giacometti M, Wade S. Growth hormone and insulin-like growth factor I in athletes performing a marathon at 4000 m of altitude. Growth Reg 1994; 4:82–86.
209. Sutton JR, Jones NL, Toews CJ. Growth hormone secretion in acid-base alterations at rest and during exercise. Clin Sci Mol Med 1976; 50:241–247.
210. Giustina A, Wehrenberg WB. Influence of thyroid hormones on the regulation of growth hormone secretion. A review. Eur J Endocrinol 1995; 133:646–653.
211. Boyd AE, Lebovitz HE, Pfeiffer JB. Stimulation of human growth hormone secretion by L-DOPA. N Engl J Med 1970; 283:1425–1429.
212. Sutton J, Lazarus L. Effect of adrenergic blocking agents on growth hormone responses to physical stress. Horm Metab Res 1974; 6:428–429.
213. Hackney AC, Premo MC, McMurray RG. Influence of aerobic versus anaerobic exercise on the relationship between reproductive hormones in men. J Sports Sci 1995; 13:305–311.
214. Vasankari TJ, Rusko H, Kujala UM, Huhtaniemi IT. The effects of ski training at altitude and racing on pituitary, adrenal and testicular function in men. Eur J Appl Physiol 1993; 66:221–225.
215. Vaernes RJ, Owe JO, Myking O. Central nervous reactions to a 6,5-hour altitude exposure at 3048 meters. Aviat Space Environ Med 1984; 55:921–926.
216. Friedl KE, Plymate SR, Bernhard WN, Mohr LC. Elevation of plasma estradiol in healthy men during a mountaineering expedition. Horm Metabol Res 1988; 20:239–242.
217. Vander AJ, Moore LG, Brewer G, Menon KMJ, England BG. Effects of high altitude on plasma concentrations of testosterone and pituitary gonadotropins in man. Aviat Space Environ Med 1978; 49:356–357.
218. Guerra-Garcia R. Testosterone metabolism in men exposed to high altitude. Acta Endocrin Panam 1971; 2:55–59.

219. Garmendia F, Valdivia H, Castillo O, Ugarte N, Garmendia A. Hypothalamo-hypophyso-gonadal response to clomiphene citrate at median high altitude. Horm Metab Res 1982; 14:679–680.
220. Fellmann N, Bedu M, Spielvogel H, Falgairette G, van Praagh E, Jarridge JF, Coudert J. Anaerobic metabolism during pubertal development at high altitude. J Appl Physiol 1988; 64:1382–1386.
221. Beall CM, Worthman CM, Stallings J, Strohl KP, Brittenham GM, Barragan M. Salivary testosterone concentration of Aymara men native to 3600m. Ann Human Biol 1992; 19:67–78.
222. Bangham CRM, Hackett PH. Effects of high altitude on endocrine function in the sherpas of Nepal. J Endocrinol 1978; 79:147–148.
223. Ou LC, Sardella GL, Leiter JC, Brinck-Johnsen T, Smith RP. Role of sex hormones in development of chronic mountain sickness in rats. J Appl Physiol 1994; 77:427–433.
224. Wetzel RC, Zacur HA, Sylvester JT. Effect of puberty and estradiol on hypoxic vasomotor response in isolated sheep lungs. J Appl Physiol Respir Environ Exercise Physiol 1984; 56:1199–1203.
225. Kryger M, Glas R, Jackson D, McCullough RE, Scoggin C, Grover RF, Weil JV. Impaired oxygenation during sleep in excessive polycythemia of high altitude: improvement with respiratory stimulation. Sleep 1978; 1:3–17.
226. Moore LG, McCullough RE, Weil JV. Increased HVR in pregnancy: relationship to hormonal and metabolic changes. J Appl Physiol 1987; 62:158–163.

19

High Altitude and Human Immune Responsiveness

RICHARD T. MEEHAN

National Jewish Medical Research Center
Denver, Colorado

PETER N. UCHAKIN

Mercer University School of Medicine
Macon, Georgia

CLARENCE F. SAMS

NASA–Johnson Space Center
Houston, Texas

I. Introduction

This chapter constitutes a review of published results on the effects of altitude exposure on infectious illness in humans. In addition to various epidemiological investigations, nine studies have been conducted at high altitude for the specific purpose of monitoring indicators of immune function in humans (Table 1). Variables measured in those studies included peripheral blood counts and subpopulations of mononuclear leukocytes; various serum proteins, immunoglobulins, complement and lysozyme levels; in vitro T-cell function (i.e., mitogen-stimulated proliferation), B-cell function (i.e., antibody production), and cytokine production. Some investigators also have measured T-cell function in vivo by assessing delayed hypersensitivity responses to intradermal recall antigens and B-cell function in vivo by measuring antibodies produced in response to a vaccine. Results from immunological studies involving decompression chambers, which are used to mimic high-altitude exposures, are also discussed. Similarities between immunological findings obtained during hypobaric hypoxia exposure and those from other diverse environmental stressors are addressed. Finally, the practical implications of these studies, the potential

Table 1 Summary of Human Immune Findings at High Altitudes

Altitude (mountain)	No. of subjects	Cellular	Humoral	Other	Ref.
2300 m (Tien Shan)	15 (men)	WBC ↑	α_1-globulins ↑ α_2-globulins ↑ β-globulins ↓ γ-globulins ↓	Lysozyme ↑ Properdin ↓ Phagocytosis ↑	11
3700–5600 m (Mt. Elbrus)	16		Influenza A & B ↓	Phagocytic activity ↑/↓, autoimmune rxn ↑, plaque-producing cells ↑/↓	21
3200–3800 m (Tien Shan)	47	PHA response ↓ at day 5, normal at day 25	IgA, IgM, IgG ↓ at day 5, normal at day 25 Normal Ab to TABT vaccine	Lysozyme, complement ↓ at day 5, normal at day 25	22
3600 m (Pamir)	57 (men)	At 30 days, # T cells ↓, # B cells ↑, PHA response ↓ ConA response normal	At 5 days, IgG, IgM, IgA normal; at 30 days, IgM ↑		23

3600 m (Tien Shan and Pamir)	38 (men)	At 3–5 days, #, % T cells ↓, PHA response ↓, T_{helper} ↓, $T_{suppressor}$ ↑			24
		At 25–30 days, # B cells ↑	At 25–30 days, IgM ↑		
3200 m (Tien Shan)	18	At 3–5 days, # T cells ↓, # B cells ↑, PHA response ↓, spont. prolif. ↓, ConA response normal	At 3–5 days, IgM, IgG ↓, IgA normal		25
		At 25–30 days, # B cells ↑, PhA, ConA responses normal, spont. prolif. ↓	At 25–30 days, IgM ↑, IgG, IgA normal		
1559 m (Davos)	56 (22 M, 34 F) with atopic dermatitis			↑ dermal response to 7 recall Ag (22% for women, (29% in men)	26
4930 m (Mt. Poumori)	18 (men)		Normal Ab response to *N. meningitidis* polyclonal A & C		27
2700 m (Park City)	9 (men)	IL-2 ↑, TNF-α ↓ no change in IFN-γ, IL-10, sIL-2Rα		No change in ACTH, cortisol	28

mechanisms responsible for the observed findings, and some suggestions for future studies are presented.

For readers unfamiliar with immunology, we include here a brief overview of the human immune system and describe some of the methods used to assess its functional components. The immune system comprises both fixed and circulating components. The fixed component includes immune effector cells that reside within multiple lymphoid organs, including the lymph nodes, thymus, spleen, tonsils, bone marrow, and gut-associated lymphoid tissue, as well as in the skin and mucosal surfaces. The circulating components include both immune cells [granulocytes, monocytes, T cells, natural killer (NK) cells, and B cells] and proteins such as immunoglobulins, lysozyme, and complement. Together, the fixed and circulating immune components constitute a formidable defense for targeting and eliminating pathogenic organisms. Other historical classifications of immunity refer to ''innate'' or ''nonspecific'' immunity, which collectively consists of the skin, mucosal surfaces, complement, macrophages, and NK cells, and ''specific'' immunity, which consists of immunoglobulins and T cells, which can recognize large numbers of foreign (''nonself'') molecules expressed on the surface of invading pathogens or foreign tissues. In addition, numerous cytokines, released from activated immune cells, serve first to amplify the immune response by activating and recruiting additional cells to the site of the immune response and then later to downregulate the immune response in order to minimize local tissue damage and restrict the response to the site of infection. Disturbances in the delicate balance of this highly regulated system can result in allergies, life-threatening infections, autoimmune disease, or cancer.

With regard to assessments of immune function, standard clinical methods can be used to count the number of circulating white blood cells, but accurate identification of T-cell subsets, NK cells, B cells, and even monocytes requires the use of monoclonal antibodies and flow cytometry. Indeed, before these technologies became available, reports of reductions in T-cell, B-cell, or NK-cell function in response to environmental stress could well have reflected artificial reductions in cell numbers rather than any functional changes. These phenotypically and functionally diverse mononuclear cells can now be identified precisely by using monoclonal antibodies against cell-surface proteins that are specific for each type of cell. Current nomenclature for these proteins includes the initials CD (for cluster of differentiation) and a number; for example, the presence of CD4 vs. CD8 proteins is used to distinguish between two subtypes of T cells.

B cells, which produce antibodies, can be identified easily by their expression of surface immunoglobulins. In vitro, circulating B cells must be activated in order for them to produce immunoglobulins. Since very few cells in culture can be activated with a specific antigen, a polyclonal activator such as pokeweed mitogen (PWM) is used, which stimulates large numbers of cells. Large interassay variations are typical, which can make detecting stress-induced immune suppression difficult. The amounts and types of specific immunoglobulins produced are usually quantified by ELISA (enzyme-linked immunosorbent assay). An earlier, less precise technique is the plaque assay, which involves identifying zones of lysis around B cells as they secrete antibody against antigen-coated red blood cells. In vivo B-cell function is best assessed by measuring the IgM antibody response to a novel antigen after the subject has been vaccinated with that antigen. Standard clinical methods (e.g., radial immunodiffusion or nephelometry) can be used to measure the amounts of circulating immunoglobulins, but these methods are not very sensitive for detecting impairment of B-cell function, except for conditions involving long-standing suppression of B-cell function, e.g., congenital defects or the use of cytotoxic agents to treat cancer or autoimmune disease. Moreover, polyclonal expansion and subsequent elevation of circulating immunoglobulins are common after infections. Thus, finding an increase in levels of ''autoantibodies'' (e.g., antinuclear antibodies) may reflect nonspecific polyclonal activation rather than any autoimmune reaction induced by the stress.

T cells and their subsets also can be identified and quantified by the expression of specific cell-surface markers. Like B cells, very few T cells can be activated in vitro with a specific antigen, so T-cell function typically is assessed by using mitogenic activators such as phytohemagglutinin (PHA) or concanavalin A (ConA) and measuring the subsequent biochemical events, e.g., increases in protein synthesis, release of regulatory cytokines, expression of specific markers on the cell surface, or increases in cell size (blastogenic transformation). The most common method of assessing T-cell activation is by following DNA synthesis through cellular incorporation of radiolabeled thymidine. In addition, since many cytokines themselves modulate immune function by suppressing or enhancing the immune response, identifying specific cytokines released during stress can help in clarifying the mechanisms responsible for T-cell dysfunction induced by hypoxia or environmental stressors. In vivo T-cell function typically is determined with the delayed hypersensitivity skin response, which involves measuring the local tissue response (induration) to an antigen given intradermally after the subject has already been exposed to that antigen.

Granulocyte function can be tested in vitro with histochemical techniques, e.g., by following the metabolic reduction of nitroblue tetrazolium to blue formazan. Phagocytic function can be assessed by measuring the luminescence of the ''respiratory burst'' reaction that takes place during phagocytosis. The cytotoxicity of NK or other cells can be quantified by labeling a target cell (e.g., virally infected cells or K562 tumor cells) with a radioisotope (e.g., ^{51}Cr) and measuring the release of that isotope into the supernatant as the target cells are killed by the cytotoxic cells.

These assays have proven to be useful in clarifying which cell types are involved in certain disease states (e.g., X-linked agammaglobulinemia, AIDS, chronic granulomatous disease, or severe combined immunodeficiency) characterized by defects or deficiencies in T cells, B cells, or the serum protein complex called complement. However, the role of these assays in detecting stress-induced alterations in B- and T-cell function and subsequent impairment of immunocompetence is less clear.

II. Infections at High Altitude

Travelers to high altitudes frequently experience infectious illness during high-altitude excursions, as well as failure to improve until descent (1,2). Of greater concern is the likelihood that preexisting respiratory infections increase the risk of developing high-altitude pulmonary edema (HAPE) or cerebral edema (HACE), both of which are potentially life-threatening conditions. The release of vasoactive inflammatory mediators during systemic or pulmonary infections has been postulated to increase capillary permeability, which might worsen hypoxia by impairing gas exchange at the level of the alveolus (3–6). For example, rats exposed to normobaric hypoxic conditions after being infected with a parainfluenza virus demonstrated an increase in lung water content and had elevated cell counts and protein content in lavage fluid compared to normoxic-control infected rats (7). Urinary leukotriene E_4 levels among 71 people with HAPE at altitudes above 2727 m were elevated relative to those of control subjects, which suggests that an inflammatory mechanism may participate in aggravating pulmonary edema (8). One recent retrospective study of children with HAPE in Colorado at 2800 m revealed that 79% had evidence of preexisting respiratory illnesses, with 53% showing upper respiratory infection (URI), 21% bronchitis, and 16% otitis media (9). However, among 127 adults with HAPE, only 13% had preexisting viral URIs (9).

Several studies have documented higher incidences of infections in people moving to or living at high altitudes. In one such study, the incidence of lobar pneumonia and amoebic hepatitis was higher among 20,000 soldiers who had been stationed at altitudes above 3692 m than among soldiers living below 760 m (10). However, tuberculosis, dysentery, varicella, and ''infectious hepatitis'' occurred less often in the high-altitude group. In a Russian study, 6 of 15 men who spent 30 days at 2100–2300 m with occasional excursions to 3200 m developed mild respiratory infections (11).

Questionnaires given to 283 hikers in the Mt. Everest region revealed that 87% had symptoms of infectious illness (12). Earache, productive cough (although not characterized as purulent), and ''sinusitis'' may well have been due to infectious agents among 36% of subjects. However, the authors of this study astutely noted that many of the symptoms, especially cough, rhinitis, and sore throat, could have resulted from nonspecific irritation of the respiratory tract caused by inhalation of cold dry air. The prevalence of diarrhea among that group (36%) is similar to that experienced by new arrivals to underdeveloped countries at sea level. Among the 40 deaths that took place among trekkers in Nepal from 1987 to 1991, two were attributed to infections (13). High incidences of lung infections also have been reported among high-altitude natives (14,15). In another group, indigenous natives had a higher incidence of otitis media than did lowland dwellers (16).

Although it is tempting to speculate that hypoxia-impaired host defenses underlie the reported increases in frequency and severity of some infectious illnesses in mountaineering expeditions, other factors can facilitate the transmission of infectious agents (Fig. 1). Maintaining personal hygiene and keeping food and water

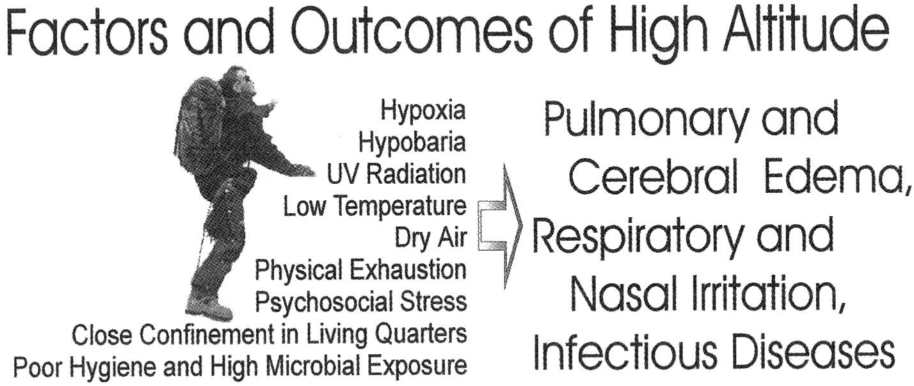

Figure 1 Factors associated with mountaineering that can facilitate transmission of infectious agents.

safe for consumption during expeditions are often difficult. Exposure to aerosolized infectious particles from coughing, whether from inhalation of cold dry air or from infection, is increased in poorly ventilated sleeping quarters. Interpersonal transfer of novel pathogens is likely while expedition members from geographically separate regions live together. Outbreaks of infections are common among other isolated populations, such as cadets during military training (17), submarine (18), space flight crews (19), or Antarctic research crews after the arrival of new members (20). The following sections provide a brief review of immunological findings from humans at high altitudes.

III. Survey of Human High-Altitude Immunology Studies

A. Altitude Studies

In the early 1970s, Krupina and colleagues studied 15 men during a 30-day expedition to 2100–2300 m, with occasional ascensions to 3200 m, in the Tien Shan mountains (11). As noted previously, 6 of these men reported having symptoms of mild respiratory infections. Other findings included increases in the absolute numbers of leukocytes, numbers of phagocytes, and the phagocytic index (i.e., neutrophilic activity against *Staphylococcus aureus* strain #209) during the 30-day expedition. Serum lysozyme activity (measured against *Micrococcus lysodeicticus*) was elevated, and properdin proteins were reduced on day 25 of the expedition relative to baseline values. Total blood proteins exceeded baseline values on days 3 and 25 of the expedition: the albumin fraction was higher and the β-globulin level lower on the 25th expedition day relative to their baseline levels. In another 30-day study in which 16 men were evaluated at 3700–5600 m, serum antibodies to influenza A2 and B were lower than the corresponding measurements made before the expedition; however, none of those subjects had evidence of influenza or respiratory infections (21). The number of plaque-forming (antibody-producing) cells was higher at altitude, as were the phagocytic activity of neutrophils and the intensity of autoimmune reactions, the latter measured by the percentage of cells that produced antibodies to autologous erythrocytes.

T- and B-cell functions were assessed in a series of three studies in the Tien Shan and Pamir mountains (22–24). In the first of these studies, 47 subjects who lived at elevations of 760–980 m were brought to 3200–3800 m, where they remained for 40 days (22). Lymphocyte response to PHA stimulation in these subjects was reduced after 5 days at high altitude, as were serum IgG, IgA, and IgM, complement, and lysozyme levels. All of these measurements had returned to baseline values by the 25th day at altitude (22). Interestingly, active immunization of 23 subjects with TABT vaccine on day 6 at 3200 m led to a fourfold increase in Vi antibody titers on day 26 at altitude (22).

In a study by Mirrakhimov and colleagues (23), T- and B-cell functions were assessed with regard to whether the subjects did or did not have acute mountain sickness (AMS). (This is the only study in which immunology results were assessed on this basis.) Of the 25 men who did not have AMS, lymphocyte response to PHA was reduced at day 3 at altitude, but the response to ConA was unchanged. IgG, IgA, and IgM levels were normal at day 3 at altitude, but IgM was elevated after 30 days at altitude. Both the percentage and absolute numbers of null cells increased in both groups upon exposure to altitude. Interestingly, the 32 men who had AMS had fewer T cells than the non-AMS group before the ascent. By 3 days at altitude, lymphocyte reaction to either PHA or ConA in the AMS group was greatly reduced and had not recovered by day 30 at altitude (23). As for B-cell function, the numbers of B cells per volume of blood were different in the AMS and non-AMS groups. The percentage of B cells in the AMS group was no different from baseline at day 3 at altitude, but both the baseline and the day 3 measurements were less that those of the group without AMS. After 30 days at altitude, IgM production in the AMS group was considerably less than that of the non-AMS group. The third study (24), which involved 38 men, revealed similar findings, i.e., reduced numbers of T cells and concomitant depression in PHA-stimulated lymphocyte proliferation. Lymphocyte subset analysis revealed a reduced percentage of T-helper cells (TH) and an increased percentage of T-suppressor cells (Ts), which may have contributed to the observed reduction in T-cell proliferation in vitro.

Depressed T-cell proliferation in vitro also was documented in 18 men who ascended from 760 to 3200 m in the Tien Shan mountains (25). By 3–5 days at 3200 m, both the absolute numbers and percentages of T cells were reduced relative to measurements made at 760 m. Both spontaneous and PHA-stimulated transformations also were reduced at 3–5 days at high altitude, but ConA-stimulated cultures were no different from baseline values. All of these assays, except for unstimulated transformations, had returned to baseline values by days 25–30 at altitude. Plasma IgM and IgG levels dropped below baseline values at 3–5 days at altitude, but rebounded by 25–30 days at altitude, with IgM values exceeding baseline values at that time.

In another study, cell-mediated immune responses in 56 subjects with atopic dermatitis were evaluated before and after 4.5 weeks of exposure to 1560 m to evaluate the potentially beneficial effect of "climatotherapy" (26). Before treatment, anergy was present in 9.1% of the 22 men (normal prevalence 0.4%) and 8.8% of the 34 women (normal prevalence 4%); mean induration scores also were lower in the test group than in normal controls (7.5 vs. 21.0 mm for men and 8.2 vs. 15.4 mm for women). After treatment, the number of positive skin responses to 7 recall antigens (tetanus, diphtheria, streptococci, tuberculin, candida, trichophyton, and proteus) increased from 2.0 to 2.6 ($p < 0.001$), as did the induration scores (29% increase for the men and 22% for the women).

Biselli et al. immunized two groups of men with meningococcal vaccine A & C: 18 control subjects at sea level, and 18 men who had spent 20 days at 4930 m (27). Titers of antibodies to eight polysaccharide antigens measured by ELISA 18 days after the vaccination revealed no difference in antibody titers between groups.

Other more recent studies have involved assessment of cytokine production and lymphocyte subpopulations in association with hypoxia and exercise. Uchakin and others (28) measured IL-2, IFN-gamma, IL-10, TNF-alpha, and sIL-2R production in nine male runners at sea level and at 2500 m. Whole blood was used to assay mitogen-stimulated cytokine production and mitogen-stimulated release of sIL-2R by white blood cells. IL-2 was increased and TNF-alpha decreased after 4 weeks at altitude; IFN-gamma, IL-10, and sIL-2R were unchanged. Gabriel and others (29) used flow cytometry to assess peripheral-blood mononuclear cell subsets before and after exercise under normoxic and hypoxic conditions. (Hypoxic conditions were created by having the subjects breathe an F_{IO_2} of 14% to simulate a 3000 m altitude.) During exercise in hypoxic conditions, granulocyte numbers did not increase, nor were differences found in the numbers of monocytes, total lymphocytes, or the subpopulations CD4+CD45RO−, CD4+CD45RO+, CD8+CD45RO−, CD8+CD45RO+, CD3HLADR+, CD3−CD16/CD56+, CD3+CD16/CD56+, or CD19+. A summary of the results of studies conducted at high altitude is provided in Table 1.

B. Decompression Chamber Studies

Many other investigations have involved using decompression chambers to simulate high-altitude hypoxia. Novikov et al. found that several aspects of immune function declined among 12 men who were exposed daily to 30 minutes of decompression to simulate an altitude of 3500 m (30). Reductions in ConA-activated lymphocyte proliferation were accompanied by decreases in absolute numbers of circulating total leukocytes, neutrophils, monocytes, and CD4+ and CD8+ lymphocytes. The numbers of B cells expressing surface IgM were elevated, serum immunoglobulin levels were unchanged, and C3 complement levels were increased.

Klokker et al. exposed eight men to 20 minutes at 380 torr (5486 m) in a crossover study design with and without supplemental oxygen (31). This brief hypoxic exposure was sufficient to elevate the numbers of NK cells, monocytes, and CD8+ lymphocytes. Substantial increases also were noted in spontaneous and augmented NK-cell activity and neutrophil chemiluminescence response during hypoxia. On the other hand, neither T-cell proliferation nor IL-2 or IL-1β secretion in response to PHA or PPD stimulation were affected by hypoxia.

In another short-duration chamber study, 5 hours of exposure to 5100 m was sufficient to increase the numbers of circulating granulocytes, T cells, and B cells; the numbers of ''null'' cells, however, were reduced (32). Seven days after the exposure, all values had returned to baseline except for spontaneous lymphocyte proliferation and numbers of T cells, both of which remained elevated.

Operation Everest II provided the opportunity to measure several immune variables in seven men during a 6-week period of progressive hypobaric hypoxia in a decompression chamber (33,34). Findings at sea level were compared to those at simulated altitudes of 2286 and 7620 m. For the most part, immunological findings were similar to results from the Russian high-altitude studies. Relative to sea level and 2886 m values, the T-cell proliferation response at 7620 m was reduced, as measured by PHA-stimulated thymidine uptake at 72 hours and by protein synthesis at 24 hours of culture. B-cell function, assessed in vitro by the production of IgG, IgA, and IgM during PWM stimulation, was unaffected. However, plasma levels of IgM and IgA—but not IgG—were increased at 7620 m. No increases were seen at high altitude in plasma levels of autoantibodies against nuclear antigens, in vitro interferon production, or NK cytotoxicity. Mucosal immunity also was unimpaired at high altitude, as deduced from normal levels of nasal-wash IgA and lysozyme levels. Flow cytometry analysis of peripheral-blood mononuclear cells revealed an increase in the number of monocytes but no changes in the numbers of B cells, T cells, or T-cell subsets.

IV. Discussion

Inconsistencies in the findings from the studies described above undoubtedly reflect the considerable variation among study protocols as well as the difficulties inherent in conducting immunological assessments under field conditions (35). Sources of variation among studies include the duration of hypoxia exposure and the altitude studied, interassay and intersubject variability, lack of suitable control subjects, differences in immunological assays and methods, timing of sample collections, and the presence of concomitant environmental stressors such as altitude illness or physical exertion. Moreover, nearly all of the studies conducted thus far involved only adult male subjects; thus, caution is needed in extrapolating these results to women or children.

Nonetheless, some consistent trends have emerged from these diverse studies. For one, the human immunological response to altitude-induced hypoxia is quite similar to the response to other environmental stressors, including military training, space flight, and Antarctic isolation (17,36–39). Most of the stress-immunology studies indicate that B-cell functions are relatively unaffected; for example, the immune response to immunization has been largely normal (22,27), as are numbers of circulating B cells (23–25,33) and B-cell function in vitro (21,33). Variable changes in immunoglobulin levels (22,23,25,30,33) may reflect a dilution effect caused by intravascular fluid shifts, which are common at high altitudes (6). T-cell function, in contrast, seems to be impaired more readily during hypoxic stress than does B-cell function, especially as reflected by in vitro T-cell assays (22–25,33). Impaired T-cell function in vivo (as determined by delayed hypersensitivity skin responses) has been reported during space flight (39) and during winter over-isolation in Antarctica (36). Unfortunately, Drosner's study (26) did not include a control group without preexisting immunomodulatory defects. Impaired mitogenic responses of T cells also have been found during military training and after space flight (17,37). More variable changes among other peripheral-blood mononuclear cell populations have been reported under these circumstances as well (17,38).

The significance of reductions in in vitro T-cell function in healthy humans is unclear. In disease states, depressed in vitro T-cell function correlates with adverse clinical outcomes due to infectious complications (40,41). However, an assessment of 96 U.S. Air Force Academy cadets during basic training revealed no significant associations between the magnitude of depressed in vitro T-cell proliferation and susceptibility to viral illnesses (17). In our opinion, a large, well-controlled prospective study is needed if correlations are to be established between depressions of in vivo or in vitro immune responses and susceptibility to infectious illness. Such a study could involve introducing an infectious agent into an environment at high altitude and assessing the immune response of large numbers of susceptible subjects.

The observed immunomodulatory in vivo and in vitro effects of hypoxic stress may well be mediated by the neuroendocrine system (Fig. 2). As reviewed elsewhere in this volume, several stress hormones that influence the immune system are elevated during hypoxia stress. However, the observed immunological findings probably are not mediated exclusively by a glucocorticoid effect, since similar changes have been observed in the absence of elevated cortisol levels after space flight and among Air Force Academy cadets (17,33,38,42). Catecholamine levels, on the other hand, vary widely during and after space flight and may remain elevated for up to 3 days after landing (38,42).

Numerous well-designed murine in vivo and human in vitro studies have established that the central nervous system and immune system communicate via shared effector molecules (cytokines and neuropeptides) and common receptors (43–47). Physiological concentrations of glucocorticoid, catecholamine, and neuropeptide hormones all exhibit immunomodulatory activity both in vivo and in vitro. The specific immunological response depends on the biochemical activator and the target cell being studied (46,48–50). When stimulated in vitro, lymphocytes produce a variety of neuropeptides, including ACTH, GH, prolactin, TSH, and HCGH, all

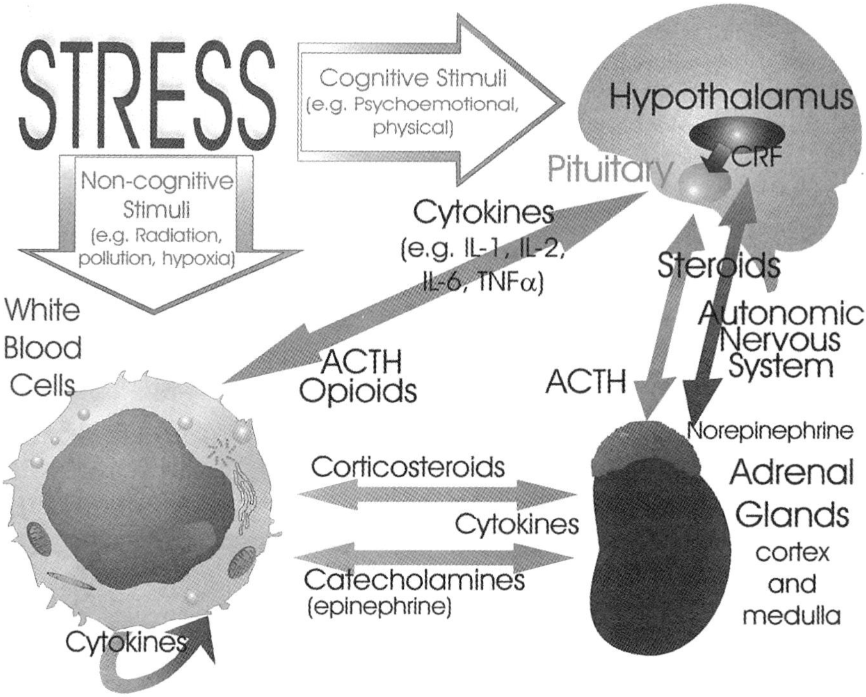

Figure 2 Interactions among the neuroendocrine and immune systems under stress.

of which provide autocrine and paracrine influence on immune cells within the microenvironment of an immune response (51).

V. Conclusions

This review represents an attempt to identify the possible influence of hypobaric hypoxia on human immune function. Studies of humans, of course, will always involve constraints (35), which may well confound interpretation of results. Epidemiological studies of infectious illness in the field, which often include symptom surveys, can be misleading because of the difficulty in differentiating symptoms caused by infectious agents from symptoms caused by nonspecific irritation of mucous membranes due to environmental factors. Moreover, symptoms of altitude illness can overlap those of infectious illness. Many of the nonhypoxic stressors that exist in the mountaineering environment also can modulate the immune system, possibly via neuroendocrine-mediated mechanisms such as disrupted circadian cycles (52), sleep deprivation (53), exhaustion (54), fear, isolation, interpersonal conflicts, dehydration, and inadequate caloric intake (55–59).

Future research protocols designed to isolate the effects of hypoxia per se from the other environmental stressors that can modulate immune responses are best conducted in a controlled environment such as that used during Operation Everest II (33,34). That kind of experimental environment also allows sophisticated in vivo and in vitro immunology studies to be conducted without delays in sample processing. Crossover designs such as that used by Klokker et al. (31) are especially valuable for studies with small numbers of subjects, particularly because in vitro cell-function assays involve considerable interassay and intersubject variability. Improved study designs in the future are needed to definitively answer the following questions:

1. Does hypoxia per se directly modulate the immune system, or do all of the observed changes result exclusively from neuroendocrine-mediated effects?
2. Are the observed alterations in immune effector-cell function clinically significant, i.e., is host resistance compromised?
3. Do immune mechanisms directly increase susceptibility to or aggravate AMS, HAPE, or HACE?

Future progress in cryopreservation techniques for peripheral-blood mononuclear cells and the use of DNA probes to identify infectious agents and specific markers of cell function will allow more precise studies to be conducted in the future. Results from these types of studies will be useful in elucidating the precise mechanisms for neuroendocrine-mediated, stress-induced immune modulation. As additional specific hematopoietic growth factors become available for clinical use, it should be possible to restore depressed immune competence caused by environmental stressors such as hypoxia, provided that specific cellular defects are identified (60).

From a practical standpoint, the studies conducted to date indicate that exposure to high altitudes does not severely depress the human immune response. Indeed, some findings suggest that high altitude may have beneficial effects on some immunological disorders, including reactive airway disease, immune thrombocytopenia, and atopic dermatitis (26,61). Nevertheless, precautions are still appropriate for those who travel to high altitudes. Immunizations before ascent and attention to personal hygiene in the field would do much to reduce the spread of infectious agents (62,63). Since respiratory infection may increase susceptibility to more severe forms of altitude illness, climbers with infectious illness should exercise caution if they elect to continue ascending. The growth in the trekking industry (60,000 people visit Nepal every year) means that the trekker population will expand to include older individuals with age-related immune suppression and people with preexisting diseases that compromise the immune system. Persons infected with HIV, those who have organ transplants or malignancies, and those who receive corticosteroids or other immunosuppressive agents may be especially vulnerable to infectious complications at high altitudes. People who take high doses of dexamethasone to prevent or treat AMS may also be immune-compromised, especially with regard to novel infectious agents (64,65). Finally, the emergence of antibiotic-resistant microorganisms such as *Enterococcus*, *Mycobacterium tuberculosis*, *Haemophilus influenzae*, *S. aureus*, *Streptococcus pneumoniae*, and *Shigella* sp. requires particular caution for potentially immune-compromised individuals (66).

Acknowledgments

The authors thank Ms. Christine Wogan of Wyle Life Sciences for her significant contributions to the style and content of this chapter.

References

1. Ward M. Man and the mountain environment. In: Mountain Medicine: A Clinical Study in Cold and High Altitude. London: Crosby Lockwood Staples Ltd, 1975:28–47.
2. Hultgren H, Spickard W, Lopez E. Further studies of high altitude pulmonary edema. Br Heart J 1962; 24:95–102.
3. Hultgren HN. High altitude medical problems. West J Med 1979; 131:8–23.
4. Schoene RB, Hackett PH, Henderson WR, Sage EH, Chow M, Roach RC, Mills WJ, Martin TR. High-altitude pulmonary edema characteristics of lung lavage fluid. JAMA 1986; 256:63–69.
5. Dinarello CA, Mier JW. Lymphokines. N Engl J Med 1987; 317:940–945.
6. Meehan RT, Zavala DC. The pathophysiology of acute high altitude illness. Am J Med 1982; 73:108–111.
7. Carpenter TD, Reeves JT, Durmowicz AG. Viral respiratory infection increases susceptibility of young rats to hypoxia-induced pulmonary edema. J Appl Physiol 1998; 84(3): 1048–1054.
8. Kaminisky DA, Jones BS, Schoene RD, Voelkel NF. Urinary leukotriene E4 levels in high-altitude pulmonary edema: a possible role for inflammation. Chest 1996; 110:939–945.
9. Durmowicz AG, Noordewein E, Nicholas R, Reeves JT. Inflammatory process may predispose children to high-altitude pulmonary edema. J Pediatr 1997; 130:838–840.
10. Singh I, Chohan IS, Lal M, Khanna PK, Srivastava MC, Nanda RB, Lamba JS, Malhotra MS. Effects of high altitude stay on the incidence of common diseases in man. Int J Biometeorol 1977; 21:93–122.
11. Krupina TK, Pukhova MM, Tsyganova IaI, Reutova MB. [Characteristics of the human immunological state during hypoxic hypoxia.] Kosm Biol Aviakosm Med 1974; 8:56–60 (in Russian).
12. Murdoch DR. Symptoms of infection on altitude illness among hikers in the Mount Everest region of Nepal. Aviat Space Environ Med 1995; 66:148–151.
13. Shlim DR, Gallie J. The causes of death among trekkers in Nepal. Int J Sports Med 1992; 13:374–S76.
14. Mirrakhimov MM, Kitayev MI. Problems and prospects of high altitude immunology. Vestn Akad Med Nauk SSSR 1979; 4:64–69 (in Russian).
15. Clegg EJ, Harrison GA, Baker PT. The impact of high altitude on human populations. Hum Biol 1970; 42:486–518.

16. de Meer K, Heymans HSA, Zijlstra WG. Physical adaptation of children to life at high altitude. Eur J Pediatr 1995; 154:263–272.
17. Lee DJ, Meehan RT, Robinson C, Mabry TR, Smith ML. Immune responsiveness and risk of illness in U.S. Air Force Academy cadets during basic cadet training. Aviat Space Environ Med 1991; 63:517–523.
18. Sawyer R, Sommerville RG. An outbreak of mycoplasma pneumoniae infection in a nuclear submarine. JAMA 1966; 195:958–959.
19. Nicogossian AE, Pool SL. Medical care and health maintenance in flight. In: Nicogossian AE, Huntoon CL, Pool SL, eds. Space Physiology and Medicine, 2d ed. Philadelphia: Lea and Febiger, 1989:349–363.
20. Muchmoore HG, Parkinson AJ, Scott EN, Scott LV, Miles JAR. Respiratory virus infection late in isolation at the South Pole. Antarct J 1978; 13:171–172.
21. Krupina TK, Kerotaev MM, Pukhova VaI, Tsyganova NI, Lichacheva NP, Reutova MB. [Comparative evaluation of studies of the action of different levels of hypoxia on the human immunobiological status.] Kosm Biol Aviakosm Med 1977; 11:38–43 (in Russian).
22. Mirrakhimov MM, Tulebekov BT, Kitaev MI, Amanturova KA, Saburov KA. Immunophysiological aspects of human adaptation to life at high altitude. Fiziologia Cheloveka 1979; 5:300–305.
23. Mirrakhimov MK, Kitaev MI, Khamzamulin Tokhtabaev AG, Pogrebitskiy SM. [T and B components of immunity in the presence of acute mountain sickness.] Kosm Biol Aviakosm Med 1986; 29(1):53–56 (in Russian).
24. Mirrakhimov MK, Kitaev MI, Tokhtabaev AG. [The human immunocompetent system in adaptation to high-altitude hypoxia.] Fiziologia Cheloveka 1987; 13:265–69.
25. Kitaev MI, Tokhtabaev AG. [T and B lymphocytes in adaptation to high altitude.] Kosm Biol Aviakosm Med 1981; 15(3):87–89.
26. Drossner M. Klimatherapie und Änderung der zeilularen Immunität [Climate therapy and change in cellular immunity]. Z Hautkr 1988; 63:104–107 (in German).
27. Biselli RL, Le Moli S, Matricardi PM, Farrace S, Fattorossi A, Nisini A, D'Amelio R. The effects of hypobaric hypoxia on specific B cell responses following immunization in mice and humans. Aviat Space Environ Med 1991; 62:870–874.
28. Uchakin PN, Gotovtseva EP, Levine B, Stray-Gundersen J. Neuroimmunohumoral changes associated with altitude training (abstr). Med Sci Sports Exercise 1995; 27(suppl):174.
29. Gabriel H, Kullmer T, Schwarz L, Urhausen A, Weiler B, Born P, Kindermann W. Circulating leucocyte subpopulations in sedantary subjects following graded maximal exercise with hypoxia. Eur J Appl Physiol 1993; 67:348–353.
30. Novikov VS, Goranchuk VV, Smirnov VS, Lustin SI. [Immunophysiologic mechanisms of hypoxia.] Fiziologia Cheloveka 1993; 19:104–113 (in Russian).
31. Klokker M, Kharazmi A, Galbo H, Bygbjerg I, Pedersen BK. Influence of in vivo hypobaric hypoxia on function of lymphocytes, neutrocytes, natural killer cells and cytokines. J Appl Physiol 1993; 74:1100–1106.
32. Tulebekov BP, Vorob'ev TA, Iseev GF, Chadov VI. [Effect of acute hypoxic hypoxia on the immune system, hemostasis and acid-base state of the blood.] Fiziologia Cheloveka 1988; 14:115–122.
33. Meehan RT, Duncan U, Neale L, Taylor G, Muchmore H, Scott N, Ramsey K, Smith E, Rock P, Goldblum R, Houston C. Operation Everest II: alterations in the immune system at high altitudes. J Clin Immunol 1988; 8:397–406.

34. Houston CS, Sutton JR, Cymerman A, Reeves JT. Operation Everest II: man at extreme altitude. J Appl Physiol 1987; 63:877–882.
35. Meehan RT. Immune suppression at high altitude. Ann Emerg Med 1987; 16:974–979.
36. Williams DL, Climie A, Muller HK. Cell-mediated immunity in healthy adults in Antarctica and the sub-Antarctic. J Clin Lab Immunol 1986; 20:43–49.
37. Taylor GT, Dardano JR. Human cellular immune responsiveness following space flight. Aviat Space Environ Med 1983; 54:S55–S59.
38. Meehan RT, Whitson P, Sams C. The role of psychoneuroendocrine factors on spaceflight-induced immunological alterations. J Leukoc Biol 1993; 54:236–244.
39. Taylor GR, Janney RP. In vivo testing confirms a blunting of the human cell-mediated immune mechanism during space flight. J Leukoc Biol 1992; 51:129–132.
40. Faist E, Kupper TS, Baker CC, Chaudry IH, Dwyer J, Baue AE. Depression of cellular immunity after major injury. Arch Surg 1986; 121:1000–1005.
41. O'Gorman RB, Feliciano DV, Mathews KS, Matthew R, Bitondo CG, Mattox KL, Jordan GL. Correlation of immunologic and nutritional states with infectious complications after major abdominal surgery. Surgery 1986; 99:549–555.
42. Leach CS. Biochemical and hematological changes after short term space flight. Microgravity Q 1992; 2:69–75.
43. Besedovsky HO, del Rey A, Sorkin E, Da Prada M, Miller HH. Immunoregulation mediated by the sympathetic nervous system. Cell Immunol 1979; 48:346–355.
44. Besedovsky HO, Sorkin E, Keller M, Mueller J. Changes in blood hormone levels during the immune response. Proc Soc Exp Biol Med 1975; 150:466–470.
45. Besedovsky HO, del Rey A, Sorkin E. Regulatory immune-neuroendocrine feedback signals. In: Berczi I, ed. Pituitary Function and Immunity. Boca Raton, FL: CRC Press, 1986:242–247.
46. Blalock JE, Harbour-McMenaman D, Smith EM. Peptide hormones shared by the neuroendocrine and immunologic systems. J Immunol 1985; 135(suppl):858–861.
47. Carr DJ. Neuroendocrine peptide receptors on cells of the immune system. Chem Immunol 1992; 52:84–105.
48. Cupps TR, Fauci AS. Corticosteroid-mediated immunoregulation in man. Immunol Rev 1982; 65:133–155.
49. Kroc CJ, Nguyen TT, Go VL. Modulation of T cell functions by peptide hormones. Adv Exp Med Biol 1987; 225:183–188.
50. Lotz M, Vaughan JH, Carson DA. Effect of neuropeptides on production of inflammatory cytokines by human monocytes. Science 1988; 241:1218–1220.
51. Blalock JE, Bost KL, Smith EM. Neuroendocrine peptide hormones and their receptors in the immune system, production, processing and action. J Neuroimmunol 1985; 10: 31–40.
52. Abo TR, Kawate T, Itoh K, Kumagai K. Studies on the bioperiodicity of the immune response. J Immunol 1981; 126:1360–1361.
53. Palmblad J, Petrini B, Wasserman J, Akerstedt T. Lymphocyte and granulocyte reactions during sleep deprivation. Psychosom Med 1979; 41:273–278.
54. Gmunder FK, Lorenzi G, Bechler B, Joller P, Muller J, Ziegler W, Cogoli A. Effect of long-term physical exercise on lymphocyte reactivity: similarity to space flight reactions. Aviat Space Environ Med 1988; 59:146–151.
55. Ader R, Cohen N. Conditioned immunopharmacologic responses. In: Ader R, ed. Psychoneuroimmunology. New York: Academic Press, 1981:281–319.
56. Hillhouse JE, Kiecolt-Glaser JK, Glaser R. Stress-associated modulation of the immune

response in humans. In: Plotnikoff N, Murgo A, Faith R Wybran J, eds. Stress and Immunity. Boca Raton, FL: CRC Press, 1991:3–28.
57. Landmann RM, Muller FB, Perini CH, Wesp M, Erne P, Buhler FR. Changes of immunoregulatory cells induced by psychological and physical stress: relationship to plasma catecholamines. Clin Exp Immunol 1984; 58:127–135.
58. Calabrese JR, Kling MA, Gola PW. Alterations in immunocompetence during stress, bereavement and depression: focus on neuroendocrine regulation. Am J Psychiatry 1987; 144:1123–1134.
59. Selye H. Stress in Health and Disease. London: Butterworths, 1975.
60. Naparstek E. Granulocyte colony-stimulating factor, congenital neuropenia, and acute myeloid leukemia. N Engl J Med 1993; 333:487–493.
61. Gippenreiter E, West JB. High altitude medicine and physiology in the former Soviet Union. Aviat Space Environ Med 1996; 67:576–584.
62. Gardner P, Schaffner W. Immunization of adults. N Engl J Med 1993; 328:1252–1258.
63. Glaser R, Kiecolt-Glaser J, Bonneau RH, Malarkey W, Kennedy S, Hughes J. Stress-induced modulation of the immune response to recombinant hepatitis B vaccine. Psychosom Med 1992; 54:22–29.
64. Johnson TS, Rock PB. Current concepts in acute mountain sickness. N Engl J Med 1988; 319:841–845.
65. Levine BD, Yoshimura K, Kobayashi T, Fukushima M, Shibamoto T, Ueda G. Dexamethasone in the treatment of acute mountain sickness. N Engl J Med 1989; 1321: 1707–1713.
66. Iseman MD. Tailoring a time bomb: inadvertent genetic engineering. Am Rev Respir Dis 1985; 132(4):735–736.

20

Exercise and Hypoxia

Performance, Limits, and Training

ROBERT ROACH

New Mexico Highlands University
Las Vegas, New Mexico

BENGT KAYSER

University of Geneva
Geneva, Switzerland

Once more I must pull myself together. I can scarcely go on. . . . I consist only of will. After each few metres [of upward progress] this too fizzles out in an unending tiredness.
Reinhold Messner writing about his experiences a few meters from the top of Mt. Everest during his solo ascent without supplementary oxygen (1).

I. Introduction

On May 8, 1978, Reinhold Messner and Peter Habeler became the first humans to summit Mt. Everest without the use of supplementary oxygen, thus achieving what many physiologists had believed to be impossible because sudden exposure to air with an oxygen concentration as on the summit of Mt. Everest results in unconsciousness within minutes, and if continued, death (2). Indeed, it was well known that capacity for exercise performance is progressively limited with increasing altitude (Fig. 1). Since the first ascent "by fair means" in 1978, several others, includ-

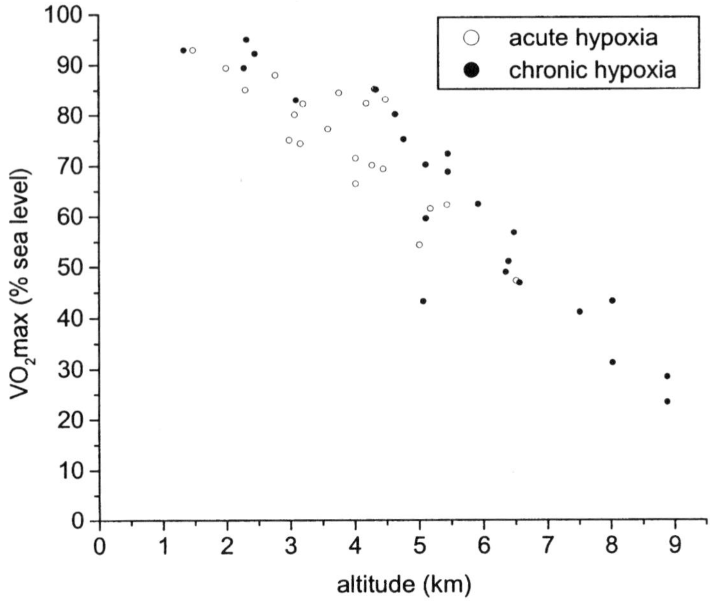

Figure 1 The decrease in $\dot{V}O_2max$ as a function of altitude expressed in % of sea level control. Open symbols are acute hypoxic exposure and closed symbols chronic exposure.

ing Messner himself in 1980, have repeated this extraordinary feat, but more have failed, and many have died while trying. Thus, with or without bottled oxygen, it remains extremely demanding to climb as high as Mt. Everest. We will consider the physiological mechanisms that allow lowlanders to climb to extreme altitude and describe the exercise responses of the millions of people who live and work permanently at moderate to high altitudes (see Chapter 3). We define moderate altitude as 1500–3500 m above sea level (P_{IO_2} 124–96 torr); high altitude as 3500–5500 m above sea level (P_{IO_2} 96–73 torr) and extreme altitude as 5500–8848 m or more above sea level (P_{IO_2} 73–43 torr).

In this chapter, we first present what is known about work in hypoxia, with an emphasis on factors that may contribute to limiting work at very high altitude. Next, we explore the complex metabolism of lactate and how it may be altered in hypoxia. These two sections culminate in a discussion of how humans can climb Mt. Everest without supplementary oxygen. Brief mention is then made of exercise performance in high-altitude native residents and of the use of altitude training to improve sea level athletic performance. The chapter finishes with a presentation of several areas rich with potential for future investigations.

II. Work in Hypoxia

Of the three major sources of energy for work (aerobic, alactic, and lactic), the aerobic pathway is the most important because on a long-term basis all alactic and lactic anaerobic metabolic oxygen deficits are always followed by an aerobic metabolic debt payment. While walking, running, or climbing virtually all energy necessary for work comes from the full oxidation of substrate. Thus, the performance capacity for endurance activities depends heavily on the maximum oxygen flux ($\dot{V}O_2max$) through the system from the mouth to the mitochondria. The $\dot{V}O_2max$ and the fraction of $\dot{V}O_2max$ that can be sustained for one or more hours (maximal sustainable $\dot{V}O_2$) are therefore important determinants of endurance exercise capacity (3).

A. Oxygen Transport

Oxygen transport from the inspired air to the site of oxidation in the mitochondria occurs along a series of steps known as the oxygen transport cascade. The first step is convective and involves alveolar ventilation, the second is diffusional from the alveolar space through the gas-blood barrier and intererythrocyte plasma to the intraerythrocyte hemoglobin, and the third is again convective and involves cardiovascular blood transport. The fourth and last step is diffusion of oxygen from the capillary to the myocyte and into the oxygen sink of the mitochondria.

Lungs

The first step in the oxygen cascade, transport from ambient air to the alveolar space, depends on ventilation. At sea level exercise hyperpnea allows maintenance of a sufficiently high alveolar oxygen tension (P_{AO_2}) for the next diffusive step over the alveolar membrane. At high altitude, the partial pressure of inspired oxygen (P_{IO_2}) is lower than at sea level so that P_{AO_2} falls, leading to a drop in arterial P_{O_2} (Pa_{O_2}). This hypoxemia triggers carotid chemoreceptor firing, leading to a rise of ventilation in an attempt to elevate P_{AO_2} and Pa_{O_2}. At moderate altitude, during rest through moderate exercise a rise in alveolar ventilation can maintain Pa_{O_2} despite the drop in P_{IO_2}. However, at heavier workloads and higher altitudes a dilemma may develop. Although one would expect that *any* increase in alveolar ventilation would be beneficial, the energy costs of the increased work of breathing above 60–70% $\dot{V}O_2max$ may balance any gain in oxygen availability. If at sea level this does not seem to have a relevant effect (4), this trade-off may occur as low as 500 m (5,6). Another interesting basic physiological phenomenon occurs at the pulmonary blood-gas barrier (7). Since diffusion depends on the thickness of the blood-gas barrier, the barrier needs to be as thin as possible to allow for efficient gas exchange. On the other hand, it must also be able to withstand the high stresses that ensue when the pulmonary capillary pressure rises on exercise and when flow rates are high. An increase in thickness of the blood-gas barrier will decrease diffusional oxygen transport. A third dilemma involves maximum cardiac output during exercise at high altitude. The Fick equation [$\dot{V}O_2$ = cardiac output (CaO_2—$C\bar{v}O_2$)] says that at a given arterial oxygen content (CaO_2), in order to keep $\dot{V}O_2$max as close as possible to that achieved at sea level, one has to keep cardiac output as high as possible (in the face of unchanging oxygen extraction). However, since the pressure gradient across the blood-gas barrier is reduced, a shortened blood transit time as a consequence of increased cardiac output may not allow time for equilibration of the pulmonary capillary blood oxygen partial pressure with that in the alveoli. Highly trained athletes who have high cardiac outputs and therefore short pulmonary capillary transit times show considerably more arterial desaturation upon hypoxic exercise compared to untrained subjects (5,8,9). With acclimatization hemoglobin concentration typically rises, increasing CaO_2 and allowing a lower cardiac output for a given $\dot{V}O_2$, which ameliorates to some extent the effects of arterial oxygen desaturation. Hemoglobin concentration, blood viscosity, and O_2 affinity of hemoglobin all can affect oxygen uptake (see Chapter 16).

The Cardiovascular System

Perhaps the most important observation about the heart at high altitude is that its function is well preserved (10,11). Cardiovascular regulation certainly changes with exposure to hypoxia, but contrary to what one would expect for such an aerobic organ, known to suffer greatly from acute focal ischemia, the heart performs well even at the summit of Mt. Everest, as was shown by the invasive measurements obtained during Operation Everest 2 (OEII) (10). Altered cardiovascular regulation during hypoxic exercise was first described by Christensen and Forbes in 1936 (12). They observed a decline of heart rate during maximal exercise that was more pronounced at higher altitudes or with more prolonged exposure (12). The finding of a lower peak heart rate in chronic hypoxia has been consistently confirmed (13–18). In contrast, during acute hypoxia some studies show no depression of peak heart rate (13,19–21), while others show a significant decrease (22–25). Why peak heart rate falls with sustained hypoxia and why the response to acute hypoxia is so variable is not known. One explanation for the individual variability may be different training status between studies. A high degree of physical fitness at sea level is associated with a larger decrement in peak heart rate and $\dot{V}O_2$max with hypoxia (22,25–27). Another possible contributor to the fall in peak heart rate during exercise in hypoxia is an alteration in autonomic balance as discussed below. During submaximal exercise in hypoxia the heart rate is elevated above levels seen for the same workload at sea level (10).

Initially, stroke volume is slightly lower during submaximal and maximal exercise at altitudes from 3000 to 5000 m (PIO_2 103–78 torr) and returns toward but not to sea level values with acclimatization (28). The decrease in myocardial work as a consequence of the drop in peak heart rate and stroke volume could, in theory, protect the myocardium from extreme hypoxemia (29). However, myocardial blood flow rises markedly during exercise in severe, acute hypoxia and the rise resulting in an increase in oxygen delivery that nearly compensates for the hypoxemia (at least down to an F_{IO_2} of 0.10), such that the myocardium appears well protected (30). Another possible mechanism for the reduction of stroke volume early in acclimatization is the reduction in total blood volume due to the contraction of the plasma volume. However, acute expansion of plasma volume with dextran infusion in two subjects exercising at 3100 m resulted in no increase in stroke volume (31). Subsequent echocardiographic studies done at the same altitude ruled out left ventricular dysfunction as a possible cause of the lower stroke volume (32). β-Receptor densensitization or downregulation is another possible explanation for the initial drop in stroke volume with hypoxia and has been invoked to explain the drop in peak heart rate. This possibility will be discussed subsequently.

During submaximal exercise at altitude, cardiac output initially may be similar to sea level values (33) or increased (18,34,35), but all reports confirm a fall in cardiac output from normoxic values during peak exercise (13,18,36). After 2 weeks at 4300 m, peak cardiac output was 22% lower than sea level values (13). Similar findings were reported at higher altitudes from field (18) and chamber (36) studies. In four men who spent 4–7 months at 4600 m or above, including 2–3 months at 5800 m, Pugh et al. observed a 30% decrease in peak cardiac output from 23 L/min at sea level to 16 L/min (18). In OEII, near peak cardiac output was 24 L/min at sea level, 20 L/min at a P_{IO_2} of 63 torr, and 16 L/min at the ''summit'' (P_{IO_2} = 43 torr) (36). The cardiac output response to hypoxia during submaximal exercise is largely determined by increased heart rate and a decrease at peak effort, with only a minor influence of stroke volume.

Mean arterial pressure during exercise with acute hypoxia is similar to normoxia control values (37). In contrast, after 3 weeks of acclimatization to 4300 m, mean arterial pressure at rest and exercise was about 20 torr higher than sea level or acute hypoxia values. Whether mean arterial pressure is maintained at maximal exhaustive work with chronic hypoxia is not settled. Some authors report a decrease in mean arterial pressure at maximal exercise (15,19), while others report a slight increase (38,39). In OEII, mean arterial pressure was slightly elevated from 112 mmHg at sea level to 128 mmHg at the ''summit'' (P_{IO_2} = 43 torr) during near maximal effort (97% $\dot{V}O_2max$) (38,39).

Sympathetic Regulation

During exercise in acute hypoxia, epinephrine levels are uniformly elevated (22,24,40). After several days to weeks of hypoxia, adrenal medullary secretion of epinephrine is attenuated (40). This attenuation of epinephrine secretion is coincident with a rise in norepinephrine to levels significantly above sea level, beginning after several days and continuing through at least several weeks of hypoxia (40–42). Cunningham et al. were the first to report adrenergic upregulation by hypoxia in humans (41). They noted a 22% increase in plasma norepinephrine from sea level after 12 days at 4560 m (P_{IO_2} = 83). In another study in five men studied during steady-state submaximal exercise after 21 days at 4300 m, arterial norepinephrine rose 174% from sea level values (40). After 7 days at 6542 m (P_{IO_2} = 62 torr) norepinephrine levels increased three- to fourfold during rest and submaximal exercise (43). The picture is less clear when considering more prolonged exposure to severe hypoxia. For example, 3 weeks at 6542 m resulted in a fall in norepinephrine values, particularly during exercise, although the values were still above those seen in normoxia (43). In addition, OEII subjects spent almost a week at a pressure of about 280 torr (P_{IO_2} = 49 torr) while intermittently going to higher altitudes. Their resting norepinephrine values at the start of their stay rose threefold from sea level values to 630 pg/mL (44) but dropped to 351 pg/mL after several days at 280 torr. Together these data suggest an initial increase followed by a marked attenuation of sympathetic activation during hypoxic exposure lasting for several weeks (43–45). Whether these changes would be sustained after yet more prolonged exposure to hypoxia is not known.

β-Receptor downregulation is well documented at sea level in response to sustained sympathetic activation (46,47). When subjects exercise for 90 minutes at a high submaximal workload and then undergo a β-receptor agonist challenge, significant β-receptor desensitization is revealed (48), presumably due to the elevation of circulating norepinephrine levels. Similar mechanisms seem to be at play during acclimatization to hypoxia. β-Receptor downregulation has been noted after several days to weeks of hypoxia in animals (49–52) and humans (43). In humans, marked insensitivity to β-agonist has also been described (29,43,53–55). The picture during chronic hypoxia could then be a combination of β-receptor downregulation and a fall in sympathetic outflow, resulting in the observed drop in peak heart rate and the slight fall or lack of change in stroke volume.

Parasympathetic Regulation

Evidence in humans of parasympathetic withdrawal during normoxic exercise is indirect and largely based on the absence of a change in peak heart rate when atropine, a muscarinic acetylcholine antagonist, is given. Marked sympathetic activation in normoxic exercise only occurs after heart rate has reached ~150 beats per minute. Thus, at sea level, parasympathetic withdrawal and sympathetic activation appear as distinct phenomena. By contrast, marked elevations in peak heart rate at high altitude after atropine administration have been consistently observed (14,56–58). In five men studied before and after several days acclimatization to 4600 m, peak heart rate fell 14 bpm from sea level values. After parasympathetic blockade with atropine peak heart rate increased an average of 11 bpm, while $\dot{V}O_2$ was unchanged (59) (see Fig. 2). Similar results were obtained in climbers tested at 5300 m on Mt. Everest (57). These data suggest a high parasympathetic tone during exercise in acclimatized individuals. On the other hand, elevated adrenergic tone in chronic hypoxia might augment the effect of parasympathetic activation. With increased sympathetic tone, the same degree of vagal nerve stimulation caused a much larger drop in heart rate than in the absence of such tone (60,61). In long-term hypoxia when sympathetic tone has returned to near sea level values, cardiac parasympathetic activation may not be altered; rather its effects may be unveiled because of the reduced sympathetic tone and lower absolute heart rate compared to normoxia (57).

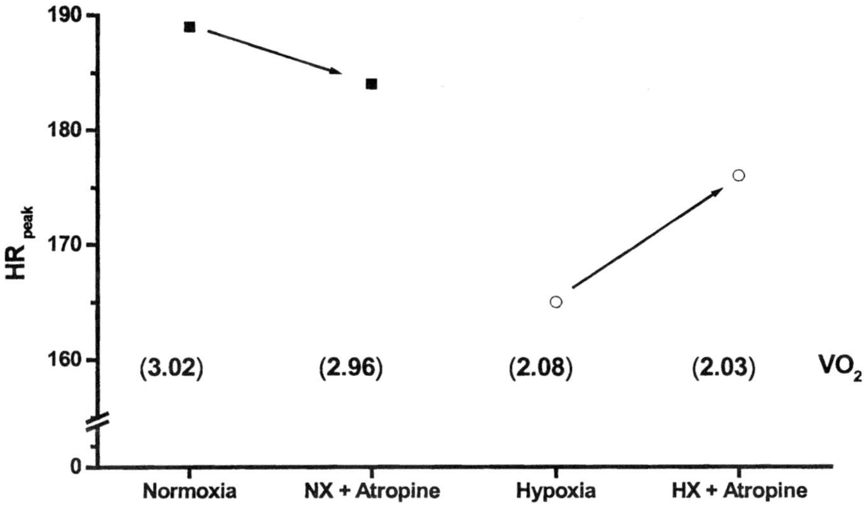

Figure 2 Increased peak exercise heart rate after atropine infusion at high altitude in five men in normoxia and ambient hypoxia at 4600 m. (Adapted from Ref. 14.)

In summary, during the early days of altitude acclimatization, elevated cardiac parasympathetic and sympathetic tone determine the cardiovascular responses to exercise. The picture during chronic hypoxia, lasting several weeks to months, is less clear. In this setting, sympathetic tone may drop to near sea level values, while the state of parasympathetic tone in acclimatizing lowlanders is not known.

Tissue

The last step in the oxygen transport chain involves the delivery of oxygen to cells. Oxygen, unloaded from hemoglobin, diffuses through plasma and the capillary wall into the myocyte, then via myoglobin to the mitochondria. Since oxygen delivery is the product of Cao_2 and blood flow, any strategy that increases Cao_2 with the same blood flow will increase the oxygen available for diffusion into the muscle cell. Initially, plasma volume contracts in response to hypoxia (31,33,62,63), with only a minimal rise over 2 months at the same altitude [4300 m (63)]. This initial 10–15% reduction in plasma volume results in an increased hemoglobin concentration and hence Cao_2 (64). An actual increase in total red cell mass occurs after several weeks of hypoxia, secondary to the stimulation of erythropoiesis by hypoxia (65,66). Finally, muscle oxidative capacity and structure play a role in delivery of oxygen for muscular exercise.

Muscle Oxidative Capacity

In animals, chronic hypoxia increases the capacity for muscle to oxidize substrate used for fuel, known as the muscle oxidative capacity (67). In humans, the opposite seems to occur. During OEII, after 33 days of progressive decompression to 282 torr, no changes in aerobic enzyme activity in vastus lateralis muscle biopsies were seen (68). In fact, a significant decrease was found (−25%) during the subsequent 7 days of exposure to 282 torr. This finding was confirmed on subjects exposed to ''real'' altitudes between 5000 and 8000 m for 2 months (69,70). The tricarboxylic acid cycle enzyme citrate synthase and the respiratory chain enzyme cytochrome *c* oxidase were reduced by −23%, and these reductions correlated to a morphometrically measured drop in total mitochondrial volume density of the muscle fibers (−19%). The observed changes in mitochondrial enzyme activities were therefore interpreted to be due to a loss of mitochondrial volume rather than to qualitative changes of the mitochondria per se.

Muscle Loss, Body Composition, and Nutrition

Prolonged altitude exposure often leads to a marked drop in body mass (71–75), including a performance diminishing fall in muscle mass (up to 70% of total body mass lost may be due to muscle wasting) (76–79). Initially, inhospitality of life in high mountains was blamed for the cachexia of high altitude, but recent studies with ad lib food intake in an environmental chamber showed losses similar to those seen in the field (79). Similarly, chronic hypoxia secondary to pulmonary disease is associated with loss of body mass (80). Thus, hypoxia per se plays a major role in weight loss at high altitude. After 8 months living above 5000 m, members of the Silver Hut expedition reported weight losses of 6–9 kg (75). Similar losses of body weight were reported from OEII (~7 kg) (79). The loss of body mass is explained largely by increased energy expenditure coupled with anorexia (74). The cause of greater energy expenditure at high altitude is a rise in basal metabolic rate (81) and, in the field setting, an increase in activity. The reasons for loss of appetite at high altitude are unknown. Leptin plays a key role in the neuroendocrine regulation of energy metabolism and appetite at sea level (82). A recent preliminary report showing a corresponding rise in serum leptin levels and symptoms of anorexia during acute exposure to high altitude (~4500 m) suggests a viable explanation for the anorexia at high altitude (83). Moreover, the suggestion that elevated leptin levels in some individuals prior to ascent correlated with those who subsequently became anorexic suggests that an inherent characteristic may determine anorexia, and perhaps weight loss, at high altitude. A reduction in body water content and possibly some malabsorption play minor roles (84,85). The reduced intake can be partly explained by changes in food availability and living conditions. At least up to 5050 m, the loss of body mass can be largely avoided, provided there is free access to a large choice of palatable food and that comfortable, warm lodging is available (86). Using this regime in a field setting in very active young adults, field measures of body composition and muscular strength suggest no loss of muscle mass. In addition, a diet with an energy content known to be adequate for the expenditure (81) will also contribute to maintenance of body mass. Whether muscle mass is also maintained using the latter regime, in relatively sedentary research volunteers, is not known.

The muscle loss is characterized by a ~25% drop in muscle fiber size (76,77), which is of similar magnitude for type I and II fibers (78). It may be accompanied by a decrease in thigh muscle cross-sectional area of up to 15%, as seen after a 2-month period at altitudes above 5500 m (76–78). This reduction in muscle fiber size will cause a drop in the force-generating capacity of the muscles involved (87), without affecting the force per cross-sectional area ratio. Although the loss in muscle mass at altitude mainly results from malnutrition, inactivity or detraining may also play a role. Maximal exercise capacity and therefore aerobic training intensity is reduced in hypoxia. On typical climbing expeditions to the Himalaya, a considerable amount of time is spent at base camp with little physical activity. Climbers, usually well trained before leaving for a mountaineering expedition, may thus undergo muscle atrophy at altitude due to a relative lack of exercise. Nevertheless, such an effect of detraining would not explain all of the observed muscle atrophy since only after 6 weeks of complete bed rest does thigh muscle cross-sectional area decrease by a similar extent (14%) (88) as seen in chronic hypoxia (76–78). It therefore appears that hypoxia influences muscle mass homeostasis. Whether this effect is direct or mediated via an altered hormone status is not yet clear (see also Chapter 14). Acute hypoxia depresses muscle protein synthesis in humans (89) as well as in the rat (90). At sea level an increase in muscle mass from strength training results from an increase in protein synthesis coupled to an unchanged protein catabolism (91). In chronic normobaric hypoxia, as in patients suffering from pulmonary emphysema (80), body mass drops and protein synthesis is depressed. Whether chronic hypoxia reduces the hypertrophic response of human skeletal muscle when solicited during strength training was studied as subjects strength trained their elbow flexors at equal intensity at 5050 m and at sea level (92). In spite of a nearly maintained energy balance as well as an appropriate protein intake at altitude, chronic hypoxia reduced the potential for hypertrophy of human skeletal muscle by 30%. Bigard et al. (93) convincingly showed, in a rat model, that chronic exposure to hypobaric hypoxia specifically decreases (muscle) growth rate as compared to normobaric normoxia. Increasing the dietary protein intake in the hypoxic group of rats had no effect on the depression of muscle growth excluding an effect of protein content of the diet.

B. Limitation to $\dot{V}O_2$max in Hypoxia

Endurance performance capacity depends on aerobic metabolism and hence on the maximum sustainable oxygen uptake. In a moderately trained subject at sea level, the maximum sustainable oxygen uptake is typically located at about 80% $\dot{V}O_2$max, midway between the lactate threshold and $\dot{V}O_2$max. The combination of a high $\dot{V}O_2$max and a high maximum sustainable oxygen uptake is thus a good physiological basis for endurance exercise at sea level. It is well known that $\dot{V}O_2$max decreases as P_{IO_2} falls. As shown in Figure 1, the drop in $\dot{V}O_2$max becomes progressively steeper at lower P_{IO_2}. The $\dot{V}O_2$max declines ~1% per 100 m elevation gain above 1500 m (94).

After the initial decrease of $\dot{V}O_2$max with acute hypoxia, further exposure has a small influence on $\dot{V}O_2$max. For example, during a 5-week sojourn at 5050 m, $\dot{V}O_2$max initially fell 47%, followed by partial recovery of 4–8% after 15 and 35 days, respectively (95). Upon descent to sea level, $\dot{V}O_2$max was 92% of the preexpedition level and did not recover to preexpedition values for at least 5 weeks. However, endurance time at a given submaximal workload rises with acclimatization (96). The elevation of endurance time is not strictly due to enhanced oxygen delivery. The rise in CaO_2 (secondary to elevated hemoglobin) is balanced by a drop in cardiac output, the net effect being little change in oxygen delivery. The mechanisms for the rise in endurance time with acclimatization during a prolonged stay in chronic hypoxia or for the rise in $\dot{V}O_2$max during prolonged hypoxia (95) are not known.

What limits $\dot{V}O_2$max at sea level has received considerable attention in the past and continues to be a field of vivid debate. Brief consideration is now given to theoretical explanations of what limits $\dot{V}O_2$max in hypoxia. It is now generally accepted that limitation to $\dot{V}O_2$max in hypoxia is multifactorial. Today's existing models approach the problem from different viewpoints and necessarily yield different insights.

CNS Limitations to $\dot{V}O_2$max

Exercise begins and ends in the brain. Brain function and performance are profoundly influenced by acute and chronic hypoxia (97–99). However, the role that the brain plays during exercise at altitude is a topic that has received very little attention. All accounts of climbing at extreme altitude mention an intense sensation of effort and a very high level of exertional discomfort in spite of very low absolute levels of mechanical power output. Both at sea level and at altitude, the ''real'' limiting factor to endurance exercise performance is the brain. The cessation of heavy exercise occurs when the subjective exertional discomfort exceeds a threshold and a conscious decision is made to stop exercising. This threshold will be different for an athlete and a patient with cardiovascular or pulmonary disease. Furthermore, the threshold will be determined by limb fatigue and breathlessness (100). The perception of exertional discomfort is certainly multifactorial, with input from various sources. Once a threshold of discomfort is reached, exercise is stopped by a reduction of motor command. The various inputs during exercise at sea level and in hypoxia are probably the same, but their relative importance and central processing may be altered by hypoxia. For example, the sensation of breathlessness felt at a ventilation of 75 L/min during a standard exercise challenge at sea level was reduced by 35% after return from a 4-week sojourn at 4000 m, an effect that lasted at least 6 weeks (101). As another example, in normoxia, but not in hypoxia, signs of peripheral metabolic fatigue of the locomotor muscles and the diaphragm are seen on cessation of exhaustive large muscle group exercise (102).

A hint that central drive to exercise may be diminished in chronic hypoxia comes from OEII. Experimental subjects were asked to perform maximum voluntary contractions with the dorsal foot flexors. At extreme simulated altitudes, (PB =

282 torr, P_{IO_2} = 49 torr), using the twitch interpolation technique by electrically stimulating the muscle during a maximum voluntary contraction, twitch effects superimposed randomly on top of the voluntary torque trace were seen. Such twitch effects suggest that voluntary activation was not maximal and therefore that central drive had slackened. However, after verbal encouragement a greater effort invariably increased voluntary torque and superceded the twitch effects. Thus, in conditions of extreme hypoxia a minor reduction in motor drive was seen that could lead to a diminished power during such explosive contractions, but this was overcome by additional volition (103).

Further support for a central limitation to exercise comes from whole-body exercise studies carried out during OEII. During peak exercise at the highest altitudes, arterial oxygen saturation fell to 40%, and Pa_{O_2} dropped as low as 26 torr (104). The clinical findings of near unconsciousness at the end of exercise, coupled with the extreme hypoxemia, suggests that the brain may well have been the critical limiting organ to exercise performance. During such severe hypoxemia breathlessness was expected to be an exercise limiting symptom, but both breathlessness and leg fatigue were graded equal and maximal (105). These symptomatic observations, taken in conjunction with the absence of metabolic fatigue in the exercising legs, suggest that a failure of the central nervous system (CNS) to drive the locomotor apparatus rather than any muscular metabolic inhibition is what limits performance. Whether at these extreme altitudes a lack of motor drive also limits $\dot{V}_{O_2}$max is, however, still an open question.

Based on data from the OEII study, Bigland-Ritchie and Vøllestad (106) hypothesized that a maximally stressed respiratory system could, via the central nervous system, limit the motor drive to active locomotor muscle groups before their full potential was reached. For a given %$\dot{V}_{O_2}$max the respiratory system is stressed more during heavy exercise in hypoxia than in normoxia. A subsequent report supports the notion of an interaction between respiratory muscle fatigue and limitation to exercise in hypoxia. Five subjects cycled to exhaustion at 75% of $\dot{V}_{O_2}$max in normoxia and after one month at 5050 m (107). At altitude, the absolute work load was 24% below sea level values. In spite of the reduced load, time to exhaustion was only 55% of the sea level value. Although mechanical power output of the locomotor muscles on the cycle ergometer and $\dot{V}_{O_2}$ were lower at altitude, the ventilatory requirement rose (see Fig. 3). The authors argued that there were indirect indications pointing to possible respiratory muscle fatigue, including a drop in tidal volume coupled to a rise in breathing frequency. In addition, they observed negative inspiratory gastric pressure swings that are consistent with a paradoxical upward movement of the diaphragm during inspiration, which supports the presence of diaphragm fatigue. In one subject, just prior to exhaustion, a decrease in the tidal volume–to–inspiratory time ratio suggested a drop in motor drive to the respiratory muscles. Fatigue, or the dramatic continuous increase in ventilatory drive during cycle exercise at 75% of $\dot{V}_{O_2}$max, could thus provide an input to the central nervous system leading to a decrease in central drive to the active locomotor muscles before those muscles develop exercise-terminating peripheral metabolic fatigue. In hyp-

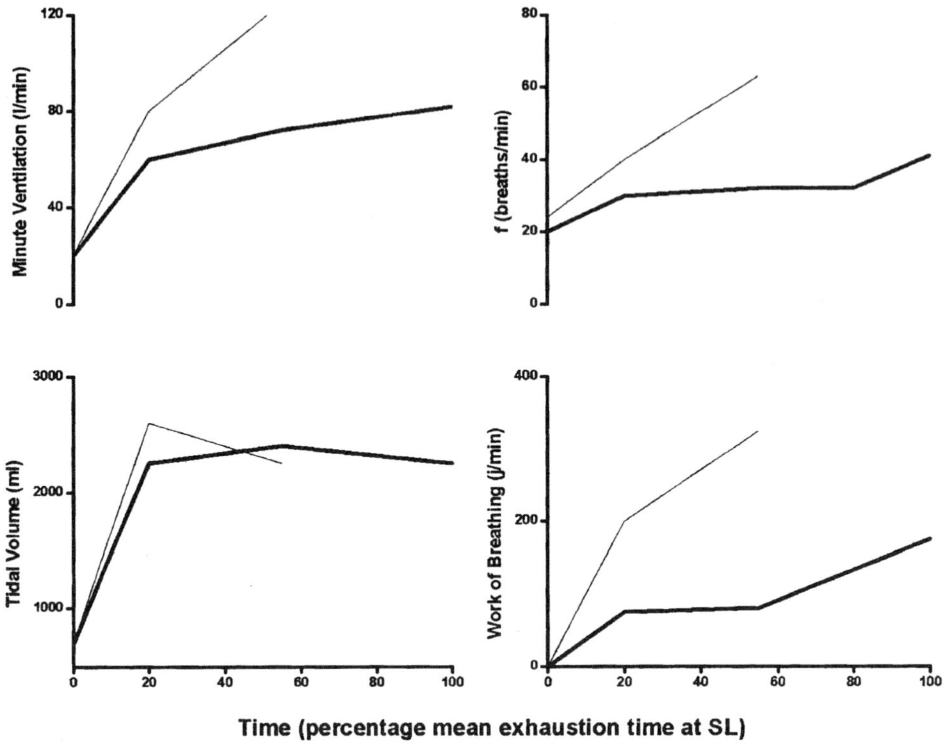

Figure 3 Ventilation (Ve), tidal volume (Vt), breathing frequency (f), and respiratory power (w) as a function of sea level exhaustion time, at sea level (thin line) and at altitude (thick line). Exercise intensity of 75% $\dot{V}O_2$max was 24% lower at altitude than at sea level. In spite of the lower load, the exhaustion time was reduced by 55% and ventilatory requirements increased greatly. (Adapted from Ref. 107.)

oxia, the legs do not show signs of peripheral metabolic fatigue. In fact, when working muscle volume is reduced (e.g., one-legged exercise), the maximum power output of the working muscles can reach sea level values, and only then does peripheral metabolic fatigue occur (57,108,109).

However, a recent experiment performed at 5200 m does not seem to support a role for respiratory muscle fatigue leading to a central limitation of exercise in hypoxia (57). Acclimatized subjects at 5200 m performed repeated maximal isometric voluntary contractions with the forearm flexors before and during exhausting bicycle exercise with and without 4% inspiratory carbon dioxide. The addition of carbon dioxide caused an elevation in ventilation at peak effort. The additional respiratory muscle work did not alter the force generated by a maximal voluntary contraction of the forearm muscles during peak effort. The authors concluded that central drive to the respiratory muscles or locomotor muscles was intact and had not changed with the increased work of breathing during inhalation of carbon dioxide.

Another hypothetical route to a ventilation-engendered central limitation to exercise emerged from observations made on soldiers suffering from high-altitude pulmonary edema. Paintal (110) suggested that hypoxic pulmonary arterial constriction at high altitude, combined with an exercise-induced additional rise in pulmonary arterial pressure and flow, might activate pulmonary juxta-capillary receptors. In turn, the juxta-capillary receptors, which are sensitive to pressure as well as interstitial edema, would via the vagus give rise to dyspnea and could by a reflex mechanism limit central motor drive. To date there are, however, no data to support this hypothesis.

Thus, the origin of the signals leading to the cessation of the central motor drive at exhaustion during heavy large muscle group exercise in hypoxia remains unclear. Possible candidates controlling the signal to stop exercise include arterial O_2 desaturation with exercise causing marked central nervous system hypoxia (111), other factors acting on the respiratory and/or other higher nervous centers, with or without contribution of fatigued respiratory muscles, or the effects of pulmonary hypertension. The proposed mechanism of early central nervous limitation to exercise in hypoxia is compatible with the absence of signs of peripheral metabolic fatigue at the end of exhausting maximum exercise with large muscle groups in hypoxia and with the lower levels of maximum lactate ([La]max) reached after acclimatization to chronic hypoxia (see Sec. III.B).

Central Cardiovascular Limitation

This model is based on the observation that human maximal cardiac output ($\dot{Q}$max) is limited to 25–30 L/min (112). This $\dot{Q}$max will not change whether they exercise with just their legs or with both arms and legs. An important observation is that if cardiac output serves only a small volume of muscle mass, such as during quadriceps-only exercise, the muscle maximal blood flow and $\dot{V}O_2$ are much higher than observed for the same muscle mass during cycling or running exercise (113). Since cardiac output cannot supply such amounts of blood (and therefore oxygen) to all muscles active during whole body exercise, it is argued that cardiovascular oxygen transport must limit $\dot{V}O_2$max.

With chronic hypoxia, control of heart rate during exercise is certainly changed, but why heart rate and cardiac output do not reach levels comparable to sea level during intense hypoxic exercise is not clear. A lower cardiac output favors gas exchange in the lungs by keeping capillary blood transit time longer and thereby minimizing the A-aDO_2 due to diffusion limitation (112). Evidence for an altered neurogenic control of heart rate was found by Åstrand and Åstrand (17), who gave subjects 100% oxygen during intense exercise at altitude and observed a sudden increase in heart rate that occurred within seconds and before mass oxygen transport was improved, a finding reproduced recently (57).

Peripheral Diffusive Oxygen Transport Limitation

Wagner developed a model to explain limits to $\dot{V}O_2max$ using the convective-diffusive interaction equation (114), which postulates diffusion limitation for oxygen from muscle capillaries to muscle mitochondria. This model assumes that the mean capillary to mitochondria difference is practically expressed by the O_2 tension of the effluent venous blood from the contracting muscles (mitochondrial PO_2 assumed to be zero). The model is based on the combination of two equations: the Fick equation describing convective oxygen transport $\dot{V}O_2 = \dot{Q}\ (CaO_2 - C\overline{v}O_2)$ and the equation describing peripheral diffusive flux, $\dot{V}O_2 = DO_2(P\overline{cap}O_2 - P_{mit}O_2)$, where $\dot{V}O_2$ is oxygen uptake, $\dot{Q}$ muscle blood flow, DO_2 is overall muscle diffusing capacity. $P\overline{cap}O_2$ is mean capillary PO_2 and $P_{mit}O_2$ is mitochondrial PO_2. Assuming $P\overline{cap}O_2$ to be proportionally related to $P\overline{v}O_2$, the relationship between muscle $\dot{V}O_2max$ can be plotted against the $P\overline{v}O_2$ of both equations. The intersection of these relationships is the functional $\dot{V}O_2max$ vs. $P\overline{v}O_2$ relationship. This analysis appears to be a promising approach to understanding the interrelationship between muscular blood flow, peripheral diffusion, and muscle aerobic capacity both at sea level and in hypoxia. There is evidence that the model also applies to whole-body $\dot{V}O_2max$. However, the exact implications with regard to limitation to $\dot{V}O_2max$ in different experimental conditions at the level of the organism are, as yet, less clear (for details see Chapter 8).

The Multifactorial Model

A quantitative multifactorial model of $\dot{V}O_2max$ limitation was developed by Di-Prampero (115–117) on the assumption that each physiological resistance to the O_2 flow from ambient air to the mitochondria contributes to the overall $\dot{V}O_2max$ limitation. The oxygen transport system was considered as a flow along a pressure gradient through a resistor such that $\dot{V}O_2 = dP/Rt$. The overall resistance (Rt) to O_2 flow is equal to the sum of four physiological resistances in series, associated with alveolar ventilation (ventilatory resistance Rv), alveolar-capillary O_2 transfer (lung resistance Rl), blood O_2 transport (circulatory resistance Rq), and peripheral O_2 diffusion and utilization (peripheral resistance, Rp). At maximal exercise $\dot{V}O_2max = (PIO_2 - P_mO_2/Rt + (PAO_2 - PaO_2)/Rl + (PaO_2 - P\overline{v}O_2)/Rq + (PvO_2 - P_mO_2)/Rp$ where PIO_2, PAO_2, PaO_2, $P\overline{v}O_2$, and P_mO_2 represent the O_2 partial pressures in inspired air, alveolar gas, arterial blood, mixed venous blood, and the mitochondrial matrix, respectively. Each resistance is considered responsible for a fraction of the overall resistance to oxygen flow. The changes on $\dot{V}O_2max$ and of the ith resistance to O_2 flow (dRi) can be expressed in relative terms: $\dot{V}O_2max/(\dot{V}O_2max + d\dot{V}O_2max) = 1 + Fi * dRi/Ri$, where Fi is the fractional resistance to $\dot{V}O_2max$ imposed by the resistance in question (115). In the most recent version of the model (115), the calculated fractional limitations to $\dot{V}O_2max$ in normoxia, varying alveolar ventilation between 125 and 75% of a reference value, resulted as follows: Fv and Fl provided between 5 and 12%, Fq between 59 and 78%, and Fp between 13 and 19%, of the overall limitation to $\dot{V}O_2max$. In hypoxia ($PIO_2 = 90$ torr), Fv and Fl increased, and

Fq decreased as the steep part of the oxygen dissociation curve was approached. At P_{IO_2} = 90 torr, Fv, Fl, Fq, and Fp amounted to 35, 31, 20, and 14%, respectively. Thus, at altitude the lungs and their ventilation take over from the circulation as a critical limiting factor in exercise.

This model was recently used to explain the nonlinear fall in the $\dot{V}O_2$max to P_{IO_2} relationship (9). The slope of that relationship is equal to the overall conductance to oxygen flow allowed by the oxygen transfer system. This implies an increase in the overall conductance to oxygen flow (or decrease of its reciprocal, resistance) in hypoxia. Because of the nonlinear shape of the oxygen dissociation curve, the cardiovascular conductance (Gq) to oxygen flow will vary with P_{IO_2}. $Gq = Q * \beta b$, where the oxygen transport coefficient $\beta b = (Ca_{O_2} - C\bar{v}_{O_2})/(Pa_{O_2} - P\bar{v}_{O_2})$ or the average slope of the dissociation curve.

On the basis of the above and on reports of desaturation during maximal exercise at sea level in subjects with high but not with low $\dot{V}O_2$max values (118), it was hypothesized that the drop in $\dot{V}O_2$max in hypoxia would be more pronounced in athletes than in sedentary subjects. This is the case. Athletes ($\dot{V}O_2$max 62 ± 4 mL/min/kg) had a progressively more pronounced drop in $\dot{V}O_2$max in hypoxia than sedentary subjects ($\dot{V}O_2$max 42.1 ± 6 mL/min/kg) so that the absolute differences in $\dot{V}O_2$max and mechanical power output between the two groups decreased from 22.6 to 9.1 mL/min/kg, and 113 to 49 W, respectively (9) (Fig. 4). The drop in Sa_{O_2} and that in $\dot{V}O_2$max had identical shapes in both groups, and it was concluded that the shape of the drop in $\dot{V}O_2$max with increasing altitude is a direct consequence of the shape of the oxygen dissociation curve. These observations have some inter-

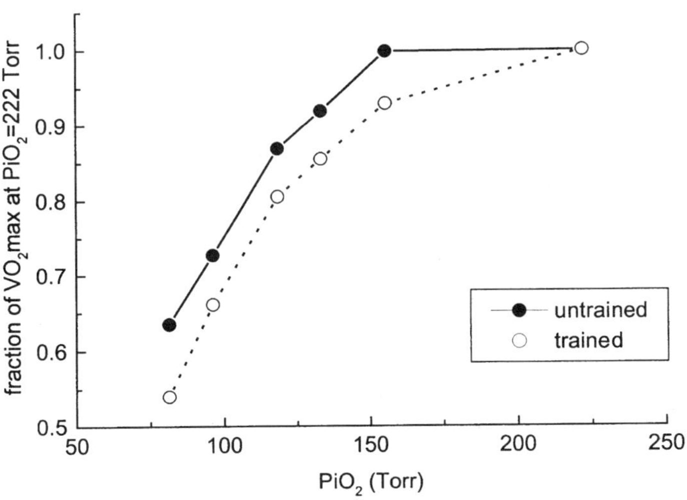

Figure 4 The difference in decrease in $\dot{V}O_2$max in acute hypoxia between untrained and trained subjects. (Adapted from Ref. 9.)

esting consequences: (1) a very high $\dot{V}O_2max$ per se is not necessarily a prerequisite to perform well at high altitude, (2) the difference in performance capacity of trained and untrained subjects should diminish at high altitude, and (3) ventilation and lung diffusion are the dominant limiting sites for overall oxygen flux in hypoxia.

III. Anaerobic Metabolism

Climbing a mountain is hard work. Obviously, it depends on mechanical power generated by muscular contraction. Ultimately all the energy expended for muscular contraction is derived aerobically by oxidation of substrate. However, for short periods (seconds for high-energy phosphate stores, minutes for lactic acid production) energy can also be made available through nonoxidative processes (anaerobic metabolism). Anaerobic metabolism is described as alactic if it concerns the high-energy phosphates ATP and phosphocreatine (PC) and lactic if the energy is made available through the production of lactic acid (for a detailed review on the energetics of muscular exercise see Refs. 3 and 119). The reasons that we address these nonoxidative pathways quite extensively in the following paragraphs are: (1) the regulation of oxidative metabolic pathways is dependent on alactic and lactic metabolism; therefore, it is important to understand the effects of oxygen deprivation on these metabolic pathways, (2) the regulation of oxidative metabolic pathways is dependent on alactic and lactic metabolism, (3) one would expect that oxygen deprivation would simply shift energy liberation necessary for work towards nonoxidative metabolism (Pasteur effect), but in humans at altitude this appears not be the case, and (4) acclimatization leads to striking changes in lactate accumulation in arterial blood during whole body exercise.

A. Alactic Anaerobic Metabolism

In theory, severe environmental hypoxia could impede ATP homeostasis. However, high-energy phosphates measured in skeletal muscle of humans show no significant decrement in hypoxia at rest or even after exhaustive exercise. A lack of an effect of hypoxia on ATP and PC homeostasis is supported by findings from biopsies of the vastus lateralis (68,120) as well as indirect measurements (maximum voluntary or electrically induced contraction force, maximum vertical jumping height, or cycling peak power (see Refs. 86,121–126). These findings imply that the maximal in vivo rate of ATP hydrolysis is not influenced by hypoxia, since $\dot{V}O_2$ consumption and lactic acid production do not contribute significantly to the energy necessary for such short explosive efforts (126). The finding after a prolonged altitude exposure of unchanged in vitro activity in muscle biopsy samples of creatine kinase, the enzyme catalyzing the transfer of a high-energy phosphate bond from PC to ADP to form ATP (70), also supports the notion that ATP and PC homeostasis are well protected, even in conditions of maximum exercise in extreme hypoxia as encountered on the top of Mt. Everest.

However, a transient effect of extreme hypoxia during exercise on alactic metabolism can be seen. At the onset of a square-wave exercise, aerobic energy supply needs time to attain a new steady state, thus creating an energy deficit that must be met from anaerobic energy sources. In hypoxia, the transient of oxygen uptake at the onset of a square wave exercise, known as the $\dot{V}O_2$-on response, is slowed compared to normoxia (127–132). During such a slowed transient, for a given rate of ATP hydrolysis, more energy is must therefore be derived through anaerobic metabolism as compared to normoxia (127–132), although no change has also been reported (133). To date there are no published studies on the effects of acclimatization to hypoxia on the $\dot{V}O_2$-on response. However, in spite of the slowing effect on the $\dot{V}O_2$-on response, likely leading to a fall in PC during heavy exercise in hypoxia but not in normoxia, exhaustive exercise with large muscle groups, be it at sea level or in hypoxia, never leads to depletion of PC or significant reductions in ATP (3).

Maximal explosive muscle power output is diminished after an expedition to very high altitudes or in some clinical conditions involving hypoxia that are linked to muscle wasting (80). This diminution is due to the loss of muscle mass, not to depletion of energy stores. For example, upon return from several months of exposure to extreme altitude (5300–8000 m), climbers had a drop in the instantaneous peak power (−9.8%) and the average power (−14.7%) during the push phase of a standing high jump off both feet on a force platform (87). The changes in power output could be fully explained by the loss of muscle mass that had occurred, since the thigh muscle cross-sectional area measured by CT scan had diminished by 11% (87). Thus, in spite of the maintained high-energy stores and maximum rates of ATP utilization, altitude exposure may lead to decreased power output simply because of loss of muscle mass (see Sec. II.A).

B. Lactate Metabolism and the Lactate Paradox

Lactate has several roles during exercise (134). Beyond being a by-product of glycolysis, lactate is an anaerobic source of energy (3), a means to shuttle carbohydrate between organs and cells (135), and its hydrogen ion helps in raising or maintaining capillary PO_2 to facilitate O_2 consumption by means of the Bohr effect (136). Two puzzling phenomena regarding lactate occur during exercise and in the course of acclimatization to hypoxia. First, less lactate accumulates in blood during maximal exercise ([La]max; known as the lactate paradox; Fig. 5), and second, the lactate produced for a given submaximal power output is diminished. The term lactate paradox was introduced by Hochachka (137) based on his observations in Quechuas Indians, high-altitude natives of the Andes in whom low maximal blood lactate levels were found both at altitude and upon exposure to normoxia. Reeves expanded the definition of the lactate paradox to include lowlanders, a situation ''in which blood lactate accumulation during [submaximal] exercise is increased on arrival at high altitude but falls with acclimatization'' (138). We will now review several factors that influence lactate metabolism during exercise in hypoxia.

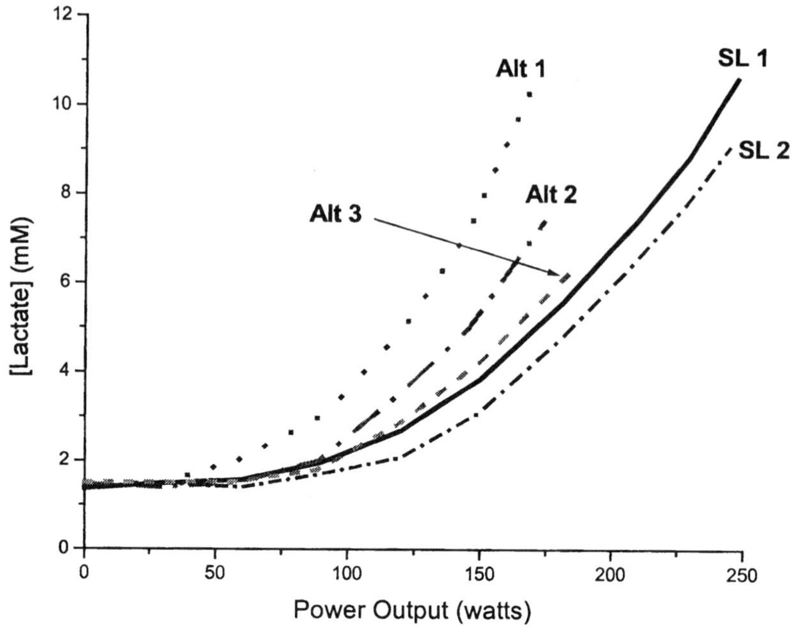

Figure 5 Blood lactate vs. power output during incremental cycle exercise in lowlanders at sea level (SL 1), at 5050 m on arrival (Alt 1) and at 3 (Alt 2) and 5 weeks (Alt 3), and one week after return to sea level (SL 2). The ends of the different lines indicate the mean maximum blood lactate values. Notice that at Alt 1 [La]max is similar to that at sea level but attained at lower work load. With acclimatization the curves (Alt 2 and Alt 3) tends towards SL 1, and [La]max is progressively decreased and attained at slightly higher work loads. On return from altitude the curve (SL 2) is initially to the right of that prior to exposure (SL 1), indicating less lactate being produced per watt, and only returns to SL over a 6-week deacclimatization period (not shown). (Adapted from Ref. 95.)

Duration of Exercise, Central Drive, and Peripheral Fatigue

During graded cycle-ergometer exercise in normoxia, blood lactate concentration ([La]) depends on the type of exercise and its duration. At low work rates no increase in [La] is found, at slightly higher levels a temporary increase is seen, and during heavy exercise lactic acidosis is a hallmark. The increase in [La] is the result of a change in the balance between lactate appearance (production and washout) and lactate removal (uptake and utilization). Lactate production during the transient phase at the onset of exercise increases with exercise intensity (139), to be compensated for later by a corresponding increase in lactate removal, so that at steady-state exercise higher stable [La] values are observed. This equilibrium is lost during high-intensity exercise when lactate production exceeds lactate removal and [La] keeps increasing with time. The increased lactate removal results from increased lactate oxidation as a substrate and from gluconeogenesis, which takes place either in organs such as the liver or in the muscles themselves (''lactate shuttle''; see Ref. 135). Under certain circumstances, contracting muscles may even remove lactate produced elsewhere (140).

The relationship between [La] and workload is nonlinear, with a higher slope at higher workloads. This particular shape with an inflection point (lactate threshold) has been interpreted as (1) a reflection of the onset of a lack of oxygen in the contracting muscle; (2) an increase in the recruitment of fast twitch fibers at high workloads; (3) an increased reliance on carbohydrates; (4) the energy state of the cell; and (5) increasing levels of circulating hormones like epinephrine. Even though severe tissue hypoxia and/or ischemia do lead to increased lactate production, there is now persuasive evidence both from animal and human studies that lactate production often occurs under strictly aerobic conditions. Therefore, the premise that an elevation of plasma [La] during exercise of increasing intensity necessarily reflects tissue hypoxia seems doubtful (119,135,141).

Another important modulator of blood lactate during exercise in hypoxia may be the metabolic rate that can be sustained for several minutes or more. Maximal power output during incremental exercise at altitude is decreased, and so is maximal metabolic rate. Lactate appearance in blood during exercise depends on the duration, the intensity, and the volume of active muscle. In Figure 6 $\dot{V}O_2$max and [La]max measured during OEII (142) suggest a parallel drop in maximum anaerobic glycolysis and in maximum oxidative phosphorylation ($\dot{V}O_2$max) with the decrease in barometric pressure. In hypoxia, if the high lactate levels in the nonacclimatized state are due to an increase in epinephrine levels, perhaps the low lactate levels in acclimatized subjects could reflect diminished muscle activation during exercise with large muscle groups.

During OEII, at low barometric pressures, cycle exercise induced exhaustion with less biochemical signs of muscle fatigue than at sea level (120). Similarly, muscle pH was higher at the end of heavy cycling exercise to exhaustion at 5050 m compared to sea level (143). The neuromuscular system seemed able to function normally. Electrical stimulation and voluntary activation studies showed that nerve conduction velocity, muscle end-plate transmission, muscle excitation-contraction coupling, and motor unit activation patterns were not altered by prolonged altitude exposure (103,108,144,145). Although explosive power output (i.e., maximum rate of ATP and PC hydrolysis) was not changed, longer duration power output with big muscle groups was reduced (108). By contrast, endurance exercise with small muscle groups is not much affected by altitude exposure. Acclimatized subjects at 5050 m sustained maximum aerobic exercise with a small muscle group (forearm flexors) at the same absolute load, for the same time and with similar signs of peripheral fatigue as at sea level, whereas exhaustion during maximum cycling exercise, although performed at a lower absolute load than sea level, was reached after similar time but with no signs of peripheral fatigue in the legs (108).

Acclimatized muscle does not seem to lack the ability to produce lactate up to the levels seen at sea level. If the muscle mass is small and does not put a burden

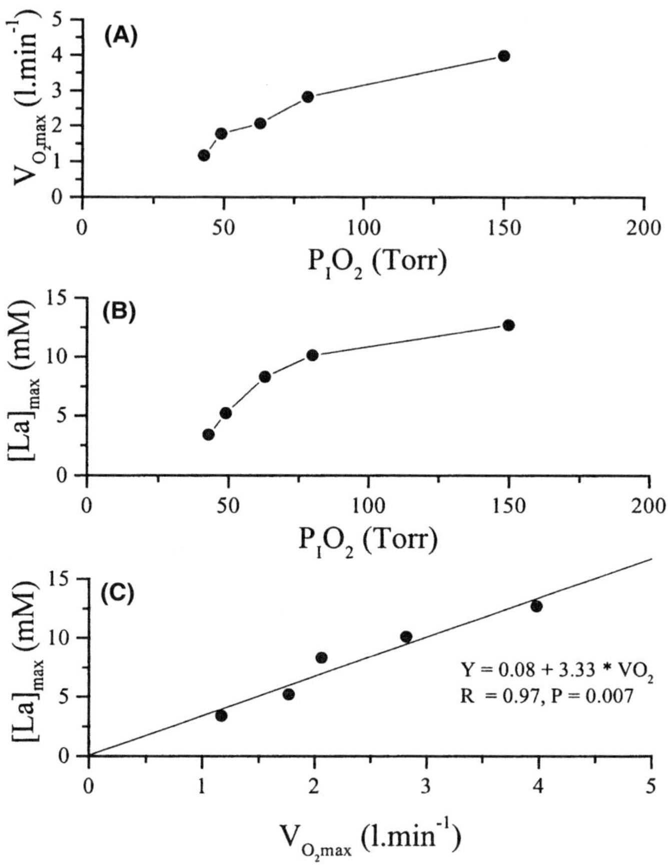

Figure 6 Average $\dot{V}o_2$max (A) and [La]max (B) vs. power output at the different inspiratory oxygen pressures reached during OEII. In (C) $\dot{V}o_2$max and [La]max are plotted against each other. The linear best fit suggests that in acclimatized subjects the reduction of [La]max occurs in parallel with the reduction in maximum rate of oxidative phosphorylation, suggesting that the reduction in muscle drive may have set the maximum blood lactate accumulation. (Adapted from Refs. 104, 142.)

on mass oxygen transport, several studies have shown that acclimatized skeletal muscle can produce lactate at or even above sea level values. Raynaud et al. sampled venous blood from the exercising forearm (146,147). Six lowlanders were studied at sea level, after 3 weeks at 3650 m and compared with six highlanders. Resting and exercise Pvo_2 (~18 torr) was not different at altitude compared to sea level and was similar to that of the altitude natives. Lactate levels during exercise were 80% higher at altitude than at sea level. Similarly, several studies have now reported that the fraction of sea level two-legged exercise maximum power output and $\dot{V}o_2$max that can be reached during exercise with just one leg is much higher at altitude than at sea level and is accompanied by much higher systemic lactate levels than reached during two-legged exercise (57,108,109). Summarizing, the studies mentioned suggest that the capacity of muscle to perform work is not decreased at high altitude. Rather, maximum work output is limited by diminished ability of the cardiorespiratory system to deliver oxygen to muscle or brain.

Reduced Maximum Power Output

The higher blood [La] during or after exercise with acute hypoxia depends on the work rate. Up to about 4000 m, blood [La] of nonacclimatized lowlanders at rest or during low-intensity exercise is about the same as at sea level, but when $\dot{V}O_2$ exceeds about 1.5 L/min, blood [La] increases beyond sea level values. At lower P_{IO_2}, even resting [La] can be elevated in nonacclimatized lowlanders. At the same $\%\dot{V}O_2max$ [La] is not different between levels of hypoxia in exercise bouts of the same relative intensity (148). In one study, muscle [La] at high altitude was not different from sea level when exercise of the same relative intensity was compared. In another study, exercise in normoxia, hypoxia ($F_{IO_2} = 0.12$) and hyperoxia ($F_{IO_2} = 1.9$) induced similar [La] at all submaximal workloads if expressed in percent of the F_{IO_2}-specific maximum (149). They also reported that in all conditions and at all loads, more than 90% of the increase in circulating [La] came from the exercising legs. Thus, when expressed in relative terms ($\%\dot{V}O_2max$), the threshold for lactate accumulation seems not to be different (138,150). During continuous incremental exercise (25 W/min) in nonacclimatized subjects at 5050 m, the relationship between [La] and workload was shifted such that higher [La] was found at any given workload, whereas maximal [La] was similar to that with normoxia (95). Maximal [La] was achieved at lower power outputs at altitude than at sea level (95) (see Fig. 5). In the same subjects after 5 weeks at 5050 m, the relationship between [La] and workload was close to that in normoxia, so that the decrease in [La]max was appropriate to the lower $\dot{V}O_2max$, a finding supported by many other studies (151–154) (see Figs. 5 and 6).

Hypoxia

For detailed review of the O_2 dependency of the increase in muscle lactate production during exercise see Wasserman (119) and Richardson et al. (155). Wasserman reasons that after the start of heavy exercise, the rate of adjustment of O_2 delivery to the capillaries, and ultimately to the mitochondria, is insufficient to fully support the rising oxidative metabolism. In response, there is greater phosphocreatinine (PC) breakdown than predicted from moderate exercise, as implied by a longer time constant of the second component of the $\dot{V}O_2$-on response. Associated with a greater PC breakdown and accumulation of inorganic phosphorus is a greater lactate production. The resulting acidosis results in a rightward shift in the HbO_2 dissociation curve, which enables the unloading of O_2 from hemoglobin and uptake by the muscles during heavy exercise. Under acute hypoxic conditions, the insufficiency of O_2

delivery after the onset of exercise is more pronounced. PC breakdown is even greater, as is lactate production, creating a condition that allows muscle O_2 uptake ultimately to rise to similar levels despite the hypoxia. In this scheme, lactate production and the resulting acidosis during heavy exercise can be seen as part of a feedback loop in the capillary microenvironment, in which lactate production continues until conditions are fulfilled for adequate O_2 delivery from the capillaries to the mitochondria to sustain the desired rate of oxidative metabolism (119). This view is in agreement with the contention that the maximum oxygen uptake of a muscle is set by a compromise between convective and diffusive oxygen transport, with the prime limitation being the diffusion of oxygen from the red cell to the mitochondria (156). Although at sea level it is likely that changes in cellular redox state play a role in controlling anaerobic glycolysis during exercise, there are several reasons to believe that the changes in lactate metabolism at altitude are determined by other mechanisms, as will be discussed below.

The effects of acclimatization on oxygen transport to the leg during cycle exercise were investigated on Pikes Peak (4300 m; P_B = 460 torr) (150,157). The investigators measured leg blood flow (by thermo-dilution) and femoral arterial-venous differences for oxygen and lactate. Subjects exercised at identical absolute power outputs, at sea level, and at the beginning and end of a prolonged sojourn at altitude. As expected, Sa_{O_2} and hemoglobin concentration rose with acclimatization, but these changes were offset by a drop in leg blood flow, such that mass oxygen transport to the leg remained unchanged and closely matched demand. Also, femoral Pv_{O_2} was not different after acclimatization when compared to acute hypoxia. If one assumes that femoral Pv_{O_2} reflects mean capillary P_{O_2}, then the driving pressure for diffusion of oxygen was not different between acute and chronic hypoxia. Thus, the lactate production and release by the muscle, which was lower in the chronic as compared to the acute state (i.e., the lactate paradox), could not be explained by improved oxygen delivery to the contracting muscle after acclimatization.

Buffer Capacity

It seems unlikely that early cellular acidification from lactate inhibiting certain key enzymes can explain the changes in lactate metabolism at high altitude. Humans hyperventilate in a hypoxic environment and develop a metabolically compensated respiratory alkalosis with losses of CO_2 and HCO_3^- and thereby develop a diminished body buffer capacity. As early as 1936 H. T. Edwards (158) hypothesized that at altitude a greater pH change because of decreased buffer capacity due to an increased proton load for a given lactate production could inhibit certain rate-limiting enzymes of glycolysis like phosphofructokinase at an earlier stage than at sea level (151,154,159). For a given amount of blood lactate, the H^+ concentration is increased in acclimatized lowlanders as well as in natives at altitude more than in nonacclimatized subjects (159). The H^+ transmembrane transport is an active process and involves an exchange of Na^+-H^+ and HCO_3^--Cl^-. Extracellular [HCO_3^-] is therefore important for proton transport, and in theory could become a limiting factor to anaerobic metabolism in hypoxia. This hypothesis was tested in six subjects after one month at 5050 m during exhaustive exercise at $\dot{V}O_2max$, with and without prior oral sodium-bicarbonate ingestion (0.3 g/kg) (152). The sodium-bicarbonate ingestion brought the subjects back to a normal sea level blood base excess, hence they were again in a state of acute noncompensated respiratory alkalosis, as if nonacclimatized. In acute hypoxia, when the respiratory alkalosis is not yet compensated, lactate concentration is usually the same as at sea level. However, despite the normalized buffer stores in blood, peak [La] at exhaustion did not increase.

Glycolytic Flux: Substrate Availability and Enzyme Activity

Lactate is produced from glucose. A major source of muscle glucose is glycogen. A decrease in glycogen availability could therefore potentially cause the decrease in lactate at altitude (158,160). However, the glycogen levels in resting muscles seem largely unaffected by hypoxia and even in chronic extreme hypoxia are well preserved (120,161). Upon arrival at altitude, when lactate appearance rates are higher then at sea level, a standardized exercise (identical absolute submaximal work rate) causes a drop in glycogen somewhat greater than at sea level. After acclimatization, the depletion is again the same as at sea level, and lactate appearance rates drop (162). At the end of exhausting exercise at altitude, depletion of glycogen has never been reported. A lack of muscle glycogen stores can therefore not be causing lower plasma [La] after acclimatization to altitude. This muscle glycogen-sparing effect upon acclimatization was attributed to a lesser β-adrenergic stimulation of glycogenolysis and to a lesser dependence on intramuscular glucose sources (162,163) compared to acute hypoxia.

Diminished activity of glycolytic enzymes in hypoxia in theory could explain the lactate paradox. However, this idea is not supported by the experimental data (70,164–166). Several studies have looked at the activity of different glycolytic enzymes, but none reported major changes. Sea level natives residing at 4300 m for 4 weeks showed no change in the activities of either glycolytic or citric acid cycle enzymes (166). Furthermore, lowlanders residing at 3450 m for 4–6 weeks had reductions in the activities of both citric acid cycle and fatty acid oxidation enzymes. But the enzymes and metabolites of alactic and lactic pathways showed no change in activity (165). Similarly, individuals in OEII exhibited a reduction in the activities of citric acid cycle enzymes only (68). Therefore, glycolytic enzyme activity is either unchanged or increased at altitude compared to sea level. Since enzyme activity is measured on homogenized muscle tissue in vitro, the hypothesis that an in vivo modulation of enzymatic metabolic control of glycogenolysis or glycolysis may be at the basis of the ''lactate paradox'' remains possible.

In a Pikes Peak study (4300 m; P_B = 460 torr) (150,157), investigators measured leg blood flow (by thermo-dilution) and femoral arterial-venous differences for lactate. Subjects exercised at identical absolute power outputs, at sea level, and at the beginning and end of a prolonged sojourn at altitude. β-Blockade in acute hypoxia largely prevented the initial rise in lactate (138,150,167). Since epinephrine is one potential modulator of circulating lactate during an altitude sojourn, the β-blockade study mentioned above is potentially revealing. As described earlier, epinephrine levels rise in humans upon acute exposure to hypoxia. After several days, epinephrine levels fall back to near sea level values, but norepinephrine levels climb to two to three times those seen at sea level. Adrenergic activation modulates glycogenolysis and thus the influx of glucose into glycolysis. Epinephrine binds to β-receptors in the muscle, thereby, through a second messenger system (cAMP), activating a kinase. The latter subsequently transforms phosphorylase *b* to its more active form *a*, which is the enzyme controlling the rate of glycogen to glucose transformation. At sea level, infusion of epinephrine at rest increases the activated fraction of phosphorylase from 28% to 88% (145). The apparent consequence of phosphorylase activation by infused epinephrine is a rise in lactate levels during exercise with an unchanged $\dot{V}O_2$ (168–170). Since acute hypoxia increases arterial epinephrine levels during exercise (40), glycolysis might be stimulated. During sustained submaximal bicycle exercise, both at sea level and at high altitude, the rate of lactate appearance in blood, as well as arterial [La], are closely related to the epinephrine blood level, suggesting a relationship between adrenergic drive and [La] (40,138,140). Unacclimatized subjects with β-blockade by popranalol had low blood [La] levels upon arrival at 4300 m, similar to those of acclimatized subjects (171). Thus, the observed higher [La] values during exercise at submaximal work loads during acute hypoxia result, at least in part, from increased β-adrenergic stimulation. Unfortunately, the subjects at sea level and high altitude worked at the same relative power output (80% of $\dot{V}O_2max$); thus, workload-dependent effects could not be excluded. In another study the power output was standardized (~50% of sea level $\dot{V}O_2max$) and a close relationship was again found between epinephrine concentration and [La] (172). β-Blockade attenuated the effect of acute altitude exposure on glycolysis, and blood [La] was lower than in the control condition, although still higher than at sea level. β-Blockade, however, did not fully prevent the progressive drop in exercise blood [La] that accompanies acclimatization to high altitude, although the amplitude of the decrease was considerably less (172). Thus, even though epinephrine appears to accelerate glycolysis in acute hypoxia, since β-blockade did not completely abolish the progressive reduction in [La], the lactate paradox cannot be fully explained by an alteration of adrenergic control of glycolysis. This is not so surprising since even if glycogenolysis is also controlled by phosphorylase *a*, activation of phosphorylase *alone* is not sufficient to increase glycogenolytic rate. Even in the presence of fully activated phosphorylase, only when the muscle contracts does glycogenolysis substantially increase (145,173).

Summarizing, the increase in blood lactate during exercise at submaximal workloads in unacclimatized humans is possibly due to increased adrenergic drive. The subsequent reduction in blood lactate is due to a decrease in adreneragic drive with acclimatization and possibly β-receptor downregulation. The reason for the reduction in the maximum blood lactate accumulation in acute and chronic hypoxia is likely due to the decrease in maximum power output during aerobic exercise. In either acute or chronic hypoxia, it seems unlikely that the Pasteur effect plays any important role.

IV. Climbing Everest Without Supplementary Oxygen

Acclimatized humans can exercise in extreme hypoxia. Mt. Everest is now climbed frequently without supplemental oxygen, and one Sherpa recently spent 21 hours on the summit! Such astounding physical feats result from a complex interaction of physiological adjustments that are not completely understood. Experienced climbers like Habeler, Messner, and those who followed their example to the top of Everest perhaps climb with high levels of biomechanical efficiency as witnessed by their remarkable ascent and descent rates. Arguing against this are laboratory tests of efficiency of these climbers during hypoxic exercise with results similar to those of untrained lowlanders (174). Overall aerobic fitness is also unremarkable in Everest summiteers; they are often only as fit as the normal recreational athlete (174). Habeler and Messner, with their relatively short stature and legs trained for climbing, are similar in physique to the Sherpas, Himalayan residents renowned for their climbing speed at very high altitudes (175). However, other climbers are tall, angular, and equally fast (or faster!). Tolerance to dehydration, practiced by Messner, who would abstain from fluids for as long as 8 hours while climbing, might be helpful at extreme altitude where fluid intake is limited and losses are high (176). Environmental and inborn factors may also influence climbing success. For example, a small increase in barometric pressure on the summit of Mt. Everest can lead to an important rise in arterial oxygen supply (177). However, a bright, clear and windless day is a winner on the summit of Mt. Everest for more reasons than barometric pressure. Inborn factors such as the control of ventilation have been invoked to explain performance at extreme altitude. Although some climbers have a brisk ventilatory response to hypoxia, others do not (174,178–180). Whether other, yet undiscovered or unexplained, genetic traits might determine climbing success at very high altitudes is a fertile area of ongoing research (181).

The brain is the final major contributor to climbing Everest without supplemental oxygen. Although a climber may have the presumed physical and physiological characteristics thought necessary for climbing at extreme altitude, few actually succeed even when afforded the opportunity. It is beyond the scope of this review to delve into the psyche of this elite group of athletes, but suffice it to say that the successful extreme altitude climber must have a mental fabric that is woven with a unique combination of purpose, goal orientation, and willingness to take extraordinary risks combined with the judgment to turn back when necessary. Success and virtuosity are measured in tasks completed, i.e., ascent *and* descent. Another point to consider is the effect of extreme hypoxemia on the brain (111) and its role in initiating and perpetrating the act of climbing for many hours. Climbers at extreme altitude emphasize a dimming of resolve to continue as the factor limiting exertion

(1), rather than leg fatigue or dyspnea, which are more common at low altitudes. This may be a reflection of the brain's role as a guardian of the body and provides a fertile area for future investigations. In summary, adaptations from the brain, lungs, heart, and tissues allow exercise to the summit of Mt. Everest, an environment that in the absence of such adjustments is deadly.

V. Altitude Training for Endurance Exercise

In 1968 the Mexico City Olympic Games focused world attention on the problems of athletic competition at high altitude (182). Exercise and sports scientists from around the world sought to identify which sports would be most affected, the optimum combination of length of altitude exposure before competition, and the ideal height for altitude training camps. Several clear lessons resulted from the intense scrutiny: one, endurance sports are adversely affected by even moderately high altitude; two, sports involving people or objects flying through the air benefited from the lower air density at altitude; three, for the endurance events, the best training regime for competition at high altitude was to train at the altitude of competition; and four, when the many athletes and test subjects returned to their sea level homes, some noted an improvement in their sport performance, while others did not. Until recently whether altitude training actually benefited performance was not certain. Yet altitude training is used routinely by competitive athletes to improve sea level performance (183); the variable individual responses to altitude training were dismissed as artifacts of poor research methodologies.

As mentioned above, $\dot{V}O_2max$ falls with increasing altitude. Therefore, an athlete's training intensity for endurance performance will be lower at altitude compared to sea level. On the other hand, an important beneficial effect of acclimatization that aids sea level performance is the rise in hemoglobin that accompanies acclimatization (184). Up to at least 5000 m, the higher one goes, the greater the stimulation of erythropoiesis and thus rise in hemoglobin (assuming intact iron stores). Thus, high altitude provides physiological blood doping while at the same time diminishing the intensity of training. The conundrum that physiologists and coaches have attempted to answer since the Mexico City Olympics is whether high-altitude exposure benefits sea level performance.

Two recent studies indicate that altitude training has benefits (185) and that variable individual responses to altitude training are due to the presence of nonresponders and responders to altitude training in groups of sea level resident athletes (186). For a review of altitude training predating these two recent studies, the reader is referred to Ref. 187.

To examine the effect of training and/or living at altitude on sea level performance, Levine and coworkers studied a group of college age, national caliber middle distance runners. Thirty-nine competitive runners (12 women, 27 men) were assigned to one of three 4-week long training regimes (13 in each regime): training and living at sea level (low-low), living at high altitude (2500 m) and training at low altitude (1250 m) (high-low), and living and training at high altitude (2500 m) (high-high) (185). A rise in red cell mass (9%) and $\dot{V}O_2$max (5%) was noted in both altitude groups but not in the low-low group. However, only in the high-low group was 5-km race time improved (~13 ± 10 seconds faster). In addition, the high-low group ran faster at the ventilatory threshold and at $\dot{V}O_2$max than either of the other groups. This study raises intriguing questions about the reasons for improvement in the high-low group. The authors speculate that the ability to train at near sea level intensity allowed the high-low group to better utilize their increased oxygen-carrying capacity compared to the high-high group. In addition, whether living at even higher altitude and training at sea level would impart greater benefits remains to be determined.

Retrospective analysis of the athletes in the Levine et al. study (185) showed that some had a marked and sustained erythropoietic response to altitude exposure that was linked to improved performance, while others did not. Chapman et al. (186) expanded on this observation in a prospective trial in 22 elite runners (14 men and 8 women). All completed 4-week living at moderate altitude (2700 m), endurance training at moderate altitude (2700 m), but with interval training at low altitude (1250 m). The authors observed that those with an early erythropoietic response and a greater increase in red cell mass had a higher $\dot{V}O_2$max than those with a poor erythropoietic response. For those with a marked erythropoietic response, at least, living high and training low seems to benefit.

VI. Effects of Life-Long Acclimatization

The preceding discussion has focused primarily on exercise by lowland residents exposed to acute or prolonged hypoxia. Natives of the world's mountain regions appear to be a breed apart, many of them exhibiting extraordinary physical prowess at high altitude. Anyone who has attempted to play football with a group of Quechua at 4000 m in the Andes or tried to keep up with a Sherpa while climbing in Nepal are certain that these high-altitude residents possess special physiological traits. In general, high-altitude natives breathe more, have larger ventilatory response to hypoxia, and tend to have larger hematocrits (188–192). We will focus on oxygen uptake and lactate metabolism to illustrate what distinguishes natives of high altitudes from lowlanders in these two key areas.

High-altitude natives, like many elite high-altitude mountaineers (174), do not exhibit a particularly high $\dot{V}O_2$max. In Tibetans and Sherpas their $\dot{V}O_2$max (~48 ± 5 mL O_2/kg/min, measured at 1300 m) (193,194) is comparable to that of untrained lowlanders, Quechua Indians studied at sea level (195), and climbers returning from the Himalaya (196). But altitude natives seem to have a smaller decrement in $\dot{V}O_2$max with ascent to high altitude than lowlanders (197). Perhaps this ability to maintain a higher fraction of normoxic $\dot{V}O_2$max when exercising in hypoxia is what distinguishes the performance of natives from lowlanders.

Growth and development at high altitude have been assumed necessary to confer performance advantages on high-altitude natives. However, in a study of 10,000 Han Chinese military occupying Tibet at ~4000 m, they had nearly regained their sea level $\dot{V}O_2max$ after 15 months of residence (198). This observation raises the possibility that if sea level residents have sufficient time to acclimatize, they can reach levels of oxygen transport similar to high-altitude natives, suggesting that this is more acquired than inherited. Support for this view comes from studies on Tibetans born at low altitude to parents who came originally from the Tibetan Plateau (3500–4500 m). These lowland-born but genetic highlanders have the same $\dot{V}O_2max$ as controls born to and raised by lowland parents (199).

Lactate levels reached during maximal exercise at ~4000 m in high-altitude natives in the Himalaya and Andes at 6 mM are about half the values seen at sea level (151). In addition, Andean natives had lower lactate levels than acclimatized lowlanders when exercising only a small muscle mass (forearm) (146,147). In Andean natives, low lactate levels seem to be a persistent characteristic. Six Quechua Indians, born and living at altitudes ranging from 3600 to 4500 m, when studied at sea level had lower [La]max than lowlanders (137). In lowlanders acclimatizing to 5050 m, the drop from sea level in [La]max stabilizes after about 10 days (95). Upon return to sea level, preexposure [La]max values are attained only after a few weeks. By contrast, even after residence at sea level for six weeks the Quechua were still unable to accumulate as much [La] as lowlanders during exhausting cycle exercise (195,200). Tibetans born at low altitude have [La]max levels similar to those of lowlanders (201) and have slightly higher [La]max levels during a sojourn at 5050 m as compared to lowlanders (Kayser, unpublished observation). The finding that lactate metabolism in genetic highlanders born at low altitude was similar to that seen for lowlanders sheds doubt on the proposed inborn tight coupling between glycolysis and oxidative phosphorylation of altitude natives (137). The fact that in the Quechua the lactate paradox does not disappear within 6 weeks of sojourning at sea level is intriguing but does not necessarily imply that it is a permanent feature. In lowlanders acclimatized to 5050 m for one month, [La]max only reached prealtitude levels 3 weeks after return (95). Therefore, it is possible that sea level residence for longer than 6 weeks would lead to a greater [La]max in altitude-native Quechua.

Endurance training at altitude of altitude natives increased aerobic enzyme activity but did not change glycolytic enzyme activity (202). Whereas lowland born Tibetans have similar glycolytic enzyme activities when compared to unacclimatized lowlanders (201), altitude-born Quechuas and Sherpas have reduced aerobic and increased glycolytic enzyme activities (203,204). Muscle biopsies from the vastus lateralis muscle of altitude native Sherpas showed that muscle fiber size was relatively small and similar to that observed in elite high-altitude climbers as well as in climbers upon return from a prolonged sojourn in the Himalaya, but significantly smaller than that usually observed in trained or untrained lowlanders (77,174,194). As Sherpas do not seem to lose much weight while climbing (73), a decrease in fiber size could be a useful adaptation to altitude as it decreases the diffusion distance for oxygen from the capillary to the mitochondrion (194).

Interestingly, the total mitochondrial volume density of 3.96 ± 0.54% in the Sherpa is the lowest number found in any population so far (194). This coupled with normal $\dot{V}O_2max$ in the Sherpa results in a ratio of total mitochondrial volume density and $\dot{V}O_2max$ that is lower than that of lowlanders (194). That this low ratio of mitochondrial volume density coupled to normal $\dot{V}O_2max$ may be an ubiquitious feature among high-altitude natives is supported by its finding in Tibetans born at low altitude (201) and in altitude natives from La Paz, Bolivia (3600 m) (202). As a possible cause for the observed higher ratio between maximal oxygen uptake and mitochondrial volume density in altitude natives, it may be argued that the mitochondrial function of the altitude native is perhaps different from that of lowlanders. Although the possibility cannot be ruled out, it seems to be quite unlikely for the following reasons: (1) recent analysis of mitochondrial DNA (mtDNA) suggests that the adaptation of Tibetans to high altitude is likely not the result of mtDNA mutations (205), and (2) the observed reduction of mitochondrial volume density is accompanied by a proportional decrease in the activity of mitochondrial enzymes, such as citrate synthase (70). Summarizing, even though a lot of anecdotal information suggests that altitude natives may be better adapted for exercise at altitude, to date little physiological explicatory evidence exists. Both altitude natives and lowlanders chronically exposed to altitude are characterized by low peripheral aerobic capacity. It may be that this is part of a strategy optimizing tissue oxygen transport.

VII. Where To Now?

We believe that important new insights into how humans exercise in hypoxia will come from the following five areas: women exercising in hypoxia, the genetics of high-altitude exposure over many generations, the regulation of body mass in chronic, severe hypoxia, oxygen flow and sensing, and deacclimatization.

A. Women

Of the hundreds of studies that examine exercise responses to hypoxia, few have included women as subjects. Essentially the entire realm of questions regarding acclimatization and exercise at altitude in women is open to investigation. Women have several physiological traits that in theory could profoundly influence their responses to exercise in hypoxia. Women may have altered ventilatory control in hypoxia during rest (206,207) and exercise (208–211), particularly in the luteal phase of the menstrual cycle when progesterone levels are high (see Chapter 7). In addition, marginal iron stores are common in women of menstruating age (212), a factor that during altitude acclimatization could hamper the normal and necessary increase in red cell mass and be detrimental to exercise performance (213). However, evidence to date supports the ability of women and men to perform equally well in hypoxia (214,215).

B. Natives and Genetic Endowment

Although considerable debate exists on possible genetic advantages of altitude natives compared to lowlanders with regard to exercise performance at altitude, the overall picture is still far from clear (see also Chapter 4). The recent work involving lowland-born altitude natives (194,201), altitude natives of mixed genetic origin (216), deacclimatization of altitude natives (137), or altitude-born children (217) is promising and worth pursuing. Such work may reveal genetic control of exercise capacity and oxygen transport applicable at sea level as well as at altitude.

C. Muscle Mass and Nutrition

The recent preliminary work on a connection between raised serum leptin levels and the anorexia of acute altitude exposure (82,83) holds promise for unraveling one of the most enduring mysteries of human acclimatization to high altitude: How does chronic hypoxia suppress appetite? Furthermore, the mechanisms responsible for the loss of muscle mass in hypoxia in the face of falling or even maintained body mass are yet to be identified. Hypoxia-induced muscle wasting may provide a model for advancing knowledge of the basic mechanisms of the control of appetite and body mass at sea level, with application to muscle-wasting hypoxic disease states as well as in the healthy human.

D. Oxygen Flow and Oxygen Sensing

The puzzle of the reduced maximum heart rate and cardiac output at high altitude warrants further investigation. Saltin (112) argues that the maximal heart rate is lowered to prevent a further reduction in arterial saturation that would follow an elevation in pulmonary blood flow and shortened transcapillary transit time. Thus, the beneficial effects of an elevated cardiac output could be offset by a reduced arterial oxygen partial pressure and content. This concept would favor an adaptation resulting in the highest possible pulmonary blood flow still enabling optimal oxygenation of the blood to occur. No mechanisms are known that could account for such a regulation so far, but Saltin proposes hyperventilation and as-yet-unidentified sensors for critically low P_{CO_2} and P_{O_2} altering the autonomic outflow to the heart as possible candidates (112).

During exercise at sea level the relationship between oxygen uptake ($\dot{V}_{O_2}$) and oxygen supply ($\dot{Q}_{a}O_2 = Ca_{O_2} * \dot{Q}$) is linear with a slope of 1 and an intercept about 1 L/min (218). The intercept is equal to the mixed venous flow of oxygen, or the amount of oxygen that returns to the heart, independent of $\dot{V}_{O_2}$. The fact that $\dot{Q}\bar{v}_{O_2}$ is a constant makes it a good candidate to be a regulated variable, but so far there is no experimental evidence for such a role. A constant $\dot{Q}\bar{v}_{O_2}$ may also very well be a result of other mechanisms of homeostasis. In Figure 7 are shown the results of several studies from the literature on humans during exercise, including OEII and measurements on Tibetans at altitude, where $\dot{Q}$ and Ca_{O_2} were measured along with $\dot{V}_{O_2}$. $\dot{Q}\bar{v}_{O_2}$ appears to be constant in normoxia, anemia, and acute as

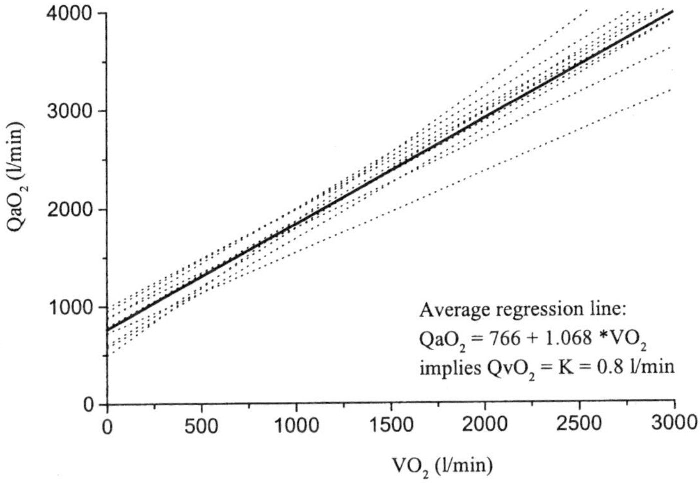

Figure 7 Relationship between mass oxygen flux (cardiac output × CaO_2) and $\dot{V}O_2$ during exercise suggesting constant mass oxygen return to the right heart.

well as chronic hypoxia and therefore merits further investigation. Questions include what factors regulate $\dot{Q}\bar{v}O_2$, and how and where is it sensed.

Below maximum cardiovascular oxygen transport, peripheral oxygen flow is regulated to match oxygen consumption. A change in CaO_2 is offset by a change in perfusion. A recent study suggested that C_aO_2 may be a key regulatory variable as limb blood flow and cardiac output rose as C_aO_2 fell, independent of PaO_2 (219). Moreover, similar C_aO_2-dependent perfusion changes have been noted in the cerebral circulation (220), suggesting that changes in CaO_2 may be a universal signaling mechanism. Several compounds await further investigation as likely regulators of blood flow according to changes in C_aO_2 or hemoglobin concentration. These include red cell ATP release (221), arachidonic acid metabolites released in response to changes in oxygen levels (222), and a hemoglobin concentration specific effect on scavenging of nitric oxide to effect vasodilation in the face of lowered hemoglobin concentration (223).

E. Loss of Acclimatization

The loss of acclimatization that occurs after return from a prolonged stay at high altitude has been largely unexplored (224). Studies of deacclimatization may reveal mechanisms associated with acclimatization that result in structural changes, such as altered enzyme kinetics or mRNA expression, and other nonstructural adjustments that resolve immediately upon cessation of the hypoxic challenge. Such findings, in turn, may open new avenues of investigation into and understanding of the process of acclimatization.

References

1. Messner R. The Crystal Horizon. Everest: The First Solo Ascent. Seattle: The Mountaineers, 1989:322.
2. West JB, Wagner PD. Predicted gas exchange on the summit of Mt. Everest. Respir Physiol 1980; 42:1–16.
3. di Prampero PE. Energetics of muscular exercise. Rev Physiol Biochem Pharmacol 1981; 89:143–222.
4. Aaron EA, Seow KC, Johnson BD, Dempsey JA. Oxygen cost of exercise hyperpnea: implications for performance. J Appl Physiol 1992; 72:1818–1825.
5. Gore CJ, Hahn AG, Scroop GC, et al. Increased arterial desaturation in trained cyclists during maximal exercise at 580 m altitude. J Applied Physiol 1996; 80:2204–2210.
6. Cibella F, Cuttitta G, Romano S, Grassi B, Bonsignore G, Milic-Emili J. Respiratory energetics during exercise at high altitude. J Appl Physiol 1999; 86:1785–1792.
7. West JB, Mathieu-Costello O. Stress failure of pulmonary capillaries as a limiting factor for maximal exercise. Eur J Appl Physiol 1995; 70:99–108.
8. Gore CJ, Little SC, Hahn AG, et al. Reduced performance of male and female athletes at 580 m altitude. Eur J Appl Physiol Occup Physiol 1997; 75:136–143.
9. Ferretti G, Moia C, Thomet JM, Kayser B. The decrease of maximal oxygen consumption during hypoxia in man: a mirror image of the oxygen equilibrium curve. J Physiol Lond 1997; 498:231–237.
10. Reeves JT, Groves BM, Sutton JR, et al. Operation Everest II: preservation of cardiac function at extreme altitude. J Appl Physiol 1987; 63:531–539.
11. Roach RC. Cardiovascular adjustments to high altitude. In: Saltin B, Secher N, Mitchell J, eds. Cardiovascular Regulation During Exercise. Champaign, IL: Human Kinetics Press, 1999.
12. Christensen EH, Forbes WH. Der Kreislauf in grossen Höhen. Skand Arch Physiol 1937; 76:75–89.
13. Saltin B, Grover RF, Blomqvist CG, Hartley LH, Johnson RL. Maximal oxygen uptake and cardiac output after 2 weeks at 4,300 m. J Appl Physiol 1968; 25:400–409.
14. Hartley LH, Vogel JA, Cruz JC. Reduction of maximal exercise heart rate at altitude and its reversal with atropine. J Appl Physiol 1974; 36:362–365.
15. Vogel JA, Hansen JE, Harris CW. Cardiovascular responses in man during exhaustive work at sea level and high altitude. J Appl Physiol 1967; 23:531–539.
16. West JB, Boyer SJ, Graber DJ, et al. Maximal exercise at extreme altitudes on Mount Everest. J Appl Physiol 1983; 55:688–698.
17. Åstrand PO, Åstrand I. Heart rate during muscular work in man exposed to prolonged hypoxia. J Appl Physiol 1958; 13:75–80.
18. Pugh LG. Cardiac output in muscular exercise at 5,800 m (19,000 ft). J Appl Physiol 1964; 19:441–447.
19. Stenberg J, Ekblom B, Messin R. Hemodynamic response to work at simulated altitude, 4000m. J Appl Physiol 1966; 21:1589–1594.
20. Fagraeus L, Karlsson J, Linnarsson D, Saltin B. Oxygen uptake during maximal work at lowered and raised ambient air pressure. Acta Physiol Scand 1973; 87:411–421.
21. Lawler J, Powers SK, Thompson D. Linear relationship between VO2 max and VO2 max decrement during exposure to acute hypoxia. J Appl Physiol 1988; 64:1486–1492.

22. Kjaer M, Bangsbo J, Lortie G, Galbo H. Hormonal response to exercise in humans: influence of hypoxia and physical training. Am J Physiol 1988; 23:R197–R203.
23. Dill DB, Myhre LG, Phillips EE, Brown DK. Work capacity in acute exposures to altitude. J Appl Physiol 1966; 21:1168–1176.
24. Escourrou P, Johnson DG, Rowell LB. Hypoxemia increases plasma catecholamine concentrations in humans. J Appl Physiol 1984; 57:1507–1511.
25. Shephard RJ, Bouhlel E, Vendewalle H, Monod H. Peak oxygen uptake and hypoxia: Influence of physical fitness. Int J Sport Med 1988; 9:279–283.
26. Martin D, O'Kroy J. Effects of acute hypoxia on the Vo_2max of trained and untrained subjects. J Sports Sci 1993; 11:37–42.
27. Grover RF, Reeves JT, Grover EB, Leathers JE. Muscular exercise in young men native to 3100 meter altitude. J Appl Physiol 1967; 22:555–564.
28. Grover RF, Weil JV, Reeves JT. Cardiovascular adaptation to exercise at high altitude. In: Pandolf KB, ed. Exerc Sport Sci Rev Vol. 14. New York: Macmillian Publishing Co., 1986:269–302.
29. Richalet JP. The heart and adrenergic system in hypoxia. In: Sutton JR, Coates G, Remmers JE, eds. Hypoxia: The Adaptations. Philadelphia: B.C. Decker, 1990:231–240.
30. Kaijser L, Grubbström J, Berglund B. Myocardial lactate release during prolonged exercise under hypoxemia. Acta Physiol Scand 1993; 149:427–433.
31. Alexander JK, Hartley LH, Modelski M, Grover RF. Reduction of stroke volume during exercise in man following ascent to 3,100 m altitude. J Appl Physiol 1967; 23: 849–858.
32. Alexander JK, Grover RF. Mechanism of reduced cardiac stroke volume at high altitude. Clin Cardiol 1983; 6:301–303.
33. Hartley LH, Alexander JK, Modelski M, Grover RF. Subnormal cardiac output at rest and during exercise in residents at 3100m altitude. J Appl Physiol 1967; 23:839–848.
34. Asmussen E, Nielsen M. The cardiac output in rest and work at low and high oxygen pressure. Acta Physiol Scand 1955; 35:73–83.
35. Asmussen E, Consolazio CF. The circulation in rest and work on Mount Evans (4,300 m). Am J Physiol 1941; 132:555–563.
36. Sutton JR, Reeves JT, Wagner PD, et al. Operation Everest II. Oxygen transport during exercise at extreme simulated altitude. J Appl Physiol 1988; 64:1309–1321.
37. Bender PR, Groves BM, McCullough RE, et al. Oxygen transport to exercising leg in chronic hypoxia. J Appl Physiol 1988; 65:2592–2597.
38. Cymerman A, Reeves JT, Sutton JR, et al. Operation Everest II: maximal oxygen uptake at extreme altitude. J Appl Physiol 1989; 66:2446–2453.
39. Reeves JT, Groves BM, Sutton JR, et al. Operation Everest II: preservation of cardiac function at extreme altitude. J Appl Physiol 1987; 63:531–539.
40. Mazzeo RS, Bender PR, Brooks GA, et al. Arterial catecholamine response during exercise with acute and chronic high altitude exposure. Am J Physiol 1991; 261:E419–E424.
41. Cunningham WL, Becker EJ, Kruezer F. Catecholamines in plasma and urine at high altitude. J Appl Physiol 1965; 20:607–610.
42. Becker EJ, Kreuzer F. Sympathoadrenal response to hypoxia. Pflugers Arch 1968; 304: 1–10.
43. Antezana AM, Kacimi R, Le Trong JL, et al. Adrenergic status of humans during prolonged exposure to the altitude of 6,542 m. J Appl Physiol 1994; 76:3001–3005.

44. Young PM, Rose MS, Sutton JR, Green HJ, Cymerman A, Houston CS. Operation Everest II: plasma lipids and hormonal responses during a simulated ascent of Mount Everest. J Appl Physiol 1989; 66:1430–1435.
45. Roach RC. Cardiovascular regulation during hypoxia. In: Ohno H, Kobayashi T, Masuyama S, Nakashima M, eds. Progress in Mountain Medicine and High Altitude Physiology. Matsumoto: Japanese Society of Mountain Medicine, 1998:264–270.
46. Choudhury L, Rosen SD, Lefroy DC, Nihoyannopoulos P, Oakley CM, Camici PG. Myocardial beta adrenoceptor density in primary and secondary left ventricular hypertrophy. Eur Heart J 1996; 17:1703–1709.
47. Eysmann SB, Gervino E, Vatner DE, Katz SE, Decker L, Douglas PS. Prolonged exercise alters β-adrenergic responsiveness in healthy sedentary humans. J Appl Physiol 1996; 80:616–622.
48. Eysmann SB, Gervino E, Vatner DE, Katz SE, Decker L, Douglas PS. Prolonged exercise alters β-adrenergic responsiveness in healthy sedentary humans. J Appl Physiol 1996; 80:616–622.
49. Kacimi R, Moalic JM, Aldashev A, Vatner DE, Richalet JP, Crozatier B. Differential regulation of G protein expression in rat hearts exposed to chronic hypoxia. Am J Physiol 1995; 38:H1865–H1873.
50. Kacimi R, Richalet JP, Crozatier B. Hypoxia-induced differential modulation of adenosinergic and muscarinic receptors in rat heart. J Appl Physiol 1993; 75:1123–1128.
51. Maher JT, Manchanda SC, Cymerman A, Wolfe DL, Hartley LH. Cardiovascular responsiveness to b-adrenergic stimulation and blockade in chronic hypoxia. Am J Physiol 1975; 228:477–481.
52. Voelkel NF, Hegstrand L, Reeves JT, McMurtry IF, Molinoff PB. Effects of hypoxia on density of beta-adrenergic receptors. J Appl Physiol 1981; 50:363–366.
53. Richalet J-P, Le-Trong J-L, Rathat C, et al. Reversal of hypoxia-induced decrease in human cardiac response to isoproterenol infusion. J Appl Physiol 1989; 67:523–527.
54. Richalet JP, Kacimi R, Moalic JM, Aldashev A, Vatner D, Crozatier B. Role of G proteins in the hypoxia-induced desensitization of cardiac beta-receptors (abst). In: Sutton JR, Houston CS, Coates G, eds. Hypoxia and the Brain. Burlington, VT: Queen City Press, 1995.
55. Richalet J-P, Mehdioui H, Rathat C, et al. Acute hypoxia decreases cardiac response to catecholamines in exercising humans. Int J Sports Med 1988; 9:157–162.
56. Richalet JP, Rathat C, Keromes A, Larmignant P. Effects de l'atropine sur la réponse adrenergique a l'exercise en hypoxia d'altitude (4350 m). Sci Sport 1990; 5:77–82.
57. Savard GK, Areskog NH, Saltin B. Cardiovascular response to exercise in humans following acclimatization to extreme altitude. Acta Physiol Scand 1995; 154:499–509.
58. Zhuang J, Droma T, Sutton JR, et al. Autonomic regulation of heart rate response to exercise in Tibetan and Han residents of Lhasa (3658 m). J Appl Physiol 1993; 75:1968–1973.
59. Hartley LH, Vogel JA, Cruz JC. Reduction of maximal exercise heart rate at altitude and its reversal with atropine. J Appl Physiol 1974; 36:362–365.
60. Furukawa Y, Hoyano Y, Chiba S. Parasympathetic inhibition of sympathetic effects on sinus rate in anesthetized dogs. Am J Physiol 1996; 271:H44–H50.
61. Levy MN. Sympathetic-parasympathetic interactions in the heart. Circ Res 1971; 29:437–445.

62. Myhre LG, Dill DB, Hall FG, Brown DK. Blood volume changes during three-week residence at high altitude. Clin Chem 1970; 16:7–14.
63. Hannon JP, Vogel JA. Oxygen transport during early altitude acclimatization: a perspective study. Eur J Appl Physiol 1977; 36:285–297.
64. Horstman DW, Weiskopf R, Jackson RD. Work capacity during 3-week sojourn at 4300 meters: effects of relative polycythemia. J Appl Physiol 1980; 49:311–318.
65. Jelkman W. Erythropoietin: Structure, control of production and function. Physiol Rev 1992; 72:449–489.
66. Semenza GL, Roth PH, Fang HM, Wang GL. Transcriptional regulation of genes encoding glycolytic enzymes by hypoxia-inducible factor 1. J Biol Chem 1994; 269: 23757–23763.
67. Lepawsky M. Hypoxia as a major pathogenetic event in catastrophic human anaerobic infections requiring multiple therapeutic modalities including hyperbaric oxygen to salvage life and limb (abstr). In: Sutton JR, Houston CS, Jones NL, eds. Hypoxia, Exercise and Altitude. New York: A.R. Liss, 1983:463–464.
68. Green HJ, Sutton JR, Cymerman A, Young PM, Houston CS. Operation Everest II: adaptations in human skeletal muscle. J Appl Physiol 1989; 66:2454–2461.
69. Hoppeler H, Desplanches D. Muscle structural modifications in hypoxia. Int J Sport Med 1992; 13:S166–S168.
70. Howald H, Pette D, Simoneau JA, Hoppeler H, Cerretelli P. Effects of chronic hypoxia on muscle enzyme activities. Int J Sport Med 1990; 11:S10–S14.
71. Consolazio CF, Matoush LO, Johnson HL, Daws TA. Protein and water balances of young adults during prolonged exposure to high altitude (4,300 meters). Am J Clin Nutr 1968; 21:154–161.
72. Cerretelli P. Limiting factors to oxygen transport on Mount Everest. J Appl Physiol 1976; 40:658–667.
73. Boyer SJ, Blume FD. Weight loss and changes in body composition at high altitude. J Appl Physiol 1984; 57:1580–1585.
74. Westerterp KR, Kayser B, Brouns F, Herry JP, Saris WH. Energy expenditure climbing Mt. Everest. J Appl Physiol 1992; 73:1815–1819.
75. Pugh LG. Physiological and medical aspects of the Himalayan Scientific and Mountaineering Expedition, 1960–61. Br Med J 1962; 2:621–627.
76. Hoppeler H, Howald H, Cerretelli P. Human muscle structure after exposure to extreme altitude. Experentia 1990; 46:1185–1187.
77. Hoppeler H, Kleinert E, Schlegel C, et al. Morphological adaptations of human skeletal muscle to chronic hypoxia. Int J Sport Med 1990; 11:S3–S9.
78. MacDougall JD, Green HJ, Sutton JR, et al. Operation Everest II: structural adaptations in skeletal muscle in response to extreme simulated altitude. Acta Physiol Scand 1991; 142:421–427.
79. Rose MS, Houston CS, Fulco CS, Coates G, Sutton JR, Cymerman A. Operation Everest II: nutrition and body composition. J Appl Physiol 1988; 65:2545–2551.
80. Morrison WL, Gibson JN, Scrimgeour C, Rennie MJ. Muscle wasting in emphysema. Clin Sci 1988; 75:415–420.
81. Butterfield GE, Gates J, Fleming S, Brooks GA, Sutton JR, Reeves JT. Increased energy intake minimizes weight loss in men at high altitude. J Appl Physiol 1992; 72: 1741–1748.
82. Auwerx J, Staels B. Leptin. Lancet 1998; 351:737–742.
83. Tschöp M, Strasburger J, Hartmann G, Biollaz J, Bärtsch P. Raised leptin concentra-

tions at high altitude associated with loss of appetite (letter). Lancet 1998; 352:1119–1120.
84. Blume FD. Metabolism and nutrition at altitude. In: Sutton JR, Houston CS, Jones NL, eds. Hypoxia, Exercise and Altitude. New York: A.R. Liss, 1983:311–316.
85. Kayser B. Nutrition and energetics of exercise at altitude: Theory and possible practical implications. Sports Med 1994; 17:309–323.
86. Kayser B, Narici M, Milesin S, Grassi B, Cerretelli P. Body composition and maximum lactic anaerobic performance during a one month stay at high altitude. Int J Sport Med 1993; 14:244–247.
87. Ferretti G, Hauser H, di Prampero PE. Maximal muscular power before and after exposure to chronic hypoxia. Int J Sport Med 1990; 11:S31–S34.
88. Berg H, Larsson L, Tesch P. Lower limb skeletal muscle function after 6 wk of bed rest. J Appl Physiol 1997; 82:182–188.
89. Rennie MJ. How can oxygen availability affect metabolism and how does it affect sugar transport? In: Sutton JR, Houston CS, Jones NL, eds. Hypoxia, Exercise and Altitude. New York: A.R. Liss, 1983:289–296.
90. Preedy VR, Smith DM, Sugden PH. The effects of 6 hours of hypoxia on protein synthesis in rat tissue in vivo and in vitro. Biochem J 1985; 228:179–185.
91. Goldberg AL, Etlinger JD, Goldspink DF, Jablecki C. Mechanism of work-induced hypertrophy of skeletal muscle. Med Sci Sports Exerc 1975; 7:185–198.
92. Narici MV, Kayser B. Hypertrophic response of human skeletal muscle to strength training in hypoxia and normoxia. Eur J Appl Physiol 1995; 70:213–219.
93. Bigard AX, Lavier P, Ullmann L, Legrand H, Douce P, Guezennec CY. Branched-chain amino acid supplementation during repeated prolonged skiing exercises at altitude. Int J Sport Nutr 1996; 6:295–306.
94. Buskirk ER, Kollias J, Akers RF, Prokop EK, Reategui EP. Maximal performance at altitude and return from altitude in conditioned runners. J Appl Physiol 1967; 23:259–266.
95. Grassi B, Marzorati M, Kayser B, et al. Peak blood lactate and blood lactate vs workload during acclimatization to 5,050 m and in deacclimatization. J Appl Physiol 1996; 80:685–692.
96. Maher JT, Jones LG, Hartley LH. Effects of high altitude exposure on submaximal endurance capacity of men. J Appl Physiol 1974; 37:895–898.
97. Siesjo BK, Johannsson H, Ljunggren B, Norberg K. Brain dysfunction in cerebral hypoxia and ischemia. Res Publ Assoc Res Nerv Ment Dis 1974; 53:75–112.
98. Gibson GE, Pulsinelli W, Blass JP, Duffy TE. Brain dysfunction in mild to moderate hypoxia. Am J Med 1981; 70:1247–1254.
99. Luft UC. Aviation physiology-the effects of altitude. In: Dill DB, ed. Handbook of Physiology: Adaptation to the Environment. Washington, DC: American Physiological Society, 1964:1099–1145.
100. Killian KJ, Jones NL. Respiratory muscles and dyspnea. Clin Chest Med 1988; 9:237–248.
101. Wilson RC, Oldfield WL, Jones PW. Effect of residence at altitude on the perception of breathlessness on return to sea level in normal subjects. Clin Sci 1993; 84:159–167.
102. Johnson BD, Babcock MA, Suman OE, Dempsey JA. Exercise-induced diaphragmatic fatigue in healthy humans. J Physiol 1993; 460:385–405.
103. Garner SH, Sutton JR, Burse RL, McComas AJ, Cymerman A, Houston CS. Operation

Everest II: neuromuscular performance under conditions of extreme simulated altitude. J Appl Physiol 1990; 68:1167–1172.

104. Sutton JR, Reeves JT, Wagner PD, et al. Operation Everest II. Oxygen transport during exercise at extreme simulated altitude. J Appl Physiol 1988; 64:1309–1321.
105. Sutton JR, Balcomb A, Green HJ, Young PM, Cymerman A, Reeves JT. Breathlessness at altitude. In: Breathlessness. Hamilton, Ontario: Boehringer-Ingelheim, 1992:143–148.
106. Bigland-Ritchie B, Vøllestad NK. Hypoxia and fatigue: How are they related? In: Sutton JR, Houston CS, Coates G, eds. Hypoxia. The Tolerable Limits. Indianapolis: Benchmark Press, 1988:315–328.
107. Cibella F, Cuttitta G, Kayser B, Narici M, Romano S, Saibene F. Respiratory mechanics during exhaustive submaximal exercise at high altitude in healthy humans. J Physiol 1996; 494:881–890.
108. Kayser B, Narici M, Einzoni T, Grassi B, Cerretelli P. Fatigue and exhaustion in chronic hypobaric hypoxia: influence of exercising muscle mass. J Appl Physiol 1994; 76:635–640.
109. Fulco CS, Rock PB, Trad L, Forte VJ, Cymerman A. Maximal cardiorespiratory responses to one- and two-legged cycling during acute and long-term exposure to 4300 meters altitude. Eur J Appl Physiol 1988; 57:761–766.
110. Paintal AS. Sensations from J-receptors. News Physiol Sci 1995; 10:238–243.
111. Colier WNJM, Hornbein TF, Paulson O, Roach RC. Cerebral de-oxygenation during peak exercise at 5260 m in well acclimatized sea level subjects. FASEB J 2000.
112. Saltin B. Limitations to performance at altitude. In: Sutton JR, Houston CS, Coates G, eds. Hypoxia. The Tolerable Limits. Indianapolis: Benchmark Press, 1988:9–34.
113. Andersen P, Saltin B. Maximal perfusion of skeletal muscle in man. J Physiol 1985; 366:233–249.
114. Piiper J, Scheid P. Comparison of diffusion and perfusion limitations in alveolar gas exchange. Respir Physiol 1983; 51:287–290.
115. Ferretti G, di Prampero PE. Factors limiting maximal O_2 consumption: effects of acute changes in ventilation. Respir Physiol 1995; 99:259–71.
116. di Prampero PE. Metabolic and circulatory limitations to VO_2 max at the whole animal level. J Exp Biol 1985; 115:319–331.
117. di Prampero PE, Ferretti G. Factors limiting maximal oxygen consumption in humans. Respir Physiol 1990; 80:113–127.
118. Dempsey JA, Hanson PG, Hendersen KS. Exercise-induced arterial hypoxemia in healthy human subjects at sea level. J Physiol (London) 1984; 355:161–175.
119. Wasserman K. Coupling of external to cellular respiration during exercise: the wisdom of the body revisited. Am J Physiol 1994; 266:519–539.
120. Green HJ, Sutton J, Young P, Cymerman A, Houston CS. Operation Everest II: muscle energetics during maximal exhaustive exercise. J Appl Physiol 1989; 66:142–150.
121. Robach P, Biou D, Herry JP, et al. Recovery processes after repeated supramaximal exercise at the altitude of 4,350 m. J Appl Physiol 1997; 82:1897–1904.
122. Bedu M, Fellmann N, Spielvogel H, Falgairette G, Van Praagh E, Coudert J. Force-velocity and 30-s Wingate tests in boys at high and low altitudes. J Appl Physiol 1991; 70:1031–1037.
123. Bedu M, Falgairette G, Van Praagh E, Coudert J. Effect of chronic hypoxia and socio-economic status on anaerobic power of 10- to 12-year-old Bolivian boys. Int J Sports Med 1994; 15:S84–S89.

124. Coudert J. Anaerobic performance at altitude. Int J Sport Med 1992; 13:S82–S85.
125. McLellan TM, Kavanagh MF, Jacobs I. The effect of hypoxia on performance during 30 s or 45 s of supramaximal exercise. Eur J Appl Physiol 1990; 60:155–161.
126. di Prampero PE, Mognomi P, Veicsteinas A. The effects of hypoxia on maximal anaerobic lactic power in man. In: Brendel W, Zink RA, eds. High Altitude Physiology and Medicine. New York: Springer-Verlag, 1982:88–93.
127. Linnarsson D, Karlsson J, Fagraeus L, Saltin B. Muscle metabolites and oxygen deficit with exercise in hypoxia and hyperoxia. J Appl Physiol 1974; 36:399–402.
128. Hughson RL, Green HJ, Sharratt MT. Gas exchange, blood lactate and plasma catecholamines during incremental exercise in hypoxia and normoxia. J Appl Physiol 1995; 79:1134–1141.
129. Xing HC, Cochrane JE, Yamamoto Y, Hughson RL. Frequency domain analysis of ventilation and gas exchange kinetics in hypoxic exercise. J Appl Physiol 1991; 71: 2394–2401.
130. Springer C, Barstow TJ, Wasserman K, Cooper DM. Oxygen uptake and heart rate responses during hypoxic exercise in children and adults. Med Sci Sport Exerc 1991; 23:71–79.
131. Murphy PC, Cuervo LA, Hughson RL. A study of cardiorespiratory dynamics with step and ramp exercise tests in normoxia and hypoxia. Cardiovasc Res 1989; 23:825–832.
132. Engelen M, Porszasz J, Riley M, Wasserman K, Maehara K, Barstow TJ. Effects of hypoxic hypoxia on O2 uptake and heart rate kinetics during heavy exercise. J Appl Physiol 1996; 81:2500–2508.
133. Griffiths TL, Henson LC, Whipp BJ. Influence of inspired oxygen concentration on the dynamics of the exercise hyperpnoea in man. J Physiol (Lond) 1986; 380:387–403.
134. Gladden LB. Lactate transport and exchange during exercise. In: Rowell LB, Shephard JT, eds. Section 12: Exercise: Regulation and Integration of Multiple Systems. New York: Oxford University Press, 1996:614–648.
135. Brooks GA. Current concepts in lactate exchange. Med Sci Sports Exerc 1991; 23: 895–906.
136. Wasserman K, Hansen JE, Sue DY. Facilitation of oxygen consumption by lactic acidosis during exercise. NIPS 1991; 6:29–34.
137. Hochachka PW. The lactate paradox: analysis of underlying mechanisms. Ann Sports Med 1989:184–188.
138. Reeves JT, Wolfel EE, Green HJ, et al. Oxygen transport during exercise at altitude and the lactate paradox: lessons from Operation Everest II and Pikes Peak. Exerc Sport Sci Rev 1992:275–296.
139. Cerretelli P, Pendergast D, Paganelli WC, Rennie DW. Effects of specific muscle training on VO2 on-response and early blood lactate. J Appl Physiol 1979; 47:761–769.
140. Brooks GA, Wolfel EE, Groves BM, et al. Muscle accounts for glucose disposal but not blood lactate appearance during exercise after acclimatization to 4,300 m. J Appl Physiol 1992; 72:2435–2445.
141. Connett RJ, Honig CR, Gayeski TE, Brooks GA. Defining hypoxia: a systems view of VO2, glycolysis, energetics, and intracellular PO2. J Appl Physiol 1990; 68:833–842.
142. Young PM, Sutton JR, Green H, et al. Operation Everest II: metabolic and hormonal responses to incremental exercise to exhaustion. J Appl Physiol 1992; 73:2574–2579.

143. Kayser B, Ferretti G, Grassi B, Binzoni T, Cerretelli P. Maximal lactic capacity at altitude: effect of bicarbonate loading. J Appl Physiol 1993; 3:1070–1074.
144. Orizio C, Esposito F, Veicsteinas A. Effect of acclimatization to high altitude (5050 m) on motor unit activation pattern and muscle performance. J Appl Physiol 1994; 77:2840–2844.
145. Ren JM, Hultman E. Regulation of phosphorylase a activity in human skeletal muscle. J Appl Physiol 1990; 69:919–923.
146. Raynaud J, Douget D, Legros P, Capderou A, Raffestin B, Durand J. Time course of muscular blood metabolites during forearm rhythmic exercise in hypoxia. J Appl Physiol 1986; 60:1203–1208.
147. Raynaud J, Bailliart O, Duran JC, Marotte H, Durand J. PO_2, PCO_2, pH and lactates in muscular venous blood during rhythmic forearm exercise at sea level and high altitude (3,850 m) (abstr). In: Sutton JR, Houston CS, Coates G, eds. Hypoxia. The Tolerable Limits. Indianapolis: Benchmark Press, 1988:378.
148. Hermansen L, Saltin B. Blood lactate concentration during exercise at acute exposure to high altitude. In: Margaria R, ed. Exercise at High Altitude. New York: Excerpta Medica, 1967:48–53.
149. Knight DR, Poole DC, Hogan MC, Bebout DE, Wagner PD. Effect of inspired O_2 concentration on leg lactate release during incremental exercise. J Appl Physiol 1996; 81:246–251.
150. Wolfel EE, Groves BM, Brooks GA, et al. Oxygen transport during steady-state submaximal exercise in chronic hypoxia. J Appl Physiol 1991; 70:1129–1136.
151. Cerretelli P, Veicsteinas A, Marconi C. Anaerobic metabolism at high altitude: The lactacid mechanism. In: Brendel W, Zink RA, eds. High Altitude Physiology and Medicine. New York: Springer-Verlag, 1982:94–102.
152. Kayser B, Ferretti G, Grassi B, Binzoni T, Cerretelli P. Maximal lactic capacity at altitude: effect of bicarbonate loading. J Appl Physiol 1993; 75:1070–1074.
153. Bender PR, Groves BM, McCullough RE, et al. Decreased exercise muscle lactate release after high altitude acclimatization. J Appl Physiol 1989; 67:1456–1462.
154. West JB. Lactate during exercise at extreme altitude. Fed Proc 1986; 45:2953–2957.
155. Richardson RS, Noyszewski EA, Leigh JS, Wagner PD. Lactate efflux from exercising human skeletal muscle: role of intracellular PO_2. J Appl Physiol 1998; 85:627–634.
156. Wagner PD. Muscle O_2 transport and O_2 dependent control of metabolism. Med Sci Sports Exerc 1995; 27:47–53.
157. Bender PR, Groves BM, McCullough RE, et al. Oxygen transport to exercising leg in chronic hypoxia. J Appl Physiol 1988; 65:2592–2597.
158. Edwards HT. Lactic acid in rest and work at high altitude. Am J Physiol 1936; 116: 367–375.
159. Cerretelli P. Lactacid O_2 debt in acute and chronic hypoxia. In: Margaria R, ed. Exercise at High Altitude. New York: Excerpta Medica, 1967:58–64.
160. Christensen EH. Oxygen uptake and respiratory functions at high altitudes. Scand Arch Physiol 1937; 76:88–100.
161. Young AJ, Evans WJ, Cymerman A, Pandolf KB, Knapik JJ, Maher JT. Sparing effect of chronic high-altitude exposure on muscle glycogen utilization. J Appl Physiol 1982; 52:857–862.
162. Green H, Sutton J, Wolfel E, Reeves JT, Butterfield GE, Brooks G. Altitude acclimatization and energy metabolic adaptations in skeletal muscle during exercise. J Appl Physiol 1992; 73:2701–2708.

163. Brooks GA, Butterfield GE, Wolfe RR, et al. Decreased reliance on lactate during exercise after acclimatization to 4,300 m. J Appl Physiol 1991; 71:333–341.
164. Mizuno M, Juel C, Bro-Rasmussen T, et al. Limb skeletal muscle adaptation in athletes after training at altitude. J Appl Physiol 1990; 68:496–502.
165. Saltin B, Nygaard E, Rasmussen B. Skeletal muscle adaptation in man following prolonged exposure to high altitude (abstr). Acta Physiol Scand 1980; 109:31A.
166. Young AJ, Evans WJ, Fisher EC, Sharp RL, Costill DL, Maher JT. Skeletal muscle metabolism of sea-level natives following short-term high altitude residence. Eur J Appl Physiol 1984; 52:463–466.
167. Young AJ, Young PM, Rock PB, Cymerman A, Moore LG, Reeves JT. Beta-adrenergic blockade during altitude acclimatization fails to prevent glycogen-sparing adaptations in muscle metabolism (abstr). In: Sutton JR, Houston CS, Coates G, eds. Hypoxia. The Tolerable Limits. Indianapolis: Benchmark Press, 1988:386.
168. Jansson E, Hjemadahl P, Kaijser L. Epinephrine-induced changes in muscle carbohydrate metabolism during exercise in male subjects. J Appl Physiol 1986; 60:1466–1470.
169. Turner MJ, Howley ET, Tanaka H, Ashraf M, Basett DR, Keefer DJ. Effect of graded epinephrine infusion on blood lactate response to exercise. J Appl Physiol 1995; 79: 1206–1211.
170. Gaesser GA, Ward SA, Baum VC, Whipp BJ. Effects of infused epinephrine on slow phase of O2 uptake kinetics during heavy exercise in humans. J Appl Physiol 1994; 77:2413–2419.
171. Young AJ, Young PM, McCullough RE, Moore LG, Cymerman A, Reeves JT. Effect of beta-adrenergic blockade on plasma lactate concentration during exercise at high altitude. Eur J Appl Physiol 1991; 63:315–322.
172. Mazzeo RS, Brooks GA, Butterfield GE, et al. β-Adrenergic blockade does not prevent the lactate response to exercise after acclimatization to high altitude. J Appl Physiol 1994; 76:610–615.
173. Ren JM, Gulve EA, Cartee GD, Holloszy JO. Hypoxia causes glycogenolysis without an increase in percent phosphorylase a in rat skeletal muscle. Am J Physiol 1992; 263: E1086–E1091.
174. Oelz O, Howald H, di Prampero PE, et al. Physiological profile of world-class high altitude climbers. J Appl Physiol 1986; 60:1734–1742.
175. Saha H. Studies of the oxygen uptake and efficiency of climbing of Tensing Norgay and other subjects. Q J Exp Physiol 1958; 43:295–299.
176. Messner R. Everest: Expedition to the Ultimate. London: Kay and Ward, 1979.
177. West JB, Lahiri S, Maret KH, Peters Jr RM, Pizzo CJ. Barometric pressures at extreme altitudes on Mt. Everest: Physiological significance. J Appl Physiol 1983; 54:1188–1194.
178. Schoene RB, Hackett PH, Roach RC. Blunted hypoxic chemosensitivity at altitude and sea level in an elite high altitude climber (abst). In: Sutton JR, Houston CS, Coates G, eds. Hypoxia and Cold. New York: Praeger, 1987:532.
179. Schoene RB, Lahiri S, Hackett PH, et al. The relationship of hypoxic ventilatory response to exercise performance on Mount Everest. J Appl Physiol 1984; 56:1478–1483.
180. Lahiri S. Respiratory control in Andean and Himalayan high-altitude natives. In: West JB, Lahiri S, eds. High Altitude and Man. Bethesda, MD: American Physiological Society, 1984:147–162.

181. Montogomery HE, Marshall R, Hemongway H, et al. Human gene for physical performance. Nature 1999; 393:221–222.
182. Pugh LG. Altitude and athletic performance. Nature 1965; 207:1397–1398.
183. Dick FW. Training at altitude in practice. Int J Sport Med 1992; 13:S203–S206.
184. Weil JV, Jamieson G, Brown DW, Grover RF. The red cell mass-arterial oxygen relationship in normal man. J Clin Invest 1968; 47:1627–1639.
185. Levine BD, Stray-Gundersen J. "Living high-training low": effect of moderate-altitude acclimatization with low-altitude training on performance. J Appl Physiol 1997; 83:102–112.
186. Chapman RF, Stray-Gundersen J, Levine BD. Individual variation in response to altitude training. J Appl Physiol 1998; 85:1448–1456.
187. Levine BD, Roach RC, Houston CS. Work and training at altitude. In: Sutton JR, Coates G, Houston CS, eds. Hypoxia and Mountain Medicine. Burlington, VT: Queen City Press, 1992.
188. Beall CM, Goldstein MC. Hemoglobin concentration, percent oxygen saturation and arterial oxygen content of Tibetan nomads at 4,850 to 5,450 meters. In: Sutton JR, Coates G, Remmers JE, eds. Hypoxia: The Adaptations. Philadelphia: B.C. Decker, 1990:59–65.
189. Beall CM, Brittenham GM, Strohl KP, et al. Hemoglobin concentration of high-altitude Tibetans and Bolivian Aymara. Am J Phys Anthropol 1998; 106:385–400.
190. Chiodi H. Respiratory adaptations to chronic high altitude hypoxia. J Appl Physiol 1957; 10:81–87.
191. Lahiri S, Milledge JS. Sherpa physiology. Nature 1965; 207:610–613.
192. Kobayashi T, Masuyama S, Masuda A, et al. Control of breathing and metabolism in Tibetans. In: Ueda G, ed. High Altitude Medicine. Matsumoto, Japan: Shinshu University Press, 1992.
193. Curran LS, Zhuang J, Droma T, Moore LG. Superior exercise performance in lifelong Tibetan residents of 4,400 m compared with Tibetan residents of 3,658 m. Am J Phys Anthropol 1998; 105:21–31.
194. Kayser B, Hoppeler H, Claassen H, Cerretelli P. Muscle structure and performance capacity of Himalayan Sherpas. J Appl Physiol 1991; 70:1938–1942.
195. Hochachka PW, Stanley C, Matheson GO, McKenzie DC, Allen PS, Parkhouse WS. Metabolic and work efficiencies during exercise in Andean natives. J Appl Physiol 1991; 70:1720–1730.
196. Ferretti G, Boutellier U, Pendergast DR, et al. Oxygen transport system before and after exposure to chronic hypoxia. Int J Sport Med 1990; 11:S15–S20.
197. Favier R, Spielvogel H, Desplanches D, Ferretti G, Kayser B, Hoppeler H. Maximal exercise performance in chronic hypoxia and acute normoxia in high-altitude natives. J Appl Physiol 1995; 78:1868–1874.
198. Niu W, Wu Y, Li B, Chen N, Song S. Effects of long-term acclimatization in lowlanders migrating to high altitude: comparison with high altitude residents. Eur J Appl Physiol 1995:543–548.
199. Kayser B, Marconi C, Amatya T, et al. The metabolic and ventilatory response to exercise in Tibetans born at low altitude. Respir Physiol 1994; 98:15–26.
200. Matheson GO, Allen PS, Ellinger DC, et al. Skeletal muscle metabolism and work capacity: a P-NMR study of Andean natives and lowlanders. J Appl Physiol 1991; 70: 1963–1976.
201. Kayser B, Hoppeler H, Desplanches D, Marconi C, Broers B, Cerretelli P. Muscle

ultrastructure and biochemistry of lowland Tibetans. J Appl Physiol 1996; 81:419–425.

202. Desplanches D, Hoppeler H, Tuscher L, et al. Muscle tissue adaptations of high-altitude natives to training in chronic hypoxia or acute normoxia. J Appl Physiol 1996; 81:1946–1951.
203. Rosser BWC, Hochachka PW. Metabolic capacity of muscle fibers from high-altitude natives. Eur J Appl Physiol 1993; 67:513–517.
204. Hochachka PW. Muscle enzymatic composition and metabolic regulation in high altitude adapted natives. Int J Sport Med 1992; 13:S89–S91.
205. Torroni A, Miller JA, Moore LG, et al. Mitochondrial DNA analysis in Tibet: implications for the origin of the Tibetan population and its adaptation to high altitude. Am J Phys Anthropol 1994; 93:189–199.
206. Schoene RB, Robertson HT, Pierson DJ, Peterson AP. Respiratory drives and exercise in menstrual cycles of athletic and nonathletic women. J Appl Physiol 1981; 50:1300–1305.
207. White DP, Douglas NJ, Pickett CK, Weil JV, Zwillich CW. Sexual influence on the control of breathing. J Appl Physiol 1983; 54:874–879.
208. Drinkwater BL, Kramar PO, Bedi JF, Folinsbee LJ. Women at altitude: cardiovascular response to hypoxia. Aviat Space Environ Med 1982; 53:472–477.
209. Drinkwater B. Responses of women mountaineers to maximal exercise during hypoxia. Aviat Space Environ Med 1979; 50:657–662.
210. Shephard RJ, Vandewalle H, Bouhlel E, Monod H. Sex differences of physical working capacity in normoxia and hypoxia. Ergonomics 1988; 31:1177–1192.
211. Miles DS, Wagner JA, Horvath SM, Reyburn JA. Absolute and relative work capacity in women at 758, 586, and 523 torr barometric pressure. Aviat Space Environ Med 1980; 51:439–444.
212. Galan P, Yoon HC, Preziosi P, et al. Determining factors in the iron status of adult women. Eur J Clin Nutr 1998; 52:383–388.
213. Mairbaurl H. Red blood cell functions at high altitude. Ann Sport Med 1989; 4:189–195.
214. Braun B, Mawson JT, Muza SR, et al. Women at altitude: carbohydrate utilization during exercise at 4300 m. J Appl Physiol 2000;
215. Hannon JP. Comparative altitude adaptability of young men and women. In: Folinsbee LJ, Wagner JA, Borgia JF, Drinkwater BL, Gliner JA, Bedi JF, eds. Environmental Stress: Individual Human Adaptations. New York: Academic Press, 1978:335–350.
216. Moore LG, Curran-Everett L, Droma TS, et al. Are Tibetans better adapted? Int J Sport Med 1992; 13:S86–S88.
217. Saltin B, Larsen H, Terrados N, et al. Aerobic exercise capacity at sea level and at altitude in Kenyan boys, junior and senior runners compared with Scandinavian runners. Scand J Med Sci Sport 1995; 5:209–221.
218. Ferretti G, Kayser B, Schena F, Turner DL, Hoppeler H. Regulation of perfusive O_2 transport during exercise in humans: effects of changes in haemoglobin concentration. J Physiol 1992; 455:679–688.
219. Roach RC, Koskolou M, Calbet J, Saltin B. Arterial O_2 content and tension in regulation of cardiac output and leg blood flow during exercise in humans. Am J Physiol: Heart Circ Physiol 1999; 276:H1–H8.
220. Jones MD, Traysman RJ, Simmons MA, Molteni RA. Effects of changes in arterial O_2 content on cerebral blood flow in the lamb. Am J Physiol 1981; 240:H209–H215.

221. Ellsworth ML, Forrester T, Ellis CG, Dietrich HH. The erythrocyte as a regulator of vascular tone. Am J Physiol 1995; 269:H2155–H2161.
222. Harder DR, Narayanan J, Birks EK, et al. Identification of a putative microvascular oxygen sensor. Circ Res 1996; 79:54–61.
223. Stamler JS, Jia L, Eu JP, et al. Blood flow regulation by S-nitrosohemoglobin in the physiological oxygen gradient. Science 1997; 276:2034–2037.
224. Beidleman BA, Muza SR, Rock PB, et al. Exercise responses after altitude acclimatization are retained during reintroduction to altitude. Med Sci Sports Exerc 1997; 29: 1588–1595.

21

Sleep

JOHN V. WEIL

University of Colorado Health Sciences Center
Denver, Colorado

DAVID P. WHITE

Brigham and Women's Hospital
Boston, Massachusetts

I. Introduction

Serious climbers, trekkers, and casual travelers to high altitude are commonly impressed with the extent to which the experience is marred by restless and nonrefreshing sleep. The subjective assessment is usually of frequent awakenings with a sense of suffocation. There is also a subjective impression of very little time spent asleep, which it turns out is only partially correct. Although once viewed as an indication of altitude maladaptation, current evidence suggests that this sleep disturbance may instead be an unfortunate byproduct of effective general adaptation. In this chapter we review the nature of disturbed sleep at altitude with attention to mechanisms, functional consequences, and therapeutic aspects.

II. Physiological Setting

The physics of altitude ascent is the decrease in atmospheric pressure with the result that, despite constancy of the content (fractional concentration) of oxygen, the partial pressure or driving force for oxygen (the product of atmospheric pressure × fraction) falls. This reduction in ambient oxygen tension leads to arterial hypoxemia, which is substantially, but not completely, mitigated by an immediate increase in ventilation attributed largely to the action of peripheral chemoreceptors, mainly the carotid bodies. This ventilatory response supports oxygenation but comes at an important price—the development of hypocapnic alkalosis, a potent inhibitor of ventilation and of the ventilatory response to hypoxia. This combination of respiratory stimulation by hypoxia and inhibition by hypocapnic alkalosis constitutes the respiratory dilemma of altitude of which disturbed sleep is a prime consequence. The adaptive solution to this dilemma is the phenomenon of ventilatory acclimatization to high altitude (Chapter 7), a central feature of which is a net augmentation of ventilation due largely to a progressive enhancement of the peripheral chemosensitivity to hypoxia. This increase in ventilation leads to improved oxygenation, progressive hypocapnia, and a less well understood but likely important diminution in the inhibitory action of hypocapnic alkalosis.

III. Sleep Disturbance at Altitude

A. Acute Ascent

Sleep disruption at altitude was described as early as the middle of the nineteenth century by Tyndall, Egli-Sinclair and Mosso (1). However, systematic study of its origins has occurred only relatively recently.

Subjective and objective sleep disturbance is nearly invariable with ascent if the altitude increment is sufficiently great and the ascent is rapid. Altitudes of 7600 m tend to produce disruption in all subjects, even with gradual ascent, while lesser altitudes of <5000 m do this with rapid ascent. Lesser, slower ascents produce less consistent disruption (2–10).

Typical subjective assessments are, as mentioned, those of restless, sparse sleep punctuated by periodic awakening with a sense of suffocation, deep breathing, and subsequent return to sleep (4,8,11,12). Much of this is echoed by objective findings that show frequent awakenings, a shift from deeper (stages 3 and 4) to lighter (stages 1 and 2) sleep, with decreased rapid eye movement (REM) sleep. However, despite subjective impression of sleeplessness, there is usually little or no decrement in total sleep time (2,6–8,13,14). Studies over a range of altitudes suggest that the decrement in subjective sleep quality is more clearly related to fragmentation of sleep by arousals than to alterations in sleep duration or stage distribution (2).

B. Breathing During Sleep at High Altitude

The most striking feature of sleep at altitude is the frequent and persistent change in respiratory pattern, with clusters of breaths of increasing depth, followed by a progressive decline eventuating in hypopnea or apnea (2,4,8,9,15–17) (Fig. 1). Various terms are used to describe this pattern, the most common being periodic breathing. Some authors prefer the designation Cheyne-Stokes respiration, which in accordance with early descriptions of patients in heart failure (18) specifically designates

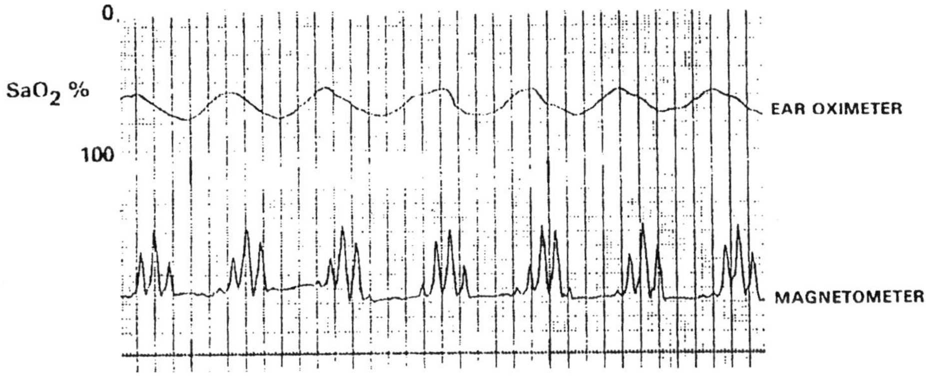

Figure 1 Breathing pattern and arterial oxygenation in a normal subject during sleep at high altitude (4300 m). There is characteristic monotonously repetitive, machinery-like periodic alternation of apnea and repetitive clusters of hyperpneic breaths, which continues throughout much of sleep in most individuals early after ascent. (From Ref. 13.)

a crescendo-decrescendo respiratory pattern. However, because there are differences between the respiratory pattern at altitude and the Cheyne-Stokes breathing of heart failure (discussed below) and because the term periodic breathing is the most common parlance, we shall use it here. However, this is with the intent to indicate a waxing-waning pattern similar to that of the Cheyne-Stokes variety.

The temporal pattern of periodic breathing and its linkage to sleep stages show some night-to-night and considerable intersubject variation (2,3,5–8,13,19,20), with periodicity most evident early in sleep and during light sleep stages. Indeed, in some subjects periodic breathing occurs in wakefulness, especially during periods of drowsiness (21,22). The pattern, as mentioned, almost invariably has the classic crescendo-decrescendo, Cheyne-Stokes, morphology, but with a relatively short cycle length (12–34 seconds), which shortens with increasing altitude (9,23,24) and differs from the longer 40- to 90-second cycles seen in patients with heart failure (25). The major contributor to the periodic oscillation of ventilation is the fluctuation in tidal volume with less obvious cycling of breathing frequency. However, subtle parallel variations in frequency contribute to changes in ventilation (23). Periodicity at altitude, as at sea level, is often initiated by movement and/or a deeper breath with a transient increase in hypocapnia (26). As mentioned, periodic breathing at altitude is also most prevalent in light, stage 1 and 2 sleep and less common in deeper stages of NREM sleep, although these stages are relatively rare after acute ascent. The most striking influence of sleep stage on periodic breathing is the observation of many (7,8,13) but not all investigators (4,6) that REM sleep promptly reverts periodicity to regular breathing.

As at sea level, sleep at altitude is associated with a reduction of ventilation and oxygenation below waking values and to similar relative extent as at sea level

(10). However, at altitude, awake oxygenation and P_{CO_2} are lower and thus closer to critical thresholds (see Fig. 2). Oxygen tensions are closer to the thresholds for hemoglobin desaturation and for stimulation of ventilation, and carbon dioxide tensions are closer to the dog-leg of the hypercapnic response curve where CO_2 loses its ventilatory stimulus potential. Under such conditions variation in ventilation produce changes in blood gases, which have enhanced ventilatory impact, as will be discussed below.

Although it might seem that periodic breathing would lower average ventilation, most studies show little effect, and when a difference is found there appears to be, on average, higher ventilation and improved oxygenation during periodic than during nonperiodic breathing (5,7). Potential reasons for this are discussed below.

C. Sleep Architecture

As mentioned, frequent arousal is a common subjective complaint of sleep early after ascent and is an especially prominent feature of polysomnographic recordings, which commonly show a marked increase in arousals (2,6–8,13,14) (see Fig. 3). Many are very brief micro-arousals, lasting but a few seconds, while others are longer awakenings, which persist for several minutes. The brief arousals are most commonly synchronous with the transition from the end of an apnea or hypopnea to the onset of hyperpnea. As will be discussed below, the precise sequence of these events and the mechanisms of arousal remain uncertain. However, although arousals and periodic breathing are linked, this linkage is only partial, as seen, for example,

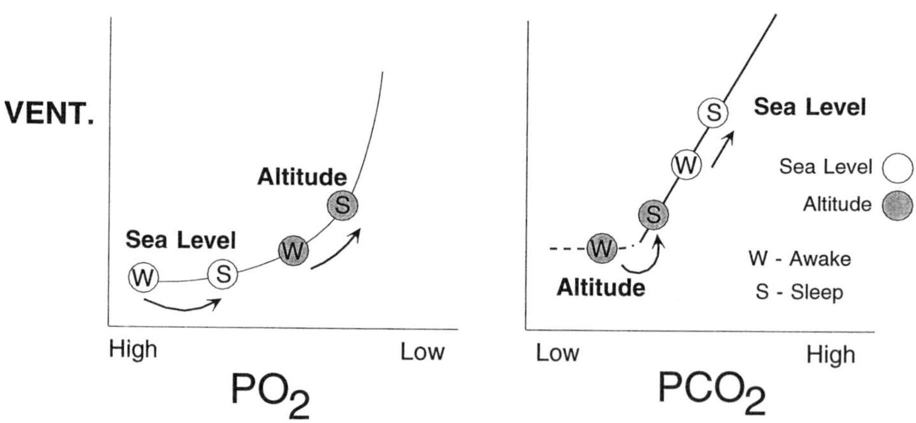

Figure 2 A schematic illustration of the relationship of blood gas set points to ventilatory responses during wakefulness and sleep at low and high altitude. At altitude, oxygen tensions move to the steeper portion of the response curve and P_{CO_2} shifts toward the dog-leg of the response. The result is that small changes in either P_{O_2} or P_{CO_2} have greatly enhanced effects on ventilation than at low altitude.

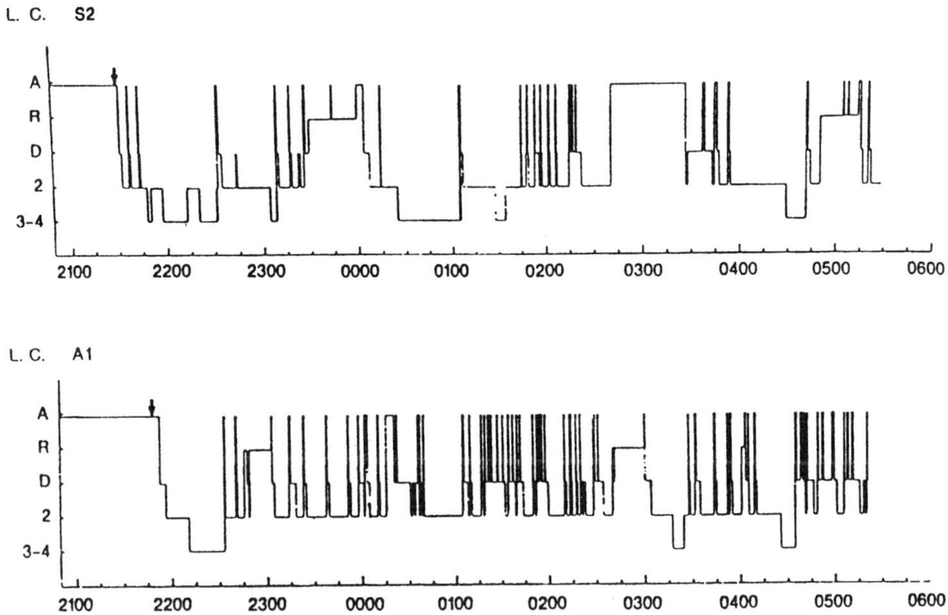

Figure 3 All-night sleep plots for a single subject during a night at sea level (top) and the first night at altitude (bottom). Time in hours is plotted on the horizontal axis—lights out occurred at the small vertical arrow in each plot. Sleep stage is shown on the vertical axis, with A (awake), R (REM) sleep; D (drowsiness, stage 1) sleep; 2, stage 2 sleep; 3–4, stages 3 and 4 sleep combined. Sleep at high altitude was associated with increased fragmentation by frequent awakenings and with a reduction in stage 3/4 sleep. (From Ref. 8.)

with oxygen administration, which abolishes periodic breathing but not the increased frequency of arousal (4–8).

The other relatively consistent alteration in sleep after ascent is the relative paucity of deeper NREM sleep (stages 3 and 4) with an increase in light, stage 1, sleep (4,7,8). There seems to be no consistent alteration in the amount of REM sleep reported to be either unchanged (7,8) or decreased (2,6). A study in rats showed that the stage shifts induced by ambient hypoxia can be reproduced by administration of carbon monoxide (27), which induces central hypoxia without respiratory stimulation. This observation suggests a role for brain hypoxia in the altered distribution of sleep stages.

Remarkably, despite the common subjective complaint of sleeplessness; total sleep time, as mentioned, is generally found to be well maintained at altitude. This disparity provided early evidence pointing to the importance of sleep continuity in the apparent subjective quality of sleep and suggested that sleep fragmentation despite normal cumulative sleep duration might produce the impression of sleeplessness (8,28). Reasons for preservation of total sleep time despite the disruptive effect of periodic breathing are unknown. However, it is commonly observed (although largely unstudied) that ascent to altitude induces sleepiness that is consistent, prompt, and often profound. Although this is often presumed to be due to hypoxia, a recent study suggests a marked apparent proclivity for human subjects to fall asleep quickly following brief voluntary hyperventilation at sea level pointing to a potential role of hypocapnia (29). In either case, the hypnotic effect of altitude could contribute to the maintenance of total sleep duration.

D. Relationship of Sleep Disturbance and Periodic Breathing to Altitude Syndromes

Rapid ascent in susceptible individuals can lead to two well-recognized syndromes of acute altitude maladaptation: acute mountain sickness (AMS) and high-altitude pulmonary edema (HAPE), which are discussed in Chapters 22 and 23. These syndromes are often most severe after awakening from sleep and are most common in the early postascent period when sleep disturbance and respiratory periodicity are also most pronounced, suggesting an exacerbation by events in sleep. However, it seems that they are un- or even inversely related (3,4). In subjects with these syndromes periodic breathing is commonly replaced by an irregular pattern (see Fig. 4). For reasons to be discussed, periodic breathing may be marker of strong hypoxic chemosensitivity, which may promote better oxygenation at altitude and relative freedom from these maladaptive syndromes.

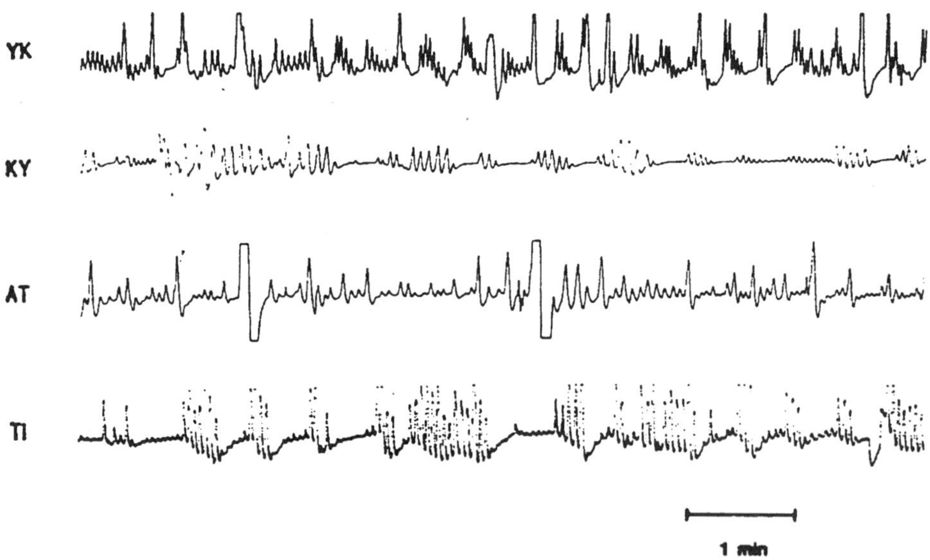

Figure 4 Irregular, nonperiodic, breathing during sleep at 2850 m in four subjects susceptible to high-altitude pulmonary edema. (From Ref. 3.)

E. Influence of Acclimatization

The greatest decrements in subjective quality of sleep and largest increments in respiratory periodicity and arousals occur during the first few nights following ascent. With time and likely with progressive ventilatory acclimatization, there is generally progressive improvement in sleep quality and lessening of respiratory periodicity (4,8,10,15). However, at very high altitudes (>5000 m) periodicity tends to persist (9,17), with one report describing periodic breathing during sleep after several days at 4150–4846 m (30) and others suggesting that periodicity may persist indefinitely at very high altitudes (9,17).

F. Sleep in Chronic High-Altitude Residents

Sleep in natives or long-term residents of altitude is generally characterized by few of the subjective symptoms so prevalent in sojourners. However, objective findings vary, in part as a reflection of interindividual differences and because of differing altitudes at which measurements are made. Most studies show that sleep architecture is generally similar to that of normal subjects at low altitude (31–34). Breathing during sleep often shows periodic oscillations, but unlike findings in sojourners, apneas and arousals are typically absent (17,31) (see Fig. 5). Some exceptions include the occurrence of nonperiodic central apneas in polycythemic altitude natives (32) and periodic breathing in children native to the Andes with relatively high hypoxic ventilatory responses, but not in Himalayan children in whom hypoxic responses were lower (20). Periodic breathing was noted in some natives of 4300 m (34) in whom there was a broad range of hematocrit (48–64), which was negatively correlated with apnea and arousal. Polycythemia at altitude is related to the extent of hypoxemia (35) and likely to low hypoxic chemosensitivity. Thus, a low ventilatory

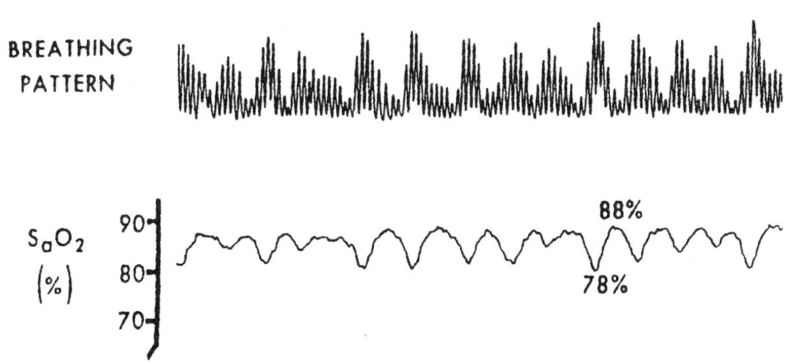

Figure 5 Breathing pattern and arterial oxygenation in a normal subject during sleep at high altitude (4300 m). Such a pattern is seen throughout much of sleep in most individuals after ascent. There is characteristic monotonously repetitive, machinery-like periodic alternation of apnea and repetitive clusters of hyperpneic breaths. (From Ref. 13.)

response to hypoxia may underlie the linkage of polycythemia and absence of periodic breathing in long-term residents of altitude.

Depending on length of stay and extent of altitude, long-term exposure leads to progressive, often nearly complete blunting of ventilatory sensitivity to hypoxia (16), which is more pronounced in adults than in children (36) (Chapter 6). The improvement of arterial oxygenation and polycythemia produced by medroxyprogesterone acetate, which augments the hypoxic ventilatory response in long-term altitude residents, may reflect the underlying role of decreased hypoxic chemosensitivity in the development of polycythemia in these individuals.

IV. Mechanisms of Periodic Breathing at Altitude

A. Historical Aspects

As pointed out by Ward (1), the earliest observation of Cheyne-Stokes breathing at low altitude may have been that of Hippocrates followed by that of Hunter in 1781. Later, Cheyne (1818) and Stokes (1854) clearly described in cardiac patients the typical decrescendo-crescendo breathing pattern, which now carries their names. Similar periodic breathing at high altitude in normal subjects was observed by Tyndall in 1857 and subsequently by Egli-Sinclair and Mosso in 1893–94.

Subsequent deceptively simple experiments by Douglas and Haldane explored physiological mechanisms and highlighted the central importance of both hypoxia and hypocapnia in the evolution of Cheyne-Stokes respiration (21,37). They were aware that during wakefulness, breathing was characterized by intrinsic momentum, a "flywheel" effect, which caused breathing to continue despite brief episodes of hypocapnia. However, they also showed that respiratory periodicity could be readily induced in awake normal sea level subjects by intense sustained voluntary hyperventilation with room air which elicited apnea followed by Cheyne-Stokes breathing. In contrast hyperventilation with an oxygen-enriched gas produced only apnea, without periodic breathing, indicating that the development of hypoxia during apnea was important in the subsequent induction of Cheyne-Stokes periodicity (see Fig. 6). In an ingeniously simple experiment, they showed the specific role of phasic hypoxia by the application of dead space containing the CO_2 absorber, soda lime. This produced rebreathing-induced hypoxia without the usual attendant hypercapnia. When the resulting progressive increase in tidal volume exceeded the dead space and intrained "fresh air" replete with oxygen, apnea occurred leading to recurrent hypoxia and perpetuation of the cycle. Similarly, application of a simple tubular dead space which produced intermittent hypercapnia also induced periodicity indicating that phasic hypoxia and hypercapnia were each capable of inducing periodic breathing.

B. Theoretical Models

Thus, early on, oxygen and carbon dioxide were ascribed important roles in the respiratory periodicity typical of sleep at altitude. These ideas, buttressed by much recent information, have led to unified, feedback-based views of respiratory period-

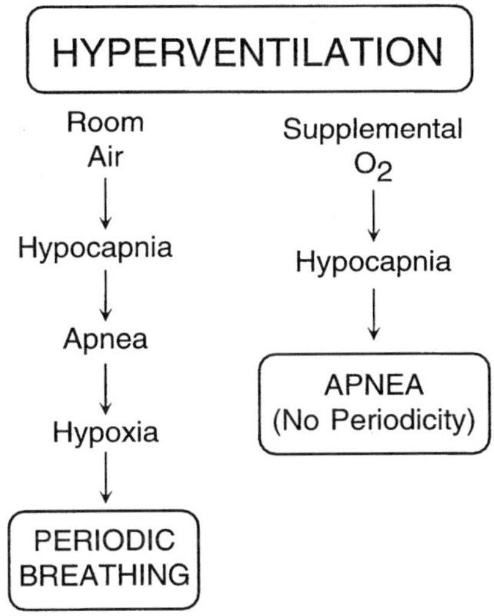

Figure 6 Summary of findings of Haldane and Douglas, who showed that hyperventilation with room air induced hypocapnia followed by apnea and hypoxia with subsequent periodic breathing (left). However, when apnea-induced hypoxia was prevented by inspired oxygen supplementation, apnea again occurred, but without subsequent periodic breathing (right). These findings pointed to the combined importance of ventilatory inhibition by hypocapnia and stimulation by hypoxia in the development of periodicity.

icity. These constructs emphasize the roles of increased controller gain, decreased system damping, and delayed information transfer (38,39). At high altitude increased controller gain is probably most relevant where a low basal Po_2 moves the resting arterial tension to the steeper, curvilinear portion of chemoreceptor hypoxic response curve (Fig. 2). Under these conditions small changes in Po_2 have large effects on ventilation and the slope (gain) of the response curve increases with further hypoxia. In contrast the contribution of the stable, linear gain of the CO_2-pH response is diminished by hypocapnic alkalosis. Taken together these alterations in respiratory control set the stage for ventilatory over- and undershoot and respiratory periodicity.

These concepts suggest a mechanism for altitude-induced periodic breathing during sleep shown in Figure 7 that serves as a framework for the discussion that follows.

C. Altitude Ascent

As discussed, ascent moves the basal arterial oxygen tension to steeper portion of the hypoxic ventilatory response curve (Fig. 2), which augments the ventilatory effects of further decrements in Po_2. This, in turn, promotes an hypoxia-induced ventilatory overshoot. The associated hypocapnia moves resting arterial Pco_2 toward and eventually below the apneic threshold and ventilation ceases.

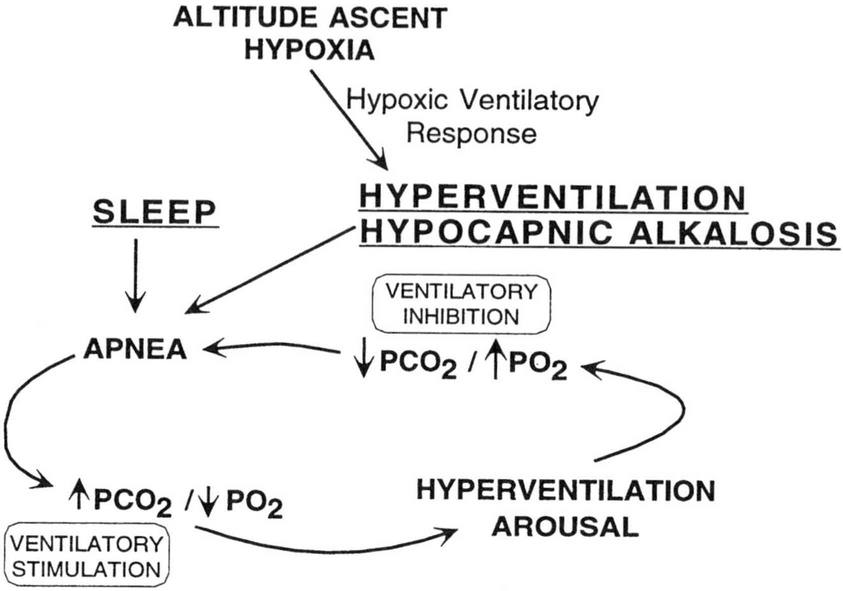

Figure 7 Schematic summary of mechanisms responsible for periodic breathing during sleep at altitude. Altitude-induced hypoxia stimulates increased ventilation. Ventilatory inhibition by hypocapnic alkalosis together with the decreased ''wakefulness drive'' of sleep promotes apnea. During apnea, increasing hypoxemia and lessening of hypocapnic alkalosis stimulates resumption of ventilation and arousal. This augments oxygenation and increases hypocapnic alkalosis permitting recurrence of apnea.

Hypoxemia reflects the net balance of decreased partial pressure of oxygen and the partially compensatory hypoxemia-induced hyperventilation. The importance of hypoxemia in periodic breathing during sleep at altitude is indicated by the prompt restoration of regular breathing produced by oxygen administration (8,10).

The preascent hypoxic ventilatory response, or drive, varies considerably among individuals due in part to genetic effects (40–42). At altitude, the initial postascent response is progressively increased by acclimatization (10) but later is blunted in long-term altitude residents by chronic hypoxic exposure (16) (see Chapter 5). Preascent differences in this response are predictors of blood gases after ascent-persons with high responses generally have higher Po_2s and lower Pco_2s upon arrival at altitude (43) and greater extent of periodic breathing (10). These drives also have substantial implications for function at and adaptation to altitude. Preascent hypoxic ventilatory responses tend to be high in individuals capable of very high-altitude climbs (44,45), while low responses are common in persons who experience acute mountain sickness and high-altitude pulmonary edema (43,46–49). As mentioned, both hypoxic and hypercapnic ventilatory responses increase over time at high altitude. The hypoxic response shows little change in the first 24 hours, but thereafter increases progressively and markedly over the next several days to weeks during acclimatization (10,50,51). The hypercapnic response shows an immediate increase in slope followed by a progressive left shift (Chapter 5).

That the hypoxic response contributes to the genesis of sleep-associated periodic breathing is suggested by the correlation of the extent of periodicity during sleep on the first few nights at altitude with the slope of pre-ascent hypoxic ventilatory response (4,5,10,19,52) (see Fig. 8). Further, the carotid body stimulant almitrine, which augments hypoxic ventilatory responses, increases respiratory periodicity (53). However, some studies find no clear relationship of periodicity to hypoxic ventilatory response (9,54).

A less consistent relationship of periodic breathing to the hypercapnic ventilatory response is observed in some studies (5,10).

With acclimatization the relationship of periodicity to increased hypoxic response seen early after ascent is if anything reversed—acclimatization increases the hypoxic response and decreases periodic breathing (4,8,10,15) (see Fig. 9). The likely explanation is that acclimatization provides an escape from the ventilatory dilemma of acute ascent—ventilatory stimulation by hypoxia versus inhibition by hypocapnic alkalosis. Acclimatization seems to allow greater ventilatory correction of hypoxia, while blunting the inhibition by hypocapnic alkalosis. In other words,

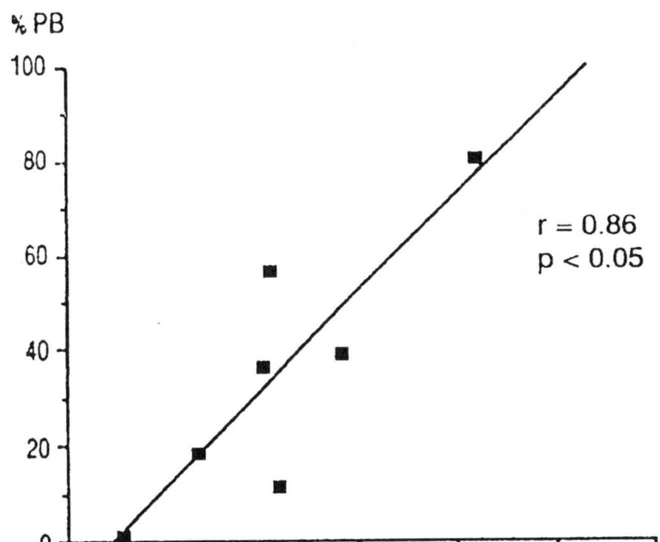

Figure 8 Relationship of hypoxic chemosensitivity to the development of periodic breathing during sleep is illustrated by the positive correlation of prevalence of periodic breathing in sleep at 6542 m with the hypoxic ventilatory response. (From Ref. 4.)

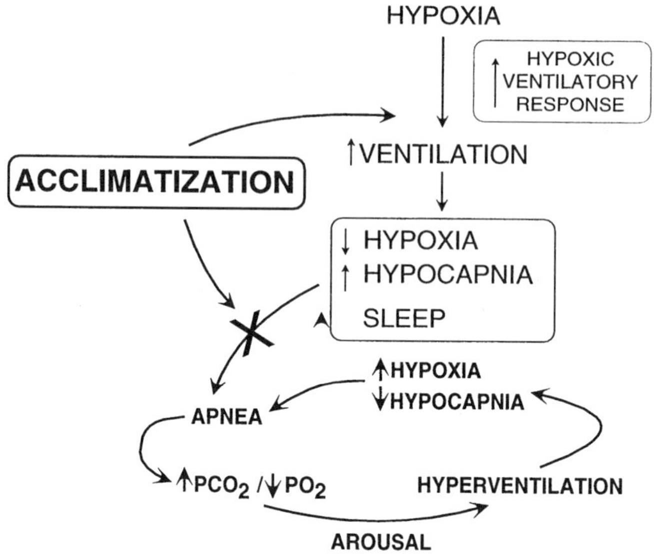

Figure 9 Schematic representation of potential mechanisms by which acclimatization decreases periodic breathing during sleep at high altitude. Despite an increase in hypoxic ventilatory response, which in newcomers is associated with increased periodicity, acclimatization is accompanied by decreased periodic breathing, which may reflect a decrease in the tendency of hypocapnia to induce apnea (a lowering of the apneic threshold).

the threshold at which P_{CO_2} triggers apnea appears to fall with acclimatization. The net result is increased hyperventilation with reduced phasic ventilatory inhibition and decreased apnea and periodic breathing (10).

The locus and mechanism of this lowering of apneic threshold are largely unknown. Specifically it is unclear whether loss of ventilatory inhibition by hypocapnia occurs at peripheral or central chemoreceptors, nor is the molecular nature of this effect understood. It is noteworthy that decreased hypocapnic inhibition is evident despite increasing hypocapnia. It is possible that the decreased inhibitory action of hypocapnia reflects a correction of pH due perhaps to the generation of a relative metabolic acidosis. However, arterial blood and cerebrospinal fluid generally show little or no evidence of correction of respiratory alkalosis during early acclimatization (Chapter 5). However, local acidification within the central nervous system, perhaps by lactate generation, might be involved (55), but a study using NMR spectroscopy showed no decline in brain tissue pH during a 7-day exposure to simulated altitude (56). Nonetheless, several observations suggest an important role for the central chemoreceptor in setting the apneic threshold. A central action is supported by the observation that hypocapnia must last at least 1 minute to induce apnea, and the more prolonged the hypocapnia the longer the apnea (57,58). This is consistent with a role for the slower-responding central chemoreceptor. Further evidence in support of a central action comes from a study in dogs with extracorporeal perfusion isolated to the carotid body, which showed that selective peripheral chemoreceptor hypocapnia during sleep produced hypoventilation, but not apnea (59).

Hypocapnia, a consequence of ventilatory stimulation by hypoxemia, plays a critical role in generation of periodic breathing at altitude. As mentioned earlier, P_{CO_2} seems to provide the principal drive to resting ventilation, particularly during sleep and withdrawal of this critical stimulus likely is a major contributor to the genesis of apnea. The early studies by Douglas and Haldane showed that hyperventilation induces apnea in normal subjects and is prevented by addition of carbon dioxide to the inspired air. They also showed that hyperventilation-induced apnea was followed by Cheyne-Stokes breathing, but hyperventilation with added oxygen produced only apnea followed by resumption of regular breathing. Thus the critical contributions of both hypoxia and hypercapnia were evident. Further indication of the important role of hypocapnia is evident at sea level in normal subjects in whom periodic breathing at sleep onset is often preceded by transient increases in ventilation (26,52). Linkage of phasic hyperventilation to onset of apnea is also evident in patients with idiopathic central apnea and with Cheyne-Stokes breathing in heart failure (60). Indeed, hypocapnia during wakefulness is a common feature of patients with Cheyne-Stokes breathing which, as mentioned, likely brings P_{CO_2}s closer to the apneic threshold (25,60). The critical role of hypocapnia in periodic breathing is suggested by the regularization of breathing produced by CO_2 administration during central apnea at sea level and at altitude (8,10,61). Although in one study administration of CO_2 prevented apnea, periodic breathing continued (2). Hypocapnia could induce phasic respiratory inhibition with periodicity by actions at the peripheral and/or central chemoreceptors.

The importance of sleep in the genesis of periodic breathing was evident to Douglas and Haldane through its weakening of the ''flywheel'' inertial feature of breathing during wakefulness (22) wherein breathing continues despite hypocapnia, now commonly referred to as the ''wakefulness stimulus'' (62). Bulow also noted the frequent onset of periodic breathing in normal subjects with sleep onset (26). An important advance was the finding by Sullivan et al. that in dogs withdrawal of respiratory stimuli by induction of hyperoxia or alkalosis, which produces no apnea during wakefulness, did so readily during sleep (63). Collectively these findings demonstrate that maintenance of regular breathing is much more dependent on chemical (P_{O_2}, P_{CO_2}, pH) stimulation during sleep than in wakefulness. Put another way the flywheel effect or respiratory momentum of wakefulness is lost during sleep. This sleep-induced requirement for respiratory chemical drive is evident in normal humans in whom hypocapnia, induced by positive pressure ventilation, induces apnea during NREM sleep when end-tidal P_{CO_2} is lowered by as little as 1–3 mmHg below awake resting levels (64). This was insufficient to induce apnea during wakefulness. Conversely, periodic breathing induced by simulated altitude exposure is readily reversed by selective lessening of hypocapnia by administration of CO_2 with addition of nitrogen to prevent amelioration of hypoxemia (52). Thus, exquisite sensitivity to respiratory inhibition by hypocapnia is characteristic of NREM sleep such that very slight decrements in P_{CO_2} below resting values are sufficient to trigger apnea or periodic breathing. This has led to the concept of an apneic P_{CO_2} threshold set only slightly below the awake setpoint, which may explain the often observed onset of periodic breathing after a deep breath (26). It is possible that hyperpnea through increased tidal volume per se may also contribute to development of apnea (57).

Hypocapnia and hypoxemia interact reciprocally in induction of periodic breathing. Both are important determinants of the net level of chemosensory stimulation. Thus, the reduced stimulus potential of P_{CO_2} in hyperoxia may account for the finding that regularization of breathing during sleep at altitude requires greater elevation of P_{CO_2} during oxygen administration than when breathing ambient air (10,52) (see Fig. 10).

Thus the ventilatory response to the ambient hypoxia of altitude produces hypocapnia, which, during sleep, becomes a major inhibitor of respiratory rhythm triggering apnea. The ensuing apnea induces ventilatory stimulation by lessening hypocapnia and lowering oxygen tension, moving it to a steeper portion of the hypoxic response curve and enhancing its stimulus potential.

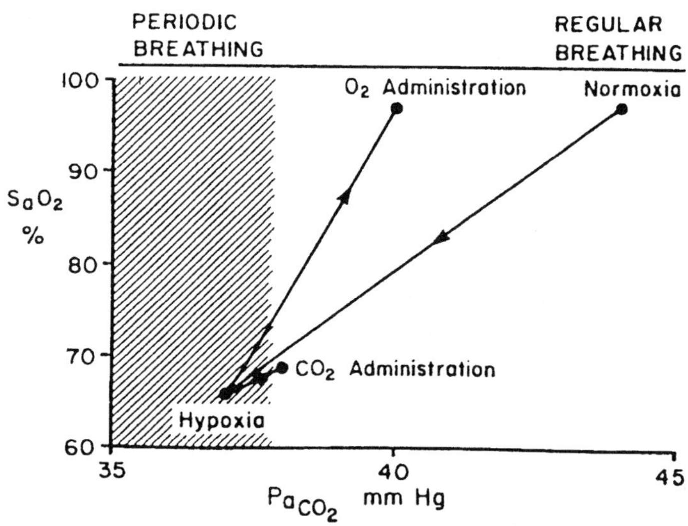

Figure 10 Relationship of respiratory pattern to Pa_{CO_2} and Sa_{O_2} during sleep. This figure is a schematic representation of data of Berssenbrugge et al. (52) and demonstrates that reduction of P_{CO_2} is a critical determinant of periodic breathing in normal human subjects during hypoxia—even a small increase in P_{CO_2} after restores normal breathing. Oxygen administration also reverses periodicity, which may reflect dual effects of removal of ventilatory inhibition by decreased hypocapnia and reduction of stimulation by reversal of hypoxia. (From Ref. 52.)

Apneas are frequently followed by arousal, which, as mentioned earlier, occur in close synchrony to apnea termination (so closely that it is often unclear whether arousal stimulates resumption of breathing) or the reverse, that it is the resumption of breathing that causes arousal. Consequently, it is uncertain whether arousal is triggered by apnea-induced asphyxia via chemosensitivity or by mechanical stimuli. It is also possible that CNS command signals responsible for the resumption of breathing act centrally to stimulate arousal. Observations in low-altitude patients suggest that apnea-associated arousals correlate poorly with chemical variables but have a consistent relationship to respiratory effort (65), suggesting a mechanistic linkage to resumption of ventilation rather than to blood chemistry.

Whatever the precise cause, it seems likely that arousal, or a lightening of sleep stage, makes an important contribution to periodicity by restoring the wakefulness drive and promoting restoration of breathing. This might reflect both a lessening of sleep-induced decrement in chemosensitivity and a lowering, or weakening, of the sleep-associated hypocapnic "apneic threshold."

Although most arousals occur as part of the periodic breathing cycle, this relationship is variable and probably accounts for only a portion of the excess arousal frequency at altitude. Indeed, a clear dissociation is seen with oxygen administration, which abolishes periodic breathing but fails to prevent arousals (8). One study suggests that, at low altitude, arousals from stage 2 sleep are preceded by changes in heart rate variability, which indicate central sympathetic activation (66). It might be that some aspect of altitude exposure, perhaps hypoxia, acting peripherally or centrally, triggers arousal through such a mechanism.

Whatever their cause, fragmentation of sleep by frequent arousals probably contributes substantially to decreased function during wakefulness at altitude. Studies of periodic arousal in healthy subjects at sea level show that frequent nocturnal arousals induce profound deterioration in mood (67) with marked impairment of problem solving during subsequent wakefulness (68). Frequent arousals also produce the perception of poor quality sleep, despite the fact that subjects recall only a small fraction of total awakenings (68). These findings resemble those at altitude where a decline in perceived sleep quality is proportional to frequency of arousal (2). The extent to which these effects of sleep fragmentation interact with those of hypoxia on intellectual function and mood are largely unknown, but it seems likely that together they contribute to impaired judgment and consequent dangers of high altitude climbs.

Therefore, periodic breathing is comprised of a cycle in which hyperpneic phase is stimulated by apnea-associated hypoxemia, with attendant shift to steeper hypoxic drive, augmented by the associated rise of P_{CO_2} above its apneic threshold, and likely further increased by arousal or a lightening of sleep. This is followed by a hyperventilation-induced stimulus withdrawal, a consequence of lessened hypoxia, increased hypocapnia, and restoration, or deepening, of sleep with withdrawal of wakefulness drive leading to apnea and periodicity.

Just as cycling is initiated by sleep onset and transient increases in ventilation; several factors are known to terminate this pattern. The most frequent spontaneous factor is sustained awakening, which likely acts by lowering the sleep-associated hypocapnic apneic threshold by relieving hypoxia and restoring the wakefulness effect. Another is REM sleep, which usually produces near-immediate onset of regular breathing (7,8,13), although some reports indicate continued periodicity during REM periods (2). Mechanisms of this regularizing effect of REM sleep are uncertain but may reflect the profound differences in ventilatory control in REM versus NREM sleep. In REM sleep ventilation is much less responsive to chemosensory regulation and dominated by other input. Thus, the impact of enhanced ventilatory drive at end-apnea responsible for hyperventilation and impact of subsequent withdrawal of respiratory stimuli, which promote apnea, would be greatly blunted during REM periods. Alternatively, a REM-associated decrease in alveolar ventilation with lessening of hypocapnia might prevent apnea, but the prompt regularization with REM onset seems to provide insufficient time for resolution of hypocapnic alkalosis.

D. Other Contributors to Periodic Breathing

The above scheme is a deliberate simplification focused on the best documented determinants of periodic breathing in sleep at altitude. However, it is clear that this is not likely the whole story and there may be several other contributors (69).

It is possible that increased upper airway resistance contributes to the development of apnea. However, major obstruction seems unlikely. There is no association with usual risk factors for obstruction such as obesity, snoring is usually minimal or absent, and measures of respiratory effort indicate that altitude apnea is usually central (2,4,5,7). Yet some narrowing of the upper airway is probably common with all apneas, most of which are ''mixed'' with associated central and obstructive features. Indeed even obstructive apneas often begin as central events with obstruction becoming evident only after establishment of apnea with resumption of respiratory effort and relief of obstruction may uncover residual central apneas (70). Airway narrowing is also evident with central apneas (71), which may account for the improvement of central apneas with continuous positive airway pressure (72,73). This is likely a reflection of the phasic ''respiratory'' nature of the activation pattern of many upper airway muscles such that loss of respiratory drive reduces activation of both classical inspiratory muscles and those responsible for maintenance of upper airway patency. Still, the extent to which increased upper airway resistance contributes to periodic breathing at altitude remains unknown.

Because of the rapidity with which apnea follows hyperpnea, it has been suggested that reflexes other than those linked directly to chemosensory stimuli may be involved (69). Some potential contributors include inhibitory vagal feedback stimulated perhaps by the high tidal volumes of hyperpnea or inhibitory consequences of baroreceptor reflexes induced by phasic hypertension induced by apnea and/or arousal.

Finally, delayed lung-to-chemoreceptor transit is a well-recognized theoretical cause of periodic breathing with circulatory delay and has a time-honored linkage to the Cheyne-Stokes respiration of heart failure. However, prolonged circulation time tends to produce periodic breathing with long cycle lengths (40–90 s), which are proportional to circulatory delay (74), but at altitude, cycle lengths are relatively short (~20 s) (9,17,24). Estimated circulation times are brief, comparable to those at sea level (23), and resting cardiac output is variable but only slightly reduced at moderate elevation and increased at higher altitudes (75). Thus it seems unlikely that circulatory delay is important in the genesis of periodicity at altitude. It is, however, possible that the exaggerated oscillation of heart rate associated with periodic breathing (76,77) may indicate complex dynamic effects on circulation time and baroreceptor responses contributing to respiratory cycling.

E. Influence of Periodic Breathing on Average Ventilation

Although it would seem intuitively that periodic breathing is likely to result in net hypoventilation, as mentioned above, most evidence suggests that periodic breathing is associated with increased ventilation higher Sa_{O_2} (5) and a lesser incidence of syndromes of altitude maladaption. There are at least two potential reasons for this. First, it may reflect the association of periodicity with increased chemosensitivity with the latter setting the level of overall ventilation and periodic breathing appearing as a byproduct. In addition it may be that periodic breathing is mechanically and energetically efficient. Theoretical analysis suggests that the high tidal volumes of the hyperpneic phase enhance relative alveolar, Vs dead space, and ventilation and that apneas actually conserve respiratory muscle work. As a result, clusters of two to four large breaths separated by apneas are estimated to produce optimal oxygenation with lowest ''pressure costs'' (23,78). Such a pattern might be especially efficient at altitude where decreased air density would further reduce the respiratory pressure cost.

V. Treatment

Staged, gradual ascent to altitude is an effective way to prevent or blunt sleep-related symptoms but may be inconvenient. As a result, pharmacological approaches have evolved.

A. Carbonic Anhydrase Inhibitors

The most common and best-studied approach to the treatment of disturbed sleep and periodic breathing at altitude is the preascent administration of a carbonic anhydrase inhibitor, usually acetazolamide. This agent has the added advantage of also reducing symptoms of acute mountain sickness (Chapter 22). Several studies show that these agents raise ventilation during wakefulness and sleep, with improved oxygenation and greater hypocapnia. Hypoxic ventilatory responses are shifted upward (higher ventilation at all points), but the steepness of relationship is unchanged, indicating no potentiation of the response (13,79–81). The effects on the hypercapnic response are variable, and, for uncertain reasons, findings differ for progressive Vs steady-state measurements (80).

Carbonic anhydrase inhibitors produce marked reductions in periodic breathing during sleep with higher and less oscillatory arterial oxygen saturation (13,79) (see Fig. 11). Although not well studied, it appears that these agents also reduce arousals and improve subjective and objective sleep quality with enhancement of stage 2 sleep and decreased wakefulness (30).

The primary action of these agents is to block enzymatic hydration of carbon dioxide to carbonic acid with consequent induction of a bicarbonate diuresis and metabolic acidosis. Blockade of carbonic anhydrase could also promote accumulation of CO_2 in tissue. This might be the mechanism of the increase of cerebral blood flow induced by intravenous administration of acetazolamide (82). This could also stimulate central chemoreceptors (81). However, these central effects on blood flow are minimal or absent with the relatively low doses used for symptom reduction at altitude (83). Further, selective agents with action limited to the kidney seem to be as effective as acetazolamide (80). Thus, the mechanism of action is likely the renal tubular effect with induction of systemic metabolic acidosis with stimulation of ventilation and lessening of hypoxemia. However, carbonic anhydrase inhibitors have similar utility in eliminating central apneas at low altitude (84,85), where hypoxemia is not a likely factor, suggesting that induction of acidosis is the major contribution.

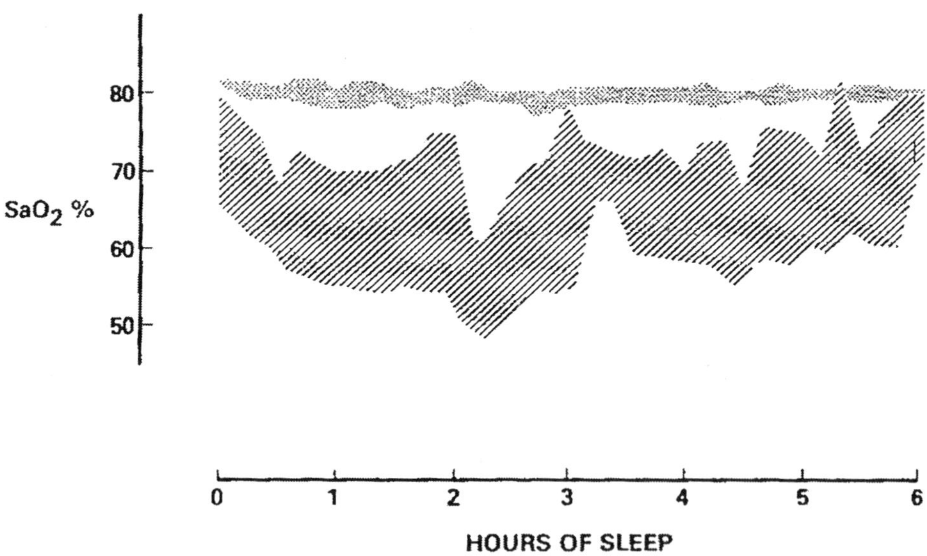

Figure 11 Average arterial oxygenation in a sleeping subject at altitude (5360 m) without (lower hatched area) and with (upper) acetazolamide. Treatment raised and stabilized arterial oxygen saturation. (From Ref. 93.)

That these agents reduce respiratory periodicity provides useful insights into pathophysiological mechanisms of periodic breathing, which improves despite worsening hypocapnia, which is so heavily implicated in the genesis of this respiratory dysrhythmia. The explanation of this apparent paradox may be that the apneic threshold is in fact a pH rather than a P_{CO_2} threshold. Although alkalosis and hypocapnia are typically coexistent at altitude, their dissociation by carbonic anhydrase inhibition points to the dominant role of pH. The utility of low-dose or renal-selective inhibition suggests that it may be blood pH rather that local tissue pH that is most important in hyperventilation-induced apnea. Taken together, the failure of these agents to augment hypoxic ventilatory response and the evidence, mentioned earlier, that periodic breathing does not result from hypocapnia confined to the carotid body (59) suggest the importance of central, rather than peripheral chemoreceptor pH in the genesis of periodicity and the response to carbonic anhydrase inhibitors.

B. Benzodiazepines

These agents constitute another potentially useful approach to the reduction of sleep difficulties at altitude. Benzodiazepines are able to substantially reduce the hypoxic ventilatory response (86) and were once thought to be hazardous in respiratory disorders of sleep. However, recent evidence suggests that in low doses they are relatively safe in such situations. Studies indicate that these agents do not increase sleep disordered breathing in the elderly (87) in nonselected apnea patients (88). In patients with chronic obstructive respiratory disease, midazolam, zolpidem, and triazolam produced no clear increase in apnea, hypopnea, or hypoxemia during sleep (89,90). Little or no adverse effect is seen with triazolam in patients with obstructive apnea (91). In patients with heart failure, temazepam decreased microarousals and improved daytime alertness with no change in the extent of periodic breathing or oxygenation in sleep (92).

Two studies indicate the safety and utility of these agents in the sleep disturbance of high altitude. Low-dose temazepam shortened sleep onset latency, decreased arousals, increased sleep efficiency, increased REM, and produced subjectively better sleep. These effects were unaccompanied by any clear changes in prevalence of periodic breathing and they augmented the favorable action of acetazolamide (30). Loprazolam is said to slightly reduce periodicity, augment slow wave sleep, and reduce wakefulness during acclimatization, but not in early nights after ascent (4).

C. Almitrine

This agent, which stimulates the carotid body and augments hypoxic ventilatory responses, improves arterial oxygenation during sleep at altitude but increases respiratory periodicity (53). That an agent that augments hypoxic chemosensitivity also increases respiratory periodicity illustrates the importance of the hypoxic ventilatory response in the development of periodic breathing at altitude. This effect would be expected to decrease the continuity and subjective quality of sleep, but this has not been directly studied.

D. Other Agents

Progestational agents such as medroxyprogesterone acetate substantially reduce periodic breathing with little change in oxygenation in sojouners (13) but have greater effects on oxygenation in long-term residents with chronic mountain sickness (31). Whether other drugs with major utility in the treatment of high-altitude pulmonary edema such as calcium channel blockers and glucocorticoids effect breathing, oxygenation and symptoms related to sleep remains unknown.

VI. Functional Significance

Thus, sleep disturbance at high altitude is largely but not entirely linked to periodic cycling of breathing with consequent sleep fractionation. These events reflect the physiological dilemma of altitude—the stimulation of ventilation that addresses the vital issue of oxygenation, at the expense of hypocapnia with its potent ventilatory inhibitory effects. The adaptive aspect is evident in the association of periodic breathing with high preascent hypoxic responsiveness and a relative freedom from syndromes of altitude maladaptation. The unfortunate price is sleep disruption with subjective decrements in daytime mood and likely intellectual function—recognized consequences of sleep fragmentation.

References

1. Ward M. Periodic respiration. A short historical note. Ann R Coll Surg Engl 1973; 52: 330–334.
2. Anholm JD, Powles AC, Downey Rd, et al. Operation Everest II: arterial oxygen saturation and sleep at extreme simulated altitude. Am Rev Respir Dis 1992; 145:817–826.
3. Fujimoto K, Matsuzawa Y, Hirai K, et al. Irregular nocturnal breathing patterns high altitude in subjects susceptible to high-altitude pulmonary edema (HAPE): a preliminary study. Aviation Space Environ Med 1989; 60:786–791.
4. Goldenberg F, Richalet JP, Onnen I, Antezana AM. Sleep apneas and high altitude newcomers. Int J Sports Med 1992; 13:S34–36.
5. Masuyama S, Kohchiyama S, Shinozaki T, et al. Periodic breathing at high altitude and ventilatory responses to O_2 and CO_2. Jpn J Physiol 1989; 39:523–535.
6. Mizuno K, Asano K, Okudaira N. Sleep and respiration under acute hypobaric hypoxia. Jpn J Physiol 1993; 43:161–175.
7. Normand H, Barragan M, Benoit O, Bailliart O, Raynaud J. Periodic breathing and O2 saturation in relation to sleep stages at high altitude. Aviation Space Environ Med 1990; 61:229–235.
8. Reite M, Jackson D, Cahoon RL, Weil JV. Sleep physiology at high altitude. Electroencephalogr Clin Neurophysiol 1975; 38:463–471.

9. West JB, Peters R, Jr., Aksnes G, Maret KH, Milledge JS, Schoene RB. Nocturnal periodic breathing at altitudes of 6,300 and 8,050 m. J Appl Physiol 1986; 61:280–287.
10. White DP, Gleeson K, Pickett CK, Rannels AM, Cymerman A, Weil JV. Altitude acclimatization: influence on periodic breathing and chemoresponsiveness during sleep. J Appl Physiol 1987; 63:401–412.
11. Ravenhill TH. Some experience of mountain sickness in the Andes. J Trop Med Hygiene 1913; 16:313–320.
12. Nicholson AN, Smith PA, Stone BM, Bradwell AR, Coote JH. Altitude insomnia: studies during an expedition to the Himalayas. Sleep 1988; 11:354–361.
13. Weil JV, Kryger MH, Scoggin CH. Sleep and breathing at high altitude. In: Guilleminault C, Dement WC, eds. Sleep Apnea Syndromes. New York: Liss, 1978:119–136.
14. Miller JC, Horvath SM. Sleep at altitude. Aviation Space Environ Med 1977; 48:615–620.
15. Douglas C, Haldane J, Henderson Y, Schneider E. Physiological observations made of Pikes Peak, Colorado, with special reference to adaptation to low barometric pressures. Phil Trans R Soc Lond 1913; 203:185–381.
16. Weil JV, Byrne-Quinn E, Sodal IE, Filley GF, Grover RF. Acquired attenuation of chemoreceptor function in chronically hypoxic man at high altitude. J Clin Invest 1971; 50:186–195.
17. Lahiri S, Maret K, Sherpa MG. Dependence of high altitude sleep apnea on ventilatory sensitivity to hypoxia. Respir Physiol 1983; 52:281–301.
18. Ancoli-Israel S, Engler RL, Friedman PJ, Klauber MR, Ross PA, Kripke DF. Comparison of patients with central sleep apnea. With and without Cheyne-Stokes respiration. Chest 1994; 106:780–786.
19. Lahiri S, Barnard P. Role of arterial chemoreflex in breathing during sleep at high altitude. Progr Clin Biol Res 1983; 136:75–85.
20. Lahiri S, Data PG. Chemosensitivity and regulation of ventilation during sleep at high altitudes. Int J Sports Med 1992; 13:S31–33.
21. Douglas C, Haldane J. The causes of periodic or Cheyne-Stokes breathing. J Physiol 1909; 38:401–419.
22. Douglas CG, Haldane JS. The regulation of normal breathing. J Physiol 1909; 38:420–440.
23. Brusil PJ, Waggener TB, Kronauer RE, Gulesian P, Jr. Methods for identifying respiratory oscillations disclose altitude effects. J Appl Physiol Respir Environ Exercise Physiol 1980; 48:545–556.
24. Waggener TB, Brusil PJ, Kronauer RE, Gabel RA, Inbar GF. Strength and cycle time of high-altitude ventilatory patterns in unacclimatized humans. J Appl Physiol Respir Environ Exercise Physiol 1984; 56:576–581.
25. Naughton M, Benard D, Tam A, Rutherford R, Bradley TD. Role of hyperventilation in the pathogenesis of central sleep apneas in patients with congestive heart failure. Am Rev Respir Dis 1993; 148:330–338.
26. Bulow K. Respiration and wakefulness in man. Acta Physiol Scand 1963; 59:1–110.
27. Pappenheimer JR. Hypoxic insomnia: effects of carbon monoxide and acclimatization. J Appl Physiol Respir Environ Exercise Physiol 1984; 57:1696–1703.
28. Bonnet MH. Performance and sleepiness as a function of frequency and placement of sleep disruption. Psychophysiology 1986; 23:263–271.

29. Ohi M, Chin K, Hirai M, et al. Oxygen desaturation following voluntary hyperventilation in normal subjects. Am J Respir Crit Care Med 1994; 149:731–738.
30. Nicholson A, Smith P, Stone B, et al. Altitude insomnia: studies during an expedition to the Himalayas. Sleep 1988; 11:354–361.
31. Kryger M, Glas R, Jackson D, et al. Impaired oxygenation during sleep in excessive polycythemia of high altitude: improvement with respiratory stimulation. Sleep 1978; 1:3–17.
32. Normand H, Vargas E, Bordachar J, Benoit O, Raynaud J. Sleep apneas in high altitude residents (3,800 m). Int J Sports Med 1992; 13:S40–42.
33. Coote JH, Stone BM, Tsang G. Sleep of Andean high altitude natives. Eur J Appl Physiol Occup Physiol 1992; 64:178–181.
34. Coote JH, Tsang G, Baker A, Stone B. Respiratory changes and structure of sleep in young high-altitude dwellers in the Andes of Peru. Eur J Appl Physiol Occup Physiol 1993; 66:249–253.
35. Weil JV, Jamieson G, Brown DW, Grover RF. The red cell mass–arterial oxygen relationship in normal man. J Clin Invest 1968; 47:1627–1639.
36. Byrne-Quinn E, Sodal IE, Weil JV. Hypoxic and hypercapnic ventilatory drives in children native to high altitude. J Appl Physiol 1972; 32:44–46.
37. Douglas CG, Haldane JS, Henderson Y, Schneider E. The physiological effects of low atmospheric pressures as observed on Pikes Peak, Colorado. J Physiol 1912; 85:65–67.
38. Khoo MC, Kronauer RE, Strohl KP, Slutsky AS. Factors inducing periodic breathing in humans: a general model. J Appl Physiol Respir Environ Exercise Physiol 1982; 53: 644–659.
39. Cherniack N, Gothe B, Strohl K. Mechanisms for recurrent apneas at altitude. High Altitude Man 1984; 129–140.
40. Collins DD, Scoggin CH, Zwillich CW, Weil JV. Hereditary aspects of decreased hypoxic response. J Clin Invest 1978; 62:105–110.
41. Fleetham JA, Arnup ME, Anthonisen NR. Familial aspects of ventilatory control in patients with chronic. Am Rev Respir Dis 1984; 129:3–7.
42. Kawakami Y, Yoshikawa T, Shida A, Asanuma Y. Genetic aspects of serum immunoglobulins and respiratory chemosensitivity. Respiration 1982; 43:101–107.
43. Moore LG, Harrison GL, McCullough RE, et al. Low acute hypoxic ventilatory response and hypoxic depression in. J Appl Physiol 1986; 60:1407–1412.
44. Schoene RB. Control of ventilation in climbers to extreme altitude. J Appl Physiol 1982; 53:886–890.
45. Masuyama S, Kimura H, Sugita T, et al. Control of ventilation in extreme-altitude climbers. J Appl Physiol 1986; 61:500–506.
46. Matsuzawa Y, Kobayashi T. [Exposure to high altitude: ventilatory control in relation to syndromes of high altitude]. [Japanese]. Nippon Kyobu Shikkan Gakkai Zasshi Jpn J Thoracic Dis 1992; 30:139–146.
47. Hackett PH, Roach RC, Schoene RB, Harrison GL, Mills WJ, Jr. Abnormal control of ventilation in high-altitude pulmonary edema. J Appl Physiol 1988; 64:1268–1272.
48. Hohenhaus E, Paul A, McCullough RE, Kucherer H, Bartsch P. Ventilatory and pulmonary vascular response to hypoxia and susceptibility to high altitude pulmonary oedema. Eur Respir J 1995; 8:1825–1833.
49. Matsuzawa Y, Fujimoto K, Kobayashi T, et al. Blunted hypoxic ventilatory drive in

subjects susceptible to high-altitude pulmonary edema. J Appl Physiol 1989; 66:1152–1157.

50. Sato M, Severinghaus JW, Bickler P. Time course of augmentation and depression of hypoxic ventilatory responses at altitude. J Appl Physiol 1994; 77:313–316.
51. Schoene RB, Roach RC, Hackett PH, Sutton JR, Cymerman A, Houston CS. Operation Everest II: ventilatory adaptation during gradual decompression to extreme altitude. Med Sci Sports Exerc 1990; 22:804–810.
52. Berssenbrugge A, Dempsey J, Iber C, Skatrud J, Wilson P. Mechanisms of hypoxia-induced periodic breathing during sleep in humans. J Physiol 1983; 343:507–526.
53. Hackett PH, Roach RC, Harrison GL, Schoene RB, Mills W, Jr. Respiratory stimulants and sleep periodic breathing at high altitude. Almitrine versus acetazolamide. Am Rev Respir Dis 1987; 135:896–898.
54. Powles AC, Sutton JR, Gray GW, Mansell AL, McFadden M, Houston CS. Sleep hypoxemia at altitude: its relationship to acute mountain sickness and ventilatory responsiveness to hypoxia and hypercapnia. In: Folinsbee LJ, et al., eds. Environmental Stress. New York: Academic Press, 1978:140.
55. Weil J. Ventilatory Control at High Altitude. Handbook of Physiology: The Respiratory System II, 1986.
56. Goldberg SV, Schoene RB, Haynor D, et al. Brain tissue pH and ventilatory acclimatization to high altitude. J Appl Physiol 1992; 72:58–63.
57. Chow CM, Xi L, Smith CA, Saupe KW, Dempsey JA. A volume-dependent apneic threshold during NREM sleep in the dog. J Appl Physiol 1994; 76:2315–2325.
58. Badr MS, Skatrud JB, Dempsey JA. Determinants of poststimulus potentiation in humans during NREM sleep. J Appl Physiol 1992; 73:1958–1971.
59. Smith CA, Saupe KW, Henderson KS, Dempsey JA. Ventilatory effects of specific carotid body hypocapnia in dogs during wakefulness and sleep. J Appl Physiol 1995; 79:689–699.
60. Xie A, Wong B, Phillipson EA, Slutsky AS, Bradley TD. Interaction of hyperventilation and arousal in the pathogenesis of idiopathic central sleep apnea. Am J Respir Crit Care Med 1994; 150:489–495.
61. Badr MS, Grossman JE, Weber SA. Treatment of refractory sleep apnea with supplemental carbon dioxide. Am J Respir Crit Care Med 1994; 150:561–564.
62. Fink BR, Hanks EC, Ngai SH, Papper EM. Central regulation of respiration during anesthesia and wakefulness. Ann NY Acad Sci 1963; 109:892–900.
63. Sullivan C, Kosar L, Murphy E, Phillipson E. Primary role of respiratory afferents in sustaining breathing rhythm. J Appl Physiol 1978; 45:11–17.
64. Skatrud JB, Dempsey JA. Interaction of sleep state and chemical stimuli in sustaining rhythmic ventilation. J Appl Physiol Respir Environ Exercise Physiol 1983; 55:813–822.
65. Berry RB, Gleeson K. Respiratory arousal from sleep: mechanisms and significance. Sleep 1997; 20:654–675.
66. Bonnet MH, Arand DL. Heart rate variability: sleep stage, time of night, and arousal influences. Electroencephalogr Clin Neurophysiol 1997; 102:390–396.
67. Bonnet MH. Effect of sleep disruption on sleep, performance, and mood. Sleep 1985; 8:11–19.
68. Bonnet MH. Cognitive effects of sleep and sleep fragmentation. Sleep 1993; 16:S65–S67.

69. Dempsey JA, Smith CA, Harms CA, Chow C, Saupe KW. Sleep-induced breathing instability. Sleep 1996; 19:236–247.
70. Fletcher EC. Recurrence of sleep apnea syndrome following tracheostomy. A shift from obstructive to central apnea. Chest 1989; 95:205–209.
71. Badr MS, Toiber F, Skatrud JB, Dempsey J. Pharyngeal narrowing/occlusion during central sleep apnea. J Appl Physiol 1995; 78:1806–1815.
72. Issa FG, Sullivan CE. Reversal of sleep apnea using nasal CPAP. Chest 1986; 90:165–171.
73. Hoffstein V, Slutsky AS. Central sleep apnea reversed by continuous positive airway pressure. Am Rev Respir Dis 1987; 135:1210–1212.
74. Naughton MT, Liu PP, Bernard DC, Goldstein RS, Bradley TD. Treatment of congestive heart failure and Cheyne-Stokes respiration during sleep by continuous positive airway pressure. Am J Respir Crit Care Med 1995; 151:92–97.

22

Acute Mountain Sickness and High-Altitude Cerebral Edema

PETER BÄRTSCH*

University Hospital
Heidelberg, Germany

ROBERT ROACH*

New Mexico Highlands University
Las Vegas, New Mexico

I. Introduction

Acute mountain sickness (AMS) and high-altitude cerebral edema (HACE) are illnesses caused by rapid ascent to high altitudes. Early reports of symptoms of AMS date back to Chinese travelers crossing the Great and Little Headache Mountains (32 BC). A vivid description of AMS was given in 1604 by a Spanish priest, Father Acosta, traveling in the Peruvian Andes (1). In 1913 Ravenhill (2) characterized three forms of illnesses associated with acute exposure to high altitude. He distinguished between *puna*, the word used in the Andes for an illness caused by acute exposure to high altitude, of the (1) normal type, with headache, insomnia, vomiting, and lassitude, (2) nervous type, which he considered a ''rare'' divergence from the normal with ataxia and decreased consciousness, and (3) cardiac type with râles and severe dyspnea, which he attributed to heart failure. Today we refer to these three types of *puna* as AMS, HACE, and high-altitude pulmonary edema (HAPE). In this chapter we will discuss AMS and HACE. AMS is a benign, self-limiting illness, which may, however, progress to HACE, a severe life-threatening illness with distinct neurological impairment.

---

* With Diana Depla, Newcastle Upon Tyne, England.

II. AMS

A. Definition of AMS

AMS is defined as the presence of a combination of nonspecific symptoms such as headache, loss of appetite, nausea, vomiting, weakness, dizziness, and difficulty sleeping, but without abnormal neurological findings. AMS appears within 4–36 hours after ascent to high altitudes.

B. Clinical Aspects

Symptoms

During the first few days of acute exposure to high altitude, a number of sensations can occur, which may be related to hypoxia or to other factors such as dry and cold air, exertion, or changes from one's usual lifestyle (food, beverages, housing). None of the most frequent AMS symptoms listed in Table 1 are specific for high-altitude

Table 1 Symptoms Reported in 21 Studies Dealing with AMS

Symptoms	Studies reporting symptom (No.)	Studies reporting symptom (%)
Headache	20	95.2
Nausea/vomiting	17	81.4
Tired/sleepy	15	71.4
Difficulty breathing	14	66.7
Dizzy/lightheaded	11	52.4
Loss of appetite	10	47.6
Rapid heart rate	9	42.9
Trouble sleeping	9	42.9
Difficulty concentrating	8	38.1
Chilly/cold/cyanotic	8	38.1
Weakness	7	33.3
Depressed	6	28.6
Irritable/moody	6	28.6
Elated/alert/giddy	4	19.0
Stomach/abdominal pain	4	19.0
Chest pain/pressure	4	19.0
Dry nose/mouth	2	9.5
Eyes irritated	2	9.5
Constipated	2	9.5
Loss of taste	2	9.5
Loss of smell	1	4.8
Frequent urination	1	4.8
Feverish	1	4.8

Source: Ref. 22.

exposure. Thus, distinguishing mild AMS from other causes of symptoms can be difficult.

Clinical Signs

Diagnostic clinical signs are absent in AMS. Sometimes changes in resting heart rate, body temperature, or presence of peripheral edema or râles have been noted. Upon arrival at altitudes up to 4500 m, systemic blood pressure and respiratory rate are not significantly different between subjects with and without AMS (3,4). Resting heart rate is initially higher by about 20 beats per minute in subjects with AMS (3), although Singh et al. (5) reported low heart rates in severely ill subjects with incipient cerebral edema. Axillary temperature (6) was slightly but not diagnostically higher when AMS was present (37.2 ± 0.5 vs. 36.8 ± 0.5°C in controls, $p < 0.01$).

Edema of the face, wrists, and ankles was found in 18% of 200 Himalayan trekkers (7) and in 25% of 466 mountaineers (4). However, the association between AMS and peripheral edema is not strong. One-third of the cases of peripheral edema occurred in the absence of AMS (7). Furthermore, the incidence of peripheral edema in mountaineers did not increase at higher altitudes (4), suggesting that exercise may contribute to peripheral edema, for studies of prolonged, submaximal exercise at sea level (8) and moderate altitude (9) show sodium retention and edema after several days exercise.

The prevalence of râles reported from comparable altitudes by different observers (4,7,10) ranges from 5 to 23% of the examined population. Râles may occur with (4,10) or without AMS (7). This variability between studies suggests that the interpretation of findings from pulmonary auscultation may be observer dependent and that clearly abnormal findings with auscultation linked to AMS may be rare.

AMS is, by definition, not accompanied by abnormal neurological findings. Decreased consciousness, ataxia, or focal neurological signs indicate progression of AMS to HACE (see below). More sensitive tests indicate that at altitudes between 3600 and 4700 m AMS may be associated with impairment of motor (11) and cognitive function (11–13) and color vision (14). Postural ataxia detected by posturography was, however, not different between AMS and control subjects on the first 3 days at 4559 m (Baumgartner, unpublished data).

Additional Examinations

CSF pressure is not increased in AMS (15,16). After 48 hours at a simulated altitude of 3700 m, one of six subjects with AMS showed mild cerebral edema on CT scans (17) and in a similar exposure over 24 hours T_2-weighted MRI revealed a slightly increased signal intensity in four of seven subjects with AMS, suggesting that mild vasogenic edema may be present (18). Recent studies using MRI found, however, similar mild increase of brain volume (19) and similar reduction of cerebral spinal fluid (CSF) volume (20) in subjects with and without AMS after 12–32 hours of exposure to a simulated altitude of 4800 m. These findings, which appeared to occur in all individuals exposed to high altitude, can be explained by increased cerebral blood flow and, possibly, mild brain edema. Chest radiographs do not usually show signs of pulmonary edema or other abnormalities (3,21).

Assessment of AMS

Two systems for assessing the severity and incidence of AMS are in wide use today: the Environmental Symptom Questionnaire (ESQ) (22) and the Lake Louise scoring system (23). The ESQ was developed to assess symptoms due to a variety of environmental challenges. Its advantage is that a large body of literature exists comparing responses between various groups of subjects at different altitudes, which were reached with different rates of active or passive ascent. Major weaknesses of the ESQ include poor identification of those ill with AMS (about 40% may be missed), a demanding 67-item questionnaire, and a complicated algorithm for calculating the final AMS-C score (C stands for cerebral symptoms). Furthermore, a recent study (24) demonstrated that the ESQ lacked specificity in distinguishing AMS from others stressors such as heat, exercise, or cold. The Lake Louise scoring system consists of a self-report questionnaire, which may be used by itself or in combination with clinical assessment (Table 2). It was designed to provide a briefer, simpler set of questions that would be at least as sensitive as the ESQ in identifying those with AMS as well as correctly identifying those who did not have AMS. Comparable cut-off levels for the two scoring systems are: AMS-C score of ESQ > 0.70, Lake Louise score (self-report questionnaire) > 4, and Lake Louise score (self-report questionnaire and clinical assessment) > 5 (24,25).

Clinical Course

Symptoms of AMS usually develop between 4 and 36 hours after reaching a given altitude depending on the altitude and speed of ascent. After an airlift from 500 to 4559 m, symptoms developed within 4–8 hours (26). Honigman et al. (27) found that at 2000–3000 m in those contracting AMS, the onset of symptoms occurred within 12 hours of arrival in 65%, between 12 and 36 hours in 34%, and after 36 hours in the remaining 1%. AMS is most prominent after the first night spent at a given altitude and resolves spontaneously over the next 2 or 3 days if no additional gain in altitude occurs (3,28). In accordance with these observations, Singh et al. (5) report a duration of 2–5 days of incapacitating illness in 840 soldiers with AMS at altitudes of 3350–5500 m. AMS can progress to HACE, or it may persist for many days without progression to more severe illness in a few individuals (5).

C. Determinants of Incidence

Altitude, rate of ascent, and extent of prior acclimatization are major determinants of AMS susceptibility, while fitness and general health, age, gender, and obesity play a minor role. The reasons for the considerable variation in individual susceptibility remain unresolved.

Table 2 Lake Louise Score

A. Questionnaire		
1. Headache	0	No headache
	1	Mild headache
	2	Moderate headache
	3	Severe headache, incapacitating
2. Gastrointestinal symptoms	0	No gastrointestinal symptoms
	1	Poor appetite or nausea
	2	Moderate nausea or vomiting
	3	Severe nausea and vomiting, incapacitating
3. Fatigue and/or weakness	0	Not tired or weak
	1	Mild fatigue/weakness
	2	Moderate fatigue/weakness
	3	Severe fatigue/weakness, incapacitating
4. Dizziness/lightheadedness	0	Not dizzy
	1	Mild dizziness
	2	Moderate dizziness
	3	Severe dizziness, incapacitating
5. Difficulty sleeping	0	Slept as well as usual
	1	Did not sleep as well as usual
	2	Woke many times, poor night's sleep
	3	Could not sleep at all
B. Clinical Assessment		
6. Change in mental status	0	No change in mental status
	1	Lethargy/lassitude
	2	Disoriented/confused
	3	Stupor/semiconsciousness
	4	Coma
7. Ataxia (heel-to-toe walking)	0	No ataxia
	1	Maneuvers to maintain balance
	2	Steps off line
	3	Falls down
	4	Can't stand
8. Peripheral edema	0	No peripheral edema
	1	Peripheral edema at one location
	2	Peripheral edema at two or more locations
	3	Severe dizziness, incapacitating
9. Difficulty sleeping	0	Slept as well as usual
	1	Did not sleep as well as usual
	2	Woke many times, poor night's sleep
	3	Could not sleep at all
C. Individual Score		
Overall, if you had any symptoms, how did they affect your activity?		
0 No reduction in activity		
1 Mild reduction in activity		
2 Moderate reduction in activity		
3 Severe reduction in activity (e.g., bedrest)		

Altitude, Rate of Ascent, and Preacclimatization

The prevalence of AMS increases with increasing altitude (Table 3) and with a faster rate of active or passive ascent (Table 3). This was shown in Alpine mountaineers (4,27), in visitors of the Rocky Mountains (27), in trekkers in Nepal (7,29,30), and in mountaineers in the Cascades (31). Differences in prevalence between studies may be explained by varying definitions of AMS, differences in study population, mode of ascent, altitude, and activity at altitude.

Ascending from 2900 to 5400 m over 6 instead of 4 days is associated with a 26% reduction in the incidence of AMS (32). A reduction of severity and prevalence of symptoms of AMS occurred with a one week pause at 1600 m before ascending to 4300 m (33) and with gradual passive ascent over 4 days to 3500 m (34) compared to rapid ascent to these altitudes. No symptoms of AMS were observed in sea level residents going within one day to 3500 after becoming acclimatized for 2–8 months at 1800 m (33). Furthermore, permanent residency above 900 m compared to residency below 900 m reduces the incidence of AMS from 27% to 8% when going quickly to altitudes between 2000 and 3000 m (27).

Prior altitude exposure prevents or ameliorates AMS symptoms on reascent for up to at least 8 days after return to sea level (35,36). Arterial oxygen saturation on reascent reached values intermediate between sea level and the highest value obtained on initial ascent, suggesting a carry-over of ventilatory acclimatization (35). Another study suggests a carry-over effect of acclimatization of at least 5 days after working at 4200 m and sleeping at 2800 m for 5 days (37). Subjects who had more frequent or more recent previous exposures to altitudes above 3000 m were less likely to get AMS at 4559 m (28). Taken together, these data demonstrate that the level of altitude, the rate of ascent, and preceding altitude exposure are major determinants of the prevalence of AMS.

Fitness and Physical Activity

Physical fitness has minimal influence on the prevalence of AMS (28,30,33,38). One aspect of fitness, the maximal oxygen uptake before ascent, bore no relationship to subsequent AMS in 128 participants in expeditions to Himalayan peaks between 6200 and 8848 m (38) or in 24 soldiers exposed to 4300 m (33), a finding supported by studies in mountaineers in the Alps (28) and trekkers in Nepal (30). In contrast, conference attendees who judged themselves as having poor physical fitness were twice as likely to get AMS at moderate altitudes as those who judged themselves to be fit (27). However, a follow-up study using a more detailed self-assessment of physical fitness found no relationship between self-report physical fitness and AMS in 205 conference attendees at 3000 m (39). Only one controlled study specifically addressed the effect of endurance training before altitude exposure on susceptibility to AMS (40). Eight weeks of endurance training at sea level significantly reduced the total AMS score obtained during the first week after being airlifted to 3500 m. A subset of well-trained athletes had even lower symptom scores. Except for the results of the last study, which needs to be confirmed, the cited data indicate that

Table 3 Factors Affecting Prevalence of AMS

Study population	Age (years and range)	n	Definition AMS	Altitude (lodging)	Way and rate of ascent	Incidence or prevalence (%)	Ref.
Trekkers	32 (19–62)	200	> 1 symptoms or signs	5336 m	By foot from 2800 m within 5–7 days	43	7
Trekkers	35 (19–71)	158	> 1 symptoms or signs	5336 m	By foot from 2800 m within 4–6 days	69	29
		71		5336 m		24[a]	29
Mountaineers		47	> 2 symptoms or signs[b]	2850 m	Within 1 or more day by foot + passive transport to intermediate altitude	9	4
		128		3050 m		13	
		82		3650 m		34	
		209		4559 m		52	
Conference attendees	37 (22–65)	454	> 2 symptoms	2100 m	Car within 1 day	25	221
	44 (26–69)	96		0 m		5	
Conference attendees	42 (30–62)	101	> 2 symptoms	2987 m	Car mostly within 1 day	42	227
Conference attendees	44 (16–87)	3158	> 3 symptoms	1828-2134	Car within 1 or 2 days	18	27
		?		2135-2743		22	
		?		2744-2957		27	
		2023					

[a] All on acetazolamide.
[b] Including peripheral edema.

the level of physical fitness has only a minor impact on susceptibility to AMS when compared to the effects of the rate of ascent, the level of altitude, and the degree of preacclimatization.

The concept of the pathophysiology of AMS (see Sec. IV) suggests that exercise at altitude is likely to enhance AMS because it increases hypoxemia (41) and sodium retention (9). There is not sufficient experimental data supporting this assumption. Two preliminary studies yielded controversial results. Exercise in a chamber enhanced symptoms of AMS occurring during a 10-hour exposure (42), while active and passive ascent to 4559 m lead to similar prevalences of AMS (43).

Age

In adults, Hackett and colleagues (29) were the first to note a decrease in AMS with increasing age in Himalayan trekkers. In conference attendees at moderate altitude (2000–3000 m), those less than 60 years old were twice as likely to get AMS as those 60 years old and above (27). These findings were confirmed in a group of elderly visitors (mean age 69.8 years, range 59–83 years) at 2500 m who had an incidence of AMS of only 16% (44), compared to an overall incidence of 27% in a population with a mean age of 44 years at a comparable altitude (45). In mountaineers (4) the AMS score of subjects between 20 and 40 years was significantly higher than for those above 40 years. In contrast, one study on 63 trekkers (age 23–58, mean 35 years) there was no influence of age on AMS (46). In summary, these data suggest that age above 60 years may protect somewhat from AMS. Whether the lower AMS incidence in the elderly has an intrinsic physiological or a psychological basis is not known. The perception of presence and severity of symptoms is subjective, and psychological make-up may undoubtedly influence individual ratings. Age-related behavioral differences could also account for the lesser problem with AMS in the elderly. Another possible contribution is a lesser likelihood of indulging in strenuous physical activity soon after ascent.

At the other end of the age spectrum, it seems that preverbal and prepubertal children have a similar incidence and severity of AMS as young adults (<40 years). In 23 subjects between 3 and 36 months (mean age 21 months), the incidence at 3488 m of 22% was similar to that in adults at the same altitude (20%) (47). In older children (9–14 years) living at an altitude of 1600 m and participating in a science camp at 2835 m, 28% were ill with symptoms characteristic of AMS (48), not significantly different from the 21% incidence of the same symptoms in children attending a camp at sea level. Attributing the 7% higher incidence of AMS-like symptoms in the summer camp at 2835 m to high altitude gives a similar incidence as reported for adults living above 900 m when going to comparable altitudes (27). These studies underscore the challenge of assigning cause to nondiagnostic symptoms, perhaps a more significant problem in children than in adults (45). In summary, the presented data indicate that children compared with adults below 60 years are not at an increased risk for AMS.

Gender

Men and women seem to suffer from AMS with comparable frequency and severity. The incidence in men and women trekking to Everest base camp is 51% vs. 53%, respectively (29). In contrast, AMS was significantly more frequent in male than in female conference attendees (27.9% vs. 23.6%) at intermediate altitude (27). A substantially higher incidence of AMS in women during trekking and expeditions (38) could not be confirmed in a larger group of over 400 individuals (Richalet, personal communication). A recent study of 18 women suggests no effect of menstrual cycle status on the symptoms of AMS (49). In summary, the presented data indicate that there are most likely no substantial gender differences in susceptibility to AMS.

State of Health

Even mild obesity appears to increase susceptibility to AMS. The prevalence of AMS was 31% in obese vs. 24% in "nonobese" visitors to intermediate altitude when obesity was defined as a body mass index [BMI: weight/(height in m)^2] $>$ 27.3 in women or $>$ 27.8 in men (27). Participants in a Himalayan expedition with a BMI $>$ 24 had a minor but significant increase in prevalence of AMS (46), as did male trekkers in Nepal (30). In contrast, body weight alone was not significantly different between trekkers with and without AMS in another investigation (29).

Three studies reported no effect of cigarette smoking (27,30,38) and one study no effect of oral contraceptives (30) on the prevalence of AMS. The risk for AMS associated with preexisting illnesses has only been examined at intermediate altitude. While heart disease, hypertension, and diabetes had no influence on the prevalence of AMS at intermediate altitudes, preexisting lung diseases increased the odds ratio for AMS to 1.6 (95% confidence interval 1.05–2.5) (27). The independence of AMS prevalence from preexisting heart disease, diabetes, and hypertension was confirmed in an elderly group at 2500 m (44).

Individual Susceptibility

AMS is likely to recur (27,38,50,51). Recurrence, however, can be avoided in most cases by slow ascent or preacclimatization, as was shown for individuals with susceptibility to HAPE (52). In spite of numerous investigations over more than 100 years, it is not clear which particular physiological or pathophysiological features are associated with an increased individual susceptibility to AMS. The possible contributions of hypoxic ventilatory response (HVR) and hypercapnic ventilatory response (HCVR) will be discussed in Sec. IV.

Whether other parameters such as diuretic response to hypoxia (53), ventilatory and cardiac response to hypoxic exercise (38), end-tidal P_{O_2} in normoxia (54), breath hold time, response to maximal voluntary hyperventilation or pharyngeal stimulation (55) either predict or point to mechanisms of individual susceptibility needs to be examined and confirmed in larger groups.

III. High-Altitude Cerebral Edema

A. Definition

High-altitude cerebral edema (HACE) is a potentially fatal neurological syndrome, which typically presents with ataxia and progressive deterioration of consciousness in persons with AMS or HAPE.

B. Epidemiology

HACE generally occurs at altitudes above 4000 m and particularly with rapid ascent (56–58). Hackett et al. found an incidence of 1.0% for HACE and 1.5% for HAPE in 522 trekkers at 4200–5500 m (29). The incidence was 0.5% for HACE and 1.3% for HAPE (Wu, personal communication) in 5355 visitors to 4555 m in Tibet. Five of 45 cases (11%) of early HAPE occurring during prospective studies in the Alps at 4559 m had concomitant HACE (P. Bartsch, unpublished observation). Ten of 50 cases (20%) of HAPE rescued by helicopter in the Swiss Alps also had HACE (59), while the combination of HACE with HAPE appears to be much more frequent in severe cases of HACE reported in the literature (for review, see Ref. 60). These observations indicate that HACE is less frequent than HAPE and that severe HACE and HAPE are likely in the same individual.

C. Clinical Aspects

Early symptoms of HACE are truncal ataxia and clouded consciousness usually preceded by symptoms of AMS. The neurological findings are accompanied by a progressive worsening of headache, malaise or weakness, vomiting, and irritability (5,56,58). Headache, however, is not always present. Hallucinations and seizures may occur. Often HACE is preceded or accompanied by HAPE (56,57,60). Deterioration may be very rapid. In one case coma occurred within 24 hours and death within 48 hours after high-altitude exposure (58). Due to dissimulation or neglect of symptoms because of early mental impairment, prodromal symptoms may not be heeded and patients appear to fall suddenly severely ill.

Clinical examination often reveals somnolence, stupor or coma, and elevated body temperature (>38.0°C) (6). Papilledema, retinal hemorrhages, truncal ataxia, pyramidal signs, and sometimes focal neurological signs can be observed. Focal deficits include cranial nerve palsies, which may result in blurred speech and double vision (5,56–59). When focal signs dominate, other illnesses like stroke or transient ischemic attacks (61), cortical blindness (62), or migraine with aura (63,64) need to be considered. HACE and these other illnesses can occur in acclimatized mountaineers without signs of AMS or HAPE (58).

Due to the remote areas in which HACE usually occurs, detailed clinical or laboratory examinations are only available upon return to low altitude. There are no diagnostic laboratory findings. Increased CSF pressure (60–300 cm H_2O above normal) was reported in advanced cases (5,15,56). The increase in intracranial pressure is reflected in MR and CT images. CT scans showed compression of ventricles, disappearance of sulci, low density appearance of the white matter or in the whole cerebrum in mountaineers with combined HACE and HAPE (65–67). Evidence of edema in the white matter, especially in the splenium of the corpus callosum (68), and no abnormal findings in the gray matter have been reported from MRI studies (Fig. 1). These abnormalities may persist for several days after clinical recovery (67,68).

Clinical course and prognosis can only be extrapolated from the few published case histories. These suggest that patients with HACE do not improve at low altitude as rapidly as patients with HAPE. Once coma has developed, without treatment death is likely. Three of 24 patients hospitalized with serious neurological manifestations died despite aggressive treatment (5). Several case histories demonstrate that a delayed descent considerably increases mortality. Long-term recovery after HACE has not been studied. Residual deficits have, however, persisted in some patients (P. Hackett, personal communication).

D. Postmortem Findings

Findings at autopsy after death from HACE (57,69) include generalized cerebral edema, petechial and gross intracerebral hemorrhages, thrombosis, and infarction. Many of these findings may only occur terminally. Rarely, only cerebral edema without bleeding or thrombosis is found. Often there is simultaneous pulmonary edema with intra-alveolar fibrin deposition and arteriolar thrombi. Increased markers of fibrin formation and fibrin degradation found in a mountaineer suffering from cerebral and advanced pulmonary edema suggest that thrombosis found at autopsy may be present in severe disease (70).

IV. Pathophysiology of AMS and HACE

The signs and symptoms of AMS and HACE point to the central nervous system as the target organ for these illnesses. One current view is that AMS is simply caused by early stages of brain swelling, which becomes more apparent with HACE. In this section we will discuss a number of questions:

How does the hypoxia of high altitude cause the brain to become edematous?
What is the evidence for increased capillary permeability at high altitude?
Can mild brain edema explain the symptoms of AMS?
Is individual susceptibility to AMS/HACE linked to the control of ventilation in hypoxia?

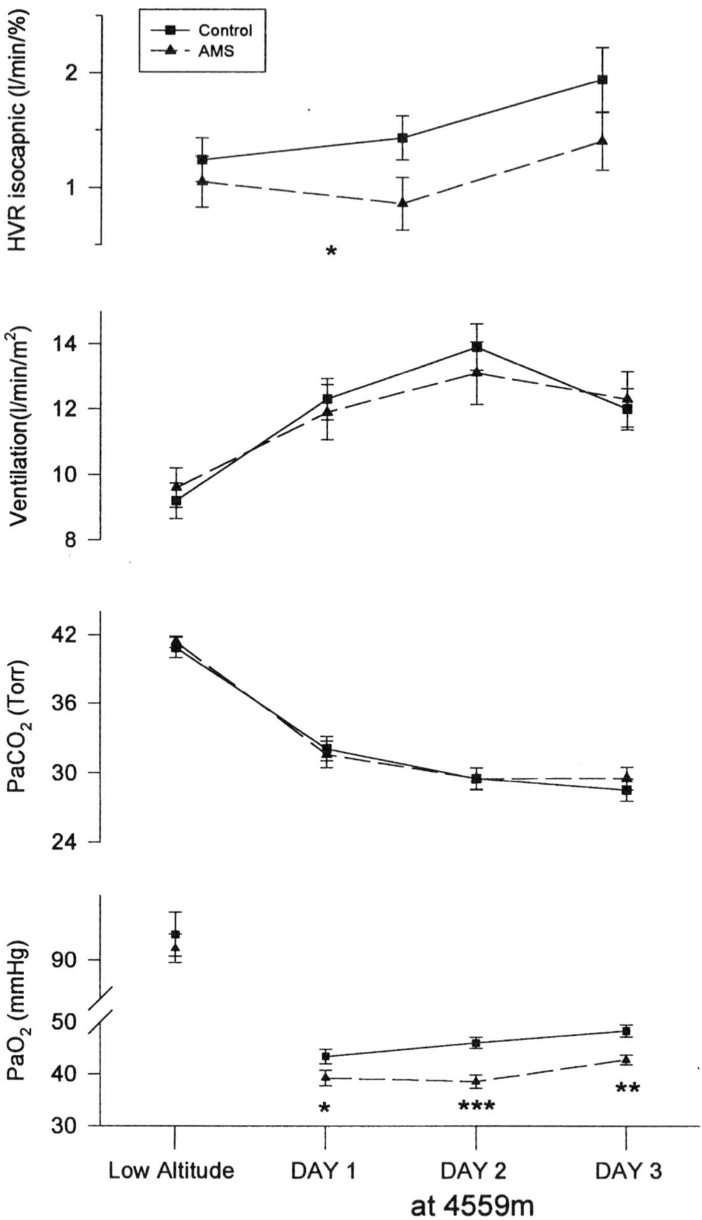

Figure 1 CT scan and MRI of HACE. (Courtesy of Peter Hackett.)

A. Hypoxemia—The Essential Cause of AMS/HACE

Hypoxemia is essential to the production of AMS and HACE. The lower the arterial oxygen saturation (Sa_{O_2}), the more rapid the onset (71) (Baumgartner, unpublished observations) and the more severe the illness (10,72). Supplemental oxygen rapidly ameliorates symptoms (73), while further desaturation during sleep makes it worse (74,75). Recovery occurs usually over a few days coincident with a rising Sa_{O_2} as a consequence of ventilatory acclimatization. Differences of Sa_{O_2} in individuals with and without AMS are rather small, on the order of 5–10 Sa_{O_2}%. An individual getting AMS at an altitude of 2500 m may have an Sa_{O_2} that is measured in an individual without AMS at 4500 m. These data suggest that hypoxemia alone does not account for individual differences in susceptibility.

In 1898, Mosso suggested that the hypocapnia consequent to hypoxia-induced hyperventilation, not hypoxemia per se, was the cause of AMS (76), and though others subsequently have been attracted to this possibility (77,78), a variety of observations suggest that a low Pa_{CO_2} does not explain the signs and symptoms of AMS. First, sustained hypocapnia without hypoxia does not result in AMS in sheep (79). With ventilatory acclimatization, symptoms lessen even when Pa_{CO_2} continues to decline (3,21). Acetazolamide, an effective therapy for AMS, concurrently decreases Pa_{CO_2} and increases Pa_{O_2} (72), and raising Pa_{CO_2} by inhalation of carbon dioxide worsens symptoms if Sa_{O_2} is prevented from rising (80). Improvement of AMS by addition of 3% CO_2 to the inspired air (78) could not be confirmed in a double-blind, placebo-controlled trial (73).

These data indicate that hypoxemia is the principal cause of AMS. That onset of symptoms is delayed by 4–8 hours or more after initial exposure implies that hypoxia triggers processes accounting for the illness, ultimately resulting in cerebral edema. The level of hypoxemia at which such processes are triggered or the magnitude of hypoxia-induced processes may be different between individuals and contribute to the extent of susceptibility to AMS.

B. Mechanisms Causing Brain Edema at High Altitude

With currently available methods, increases in brain water are not apparent in those acclimatizing uneventfully at high altitude or in those with early symptoms of AMS. But the symptoms and signs of HACE, including obtundation and ataxia, exist in concert with clear evidence of brain swelling. We presume, therefore, that these neurological impairments are a consequence of the edema.

How does hypoxia cause brain water to increase? Fluid may be pulled into the brain, as by an increase in brain osmolarity, or pushed into the brain as a consequence of increased vascular pressure or brain blood vessel permeability. These two modes are commonly referred to as cytotoxic and vasogenic edema. Cytotoxic edema as a cause of HACE was proposed to result from hypoxia-induced failure of cellular ion pumps, resulting in increased intracellular osmolarity and consequent cell swelling (56). While such changes may contribute late in the illness, it seems unlikely that early symptoms, sometimes occurring as low as 2500 m, could be associated with hypoxemia sufficient to impair cell ion homeostasis. In advanced cases of HACE, increased CSF pressure and edema might lead to a decrease of perfusion that results in focal ischemia and cytotoxic edema. Recent investigations demonstrate a 20% reduction of membrane-associated NaK-ATPase activity and ATP concentration in tissues or cell cultures (lung and kidney) with acute, extreme hypoxia ($F_{IO_2} < 6\%$) (81; Mairbäurl, unpublished data). If the brain is similarly affected, outward sodium and water transport from neurons and across the blood brain barrier could contribute to edema. But current thinking is that brain edema is more likely of vasogenic than cytotoxic origin.

Vasogenic edema may result from increased hydrostatic pressure across blood vessels or an altered permeability of the blood-brain barrier. MRI findings in HACE suggest that brain edema is vasogenic (82). Thus, one possibility is that AMS/HACE is caused by hypoxic cerebral vasodilation resulting in increased cerebral blood flow, intravascular pressures, and consequent edema (83). This hypothesis implies that on average those developing AMS should have a higher CBF than those without. This was found after 24 hours at 4559 m (84), while others reported equally increased CBF with and without AMS on two expeditions (85). Furthermore, Krasney et al. showed that even a doubling of CBF by normoxic hypercapnia does not cause brain edema in sheep (86), suggesting that other mechanisms elicited by hypoxia, perhaps in conjunction with increased CBF, must account for the development of HACE.

Increased vascular permeability is an attractive pathogenic mechanism for AMS/HACE. Hypoxia can increase vascular permeability (87) and vascular endothelial growth factor (VEGF) expression (88) in animal models, and endothelial cell cultures express various proteins that can trigger or enhance inflammatory response such as cytokines and adhesion molecules (89), procoagulatory proteins (90) and VEGF (91). Increased expression of VEGF, which was demonstrated in the brain of hypoxic rats (68), might point to hypoxia-induced endothelial activation and contribute to increased permeability of brain vessels. One has to consider, however, that these experiments with animals and cell cultures were often performed at considerably lower P_{O_2} than exists between 3000 and 5000 m, where most field studies are carried out. Therefore, we look at the evidence for increased vascular permeability in data obtained in humans at high altitude.

C. Vascular Permeability at High Altitude

Direct Measurements

An inflammatory response due to endothelial activation would cause a generalized increased capillary leakage for large molecules. However, as summarized in Table 4, most studies thus far suggest that inflammatory activation of endothelial cells at high altitude is minimal. Furthermore, activation appears to be similar for those with and without AMS. One has to consider that these substances act predominantly locally and have a short half-life in plasma. Therefore, normal plasma levels do not rule out local production, and increased plasma levels do not show which vascular

Table 4 Plasma Levels of Mediators of Inflammation and Markers of Endothelial Activation in AMS

	Eicosanoids	Cytokines	Endothelial activation
Not changed compared to low altitude	F_2 isoprostanes (228) Urinary leukotriene E_2 (229)	Interleukins (IL)-1β, 2 (228,230) IL-6 (231) IL-8 (188) Tumor necrosis factor α (TNFα) and/or soluble receptors of IL-2 and TNF-α (188) C-reactive protein (230,232) Cleaved HMWK (233)	P-selectin (234) L-selectin (235) VEGF (232) von Willebrand factor (234) Thrombomodulin (14)
Increase at high altitude independent of AMS		IL-6 (230,235)	E-selectin (234–236) Endothelin (236,237) von Willebrand factor (HAPE, susceptibles included (70)
Increase in AMS versus control	Index composed of prostaglandin (E_2) and leukotrienes (LTB_4 and LTC_4) (no statistics provided) (238) Serum phospholipid bound arachidonic acid (239) Urinary leukotriene E_4 (240)		

pathways were involved in their production. Nevertheless, arachidonic acid and its metabolites appear to be increased in plasma and urine at high altitude, especially in AMS (see Table 4).

In vivo, overall capillary permeability was examined by measuring the transcapillary escape rate (TER) of I-labeled albumin. Two investigations, one involving 9 subjects after 24 hours at 4350 m (92), and the other involving 8 subjects after 4–8 hours at simulated altitude of 4268 m (93), showed a slight increase in TER. In contrast, in a large group of 29 mountaineers including 15 with AMS, TER of albumin was unchanged after 18 hours at 4559 m compared to values obtained at low altitude (Figs. 2, 3) (94). Furthermore AMS scores and TER of albumin were not correlated. A further study performed at simulated altitude of 3454 m also found no significant changes of TER of albumin (95). Recently the group reporting increased TER of albumin (92) could not confirm their findings in a further study (96). These data indicate that increased capillary permeability, if present, may be difficult to assess with this method. Furthermore, these studies do not rule out the possibility of increased permeability of specific capillary beds that have been examined in AMS in the kidney, the lung, and the leg.

Kidney

Proteinuria of 200–500 μg/mL, which could not be attributed to the effects of exercise, was found after acute and chronic exposure to altitudes up to 5500 m (97–99). Changes in tubular reabsorption and glomerular sieving account for the proteinuria (99–101). A reduction of proteinuria and increase of Sao_2 when acetazolamide is given for prophylaxis of AMS (102) and lower protein excretion when high-altitude residents were taken to sea level (98) suggest that proteinuria may be a normal response of the kidney to hypoxia. A recent study reporting no correlation between albuminuria and AMS is compatible with this explanation (103).

Lung

Impairment of gas exchange in AMS could be explained by interstitial pulmonary edema. In contrast to those with HAPE, broncho-alveolar lavage in subjects with AMS does not reveal differences from healthy individuals with regard to concentration of proteins or mediators of inflammatory response (104) and thus does not provide evidence for an increased leak across pulmonary capillaries in those with AMS.

Leg

An increased filtration coefficient in leg muscle capillary bed, found after 24 hours but not after 5 hours at 4559 m, did not correlate with AMS scores (103). Changes in capillary filtration coefficient are difficult to interpret since its important determinants such as arterial and venous pressure and functional capillary surface area may all change when going to high altitude.

In summary, investigations of localized permeability in kidney, lung, and leg do not provide evidence that AMS is associated with an increased vascular leak for proteins. However, permeability in the target organ for AMS has not been examined.

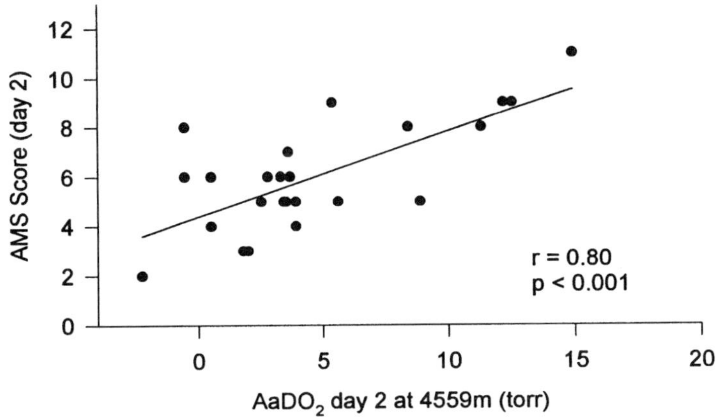

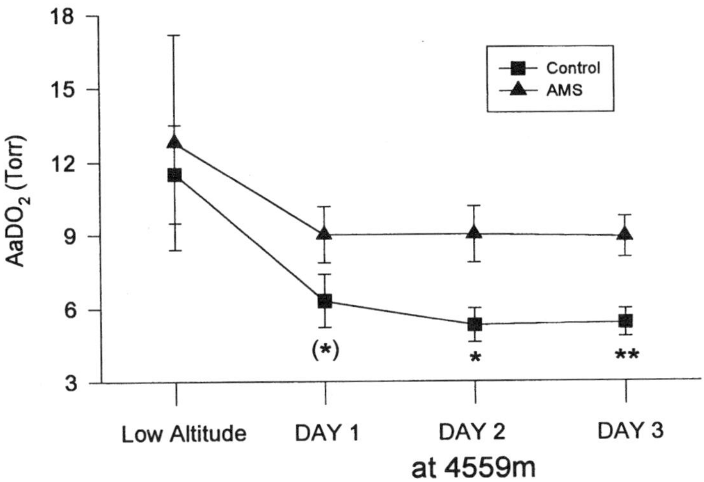

Figure 2 Mean values ± of isocapnic hypoxic ventilatory response (HVR), resting ventilation, arterial P_{CO_2} and P_{O_2} in 11 subjects developing AMS at 4559 m and in 12 controls. Note that prior to measurements on day 1 subjects had been above 3200 m for the preceding 20 hours and had exercised to reach the final altitude; $*p < 0.05$, $**p \leq 0.01$, $***p < 0.0001$ compared to corresponding values of control group. (Data from Ref. 94.)

Effects of various mediators, such as VEGF, on permeability of the blood-brain barrier in those with AMS/HACE need further investigation.

Water and Electrolyte Balance

While acute short-term exposure of humans to hypoxia is normally associated with increased water and salt excretion (105) with considerable individual variability

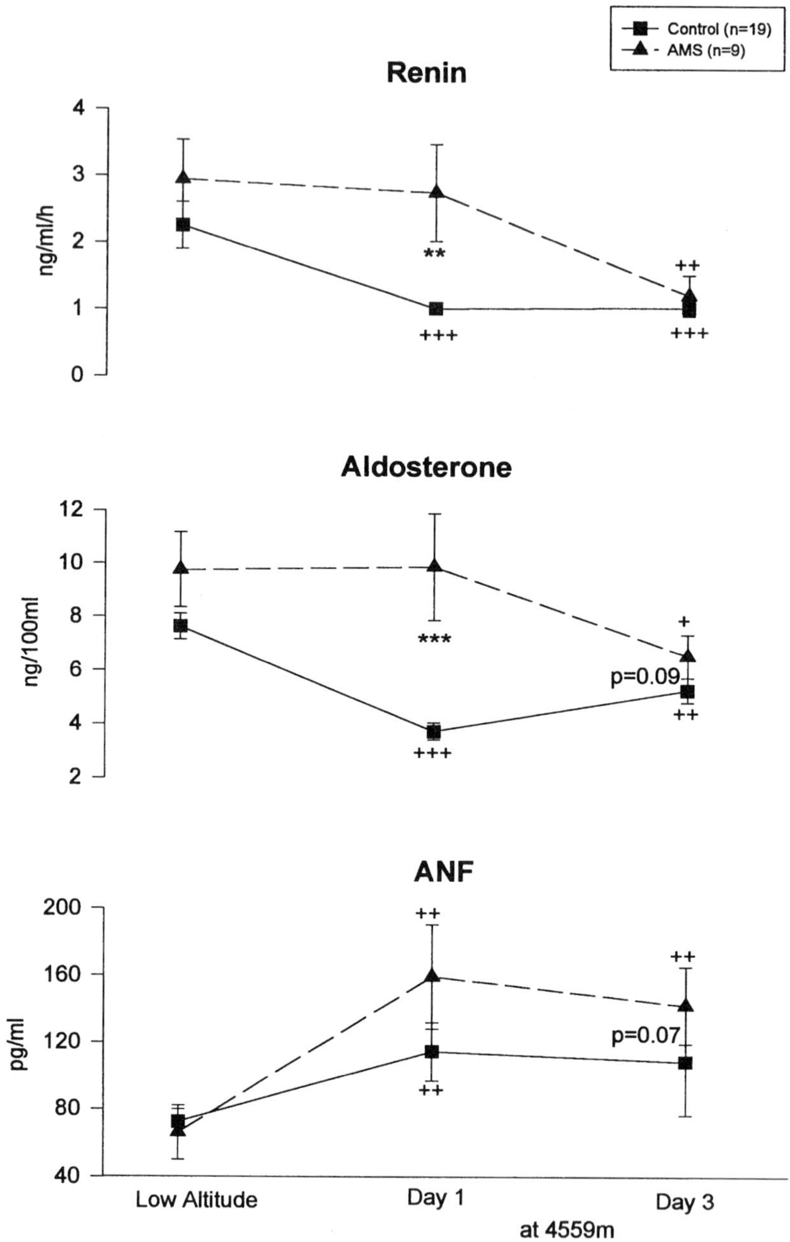

Figure 3 Correlation of Lake Louise score and $AaDo_2$ in 23 subjects. One day after arrival at 4559 m and mean values ± SE of $AaDo_2$ at low altitude and days 1–3 at 4559 m in 11 subjects with AMS and 12 controls; $*p < 0.05$, $**p < 0.01$ compared to corresponding values of control group. (Data from Ref. 94.)

(106,107), those with AMS (Fig. 4) (3,5,108–110) commonly exhibit less diuresis and natriuresis, resulting in weight gain when compared to healthy controls (3,111–114). Research in animals and humans suggests that this response may be linked to peripheral chemoreceptor responsiveness to hypoxia (53,115); although one field study did not discern a difference between those with and those without AMS (116).

The most likely mechanism for sodium and water retention in those with AMS is increased aldosterone and antidiuretic hormone (ADH) action, both at rest and induced by exercise, as compared to healthy individuals. Plasma renin activity and aldosterone levels decreased with initial exposure to high altitude, but the decrease was less in those with AMS (3,108,109), though one study reported lower renin activity in those with AMS (117). While plasma levels of ADH were not higher at rest in trekkers or mountaineers with AMS (3,109,118), with exercise on the first day after ascent mountaineers developing AMS had significantly higher plasma ADH as well as aldosterone levels than those who remained well (109).

Higher plasma norepinephrine and epinephrine in subjects with AMS suggests greater sympathetic activation, which may be attributed to more severe hypoxemia (Fig. 5) (3,109). Plasma levels of atrial natriuretic factor (ANF) were slightly higher in those with AMS, but not enough to have much effect on diuresis and natriuresis (3,108,109,119). Whether ANF contributes to edema formation by increasing capillary permeability remains to be determined (120).

Whether or how increased water and salt retention are related to the signs and symptoms of AMS is unclear. Generalized edema can be seen in totally asymptomatic individuals. Additionally, fluid retention may not be essential for AMS. A recent investigation where diet was controlled and in the absence of exertion found that subjects developing AMS and those that remained well 24 hours after airlift to 4559 m had comparable urine and sodium excretion (116). The fluid retention consistently reported from less well-controlled field studies could be attributable in part to the confounding factors of exercise and uncontrolled diet. Given the hypothesis of increased vascular permeability in AMS, fluid retention and weight gain may be the consequence of a volume shift from the intra- to the extravascular compartment.

Impaired Gas Exchange

Several studies have demonstrated that AMS is associated with impaired gas exchange and a widened $AaDo_2$ (10,21,121–123). The increase of $AaDo_2$ is not apparent within the first few hours after ascent and is greatest after 2 days at the time when altitude illness is most severe (21,26,122). The widened $AaDo_2$ may be due to interstitial pulmonary edema for its time course parallels that of the fluid retention observed in AMS (21,110). The increased slope of the phase III of the nitrogen washout curve, denoting increased V/Q variance, correlates temporally with widening of the $AaDo_2$ (124), and hypoxemia with a decrease in the lung-diffusing capacity (125). A lower vital capacity and FEV_1 in those with AMS compared to controls at high altitude may also reflect interstitial edema (126), although not all investiga-

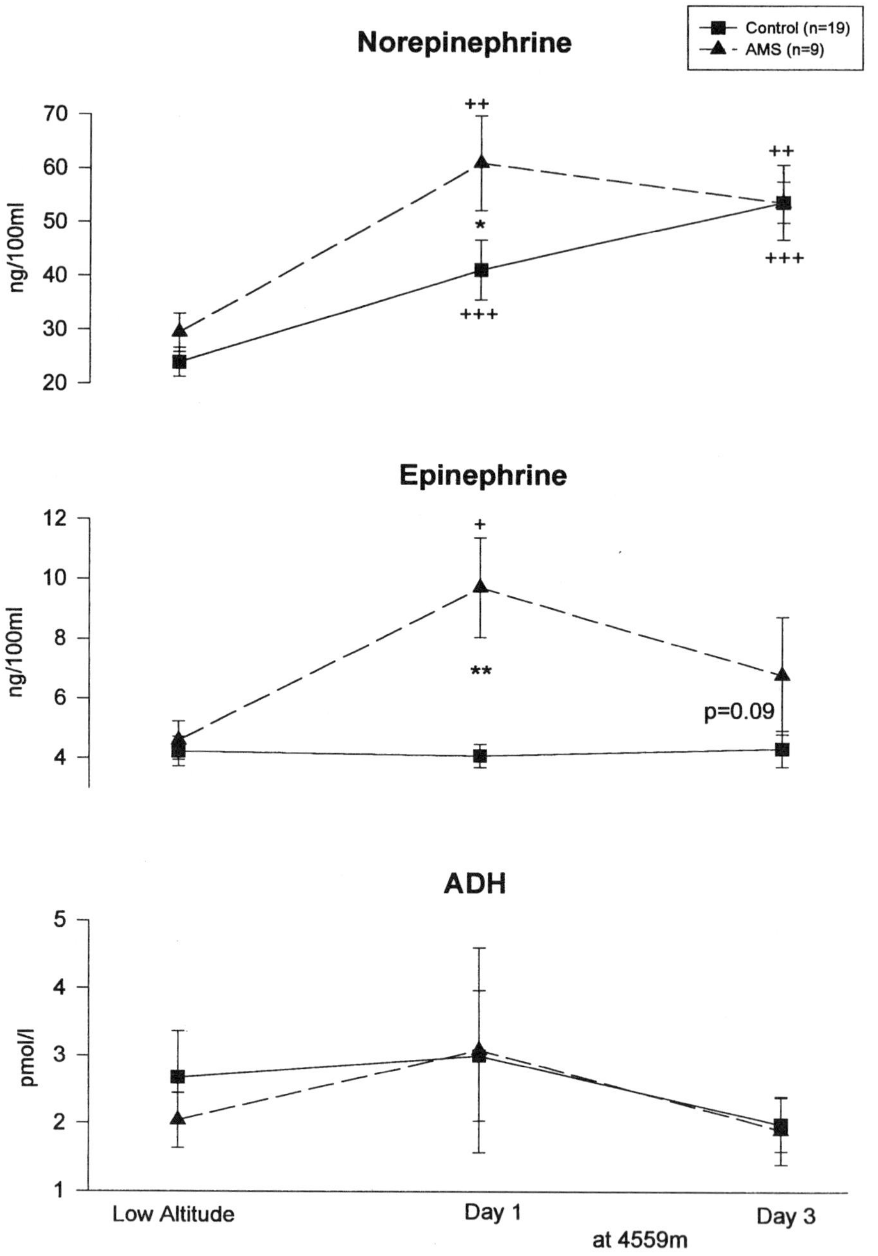

Figure 4 Mean values ± SE of changes in body weight and urine output (expressed in % of fluid intake) in 11 subjects with AMS and 12 controls; $*p < 0.05$, $**p < 0.01$ compared to corresponding values of controls (110).

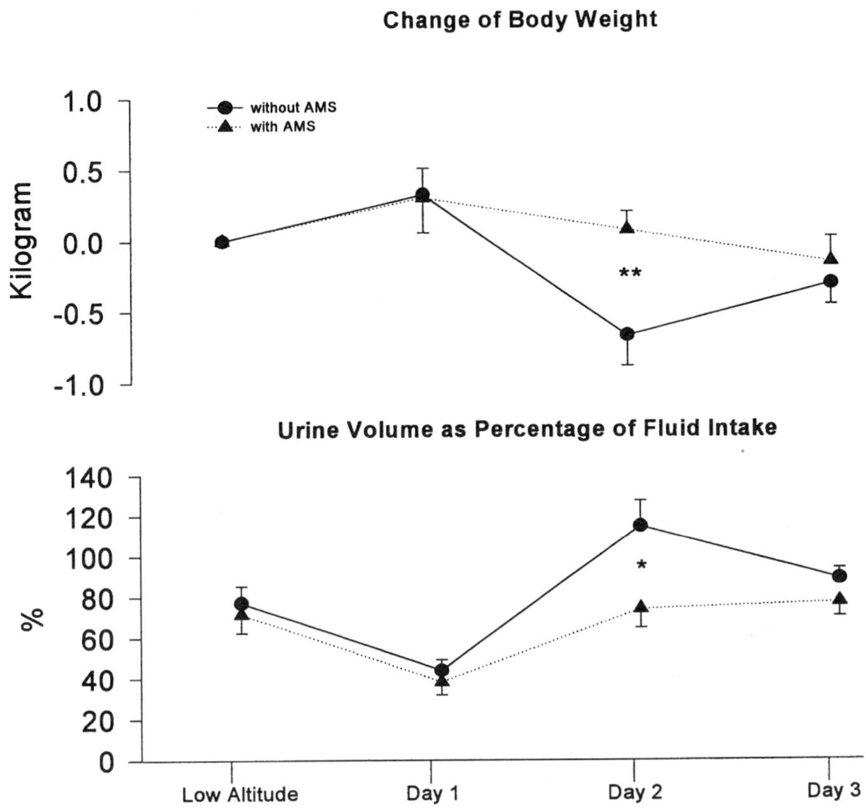

Figure 5 Mean values ± SE in 9 subjects with AMS and 19 controls at low altitude and on day 1 and 3 at high altitude. $*p < 0.05$, $**p < 0.01$, $***p < 0.001$ and stated p-values for comparison between groups (Kruskal-Wallis-Test); $^{+}p < 0.085$, $^{++}p < 0.01$, $^{+++}p < 0.001$ compared to corresponding value at low altitude (Wilcoxon-Sign-Rank-Test). Data were combined from two studies excluding 13 subjects with HAPE (3,109).

tors have found a correlation between AMS scores and changes in vital capacity or FEV_1 (17,127).

The assumption that impaired gas exchange is due to interstitial pulmonary edema fits well with a postulated generalized increase in capillary permeability in AMS. But, unlike HAPE, bronchoalveolar lavage of individuals with AMS alone does not show an inflammatory response (104). Nor is the impaired gas exchange of AMS prevented by nifedipine (121) or improved by inhalation of NO (128). Whether these differences between AMS and HAPE are due to the location or quantity of the edema fluid or whether they point to a different mechanism for leak in HAPE is unclear.

D. Is AMS Early HACE?

It is enticing to think that the entire spectrum of symptoms from early AMS through the end stage of HACE derives from a common cause, namely, increasingly severe brain edema. But for a couple of reasons, AMS, at least until its more severe stages, may be hard to reconcile with a progressive edema hypothesis.

Evidence that edema is present with AMS in meager at best. Cerebrospinal fluid pressure is not elevated in individuals with AMS (15,16). MRI (18) or CT (17) scans show mild cerebral edema only in a few advanced cases of AMS (AMS-C scores >1.6), and whether signs indicative of early HACE were present in those cases was not well documented. Conceivably, current methodologies could lack sufficient sensitivity to detect the presence of early edema, but even if incipient edema exists, can it account for the early signs and symptoms? Preliminary reports of recent studies using MRI suggest that this is not the case. There was no difference between those with and without AMS regarding signs that could point to mild cerebral edema like increased brain volume (19), reduced CSF volume (20), and minor enhancement of T_2-weighted MRI signal in the corpus callosum (129).

Headache, the most common and at times the most debilitating of AMS symptoms, characteristically occurs early in the illness, at a time when its causation is difficult to attribute to brain edema. The only known loci of pain sensation in the brain are the meninges and large blood vessels, afferent innervation from which is via C-fibers to the trigeminal ganglion (130). With no increase in intracranial pressure in early stages of AMS, stretch of meninges seems an improbable cause for head pain. Blood flow velocity in the internal carotid and vertebral arteries (131,132) and in the middle cerebral arteries (133) does not increase significantly within the first few hours after acute exposure to 4300–4800 m, even though headache (131) and other symptoms of AMS may be well developed (133). Furthermore, an increase in cerebral blood flow by 20% observed during treatment of AMS with 1.5 g acetazolamide (134) and a 37% increase of CBF with inhalation of air enriched with 7% CO_2 (135) did not worsen symptoms of AMS. Even so, lack of dilation of resistance vessels does not preclude the possibility that mechanical stretch of larger vessels that contain pain C-fiber innervation is the cause of headache.

Nevertheless, stretch of these structures is unlikely with normal pressure and at best minimal cerebral edema. One might postulate that other mechanisms account for early symptoms of AMS. Hypoxia may trigger release of neuromodulators such as substance P, calcitonin gene-related peptide, or bradykinin, which are known modulators of pain sensitivity at nerve endings. This mechanism is believed to be involved in the pathophysiology of migraine (130). Sumatriptan, a selective agonist of 5-hydroxytryptamine 1 receptors used for treatment of migraine, caused a significant amelioration of headache for 1–3 hours in a placebo-controlled trial at 4559 m (136) but was not effective in a double-blind study during early exposure at 3480 m. This observation suggests that mechanisms similar to those of migraine may be involved in the pathophysiology of headache at high altitude. Ibuprofen appears to be more effective therapy for high-altitude headache than sumatriptan, implying a role for cyclo-oxygenase in causing the headache (137,138).

In summary, at this time we do not understand how hypoxia causes AMS or HACE. Nor is it clear that these entities, which we distinguish clinically, represent different stages of a common mechanism. Changes in blood-brain barrier permeability seem a fruitful area for exploration of how hypoxia can cause brain edema. Possibly symptoms of AMS such as headache are a consequence of the same mechanism or an independent process triggered by release of some of the same neuromodulators that alter brain vascular permeability. Future inquiry might well include a search for these substances, how hypoxia effects their release, and how their presence alters vascular permeability or stimulates nociceptive nerve endings.

E. Individual Susceptibility to AMS/HACE and Control of Ventilation

Although we all will succumb to acute altitude illness if we go high enough fast enough, as noted above, individual differences in susceptibility are striking. Understanding the reason for these differences between individuals has been a major interest of researchers, both to predict how a given individual will fare at altitude and also in the hope that understanding such variability might provide clues to the mechanism causing AMS/HACE. Certain differences between those more and less susceptible have been noted, and one hypothesis in particular has been fairly extensively explored, namely, that those more susceptible exhibit hypoventilation and decreased ventilatory response to hypoxia.

Lower ventilation appears to be an important factor that contributes to hypoxemia in AMS since most studies discussed below demonstrate direct or indirect evidence of hypoventilation in subjects with compared to those without AMS. There are two studies performed in hypobaric chambers (139,140) and three field studies (111,122,141), which show lower ventilation in AMS subjects. Three of these studies (122,139,140) demonstrated hypoventilation only within the first 6 hours of exposure. In most field studies ventilation was assessed 24 or more hours after exposure to high altitude, and it was not found to be significantly different between AMS and control groups (10,21,31,75,142,143). These data, nevertheless, indicate that subjects with AMS ventilate less than control subjects when considering the difference of arterial Po_2 between these groups. Due to their more pronounced hypoxemia AMS subjects had a greater stimulus to ventilate and should have had a higher ventilation than control subjects if similar ventilatory drives were present.

Controversial data are reported from investigations examining whether hypoventilation at high altitude may be explained by a lower hypoxic ventilatory response (HVR) already present at low altitude. Two chamber studies (139,140) found a lower isocapnic HVR in those developing AMS. One of these studies (139) looked only at the five most and five least ill out of 24 subjects. Furthermore, it is possible that individuals with susceptibility for HAPE, who have consistently lower HVR (144–147), may have been included in these studies. Six field studies (21,71,122,145,

148,149) involving between 7 and 32 subjects per study failed to show a significant relationship between HVR measured at low altitude and occurrence of AMS. However, a trend toward lower HVR in the AMS group was observed in these studies (21,145,149). Measuring HVR during exercise did not increase differences between AMS susceptible and nonsusceptible individuals (21,145). In 128 mountaineers, however, a smaller increase of ventilation after 5 minutes of hypoxic exercise (50% Vo_2max, $Fio_2 = 0.115$) was associated with a higher incidence of AMS on expeditions (38). The differences between the AMS and non-AMS groups were small, and the big overlap between these groups does not allow reliable identification of AMS-susceptible subjects by this method. In summary, HVR at low altitude is slightly lower in individuals with susceptibility to AMS. Because of large interindividual variability, it can only be shown in studies with sufficient statistical power (e.g., studies involving a large number of subjects).

Since HVR measured at low altitude is only weakly associated with AMS, other mechanisms may contribute to hypoventilation associated with AMS. These include: ventilatory response to CO_2 (HCVR), changes in ventilatory control during prolonged hypoxia and periodic breathing during sleep. HCVR measured at low altitude is not different between susceptible and nonsusceptible subjects (21,145, 149). One study suggests that the increase of HVR that occurs with acute exposure to altitude (150) is somewhat delayed in individuals with AMS (21). Furthermore, the decline of ventilation occurring between 10 and 30 minutes of hypoxia (151) may be more pronounced in subjects developing AMS (140). Periodic breathing increases with altitude (76) and accounts for oscillations of Sao_2 (152) and a decrease in mean Sao_2 during sleep (74,153). However, at a simulated altitude of 3700 m (154) and in a field study at 4559 m (75), AMS was not correlated with periodic breathing during sleep. These data suggest that differences in HCVR and periodic breathing during sleep cannot account for susceptibility to AMS, while differences in ventilatory control with prolonged hypoxia, which cannot be assessed by measuring HVR, may contribute to susceptibility for AMS and account for the insufficient increase of ventilation compared with the degree of hypoxemia.

F. A Role for Hypobaria?

Conceivably, the low barometric pressure at high altitude may influence AMS symptoms, fluid balance, and ventilation. The severity of AMS symptoms rose when altitude responses were compared to responses to normobaric hypoxia (155), while no symptoms of AMS were noted during hypobaria without hypoxia. Several other studies support a higher severity of AMS with hypobaric compared to normobaric hypoxia (156–158). In addition, hypobaria may contribute to the fluid balance abnormalities observed during acute altitude exposure. Epstein and Suruta reported that prolonged normoxic hypobaria (258 mmHg) caused a 40% decrease in creatinine clearance, a tendency to sodium retention, and a smaller decrement in body weight (159). Also, normoxic hypobaria attenuated the normal contraction of plasma volume seen with hypobaric hypoxia (160). The first intimation that hypobaria per se may influence ventilation came from the observation that ventilation is blunted by altitude compared to hypoxia alone. In nine healthy young men exposed to normobaric hypoxia, resting minute ventilation increased about 60% above control values, a value twice that during a comparable altitude exposure, suggesting a blunting of ventilation by low pressure (161). A recent report found a similar degree of depression of resting minute ventilation by hypobaria when comparing hypobaric hypoxia, normobaric hypoxia, and normoxic hypobaria (162). An explanation for the apparent blunting of ventilation during exposure to hypobaric hypoxia is not immediately apparent. In summary, hypobaria increases AMS and fluid retention and reduces ventilatory response to hypoxia. Further investigations are needed to clarify the importance of these factors in the pathophysiology of AMS.

V. Retinal Hemorrhages

A. Retinopathy at High Altitude

High-altitude retinal hemorrhages (HARH) are hemorrhages of the inner and middle layers of the retina occurring in lowland dwellers when they ascend to altitudes of over 5000 m. They are not observed in permanent inhabitants of high altitude (163,164). They may occur in association with other acute altitude illnesses (AMS, HACE, HAPE), but frequently they are seen without other signs of high-altitude maladaptation. Many commentators use the term ''altitude retinopathy'' (165,166) to describe the picture of high-altitude retinal hemorrhages with the expected physiological changes of vascular dilatation and tortuosity (both arteries and veins) and optic disc hyperemia found in hypoxemia. The hemorrhages have been documented clinically and histologically and can be found in any part of the fundus of the eye, but they are most frequently found adjacent to the superior and inferior temporal arcades, usually sparing the macula. When in the superficial (axonal) layer of the retina they are red and flame shaped, and when in the compact middle layers they are dark and rounded ''dot and blot'' hemorrhages.

Cotton wool spots (''soft exudates'') are lesions of the nerve fiber layer of the retina produced by interruption of orthograde and retrograde axoplasmic transport in ganglion cell axons (167). Their fluffy appearance is due to intracellular accumulation of organelles. Retinal cotton wool spots have been reported as isolated findings at altitude (166) and in association with hemorrhages (168). They have also been found overlying hemorrhages, especially the larger, deeper hemorrhages. Cotton wool spots are seen predominantly in climbers ascending over 7000 m (21,000 ft) but have also been observed in trekkers at 5400 m (169). These cotton wool spots would not necessarily produce scotomata as some axons can survive within small retinal infarcts.

B. Clinical Aspects

Most hemorrhages are asymptomatic, and even macular (but not foveal) hemorrhages may go unnoticed by the subject. The hemorrhages clear quickly, often within days, but the closer they are to the fovea, the more likely they are to produce a positive scotoma (dark patch in the vision). These scotomata have been reported to last for a year or more (165), though most resolve within days to weeks. Small macula hemorrhages have been seen to resolve over a matter of days in subjects remaining at an altitude of 5300 m (personal observation), but if a hemorrhage is seen to be enlarging (on daily examination), descent is advisable.

Singh (5) found vitreous hemorrhages in 3 of 1925 physically fit men between the ages of 18 and 53 staying at altitudes of 11,000–18,000 ft—an incidence more than 22 times the number of hemorrhages one would expect in a similar population at sea level (170). While the hemorrhages seem to be altitude related, the cause remains unknown. One case of vitreous hemorrhage is reported in the literature (166).

C. Determinants of Incidence

HARHs are common, the incidence above 5300 m averaging 34% (0–66%) (171–174). No association has been found with age or gender. Although no clear association with rate of ascent has been documented, it has been suggested that slow ascent may be protective (168,172). The variability in prevalence of retinal hemorrhages between these studies on high-altitude sojourners might be attributed to differences in speed of ascent, duration of exposure and altitude attained, number of previous exposures to high altitude, and diagnostics tools. Because HARH is not seen in sherpas, repeated exposure could be protective. However, this hypothesis could not be confirmed in Himalayan climbers (175). This study showed also no association between retinal hemorrhages and maximum altitude reached during the expedition, number of nights spent at extreme altitude, weight loss (used as an indicator of tolerance to chronic exposure to altitude), hemoglobin, cough, Valsalva maneuvers, severity of AMS, or drug use (aspirin, acetazolamide, dexamethasone, or nifedipine). Schumacher and Petajan (172) found in a group of 39 subjects an association between a history of vascular headaches (''common migraine''), altitude-related headache, and retinal hemorrhages, which was not observed in another study (175). This discrepancy may be due to ethnic differences (Japanese vs. Caucasian), selection bias, and small sample sizes.

The altitude at which a particular individual will develop hemorrhages is unknown. Many subjects will have developed HARHs before reaching 5000 m. Most people going to extreme altitude (8000 m) will probably develop HARH. A few subjects, however, had no retinal hemorrhages at 5000 m before and after return from altitudes of 8000 m. This absence may, of course, simply reflect the limitations of fundal examination at altitude. Reports of altitude retinopathy at altitudes below 3000 m are rare and tend to be associated with coexisting ocular or systemic disease (176).

D. Pathophysiology

The eye is embryologically an outgrowth of the brain, and retinal blood flow responds to changes in arterial Pa_{CO_2} and Pa_{O_2} in the same way as does the cerebral circulation. At 5300 meters (17,500 feet) retinal blood flow is more than double that at sea level (171) in spite of the hypoxia-induced hypocapnia ($Paco_2$ about 24 mmHg), suggesting that retinal vascular flow is controlled by the metabolic needs of the retina. This increased blood flow is reflected in vascular dilatation and tortuosity (both arteries and veins) and optic disc hyperemia, similar to that seen at sea level in infants with hypoxia secondary to congenital heart disease. Increased blood flow is likely an important pathophysiological factor for HARH. Retinal venous hypertension in high-altitude climbers, detected by diastolic ophthalmodynamometry (174), might diminish retinal vascular reactivity leading to impaired autoregulation in the face of sudden intravascular or intraocular pressure changes.

The following observations suggest that increased endothelial permeability due to hypoxia (87,97) may also contribute to HARH. HARHs are most commonly seen in the areas around the vascular arcades with a relative sparing of the macula, which is the area with the highest oxygen delivery (177,178). At low altitude, retinal hemorrhages are observed in illnesses with acute or chronic tissue hypoxia like cystic fibrosis (176), carbon monoxide poisoning, local hypoxia due to venous occlusion, or diabetes (179). However, retinal hemorrhages with the same appearance as at high altitude can also occur after prolonged Valsalva maneuvers (180). These observations suggest that increased blood flow and pressure as well as increased endothelial permeability due to hypoxia may play a role in the pathophysiology of HARH.

Fluorescein angiography indicates that hypoxia alone does not increase retinal vascular permeability at high altitude. There was no evidence of either leaking blood vessels or papilledema in subjects at 17,500 ft at rest, even though 36% exhibited retinal hemorrhages. At the same altitude after exercise, however, fluorescein leakage from around the optic disk was found in 40% of all subjects and in 100% of those with retinal hemorrhages. Findings with fluorescein angiography indicate that exercise in hypoxia transiently increases retinal vascular permeability. This increase may facilitate the occurrence of retinal hemorrhages perhaps in combination with decompensation of retinal vascular autoregulation and increased venous pressure, exacerbated, for example, by coughing and straining or increased intracranial pressure.

At sea level cotton wool spots most often develop at the edges of areas of ischemic retina, suggesting microvascular occlusion. Local hypoxia leads to failure of the sodium pump, accumulation of the products of glycolysis, and intracellular acidosis. It is conceivable that HARH may act as a barrier to diffusion of oxygen to the overlying axons, the decreased vitreal oxygen tension unable to meet the demands of the inner retinal layers. Alternatively, microvascular occlusion could occur at altitude as a result of the increase in serum viscosity or endothelial failure resulting in blood cell adhesion to the vessel wall.

E. Do HARHs Relate to AMS and HACE?

We are tempted to look upon the retina as the window to the brain, perhaps in part because the retinal blood supply and its control seem to be under the same governance as that of the brain. Retinal hemorrhages can occur in the absence of AMS and HAPE, and they do not necessarily herald impending AMS. They are often seen with HACE, along with disc hemorrhages and papilledema. Lubin et al. (181) found many small hemorrhages in the brain of a subject who died of HACE along with clinical and histological evidence of retinal hemorrhages, suggesting that similar processes might be occurring in both the retinal and cerebral circulations. Nevertheless, at present it is unclear whether retinal and cerebral hemorrhages of high altitude are caused by the same mechanism.

Retinal and cerebral hemorrhages of similar appearance to those occurring at high altitude are seen at sea level in victims of carbon monoxide (182). Morphologically these cerebral petechiae are also similar to those of hypoglycemia where the tissues suffer ischemic cell death due to loss of their principal metabolic substrate (183), supporting cellular failure as the cause of the hemorrhages rather than a mechanical failure of the vessel wall. Endothelial failure of altitude hypoxia may increase the permeability of the blood-retinal and blood-brain barriers, causing cerebral edema and increased CSF pressure along with hemorrhages in the retina and brain.

F. Further Research

A fair knowledge of the epidemiology of HARH currently exists, though reasons for the variable incidence in different studies and presumably between individuals have yet to be elucidated. Retinal hemorrhages at altitude may be caused by hypoxic capillary endothelial failure allowing extravasation of blood, perhaps accompanied by decompensation of retinal vascular autoregulation and increased venous pressure. Mechanical factors, such as coughing and straining or increased intracranial pressure or microinfarcts secondary to hematological and vascular changes, may exacerbate the condition. No work has yet been done to test this or any other hypothesis concerning why HARH occurs.

VI. Prevention and Treatment

A. General Measures

The likelihood of experiencing AMS can be decreased by slower, staged ascent to high altitude. Beneficial effects of acclimatization likely last several days to perhaps weeks, depending on duration and degree of altitude exposure (36). Residence at an altitude above 900 m reduces AMS at intermediate altitude (27). An acclimatization to 4350 m over 7 days followed by 4 days of exposure to altitudes of 6000–8000 m for several hours in a hypobaric chamber improved the climbing rate on Mount Everest (184), while intermittent exposure 8 hours daily to an F_{IO_2} of 13% (equivalent to 3300 m) over 10 days did not benefit individuals subsequently exposed to 4500 m (185). Endurance training may reduce the incidence of AMS (27,40), but this seems unlikely since many studies find no association between

V_{O_2}max and AMS (28,30,33,38). Diet high in carbohydrates was protective in one study (186) but not in others (187,188).

Preferred treatment for AMS consists of rest, analgesics for headache in the case of mild to moderate illness, and descent when symptoms worsen despite a day of rest or treatment or with severe AMS or HACE. Controlled trials have shown significant benefit from rest and placebo effect (73,189) compared with no improvement over 12 hours in control groups (190,191). The rationale for a day of rest is to allow spontaneous improvement of AMS by avoiding additional hypoxemia due to exercise or further ascent.

Headache can be successfully treated with nonsteroidal analgesics and acetaminophen. Ibuprofen (400 mg) is the only such medication that has been tested in a placebo-controlled randomized, double-blind trial (138). Sumatriptan, a 5-hydroxytryptamine$_1$-receptor agonist used for migraine, was either not effective (137) or relieved headache for only 1–3 hours (136,192). Other measures reported to relieve high-altitude headache, like hyperventilation (72) and digital compression of temporal arteries (193), have not been adequately evaluated.

B. Increase in Partial Pressure for Oxygen

Increasing inspiratory P_{O_2} by inspiring an increased F_{IO_2} (73) or by increasing barometric pressure by descent or artificial pressurization in a hyperbaric device rapidly improves AMS (189,194,195). AMS recurs within a few hours when the victim is repressurized (189,194–196). Several cases of HACE were successfully treated in the bag for 4–6 hours (197,198). Additional concomitant treatments, such as dexamethasone, make it difficult to quantify the efficacy of the bag by itself for treatment of HACE. Thus, portable hyperbaric devices may be useful in remote areas and valuable support in conjunction with other treatments because of their rapid action. Due caution should be taken since delay of descent and loss of clinical control over the patients may contribute to fatal outcomes (199).

C. Acetazolamide

Prophylaxis

Acetazolamide, a sulfonamide carbonic anhydrase inhibitor, is effective for prevention and treatment of AMS. Five randomized, double-blind, placebo-controlled trials of acetazolamide for prevention of AMS (29,31,72,141,200) involving a total of about 270 subjects demonstrated a significant reduction of symptoms of AMS at altitudes between 3900 and 5900 m with prophylactic administration of 250 mg b.i.d. or t.i.d. or 500 mg slow-release acetazolamide qd. Medication was started 5 to 0 days prior to reaching the altitude of 3000 m and taken for a total time of 2–14 days. The drug was well tolerated; side effects were paresthesias and an unpleasant taste of carbonated beverages.

Mode of Action

The beneficial effects of acetazolamide and other carbonic anhydrase inhibitors are conventionally attributed to increased ventilation and improved oxygenation documented in several studies (31,72,141,200). Increased ventilation results from metabolic acidosis because of renal sodium bicarbonate excretion and perhaps tissue respiratory acidosis due to inhibition of carbonic anhydrase in brain and peripheral chemoreceptors (201). Furthermore, reduction of CSF production has also been suggested to contribute to the beneficial effect of acetazolamide (141). Selective inhibition of renal carbonic anhydrase appears to be sufficient for prevention of AMS, as was demonstrated by the use of benzolamide, a selective inhibitor of renal carboanhydrase (202). Since 5 mg/kg body weight or less produces sufficient renal carbonic anhydrase inhibition (203), lower doses of acetazolamide deserve to be tested for prevention of AMS because they might maintain efficacy with fewer gastrointestinal side effects.

Another way in which carbonic anhydrase inhibitors may benefit is the reduction in periodic breathing and hypoxemia during sleep at high altitude (74,204), associated with improved sleep quality (205). This benefit, too, may result from the influence of metabolic acidosis on ventilation (206,207). although high doses of intravenous acetazolamide increase CBF, Doppler measurements suggest that cerebral blood flow is not changed by oral doses of acetazolamide used for prevention of AMS (208). Increased arterial Po_2 and improved well-being most likely also account for better performance in laboratory tests (209) and during climbing (200) as well as for less weight and muscle mass loss during an expedition (209).

Further support for the importance of acetazolamide's stimulation of ventilation comes from studies of furosemide, spironolactone, and progesterone. Furosemide, a diuretic with no ventilatory stimulant effect, was no effective in treating AMS (210) and even had detrimental effects at 5400 m (211), although Singh et al. suggested that furosemide was effective for both prevention and treatment of AMS (5). Spironolactone, although it induces a mild metabolic acidosis and increases diuresis, has proven largely ineffective for prevention of AMS (212–214). Progesterone, which stimulates ventilation, appears to have some effect for prevention of AMS (215), while preliminary reports on preventive effects of theophylline (216) need to be confirmed and mechanisms accounting for this effect need to be elucidated.

The results of these interventional trials also give clues about the pathophysiology of AMS. Effective medications like acetazolamide or progesterone suggest that drugs increasing oxygenation by increasing ventilation help to lessen symptoms of AMS. They support a concept that low ventilation is associated with increased susceptibility to AMS. Drugs that enhance diuresis and natriuresis like furosemide or spironolactone are not effective, suggesting that fluid retention is not a primary pathophysiological mechanism of AMS. This notion is supported by measurements of sodium and urine output under a controlled diet during the early hours after passive exposure to 4559 m (116).

Treatment

Acetazolamide appears useful for treating as well as preventing AMS. High dosage for treatment may, however, do more harm than good. At altitudes between 3500 and 5500 m, a study of 37 subjects given acetazolamide (1–1.5 g) showed significant a improvement in overall AMS scores, although headache worsened in 28% (134). Acetazolamide (250 mg every 8 h) reduced AMS in 6 subjects at 4200 m over 24 hours compared to 6 controls in whom $AaDo_2$ worsened (123). One dose of 250 mg acetazolamide in 14 subjects at 4559 m reduced symptoms of AMS within 12 hours, significantly increased P_ao_2, and decreased $AaDo_2$ (191).

D. Steroids

Prophylaxis

Four mg of dexamethasone given every 6 hours unequivocally reduces severity and incidence of AMS after rapid exposure to altitudes of 4000 m or above (217–223). Administration of 4 mg dexamethasone every 6 hours starting 2 days prior to acute exposure to a simulated altitude of 4570 m (217) and passive ascent to an altitude of 4300 m (218) significantly reduced symptoms of AMS. Four mg of dexamethasone given every 12 hours (223) or 2 mg given every 4 hours (220), however, was not sufficient at altitudes above 4000 m when exercise was involved in the exposure. Starting medication a few hours before ascent appears to be sufficient given the rapid effects observed with therapeutic administration of dexamethasone (196,220). Two studies looked at the effect of discontinuing dexamethasone and found an increase of AMS symptoms, suggesting that dexamethasone could delay acclimatization (218,220).

Dexamethasone should not be considered as first choice for prevention of AMS since measures like preacclimatization, slow ascent, or acetazolamide, which have less potentially harmful side effects, are available. Also meta-analysis of 20 studies suggests that acetazolamide may be slightly more effective than dexamethasone for prevention of AMS (224), despite the opposite findings in one limited direct comparison (219).

Treatment

Controlled trials in mountaineers with AMS at an altitude of 4559 m reported the beneficial therapeutic effects of dexamethasone (8 mg i.v. or p.o. followed by 4 mg p.o. every 6 h) comparing the drug against placebo treatment (190), simulated descent in a portable hyperbaric chamber (196), or acetazolamide, almitrine, and placebo in a four-arm trial with 55 subjects (191). In the last study the decrease of AMS scores over 12 hours was more pronounced with dexamethasone than with acetazolamide.

The data demonstrate the beneficial effect of dexamethasone for treatment of AMS. None of these studies, however, looked at the long-term effects and at the possibility of rebound after discontinuing treatment. The fact that AMS might recur after treatment with dexamethasone, as happens after stopping of prophylactic administration of dexamethasone (218,220), points to the likelihood of rebound and mandate descent or delay of further ascent for 1 or 2 days after treatment with dexamethasone.

Mode of Action

In a preventive trial, administration of dexamethasone was not associated with significant changes in periodic breathing, oxygen saturation, activation of the sympatho-adrenal axis, or fluid shifts (17). The authors suggested that dexamethasone primarily acts by reducing nausea and by providing euphoria without significantly interfering with the pathophysiological process causing AMS. CT scan findings in one subject compatible with mild cerebral edema (17) after treatment with dexamethasone and the observation of rebound of AMS after discontinuation of dexamethasone (218,220) are compatible with this hypothesis. While unchanged oxygen saturation (223) and no effect on fluid shifts (222) were also observed in other preventive trials, reduced diuresis and decreased diameter of retinal arteries, suggesting decrease of cerebral blood flow, were found in another investigation (217). Controlled field studies (190,191,196) revealed a significant increase of Sao_2 after 12 hours associated with a slight increase in lung function in one (190) and a significant decrease of $AaDo_2$ in another study (191). These observations suggest that dexamethasone may improve gas exchange. A possible mechanism of action may involve capillary permeability: interaction of leukocytes and endothelial cells and the associated release of cytokines is inhibited by dexamethasone. By this mechanism, dexamethasone could also improve vascular integrity in the brain, as shown in rat lungs (87), and thus improve cerebral symptoms.

E. Drugs Without Effects

The cyclooxygenase inhibitor naproxen (157); calcium channel blockade, (121,225), and antacids (142) were all tested and found to have no effect on preventing AMS.

VII. Research Directions

Questions about research directions regarding AMS and HACE for the near future can be divided into three major areas: pathophysiology, epidemiology, and prevention/treatment. When analyzing studies about AMS, we should bear in mind that research in AMS will always be complicated by the lack of a clear, objective definition of the illness. AMS is a combination of various, nonspecific symptoms the perception and rating of which are influenced by personality traits. Thus, higher symptom scores, especially in mild disease, may not necessarily be caused by pathophysiological processes—they might also be attributed to psychological factors like increased trait anxiety and anxious self-attention (226).

Investigations about the pathophysiology of AMS and HACE should focus on the mechanisms that lead to edema and the significance of edema for early AMS. Is the widened $AaDo_2$ in AMS due to interstitial pulmonary edema, and is it a cause or consequence of AMS? As it appears that brain edema is at best minimal in AMS and may not account for the symptoms, we need to consider other mechanisms such as those discussed for headache at low altitude. Another question that needs to be addressed is: What are the processes that account for progression of edema to HACE? We need to examine the significance of fluid retention and increased permeability for edema found in AMS and HACE. Recent observations suggest that net fluid retention may not be necessary for the development of AMS. Evidence from cell cultures and animal studies and indirectly from human studies suggest that hypoxia triggers an inflammatory response and enhances capillary permeability. Manipulation of the inflammatory pathway with specific inhibitors or receptor blockers like those interfering with the lipoxygenase pathway may help to determine the role of inflammation in AMS or HACE.

To answer these questions it is essential to perform investigations during the early hours of exposure to high altitude. In addition, many of these questions demand the use of sophisticated technologies not available at altitudes of 4000 m or higher. For these reasons, investigations under artificial conditions are likely to become more frequent. This leads to the question of how exposure in the field and in a chamber compare and whether hypobaria is an essential component for such studies. Preliminary work suggesting that normobaric and hyperbaric hypoxia do not cause similar AMS incidence and severity needs to be confirmed and responsible mechanisms identified.

The epidemiology of AMS has been explored in considerable detail. There remain, however, some questions about the role of age, preexisting illnesses, exercise, and physical fitness in susceptibility to AMS. A major, mostly untouched area is the role of genetic factors in susceptibility. We need to know whether susceptibility or absence of susceptibility to AMS runs in families and whether genetic markers are associated with susceptibility. In the area of prevention and treatment, questions that focus on the mechanisms of action of effective agents, like acetazolamide and dexamethasone, may help to better understand the pathophysiology of AMS and HACE. New drugs like inhibitors of specific carboanhydrases, 5HT1-receptor blockers, leukotriene antagonists, or theophylline may offer additional insight into the pathophysiology of AMS and HACE.

References

1. Bert P. Barometric Pressure. 1978.
2. Ravenhill TH. Some experiences of mountain sickness in the Andes. J Trop Med Hyg 1913; 16:313–320.
3. Bärtsch P, Shaw S, Franciolli M, Gnädinger MP, Weidmann P. Atrial natriuretic peptide in acute mountain sickness. J Appl Physiol 1988; 65:1929–1937.

4. Maggiorini M, Bühler B, Walter M, Oelz O. Prevalence of acute mountain sickness in the Swiss Alps. Br Med J 1990; 301:853–855.
5. Singh I, Khanna PK, Srivastava MC, Lal M, Roy SB, Subramanyam CSV. Acute mountain sickness. N Engl J Med 1969; 280:175–184.
6. Maggiorini M, Bärtsch P, Oelz O. Association between raised body temperature and acute mountain sickness: cross sectional study. Br Med J 1997; 315:403–404.
7. Hackett PH, Rennie D. Rales, peripheral edema, retinal hemorrhage and acute mountain sickness. Am J Med 1979; 67:214–218.
8. Williams ES, Ward MP, Milledge JS, Withey WR, Older MWJ, Forsling ML. Effect of the exercise of seven consecutive days hill-walking on fluid homeostasis. Clin Sci 1979; 56:305–316.
9. Withey WR, Milledge JS, Williams ES, Minty BD, Bryson EI. Fluid and electrolyte homeostasis during prolonged exercise at altitude. J Appl Physiol 1983; 55:409–412.
10. Bärtsch P, Vock P, Maggiorini M, Franciolli M, Fretz C, Schobersberger W, Mairbäurl H, Oelz O. Respiratory symptoms, radiographic and physiologic correlations at high altitude. In: Sutton JR, Coates G, Remmers JE, eds. Hypoxia: The Adaptations. Philadelphia: B.C. Decker, Inc., 1990:241–245.
11. Hamilton AJ, Trad LA, Cymerman A. Alterations in human upper extremity motor function during acute exposure to simulated altitude. Aviat Space Environ Med 1991; 62:759–764.
12. White AJ. Cognitive impairment of acute mountain sickness and acetazolamide. Aviat Space Environ Med 1984; 55:598–603.
13. Regard M, Landis T, Casey J, Maggiorini M, Bärtsch P, Oelz O. Cognitive changes at high altitude in healthy climbers and in climbers developing acute mountain sickness. Aviat Space Environ Med 1991; 62:291–295.
14. Richalet J-P, Rutgers V, Bouchet P, Rymer J-C, Kéromès A, Duval-Arnould G, Rathat C. Diurnal variations of acute mountain sickness, colour vision, and plasma cortisol and ACTH at high altitude. Aviat Space Environ Med 1989; 60:105–111.
15. Hartig GS, Hackett PH. Cerebral spinal fluid pressure and cerebral blood velocity in acute mountain sickness. In: Sutton JR, Coates G, Houston CS, eds. Hypoxia and Mountain Medicine. New York: Pergamon Press, 1992:260–265.
16. Wright AD, Imray CHE, Morissey MSC, Marchbanks RJ, Bradwell AR. Intracranial pressure at high altitude and acute mountain sickness. Clin Sci 1995; 89:201–204.
17. Levine BD, Yoshimura K, Kobayashi T, Fukushima M, Shibamoto T, Ueda G. Dexamethasone in the treatment of acute mountain sickness. N Engl J Med 1989; 321: 1707–1713.
18. Matsuzawa Y, Kobayashi T, Fujimoto K, Shinozaki S, Yoshikawa S, Yamaguchi S, Kubo K, Sekiguchi M, Hayashi R, Sakai A, Ueda G. Cerebral edema in acute mountain sickness. In: Ueda G, Reeves JT, Sekiguchi M, eds. High-Altitude Medicine. Matsumoto: Shinshu University Press, 1992:300–304.
19. Muza SR, Lyons TP, Rock PB, Fulco CS, Beidleman BA, Smith S, Morocz IA, Zientaral GP, Cymerman A. Effect of altitude exposure on brain volume and development of acute mountain sickness (AMS) (abstr). In: Roach RC, Hackett PH, Wagner PD, eds. Hypoxia: Into the Next Millenium. New York: Plenum/Kluwer Academic Publishing, 1999:414.
20. Icenogle M, Kilgore D, Sanders J, Caprihan A, Roach RC. Cranial CSF volume (cCSF) is reduced by altitude exposure but is not related to early acute mountain sickness

(AMS) (abstr). In: Roach RC, Hackett PH, Wagner PD, eds. Hypoxia: Into the Next Millenium. New York: Plenum/Kluwer Academic Publishing, 1999:392.

21. Hohenhaus E, Paul A, Bärtsch P. Hypoxic ventilatory response and gas exchange in acute mountain sickness. Eur Respir J 1994; 7(18):200s.
22. Sampson JB, Cymerman A, Burse RL, Maher JT, Rock PB. Procedures for the measurement of acute mountain sickness. Aviat Space Environ Med 1983; 54:1063–1073.
23. Roach RC, Bärtsch P, Hackett PH, Oelz O. The Lake Louise acute mountain sickness scoring system. In: Sutton JR, Houston CS, Coates G, eds. Hypoxia and Molecular Medicine. Burlington, VT: Queen City Press, 1993:272–274.
24. Bärtsch P, Müller A, Hofstetter D, Maggiorini M, Vock P, Oelz O. AMS and HAPE scoring in the Alps. In: Sutton JR, Houston CS, Coates G, eds. Hypoxia and Molecular Medicine. Burlington: Queen City Press, 1993:265–271.
25. Maggiorini M, Müller A, Hofstetter D, Bärtsch P, Oelz O. Assessment of acute mountain sickness by different score protocols in the Swiss alps. Aviat Space Environ Med 1998; 69(12):1186–1192.
26. Hildebrandt W, Buclin T, Biollaz J, Swenson E, Bärtsch P. Accentuated hypoxemia precedes impairment of gas exchange in subjects with AMS (abstr). In: Houston CS, Coates G, eds. Hypoxia and Mountain Medicine: Women at Altitude. Proceedings of the 10th International Hypoxia Symposium at Lake Louise, Feb. 18–22, 1997.
27. Honigman B, Theis MK, Koziol-McLain J, Roach R, Yip R, Houston C, Moore LG. Acute mountain sickness in a general tourist population at moderate altitudes. Ann Intern Med 1993; 118:587–592.
28. Bircher HP, Eichenberger U, Maggiorini M, Oelz O, Bärtsch P. Relationship of mountain sickness to physical fitness and exercise intensity during ascent. J Wilderness Med 1994; 5:302–311.
29. Hackett PH, Rennie D, Levine HD. The incidence, importance, and prophylaxis of acute mountain sickness. Lancet 1976; II(Nov. 27):1149–1154.
30. Kayser B. Acute mountain sickness in western tourists around the Thorong pass (5400 m) in Nepal. J Wilderness Med 1991; 2:110–117.
31. Larson EB, Roach RC, Schoene RB, Hornbein TF. Acute mountain sickness and acetazolamide. JAMA 1982; 248:328–332.
32. Hackett PH, Rennie D. Avoiding mountain sickness. Lancet 1978; II(Oct. 28):938.
33. Hansen JE, Harris CW, Evans WO. Influence of elevation of origin, rate of ascent and a physical conditioning program on symptoms of acute mountain sickness. Military Med 1967; 132:585–593.
34. Purkayastha SS, Ray US, Arora BS, Chhabra PC, Thakur L, Bandopadhyay P, Selvamurthy W. Acclimatization at high altitude in gradual and acute induction. J Appl Physiol 1995; 79:487–492.
35. Muza SR, Fulco CS, Beidleman BA, Rock PB, Lyons T, Cymerman A. Ventilatory acclimatization to 4300m is retained eight days after return to sea level (abstr). In: Sutton JR, Coates G, Houston CS, eds. Hypoxia and Molecular Medicine. Burlington, VT: Queen City Press, 1993.
36. Lyons TP, Muza SR, Rock PB, Cymerman A. The effect of altitude preacclimatization on acute mountain sickness during reexposure. Aviat Space Environ Med 1995; 66:957–962.
37. Brown DE. Acute mountain sickness and physiological stress during intermittent exposure to high altitude. Ann Hum Biol 1989; 16:15–23.

38. Richalet J-P, Keromes A, Dersch B, Corizzi F, Mehdioui H, Pophillat B, Chardonnet H, Tassery F, Herry J-P, Rathat C, et al. Caractéristiques physiologiques des alpinistes de haute altitude. Sci Sports 1988; 8:89–108.
39. Honigman B, Read M, Lezotte D, Roach RC. Sea-level physical activity and acute mountain sickness at moderate altitude. West J Med 1995; 163:117–121.
40. Sen Gupta J, Joseph NT, Malhotra MS. Physical fitness status and adaptation to high altitude. Indian J Med Res 1978; 68:312–321.
41. Wagner PD, Sutton JR, Reeves JT, Cymerman A, Groves BM, Malconian MK. Operation Everest II: pulmonary gas exchange during a simulated ascent of Mt. Everest. J Appl Physiol 1987; 63:2348–2359.
42. Roach RC, Maes D, Sandoval D, Robergs RA, Icenogle M, Hinghofer-Szalkay H, Lium D, Loeppky JA. Exercise exacerbates acute mountain sickness at simulated altitude. Med Sci Sports Exerc 1999; 81:S182.
43. Bärtsch P, Hildebrandt W, Buclin T, Hohenhaus E, Biollaz J. Comparable incidence and severity of acute mountain sickness after active and passive ascent to an altitude of 4559 m (abstr). Med Sci Sports Exerc 1998; 30(5):S265.
44. Roach RC, Houston CS, Honigman B, Nicholas R, Yaron M, Grissom CK, Alexander JK, Hultgren HN. How well do older persons tolerate moderate altitude? West J Med 1995; 162:32–36.
45. Montgomery AB, Mills J, Luce JM. Incidence of acute mountain sickness at intermediate altitude. JAMA 1989; 261:732–734.
46. Hirata K, Masuyama S, Saito A. Obesity as risk factor for acute mountain sickness. Lancet 1989; 2(Oct. 28):1040–1041.
47. Yaron M, Waldman N, Niermeyer S, Nicholas R, Honigman B. The diagnosis of acute mountain sickness in preverbal children. Arch Pediatr Adolesc Med 1998; 152:683–687.
48. Theis MK, Honigman B, Yip R, McBride D, Houston CS, Moore LG. Acute mountain sickness in children at 2835 meters. AJDC 1993; 147:143–145.
49. Riboni K, Maes DP, Loeppky JA, Icenogle M, Roach RC. Women at altitude: No gender or menstrual cycele effects on acute mountain sickness (AMS) (abstr). In: Roach RC, Hackett PH, Wagner PD, eds. Hypoxia: Into the Next Millenium. New York: Plenum/Kluwer Academic Publishing, 1999:423.
50. Robinson SM, King AB, Aoki V. Acute mountain sickness: reproducibility of its severity and duration in an individual. Aerospace Med 1971; 42:706–708.
51. Forster P. Reproducibility of individual response to exposure to high altitude. Br Med J [Clin Res] 1984; 289:1269.
52. Bärtsch P. Wer wird bergkrank? Schweiz Med Wochenschr 1992; 122:307–314.
53. Swenson ER, Duncan TB, Goldberg SV, Ramirez G, Ahmad S, Schoene RB. Diuretic effect of acute hypoxia in humans: relationship to hypoxic ventilatory responsiveness and renal hormones. J Appl Physiol 1995; 78:377–383.
54. Savourey G, Moirant C, Eterradossi J, Bittel J. Acute mountain sickness relates to sea-level partial pressure of oxygen. Eur J Appl Physiol 1995; 70:469–476.
55. Austin D, Sleigh J. Prediction of acute mountain sickness. Br Med J 1995; 811:989–990.
56. Houston CS, Dickinson J. Cerebral form of high-altitude illness. Lancet 1975; II(Oct. 18):758–761.
57. Dickinson J, Heath D, Gosney J, Williams D. Altitude-related deaths in seven trekkers in the Himalayas. Thorax 1983; 38:646–656.

58. Clarke C. High altitude cerebral oedema. Int J Sports Med 1988; 9:170–174.
59. Hochstrasser J, Nanzer A, Oelz O. Das Höhenödem in den Schweizer Alpen. Beobachtungen über Inzidenz, Klinik und Verlauf bei 50 Patienten der Jahre 1980–1984. Schweiz Med Wochenschr 1986; 116:866–873.
60. Hultgren HN. High altitude pulmonary edema: hemodynamic aspects. Int J Sports Med 1997; 18(1):20–25.
61. Song S-Y, Asaji T, Tanizaki Y, Fujimaki T, Matsutani M, Okeda R. Cerebral thrombosis at altitude: its pathogenesis and the problems of prevention and treatment. Aviat Space Environ Med 1986; 57:71–76.
62. Hackett PH, Hollingshead KF, Roach R, Schoene RB, Mills WJ. Cortical blindness in high altitude climbers and trekkers—a report on six cases. In: Sutton JR, Houston CS, Coates G, eds. Hypoxia and Cold. New York: Praeger Publishers, 1987:536.
63. Jenzer G, Bärtsch P. Migraine with aura at high altitude. J Wilderness Med 1993; 4: 412–415.
64. Murdoch DR. Focal neurological deficits and migraine at high altitude. J Neurol Neurosurg Psychiatry 1995; 58:637.
65. Kobayashi T, Koyama S, Kubo K, Fukushima M, Kusama S. Clinical features of patients with high-altitude pulmonary edema in Japan. Chest 1987; 92:814–821.
66. Fukushima M, Kobayashi T, Kubo K, Yoshimura K, Shibamoto T. A case of high altitude pulmonary edema followed by brain computerized tomography and electroencephalogram. Aviat Space Environ Med 1988; 59:1076–1079.
67. Yamaguchi S, Kobayashi T, Matsuzawa Y, Koizumi T, Shinozaki S, Fujimoto K, Fukushima M, Kubo K, Sekiguchi M. MRI findings in patients with high altitude pulmonary edema. In: Ueda G, Reeves JT, Sekiguchi M, eds. High-Altitude Medicine. Matsumoto: Shinshu University Press, 1992:305–308.
68. Xu F, Severinghaus JW. Rat brain VEGF expression in alveolar hypoxia: possible role in high-altitude cerebral edema. J Appl Physiol 1998; 85(1):53–37.
69. Nayak NC, Roy S, Narayanan TK. Pathologic features of altitude sickness. Am J Pathol 1964; 45:381–391.
70. Bärtsch P, Haeberli A, Franciolli M, Kruithof EKO, Straub PW. Coagulation and fibrinolysis in acute mountain sickness and beginning pulmonary edema. J Appl Physiol 1989; 66:2136–2144.
71. Hildebrandt W, Biollaz J, Buclin T, Swenson E, Bärtsch P. Accentuated hypoxemia in acute mountain sickness: ventilatory and metabolic factors (abstr). Int J Sports Med 1998; 19:S16.
72. Birmingham Medical Research Expeditionary Society Mountain Sickness Study Group. Acetazolamide in control of acute mountain sickness. Lancet 1981; I(Jan. 24): 180–183.
73. Bärtsch P, Baumgartner RW, Waber U, Maggiorini M, Oelz O. Comparison of carbon-dioxide-enriched, oxygen-enriched, and normal air in treatment of acute mountain sickness. Lancet 1990; 336:772–775.
74. Sutton JR, Houston CS, Mansell AL, McFadden MD, Hackett PH, Rigg JRA, Powles ACP. Effect of acetazolamide on hypoxemia during sleep at high altitude. N Engl J Med 1979; 301:1329–1331.
75. Eichenberger U, Weiss E, Riemann D, Oelz O, Bärtsch P. Nocturnal periodic breathing and the development of acute high altitude illness. Am J Respir Crit Care Med 1996; 154:1748–1754.
76. London: Fischer Unwin; 1898.

77. Milledge JS. The great oxygen secretion controversy. Lancet 1985; 2(Dec. 21/28): 1408–1411.
78. Harvey TC, Raichle ME, Winterborn MH, Jensen J, Lassen NA, Richardson NV, Bradwell AR. Effect of carbon dioxide in acute mountain sickness: a rediscovery. Lancet 1988; II:639–641.
79. Krasney JA. A neurogenic basis for acute altitude illness. Med Sci Sports Exerc 1994; 26:195–208.
80. Maher JT, Cymerman A, Reeves JT, Cruz JC, Denniston JC, Grover RF. Acute mountain sickness: Increased severity in eucapnic hypoxia. Aviat Space Environ Med 1975; 46:826–829.
81. Mairbäurl H, Wodopia R, Eckes S, Schulz S, Bärtsch P. Impairment of cation transport in A549 cells and ral alveolar epithelial cells by hypoxia. Am J Physiol (Lung Cell Mol Physiol) 1997; 273 (17):L797–L806.
82. Hackett PH, Yarnell PR, Hill R, Reynard K, Heit J, McCormick J. High-altitude cerebral edema evaluated with magnetic resonance imaging. JAMA 1998; 280(22):1920–1925.
83. Lassen NA, Harper AM. High-altitude cerebral oedema. Lancet 1975; (Dec. 6): 1154.
84. Baumgartner RW, Bärtsch P, Maggiorini M, Waber U, Oelz O. Enhanced cerebral blood flow in acute mountain sickness. Aviat Space Environ Med 1994; 65:726–729.
85. Jensen JB, Wright AD, Lassen NA, Harvey TC, Winterborn MH, Raichle ME, Bradwell AR. Cerebral blood flow in acute mountain sickness. J Appl Physiol 1990; 69: 430–433.
86. Yang S-P, Bergö GW, Krasney E, Krasney JA. Cerebral pressure-flow and metabolic responses to sustained hypoxia: effect of CO_2. J Appl Physiol 1994; 76:303–313.
87. Stelzner TJ, O'Brien RF, Sato K, Weil JV. Hypoxia-induced increases in pulmonary transvascular protein escape in rats; modulation by glucocorticoids. J Clin Invest 1988; 82:1840–1847.
88. Tuder RM, Flook BE, Voelkel NF. Increased gene expression for VEGF and the VEGF receptors KDR/Flk and Flt in lungs exposed to acute or to chronic hypoxia. J Clin Invest 1995; 95:1798–1807.
89. Shreeniwas R, Koga S, Karakurum M, Pinsky D, Kaiser E, Brett J, Wolitzky BA, Norton C, Plocinski J, Benjamin W, et al. Hypoxia-mediated induction of endothelial cell interleukin-1a—an autocrine mechanism promoting expression of leukocyte adhesion molecules on the vessel surface. J Clin Invest 1992; 90:2333–2339.
90. Lawson CA, Yan SD, Yan SF, Liao H, Zhou YS, Sobel J, Kisiel W, Stern DM, Pinsky DJ. Monocytes and tissue factor promote thrombosis in a murine model of oxygen deprivation. J Clin Invest 1997; 99:1729–1738.
91. Plate KH, Breier G, Millauer B, Ullrich A, Risau W. Up-regulation of vascular endothelial growth factor and its cognate receptors in a rat glioma model of tumor angiogensis. Cancer Research 1993; 53:5822–5827.
92. Hansen JM, Olsen NV, Feldt-Rasmussen B, Kanstrup I-L, Déchaux M, Dubray C, Richalet J-P. Albuminuria and overall capillary permeability of albumin in acute altitude hypoxia. J Appl Physiol 1994; 76:1922–1927.
93. Coates G, Gray GW, Nahmias C, Powles AC, Sutton JR. Transcapillary escape rate of albumin after exposure to 4300 m. In: Brendel W, Zink RA, eds. High Altitude Physiology and Medicine. New York: Springer-Verlag, 1982:176–178.

94. Ballmer PE, Kleger G-R, Imoberdorf R, Manke HG, Geiser T, Bärtsch P. Transcapillary escape rate of albumin at high altitude. In: Sutton JR, Houston CS, Coates G, eds. Hypoxia and the Brain. Burlington, VT: Queen City Printers, 1995:235–242.
95. Siggaard-Andersen J, Petersen FB, Hansen TI, Mellemgaard K. Plasma volume and vascular permeability during hypoxia and carbon monoxide exposure. Scand J Clin Lab Invest 1968; (suppl)103:39–48.
96. Poulsen TD, Klausen T, Richalet J-P, Kanstrup I-L, Fogh-Andersen N, Olsen NV. Plasma volume in acute hypoxia: comparison of a carbon monoxide rebreathing method and dye dilution with Evans' blue. Eur J Appl Physiol 1998; 77:457–461.
97. Rennie IDB, Joseph BJ. Urinary protein excretion in climbers at high altitudes. Lancet 1970; I1247–1251.
98. Rennie D, Marticorena E, Monge C, Sirotzky L. Urinary protein excretion in high-altitude residents. J Appl Physiol 1972; 31:257–259.
99. Rennie D, Frayser R, Gray G, Houston C. Urine and plasma proteins in men at 5,400 m. J Appl Physiol 1972; 32:369–373.
100. Winterborn MH, Bradwell AR, Chesner IM, Jones GT. The origin of proteinuria at high altitude. Postgrad Med J 1987; 63:179–181.
101. Bradwell AR, Imray C. Generalised protein leakage in acute mountain sickness. In: Sutton JR, Houston CS, Coates G, eds. Hypoxia and the Brain. Burlington, VT: Queen City Printers, 1995:315.
102. Bradwell AR, Delamere JP. The effect of acetazolamide on the proteinuria of altitude. Aviat Space Environ Med 1982; 53:40–43.
103. Lewis DM, Bradwell AR, Shore AC, Beaman M, Tooke JE. Capillary filtration coefficient and urinary albumin leak at altitude. Eur J Clin Invest 1997; 27:64–68.
104. Schoene RB, Swenson ER, Pizzo CJ, Hackett PH, Roach RC, Mills WJ, Henderson WR, Martin TR. The lung at high altitude: bronchoalveolar lavage in acute mountain sickness and pulmonary edema. J Appl Physiol 1988; 64:2605–2613.
105. Ullman E. Acute anoxia and the excretion of water and electrolyte. J Physiol Lond 1961; 155:417–437.
106. Heyes MP, Farber MO, Manfredi F, Robertshaw D, Weinberger M, Fineberg N, Robertson G. Acute effects of hypoxia on renal and endocrine function in normal humans. Am J Physiol 1982; 243:R265–R270.
107. Koller EA, Bührer A, Felder L, Schopen M, Vallotton MB. Altitude diuresis: endocrine and renal responses to acute hypoxia of acclimatized and non-acclimatized subjects. Eur J Appl Physiol 1991; 62:228–234.
108. Milledge JS, Beeley JM, McArthur S, Morice AH. Atrial natriuretic peptide, altitude and acute mountain sickness. Clin Sci 1989; 77:509–514.
109. Bärtsch P, Maggiorini M, Schobersberger W, Shaw S, Rascher W, Girard J, Weidmann P, Oelz O. Enhanced exercise-induced rise of aldosterone and vasopressin preceding mountain sickness. J Appl Physiol 1991a; 71:136–143.
110. Bärtsch P, Jülg B, Hohenhaus E. Urine volume in acute mountain sickness is not related to hypoxic ventilatory response (HVR) (abstr). Eur Respir J 1995; 8:62s.
111. Hackett PH, Rennie D, Hofmeister SE, Grover RF, Grover EB, Reeves JT. Fluid retention and relative hypoventilation in acute mountain sickness. Respiration 1982; 43: 321–329.
112. Currie TT, Carter PH, Champion WL, Fong G, Francis JK, McDonald IH, Newing RK, Nunn IN, Sisson RN, Sussex M, et al. Spironolactone and acute mountain sickness. Med J Aust 1976; 2:168–170.

113. Hackett PH, Rennie D, Grover RF, Reeves JT. Acute mountain sickness and the edemas of high altitude: a common pathogenesis? Respir Physiol 1981; 46:383–390.
114. Richalet J-P, Rathat C, Kéromès A, Herry J-P, Larmignat P, Garnier M, Pilardeau P. Plasma volume, body weight, and acute mountain sickness. Lancet 1983; i(Mar. 5): 525.
115. Honig A. Peripheral arterial chemoreceptors and reflex control of sodium and water homeostasis. Am J Physiol 1989; 257:R1282–R1302.
116. Hildebrandt W, Buclin T, Swenson E, Bärtsch P, Biollaz J. Development of acute mountain sickness without sodium and fluid retention (abstr). Int J Sports Med 1998; 19:S16.
117. Hogan RPI, Kotchen TA, Boyd AEI, Hartley LH. Effect of altitude on renin-aldosterone system and metabolism of water and electrolytes. J Appl Physiol 1973; 35:385–390.
118. Hackett PH, Forsling ML, Milledge J, Rennie D. Release of vasopressin in man at altitude. Horm Metab Res 1978; 10:571.
119. Rock PB, Kraemer WJ, Fulco CS, Trad LA, Malconian MK, Rose MS, Young PM, Cymerman A. Effects of altitude acclimatization on fluid regulatory hormone response to submaximal exercise. J Appl Physiol 1993; 75:1208–1215.
120. Weidmann P, Hasler L, Gnädinger MP, Lang RE, Uehlinger DE, Shaw S, Rascher W, Reubi FC. Blood levels and renal effects of atrial natriuretic peptide in normal man. J Clin Invest 1986; 77:734–742.
121. Hohenhaus E, Niroomand F, Goerre S, Vock P, Oelz O, Bärtsch P. Nifedipine does not prevent acute mountain sickness. Am J Respir Crit Care Med 1994; 150:857–860.
122. Sutton JR, Bryan AC, Gray GW, Horton ES, Rebuck AS, Woodley W, Rennie ID, Houston CS. Pulmonary gas exchange in acute mountain sickness. Aviat Space Environ Med 1976; 47:1032–1037.
123. Grissom CK, Roach RC, Samquist FH, Hackett PH. Acetazolamide in the treatment of acute mountain sickness: clinical efficacy and effect on gas exchange. Ann Intern Med 1992; 116:461–465.
124. Sutton JR, Gray GW, McFadden MD, Bryan AC, Horton ES, Houston CS. Nitrogen washout studies in acute mountain sickness. Aviat Space Environ Med 1977; 48:108–110.
125. Ri-Li G, Gen-Den, Zhang SQ, Matsuzawa Y, Kubo K, Takeoka M, Kobayashi T. Low pulmonary diffusion capacity in subjects with acute mountain sickness. In: Sutton JR, Houston CS, Coates G, eds. Hypoxia and the Brain. Burlington, VT: Queen City Printers, 1995:327.
126. Anholm JD, Houston CS, Hyers TM. The relationship between acute mountain sickness and pulmonary ventilation at 2,835 meters (9,300 ft). Chest 1979; 75:33–36.
127. Kriemler S, Maggiorini M, Bärtsch P, Oelz O. Gas exchange in acute mountain sickness (AMS) (abstr). In: Sutton JR, Coates G, Houston CS, eds. Hypoxia and Mountain Medicine. New York: Pergamon Press, 1992:309.
128. Scherrer U, Vollenweider L, Delabays A, Savcic M, Eichenberger U, Kleger G-R, Firkrle A, Ballmer P, Nicod P, Bärtsch P. Inhaled nitric oxide for high-altitude pulmonary edema. N Engl J Med 1996; 334:624–629.
129. Kilgore D, Loeppky J, Sanders J, Caprihan A, Icenogle M, Roach RC. Corpus callosum (CC) MRI: early altitude exposure. In: Roach RC, Wagner PD, Hackett PH, eds. Hypoxia: Into The Next Millenium. New York: Plenum/Kluwer Academic Publishing, 1999:396.

130. Moskowitz M. Basic mechanisms in vascular headache. Neurol Clin 1990; 8:801–815.
131. Reeves JT, Moore LG, McCullough RE, McCullough RG, Harrison G, Tranmer BI, Micco AJ, Tucker A, Weil JV. Headache at high altitude is not related to internal carotid arterial blood velocity. J Appl Physiol 1985; 59:909–915.
132. Huang SY, Moore LG, McCullough RE, McCullough RG, Micco AJ, Fulco C, Cymerman A, Manco-Johnson M, Weil JV, Reeves J. T. Internal carotid and vertebral arterial flow velocity in men at high altitude. J Appl Physiol 1987; 63:395–400.
133. Baumgartner RW, Bärtsch P, Maggiorini M, et al. The role of cerebral blood flow in acute mountain sickness. In: Sutton JR, Coates G, Houston CS, eds. Hypoxia and Mountain Medicine. New York: Pergamon Press, 1992:252–259.
134. Wright AD, Winterborn MH, Forster PJ, Delamere JP, Harrison GL, Bradwell AR. Carbonic anhydrase inhibition in the immediate therapy of acute mountain sickness. J Wilderness Med 1994; 5:49–55.
135. Hackett PH, Roach RC, Greene ER. Oxygenation, but not increased cerebral blood flow, improves high altitude headache (abstr). In: Sutton JR, Coates G, Remmers JE, eds. Hypoxia: The Adaptions. Toronto: B. C. Decker Inc., 1990:295.
136. Utiger D, Bernasch D, Eichenberger U, Bärtsch P. Transient improvement of high altitude headache by sumatriptan in a placebo-controlled trial. Int J Sports Med 1999; 20:S8.
137. Burtscher M, Likar R, Nachbauer W, Schaffert W, Philadelphy M. Ibuprofen versus sumatriptan for high-altitude headache. Lancet 1995; 346 (July 27):254–255.
138. Broome JR, Stoneham MD, Beeley JM, Milledge JS, Hughes AS. High altitude headache: Treatment with Ibuprofen. Aviat Space Environ Med 1994; 65:19–20.
139. King AB, Robinson SM. Ventilation response to hypoxia and acute mountain sickness. Aerospace Med 1972; 43:419–421.
140. Moore LG, Harrison GL, McCullough RE, McCullough RG, Micco AJ, Tucker A, Weil JV, Reeves JT. Low acute hypoxic ventilatory response and hypoxic depression in acute altitude sickness. J Appl Physiol 1986; 60:1407–1412.
141. Forwand SA, Landowne M, Follansbee JN, Hansen JE. Effect of acetazolamide on acute mountain sickness. N Engl J Med 1968; 279:839–845.
142. Roach RC, Larson EB, Hornbein TF, Houston CS, Bartlett S, Hardesty J, Johnson D, Perkins M. Acute mountain sickness, antacids, and ventilation during rapid, active ascent of Mount Rainier. Aviat Space Environ Med 1983; 54:397–401.
143. Bärtsch P, Waber U, Haeberli A, Maggiorini M, Kriemler S, Oelz O, Straub WP. Enhanced fibrin formation in high-altitude pulmonary edema. J Appl Physiol 1987; 63:752–757.
144. Mathew L, Gopinathan PM, Purkayashta SS, Sen Gupta J, Nayar HS. Chemoreceptor sensitivity and maladaptation to high altitude in man. Eur J Appl Physiol 1983; 51: 137–144.
145. Hohenhaus E, Paul A, McCullough RE, Kücherer H, Bärtsch P. Ventilatory and pulmonary vascular response to hypoxia and susceptibility to high altitude pulmonary oedema. Eur Respir J 1995; 813:1825–1833.
146. Hackett PH, Roach RC, Schoene RB, Harrison GL, Mills WJ. Abnormal control of ventilation in high-altitude pulmonary edema. J Appl Physiol 1988; 64:1268–1272.
147. Matsuzawa Y, Fujimoto K, Kobayashi T, Namushi NR, Harada K, Kohno H, Fukushima M, Kusama S. Blunted hypoxic ventilatory drive in subjects susceptible to high-altitude pulmonary edema. J Appl Physiol 1989; 66(3):1152–1157.

148. Milledge JS, Thomas PS, Beeley JM, English JSC. Hypoxic ventilatory response and acute mountain sickness. Clin Sci 1988; 75(suppl 19):26.
149. Milledge JS, Beeley JM, Broome J, Luff N, Pelling M, Smith D. Acute mountain sickness susceptibility, fitness and hypoxic ventilatory response. Eur Respir J 1991; 4:1000–1003.
150. Sato M, Severinghaus JW, Powell FL, Xu F-D, Spellman MJ. Augmented hypoxic ventilatory response in men at altitude. J Appl Physiol 1992; 73:101–107.
151. Easton PA, Slykerman LJ, Anthonisen NR. Ventilatory response to sustained hypoxia in normal adults. J Appl Physiol 1986; 61:906–911.
152. Anholm JD, Powles ACP, Downey R, Houston CS, Sutton JR, Bonnet MH, Cymerman A. Operation Everest II: arterial oxygen saturation and sleep at extreme simulated altitude. Am Rev Respir Dis 1992; 145:817–826.
153. West JB, Peters RM, Aksnes G, Maret KH, Milledge JS, Schoene RB. Nocturnal periodic breathing at altitudes of 6300 and 8050 m. J Appl Physiol 1986; 61:280–287.
154. Matsuzawa Y, Kobayashi T, Fujimoto YS, Shinozaki S, Kubo K, Sekiguchi M, Hayashi R, Sakai A, Ueda G. Nocturnal periodic breathing and arterial oxygen desaturation in acute mountain sickness. J Wilderness Med 1994; 5:269–281.
155. Roach RC, Loeppky JA, Icenogle MV. Acute mountain sickness: increased severity during simulated altitude compared with normobaric hypoxia. J Appl Physiol 1996; 81(5):1908–1910.
156. Tucker A, Reeves JT, robertshaw D, Grover RF. Cardiopulmonary response to acute altitude exposure: water loading and denitrogenation. Respir Physiol 1983; 54:363–380.
157. Meehan RT, Cymerman A, Rock P, Fulco CS, Hoffman J, Abernathy C, Needleman S, Maher JT. The effect of naproxen on acute mountain sickness and vascular responses to hypoxia. Am J Med Sci 1986; 292:15–20.
158. Knight DR, Cymerman A, Devine JA. Symptomatology during hypoxic exposure to flame-retardant chamber atmospheres. Undersea Biomed Res 1990; 17:33–44.
159. Epstein M, Saruta T. Effects of simulated high altitude on renin-aldosterone and Na homeostasis in normal man. J Appl Physiol 1972; 33:204–210.
160. Roach RC, Loeppky JA, Robergs R. Fluid balance in humans at high altitude: Does hypobaria play a role. FASEB J 1994; 8:A553.
161. Grover RF, Tucker A, Reeves JT. Hypobaria: an etiologic factor in acute mountain sickness. In: Loeppky JA, Riedesel ML, eds. Oxygen Transport to Human Tissue. New York: Elsevier/North Holland, 1982:223–230.
162. Loeppky JA, Icenogle M, Scotto P, Robergs R, Hinghoferszalkay H, Roach RC. Ventilation during simulated altitude, normobaric hypoxia and normoxic hypobaria. Respir Physiol 1997; 107:231–239.
163. Clarke C, Duff J. Mountain sickness, retinal haemorrhages, and acclimatisation on Mount Everest in 1975. Br Med J 1975; 2:495–497.
164. Rennie D, Morrissey J. Retinal changes in Himalayan climbers. Arch Ophthalmol 1975; 93:395–400.
165. Shults WT, Swan KC. High altitude retinopathy in mountain climbers. Arch Ophthalmol 1975; 93:404–408.
166. Butler FK, Harris DJ, Reynolds RD. Altitude Retinopathy on Mount Everest, 1989. Ophthalmology 1992; 99:739–746.
167. McLeod D, Marshall J, Kohner E, Bird A. The role of axoplasmic transport in the pathogenesis of retinal cotton-wool spots. Br J Ophthalmol 1977; 61:177–191.

168. McFadden DM, Houston CS, Sutton JR, Powles AP, Gray GW, Roberts R. High altitude retinopathy. JAMA 1981; 245:581–586.
169. Depla DN, Green W. Retinopathy at altitude—a new theory. Congress of the Royal College of Ophthalmologists, 1996.
170. Lindgren G, Sjodell L, Lindblom B. A prospective study of dense spontaneous vitreous hemorrhage. Am J Ophthalmol 1995; 119:458–465.
171. Frayser R, Houston CS, Gray GW, Bryan AC, Rennie ID. The response of the retinal circulation to altitude. Arch Intern Med 1971; 127:708–711.
172. Schumacher GA, Petajan JH. High altitude stress and retinal hemorrhage: relation to vascular headache mechanisms. Arch Environ Health 1975; 30:217–221.
173. Steele P. Medicine on Mount Everest 1971. Lancet 1971; II(July 3):32–39.
174. Wiedman M. High altitude retinal hemorrhage. Arch Ophthalmol 1975; 93:401–403.
175. Botella de Maglia J, Martinez-Costa R. High altitude retinal hemorrhages in the expeditions to 8,000 meter peaks. A study of ten cases. Med Clin Barc 1998; 110:457–461.
176. Rimsza ME, Hernried LS, Kaplan AM. Hemorrhagic retinopathy in a patient with cystic fibrosis. Pediatrics 1978; 62:336–338.
177. Hickam JB, Frayser R. Studies of the retinal circulation in man. Observation on vessel diameter, arteriovenous oxygen difference, and mean circulation time. Circulation 1966; 33:302–316.
178. Saari JC. Adler's Physiology of the Eye. Clinical Application. Moses and Hart, 1987.
179. Kanski JJ. Clinical Ophthalmology. London: Butterworths, 1989.
180. Duane TD. Valsalva haemorrhagic retinopathy. Am J Ophthalmol 1973; 75:637–642.
181. Lubin JR, Rennie ID, Hackett PH, Albert DM. High altitude retinal hemorrhage: a clinical and pathological case report. Ann Ophthalmol 1982; 14:1071–1076.
182. Wilmer WH. Effects of carbon monoxide upon the eye. Am J Ophthalmol 1921; 4: 73.
183. Weller RO. Symmers Systemic Pathology (4: Nervous System, Muscle and Eyes). 3rd ed. Churchill Livingstone, 1990.
184. Richalet J-P, Bittel J, Herry J-P, Janin C, Savourey G, Le Trong J-L, Auvert J-F. Preacclimatization to high altitude in a hypobaric chamber: Everest Turbo. In: Sutton JR, Coates G, Houston CS, eds. Hypoxia and Mountain Medicine. New York: Pergamon Press, 1992:202–212.
185. Burse RL, Forte VA. Acute mountain sickness at 4500 m is not altered by repeated eight-hour exposures to 3200-3550 m normobaric hypoxic equivalent. Aviat Space Environ Med 1988; 59:942–949.
186. Consolazio CF, Matoush LO, Johnson HL, Krzywicki HJ, Daws TA, Isaac GJ. Effects of high carbohydrate diets on performance and clinical symptomatology after rapid ascent to high altitude. Fed Proc 1969; 28:937–943.
187. Gray D, Milne D. Effect of dietary supplements on acute mountain sickness. Percept Mot Skills 1986; 63:873–874.
188. Swenson ER, MacDonald A, Vatheuer M, Maks C, Treadwell A, Allen R, Schoene RB. Acute mountain sickness is not altered by a high carbohydrate diet nor associated with elevated circulating cytokines. Aviat Space Environ Med 1997; 68:001–005.
189. Bärtsch P, Merki B, Hofstetter D, Maggiorinni M, Kayser B, Oelz O. Treatment of acute mountain sickness by simulated descent: a randomised controlled trial. Br Med J 1993; 306:1098–1101.
190. Ferrazzini G, Maggiorini M, Kriemler S, Bärtsch P, Oelz O. Successful treatment of acute mountain sickness with dexamethasone. Br Med J 1987; 294:1380–1382.

191. Maggiorini M, Merki B, Pallavicini E, Bärtsch P, Oelz O. Acetazolamide and almitrine in acute mountain sickness (AMS) treatment (abstr). In: Sutton JR, Houston CS, Coates G, eds. Hypoxia and the Brain. Burlington, VT: Queen City Press, 1995:330.
192. Bärtsch P, Maggi S, Kleger G-R, Ballmer PE, Baumgartner RW. Sumatriptan for high-altitude headache. Lancet 1994; 344:1445.
193. King AB, Robinson SM. Vascular headache of acute mountain sickness. Aerospace Med 1972; 43:849–851.
194. Kasic JF, Yaron M, Nicholas RA, Lickteig JA, Roach RC. Treatment of acute mountain sickness: hyperbaric versus oxygen therapy. Ann Emerg Med 1991; 20:1109–1112.
195. Kayser B, Jean D, Herry J-P, Bärtsch P. Pressurization and acute mountain sickness. Aviat Space Environ Med 1993; 64:928–931.
196. Keller H-R, Maggiorini M, Bärtsch P, Oelz O. Simulated descent v. dexamethasone in treatment of acute mountain sickness: a randomised trial. Br Med J 1995; 310:1232–1235.
197. Taber RL. Protocols for the use of portable hyperbaric chamber for the treatment of high altitude disorders. J Wildern Med 1990; 1:181–192.
198. King SJ, Greenlee RR. Successful use of the Gamow Hyperbaric bag in the treatment of altitude illness at Mount Everest. J Wildern Med 1990; 1:193–202.
199. Pollard AJ. Treatment of acute mountain sickness. Br Med J 1995; 311:629.
200. Greene MK, Kerr AM, McIntosh IB, Prescott RJ. Acetazolamide in prevention of acute mountain sickness: a double-blind controlled cross-over study. Br Med J 1981; 283: 811–813.
201. Swenson ER. Carbonic anhydrase inhibitors and ventilation: a complex interplay of stimulation and suppression. Eur Respir J 1998; 12:1242–1247.
202. Kronenberg RS, Cain SM. Hastening respiratory acclimatization to altitude with benzolamide (CL 11,366). Aerospace Med 1968; 39:296–300.
203. Swenson ER, Maren TH. Acute mountain sickness. N Engl J Med 1989; 320:1492–1493.
204. Hackett PH, Roach RC, Harrison GL, Schoene RB, Mills WJ. Respiratory stimulants and sleep periodic breathing at high altitude. Am Rev Respir Dis 1987; 135:896–898.
205. Nicholson AN, Smith PA, Stone BM, Bradwell AR, Coote JH. Altitude insomnia: studies during an expedition to the Himalayas. Sleep 1988; 11:354–361.
206. Swenson ER, Leatham KL, Roach RC, Schoene RB, Mills WJ, Hackett PH. Renal carbonic anhydrase inhibition reduces high altitude sleep periodic breathing. Respir Physiol 1991; 86:333–343.
207. Swenson ER, Hughes JMB. Effects of acute and chronic acetazolamide on resting ventilation and ventilatory responses in men. J Appl Physiol 1993; 73:230–237.
208. Huang SY, McCullough RE, McCullough RG, Micco AJ, Manco-Johnson M, Weil JV, Reeves JT. Usual clinical dose of acetazolamide does not alter cerebral blood flow velocity. Respir Physiol 1988; 72:315–326.
209. Bradwell AR, Dykes PW, Coote JH, Forster PJE, Milles JJ, Chesner I, Richardson NV. Effect of acetazolamide on exercise performance and muscle mass at high altitude. Lancet 1986; I(May 3):1001–1005.
210. Aoki VS, Robinson SM. Body hydration and the incidence and severity of acute mountain sickness. J Appl Physiol 1971; 31:363–367.
211. Gray GW, Bryan AC, Frayser R, Houston CS, Rennie IDB. Control of acute mountain sickness. Aerospace Med 1971; 42:81–84.

212. Turnbull G. Spironolactone prophylaxis in mountain sickness. Br Med J 1980; 280: 1453.
213. Brookfield DSK, Liston WA, Brown GV. Use of spironolactone in the prevention of acute mountain sickness on Kilimanjaro. East Afr Med J 1977; 64:689–691.
214. Larsen RF, Rock PB, Fulco CS, Edelman B, Young AJ, Cymerman A. Effect of spironolactone on acute mountain sickness. Aviat Space Environ Med 1986; 57:543–547.
215. Hillenbrand P, Beazley M, Forster PJG, Milles JJ, Clayton RN, Wright AD, Imray E. A randomised controlled trial of progesterone in preventing acute mountain sickness (AMS). In: Roach RC, Wagner PD, Hackett PH, eds. Hypoxia: Into The Next Millenium. New York: Plenum/Kluwer Academic Publishing, 1999:388.
216. Kuepper T, Hoefer M, Gieseler U, Netzer N. Prevention of acute mountain sickness with theophylline (abstr). In: Roach RC, Hackett PH, Wagner PD, eds. Hypoxia: Into the Next Millenium. New York: Plenum/Kluwer Academic Publishing, 1999:400.
217. Johnson TS, Rock PB, Fulco CS, Trad LA, Spark RF, Maher JT. Prevention of acute mountain sickness by dexamethasone. N Engl J Med 1984; 310:683–686.
218. Rock PB, Johnson TS, Cymerman A, Burse RL, Falk LJ, Fulco CS. Effect of dexamethasone on symptoms of acute mountain sickness at Pikes Peak, Colorado (4300 m). Aviat Space Environ Med 1987; 58:668–672.
219. Ellsworth AJ, Larson EB, Strickland D. A randomized trial of dexamethasone and acetazolamide for acute mountain sickness prophylaxis. Am J Med 1987; 83:1024–1030.
220. Hackett PH, Roach RC, Wood RA, Foutch RG, Meehan RT, Rennie D, Mills WJJr. Dexamethasone for prevention and treatment of acute mountain sickness. Aviat Space Environ Med 1988; 59:950–954.
221. Montgomery AB, Luce JM, Michael P, Mills J. Effects of dexamethasone on the incidence of acute mountain sickness at two intermediate altitudes. JAMA 1989; 261:734–736.
222. Rock PB, Johnson TS, Larsen RF, Fulco CS, Trad LA, Cymerman A. Dexamethasone as prophylaxis for acute mountain sickness. Effect of dose level. Chest 1989; 95:568–573.
223. Bernhard WN, Miller Schalick L, Gittelsohn A. Dexamethasone for prophylaxis against acute mountain sickness during rapid ascent to 5334 m. J Wilderness Med 1994; 5:331–338.
224. Ried LD, Carter KA, Ellsworth A. Acetazolamide or dexamethasone for prevention of acute mountain sickness: a meta-analysis. J Wilderness Med 1994; 5:34–48.
225. Dugas L, Dubray C, Herry JP, Olsen NV, Court-Payen M, Hansen JM, Robach P, Ter-Minassian A, Richalet JP. Effects cardiovasculaires d'un bloqueur calcique en hypoxie d'altitude. Presse Med 1995; 24:763–768.
226. Hildebrandt W, Schuster M, Hartmann M, Herzog W, Bärtsch P. Relation of symptoms of acute mountain sickness to personality traits. Int J Sports Med 1999; 20:S7.
227. Dean AG, Yip R, Hormann RE. High incidence of mild acute mountain sickness in conference attendees at 10000 foot altitude. J Wilderness Med 1990; 1:86–92.
228. Kleger G-R, Bärtsch P, Vock P, Heilig B, Roberts LJI, Ballmer PE. Evidence against an increase of capillary permeability in subjects exposed to high altitude. J Appl Physiol 1996; 81:1917–1923.
229. Bärtsch P. Urinary leukotriene E4 levels are not increased in high altitude pulmonary edema. In: Sutton JR, Houston CS, Coates G, eds. Hypoxia and Mountain Medicine: Women at Altitude. New York: Pergamon, 1997.

230. Klausen T, Olsen NV, Poulsen TD, Richalet J-P, Pedersen BK. Hypoxemia increases serum interleukin-6 in humans. Eur J Appl Physiol 1997; 76:480–482.
231. Eldridge MW, Johnson DH, Hill HR. Cytokines and adhesion molecules in high altitude pulmonary edema. In: Sutton JR, Houston CS, Coates G, eds. Hypoxia and the Brain. Proceedings of the 9th International Hypoxia Symposium at Lake Louise, Canada, February 14–18, 1995. Burlington, VT: Queen City Printers, 1995:243–250.
232. Pavlicek V, Schirlo C, Gibbs S, Marti H, Wenger R, Kohl J, Maly F, Oelz O. No evidence for inflammatory reactions during the first 24 hours of hypobaric hypoxia equivalent to 4000 m in subjects susceptible to high altitude pulmonary edema (abstr). In: Sutton JR, Coates G, Houston CS, eds. Hypoxia and Mountain Medicine: Women at Altitude. New York: Pergamon, 1997.
233. Bärtsch P, Lämmle B, Huber I, Haeberli A, Vock P, Oelz O, Straub PW. Contact phase of blood coagulation is not activated in edema of high altitude. J Appl Physiol 1989; 67:1336–1340.
234. Grissom CK, Zimmerman GA, Whatley RE. P-selectin, E-selectin and von Willebrand factor in acute mountain sickness and high altitude pulmonary edema (abstr). In: Sutton JR, Houston CS, Coates G, eds. Hypoxia and the Brain. Burlington, VT: Queen City Printers, 1995:322.
235. Eldridge MW, Johnson DH, Hill HR. Cytokines and adhesion molecules in high altitude pulmonary edema. In: Sutton JR, Houston CS, Coates G, eds. Hypoxia and the Brain. Burlington, VT: Queen City Printers, 1995:243–250.
236. Déchaux M, Blazy I, Khayal S, Souberbielle J-C, Letournel M, Richalet J-P. Hypoxia-induced activation of endothelial cells. In: Sutton JR, Houston CS, Coates G, eds. Hypoxia and the Brain. Burlington, VT: Queen City Printers, 1995:335.
237. Goerre S, Wenk M, Bärtsch P, Lüscher TF, Niroomand F, Hohenhaus E, Oelz O, Reinhart WH. Endothelin-1 in pulmonary hypertension associated with high-altitude exposure. Circulation 1995; 90:359–364.
238. Richalet J-P, Hornych A, Rathat C, Aumont J, Larmignat P, Rémy P. Plasma prostaglandins, leukotrienes and thromboxane in acute high altitude hypoxia. Respir Physiol 1991; 85:205–215.
239. Imray CHE, Cooper M, Mead G, Wright A, Bradwell AR, Dykes P. Serum phospholipid bound arachidonic acid (22:4) at altitude and in acute mountain sickness. In: Sutton JR, Houston CS, Coates G, eds. Hypoxia and the Brain. Burlington, VT: Queen City Printers, 1995:325.
240. Roach JM, Muza SR, Rock PB, Lyons TP, Lilly CM, Drazen JM, Cymerman A. Urinary leukotriene E (4) levels increase upon exposure to hypobaric hypoxia. Chest 1996; 110:946–951.

23

High-Altitude Pulmonary Edema

ROBERT B. SCHOENE and ERIK R. SWENSON

University of Washington School of Medicine
Seattle, Washington

HERBERT N. HULTGREN†

Stanford University School of Medicine
Palo Alto, California

I. Introduction

Many of the pathophysiological manifestations of acute altitude illness are associated with extravasation of fluid from the intra- to extravascular space in the brain, lung, and peripheral tissues. Nowhere is this leak more apparent than in the lungs where fluid in the interstitial and alveolar space impairs gas exchange and can lead to life-threatening hypoxemia. The mechanism causing the leak, however, remains elusive.

This chapter will review what is known about high-altitude pulmonary edema (HAPE) from historical, clinical, and pathophysiological perspectives in order to chart a course for future investigation, the ultimate goal of which is to gain an understanding of the cause of HAPE so that scientists and clinicians can develop a targeted approach to prediction, prevention, and treatment of HAPE, which strikes both young and old.

† Deceased.

II. History

For centuries, descriptions of human responses to high altitude failed to unveil HAPE as a discrete entity, unique to rapid ascent in otherwise healthy sojourners. Fulminant pneumonia and congestive heart failure were diagnoses that clinicians gave to the constellation of deleterious pulmonary responses seen at high altitude in some individuals. Retrospective analysis of some of the early descriptions strongly suggests that HAPE was the actual problem.

In 1894 the prominent Italian physiologist, Angelo Mosso, in his seminal volume, *Life of Man in the High Alps*, describes a young soldier who, upon ascending to the Capana Regina Margherita hut on the summit of Monta Rosa (4559 m), developed a severe headache, then cyanosis, dyspnea, tachycardia, riles, and pink frothy sputum, but he did not have a high fever or chills. In spite of the puzzling findings, he was thought to have pneumonia and recovered after several days there without treatment. In retrospect, this account is, in fact, a perfect description of HAPE (1).

In 1913, Ravenhill, the British medical officer at a mining camp in the Chilean Andes, described three cases of what was probably HAPE (2). Physicians in the Peruvian Andes in the 1920s and 1930s encountered patients with severe cough and dyspnea after arrival from sea level, also most likely HAPE. All of these cases resolved with descent. Lizarraga's 1954 medical thesis describes 14 cases of HAPE observed at the Chulec Hospital in the Peruvian Andes. He made the distinction between what had always been thought to be congestive heart failure or pneumonia and this noncardiogenic form of pulmonary edema (3). Subsequently, Peruvian and Bolivian physicians published several similar reports (4,5). In 1956 Vega provided an English summary of patients with HAPE in Peru (6), and Hultgren published a report in May 1960 of HAPE cases (7), which was followed by Houston's report of a skier in Colorado (8) and a more detailed report by Hultgren and Spickard in 1961 (9). Over the next two decades reports emanated from other high mountain areas around the world (10–16). Thus, a new clinical entity emerged, and the scene was ripe for clinical and basic scientific investigation.

III. Hypothetical Model

Observations of clinical and physiological data allow one to speculate about the characteristics, which predispose healthy humans to HAPE. Although not all individuals with these certain physiological characteristics get HAPE upon each ascent, the following model may explain the evolution of the disease in many and will act as a framework upon which to base the discussion in the rest of this paper (Fig. 1).

Those individuals with a blunted hypoxic ventilatory response (HVR) have more profound alveolar hypoxia and are, therefore, more hypoxemic than others with an intact HVR. They also tend to have a more brisk pulmonary vascular response to hypoxia (HPVR), which is perhaps mediated by a lower nitric oxide (NO) synthesis than normal individuals. The marked increase in pressure in the pulmonary vascula-

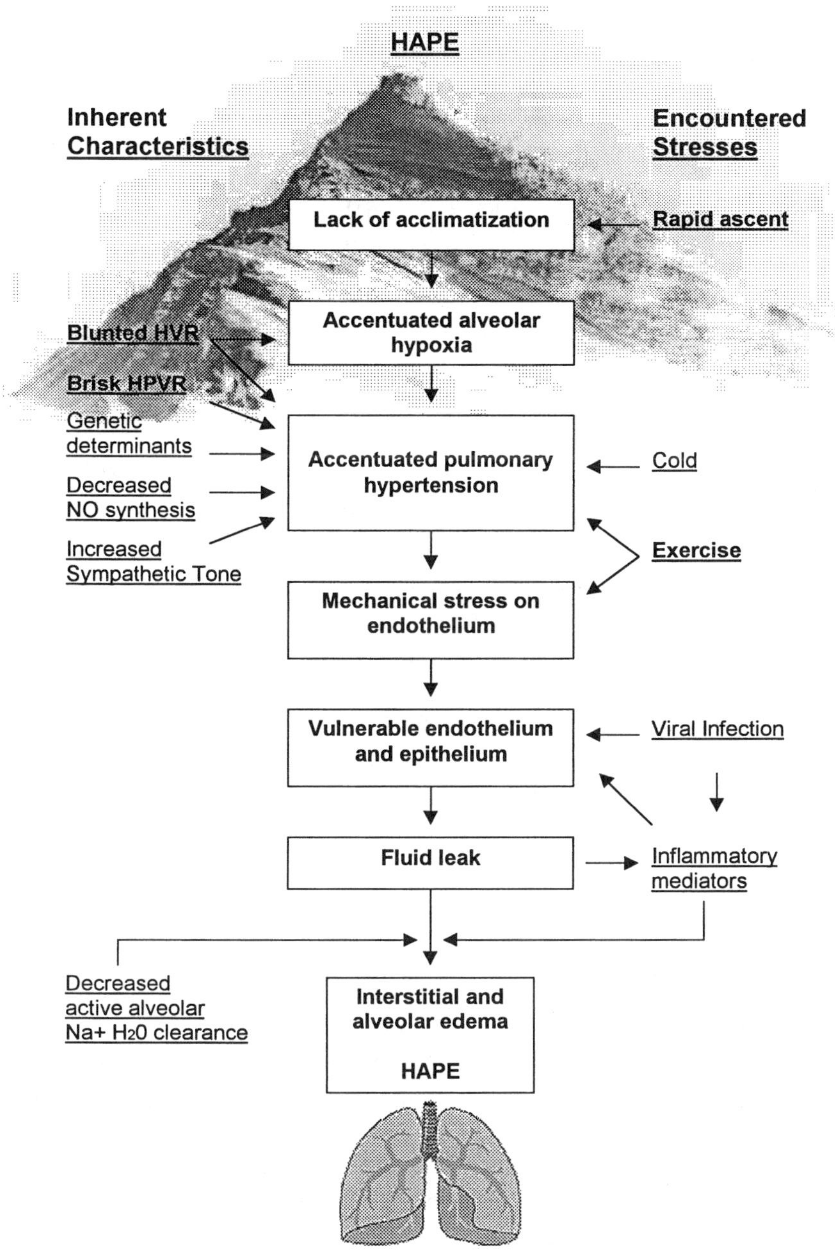

Figure 1 Schema for HAPE mechanism. A proposed mechanism for both inherent characteristics and encountered stresses, some or all of which a subject with HAPE may have during an initial or repeat bout of HAPE (see text). HVR = Hypoxic ventilatory response; NO = nitric oxide; HPVR = hypoxic pulmonary vascular response.

ture is magnified with cold and exercise. The fragile endothelial layer cannot withstand these stresses, stretches and allows extravasation of fluid from the intravascular space. Exposure of the basement membrane by endothelial stretching elicits an inflammatory response with chemotaxis of alveolar macrophages (AM) and neutrophils (PMN), which release inflammatory mediators that augment the leak until the stress of hypoxia is removed, at which point alveolar fluid reabsorption and interstitial clearance by lymphatics leads to resolution. It is conceivable that certain genetic factors or preexisting or concomitant inflammatory responses from either infection or hypoxia per se make the endothelium more vulnerable to the stresses of increased pressures. Further speculation about the Na^+/K^+ ATPase pump in the epithelial layer, which helps to mediate alveolar fluid clearance, offers the intriguing possibility that impairment of clearance may contribute to HAPE susceptibility in some individuals.

IV. Clinical and Physiological Correlation

Several excellent reviews on HAPE are available (17–24), so we will only briefly describe the clinical presentation.

HAPE occurs rarely below 2450 m, but the incidence increases progressively above this altitude. HAPE is much less common than acute mountain sickness (Table 1). In Keystone, Colorado, at 2928 m, Hultgren and colleagues reported 150 cases over 39 months (25). The incidence of HAPE depends on altitude achieved, rapidity of ascent, physical activity during and after ascent, gender, age, and individual susceptibility. At 3782 m sojourners experience a 3% incidence (13) while Indian troops transported to 3500 m had an incidence from 2.3 to 15.5%, the more rapidly transported soldiers having the higher rate of occurrence (26). Bartsch and colleagues found that 60% of climbers who had radiographically proven HAPE in the past and who ascended within one day to 4559 m developed the disease (27–29). Men appear to be more susceptible than women are. A review of 229 cases found 87% of them to be male (25,30–33), possibly in part related to the advantageous respiratory stimulation in women from higher progesterone levels, but undoubtedly because fewer women were included in the sample (34).

Symptoms most commonly evolve in 2–4 days after ascent, often in those experiencing acute mountain sickness (see Chapter 22), but can occur much later, particularly upon ascent to higher altitudes (Table 2). Dyspnea, especially with exercise, tachypnea, marked decrease in exercise tolerance, a dry cough, and weakness, are early symptoms, associated with signs of progressive cyanosis, tachypnea, tachycardia, and low-grade fever. The tachypnea might be due to hypoxemia and/or decreased pulmonary compliance. As the disease progresses, hypoxemia worsens and can be profound (14,35–37). Symptoms increase to severe dyspnea, a cough productive of pink, frothy sputum, and neurological changes compatible with concomitant high-altitude cerebral edema or hypoxic cerebral depression. Obtundation, coma, and death may follow (Table 2).

Table 1 Incidence of Altitude Illness in Various Groups

Study group	Number at risk per year	Sleeping altitude (m)	Maximum altitude reached (m)	Average rate of ascent[a]	Percent with AMS	Percent with HAPE and/or HACE
Western state visitors	30 million	~2000	3500	1–2	18–20	0.01
		~2500			22	
		~3000			27–42	
Mount Everest trekkers	6000	3000–5200	5500	1–2 (fly in)	47	1.6
				10–13 (walk in)	23	0.05
Mount McKinley climbers	800	3000–5300	6194	3–7	30	2–3
Mount Rainier climbers	6000	3000	4392	1–2	67	—
Indian soldiers	Unknown	3000–5500	5500	1–2	[b]	2.3–15.5

[a] Days to sleeping altitude from low altitude.
[b] Reliable estimate unavailable.
AMS, Acute mountain sickness; HAPE, high-altitude pulmonary edema; HACE, high-altitude cerebral edema.
Source: Ref. 144.

Table 2 Severity Classification of High-Altitude Pulmonary Edema

Grade	Clinical symptoms	Heart rate (beats/min)	Respiratory rate (breaths/min)	Chest x-ray film
1. Mild	Minor symptoms with dyspnea on moderate exertion May be able to perform light activity	<110	<20	Minor opacities involving less than 1/4 of one lung field
2. Moderate	Symptoms of dyspnea, weakness, fatigue on slight effort Cannot perform light activity Headache with cough, dyspnea at rest	110–120	20–30	Opacities involving at least 1/2 of one lung field
3. Serious	Severe dyspnea, headache, weakness, nausea at rest Loose recurrent productive cough Wheezy, difficult respirations with obvious cyanosis	121–140	31–40	Opacities involving at least 1/2 of each lung field or unilateral exudate involving all of one lung field
4. Severe	Clouded consciousness, stupor or coma Unable to stand or walk Sever cyanosis Bubbling rales present with copious sputum, usually bloody Severe respiratory distress	>140	>40	Bilateral opacities involving more than 1/2 of each lung field

Individuals with high vital capacities may be less predisposed to HAPE (38–40); lower lung volumes and lower diffusion capacities both before and during HAPE (41) have been documented. Speculation about this association of lung or thoracic volume with HAPE is that larger lungs may have a larger pulmonary vascular bed, which could accommodate greater pulmonary blood flow with less pulmonary hypertension and less microvasculature endothelial injury. Arterial oxygen saturations in climbers with HAPE at 4300 m were very low (mean = 52%) (35,36), while at 4559 m four HAPE victims had mean Pa_{O_2} of 23 torr and Sa_{O_2} of 48% compared to a mean of 40 torr and 78% in healthy individuals (27,29). In the Colorado Rockies at 2928 m, Hultgren et al. studied 45 patients with HAPE and found great variability of Sa_{O_2} with values ranging from 38 to 93% (mean = 74%) (25).

Radiographic findings (Fig. 2) are variable, with early linear opacities, followed by patchy, more confluent ones, which are more commonly found in the mid-lung and lower zones but can be present in the upper lobes. The infiltrates

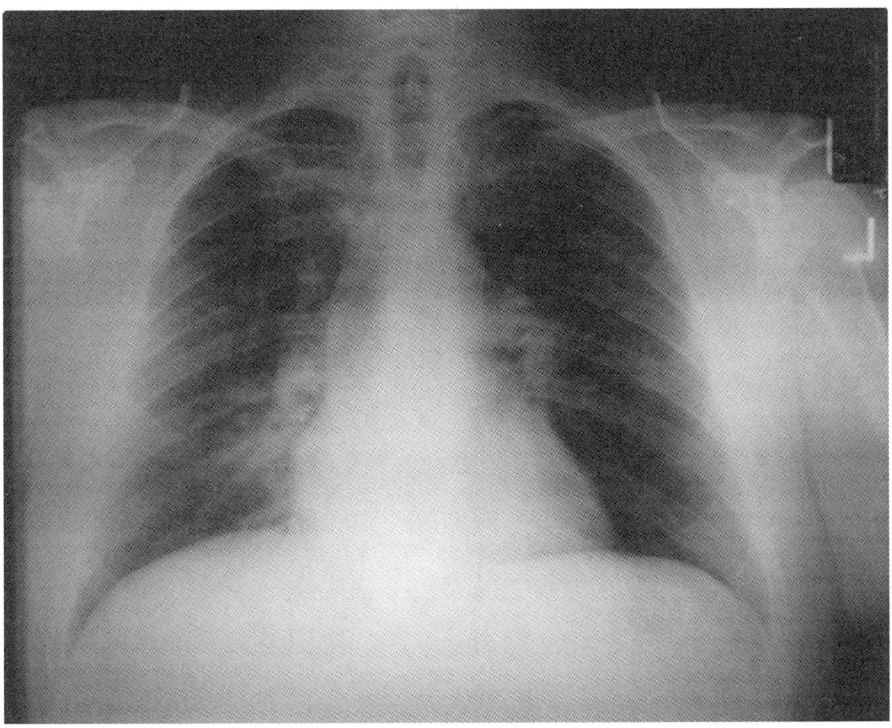

Figure 2 Chest radiograph: of a 38-year-old man with history of four episodes of HAPE encountered at 2500–3000 m. The cardiac silhouette is normal, and there are diffuse patchy opacities throughout both lung fields as well as prominent pulmonary artery shadows. The chest radiograph returned to normal between each bout of HAPE.

are commonly bilateral but if unilateral usually begin in the right mid-lung fields, progressing to bilateral diffuse densities (9,10,28,37,42,43). Patients with repeat occurrences of HAPE do not necessarily develop infiltrates in the same areas, which suggests that fixed structural abnormalities in the lung parenchyma do not account for development of leak in the lung. The pattern of the infiltrates is not ''bat-wing,'' such as in cardiogenic pulmonary edema, and the cardiac silhouette is normal. There is usually dilatation of the pulmonary arteries, but this finding is not diagnostic of HAPE since it occurs in most people ascending to high altitude. CT scans confirm the patchy nature of the infiltrates (28,29). Resolution of the infiltrates occurs quickly if the patients improve, lagging only briefly behind signs of clinical improvement.

These observations are important for reasons that will be discussed more fully later. First, the diffuse nature of the edema helps to explain the impairment in lung mechanics. Second, this diffuse pattern correlates with the gas exchange abnormalities, i.e., patchy infiltrates in the airspaces compatible with right-to-left intrapulmonary shunt. Third, the patchy nature suggests leak, secondary to uneven hypoxic pulmonary vasoconstriction. This possibility has important implications for understanding the mechanisms causing leak.

Observations in three particular groups of individuals offer clues to the underlying pathophysiology:

1. HAPE susceptibles
2. Reentry HAPE
3. Congenital or acquired absence of one pulmonary artery

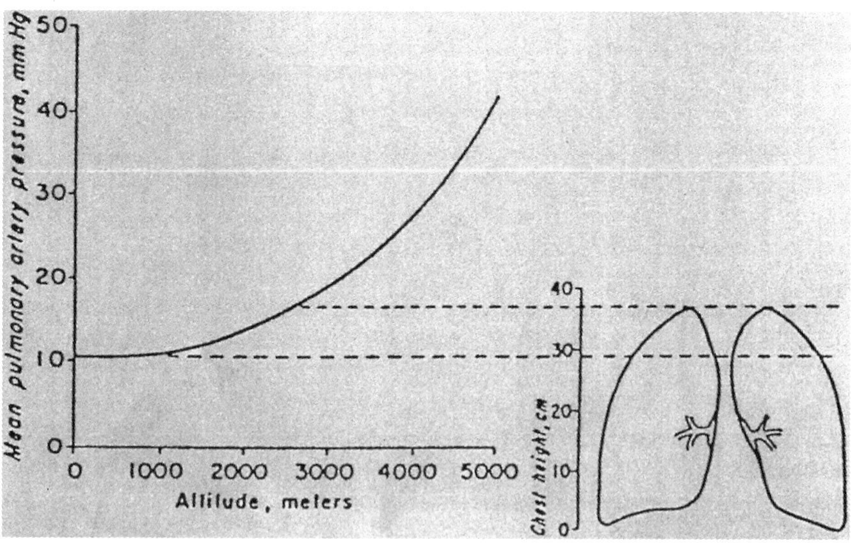

Figure 3 Characteristic increase of pulmonary artery pressures with greater degrees of hypoxia.

HAPE-susceptible individuals are defined as those who have had one or more episodes of HAPE upon acute ascent to high altitude. These individuals characteristically become symptomatic at a lower altitude upon repeated exposures. In this group, two distinct physiological features stand out:

1. Accentuated hypoxic pulmonary vasoconstrictor response (HPVR) and pulmonary hypertension (13,15,16,44–47)
2. Blunted hypoxic ventilatory responsiveness (HVR) (40,48–50)

The HPVR (Fig. 3) is generally, but not always, more intense in those who develop HAPE than in those who do not. These measurements have been made invasively (12,13,16,44) and noninvasively (45–47,51), both during the illness at high altitude and at low altitude after recovery. The persistently elevated pulmonary pressures and accentuated HPVR suggest a genetic predisposition. It is not clear whether indi-

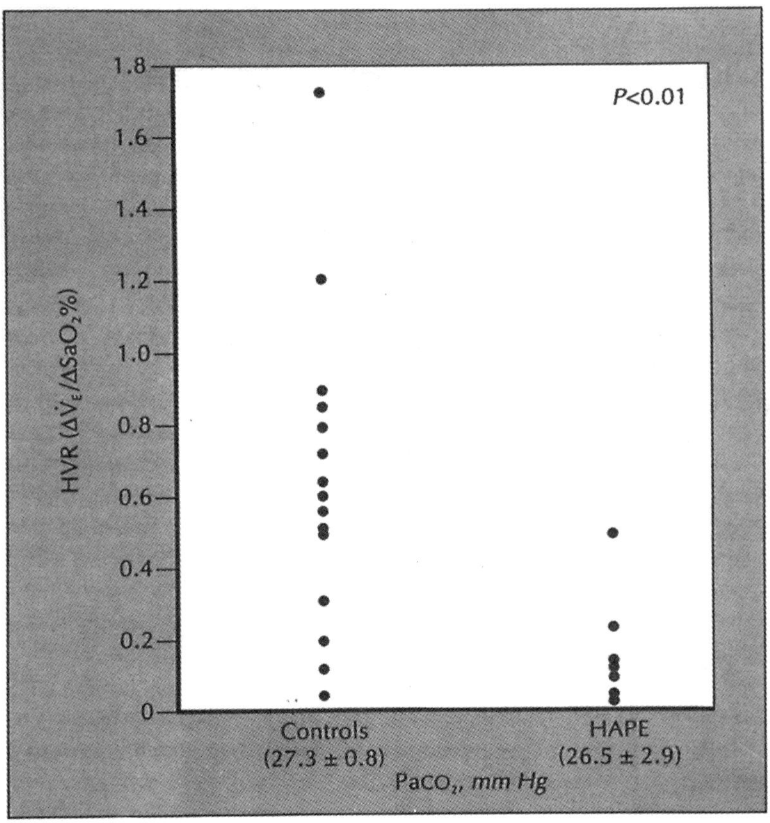

Figure 4 Hypoxic ventilatory response (HVR) is lower in climbers on Mount McKinley than in healthy controls at 4300 m. (From Ref. 50.)

viduals with a low HVR (another inherent characteristic discussed below) have a more intense HPVR for any given level of alveolar hypoxia.

It should be stressed that a strong HPVR may only be one permissive characteristic, since not all individuals with high HPVR develop HAPE (52). Conversely, a low HPVR does not necessarily prevent HAPE (53).

HAPE has been observed in permanent high-altitude dwellers upon returning home from brief forays to low altitude (as short as 24 hours, usually 3–14 days). This reentry HAPE has a predisposition for children (9,54). A possible explanation is that since high-altitude dwellers have long-standing pulmonary hypertension and hypertrophied pulmonary vascular smooth musculature, upon reascent they may have a hyperresponsive HPVR and accentuated pulmonary hypertension with further stress on the microvascular endothelium. This mechanism is speculative since no measurements of pulmonary artery pressures have been made during the course of the illness when these patients are ascending. Furthermore, its occurrence is sporadic and unpredictable.

The third group, those with a congenitally absent pulmonary artery, also has accentuated stresses on the pulmonary vasculature as all of the cardiac output, especially during exercise, must pass through a smaller pulmonary vascular bed. These

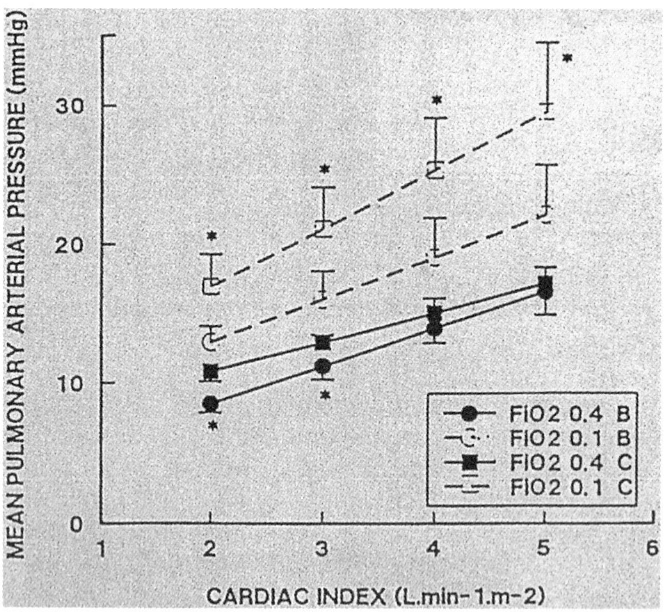

Figure 5 Composite plots of mean pulmonary artery pressure (PAP) vs. cardiac index (Q) in hyperoxia (inspired fraction of oxygen, $FIO_2 = 0.4$) and hypoxia ($FIO_2 = 0.1$) before (B) and after surgical chemodenervation (C) in nine dogs. Chemodenervation resulted in an increase in PAP at the lowest Q in hyperoxia and at all levels of Q in hypoxia. (From Ref. 61.)

cases of HAPE characteristically occur at moderate altitudes and have resulted in at least one death (55,56). Staub considered this form of permeability leak secondary to shear stress from overperfusion of the pulmonary vascular bed (57).

HAPE susceptibility is associated with a lower than normal ventilatory response to hypoxia (HVR) (40,48–50) (Fig. 4), although not all individuals with a low HVR develop HAPE. A low HVR is likely only a permissive characteristic, which upon ascent results in a lower alveolar and arterial oxygen level, which may then accentuate the cascade of events that lead to HAPE.

Is there a link other than a lower P_{AO_2} between a low HVR and high HPVR in HAPE-susceptible individuals? Although not generally accorded any direct influence on pulmonary vascular responses, perhaps greater peripheral chemoreceptor activity might affect pulmonary vascular tone. Several studies in dogs (58–60) and one in rats (61) with manipulation of carotid body innervation suggest that peripheral arterial chemoreceptor stimulation blunts HPVR, likely via efferent lung vasodilator vagal pathways (Fig. 5). Thus, a weak HVR may increase HPVR in HAPE susceptibles.

V. Pathological Findings

A number of descriptions of the appearance of lungs of those dying from HAPE have been published (10,14,62–65). The consistent findings at autopsy are severe pulmonary edema with bloody, frothy, proteinaceous fluid in the airspace, and right atrial and ventricular distension and hypertrophy. The pulmonary vessels are engorged, and thrombi are present in the small pulmonary arteries and septal capillaries. This observation has led to investigation of the role of enhanced fibrin formation in HAPE (66). Hyaline membranes, compatible with a high protein content in the edema fluid, are found in about half the cases (11,63). In some cases, there are small areas of patchy bronchopneumonia. It is difficult to ascertain how many of the findings are secondary to peri-mortem changes, especially the thrombi formation, and how much of the inflammatory response (i.e., the bronchopneumonia) represents a concomitant, pre-mortem process that contributes to a fatal outcome in a disease that usually resolves without sequelae.

VI. Pathophysiology

Hypotheses about the pathophysiological mechanisms causing HAPE emerge mostly from clinical studies plus a few as yet not totally successful attempts at creating an animal model of HAPE.

This section will deal with the following areas:

Hemodynamics in patients with HAPE and those predisposed to HAPE
Nature of the high-protein edema fluid
Inflammation in blood and alveolar fluid
Potential sites of leak in the pulmonary microvasculature

For years HAPE was thought to be a result of a filtration of low-protein fluid from the intra- to the extravascular space. Results of broncho-alveolar lavage (BAL) have shown that the alveolar fluid has a high protein concentration, which suggests a permeability leak, classically attributed to inflammation, but these findings are not incompatible with violation of the endothelial barrier from high pressures. The presence of inflammatory mediators in the BAL fluid suggests that inflammation contributes, at least in part, to the etiology of the pulmonary edema. Whether it initiates or helps to perpetuate the leak is not known.

VII. Hemodynamics

The 1960s and 1970s were important years for the understanding of the hemodynamic characteristics associated with HAPE. Several investigators performed invasive pulmonary artery catheterizations, which demonstrated two important points:

1. Pulmonary artery pressures and pulmonary vascular resistances were high.
2. HAPE was not secondary to left ventricular failure.

In 1962 Fred et al. (17) first reported a markedly elevated pulmonary artery pressure of 68/39 mmHg, which decreased substantially with oxygen breathing. This individual had a normal pulmonary arterial wedge pressure. Hultgren and coworkers in 1964 and 1971 (15,16) also catheterized the pulmonary artery of HAPE victims, both during and after their episodes of HAPE, reporting substantially elevated pulmonary artery pressures, normal wedge pressures, and accentuated HPVR after recovery from HAPE. Roy et al. in 1969 (44) confirmed these observations. Most of these studies found that oxygen administration lowered PAP but not to sea level values, suggesting preexisting or residual pulmonary hypertension. A recent study from Japan noted improved gas exchange and diminished infiltrates radiographically with normalization of pulmonary artery pressures (37), suggesting a relationship between severity of leak and elevation of pressures, but the study could not attribute causality between pressure and leak.

Several investigators have used noninvasive echocardiography to assess pulmonary vascular responses in patients who are HAPE susceptible (45–47,51,67). Most, but not all, of the subjects had accentuated pulmonary vascular responses to hypoxia. These studies lend credence to the hypothesis that increased pressures in the pulmonary vasculature play a role in the leak (68,69). Using ultrasound to assess cardiac function and estimate PAP in individuals with HAPE, Hackett et al. (46) found that alpha-blockade with phentolamine improved gas exchange and cardiac function, while beta-blockade with esmolol led to worsening. These findings imply that catecholamines and/or autonomic tone influence HPVR and HAPE.

In an elegant field study at the Campana Regina Margherita Hut on the summit of Monta Rosa (4559 m), Bartsch and associates obtained further evidence that high pulmonary intravascular pressures play a role in the etiology of HAPE (27). They tested the hypothesis that upon ascent to high altitude attenuating the rise in pulmonary artery pressure with a calcium-channel blocker, nifedipine, would prevent HAPE in HAPE-susceptible subjects. Using Doppler techniques to estimate PAP, chest radiographs, measurements of gas exchange, and clinical evaluations, they demonstrated that nifedipine effectively lowered PAP and prevented the development of HAPE in these subjects. Any putative anti-inflammatory or other beneficial actions that nifedipine (70–72) might have had, other than on PAP, could not be assessed.

In a subsequent study on Monta Rosa, these investigators administered a specific pulmonary vasodilator, inhaled nitric oxide (NO), to HAPE-susceptible subjects (52). All subjects had elevations of PAP, which were inversely associated with arterial oxygenation. With NO inhalation the PAP of these subjects decreased three times more than that of control subjects. Their Sa_{O_2} improved, while Sa_{O_2} of controls actually decreased. Perfusion scans in the HAPE subjects during NO showed redistribution of blood flow to the nonedematous areas, which would explain the improvement in oxygenation.

In 14 patients with severe HAPE (Pa_{O_2}35 ± 3.1 torr) in the Himalayas, Anand et al. (73) showed that nitric oxide and oxygen (F_{IO_2} = 0.5) separately decreased pulmonary vascular resistance 36% and 23%, respectively; both together caused a 54% decrease. Systemic hemodynamics were unaffected while gas exchange improved with both interventions, individually as well as together (Pa_{O_2} increased 21% and (A-a)D_{O_2} decreased 18%). The authors wondered whether oxygen and nitric oxide might be acting at different sites, but the design of the study could not address that possibility.

Using 16 HAPE-susceptible individuals and 14 controls, Maggiorini et al. (74) reconfirmed the accentuated pulmonary artery pressures in the former and then showed that inhaled nitric oxide and inhaled prostaglandin, iloprost, both decreased pulmonary artery pressures more than 33%, a response greater than that of oxygen (F_{IO_2} = 0.33) alone. All of these studies have important physiologic and therapeutic implications in that lowering pulmonary artery pressures may lead to improvement of HAPE.

In summary, these studies demonstrate a strong association between high pulmonary artery pressures and HAPE. The factors underlying the enhanced pulmonary vasoreactivity include a greater hypoxic activation of the sympathetic nervous system, differences in local vasoactive mediator formation, perhaps smaller lung volumes with a reduced vascular bed leading to higher pulmonary vascular resistance and lower HVR. The higher pulmonary vascular reactivity may be secondary to higher levels of plasma norepinephrine (75) prior to and during HAPE compared to normal subjects and may, therefore, be similar to neurogenic pulmonary edema (76).

An excess or imbalance of vasoactive mediators may also be responsible for an accentuated HPVR (35,36,77,78). Increased production of endothelin-1 (79) or a decreased clearance of endothelin-1 (80,81) and decreased NO synthesis (82) may contribute to higher PAP in individuals susceptible to HAPE, but other studies (78,83) have not found increased plasma endothelin levels in HAPE-susceptible patients with a brisk HPVR.

Surface area of the pulmonary vascular bed may play a role in the resistance that is encountered when both HPVR and an increased cardiac output combine to stress the pulmonary vasculature. In that regard, smaller lung volumes have been reported in those susceptible to HAPE (84–87). Maximal lung volumes, such as total lung capacity and vital capacity, corrected for height, are about 10–15% lower in HAPE-susceptible victims than in subjects with no history of HAPE, although their volumes still remain within the predicted normal values. Recently, Steinacker et al. (41) found lower values of functional residual capacity (~30%), intimating larger differences in pulmonary function at normal operating lung volumes than suggested by the maximal volume differences (Table 3). Why and how such differences in lung volumes could relate to disparate vascular resistances are not certain, but a smaller functional vascular bed at a lower lung volume may be one consequence. Furthermore, they found that HAPE-susceptible individuals had smaller increases in carbon monoxide–diffusing capacity during exercise, suggesting a diminished area for exchange between gas and blood.

Table 3 Lung Function and Maximal Oxygen Consumption $\dot{V}O_2max$

	HAPE (n = 8)	Control (n = 5)	p-value
Age, years	45 ± 8.4	38.2 ± 11.4	0.238
Weight, kg	74.1 ± 5.0	77.8 ± 8.5	0.173
Height, cm	171.3 ± 3.0	183.8 ± 6.6	0.030
BSA, cm^2	1.86 ± 0.07	2.00 ± 0.13	0.046
BMI, $kg \cdot cm^{-2}$	23.1 ± 2.37	25.3 ± 2.10	0.110
TLC, L	6.88 ± 0.73	8.52 ± 0.52	0.005
TLC, % pred	102.9 ± 11.2	111.8 ± 7.9	0.126
FVC, L	4.85 ± 0.44	6.26 ± 0.29	0.001
FVC, % pred	106.4 ± 8.1	113.4 ± 12.4	0.135
FRC, L	2.96 ± 0.63	4.46 ± 0.41	0.001
FRC, % pred	90.1 ± 18.1	124.0 ± 9.8	0.001
*D*L, CO, sb, % pred	35.1 ± 5.3	41.0 ± 7.2	0.125
*D*L, CO, sb, $mL \cdot mmHg^{-1} \cdot min^{-1}$	115.9 ± 14.3	114.6 ± 14.6	0.880
FEV_1/FVC, %	80.2 ± 4.1	75.7 ± 3.9	0.172
$\dot{V}O_2max$, $mL \cdot min^{-1} \cdot kg^{-1}$	45.1 ± 5.0	50.1 ± 8.4	0.374

Values are presented as mean ± SD. HAPE: High-altitude pulmonary edema; BSA: body surface area; BMI: body mass index; TLC: total lung capacity; % pred: percentage of predicted value, corrected for age, gender, height, and weight for a Middle European population; FVC: forced vital capacity; FRC: functional residual capacity; *D*L,CO,sb: single-breath diffusing capacity of the lung for carbon monoxide; FEV_1: forced expiratory volume in one second; FVC: forced vital capacity.
Source: Ref. 88.

During hypoxia and high pulmonary blood flow from exercise, the pressure in the pulmonary veins and venules rises. Although conventionally we think that the major effect of hypoxia is precapillary vasoconstriction, some work suggests that more prolonged hypoxia may cause pulmonary venoconstriction (88). Studies of dog lungs perfused in vivo have shown that when blood flow to a pulmonary lobe is quadrupled, the pressure drop in the precapillary vessels is diminished, capillary pressure rises, and the site of resistance is shifted to the pulmonary veins (88). These findings suggest a role for pulmonary venous resistance in increasing pressure within pulmonary capillaries.

Because normal pulmonary capillary wedge pressures are observed in HAPE victims at rest, left ventricular failure has been thought not to contribute to HAPE, but several avenues of investigation suggest otherwise. During exercise under hypoxic conditions pulmonary capillary wedge pressures are elevated in normal subjects (30 mmHg) (87,89) and even higher pressures are seen in HAPE-susceptible individuals (45 mmHg) with accompanying left ventricular dysfunction. Even at rest, acute hypoxia is associated with changes in left ventricular geometry indicative of leftward shift of the intraventricular septum and diastolic dysfunction (90). Marked pulmonary and right ventricular hypertension is presumably the cause of the ventricular septal shift. If the ventricular shift impairs left ventricular function and filling, one could then invoke increased left atrial and pulmonary venous pressures, leading to increased pulmonary capillary pressures, resulting in pulmonary edema.

Left ventricular dysfunction could also result from a decreased left ventricular compliance. In dogs with banded pulmonary arteries, pulmonary hypertension of modest degree over several hours was observed (91). Septal deviation and left ventricular edema and impaired ventricular wall motion were demonstrated. The edema was attributed to impaired lymphatic drainage of the left ventricle since the lymphatic drainage in the myocardium goes from the left to right ventricle. This idea is preliminary and deserves follow-up. As yet, evidence that conclusively links left ventricular dysfunction with HAPE does not exist.

Researchers have attempted to develop animal models of HAPE to explore pulmonary hemodynamics. One of the earliest studies was in rats. Whayne and Severinghaus exposed rats to severe hypoxia ($F_{IO_2} = 0.08$) for 3 hours and then stressed them with swimming for 10 minutes. Lung morphology showed microatelectasis and scattered alveolar hemorrhage and edema (92). Bartlett and Remmers exposed rats to an altitude of 4300 m for 3 weeks and found increased lung water at one week but not at 3 weeks (93). Colice et al. (94) exposed two strains of rats to hypoxia and found patchy, high-protein pulmonary edema. One of the most intriguing studies was by Stelzner et al. (95), who infused radiolabeled 125 I albumin intravascularly and measured alveolar levels of the tagged albumin to assess the leak of fluid and protein from the intra- to the extravascular space. Hypoxic exposure of 24–48 hours caused a significant albumin leak. Pretreatment with dexamethasone prevented the leak, which suggests that the drug maintained pulmonary endothelial integrity. An inhibition of inflammation is a possible explanation. Another study combined exercise and severe hypoxia in rats (96) and found scattered pulmonary hemorrhage, modestly elevated alveolar protein, and expression of von Willebrand factor, a protein released by injured endothelial cells and platelets.

In humans and other mammals high levels of exercise with increased intravascular pressures may induce similar violation of the pulmonary-capillary barrier. Rugby players, marathon runners, and cyclists (97–103) have had hemorrhage in the airspaces after intense exercise. Thoroughbred racehorses and greyhounds are perhaps the best example of reproducible high-protein pulmonary edema and hemorrhage after high levels of exercise (97,101–103). These extraordinary athletes, especially the horse, develop very high cardiac output and marked pulmonary hypertension with exercise, presumably resulting in stress on the pulmonary microvasculature. Stress failure and permeability leak are thought to be similar to HAPE.

Earlier studies tested the concept of "overperfusion" and the ability of the pulmonary vasculature to accommodate high flows and pressures. Hultgren gradually ligated segments of the pulmonary blood flow in dog lungs until all of the flow was directed to the left upper lobe, which resulted in a 600% increase in flow to the lobe (104). This procedure resulted in edema of the hyperperfused lobe with a high protein content in the alveolar fluid.

Mitzner and Sylvester (105) stressed pigs with moderate hypoxia and found a correlation between fluid filtration in the lymphatics of the lung and pulmonary blood flow, presumably from increased filtration pressure. In subsequent studies anesthetized pigs exposed to moderate hypoxia ($F_{IO_2} = 0.11$) developed a shunt fraction of 15%, which returned to 6% upon return to normoxia. Broncho-alveolar lavage (BAI) fluid showed a modest increase in protein, while lung morphology revealed some interstitial edema and perivascular cuffing. Similar findings were present in unanesthetized piglets exposed to an F_{IO_2} of 0.10 for 72 hours (96).

Using a monolayer cell culture of bovine pulmonary artery endothelial cells exposed to a hypoxic media, Ogawa et al. (106) induced a permeability leak across the cell layer. cAMP analogs prevented the leak, as did pretreatment with dexamethasone. The results of both interventions suggest a reestablishment of membrane integrity that is otherwise violated by a hypoxic environment, but the exact mechanism is not clear.

The choice of species makes a difference in outcome and interpretation of the data. For instance, sheep have not been a promising model in that they do not have a brisk HPVR, and attempts to create a permeability leak with hypoxia have not been consistently successful (92,96,105,107–109). Animals that deserve further investigation are the pig and the rat, whose hypoxic pulmonary vasoconstrictive responses are more brisk than that of the sheep. In summary, understanding the pulmonary hemodynamic response to hypoxia and exercise is critical to unraveling the mechanism of HAPE. It is clear that high intravascular pressures predispose one to HAPE, and the ability to reproduce this phenomenon in a consistent animal model would be crucial, but as yet such a model is not available.

VIII. Site of Leak

If pressure is an important force initiating leakage from blood to the interstitium and alveolar space in HAPE, where in the vascular bed does the leak occur? Is the pulmonary hypertension a consequence of homogeneous, uniform constriction of precapillary arterioles, or is the constriction uneven, resulting in areas of variable ventilation-perfusion (V/Q) and differing alveolar Po_2? Since the alveolar PAo_2 is the major determinant of pulmonary vasoconstriction, the heterogeneity of PAo_2 would result in patchy vasoconstriction and leak.

Consider the following scenario (Fig. 6). If alveolar hypoxia causes HPVR, then global alveolar hypoxia in the normal lung should cause uniform vasoconstriction of the precapillary arterioles and increase pressures, especially with exercise, resulting in a precapillary or extra-alveolar vessel leak. If HPVR is uniform, then all sites distal to the small arteries should be uniformly protected against the raised mean PA pressure, assuming no venoconstriction or left atrial pressure elevation. The site of leak could then only be precapillary with fluid accumulation first in the broncho-vascular bundles followed by movement into the interstitium and eventually into air spaces. The problem with this hypothesis is that even in the normal lung there is heterogeneity of regional gas exchange, which will result in some areas being more hypoxic than others, thus leading to areas with more intense vasoconstriction. Furthermore, it ignores the possible impact of active pulmonary venoconstriction and consequent increased pulmonary capillary pressures.

If the most critical stress is downstream from the pulmonary arterioles, the other potential sites of leak are capillaries or venules. If regional HPVR were uneven, the flow and/or shear forces would be greater in those areas with less HPVR. In this scenario, some lung regions could bear the brunt of both higher flow and higher capillary pressures. Conventional radiographic and CT imaging, albeit crude, show that HAPE is indeed patchy (28,29), suggesting nonuniform vasoconstriction and heterogeneity of blood flow, which could impose varying degrees of stress on the vasculature and thus result in patchy areas of edema. With respect to distribution of regional blood flow with hypoxia, data in animals are conflicting, with factors such as measurement technique, severity of hypoxia, in vitro vs. in vivo hypoxia, region size, and species differences contributing to differing results. With mild hypoxia ($FIo_2 \sim 0.15$), better VA/Q matching suggests more uniform pulmonary perfusion; however, with more severe hypoxia ($FIo_2 < 0.1$), there is greater regional blood flow dispersion consistent with uneven HPVR (110,111).

In humans, information is limited. VA/Q matching with mild hypoxia at rest is possibly improved (112) but may become less so with exercise and more severe hypoxia (113). Viswanathan and colleagues used external planar radionuclide scanning with limited spatial resolution in HAPE-susceptible and control subjects who were exposed to hypoxia. The results showed patchy perfusion in the HAPE-susceptible subjects and more uniform flow in the controls (114,115). More recently, evidence with SPECT scanning suggests that HAPE-resistant subjects show more

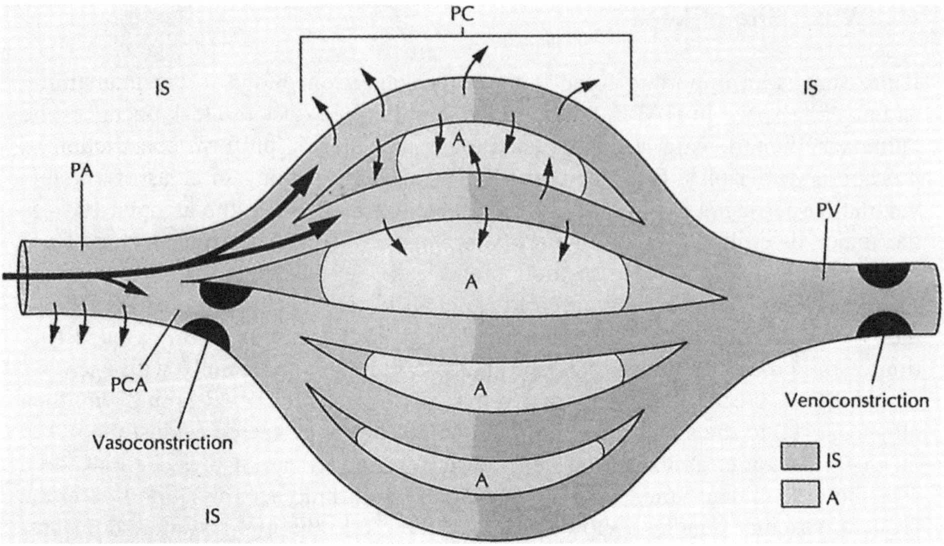

Figure 6 Schematic representation of the pulmonary microvasculature and alveolar/capillary interface and potential sites of leak in HAPE. Patchy precapillary arteriolar (PCA) vasoconstriction can have two effects on pressures and flow. First, there could be an increase in pressures upstream from the constriction leading to leak of fluid into the interstitial space (IS). Second, there would be diversion of flow to pulmonary capillaries (PC), potentially leading to stress on the endothelial layer and extravasation of fluid into alveolar space (A). Third, constriction of the pulmonary venules (PV) could lead to increases in the PC pressures.

homogeneous blood flow in the lungs after 5 hours of breathing 11% oxygen. This is in contrast to HAPE-susceptible subjects, who demonstrated heterogeneity of flow, with some areas receiving 300% greater flow with the hypoxic stress. These findings suggest that there may be a threshold level of hypoxia in the susceptible subjects which leads to a more heterogeneous distribution of flow with overperfusion in some areas, consistent with uneven HPVR. The point at which this patchy vasoconstriction occurs may be quite variable not only between individuals but possibly within a single individual at different times.

If HPVR is heterogeneous, then there are areas of accentuated pressures in the precapillary arterioles with decreased flows to the pulmonary capillaries as well as areas of high flow in the arterioles and capillaries to which the blood flow is diverted. The sites of leak, therefore, could be into the interstitium proximal to precapillary arterioles as well as from the pulmonary capillaries directly into the alveolar space. As mentioned earlier, hypoxic pulmonary venoconstriction might also lead to increased pulmonary capillary pressure.

Recently, researchers have entertained the concept of pulmonary capillary fragility, i.e., a microvascular bed that is susceptible to disruption from pressures alone

(97,116–121). With an imposed abrupt increase in left atrial pressure (52.5 cm H_2O) in rabbit lungs or in rats placed in a hypobaric chamber and immediately decompressed to 1/2 atmosphere pressure, these investigators have demonstrated damage to the endothelium and widening of the pores at the endothelial junctions, associated with leak of high molecular weight proteins, red blood cells, and LTB4 (Fig. 7). These disruptions were reversed with a decrease in pressure, which corresponds to the clinical course of HAPE. The concept of ''stress failure of the pulmonary capillaries'' could explain the course of HAPE.

The tools available in the past have not allowed adequate resolution of the pattern of perfusion in the pulmonary microvasculature to answer the questions posited above, especially in humans. Investigators, therefore, have turned to animal experiments to try to clarify the situation.

IX. Alveolar Fluid

Classically, pulmonary edemas are divided into those caused by hydrostatic and permeability factors. Starling forces, the balance of oncotic and hydrostatic forces across semi-permeable endothelial-epithelial barriers, define a hydrostatic edema. A permeability edema refers to conditions where factors such as inflammation result in loss of selectivity of the endothelial barrier, so that large protein molecules and sometimes red blood cells and plasma gain access to the pulmonary extravascular space. We now realize that this categorization of pulmonary edema is too simple and obfuscates the possibility that both hydrostatic and permeability leaks may occur together. In a rabbit model of hydrostatic pulmonary edema, Bachofen et al. (122) found marked heterogeneity of alveolar protein concentrations, indicating that the conditions that result in extravasation of fluid are nonuniform.

Fluid flux across the lung vasculature is a normal, dynamic process. An imbalance may be occur when (1) hydrostatic gradients become too great, (2) the normal oncotic gradient becomes too small, (3) selective impermeability of the endothelial-epithelial barriers is lost, or (4) the normal clearance of interstitial fluid of the interstitium by lymphatic drainage or alveolar fluid by active sodium reabsorption is impaired.

For many years, investigators presumed that high pulmonary artery pressures led to a transudative leak of low-protein fluid. To clarify this issue investigators have used broncho-alveolar lavage with normal saline in individuals with HAPE to examine the edema fluid. In two field studies at 4300 m on Mount McKinley in Alaska, the BAL fluid of climbers acutely ill with HAPE contained higher concentrations of low and high molecular weight proteins than from low and high-altitude controls as well as patients with the adult respiratory distress syndrome (ARDS) at sea level (35,36) (Fig. 8). Also, the composition of the BAL fluid in climbers with acute mountain sickness but without HAPE was comparable to that of healthy controls in spite of relatively lower arterial oxygen saturations.

A more recent study in four hospitalized patients recovering from HAPE also

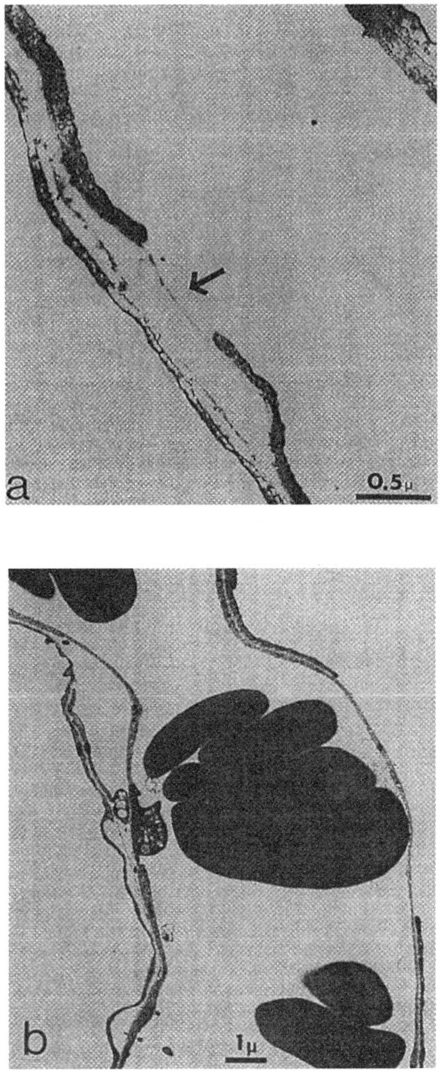

Figure 7 Electron micrographs showing stress failure in pulmonary capillaries of rabbits with acute increase in capillary transmural pressures to 52.5 cm H_2O. (a) Disruption of the capillary endothelium but intact basement membrane and alveolar epithelium. (b) Disruption of both the capillary endothelium and alveolar epithelium. (c) Passage of red blood cells through the capillary wall. (From Ref. 93.)

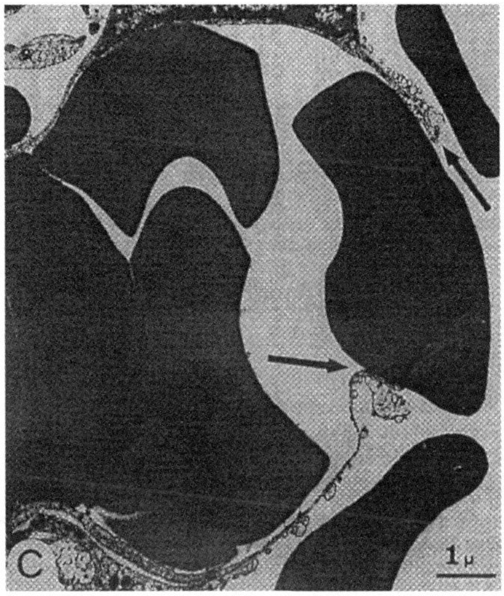

Figure 7 Continued

documented BAL protein concentrations that were much higher than in control subjects (123). Swenson et al. (67) found very high protein levels as well as very high pulmonary artery pressures in HAPE-susceptible subjects within the first hours of illness at 4559 m. The important finding in these studies was that high alveolar protein concentrations are present early on and throughout the course of HAPE.

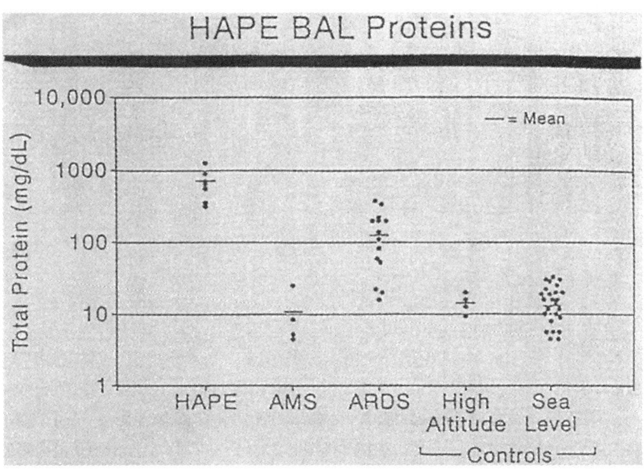

Figure 8 Protein concentrations in bronchoalveolar lavage fluid in subjects with HAPE, acute mountain sickness (AMS) at 4300 m on Mount McKinley, adult respiratory distress syndrome (ARDS), and low- and high-altitude controls. (Data from Refs. 35,36.)

Although clearance of alveolar water may have resulted in concentration of proteins, contributing to their high values, it seems apparent that HAPE results in a high protein permeability leak into the alveolar space. If so, what factors contribute to violation of the endothelial barrier?

X. Inflammation

As mentioned earlier, the classic concept of a permeability leak is that inflammation, such as in sepsis, leads to a loss of endothelial integrity and leak of high protein fluid into the extravascular space (Fig. 9). Certain markers of inflammation have been found in HAPE, which suggest that an inflammatory response plays either a primary or secondary role in the permeability leak, and it is, therefore, important to try to incorporate this information into an understanding of the disease. The questions, therefore, are: (1) Does inflammation trigger the leak at high altitude? (2) Can preexisting or concomitant viral infections prime the endothelium for the increase in pressures and thus lower the threshold for leak? (3) At what time in the course of the disease do inflammatory mediators appear?

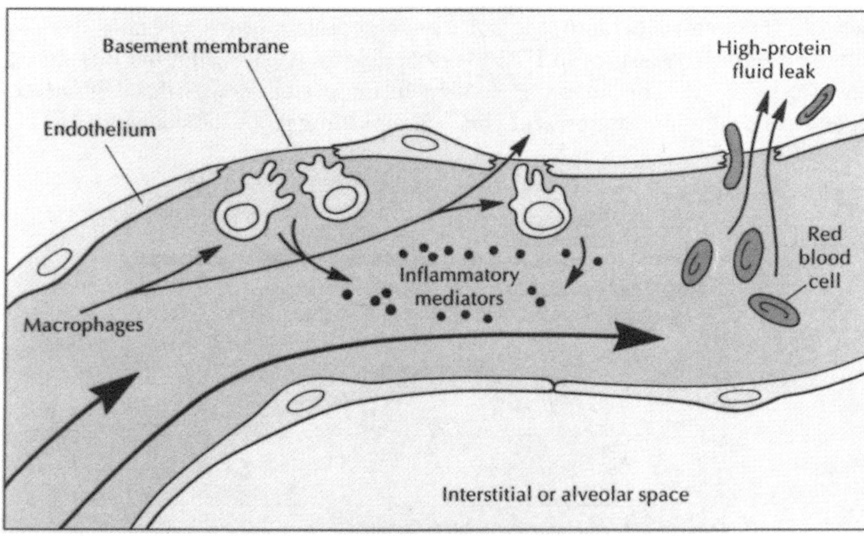

Figure 9 Schematic representation of the pulmonary microvasculature during HAPE. Increased intravascular pressures presumably cause stretching of the endothelial layer and baring of the basement membrane. This endothelial gap results in leak of protein-rich fluid and red blood cells from the intra- to extravascular space while baring of the basement results in chemotaxis of neutrophils and macrophages with release of cytokines and other inflammatory mediators.

Evidence for inflammation has come from several clinical arenas. First, many patients with HAPE have elevated temperatures, leukocytosis, sedimentation rates, and serum enzymes, such as creatine phosphokinase. Further evidence can be found in the lung itself. The BAL fluid of climbers with HAPE on Mount McKinley at 4300 m showed a markedly increased cellular response in the HAPE subjects, with a predominance of alveolar macrophages (AM) and a modestly increased absolute number of neutrophils (PMN) compared to controls (35,36). These findings contrasted with patients with ARDS, who also had high cell counts but with a marked predominance of neutrophils (Fig. 10).

A study of patients hospitalized with HAPE also showed increased numbers of cells but with a more even distribution between AM and PMN than was seen in the climbers on Mount McKinley (124). This difference may be a function of the timing of the lavages. Swenson et al. in a recent study at the Campana Margherita Hut performed BAL on HAPE-susceptible and control subjects before and within 2 hours of ascent to 4559 m. Nine of the 11 HAPE-susceptible subjects developed HAPE. They had a high protein concentration in the BAL but no neutrophils or IL-8, a neutrophil chemotactic mediator (67). This study is important in that, unlike the studies from Mount McKinley where climbers were very ill and had had symptoms of HAPE for at least 24 hours, Swenson et al. studied subjects early in their illness. While high concentrations of proteins were found in the BAL at this stage, inflammatory mediators were lacking. This difference from the Mount McKinley studies suggests that inflammation is not necessary for the initiation of a permeability leak but may just be a post facto phenomenon triggered by high pressures and/or flow. The exceptional rise in pulmonary pressures, therefore, appears to be the critical factor in initiation of the leak. The inflammation found in other studies probably occurs later and may enhance the permeability leak. This possibility lends more credence to the deleterious effect of high pressures on a fragile pulmonary vasculature.

In the Mount McKinley studies there was evidence of leukotriene B_4 (LTB4) and complement, both inflammatory mediators, and thromboxane B_2, a vasoactive mediator from the arachadonic acid cascade (35,36) (Fig. 10). LTB4 and complement are both chemotactically active as are some of the cytokines, which along with adhesion molecules may contribute to the insult to the endothelium. Kubo et al. (124), using BAL, measured a number of the chemotactically active cytokines, which also play a role in endothelium adhesion and found elevated levels of IL-1B, IL-8, TNF-α, and IL-6, all of which are elevated in ARDS. These molecules returned to normal after recovery. Kleger et al. (125) prospectively studied mountaineers at 4559 m in the Alps and found no increase in total body systemic transcapillary leak of albumin or plasma levels of cytokines prior to AMS or HAPE, but some of these markers of inflammation (acute phase and C-reactive proteins and IL-6) were elevated if subjects developed HAPE. These data suggest that inflammation may follow the initial leak, which is presumably caused by high pressures. The presence of high proteins, chemotaxis of cells, and the exposure of the basement membrane, each

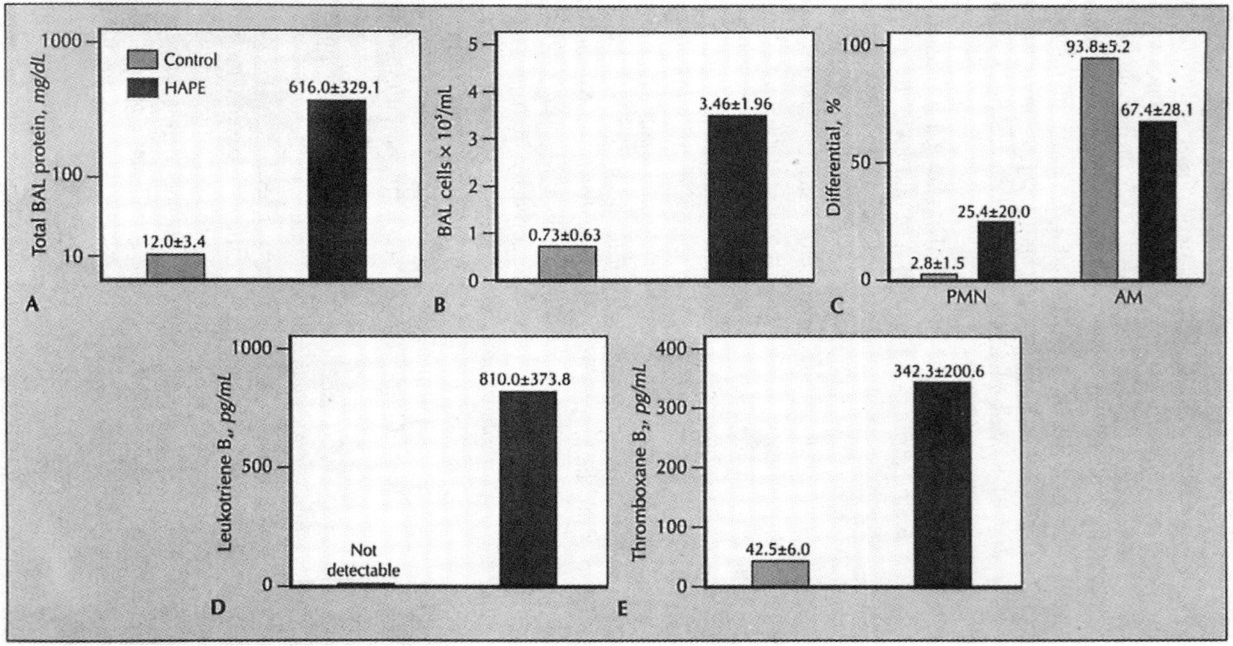

Figure 10 Cell counts and biochemical mediators in subjects with HAPE, AMS at 4300 m on Mount McKinley, compared to ARDS and high- and low-altitude controls. (From Ref. 36.)

and together, may invoke an element of inflammation as a secondary occurrence. This sequence is distinctly different from ARDS, which is thought to be a consequence of acute and severe inflammation without a comparable rise in pulmonary artery pressures.

The endothelium may also be affected by inflammatory precursors that may prime the cellular layer for insult. A clinical study at an altitude of slightly less than 3000 m in 38 patients with HAPE and 10 controls showed elevated levels of urinary LTE4, the stable metabolite of the arachadonic acid cascade, in the HAPE patients, almost three-quarters of whom had clinical histories of a preceding or concomitant viral infections (126). Similarly, in children, recent or concurrent upper respiratory tract infections were associated with 79% of paediatric cases of HAPE in the Colorado Rockies (127). These studies document the presence of inflammation and suggest that even mild inflammation from viral infections may play a role in priming the endothelium for the subsequent high pressures upon ascent to high altitude. Unfortunately, because of the limitations of such clinical studies, none could document the sequence of events, i.e., pressure versus inflammation, in the evolution of the pulmonary edema, which makes understanding of a mechanism difficult. It is also conceivable that these factors interact in varying ways depending on the magnitude of each response.

Bartsch (23) has aptly pointed out that many of the mediators that are measured act in local tissue beds and have a short half-life, thus making estimations of cause and effect difficult. Bartsch suggests that the disease, its inception, and patients' vulnerability to it may vary depending on the altitude, degree of acclimatization, presence of viral infection, and many other factors.

The presence in alveolar fluid and blood of several pro-inflammatory mediators and cytokines in patients with HAPE makes one wonder whether hypoxia itself may stimulate immune reactions, cellular or humoral, to generate these substances. In vitro studies of macrophages and neutrophils (128–130) find that until the Po_2 is considerably more hypoxic (<15 mmHg) than levels more relevant to those preceding HAPE in humans, there is no stimulation of these cells. It is possible, on the other hand, that certain regions of the lung are, in fact, that hypoxic, thus making initiation of an inflammatory cascade by low Po_2 possible. On the whole, however, the evidence appears to favor cytokine and inflammatory mediator release as a response to injury (either hemodynamic or infectious) rather than hypoxia per se. E-selection and P-selection are transmembrane glycoproteins expressed on endothelial cells in response to certain activating stimuli or injury. They act as adhesion molecules for leukocytes and have been found in the blood of patients with a number of inflammatory processes. Leukocyte adherence is the first step in the migration of these cells from the intra- to extravascular space in areas of inflammation. Leukocyte adherence is preceded by markers of inflammation and chemotaxis, such as leukotrienes, which, if present, imply that chemotaxis has been followed by adhesion and the consequences of neutrophil activation on the endothelium. E-selectin is made by endothelial cells shortly after activation by inflammatory cytokines such as IL-1, TNF-α, or lipopolysaccharide (130,131). E-selectin is elevated in hypoxic patients with HAPE but not in individuals with AMS and no HAPE, nor in controls (132), which suggests that the stress of hypoxia is not the trigger of the cascade of events that affects the endothelium and subsequent leukocyte adhesion. The expression of P-selectin differs from that of E-selectin in that its movement from within the endothelial cell to the membrane and back is very brief and has not been found in a number of studies either in the plasma or alveolar fluid of patients with HAPE or AMS (132). Presently, whether E- and P-selectins play a role in altering endothelial membrane permeability in HAPE is no more than intriguing speculation.

Neutrophils may also play a role in increasing endothelial permeability. Sheep were exposed to hypoxia, with both induced leukopenia and normal neutrophil counts (107,109). Pulmonary edema occurred only in those with repleted white cells. This study suggests some role for both cellular and humoral inflammatory mediators in the facilitation of permeability leak.

Wood et al. (133) recently examined the effect of NO on leukocyte adhesion in the mesenteric circulation of rats both acclimatized and nonacclimatized to hypo-

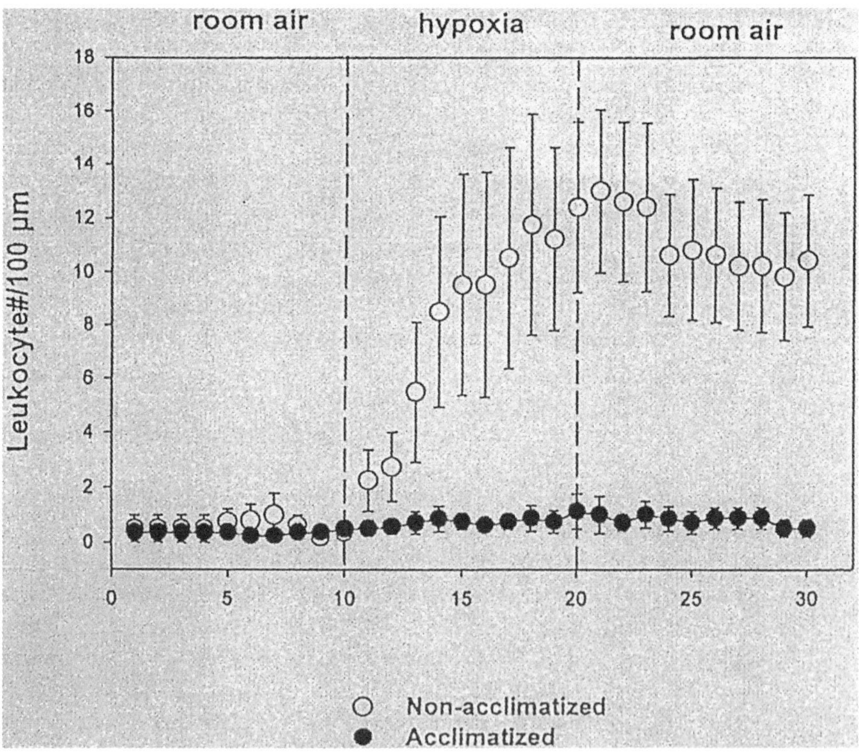

Figure 11 Cumulative results for changes in number of adherent leukocytes and shear rate during control period (breathing room air, hypoxia, an recovery period in acclimatized (full circles) and unacclimatized (empty circles). (From Ref. 134.)

baria comparable to altitudes over 4000 m for 3 weeks. Both groups were then acutely exposed to barometric pressure of 370 torr, a pressure equivalent to an altitude of approximately 6000 m. The acutely exposed, nonacclimatized rats demonstrated a marked increase in leukocyte adherence, whereas the acclimatized rats showed no increase from sea level values (Fig. 11). The investigators attributed these findings, in part, to a suppression of NO synthesis with acute exposure and/or an upregulation of NO synthesis with acclimatization. No, thus, may play a role in mediating leukocyte adherence. Even though this study was done in the mesenteric microcirculation, the relevance to the endothelium in the pulmonary and other vascular beds is intriguing.

Support for a possible role for genetic predisposition in the inflammatory component of HAPE comes from a recent study of HAPE-susceptible subjects in Japan (134). These investigators studied the frequencies of human leukocyte antigen (HLA) alleles in 30 HAPE-susceptible individuals and 100 healthy controls whose altitude tolerance was not known. They found that there were significant associations between HAPE susceptibility and the major histocompatibility complex alleles, HLA-DR6 and HLA-DQ4, and HLA-DR. These findings require confirmation in other populations and most importantly in known HAPE-resistant individuals. Susceptibility to many diseases has been linked with the histocompatibility alleles, including some types of pulmonary hypertension, but it remains highly conjectural whether these antigens or other tightly linked, nonimmunological genes are responsible (135).

Because of the presence of thrombi in postmortem specimens, some investigators have tried to link coagulation abnormalities with the pathogenesis of HAPE. Bartsch and colleagues (136) have provided hematological data in early and late HAPE which suggest that HAPE is well established clinically and radiographically before there is any activation of coagulation cascades indicative of endothelial cell disruption and significant exposure of the thrombogenic subcapillary basement membrane. Platelet aggregation and hypercoagulability, therefore, is felt to be a phenomenon occurring later and not an initiating event.

XI. Alveolar Epithelial Fluid Clearance

The alveolar epithelium plays an important role in the fluid balance of the lung. Sodium is transferred actively from the interstitium and alveolar space across the epithelial membrane by a Na^+/K^+ ATPase pump (Fig. 12). There appears to be individual variability in the capacity of this pump to handle sodium flux. Fluid follows the flux of sodium. In a quantitative sense, it is not clear how much this mechanism contributes to the susceptibility to and resolution of HAPE, but perhaps some individuals may have a blunted ability to transport sodium and fluid, thus impairing the clearance of fluid and prolonging the presence of alveolar edema. It is an intriguing possibility that may play an important role in patients who are predisposed to HAPE.

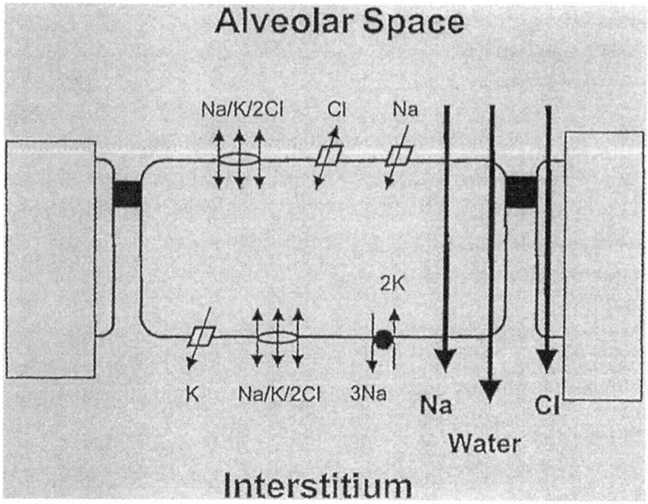

Figure 12 Alveolar epithelial apical and baso-lateral membrane ion channels and exchangers involved in active transepithelial sodium and water absorption. Acute hypoxia may reduce alveolar fluid clearance by its own inhibition of apical sodium entry pathways and baso-lateral Na^+/K^+ ATPase activity. (Figure kindly provided by Heimo Mairbaeurl, University of Heidelberg.)

Richalet proposed this possibility (22), stating that fluid flux is a normal process and that the alveolar epithelium is another important barrier across which fluid gains both access and egress from the alveolar space. Since the epithelium imposes a barrier to fluid flux, egress of any alveolar fluid to lymphatic drainage depends on an intact Na^+/K^+ ATPase pump operating at the basement membrane of the alveolar epithelium. Sodium can enter the cell on the alveolar side, via the Na-K-2Cl co-transporter and the recently cloned epithelial Na channel protein (ENaC). Rapid clearance of alveolar fluid implies an intact pump and sodium entry mechanism. Preliminary evidence showing that HAPE-prone individuals may have less alveolar ENaC, as evidenced by amiloride-sensitive potential difference changes in the nasal epithelium (137), and that a mouse strain partially deficient in ENaC shows greater lung water accumulation with hypoxia (138) point in this direction. A healthy epithelial barrier may explain the usually good recovery of HAPE victims as well as the good prognosis noted by Matthay and Wiener-Kronish in patients with high permeability lung leak who had more rapid clearance of alveolar fluid (139).

In concurrence with this idea are recent studies in rat type II pneumocytes and transformed lung epithelial cell lines exposed to profound hypoxia ($FIO_2 < 0.05$) for up to 48 hours. These studies show a time-dependent decrease in Na^+/K^+ ATPase function, downregulation of apical ENaC and Na-K-2Cl co-transport with hyp-

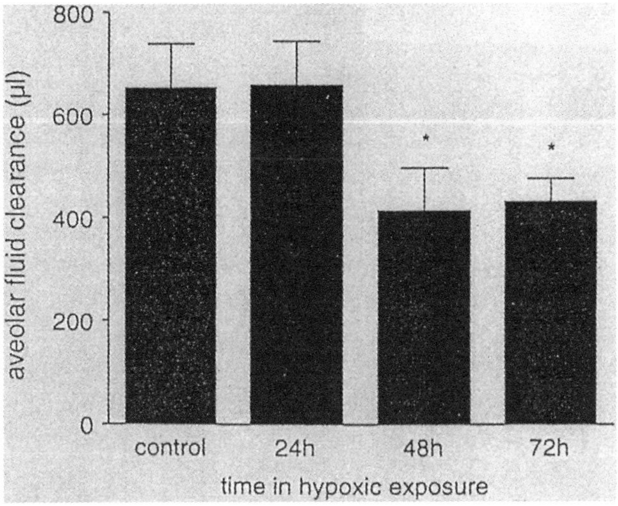

Figure 13 Time course of effect of hypoxia ($FIO_2 = 0.10$) on alveolar fluid clearance (AFC). There is a significant decrease in AFC of rats exposed for 48 hours or longer. (From Ref. 144.)

oxia, that is entirely reversible when the cells are returned to normoxic conditions (139–142). Suzuki et al. (143) extended these studies to intact rats exposed to 10% oxygen for 48 hours (Fig. 13). They found a decrease in clearance of alveolar fluid as well as a decrease in Na^+/K^+ ATPase activity after 40 hours of hypoxia that supports the in vitro findings (140–142).

The reasons for reduced energy requirements of a decreased sodium and fluid reabsorption in hypoxia by the alveolar epithelium are not known, but one might speculate that this response represents an attempt at energy conservation by the epithelium to reduce its own O_2 consumption in an hypoxic environment. Whatever the explanation, any reduction in active sodium and water reabsorption by the alveolar epithelium during hypoxia would seem to facilitate the evolution of HAPE and could contribute to impaired fluid clearance from the alveolar space. The levels of hypoxia necessary to elicit these changes in ion transport by the alveolar epithelium, on the other hand, occur at hypoxic levels much lower than those at which HAPE develops; however, once edema begins to form, the rate at which active fluid reabsorption takes place could become a critical factor in maintaining regional ventilation for local O_2 availability. As more edema fluid accumulates and ventilation is compromised in certain regions, the local PO_2 may fall to a level that ultimately impairs active fluid reabsorption. Whether differences during hypoxia exist among individuals in the capacity for alveolar fluid reabsorption is not known, but such differences could help to explain susceptibility to HAPE.

XII. Prevention and Therapy

The best way to prevent HAPE is to allow sufficient time for acclimatization. HAPE-susceptible individuals can dramatically reduce the incidence of HAPE by ascending gradually over days. It is not always possible, however, for climbers, trekkers, or tourists to ascend at a pace that is slow enough to prevent HAPE. Therefore, HAPE will remain an important clinical problem for physicians to recognize and treat.

Because of the low incidence and unpredictability of HAPE, controlled clinical and pharmacological trials have been difficult to conduct. Are drugs, such as acetazolamide and dexamethasone, which are efficacious for the prevention and treatment of acute mountain sickness (see Chapter 22), helpful in prevention and treatment of HAPE? No studies to answer this question have yet been performed. On the other hand, as was mentioned earlier, Bärtsch and coworkers in an elegant field study (27) on HAPE-susceptible subjects tested the hypothesis that high pulmonary vascular pressures are associated with the development of HAPE. By attenuating the pulmonary hypertension with a pulmonary vasodilator (nifedipine), they prevented HAPE. In two other studies investigators used nifedipine in the treatment of HAPE and found that sublingual administration of the drug improved gas exchange in some but not all of the subjects.

In light of these studies, HAPE-susceptible individuals who have to ascend quickly to high altitude should take an extended-release preparation of nifedipine. Adverse side effects have not been reported. In individuals with HAPE who are not near medical care, careful administration of nifedipine with monitoring of arterial oxygen saturation and blood pressure may be helpful.

The most important step in the treatment of HAPE is its early recognition. In remote settings, individuals should descend while they are still able to walk. A descent of 500–1000 m may be all that is necessary to prevent progression of this potentially fatal disease and actually result in resolution. Oxygen administration, if available, and rest are also helpful but should not delay descent. In areas where medical help is available, such as recreational ski locations, patients whose clinical presentation is of the mild or moderate degree and whose oxygen saturation can be improved to >90% by the administration of oxygen and who have family or friends to watch them can stay at high altitude using low-flow oxygen therapy, rest, and with daily follow-up thereafter. Pharmacotherapy to allow ascent while preventing HAPE or to reliably reverse its course once begun is still a challenge for future investigation.

XIII. Directions

HAPE is a condition that affects only humans. Investigators have made important advances with observations of physiological, cellular, and biochemical responses that have led to a better understanding of the mechanism of the disease. Recognition of the disease in the field setting is essential to prevent mortality since treatment even in the worst of situations is easy and usually successful. But a full understanding of the disease is important not only for its own sake but also for the knowledge it might provide to an understanding of other disorders where fluid leakage is a major component.

Unraveling these mysteries will be greatly abetted by an animal model that could answer the following questions:

What is the sequence of events that leads to a permeability leak?
Do high pressures precede leak of proteinaceous fluid into the alveolar space?
Can prevention of excessive pulmonary hypertension prevent HAPE?
Are markers of inflammation present in the early phase of HAPE, or do they appear only after vascular integrity is violated?
Does active alveolar salt and water reabsorption across the epithelium play a role in development and resolution of HAPE?
Are there genetic markers of alveolar salt transport, pulmonary vascular reactivity, and inflammatory responsiveness that predict whether certain individuals are more susceptible to HAPE?
How does acclimatization decrease susceptibility to HAPE? Are changes in NO production involved?

References

1. Mosso A. Life of Man in the High Alps. London: T. F. Unwin, 1898.
2. Ravenhill T. Some experiences of mountain sickness in the Andes. J Trop Med Hygiene 1913; 1620:313–320.
3. Lizarraga L, Soroche A. Edema agudo de pulmon. An Fac Med (Lima) 1955;38:244.
4. Hurtado A. Pathological aspects of life at high altitudes. Milit Med 1955; 117:272–284.
5. Alzamora-Castro V, Garrido-Lecca G, Battilana G. Pulmonary edema of high altitude. Am J Cardiol 1961; 7:769–778.
6. Vega A. Algunos casos de edema pulmonar agudo por soroche grave. An Facul Med Lima 1955; 38:233–240.
7. Hultgren H, Spickard W. Medical experiences in Peru. Stanford Med Bull 1960; 263: 478–480.
8. Houston C. Acute pulmonary edema of high altitude. N Engl J Med 1960; 263:478–480.
9. Hultgren H, Spickard W. Medical experiences in Peru. Stanford Med Bull 1960; 18: 76–95.
10. Singh I, et al. High altitude pulmonary edema. Lancet 1965; 1:229–234.
11. Menon N. High altitude pulmonary edema: a clinical study. N Engl J Med 1965; 273: 66–73.
12. Marticorena E, Tapia F, Dyer J, et al. Pulmonary edema by ascending to high altitudes. J Dis Chest 1964; 45:273–283.
13. Hultgren H, Marticorena E. High altitude pulmonary edema: epidemiologic observations in Peru. Chest 1978; 74:372–376.

14. Hultgren H, Spickard W, Lopez C. Further studies of high altitude pulmonary edema. Br Heart J 1962; 24:95–102.
15. Hultgren H. Rapid ascent, too early exercise may cause acute pulmonary edema at high altitude. Circulation 1964; 29:393–408.
16. Hultgren H, Grover R, Hartley L. Abnormal circulatory responses to high altitude in subjects with a previous history of high altitude pulmonary edema. Circulation 1971; 44:759–770.
17. Fred H, et al. Acute pulmonary edema at altitude. Clinical and psychological observations. Circulation 1962; 25:929–937.
18. Hultgren H. High altitude pulmonary edema. In: Staub NC, ed. Lung Water and Solute Exchange. New York: Marcel Dekker, 1978:437–469.
19. Viswanathan R, Jain S, Subramanian S. Pulmonary edema of high altitude III. Pathogenesis. Am Rev Respir Dis 1969; 100:342–349.
20. Schoene R. High altitude pulmonary edema: pathophysiology and clinical review. Ann Emerg Med 1987; 16:987–992.
21. Hackett P, Roach R. High altitude pulmonary edema. J Wilderness Med 1990; 1:3–26.
22. Richalet J. High altitude pulmonary edema: still a place for controversy? (editorial). Thorax 1995; 50:923–929.
23. Bärtsch P. High altitude pulmonary edema. Respiration 1997; 64:435–443.
24. Hultgren H. High altitude pulmonary edema: current concepts. Ann Rev Med 1996; 47:267–284.
25. Hultgren H, et al. High-altitude pulmonary edema at a ski resort. West J Med 1996; 164:222–227.
26. Singh I, et al. Acute mountain sickness. N Engl J Med 1969; 280:175–184.
27. Bärtsch P, et al. Prevention of high altitude pulmonary edema by nifedipine. N Engl J Med 1991; 325:1284–1289.
28. Vock P, et al. High-altitude pulmonary edema: findings at high altitude chest radiography and physical examination. Radiology 1989; 170:661–666.
29. Vock P, et al. Variable radiomorphologic data of high altitude pulmonary edema. Features from 60 patients. Chest 1991; 100:1306–1311.
30. Sophocles A. High-altitude pulmonary edema in Vail, Colorado, 1975–1982. West J Med 1986; 144:569–573.
31. Sophocles A, Bachman J. High altitude pulmonary edema among visitors to Summit County, Colorado. J Fam Practice 1983; 17:1015–1017.
32. Hochstrasser J, Nanzer A, Oelz O. Das Hoehenoedem in den Schweizer Alpen. Schweiz Med Wschr 1986; 116:866–873.
33. Lobenhoffer H, Zink R, Brendel W. High altitude pulmonary edema: analysis of 166 cases. In: Brendel W, Zink R, eds. High Altitude Physiology and Medicine. New York: Springer-Verlag, 1982:219–231.
34. Regensteiner J, Woodward W, Hagerman D, et al. Combined effects of female hormones and metabolic rate on ventilatory drives in women. J Appl Physiol 1989; 66: 808–813.
35. Schoene R, Hackett P, Henderson W, et al. High altitude pulmonary edema. Characteristics of lung lavage fluid. JAMA 1986; 256:63–69.
36. Schoene R, Swenson E, Pizzo C, et al., The lung at high altitude bronchoalveolar lavage in acute mountain sickness and pulmonary edema. J Appl Physiol 1988; 64: 2605–2613.

37. Kobayashi T, Koyama S, Kubo K, et al. Clinical features of patients with high altitude pulmonary edema in Japan. Chest 1987; 92:814–821.
38. Mansell A, Powles A, Sutton J. Changes in pulmonary PV characteristics of human subjects at an altitude of 5,366m. J Appl Physiol 1980; 49:79–83.
39. Gautier H, Peslin R, Grassino A, et al. Mechanical properties of the lungs during acclimatization to altitude. J Appl Physiol 1982; 52:1407–1415.
40. Selland M, Stelzner T, Stevens T, et al. Pulmonary function and hypoxic ventilatory response in subjects susceptible to high altitude pulmonary edema. Chest 1993; 103: 111–116.
41. Steinacker J, Tobias P, Menold E, et al. Lung diffusing capacity and exercise in subjects with previous high altitude pulmonary edema. Eur Respir J 1998; 11:643–650.
42. Maldonado D. High altitude pulmonary edema. Rad Clin North Am 1978; 16:537–549.
43. Koizumi T, et al. Radiographic and hemodynamic changes during recover from high altitude pulmonary edema. Intern Med 1994; 33:525–528.
44. Roy S, Guleria J, Khanna P, et al. Haemodynamic studies in high altitude pulmonary edema. Br Heart J 1969; 31:52–58.
45. Kawashima A, et al. Hemodynamic responses to acute hypoxia, hypobaria, and exercise in subjects susceptible to high altitude pulmonary edema. J Appl Physiol 1989; 67:1982–1989.
46. Hackett P, Roach R, Hartig G, et al. The effect of vasodilators on pulmonary hemodynamics in high altitude pulmonary edema: a comparison. Int J Sports Med 1992; 13: 268–271.
47. Yagi H, et al. Doppler assessment of pulmonary hypertension induced by hypoxic breathing in subjects susceptible to high altitude pulmonary edema. Am Rev Respir Dis 1990; 142:796–801.
48. Hyers T, et al. Accentuated hypoxemia at high altitude in subjects susceptible to high-altitude pulmonary edema. J Appl Physiol 1979; 46:41–46.
49. Hackett P, Roach R, Schoene R, et al. Abnormal control of ventilation in high-altitude pulmonary edema. J Appl Physiol 1988; 64:1268–1272.
50. Matsuzawa Y, Fujimoto K, Kobayashi T, et al. Blunted hypoxic ventilatory drive in subjects susceptible to high altitude pulmonary edema. J Appl Physiol 1989; 66:1152–1157.
51. Vachiery J, McDonagh T, Moraine J, et al. Doppler assessment of hypoxic pulmonary vasoconstriction and susceptibility to high altitude pulmonary edema. Thorax 1995; 50:22–27.
52. Scherrer U, Vollen weider L, Delabays A, et al. Inhaled nitric oxide for high-altitude pulmonary edema {see comments}. N Engl J Med 1996; 334:624–629.
53. Naeije R, Melot C. Acute pulmonary edema on the Ruwenzori mountain range. Br Heart J 1990; 64(6):400–402.
54. Scoggin C, et al. High-altitude pulmonary edema in the children and young adults of Leadville, Colorado. N Engl J Med 1977; 297:1269–1272.
55. Hackett P, Creagh C, Grover R, et al. High altitude pulmonary edema in persons without the right pulmonary artery. N Engl J Med 1980; 302:1070–1073.
56. Fiorenzano G, Rastelli V, Greco V, et al. Unilateral high altitude pulmonary edema in a subject with right pulmonary artery hypoplasia. Respiration 1994; 61:51–54.
57. Staub N. Overperfusion edema. N Engl J Med 1980; 302:1085–1086.

58. Wilson L, Levitzky M. Chemoreflex blunting of hypoxic pulmonary vasoconstriction is vagally mediated. J Apply Physiol 1989; 66(2):782–791.
59. Chapleau M, et al. Chemoreflex stimulation interferes with regional hypoxic pulmonary vasoconstriction. Respir Physiol 1988; 71(2):185–200.
60. Naeije R, Lejeune P, Leeman M, et al. Pulmonary vascular responses to surgical chemodenervation and chemical sympathectomy in dogs. J Appl Physiol 1989; 66: 42–50.
61. Sugito K, Tatsumi K, Igari H, et al. Role of carotid body in pressure response of pulmonary circulation in rats. Respir Physiol 1998; 111(3):283–293.
62. Navak N, Roy S, Narayaran T. Pathologic features of altitude sickness. Am J Pathol 1964; 45:381–391.
63. Arias-Stella J, Kruger H. Pathology of high altitude pulmonary edema. Arch Pathol 1963; 76:147–157.
64. Heath D, Moosavi H, Smith P. Ultrastructure of high altitude pulmonary edema. Thorax 1973; 28:694–700.
65. Fukushima M, et al. A case of high altitude pulmonary edema. Nihon Kyobu Shikkan Gakkai Zasshi 1980; 18:753–757.
66. Bartsch P, Waber U, Haeberli A, et al. Enhanced fibrin formation in high-altitude pulmonary edema. J Apply Physiol 1987; 63:752–757.
67. Swenson E, Mongovin S, Gibbs S, et al. Stress failure in high altitude pulmonary edema (abstr). Am Rev Respir Crit Care Med 2000; 161:A418.
68. Hoon R, Balasubramanian V, Tiwari S, et al. Changes in transthoracic electrical impedance at high altitude. Br Heart J 1977; 39:61–66.
69. Lockhart A, Saiag B. Altitude and human pulmonary circulation. Clin Sci 1981; 60: 599–605.
70. Ahmed T, D'Brot J, Wasserman M, et al. Effect of verapamil on pulmonary and eicosanoid responses to endotoxin in awake sheep. J Appl Physiol 1988; 64(4):1700–1708.
71. Fedorak R, Empey L, Walker K. Verapamil alters eicosanoid synthesis and accelerates healing during experimental colitis in rats. Gastroenterology 1992; 102:1229–1235.
72. Ono S, et al. Endotoxin priming followed by high altitude causes pulmonary edema in rats. J Appl Physiol 1993; 24:1534–1542.
73. Anand I, Prasad B, Chugh S, et al. Effects of inhaled nitric oxide and oxygen in high altitude pulmonary edema. Circulation 1998; 98:2441–2450.
74. Maggiorini M, Melot C, Pierre S, et al. Effects of inhaled nitric oxide and prostaglandin on pulmonary hemodynamics in high altitude pulmonary edema resistant and susceptible climbers. In: Roach R, Wagner P, Hackett P, eds. Hypoxia into the Next Millenium. New York: Kluwer Academic/Plenum Publishers, 1999:408.
75. Hoon R, et al. Effects of high altitude exposure for 10 days on stroke volume and cardiac output. J Appl Physiol 1977; 42:722–727.
76. Theodore J, Robin E. Pathogenesis of neurogenic pulmonary edema. Lancet 1975; 2: 749–751.
77. Hanaoka M, Tanaka M, Ge RL, et al. Hypoxia-induced pulmonary blood redistribution in subjects with a history of high-altitude pulmonary edema. Circulation 2000; 101(12): 1418–1422.
78. Droma Y, Ge R, Tanaka M, et al. Acute hypoxic pulmonary vascular response does not accompany plasma endothelin-1 elevation in subjects susceptible to high altitude pulmonary edema. Intern Med 1996; 35:257–260.
79. Dechaux M, et al. Hypoxia-induced activation of endothelial cells. In: Sutton J, Hous-

ton C, Coates G, eds. Hypoxia and the Brain. Burlington, VT: Queen City Printers, 1995:10.

80. Loffler B, Schirlo C, Davlick V, et al. Impaired clearing function of the lung for endothelin-1 in subjects susceptible to HAPE. In: Roach R, Wagner P, Hackett P, eds. Hypoxia into the Next Millenium. New York: Kluwer Academic/Plenum, 1999:406.
81. Maggiorini M, Zaccaria M, Greve I, et al. Effects of high altitude on serotonin, angiotensin, and endothelin-1 plasma levels in HAPE susceptible and central subjects. In: Roach R, Wagner P, Hackett P, eds. Hypoxia into the Next Millenium. New York: Kluwer Academic/Plenum, 1999:408.
82. Busch T, Swenson E, Gruenig E, et al. Exhaled nitric oxide during normoxia and prolonged hypoxia. Am J Respir Crit Care Med 1998; 157:A228.
83. Droma Y, Hayano T, Takabayashi Y, et al. Endothelin-1 and interleukin-8 in high altitude pulmonary edema. Eur Respir J 1996; 9:1947–1949.
84. Viswanathan R, et al. Pulmonary edema of high altitude I. Production of pulmonary edema in animals under conditions of simulated high altitude. Am Rev Respir Dis 1969; 100:327–333.
85. Viswanathan R, et al. Pulmonary edema of high altitude II. Clinical, aerohemodynamic, and biochemical studies in a group with history of pulmonary edema of high altitude. Am Rev Respir Dis 1969; 100:334–341.
86. Hohenhaus E, et al. Ventilatory and pulmonary vascular response to hypoxia and susceptibility to high altitude pulmonary edema. Eur Respir J 1995; 8:1825–1833.
87. Eldridge M, Podolsky A, Richardson R, et al. Pulmonary hemodynamic response to exercise in subjects with prior high-altitude pulmonary edema. J Appl Physiol 1996; 81:911–921.
88. Younes M, Bshouty Z. Effect of high blood flow, ventilation, breathing pattern and alveolar hypoxia on lung fluid flux. Hypoxia: The Adaptations. Philadelphia: BC Decker, 1990:155–162.
89. Podolsky A, Eldridge M, Richardson R, et al. Exercise-induced VA/Q inequality in subjects with prior high altitude pulmonary edema. J Appl Physiol 1996; 81:922–932.
90. Ritter M, Jenni R, Maggiorini M, et al. Abnormal left ventricular diastolic filling patterns in acute hypoxic pulmonary hypertension at high altitude. Am J Noninvas Cardiol 1993; 7:33–38.
91. Davis K, et al. Myocardial edema, left ventricular function, and pulmonary hypertension. JAP 1995; 78:132–137.
92. Whayne T, Severinghaus J. Experimental hypoxic pulmonary edema in the rat. J Appl Physiol 1968; 25:729–732.
93. Bartlett DJ, Remmers J. Effects on high altitude exposure on the lungs of young rats. Respir Physiol 1971; 13:116–125.
94. Colice G, Lee Y, Chen J, et al. Susceptibility to high altitude pulmonary edema in Madison and Hilltop rats. I Ventilation and fluid balance. J Appl Physiol 1995; 78: 2279–2285.
95. Stelzner T, et al. Hypoxia-induced increases in pulmonary transvascular protein escape in rats. J Clin Invest 1988; 82:1840–1847.
96. Schoene R, Goldberg S. The quest for an animal model of high altitude pulmonary edema. Int J Sports Med 1992; 13:S59–61.
97. West J, Mathieu-Costello O, Jones J, et al. Stress failure of pulmonary capillaries in racehorses with exercise-induced pulmonary hemorrhage. J Appl Physiol 1993; 75: 7097–1109.

98. McKechnie J, Leary W, Noakes T, et al., Acute pulmonary edema in two athletes during a 90-km running race. S Afr Med J 1979; 56:261–265.
99. Everson R, et al. Host determinants of cellular and biochemical constituents of bronchoalveolar lavage fluids. Implications for design of epidemiologic studies. Am J Respir Crit Care Med 1994; 149:899–904.
100. Hopkins S, et al. Intense exercise impairs the integrity of the pulmonary blood-gas barrier in elite athletes. Am J Respir Crit Care Med 1997; 155:1090–1094.
101. Pascoe J, et al. Exercise-induced pulmonary hemorrhage in racing thoroughbreds: a preliminary study. Am J Vet Res 1981; 42:703–707.
102. Whitwell K, Greet T. Collection and evaluation of tracheo-bronchial washes in the horse. Equine Vet J 1984; 16:499–508.
103. King R, Raskin R, Rosbelt J. Exercise-induced pulmonary hemorrhage in the racing greyhound dog. J Vet Int Med 1990; 4:130.
104. Hultgren H. Pulmonary hypertension and pulmonary edema. In: Loeppky J, Riedesel M, eds. Oxygen Transport to Human Tissue. New York: Elsevier/North Holland, 1982: 243–254.
105. Mitzner W, Sylvester J. Hypoxic vasoconstriction and fluid filtration in pig lungs. J Appl Physiol 1981; 51:1065–1071.
106. Ogawa S, Koga S, et al. Hypoxia-induced increased permeability of endothelial monolayers occurs through lowering of cellular cAMP levels. Am J Physiol 1992; 262(31): C546–C554.
107. Landolt C, Matthay M, Albertine K, et al. Overperfusion, hypoxia and increased pressure cause only hydrostatic pulmonary edema in anesthetized sheep. Circ Res 1983; 52:335–341.
108. Bland R, et al. Effects of alveolar hypoxia on lung fluid and protein transport in unanesthetized sheep. Circ Res 1977; 40:269–274.
109. Flick M, Perel A, Staub N. Leukocytes are required for increased lung microvascular permeability after microembolization in sheep. Circ Res 1981; 48:344–351.
110. Dawson C, et al. Influence of pulmonary vasoconstriction on lung water and perfusion heterogeneity. J Appl Physiol 1983; 54:654–660.
111. Oyamada Y, Mori M, Kuwahira I, et al. Effects of active vasoconstriction and total blood flow on perfusion distribution in the rabbit lung. Am J Physiol 1997; 273: R1465–1473.
112. Melot C, Naeije R, Hallemans R, et al. Hypoxic pulmonary vasoconstriction and pulmonary gas exchange in normal man. Respir Physiol 1987; 68:11–27.
113. Gales G, Torre-Bueno J, Moon R, et al. Ventilation-perfusion inequality in normal humans during exercise at sea level and simulated altitude. J Appl Physiol 1985; 58: 978–988.
114. Viswanathan R, Subramanian S, Radha T. Effect of hypoxia on regional lung perfusion, by scanning. Respiration 1979; 37:142–147.
115. Elser H, Swenson E, Hildebrandt J, et al. Regional distribution of pulmonary perfusion after five hours of normobaric hypoxia in subjects susceptible to high altitude pulmonary edema. Eur Respir J 1998; 12:A2328.
116. West J, et al. Stress failure in pulmonary capillaries. J Appl Physiol 1991; 70:1731–1732.
117. Tsukimoto K, Mathieu-Costello O, Prediletto R, et al. Ultrastructural appearances of pulmonary capillaries at high transmural pressures. J Appl Physiol 1991; 71:573–582.

118. West J, Mathieu-Costello O. High altitude pulmonary edema is caused by stress failure of pulmonary capillaries. Int J Sports Med 1992; 13:S54–58.
119. West J, Mathieu Costello O. Stress failure of pulmonary capillaries in the intensive care setting. Schweiz Med Wochenschr 1992; 122:751–757.
120. West J, Mathieu-Costello O. Pulmonary blood-gas barrier: a physiological dilemma. NIPS 1993; 8:249–253.
121. West J, Mathieu-Costello O. Stress failure of pulmonary capillaries as a limiting factor for maximal exercise. Eur J Appl Physiol 1995; 70:99–108.
122. Bachofen H, et al. Experimental hydrostatic pulmonary edema in rabbit lungs: morphology. Am Rev Respir Dis 1993; 147:989–996.
123. Tsukimoto K, Yoshimura N, Ichioka M. Protein, cell, and LTB4 concentrations of lung edema fluid produced by high capillary pressure in rabbit. J Appl Physiol 1994; 76:321–327.
124. Kubo K, Hanaoka M, Yamaguchi S, et al. Cytokines in bronchoalveolar lavage fluid in patients with high altitude pulmonary edema at moderate altitude in Japan. Thorax 1996; 51:739–742.
125. Kleger G, Bartsch P, Vock P, et al. Evidence against an increase in capillary permeability in subjects exposed to high altitude. J Appl Physiol 1996; 81:1917–1923.
126. Kaminsky D, et al. Urinary leukotriene e (4) levels in high altitude pulmonary edema: a possible role for inflammation. Chest 1996; 110:939–945.
127. Durmowicz A, et al. Inflammatory processes may predispose children to high-altitude pulmonary edema. J Pediatr 1997; 130:838–840.
128. Shreeniwas R, Koga S, Karakurum M, et al. Hypoxia-mediated induction of endothelial cell interleukin-1. J Clin Invest 1992; 90:2333–2339.
129. Ertel W, et al. Hypoxemia in the absence of blood loss or significant hypotension causes inflammatory cytokine release. Am J Physiol 1995; 269:R160–R166.
130. Zund G, Dzus A, McGuirk D, et al. Hypoxic stress alone does not modulate endothelial surface expression of bovine E-selectin and intercellular adhesion molecule-1 (ICAM-1). Swiss Surg Suppl 1996; 1:41–45.
131. Zund G, Nelson D, Neufield E, et al. Hypoxia enhances stimulus-dependent induction of E-selectin on aortic entothelial cells. Proc Natl Acad Sci USA 1996; 93:7075–7080.
132. Grissom C, Whatley R, Zimmerman G. P-selectin, E-selectin, and von Willebrand factor in acute mountain sickness and high altitude. Am J Respir Crit Care Med 1996; 153:A195.
133. Wood J, Mattioli L, Gonzalez N. Hypoxia causes leukocyte adherence to mesenteric venules in non-acclimatized, but not in acclimatized rats. J Appl Physiol 1999; 87: 873–881.
134. Hanaoka M, Kubo K, Yamazaki Y, et al. Association of high altitude pulmonary edema with major histocompatibility complex. Circulation 1998; 97:1124–1128.
135. Arnett F. High-altitude pulmonary edema—an immunogenetically mediated disease? Am Heart Assoc 1998; 97:1111–1113.
136. Bartsch P, Lammie B, Huber I, et al. Contact phase of blood coagulation is not activated in edema of high altitude. J Appl Physiol 1989; 67:1336–1340.
137. Sartori C, Lepori M, Maggiorini M, et al. Impairment of amiloride-sensitive sodium transport in individuals susceptible to HAPE. In: Roach R, Wagner P, Hackett P, eds. Hypoxia into the Next Millenium. New York: Kluwer Academic/Plenum, 1999:426.
138. Lepori M, Hummler E, Feihl F, et al. Amiloride-sensitive sodium transport dysfunction augments susceptibility to hypoxia-induced lung edema. In: Roach R, Wagner P, Hack-

ett P, eds. Hypoxia into the Next Millenium. New York: Kluwer Academic/Plenum, 1999:403.

139. Matthay M, Wiener-Kronish J. Intact epithelial barrier function is critical for the resolution of alveolar edema in humans. Am Rev Respir Dis, 1990; 142:1250–1257.
140. Mairbaurl H, Wodopia R, Eckes S, et al. Impairment of cation transport in A549 cells and rat alveolar cells by hypoxia. Am J Physiol 1997; 273:L797–L800.
141. Planes C, et al. Inhibition of Na-K-ATPase activity after prolonged hypoxia in an alveolar epithelial cell line. Am J Physiol 1996; 271:L70–L78.
142. Planes C, et al. Hypoxia downregulates expression and activity of epithelial sodium channels in rat alveolar epithelial cells. Am J Respir Cell Mol Biol 1997; 17:508–518.
143. Suzuki S, Noda M, Sugita M, et al. Impairment of trans-alveolar fluid transport and lung Na^+-K^+-ATPase function by hypoxia in rats. J Appl Physiol 1999; 87:963–968.
144. Hackett PH, Roach RC. High altitude medicine. In: Auerbach PS, ed. Wilderness Medicine. 3rd ed. Boston: Mosby Press, 1995:1–37.

24

Chronic Mountain Sickness in Andeans

CARLOS C. MONGE, FABIOLA LEÓN-VELARDE, and ALBERTO ARREGUI

Cayetano Heredia University
Lima, Peru

I. Introduction

Chronic mountain sickness (CMS) afflicts people who are native or long-time residents of high altitude. It is characterized by excessive erythrocytosis. It can be classified as primary (without identified cause) or secondary (due to underlying conditions). CMS usually begins insidiously in adult life, often during the fourth decade, as progressive hypoxemia stimulates erythrocytosis. The clinical picture disappears when the patient moves to lower altitudes. Therefore, the primary etiology of CMS is presumed to be hypoxemia.

Although the clinical and laboratory findings in CMS represent a continuum of variation from normal (for the altitude) to severe, patients who show only mild disturbances and symptoms or who develop symptoms of congestive heart failure after a relatively short exposure to altitude (1) are sometimes said to suffer with "subacute mountain sickness." This entity, more often described in the Himalayas than in the Andes, may be a variant of CMS, but this classification must be considered tentative.

This chapter presents the findings on CMS that have been a matter of research mainly in the Andes. However, allusions to research in other mountainous areas of the world extend and illustrate the concepts and factors involved in the development of CMS. The chapter will cover a brief historical background of CMS, its clinical and epidemiological expressions, its physiological mechanisms and pathophysiology, the organ effects of hypoxia, the prevention and treatment of CMS, and its biological basis.

The clinical and epidemiological aspects of CMS include the clinical description of the disease, symptoms, signs, laboratory findings, and the geographical distribution and altitude relationships. This section also examines the physiological mechanisms by which other pathologies, as well as age and gender-specific factors are likely to influence the development of CMS. We also present an integrative approach encompassing the clinical, pathophysiological, and epidemiological aspects of the disease. The section on mechanisms and pathophysiology describes the evolution of erythrocytosis and the role of the ventilatory function in the development of hypoxemia. In the section on organ effects of hypoxia, we examine how hypoxia and excessive erythrocytosis affect the plasma volume, the pulmonary hemodynamics, and kidney and endocrine function. The section on prevention and treatment discusses the different treatments available and future therapeutic possibilities. Finally, the section that presents the biological basis of CMS deals with the genetic adaptation in high-altitude native animals and puts CMS in the context of comparative pathophysiology.

II. Brief Historical Background

Carlos Monge presented the first description of CMS in 1925 at the Academy of Medicine of Lima (2). The patient was a native of the mining town of Cerro de Pasco located at 4300 m in the Peruvian Andes. He worked as an office employee, not in the mining environment. His symptoms disappeared with descent to Lima (sea level) but recurred upon returning to 4300 m. In 1928 Monge published an extensive article in Spanish (3) on cases of CMS from Cerro de Pasco and Puno (3800 m), an agricultural town. The purpose of this study was to show that CMS could occur in people not exposed to environmental contamination. The dean of the Medical School of Paris suggested the name of Monge's disease for this clinical picture. The first case of Monge's disease reported in the international scientific literature was in 1936 (4). Monge considered CMS to be a ''loss of acclimatization'' because it developed only after prolonged exposure to altitude in previously well-acclimatized subjects.

III. Clinical and Epidemiological Aspects

A. Symptoms and Signs of CMS

The most common symptoms of CMS are headaches, dizziness, dyspnea, sleep disturbances (insomnia, hypersomnia), tinnitus, physical and mental fatigue, alterations of memory, loss of appetite, and bone and muscle pains. Patients with CMS usually have signs of intermittent or permanent cyanosis, venous dilatation in the hands and feet, and clubbing of the fingers and toes. In advanced stages of the disease right heart failure secondary to excessive pulmonary hypertension can appear. The laboratory findings show hemoglobin concentration, hematocrit, and red cell counts above the normal level for the altitude of residence. The arterial O_2 saturation is lower and CO_2 partial pressure ($Paco_2$) higher than normal for the altitude. The relatively high $Paco_2$, an indication of chronic hypoventilation, is not necessarily the result of a blunted hypoxic ventilatory response. The pH is slightly elevated, and the bicarbonate concentration is higher than the normal for the altitude, indicating mild respiratory alkalosis with partial renal compensation (Table 1).

B. Geographical Distribution and Altitude Relationship

CMS is more common in Andean natives than in Tibetans. Contrary to what occurs in the Andes, subacute mountain sickness (called CMS by Chinese doctors) is more common in Han (low-altitude native) Chinese sojourners than in Tibetan (high-altitude native) residents (5,6). The disease can be considered as a loss of acclimatization in natives (either Tibetan or Andean) and as a lack of acclimatization in the Han Chinese sojourners. Cases of subacute mountain sickness may have a different etiological and, perhaps, racial basis than the typical cases of CMS.

Erythrocytosis, the main pathophysiological sign of CMS, can be excessive even at moderate altitudes. Because this concept is central for the understanding of the development of the disease in different parts of the world, it is important to analyze it from a physiological point of view. In the well-acclimatized individual the arterial O_2 content (Cao_2), the product of hemoglobin concentration and saturation, increases exponentially with altitude (7), while the decrease in Po_2 is almost linear, suggesting an overcompensation by the erythropoietic system.

C. CMS and Age

Ventilation is the first step in oxygen transport from air to mitochondria. In mammals and birds that respond to hypoxemia with erythrocytosis, decreased ventilation in-

Table 1 Blood Parameters in High-Altitude Normal Andeans and Andeans with CMS

Blood	Natives, 4540 m	CMS, 4540 m
RBC, cells $\times 10^6/mm^3$	6.14	6.54–10.0
Hemoglobin, g/dL	20.8	20.8–28.4
Hematocrit, %	59.9	55.0–93.8
Sao_2, %	81.4	59.6–90.0
Pco_2, torr	32.5	35.0–45.6
HCO_3^-, mmol/L	20.9	23.4–28.4
pH	7.431	7.393–7.457

Source: Ref. 28.

creases red blood cell number, hematocrit, and hemoglobin concentration. With normal aging, Pa_{O_2} decreases approximately linearly (8,9). The Pa_{O_2} diminishes with age from approximately 95 torr at the age of 20 to about 75 torr at the age of 75 (9). The hemoglobin-oxygen equilibrium curve (OEC) is sigmoid in shape and varies little with altitude (10). Thus, at sea level, a saturation of 85% or more falls on a flat part of the OEC such that small changes in Pa_{O_2} have little effect on saturation, which is protected with age. However, at altitude, where Pa_{O_2} is lower, small changes in P_{O_2} can have large effects on saturation, and the same changes that accompany aging may have significant effects on arterial hemoglobin saturation and, therefore, Ca_{O_2}.

Monge and Whittembury have shown, by combining the equation describing the Pa_{O_2} drop with age with the one describing the hemoglobin or hematocrit response to Pa_{O_2} in high-altitude natives, that the increase in hematocrit at high altitude as a function of age can be predicted (7). The predicted equation fits the increase in hematocrit described by Sime in high-altitude natives at 4500 m (11,12). Based on the above physiological observations and recent epidemiological studies (13,14), we have proposed that at high altitude, as age increases, CMS will increase its prevalence. Any additional impairment of pulmonary function will accentuate this physiological erythrocytosis.

Figure 1 shows indirectly that age is an important contributory factor for the development of excessive erythrocytosis and CMS. It confirms previous studies

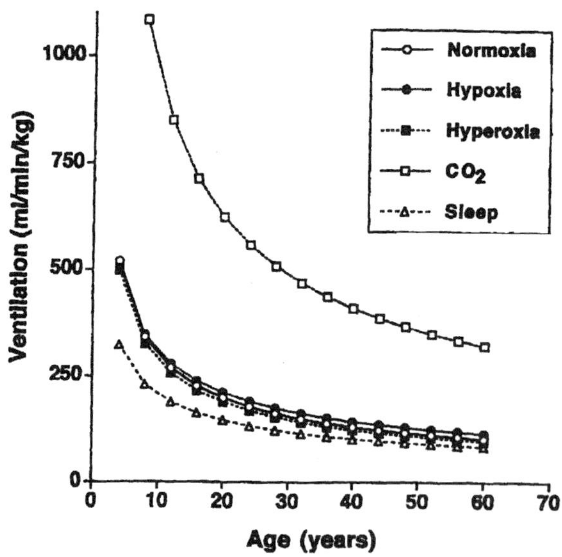

Figure 1 Age-related ventilation (V_e) in high-altitude natives residing at 4500 m. V_e was measured under conditions of normoxia, hypoxia, hyperoxia, with CO_2 added to the inspired air, and during sleep. (From Ref. 11.)

showing that the ventilation of Andean high-altitude natives responds poorly to acute hypoxia and hyperoxia, but does respond to CO_2 added to the inspired air. However, all these responses fall with age, and, therefore, a robust response to CO_2 does not mean that the ventilation will be independent of age.

As a confirmation that ventilation diminishes with age, Arai et al. (15) have shown that also the $Paco_2$ increases with age at an altitude between 3450 and 3850 m. Additionally, vital capacity of high-altitude men decreases more with age than that of sea level men, together with a decrease in oxygen saturation (16). Some authors, however, have not been able to find an association between hematocrit and hemoglobin with age and have rejected the hypothesis that age is a risk factor for CMS (17–20). Unfortunately, while some of them did longitudinal studies relating hematocrit and age, they used a small number of subjects, making the results difficult to interpret (17,18). Conflicting views on this interesting question need to be resolved by further studies.

Work carried out in La Paz, Bolivia, (19) has also found no association between age and CMS. This is not surprising since, according to Whittembury and Monge (21), at an altitude of 3600 m the rise of hematocrit with age should be very small. In fact, hemoglobin concentration at high altitude is so variable that at moderate altitudes it is difficult to obtain a high correlation of hematocrit with age. The comparison of prevalences of people with CMS at different ages should be more likely to reveal the effect of aging on the prevalence of CMS in a certain population.

D. CMS in Women

Until recently CMS was considered an almost exclusively male malady, since little is known about its prevalence and associated risk factors in women. Premenstrual women have been thought to be protected from CMS because the female hormones, progestin and estrogen, increase alveolar ventilation and the hypoxic ventilatory response (22). In addition, having low concentrations of male hormones may lessen erythropoiesis (23). Menstruation might act like phlebotomy to protect premenopausal women from developing excessive erythrocytosis. Unfortunately, no data have been available to shed light on these possibilities.

León-Velarde et al. (24) measured hematocrit, Sao_2 and peak expiratory flow rates (PEFR) in pre- and postmenopausal high-altitude women. After menopause women have higher hematocrit, lower Sao_2, and lower PEFR. In addition, postmenopausal women have a significantly higher frequency of symptoms associated with CMS. Progesterone increases alveolar ventilation during pregnancy and the luteal phase of the menstrual cycle, and this effect is enhanced when combined with estrogen (25,26). In rats, female sex hormones suppress the erythremic and cardiopulmonary responses seen during chronic exposure to hypoxia (23). The postmenopausal decrease of hormones could depress alveolar ventilation and Pao_2, stimulate erythropoiesis, increase viscosity, decrease tissue perfusion, and lead to further erythropoiesis and, ultimately, CMS.

Santolaya et al. (27) measured Pa_{O_2} and Pa_{CO_2} in 162 women living at 2800 m. In those 55 years of age or older, Pa_{O_2} values were below those of age-matched men living at the same altitude. Arterial P_{CO_2} showed a sharp rise after age 40, reaching values similar to or greater than those of men. Although these authors did not relate their findings to the appearance of menopause, their data coincide well with our findings related to the increase of hematocrit with menopause in high-altitude women.

The above observations support the notion that ventilatory drive in women has an important hormonal component and that this stimulation decreases after menopause. Interruption of the hormonal dependence of ventilation would be expected to elevate the hematocrit and to provoke the appearance of signs and symptoms of CMS in postmenopausal women living at high altitude.

E. Secondary CMS

CMS refers to the presence of excessive erythrocytosis in normal natives who are assumed to have normal respiratory function. When there is an obvious respiratory disease (or another underlying condition) that triggers the excessive erythrocytosis, the disease can be called secondary CMS (28). However, it seems difficult to separate cases of CMS with a purely diminished ventilation from those having additional pulmonary dysfunction, which aggravates the condition. It has been shown that even common chronic lower respiratory tract disorders may be additional risk factors for CMS (29).

The blood oxygen desaturation produced during sleep at high altitude has been considered an additional factor that contributes to excessive erythrocytosis (30,31). Figure 1 shows that during sleep ventilation falls below that of awake high-altitude natives in the younger age group. However, as age increases, the difference between asleep and awake ventilation disappears (11,12).

In addition to the role played by the age-related decrease in ventilation in the etiology of CMS (12), other additional factors may be involved in the development of CMS: an increase in the alveolar-arterial O_2 gradient with possible venous-arterial shunting and a reduction of the ventilation-perfusion ratio (32–34), and obesity (13,19) may aggravate the hypoxemia.

F. Epidemiological Studies

We have surveyed adult men living in Cerro de Pasco (Central Andes, Peru, total population: 70,000; altitude: 4300 m). In the population studied, 84% were natives of the mining town and the rest had lived there for more than 10 years. Most men answered questions related to the presence of symptoms and signs associated with CMS [from which we developed a CMS score (13,29)] and had Sa_{O_2}, hemoglobin, and PEFR measured. The analysis of the epidemiological data is based on the assumption that the average hemoglobin concentration of acclimatized young high-altitude populations defines normality and that values more than two standard deviations above the average for the altitude of residence are considered excessive. If symptoms appear, CMS is diagnosed based on the correlation between the appearance of excessive erythrocytosis and the variables involved in the pathophysiological sequence that leads to CMS (14,16). With this criterion, the prevalence of CMS in Peruvian Quechuas at Cerro de Pasco has been found to be 15.6%. In addition, we also showed that there are age-related increases in the frequency of men with excessive erythrocytosis (hemoglobin > 21.3 g/dL), $Sa_{O_2} < 81\%$, PEFR < 276 L/min/m, and high CMS scores (13,48).

These studies also show that the average hemoglobin for men with chronic pulmonary diseases (20.2 g/dL) is significantly higher than that of men without respiratory complaints (18.8 g/dL) (29). Among the former the frequency of excessive erythrocytosis is 32.4% compared to 11.3% among men without respiratory problems, thus providing evidence that having chronic lower respiratory diseases aggravates excessive erythrocytosis (29).

Epidemiological studies suggest the identification of risk factors for development of CMS. For example, men with high hemoglobin values (hemoglobin > 21.3 g/dL) have lower Sa_{O_2} and higher body weights than men with normal hemoglobin for the altitude of residence or men with lower hemoglobin (Table 2). Men who have high CMS scores (Table 3) have higher hemoglobin, lower Sa_{O_2}, and lower PEFR than men with intermediate or low CMS scores. It thus appears that risk factors for development of CMS include age, obesity, low Sa_{O_2}, and reduced PEFR.

The epidemiological data in Tibet reveal considerable variation between world regions in the prevalence of CMS. The prevalence of CMS in natives of Qinghai, Tibetan plateau, is 1.21% as compared with 5.59% in Han immigrants. Using the same criteria as Leon-Velarde et al. (29), and at a comparable altitude, a CMS prevalence of 0.91% of CMS was found in the Tibetans (20). These data suggest that Tibetans might be protected from CMS due to genetic factors. In fact, several physiological differences in their capacity to acclimatize to life at high altitude, when compared with Andeans, support this proposition (see Ref. 37 for review; see also

Table 2 Age, Oxygen Saturation (Sa_{O_2}), Body Weight, Peak Expiratory Flow Rates (PEFR), and Chronic Mountain Sickness (CMS) Scores Among High-Altitude Men with Different Hemoglobin Concentrations

	Hemoglobin > 21.3	Hemoglobin (*n*) 17.0–21.3	Hemoglobin < 17.0 g/dL
Hemoglobin (g/dL)	23.6 ± 1.1 (39)	19.1 ± 1.3 (265)	15.8 ± 1.1 (72)
Age (years)	41.3 ± 10.7 (39)	42.3 ± 11.6 (265)	42.1 ± 10.4 (72)
Sa_{O_2} (%)	81.7 ± 4.3[bc] (39)	84.4 ± 4.3 (264)	85.5 ± 4.9 (71)
Body weight (kg)	67.9 ± 7.3[bc] (37)	63.9 ± 7.3[b] (263)	61.2 ± 8.6 (67)
PEFR (L/min/m)	335.3 ± 83.4 (25)	325.1 ± 57.1 (151)	327.1 ± 56.7 (39)
CMS score[a]	7.1 ± 3.8 (39)	6.3 ± 3.7 (265)	6.5 ± 3.4 (72)

[a] Developed from 10 symptoms and signs usually associated with chronic mountain sickness (13,29).
[b] $p < 0.05$ when compared to hemoglobin < 17.0 g/dL.
[c] $p < 0.05$ when compared to hemoglobin 17.0–21.3 g/dL.

Table 3 Age, Hemoglobin, Body Weight, Oxygen Saturation (Sa_{O_2}), and Peak Expiratory Flow Rates (PEFR) in High-Altitude Men with Low, Intermediate, and High Chronic Mountain Sickness Scores

	Chronic mountain sickness scores (*n*)		
	Low	Intermediate	High
Age (years)	39.1 ± 10.3 (160)	42.4 ± 11.5[a] (242)	47.4 ± 11.4[ab] (37)
Hemoglobin (g/dL)	18.6 ± 2.2 (123)	18.9 ± 2.4 (217)	19.9 ± 2.2[ab] (36)
Body weight (kg)	63.3 ± 7.6 (122)	63.9 ± 7.8 (212)	64.5 ± 8.6 (35)
Sa_{O_2} (%)	84.5 ± 4.3 (123)	84.7 ± 4.3 (218)	82.6 ± 5.5[ab] (36)
PEFR (L/min/m)	334.6 ± 65.9 (54)	332.0 ± 51.3 (147)	285.3 ± 70.5[ab] (30)

[a] $p < 0.05$ when compared to low.
[b] $p < 0.05$ when compared to intermediate.

Chapter 3). However, it is noteworthy that although there is a very low prevalence of CMS, the disease does exist in the Tibetan population.

IV. Mechanisms of CMS

A. Ventilatory Function

Because high-altitude natives have a reduced hypoxic ventilatory response, it is tempting to propose that this may be the cause of CMS. Although reduced hypoxic ventilatory response may be contributory, evidence is contradictory. For example, Kryger and Grover (38) did not find differences between the hypoxic ventilatory responses of healthy high-altitude residents and patients with CMS studied in Leadville, Colorado (3100 m). However, Severinghaus et al. (39) found that the ventilatory response to acute hypoxia, already low in healthy high-altitude natives, was even lower in those with CMS. Because the response to CO_2 was normal in both groups, they considered that the desensitization of the carotid bodies was the etiology of CMS. The data of Vargas and Villena support this hypothesis (19,34).

Against the role of the hypoxic ventilatory response in causing CMS, Sime et al. (11,12) showed at 4500 m a lack of ventilatory response even in healthy high-altitude natives from ages 4 to 60. Sørensen and Severinghaus (40) also pointed out that during the first 2 years of life, chronic hypoxia desensitizes irreversibly the reflex response to acute hypoxia mediated by peripheral chemoreceptors. Lahiri et al. (41) found a blunted response to hypoxia in 40 adult high-altitude natives over the age of 22. Since cases of CMS are more frequent after the age of 40 years, the etiological role of a diminished ventilatory response to hypoxia should be accepted with caution. However, Arias-Stella and Valcárcel (42) showed that in high altitude natives the frequency of carotid body hyperplasia increases with age and related these changes to the functional ventilatory decay with age described by Sime et al. (12).

Although it is difficult to separate the possible role played by the carotid bodies from that played by the central brain chemoreceptors in the genesis of CMS, it is clear that the aging process seems to be involved in the functional and anatomical changes found in high altitude natives. Finally, while a low hypoxic ventilatory response may be a risk factor in the development of CMS, it is not as important an etiologic factor as was once presumed.

B. Hypoxemic Erythrocytosis and CMS

Erythrocytosis, and the consequent rise in hemoglobin concentration, increases the O_2 content of blood but not necessarily the O_2 delivery to the tissues. Oxygen is transported from the tissue capillaries to the mitochondria by diffusion down a Po_2 gradient. At rest, the total blood flow (cardiac output), the oxygen consumption and the oxygen-hemoglobin affinity of high altitude Andeans is similar to sea level natives.

Tenney (43), using a mathematical model, concluded that an Sao_2 of 70% corresponds to a maximal altitude at which humans can make permanent residence. Since at that altitude his model predicts a hemoglobin of 25 g/dL (hematocrit 75%), we believe that the limiting factor is not the Sao_2 but excessive erythrocytosis. Moreover, the optimal hematocrit, above and below which Pvo_2 declines, is close to the normal hematocrit (44,45). Figure 2 shows that high-altitude natives with normal erythrocytosis exhibit marked variability in Sao_2 and that the values are much above 70%. In subjects with excessive erythrocytosis, the average Sao_2 is also much higher than 70%. The shape of the oxygen equilibrium curve of hemoglobin favors the

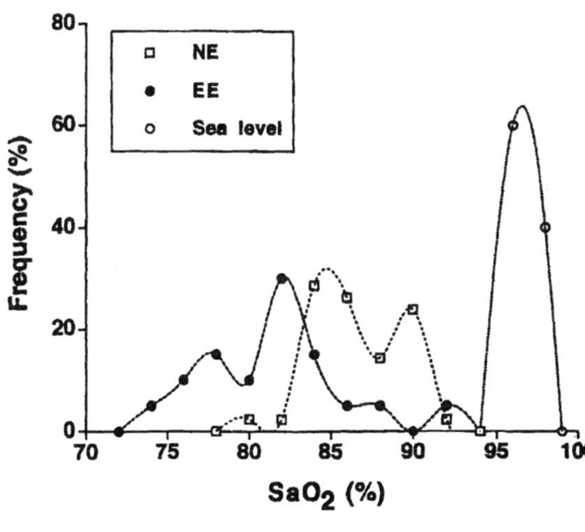

Figure 2 Distribution curves of arterial O_2 saturation (Sao_2) in high-altitude natives with excessive erythrocytosis (EE), normal erythrocytosis (NE), and in sea level natives (SL). (From Ref. 39.)

scattering of the high-altitude data, because at reduced Sao_2, small changes in Po_2 result in large changes in Sao_2. It is clear that the situation is too complex for a simple model.

A more comprehensive mathematical model allows the analysis of the relationship of hemoglobin to the Pvo_2 (7,46,47). The model combines several equations including the Hill equation, the convection equation and the equation relating hemoglobin and Pao_2 at different altitudes. The model fixes the invariant parameters and because the hemoglobin concentration as a function of Pao_2 at different altitudes is known in Peruvian high-altitude natives, the model permits the study of the effect of hemoglobin and/or Pao_2 on the Pvo_2. Figure 3 has been constructed using the model and plotting Pvo_2 as a function of the average hemoglobin found in high-altitude natives at different altitudes. The Pvo_2 varies little until hemoglobin values exceed 17 g/dL (corresponding to an altitude of 3200 m) and then declines despite a continuous elevation of hemoglobin. It could be argued that for a fixed level of Pao_2, an increase of hemoglobin above the native's normal hemoglobin (excessive erythrocytosis) could help maintain the Pvo_2 at higher levels and be adaptive. The fact is, however, that there is an inverse relationship between Pao_2 and hemoglobin, and an increase in hemoglobin is the result of excessive hypoxemia, which will keep the Pvo_2 low even in the presence of high hemoglobin levels (46,47). Figure 3 also

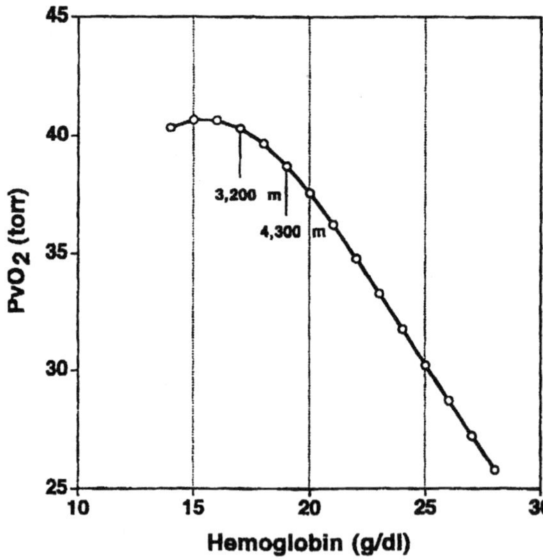

Figure 3 Central venous O_2 pressure (Pvo_2) as a function of hemoglobin concentration at different altitudes. Pvo_2 remains at sea level values up to an altitude of about 3200 m and a hemoglobin close to 17 g/dL. At higher altitudes the Pvo_2 declines despite an increase in hemoglobin. At 4300 m the hemoglobin is close to 19 g/dL but the Pvo_2 is low. (From Ref. 45.)

shows that an altitude of 4300 m corresponds to a hemoglobin of 19 g/dL, which is the average hemoglobin found in the Cerro de Pasco (4300 m) population (29,48,49). At this altitude polycythemia becomes ineffective in adaptation in Andeans.

C. Erythropoietin and Excessive Erythrocytosis

The excessive erythrocytosis seen in patients with CMS has made several researchers suspect that the erythropoietin (EPO) axis may be implicated in the pathogenesis of the disease. The findings, however, are not convincing and more work needs to be done to test the hypothesis of a primary abnormality of the EPO response as a cause of CMS. The data seem to indicate that hematocrit and hemoglobin levels correlate with circulating EPO levels. Excessive erythrocytosis, which arises from the hypoxic stimulus, can become self-perpetuating if decreased flow resulting from increased viscosity causes further hypoxia and the subsequent production of EPO. It seems that at high altitude, the simple negative feedback model, which operates at sea level, has been reversed to a positive feedback mechanism that counteracts the hemoglobin production arrest. The EPO concentration in the blood of high-altitude natives when compared with sea level controls has been found elevated by several workers (50–53). Furthermore, the higher titers of serum levels of EPO in

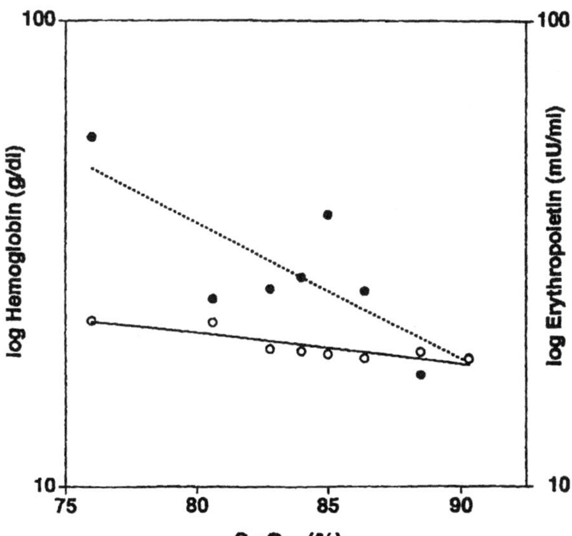

Figure 4 Log of hemoglobin concentration (○) and log of erythropoietin concentration (●) are plotted against different levels of arterial blood O_2 saturation (Sao_2). The points represent averages of several determinations. The linear correlation coefficients are $r = 0.875$ and $r = 0.81$, respectively, and are significant. (From Ref. 51.)

the presence of excessive erythrocytosis is an indication of the failure of erythrocytosis to prevent tissue hypoxia at high altitude.

Figure 4 shows the correlation that exists between Sao_2 and both hemoglobin and EPO. In order to compare the slopes of functional relationships for a common independent variable (Sao_2) but with different units of the dependent variables (hemoglobin and EPO), a semi-logarithmic plot is used. The plot shows that EPO response is disproportionate to hemoglobin response. This is why it seems that EPO response does not maintain a negative feedback with erythrocytosis. This would imply higher EPO production as hemoglobin rises, rather than the contrary. The EPO concentration shows, however, more dispersion and a lower correlation, which agrees with the concept of a more complicated degree of regulation between Sao_2 and EPO than between Sao_2 and hemoglobin (54,55).

Based on data of two subjects having elevated EPO, Dainiak et al. (50) suggested that two subgroups of CMS may exist: those with elevated EPO, and those with ''normal'' EPO levels. Winslow et al. (53) and León-Velarde et al. (51) also found one and two subjects among their samples with markedly elevated EPO levels, but they postulated that these overresponders might represent the extremes of the reversed feedback mechanisms relating hypoxia, EPO production, and excessive erythrocytosis. (See also Refs. 52 and 55 for reviews.)

V. Organ Effects

A. Plasma Volume and Erythrocytosis

The erythrocytosis of high altitude results in elevated red cell volume and hematocrit, but plasma volume is apparently contracted (44,56). This contraction produces a prolonged plasma circulation time (44), and, in the case of the kidney, there is a negative correlation between hematocrit and kidney plasma flow (57). High-altitude natives with high hematocrit may have a plasma flow reduced to one-third that found at sea level (58).

Sánchez et al. (58) measured body hematocrit in sea level and high-altitude natives using ^{55}Cr-labeled red cells and determined simultaneous plasma volume using Evans blue. They reported a modest correlation between peripheral hematocrit and plasma volume in high-altitude natives. Using their data we have now studied the correlation between plasma volume and central (instead of peripheral) hematocrit. Figure 5 shows that the correlation is very good in high-altitude natives (r = 0.939); the same applies to sea level subjects. This finding reinforces our view that the reduction of plasma volume is functionally related to the normal or excessive increase in hematocrit.

Comparing Figure 5 with Figure 6 it may be seen that for the same range of hematocrits the drop in plasma volume is of the same magnitude as the drop in kidney plasma flow. It would be difficult to explain this marked reduction in kidney plasma flow without a reduction in plasma volume. What, then, is the physiological significance of a contracted plasma volume and prolonged plasma circulation time?

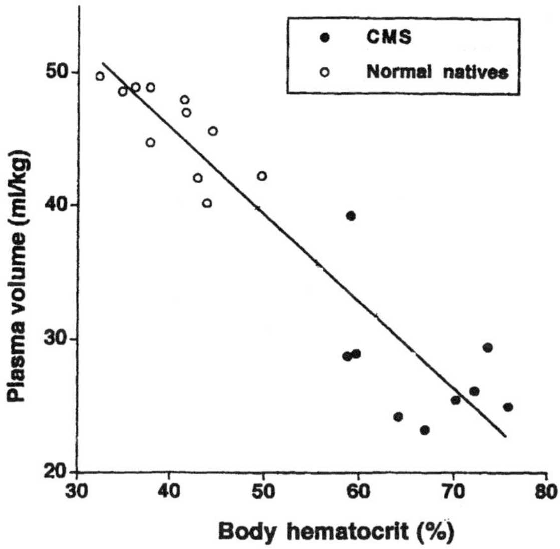

Figure 5 Blood plasma volume as a function of the ratio red cell mass/plasma volume (body hematocrit) in normal high-altitude natives (○) and patients with chronic mountain sickness (CMS) (●). At very high body hematocrits the plasma volume may be close to 50% of the sea level value. (From Ref. 58.)

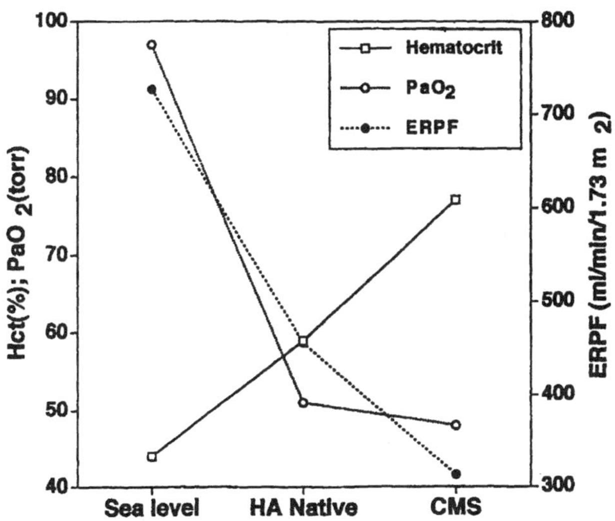

Figure 6 The graph contains data on sea level (SL), high-altitude (HA), and HA natives with CMS. The inverse relationship of the hematocrit with the Pa_{O_2} and with the effective renal plasma flow (ERPF) is evident. (From Ref. 28.)

If plasma solutes normally cleared in the kidney were to keep the clearance rates unmodified in the presence of contracted plasma volume, then their plasma concentration should increase. This does not happen, leaving in question how the kidney adapts to reduced plasma flow.

B. Pulmonary Hemodynamics

The high-altitude native has, in resting conditions, normal blood flow and increased mean pulmonary artery pressure (MPAP). The systemic arterial pressure is usually slightly below that of sea level controls. The blood viscosity is high (21), and in CMS, MPAP is even more elevated (34,59); we might consider a population with excessive pulmonary hypertension much as we have done in the case of excessive erythrocytosis. Figure 7 shows MPAP in CMS and normal subjects from Cerro de Pasco (4300 m) plotted as a function of age. Although the two populations are not matched for age, there is a clear correlation between MPAP and age in these subjects with CMS.

Peñaloza and Sime (59) considered their patients with CMS as special cases of high-altitude cor pulmonale. We prefer to regard these individuals as having excessive MPAP leading to circulatory insufficiency during the development of CMS. Like excessive erythrocytosis, the pulmonary hypertension seen in CMS is excessive compared with the pressures found in normal natives, which is the result of hypoxic

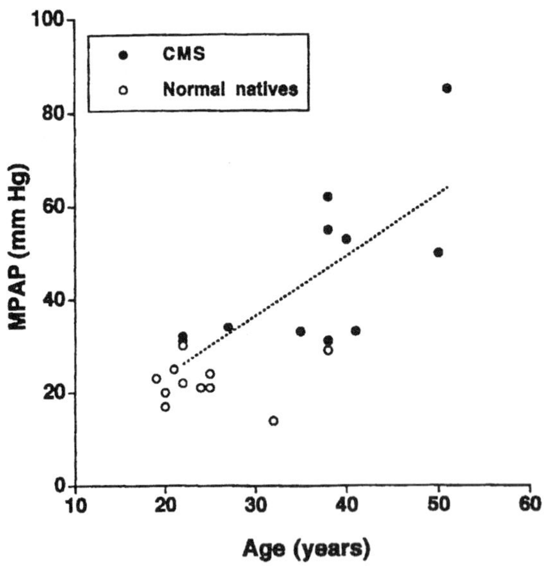

Figure 7 Age-related increments of mean pulmonary artery pressure (MPAP). (From Ref. 59.)

vasoconstriction of the pulmonary tree. Pulmonary hypertension may be associated with and/or aggravated by excessive erythrocytosis because of the increased blood viscosity, but in some cases excessive MPAP can be found with no excessive erythrocytosis. This is not surprising since the large variations in each condition do not necessarily coincide, as happens in other clinical conditions at sea level.

The increased blood viscosity due to the high hematocrit may contribute to pulmonary hypertension by increasing pulmonary capillary resistance. However, pulmonary hypertension also occurs in response to hypoxia in the absence of excessive erythrocytosis (80). In order to compare the increase in MPAP with changes in hematocrit and viscosity, we have plotted the age-related increase of the logarithms of these parameters (Fig. 8). It can be seen that when the viscosity is plotted as a function of age, MPAP has a steeper slope than the corresponding slope for hematocrit and viscosity. We believe this indicates that viscosity is a minor contributor to the pulmonary hypertension of CMS.

C. Kidney Function

Because the kidney has small dimensions and a large blood flow, its arterial-venous O_2 concentration difference is small, and its Pvo_2 is higher than the Pvo_2 of the systemic venous circulation. At high altitude the kidney Pvo_2 decreases more than

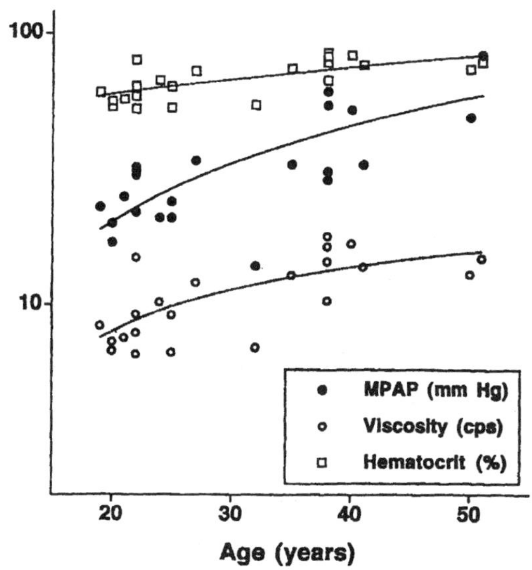

Figure 8 Age-related increments of hematocrit (Hct), blood viscosity, and mean pulmonary artery pressure (MPAP) with age. To compare the corresponding slopes, the ordinate is logarithmic. The MPAP slope is greater than the one corresponding to the hematocrit or to the viscosity. (From Refs. 44, 59.)

the corresponding drop in the systemic one. Our previous studies show that in CMS kidney plasma flow falls as hematocrit rises (Fig. 6) (57,60).

Despite the marked alterations in kidney hemodynamics at high altitude, oxygen consumption (V_{O_2}) and tubular functions related to water and acid-base regulation remain intact and similar to those of sea level. So far no one has been able to explain how the kidney can adapt not only to high altitude but also to its own hemodynamic alterations. Understanding the function of the kidney at high altitude may prove valuable to understanding kidney function at sea level.

D. Endocrine Function and CMS

Pretell et al. (61) showed that thyroid function (T4) in healthy young high-altitude males was the same as that at sea level but declined with age with an inverse relationship with hematocrit. In CMS, T4 was low and TSH and the T3/T4 ratio were high. Although iodine deficiency was found in a large percentage of the population, these findings were also seen in high-altitude natives without iodine deficiency. Pretell (62) concluded that thyroid function is lower at high altitude and more severely depressed in cases of CMS. These changes, however, were reversible upon descent to sea level (63).

Patients with CMS studied at 4300 m have a lower urinary excretion of testicular hormones after the administration of human chorionic gonadotropin, a finding not seen in normal high-altitude natives (64). Aldosterone has a normal increase after orthostasis in high-altitude natives, but the response is impaired in patients with CMS (65). These patients also have a decreased plasma cortisol response to ACTH when compared to normal natives (66). Sex hormones seem to play a clear role in the development of excessive erythrocytosis in women (see CMS and women).

VI. Prevention and Treatment

Hypoxemia improves on descent to sea level, and after 2 or 3 weeks there is a gradual increase in Pa_{CO_2} while the static pulmonary volumes remain elevated (67) and hematocrits return to sea level values (28,67). Right heart hypertrophy and pulmonary hypertension reverse more slowly and return to normal values after 2 years (68,88). CMS is, therefore, a condition whose symptoms and signs are reversed by normoxia. The ideal treatment is migration to low altitude, but alternative proposals have directed attention to reduction of red blood cells by phlebotomy.

Bloodletting can be done alone or with volume replacement (isovolemic hemodilution). Sedano et al. (33) and Sedano and Zaravia (69) have suggested that the latter is the better choice because the improvement of symptoms is long-lasting (up to a year). Others have found that bloodletting improves respiratory function, including ventilation/perfusion ratios (45,70).

Pharmacological attempts to reduce erythrocytosis have been carried out with methylxanthines (71–73), adrenergic (74–77) and angiotensin-converting enzyme inhibitor (78,79). In cases of pulmonary hypertension, calcium blockers have been shown to be useful in experimental settings (80,81). None of these drugs, however, have been used in the treatment of CMS.

Kryger et al. (38,82) have used medroxyprogesterone (20–60 mg/day for 10 weeks) in CMS sufferers living at 3100 m. This drug increases ventilation and normalizes Pa_{O_2} and PA_{O_2} with a parallel drop in hematocrit and subsequent reduction of symptoms for up to 5 years. Since medroxyprogesterone is a female hormone, its use in men with CMS may be limited. However, it can be used in postmenopausal women who are at risk for CMS. Other attempts to improve blood oxygenation include the use of drugs that stimulate peripheral chemoreceptors. Almitrine (3 mg/kg) increases Pa_{O_2}, pH, and respiratory frequency but at a lower dose (1.5 mg/kg for 4 weeks) has virtually no effect (83).

From the above we can conclude that the best available treatment for CMS is residence at sea level. Unfortunately, this is seldom possible, and bloodletting could be the second treatment of choice for men or women and possibly medroxyprogesterone for postmenopausal women.

VII. Biological Basis of CMS

The comparative physiology of the loss of adaptation to high altitude has shown that this process affects both humans and some domestic animals but not animals that are native to high altitude, which are presumed to be genetically adapted (84). The absence of CMS in the latter is related to several adaptations not only of the oxygen transport chain but also of the pulmonary circulation. Species genotypically adapted to high altitude do not have attenuation of the respiratory sensitivity to acute hypoxia, enlarged peripheral chemoreceptors, pulmonary vasoconstriction, right ventricular hypertrophy, or increased hematocrit. High hemoglobin-oxygen affinity is the most typical mark of genetic adaptation present in high-altitude mammals, birds, and amphibians (44,84).

The absence of CMS in genotypically adapted animals confirms that a different physiological design is needed in the oxygen transport system for the animal to be considered adapted. On the contrary, a disease of acclimatization, like CMS, is an indication of the limitations of the use of an extended phenotypic capacity beyond the limits of tolerance during life at high altitude. Tibetans are considered genetically adapted humans, but they also can develop CMS (20,85–87), so their adaptation capacity is also limited. High-altitude humans, Andeans or Tibetans, lack the generalized genotypic hypoxic characteristic of hemoglobin with high affinity for oxygen.

VIII. Conclusions

CMS is a multifactorial condition characterized by the presence of excessive hypoxemia and excessive erythrocytosis. Studies done among high-altitude Andean natives show an age-related decline in ventilation. Therefore, a failure of the hyperventilatory capacity and/or inefficient oxygenation of the blood may be the underlying pathophysiological cause of excessive hypoxemia. The hypoxic condition is usually accompanied by pulmonary arterial hypertension, leading in advanced cases to right heart failure. Epidemiological studies confirm the presence of the underlying pathophysiological factors at work in CMS. Most of the pathophysiological characteristics are present in lesser degrees in apparently normal high-altitude Andean natives and are more prominent in those with the clinical picture of CMS. The appearance of this condition, as well as the fact that genotypically adapted species to high altitude do not develop CMS, shows the limited capacity of humans to acclimatize to life at high altitudes.

Overlapping cause and effect confounds a clear understanding of the pathophysiological mechanisms in CMS. For example, polycythemia per se, presumed to be the consequence of hypoxemia, can be itself impair lung function. Similarly, the polycythemia resulting from increased EPO secretion, presumed to result from hypoxemia, can very likely itself lead to reduced kidney perfusion, which can further stimulate EPO secretion. Thus, one of the intriguing aspects of this fascinating clinical syndrome is an analysis of what is "adaptive" versus what is pathological.

IX. Future Directions

Several kinds of studies should address some of the questions that have come out in this chapter. In regard to the erythrocyte response at high altitude, it seems that the simple negative feedback model that operates at sea level has been reversed to a positive feedback mechanism that counteracts the hemoglobin production arrest: EPO seems to increase with hemoglobin, rather than the reverse. It would be useful to document the change in EPO levels when patients with different hematocrits are subjected to phlebotomy. This would test whether the normal feedback mechanism is intact or, as suggested, reversed.

There are no actual plasma clearance or pharmacokinetic measurements in polycythemic subjects. However, plasma distribution of Evans blue dye confirms a prolonged plasma circulation time with normal total blood flow in high-altitude polycythemia (84), which means reduced plasma volume. It seems, therefore, important to explore this problem experimentally. It is surprising how little attention has been given to plasma transport functions compared to the oxygen transport function of red cells.

Chronic mountain sickness is a loss of normal adaptation in natives and long-term residents at high altitudes. There is great variability in susceptibility, so one of the challenges is to explain not only the causes for its signs and symptoms, but also why at any given altitude some individuals are more likely to develop CMS than others. In addition to the physiological variables that lead to the pathophysiological sequence of CMS, other etiological factors of genetic origin must also be considered. Studies at the molecular level should follow, looking for the genetic and molecular basis of the disease, i.e., the variability in HVR and EPO response, hemoglobin levels, etc. Furthermore, studies should be undertaken at the microvascular level to understand the patterns of O_2 distribution in vessels of various sizes and the distribution of hematocrits and viscosity in these vessels. One area of glaringly inadequate knowledge is in the patterns of capillary density in animals and humans adapted to high altitude and hypoxia.

In general, CMS provides the opportunity to expand our knowledge about the mechanisms found at different levels of acclimatization to high altitude and, based on the latter, the opportunity to find a promising treatment for CMS.

References

1. Anand IS, Malhotra RM, Chandrashekhar Y, Bali HK, Chauhan SS, Jindal SK, Bhandari RK, Wahi PL. Adult subacute mountain sickness—a syndrome of congestive heart failure in man at very high altitude [see comments]. Lancet 1990; 335:561–565.
2. Monge M. Sobre un caso de Enfermedad de Vaquez. Communicación presentada a la Academia Nacional de Medicina, Lima, 1925:1–6.
3. Monge M. La Enfermedad de los Andes. Anales de la Facultad de Medicina, Universidad de Lima, 1928:1–309.
4. Talbott J, Dill D. Clinical observations at high altitude: observations on six healthy persons living at 17,500 feet and a report of one case of chronic mountain sickness. Am J Med Sci 1936; 192:626–629.
5. Huang SY, Ning XH, Zhou ZN, Gu ZZ, Hu ST. Ventilatory function in adaptation to high altitude. In: West JB, Lahiri S, eds. Studies in Tibet: High Altitude and Man. Bethesda, MD: American Physiological Society, 1984:173–177.
6. Xu-Chu HG, Zheng-zhong N, Xue-han Z, Chang-fu L, Hua-ying F, Zhong-ming C, Zheng-zheng C, Tie-chen P. The role of respiratory function in the pathogenesis of severe hypoxiemia in chronic mountain sickness. Symposium on Quinghai-Xizang (Tibet) Plateau [Beijing, China], Vol. II: Environment and Ecology. Beijing: Science Press, 1981.
7. Monge CC, Whittembury J. Chronic mountain sickness and the physiopathology of hypoxemic polycythema. In: Sutton JR, Jones NL, Houston CS, eds. Hypoxia: Man at Altitude. New York: Thieme-Stratton Inc., 1982:51–56.
8. Loew PG, Thews G. Die Altersabhängigkeit des arteriellen Saurstoffdruckes bei der berufstätigen Bevölkerung (The dependency of age for arterial oxygen pressure in a working population). Klin Wochensch 1962; 40:1093–1098.
9. Sorbini CA, Grassi V, Solinas E, Muiesan G. Arterial oxygen tension in relation to age in healthy subjects. Respiration 1968; 25:3–13.
10. Winslow RM, Monge CC, Statham NJ, Gibson CG, Charache S, Whittembury J, Moran O, Berger RL. Variability of oxygen affinity of blood: human subjects native to high altitude. J Appl Physiol 1981; 51:1411–1416.
11. Sime F. Ventilación humana en hipoxia crónica etiopatogenia de la enfermadad de Monge o desadaptación cronica a la altura. Doctoral thesis, Universidad Peruana Cayetano Heredia, Instituto de Investigaciones de la Altura, Lima, Peru, 1973.
12. Sime F, Monge C, Whittembury J. Age as a cause of chronic mountain sickness (Monge's disease). Int J Biometr 1975; 19:93–98.
13. León-Velarde F, Arregui A. Desadaptación a la vida en las grandes alturas. Travaux del

Institut Français d'Etudes Andines a (IFEA) Tomo 85. Eds IFEA/Universidad Cayetano Heredia. Lima, 1994:1–145.

14. Monge CC, León-Velarde F, Arregui A. Pathophysiology and epidemiology of chronic mountain sickness. Int J Sports Med 1992; Suppl 1:S79–S81.
15. Arai Y, Sherpa NK, Horie Y, Furuyama N, Masuyama S, Kuriyama T. Arterial blood gas change with aging in Sherpa. In: Ohno H, Kobayashi T, Nakashima M, Matsumoto M, eds. Progress in Mountain Medicine and High Altitude Physiology. Press Committee of the 3rd World Congress in Mountain Medicine and High Altitude Physiology, 1998.
16. León-Velarde F, Arregui A, Monge CC, Ruiz y Ruiz H. Aging at high altitudes and the risk of chronic mountain sickness. J Wild Med 1993; 4:183–188.
17. Gonzales GF. Guerra-García R. Niveles de hemoglobina en residentes de Cerro de Pasco (4200 m) según procedencia, actividad y edad, La Oroya, Peru, Centromín Peru, 1978: 166–171.
18. Gonzales GG, Hum N, Guerra-García R. Relación del hematocrito con la edad en varones de Huancayo (3200 m) y Cerro de Pasco (4200 m). Acta de las Primeras Jornadas de Medicina y Cirugía de la Altura. Centromín Peru: La Oroya, 1978:82–88.
19. Vargas E, Villena M. Factores predominantes en la etiopatogenia de la enfermedad de Monge (EPA) in La Paz, Bolivia (3600–4000 m). In: León-Velarde F, Arregui A, eds. Hipoxia: Investigaciones Básicas y Clínicas Homenaje a Carlos Monge Cassinelli. Lima: IFEA/UPCH, 1993: 263–282.
20. Wu TY, Zhang Q, Jin B, Xu F, Cheng Q, Wan X. Chronic mountain sickness (Monge's disease): an observation in Quingai-Tibet plateau. In: Ueda G, ed. High Altitude Medicine. Matsumoto, Japan: Shinshu University Press, 1992:314–324.
21. Whittembury J, Monge CC. High altitude, hematocrit, and age. Nature (London) 1972; 238:278–279.
22. Tatsumi K, Hannhart B, Moore LG. Hormonal influences on ventilatory control. In: Dempsey JA, Pack AI, eds. Regulation of Breathing. New York: Marcel Dekker, 1995: 829–864.
23. Ou LC, Sardella GL, Leitere JC, Brinck-Johnson T, Smith RP. role of sex hormones in development of chronic mountain sickness in rats. J Appl Physiol 1994; 77:427–433.
24. León-Velarde F, Ramos MA, Hernández JA, De Idiáquez D, Muños LS, Gaffo A, Córdova S, Durand D, Monge CC. The role of menopause in the development of chronic mountain sickness. Am J Physiol (Regul Integr Comp Physiol 41) 1997; 272:R90–R94.
25. Goodland RL, Reynolds JG, Pommerenke WT. Alveolar carbon dioxide tension levels during pregnancy and early puerperium. J Clin Endocrinol 1954; 14:522–530.
26. Takano N, Sakai A, Iida Y. Analysis of alveolar pCO_2 control during the menstrual cycle. Pflugers Arch 1981; 390:56–62.
27. Santolaya BR, Arraya CJ, Vecchiola DA, Fabres CH, Prieto PR, Vergara R. Gases y pH en sangre arterial en 176 hombres y 162 mujeres sanas trabajadores no mineros residentes a 2899 mts de altura. Rev Hosp R H Glover (Chile) 1982; 2:7–18.
28. Monge M, Monge C. High Altitude Diseases: Mechanism and Management. Springfield, IL: Charles C Thomas, 1966.
29. León-Velarde F, Arregui A, Vargas M, Huicho L, Acosta R. Chronic mountain sickness and the effect of chronic lower respiratory disorders. Chest 1994; 106:151–155.
30. Kryger M, Weil J, Grover R. Chronic mountain polycythemia: a disorder of the regulation of breathing during sleep? Chest 1978; 73(2 suppl):303–304.

31. Normand H, Vargas E, Bordacher J, Benoit O, Raynaud J. Sleep apnea in high altitude residents (3800 m). Int J Sports Med 1992; Suppl 1:540–542.
32. Kreuzer F, Tenney M, Mithoefer JC, Remmers J. Alveolar arterial oxygen gradient in Andean natives at high altitude. J Appl Physiol 1964; 19:13–16.
33. Sedano O, Pastorelli J, Gómez A, Flores V. ‘‘Sangría roja’’ aislada vs. hemodilución isovolémica inducida en mal de motaña crónico. Resúmenes de trabajos libres. V Congreso Nacional. X Curso Internacional de Medicina Interna (Resumen 249), Lima, Sociedad de Medicina Interna, 1988.
34. Vargas E, Villena M. Intercambio gaseoso y relación ventilación-perfusión en el mal de montaña crónico (Resumen 34). Acta Andina, 1992.
35. Guenard H, Vargas E, Villena M, Caras PM. Hypoxemie et hematocrite dan la polyglobulie pathologique d’altitude. Bull Eur Physiopathol Respir 1984; 20:319–324.
36. Reategui LL. Soroche crónico: observaciones realizadas en el Cuzco en 30 casos. Rev Agrup Med Amauta (Cusco) 1965; 1:7–15.
37. Moore LG, Asmus I, Curran L. Chronic mountain sickness: gender and geographic variation. In: Ohno H, Kobayashi T, Masuyama S, Nakashima M, eds. Progress in Mountain Medicine and High Altitude Physiology. Matsumoto: Press Committee of the Third World Congress, 1998:114–119.
38. Kryger MH, Grover RF. Chronic mountain sickness. Semin Respir Med 1983; 5:164–168.
39. Severinghaus JW, Bainton CR, Carcelen A. Respiratory insensitivity to hypoxia in chronically hypoxic man. Respir Physiol 1966; 1:308–334.
40. Sørensen SC, Severinghaus JW. Irreversible respiratory insensitivity to acute hypoxia in man born at high altitude. J Appl Physiol 1968; 25:217–220.
41. Lahiri S, Delaney RG, Brody JS, Simpser M, Velasquez T, Motoyama EK, Polgar C. Relative role of environmental and genetic factors in respiratory adaptation to high altitude. Nature 1976; 261:133–135.
42. Arias-Stella J, Valcárcel J. Chief cell hyperplasia in the human carotid body at high altitudes: physiologic and pathologic significance. Human Pathol 1976; 7:361–373.
43. Tenney SM. Maximal oxygen uptake rate at high altitude: a graphical analysis. In: Leon-Velarde F, Arregui A, eds. Hipoxia: Investigaciones basicas y clinicas. Tomo 76, serie Travaux de L’institut Francais d’Edudes Andines. Lima: IFEA/UPCH, 1993:127–140.
44. Winslow RM, Monge CC. Hypoxia, Polycythemia, and Chronic Mountain Sickness. Baltimore: Johns Hopkins University Press, 1986.
45. Winslow RM, Monge CC, Brown EG, Klein HG, Sarnquist F, Winslow NJ. The effect of hemodilution on O_2 transport in high-altitude polycythemia. J Appl Physiol 1985; 59:1495–1502.
46. Monge CC. Hemoglobin regulation in hypoxemic polycythemia: adjustments to high altitude. In: Chamberlayne EC, Condliffe P, eds. Proceedings of the International Symposium on Acclimatization, Adaptation, and Tolerance to HA. Bethesda, MD: NIH, 1983:53–56.
47. Monge CC. Regulación de la concentración de hemoglobina en la policitemia de altura: modelo matemático. Bull Inst Etudes Andines 1990; 19:455–467.
48. Arregui A, León-Velarde F, Valcarcel M. Salud y Minería. El Riesgo del Mal de Montaña Crónico entre Mineros de Cerro de Pasco. Lima: ADEC-ATC/Mosca Azul, 1990.
49. Monge CC, León-Velarde F, Arregui A. Increasing prevalence of excessive erythrocytosis with age among healthy high-altitude miners (letter). N Engl J Med 1989; 321: 1271.

50. Dainiak N, Spielvogel H, Sorba S, Cudcowicks L. Erythropoietin and the polycythemia of high altitude dwellers. In: Ascensao JL, ed. Molecular Biology of Erythropoiesis. New York: Plenum Press, 1989:17–21.
51. León-Velarde F, Monge CC, Vidal A, Carcagno M, Criscuolo M, Bozzini C. Serum immunoreactive erythropoietin in high altitude natives with and without erythrocytosis. Exp Hematol 1991; 19:257–260.
52. Spivak JL. Erythropoietin: a brief review. Nephron 1989; 52:289–294.
53. Winslow RM, Chapman KW, Gibson CC, Samaja M, Monge CC, Goldwasser E, Sherpa M, Blume D. Different hematologic responses to hypoxia in Sherpas and Quechua Indians. J Appl Physiol 1989; 66:1561–1569.
54. Antezana G, Villena M, Aparicio O, Noriega I, Ugarte H, Valer R. Estudio hemodinámico de la eritrocitosis de altura (Resumen 36). Acta Andina 1993; 2:41–42.
55. Bozzini CE, Alippi RM, Barcelo C, Conti MI, Bozzini C, Lezon CE, Olivera MI. The biology of stress erythrocytosis and erythropoietin production. In: Rich IN, Lappin TRJ, New York Academy of Sciences, eds. Molecular, Cellular, and Developmental Biology of Erythropoietin and Erythropoiesis. New York: New York Academy of Sciences, 1994:83–93.
56. Monge C, Cazorla A, Whittembury G, Sakata Y, Rizo-Patron C. A description of the circulatory dynamics in the heart and lungs of people at sea level and at high altitude by means of the dye dilution technique. Acta Physiol Latinoam 1955; 5:189–210.
57. Monge CC, Lozano R, Marchena C, Whittembury J, Torres C. Kidney function in the high altitude native. Fed Proc 1969; 28:1199–1203.
58. Sánchez C, Merino C, Figallo M. Simultaneous measurement of plasma volume and red cell mass in polycythemia of high altitude. J Appl Physiol 1970; 28:775–778.
59. Peñaloza D, Sime F. Chronic cor pulmonale due to loss of altitude acclimatization (chronic mountain sickness). Am J Med 1971; 50:728–743.
60. Lozano R, Monge CC. Renal function in high-altitude natives and in natives with chronic mountain sickness. J Appl Physiol 1965; 20:1026–1027.
61. Pretell EA. Cambios en la función tiroidea en nativos de altura y en pacientes con de montaña crónico (Resumen). IV Congresso Nacional de Medicina: Libro de Resúmenes. Lima: Associación Medica Peruana ''Daniel Alcides Carrión,'' 1989.
62. Pretell EA. Deficiencia de iodo y función tiroidea en nativos de altura (Resumen 15). In: Universidad Peruana Cayetano Heredia, ed. Jornadas de Cientificas Estudiantiles. Lima, 1986:65.
63. Guerra-García R, Llaque WR, Crandall ED. Observaciones sobre la función endocrina de pacientes con mal de montaña cronico (MMC) estudiados a nivel del mar (Resumen 52). Ica: Sociedad Peruana de Endocrinología, 1977:80.
64. Guerra-García R, Llerena LA, Garayar D, Ames R. Función endocrino hipófiso testicular en nativos de altura y en pacientes con mal de montaña crónico. Cusco: Sociedad Peruana de Endocrinologia, 1973:40.
65. Villena A, Zorrilla R, Guerra-García R. Respuesta ortostática de aldosterona sérica en nativos normales y residentes de la altura y en pacientes con ''mal de montaña crónico'' (Resumen 9). II Congresso Peruano de Endocrinología: communicaciones cortas. Lima: Sociedad Peruana de Endocrinología, 1987.
66. Guerra-García R, Goñez C, Zubiate M, Garmedia F. Función suprerrenal en nativos de altura y en pacientes con mal de montaña crónico (Resumen 24). Cusco: Sociedad Peruana de Endocrinología, 1973:42.
67. Coudert J, Paz-Zamora M, Vargas E. Volúmenes pulmonares, ventilación y presión de

los gases en sangre, en residentes de la altura transportados a baja altitud (Communicación preliminar), Annuario 1971–1973. La Paz: Instituto Boliviano de Biología de Altura, 1975:58.
68. Sime F, Peñaloza D, Ruiz L. Bradycardia, increased cardiac output, and reversal of pulmonary hypertension in altitude natives living at sea level. Br Heart J 1971; 33:647–657.
69. Sedano O, Zaravia A. Hemodilución isovolémica inducida en mal de montaña crónico. Resúmenes de Trabajos LIbres. V Congreso Nacional. X Curso Internacional de Medicina Interna (Resumen 250). Lima: Sociedad Peruana de Medicina Interna, 1988.
70. Cruz JC, Diaz C, Marticorena E, Hilario V. Phlebotomy improves pulmonary gas exchange in chronic mountain sickness. Respiration 1979; 38:305–313.
71. Paul P, Rothmann SA, Meagher RC. Modulation of erythropoietin production by adenosine. J Lab Clin Med 1988; 112:168–173.
72. Schooley JC, Mahlmann and LJ. Adenosine, AMP, cyclic AMP, theophylline and the action and production of erythropoietin. Proc Soc Exp Biol 1975; 150:215–219.
73. Ueno M, Brookins J, Fisher JW. A1 and A2 adenosine receptor regulation of erythropoietin production. Life Sci 1988; 43:229–237.
74. Fink GD, Fisher JW. Erythropoietin production after denervation or beta-adrenergic blockade. Am J Physiol 1976; 230:508–513.
75. Fink GD, Fisher JW. Characterization of activity in polycythemic mice. J Pharmacol Exp Ther 1977; 202:192–198.
76. Fink GD, Fisher JW. Role of the sympathetic nervous system in the control of erythropoietin production. In: Fisher JW, ed. Kidney Hormones: Erythropoietin. New York: Academic Press, 1977:387–413.
77. Izaguirre V, Vargas M, León-Velarde F, Huicho L, Barcelo AC, Alippi RM, Monge CC. Inhibitory effect of an alfa-1 adrenergic antagonist on erythropoiesis in normoxic and hypoxic mice. Int J Clin Lab Res 1994; 24:213–216.
78. Gaston RS, Julian BA, Diethelm AG, Curtis JJ. Effects of enalapril on erythrocytosis after renal transplantation. Ann Intern Med 1991; 115:954–955.
79. Islam MS, Bourboigt B, Codet JP, Songy B, Fournier G, Cledes J. Captoril induces correction of post-renal transplant erythremia. Transpl Int 1990; 3:222–325.
80. Antezana A-M, Antezana G, Aparicio O, Noriega I, León-Velarde F, Richalet J-P. Pulmonary hypertension in high altitude chronic hypoxia: response to dihydropyridones. Eur Respir J 1998; 12:1181–1185.
81. Stanbrook HS, Morris KG, McMurtry IF. Prevention and reversal of hypoxic pulmonary hypertension by calcium antagonists. Am Rev Respir Dis 1984; 130:81–85.
82. Kryger MH, Glas RD, Jackson RD, McCullough RE, Scoggin CH, Grover RF, Weil JV. Impaired oxygenation during sleep in excessive polycythemia of high altitude: improvement with respiratory stimulation. Sleep 1978; 1:3–17.
83. Villena M, Vargas E, Guenard H, Nallar N, Tellez W, Spielvogel H. Etude en double insu de l'effet de l'almitrine sur les malades porteus de polyglobulie d'altitude. Bull Eur Physiopathol Respir 2000; 21:165–170.
84. Monge CC, León-Velarde F. Physiological adaptation to high altitude: oxygen transport in mammals and birds. Physiol Rev 1991; 71:1135–1172.
85. Sun SF, Huang SY, Zhaung JG, Droma TS, Banden G, McCullough RE, McCullough RG, Cymerman A, Reeves JT, Moore and LG. Decreased ventilation and hypoxic ventilatory responsiveness are not reversed by naloxone in Lhasa residents with chronic mountain sickness. Am Rev Respir Dis 1990; 142:1294–1300.

86. Wu TY, Ki WS, Li Y, et al. High altitude disease on the Tibetan Plateau. First World Congress of High Altitude and Physiology. La Paz, Bolivia, 1994:55.
87. Wu TY, Li W, Li Y, Ge RL, Cheng Q, Wang S, Zhao G, Wei L, Jin Y, Don G. Epidemiology of chronic mountain sickness: ten years study in Kinghai, Tibet. In: Ohno H, Kobayashi K, Masuyama S, Nakashima M, Matsumoto M, eds. Progress in Mountain Medicine and High Altitude Physiology. Press Committee of the Third World Congress, 1998:120–125.
88. Corone P, DRouet I, Escourrou P, Antezana G. Epidemiologie cardio-vasculaire des sujects boliviens resident en haute et basse altitude. In: Ruffie J, Quilici JC, Lacoste MC, eds. Antropologie des Populations Andines. Paris: INSERM, 1976:441–451.

25

High Altitude and Common Medical Conditions

PETER H. HACKETT

International Society for Mountain Medicine
Ridgway, Colorado

This study addresses the important topic of response to a disease in a situation that patients may choose for themselves, because it adds to their enjoyment or quality of life. Too little research has been done in this area of medical care (1).

I. Introduction

Dr. William Barclay made the above statement nearly 20 years ago in an editorial accompanying an article by Graham and Houston that investigated ascent to altitude by patients with chronic obstructive pulmonary disease (COPD) (2). Since then, more and more people with various medical conditions have been traveling to high-altitude areas for recreation or work, and many are choosing to live there, including older retirees. In response to the increased need for knowledge about how illness and altitude mix, the nascent literature is starting to grow. Overall, however, for many conditions, few conclusions can be drawn from the available work that will aid the clinician in advising his or her patients about their potential risks of ascent to high altitude. Hopefully, this attempt to present and critique this largely anecdotal literature will encourage scientists and clinicians to pursue further investigations, as Barclay beseeched. This chapter will focus on the available research, what conclusions may or may not be drawn from these studies, and what important questions remain to be answered. The scope includes the effect of altitude on preexisting illnesses, the effect of illnesses on altitude acclimatization, and altitude effects on special states such as pregnancy and the elderly.

II. Stresses of the High-Altitude Environment

Ascent to high altitude may affect people in various ways. The gas expansion from hypobaria can cause problems in sinuses, ears, lungs, the gut, and any space that has air or gas, such as a pneumothorax. Problems of dysbarism are more relevant with rapid ascent as in airplanes or balloons, but even more gradual ascent to terrestrial elevations can be associated with such things as aerodentalgia and increased bowel gas. On the other hand, less dense air may improve lung mechanics. The decreased air resistance will improve performance in events such as jumping, vaulting, and sprinting.

Temperature decreases at a rate of roughly 6.5°C per 1000 m gain in altitude. Cold and hypoxia are generally interactive in contributing to hypothermia and frostbite, and perhaps, via the sympathetic nervous system, cold contributes to high-altitude pulmonary edema (3–6). For recent reviews of interaction of altitude and cold, see Wood (7) and Sutton et al. (4).

Because particulate matter, water vapor, and cloud cover diminish with increasing altitude, ultraviolet radiation reaching the ground increases by approximately 4% per 300 m gain in altitude, increasing the risk of sunburn, dermal photoallergy or phototoxicity, drug photosensitivity, snow blindness (UV keratitis), cataracts, and skin cancer. On windless days, reflection of sunlight from snow can cause intense heat, and heat-related illnesses are more common than generally appreciated. Due to the risks of impaired sweating from anticholinergics, dehydration from diuretics, and impaired central temperature control from phenothiazines, patients taking these medications must exercise particular caution in hot environments.

Humidity is appreciably decreased at high altitude, primarily because of air cooling, and above the snow line water must be made from snow and ice. Difficulty obtaining water, combined with the increased insensible water loss from the skin and lungs, may lead to dehydration, especially for those with compromised cardiovascular systems, patients on diuretics, and perhaps in pregnancy. Drying of the airways increases the risk of tracheal and bronchial plugging, especially in those with lung disease. Due to less particulate matter in the air, high altitude can be beneficial to those with allergic asthma or other allergic conditions and to those whose reactive airways disease is aggravated by pollutants, although air pollution is now increasing in high-altitude locations throughout the world (8).

The major physiological stress of the high altitude environment is, of course, hypoxia. As barometric pressure falls with increasing altitude, the partial pressure of oxygen declines. Compared to sea level, Denver has 17% less atmospheric oxygen pressure, Aspen 25% less, and half the sea level pressure of oxygen is present at approximately 5500 meters (18,000 ft.). In addition, weather, season, and latitude all affect barometric pressure and therefore the partial pressure of oxygen in ambient air (see Chap. 2). Pressure at high altitude is lower in winter, and lower the further away from the equator for any given altitude. The degree of environmental hypoxia is directly related to the barometric pressure, not the altitude per se.

III. Effects of High Altitude on Common Illnesses

Many individuals experience some hypoxemia in everyday life, e.g., during sleep, airline travel, or because of lung disease. Commercial aircraft may have cabin altitudes as high as 2600 m, so that even airline crew members exhibit hypoxemia; a mean nadir Sa_{O_2} of 88.6% (range 80–93%) was found in a recent study (9). The effects of such hypoxemia are thought to be inconsequential in the healthy. Persons with preexisting illnesses affecting oxygen transport, however, could be at greater risk for adverse effects, especially during the more prolonged hypoxia of traveling to a high-altitude destination. This section presents what studies are available and also considers some theoretical issues concerning the impact on patients with common illnesses of both acute and longer-lasting hypoxia.

IV. Ventilatory Conditions

Transfer of oxygen from air to blood is the first stage of oxygen transport and the function of the respiratory system. Components of this system include control of ventilation, pulmonary mechanics, matching of ventilation and perfusion in the lung, and gas exchange across the alveolar capillary membrane. Medical conditions affecting any of these factors may impair oxygenation, especially at high altitude, where adjustments in the respiratory system are crucial for well-being.

A. Disorders of Ventilatory Control

Carotid body responsiveness to hypoxia is generally considered necessary for optimal acclimatization to altitude, and a lesser ventilatory response to hypoxia has been linked in some studies to susceptibility to altitude illness (see Chapter 5 and 21). Animals with denervated carotid bodies have been shown to have slower and less complete ventilatory acclimatization (10–12). Persons without carotid bodies might thus be expected to suffer greater hypoxemia at altitude and consequently be more susceptible to altitude effects (13) (see Chap. 5). Any disorder diminishing ventilation or ventilatory response, such as stroke, central nervous system (CNS) trauma, neuromuscular diseases, primary alveolar hypoventilation syndrome, obesity-hypoventilation syndrome, COPD, or sleep apnea might pose greater risks at high altitude. Medications causing ventilatory depression may have the same effect. Long-acting benzodiazepines reduced nocturnal arterial oxygen saturation at high altitude (14), while small doses (10 mg) of short-acting temazepam taken for sleep at 5300 m maintained oxygen saturation and improved quality of sleep (15).

Carotid Surgery

Since the carotid body is commonly damaged or ablated during carotid artery surgery, these patients might be expected to have problems upon subsequent ascent to altitude. To investigate this possibility, Roeggla et al. (16) exposed patients to 1600 m altitude before and after carotid artery surgery. The ventilatory response to hypoxia observed prior to surgery was abolished by the procedure, and the authors caution that altitude acclimatization might be impaired in such patients. Similarly, patients with bilateral carotid body resection for treatment of asthma or emphysema also lack an adequate ventilatory response to acute altitude hypoxia (13). Whether the relative hypoventilation observed at altitude might lead to altitude illness in these patients remains to be determined. Unlike persons with inherently low carotid body function, these patients can be easily identified postoperatively and may need to be cautioned about altitude exposure.

Sleep-Disordered Breathing and Apnea

Patients with breathing dysrhythmia during sleep at sea level [snoring, obstructive apnea, sleep-disordered breathing (SDB)], especially the elderly, might have greater nocturnal hypoxemia at high altitude, with consequent increased cardiac arrhythmia and increased pulmonary artery pressure or susceptibility to altitude illness. Certainly a given period of apnea at high altitude would produce greater hypoxemia than at sea level (Fig. 1). However, altitude-induced changes in ventilatory control and breathing might conceivably improve certain apnea syndromes. Unfortunately, virtually nothing is known about the interaction of sleep-disordered breathing with high altitude. Such important questions as the relationship of SDB to high-altitude periodic breathing, or whether the elderly with SDB may have more problems at altitude remain to be explored. For individuals living at high altitude, nocturnal hypoxemia due to SDB may contribute to chronic mountain polycythemia (17). More relevant to sojourners, sleep-disordered breathing at high altitude has been invoked in the pathogenesis of acute altitude illnesses, including high-altitude pulmonary edema (18). In this respect SDB may contrast to the periodic breathing of high altitude, which has not been related to development of acute mountain sickness (AMS), and, in fact, is associated with a brisk hypoxic ventilatory response, which is generally considered beneficial at altitude (19) (see Chapters 20 and 21). At high altitude, patients using a CPAP machine without pressure-compensating features as treatment for SDB need to adjust it, since altitude will decrease the delivered pressure (20). The error is greater the higher the altitude and the higher the initial pressure setting. At 8000 ft, for example, error was -1.5 cm H_2O for the 5 cm setting, -2.6 cm H_2O for the 10 cm setting, and -3.3 cm H_2O at the 12 cm setting (20). At 10,000 and 12,000 ft, the error was even greater.

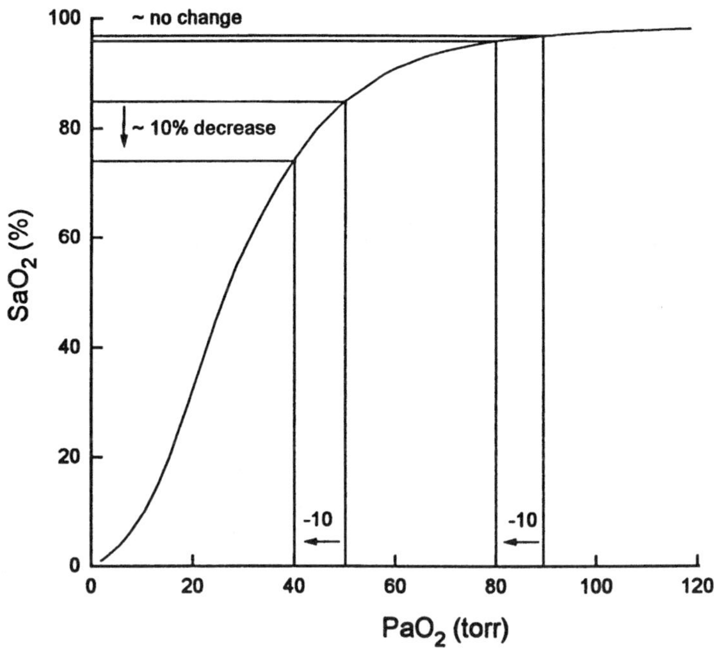

Figure 1 The oxygen-hemoglobin dissociation curve: note a much greater effect on oxygen saturation of a 10 torr decrement in Pao_2 at 4400 m as compared to sea level. (From Ref. 120.)

B. Lung Disorders

Impairment of gas exchange consequent to hypoventilation or V/Q shunt leave those afflicted with a lower Pao_2 at any given altitude than that of their healthy counterparts, i.e., physiologically those with lung disease are at a higher altitude and may thus suffer greater effects of the altitude. In addition, greater hypoxemia may exacerbate their underlying lung disease, for example, through the increase in pulmonary artery pressure due to greater hypoxic pulmonary vasoconstriction (see Chapter 8).

COPD in Altitude Residents

Lung disease is the most common cause of impaired oxygen transport and thus would be expected to impact those living at high altitude. Studies of high-altitude residents suggest increased morbidity from COPD as well as a propensity to migration to lower altitude. Renzetti et al. compared deaths in a COPD group at high altitude in New Mexico and Utah to a COPD group at low altitude and found a 50% versus 35% death rate at follow-up 2.5 years later (21). By examining regional death statistics, Sauer observed increased death rates at high altitude from all chronic respiratory diseases (22). Moore et al. reported increased mortality from emphysema in high-altitude areas of Colorado (23). That this finding was not substantiated in a similar study in New Mexico (24) was likely because of out-migration from altitude of those with COPD in the New Mexico study, a phenomenon noted in a Colorado population of those with COPD by Regensteiner and Moore (25).

In South America and Asia, the problem of acquired lung disease is greater. The recent report of an extremely high incidence of pneumoconiosis at high altitude in central Ladakh, for example, was attributed to high-silica dust storms and lack of chimney use in homes (26). Smoking is another important factor. Altitude and smoking were both found to be independently associated with mortality from COPD in the United States. In an analysis by states, COPD mortality rose by 1 per 100,000 population for every 5.4 increase in mean packs consumed per person per year and for each 95 m increase in resident altitude (27). The author suggested that persons with COPD might benefit from migration to a lower altitude. Supporting this suggestion was an interesting report of sea level patients with hypoxemic, oxygen-dependent COPD who markedly improved when taken to the Dead Sea, 402 m below sea level (28). In summary, the data indicate that persons living at high altitude with lung disease appear to do poorly compared to their sea level counterparts. (For a discussion of the role of lung disease in chronic mountain sickness, see Chapter 24.)

Lowlanders with Lung Disease

Stable Hypoxemic Disease

One would expect lowlanders with lung disease to tolerate high altitude less well. Also, altitude might unmask lung disease that has gone undetected at low altitude. While the issue of acute hypoxic exposure with air travel has received considerable attention, much less is known about more prolonged hypoxia, as with a stay at a ski resort. Graham and Houston (2) evaluated eight patients with moderate COPD with hypoxemia who ascended to an altitude of 1920 m for 4 days. Persons with pulmonary hypertension, a history of respiratory failure with CO_2 retention, or angina were excluded. The subjects had only mild symptoms on ascent, primarily mild fatigue and insomnia, despite the fact that mean Pa_{O_2} declined from 66 at sea level to 51 mm Hg while at rest and from 63 to 47 mm Hg with exercise (Table 1). The patients exhibited ventilatory acclimatization, with an abrupt drop in P_{CO_2} and further decline over 4 days, and a corresponding increase in Pa_{O_2}, the same response as seen in healthy persons. The authors concluded that travel to this moderate altitude is safe for such patients and speculated that their preexisting hypoxemia may have anticipated adaptation, decreasing the likelihood of developing AMS. Despite the authors' plea for further investigations with sicker patients or at higher altitudes, subsequent reports are lacking.

In the absence of other studies, examining the literature on acute hypoxia and air travel in persons with COPD may be instructive. Finkelstein et al. found that hypobaria per se reduced dyspnea in COPD patients at high altitude because airflow improved (29). Reassuringly, emphysematous blebs and bullae did not enlarge or

Table 1 Spirometric and Blood Gas Data for 8 COPD Patients after 3 Days of Acclimatization to Moderate Altitude (1920 m)

Measurement	Sea level, rest	Sea level, exercise	Altitude, rest	Altitude, exercise
V_E L/min	7.85 ± 1.95	26.6 ± 6.64	7.85 ± 2.28	31.2 ± 8.39*
Pao_2 mmHg	66.0 ± 7.1	63.0 ± 9.33	54.0 ± 5.96*	46.6 ± 8.55*
$Paco_2$ mmHg	37.6 ± 3.03	38.8 ± 3.88	32.9 ± 2.1*	33.7 ± 3.07*
A-a Do_2	35.4 ± 6.65	37.6 ± 8.43	24.9 ± 3.76*	28.9 ± 8.36*
pH	7.42 ± 0.02	7.39 ± 0.02	7.45 ± 0.04	7.43 ± 0.04*

* $p < 0.05$ between sea level and altitude.
Mean FVC and FEV_1 were 75 and 41.8% of predicted, respectively.
V_E, minute ventilation; A-a Do_2, alveolar-arterial oxygen tension gradient.
Values for Pao_2, pH, and $Paco_2$, taken 3 hours after ascent and at rest, were 51.5 ± 6.7 mmHg, 7.439 ± 0.01, and 33.9 ± 3.14 mmHg, respectively.
Source: Ref. 2.

rupture on ascent to altitude, most likely because of communication with the surrounding lung tissue (30). Shillito et al. (31) subjected 20 patients with lung disease more severe than those in the Graham and Houston study to a simulated altitude of 2440 m. Thirteen had prior pneumonectomy. Eight developed ''undesirable'' side effects, and those with Sao_2 less than 90% at sea level and/or with maximal voluntary ventilation less than 40 L/min tolerated altitude poorly, with uncomfortable dyspnea, cyanosis, and occasional arrhythmia. As simple screening tests, they recommended the ''match test'' (blowing out a match) to assure adequate expiratory flow, and the ability to walk up a flight of stairs without respiratory distress. They suggested supplemental oxygen during flight for those who ''failed'' (31). Since this early report, the relationship of low-altitude Pao_2 to Pao_2 during hypoxic exposure has been evaluated with more sophisticated techniques (32–36), and Gong et al. developed a nomogram for predicting the Pao_2 at altitude from that at sea level (33). Dillard et al. subsequently refined the prediction by adding spirometry values (37). These hypoxic challenge tests may be useful for predicting oxygenation during transient (up to 4 hours) modest hypoxia, but clinical utility is limited, since sea level Pao_2 or Sao_2 did not correlate well with how people fared at altitude. These studies also did not address how patients might do with longer stay at altitude and with associated stresses such as sleeping, activity, smoking, concomitant angina, hypertension, or other problems. Despite these limitations, a physician attempting to evaluate the risk of high altitude for a patient with COPD may want to use the predicted initial resting Pao_2 at altitude as a way to determine the need for supplemental oxygen.

Efforts have also been made to predict oxygenation at high altitude in children with hypoxemic lung disease. A paper describing exacerbations of cystic fibrosis at high altitude (38) provoked further work to determine whether a hypoxic challenge test at sea level could predict hypoxemia on ascent to altitude (39). Spirometry and pulse oximetry were performed on 22 children with cystic fibrosis, during room air and 15% oxygen breathing. Sao_2% was subsequently measured during commercial air flight and at 1800 m in the Alps. Hypoxic challenge with 15% oxygen was a better predictor of Sao_2% at altitude than spirometry and baseline oxygen saturation. The authors advised using the provocative test to determine which patients might require supplemental oxygen during flight or in the mountains (39). As in the previous studies with COPD, these measurements done only during rest probably underestimate supplemental oxygen requirements during exercise and perhaps during sleep.

Oxygen therapy may be required for those who are symptomatic at altitude or for those predicted to become severely hypoxemic (40). For persons already on supplemental oxygen, F_{IO_2} can be increased by the ratio of higher to lower barometric pressure. Oxygen administration to COPD patients going to altitude helped relieve cardiovascular stress: at a simulated altitude of 2438 m, oxygen improved hemodynamics (lowered BP), and decreased pulsus paradoxus and pulse pressure (41). Whether acetazolamide and medroxyprogesterone acetate, both of which have been shown to aid acclimatization by stimulating ventilation, might be useful (or harmful) in patients with COPD, especially with hypercapnia, is unknown. Controversy surrounds the question of whether a high-carbohydrate diet may benefit patients with COPD on ascent to altitude. Because a high-carbohydrate diet increases RQ, and hence P_{AO_2} and Pao_2, Hansen recommended a high-carbohydrate diet to improve Sao_2 during hypoxic exposure of those with COPD (42). The value of this strategy was questioned by Berger, who warned that some patients with pulmonary disease may not be able to increase minute ventilation to keep $Paco_2$ constant, which is required to improve the Pao_2 in the face of an increased RQ value (43); his concern may be most applicable to patients with severe COPD. This question can only be resolved by an appropriate study. Individuals should have pulmonary function optimized prior to ascent to altitude and receive instructions on medication use (including oxygen) should problems develop.

In summary, information regarding the effect of high altitude on chronic lung disease (or vice versa) is extremely limited. The only study to date suggested that COPD patients did well at the modest altitude of 1920 m because they were already "acclimatized" to hypoxia. Perhaps such persons can be viewed as already living high and then going higher, but altered pulmonary mechanics and impaired gas exchange in these individuals might render perceived or real physiological advantages inadequate at higher altitudes. Predicting altitude Pao_2 from that at sea level appears feasible, but correlation of the acute or predicted Pao_2 with ability to acclimatize and susceptibility to altitude illness remains unexplored. Clearly, much more research evaluating clinical and physiological outcomes of altitude exposure is required in order to advise those with varying types and severity of chronic lung disease regarding risks of altitude exposure. At a simplistic level, the hypoxemia of impaired gas exchange due to lung disease represents, much as does the progressive fall in Pao_2 with normal aging, a high-altitude equivalent that will in itself limit the range of capability of the affected individual. Obtaining a capacity to predict who will tolerate and who may be harmed by the additional hypoxia of high altitude will be a difficult and possibly futile proposition, leaving old-fashioned clinical judgment based upon the overall health and capacity to function of each individual as the best guidance available.

Asthma

Data on mortality and morbidity from asthma in high-altitude residents are scant. What literature is available suggests that asthma may be less of a problem at high altitude than at sea level, both for residents and sojourners, primarily because of decreased allergens and pollution. Studies worldwide have reported a lower number of mites and low mite antigenic load (mostly due to lower indoor relative humidity) and improvement in childhood asthma with time spent at high altitude (44,45). High altitude as a treatment for asthma has been popular in Europe for many decades (46). However, children at high altitude in New Mexico still had significant morbidity from asthma despite a mite-free and low air pollution environment; many were sensitized to household pets (47,48). A study in 1980 found no influence of altitude on asthma mortality in the age group of 5–34 years in the United States (49). That asthmatics living at high altitude may become worse during a stay at sea level (50) and that those living at sea level do better at altitude and then become worse upon returning home adds to the impression that the high-altitude environment may be more hospitable for patients with allergic asthma (51).

Given the possibly of greater hypoxemia during bronchoconstriction in asthmatics at high altitude, it is surprising that there are no reports of an association between asthma and high-altitude pulmonary edema (HAPE) or AMS. One investigation of 16 asthmatic patients tested the effect of acetazolamide upon ascent to 3200 m. Like nonasthmatics, the treated group had higher oxygen saturation and fewer AMS symptoms compared to the placebo control group. Seven of the eight asthmatics in the control group developed symptoms of acute mountain sickness, a rather high incidence, but whether this incidence would be higher than for a nonasthmatic control group under the same conditions was not determined (52).

Nor are there reports in the literature of unexpected asthma exacerbation in lowlanders at high altitude. Matsuda et al. (53) investigated the effect of altitude on 20 asthmatic children with exercise-induced bronchospasm in a hypobaric chamber simulating 1500 m; temperature and humidity were held constant. Respiratory rate during exercise at altitude was increased, as expected, but mean fall in FEV_1, exercise time, $\dot{V}O_2$, and heart rates were unchanged compared to sea level. Although one subject developed wheezing while at altitude, the authors concluded that the reduced barometric pressure and mild hypoxia of 1500 m do not exacerbate asthma. Future studies will hopefully compare sea level responses with those at higher altitude in the field, where humidity and temperature are lower and hypoxia is greater.

Although airflow obstruction would be expected to produce more hypoxemia at altitude than at sea level, neither data nor anecdotal reports validate this possibility. Likewise, no information exists to document the value of a whole variety of logically helpful clinical aids for prevention, including increased hydration or the use of an airway warming mask (54,55). In summary, limited data suggest that high altitude appears not to exacerbate asthma and actually improves allergic asthma. The important question of whether a severe asthma attack at high altitude puts a person at more risk for harm than at low altitude is unanswered.

V. Cardiovascular Problems

Over the first few days of ascent to high altitude, the cardiovascular system of healthy individuals makes important adjustments. These physiological changes may be limited in those with underlying disease, or these adjustments may stress the compromised cardiovascular system and aggravate illness. The principal initial adjustments are increases in resting and exercise heart rate, increase in blood pressure and systemic vascular resistance, increase in cardiac output and velocity of contraction, and contraction of the veins in the skin, muscle, and viscera resulting in increased central blood volume. Largely mediated by increased sympathetic nerve activity, the net effect is an increase in cardiac work, myocardial oxygen consumption, and coronary blood flow. After 4–8 days at high altitude most of these changes diminish, and with prolonged exposure, cardiac output, coronary blood flow, and eventually blood pressure may fall to below sea level values. Pulmonary hypertension, however, persists. The cardiovascular adjustments to altitude are reviewed elsewhere (56) (see Chapter 9). This section focuses on the impact of high altitude on those with cardiovascular illnesses.

A. Arterial Hypertension

What happens to blood pressure depends on many things: length of exposure, degree of hypoxemia, genetic and dietary variables, and cold exposure. Long-term altitude residence has been associated with lower systemic arterial pressure compared to sea level values in many reports from South and North America, Africa, and Asia (57–63). Chronic altitude exposure has also been shown to inhibit progression of hypertension (60,63,64). A preliminary survey in Tibet, however, found an incidence of hypertension higher (9.4%) than that reported from South American high-altitude natives (4.5%) (61). This difference was attributed to ingestion of large amounts of salted tea (65).

The effect of acute exposure to altitude on blood pressure is not entirely clear, in part due to the considerable differences in study methodologies, such as bedrest versus ad lib exercise, salt and diet control, and duration and degree of hypoxic stress. Many studies indicate that blood pressure increases during the first week or two at high altitude, the more so the higher the altitude (66–72). Systemic vascular resistance is increased (up to 65%), as are catecholamines (73,74). A few investigations have reported no significant change in blood pressure, or even a slight decrease (75–80) (see Chapter 9).

Whether individuals who have high blood pressure at sea level have an exaggerated hypertensive response on initial ascent to altitude has had little systematic evaluation. Anecdotally, some individuals develop severe hypertension (81). As Hultgren points out, such occurrences, whatever their frequency, are not a trivial problem, given the large number of hypertensive patients visiting ski resorts and trekking at high altitude (81).

The magnitude of the blood pressure change in hypertensives may depend upon the hypoxic stress. At low altitudes, the increase on average appears to be minimal. Palatini et al. studied 12 normotensives and 12 untreated mild hypertensives with 24-hour ambulatory blood pressure monitoring at sea level, after 12 hours at 1210 m, and after 1.5–3 hours at 3000 m (67). Blood pressure was higher at 1210 m in all subjects during the day, but not at night, and the increase over sea level values was similar in the two groups (6.1 mmHg systolic/1.5 mmHg diastolic in the normotensives and 5.5 mmHg systolic/4.3 mmHg diastolic in the hypertensives). However, individual variability was great; the maximum change was 17.4 mmHg for systolic and 16.3 mmHg for diastolic blood pressures. The brief exposure to 3000 m caused another small increase in blood pressure (BP) in both groups. Heart rates also increased. Mean plasma catecholamines increased substantially, but without statistical significance due to the variability. The authors concluded that the increase in BP in both normotensives and hypertensives was not important at 1200 m, but could become so at 3000 m.

With acute exposure to higher altitudes, hypertensives did demonstrate a greater blood pressure response than normotensives. D'Este et al. (79) found systolic pressure at rest was slightly increased in 13 hypertensives upon acute ascent to 2572 m and was unchanged in 10 normotensive subjects. During submaximal exercise, hypertensives also had a slightly greater systolic blood pressure response. Importantly, individual variation was considerable, with a few hypertensives showing an exaggerated pressure increase (79). Savonitto et al. examined 11 patients with mild to moderate untreated hypertension within 2 hours of arrival, at an even higher altitude of 3460 m, at rest and with maximal isometric and aerobic exercise (82). Systolic pressure, but not diastolic was elevated at rest and during aerobic exercise, while the slope of the increment in BP per minute of exercise was unchanged from that observed in these individuals at low altitude. Blood pressure responses to isometric exercise were the same at low and high altitude. They too concluded that the mean increase in BP was minor, but warned that individual variation was high: one individual had an increase of 25 mmHg in systolic pressure at rest and 40 mmHg at maximal exercise. Unfortunately, no measurements were made after the first hours at altitude in either of these studies, so whether blood pressure continues to increase over a period of days, as it does in normotensives (73), is an important unanswered question. Also unknown is whether more prolonged stay at altitude might reliably reduce arterial pressure to below sea level values, as has been reported in normotensives after 5–10 years (60,83) and also suggested in uncontrolled study of hypertensives (69,84) (Fig. 2). Halhuber et al. claimed a significant reduction in the blood pressure of 593 persons with hypertension after 14 days at 1700–2000 m in the Alps (84). The mechanism of a reduction in blood pressure with prolonged hypoxia may well be downregulation of adrenergic receptors (85).

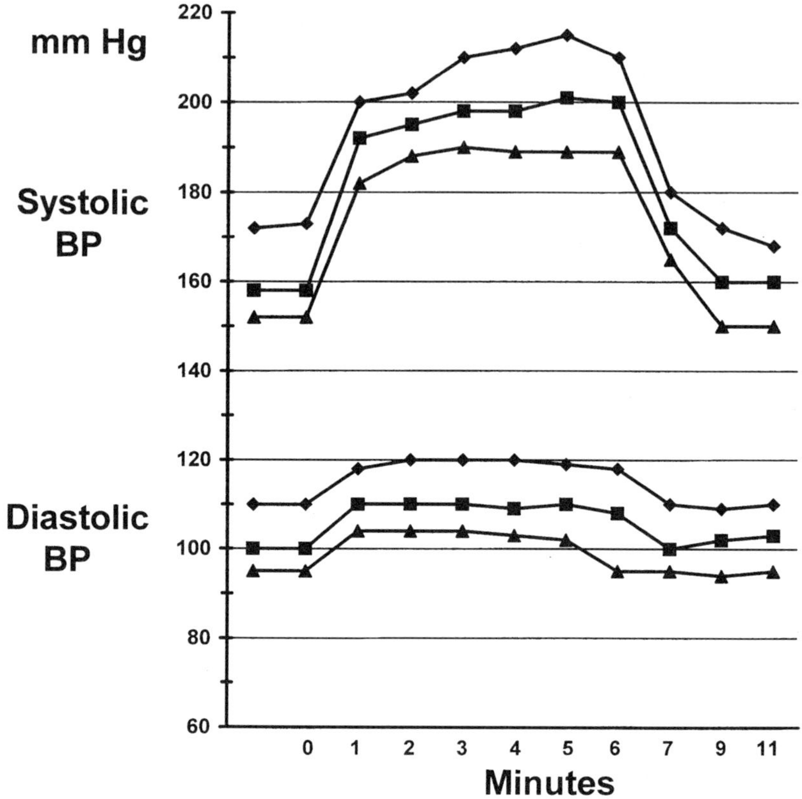

Figure 2 Systolic and diastolic blood pressure in 25 hypertensives during 70 W workload before (◆) and after 2 days (■) and 25 days (▲) at 2500 m. (From Ref. 84.)

Thus the scientific basis for advising individuals with hypertension about ascent to altitude is limited. Because some patients may become markedly hypertensive acutely, and because we have no current means to predict an individual's response, blood pressure monitoring may be prudent. Which medication is best for altitude-aggravated or -induced hypertension is a common question not yet answered. Given that the mechanism appears to be primarily α-adrenergic stimulation, an α-blocker would be reasonable, but this therapy remains untested. Calcium channel blockers may have value. In a preliminary study, Deuber had 24 hypertensive patients exercise after acute ascent to a simulated altitude of 3000 m under three conditions: untreated, treated with nifedipine, and treated with atenolol. Systolic blood pressure during exercise was reduced similarly by the calcium channel blocker and the β-blocker (from 230 mm Hg to 181 and 182 mmHg, respectively), while work capacity was substantially greater with nifedipine than atenolol (86). Clonidine, since it reduces central sympathetic outflow, might also be considered. Hultgren reported on a patient taking enalapril (an ACE inhibitor) at sea level who became severely hypertensive in the Rocky Mountains and was treated successfully on multiple visits with clonidine and nifedipine (81). His report raises the question of whether antihypertensive medication regimens effective at low altitude may be insufficient upon ascent to high altitude.

Those with hypertension may experience other untoward effects of altitude in addition to increased blood pressure. Hypertensives have a greater pulmonary pressor response to hypoxia, which could predispose to HAPE (87), a possible relationship yet to be explored. Hypertensives also have an exaggerated sympathetic response to hypoxia. Somers et al. (88) found that sympathetic nerve activity (measured by direct peroneal nerve recordings) in a hypertensive group was double that of the control group with acute hypoxia, and when voluntary apnea (breath hold) was performed during hypoxia, it was 12 times greater. The prevalence of periodic breathing at altitude could cause hypertensives to experience very high levels of sympathetic stimulation during sleep, with increased risk of exaggerated systemic hypertension, pulmonary hypertension, and arrhythmia. In fact, even normotensives were recently found to have elevated nocturnal blood pressure at high altitude (89). Although these concerns about hypertensive patients are hypothetical, the issues of untoward events during sleep and susceptibility to HAPE in these people certainly warrant further attention.

In summary, the question of how hypertensive patients fare upon ascent to high altitude deserves considerable more study. Given the demonstrated great individual variability, the susceptible person needs to be identified and the mechanism of the labile blood pressure determined. The question needs to be answered as to whether hypertensive patients (perhaps the labile subset) ascending to altitude are subject to increased morbidity or mortality. If drugs that generally suffice at low altitude are not as effective at high altitude, then the mechanistic basis for this and the ideal agents need to be elucidated. Finally, the apparent existence of a reduced blood pressure with a prolonged stay at altitude needs to be confirmed in controlled studies of hypertensive patients.

B. Coronary Artery Disease

High-Altitude Residents

Various avenues of research suggest that the incidence of coronary artery disease and consequent death rates are lower in persons with life-long residence at high altitude (56). Firm conclusions are difficult for lack of control in some studies of such variables as age, genetics, smoking, exercise, diet and downmigration due to illness, but the evidence is not to be discounted. No study has ever suggested an increased incidence of ischemic heart disease at high altitude. Two of three epidemiological studies in North America controlled for age found a lower death rate at high altitude (90,91), while one report claimed there was no effect (92). In South America, a series of 300 routine autopsies of Andean highlanders found not a single case of myocardial infarction (93). Consistent with this observation was the work of Ruiz et al. in Peru showing that ischemic ECG abnormalities were fewer in high-altitude natives compared to matched controls at sea level (94). Arias-Stella and Topilsky followed up on this work by showing increased myocardial vascularity in postmortem studies of high-altitude dwellers compared to sea level controls and offered this as a possible explanation for the lower incidence of ischemic heart disease (95). Subsequent stereoangiographic studies in 53 altitude natives and 30 sea level controls also demonstrated a relatively high density of peripheral coronary artery branches (96). While this work may seem compelling, whether high-altitude residence truly confers protection from death due to coronary artery disease in various populations still requires further investigation (see Chapter 3).

Lowlanders Going to High Altitude

With acclimatization, the normal heart can tolerate even extreme hypoxemia ($Pao_2 <$ 30 mmHg) without evidence of ischemic changes or impaired function (97) (see Chapter 9). A theoretical basis for exacerbation of coronary artery disease at high altitude, however, exists. Cardiac work is slightly increased, and the usual compensatory increase in coronary blood flow may not be attained in those with CAD. In fact, atheromatous coronary vessels without the ability to vasodilate because of abnormal endothelial vasomotor control may actually vasoconstrict due to unopposed sympathetic activation, resulting in ischemia (98). To assess whether these concerns are clinically relevant, the questions I will examine are whether altitude might (1) unmask previously unknown coronary artery disease, (2) exacerbate heretofore stable ischemia, (3) increase risk for myocardial infarction or sudden death, or (4) provoke arrhythmia. Anecdotal reports make it clear that these events may all occur at high altitude (81,99), as they do at sea level, but whether altitude is a contributor is unclear. Additional issues to consider are the possible influence of cold and the risk of ischemia in those without signs or symptoms suggesting CAD, especially in the elderly population. Finally, what recommendations can be offered that may make ascent to high altitude safer for those with known or unknown CAD?

Does High Altitude Unmask Coronary Artery Disease?

Investigations from decades ago suggested that acute, severe hypoxia is indeed a stress on the heart of those with coronary artery disease, but less so than is exercise (100). In a direct comparison, normoxic exercise produced more ischemia-related ECG changes in patients with known coronary artery disease than did inhalation of 10% oxygen while at rest (the anoxemia or Levy test) (101). As a result, exercise testing has become the preferred way to screen. Reassuringly, of the several thousand patients with suspected coronary disease who were given the anoxemia test for 20 minutes, none developed serious cardiac events or died (102,103). When hypoxia and exercise were combined, however, the results were different. Khanna et al. sought to identify coronary artery disease in soldiers being considered for high-altitude duty with the use of either simulated altitude (4592 m), an exercise test, or the combination of hypoxia and exercise (104). Seventy asymptomatic soldiers (ages 30–50) but with abnormal ECGs (nonspecific ST segment or T-wave changes) were evaluated. Hypoxia at rest resulted in positive ECG changes (ischemic-looking ST segment depression) in 4 subjects, 10 were positive with exercise at sea level, and 20 were positive with the combination of hypoxia and exercise; the higher yield in the latter condition may be explained by a considerably higher heart rate compared to the other test conditions. Those with negative hypoxic exercise tests showed no symptoms or signs of coronary disease during 3 years of high-altitude duty, those with positive tests apparently did not go to altitude. I could find no other reports of acute hypoxic exercise unmasking coronary artery disease. While these studies are interesting, they assessed only ECG changes (therefore, the false-positive rate is unknown) and only in persons with pretesting ECG abnormalities. Nonetheless, hypoxic exercise appears better able to unmask coronary artery disease (by ECG) than hypoxia at rest or normoxic exercise. If acute hypoxic exercise can unmask ECG evidence of CAD, then more sustained altitude exposure may do so as well.

Because of the increase in catecholamines over the first few days at high altitude, varying exertion, sleep-exaggerated hypoxemia, cold, and other factors, a stay at high altitude might stress the coronary circulation as much or even more than acute hypoxic exercise. One of the earliest field attempts to assess the effect of exertion at high altitude on unmasking coronary disease occurred in 1970: 149 middle-aged generally fit male skiers (ages 20–70, with 70% > 40) were assessed by ECG telemetry during a ski descent from 3350 to 3100 m (105). Maximal heart rates were achieved by 67% of subjects, indicating that downhill skiing is indeed a cardiovascular stress. Five individuals (5.6%) of those over age 40 developed ST segment depression of 1–2 mm at heart rates of 120–150, an incidence similar to that of asymptomatic sedentary men over 40 during submaximal exercise at sea level (105). Only one subject had known coronary artery disease, a 49-year-old physician who had sustained a myocardial infarction 3 months earlier; he had effort-induced angina at low altitude and while skiing. Although admittedly a rather limited study, Grover surmised that for physically fit middle-aged men, skiing at altitude was probably no greater a coronary stress than comparable exercise at sea level in less fit asymptomatic men. The chance of unmasking coronary artery disease (by ECG) at altitude depended more on the exercise stress than the hypoxia. This is at variance with the Khanna study, but since Khanna's group was selected on the basis of abnormal ECGs, the incidence of CAD in his group was very likely higher.

Does Altitude Exacerbate Preexisting Stable Cardiac Ischemia?

Small increases in heart rate and blood pressure on initial ascent to altitude may cause angina or induce angina in those with coronary artery disease, as has been reported anecdotally (81). How people with symptomatic coronary artery disease fare at high altitude was initially addressed in a pilot study by Okin (106). He subjected 11 individuals with known CAD and 16 controls to a Master's step test at 1600 m (Denver) and again on the first day at 2440 m and 3170 m. In addition, the subjects breathed 16% oxygen during exercise in Denver, simulating 3350 m. No one who had a normal Master's test in Denver had a positive test at altitude, although one person had a positive test with the hypoxic gas breathing. Although acknowledging the limitations of this small study, Okin concluded that hypoxia was not a major contributor to exercise-induced angina in these subjects. Khanna et al. studied 30 individuals with known coronary artery disease at sea level and 40 minutes after exposure to a simulated altitude of 4599 m (104). Seventeen had ischemic ECG changes with exercise at sea level, and an additional three were positive at altitude, supporting the notion that hypoxia if severe enough may add a coronary stress in addition to exercise.

Subsequently, more sophisticated studies using exercise treadmill tests in symptomatic patients with coronary artery disease have shown that altitude might exacerbate angina, at least with acute exposure. Morgan et al. evaluated nine men with stable exercise-induced angina by treadmill test at 1600 m (Denver) and again within the first hour of arrival at 3100 m. The angina threshold was at a lower work load at altitude, but at about the same double product (systolic blood pressure times heart rate) as at sea level. The effect of altitude was thus to increase the work of the heart, rather than to decrease oxygen delivery, since angina did not appear at a lower double product (107) (Fig. 3). The authors proposed that the activity level for angina patients at altitude should be based on heart rate rather than workload, at least on the day of arrival: heart rate of 70–85% of the rate that produced ischemia at low altitude was associated with angina-free exercise at 3100 m. Brammel et al. reported similar results, with angina patients from Denver needing to reduce their activity at higher altitude to avoid angina episodes (108). In a more recent study by Levine et al. of 20 men who were much older than those in the previous investigations (mean age 68 ±3 years), 10 had positive treadmill tests at sea level, and 9 of these plus 3 others had ischemic changes with acute simulated altitude of 2500 m (98). With acute exposure the double product required to induce 1 mm ST depression was decreased about 5%, but after 5 days of acclimatization at 2500 m, this value was unchanged from sea level and incidence of ischemia was the same as at low altitude (Fig. 4). The authors also noted that maximal ST segment depression was the same at sea level, acutely and after 5 days at 2500 m. Also, no new wall motion abnormalities on echocardiography were seen at high altitude. Only one subject exhibited increased angina at altitude. A single individual with severe coronary artery disease sustained a myocardial infarction after maximal exercise at 2500 m. Levine et al. concluded that coronary artery disease patients who are well compensated at sea level do well at a moderate altitude after a few days of acclimatization, but that acutely angina threshold may be lower and activity should be reduced (98). That 2 of the 20 had untoward events (one with increased angina and one with MI) indicates the need to better define those with CAD who are at risk for progression of ischemia from stable to unstable coronary syndromes upon ascent to altitude.

In Vail, Colorado (2500 m), a reunion of the U.S. Army 10th Mountain Division provided an opportunity to observe a large number of elderly (mean age 69.8 ± 4.4, range 59–83 years) over 4 days after ascent from near sea level (109). Of 77

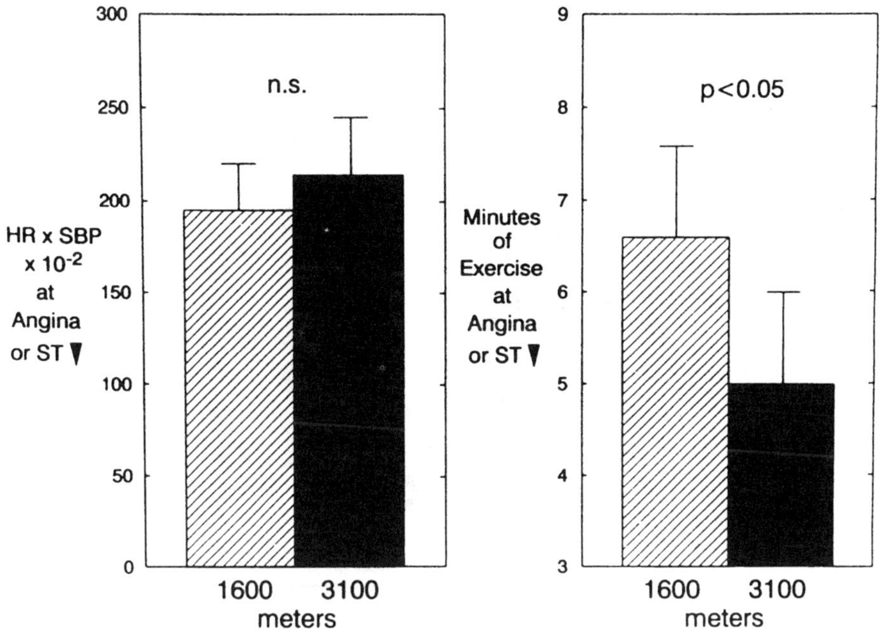

Figure 3 Mean and standard deviation for double product (systolic blood pressure × heart rate) and minutes of exercise at the ischemic threshold at 1600 and after 1 hour at 3100 m. Angina appeared at the same double product but at a lower workload (shorter time) at high altitude compared to sea level. (From Ref. 107.)

men and 20 women, 20 had coronary artery disease, evidenced by past MI, bypass surgery, and use of anti-angina medications. Thirty-eight had abnormal ECGs on arrival at high altitude (lowland ECGs just prior to ascent were not available). Cardiac work as estimated by the heart rate–blood pressure double product was increased 13% at rest, but no new ST-T changes indicative of ischemia developed in this population over the 4 days, nor did any clinical events suggestive of ischemia. Thus, in these elderly subjects with both symptomatic and presumed asymptomatic coronary artery disease (and frequently hypertension), moderate altitude was well tolerated, without evidence of exacerbation of stable ischemia.

Does High Altitude Increase Risk of Myocardial Infarction and Sudden Death?

Halhuber et al. attempted to answer the question of whether high altitude increases the incidence or severity of cardiac events by observing 434 men and women aged 60–85 years who were taken to altitudes of either 1700–2500 or 3200 m for 4 weeks (84). This cohort was selected from 1273 cardiac patients because they could tolerate 4 minutes exercise at 50 W without ECG changes at sea level. Of the 434, 139 had

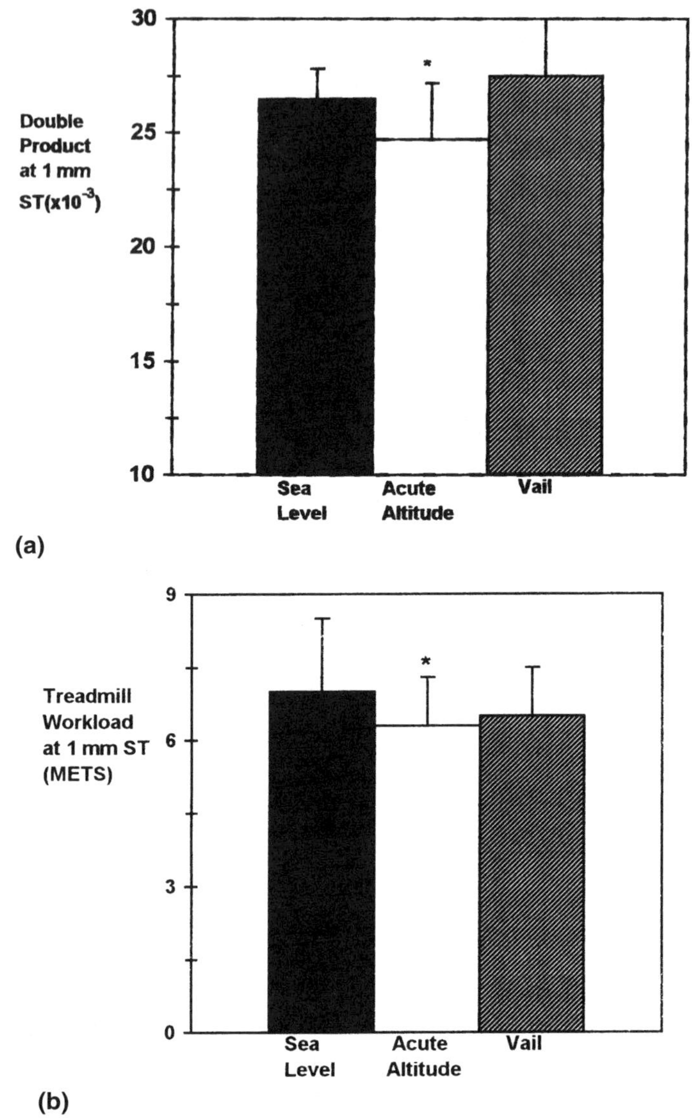

Figure 4 Values represent mean ± SEM for the ischemic threshold of the (a) double product (heart rate × systolic blood pressure) and (b) treadmill workload (in METS) at which 1 mm flat, downsloping, or slowly upsloping ST depression was first identified. Solid bars, sea level. Open bars, acute simulated altitude of 2500 m. Hatched bars, 5 days of acclimatization to 2500 m in the field. In contrast to the data in Figure 3, the double-product ischemic threshold with acute exposure was lower than at sea level (5% less). (From Ref. 98.)

arrhythmias or conduction disturbances, and 141 had previous myocardial infarction. (Whether these groups overlapped is not clear.) On ascent to altitude, and with ad lib activity, one patient had a myocardial infarction, an incidence of 0.2%, which the authors considered low for this population.

In terms of the risk of sudden cardiac death, Halhuber mentioned that among 151,000 vacationers of both sexes who went to high altitudes in the Alps, including 69,640 over the age of 40, there were only 6 deaths. Of these, 1 person had a myocardial infarction in a low-altitude valley (1500 m), and 2 had myocardial infarctions above 2500 m. Two persons died of unknown causes. In contrast to this seemingly low incidence of cardiac events in this age group are data from Austria claiming a higher rate of sudden cardiac death in the mountains compared to the overall risk of sudden cardiac death (110). Analyzing 416 such deaths in the modest altitude range of 1000–2100 m, this study reported an increased risk ratio for sudden death of 4.3 for mountain hiking and 2.1 for skiing, in men over age 34, and no increased sudden death in women of any age. No increased risk, however, was evident in men who participated regularly in sports. The authors suggest that abrupt onset of exercise in sedentary men combined with altitude stress might induce cardiac sudden death. But whether altitude contributed at all to this mortality is uncertain.

Data on the risk of sudden cardiac death at much higher altitudes are even more sparse. Shlim and Gallie reported a low number of cardiac deaths among 275,900 trekkers in Nepal, although many trekking were middle-aged and older males (111). They attributed 4 of 40 deaths to cardiac causes. Dickinson et al. presented autopsy results of seven trekkers aged 27–62 who died of altitude illness in Nepal (112). Although one, aged 38, who had just climbed Kilimanjaro (5900 m) had severe occlusive disease of the anterior descending and circumflex arteries, none of the seven sustained a cardiac death. In summary, data are inadequate to draw firm conclusions regarding the risk of sudden death and myocardial infarction at high altitude in those with coronary artery disease. The limited available evidence to date does not identify an increased risk, at least up to the modest altitude of 2500 m.

Cold, Altitude, and Ischemia

Whether cold provokes myocardial ischemia at high altitude, as it is known to do at low altitude (113), has not been studied, but seems likely. At sea level Lassvik and Areskog observed the effect of cold air inhalation and/or a cold room on individuals with effort angina (114). The 12 men performed submaximal bicycle exercise. Skin cooling (room at −10°C) augmented sympathetic response and decreased work capacity due to onset of angina. Inhalation of extremely cold air (−35°C) had a similar effect. The cold-induced increase in double product at rest (14%) and exercise (8%) was similar to that reported at moderate altitude without cold. Therefore, cold and altitude might interact to incite effort-induced angina. Because high-altitude locations are commonly cold, this question of interaction deserves pursuit.

The Risk of Acute Ischemic Events in Those Without CAD

In addition to atherosclerosis, coronary artery vasospasm has also been invoked as a possible mechanism for ischemia at altitude. One patient of Hultgren's with T-wave inversion in the precordial leads during chest pain was subsequently found to have a normal coronary angiogram, and he was assumed to have coronary artery spasm perhaps induced by the altitude. Hultgren suggested this would not be surprising at altitude, given that sympathetic stimulation and respiratory alkalosis are known coronary vasoconstrictors (81).

What to Advise Persons with CAD Regarding Altitude Exposure?

The stress of high altitude on the coronary circulation appears to be minimal at rest, but significant in conjunction with exercise. The European studies mentioned above (110) imply that an exercise program at sea level prior to exercising at altitude may reduce the risk of sudden death. Ideally, no one with known CAD or even risk factors for CAD should undertake unaccustomed exercise at high altitude. In the absence of substantive information on outcome for high-altitude exposure of those with CAD, Hultgren applied the technique of risk stratification that is commonly used at sea level as a basis for providing advice (115). For asymptomatic males over age 50 with risk factors, he suggested an exercise test to determine risk status prior to altitude exposure and then further evaluation as indicated (Table 2). Low-risk patients as determined by exercise test can be reassured and given medical management as necessary. Note that patients with previous MI, bypass surgery, or angioplasty are considered high risk only if they have a strongly positive exercise treadmill test. High-risk patients may require coronary angiography to establish appropriate management. Alexander has defined somewhat differently the subset of patients he considers at high risk during altitude exposure: those with ejection fraction less than 35% at rest, a fall in exercise systolic blood pressure, ST segment depression greater than 2 mm at peak heart rate, and high-grade ventricular ectopy (103). For such patients, he recommends ascent to no more than 2500 m and proximity to medical care. Such recommendations, while reasonable, are unsubstantiated by outcome studies, falling into that time-honored realm of medical practice known as clinical judgment.

An increasingly frequent question these days is whether individuals after coronary artery bypass grafting or angioplasty are at increased risk at altitude. Cognizant of the paucity of outcome data available, a number of experienced physician-scientists have put into print their judgment as to whether and under what circumstances such individuals should or should not venture to high altitude (115–117).

Possible Beneficial Effects of Altitude on Coronary Artery Disease

Some work has suggested benefits of high altitude exposure for those with CAD. Meerson, for example, using a rat model of cardiac ischemia, was able to raise the fibrillation threshold two- to threefold in rats with intermittent exposure to hypoxia

Table 2 Suggested Approach to Evaluation of Coronary Artery Disease in Individuals Over Age 50 Who Wish to Go Trekking or Travel to High Altitude

Symptom	Evidence for coronary artery disease	Risk factors	Testing	Risk category
None	None	None	ETT optional	Low
None	None	1 or more	ETT −	Low
			ETT +, thallium −	Low
			ETT +, thallium +	High, needs angiogram
None, or stable angina	Past MI, old MI on ECG, history CABG, PTCA		ETT − or ±	Low
			ETT strongly +	High
Age < 50	None	None, 1 or more	ETT not indicated	Low
			ETT optional	Low

Risk factors: family history of coronary artery disease before age 55, history or presence of hypertension, ST depression of any magnitude in the resting ECG, and prior episode of chest pain.
Treadmill test negative or minimally positive: patient can walk 9 minutes or more (through Stage 3 Bruce protocol) without chest pain and with ≤1 mm ST segment depression. Treadmill test strongly positive: patient walks ≤6 minutes (Bruce protocol) and has either chest pain or ST depression ≥2 mm.
Risk categories: (a) low: risk for a coronary event (angina, infarct, sudden death) over 5 years of 2–4%; (b) high: risk for a coronary event in 5 years of 10–20%; requires coronary arteriography.

(118). In the same study, muscle healing and scar reduction after induced myocardial infarction were also improved with intermittent hypoxia. Researchers from Russia and Kyrgyzstan claim hypoxic training improves the functional capacity of the cardiovascular system, especially endurance, and resistance to severe hypoxia (56). In Peru, a cardiac rehabilitation program exists at 3750 m for sea level dwellers with heart disease. No acute MI, unstable angina, or sudden death in a previously known coronary patient has occurred within 2 weeks of arrival at 3750 m (81). Although detailed outcome data are not yet available, the clinicians consider the moderate hypoxic stimulus ''beneficial'' in a controlled cardiac rehabilitation program. Similarly, a French group has suggested skiing at moderate altitude as part of a program to lower cardiovascular risk factors but has not yet confirmed a benefit (119). Given the physiological and biochemical similarities of altitude acclimatization and aerobic conditioning, these proposals merit attention. Clearly, outcome data are needed to substantiate these clinical impressions.

C. Does Altitude Provoke Cardiac Arrhythmia?

Levine et al., in a study of older men (68 ± 3 years), many with coronary artery disease, found that PVCs increased 63% on acute ascent but returned to baseline

Table 3 Late Potentials and Inducible Ectopy

Altitude	QRS duration (ms)	Root-mean-square 40 (μV)	LAHFD (ms)	PVCs (beats)	Repetitive forms (beats)
Sea level	94.3 ± 3.2	66.7 ± 16.2	26.3 ± 1.8	7.5 ± 3.6	5.5 ± 2.8
Acute altitude	94.2 ± 3.1	68.7 ± 16.7	26.1 ± 2.0	12.6 ± 4.7*	4.6 ± 2.6
Acclimatization	98.2 ± 3.2	56.4 ± 12.0	34.4 ± 5.4	6.7 + 2.5	4.1 + 3.6

* $p < 0.05$ compared to sea level.
LAHFD: Low-amplitude, high-frequency duration.
Twenty elderly (17 with or at high risk for coronary artery disease) had signal-averaged ECG at sea level, during acute exposure to 2500 m, and after 5 days acclimatization to 2500 m. PVCs increased with acute exposure but without an increase in higher-grade ectopy (repetitive forms). Arrhythmia risk assessed by QRS duration, root-mean-square of the terminal 40 ms of the QRS, and duration of low-amplitude signals was unchanged. High altitude appeared not to increase risk of arrhythmia.
Source: Ref. 98.

after 5 days of acclimatization (98) (Table 3). Since urine norepinephrine increased 67% in the first 24 hours in these subjects, sympathetic activation seemed the most likely cause of the increased ectopy. They observed no increase in higher-grade ectopy, however, and no changes in signal-averaged ECG suggestive of a change in arrhythmia substrate or fibrillation threshold; in other words, the PVCs appeared benign, without suggestion of propensity to life-threatening arrhythmia. A number of other studies have noted increased ectopy in patients with coronary artery disease (84), but no dire events have been reported. Alexander recently described onset of asymptomatic PVCs and ventricular bigeminy, mostly of left ventricle origin, in himself while hiking up to 5900 m; interestingly, he had no evidence of heart disease on subsequent evaluation. He went on to thoroughly review the subject of altitude, age, and arrhythmia (99). Patients with preexisting troublesome or high-grade arrhythmia have not been evaluated systematically at high altitude, and anecdotal reports of exacerbations of supraventricular tachycardia or other common arrhythmia are largely word of mouth. No information exists to determine whether or to what extent ascent to high altitude might increase risk for those with specific rhythm disorders at sea level.

D. Heart Failure

Clinicians have noted a tendency toward acute decompensation upon arrival to high altitude in those with a history of heart failure (120), but systematic study is limited. Erdmann et al. recently performed exercise tests at 1000 m (baseline) and then upon arrival at 2500 m by cable car in 23 patients with coronary artery disease and ejection fraction less than 45% (121). None were in active failure. Compared to 23 control subjects, the decrement in exercise performance was similar, and no complications or signs of ischemia developed. They concluded that ascent to 2500 m resulted in no increased risk for an adverse effect in these individuals. While these results are encouraging for such patients, they made no observations past the first few hours at altitude. Although the hypoxemia of moderate altitude does not depress myocardial contractility (see Chapter 9), the tendency toward fluid retention in some at altitude, especially with acute mountain sickness, could aggravate heart failure. Alexander et al. noted during an exercise study in patients with angina that left ventricular ejection fraction declined at altitude, with an increase in end-diastolic and systolic volume as measured by two-dimensional echo (122). These changes were attributed to fluid overload rather than depressed ventricular contractility, which was unchanged from low altitude. Because those with impaired contractility are particularly sensitive to fluid overload, patients with heart failure may need to be cautious about high-altitude exposure. They especially need to avoid acute mountain sickness, which is associated with fluid retention. However, no information concerning this possibility exists. Outcome studies are needed to determine if patients with a known history of heart failure are at increased risk of developing failure on ascent to altitude. Presumably, prevention and treatment at altitude would be the same as at sea level, except for a probable greater benefit of low-flow oxygen to reduce the hypoxic stress. Patients may need to increase their diuretic medication at altitude and should monitor their blood pressure as well as their weight, peripheral edema, and chest congestion. Acetazolamide prophylaxis may be useful to consider in terms of speeding acclimatization, inducing a diuresis, and preventing AMS.

E. Congenital Heart Disease

The hypoxia of high altitude elevates pulmonary arterial pressure and resistance (see Chapter 10). As a result, conditions such as atrial and ventricular septal defects (ASD and VSD), patent ductus arteriosus, and partially corrected tetralogy of Fallot might have increased right-to-left shunting at high altitude. For example, children born at altitude with an ASD have pulmonary hypertension greater than those without ASD at high altitude and greater than those with ASD born at lower altitude (123). Children born in Denver with VSD have twice the pulmonary vascular resistance of those born with VSD at sea level (124). Whether patients with ASD and VSD suffer greater arterial oxygen desaturation at high altitude secondary to right-to-left shunting depends on the degree of pulmonary hypertension, both preexisting and consequent to altitude hypoxia. Such shunting would be expected to contribute to altitude illness, including high-altitude pulmonary edema, as well as to dyspnea, although these individuals may have lower HVR secondary to their chronic hypoxia (125). An interesting related question is whether defects resulting in reduced pulmonary blood flow, such as tetralogy of Fallot, might be protective for HAPE. Remarkably, children without a pulmonary pumping chamber (after a Fontan operation for palliation of tricuspid atresia) were able to tolerate exposure to 3000 m, even with some exercise, although exercise capacity appeared limited by pulmonary blood flow (126). Whether patients with persistent pulmonary hypertension after corrective surgery for ASD and VSD will have problems at high altitude is unknown. When patent foramen ovale was noticed in persons with high-altitude pulmonary edema, it raised the suspicion that shunting and hypoxemia may have contributed to the illness; further investigation revealed others with patent foramen ovale who did not develop HAPE, however, and cause and effect could not be established (127).

F. Pulmonary Vascular Disorders

In addition to congenital cardiac defects, any illness resulting in pulmonary hypertension may be a relative contraindication to high-altitude exposure. Hypoxic pulmonary vasoconstriction will most likely exaggerate preexisting pulmonary hypertension. Given the prominent role of pulmonary hypertension in the pathophysiology of HAPE, any such illness may predispose to HAPE. This caution applies to unilateral absent pulmonary artery, granulomatous mediastinitis, and restrictive lung diseases, all of which have been associated with high-altitude pulmonary edema (128–130). One would expect persons with primary pulmonary hypertension (PPH) to be more symptomatic, with increased dyspnea, more frequent syncope, and weakness on exertion. As Hultgren has observed, however, some patients with this disease are able to tolerate high altitude, and hypoxic gas breathing can be used to identify an individual's response to hypoxia if clinically indicated. Persons with PPH who must travel to high altitude may benefit from calcium channel blockers, isoproterenol, and/or low-flow oxygen (81). These individuals must also be aware of a probable increased susceptibility to HAPE, as was recently demonstrated by a lowland woman with PPH secondary to fenfluramine who developed two episodes of HAPE. The first episode was at 2300 m and the second at only 1850 m, with skiing up to 2350 m (131). Other conditions include bronchopulmonary dysplasia, recurrent pulmonary emboli, and mitral stenosis. Whether pulmonary hypertension is primary or secondary, patients should be made aware of the potential hazards of high altitude, including right heart failure and HAPE. Pulmonary hypertension should be considered at least a qualified contraindication to ascent to high altitude.

VI. Hematological Problems

A. Abnormal Hemoglobin

The hypoxia of high altitude is a well-recognized problem for those with sickle cell disease (132). Even pressurized aircraft cabin altitudes (up to 2500 m) will cause 20% of those with hemoglobin SS, SC, and sickle-thalassemia to have a vaso-occlusive crisis (133). Indeed, for some, ascent to altitude or flying triggers their first awareness of their condition. People with sickle cell disease living at high altitude (3000 m) in Saudi Arabia had twice the incidence of crises, hospitalizations, and complications as their counterparts at low altitude (134). Splenic infarction at altitude may be more common in those with the trait (Hb AS) than those with homozygous disease, who totally infarct their spleens early in life (135). The differential diagnosis of left upper quadrant pain at altitude (even as low as 1500 m) should include splenic infarction in Caucasians as well as those of African ancestry (133,135,136). The incidence of problems in those with sickle cell trait, however, is quite low, and ascent to altitude, even with vigorous exercise, is not contraindicated, in contrast to those with sickle cell disease. The U.S. Army, for example, does not consider soldiers with the trait unfit for duty at high altitude (137).

Hemoglobin with altered oxygen affinity may affect acclimatization to high altitude, as suggested by the interesting experiment of Eaton et al. (138,139). When individuals with a genetically left-shifted oxygen dissociation curve were taken to 3100 m, they evidenced a surprising lack of the usual tachycardia and tachypnea and superior exercise performance. At least at moderate altitude, increased loading of the hemoglobin in the lung presumably resulted in improved oxygen transport. Whether a decreased oxygen affinity would be detrimental has not been addressed.

B. Anemia

Given the central role of hemoglobin in oxygen transport, a low hemoglobin level might be expected to affect tolerance to high altitude. However, neither the extent of ventilatory acclimatization nor incidence or severity of AMS has been found to correlate with hemoglobin concentration. But persons with iron deficiency, even without frank anemia, have reduced exercise performance at high altitude (as at sea level) as well as a reduced erythropoietic response (140). Iron deficiency is especially common in women of menstruating age. No data are available on the role of anemia in susceptibility to high-altitude illness. Wilson described a mountaineer on Denali with a severe GI bleed from an ulcer. Incredibly, she survived 16 days, many of them above 5200 m, with an arterial O_2 content of less than 5 vol%. Hemoglobin concentration upon presentation to the hospital was 3.8 g/100 mL (141)! She developed no altitude illness, although she was weak and breathless.

VII. Neurological Conditions

A. Migraine

Whether high altitude aggravates migraine is not clear. That the headache of acute mountain sickness and that of migraine can be nearly indistinguishable presents obvious problems with diagnosis and determination of incidence. In fact, the mechanism could be similar for migraine and the headache of AMS. To test this hypothesis, sumatriptan, a $5HT_1$-receptor agonist effective against migraine, was evaluated as therapy for the headache of AMS; results were mixed. Bärtsch reported marked effectiveness of sumatriptan in an uncontrolled trial (142), but Burtscher et al., in a randomized, double blind design, found sumatriptan had no effect, while ibuprofen was uniformly successful (143). (For migraine headache at low altitude, sumatriptan is generally superior to ibuprofen.) These results suggested that the mechanism of headache in the two entities may differ. The subjects in the Bärtsch study had more severe headache of longer duration, however, and before the hypothesis of a similar mechanism is rejected, further study is required.

One population in which migraine is easier to identify at altitude is the high-altitude native, a population that does not suffer from AMS. Studies from Peru show a higher incidence of migraine in the high-altitude natives than in Peruvian lowlanders: 12.4% of a surveyed general population at 4300 m, compared to 3.6% at sea level (144). In a group of 379 adult men subsequently studied at 4300 m by the same investigators, the incidence of migraine headaches was 32.2%, with a positive correlation to increasing age, hemoglobin values, and chronic mountain sickness scores (145). For all types of headaches, incidence was 54.6%, far higher than reported in Peruvian and other populations living at low altitude. The authors postulated that migraine in this population might be higher due to the combined effects of hypoxia, polycythemia, and reduced cerebral blood flow. Since migraine in some of this population may be related to chronic mountain sickness, the implications for high-altitude sojourners are unclear.

Whether ascent to high altitude will increase the frequency or severity of headaches in lowlanders with migraine is not yet determined, but clearly, ascent can trigger migraine both in those with and those without a prior history of migraine at sea level (146,147). For some migraineurs, ascent to altitude may be a reliable trigger: migraine developed repeatedly in susceptible subjects during simulated altitude exposure between 2750 and 3400 m (147). For others, while frequency may remain stable, the severity may change. A physician reported his usual migraine frequency during 2 years at 3700 m, but with marked focal neurological deficits he did not usually experience at low altitude (148). Upon relocating to low altitude, the focal neurological deficits did not recur. The mountain climber reported by Jenzer and Bärtsch also had hemi-symptomatology, with two episodes of right-sided hemiparesis and visual and speech problems (146). Subsequent neurological evaluation each time was negative. Other authors have also observed focal, one-sided neurological symptoms or visual symptoms consistent with a migraine event, but perhaps mistakenly attributed to a transient ischemic attack (TIA), another possible but unlikely diagnosis in these patients, given their young age and good health (120,149–151). Therapy with oxygen breathing, CO_2-enriched air breathing, or descent was effective, which points more toward a cerebrovascular migraine-like phenomenon than a TIA (120,149). Migraine must be included in the differential diagnosis of a severe headache at altitude, even with apparent AMS, especially when associated with visual or other focal neurological deficits. Whether hypoxic gas breathing at sea level may be useful to identify those migraineurs for whom high altitude may trigger an attack has not been investigated.

B. Stroke, TIA, and Cerebrovascular Disorders

Whether the incidence or severity of cerebrovascular disease (CVD) and stroke is different in natives of high altitude than in sea level populations is unclear. Surveys conducted in rural areas of South America and Asia over a range of 1550–4300 m have suggested a lower incidence of CVD (152–154). A recent epidemiological study reported that the prevalence ratio for CVD at 3400 m in Peru (5.74 per 100 when age-adjusted to the WHO point population, and 8.58 per 1000 adjusted to the U.S. point population) was close to the worldwide average (155). Polycythemia, age, and alcohol consumption, all of which are known independent risk factors for stroke at low altitude, were also correlated with stroke at high altitude. In addition, urban residence was a major risk factor (155). Discerning the true relationship between altitude and cerebrovascular disease, if any, will require careful control for racial characteristics and lifestyle while comparing low- and high-altitude populations.

Rarely, stroke has been reported in generally healthy sojourners at altitude and typically ascribed to dehydration, forced inactivity, and polycythemia (120,156,157). Stroke may also be a complication of high-altitude cerebral edema; cerebral thrombosis is a common finding at autopsy (112,158). Clinically, cerebral venous thrombosis is easily confused with HACE. A number of reports emphasize the need to consider this diagnosis in patients with persistent neurological symptoms after descent from altitude (112,159–161). In a recent series of patients who were diagnosed with cerebral venous thrombosis, all had been on long air flights, and one had been mountaineering (162). An accompanying editorial suggested hypercoagulability or viscosity may be factors, and, being relatively easy to treat, this condition ought not be missed (163). Whether acute high-altitude exposure might contribute to a hypercoagulable state that may contribute to stroke is debated (see Chapter 15). The interesting question of whether high altitude might unmask cerebral vascular insufficiency in those with asymptomatic cerebral arteriosclerotic disease has not been addressed.

Transient, focal neurological signs in the absence of high-altitude cerebral edema have been reported and are usually attributed to transient ischemic attacks. However, other possible etiologies include cerebrovascular spasm or migraine equivalent, focal edema, and hypoxia in the watershed zones of minimal cerebral blood flow. Most of these individuals have been healthy, young mountaineers, not the age group expected for atherosclerotic vascular disease, the usual cause of TIA. Hackett et al. described six cases of transient blindness that appeared to be due to a cortical process, since pupillary reflexes were intact (149). Supplemental oxygen or carbon dioxide breathing promptly reversed the condition, which suggested vascular spasm rather than occlusion. Descent provided slower relief. Houston described similar incidents (50), and Wohns reported other problems consistent with TIA: transient hemiplegia and hemiparesis, aphasia, and scotoma (164). In all cases, recovery was complete, and in the few subsequently evaluated by MRI and MR angiography, no pathology was detected. The etiology of these focal neurological events remains a mystery and certainly warrants further study.

Because of the significant cerebral vasodilation on ascent to altitude, persons with cerebrovascular structural abnormalities, such as arteriovenous malformation or aneurysm, may be at risk for an untoward event. I cared for two patients in Nepal with focal neurological deficits, who upon subsequent evaluation were found to have arteriovenous malformations. A third trekker thought to have a TIA was later diagnosed with an aneurysm, which the consultants in retrospect thought had leaked at high altitude. These anecdotes cause one to wonder whether high altitude might put susceptible individuals at increased risk.

C. Brain Tumors

Space-occupying lesions can be unmasked upon ascent to high altitude, presumably because of the change in intracranial dynamics, with an increase in brain volume. Four such cases, two with meningioma and two with malignancies, have now been reported. (165) None of the patients had their tumors previously diagnosed.

D. Seizures

Acute, severe hypoxia may cause seizure, a fact more relevant and well known in aviation medicine. Incredibly, Russian investigators using rapid, stepwise ascent to over 9000 m in altitude chambers, considered the onset of seizures an index of CNS intolerance to hypoxia when selecting persons for a national high-altitude climbing team to Mt. Everest (166). These observations are of questionable relevance for sojourners who take time to acclimatize and go to more moderate altitudes. In fact, de novo seizure in sojourners at altitude, without an underlying seizure focus, and with or without HACE, appears to be rare.

In persons with seizure disorders, exacerbation possibly due to altitude has been anecdotally observed, at least in those not on medication. Shlim (personal communication) noted two persons with new onset seizure within hours after flying to Lhasa (3700 m). On subsequent work-up, one was found to have epilepsy, while the cause for the other was unexplained. On Mt. McKinley, a camp worker had a grand mal seizure within 12 hours of abrupt ascent to 4300 m. She had a childhood history of epilepsy but had discontinued her medications years earlier, since she had been without seizures. She remained at altitude, taking phenytoin, and had no further problems. An 11-year-old girl suffered a seizure on a chair lift at 3000 m, and subsequent evaluation revealed a previously unknown epileptic focus. The mechanism of exacerbation might be related to the respiratory alkalosis of altitude, similar to the acute hyperventilation used in EEG labs to trigger seizure events, or to the hypoxia. In contrast to these apparently unusual events, syncope is common at high altitude (167). Since syncope of any cause may have attendant seizure activity (convulsive syncope), confusion of convulsive syncope with a true seizure is inevitable. Systematic studies of patients with seizure disorders are sorely needed to determine whether high altitude is truly a risk for these individuals. Anecdotally, those with seizure disorders have had no increase in frequency or severity of seizures when medications were continued at high altitude. Another unanswered question is whether an individual with a grand mal seizure at very high altitude might be more at risk for residual brain injury.

VIII. Diabetes Mellitus

Do diabetics have more problems at high altitude? Shlim (personal communication) observed onset of severe diabetic ketoacidosis in five trekkers at high altitude, three of whom died: none had previous episodes and all were accustomed to exercise.

While these episodes raise the suspicion of a problem, whether incidence or severity is increased at high altitude is unknown. For diabetics assessing hyperglycemia, recent reports have questioned the reliability of self-monitoring blood glucose devices at high altitude (168,169), one study reporting a response as low as minus 34% from the reference standard. The cause of such error was not addressed. A recent study, however, showed good reliability of one system at 2668 and 3665 m in Colorado, with a clinically insignificant increase in imprecision of 1–3% from sea level (170).

IX. Ophthalmological Conditions

The parts of the eye most affected by altitude are the cornea and the retina. The most common corneal problem at high altitude is UV keratitis, or photokeratoconjunctivitis, also known as snow blindness or corneal sunburn, due to the increased ultraviolet light and increased reflection from snow at altitude. Hypoxia also affects the cornea. Since the cornea nearly exclusively obtains its oxygen by diffusion from the atmosphere, the decline in partial pressure of oxygen at high altitude causes a decline in corneal oxygen tension. This hypoxia disturbs the homeostasis of corneal fluid balance, and the cornea diffusely swells, especially after lid closure during sleep, which represents an additional hypoxic stress. Since the swelling is uniform, there is no change in corneal curvature and no visual abnormality occurs (171).

Persons who have had radial keratotomy for refraction correction, however, no longer have structurally normal corneas. Deep radial incisions have been made that interrupt stromal fibers in the area of the incision, allowing the central cornea to become flattened, improving far-sighted vision. On ascent to altitude, however, the swelling of the hypoxic cornea is not uniform, and significant visual changes may result (Fig. 5). A case report suggesting such an effect of high altitude in a post–radial keratotomy (RK) person appeared in 1988 (172). A subsequent case report (173) led to a study of two subjects after RK and two controls at sea level—12,000 and 17,000 ft—which confirmed major hyperopic shifts (greater far-sightedness) in the four operated corneas (174). The degree of refractive change was related to baseline postoperative refractive error (indicating it may be somewhat

Figure 5 A schematic diagram of the change in the cornea upon ascent to high altitude. The normal and the PRK cornea swell uniformly. The periphery of the RK cornea swells more than the center, resulting in central corneal flattening that affects vision. (Courtesy of Tom Mader, M.D.)

predictable) and was progressive with increasing altitude. The authors ascribed the changes to the observed flattening of the cornea at high altitude and speculated that hypoxia caused increased hydration of the corneal stroma with resultant preferential peripheral expansion of the cornea in the area of the keratotomy incisions. A follow-up study by Ng et al. (175) found no effect on refraction in RK patients with a 6-hour normoxic hypobaric exposure and concluded that the effect must not be due to hypobaria per se, but to a hypoxic-induced metabolic process. This interpretation was confirmed by Winkle et al., who were able to produce the same corneal flattening and hyperopic shift in post RK eyes with the use of an air-tight goggle system and a nitrogen bleed-in that created ocular surface hypoxia at sea level (176). The clinical implications are significant as dramatically documented by the problems of a post RK climber on a recent expedition to Mt. Everest. At 8300 m, over a few hours, he had increasing difficulty seeing, to the point of near-blindness, which he attributed to his prior bilateral RK procedure. (B. Weathers, personal communication). He suffered severe frostbite, nearly died, and required an enormous rescue effort. Similar visual changes have been reported in other climbers to high altitude (177,178) and may occur even below 3000 m. The problem is likely to become worse with decreasing accommodation as one ages. In contrast, photorefractive keratectomy (PRK) is a laser technique that shaves the anterior cornea uniformly, without incisions. Mader et al. compared 11 post-RK eyes with 12 post-PRK eyes and 17 untreated myopic eyes at sea level, on 3 consecutive days at 4200 m, and on return to sea level. Peripheral corneal swelling was observed in all the groups at high altitude, but only in the RK eyes were refractive changes noted; PRK caused no change in vision at high altitude (179). For individuals who may travel to high altitude, PRK is preferable to RK for surgical correction of myopia. Those with past RK can correct their vision by bringing with them eyeglass of increasing plus power (175,177).

High-altitude retinal hemorrhages are discussed in Chapter 22. No information exists as to whether high altitude may contribute to or aggravate the retinal microangiopathy of hypertension, diabetes, or other diseases, and no research has addressed the effect of ascent to altitude on such patients. Even if these populations exhibited a ''normal'' incidence of retinal hemorrhage at altitude, such an event could increase the likelihood of visual impairment. While retinal hemorrhages are rare below 4500 m, except in those with AMS or HAPE (180), whether the threshold altitude is the same in the diseased retina is unknown.

X. Altitude and Pregnancy

Pregnancy and reproduction of high-altitude dwellers are discussed in Chapter 3. Here I will deal with the effects of ascent to altitude on the pregnant lowlander and whether altitude poses any risk to the well being of mother and fetus. The discussion draws from what has been learned from research on altitude residents and what is known of the effects of acute hypoxia or ascent to high altitude in pregnant animals and humans.

First, consideration of the fetal milieu is in order. A normal mother breathing room air at sea level has an arterial Po_2 of 85–100 mmHg and a uterine venous Po_2 of approximately 40 mmHg. The Po_2 in the umbilical vein (analogous to the pulmonary vein or systemic artery in the adult) will be 30–35 mmHg and in the umbilical artery (similar to the pulmonary artery) 10–15 mmHg. Breathnach has argued that this situation should be considered fetal normoxia and that applying the term "fetal hypoxia" to it is inappropriate (181). This "low-oxygen" fetal environment is remarkably stable. Due to the shape of the oxygen dissociation curve and the high affinity of fetal hemoglobin for oxygen, changes in maternal arterial oxygenation are minimized in the fetus (182). The placenta acts as a buffer in multiple ways: it maintains a constant Pco_2 gradient (10 mmHg), is relatively impermeable to bicarbonate ions (to protect the fetus from maternal changes in hydrogen ion concentration), and maintains the fetal oxygen environment (181). Additional strategies to maintain fetal stability include increased maternal ventilation during pregnancy (even in high altitude residents who may normally be insensitive to the chronic hypoxia; see below) and increased oxygen extraction by the fetus. As a result, oxygen consumption remains stable even when stressed with a 50% reduction in either placental blood flow or blood oxygen content (183). In a mother with cyanotic congenital heart disease, e.g., a Sao_2 of only 82%, the fetal umbilical oxygen saturations are 40% venous and 7% arterial, and oxygen consumption is maintained. Further, the fetus must avoid a high Po_2 as well as a severely low Po_2, for raised Po_2 is the trigger for the drop in pulmonary vascular resistance and closure of the ductus arteriosus (181,182). Given a stable fetal environment, what degree of maternal hypoxia can be detrimental, and what evidence is there for compromise of the lowland fetus or mother upon ascent to high altitude?

A. Studies of High-Altitude Residents: Implications for Pregnant Lowlanders

In native residents at an altitude of 6000 ft (1830 m), umbilical cord arterial and venous oxygen tensions are the same as at sea level, while a slightly lower Pco_2 reflects the mild maternal hyperventilation. At altitudes over 3000 m, fetal response to hypoxia is evidenced by increased hematocrit (2–3% higher), increased fraction that is fetal hemoglobin, and increased erythropoietin in the cord blood (184,185).

Research from Colorado, South America, Africa, and Asia has suggested various effects of high-altitude residence on pregnancy. The effect best documented is intrauterine growth retardation, which leads to healthy, full-term infants that are small for gestational age (see Chapter 3). In high-altitude areas of the world with accessible medical care intrauterine growth retardation is not associated with increased morbidity or mortality unless the infants are also premature. Lower birth weight has even been suggested as a possible survival advantage at high altitude (186), a concept refuted by data from Unger et al. in Colorado (187). Intrauterine growth retardation, however, does indicate an apparent regulatory effect of oxygen on the developing fetus (182). Similar reductions in birth weight are seen with mothers who smoke, but unlike the altitude infants, the infants of mothers who smoke have higher perinatal mortality at every birth weight. This important difference indicates that effects on fetal growth can occur independently of effects on mortality (188) and also that smoking may involve other mechanisms than oxygen transport affecting fetal growth and well-being. Small size for gestational age at sea level is also associated with preeclampsia, maternal hypoxic lung disease (189), maternal cyanotic congenital heart disease (190), and various anemias; all have diminished fetal oxygen/nutrient delivery as a common pathway. Unlike some of these other conditions, particularly preeclampsia, low birth weight due to high altitude (in full-term infants) may not be associated with increased morbidity and mortality.

In a study at 4300 m, Moore et al. observed a positive relationship between birth weight at high altitude and maternal oxygen transport (191). In contrast, at low altitude the reverse is true: the "physiological anemia" of pregnancy, with a reduced arterial oxygen content, is correlated with higher birth weight than those who do not develop this "anemia." To help clarify the role of maternal arterial oxygen content, Moore and colleagues compared infant birth weights in women from low and high altitude who had the same maternal arterial O_2 content values (Hgb $\times$ $Sao_2\%$ $\times$ 1.34). For the same oxygen content, the high-altitude women still had smaller infants (191). Further investigations by Zamudio et al. suggested that the mechanism of intrauterine growth retardation at high altitude was perhaps in larger part due to lower uteroplacental blood flow and "relative ischemia" of the placenta (192). Three factors seemed to explain the decreased blood flow at high altitude: a lesser maternal blood volume increase, less uterine artery dilatation, and lack of appropriate redistribution of blood flow to the uteroplacental circulation. How hypoxia might produce these changes in unknown. The authors speculated that a common explanation for these findings could be impaired placentation, the process by which the placenta "invades" the uterus to establish its blood supply. Indeed, recent in vitro work has suggested a role for oxygen tension in the differentiation and invasion of cytotrophoblasts in the human placenta; the result of hypoxia was shallower placentation and higher vascular resistance, like that seen in preeclampsia (193). Although this finding needs to be confirmed anatomically in normal high-altitude placentas, this work makes one wonder whether during placentation (9–12 weeks) a hypoxic exposure could be detrimental. A corollary question to consider is whether altitude exposure is advisable for lowland women with any suggestion of impaired placentation and/or fetal compromise, such as hypertension or pre-eclampsia, at any time during pregnancy.

A second lesson to be learned from high-altitude residents is that intrauterine growth retardation is not linear throughout pregnancy. Only after 32 weeks gestation does fetal growth at altitude become appreciably slowed compared to sea level (187). Does this observation mean that the otherwise healthy pregnant lowlander need not worry about the possibility of impaired fetal growth at altitude until after 32 weeks gestation? Or is the stage set for intrauterine growth retardation earlier in the pregnancy, such as during placentation, so that altitude exposure afterwards has no effect on growth? Is the intrauterine growth retardation of altitude residents even relevant to lowlanders? Without answers to these questions, it seems premature to admonish all women to avoid altitude exposure throughout pregnancy, as some have done because of concerns about intrauterine growth retardation (194). At the present time, we have no evidence that exposure, at least to moderate altitudes, increases risk to the healthy pregnant lowlander or her fetus.

B. Acute Ascent—The Altitude Sojourner

Clinical and physiological investigations of pregnant lowlanders ascending to altitude are conspicuously lacking, especially to altitudes over 2500 m. Thousands of pregnant women travel to the moderate altitude of ski resorts for recreation, and the lack of any reported adverse effects could be reassuring. On the other hand, the safety of altitude exposure during pregnancy has not been systematically evaluated. Is adding the stress of exercise (e.g., skiing) to that of hypoxia a cause for concern? The few studies available are reassuring. Artal et al. studied seven sedentary women at 34 weeks gestation. Maximal and submaximal exercise tests were completed at sea level and 6000 ft. (1830 m) after 2–4 days of acclimatization (195). They reported the expected decrease in maximal aerobic work, but found no difference from sea level in submaximal endurance. Fetal heart rate responses and maternal lactate, epinephrine, and norepinephrine levels were not changed from sea level. The reduction in peak $V\text{o}_2$ was 13%, which is large for that altitude and possibly related to poor conditioning. Acknowledging the small number of subjects, the authors concluded that it was safe for third-trimester women to engage in brief bouts of submaximal exercise at moderate altitude. Drawing similar conclusions was a study of 12 pregnant subjects who exercised after ascent to 2225 m. Finding no abnormal fetal heart rate responses, the authors considered the exercise at altitude benign for mother and fetus (196). The experience of the airline industry may have some bearing on altitude exposure during pregnancy. Although the exposures are brief, cumulative time at cabin altitudes up to 2500 m is high. Pregnant cabin crew are generally permitted to fly until 7 months gestation with some variance among airline companies. Untoward effects have not yet been demonstrated in this large population, though studies are continuing. In summary, the only available data, though rather inadequate, suggest that short-term exposure to altitudes up to 2500 m, with exercise, is safe for a lowland woman with a normal pregnancy. To ascertain whether and to what extent fetal morbidity or mortality might be affected by increasing altitude would require investigation of outcomes in a large number of pregnant women.

C. Acute Hypoxia

Acute hypoxic challenge may provide information on the response of mother and fetus to altitude exposure. Human investigations have generally used hypoxic gas mixtures of known concentration of oxygen, but without control of $Sa\text{o}_2\%$, $Pa\text{o}_2$, or $Pa\text{co}_2$, and the effect on the fetus has been limited to heart rate response. Hypoxic effects have been variable, with only slight tachycardia as the usual response to moderate maternal hypoxia (12% or 15% inhaled oxygen) and a bradycardic response to more severe hypoxia (10% oxygen). In sheep, acute, severe hypoxia (10% oxygen), with an average maternal Pao_2 of 40 mmHg resulted in no change in uterine or umbilical blood flow but substantial increases in fetal heart rate (197,198). Researchers have suggested acute hypoxic gas breathing as a tool for detecting placental insufficiency and potential labor and delivery problems; fetuses with abnormal responses, such as prolonged recovery from tachycardia, were much more likely to have fetal distress in labor (199,200). Another way to assess relative hypoxia/ischemia of the uteroplacental unit is to determine fetal response to oxygen breathing; improved physiological function with oxygen implies correction of a deficit. While no reports of this intervention could be found at high altitude, maternal hyperoxia at sea level had no effect in normal pregnancy or mild preeclampsia, but caused observable physiological changes (increased heart rate, increased variability, and fetal breathing movements) in severe preeclampsia, in fetal growth retardation, and in small-for-gestational-age infants (200–202). The importance of these studies for lowland women is that they suggest that a compromised placental-fetal circulation could be unmasked at high altitude. On the other hand, in the absence of such complications, the fetus seems to tolerate a level of acute hypoxia far exceeding a moderate altitude exposure.

Given the available data, it seems prudent to recommend that women with any complication of pregnancy avoid unnecessary altitude exposure. An ultrasound or other assessment may help to reassure the clinician and mother about the absence of the more common complications. For women with no known abnormalities, there appears to be little risk to the fetus or the mother undertaking a sojourn to an altitude at which Sao_2 will remain above 85% most of the time (up to 3000 m altitude), but the data are very limited. It is not the altitude per se that determines whether the fetus becomes stressed, of course, but the maternal (and fetal) Pao_2 and Sao_2. A woman with high-altitude pulmonary edema at 2500 m, for example, is much more hypoxemic than a healthy woman at 5000 m. Altitude illness, especially pulmonary edema, must be carefully avoided. Similarly, carboxyhemoglobin from smoking, lung disease, and other problems of oxygen transport will render the pregnant patient at altitude more hypoxemic and physiologically at a higher altitude. Whether breathing a hypoxic gas mixture could be a useful challenge test has not been evaluated. The question of the safety of modest hypoxia to the fetus and mother at different stages of pregnancy and at different altitudes (levels of hypoxia), the issue of whether persons at risk for any possible untoward effects can be identified prior to exposure, and the interaction of hypoxia with other stresses, such as exercise, clearly require much more investigation. Basic science and clinical and physiological research are all necessary. Outcome studies comparing large populations of women with and without high-altitude exposure during pregnancy would be especially useful to help the pregnant lowlander make informed decisions about potential risks of high altitude.

XI. Alcohol at Altitude

Two questions regarding alcohol are frequently asked: (1) Does alcohol affect acclimatization, and (2) does altitude potentiate the effects of alcohol? A recent epidemiological study indicated that 64% of tourists ingested alcohol during the first few days at 2800 m (203). The effect of alcohol on altitude tolerance and acclimatization might therefore be of considerable relevance. Roeggla et al. determined blood gases 1 hour after ingestion of 50 g of alcohol (equivalent to 1 L of beer) at 171 m and again after 4 hours at 3000 m. A placebo-controlled, double-blind paired design was used. For the 10 subjects, alcohol had no effect on ventilation at the low altitude, but at high altitude it depressed ventilation, as gauged by a decreased arterial Po_2 (from 69 to 64 mmHg) and increased Pco_2 (from 32.5 to 34 mmHg) (204). Whether this degree of ventilatory depression would contribute to acute mountain sickness and whether repeated doses would have greater effect was not tested. Nonetheless, the authors argue that alcohol might impede ventilatory acclimatization and should be used with caution at high altitude.

Conventional wisdom proffers an additive effect of altitude and alcohol on brain function. McFarland, who was concerned about the interaction in aviators, wrote "the alcohol in two or three cocktails would have the physiological action of four or five drinks at altitudes of approximately 10,000 to 12,000 ft." (205). Also, "Airmen should be informed that the effects of alcohol are similar to those of oxygen want and that the combined effects on the brain and the CNS are significant at altitudes even as low as 8,000 to 10,000 ft." (205). His original observations were made on two subjects in the Andes in 1936. He found that blood alcohol levels rose more rapidly and reached higher values at altitude but noted no interactive effect of alcohol and altitudes of 3810 and 5335 m (206). Most subsequent studies refuted the increased blood alcohol concentration data except at the highest altitudes, over 5450 m. Higgins et al., in a series of chamber studies (207,208), found that blood alcohol levels were similar at 392 m and 3660 m, and they noted no synergistic effects of alcohol and altitude. Lategola et al. (209) found that blood alcohol uptake curves were the same at sea level and 3660 m, and performance on math tests showed no interaction between alcohol and altitude. In another study of 25 men, performance scores were similar at sea level and at a simulated 3810 m altitude, with blood alcohol level of 88 mg% (210) (Fig. 6). Performance was not affected by hypoxia, only by alcohol, and older subjects were more affected. When more demanding tasks were tested, Collins found that a blood alcohol level of 91 mg% affected performance, as did an altitude of 3660 m during night sessions when the subjects were sleep deprived, but there was no significant altitude-alcohol interaction (211). In the one study in which Collins et al. were able to discern some altitude effect, there was a simple additive interaction of altitude (hypoxic gas breathing) and alcohol (212). They concluded that performance decrements due to alcohol may be increased by altitudes of 3660 m (12,000 ft) if subjects are negatively affected by that altitude without alcohol. All of these aviation-oriented studies used acute hypoxia

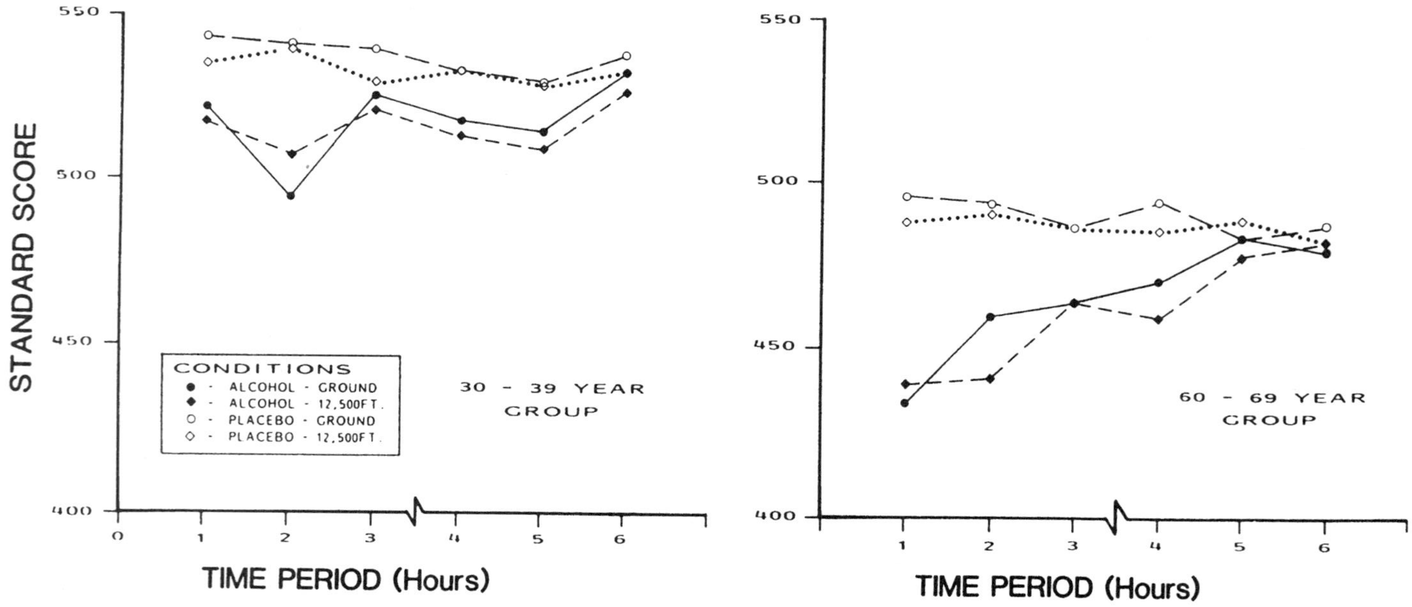

Figure 6 Mean overall composite scores from Multiple Task Performance Battery of groups of younger and older subjects at sea level and at a simulated altitude of 3810 m for 6-hourly work periods (a 1-hour lunch break preceded the fourth work period). Note an effect due to alcohol, but not due to high altitude, and no interaction between alcohol and altitude. (From Ref. 210.)

equivalent to no more than 3500 m. Perhaps the highest altitude (without supplemental oxygen) at which alcohol was studied was 4350 m, on the summit of Mt. Evans in Colorado. Freedman et al. found that alcohol affected auditory evoked potentials the same as in Denver; i.e., no influence of altitude was detectable (213).

In summary, the possibility of interactions between alcohol and altitude deserves study. The limited data on blood gases at altitude after alcohol ingestion support the popular notion that alcohol could slow ventilatory acclimatization. Considerable data, however, refute the belief that at least up to 3660 m, altitude potentiates the effect of alcohol. How altitude and alcohol might interact during various stages of acclimatization in individuals at higher altitudes is still unknown.

XII. Future Directions

To assess successfully the effect of high altitude on myriad medical conditions is a difficult task requiring multiple approaches. The most fruitful avenue of research would be development of a large, ongoing epidemiological data system, similar to the surveillance systems used for detecting adverse drug reactions or environmental toxicity. Many of the questions regarding the interaction of altitude with age, disease, and pregnancy require such a large-scale approach. Practitioners and public health workers in high-altitude counties in North America and other high-altitude locations that draw tourists with their various medical problems are ideally suited for this type of investigation. In addition to being alert for possible impacts of altitude on preexisting conditions in persons seeking medical care, focused surveys of hotel and resort guests can yield data on prevalence of medical conditions in a general population. Occupational health personnel as well as local health care providers could make valuable contributions to an epidemiological survey assessing the health of individuals who spend time at high altitude locations for variable periods, especially in the higher-altitude locations of South America and Asia. Focused, detailed studies of selected medical conditions or populations, such as the elderly and those with coronary artery disease investigated in the 10th Mountain Division Study, play an important role in understanding how specific conditions are affected by altitude hypoxia. This model can be applied to the neonatal and pediatric age groups, pregnancy, and migraine, to name but a few of the more pressing areas needing research. Such groups with their special problems are also well suited to outcome studies. Acute hypoxic stress tests need to be correlated with the effects of intermediate and long-term acclimatization. The study of high-altitude natives will continue to be important for the sojourner. For example, intrauterine growth retardation in high-altitude natives alerts us to the possible effects of hypoxia on the fetus of the sojourner as well, and understanding its mechanism will help assess its relevance for the altitude visitor and whether such effects can be prevented or minimized. Finally, as the molecular bases of adaptation to hypoxia and to disease unfold, new breakthroughs might be available to minimize hypoxic stress and maximize wellness in those persons with medical conditions who choose to visit high altitude ''for themselves, because it adds to their enjoyment or quality of life'' (1).

References

1. Barclay W. Obstructive lung disease and altitude adaptation (editorial). JAMA 1978; 240:1520.
2. Graham WG, Houston CS. Short-term adaptation to moderate altitude. Patients with chronic obstructive pulmonary disease. J Am Med Assoc 1978; 240:1491–1494.
3. Reeves JT, Wagner J, Zafren K, Honigman B, Schoene RB. Seasonal variation in barometric pressure and temperature in Summit County: effect on altitude illness. In: Sutton JR, Houston CS, Coates G, eds. Hypoxia and Molecular Medicine. Burlington, VT: Queen City Press, 1993:275–281.
4. Sutton JR, Houston CS, Coates G. Hypoxia and Cold. New York: Praeger, 1987.
5. Bligh J. The additive stresses of hypoxia and cold exposure: brisket disease in cattle. In: Sutton JR, Houston CS, Coates G, eds. Hypoxia and Cold. New York: Praeger, 1987:178–186.
6. Reeves JT. Sympathetics and hypoxia. A brief overview. In: Sutton JR, Houston CS, Coates G, eds. Hypoxia and Molecular Medicine. Burlington, VT: Queen City Press, 1993:1–6.
7. Wood SC. Interactions between hypoxia and hypothermia. Ann Rev Physiol 1991; 53: 71–85.
8. Denniston D. High Priorities: Conserving Mountain Ecosystems and Cultures. Washington, DC: Worldwatch Institute, 1995.
9. Cottrell J, Lebovitz B, Fennell R, Kohn G. Inflight arterial saturation: continuous monitoring by pulse oximetry. Aviat Space Environ Med 1995; 66:126–130.
10. Ryan ML, Hedrick MS, Pizarro J, Bisgard GE. Effects of carotid body sympathetic denervation on ventilatory acclimatization to hypoxia in the goat. Respir Physiol 1995; 99:215–224.
11. Forster HV, Bisgard GE, Klein JP. Effect of pheripheral chemoreceptor denervation on acclimatization of goats during hypoxia. J Appl Physiol 1981; 40:392–398.
12. Smith CA, Bisgard GE, Nielsen AM, et al. Carotid bodies are required for ventilatory acclimatization to chronic hypoxia. J Appl Physiol 1986; 60:1003–1010.
13. Honda Y. Respiratory and circulatory activities in carotid body-resected humans. J Appl Physiol 1992; 73:1–8.
14. Powles AP, Sutton JR. Sleep at altitude. Semin Respir Med 1983; 5:175–180.
15. Dubowitz G. Effect of temazepam on oxygen saturation and sleep quality at high altitude: randomised placebo controlled crossover study. BMJ 1998; 316:587–589.
16. Roeggla G, Roeggla M, Wagner A, Laggner AN. Poor ventilatory response to mild hypoxia may inhibit acclimatization at moderate altitude in elderly patients after carotid surgery. Br J Sport Med 1995; 29:110–112.
17. Kryger M, Weil JV, Grover RF. Chronic mountain polycythemia: a disorder of the regulation of breathing during sleep? Chest 1978; 73:303–304.
18. Fujimoto K, Matsuzawa Y, Hirai K, et al. Irregular nocturnal breathing patterns at high altitude in subjects susceptible to high-altitude pulmonary edema (HAPE): a preliminary study. Aviat Space Environ Med 1989; 60:786–791.
19. Hackett PH, Roach RC, Harrison GL, Schoene RB, Mills Jr WJ. Respiratory stimulants and sleep periodic breathing at high altitude. Almitrine versus acetazolamide. Am Rev Respir Dis 1987; 135:896–898.

20. Fromm R, Varon J, Lechin A, Hirshkowitz M. CPAP machine performance and altitude. Chest 1995; 108:1577–1580.
21. Renzetti AD, McClement JH, Litt BD. The Veterans Administration Cooperative Study of Pulmonary Function. Am J Med 1966; 41:115–130.
22. Sauer H. Geographic patterns in the risk of dying and associated factors, ages 35–74 years. United States, 1968–1972. Washington, DC: DHHS, 1980.
23. Moore L, Rohr A, Maisenbach J, Reeves J. Emphysema mortality is increased in Colorado residents at high altitude. Am Rev Respir Dis 1982; 126:225–228.
24. Coultas DB, Samet JM, Wiggins CL. Altitude and mortality from chronic obstructive lung disease in New Mexico. Arch Environ Health 1984; 39:355–359.
25. Regensteiner JG, Moore LG. Migration of the elderly from high altitudes in Colorado. J Am Med Assoc 1985; 253:3124–3128.
26. Saiyed H, Sharma Y, Sadhu H, et al. Non-occupational pneumoconiosis at high altitude villages in central Ladakh. Br J Ind Med 1991; 48:825–829.
27. Cote T, Stroup D, Dwyer D, Horan J, Peterson D. Chronic obstructive pulmonary disease mortality: a role for altitude. Chest 1993; 103:1194–1197.
28. Kramer M, Springer C, Berkman N, et al. Effect of natural oxygen enrichment at low altitude on oxygen-dependent patients with end-stage lung disease. Ann Intern Med 1994; 121:658–662.
29. Finkelstein S, Tomashefski JF, Shillito FH. Pulmonary mechanics at altitude in normal subjects and in obstructive lung disease patients. Aerospace Med 1965; 36:880–884.
30. Tomashefski J, Feeley D, Shillito F. Effects of altitude on emphysematous blebs and bullae. Aerospace Med 1966:1158–1162.
31. Shillito F, Tomashefski J, Ashe W. The exposure of ambulatory patients to moderate altitudes. Aerospace Med 1963:850–857.
32. Schwartz JS, Beucowitz HZ, Moser KM. Air travel hypoxemia with chronic obstructive pulmonary disease. Ann Intern Med 1984; 100:473–477.
33. Gong H, Tashkin DP, Lee EY, Simmons MS. Hypoxia-altitude simulation test. Evaluation of patients with chronic airway obstruction. Am Rev Respir Dis 1984; 130:980–986.
34. Gong H. Air travel and oxygen therapy in cardiopulmonary patients. Chest 1992; 101: 1104–1113.
35. Dillard TA, Berg BW, Rajagopal KR, Dooley JW, Mehm WJ. Hypoxemia during air travel in patients with chronic obstructive pulmonary disease. Ann Intern Med 1989; 111:362–367.
36. Dillard T. Hypoxemia during altitude exposure. A meta-analysis of chronic obstructive pulmonary disease. Chest 1993; 103:422–425.
37. Dillard T, Moores L, Bilello K, Phillips Y. The preflight evaluation—a comparison of the hypoxia inhalation test with hypobaric exposure. Chest 1995; 107:352–357.
38. Speechly-Dick M, Rimmer S, Hodson M. Exacerbations of cystic fibrosis after holidays at high altitude—a cautionary tale. Respir Med 1992; 86:55–56.
39. Oades P, Buchdahl R, Bush A. Prediction of hypoxaemia at high altitude in children with cystic fibrosis. BMJ 1994; 308:15–18.
40. Berg B, Dillard T, Rajagopal K, Mehm W. Oxygen supplementation during air travel in patients with chronic obstructive lung disease. Chest 1992; 101:638–641.
41. Berg BW, Dillard TA, Derderian SS, Rajagopal KR. Hemodynamic effects of altitude

exposure and oxygen administration in chronic obstructive pulmonary disease. Am J Med 1993; 94:407–412.
42. Hansen JE. Diet and flight hypoxemia (letter). Ann Intern Med 1989; 111:859–860.
43. Berger R. Oxygen, food, and air travel (letter). Ann Intern Med 1990; 112:236.
44. Vervolet D, Charpin D, Magnan A, Birnbaum J. Asthma-allergy. Altitude: a study model (editorial) (French). Presse Medicale 1994; 23:1684–1686.
45. Simon H, Grotzer M, Nikolaizik W, Blaser K, Schoni M. High altitude climate therapy reduces peripheral blood T lymphocyte activation, eosinophilia, and bronchial obstruction in children with house-dust mite allergic asthma. Pediatr Pulmonol 1994; 17:304–311.
46. Smith JM. The use of high altitude treatment for childhood asthma. Practitioner 1981; 225:1663–1666.
47. Sporik R, Ingram J, Price W, Sussman J, Honsinger R, Platts-Mills T. Association of asthma with serum IgE and skin test reactivity to allergens among children living at high altitude. Tickling the dragon's breath. Am J Resp Crit Care 1995; 151:1388–1392.
48. Platts-Mills T, Sporik R, Ingram J, Honsinger R. Dog and cat allergens and asthma among school children in Los Alamos, New Mexico, USA: altitude 7,200 feet. Int Arch Allergy Immunol 1995; 107:301–303.
49. Sly R, O'Donnell R. Lack of effect of geographic elevation on mortality from asthma. Ann Allergy 1989; 63:495–497.
50. Christie P, Yntema J, Tagari P, Ysselstijn H, Ford-Hutchinson A, Lee T. Effect of altitude on urinary leukotriene (LT) E4 excretion and airway resposiveness to histamine in children with atopic asthma. Eur Respir J 1995; 8:357–363.
51. Boner A, Comis A, Schiassi M, Venge P, Piacentini G. Bronchial reactivity in asthmatic children at high and low altitude. Effect of budesonide. Am J Respir Crit Care 1995; 151:1194–1200.
52. Mirrakhimov M, Brimkulov N, Cieslicki J, et al. Effects of acetazolamide on overnight oxygenation and acute mountain sickness in patients with asthma. Eur Respir J 1993; 6:536–540.
53. Matsuda S, Onda T, Iikura Y. Bronchial responses of asthmatic patients in an atmosphere-changing chamber. Int Arch Allergy Immunol 1995; 107:402–405.
54. Rosen A, Rosen J. Effect of a face mask on respiratory water loss during sleep in cold conditions. Wilderness Environ Med 1995; 6:189–196.
55. McFadden ER. Respiratory heat and water exchange: Physiological and clinical implications. J Appl Physiol 1983; 54:331–336.
56. Mirrakhimov M, Winslow R. The cardiovascular system at high altitude. In: Fregly MJ, Blatteis CM, eds. Handbook of Physiology; Section 4: Environmenal Physiology. Vol. II. Oxford: Oxford University Press, 1996:1241–1257.
57. Hultgren HN. Reduction of systemic arterial blood pressure at high altitude. Adv Cardiol 1979; 5:49–55.
58. Roca Cusachs A. Pattern of blood pressure among high and low altitude residents of southern Arabia [letter]. J Hum Hypertens 1995; 9:293.
59. Marticorena E, Severino J, Chavez A. Presion arterial sistemica en al nativo de altura. Arch Inst Biol Andina 1967; 2:18–26.
60. Marticorena E, Ruiz L, Severino J, Galvez J, Penaloza D. Systemic blood pressure in white men born at sea level: Changes after long residence at high altitudes. Am J Cardiol 1969; 23:364–368.

61. Ruiz L, Penaloza D. Altitude and hypertension. Mayo Clin Proc 1977; 52:442–445.
62. Appleton F. Possible influence of altitude on blood pressure (abst). Circulation 1967; 36:55.
63. Mirrakhimov MM. Biological and physiological characteristics of the high-altitude natives of Tien Shan and the Pamirs. In: Baker PT, ed. The Biology of High-Altitude Peoples. London: Cambridge University Press, 1978:299–315.
64. Hultgren HN, Grover RF. Circulatory adaptations to high altitude. Ann Rev Med 1968; 19:119–152.
65. Sun S. Epidemiology of hypertension on the Tibetan plateau. Human Biol 1986; 58: 507–515.
66. Hannon J, Sudran D. Basal metabolic and cardiovascular function of women during altitude acclimatization. J Appl Phsiol 1973; 34:4–11.
67. Palatini P, Businaro R, Berton G, et al. Effects of low altitude exposure on 24-hour blood pressure and adrenergic activity. Am J Cardiol 1989; 64:1379–1382.
68. Vogel J, Harris C. Cardiopulmonary responses of resting man during early exposure to high altitude. J Appl Physiol 1967; 22:1124–1129.
69. Scholing W. High altitude climate: adaptation of healthy persons and hypertensives. Med Klin 1985; 76:518–523.
70. Klausen K. Cardiac output in man at rest and work during and after acclimatization to 3800m. J Appl Physiol 1966; 21:609–616.
71. Kamat SR, Banerji BC. Study of cardiopulmonary function on exposure to high altitude. Acute acclimatization to an altitude of 3500 to 4000 meters. Am Rev Respir Dis 1972; 106:404–413.
72. Malhotra MS, Selvamurthy W, Purkayastha SS, Mukherjee AK, Mathew L, Dua GL. Responses of the autonomic nervous system during acclimatization to high altitude in man. Aviat Space Environ Med 1976; 47:1076–1079.
73. Reeves JT, Mazzeo RS, Wolfel EE, Young AJ. Increased arterial pressure after acclimatization to 4300 m: possible role of norepinephrine. Int J Sport Med 1992; 13:S18–S21.
74. Wolfel EE, Groves BM, Brooks GA, et al. Oxygen transport during steady-state submaximal exercise in chronic hypoxia. J Appl Physiol 1991; 70:1129–1136.
75. Korner PI. Circulatory adaptations in hypoxia. Physiol Rev 1959; 39:687–730.
76. Balke B. Cardiac performance in relation to altitude. Am J Cardiol 1964; 14:796–810.
77. Pugh LG. Cardiac output in muscular exercise at 5,800 m (19,000 ft). J Appl Physiol 1964; 19:441–447.
78. Stenberg J, Ekblom B, Messin R. Hemodynamic response to work at simulated altitude, 4000 m. J Appl Physiol 1966; 21:1589–1594.
79. D'Este D, Mantovan R, Martino A, et al. Blood pressure changes at rest and during effort in normotensive and hypertensive subjects in response to altitude acute hypoxia. G Ital Cardiol 1991; 21:643–649.
80. Knudtzon J, Myhre K, Rasch W, Neslin IL, Bognes A, Opstad PK. Reduced systolic blood pressure elevations during maximum exercise at simulated altitudes. Aviat Space Environ Med 1989; 60:1072–1076.
81. Hultgren HN. Effects of altitude upon cardiovascular diseases. J Wilderness Med 1992; 3:301–308.
82. Savonitto S, Giovanni C, Doveri G, et al. Effects of acute exposure to altitude (3,460 m) on blood pressure response to dynamic and isometric exercise in men with systemic hypertension. Am J Cardiol 1992; 70:1493–1497.

83. Hultgren H. Reduction of systemic arterial blood pressure at high altitude. Adv Card 1970; 5:49–55.
84. Halhuber M, Humpeler E, Inama K, Jungman H. Does altitude cause exhaustion of the heart and circulatory system? Med Sport Sci 1985; 19:192–202.
85. Voelkel NF, Hegstrand L, Reeves JT, McMurty IF, Molinoff PB. Effects of hypoxia on density of beta-adrenergic receptors. J Appl Physiol 1981; 50:363–366.
86. Deuber HJ. Treatment of hypertension and coronary heart disease during stays at high altitude (abstr). Aviat Space Environ Med 1989; 60:119.
87. Guazzi MDA, Berti M, Fiorentini C, Galli C, Tamborini G. Enhanced hypoxic pulmonary vasoconstriction in hypertension. Circulation 1989; 79:337–343.
88. Somers V, Mark A, Abboud F. Potentiation of sympathetic nerve responses to hypoxia in borderline hypertensive subjects. Hypertension 1988; 11:608–612.
89. Barthelemy JC, lacour JR, Roche F, et al. Elevated nocturnal blood pressure assessed by ambulatory automatic monitoring during a stay at high altitude. Eur J Appl Physiol 1995; 70:258–262.
90. Mortimer EA, Monson RR, MacMahon B. Reduction in mortality from coronary heart disease in men residing at high altitude. N Engl J Med 1977; 296:581–585.
91. Voors A, Johnson W. Altitude and arteriosclerotic heart disease mortality in white residents of 99 of the 100 largest cities in the United States. J Chron Dis 1979; 32: 157–162.
92. Morton W, Davids D, Lichty J. Mortality from heart disease at high altitude: the effect of high altitude on mortality from arteriosclerotic and hypertensive heart disease. Arch Environ Health 1964; 9.
93. Ramos A, Kruger H, Muro M, Arias-Stella J. Untitled. Bol San Pan Am 1967; 62: 496–502.
94. Ruiz L, Figueroa C, Horna C, Penaloza D. Prevalence of arterial hypertension and cardiac ischemia at high altitudes. Am J Cardiol 1969; 39:474–489.
95. Arias-Stella J, Topilsky M. Anatomy of the coronary circulation at high altitude. In: Porter R, Knight J, eds. High Altitude Physiology: Cardiac and Respiratory Aspects. Edinburgh: Churchill Livingstone, 1971:149–154.
96. Carmelino M. Man at High Altitude. Edinburgh: Churchill Livingstone, 1981:347.
97. Suarez J, Alexander JK, Houston CS. Enhanced left ventricular systolic performance at high altitude during Operation Everest II. Am J Cardiol 1987; 60:137–142.
98. Levine B, Zuckerman J, deFilippi C. Effect of high altitude exposure in the elderly: the 10th Mountain Division Study. Circulation 1997; 96:1224–1232.
99. Alexander J. Age, altitude and arrhythmia. Texas Heart Institute 1995; 22:308–316.
100. Stewart H, Carr H. The anoxemia test. American Heart J 1954; 48:293.
101. Kassebaum DG, Sutherland KI, Judkins MP. A comparison of hypoxemia and exercise electrocardiography in coronary artery disease. Am Heart J 1968; 75:759–776.
102. Burchell HB, Pruitt RD, Barnes AR. The stress and the electrocardiogram in the induced hypoxemia test for coronary insufficiency. Am Heart J 1948; 36:373–389.
103. Alexander J. Coronary heart disease at altitude. Texas Heart Inst J 1994; 21:261–266.
104. Khanna P, Dham S, Hoon R. Exercise in an hypoxic environment as a screening test for ischaemic heart disease. Aviation Space Environ Med 1976; 47:1114–1117.
105. Grover RF, Tucker CE, McGroarty SR, Travis RR. The coronary stress of skiing at high altitude. Arch Intern Med 1990; 150:1205–1208.

106. Okin JT. Response of patients with coronary heart disease to exercise at varying altitudes. Adv Cardiol 1970; 5:92–96.
107. Morgan BJ, Alexander JK, Nicoli SA, Brammell HL. The patient with coronary heart disease at altitude: observations during acute exposure to 3100 meters. J Wilderness Med 1990; 1:147–153.
108. Brammell HL, Morgan BJ, Niccoli SA, Alexander JK. Exercise tolerance is reduced at altitude in patients with coronary artery disease. Circulation 1982; 66:II-371.
109. Roach RC, Houston CS, Honigman B, et al. How well do older persons tolerate moderate altitude? West J Med 1995; 162:32–36.
110. Burtscher M, Philadelphy M, Likar R. Sudden cardiac death during mountain hiking and downhill skiing. N Engl J Med 1993; 329:1738–1739.
111. Shlim DR, Gallie J. The causes of death among trekkers in Nepal. Int J Sport Med 1992; 13:S74–S75.
112. Dickinson JG, Heath J, Gosney J, Williams D. Altitude related deaths in seven trekkers in the Himalayas. Thorax 1983; 38:646–656.
113. Klassen GA, Armour JA, Hopkins DA. Coronary circulation: responses to cold and hypoxemia. In: Sutton JR, Houston CS, Coates G, eds. Hypoxia and Cold. New York: Praeger, 1987:114–123.
114. Lassvik C, Areskog NH. Angina pectoris during inhalation of cold air. Reactions to exercise. Br Heart J 1980; 43:661–667.
115. Hultgren H. Coronary heart disease and trekking. J Wilderness Med 1990; 1:154–161.
116. Hultgren HN. The safety of trekking at high altitude after coronary bypass surgery (letter). J Am Med Assoc 1988; 260:2218.
117. Rennie ID. Will mountain trekkers have heart attacks? (editorial). J Am Med Assoc 1989; 261:1045–1046.
118. Meerson FZ, Ustinova EE, Orlova EH. Prevention and elimination of heart arrhythmias by adaptation to intermittent high altitude hypoxia. Clin Cardiol 1987; 10:783–789.
119. Kahn J, Jouanin J, Bruckert E, Guezennec C, Monod H. Physiological effects of downhill skiing at moderate altitude in untrained middle-aged men. Wilderness Environ Med 1996; 3:199–207.
120. Hackett PH, Roach RC. High-altitude medicine. In: Auerbach PA, ed. Wilderness Medicine. St. Louis: Mosby, 2001:1–37.
121. Erdmann J, Sun K, Masar P, Niederhauser H. Effects of exposure to altitude on men with coronary artery disease and impaired left ventricular function. Am J Cardiol 1998; 81:266–270.
122. Alexander J, Abinader E, Sharif D, Morgan B, Brammell H. Left ventricular function in coronary heart disease at high altitude (abstract #24). Circulation 1988; 78:II-6.
123. Dalen J, Bruce R, Cobb L. Interaction of chronic hypoxia of moderate altitude on pulmonary hypertension complicating defect of the atrial septum. N Engl J Med 1962; 266:272–277.
124. Vogel JH, McNamara DG, Blount SG. Role of hypoxia in determining pulmonary vascular resistance in infants with ventricular septal defects. Am J Cardiol 1967; 20: 346–349.
125. Sorensen SC, Severinghaus JW. Respiratory insensitivity to acute hypoxia persisting after correction of tetralogy of Fallot. J Appl Physiol 1968; 25:221–223.
126. Garcia J, MCMinn S, Zuckerman J, Fixler D, Levine B. Exercise responses at high altitude after the Fontan operation (abstr). Circulation 1993; 88:1548.
127. Levine BD, Grayburn PA, Voyles WF, Greene ER, Roach RC, Hackett PH. Intracar-

diac shunting across a patent foramen ovale may exacerbate hypoxemia in high altitude pulmonary edema. Ann Intern Med 1991; 114:569–570.
128. Hackett PH, Creagh CE, Grover RF, Honigman B, Houston CS. High altitude pulmonary edema in persons without the right pulmonary artery. N Engl J Med 1980; 302: 1070–1073.
129. Torrington KG. Recurrent high-altitude illness associated with right pulmonary artery occlusion from granulomatous mediastinitis. Chest 1989; 96:1422–1423.
130. Rios B, Driscoll DJ, McNamara DG. High-altitude pulmonary edema with absent right pulmonary artery. Pediatrics 1985; 75:314–317.
131. Naeije R, DeBacker D, Vachiery J, DeVuyst P. High-altitude pulmonary edema with primary pulmonary hypertension. Chest 1996; 110:286–289.
132. Green RL, Huntsman RG, Serjeant GR. The sickle-cell and altitude. Br Med J 1971; 4:593–595.
133. Mahoney BS, Githens JH. Sickling crisis and altitude: occurrence in the Colorado patient population. Clin Pediatr 1979; 18:431–438.
134. Adzaku F, Addae S, Annobil S, Mohammed S. Clinical features of sickle cell disease at altitude. J Wilderness Med 1992; 3:260–268.
135. Lane PA, Githens JH. Splenic syndrome at mountain altitudes in sickle cell trait: Its occurrence in nonblack persons. J Am Med Assoc 1985; 253:2252–2254.
136. Goldberg NM, Dorman JP, Riley CA, Armbruster EJ. Altitude-related splenic infarction in sickle cell trait—case reports of a father and son. West J Med 1985; 143: 670–672.
137. Diggs L. The sickle cell trait in relation to the training and assignment of duties in the Armed Forces: IV. Considerations and recommendations. Aviation Space Environ Med 1984; 55:487–492.
138. Hebbel RP, Eaton JW, Kronenberg RS, Zanjani ED, Moore LG, Berger EM. Human llamas: adaptation to altitude in subjects with high hemoglobin oxygen affinity. J Clin Invest 1978; 62:593–600.
139. Eaton JW. Low altitude hominids at high altitude. Blood Cells 1981; 7:509–511.
140. Beard JL, Haas JD, Tufts DA, Spielvogel H, Vargas E, Rodriguez C. Iron deficiency anemia and steady-state work performance at high altitude. J Appl Physiol 1988; 64: 1878–1884.
141. Wilson R. Anemia at high altitude. Alaska Med 1977; 19:49–52.
142. Bärtsch P, Maggi S, Kleger GR, Ballmer PE, Baumgartner RW. Sumatriptan for high-altitude headache. Lancet 1994; 344:1445.
143. Burtscher M, Likar R, Nachbauer W, Schaffert W, Philadelphy M. Ibuprofen versus sumatriptan for high-altitude headache (letter). Lancet 1995; 346:254–255.
144. Arregui A, Cabrera J, Leon-Velarde F, Paredes S, Viscarra D, Arbaiza D. High prevalence of migraine in a high-altitude population. Neurology 1991; 41:1668–1670.
145. Arregui A, Leon-Velarde F, Cabrera J, Paredes S, Vizcarra D, Umeres H. Migraine, polycythemia and chronic mountain sickness. Cephalgia 1994; 14:339–341.
146. Jenzer G, Bärtsch P. Migraine with aura at high altitude: case report. J Wilderness Med 1993; 4:412–415.
147. Engel RL, Webb JP, Ferris EB, Romano J, Ryder H, Blankenhorn MA. A migraine-like syndrome complicating decompression sickness. War Med 1944:304–314.
148. Murdoch DR. Focal neurological deficits and migraine at high altitude (letter). J Neurol Neurosurg Psychiatr 1995; 58:637.
149. Hackett PH, Roach RC, Hollingshead KF, Schoene RB, Mills WJ. Cortical blindness

in high altitude climbers and trekkers—a report on six cases (abstr). In: Sutton JR, Houston CS, Coates G, eds. Hypoxia and Cold. New York: Praeger, 1987:536.

150. Houston CS. Transient visual disturbance at high altitude (Abstract). In: Sutton JR, Houston CS, Coates G, eds. Hypoxia and Cold. New York: Praeger, 1987:536.
151. Wohns RN. Transient ischemic attacks at high altitude (abstr). In: Sutton JR, Houston CS, Coates G, eds. Hypoxia and Cold. New York: Praeger, 1987:536.
152. Razdan S, Koul R, Motta A, Kaul S. Cerebrovascular disease in rural Kashmir, India. Stroke 1989; 20:1691–1693.
153. Cruz M, Schoenberg B, Ruales J, et al. Pilot study to detect neurologic disease in Ecuador among a population with a high prevalence of endemic goiter. Neuroepidemiology 1985; 4:108–116.
154. Bancaleri E. Internacional de Post-Grado: neurologia de altura, Latin America Regional Satellite Symposium on Cerbrovascular Disease, 1983. Vol. 4.
155. Jaillard AS, Hommel M, Mazetti P. Prevalence of stroke at high altitude (3380 m) in Cuzco, a town of Peru. A population-based study. Stroke 1995; 26:562–568.
156. Clarke CR. Cerebral infarction at extreme altitude (Abstract). In: Sutton JR, Houston CS, Jones NL, eds. Hypoxia, Exercise and Altitude. New York: AR Liss, 1983:453–454.
157. Houston CS. Going Higher: The Story of Man at High Altitude. Boston: Little, Brown, 1987.
158. Dickinson J. Acute mountain sickness: a dissertation based on 11 years experience in the Nepal Himalaya. Oxford: Oxford University, 1981:98.
159. Song S, Asaji T, Tanizaki Y, Fujimaki T, Matsutani M, Okeda R. Cerebral thrombosis at altitude: its pathogenesis and the problems of prevention and treatment. Aviat Space Environ Med 1986; 57:71–76.
160. Fujimaki T, Matsutani M, Asai A, Kohno T, Koike M. Cerebral venous thrombosis due to high altitude polycythemia. J Neurosurg 1986; 64:148–150.
161. Hackett P. Cerebral venous thrombosis at altitude. Wilderness Med Newsletter 1987; 4:8–9.
162. Pfausler B, Vollert H, Bosch S, Schmutzhard E. Cerebral venous thrombosis—a new diagnosis in travel medicine? J Travel Med 1996; 3:165–167.
163. Grotta J. Cerebral venous thrombosis—a new diagnosis in travel medicine (editorial). J Travel Med 1996; 3:137.
164. Wohns RN. Transient ischemic attacks at high altitude. Crit Care Med 1986; 14:517–518.
165. Shlim DR, Meijer H. Suddenly symptomatic brain tumors at altitude. Ann Emerg Med 1991; 20:315–316.
166. Gazenko OG, Gippenreiter EB, Malkin VB. Expert evaluation of 'altitude' tolerance in candidates for Soviet expedition 'Everest-82' (abstr). In: Sutton JR, Houston CS, Coates G, eds. Hypoxia and Cold. New York: Praeger, 1987:532.
167. Nicholas R, O'Meara P. High-altitude syncope: History repeats itself. JAMA 1993; 269:587.
168. Parker D, Hiar C. Performance of the Glucometer Elite and Glucometer Encore at altitude (8800 ft.) [abstr]. Clin Chem 1995; 41:S185.
169. Giordano B, Thrash W, Hollenbaugh L, Dube W, Hodges C, Swain A. Performance of 7 blood glucose testing systems at high altitude. Diabetes Educ 1989; 15:444–448.
170. Chmielewski S, Kurtock D, Jennings S, et al. Precision and accuracy of the Accu-

Chek Advantage blood glucose monitoring system at high altitude. Clin Chem 1996; 42:115–117.
171. Mader TH, Friedl KE, Mohr LC, Bernhard WN. Conjunctival oxygen tension at high altitude. Aviat Space Environ Med 1987; 58:76–79.
172. Snyder R, Klein P, Solomon J. The possible effect of barometric pressure on the corneas of an RK patient: a case report. Int Contact Lens Clin 1988; 15:130–132.
173. White L, Mader T. Refractive changes with increasing altitude after radial keratotomy. Am J Ophthalmol 1993; 115:821–823.
174. Mader TH, White LJ. Refractive changes at extreme altitude after radial keratotomy. Am J Ophthalmol 1995; 119:733–737.
175. Ng J, White L, Parmley V, Hubickey W, Carter J, Mader T. Effects of simulated high altitude on patients who have had radial keratotomy. Ophthalmology 1996; 103:452–457.
176. Winkle R, Mader T, Parmley V, White L, Polse K. The etiology of refractive changes at high altitude after radial keratotomy. Hypoxia versus hypobaria. Ophthalmology 1998; 105:282–286.
177. Mader T, White L. High altitude mountain climbing after radial keratotomy (letter). Wilderness Environ Med 1996; 7:77–78.
178. Geninatti M. Climbing after keratorefractive surgery (letter). Wilderness Environ Med 1995; 6:351–352.
179. Mader T, Blanton C, Gilbert B, et al. Refractive changes during 72-hour exposure to high altitude after refractive surgery. Ophthalmology 1996; 103:1188–1195.
180. Hackett PH, Rennie ID. Rales, peripheral edema, retinal hemorrhage and acute mountain sickness. Am J Med 1979; 67:214–218.
181. Breathnach C. The stability of the fetal oxygen environment. Irish J Med Sci 1991; 160:189–191.
182. Meschia G. Supply of oxygen to the fetus. Journal Reprod Med 1979; 23:160–165.
183. Carter A. Factors affecting gas transfer across the placenta and the oxygen supply to the fetus. J Develop Physiol 1989; 12:305–322.
184. Leibson C, Brown M, Thibodeau S, et al. Neonatal hyperbilirubinemia at high altitude. Am J Dis Child 1989; 143:983–987.
185. Ballew C, Haas JD. Hematologic evidence of fetal hypoxia among newborn infants at high altitude in Bolivia. Am J Obstet Gynecol 1986; 155:166–169.
186. Beall C. Optimal birth weights in Peruvian populations at high and low altitudes. Am J Phys Anthropol 1981; 56:209–216.
187. Unger C, Weiser JK, McCullough RE. Altitude, low birth weight, and infant mortality in Colorado. J Am Med Assoc 1988; 259:3427–3432.
188. Wilcox AJ. Birth weight and perinatal mortality: The effect of maternal smoking. Am J Epidemiol 1993; 137:1098–1104.
189. Templeton A. Intrauterine growth retardation associated with hypoxia due to bronchiectasis. Br J Obstet Gynaecol 1977; 84:389–390.
190. Novy M, Petersen E, Metcalf J. Respiratory characteristics of maternal and fetal blood in cyanotic congenital heart disease. Am J Obstet Gynecol 1968; 100:821–828.
191. Moore LG, Brodeur P, Chumbe O, D'Brot J, Hofmeister SE, Monge CC. Maternal hypoxic ventilatory response, ventilation, and infant birth weight at 4,300 m. J Appl Physiol 1986; 60:1401–1406.
192. Zamudio S, Palmer SK, Droma T, Stamm E, Coffin C, Moore LG. Effect of altitude on uterine artery blood flow during normal pregnancy. J Appl Physiol 1995; 79:7–14.

193. Genbacez O, Joslin R, Damsky C, Polliotti B, Fisher S. Hypoxia alters early gestational human cytotrophoblast differentiation/invasion in vitro and models the placental defects that occur in preeclampsia. J Clin Invest 1996; 97:540–550.
194. Barry M, Bia F. Pregnancy and travel. J Am Med Assoc 1989; 261:728–731.
195. Artal R, Fortunato V, Welton A, et al. A comparison of cardiopulmonary adaptations to exercise in pregnancy at sea level and altitude. Am J Obstet Gynecol 1995; 172: 1170–1178.
196. Baumann H, Bung P, Fallenstein F, Huch A, Huch R. Reaktion von Mutter und Fetus auf die Körperliche Belastung in Höhe. Geburtshilfe Frauenheilkd 1985; 45:869–876.
197. Parer JT. Effects of hypoxia on the mother and fetus with emphasis on maternal air transport. Am J Obstet Gynecol 1982; 142:957–961.
198. Greiss F. Concepts of uterine blood flow. In: Wynn R, ed. Obstetrics and Gynecology Annual. New York: Appleton-Century-Crofts, 1973:55–83.
199. Baillie P. Nonhormonal methods of antenatal monitoring. Clin Obstet Gynecol 1974; 1:103–122.
200. Copher DE, Huber CP. Heart rate response of the human fetus to induced maternal hypoxia. Am J Obstet Gynecol 1967; 98:320–335.
201. Ritchie J, Lakhani K. Fetal breathing movements and maternal hyperoxia. Br J Obstet Gynaecol 1980; 87:1084–1088.
202. Bartnicki J, Saling E. Influence of maternal oxygen administration on the computer-analysed fetal heart rate patterns in small-for-gestational-age fetuses. Gynecol Obstet Invest 1994; 37:172–175.
203. Honigman B, Theis MK, McLain J, et al. Acute mountain sickness in a general tourist population at moderate altitudes. Ann Intern Med 1993; 118:587–592.
204. Roeggla G, Roeggla H, Roeggla M, Binder M, Laggner AN. Effect of alcohol on acute ventilatory adaptation to mild hypoxia at moderate altitude. Ann Intern Med 1995; 122:925–927.
205. McFarland R. Human Factors in Air Transportation. New York: McGraw-Hill, 1953.
206. McFarland R, Forbes W. The metabolism of alcohol in man at high altitudes. Human Biol 1936; 8:387–398.
207. Higgins E, Davia A, Vaughn J, Funkhauser G, Galerston E. The Effects of Alcohol at Three Simulated Aircraft Cabin Conditions. Washington, DC: FAA Office of Aviation Medicine, 1968.
208. Higgins E, Vaughan J, Funkhouser G. Blood Alcohol Concentrations as Affected by Combinations of Alcoholic Beverage Dosages and Altitude. Washington, DC: FAA Office of Aviation Medicine, 1970.
209. Lategola M, Lyne P, Burr M. Alcohol-Induced Physiological Displacements and Their Effects on Flight-Related Functions. Washington, DC: FAA, 1982.
210. Collins W, Mertens H. Age, alcohol, and simulated altitude: effects on performance and breathalyzer scores. Aviat Space Environ Med 1988; 59:1026–1033.
211. Collins W. Performance effects of alcohol intoxication and hangover at ground level and at simulated altitude. Aviat Space Environ Med 1980; 51:327–351.
212. Collins W, Mertens H, Higgins E. Some effects of alcohol and simulated altitude on complex performance scores and breathalyzer readings. Aviat Space Environ Med 1987; 58:328–332.
213. Freedman R, Waldo M, Adler L, Baker N, Levin D, Deitrich R. Electrophysiological effects of low dose alcohol on human subjects at high altitude. Alcohol Drug Res 1986; 6:289–297.

AUTHOR INDEX

Italic numbers give the page on which the complete reference is cited.

A

B

C

D

E

F

G

H

M

N

O

P

S

T

U

V

X

Y

SUBJECT INDEX

A

B

C

F

G

H

I

J

K

L

N

Q

R

S

T

X

Y

Z